Media and Supplements Overview

STUDENT SUPPLEMENTS

Student CD

Every copy of *Microbial Life* includes a comprehensive student CD. The CD provides students with a variety of interactive tools and activities for learning the important concepts introduced in the textbook. A list of the activities is included on the facing page. Features include:

- **Intuitive Interface:** The interface has been designed to present the review material and animated activities in a logical, intuitive manner.
- **Chapter Summaries:** The major concepts from each chapter are summarized and presented for quick review.
- **Animated Tutorials:** The CD includes 15 in-depth tutorials, consisting of three parts: an Introduction that sets up the concept to be presented; an Animation that presents the concept using a clear, easy-to-follow narrative; and a Quiz that tests the student's understanding of the material.
- **Major Techniques:** "Technique" animations thoroughly describe 18 important laboratory methods used by microbiologists in the course of their work.
- **Dynamic Illustrations:** 26 key illustrations from the text are presented in step-by-step format, emphasizing the flow of the figure and making the sequence of events easier to follow.
- **Key Terms:** Extensive lists of important terminology from the textbook are included for each chapter, with textbook page references for further review.
- **Quizzes:** Self-Quizzes for each chapter test knowledge of key concepts and facts from the textbook. References to headings and pages in the textbook provide direction for additional study.
- **Glossary:** The CD includes a comprehensive glossary of all the important terms introduced in the textbook.

Study Guide

William H. Coleman, University of Hartford
This study aid for students includes chapter-by-chapter summaries of important concepts, and key terms. The Study Guide also includes a wealth of factual and conceptual review questions in a variety of formats for effective, self-paced review of the material presented in the textbook.

INSTRUCTOR SUPPLEMENTS

Instructor's Resource CD

This set of CDs includes all of the following electronic resources to enhance the lecture and facilitate course planning and assessment. Contents include:

- **Browser Interface:** A convenient way to preview all of the images and animations on the CD.
- **Images**: All of the line-art figures and tables from the book in high-resolution JPEG format.
- **PowerPoint Presentations**: Ready-to-use presentations for each chapter that include all of the line art and tables from the book.
- **Photo Collection**: A supplemental collection of top-quality micrographs and color photos illustrating a wide range of organisms and phenomena covered in the textbook.
- **Animations and Activities**: All the animations and activities from the Student CD (59 total) are included.
- **Cross-Platform Electronic Test Bank:** Provided both as Microsoft Word documents and in test-creation software (software included).
- **Chapter Outlines:** Provided as Microsoft Word documents.
- **Instructor's Manual:** Provided as PDF documents.

Instructor's Manual

William H. Coleman, University of Hartford
The Instructor's Manual provides resources for developing the lecture and for assessment. Contents include the following for each chapter:

- **Summary & Outline:** A thorough summary and outline of the important concepts in the chapter.
- **Internet Teaching Resources:** Useful tips on how to locate resources on the Internet to engage students. Up-to-date links are provided on an Instructor's website.
- **Test File:** The Test File contains more than 900 test questions. Each chapter has an extensive set of multiple-choice questions, with additional matching and true-false questions. Using the electronic test file (also on the Instructor's Resource CD), instructors can create customized tests using any combination of Test File questions and their own questions.

Overhead Transparencies

This set of 275 full-color transparencies includes all of the key figures and tables from the textbook in a relabeled and resized format. The format is optimal for projection and produces excellent image quality and readability.

Boxes

Boxes throughout the text describe **Milestones**, **Research Highlights**, or **Methods & Techniques** in the study of microbiology.

Microbial Life

Microbial Life

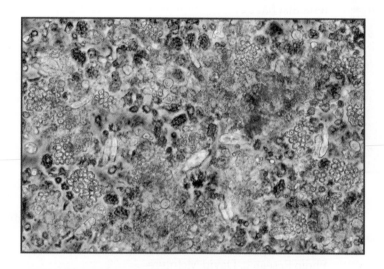

Jerome J. Perry
Raleigh, North Carolina

James T. Staley
Department of Microbiology
University of Washington

Stephen Lory
Department of Microbiology and Molecular Genetics
Harvard Medical School

 Sinauer Associates, Publishers • SUNDERLAND, MASSACHUSETTS

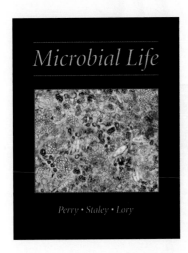

About the cover

These bacteria were collected from the anoxic (oxygen-depleted), lower depths of a lake. The red and green organisms are purple and green sulfur photosynthetic bacteria that have evolved from some of the earliest photosythetic organisms on Earth. Courtesy of J. T. Staley.

About the book

Editor: Andrew D. Sinauer
Project Editor: Chelsea D. Holabird
Market Research and Reviewing: Susan McGlew
Copy Editor: Karen Slaght
Editorial Assistant: Sydney Carroll
Production Manager: Christopher Small
Book Layout and Production: Joan Gemme, Jefferson Johnson, and Janice Holabird
Art Editing: Elizabeth Morales and Linda Strange
Illustration Program: Elizabeth Morales and Precision Graphics
Book Design: Jefferson Johnson
Cover Design: Joan Gemme
Photo Research: David McIntyre
Color Separations: Burt Russell Litho
Book and Cover Manufacture: Courier Companies, Inc.

Microbial Life

Address orders and editorial correspondence to:
Sinauer Associates, Inc., P. O. Box 407
23 Plumtree Road, Sunderland, MA, 01375 U.S.A.
Fax: 413-549-1118
Internet: www.sinauer.com; Email: publish@sinauer.com

10 9 8 7 6 5 4 3 2 1

Authors

JEROME J. PERRY, an experienced research scientist and educator, taught general microbiology to college undergraduates for more than 25 years. He received his B.S. from Pennsylvania State University and his Ph.D. from the University of Texas, then undertook postdoctoral studies at the University of Chicago. Dr. Perry's major research interests included metabolism of gaseous alkanes, co-oxidation, and bioremediation. In addition to *Microbial Life*, he has written extensively on hydrocarbon metabolism in reviewed papers, chapters, and various proceedings, and co-edited *Introduction to Environmental Toxicology*, a major textbook in its field.

JAMES T. STALEY is a Professor in the Department of Microbiology at the University of Washington, Seattle. He earned a B.A. from the University of Minnesota, an M.Sc. from Ohio State University, and a Ph.D. from the University of California, Davis. Before coming to the University of Washington, he held academic appointments at Michigan State University and the University of North Carolina. Dr. Staley's major research interests are in general microbiology, with emphasis on microbial evolution, diversity, ecology, and bacterial taxonomy. He is Director of the cross-disciplinary NSF IGERT Astrobiology Ph.D. traineeship program at the University of Washington, which uses extreme environments on Earth as models for possible life on other planets and studies the early evolution of life on Earth.

STEPHEN LORY is Professor of Microbiology and Molecular Genetics at Harvard Medical School. He earned a B.A. in Bacteriology and a Ph.D. in Microbiology, both from the University of California, Los Angeles. Following a research fellowship in the Bacterial Physiology Unit at Harvard Medical School, he taught at the University of Washington, Seattle for several years before returning to Harvard. The author of numerous journal articles and book chapters, Dr. Lory has served as a member on the editorial boards of the *Journal of Bacteriology*, *Infection and Immunity*, and *Molecular Microbiology*, and as Editor of *Microbiology and Molecular Biology Reviews*.

Contributors

CYNTHIA L. BALDWIN is a Professor in the Department of Veterinary and Animal Sciences at the University of Massachusetts, Amherst. After receiving her B.A. from Hartwick College and her Ph.D. (Immunology) from Cornell University, she was a postdoctoral fellow (Cellular Immunology) at the International Laboratory for Research on Animal Diseases in Nairobi, Kenya. Her current research interests include gamma delta T cell biology and immunity to intracellular bacteria.

SAMUEL J. BLACK is a Professor in the Department of Veterinary and Animal Sciences at the University of Massachusetts, Amherst. He received his B.S. (Zoology) from Queens University in Belfast, Northern Ireland and his Ph.D. (Immunology) at the University of Edinburgh, Scotland. His postdoctoral work in Immunogenetics was done at the University of Köln, Germany and at Stanford University Medical Center (with L. A. Herzenberg). Dr. Black's research interests include the regulation of immune function by pathogens and the molecular basis of host resistance to trypanosomiasis.

JOHN H. GUNDERSON is an Assistant Professor in the Department of Biology at Tennessee Technological University. He received his B.S. (Zoology and Russian) from the University of Nebraska and his Ph.D. from the University of California, Berkeley, where he specialized in protozoology. His postdoctoral work with Mitchell Sogin involved using ribosomal RNA sequences to study protistan phylogeny. His current research interests include: development of DNA probes for detecting parasitic dinoflagellates; detection of Legionella-like amoebal pathogens; identification of prokaryotic symbionts living in termite flagellates; and development of molecular methods for studying microbial food webs.

WILLIAM B. WHITMAN is Professor of Microbiology, Biochemistry, and Marine Sciences at the University of Georgia. He received his B.S. from the State University of New York at Stony Brook and his Ph.D. from the University of Texas at Austin. He has always been interested in unusual microorganisms. His current research focuses on carbon metabolism and the development of genetic systems in the methane-producing archaeon *Methanococcus*, prokaryotic diversity in soils and seawater, and prokaryotic evolution.

Preface

As we enter this new century and millennium, scientists are finally beginning to comprehend the full significance of microbial life. Microorganisms have inhabited Earth for about 4 billion years—more than 3 billion years longer than plants and animals— and as a consequence of their activities, microbial life formed Earth's first biosphere. Plants and animals are the evolutionary descendants of *Bacteria* and *Archaea* and could not have evolved unless microbial life evolved first. Microorganisms are a major life form: their total biomass exceeds that of animals and plants combined. The diversity of microorganisms is exemplified by their physiological tolerances and the ranges over which they grow, their many unique metabolic pathways, and the consequences of their ecological activities and geochemical transformations. They drive many of the life processes on Earth—macroscopic organisms are completely dependent on them for survival. Surprisingly, evidence supporting these fundamental concepts has only been discovered during the past 20 or so years

Microbial Life was written to introduce the reader to this remarkable field of microbiology. Microorganisms are studied, not only for their own sake, but because they are important as research tools and are of considerable consequence in animal and plant symbioses and disease. In this book we provide the reader with a balanced and broad context for the study of microorganisms, with a focus on these basic themes: the evolution of concepts in the study of microbiology; the process of scientific inquiry; and the significance of molecular approaches in the investigation of microbial life. We present microbes' cell and molecular biology, evolution, genetics, genomics, physiology, metabolism, taxonomy, phylogeny, and the impact they have had on other biological fields. Emphasis is placed on the development of concepts in microbiology, from the observations of Antony van Leeuwenhoek in the 17th century to the contributions of the great scientists in the latter part of the 19th century—Robert Koch, Louis Pasteur, Martinus Beijerinck, and Sergei Winogradsky. These scientists were the principals in the Golden Age of Bacteriology. Their discoveries laid the foundation for the contributions that followed and led to our current understanding of microbial life. In addition, we explore microbes' involvement in disease and immunity, and we investigate the various bacterial types (*Bacteria* and *Archaea*) along with the viruses and eukaryotic microorganisms. We also discuss their larger role in ecosystems and in the biosphere. Finally, we cover the importance of microorganisms in biotechnology and the environmental, food, and health industries.

FEATURES

Microbial Life provides the student with an understanding of the microbe's place in nature in a way that will encourage learning the broad aspects of the microbial world. The text focuses on the microbe itself rather than the diseases that might be caused by microorganisms.

Both because microorganisms are fascinating to behold and because many students are visual learners, we have worked especially hard on the illustration program. The line drawings and photographs in the book have been developed or chosen to emphasize directness and drama. We have combined the strengths of both text and graphics through the use of "balloon captions." These brief statements are incorporated directly into the graphics and go beyond mere labeling to describe, define, or explain key graphic elements.

Each chapter in *Microbial Life* opens with a quote from a noted writer to provide a cultural context. The first page of each chapter presents an introductory overview. At the end of each chapter, there is an extensive narrative summary with boldfaced key terms, as well as thought questions to aid in reviewing and synthesizing major points covered in the chapter, and suggested readings.

Most chapters also include one or more boxes that employ one of three major themes: "Research Highlights" focus on recent or significant research topics; "Milestones" present historical or biographical sketches on the evolution of significant ideas or concepts; "Methods and Techniques" discuss the tools used by microbiologists in the course of their work.

Our presentation of microbial diversity is phylogenetic in orientation and Part V (Microbial Evolution and Diversity) emphasizes the "Three Domain" classification of microorganisms. The diversity coverage is more evolutionary than that in other textbooks; it is based on the latest ribosomal RNA sequencing techniques. Special features of Part V include:

- In addition to the usual tables that appear in all chapters, Part V presents unique "diversity tables" that help students categorize and compare the immense variety of microorganisms.

- A balanced and contemporary chapter on the *Archaea* explores these extraordinary organisms that thrive under some of the most unusual conditions for life on Earth.

- A separate chapter on beneficial symbiosis clearly defines the role of prokaryotes in the evolution and survival of eukaryotes and describes a way of life that is now recognized as of major significance in the biosphere.

- A chapter on eukaryotic microorganisms brings the taxonomy of this major group into perspective. It is the first textbook chapter of its kind in which the presentation is organized around the phylogenetic classification of these microorganisms.

- Taxonomy is covered in a separate chapter and separate chapters are devoted to related groups, including the Gram positive bacteria, non-photosynthetic proteobacteria, and phototrophic bacteria.

Other features of note are: (1) The chapter on microbial genomics provides a very current overview of this important developing area of investigation. (2) A separate chapter on biodegradation covers various aspects of this major role played by microorganisms in perpetuating life on Earth. (3) Medical aspects are thoroughly covered in chapters devoted to immunology, clinical microbiology, and Public Health. Diseases caused by viruses and by microorganisms are presented in separate chapters.

STUDENT AUDIENCE

This text is organized so that students are introduced to topics in a sequence that allows them to develop their understanding of microbiology in a logical, reinforcing manner. We believe that the microbe is the perfect model system for studying metabolism, genetics, and molecular interactions, and this view has been adopted in the organization and presentation of the text material. We recognize that undergraduate students who use this book will have varied backgrounds. Many students who are microbiology or biology majors will have formally studied other biological areas prior to reading this text. The text is organized to accommodate different groups of students, including students with an intermediate level of understanding. For example, it may also serve as supplementary reading in a diversity or taxonomy course.

ORGANIZATION

The text uses a level-of-organization approach beginning with basic biochemistry, structure/function, nutrition, growth, metabolism, physiology, and genetics. These basic concepts are then followed by microbial taxonomy, a treatment of the diverse microbial groups, microbial ecology, microbial and viral disease, immunology, and industrial and environmental microbiology.

Microbial Life is presented in eight parts, with two to seven chapters in each part. Part I is an overview of the science of microbiology. Chapter 1 introduces the reader to the microbial world using an evolutionary perspective. Chapter 2 provides the reader with the development of concepts in the field. For many students, Chapter 3 about chemistry will be a useful review. Part I ends with Chapter 4 on the structure and function of *Bacteria* and *Archaea*.

Microbial physiology is the subject of Parts II and III, which focus on the diverse energy-sustaining strategies adapted by microorganisms for survival. Chapter 5 describes the wide variety of nutritional types of microorganisms. The growth of single-celled organisms is treated kinetically in Chapter 6, followed by the control of microbial growth in Chapter 7. Part III covers metabolism, with emphasis on the diversity of metabolic types. Chapter 8 presents the variety of pathways utilized by microorganisms in generating energy chemically, and Chapter 9 addresses the variety of photosynthetic processes used to generate energy by an array of microorganisms. Two chapters on biosynthesis follow: Chapter 10 discusses the synthesis of major monomers, including amino acids, needed for macromolecular synthesis; the synthesis of protein and peptidoglycan is treated in Chapter 11. The final chapter (12) in this Part presents information on the biodegradation of selected organic compounds.

Part IV presents an up-to-date treatment of genetics and virology. Basic genetic principles are discussed in Chapter 13. In order to thoroughly present gene transfer between bacteria through viral transduction, which is covered in Chapter 15, viruses are introduced in Chapter 14. The last chapter (16) in this Part covers the major emerging field of microbial genomics.

In Part V, we investigate microbial evolution and diversity. We provide as complete coverage of these topics as reasonable for this level of treatment. The taxonomy and phylogeny of *Bacteria* and *Archaea* is discussed in Chapter 17, which also presents the universal tree of life, derived from Carl Woese's "Three Domains" model.

The diversity chapters provide a well-balanced treatment of the various microbial groups. They can be taught in their entirety or used as reference. The chapters are organized phylogenetically according to the latest research developments. The diversity section begins in Chapter 18, with a treatment of *Archaea*. Three following chapters discuss the various bacterial groups: Chapter 19 on nonphotosynthetic *proteobacteria*; Chapter 20 on gram-positive heterotrophic bacteria; and Chapter 21 on phototrophic bacteria. Our treatment of the cyanobacteria in Chapter 21 exemplifies our belief that this important group of bacteria is often neglected, if not overlooked, in most other comparable textbooks. This chapter is followed by Chapter 22 on novel bacterial phyla. Finally, Chapter 23 closes Part V with a discussion of the evolution and remarkable diversity of eukaryotic microorganisms.

The significance of microorganisms and their diversity in ecological processes and geochemical transformations becomes clear in Part VI, "Microbial Ecology." Chapter 24 describes the role microorganisms play in ecosystems. Chapter 25 highlights some fascinating beneficial symbiotic associations among microorganisms, as well as between microorganisms and macroorganisms. Chapter 26 covers the special symbiotic relationship of host/parasite, focusing on nonspecific host resistance.

An especially important area of host/parasite associations impacts health and disease. Part VII presents immunology and medical microbiology. Chapter 27 discusses immunology and the specific strategies of host resistance. This is followed by Chapter 28 addressing microbial diseases of humans, and Chapter 29 on viral diseases of humans. The final chapter in Part VII, Chapter 30, highlights epidemiology and clinical microbiology.

Part VIII, Applied Microbiology, describes microbes as tools. Many new and important careers for microbiologists will continue to be found in these areas. The first of these chapters (31) deals with industrial microbiology and biotechnology. The final chapter (32) covers the area of environmental microbiology, including water and wastewater treatment and bioremediation.

CONTRIBUTORS

Microbiology is a rapidly growing field of science, both in depth of understanding of the genetics, taxonomy, and physiology of the microbe, and in breadth of the disciplines encompassed by the term. From virus to protozoa, or from immunology to pathogenicity, the field is too broad to be covered by three authors. Therefore, we invited a few colleagues to share their expertise in writing important chapters for this book. We thank William "Barney" Whitman for the fascinating and up-to-date chapter on the *Archaea* and John Gunderson for the timely chapter on eukaryotic microorganisms. The excellent coverage of immunology was written by Cynthia L. Baldwin and Samuel J. Black.

ACKNOWLEDGMENTS

A large corps of busy scientists agreed to make the time to review our drafts to help ensure accuracy and timeliness. We greatly appreciate their generosity. Particular thanks to Paul Dunlap, who reviewed several chapters in manuscript and then provided a thorough check of every chapter in page proofs.

The authors are indebted to Andy Sinauer and his colleagues at Sinauer Associates for it was their efforts that led to the completion of this book. Andy's guidance, leadership, and understanding were apparent throughout the process and we are most grateful. We offer a special thanks to Chelsea Holabird for her patience and kindness in keeping us focused and on schedule. It was not easy. The cooperation of Christopher Small, Susan McGlew, Dean Scudder, David McIntyre, Jason Dirks, Marie Scavotto and others at Sinauer Associates is greatly appreciated. Their friendly, cheerful professionalism was always evident.

SPECIAL THANKS

J. J. Perry expresses special thanks to Larry F. Grant, Wesley E. Kloos, and T. J. Schneeweis. Thanks also to Todd McPherson at the North Carolina Department of Public Health and Human Services for his assistance.

Jim Staley thanks his microbiology colleagues and mentors as well as his graduate students and post docs for the opportunity to collaborate and learn microbiology with them. He also extends special thanks to two current post docs, Sujatha Srinivasan and Cheryl Jenkins. Douglas Vollgraff and Katharine Norris in the departmental office provided welcome assistance with support services.

Steve Lory thanks Katie Rhodes, Tamara Will, and Matt Wolfgang for their assistance with the text and illustrations in the various chapters.

JEROME J. PERRY
JAMES T. STALEY
STEPHEN LORY
March, 2002

Reviewers

Phil Achey, University of Florida

Craig A. Almeida, Stonehill College

Vivian Lam Braciale, University of Texas Medical Branch

Jean E. Brenchley, Pennsylvania State University

Carl E. Cerniglia, National Center for Toxicological Research

James F. Curran, Wake Forest University

Paul V. Dunlap, University of Michigan

Roger Fujioka, University of Hawaii at Manoa

George Hegeman, Indiana University

John D. Helmann, Cornell University

Ken F. Jarrell, Queen's University

Laura Katz, Smith College

Wesley Kloos, North Carolina State University

Laura J. Knoll, University of Wisconsin, Madison

Edward R. Leadbetter, University of Connecticut

John A. Leigh, University of Washington

Alan C. Leonard, Florida Institute of Technology

Steven E. Lindow, University of California, Berkeley

Carol Litchfield, George Mason University

Jim McAlpine, Phytera, Inc.

Bill McCleary, Brigham Young University

Sue Merkel, Cornell University

David Mullin, Tulane University

Martin Polz, Massachusetts Institute of Technology

Anna-Louise Reysenbach, Portland State University

Erle S. Robertson, University of Michigan Medical School

Pratibha Saxena, University of Texas, Austin

Imke Schroeder, University of California, Los Angeles

Edward Simon, Purdue University

Mitchell L. Sogin, The Marine Biological Laboratory, Woods Hole

Kevin R. Sowers, Center for Marine Biotechnology

Sujatha Srinivasan, University of Washington

Valley Stewart, University of California, Davis

Dave Westenberg, University of Missouri, Rolla

Christine White-Ziegler, Smith College

Brief Contents

Contents

PART II Microbial Physiology: Nutrition and Growth

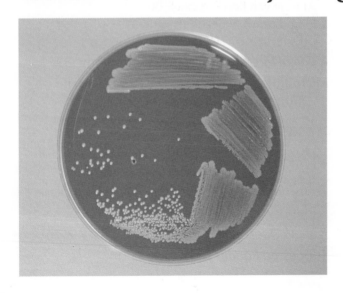

5 Isolation, Nutrition, and Cultivation of Microorganisms 103

PART III   *Microbial Physiology: Metabolism*

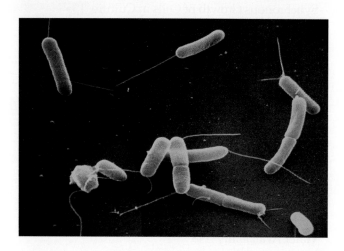

8   Cellular Energy Derived from Chemicals 165

9   Cellular Energy Derived from Light 185

10   Biosynthesis of Monomers 201

PART IV Genetics and Basic Virology

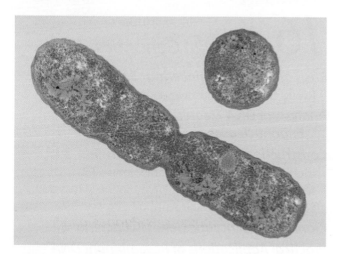

PART V   *Microbial Evolution and Diversity*

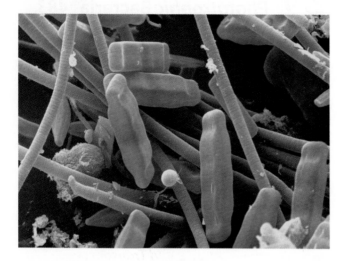

17   Taxonomy of Bacteria and Archaea   359

18   Archaea   385

19   Nonphotosynthetic Proteobacteria   409

PART VI *Microbial Ecology*

PART VII *Immunology and Medical Microbiology*

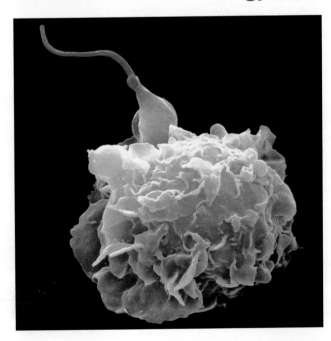

27 *Immunology* **639**

PART VIII *Applied Microbiology*

PART I

The Scope of Microbiology

Chapter 1

Evolution, Microbial Life, and the Biosphere

In biology nothing makes sense, except in the light of evolution.
— THEODOSIUS DOBZHANSKY, 1970

Our solar system originated through physical and chemical processes. After Earth formed, organisms originated and evolved. These first organisms were microorganisms, and they had a profound impact on Earth and the formation of its biosphere, the shell about Earth where life occurs. Certain bacterial groups played especially crucial roles early on in Earth's development. For example, geochemical and fossil evidence indicates that the production of oxygen in the atmosphere was due to the photosynthetic activity of cyanobacteria. The evolution of microorganisms that produced oxygen was of monumental significance, because all plant and animal life that exists today requires oxygen. Thus, plants and animals could evolve only because microorganisms evolved first. In this chapter we discuss Earth's origin, the evolution of life, and the importance of microorganisms to life on Earth.

ORIGIN OF EARTH AND LIFE

The origin of Earth and the evolution of life on our planet has been a long process. The universe, which is estimated to have an age of 18 Ga (1 Ga, a giga-annum, is 10^9 years), began with a "Big Bang" that produced two principal elements, hydrogen (^{1}H) and helium (^{4}He), with smaller amounts of other light (low atomic weight) elements. Following the Big Bang, the universe expanded, as it continues to do today. At its periphery the original light elements condensed to form clouds of gases and dust. In the clouds heavier elements evolved from the lighter ones.

Our solar system was formed by an accretional process in which micrometer-sized dust particles collided to form centimeter-sized bodies. These particles were located in a planar disk that orbited the sun. The accretional process continued as the dust and rock particles aggregated to form boulders and larger bodies that eventually attained the size of the planets. Thus, ultimately by gravitational contraction, our solar system, with the Sun, Earth, and other planets, formed about 4.5 Ga ago. The final stages of accretion involved collisions between large bodies at high velocities. A major collision between early Earth and a Mars-sized object resulted in the formation of our moon and Earth.

The 600 million years following Earth's formation is called the era of "heavy bombardment" because of the high frequency of collisions between Earth and large asteroids and comets. Some of these collisions, such as the one responsible for the formation of the moon, were so violent that they heated Earth to sterilizing temperatures. Even collisions with bodies only 100 kilometers in diameter could result in sterilization within the planet to depths of several kilometers. Furthermore, the heat from these collisions would have removed volatile substances such as water.

During its first 600 million years, Earth was not a hospitable planet for life. Water was not initially available. It was brought to Earth by comets and asteroids that came from farther out in the solar system. Once water was available and the era of heavy bombardment had ended, conditions became conducive to the evolution of living organisms.

Scientists have determined the date of Earth's formation by studying slowly decaying radioactive isotopes, whose decay occurs at a constant rate independent of temperature and pressure. The isotope most relied on for dating such ancient events is potassium (^{40}K), which decays to argon (^{40}Ar) with a half-life, the time required for half the radioactivity to decay, of 1.26 billion years. Radioisotopic methods are also used for dating strata in sedimentary rocks and therefore offer a means of dating fossilized life forms in rocks.

Fossil Evidence of Microorganisms

By the nineteenth century it was known that fossils were the remains or impressions of plants and animals that had been preserved in sedimentary rocks. Accurate dating methods were not yet available, so the estimated dates were only guesses. We now know that some of the organisms that became fossilized, such as the dinosaurs, lived and became extinct millions of years ago. By examining fossils, paleontologists came to several conclusions about the evolution of life. They noted that fossils nearest the surface, that is, in the most recently deposited sedimentary rocks, were structurally more complex than those in deeper layers. Fossils found in deeper strata were increasingly simple in structure, fossils of very simple, extinct animals such as trilobites. The gradation of complexity, from simple organisms in the most ancient rocks to more complex forms in the more recent sedimentary rocks, argued for an evolutionary process in which more complex forms of plants and animals arose from simpler organisms. Geologists and paleontologists worked hand in hand to develop time scales for sedimentary rock deposits and named the various time periods of Earth's history based upon fossil records (**Table 1.1**).

At the time of Charles Darwin (1809–1882) the fossil record was being carefully studied, but there was no means of estimating ages. Today, we realize that the fossil record for plants and animals extends back to a period of 570 million years ago (570 mya). Rocks older than that contain no plant and animal fossils. This time and all earlier times became known as the Precambrian era (Table 1.1).

In the 1950s two American scientists, Stanley Tyler and Elso Barghoorn, made a startling discovery. They reported to the scientific world that they had found fossils of microorganisms in sedimentary rocks dated to the Precambrian era (**Box 1.1**). This important discovery provided the first convincing evidence that the earliest life forms on Earth were microorganisms.

The microbial fossils were discovered in laminated sedimentary rocks called **stromatolites** (**Figure 1.1A**). Many of the multilayered stromatolite structures contain calcium carbonate along with the fossils of filamentous microorganisms (Figure 1.1B). Living stromatolites still exist on Earth today. The columnar stromatolites occur in intertidal marine areas such as Shark Bay, Western Australia (**Figure 1.2**). These living

Table 1.1	Geological timetable on Earth		
Eon/era	**Period**	**Years Before Present (millions)**	**Major Events**
Precambrian	Hadean	4,500	Heavy bombardment period
	Archaean	3,800	First sedimentary rocks
	Proterozoic	2,600	Appearance of O_2
Paleozoic	Cambrian	570	Animals evolve
	Ordovician	500	
	Silurian	440	Land colonization by plants, animals
	Devonian	395	Fish diversify
	Carboniferous	345	Reptiles evolve; large "fern" forests
	Permian	280	Mass extinction at end
Mesozoic	Triassic	245	Early dinosaurs; first mammals
	Jurassic	190	Plants and animals diversify
	Cretaceous	145	Mass extinction at end; 75% of species lost
Cenozoic	Tertiary	65	Plant and animal radiation
	Quaternary	1.8	Humans evolve

(A)

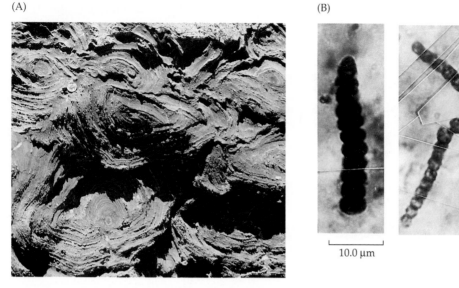

(B)

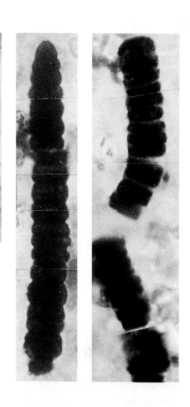

10.0 μm

Figure 1.1 Stromatolites and mat communities

(A) Fossil columnar stromatolites from Glacier National Park, shown in cross section. A U.S. quarter is shown for size. (B) Filamentous microbial fossils observed in sections of 860 million-year-old stromatolites from the Bitter Springs formation in central Australia. A, courtesy of Beverly Pierson; B, courtesy of William Schopf.

Milestones Box 1.1   The Discovery of Microbial Fossils

In the early twentieth century an American geologist, Charles Doolittle Walcott, was studying Precambrian sedimentary rocks in Glacier National Park in northwestern Montana. He noted that some had curious undulating wavelike structures (these are now called stromatolites) and postulated that they were fossilized forms of Precambrian reefs. Contemporary scientists doubted his theory, and it remained untested for many years.

American micropaleontologists Stanley Tyler from the University of Wisconsin and Elso Barghoorn from Harvard University were the first to test Walcott's hypothesis. They were studying stromatolites from 1 to 2 billion-year-old Precambrian Gunflint chert deposits from the Great Shield area in the Great Lakes vicinity of North America. When they examined sections of these stromatolites using the light microscope they discovered microbial fossils.

Microbiologists were incredulous when Tyler and Barghoorn first reported their observations in the 1950s and 1960s, as most microbiologists did not believe microbial fossils existed. However, Tyler and Barghoorn's clear photomicrographic evidence convinced a whole generation of skeptical microbiologists. More recently, micropaleontologists have discovered microbial fossils in stromatolites 3.5 Ga old, dating back to about 1 Ga after the origin of Earth.

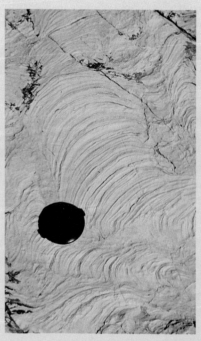

The undulating layers of this sedimentary rock of Glacier National Park are stromatolites containing fossil microorganisms. The lens cap serves as a scale marker. Courtesy of Beverly Pierson.

**Figure 1.2 Living columnar stroma-
tolites, Shark Bay, Western Australia**
The largest stromatolite shown here is
about 1 m in diameter. Courtesy of
Beverly Pierson.

stromatolites contain microorganisms
that deposit calcium carbonate and
other minerals, forming the successive
layers of the stromatolite structure.
Other precursors of fossil stromatolites
are microbial **mat communities**, which
occur extensively in intertidal marine
environments throughout the world
(**Figure 1.3A**). Photosynthetic microorganisms, including cyanobacteria, and
photosynthetic bacteria are found in
distinct layers in living stromatolites (Figure 1.3B–D).
These mat communities are flatter and broader than the
columnar-shaped classical stromatolites, but they are produced in similar saline, intertidal environments by similar microorganisms. Evidently during some major geological events living stromatolites became fossilized and
preserved in sedimentary deposits.

Fossil microorganisms have been dated at 3.5 Ga
before the present and therefore are found in some of
the earliest sedimentary deposits on Earth. Other evidence for early microbial life comes from studies of
chemicals left by microbial activities in early sedimentary rocks. These chemicals are found in organic materials, called **kerogen**, deposited in ancient rocks.

The Ishua formation in Greenland, which is more than
3.5 Ga old, is one of the oldest sedimentary deposits
known. Over the long period of time following the deposition of organic carbon by microorganisms, the organic matter was altered considerably to form the kerogen.
Geochemists who have examined the Ishua kerogen note
that it has a significantly higher ratio of ^{12}C to ^{13}C than
does the associated inorganic carbon from the same strata. This is indicative of a biological process that deposited organic material that was eventually transformed to
kerogen. This dates the biological process to 3.5 Ga ago.

How can the occurrence of high concentrations of ^{12}C
in the kerogen be attributed to biological activity? Here
is the reasoning. Some organisms, called **autotrophs**, use
carbon dioxide as a carbon source for growth and from
this produce organic cellular material called **biomass**.
These organisms selectively use $^{12}CO_2$ in preference to
its heavier, stable isotopic form, $^{13}CO_2$, which is also
present in the environment. As a result, by a process
called **isotopic fractionation**, the biomass becomes
enriched in the lighter isotope (^{12}C) leaving behind the
heavier isotope in the environment. Determination of the
relative amounts of ^{12}C and ^{13}C isotopes in the kerogen

and inorganic carbon deposits of a sample can therefore
be used to determine whether biological activity is
involved in geochemical processes (see Chapter 24).

Thus, both the fossil and the geochemical evidence
suggest that microorganisms originated on Earth within a billion years of its formation. In fact, during the 3
billion years between 3.5 to about 0.5 Ga ago, living mat
and stromatolite communities covered vast areas of
intertidal zones on the planet and were likely the dominant feature of life on Earth.

Mat communities are still common in intertidal areas,
but the columnar stromatolites are much rarer. Presumably the evolution of predatory animals led to the
selection of organisms that preyed on the microorganisms in stromatolite communities, and this led to the
demise of these microorganisms in many areas on Earth.
So, except in special environments such as Shark Bay,
with its high salt concentration that is inhibitory to predators, columnar stromatolites have disappeared.

Origin of Life on Earth

Early fossils provide evidence that microbial life existed
on Earth within a billion years of its formation, but we
have many questions about this early period. How did
life originate? What were the first forms of life? What
were the conditions on Earth that permitted the origin
of life? These are important and intriguing questions,
but they cannot be answered by direct observation.

Nonetheless, from what we know of life and the early
history of the planet, the process can be partially reconstructed. For example, we know that life cannot exist
without liquid water. This means that, at the time life
originated, the temperature somewhere on Earth must
have been between 0°C and 100°C (at atmospheric pressure). Furthermore, we know that the atmosphere was
anoxic, that is, without free oxygen gas (O_2). Oxygen
could not have formed chemically in any great amount,

(A)

(B)

(C)

Four layers of photosynthetic organisms are visible (from top to bottom): cyanobacteria, two layers of purple sulfur (of different species), and green sulfur bacteria.

(D)

Multiple years of bacterial buildup are visible. The green surface layer contains living cyanobacteria.

Figure 1.3 Microbial mat communities
(A) This marine intertidal community in Massachusetts, called Sippewisset Marsh, contains a microbial mat community. Some areas are sectioned off by ribbons for research purposes. (B) The mat community just beneath the surface is made visible by cutting through the upper layers of the sand using a razor blade, shown here to provide a size scale. (C) A vertical section of the mat showing the four layers of photosynthetic microorganisms. Each layer is about 1 mm thick. The Sippewisset Marsh mat forms during the summer months; winter storms disrupt it, and a mat re-forms the next summer season. Other mat communities remain stable for many years, such as this one (D) at Laguna Mormona (Laguna Figueroa), Baja California del Norte, Mexico. A–C courtesy of Beverly Pierson; D, courtesy of William Schopf.

and certainly it did not make up 20% of the atmosphere as it does today.

Another precondition for the origin of life is the presence of organic compounds. It is inconceivable that cells could have originated de novo in the absence of organic compounds, which are part and parcel of all living organisms and biological processes. Thus, an important question is, can organic compounds such as sugars and amino acids be produced in the absence of organisms, that is, **abiotically**? The first experiments to address this question were conducted by Stanley Miller in 1953. He constructed an apparatus for the interaction of a mixture of gases thought to be present in Earth's early atmosphere. The experimental device mimicked prebiotic conditions (**Figure 1.4**).

The sterile apparatus contained 500 ml of water, representing the "ocean," and an "atmosphere," consisting of an anoxic gas mixture of methane, hydrogen, and ammonia. The water was boiled, and steam rose into the atmosphere to mix with the gases. A condenser subsequently cooled the gases to produce liquid water, that is, "rain." Miller included as a source of energy a 60,000 volt spark discharge that represented lightning in the atmosphere. The gases and water were recirculated and the anoxic process was run continuously.

In a matter of a few days of operation, Miller's apparatus yielded a dark tarry liquid. This material was analyzed and found to contain, in addition to tarry hydrocarbons, a variety of other organic compounds such as glycine, alanine, lactate, glycolate, acetate, and formate,

(A)

Figure 1.4 Diagram of Stanley Miller's apparatus
(A) Stanley Miller shown observing his apparatus for generating organic material. (B) Using this apparatus, Miller and others produced organic compounds from inorganic sources. Photo ©Roger Ressmeyer/CORBIS.

(B)

1 Steam produced in boiler passes into the spark chamber.

Spark chamber

80°C

2 Steam joins gas mixture containing CH_4, NH_3, and H_2.

3 Spark discharge mimics the effect of lightning.

Condenser

Boiler

Heat

4 Water and gases recirculate through the apparatus; reaction products condense in the collecting chamber.

as well as smaller amounts of other organic compounds. Thus, organic materials were formed under anoxic, abiotic conditions that resembled those found on the early Earth.

However, we know that the gas mixture used by Miller does not best represent that of the early atmosphere; similar experiments have been conducted by other investigators, who used gas mixtures with compositions more closely resembling those of the atmosphere of early Earth. These are the gases, called fumarolic gases, that are released from Earth's hot mantle by volcanoes. In addition to the gases and water that Stanley Miller used, the fumarolic gases include large amounts of carbon dioxide, nitrogen, sulfur dioxide, and hydrogen sulfide. Ultraviolet light, which was intense on early Earth, has been successfully used as an alternative to Miller's spark discharge as an energy source. In addition, volcanism was more prevalent on early Earth, because the nuclear reactions in its interior core produced more heat than they now do. Thus, the heat from within Earth's crust would have influenced many of these early reactions. In all experiments in which conditions were anoxic, as they were on early Earth, organic compounds similar to those found by Miller were synthesized.

The overall results of these Miller-type experiments indicate that organic compounds can readily be synthesized from inorganic compounds under conditions that resemble Earth's prebiotic environment. However, we also know that organic compounds are synthesized in intergalactic space. These organic compounds, including amino acids and polycyclic aromatic hydrocarbons, would have been brought to Earth by comets and meteors. Thus, a large variety of organic compounds would have been present on Precambrian Earth in the so-called **primordial soup**.

We now realize that it is unlikely that life originated and evolved in shallow aquatic habitats, because these habitats would have been continually susceptible to destruction during the period of heavy bombardment. Many scientists now believe that life evolved either in deep sea environments such as hydrothermal vents (see Chapters 24 and 25) or in subterranean environments, because these environments were less likely to be destroyed by asteroid impacts.

The most difficult questions still remain unanswered. How did the first cell originate? What were its characteristics? Was the first cell a progenitor of all life.

TRACING BIOLOGICAL EVOLUTION

How can we trace biological evolution? Two approaches have been used. The first is to look at the fossil evidence for microorganisms in sedimentary deposits. This approach, discussed earlier in the chapter, requires the examination of sedimentary rocks for evidence of fossilized microorganisms or their chemical traces or for evidence of their geochemical activity.

The second approach is to construct an evolutionary tree based on knowledge about current living organisms. This is accomplished by analyzing the sequences of the monomers (smaller units) of large molecules called macromolecules such as deoxyribonucleic acid

(DNA), whose monomers are purines and pyrimidines, or protein, with amino acid monomers. The sequences in these macromolecules provide the necessary information to trace the evolutionary history of organisms, as discussed in greater detail in Chapter 17.

However, before further discussion of the evolution of life, we need to provide some background on the characteristics of organisms that live on Earth today. The first characteristic of all organisms is that they are composed of one or more cells, this is the **cell theory of life**.

Cell Theory: A Definition of Life

Microorganisms can be divided into four groups on the basis of form and function: bacteria, fungi, and algae and protozoa (protists). Like plants and animals, all microorganisms consist of one or more cells. Or to put it another way, if something is living, it must be cellular. Viruses, also discussed in this book, are not cellular and therefore they are not regarded as living organisms. Nonetheless, they are important biological agents that develop only as intracellular parasites of organisms, including microorganisms.

The **cell** is the fundamental unit of organisms and has characteristic functional and structural features. These functions include **metabolism**, the chemical reactions and physical activities by which cells obtain and transform energy and synthesize cell material for growth. Metabolism is accomplished by biochemical reactions catalyzed by proteins called **enzymes**. The other basic function of cells is **reproduction**, the process by which cells duplicate themselves to produce progeny.

Cells have three major groups of structural components (**Figure 1.5**):

- The cytoplasm, the aqueous fluid of the cell in which most of the enzymatic and metabolic activities occur. For example, **ribosomes,** small structures responsible for protein synthesis, are located in the cytoplasm.

- A central nuclear area that contains deoxyribonucleic acid (DNA), the hereditary material that is duplicated during reproduction.
- A cell membrane, or plasma membrane, the boundary between the cell's cytoplasm and its environment. The cell membrane consists of lipids and proteins.

Many microorganisms also contain a layer external to the cell membrane that is referred to as the **cell wall**, a rigid structure that confers shape on the cell. All fungi have cell walls, as do most algae and bacteria. Most protozoa lack cell walls, so their bounding structure is the cell membrane. The chemical composition and structure of the cell walls of microorganisms differ from one group to another. Chapter 4 covers these structures and their functions in greater detail.

Unlike plants and animals, which are all multicellular (containing millions of cells), many microorganisms consist of a single cell and are therefore called **unicellular**. Most but not all bacteria and protozoa are unicellular. Only one group of fungi is unicellular—the yeasts. Plants and animals are macroscopic; they contain many cells organized into tissues and organs, neither of which are found in microorganisms.

The Tree of Life

The macromolecules that have been most useful in tracing evolution are found in the ribosomes, which are responsible for protein synthesis in all organisms, microorganisms as well as plants and animals. The ribosome is a complex structure containing RNA and protein (see Chapter 4). Studies of ribosomal RNA (rRNA) molecules indicate that they have changed very slowly during evolution. Because of their highly conserved nature and universal occurrence, rRNA molecules have been used in the study of the evolutionary relatedness among organisms. The 16S and 18S rRNA molecules are the most commonly used (where S refers to the Svedberg unit, which relates to the mass and density of a molecule).

As a consequence of these studies, three major domains of organisms are now recognized by biologists: the *Bacteria*, the *Archaea*, and the *Eucarya* (**Figure 1.6**). The *Bacteria* contains many of the common microorganisms encountered in typical soil and aquatic environments and includes those that are known to cause disease. The *Archaea* comprises a separate group of microorganisms, some of which live in saturated salt environments or

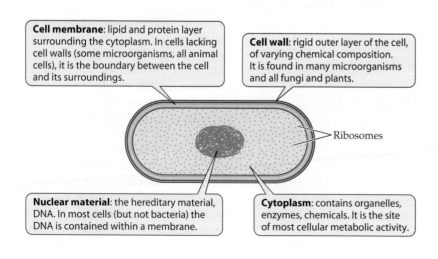

Cell membrane: lipid and protein layer surrounding the cytoplasm. In cells lacking cell walls (some microorganisms, all animal cells), it is the boundary between the cell and its surroundings.

Cell wall: rigid outer layer of the cell, of varying chemical composition. It is found in many microorganisms and all fungi and plants.

Ribosomes

Nuclear material: the hereditary material, DNA. In most cells (but not bacteria) the DNA is contained within a membrane.

Cytoplasm: contains organelles, enzymes, chemicals. It is the site of most cellular metabolic activity.

Figure 1.5 Cell structure
The diagram shows the four major components of a typical cell: cell wall, cell membrane, cytoplasm, and nuclear area.

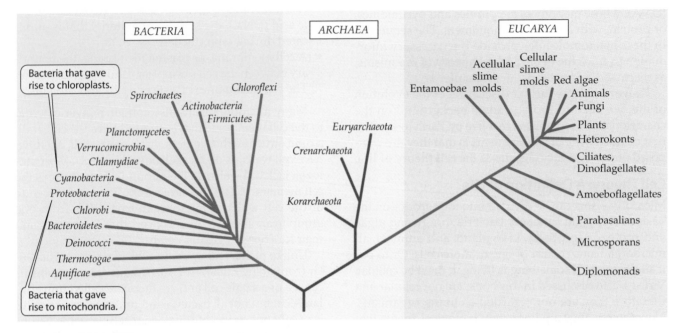

Figure 1.6 Tree of Life
This diagram shows the evolutionary tree of various groups of organisms based on 16S and 18S rRNA sequence analysis. The two prokaryotic domains are the *Bacteria* and *Archaea*. All eukaryotic microorganisms are placed in a separate domain, the *Eucarya*, along with the plant and animal "kingdoms." The *Eucarya* contains many "kingdoms" of microorganisms, including the fungi and various protists. The *Bacteria* and *Archaea* also contain many "kingdoms," which in this book we call phyla. The *Bacteria* contains at least 30 phyla, many of which have never been studied in the laboratory.

high-temperature environments. The *Eucarya* contains the microbial groups fungi, algae, and protozoa as well as plants and animals. The major differentiating characteristics among these organisms are shown in **Table 1.2**. It is noteworthy that the three-domain system is the first truly scientific classification of life (see Chapter 17).

Biologists have known for a long time that the cell types of the *Eucarya* are structurally different from those of the *Bacteria* and *Archaea*; the cells of the *Eucarya* are called eukaryotic and those of the bacteria prokaryotic. In the next section we discuss the differences between eukaryotic and prokaryotic cell structure, or **morphology**, and compare other major features of eukaryotic and prokaryotic microorganisms.

PROKARYOTIC VERSUS EUKARYOTIC MICROORGANISMS

When examined under the light microscope, bacteria appear different from eukaryotic microorganisms (**Figure 1.7**). Bacterial cells are usually very small and have no apparent nucleus. In contrast, cells of algae, protozoa, fungi, plants, and animals are typically much larger and have a distinct nucleus.

These differences noted by observations with the light microscope are borne out by more detailed examination using the electron microscope. Microorganisms can be sliced into very thin sections and examined at high magnification with the transmission electron microscope (TEM) (see Chapter 4). When viewed in this manner, the structural differences between bacteria (**Figure 1.8**) and other microorganisms (**Figure 1.9**) are striking. Bacteria have a much simpler cell structure and are referred to as **pro-**

Table 1.2	Major differentiating characteristics of the three domains of life		
	Bacteria	*Archaea*	*Eucarya*
Nuclear membrane	No	No	Yes
Plastids	No	No	Yes
Peptidoglycan cell walls	Yes[a]	No	No
Membrane lipids	Ester-linked	Ether-linked	Ester-linked
Ribosome size	70S	70S	80S

[a]Three bacterial groups, the chlamydia, planctomycetes, and mycoplasmas, lack cell wall peptidoglycan (the structure of this material is discussed in Chapter 4).

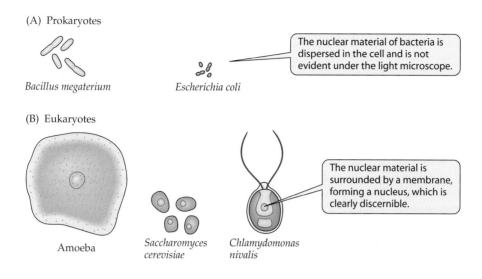

(A) Prokaryotes

Bacillus megaterium

Escherichia coli

> The nuclear material of bacteria is dispersed in the cell and is not evident under the light microscope.

(B) Eukaryotes

Amoeba

Saccharomyces cerevisiae

Chlamydomonas nivalis

> The nuclear material is surrounded by a membrane, forming a nucleus, which is clearly discernible.

Figure 1.7 Drawings of representative microorganisms, as they appear by light microscopy
The two examples of bacteria are a large rod, *Bacillus megaterium*, and a small rod, *Escherichia coli*. The eukaryotic organisms are an amoeba (a protozoan), a yeast (*Saccharomyces cerevisiae*), and an alga (*Chlamydomonas nivalis*). Note the cup-shaped chloroplast and the two flagella of *C. nivalis*.

karyotic (from the Greek meaning "before nucleus") organisms. In contrast, algae, fungi, and protozoa are called **eukaryotic** ("good nucleus" or "true nucleus"). **Table 1.3** lists the major differences between these two basic types of cellular organization. As the terms imply, the single major difference between these two cell types is related to their nuclear material (**Box 1.2**). The nucleus of the cell of a eukaryotic microorganism (as well as plants and animals) is bounded by a membrane referred to as a **nuclear envelope** or **nuclear membrane** (Figure 1.9). In bacteria or prokaryotic organisms, the nuclear material, which appears as a central fibrous mass in thin sections, is not bounded by a membrane but is in direct contact with the cytoplasm (Figure 1.8).

Other differences exist between prokaryotic and eukaryotic cells, some structural, others genetic and physiological. The nuclear material of prokaryotes typically consists of a single type of DNA molecule, called a **chromosome**. More than one copy of it may be present, depending on how fast the organism is growing. Thus,

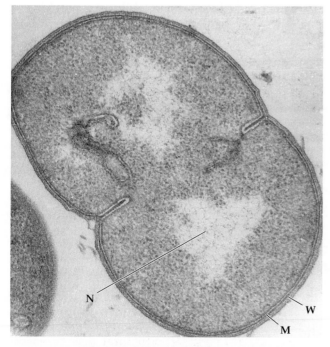

Figure 1.8 Cross section of a bacterial cell
This electron micrograph of a thin section of *Sporosarcina ureae* shows the cell wall (W), cell membrane (M), and nuclear material (N), which appears as fibrous matter dispersed in the cytoplasm. ©T. J. Beveridge/Biological Photo Service.

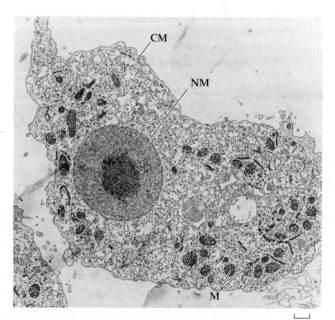

Figure 1.9 Cross section of a eukaryotic cell
A protozoan of the genus *Acanthamoeba*, showing the cell membrane (CM), nuclear membrane (NM), and mitochondria (M). Courtesy of T. Fritsche.

10.0 μm

Table 1.3   **Major differentiating characteristics of prokaryotes and eukaryotes**

Characteristic	Prokaryote	Eukaryote
Nuclear structure and function		
Nucleus with membrane	No	Yes
Chromosomes	One	Two or more
Mitosis	No	Yes
Sexual reproduction	Rare; only part of genome involved	Common; all chromosomes involved
Meiosis	No	Yes
Cytoplasmic structures		
Mitochondria	No	Yes[a]
Chloroplasts	No	Yes (if photosynthetic)
Ribosomes	70S	80S[b]
Typical cell volume	<5 μm^3	>5 μm^3

[a]A few lack mitochondria.
[b]Some rare, primitive eukaryotic microorganisms have 70S ribosomes.

rapidly growing cells might have two or four copies of the DNA molecule, but all copies are identical. Some bacteria also have nonchromosomal DNA in their cells called **plasmids**, which are discussed in greater detail in Chapter 15. Plasmids are smaller than the chromosome but contain genes that are often significant to the bacterium.

In contrast, the nucleus of eukaryotic organisms contains many separate chromosomes, each with its own genetic material. Thus, bacteria can be regarded as typically having a single chromosome and eukaryotic microorganisms as having more than one chromosome. To ensure orderly, accurate, and precise delivery of their multiple chromosomes during the process of cell division, eukaryotic organisms undergo **mitosis**. In this process, each chromosome replicates and aligns along the division axis of the cell before asexual cell division

occurs (**Figure 1.10**). This elaborate physiological and morphological orchestration does not occur in prokaryotes.

Other Morphological Differences

Ribosomes appear as granules (about 5 nm in diameter) in the cytoplasm. Prokaryotic ribosomes are called 70S ribosomes. Eukaryotes, with rare exceptions in some protozoa, have slightly larger 80S ribosomes (see Chapter 4). Molecular differences in the RNA and protein of ribosomes account for the differences in size.

One of the striking features of eukaryotes is their organelles, small structures in the cytoplasm. There are several types of organelles. All are distinct compartments surrounded by one or more membranes, which, like the cell membrane, contain both protein and lipid. The most common organelle of this type, found in almost all eukaryotic cells, is the mitochondrion (**Figure 1.11**). The **mitochondrion** is the site of respiratory activity in eukaryotes. Mitochondria have their own internal DNA, cytoplasm, and ribosomes. One exciting fact of cell biology is that the DNA of the mitochondrion is similar to prokaryotic DNA, that is, it has no nuclear envelope. Furthermore, the ribosomes of the mitochondrion are 70S in size, like those of prokaryotes. These features and other lines of evidence (see "Evolution of Eukaryotes" on page 17) suggest that the mitochondrion evolved from a bacterium that developed a close interdependence or **symbiotic association** with another cell over 1 billion years ago.

Milestones Box 1.2   **Separating the Organisms of Earth into Two Categories on the Basis of Cell Structure**

Although his views were largely ignored in the 1930s, the French biologist E. Chatton noted the differences in cellular structure between "higher" and "lower" forms of life. He coined the terms "eukaryotic" and "prokaryotic" based on his light microscopic observations of the differences between the cells of higher organisms and bacteria (see Figure 1.7). Only after the invention of the electron microscope (in the late 1930s) and the subsequent development of appropriate procedures to thin-section organisms (1950s to 1960s) did other biologists confirm the detailed differences between these two types of cellular organization. In addition to these morphological features, a number of other differences were also discovered that permitted the clear distinction of these two types of cells. The major features that distinguish prokaryotic from eukaryotic cells were eloquently stated in an important publication by Roger Stanier and C. B. van Niel in 1962.

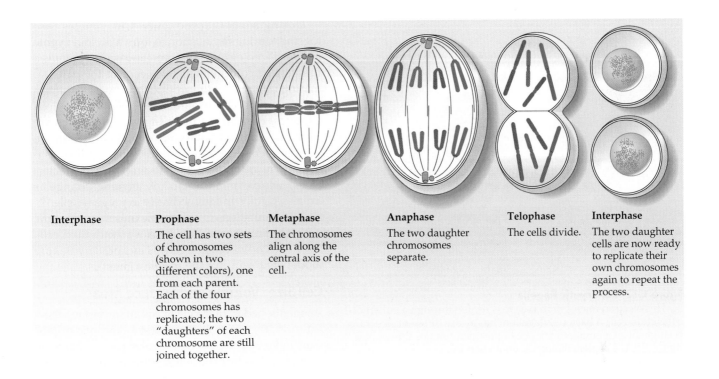

Interphase	Prophase	Metaphase	Anaphase	Telophase	Interphase
	The cell has two sets of chromosomes (shown in two different colors), one from each parent. Each of the four chromosomes has replicated; the two "daughters" of each chromosome are still joined together.	The chromosomes align along the central axis of the cell.	The two daughter chromosomes separate.	The cells divide.	The two daughter cells are now ready to replicate their own chromosomes again to repeat the process.

Figure 1.10 Mitosis
In mitosis, a dividing eukaryotic cell duplicates its chromosomes and distributes one copy to each of the newly forming daughter cells. This particular cell has two sets of chromosomes.

The chloroplast is the site of photosynthesis in eukaryotes. Like the mitochondrion, the **chloroplast** is a membrane-bounded organelle found in the cytoplasm. It also resembles the mitochondrion in that its DNA has no nuclear envelope and its ribosomes are 70S in size. However, unlike mitochondria, it also has internal membranes containing the chlorophyll pigments involved in photosynthesis. Chloroplasts are found in algal and plant cells. They, too, are thought to be derived through an evolutionary process from a prokaryotic organism, in this case an organism from the photosynthetic group called the cyanobacteria (see Chapter 21).

The organelles of motility of eukaryotic cells—the flagellum and cilium—are larger and more complex than the flagellum of prokaryotes. A cross section of the eukaryotic flagellum reveals an elaborate fibrillar system called the "9 + 2" arrangement, with nine outer doublets of fibrils called microtubules and an inner pair (**Figure 1.12**). In contrast, the prokaryotic flagellum has a single fibril when viewed in cross section; the thread is of such a fine diameter that a single flagellum cannot be seen by light microscopy. Eukaryotic flagella and cilia, in contrast, are readily observed with the light microscope (see the alga in Figure 1.7).

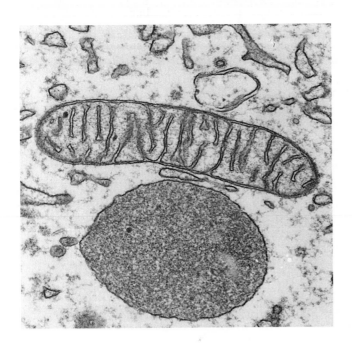

Figure 1.11 Mitochondrion
Electron micrograph of a mitochondrion from a eukaryotic microorganism, showing the outer membrane, the inner membrane folded into cristae, and the enclosed fluid, the matrix. ©Barry F. King/Biological Photo Service.

(A)   (B)

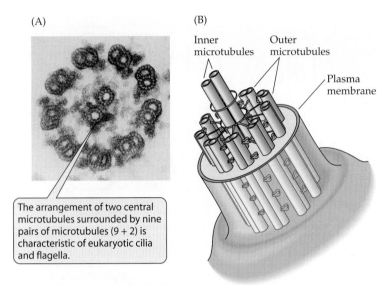

The arrangement of two central microtubules surrounded by nine pairs of microtubules (9 + 2) is characteristic of eukaryotic cilia and flagella.

Figure 1.12 Eukaryotic flagella
(A) This electron micrograph is a cross section through a eukaryotic flagellum. (B) A bacterial flagellum has a very different structure: its single fibril is smaller than one of the microtubules shown here. Photo ©W. L. Dentler/Biological Photo Service.

Reproductive Differences

All prokaryotes reproduce by asexual cell division. Cells simply enlarge in size, replicate their DNA (i.e., produce a second identical copy of their DNA), and divide to form two new cells, each containing a copy of the DNA molecule (**Figure 1.13**). Thus, prokaryotes have only one copy of DNA and are called **haploid**. Sexual reproduction is relatively rare in prokaryotes. Though many bacteria are able to exchange genetic material between mating types, this is not known to be a universal characteristic. As discussed later (see Chapter 15), this rarely results in the formation of a **diploid** cell, with one copy of the DNA molecule from each of the mating cells. A diploid cell has two copies of each chromosome, that is, two copies of each DNA molecule.

In contrast to prokaryotes, most eukaryotes exist as diploid organisms or have diploid stages in their life cycles. Thus, their cells have two sets of chromosomes, one set from the "male" and another set from the "female" mating types. For example, human body cells have 46 chromosomes. These exist as 23 paired chromosomes. Half, or one set of 23, is derived from the father and the other 23 from the mother. To generate reproductive cells, the number of chromosomes and amount of DNA are reduced by half. **Meiosis** is the process whereby, for example, the 46 human chromosomes are reduced to 23 in preparation for sexual reproduction. Meiosis results in the formation of haploid mating cells, called **gametes**—the sperm and the ovum, produced by male and female mating types, respectively (**Figure 1.14**).

During sexual reproduction the gametes fuse together during fertilization to form a diploid **zygote** (the fertilized egg). Therefore, the zygote contains a full genetic complement from each of the parental mating cells. Sexual reproduction is very common among eukaryotic organisms. Except for haploid gametes, the cells of most higher eukaryotes are diploid. Because prokaryotes contain only a single copy of each gene, genetic studies are much simplified. There are no dominant and recessive characteristics, which means that any genetic change is expressed fully and immediately in progeny cells. In contrast, mutations in eukaryotic cells may not show up in the next generation, because the diploid cells have two copies of each gene. Thus prokaryotes are model organisms for the study of genetics.

Cell Size: Volume and Surface Area

As mentioned earlier, cell size is an important characteristic for an organism. Most eukaryotic organisms have larger cells than prokaryotic organisms (Figure 1.7)—but there are some exceptions. For example, although typical bacterial cells range in diameter from 0.5 to 1.0 µm, some wider than 50 µm have been reported (**Box 1.3**). The cells of typical eukaryotes range in diameter from 5 to 20 µm, with most about 20 µm, although some species have larger ones. Specialized cells in multicellular organisms can be much larger. A human neuron can be as long as 1 m.

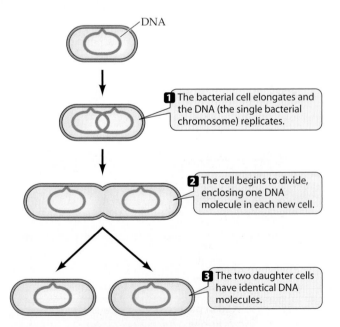

DNA

1 The bacterial cell elongates and the DNA (the single bacterial chromosome) replicates.

2 The cell begins to divide, enclosing one DNA molecule in each new cell.

3 The two daughter cells have identical DNA molecules.

Figure 1.13 Prokaryotic cell division
Though this process is analogous to mitosis in eukaryotic organisms (compare with Figure 1.10), mitosis involves complex structural features that are absent in bacteria.

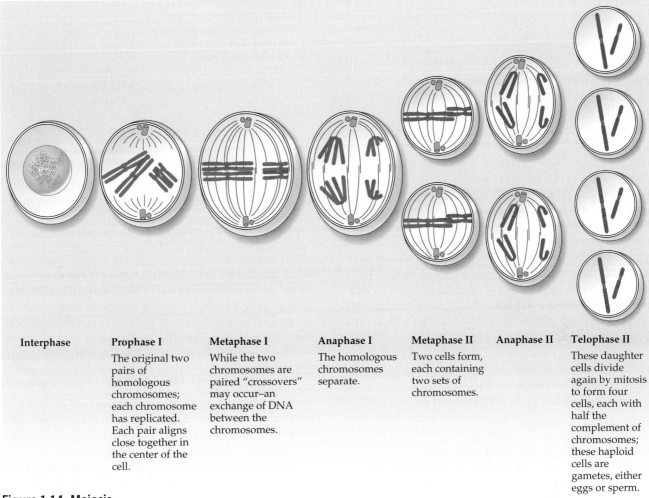

Interphase	Prophase I	Metaphase I	Anaphase I	Metaphase II	Anaphase II	Telophase II
	The original two pairs of homologous chromosomes; each chromosome has replicated. Each pair aligns close together in the center of the cell.	While the two chromosomes are paired "crossovers" may occur—an exchange of DNA between the chromosomes.	The homologous chromosomes separate.	Two cells form, each containing two sets of chromosomes.		These daughter cells divide again by mitosis to form four cells, each with half the complement of chromosomes; these haploid cells are gametes, either eggs or sperm.

Figure 1.14 Meiosis
In meiosis, a process occurring in organisms that undergo sexual reproduction, a diploid cell undergoes two rounds of division to form four haploid cells, the gametes. In this case, two pairs of chromosomes are shown.

Research Highlights Box 1.3   You Can't Tell a Bacterium by Its Size Alone!

Although most bacteria are very small, some are amazingly large. The largest bacterium we know of is *Thiomargarita*. Individual cells of this bacterium can be seen by the naked eye. The bacterium lives in the inter-tidal area off the coast of Namibia, in southwest Africa. Although it has not yet been isolated in pure culture, *Thiomargarita* is known to be a sulfur bacterium that lives by the oxidation of reduced sulfur compounds.

Three cells of *Thiomargarita*

A chain of spherical cells of *Thiomargarita* is lying next to a fruit fly, indicating their huge (for bacteria) size. Reprinted with permission from *Science*, Vol. 284, pp. 493–495 ©1999 AAAS.

6 mm

Many bacteria grow and reproduce at very rapid rates. Some can double in size or in number of cells in less than 10 minutes under optimal growth conditions. This implies that metabolic processes can be extremely rapid in these organisms. The rapid metabolic rate is due in part to the small size of bacteria. Their small size ensures that all the cytoplasm is in close proximity to the surrounding environment from which they derive their nutrients. The greatest distance between the cytoplasm and the growth environment is only 0.5 μm in a bacterium with a diameter of 1.0 μm, whereas it is 10.0 μm in a eukaryotic organism with a diameter of 20.0 μm.

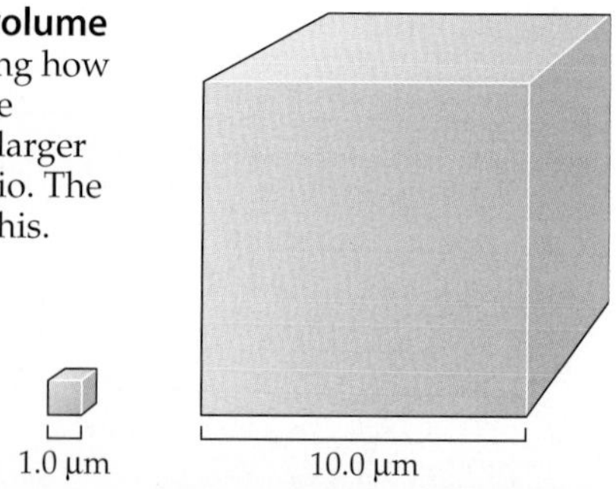

Another way to consider the close spatial relationship between the cytoplasm of a cell and its environment is to calculate the ratio of its surface area to its volume (**Figure 1.15**). Let's assume that a bacterial cell is cubical, with sides 1.0 μm in length (actually, one extreme salt-loving bacterium is a cube!); its surface area is 6.0 μm^2 and its volume is 1.0 μm^3. Thus, the ratio of its surface area to its volume (SA/V) is 6.0. In comparison, a hypothetical eukaryotic microorganism of the same shape with 10.0 μm sides has an SA/V of 0.6. This smaller value for the eukaryote indicates that it has a tenfold greater amount of cytoplasm per unit of cell membrane surface than does the smaller prokaryote. Given that nutrients for growth must enter the cell by crossing the cell membrane, more nutrients are available per unit of cytoplasm in the prokaryote (6.0) than in the eukaryote (0.6). The larger SA/V ratio enables faster metabolism and growth.

Microbial Nutrition

Algae and several groups of bacteria are **photosynthetic**, that is, like plants, they obtain their energy from sunlight. Also, like plants, they use carbon dioxide as their principal source of carbon for growth. This type of nutrition, which is based entirely on inorganic compounds, is referred to as **autotrophic** (self-nourishing or self-feeding). Algae are therefore called **photoautotrophic**, to indicate that they obtain their energy from sunlight and their carbon from carbon dioxide.

In contrast to algae, fungi obtain their energy directly from chemical compounds, not sunlight. This type of nutrition is referred to as **chemotrophic** (chemical feeding). Fungi require organic chemical compounds as their sources of energy and carbon. Such nutrition is termed **heterotrophic** (other or different feeding, as distinct from autotrophic) or **organotrophic**. Thus, fungi are **chemoheterotrophic**—they use chemical compounds as energy sources and organic compounds as carbon sources. Only dissolved organic carbon sources can pass through the cell walls of fungal cells. Thus, fungi are

Figure 1.15 Surface area and volume
Hypothetical cubical cells, showing how the ratio of surface area to volume (SA/V) varies with cell size. The larger cell has a much smaller SA/V ratio. The text explains the implications of this.

	1.0 μm	10.0 μm
Surface area (SA)	6.0 μm^2	600 μm^2
Volume (V)	1.0 μm^3	1,000 μm^3
SA/V	6	0.6

well known for their ability to use simple sugars and other dissolved substances as carbon sources. Some fungi can also degrade particulate organic materials, such as cellulose, by excreting enzymes that solubilize the organic material outside the cell; they then transport the dissolved compounds into the cell.

Like fungi, protozoa are chemoheterotrophic organisms, using organic compounds as sources of carbon and energy for growth. Typical protozoa, which lack cell walls, engulf bacteria and other microorganisms in much the same way that higher animals eat food. The protozoa's source of food is particulate organic material; this type of feeding is called **phagotrophic**.

As a group, bacteria are exceedingly diverse in their nutritional capabilities. Some are similar to algae in being photoautotrophic. Others are **photoheterotrophic**, that is, they can obtain energy from sunlight but use organic compounds as carbon sources. Most prokaryotic organisms are chemoheterotrophic, deriving both energy and carbon from organic compounds. One especially interesting group of bacteria can obtain energy by the oxidation of inorganic compounds, such as ammonia or hydrogen sulfide, and use carbon dioxide as their principal carbon source. This type of nutrition is termed **chemoautotrophic**, a nutritional category unique to these specialized bacteria. The four principal groups of microbes and their types of nutrition are shown in **Table 1.4**. Microbial nutrition is discussed in more detail in Chapter 5.

With this background in microbiology, we are ready to address more specifically the early evolution of organisms.

MICROBIAL EVOLUTION AND BIOGEOCHEMICAL CYCLES

Although we do not know which organisms were the first biological entities on Earth, various theories have been presented. Most scientists believe that the first forms of life were **anaerobic** (living in the absence of

Table 1.4	Microorganisms and their nutritional types		
Microbial Group	**Number of Cells per Organism**	**Cell Walls**	**Nutritional Type**
Algae	Usually one, some filamentous	Yes	Photoautotrophic
Protozoa	One	No	Chemoheterotrophic
Fungi	Filamentous, except yeasts (unicellular)	Yes	Chemoheterotrophic
Bacteria *Bacteria* and *Archaea*	Usually one, some multicellular	Yes[a]	Photoautotrophic, Photoheterotrophic, Chemoautotrophic, or Chemoheterotrophic

[a]A few bacteria, namely, the mycoplasmas and thermoplasmas, lack cell walls.

O_2), based on evidence that early Earth was anoxic. Many also believe that the earliest bacteria were **thermophilic** (heat-loving), living in high-temperature environments such as hydrothermal systems or deep within Earth's crust where it is hot. Evidence supporting this hypothesis is that the earliest branches in the Tree of Life contain thermophilic *Bacteria* and *Archaea* (Figure 1.6; see also Chapter 17).

Carl Woese from the University of Illinois calls the progenitor of microbial life the progenote, the prototypical precellular "organism" that gave rise to both the *Bacteria* and the *Archaea*, and ultimately the *Eucarya* as well. The progenote likely had a cell membrane that conferred the ability to concentrate important chemicals and carry out simple reactions.

Wolfram Zillig, a German biochemist, has proposed that the Progenote populations must have separated physically, possibly geographically, into two communities early in Earth's history and that this separation led to the evolution of the two main lines of descent—the *Bacteria* and the *Archaea*. However, all theories on the origin of life and the first microorganisms are speculative and will not be addressed in great detail here.

Possible Early Metabolic Types

The Russian evolutionist A. I. Oparin argued that the initial metabolic type was likely a simple heterotrophic bacterium. He reasoned that autotrophic organisms are inherently more complex, so they would not have evolved first. As he noted, although autotrophs can live on simple nutrients, they are more complex in that they need not only metabolic pathways for the generation of energy but also additional pathways to carry out carbon dioxide fixation, that is, the conversion of CO_2 into organic material. In contrast, simple fermentative heterotrophs require only a few enzymes for energy generation, and they could have lived on the organic compounds formed abiotically early in Earth's history.

Others have argued that early life forms might have been hydrogen bacteria, those that obtain energy from the oxidation of hydrogen gas. Both bacterial and archaeal hydrogen users are known. These organisms have simple nutritional requirements. Some grow autotrophically, generating energy from the oxidation of hydrogen gas and using carbon dioxide as a sole source of carbon (see Chapter 8). Furthermore, many are anaerobic and could have existed in an anoxic environment like that of early Earth.

Although photosynthetic bacteria may not have been the first organisms, it is believed that they evolved early. The ability to carry out photosynthesis using chlorophyll-type compounds probably evolved shortly after the split between the *Bacteria* and the *Archaea*. Several groups of the *Bacteria* carry out photosynthesis, whereas none of the *Archaea* produce chlorophyll compounds. The first photosynthetic organisms may have resembled the photosynthetic *Proteobacteria* or *Chlorobia*, two of the major lineages of *Bacteria*; this is consistent with recent molecular evidence on the origin of photosynthesis. These bacteria carry out photosynthesis anaerobically using hydrogen sulfide or elemental sulfur for carbon dioxide fixation (see Chapters 9 and 21):

$$(1) \ CO_2 + H_2S \rightarrow (CH_2O)_n + S^0$$

$$(2) \ CO_2 + S^0 \rightarrow (CH_2O)_n + SO_4^{2-}$$

where $(CH_2O)_n$ represents organic material (equations are not balanced).

The volcanic atmosphere of early Earth would have been ideal for these organisms, because it provided abundant quantities of carbon dioxide and hydrogen sulfide, the essential "ingredients." This type of photosynthesis is termed **anoxygenic photosynthesis**, because it proceeds in an anoxic environment without the production of oxygen gas. Most photosynthesis that occurs today, however, generates oxygen. Cyanobacteria are the only prokaryotic organisms that can carry out this oxygenic (oxygen-producing) photosynthesis.

Cyanobacteria and the Production of Oxygen

To understand the importance of cyanobacteria in the production of oxygen, we must first review their metabolism. The metabolism of the cyanobacteria is similar to that of the anoxygenic photosynthetic bacteria (see reaction (1) above). However, there is one major difference: cyanobacteria use water in place of hydrogen sulfide as the hydrogen donor. Thus, the overall equation for cyanobacterial photosynthesis is:

$$(3) \ CO_2 + H_2O \rightarrow (CH_2O)_n + O_2$$

This process is termed **oxygenic photosynthesis,** because oxygen is produced.

C. B. van Niel, a Dutch-born American microbiologist who studied photosynthetic bacteria, noted that the O_2 produced in reaction (3) must be derived from the water molecule rather than from the carbon dioxide. He concluded this based upon analogy to reaction (1) for anoxygenic photosynthesis. His hypothesis was confirmed when scientists used radiolabeled water, $H_2^{18}O$, to show that the label ended up in the oxygen produced ($^{18}O_2$); the label from $C^{18}O_2$ did not. Thus, the oxygen comes from a reaction referred to as the "water-splitting" reaction. This reaction, which is the key reaction of oxygenic photosynthesis, is found in all cyanobacteria, algae, and plants.

The cyanobacteria evolved about 2.5 to 3.0 Ga ago, when Earth's atmosphere still lacked O_2. But we know that by 2.5 to 1.5 Ga ago free oxygen was present in the atmosphere, because this was the time that iron oxides were deposited in geological strata called **banded iron formations**. The bands of these formations are alternating millimeter-thick layers of quartz and iron oxides, partially oxidized forms of iron (FeO and Fe_2O_3) that could have been produced only in the presence of atmospheric oxygen. Banded iron formations are not formed today, because the concentration of oxygen in the atmosphere and in the oceans is too high. Instead, contemporary iron deposits form red beds, so named because they contain hematites (Fe_3O_4), a more highly oxidized form of iron that gives them their red color.

From the locations of the iron oxides, it is concluded that the banded iron formations were produced during the period in which O_2 was first formed but before significant concentrations accumulated in the atmosphere, that is, between about 2.5 and 1.7 Ga ago. The O_2 needed to oxidize the iron is believed to have come from oxygenic photosynthesis first carried out by cyanobacteria.

Initially, when the cyanobacteria first began producing oxygen, its concentration in the atmosphere would have remained very low. This is because oxygen is highly reactive chemically and would have combined with the large amounts of highly reduced compounds that existed on Earth at the time. These reduced compounds, such as ferrous iron and sulfides, would have reacted with the free oxygen, preventing it from accumulating rapidly in the atmosphere. Therefore, the oxygen concentration in the atmosphere increased very gradually over the past 2 to 3 billion years to reach its present level of about 20% of atmospheric gases.

One of the recent exciting discoveries about cyanobacteria is that some of them can also carry out anoxygenic photosynthesis, as in reaction (1) above. This finding suggests that the cyanobacteria may have evolved from anoxygenic photosynthetic bacteria similar to purple or green sulfur bacteria. Indeed, evidence supporting this comes from molecular phylogenetic studies of the two photosystems of photosynthesis (see Chapter 9), one of which is thought to be derived from a member of the *Chlorobi* and the other from a member of the *Firmicutes*, a photosynthetic gram-positive group (see Chapter 21). However, some variant must have evolved that could use water in place of hydrogen sulfide as a reductant in photosynthesis and therefore could split water and carry out oxygenic photosynthesis, as in reaction (3). This important process may have evolved by natural selection when hydrogen sulfide became scarce and water for photosynthesis was abundant.

The oxygen produced by cyanobacteria would have been toxic to early life forms. Fortunately, for the reasons noted above, free oxygen would not have been available in the atmosphere for many millions of years after the first oxygenic cyanobacterium began producing it. This lengthy time provided favorable conditions for the selection and evolution of enzymes such as peroxidases, which would protect sensitive bacteria from the oxidizing effects.

Impact of *Bacteria* and *Archaea* on Biogeochemical Cycles

Bacterial groups carry out significant reactions, called **biogeochemical reactions**, that are crucial to the operation of Earth's biosphere. Examples of these reactions occur in the great cycles of elements such as the carbon cycle, in which autotrophic organisms fix carbon dioxide to form organic carbon that is recycled back to CO_2 by heterotrophic organisms. These reactions are discussed in greater detail later in the book, so we present just a brief summary here.

Because of the early evolution of microorganisms, particularly *Bacteria* and *Archaea*, they were provided with many energy sources 3 billion years before plants and animals evolved. As a result of this long period of evolution, they have diversified into many different metabolic groups that uniquely use unusual growth substrates such as methane, ammonia, hydrogen gas, sulfur, and reduced iron. In addition, several different groups of the *Bacteria* carry out photosynthesis using light energy.

All the biological transformations of the nitrogen cycle can be carried out by microorganisms. Likewise, all the biological transformations of the carbon and sulfur cycles can be carried out by microorganisms. Indeed, these cycles as we know them today were in place 1 to 2 Ga ago, at least 1 Ga before plants and animals evolved. The *Bacteria* and *Archaea* continue to carry out unique steps in these cycles, such as nitrogen fixation, that eukaryotic microorganisms, plants, and animals cannot perform. More information on these cycles and the important roles of microorganisms in them is provided in Chapter 24.

EVOLUTION OF EUKARYOTES

The origin of eukaryotic organisms is obscure at this time. The most popular hypotheses about eukaryotic evolution stem from the ideas of scientists such as Lynn Margulis and Wolfram Zillig. Zillig has proposed that eukaryotic organisms, the *Eucarya*, evolved through a fusion event between an ancestor of the *Bacteria* and an ancestor of the *Archaea* (**Figure 1.16**). According to this theory, the progenote had a permeable cell membrane (possibly protein) that allowed genetic communication with other species. As evolution proceeded, some event, perhaps geographic isolation, led to the separation of two different populations—the Prebacteria and the Prearchaea. During this period of separation, the two groups of organisms developed different metabolic patterns and genetic systems. Also at this time, the organisms developed their own characteristic lipid cell membranes, ester-linked fatty acid lipids for the Prebacteria and ether-linked isopranyl lipids for the Prearchaea. According to Zillig, it was at about this time that the fusion event occurred between these two cell types giving rise to the eukaryotic cell.

Similarities between *Eucarya* and the *Bacteria* and *Archaea* are cited as evidence supporting this hypothesis. For example, the cell membranes of *Eucarya* and *Bacteria* contain fatty acids linked to glycerol by ester linkages. In contrast, archaea do not produce long-chain fatty acids and they use ether linkages in their isopranyl cell membranes (see Chapters 4 and 18). This evidence suggests that if a fusion event occurred, it occurred when the bac-

terial ancestor had already developed its cell membrane and this feature was thus incorporated into the *Eucarya*.

Although much of this is speculative, it is interesting to note that as scientists further dissect organisms at the molecular level, they are beginning to infer likely, if not actual, evolutionary events that occurred billions of years ago.

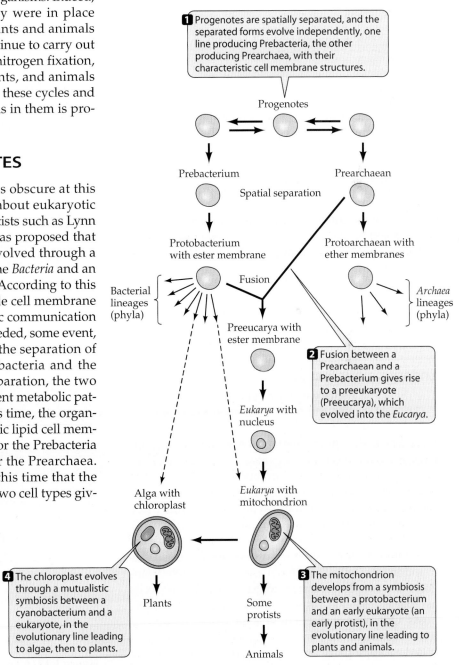

1 Progenotes are spatially separated, and the separated forms evolve independently, one line producing Prebacteria, the other producing Prearchaea, with their characteristic cell membrane structures.

Progenotes

Prebacterium

Spatial separation

Prearchaean

Protobacterium with ester membrane

Fusion

Protoarchaean with ether membranes

Bacterial lineages (phyla)

Archaea lineages (phyla)

Preeucarya with ester membrane

2 Fusion between a Prearchaean and a Prebacterium gives rise to a preeukaryote (Preeucarya), which evolved into the *Eucarya*.

Eukarya with nucleus

Alga with chloroplast

Eukarya with mitochondrion

4 The chloroplast evolves through a mutualistic symbiosis between a cyanobacterium and a eukaryote, in the evolutionary line leading to algae, then to plants.

Plants

Some protists

3 The mitochondrion develops from a symbiosis between a protobacterium and an early eukaryote (an early protist), in the evolutionary line leading to plants and animals.

Animals

Figure 1.16 Evolution of the main lines of descent
This diagram illustrates the possible origin of prokaryotic and eukaryotic organisms from a progenote. The early progenotes may have been precellular RNA forms without organic cell membranes.

Impact of Oxygen on the Evolution of Plants and Animals

The evolution of oxygenic photosynthetic organisms had a profound impact, not only on the chemistry of Earth but on the evolution of animal and plant life. Virtually all plants and animals use oxygen and carry out aerobic respiration, a process that occurs in the mitochondrion. Mitochondria are not found in *Bacteria* or *Archaea*. However, the mitochondrion is bacterial, by which we mean:

- A mitochondrion has its own DNA, but its DNA is not bounded by a nuclear membrane.
- A mitochondrion has ribosomes, which are not the 80S ribosomes of eukaryotes but the smaller, 70S ribosomes of prokaryotes.
- The size and structure of a mitochondrion are reminiscent of a gram-negative bacterium without a cell wall.
- Sequence analyses of mitochondrial 16S rRNA reveal that, regardless of the eukaryotic source, mitochondria are descended from the Proteobacteria (Figure 1.6).

Similarly, the chloroplast, the photosynthetic organelle of algae and higher plants, bears a striking resemblance to another prokaryotic group, the cyanobacteria. Thus, the chloroplast, like the mitochondrion, is prokaryotic. It, too, has DNA but no nuclear membrane, has 70S ribosomes, and lacks a cell wall, though it is descended from one of the bacterial phyla. Sequence analysis of its 16S rRNA places the chloroplast within the cyanobacterial branch of the *Bacteria* (Figure 1.6).

Endosymbiotic Evolution

The theory of **endosymbiotic evolution**, championed by Lynn Margulis, has been developed to explain the origin of mitochondria and chloroplasts. According to this theory, early in the evolution of eukaryotic organisms, certain prokaryotic organisms (the premitochondrion and prechloroplast) developed intracellular symbioses with eukaryotic cells (Figure 1.16). As time passed, the partners in these symbioses became more and more interdependent, until the bacterium became an organelle inside the eukaryotic cell.

Further support for the endosymbiotic theory is found in intracellular bacterial associations with protozoa. For example, certain protozoa harbor bacterial cells as "parasites" that provide unique features to their hosts. One type of association is the "killer paramecium." This strain of *Paramecium* contains a bacterium called a **kappa particle**. The kappa particle lives inside the protozoan and is responsible for the production of an organic compound that kills other paramecia that do not harbor these particles. The kappa particles retain their cell wall, so the symbiosis is not as highly evolved as that of the mitochondrion. Other *Paramecium*-bacterial relationships are also known (**Figure 1.17**). Another example of endosymbiosis occurs in certain flagellate and amoeboid protozoans that have a photosynthetic organelle called a **cyanelle**. The cyanelles appear to be cyanobacteria replete with all their bacterial features, including their cell wall.

Clearly, plants and animals, through evolution, have obtained many of their genes from microorganisms. In this manner genes have been available from a vast pool of different types of organisms. The macroorganism can therefore be viewed, at least in part, as a chimera of different genes, some derived directly from microorganisms and others modified through evolution.

Origin of the Ozone Layer

Crucial to the evolution of higher life forms was the development of the ozone layer in Earth's stratosphere. Ozone is produced by a photochemical oxidation reaction of oxygen in the upper stratosphere. It is believed to have first formed as a consequence of oxygen production by early cyanobacterial photosynthesis. The ozone layer is important to life on Earth in that it strongly absorbs ultraviolet light, a very active oxidiz-

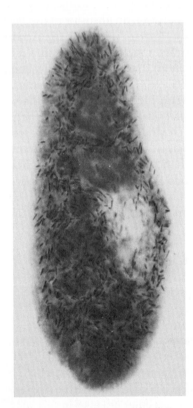

Figure 1.17 A *Paramecium*
This protist contains numerous endosymbiotic bacteria called "lambda particles," which are analogous to the "kappa particles" described in the text. Courtesy of John Preer Jr., Louise Preer, and Artur Jurand.

ing agent and mutagen. Some aquatic forms of life could have evolved and lived largely unaffected by UV radiation prior to the formation of an ozone layer, because water also strongly absorbs UV light. However, higher forms of terrestrial life would have been unprotected and could not have evolved until the ozone shield was established.

Oxygen and the development of the ozone layer had a profound impact on the evolution of higher organisms. Especially intriguing is the diversity and complexity of multicellular eukaryotic life forms, all of which evolved during the past 600 million years, particularly in terrestrial environments. In contrast, prokaryotes, which have had about 3.5 Ga to evolve, have rather simple structures. However, the simple and largely unicellular morphology of prokaryotes belies their vast genetic, metabolic, and physiological diversity, as this book so clearly demonstrates.

SEQUENCE OF MAJOR EVENTS DURING BIOLOGICAL EVOLUTION

The timetable shown in **Figure 1.18** portrays, to the best of our current knowledge, the probable major sequential events during biological evolution. The initial composi-

tion of the atmosphere was determined in large part by volcanic emissions. From these anaerobic gases and water, organic compounds could have been produced on Earth and mixed with organic compounds from intergalactic space. Within a billion years of Earth's formation, the first microorganisms appeared. These were probably thermophilic, anaerobic bacteria, including heterotrophic organisms that could live by fermentative processes. Hydrogen bacteria were also likely to have evolved early.

Anoxygenic photosynthetic bacteria likely had evolved by 3.0 to 3.5 Ga ago. These communities of organisms are thought to have colonized all aquatic habitats favorable for life. Photosynthesis had a major impact on Earth's biosphere. By producing large amounts of organic carbon, it greatly enhanced the growth of heterotrophic organisms.

The cyanobacterial branch of photosynthetic bacteria probably first appeared between 2.5 and 3.0 Ga ago. The production of oxygen would have had a major impact on many different processes in the biosphere. First, it would have caused the oxidation of reduced inorganic compounds such as iron sulfides that were present in enormous amounts early on. Indeed, it likely provided conditions that favored the evolution of new metabolic types such as iron-oxidizing bacteria. In addition,

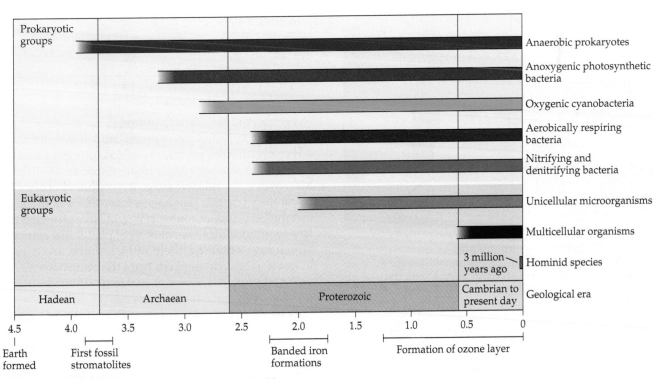

Figure 1.18 Geological and evolutionary timetable
The timetable shows the major geological events, beginning with the formation of Earth 4.5 Ga ago, and the evolution of various bacterial, archaeal, and eukaryotic groups. Note that hominids (primates) have occupied Earth for only 3 million years or so, a minute fraction (less than one-thousandth) of the time since cellular life forms arose.

(A)

(B)

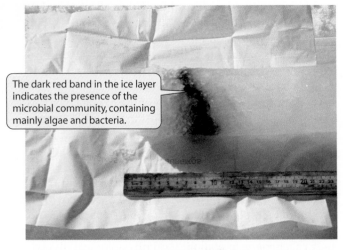

The dark red band in the ice layer indicates the presence of the microbial community, containing mainly algae and bacteria.

(D)

Bands of fungi (upper dark layer) and algae (lower greenish layer), components of a lichen, grow beneath the rock surface.

1 cm

(C)

Figure 1.19 Extreme environments
(A) Thermal hot springs, such as Mammoth Terrace in Yellowstone National Park, contain thermophilic (heat-loving) bacteria. (B) A core taken through the sea ice of Antarctica, home to sea ice microorganisms. (C) On this salt farm in Thailand, sea water is evaporated and becomes more and more concentrated in salt. Only halophilic (salt-loving) microorganisms can grow in such salt-saturated water. (D) In Antarctica's cold desert area in Victoria Land, called the Dry Valleys, microbes are the only life forms. The microbes live inside the rocks, such as this lichen revealed in a broken sandstone rock. A,B, courtesy of J. T. Staley; C, ©Claudia Adams/Dembinsky Photo Associates; D, courtesy of J. Robie Vestal.

oxygen would have been a poison to the anaerobic bacteria that had already evolved, and it would have adversely affected processes such as nitrogen fixation, which occur most efficiently under reducing conditions. Moreover, it would have led to the evolution of aerobic, respiratory bacteria that have a more efficient mode of metabolism. And, of course, it would have provided the key conditions for the evolution of plants and animals, which had obtained their mitochondria from proteobacteria and chloroplasts from the cyanobacteria.

Another consequence of oxygenic photosynthesis was the development of an ozone layer, which fostered the evolution of land plants and animals by protecting them from UV radiation. Aquatic organisms would have been less affected by UV light, because water strongly absorbs radiation of this wavelength.

It is interesting, from the standpoint of human evolution, that hominid primates date back to only about 3 mya, and the species *Homo sapiens* to a mere 100,000 years or so ago. Humans are therefore very recent participants in the biosphere and very dependent on the other processes and organisms that occur on Earth.

Because humans and other animals and the plants are so recent, they evolved in a world of microorganisms, and it is interesting to note how plants and animals made use of microorganisms as they evolved. Both groups rely on microorganisms and their activities in very intimate ways. For example, the roots of plant species have bacteria and some have special fungi, called mycorrhizae, that the plant relies on to bring in nutrients from the soil. Likewise, almost all animals have an anoxic digestive tract in which foods are broken down by microorganisms to provide nutrients, including amino acids and vitamins, for the host animal. Without these microorganisms, animals could not survive.

ASTROBIOLOGY

What we have learned about the evolution of life on Earth poses interesting questions about evolution on other planetary bodies. **Astrobiology** is the study of life in the universe. Of course, at this time, the only place where life is known to exist is Earth. Because we are just beginning to understand microbial life on Earth, much of the field of astrobiology is centered on studying microbial ecosystems on Earth.

What type of life is most likely to be found on other potentially habitable planets? Consider the following factors:

- Earth has had microbial life for more than 3.5 Ga of its 4.5 Ga history, whereas plants and animals are less than 500 million years old.
- Microbial life can survive asteroid impacts that would almost sterilize a planet.

- Microbes live everywhere that organisms live on Earth.

To amplify on the last point, it is noteworthy that some microorganisms grow at boiling temperatures in hot springs (**Figure 1.19A**) and at over 110°C in undersea volcanic hydrothermal vents. Others live in sea ice at temperatures below freezing (Figure 1.19B). Some produce sulfuric acid from sulfur compounds and live at a pH of l.0, equivalent to 0.l N H_2SO_4. A few microbes live in saturated salt brine solutions (Figure 1.19C), while others live in pristine mountain lakes as pure as distilled water. Microbes even live inside rocks in the harsh environment of Antarctica (Figure 1.19D). Thus, microorganisms can live in extreme environments not colonized by plants and animals.

All these factors are consistent with the hypothesis that if life is found elsewhere in the universe, it is less likely to be macroscopic than microscopic. This is extremely likely for our own solar system. There is now very little doubt that no "little green men" are living on Mars or any other planets in our solar system. However, extreme environments are known to be common elsewhere in the solar system, so microorganisms are their most likely inhabitants. Indeed, Mars may still be volcanically active and may contain aquatic habitats beneath its surface that are hospitable for microorganisms. Jupiter's moon, Europa, though frozen at its surface, may teem with microbial life in its ocean and possible hydrothermal vents (**Figure 1.20**). Furthermore, we know that asteroid impacts on planets can disperse rocks from one planet to another. For example, several rocks from Mars have been found on Earth.

These exciting recent developments have led the National Aeronautics and Space Administration (NASA) to launch a program on astrobiology to study early life and its evolution on Earth (http://astrobiology.arc.nasa.gov). Indeed, preparations are being made for expeditions to Mars and Europa for the exploration of past or present life. Pursuit of these lines of research may help humans understand some of the great questions of life such as, where did we come from? And are we alone in the universe?

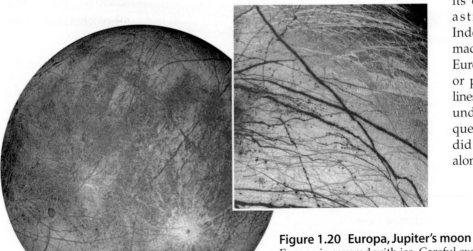

Figure 1.20 Europa, Jupiter's moon
Europa is covered with ice. Careful examination of the surface reveals that large blocks of ice have been moved, indicative of an underlying ocean (see inset). Although cold and anaerobic, conditions on Europa seem to be suitable for life, although life has not yet been found. Photos from PIRL/University of Arizona/NASA.

SUMMARY

▶ Earth and its solar system formed by physical and chemical processes about 4.5 Ga ago. Microbial fossils were first reported by micropaleontologists only recently (1950s). The oldest microbial fossils are found in stromatolites dated at about 3.5 Ga old.

▶ Stromatolites are laminated sedimentary rocks, some of which were produced by microorganisms. Intertidal mat communities are the most common type of living stromatolites currently found on Earth.

▶ Many organic molecules can be formed abiotically from inorganic compounds in an anoxic environment with an energy source such as ultraviolet light or lightning. Some organic compounds are produced in space and brought to Earth in meteors and comets.

▶ The "RNA world" theory holds that RNA was the first macromolecule and served both as a biochemical catalyst and as a template for reproduction in primitive "organisms."

▶ Microorganisms are the smallest living things. Most can be seen only under a microscope. Microorganisms are ubiquitous in the biosphere; some grow in extreme habitats such as boiling hot springs, saturated brine solutions, or sea ice; some produce sulfuric acid.

▶ Microorganisms that use light as an energy source and carbon dioxide as a carbon source are said to be photoautotrophic. Those that use light as an energy source and organic carbon as a carbon source are photoheterotrophic. Those that use organic chemical compounds as both energy and carbon sources are heterotrophic, chemoheterotrophic, or chemoorganotrophic. Microorganisms that use inorganic compounds as energy sources and carbon dioxide as a carbon source are chemoautotrophic.

▶ The term "bacteria" is synonymous with "prokaryotic organism." The two groups of prokaryotic microorganisms are *Bacteria* and *Archaea*. The three groups of eukaryotic microorganisms are fungi and the two types of protists, algae and protozoa. All organisms on Earth are classified into three domains of life and organized on a "Tree of Life," which is based on the sequence of 16S and 18S ribosomal RNAs. These domains are the *Bacteria*, *Archaea*, and *Eucarya* (the latter including all eukaryotic microorganisms and plants and animals).

▶ Viruses lack cytoplasm, and they do not produce cells. Therefore, they cannot be included in the Tree of Life.

▶ Prokaryotic organisms lack a nuclear membrane and do not have membrane-bounded organelles such as mitochondria and chloroplasts.

▶ Prokaryotes have 70S ribosomes, whereas eukaryotic organisms have larger, 80S ribosomes.

▶ Eukaryotic organisms have a nuclear membrane and most possess mitochondria and, in photosynthetic cells, chloroplasts. Mitosis is the process by which the chromosomes are duplicated and separated during asexual division of eukaryotic cells. Meiosis is the process by which the diploid chromosomes of eukaryotic cells are separated to form haploid gametes.

▶ The cyanobacteria are thought to be the first O_2-producing (oxygenic) photosynthetic organisms. The O_2 produced by cyanobacteria resulted in the banded iron formations as well as the protective ozone layer and led eventually to the high concentration of O_2 in the atmosphere. Through natural selection, O_2 in the atmosphere led to the evolution of aerobic respiration, a much more efficient type of metabolism than fermentation and anaerobic respiration.

▶ The theory of endosymbiotic evolution holds that mitochondria, chloroplasts, and possibly other organelles of higher animals and plants are derived from prokaryotic organisms that developed symbioses with higher organisms. Evidence from 16S rRNA sequence analysis of mitochondria indicates that these organelles evolved from the Proteobacteria. Evidence from 16S rRNA sequence analysis of chloroplasts indicates they evolved from cyanobacteria. The cyanelle is a cyanobacterial symbiont within some protozoa.

▶ The eukaryotic cell may have evolved from a fusion event between Prebacteria and Prearchaea coupled with endosymbiotic events with prokaryotic organisms. Plants and animals could not have evolved unless microorganisms evolved first.

▶ Microorganisms, particularly *Bacteria* and *Archaea*, carry out unique steps in the biogeochemical cycles of the biosphere.

▶ Astrobiology is the study of life in the universe. Because of their rapid evolution and hardiness, microorganisms may be a much more common form of life in the universe than macroorganisms.

REVIEW QUESTIONS

1. Describe what you believe to have been the first true bacterium on Earth. What were its properties and why were they essential?

2. Compare and contrast oxygenic and anoxygenic photosynthesis.

3. How do you explain the existence of some eukaryotes that lack mitochondria?

4. Do you believe it would be possible to place chloroplasts in human skin cells? Would humans become photosynthetic? Would they become "little green men and women"?

5. Why do bacteria grow in such unusual habitats?

6. What are viruses? Are they alive? Explain your answer.

7. Compare the nutrition of humans with that of a typical chemoheterotrophic bacterium.

8. Some bacteria can reproduce to form progeny cells in less than 10 minutes. Why is human reproduction so much slower?

9. Why don't bacteria undergo meiosis?

10. Describe Earth's biosphere before and after the evolution of cyanobacteria.

11. How much information can be obtained by the discovery and dating of microbial fossils? Can the type and metabolic activity of the microorganism be determined?

12. Is there life on Mars? On the moon? On Europa? Explain your answer.

SUGGESTED READING

Chatton, E. 1932. *Titres et Travaux Scientifiques*. Sottano, Italy: Séte.

Doolittle, W. F. 1999. "Phylogenetic Classification and the Universal Tree." *Science* 284: 2124–2128.

Gest, H. 1987. *The World of Microbes*. Madison, WI: Science Tech Publishers.

Jacobson, M. C., R. J. Charlson, H. Rodhe and G. H. Orians, eds. 2000. *Earth System Science*. New York: Academic Press.

Lwoff, A. 1957. "The Concept of Virus." *Journal of General Microbiology* 17: 239–253.

Schopf, J. W. 1978. *Evolution of Life. Scientific American* 239: 110–138.

Schopf, J. W. 1983. *Earth's Earliest Biosphere*. Princeton, NJ: Princeton University Press.

Staley, J. T., and A.-L. Reysenbach, eds. 2002. *Biodiversity of Microbial Life: Foundation of Earth's Biosphere*. New York: John Wiley & Sons.

Stanier, R. Y., and C. B. van Niel. 1962. "The Concept of a Bacterium." *Archiv für Mikrobiologie* 42: 17–35.

Woese, C. R. 1987. "Bacterial Evolution." *Microbiological Reviews* 51: 221–271.

Woese, C. R., O. Kandler and M. C. Wheelis. 1990. "Towards a Natural System of Organisms: Proposal for the Domains Archaea, Bacteria, and Eucarya." *Proceedings of the National Academy of Sciences, USA* 87: 4576–4579.

Chapter 2

Historical Overview

*The accidents of health had more to do with the march of great events
than was ordinarily suspected.*
— H. A. L. FISHER, 1865–1940

The previous chapter was devoted to a discussion of the evolution and role of microorganisms in the living world. Clearly, microbes make up a considerable part of the biosphere and are therefore studied for their own sake. However, microbiology is a multifaceted discipline that is concerned with infectious disease, agricultural practice, sanitation, and the industrial production of food, beverages, and chemicals. Microorganisms have been and remain important model organisms for studies in nutrition, metabolism, genetics, and biochemistry. Results of studies with microorganisms were instrumental in the development of biotechnology. Microorganisms have also had a profound effect—both positive and negative—on human welfare. In this chapter we will examine the contributions of pioneer scientists to our understanding of the microbe.

EFFECTS OF DISEASE ON CIVILIZATION

Microorganisms have markedly affected the course of human history. Disease-causing microorganisms often decided the well-being and morale of populations, the strength of armies, and the outcomes of battles. The mobilization of armies themselves, with the consequent concentration of young soldiers from across the country, created environments that were ripe for the spread of epidemics. Every soldier was both a potential carrier of infectious disease from his home and a potential victim of disease due to lack of previous exposure. In rapidly growing urban areas, the absence of proper sanitation under crowded conditions also contributed to the spread of infectious agents. As world trade and travel increased, infectious diseases slowly but inevitably moved from region to region. They then retraced their steps back to a new generation of susceptible individuals. This section briefly presents some examples of these and other effects of microbial disease on human populations, institutions, and wars throughout the history of civilization.

The decline of Rome under the Emperor Justinian (A.D. 565) was certainly hastened by epidemics of bubonic plague and smallpox. The inhabitants of Rome were decimated and demoralized by these massive epidemics and were left powerless against the barbarian hordes that destroyed the empire. Through the Middle Ages and beyond, each human generation was subject to renewed attacks of epidemics. Some of these epidemics spread across the continents, whereas others were more localized. Typhus, plague, smallpox, syphilis, and cholera were some of the infectious diseases that caused suffering and great loss of human life.

The populations of Europe, North Africa, and the Middle East totaled about 100 million when a bubonic plague epidemic struck in A.D. 1346 (the Black Death). The epidemic traveled west down the "Silk Road" (the main trade route from China), bringing death to Asia; the epidemic then spread

Milestones Box 2.1   The Spread of the Black Plague in Europe

The Great Plague of the fourteenth century was also referred to as the "Black Death." The epidemic originated in China in 1331 and moved slowly across Asia, reaching the outskirts of Europe in 1347. The disease spread from caravan stop to caravan stop, moving west with the rat-flea-human community associated with the caravans. The high rate of fatalities among these hosts in thinly populated areas of Asia was probably responsible for the slow movement of the epidemic across Asia. When the infected flea/rat population reached the Mediterranean area, the disease agent (*Yersinia pestis*) spread into the black rat populations. Shipping and commerce carried the black rats and their bubonic plague-infected fleas upward across Europe. The Black Death reached Sweden by late 1350, a mere three years from when it arrived in the Mediterranean area.

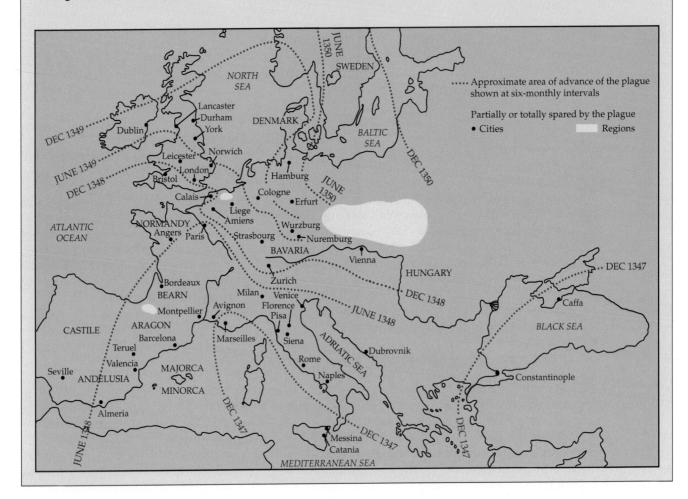

throughout Europe, resulting in the loss of 25 million people in a few short years (**Box 2.1**). Recurrences of plague through the sixteenth and seventeenth centuries kept populations in check. In 1720–1722, one last great epidemic occurred in France that killed 60% of Marseilles, 60% of Poulon, 44% of Arles, 30% of Aix, and 30% in Avignon. The most recent major plague pandemic originated in Yunnan, China, in 1892, moved across India, and arrived in Bombay in 1896. That outbreak killed an estimated 6 million people in India alone.

In wars that were fought prior to World War II, the outcome was generally decided by arms, strategy, and

pestilence. Pestilence, more often than not, played the leading role, as the following examples affirm.

By the time Napoleon began his retreat from Moscow in 1812, most of his army had fallen victim to typhus, pneumonia, dysentery, and other diseases. These diseases, the cold, and deprivation all played a role in his departure. The following year (1813), the irrepressible Napoleon recruited a new army of 500,000 young soldiers. As with the previous army, their youth and the crowded unsanitary conditions soon rendered them susceptible to infectious disease. By the time Napoleon and his new army faced the allies at Leipzig, preliminary battles and disease had reduced his army of approximately 500,000 to about 170,000. An estimated 105,000 were casualties of earlier battles, but about 220,000 were incapacitated by illness. Thus, microbial infections were a major factor in the ultimate defeat of Napoleon at Waterloo in 1815.

There were over 550,000 deaths among the soldiers in the American Civil War (1861–1865), and more were victims of infectious disease than died from battle. Of the Union Army troops lost, 93,443 were killed in action or died from wounds, whereas 210,400 succumbed from disease. Infections were a major cause of death among the wounded, as field conditions were unsanitary and care of the wounded was haphazard. Of those who died of disease, records indicate that 29,336 died from typhoid fever, 15,570 from other "fevers," 44,558 from dysentery, and 26,468 from pulmonary disease (mostly tuberculosis). The remaining deaths were from other undetermined causes. Records for the army of the Confederacy are less extensive, but an estimated 90,000 were killed in action or died from wounds. The number of deaths from disease exceeded 180,000 soldiers. It is highly probable that typhoid fever, dysentery, and pulmonary diseases were also the main causes of death among the Confederate soldiers. The approximate totals for the two armies were 183,000 deaths from combat wounds and over 390,000 victims of infectious disease.

WHY STUDY THE HISTORY OF A SCIENCE?

Microbiology, as with any field of endeavor, can be comprehended best if one has a reasonable understanding of the historical development of the field. The ingenious experimentation and insights that led scientists, such as Pasteur and Koch, to logical explanations for the observable manifestations of disease should be familiar to the modern student in microbiology. One cannot fully comprehend present theories and concepts without first understanding the logical steps that led to those ideas.

The remainder of this chapter presents the scientific contributions made by several intellectual giants in microbiology. This list of contributors is by no means complete, nor does this discussion do justice to the many early scientists who contributed to the foundations of microbiology. Suggested readings at the end of the chapter provide sources for those interested in furthering their knowledge of the historical development and significance of microbiology.

STATUS OF MICROBIAL SCIENCE PRIOR TO 1650

From the dawn of civilization until the middle of the nineteenth century, any success in combating disease, in fighting the microbe's destructive power, or in harness-

ing their fermentative capabilities came about through inexact processes of trial and error. Those occurrences we now assign to the microbe were then ascribed to a supreme being, "miasmas," spontaneous generation, magic, chemical instabilities, or other interpretations limited only by the human imagination.

Human thought was influenced by the great philosophers, whose writings indicate that they were intrigued by theories supporting **spontaneous generation**, the immediate origin of living organisms from inert organic materials. It should be emphasized that spontaneous generation is defined, for our purposes in this chapter, as the creation of identifiable living creatures from the inanimate. The evolution of the earliest viable cell was discussed in Chapter 1 and is not an example of spontaneous generation as applied here.

Aristotle (384–322 B.C.) and others wrote of the formation of frogs from damp earth and of mice from decaying grain. After the fall of Rome and the decline of civilization, free inquiry and acquisition of knowledge became severely limited. Through the centuries of the Dark Ages (A.D. 476–1000), disease epidemics and plagues were recorded, but little of scientific consequence was written. The Renaissance (1250–1550) was a period of awakening and the beginning of open inquiry into the forces that shaped human life. The first major writings on the causation of disease were by **Girolamo Fracastoro** (c. 1478–1553), who wrote extensively on the "contagions" involved in the disease process. Based on his investigations, he wrote that disease could be spread by direct contact, handling contaminated clothing, or through the air.

It is evident from the literature of that time (prior to the Age of Enlightenment) that scholars in the sixteenth and early seventeenth centuries were seeking logical explanations for the phenomena of nature. Theory and experimentation were one thing, but acceptance of concepts contrary to the dogma of the time was quite another. It was particularly difficult to gain acceptance of biological explanations for natural phenomena, such as food decay, when many renowned scientists, particularly chemists, attributed this process to chemical instabilities and clung to theories of spontaneous generation. Typical of this group was van Helmont (1580–1644), a forerunner of scientific chemistry, who published a recipe for producing mice from soiled clothing and a little wheat.

MICROBIOLOGY FROM 1650 TO 1850

During the seventeenth century there were individuals of a scientific bent whose contributions gave impetus to a further awakening of the human spirit. Biology benefitted significantly from such developments as the microscope and the realization that living matter was composed of individual cells. This was the Age of Enlightenment, a time when people questioned traditional doctrines. Science, reason, and individualism replaced obedience to accepted dogma.

Fabrication of the original microscope is generally attributed to a Dutch spectacle maker, Zacharias Janssen, and his father, Hans, between 1590 and 1610. The first to employ a microscope extensively in the examination of biological material was the Italian scientist **Marcello Malpighi** (1628–1694), and he is considered by many to be the "father of microscopic biology." Malpighi made major contributions through his discovery of capillaries and an understanding of embryonic development. **Antony van Leeuwenhoek** (1632–1723) developed a solar microscope with high resolving power that led to the first recorded observations of bacteria in 1683. **Robert Hooke** (1635–1703) examined the structure of cork and, based on these observations, suggested that all living creatures were made up of individual cells. The following section covers Leeuwenhoek and his discovery of bacteria in some detail. About 200 years passed before the nature of these organisms was generally accepted. Why 200 years? Because humans were not yet ready to disregard completely the prevailing dogma that chemical instabilities and spontaneous generation were responsible for all activities that we now ascribe to microbes.

Antony van Leeuwenhoek

Antony van Leeuwenhoek was born in Delft, Holland, in 1632 into a relatively prosperous family (**Figure 2.1**). At the age of 16, he was sent to Amsterdam to apprentice in a draper's shop and learn a useful trade. He was a bright lad and was appointed cashier in the business before returning to Delft to spend the remaining 70 years of his life. He did not attend a university, but was learned in mathematics, and he became a successful businessman, a surveyor, and the official wine gauger for the town of Delft. In the latter capacity, he assayed all of the wines and spirits entering the town, and it was his responsibility to calibrate the vessels in which they were transported. His lack of a university education was inconsequential because, at that time, such institutions were mostly devoted to theology, law, and philosophy. Scientific research was done by amateurs as an avocation, by the wealthy, or by individuals under the sponsorship of a patron.

Leeuwenhoek apparently had solar microscopes available and made extensive observations of bacteria prior to 1673 for, in that year, several of his studies were communicated to the Royal Society in London, England, by his friend, Renier de Graaf, a noted Dutch anatomist. Fortuitously, the editor of *Philosophical Transactions* (published by the Royal Society) was Henry Oldenberg, a German-born scientist who was well versed in several languages, including Dutch. Oldenberg firmly believed that scientific information should have no nationalistic restrictions, and he carried on an extensive correspon-

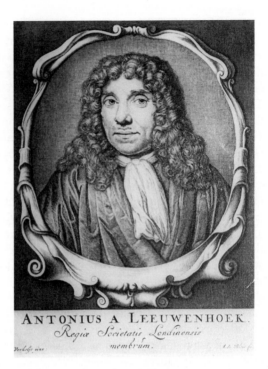

Figure 2.1 Antony van Leeuwenhoek
Antony van Leeuwenhoek (1632–1723) developed a microscope and was the first to observe and accurately describe bacteria. Courtesy of National Library of Medicine.

dence with scientists throughout Europe. Leeuwenhoek was encouraged to communicate with the Royal Society. His first report, published in the *Philosophical Transactions*, dealt with molds, mouthparts and the eye of the bee, and gross observations on the louse. Due to Oldenberg's contacts and immediate translation of Leeuwenhoek's correspondence into several languages, his work generated considerable interest in the scientific world. Leeuwenhoek continued to send letters to the Royal Society throughout his life, and in 1680, he was unanimously elected a Fellow of the Royal Society.

Over 200 of Leeuwenhoek's original letters have been preserved in the library of the Royal Society. Upon his death, Leeuwenhoek bequeathed a cabinet containing 26 microscopes to the Royal Society, and these were delivered by his daughter, Maria, with the handwritten passage, "every one of them ground by myself and mounted in silver and furthermore set in silver . . . that I extracted from the ore . . . and therewithal is writ down what object standeth before each glass." The cabinet and microscopes remained in the Royal Society collection for a century, but unfortunately have since been lost and never recovered. The microscopes were rather simple in design, and he left no description of how they were operated (**Figure 2.2**). The exactness of his drawings suggests that he had a keen eye, but most certainly he also had an unexplained method for using his microscope to examine bacteria. Because his microscopes

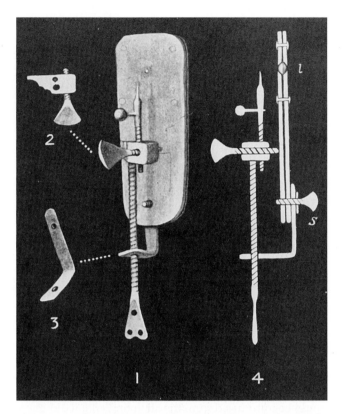

Figure 2.2 The microscope developed by Leeuwenhoek
It is very difficult to observe microbes with this crude instrument, but Leeuwenhoek held the implement to the solar light in such a way that small specimens such as bacteria could be viewed. His drawings confirm the accuracy of his observations. From *Antony van Leeuwenhoek and His "Little Animals,"* edited by Clifford Dobell, Dover Publications, 1960.

could magnify by only 300 diameters (less than one-third of what modern light microscopes can do), it is probable that he achieved his detailed observations of bacteria by careful focusing of solar light.

It is evident from Leeuwenhoek's descriptions of what he called "**animalcules**" and from his size estimates by comparison with grains of sand or human red blood corpuscles (which he measured fairly accurately at 1/300 inch) that he indeed closely observed the living forms that he depicted. His writings describe some of the major bacterial forms now known—spheres, rods, and spirals—and he described motility in rod- and spiral-shaped cells (**Figure 2.3**). Leeuwenhoek's descriptions of bacteria were superior to those made by Louis Joblot (1645–1723) and Robert Hooke (1635–1703), who used slightly more sophisticated compound microscopes. Leeuwenhoek's extensive letters described an array of protozoa, spermatozoa, and red blood cells, and he is generally considered to be the founder of protozoology and histology.

Leeuwenhoek emphasized the abundance of protozoa, yeast, bacteria, and algae in environments as diverse as the mouth and seawater. Although Leeuwen-

hoek's observations were widely known, apparently no one attributed familiar processes such as fermentation and decay to these "animalcules." He was firm in his belief that his animalcules arose from preexisting organisms of the same kind and did not arise by spontaneous generation.

From Leeuwenhoek to Pasteur

The studies by a number of experimentalists (now much ignored) between 1725 and 1850 laid the foundation for the great advances in bacteriology made in the latter half of the nineteenth century. These scientists did much to discredit the theories of spontaneous generation, provide a foundation for studies on the immune response, and suggest rational classification schemes for bacteria. A major factor in affirming the role of microbes in nature was disproving the generally held concept of spontaneous generation.

Francesco Redi (1626–1697), an Italian physician, published a book in 1688 attacking the doctrine of spontaneous generation. At that time, the presence of maggots in decaying meat was considered a prime example of spontaneous generation. Redi believed otherwise and to prove this he placed meat in beakerlike containers, covered some with fine muslin, and left others exposed to invasion by blowflies. Although fly eggs were deposited on the muslin over the covered jars, maggots did not develop on the meat. Extensive growth of maggots, however, did occur in the meat that was left uncovered.

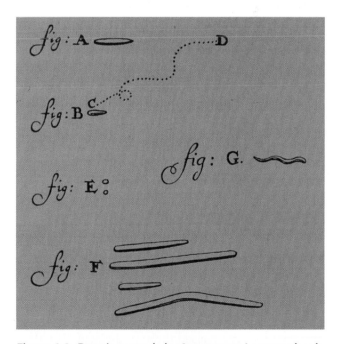

Figure 2.3 Drawings made by Antony van Leeuwenhoek
Their probable identification: (A) *Bacillus*, (B) motile *Selenomonas*, (E) *Micrococcus*, (F) *Leptothrix*, and (G) a spirochete. From *Antony van Leeuwenhoek and His "Little Animals,"* edited by Clifford Dobell, Dover Publications, 1960.

When Redi placed the eggs that were deposited on the muslin onto the surface of the meat, maggots quickly appeared. Clearly maggots came from the eggs deposited by the flies. Studies of this type led to careful experimentation and clear evidence that animals and insects, discernible by the eye, could not arise spontaneously from decaying matter. The proponents of spontaneous generation turned to phenomena whose causes were not so apparent. The inability to identify the factor responsible for fermentation and putrefaction in infusions (meat or vegetable broth) gave their theory continued life.

Lazzaro Spallanzani (1729–1799), an Italian naturalist and priest, was familiar with the studies of Redi and did a series of experiments to confirm and extend Redi's earlier work. Spallanzani hermetically (airtight) sealed meat broth in glass flasks and reported that 1 to 2 hours of heating the enclosed infusions was sufficient to render the contents incapable of supporting growth. John Needham, an English cleric and proponent of spontaneous generation attacked these studies. Needham proposed that a "vegetative force" was responsible for spontaneous generation and that hermetically sealing and heating the flasks destroyed this vital force. Spallanzani then did a series of experiments to overcome this objection, which he considered to be conclusive and from which he affirmed that to render a broth sterile, it was necessary to seal the flask and not allow unsterile air to enter. Spallanzani also concluded that boiling for a few minutes destroyed most organisms, but that others, now known to be spore formers, withstood boiling for a half hour. Spallanzani's work was quite advanced for his time, and his deductions and conclusions were similar to those reported later by Pasteur.

During the latter part of the eighteenth century, Antoine Lavoisier and others demonstrated the indispensability of oxygen (O_2) to animals. This led to the assumption that oxygen was the mysterious element (vegetative force?) necessary for spontaneous generation, and that it was this element that had been excluded in Spallanzani's experiments.

In 1836, **Franz Schulze** (1815–1873), a German chemist, performed a crucial experiment indicating that oxygen depletion was not the sole reason for sterility in Spallanzani's flasks. Schulze took a flask that was half filled with vegetable infusion and closed it with a cork through which two bent glass tubes were fitted (**Figure 2.4**). He thoroughly boiled the infusion in a sand bath, and while steam was being emitted, he attached an absorption bulb to each of the two glass tubes. One bulb contained concentrated sulfuric acid and the other a solution of potassium hydroxide (potash). Every day for several months, he drew air out through the potash bulb, and incoming air passed through the acid. The flask remained sterile, while a control flask without acid sterilization of incoming air had visible mold growth after a few days. When the sterile infusion was opened and

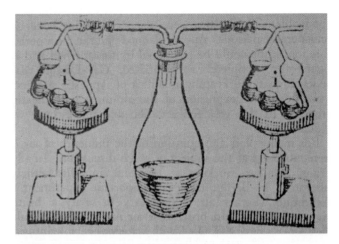

Figure 2.4 Sterilization by heat
The experiment by Franz Schulze did much to confirm that air does not contain a "vital force." From *The History of Bacteriology*, by William Bulloch, M.D., Oxford University Press, 1960.

exposed to the atmosphere, mold growth followed in a few days. This experiment demonstrated that infusions could be sterilized (no viable microbes) and that microbes could be introduced from the air. John Tyndall did many modifications of the Schulze experiment, and Theodor Schwann made further refinements, and these finally led to refutation of the doctrine of spontaneous generation.

During this period, other scientists were helping to disprove spontaneous generation by examining how microbes were involved in fermentation. The first report that gave a clear account of the yeast cell and its role in fermentation of beer and wine appeared in an 1836 report by **Charles Cagniard-Latour** (1777–1859). He asserted that **yeasts** were nonmotile organized globules capable of reproduction by budding and that they probably belonged to the vegetable kingdom. In 1837, he suggested that the vital activity of the yeast cell was responsible for converting a sugar solution to carbonic acid and alcohol.

Theodor Schwann (1810–1882), a German physiologist, independently discovered and described the yeast cell in 1837. Although his report also concerned the doctrine of spontaneous generation, he wrote that beer yeast consisted of granules arranged in rows, and that they resembled fungi. He believed them to be plants and observed their reproduction by budding. The relationship between the growth of yeast and the process of fermentation was clear to Schwann, and he called the organisms "zuckerpilz" (sugar fungus), from which the term *Saccharomyces* (a genus of common yeast) was derived. He also noted the indispensable requirement for nitrogenous compounds in the fermentation process.

A third independent worker who contributed to the discovery of the **fermentation** process was **Friedrich**

Kützing (1807–1893), a German naturalist, and his major publication was also dated 1837. He described the nucleus of the cell and developed the concept that all fermentation is caused by living organisms. He was among the first to suggest that physiologically distinct organisms brought about different types of fermentation.

The role ascribed to the yeast cell in fermentation was contemptuously attacked by the chemists of that time. J. J. Berzelius, F. Wöhler, and J. von Leibig (all influential chemists) attributed a chemical character to every vital process and suggested that chemical instabilities were responsible for fermentation. Although much rhetoric ensued, the criticisms by the chemists led to more definitive experimentation and a clearer understanding of the role of yeast and fungi in many different types of fermentation.

Classification Systems

The earliest classification scheme for living organisms that included the bacteria was formulated by the Swedish botanist **Carolus Linnaeus** (1707–1778). In *Systema Naturale* (1743), he described the "animalcules" (Leeuwenhoek) as "infusoria" and placed this group, which included the bacteria, in the genus *Chaos*. A major shortcoming of the Linnaean classification scheme was the emphasis that he placed on a few known characteristics. This emphasis has permeated classification schemes until recent times; for example, we often use terms such as *rod* or *coccus* without considering the metabolic capabilities of the microbe. There were other classification schemes suggested, and the **Adansonian classification** scheme proposed by **Michael Adanson** in 1730 gave equal weight to all characteristics of a species. The Adansonian classification scheme was more amenable to modern computerized classification systems of bacteria.

The Danish naturalist **Otto F. Muller** (1730–1784) presented several works that described, arranged, and named a number of microscopic organisms. His major study, *Animalcula infusoria et Marina*, published posthumously in 1786, was 367 pages in length and described 379 species of bacteria. In his scheme, two of five genera contained bacterial forms. The genera were named *Monas* and *Vibrio*. The term *Vibrio* has been retained to the present day.

The first attempt to define bacterial forms as distinct from complex organisms was made by **Christian G. Ehrenberg** (1795–1876) in 1838. His treatise, entitled *Die Infusionsthierchen als Volkommene Organismen* (*The Infusoria as Complete Organisms*), used some of Leeuwenhoek's drawings of microbial cells. It was 547 pages in length and recognized three families that comprised forms we now recognize as bacteria. These families— *Monadina*, *Crystomonadina*, and *Vibrionia*—encompassed several genera: *Bacterium*, *Vibrio*, *Spirochaeta*, and *Spirillum*. All of these terms are still in common usage.

MEDICAL MICROBIOLOGY AND IMMUNOLOGY

Several scientists merit mention for their contributions to our understanding of the cause of disease and immunity prior to 1850. In 1822, the Italian scientist **Enrico Acerbi** postulated that parasites existed that were capable of entering the body and that their multiplication caused typhus fever. This theory was advanced by the 1835 work of **Agostino Bassi** (1773–1856), who made classic observations on the diseases of silkworms.

A disease that was rampant at that time caused the death of the worms; the dead silkworms were covered with a hard, white, limy substance. At the time, the limy coat was considered to arise spontaneously from unknown factors. However, Bassi demonstrated that aseptic transfer of subcutaneous material from sick living worms to healthy worms resulted in the disease. He suggested that a fungus caused the disease, which was ultimately renamed *Botrytis bassiana* in his honor. In later life, although nearly blind, he developed his theory that contagion (infectious agents) in such diseases as cholera, gangrene, and plague resulted from living parasites.

During 1798, **Edward Jenner** (1749–1823) published studies on the immunization of humans against smallpox (**Figure 2.5**). Jenner and others observed that indi-

Figure 2.5 Effective vaccination
An early depiction of Edward Jenner vaccinating a child against smallpox. From *The Eradication of Smallpox from India*, by Basu, Jezek, and Ward, a World health Organization publication, 1979.

viduals who were routinely exposed to cows often developed pustules on their hands and arms that were similar to those caused by the dreaded smallpox. This "cowpox" was not fatal in humans, and apparently those infected with the cowpox did not contract the smallpox. Jenner inoculated an eight-year-old boy (James Phipps) with material from the infected cowpox pustules on the hands of a milkmaid. The boy became ill and had characteristic cowpox pustules at the site of inoculation but quickly recovered. A challenge dose later from an active case of smallpox did not result in illness. The boy was apparently immune to smallpox. Jenner then inoculated several healthy individuals with cowpox exudate and observed that they were made immune to smallpox. Jenner has often been criticized for the empirical nature of his experimentation, but given the context of the time, he deserves much credit for his contribution.

MICROBIOLOGY AFTER 1850: THE BEGINNING OF MODERN MICROBIOLOGY

During the latter part of the nineteenth century, a number of gifted scientists working independently established the disciplines that are now encompassed by the term "microbiology." Much was accomplished during this half-century, and the foundations for immunology, medical microbiology, protozoology, systematics, fermentation, and mycology were all set in place. However, up until 1860, the doctrine of spontaneous generation of microbes remained a generally accepted concept. This was largely due to the widespread acceptance that chemists were infallible—a view they did little to discourage—and they tended to be more dogmatic than were the naturalists. They would soon meet their match in a brilliant, strong-willed Frenchman named Louis Pasteur.

A notable publication appeared in 1861 written by Louis Pasteur and entitled *Mémoire sur les corpuscules organisés qui existent dans l'atmosphère: Examen de la doctrine de générations spontanées (Report on the organized bodies that live in the atmosphere: Examination of the doctrine of spontaneous generation).* It marked the beginning of a new epoch in bacteriology.

Louis Pasteur (1822–1895) was a chemist, and his approach to the study of microorganisms was markedly influenced by his chemical training and analytical mind (**Figure 2.6**). Pasteur was a genius in his thinking, argument, and experimentation, but also one of his unique qualities was the ability to communicate and convince scientists and laypeople alike of his views. He laid to rest forever the idea of spontaneous generation, established immunology as a science, developed the concepts of fermentation and anaerobiosis, and developed many microbiological techniques. In short, he contributed to every phase of microbiology.

Figure 2.6 Louis Pasteur
Louis Pasteur (1822–1895), an outstanding researcher and the developer of the rabies vaccine. Pasteur is credited with disproving "spontaneous generation." From *Life of Pasteur,* by Rene Vallery-Radot, Doubleday Page, Garden City, NY, 1923.

The Pasteur School

Louis Pasteur entered the controversy surrounding spontaneous generation at a time when the dogma of **heterogenesis** was being used to explain the origin of living matter. This theory was avidly expounded by **Felix-Archimede Pouchet** (1800–1872), a noted French physician and naturalist and honored member of many learned societies in France. Pouchet believed that life could spring de novo from a fortuitous collection of molecules and that the "vital force" came from preexisting living matter. In contrast Pasteur's studies on fermentation indicated that "ferments" (cause of fermentation) were actually organic living beings that reproduced and, by their vital activities, generated the observed chemical changes. In a series of brilliant experiments, Pasteur showed that "germs" present in the air were the cause of ferments and that such organisms were widely distributed in nature. Pasteur's "germs" are actually what we call bacteria and fungi today.

Starting in 1859, Pasteur dealt with the problem of microbes in air by designing an aspirator filter system to recover them. Pasteur was aware that H. G. F. Schröder and T. von Dusch had found spun cotton-wool to be an effective filter for airborne microbes. Employing this knowledge, Pasteur drew copious quantities of air through spun cotton and then dissolved the cotton in a mixture of alcohol and ether. Microscopic examination of the resulting sediment revealed a considerable number of small round or oval bodies that were indistinguishable from the "germs" previously described by others. Pasteur noted that the number of these organisms varied with the temperature, moisture, and movement of the air and the height above the soil that the aspirator's inlet tube was placed.

We digress briefly here to present work done by others that influenced Pasteur's experimentations.

During this era, the English physicist **John Tyndall** (1820–1893), who was also opposed to the dogma of spontaneous generation, performed a series of experiments that supported the work of Pasteur. Tyndall used optics to demonstrate that microbes were present in air and that heated infusions placed in optically clear chambers remained sterile, whereas infusions placed under an ordinary atmosphere exhibited growth. A major contribution was his empirical observation that some bacteria have phases: one is a *thermolabile* (unstable when heated) phase during which time the bacteria are destroyed at 100°C, and the other is a thermoresistant phase that renders some microbes incredibly resistant to heat. **Ferdinand Cohn** (1828–1898) confirmed these suppositions with the demonstration that hay bacilli could form the heat-resistant bodies we now call **endospores**.

To dispel the theories of spontaneous generation, Pasteur initiated a series of experiments with long-necked flasks of various types (**Figure 2.7**) to improve on experiments by earlier workers such as Schwann and Spallanzani. He fashioned one flask with a horizontal neck and placed distilled water in the flask that contained 10% sugar, 0.2% to 0.7% albuminoid (soluble proteins), and the mineral matter from beer yeast. The flask was boiled for several minutes to sterilize the contents, and the neck was attached to a platinum tube that was maintained at a red-hot temperature as the flask cooled down. The air drawn into the flask was sterilized by passage through this heated tube. Pasteur noted that flasks containing various infusions treated in this manner remained clear and free of microbial growth. He also showed that swan-necked flasks, which were long and bent down in a way that excluded passage of dust on cooling, but allowed a free exchange of air, also remained sterile. Microbes rapidly grew in all of the flasks if the neck were broken off or if its infusion were spilled into the neck and allowed to drain back into the flask.

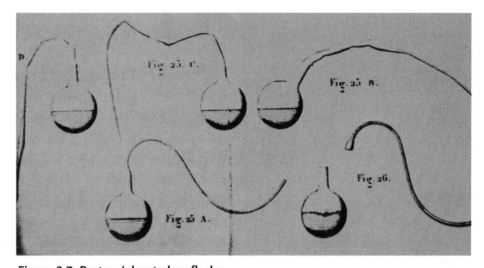

Figure 2.7 Pasteur's bent glass flasks
Pasteur's drawings of his bent neck flasks, which were employed in disproving spontaneous generation. From *The History of Bacteriology*, by William Bulloch, M.D., Oxford University Press, 1960.

In 1864 Pasteur gave a public lecture about his experiments and demonstrated some infusions that had remained unspoiled for four years. In his address, he stated:

> And, therefore, gentlemen, I could point to that liquid and say to you, I have taken my drop of water from the immensity of creation, and I have taken it full of the elements appropriated to the development of inferior beings. And I wait, I watch, I question it, begging it to recommence for me the spectacle of the first creation. But it is dumb, dumb since these experiments were begun several years ago; it is dumb because I have kept it from the only thing that man cannot produce, from the germs that float in the air, from life, for life is a germ and a germ is life. Never will the doctrine of spontaneous generation recover from the mortal blow of this simple experiment.

(From Vallery-Radot, Rene. 1920. *The Life of Pasteur.* Garden City, NY: Garden City Publishing Company, Inc., p. 108–109.)

Much heated controversy continued during the period from 1860 to 1880, and the heterogenesists continued to attack the experiments and writings of Pasteur, Tyndall, Lister, Cohn, and others. Despite this, the voice of the opposition was weakening. The Academy of Science in Paris and scientists everywhere were convinced of the correctness of Pasteur's experimental work.

Pasteur was aware of and continued the studies of Cagniard-Latour, Schwann, and Kützing (1837), who reported that the fermentation of sugar to an alcohol was due to the biological activities of a viable organism.

Pasteur initiated studies on fermentation about 1857 and published an extensive paper three years later that established a number of properties of alcohol fermentation:

- It is caused by a yeast
- The organism requires nitrogen
- Alcohol and carbon dioxide are not the sole products of sugar fermentation because yeasts utilize some of the sugar to synthesize cell protein, carbohydrate, and fat as they multiply

In 1865, Napoleon III asked Pasteur to investigate the causative agent of bad wine. Pasteur reported that one specific organism that was different from the agent of good wine caused this problem. He suggested that the juices of grapes be heated to 50° to 55°C to destroy the resident populations and that the resultant material be inoculated with a proven producer of good wine.

In addition to working with wine, Pasteur studied the fermentation of beer over several years. He felt that it was his patriotic duty to make French beer superior to the brew produced elsewhere, particularly in Germany. Whether he laid the foundation for the ultimate production of a superior brew is left to the taste of the connoisseurs. His research resulted in the publication of a book on beer fermentation in 1876.

Pasteur's studies on the butyric acid fermentation (conversion of sugar to butyric acid and other products) were very successful, as in this work he confirmed that anaerobic forms of life do exist. While microscopically examining fluid from butyric acid fermentation, he noted that organisms near the edge of the cover glass placed on the slide ceased movement whereas those in the middle swam about vigorously. He wondered whether oxygen in the air might be harmful to the organisms. To test this, he passed a stream of oxygen through a group of cells that were actively fermenting. He found that butyric acid production was halted, and the organisms died. Pasteur was the first (1863) to use the terms "aerobe" and "anaerobe" to describe the effects of air on microorganisms.

Pasteur's studies with microbes resulted in his development of new methods to work with these organisms. Many of these methods are still in use.

Knowledge of the extraordinary sterilizing effects of superheated steam came from studies in Pasteur's laboratory. He and his coworker, Charles Chamberland, noted that when hermetically sealed flasks were placed in a bath of calcium chloride and heated above 100°C, the flask solutions were free from viable organisms. Experimental work resulted in an improvement in this process so that equally effective sterilization resulted if the flasks were plugged with wool and heated in a closed container. Heating in a closed container resulted in internal temperatures that exceeded 100°C. The modern apparatus used for this purpose is the **autoclave**, which was originally manufactured in 1884 by a Parisian engineering firm under the name Chamberland's Autoclave.

In 1865, Pasteur's chemistry professor from his earlier days at the Sorbonne (who was in later life a senator from the south of France) asked his former pupil to investigate a devastating problem in silkworms. Pasteur was reluctant to go because, as a chemist, he knew nothing of silkworms. Eventually he took his entourage to Alais, France, to initiate studies on the disease. It took Pasteur five years and generous help from his able assistant, M. Gernez, to show clearly that the agent that was destroying the extensive silkworm industry in France was a transmissible microbe, specifically, the protozoan *Nosema*. Pasteur outlined a course for eradication of the disease based on isolating healthy worms, retaining the eggs produced, and examining the progenitor for a period of time. If the parent worm remained healthy, the eggs were allowed to hatch. These robust progeny were free of the disease, and following this regimen, the French silk industry was restored to its former glory.

Most of Pasteur's remaining years were dedicated to the understanding and prevention of infectious diseases in animals and humans.

Pasteur studied the disease anthrax in farm animals and suggested that immunization to this scourge was possible. Apparently, further experimentation was unsuccessful, and Pasteur abandoned the study. He was more successful in devising methods whereby immunization could prevent cholera in chickens. Pasteur obtained the chicken cholera organism and grew it in culture. Inoculation of chickens with laboratory cultures of this organism resulted in a mild illness from which they recovered. At a later date, a virulent (infectious) strain of the cholera organism was used to inoculate these chickens, but they were completely resistant to the disease. Apparently, growth in laboratory culture yielded a bacterial strain that was less virulent but could still elicit immunity. Pasteur realized the implications and potentials of this discovery and then turned to a study that was to be the crowning achievement of his career—the use of a weakened virus to prevent hydrophobia in humans bitten by rabid dogs.

During the late nineteenth century, rabies was a serious health problem in France, and Pasteur decided to devote his efforts to the eradication of this dreadful disease. Perhaps he was intrigued by the always-fatal consequence of a bite by a rabid dog and sensed the profound effect that a cure would have on the scientific world. Pasteur recognized that the infection settled in the brain and nervous system of animals. All previous efforts to isolate a microscopically visible agent (now known to be a virus) of this disease had been unsuccessful. He outlined an empirical method to develop a vaccine that would counteract the fearful effects of rabies.

In 1885, he published a method for protecting dogs after they had been exposed to rabies. A dog so exposed was protected by inoculation with an emulsion prepared from the dried spinal cord of a rabbit that had succumbed to the disease. Because he was able to prevent symptoms of rabies in animals exposed to other rabid animals through this procedure, Pasteur was convinced that the method was also applicable to unfortunate humans who were bitten by rabid animals. An opportunity to test this belief soon presented itself.

A nine-year-old boy, Joseph Meister, was brought to Pasteur in July 1885 after he had been severely bitten by a dog that was certainly rabid. Meister was injected over several days with the emulsions prepared from animal spinal cord material. After two weeks, the boy was given an injection of virus that had maximal virulence when tested in a rabbit. The boy survived, as did thousands of others treated by the same procedure, and Pasteur received worldwide acclaim.

In 1886, a commission was appointed within the Academy of Sciences in Paris to erect a scientific institute in honor of the man who had contributed so much to world health. More than 2.5 million francs were collected from throughout the world, and the Institute Pasteur was established. This institute continues today as a major scientific research center. Although Pasteur had many noble accomplishments during his lifetime, he may have erred in some of his judgments and experimentation. However, this should not detract from his great contributions to society. The well-known microbiologist A. T. Henrici summed it up well:

> It has been hinted that Pasteur was not always willing to give credit to those who had preceded him that his ideas and experiments were not always strictly original with him. It is one thing to discover a truth, another to get it established as an accepted fact. Whatever criticism may be directed towards Pasteur as regards the originality of his ideas, nothing can be said to belittle his ability to put them across. His genius for quick and accurate thinking, for keen argument, and for obtaining publicity, was not less important to the development of microbiology than his ingenious experiments. He "sold" the science to the public.

(From Henrici, A. T. 1939. *The Biology of Bacteria*, 2nd ed. New York: D. C. Heath & Co., p. 10.)

The Koch School

This school of thought concentrated on the isolation of pure cultures of pathogenic (disease causing) and saprophytic (live on dead or decaying matter) bacteria. The members of Koch's laboratory were the originators of pure culture methods and made significant contributions to many specialized branches of bacteriology (see later discussions).

Figure 2.8 Robert Koch
Robert Koch (1843–1910), the great medical microbiologist; Koch confirmed the "germ theory" of disease. From *The History of Bacteriology*, by William Bulloch, M.D., Oxford University Press, 1960.

Robert Koch (1843–1910) was trained as a physician and became a country doctor in Wollstein, East Prussia (**Figure 2.8**). His wife, seeking to allay his restless curiosity, gave him a microscope as a gift, little knowing the far-reaching consequences this small instrument would have on advancing science and in alleviating human misery. With his microscope, Koch examined many specimens, including the blood of an ox that had succumbed to **anthrax.** Anthrax is an infectious disease of warm-blooded animals. (See Box 28.4 for a discussion of anthrax in humans.) He noted the constant presence of sticklike bodies in diseased animal blood, which were absent in blood taken from healthy animals (see **Figure 2.9**). He found that the disease symptoms could be transmitted by inoculating a healthy mouse with blood from an animal that had died from anthrax. His tool for inoculation was a fire-sterilized splinter.

Koch surmised that these long, cylindrical bodies might be the viable causative disease agent and proceeded to culture the organism in fluid obtained from the eye of an ox. The organism, transferred in several passages of the fluid medium, would again cause anthrax when injected into the tail of a mouse. The transfers were done to ensure that no other agent was carried along that could cause the disease. This was the first clear experimental evidence that a bacterium was an agent of disease (see Figure 2.9).

From these experiments, Koch formulated his theories on the causal relationships between microbes and disease. From this came what we now known as **Koch's postulates:**

Figure 2.9 Causative agent of anthrax
Koch's drawings of *Bacillus anthracis* that he recognized as the causative agent of anthrax in cattle. The drawing in the lower right indicates why he described the bacterium as "sticklike." From *The History of Bacteriology*, by William Bulloch, M.D., Oxford University Press, 1960.

- A specific microorganism must be present in all cases of a disease
- The organism can be obtained in pure culture outside the host
- The organism will, when inoculated into a susceptible host, bring about symptoms equivalent to those observed in the host from which it was isolated
- The organism can be isolated in pure culture from the experimentally infected host

These postulates express a logical series of steps to be followed in identification of microorganisms obtained in the clinic and from the environment.

Koch observed that the causative agent of anthrax has a life cycle involving a dormant spore, and he theorized correctly that these were the resistant bodies responsible for survival in soil. Koch (1876) wrote to Ferdinand Cohn, a noted scientist in Breslau, Germany, telling him of his investigations and was invited to Breslau to demonstrate his work. He was received with enthusiasm, and his discoveries were acknowledged

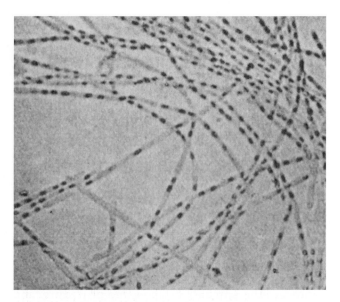

Figure 2.10 A pure culture
Koch's photomicrograph of bacteria from a single colony confirming that it was composed of one morphological type. From *The History of Bacteriology*, by William Bulloch, M.D., Oxford University Press, 1960.

with acclaim. Later he was invited to Berlin to set up a laboratory and devote his energies to studying microbes.

The origin of pure culture methods can be traced to Koch's observation that individual bacterial colonies growing on potato slices often differed in appearance (**Figure 2.10**). Microscopic examination of stained cells revealed that the organisms within a colony were similar, but were often unlike organisms in other colonies. Koch theorized that a colony arose from a single cell, and he developed streaking methods using a platinum loop (his invention) that enabled him to isolate organisms in pure culture. **Gelatin** was the solidifying agent for culture media during the early years of Koch's career (**Box 2.2**). However, gelatin had major shortcomings, and Fannie Hesse, the American-born wife of Walter Hesse, a coworker in Koch's laboratory, suggested that **agar** could be used as a solidifying agent. At that time (1882), agar was added as a jellying agent in fruit preserves. This discovery was a considerable aid in the isolation of pathogenic microorganisms. An assistant in Koch's laboratory, R. J. Petri, in 1887 developed the dish (or plate) named in his honor that is still used in culturing bacteria. The design of the **Petri dish** is virtually unchanged from his original except that glass has been replaced with plastic.

Paul Ehrlich (1854–1915) was another coworker in Koch's laboratory and made far-reaching discoveries in immunology and chemotherapy. In devising staining methods for visualizing infectious microbes in tissue, he observed that bacterial cells present often absorbed select-

Milestones Box 2.2   Discovery of Agar as a Solidifying Agent

Agar has long been the universal solidifying agent used in preparing media for growing microbes, but it was not the first employed. The first solid medium was employed by Robert Koch and was an aseptically cut slice of potato. He was able to isolate colonies on this substrate and developed his theory that all organisms in a colony were of one species. Koch assumed correctly that all organisms in a discrete colony grew from a single cell. These studies were the origin of pure culture methods. The next solidifying agent employed by Koch was gelatin. Gelatin is a protein and was unsatisfactory for two principal reasons: (1) it melts at 37°C, the favored incubation temperature for most pathogens; and (2) many bacteria can digest gelatin.

In 1882 Fannie Hesse, the American-born (New Jersey) wife of Walter Hesse, suggested that a jelling agent used in making fruit preserves might be a replacement for gelatin in bacterial media. Walter Hesse (1846–1911) was an associate

in Koch's laboratory and probably discussed the problems of gelatin with his wife. Fannie Hesse's suggestion led to the adoption of agar as the choice for a solidifying agent.

The Dutch apparently brought agar from their East Indian colonies, where it was used to improve the setting quality of jam. Chemically, agar is an extract of algae that thrive in the Pacific and Indian Oceans and Japan Sea. Agar is a complex polysaccharide containing sulfated sugars. It cannot be digested by most microbial species. Laboratory grade agar is inhibitor free, and virtually all organisms grow well in its presence.

Agar melts at 100°C and remains in the liquid state down to about 45°C. The high melting point makes agar useful for growing all organisms including thermophiles at temperatures up to almost 100°C. The low solidifying temperature permits the addition of bacteria to melted agar (45°C). They can be distributed by mixing, and the isolated colonies can be observed.

Fannie Hesse First to suggest the use of agar. Courtesy of ASM Archives, University of Maryland, Baltimore County.

Thus, an observation by Fannie Hesse on the qualities of a simple kitchen commodity became an object of worldwide utility.

---

ed dyes to a markedly greater extent than did surrounding tissue. Ehrlich reasoned that a toxic dye might destroy the bacterium without significant damage to the host tissue because so much more would adhere to the invading microbe. Ehrlich theorized that organic arsenicals might be synthesized that would be harmless to animals but toxic to invading parasites. **Arsenicals** are organic derivatives of arsenic, a toxic element. In his laboratory, scores of arsenicals were synthesized and tested on trypanosomal infections in mice. **Trypanosomes** are protozoan parasites that infect the blood of vertebrates. The 606th arsenical (Salvarsan) that he synthesized proved to be effective in curing these protozoan infections.

At the time, the spirochete that caused syphilis was reported by its discoverer, Fritz Schaudinn, to be related to the trypanosomes. Ehrlich proceeded to test 606 as a cure for syphilis. It was remarkably successful, and thus the "magic bullet" (often mentioned in the popular press) against the dreaded disease syphilis was discovered.

The magnitude of the discoveries in Koch's laboratory can be appreciated by considering that when he developed his methods in 1881, only anthrax was sus-

pected of being caused by a microbe. During the next 20 years, Koch and his coworkers confirmed that microbes were the etiological (causative) agents of 15 significant human and animal diseases (**Table 2.1**).

Chemical antiseptics originated with **Joseph Lister** (1827–1912), an English physician, who employed carbolic acid for antisepsis during surgery. Koch extended the work of Lister and devised a method for comparing the efficiency of chemical antiseptics. He dried cultures of bacteria, generally anthrax spores, on small pieces of silk thread, which were then immersed in the antiseptic solution. At intervals, a thread was removed from the antiseptic, washed in sterile water, and placed in growth medium to determine whether the organism remained viable. Koch found that carbolic acid was relatively weak in its disinfecting (killing) power, and of all the substances he tested, perchloride of mercury was most effective. It destroyed bacterial spores at a high dilution and in the shortest period of time. Mercuric compounds today remain widely used as disinfectants, although we are now concerned about the toxic effects of mercury in the environment.

Table 2.1	Diseases whose causative agents were discovered by koch and his coworkers	
Date	**Discoverer**	**Disease**
1882	Koch	Tuberculosis
1882	Loeffler and Schutz	Glanders
1884	Koch	Asiatic cholera
1884	Loeffler	Diphtheria
1884	Gaffky	Typhoid fever
1884	Rosenbach	Staphylococcal and streptococcal infections
1885	Bumm	Gonorrhea
1886	Fraenkel	Pneumonia
1887	Bruce	Malta fever
1887	Weicheselbaum	Meningococcal infections
1889	Kitasato	Tetanus
1891	Wolff and Israel	Actinomycosis
1894	Kitasato and Yersin	Plague
1897	van Ermengen	Botulism
1898	Shiga	Acute dysentery

He introduced the principles of enrichment culture, which gave clarity and rationality to microbial ecology. **Enrichment culture** is a means by which organisms that evolved to exist under any specific conditions of substrate, temperature, pH, salinity, osmolarity, and oxygen availability can be isolated. The sole limitation on isolation is the existence of an organism in the inoculum taken from a natural environment.

The enrichment culture technique provided a means for the isolation of various physiological types of microorganisms that exist in natural environments. An enrichment medium is prepared with a defined chemical composition and inoculated with soil or water rich in microbes; only those microbes capable of growth on that particular medium will grow. The microbes that thrive will be those best equipped by heredity to survive on that medium at the temperature, pH, and other conditions chosen. With this technique, Beijerinck and his followers readily obtained microbes with differing physiological capabilities, and

Microbes as Agents of Environmental Changes

The adverse effects that microbes had on humans, food, and animals were the major considerations of the Koch and Pasteur schools. The causative agents of diseases, such as tuberculosis or rabies, caused much human suffering and challenged the scholar of scientific bent. During the latter years of the nineteenth century, knowledge of disease expanded rapidly along with a broad acceptance of the germ theory of disease. Another area that caught attention during the latter part of the nineteenth century was the potential role that microorganisms might play as a component part of all life on Earth. This leads us to the next conceptual development in microbiology—the role of microbes in nature.

Ferdinand Cohn suggested in 1872 that microbes were involved in the ultimate cycling of all living matter and that the activities of organisms in the biosphere allowed for the reutilization of cellular constituents. Our knowledge of the indispensable role that diverse microorganisms play in recycling constituents of living cells has expanded greatly since this pronouncement by Cohn. Microbial ecology, the relationship between the microbe and the environment and the scope of microbiology, expanded markedly in the twentieth century.

Martinus Beijerinck An important technique developed by the great Dutch botanist **Martinus Beijerinck** (1851–1931) provided much of the foundation for the elucidation of the various functions of microbes in the cycles (carbon, nitrogen, sulfur) of matter (**Figure 2.11**).

Figure 2.11 Martinus Beijerinck
Martinus Beijerinck (1851–1931), a major contributor to our understanding of the role of microbes in nature. From *Martinus Willem Beijerinck: His Life and His Work*, by G. van Iterson Jr., L. E. den Dooren de Jong, and A. J. Kluyver, Martinus Nijholt, The Hague, 1940.

they were able to assess the potential role of that microbe under natural conditions.

Beijerinck discovered free-living, nitrogen-fixing (assimilate atmospheric nitrogen) bacteria by the application of enrichment using a medium devoid of nitrogenous compounds such as ammonia or amino acids. The aerobic microbe that he obtained was given the genus name *Azotobacter*. He also published extensively on the symbiotic nitrogen-fixing organisms (*Rhizobium*) that form nodules in the roots of legumes such as peanuts.

Beijerinck discovered and described many major groups of bacteria: the luminous organisms (*Photobacterium*), the sulfate reducers (*Desulfovibrio*), the methane-generating bacteria (now *Archaea*), and *Thiobacillus denitrificans,* an organism involved with denitrification (reduction of nitrates to nitrogen gas). Beijerinck did much of the early work on lactic acid bacteria and proposed the genus name *Lactobacillus*. He recognized that "soluble" living germs existed that he called "contagium vivum fluidum," which is generally accepted as the initial description of a virus (specifically, the tobacco mosaic virus). His contributions to microbiology are legion, and his perceptions of the great role of microbes established the foundation of modern approaches to microbial physiology and microbial ecology.

Sergei Winogradsky During the era of Koch, Pasteur, and Beijerinck, there arose another major figure in the field of general bacteriology (**Figure 2.12**). **Sergei Winogradsky** (1856–1953) was born in Russia in 1856, and during his long and fruitful life, he witnessed the origins of the science of microbiology and survived to see the Age of Antibiotics. It is noteworthy that both Winogradsky and Beijerinck spent time in the laboratory of the great mycologist and plant pathologist **Anton De Bary** (1831–1888) at Strassburg, Germany. Winogradsky arrived in Strassburg shortly after Beijerinck left. In Strassburg, Winogradsky initiated studies on sulfur-oxidizing bacteria. He concluded that the bacterium *Beggiatoa* could utilize inorganic H_2S as a source of energy and atmospheric CO_2 for carbon in the synthesis of cellular material. He named these organisms "orgoxydants" and thus opened up the entire concept of **autotrophy**, which is the ability of bacteria to manufacture their cells from CO_2 and use inorganic chemicals or light as a source of energy. Prior to his study, only chlorophyll-containing plants were believed to use CO_2 as their sole carbon source.

Following the death of De Bary, Winogradsky went to Zurich, where he isolated and clarified the role of the bacteria that convert ammonia to nitrate, the autotrophic nitrifying bacteria. During this period, he also showed that green and purple bacteria could oxidize hydrogen sulfide to sulfate, but he was uncertain whether or not this was a photosynthetic process. In

Figure 2.12 Sergei Winogradsky
Sergei Winogradsky (1856–1953), a Russian-born microbiologist. Winogradsky was the father of autotrophy. He lived from the days of Pasteur and Koch to the modern era of microbiology. From *Sergei N. Winogradsky: His Life and Work*, by S. A. Waksman, © 1953 by the Trustees of Rutgers College. Reprinted by permission of Rugers University Press.

addition, he isolated the nitrogen-fixing anaerobe *Clostridium pastorianum*.

In 1891, the noted scientist and Nobel laureate **Elie Metchnikoff** (1845–1916) carried a personal letter from Pasteur to Zurich inviting Winogradsky to work at the Pasteur Institute in Paris. Metchnikoff was the discoverer of phagocytosis and cellular immunity. After much deliberation Winogradsky decided to accept a position in St. Petersburg, Russia, thus ending the first half of his illustrious scientific career. He did go to Paris in 1892 to represent Russia at the seventieth-birthday celebration for Pasteur and met Beijerinck for the first and only time. The revolution in Russia led Winogradsky to emigrate, and in 1922 he accepted an offer from the Pasteur Institute and moved to Paris. He spent the remainder of his life there studying the broad aspects of soil bacteriology. Winogradsky compiled his life's work and published it in a monumental treatise entitled *Microbiologie du Sol* (*Microbiology of the Soil*). He survived the deprivations of World War II and died in Paris at the age of 97.

Microbes and Plant Disease

Among the first to recognize that microbes might be directly involved in plant diseases was Anton De Bary, who in 1853 suggested that *Brandpilze* (plant rust) was caused by a parasitic fungus. He later proved experimentally that a fungus, *Phytophthora infestans*, caused late blight of the potato. This disease had caused crop failures in the 1840s in Ireland, and in 1845 M. J. Berkeley demonstrated that the blight was caused by a fungus. The potato blight continued in the 1850s and 1860s and led to a widespread famine, thousands of deaths, and the immigration of 1.5 million Irish to the United States.

Although the causal relationship between bacteria and animal disease had gained widespread acceptance by 1900, few botanists were willing to believe that bacteria could cause disease in plants. The fungi were generally acknowledged to be the agents of plant disease. J. H. Wakker, a Dutch scientist working with De Bary in 1881, isolated and identified the bacterium *Xanthomonas hyacinthia* as the pathogen causing widespread damage to hyacinth bulbs in Holland. Erwin Smith, in 1895, demonstrated that a bacterium caused a wilt in cucurbits (cucumbers and other members of the gourd family), and in 1898, Beijerinck concluded that a virus was the cause of tobacco mosaic.

Despite these and other reports, considerable controversy remained regarding the role of bacteria in plant disease. It took two American scientists, Thomas Burrill and Erwin Smith, to convince the scientific world that bacteria did indeed cause plant disease. Burrill was the first to discover and demonstrate a bacterial disease of plants when he reported in 1877 that pear blight was caused by *Micrococcus* (*Erwinia*) *amylophorus*. Burrill investigated a number of diseases of corn, potatoes, and fruit, but he is probably best known as the first person in America to offer a laboratory course in bacteriology.

During the period from 1895 to 1900, Erwin Smith implicated bacteria in diseases of tomatoes, cabbage, and beans and in the wilt of maize. By applying the postulates of Koch (a hero to Smith), his group at the U.S. Department of Agriculture found definitive proof that strains of *Xanthomonas*, *Pseudomonas*, *Erwinia*, and *Corynebacterium* were the causative agents of disease in agricultural crops. Smith also demonstrated that crown gall, a plant tumor, resulted from an *Agrobacterium tumefaciens* infection. These studies established plant pathology as a field of scientific inquiry.

TRANSITION TO THE MODERN ERA

Louis Pasteur died in 1895, and Koch died in 1910. Beijerinck retired in 1921, and the first part of Winogradsky's scientific career ended at the turn of the century when he returned to Russia. Thus ended what is considered to be the "Golden Era of Microbiology," for during the 50-year period from 1870 to 1920, the discipline of microbiology was firmly established.

Albert Jan Kluyver (1888–1956) replaced Beijerinck as professor of microbiology at the Technical University at Delft in 1922. Knowledge of metabolic processes in microbes and in living cells in general was then quite limited. Most questioned whether metabolic reactions in microbes and in higher forms of life could occur in any equivalent manner. Knowledge of biochemical activities in microbes was mostly restricted to a number of unrelated transformations such as nitrification brought about by individual organisms.

During the decade following his arrival at Delft, Kluyver introduced order to chaos by presenting a simple coordinated model for metabolic events in all living cells. His experimental approach led to our modern understanding that unity exists in biochemical reactions, a concept termed **comparative biochemistry**. This concept arose from his proposal that the basic feature of virtually all metabolic processes is a transfer of hydrogen (oxidation/reduction reactions). This transfer is universal and occurs whether the organism is aerobic or anaerobic, autotrophic or heterotrophic. He also believed that biosynthetic and biodegradative pathways in cells are highly coordinated and relatively few in number. He proposed that metabolic processes are functionally equivalent in all living cells. It is interesting to note that the general ideas behind the "Unity of Biochemistry" concept have been borne out by the recent comparative study of genomes.

Kluyver and **Cornelis B. van Niel**, one of his students, proposed that aerobic and anaerobic respiration could be illustrated by the following simple formula:

$$AH_2 + B \rightarrow A + BH_2$$

in which A represents a more reduced element, such as NH_4^+, CH_4, and B represents oxygen in aerobic respiration or some less-reduced metabolic intermediate in anaerobic respiration. He was certain that oxidation/reduction reactions are basic to all life forms.

They also suggested that the general formula for all photosynthetic reactions would be as follows:

$$CO_2 + 2H_2A \xrightarrow{\text{light}} CH_2O \text{ (cell material)} + H_2O + 2A$$

In plant photosynthesis, A would represent oxygen because the hydrogen atom for CO_2 reduction is donated by H_2O. Molecular oxygen (O_2) is released in this process. Prior to their studies, it was widely considered that light was used in photosynthesis to decompose carbonic acid and generate oxygen (O_2). For photosynthesis to occur in the anaerobic photosynthetic bacteria, A could be hydrogen, sulfide, or a reduced organic compound. As a consequence of the reduction of CO_2, an oxidized product such as sulfate would be generated.

One should not underestimate the contribution that the comparative biochemistry concept made to advances in the area of cellular metabolism. The bacteria became a major model system for study only with the general acceptance of this unity by all biologists. During the decade following World War II, a fundamental understanding of the role and structure of DNA, enzymes, metabolic regulatory mechanisms, and microbial genetics was attained. These discoveries led to a broad, integrated study on cellular growth that we now call **molecular biology**, a term coined by William Astbury in 1945.

Many characteristics of *Bacteria* and *Archaea*—rapid growth, simple nutritional requirements, growth under harsh conditions, and ease in handling—make them attractive tools for physiological studies. Because researchers generally choose to progress from simpler systems to those of greater complexity, the *Bacteria* and *Archaea* have provided an attractive model system for studying animal and plant processes. The broad array of *Bacteria* and *Archaea* species available—from those that use light as an energy source to the myxobacteria, which form fruiting bodies—makes them a system of choice for investigations on regulation, biosynthesis, gene expression, and molecular interactions. Studies of their ability to grow under extreme conditions of pH, temperature, salinity, and anaerobiosis (and combinations of these extremes) have yielded many clues to the elucidation of life strategies. Other unique characteristics that allow study at the molecular level include nitrogen fixation, antibiotic synthesis and action, parasitism (for bacteria and eucaryotes), directed motility, and development of bacteriophages (bacterial viruses).

Much might be written on the growth of microbiology as a scientific discipline over the last 50 years. The following chapters give ample evidence that much has been accomplished.

SUMMARY

▶ **Epidemic diseases** played a major role in battles throughout recorded history. Napoleon was driven from Russia in 1812 by disease and deprivation and ultimately lost at Waterloo because about one-half of his army was incapacitated by illness.

▶ The first significant writings on contagious disease were those of **Girolamo Fracastoro** during the first half of the sixteenth century. He suggested that invisible organisms were the agents of disease.

▶ **Antony van Leeuwenhoek** observed and made extensive drawings of bacteria in 1683, but it was 200 years before "spontaneous generation" was disproved and the "germ theory" of disease was accepted.

▶ A number of scientists who preceded Louis Pasteur did much careful experimentation that disproved **spontaneous generation**. Among these were: Redi, Spallanzani, Schulze, Cagniard-Latour, Schwann, and Kützing.

▶ Edward Jenner developed an effective **vaccination** that protected humans against smallpox long before the germ theory of disease was established.

▶ When it became clear that animals could not spring from inanimate material, the proponents of spontaneous generation moved to changes caused by decay or putrefaction as proof that **chemical instabilities** were the basis for the observable changes in organic matter.

▶ By 1864 Pasteur had clearly proven that **spontaneous generation** did not occur, and his ideas on **fermentation** were generally accepted.

▶ Pasteur and his colleague, Chamberland, developed the **autoclave**.

▶ Pasteur was the first to observe and report that **anaerobic bacteria** existed. He also recognized the immunity to disease was possible and developed an effective rabies vaccine.

▶ The Academy of Sciences in Paris collected funds from around the world and established the renowned **Institute Pasteur** in honor of Louis Pasteur.

▶ Robert Koch was the first to clearly demonstrate that bacteria were a causative **agent of infectious disease**.

▶ Koch proposed a set of postulates (**Koch's postulates**) that could be employed to demonstrate that a culturable organism was the causative agent of disease symptoms.

▶ Koch developed **pure culture** methods. His laboratory developed the loop and the Petri dish and was the first to use agar as a solidifying agent for bacterial growth.

▶ Paul Ehrlich is considered the founder of **chemotherapy**, the use of chemicals to kill infectious microorganisms.

▶ Martinus Beijerinck was a pioneer in assessing the role of microorganisms in the **cycles of matter** (carbon, nitrogen, and sulfur) in nature. He discovered the free-living nitrogen fixing *Azotobacter* sp. and the symbiotic nitrogen fixing *Rhizobium* sp.

▶ **Sergei Winogradsky** was the first to recognize that **autotrophic** bacteria were of widespread occurrence. He spent the last 34 years of his life at the Pasteur Institute studying the activities of bacteria in soil.

▶ The American scientists Thomas Burrill and Erwin Smith elucidated the role of bacteria in plant disease.

▶ Albert Jan Kluyver, a Dutch bacteriologist, proposed that basic metabolic reactions in all living cells were equivalent, a concept known as **comparative biochemistry**.

▶ *Bacteria* and *Archaea* have proven to be an excellent **model** system for studies in metabolism, genetics, and molecular interactions.

REVIEW QUESTIONS

1. What are some of the influences the microbe has had on human history? Discuss in terms of war, food, population control, and other factors.

2. It took 200 years, from Antony van Leeuwenhoek's observations and descriptions of bacteria (1673) until the latter part of the nineteenth century, for the role of bacteria to be elucidated. Why did it take so long?

3. Define "spontaneous generation." Why did this concept have so many proponents?

4. Spallanzi and Redi made significant contributions to disproving "spontaneous generation." What were some of their contributions, and why were their experiments not accepted as definitive proof? What did Franz Schulze add that was crucial?

5. Koch and Pasteur were interested in the effects of microbes on human health and well-being, Winogradsky and Beijerinck in the role of microbes in nature. Give examples that support this reasoning.

6. How did Louis Pasteur contribute to conquering rabies? What experiment(s) led him to conclude that a treatment was possible?

7. Most of the basic techniques used in microbiological studies originated in Robert Koch's laboratory. What are some of them?

8. What are Koch's postulates? How did he use these in discovering the causative agents of disease? What are some of the common diseases for which Koch's laboratory discovered the causative agent?

9. Do bacteria play a role in plant disease? How did our understanding of the role of fungi and bacteria in plant disease evolve? What important historic development resulted from plant disease?

10. How did the philosophy of Albert Kluyver and C. B. van Niel influence modern science?

SUGGESTED READING

Beck, R. P. 2000. *A Chronology of Microbiology in Historical Context.* Herndon, VA: ASM Press.

Brock, T. D., ed. and trans. 1999. *Milestones in Microbiology 1546–1940.* Herndon, VA: ASM Press.

Brock, T. D. 1999. Robert Koch: *A Life in Medicine and Bacteriology.* Herndon, VA: ASM Press.

Bulloch, W. 1938. *The History of Bacteriology.* London: Oxford University Press.

Dobell, C., ed. and trans. 1932. *Antony van Leeuwenhoek and His "Little Animals."* New York: Harcourt, Brace and Company.

Dubos, Rene. 1998. *Pasteur and Modern Science.* Herndon, VA: ASM Press.

Joklik, W. K., L. G. Ljungdahl, A. D. O'Brien, A. von Graevenitz and C. Yanofsky. 1999. *Microbiology: A Centenary Perspective.* Herndon, VA: ASM Press.

Kamp, A. F., J. W. M. La Riviere and W. Verhoeven, eds. 1959. *Albert Jan Kluyver: His Life and Work.* Amsterdam: North-Holland Publishing Co.

Vallery-Radot, R. 1920. *The Life of Pasteur* (translated by R. L. Devonshire). New York: Garden City Publishing Co.

Van Iterson, G., L. E. Den Dooren De Jong and A. J. Kluyver. 1940. *Martinus Willem Beijerinck: His Life and Work.* The Hague, Netherlands: Martinus Nijhoff.

Waksman, S. A. 1953. *Sergei N. Winogradsky: His Life and Work.* New Brunswick, NJ: Rutgers University Press.

Chapter 3

Fundamental Chemistry of the Cell

Life, atom of that infinite space that stretcheth,
twixt the here and there.
– SIR RICHARD FRANCIS BURTON, 1821–1890

The basic physical principles that govern the chemical activities of all elements apply equally to both living and nonliving systems. The living cell differs in that it has the capacity to utilize these physical properties to generate energy, to grow, and to reproduce. To understand viable systems, we must first have a fundamental knowledge of chemical principles and how these principles relate to the structure and function of a cell. For example, we know that all matter is composed of molecules that are made up of atoms. We also know that matter has characteristics that are determined by the way in which these atoms are joined to form molecules.

A molecule is formed when two or more atoms bond to one another. They may be like atoms and form a molecule such as O_2 or N_2 or be unlike atoms and form a compound such as CO_2 or glucose. Macromolecules are composed of interconnected compounds that, working in concert, produce life. This chapter is devoted to an overview of chemical principles underlying the formation of the molecules that collectively make up a living cell.

ATOMS

The **atom** is the smallest unit that has all of the characteristics of an element. There are 92 different naturally occurring elements. An atom can exist either as a single unit or in combination with other atoms. The smallest atom is that of hydrogen, which consists of one **proton** and one **electron.** The proton is located in the dense core of the atom called the **nucleus,** and the electron orbits the nucleus at great speed. The nuclei of all elements except hydrogen also contain **neutrons.** A proton has a positive charge, an electron has a negative charge, and a neutron has no charge. The number of protons in the nucleus of an atom is equal to the number of electrons orbiting around it, which effectively renders an atom electrically neutral. The attraction between the positively charged nucleus and the negatively charged electrons keeps the atom intact. The nucleus contains 99.9% of the mass of one atom but occupies only one-hundred-trillionth (10^{-14}) of the volume.

The number of protons in the nucleus of any element is constant, and this is called the **atomic number.** The number of neutrons and electrons associated with an atom may vary. There is almost the same number of protons and neutrons in the nucleus, and the sum of these is the **atomic weight.** Electrons are inconsequential in the atomic weight of an atom because an electron is only 1/1836 of the relative mass of a proton or neutron. The structure of an element that contains an equal (or close to equal) number of protons and neutrons is considered the **elemental** form and is the form that is most abundant in nature.

Isotopes are atoms that have a greater number of neutrons than protons. Some isotopes of biological interest are listed in **Table 3.1**. The atomic number of an isotope does not differ from the

Table 3.1	Some isotopes of biological interest			
Element	**Protons**	**Neutrons**	**Isotope**	**Type^a**
Sulfur	16	16	^{32}S	Elemental
	16	18	^{34}S	Stable isotope
	16	19	^{35}S	Radioisotope
Carbon	6	6	^{12}C	Elemental
	6	7	^{13}C	Stable isotope
	6	8	^{14}C	Radioisotope
Oxygen	8	8	^{16}O	Elemental
	8	9	^{17}O	Stable isotope
	8	10	^{18}O	Stable isotope
Phosphorous	15	16	^{31}P	Elemental
	15	17	^{32}P	Radioisotope
Nitrogen	7	7	^{14}N	Elemental
	7	8	^{15}N	Stable isotope
Hydrogen	1	0	^{1}H	Elemental
	1	2	^{3}H	Radioisotope

^{a}The elemental form is most abundant in nature and has virtually the same number of protons and neutrons in the nucleus. Stable isotopes have a greater number of neutrons, but the excess neutrons do not decay. The excess neutrons in radioisotopes spontaneously decay with the release of subatomic particles.

natural element because the number of protons is invariant. The atomic weight, of course, differs because of the greater number of neutrons.

Some isotopes are stable whereas others spontaneously decay. Those that decay with the release of subatomic particles (radioactivity) are termed **radioisotopes.** Radioisotopes can occur naturally but may also be produced by physical means. Radioisotopes are generally obtained by neutron bombardment. Thermal neutrons that are activated during nuclear fission may enter the nucleus of an atom, thus creating an isotope.

The naturally occurring form of oxygen is ^{16}O, the isotopes of oxygen are ^{17}O, and ^{18}O, and their relative abundance in the environment is 99.76%, 0.04%, and 0.01%, respectively. The radioisotopes of carbon (^{14}C), phosphorous (^{32}P), sulfur (^{35}S), and the radioisotope of hydrogen, commonly called tritium, are widely employed in research. Radiolabeled atoms can be detected in exceedingly small amounts, thus allowing scientists to follow the fate of a radioisotope in a biological system. The use of a radioisotope in studying metabolic pathways is illustrated in **Box 3.1**. The stable isotopes of carbon (^{13}C), nitrogen (^{15}N), and oxygen (^{18}O) are also employed in biological research, and their presence is determined by mass spectrometry. A stable isotope would have more neutrons than

the natural form of the element and therefore a larger mass. The natural form of each of these elements is ^{12}C, ^{31}P, ^{32}S, and ^{1}H.

ELECTRONIC CONFIGURATIONS OF MOLECULES

The electrons that spin around the nucleus assume definite patterns. These patterns are termed **electronic shells** or **orbitals.** The electrons located in the shells furthest from the nucleus travel the fastest and are at the highest energy level. The first orbital (1s) is spherical in shape and has the lowest energy level. The other orbitals, which are farther from the nucleus, are termed the p, d, and f orbitals. Whereas the 1s orbital is spherical, the other orbitals can be either spherical or dumbbellshaped. The electrons can be depicted as an electron cloud that surrounds the nucleus, as presented in **Figure 3.1**. The arrangement of electrons around the nucleus may also be referred to as the **electron configuration.** An electron will always fill an orbital nearer the nucleus, and

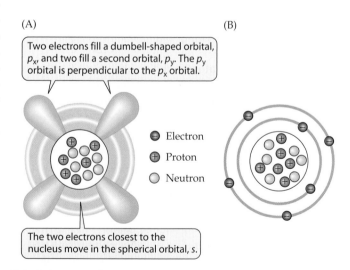

(A)

Two electrons fill a dumbell-shaped orbital, p_x, and two fill a second orbital, p_y. The p_y orbital is perpendicular to the p_x orbital.

The two electrons closest to the nucleus move in the spherical orbital, s.

(B)

⊖ Electron
⊕ Proton
○ Neutron

Figure 3.1 Structure of an atom
(A) Orbital model of a carbon atom, showing spherical and dumbbell-shaped orbitals. (B) The Bohr model (named for the Danish physicist Niels Bohr), a less-accurate representation but commonly used for convenience. The nucleus in these representations is proportionately much larger than in an actual atom, simply to accommodate display of the protons and neutrons (this applies also to Figures 3.2, 3.3, and 3.4).

Milestones Box 3.1   CO₂ Fixation and the Nuclear Age

In 1796, Jan Ingen-Housz, a Dutch scientist, wrote that plants obtain their cellular carbon by assimilating carbon dioxide from the environment. Subsequently, others discovered that selected bacteria could also obtain their cell carbon by "CO₂ fixation." Elucidating the pathway whereby CO₂ was incorporated into a cell proved to be very difficult. From a biochemical standpoint, autotrophs (CO₂ as sole carbon source) were indistinguishable from heterotrophs (organics as carbon source). Data available indicated that CO₂ entered into metabolic sequences much the same as sugars and other substrates. But how? The answer to this intriguing mystery came only when better tools were available to solve it. Ironically, the research that provided these tools occurred indirectly in the development of the atomic bomb during World War II (1941 to 1945). Research that led to atom bombs brought us into the nuclear age. The lessons learned in bomb development have been extended to provide us with nuclear energy for electrical power generation, nuclear medicine, and important research tools. In a way, the problem of CO₂ fixation ran headlong into the nuclear age.

Under proper conditions, a nuclear reactor can be manipulated to introduce extra neutrons into the nuclei of selected elements. The extra neutrons in the nucleus of an element tend to decay with the release of radioactivity (subatomic particles). This radioactivity is readily detectable and permits us to locate the site of a radioactive compound in a cell. A radioactive compound is also called a **tracer** because we can "trace" the movement of a radioactive compound in a biological system. With the knowledge that became available in 1947, it was possible to create radiolabeled carbon dioxide and follow its incorpora-tion into an organism during photosynthesis. Radiolabeled carbon dioxide is abbreviated $^{14}CO_2$, indicating

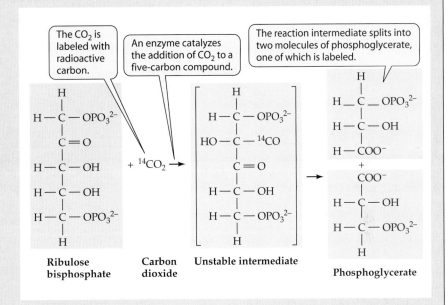

The CO₂ is labeled with radioactive carbon.

An enzyme catalyzes the addition of CO₂ to a five-carbon compound.

The reaction intermediate splits into two molecules of phosphoglycerate, one of which is labeled.

Ribulose bisphosphate   Carbon dioxide   Unstable intermediate   Phosphoglycerate

that there are 14 neutrons present in the carbon atoms rather than the normal 12.

In 1949, Melvin Calvin and Andrew Benson initiated a series of experiments employing $^{14}CO_2$, and their studies led to an understanding of the biochemistry of CO₂ fixation by exposing photosynthesizing algae to $^{14}CO_2$ for varying lengths of time. They then isolated and identified the radioactive compounds present in the cells. They reasoned that the compounds that became radioactive after brief exposure were those in which the CO₂ was initially fixed. Those compounds involved in succeeding reactions would become radioactive with time. Analysis of the pattern of $^{14}CO_2$ fixation into compounds and the distribution of the radiolabel within these compounds ultimately answered the question of how a cell can fix carbon from the atmosphere. CO₂ fixation occurs in most autotrophs as follows:

Thus, a seemingly unrelated event, the development of the atomic bomb, presented the unlikely solution to a major biological question. We now know that there are other CO₂ fixation reactions in autotrophic bacteria. These reactions will be discussed in Chapter 10.

Algae plus $^{14}CO_2$ Light

Calvin-Benson experiment Courtesy of the Lawrence Berkeley National Laboratory.

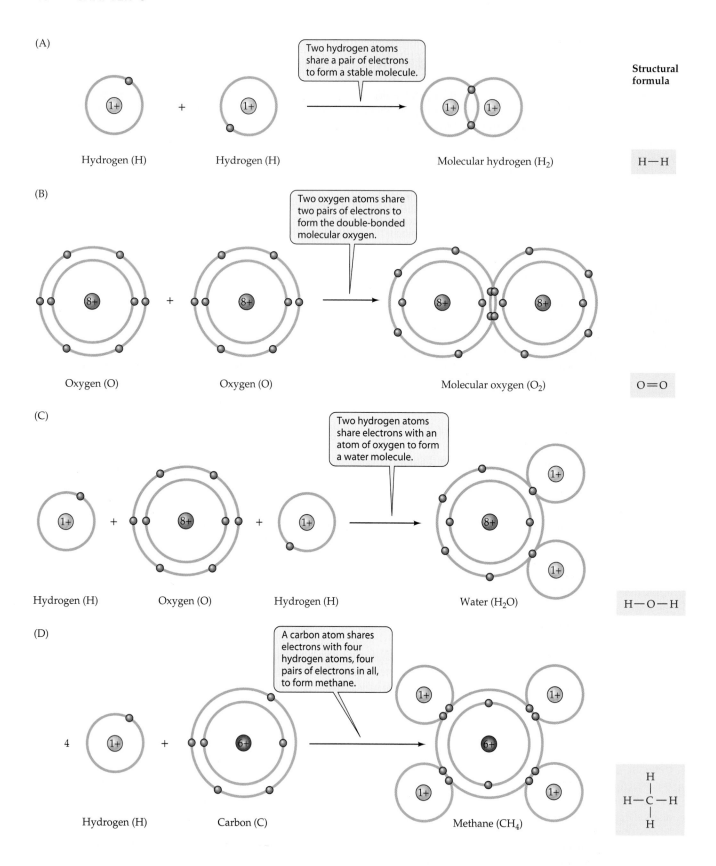

Figure 3.2 Sharing of electrons
Structures and electron-sharing (bonding) properties of
some biologically important atoms.

the total number of electrons that can occupy the first orbital is two; the second shell has four orbitals (they are spherical or dumbbell shaped, see Figure 3.1) and can contain eight electrons; the third has a maximum of 18 electrons in nine orbitals; and the fourth contains 32 electrons in 16 orbitals. Some molecules of biological interest are depicted to the left in **Figure 3.2**. The **valence** of an atom is based on the number of electrons that are needed (or may be donated) to create a full complement of electrons in the outer orbital. In the examples presented in Figure 3.2, hydrogen can fill the first orbital by gaining an electron, oxygen by gaining two electrons, and carbon by gaining or losing four electrons. Hydrogen, therefore, has a valence of +1, carbon +4, and oxygen –2. There is a tendency for atoms to interact so that the outer shell will contain the maximum allowable number of electrons. Thus, they donate, accept, or share electrons to attain this configuration. When the outer orbit is filled, a chemically stable state is attained. For example, argon, neon, and helium have filled outer shells, and these gases are mostly inert and unreactive.

MOLECULES

An atom may exist separately, in combination with dissimilar atoms, or in some cases, combined with like atoms. For example, two atoms of hydrogen may share electrons (one from each atom) to form H_2 (Figure 3.2A). Oxygen atoms are generally paired, and the Earth's atmosphere contains O_2 (Figure 3.2B). Atoms, however, most often pair with unlike atoms, and this results in the formation of **compounds.** A compound that is formed by combinations of atoms other than carbon is considered an **inorganic** compound; those that contain carbon are generally termed **organic** compounds. The configurations of other simple compounds are illustrated in Figure 3.2C and D. For example, oxygen picks up one electron from each of two hydrogen atoms to fill its outer orbital and form water. Methane is formed from the sharing of four electrons in the second orbital with four atoms of hydrogen.

When we wish to denote the chemical composition of a molecule, we use a **chemical formula.** The chemical formulas for water, methane, and carbon dioxide are H_2O, CH_4, and CO_2, respectively.

CHEMICAL BONDS

The forces that hold molecules together are called **chemical bonds.** Energy is required for the formation of chemical bonds, and as a result, each bond has some potential chemical energy. There are three main types of bonds that join atoms or molecules together: **ionic bonds, covalent bonds,** and **hydrogen bonds.** These bonds are formed because atoms seek a full complement of electrons in the outer shell or tend to achieve neu-

trality through interactions between polar (charged) molecules. A discussion of each of these bonds follows.

Ionic Bonds

Two atoms achieve a full complement of electrons in their outer orbital when one of the atoms donates an appropriate number of electrons and the other atom accepts these electrons. The donor has a **positive charge (cationic)** and the acceptor a **negative charge (anionic).** The loss of electron(s) in the donor results in an excess of protons relative to electrons, and the gain of an electron(s) results in an excess of electrons over protons. An element with an excess number of protons is referred to as positive (+). Some examples include Na^+, K^+, Ca^{2+}, Mg^{2+}, and Mn^{2+}. An element with an excess of electrons is negatively (–) charged; examples include Cl^-, I^-, and S^{2-}. An element with a positive charge can bond with an element that has a negative charge. The bond that joins them is considered an **ionic bond.**

The classical example of the formation of a compound by ionic bonding is the reaction between Na^+ and Cl^- to form sodium chloride—NaCl (table salt) (**Figure 3.3**). Elemental sodium donates an electron and achieves

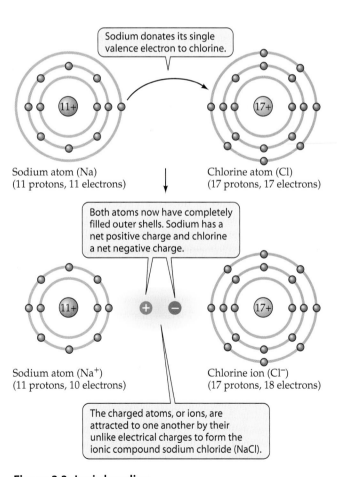

Sodium donates its single valence electron to chlorine.

Sodium atom (Na)
(11 protons, 11 electrons)

Chlorine atom (Cl)
(17 protons, 17 electrons)

Both atoms now have completely filled outer shells. Sodium has a net positive charge and chlorine a net negative charge.

Sodium atom (Na$^+$)
(11 protons, 10 electrons)

Chlorine ion (Cl$^-$)
(17 protons, 18 electrons)

The charged atoms, or ions, are attracted to one another by their unlike electrical charges to form the ionic compound sodium chloride (NaCl).

Figure 3.3 Ionic bonding
Formation of a compound by ionic bonding. Shown here is the compound sodium chloride.

Table 3.2   Types of covalent bonds that are an integral part of biomolecules. These are the major functional groups present in living cells

Carbon

$$-\overset{|}{\underset{|}{C}}-H \qquad -\overset{|}{\underset{|}{C}}-\overset{|}{\underset{|}{C}}- \qquad \overset{/}{\underset{\backslash}{C}}=\overset{\backslash}{\underset{/}{C}}$$

$$-\overset{|}{\underset{|}{C}}-O-\overset{|}{\underset{|}{C}}- \qquad \overset{\backslash}{\underset{/}{C}}=O \qquad -\overset{|}{\underset{|}{C}}-OH$$

Nitrogen

$$\overset{\backslash}{\underset{/}{N}}-H \qquad -\overset{\backslash}{\underset{/}{C}}-N= \qquad -\overset{|}{\underset{|}{C}}-N\overset{/}{\underset{\backslash}{}} \qquad \overset{\backslash}{\underset{/}{C}}=N-$$

Phosphorus

$$\overset{\backslash}{\underset{/}{P}}-O-\overset{|}{\underset{|}{P}}\overset{/}{} \qquad \overset{\backslash}{\underset{/}{}}P=O$$

Sulfur

$$-S-H \qquad -\overset{|}{\underset{|}{C}}-S- \qquad \overset{\backslash}{\underset{/}{}}\overset{|}{S}-O-$$

$$\overset{\backslash}{\underset{/}{}}S=O \qquad -S-S-$$

a full complement of electrons (8) in the outermost orbital. It is actually an ion because it carries an electrical charge due to the excess of one proton relative to electrons (Na^+). Chlorine picks up one electron, resulting in a full outer orbital, but it now has one more electron than protons in the nucleus (Cl^-). The attraction between the cation Na^+ and the anion Cl^- results in the formation of a salt.

Ionic bond formation has limited applicability to biological systems. However, it is important in the neutralization of DNA in a bacterial cell. DNA is acidic; thus, divalent cations such as Mg^{2+} form ionic bonds with and neutralize the anionic phosphate PO_4^{-2} in bacterial DNA. Neutralization of DNA in eukaryotes differs in that it is neutralized through interaction with basic (positively charged) proteins termed histones.

Covalent Bonds

The **covalent bond** is more relevant to biological systems because it is the major bond joining the component elements that form the molecules of the cell. Carbon-to-carbon bonds are examples of covalent bonds and are the basic backbone in cellular lipids, proteins, nucleic acids, and carbohydrates. Some of the covalent bonds that are important in cell structures are presented in **Table 3.2**. A covalent bond is formed when one or more pairs of electrons are shared. The simplest example of shared electrons is the hydrogen molecule where two protons share two electrons (Figure 3.2A). These electrons orbit around both nuclei. This is an example of a **single covalent bond** because one pair of electrons is shared. Other examples of the sharing of a single pair of electrons are methane and water. A **double covalent bond** is formed when two pairs of electrons are shared. An example of this is carbon dioxide (**Figure 3.4**). Carbon dioxide is a stable molecule because each oxygen molecule has a valence of –2 and carbon has a valence of +4. The four electrons in the

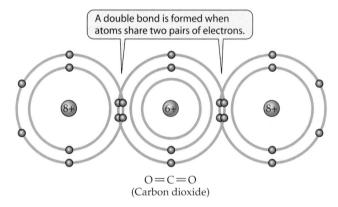

A double bond is formed when atoms share two pairs of electrons.

$$O=C=O$$
(Carbon dioxide)

Figure 3.4 Formation of double bonds
The sharing of electrons between an atom of carbon and two atoms of oxygen in carbon dioxide.

outer shell of the carbon atom are shared with two2 electrons from each of two oxygen atoms, resulting in a full complement of electrons in the outer orbital of both atoms. Double bonds, the sharing of two electron pairs, occur frequently between adjacent carbons (C=C) in many of the important compounds in cells.

In some cases, a triple bond occurs because three pairs of electrons are shared. A biomolecule of this type is N≡N (N_2). *Bacteria* and some of the *Archaea* are able to cleave triple bonds, as some species utilize molecular nitrogen gas from the atmosphere as a source of organic nitrogen. This cleaving of N_2 by these organisms requires considerable expenditure of energy, as the N≡N triple bond is one of the strongest of all chemical bonds (**Table 3.3**).

Hydrogen Bonds

When hydrogen forms a covalent bond with an electronegative atom such as nitrogen or oxygen, there is an

Table 3.3	Bond energies between atoms of biological importance—Kcal/mole consumed in breaking bonds	
Single Bonds	**Kcal/mole**	
C—C	82	
O—O	34	
S—S	51	
C—H	99	
N—H	94	
O—H	110	
Multiple Bonds		
C=C	147	
O=O	96	
C=N	147	
C=O	167	
N≡N	226	

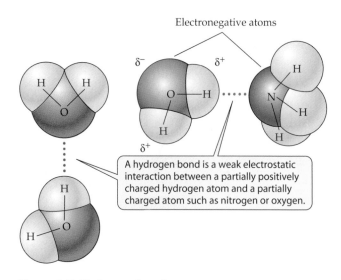

A hydrogen bond is a weak electrostatic interaction between a partially positively charged hydrogen atom and a partially charged atom such as nitrogen or oxygen.

Figure 3.5 Hydrogen bond
A hydrogen bond forms between a hydrogen atom covalently bonded to a larger electronegative atom and another electronegative atom.

uneven sharing of the electrons. The relatively large nucleus of the nitrogen or oxygen attracts the electron from hydrogen more strongly than does the small single proton in the hydrogen nucleus. In a molecule of water, the total electrons that are orbiting the two nuclei at any given moment are closer to the oxygen nucleus than to the hydrogen nucleus. This generates a partial electronegativity (δ^-) surrounding the oxygen nucleus and a positivity (δ^+) around the hydrogen nucleus. This results in an attraction between the positive part of a water molecule and the negative part of another electronegative atom such as ammonia or another water

molecule (**Figure 3.5**). The hydrogen of the water molecule is attracted to the nitrogen atom, and the resultant bond between the two molecules is called a **hydrogen bond.** A hydrogen bond possesses about 5% of the strength of a covalent bond.

Hydrogen bonds are of considerable importance in biological systems. The double helix in DNA and conformation of cellular proteins are dependent on hydrogen bonding. The role of hydrogen bonding in the double helix conformation of DNA is illustrated in **Figure 3.6**, where hydrogen bonding is involved in the planar arrangement of base pairs. Adenine and thymine are

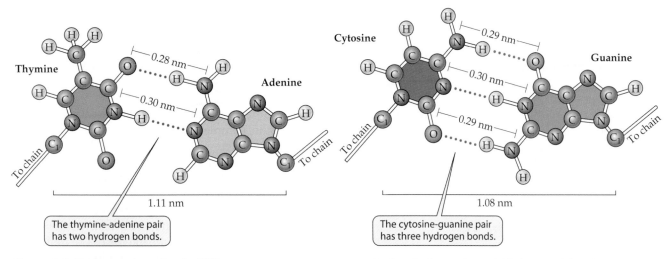

The thymine-adenine pair has two hydrogen bonds.

The cytosine-guanine pair has three hydrogen bonds.

Figure 3.6 Hydrogen bonding in DNA
Hydrogen bonding between A and T and between C and G retains the planar arrangement of purines and pyrimidines in the double helix of DNA. The AT and GC pairs are

equivalent in dimension, which is essential to the compact helical structure of DNA. Other interactions at the periphery of the helix maintain the "stacked" arrangement of base pairs one above the other (see Figure 3.7A).

(A) α-helix

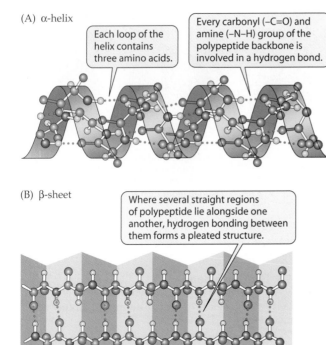

Each loop of the helix contains three amino acids.

Every carbonyl (–C=O) and amine (–N–H) group of the polypeptide backbone is involved in a hydrogen bond.

(B) β-sheet

Where several straight regions of polypeptide lie alongside one another, hydrogen bonding between them forms a pleated structure.

Figure 3.7 Hydrogen bonding in a peptide
Models of (A) α-helix and (B) β-sheet as they occur in a protein.

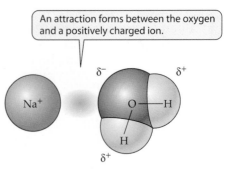

An attraction forms between the oxygen and a positively charged ion.

Figure 3.8 An ion dipole
An ion dipole between a sodium ion (Na⁺) and a polar molecule (water). As electrons orbit around the atoms of the water molecule, they can "stray" toward the oxygen atom, giving it a partial negative charge. A number of water molecules orient around the ion, a process called hydration (see Figure 3.9).

always bonded via two hydrogen bonds. Guanine is always paired with cytosine through three hydrogen bonds. The helical core is also maintained by stacking interactions between the planes of adjacent base pairs. These stacking interactions involve dipole–dipole and van der Waals forces (discussed later) that are equal in magnitude to the stabilizing energy of the hydrogen bonds between base pairs.

Hydrogen bonding is important in the conformation of functional proteins where bonding can occur either between polar constituents of amino acids on separate polypeptides chains or between polar areas within a polypeptide. Polypeptide chains with internal bonding can assume a helical shape (shown in **Figure 3.7A**). In this case, the hydrogen bonds between three amino acids form a loop (helix). Two individual polypeptide chains may form sheets by bonding portions of the same polypeptide (Figure 3.7B).

VAN DER WAALS FORCES

van der Waals forces are interactions that occur between adjacent molecules that are not due to ionic, covalent, or hydrogen bonds. These interactions can be based on either an attraction or a repulsion. The **attractions** are

due to a short-lived fluctuation in the electron charge densities that surround adjacent nonbonded atoms. This fluctuation is called a **dipole moment.** A dipole occurs when a molecule that is electrically neutral becomes polar. This polarity results when the center of negative charge (the electrons) does not center on the site of positive charge (**Figure 3.8**). As a result, there is a momentary polarity in the otherwise neutral molecule, and this can attract an opposite charge in a neighboring molecule. Some hydrophobic molecules such as hexane are liquids rather than gases because the molecules are held together by weak van der Waals forces. These van der Waals forces also play a role in enzyme substrate binding and protein interactions with nucleic acid.

Repulsion occurs when atoms that are not covalently bonded move too close together and repel one another. The electrons surrounding one molecule overlap the space occupied by the electrons of another, and the negative chrages tend to push the atoms apart.

HYDROPHOBIC FORCES

Hydrophobic forces are significant in biological systems and are quite complex. These forces essentially prevent a solute from interacting with a solvent. A hydrocarbon such as hexane is virtually insoluble in water because the water (solvent) withdraws in the area of contact with the apolar hydrophobic molecule (solute). The water molecules form a rigid hydrogen bonded network among themselves. This network effectively restricts the orientation of water molecules at the interface. The molecules of the hydrophobic compound cannot interact with the tightly oriented water molecules. This is why oil floats on water, and hydrocarbon chains in the core of cytoplasmic membranes are effective in creating a barrier that is impenetrable to most substances. Hydrophobicity is also involved in the protein/nucleic acid interactions that preserve the tertiary structure of DNA.

WATER

Water is the basic requirement of all living cells. Life evolved in water, and without water there is no life. If moisture is present in a microenvironment, it is probable that microorganisms (bacteria, cyanobacteria, or fungi) will be present as well. Water is the solvent in living cells, and a cell is 60% to 95% water. Even inert spores or seeds are 10% to 20% water by weight.

Water is essentially a neutral compound. The two constituent elements, oxygen and hydrogen, however, differ in electronegativity, and the distribution of charge around these two elements is asymmetrical. As a result, water actually has polarity (Figure 3.5). This polarity allows for hydrogen bonding and an affinity of water molecules for one another. Water molecules are dynamic and are constantly involved in arrangements where hydrogen bonds are formed and broken. This and other properties of water make it an ideal solvent.

Polarity permits water molecules to surround a solute and bring it into solution. The classical example of a solvent/solute interaction is sodium chloride in water (**Figure 3.9**). The negative area of the H_2O molecule attracts the positive parts of a solute and vice versa. These interactions permit the sodium and chloride ions to solubilize. Water has a high **dielectric constant**, which makes it an excellent solvent for ionic compounds. Organic acids and amino acids are also ionizable, and this is a major factor in their water solubility. Sugars also contain many polar hydroxyl (OH) groups and are generally quite soluble in water.

Rapid temperature changes do not occur when water is exposed to heat because of the extensive hydrogen bonding between water molecules (Figure 3.10). Generally, heat absorption by molecules increases their kinetic energy (both the rate of motion of molecules and their reactivity). Absorption of heat by water, however, results in the hydrogen bonds breaking apart rather than increasing rates of motion of the atoms. Much more heat is required to raise the temperature of water than would be required for a non-hydrogen-bonded liquid. This makes water an excellent insulator because it is slow to heat and slow to cool. This property is important to warm-blooded animals but of somewhat limited consequence to microbes that assume the temperature of their environment.

Water can form thin layers on surfaces because of its hydrogen bonds. This phenomenon is termed **surface tension** and is due to the attraction of water molecules for one another (**Figure 3.10**). Surface tension occurs on the surfaces of bodies of water and allows insects to "walk" on water because of the adhesion between water molecules.

A thin film of water protects biological membranes. Water atoms actually bond with oppositely charged atoms at or near the surface of the membrane. As a result of this surface layer of water, a membrane retains fluidity and flexibility and is slow to dry out.

Most nutrients for microbes must be water soluble because microbes are unable to ingest particles. When a microbe attacks a large molecule such as protein, lipid, or starch, it must digest the substrate to low molecular weight soluble compounds extracellularly before it can be transferred into the cell.

Water interacts with hydrophobic molecules such as lipids, and as a result, the lipids aggregate into compact molecular arrangements. This is an essential property in biological membranes where the lipid portion of the cellular membrane lines up to form a tight barrier to the flow of polar molecules in and out of the cell.

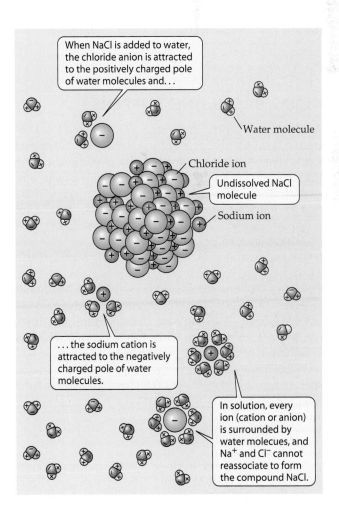

When NaCl is added to water, the chloride anion is attracted to the positively charged pole of water molecules and...

Water molecule

Chloride ion

Undissolved NaCl molecule

Sodium ion

...the sodium cation is attracted to the negatively charged pole of water molecules.

In solution, every ion (cation or anion) is surrounded by water molecules, and Na^+ and Cl^- cannot reassociate to form the compound NaCl.

Figure 3.9 Solute ionization and hydration
When an ionic compound is added to water, it dissolves as it ionizes and its component ions are hydrated.

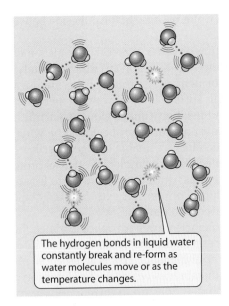

Figure 3.10 Hydrogen bonding of water
Hydrogen bonding between water molecules causes them to stick together. This property underlies the cohesive and adhesive properties of water, responsible for surface tension (cohesive forces) and capillary action (adhesive forces).

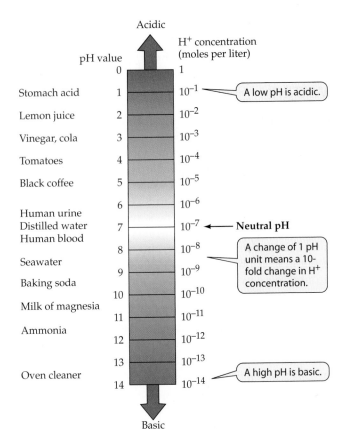

Figure 3.11 pH scale
The pH scale and the pH of some common materials.

ACIDS, BASES, AND BUFFERS

Acids are substances that ionize in water, and this ionization liberates hydrogen ions. A strong acid is one that ionizes readily in water, even in a dilute solution. Among the strong acids are hydrochloric acid, nitric acid, and sulfuric acid. Organic acids such as acetic acid or citric acid are much weaker because they ionize poorly and are only partly ionized even when in a dilute solution.

A **base** is a compound that, when ionized, releases one or more cations and negatively charged ions (OH^-). These negative ions can accept protons from solution. Thus the OH^- ions remove H^+ from solution, leading to an excess of hydroxyl groups. A solution with an excess of OH^- is basic. A strong base such as sodium hydroxide ionizes completely in dilute solution, whereas ammonium hydroxide, a weaker base, does not.

Water has a slight tendency to ionize—that is, to dissociate into hydrogen ions (H^+) and hydroxide ions (OH^-).

$$HOH \rightleftharpoons H^+ + OH^-$$

In pure water, the hydrogen ion concentration in solution is 0.0000001 M or 10^{-7}. The term *pH* is defined as the logarithm of the reciprocal of the hydrogen ion concentration. Pure water would have a pH of 7. When the concentration of H^+ is greater than 10^{-7}, the solution is acidic; if less than 10^{-7}, the solution is basic. The pH scale is logarithmic, so a solution of pH 6 contains a hydrogen ion concentration 10 times greater than a solution with pH 7. The pHs of some often-encountered materials are presented in **Figure 3.11**.

Microbial growth in laboratory media can result in the production of organic acids. These acids can lower the pH of the medium, which can result in cessation of growth or death of the cell. To prevent a significant change in pH, a **buffer** is generally added to the growth medium. A **buffer** can absorb or release hydrogen ions to maintain a favored pH. A buffering system that is often employed in growth media is composed of potassium dihydrogen phosphate (KH_2PO_4) and the salt dipotassium hydrogen phosphate (K_2HPO_4). This buffering system is effective around the neutral range (pH 6 to 8) because it can accept protons when the medium becomes acidic and donate protons if the medium becomes basic.

CHEMICAL CONSTITUENTS OF THE CELL

The number of possible reactions in living cells is limited only by the restricted range of elements present in cells. The most abundant element in cells is hydrogen, followed by carbon, oxygen, nitrogen, phosphorous, and sulfur. Cations that are present in virtually all cells are sodium, potassium, magnesium, calcium, zinc, and iron. Some organisms require trace elements (in small amounts) such as cobalt, molybdenum, copper, vanadium, or manganese.

There are a limited number of functional groups present in microbial cells, and the major groups were presented in Table 3.2. Carbon-to-carbon bonds are the backbone of the monomers that are joined to form macromolecules. These monomers are generally joined together through nitrogen, oxygen, or phosphorous molecules (or combinations of these). Phosphorous is the medium of energy exchange and also a structural component of nucleic acids. Nitrogen is a part of many cellular components including proteins and the nucleic acid bases. Sulfur is generally a component of proteins and some of the B-vitamins.

MAJOR MONOMERS

This discussion is limited to the structure of typical monomeric chemicals that make up a cell. The pathways involved in the synthesis of the major monomers (building blocks) that are present in a cell will be discussed in Chapter 10. The reactions involved in assembling these building blocks into functional structures will be presented in Chapters 4 and 11.

Carbohydrates

Carbohydrates (sugar) are a primary source of energy in many microbial cells. They are also a constituent of many structural units in the microbe, including nucleic acids, cell walls, and ATP. The general formula for a sugar is $(CH_2O)_n$. The commonly occurring sugars contain three to seven carbons. A simple sugar—and the most abundant product of living cells—is glucose (**Figure 3.12**). Glucose is also called blood sugar and is depicted along with fructose (fruit sugar). Other monosaccharides of biological interest are glyceraldehyde, erythrose, ribose, and sedoheptulose (**Figure 3.13**).

A common disaccharide (2-sugar) is sucrose or cane sugar. It is composed of one molecule of glucose and one of fructose. Lactose (milk sugar) is also a disaccharide and is composed of glucose and galactose. The structures of these two important disaccharides are presented in **Figure 3.14**. The disaccharide lactose has a **glycosidic** linkage between 1 and 4 carbons of the sugars, and this configuration is called beta (β). Lactose is galactose-β-1,4-glucose. The glycosidic linkage in sucrose is between the 1 and 2 carbons as well as in the alpha (α) configuration. Sucrose is glucose-α-1,2-fructose. Glyco-

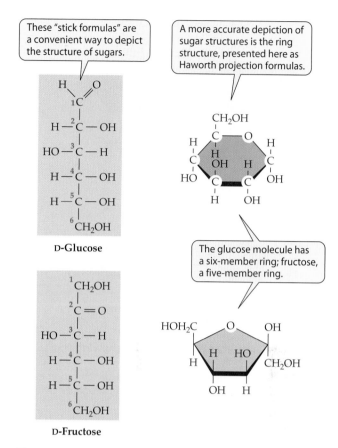

Figure 3.12 Structure of glucose and fructose
Glucose and fructose are common six-carbon sugars. They have the same chemical formula ($C_6H_{12}O_6$) and thus are structural isomers.

gen, starch, and cellulose are composed of glucose units that are polymerized into large molecules called polysaccharides. Glycogen and starch are important carbon and energy reserves in bacteria, animals, and plants.

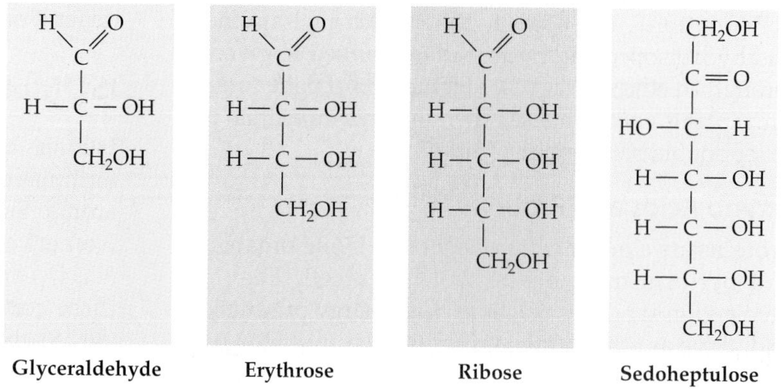

Figure 3.13 Important sugars
Examples of three-, four-, five- and seven-carbon sugars important in biological systems.

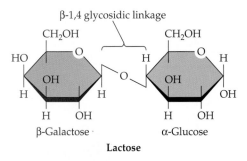

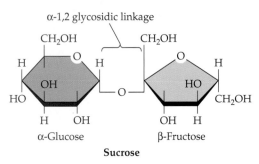

Figure 3.14 Important disaccharides, showing typical sugar linkages
The glycosidic linkages in these molecules are common in disaccharides and polysaccharides. Specific enzymes cleave these bonds.

Cellulose is present in fungi and plants as a structural component of the cell wall and is also present in a limited number of bacteria.

Lipids

Lipids are composed of fatty acids ester-linked to a sugar molecule, generally glycerol (see Figure 10.21). The fatty acids in bacteria are mostly saturated or monounsaturated fatty acids that are 16 or 18 carbons in length. A monounsaturated fatty acid has a double bond between two adjacent carbons, most often nine carbons from the carboxyl end of the long-chain fatty acid. Long-chain fatty acids that commonly occur in bacteria are depicted in Figure 10.19. The lipids in the *Archaea* are isoprenoid and hydroisoprenoid hydrocarbons linked to glycerol through an ether linkage (see Figure 4.36). Higher organisms (plants and animals) generally have more than one set of double bonds in their fatty acids.

Amino Acids and Proteins

Proteins are a major component of all living organisms and perform most of the work in a cell. They are involved in synthetic reactions, fuel energy production, and serve as structural components. About 50% (dry weight) of a cell is protein, and a bacterium, such as *Escherichia coli*, can synthesize well over 1,000 different proteins. The monomeric components of a protein are amino acids. An amino acid has one or more amino

groups ($—NH_3^+$) and one or more carboxylic acid groups ($—COO^-$). The general structure of an amino acid is as follows:

$$
\begin{array}{c}
R \\
| \\
HC{-}\overset{+}{N}H_3 \\
| \\
COO^-
\end{array}
$$

The R-group differs among the 21 amino acids. In water, the amine gains a proton and has a positive (+) charge, whereas the carboxyl loses a proton and has a negative (–) charge.

This **zwitterionic** form of the alpha amino acids occurs at physiological pH values. Molecules that bear charged groups of opposite polarity are termed zwitterions or **dipolar ions**. At near neutral pH both the carboxylic acid and amino groups are completely ionized—an amino acid can act as either an acid or a base. The structures of the 21 major amino acids are presented in **Figure 3.15**. Amino acids are joined to form an amino acid chain called a **peptide**. The bond between amino acids in a peptide chain is called a **peptide bond** (**Figure 3.16**).

Nucleic Acids

There are two major nucleic acids in all cells: **ribonucleic acid (RNA)** and **deoxyribonucleic acid (DNA),** the nucleic acid that carries the blueprint of the cell. RNA is responsible for the translation of information inherent in DNA into functional proteins. Just as amino acids are the structural units of a protein, nucleotides are the structural units of nucleic acids. There are four bases present in DNA: adenine, guanine, cytosine, and thymine. Thymine is replaced by uracil in RNA (**Figure 3.17**). The five-carbon sugar in DNA is deoxyribose, and in RNA it is ribose. The order of bases in DNA determines the sequence of amino acids in a protein, and as mentioned previously, the characteristics of a protein depend on this amino acid sequence.

ISOMERIC COMPOUNDS

Isomers are compounds that have identical molecular formulas but differ in the arrangement of the constituent atoms. Such isomers are termed **structural** isomers. Isomers composed of six carbons and 14 hydrogens (C_6H_{14}) are illustrated in **Figure 3.18**. Although these three compounds have identical molecular weights (86.18) and molecular formulas, they differ in physical properties as indicated by their boiling point. The structure would also influence biodegradability, as hexane would be the most readily degraded and 2,3-dimethylbutane the least.

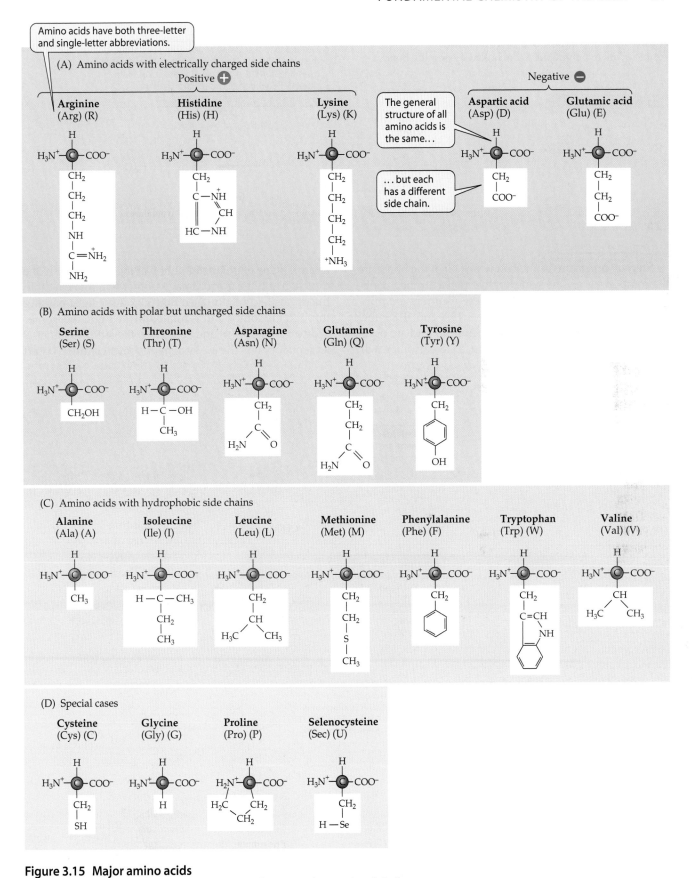

Figure 3.15 Major amino acids
Structures of the 21 amino acids most commonly present in proteins. It is the sequence of these amino acids in a protein that determines the character of that protein. All amino acids except proline exist in the zwitterion form.

Figure 3.16 Formation of a peptide bond
Amino acids in polypeptides are linked by peptide bonds.

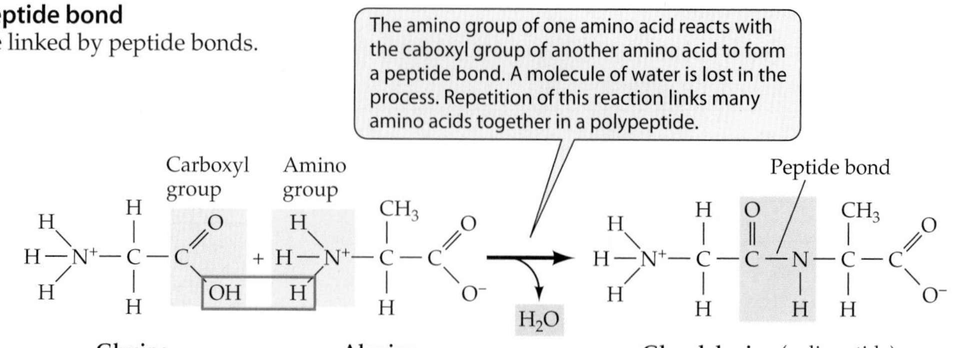

> The amino group of one amino acid reacts with the caboxyl group of another amino acid to form a peptide bond. A molecule of water is lost in the process. Repetition of this reaction links many amino acids together in a polypeptide.

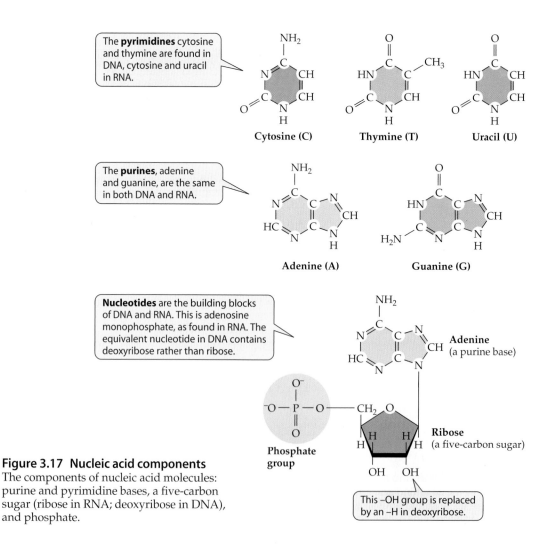

Glycine   Alanine   Glycylalanine (a dipeptide)

Stereoisomers are isomers in which all of the bonds in the compounds are the same but the spatial arrangements of the atoms differ. Essentially, stereoisomers are mirror images of one another (**Figure 3.19A** and **B**). Some stereoisomers are also termed **enantiomers** and are optical isomers. In the one configuration, a stereoisomer will rotate a plane of polarized light passing through it in one direction and the mirror image isomer will rotate light in the opposite direction (see **Box 3.2**). If the light passing through the compound is rotated to the right, the isomer is dextrorotatory (D) and rotated to the left the isomer is levorotatory (L). The D,L system of nomenclature is based

> The **pyrimidines** cytosine and thymine are found in DNA, cytosine and uracil in RNA.

Cytosine (C)   Thymine (T)   Uracil (U)

> The **purines**, adenine and guanine, are the same in both DNA and RNA.

Adenine (A)   Guanine (G)

> **Nucleotides** are the building blocks of DNA and RNA. This is adenosine monophosphate, as found in RNA. The equivalent nucleotide in DNA contains deoxyribose rather than ribose.

Adenine (a purine base)

Phosphate group

Ribose (a five-carbon sugar)

Figure 3.17 Nucleic acid components
The components of nucleic acid molecules: purine and pyrimidine bases, a five-carbon sugar (ribose in RNA; deoxyribose in DNA), and phosphate.

> This –OH group is replaced by an –H in deoxyribose.

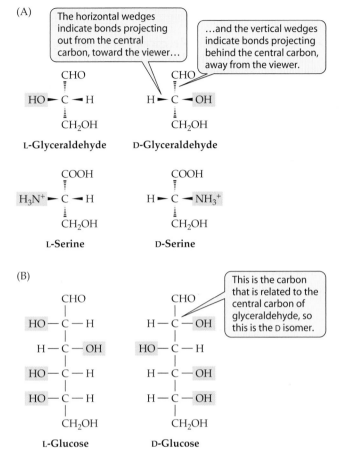

(A)

The horizontal wedges indicate bonds projecting out from the central carbon, toward the viewer…

…and the vertical wedges indicate bonds projecting behind the central carbon, away from the viewer.

L-Glyceraldehyde   D-Glyceraldehyde

L-Serine   D-Serine

Figure 3.18 Structural isomers
Isomers of the six-carbon saturated hydrocarbon, C_6H_{14}. They have the same molecular weight (86.18) but differ in physical properties, such as boiling point (BP).

n-Hexane
BP   68.95°C

2-Methylpentane
60.27°C

2,3-Dimethylbutane
58°C

(B)

This is the carbon that is related to the central carbon of glyceraldehyde, so this is the D isomer.

L-Glucose   D-Glucose

Figure 3.19 Stereoisomers
Isomers, such as those of amino acids, are denoted as D or L based on their configuration relative to D-glyceraldehyde and L-glyceraldehyde (A). D-Glucose (B) is the most common form of glucose in biological systems.

on the optical properties of D-glyceraldehyde and L-glyceraldehyde, respectively (Figure 3.19A). The absolute configurations of all other carbon-based molecules are referenced to D- and L-glyceraldehyde.

Proteins are normally composed of L-amino acids. The cell wall peptidoglycan contains D-amino acids, and the D form is also present in selected antibiotics. Many isomers of sugars exist in nature and may have the same molecular and structural formula but are mirror images of one another. These sugars are stereoisomers and designated by either D or L. Sugars are generally present in nature in the D configuration (Figure 3.19B). Enzymes that are present in bacteria (racemases) can convert the D-amino acids to the L form. Racemases are also present in bacteria that interconvert the L configuration of sugars to the D form.

Milestones Box 3.2   **Louis Pasteur, the Crystallographer**

Louis Pasteur began his scientific career as a chemist. He became intrigued with the observation that a chemical such as sulfur would crystallize in two distinct shapes. Earlier workers had reported that tartaric acid from grapes deflected the plane of polarized light, but sodium ammonium tartrate was optically neutral. When examining crystals of the tartrate salt, Pasteur noted that there were two types of crystals. One had "faces" that inclined to the right, and the faces of the second inclined to the left. He meticulously separated the two crystal types, dissolved each in water, and placed the respective solutions before polarized light. Pasteur found that the dissolved crystals with faces on one side rotated the polarized light in one direction, whereas the others rotated light in the opposite direction. An equal mixture of the two types was optically neutral. This was a clear proof that there can be a relationship between optical activity and molecular structure. That living things generally synthesized only one optical form was intriguing to Pasteur, and this ignited a lifelong interest in biological studies. Courtesy of the National Library of Medicine.

SUMMARY

▸ Essentially an atom is **electrically neutral** because the number of **negatively charged electrons** in orbit about the nucleus is equal to the **positively charged protons** present within the nucleus. **Neutrons** are present in all elements except hydrogen, and they **have no charge.**

▸ The **atomic number** is equal to the number of protons in the nucleus of an element. The sum of the number of neutrons and protons present is the **atomic weight.**

▸ The three major bonds that join atoms or molecules together are **ionic, covalent,** and **hydrogen bonds.** Carbon-to-carbon bonds are covalent bonds and vital in the formation of biological molecules.

▸ **Hydrogen bonds** are important in the conformation of nucleic acids and proteins.

▸ **Water** is essential to life, and a cell is 60% to 95% water. It is actually a polar compound and has considerable polymerization at room temperature.

▸ **Sugars** are essential components of living cells and are present in nucleic acids, cell walls, and ATP. They are also important sources of energy. Glucose is considered the most abundant product of living cells.

▸ Bacterial membranes contain **fatty acids** that are generally 16 or 18 carbons in length linked to glycerol via an ester linkage. They may be saturated fatty acids or monounsaturated. Archaeal membrane lipids are composed of **isoprenoid** chains with an ether linkage to glycerol.

▸ One-half of the **dry weight** of a **cell** is **protein,** and the constituent parts of a protein are the amino acids. The bond that joins one amino acid to another is the **peptide bond.**

▸ The major nucleic acids in cells are **deoxyribonucleic acid (DNA)** and **ribonucleic acid (RNA). DNA is the blueprint** of the cell, and **RNA** is involved in **translating** the information present in DNA to functional proteins. Purines and pyrimidines are the bases present in nucleic acids.

▸ **Structural isomers** have equivalent molecular formulas but a different arrangement of the constituent atoms. In **stereoisomers,** the bonds between atoms are in the same place but the spatial arrangement differs. They are mirror images of one another.

REVIEW QUESTIONS

1. Draw an atom and label the parts. What part retains the structure of an atom?

2. How does atomic weight differ from atomic number?

3. Define isotope. How would you design an isotope and experimentation to determine how a microbe capable of growth on methane assimilates this hydrocarbon?

4. What structural feature is involved in the reactivity of an atom? How is this related to valence?

5. Three types of bonds occur between molecules. What are they, and how do they differ? Which are most prevalent in a bacterium?

6. How are hydrogen bonds involved in the basic configuration of proteins? Nucleic acids?

7. What are van der Waals forces? Where might they play a role in biological systems?

8. Give several reasons why water is an effective solvent in biological systems.

9. How does an acid differ from a base? Why is water neutral?

10. Why are buffers often essential in bacterial culture media?

11. Name the five most abundant elements in a cell. What sort of function would they have in macromolecules?

12. What are the major monomers in polysaccharides, lipids, proteins, and nucleic acids?

13. How are the major monomers attached to one another?

14. Basically, how does one protein differ from another?

SUGGESTED READING

Brown, T. L., H. E. LeMay and B. E. Bursten. 1999. *Chemistry: The Central Science.* 8th ed. Upper Saddle River, NJ: Prentice Hall.

Garrett, R. H. and C. M. Grisham. 1998. *Biochemistry.* 2nd ed. Philadelphia: Brooks Cole Publishing.

Loewy, A. G., P. Siekevitz, J. R. Menninger and J. A. R. Gallant. 1991. *Cell Structure and Function: An Integrated Approach.* 3rd ed. Philadelphia: Saunders College Publishing.

Zubay, G. 1999. *Biochemistry.* 4th ed. Dubuque, IA: William C. Brown Publishers.

Chapter 4

Structure and Function of *Bacteria* and *Archaea*

C'est un grand progrè, monsieur.
— LOUIS PASTEUR, 1881*

In Chapter 1 we compared and contrasted prokaryotes with eukaryotes and affirmed that prokaryotes have simpler shapes and structures than eukaryotic organisms. Nonetheless, prokaryotes have all the necessary structural features for performing essential life functions, including metabolism, growth, and reproduction. And more complex prokaryotic organisms have additional structures and life cycles not found in the simpler species. This chapter introduces the various unicellular and multicellular shapes of prokaryotes and the types of cell division found among them. It also discusses the structure, chemical composition, and function of common morphological features found in prokaryotic organisms.

However, before discussing the properties of these microorganisms in greater detail, we need to recognize the special instruments, techniques, and procedures used to study them. Microbiology did not become a scientific discipline until appropriate instruments and techniques were developed to enable scientists to observe and study these small organisms (see Chapter 2). Of foremost importance was the development of quality microscopes. Indeed, it is not surprising that microorganisms were discovered by a lens maker, Antony van Leeuwenhoek, rather than by a biologist—an early illustration of the importance of instrumentation in microbiology. In this chapter, then, we focus first on microscopes, the instruments routinely used by microbiologists, then discuss the structure and function of prokaryotic organisms.

MICROSCOPY

The human eye has some ability to magnify objects. Natural magnification is simply achieved by bringing the object closer to the eye. The closer the object, the larger it appears, because the image on the retina is larger (**Figure 4.1**). Maximum enlargement depends on how close the object can be brought to the eye and still remain in focus. This **near point** is typically 250 mm for an adult human, and the distance increases as people age.

Objects smaller than about 0.1 mm (the "eye" of a needle is about 1.0 mm wide) cannot be seen distinctly because their image does not occupy a sufficiently large area on the retinal surface. Therefore, the unaided human eye cannot see small organisms such as prokaryotes, which typically have a cell diameter of only 0.001 mm (1.0 μm, or 10^{-6} m). Microscopes have been developed

---

*"That is important progress, sir." Remark made by Louis Pasteur to Robert Koch at the International Medical Congress in London, 1881, after Koch demonstrated his pure culture methods.

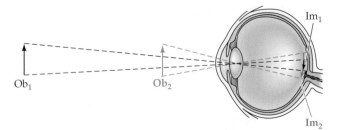

Figure 4.1 How the human eye magnifies
Diagram illustrating the magnifying capacity of the human eye. The object (Ob) is shown at two locations, Ob_1 and Ob_2. The image formed on the retina is smaller when the object is farther away from the eye (Im_1) than when it is closer (Im_2).

to aid the eye by increasing its ability to magnify. We discuss here several different types of microscopes commonly used by microbiologists.

Simple Microscope

The simplest optical device that can be used to assist the eye in enlarging objects is appropriately called the **simple microscope,** which in principle is a magnifying glass. Antoni van Leeuwenhoek constructed his own simple microscopes specifically to observe microorganisms (see Figure 2.2).

Magnifying glasses are not particularly powerful. They usually magnify objects about fivefold. It is a tribute to Leeuwenhoek that his simple microscopes not only could magnify 200- to 300-fold but did so with great clarity. Indeed, the images Leeuwenhoek produced were comparable to those obtained with light microscopes of similar magnification that are used today.

Compound Microscope

Another Dutchman, Zacharias Jansen (late sixteenth and early seventeenth century), is generally given credit for the development of the compound light microscope, although the origins of this instrument are somewhat obscure. It is interesting to note that compound microscopes were available at the time Leeuwenhoek made his momentous discoveries. However, the early compound microscopes were inferior to Leeuwenhoek's simple microscope—Robert Hooke, the English scientist who coined the term "cell," could not confirm Leeuwenhoek's reports of microorganisms with the early compound microscopes available to him.

The **compound light microscope** is named for the two lenses that separate the object from the eye. The **objective lens** is placed next to the object or specimen to be viewed; the **eyepiece** or **ocular lens** is located next to the eye. The object to be viewed is normally placed on a glass slide and illuminated with a light source.

The viewer uses the microscope to form an image by "focusing" the specimen, moving the objective lens and ocular (eyepiece) lens together relative to the specimen until the image is clear. When the specimen has been properly focused, the objective lens produces a **real image** (one that can be displayed on a screen) within the ocular diaphragm of the microscope. The viewer looking through the ocular lens sees a **virtual image**, which cannot be displayed on a screen (**Figure 4.2A**).

The magnification of a compound light microscope is determined by multiplying the magnification of the objective lens by that of the eyepiece lens. Usually the eyepiece magnification is 10×. Most microscopes of this type have several separate objective lenses, mounted on a rotating nosepiece, that typically give magnifications of 10× (low power), 60× (high, dry power), and 100× (oil immersion). The resulting magnifications attainable from this microscope would be 100×, 600×, and 1,000×, respectively.

The typical compound microscope (Figure 4.2B) also has a third lens system. Light from a lamp or other source is focused on the specimen by a **condenser lens**. The condenser lens, which is not directly involved in image formation, is necessary to provide high-intensity light, because the brightness of the object becomes a limiting factor as the specimen is increasingly magnified. The light intensity is adjusted by opening or closing a diaphragm called the **iris diaphragm**, located in the condenser.

It might seem reasonable to assume that one could increase the magnification indefinitely using the compound light microscope simply by constructing increasingly powerful objective and eyepiece lenses. In reality, the light microscope has a useful magnification maximum of only 1,000× to 2,000×. Although magnification beyond that level is attainable, it is referred to as **empty magnification** because it does not provide any greater detail of the specimen.

Improving Image Formation in Light Microscopy The ability of an optical system to produce a detailed image is termed its **resolving power**. The resolving power, d, is defined as the distance between two closely spaced points in an object that can be separated by the lens in the formation of an image (**Figure 4.3**). The equation for resolving power is:

$$d = \frac{0.61\,\lambda}{\eta \sin \theta}$$

where λ is the wavelength of light, η is refractive index, and $\sin \theta$ is the angular aperture. The lower the value of d, the greater is the resolving power. Therefore, to attain the best resolving power, or **resolution,** the following conditions are required:

(A)

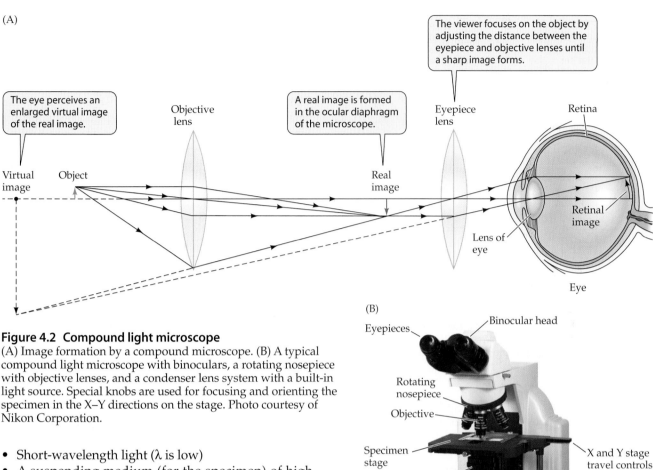

The viewer focuses on the object by adjusting the distance between the eyepiece and objective lenses until a sharp image forms.

The eye perceives an enlarged virtual image of the real image.

Objective lens

A real image is formed in the ocular diaphragm of the microscope.

Eyepiece lens

Retina

Virtual image

Object

Real image

Lens of eye

Retinal image

Eye

Figure 4.2 Compound light microscope
(A) Image formation by a compound microscope. (B) A typical compound light microscope with binoculars, a rotating nosepiece with objective lenses, and a condenser lens system with a built-in light source. Special knobs are used for focusing and orienting the specimen in the X–Y directions on the stage. Photo courtesy of Nikon Corporation.

- Short-wavelength light (λ is low)
- A suspending medium (for the specimen) of high refractive index (η is high)
- A high angular aperture, the angle at which light enters the objective (θ, and thus sin θ, is high)

The light microscope cannot be operated with wavelengths of less than 400 to 500 nm (violet light), as the eye cannot see shorter wavelengths. For maximum resolution, then, the shortest wavelength that can be used is about 500 nm (0.5 μm).

By increasing the **refractive index**, a measure of the ability of a material to bend light, more light from the specimen enters the objective lens, resulting in greater resolution. The crown glass used in lenses has a high refractive index: air has a refractive index of 1.0 (essen-

(B)

Eyepieces

Binocular head

Rotating nosepiece

Objective

Specimen stage

Condenser

X and Y stage travel controls

Fine and coarse focusing knobs

tially identical to that of a vacuum), water 1.33, and crown glass 1.5. To permit a continuous, high refractive index between the condenser and the objective, immersion oils of high refractive index are placed on the specimen and on the condenser lens system. Such **homogeneous immersion** systems allow the objective to be brought closer to the specimen, further increasing resolution (**Figure 4.4**).

The final factor affecting resolution is the aperture or opening of the objective itself. Because the objective lens has a finite aperture, a point on the specimen is not imaged as a point on the image but as a disk (called an "Airy disk") with alternate dark and light rings. The size of this disk can be decreased by increasing the aperture of the lens, but there is a limit to this. As a result, two closely spaced points on

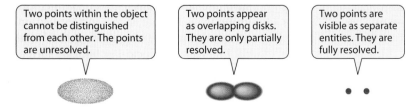

Two points within the object cannot be distinguished from each other. The points are unresolved.

Two points appear as overlapping disks. They are only partially resolved.

Two points are visible as separate entities. They are fully resolved.

Figure 4.3 Resolving power or resolution
Resolution is the degree to which two separate points in an object can be distinguished. Improved lenses allow increased resolution.

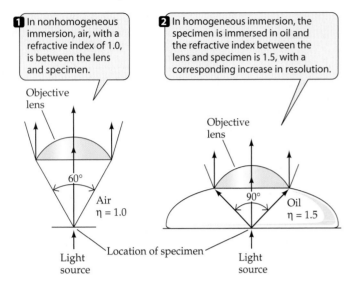

> **1** In nonhomogeneous immersion, air, with a refractive index of 1.0, is between the lens and specimen.

> **2** In homogeneous immersion, the specimen is immersed in oil and the refractive index between the lens and specimen is 1.5, with a corresponding increase in resolution.

Objective lens

60°

Air η = 1.0

Light source

Location of specimen

Objective lens

90°

Oil η = 1.5

Light source

Figure 4.4 Nonhomogeneous and homogeneous immersion
Immersion in oil increases resolving power, not only by increasing refractive index but also by allowing the specimen to be brought closer to the lens, thus increasing the angular aperture, θ.

the object are seen as fuzzy disks on the image, and if they are too close may actually overlap and not be separate points (Figure 4.3). The theoretical maximum value for the sine of the angular aperture is 1.0.

When these ideal values of the shortest possible wavelength ($\eta = 0.5$ μm), homogeneous oil immersion ($\lambda = 1.5$), and the angular aperture ($\sin \theta = 1.0$) are substituted into the equation for resolving power, then $d = 0.61(0.5$ μm$)/1.5$, which is about 0.2 μm. Since typical prokaryotic cells are about 1.0 μm in diameter, the compound light microscope has a satisfactory resolution for observation of these cells. It is not well suited, however, for observing internal cell structures, which are much smaller.

Early versions of the compound microscope were plagued by lens aberrations, which are of two types:

- **Chromatic aberration,** in which light of different wavelengths entering the objective from the specimen is focused at different planes in the formation of the image
- **Spherical aberration,** in which light rays from the specimen that enter the periphery of the objective lens are focused at a different place from those that enter the center of the lens

Correction of these aberrations was made possible by the use of multicompo-

nent lens systems (**Figure 4.5**) in the objective lens. These developments, and the introduction of the condenser, transpired over a period of about two centuries. The compound light microscope was finally perfected in the late nineteenth century by opticians including Ernst Abbe in Germany (1840–1905).

Because the refractive index of prokaryotic cells is similar to that of water, prokaryotes are almost invisible when viewed with an ordinary compound light microscope. Two approaches have been taken to overcome this problem:

- Staining the cells with dyes to produce higher contrast
- Using a modified compound microscope, such as the phase contrast microscope or the darkfield microscope

Dyes and Stains Microorganisms to be observed in the light microscope are often stained with dyes, because dyes increase the contrast between the cells and their environment. Most dyes used to stain microorganisms are aniline dyes, intensely pigmented organic salts derived from coal tar. They are called **basic dyes** if the **chromophore** (pigmented portion) of the molecule is positively charged. For example, crystal violet and methylene blue are basic dyes (**Figure 4.6**). Other basic dyes commonly used to stain microorganisms are basic fuchsin, malachite green, and safranin. Under normal growth conditions, most prokaryotes have an internal pH near neutrality (pH 7.0) and a negatively charged cell surface, so basic dyes are generally the most effective staining agents. Acid dyes such as nigrosin, Congo red, eosin, and acid fuchsin have a negatively charged chromophore and are useful in staining positively charged cell components such as protein.

Simple stains of microorganisms are made by spreading a suspension of the organism on a glass slide, allowing it to dry, and then gently heating it to fix it to the slide (a **fixative**, in this case heat, allows the specimen to adhere—like frying an egg in a pan without oil). Such a preparation is called a **smear**. The stain is added, and after a brief period of exposure the excess dye is removed by gentle rinsing. The slide is then viewed with the microscope.

Differential stains, such as the Gram stain (see Box 4.4), distinguish one microbial group from another, in this instance, gram-positive bacteria from gram-negative bacteria. Likewise, only acid-fast bacteria such as those in the genus *Mycobacterium* retain the

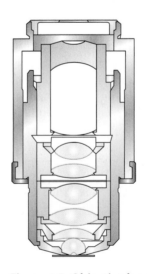

Figure 4.5 Objective lens
Cross-sectional view of a microscope objective lens showing its multicomponent lens system. Courtesy of Carl Zeiss, Inc.

$[(CH_3)_2NC_6H_4]_2C$ ⟶ $\overset{+}{N}(CH_3)_2Cl^-$

Crystal violet

$(CH_3)_2N$ ⟶ $N(CH_3)_2$

Cl^-

Methylene blue

Figure 4.6 Dyes
Chemical structures of crystal violet and methylene blue, two basic (positively charged) dyes, shown here as their chloride salts.

color of a dye when the stained preparation is rinsed in a solution of ethanol containing hydrochloric acid at a final concentration of 3%.

Some staining procedures allow the identification of structures within the cell. Thus, specific stains exist for bacterial endospores, flagella, capsules, and other cell structures. Lipophilic dyes, such as Sudan black, can be used to specifically stain lipid inclusions such as poly-β-hydroxybutyric acid (see below).

Brightfield and Darkfield Microscopes

An ordinary light microscope is called a **brightfield microscope,** because the entire field of view containing the specimen and the background is illuminated and appears bright. The brightfield microscope can be modified to produce a **darkfield microscope,** so named because the cells appear bright against a dark background (**Box 4.1**). One advantage of the darkfield microscope is that it can be used to view living cells. Cell suspensions

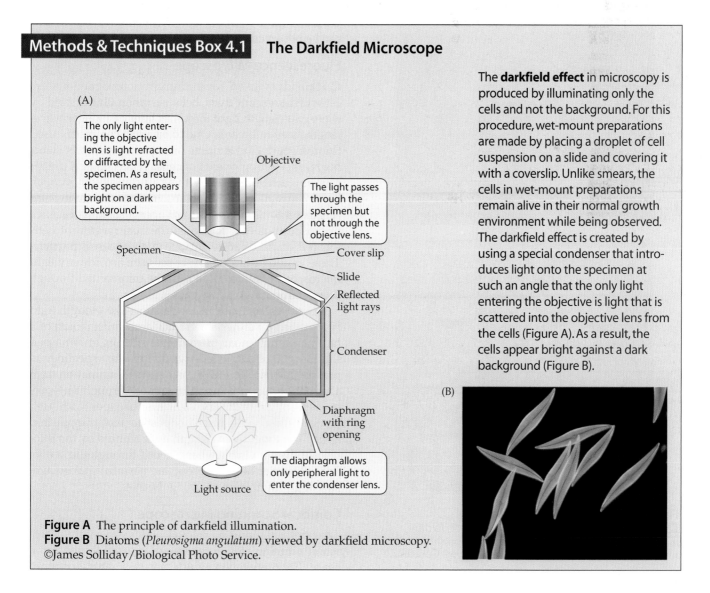

Methods & Techniques Box 4.1   The Darkfield Microscope

(A)

The only light entering the objective lens is light refracted or diffracted by the specimen. As a result, the specimen appears bright on a dark background.

The light passes through the specimen but not through the objective lens.

Objective

Specimen

Cover slip

Slide

Reflected light rays

Condenser

Diaphragm with ring opening

The diaphragm allows only peripheral light to enter the condenser lens.

Light source

The **darkfield effect** in microscopy is produced by illuminating only the cells and not the background. For this procedure, wet-mount preparations are made by placing a droplet of cell suspension on a slide and covering it with a coverslip. Unlike smears, the cells in wet-mount preparations remain alive in their normal growth environment while being observed. The darkfield effect is created by using a special condenser that introduces light onto the specimen at such an angle that the only light entering the objective is light that is scattered into the objective lens from the cells (Figure A). As a result, the cells appear bright against a dark background (Figure B).

(B)

Figure A The principle of darkfield illumination.
Figure B Diatoms (*Pleurosigma angulatum*) viewed by darkfield microscopy.
©James Solliday/Biological Photo Service.

are observed in **wet mounts**, prepared by placing a coverslip over the live suspension. The movement of motile cells, for example, is readily visible in these preparations.

Phase Contrast Microscope

As mentioned earlier, most microbial cells appear to be colorless, transparent objects when observed by ordinary brightfield microscopy (**Figure 4.7A**). A slight difference exists, however, between the refractive index of the cell (η is about 1.35) and that of its aqueous environment (η is 1.33). The **phase contrast microscope**, or **phase microscope**, amplifies this slight difference in refractive index and converts it to a difference in contrast. The result is that the cells appear very dark against a bright back-

(A)

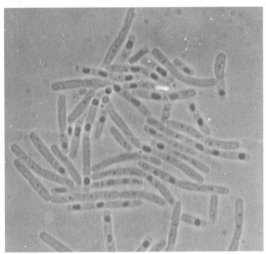

(B)

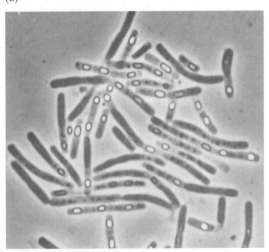

Figure 4.7   Phase contrast microscopy
Photomicrographs of an unstained bacterium, *Bacillus megaterium*, showing its appearance by (A) brightfield microscopy and (B) phase contrast microscopy. Courtesy of J. T. Staley.

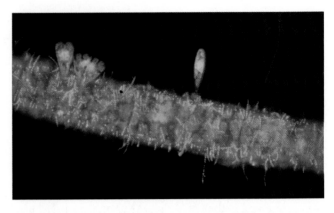

Figure 4.8   Fluorescence microscopy
Bacteria and diatoms stained with the fluorescent dye acridine orange and illuminated with ultraviolet light appear green and yellow to the eye. ©Paul W. Johnson/Biological Photo Service.

ground (Figure 4.7B). As with darkfield microscopy, cells can be observed while alive, in wet mounts.

Fluorescence Microscope

Certain dyes used for staining microorganisms are called **fluorescent dyes**, because when illuminated by short-wavelength light they emit light of a longer wavelength (they fluoresce). One of the most commonly used fluorescent dyes is acridine orange. It specifically stains nucleic acid components of cells. When stained preparations are illuminated with an ultraviolet or halogen light source, the cells fluoresce green to orange in color (**Figure 4.8**). One special advantage of fluorescence microscopy is that it allows the observation of cells located on an opaque surface such as a soil particle. These would be impossible to see with an ordinary light microscope in which light must be transmitted through the specimen.

In fluorescence microscopes, short-wavelength light is provided by either a mercury lamp (ultraviolet) or a halogen lamp (near ultraviolet). The light is passed through the objective lens system onto the specimen to provide incident illumination, that is, illumination from above the specimen (not transmitted through the specimen as in an ordinary light microscope—although some fluorescence microscopes do use transmitted light). The longer-wavelength light emitted by the fluorescent dyes is visible when viewed through the ocular. Special barrier filters prevent any harmful short-wavelength UV light from reaching the eyes.

Confocal Scanning Microscope

The confocal scanning microscope is especially useful when viewing microorganisms in three-dimensional space. The preparation, which may be a natural community containing microorganisms, is illuminated with a laser beam, which is focused on one point of the specimen using an objective lens mounted between the condenser lens and the specimen. Mirrors are used to pass (scan) the laser beam across the specimen in the x and y directions. The objective lens used for viewing the specimen magnifies the image, which is free from diffracted light, and the image is reconstructed on a video display screen.

Epifluorescence scanning microscopy uses a laser beam to illuminate the specimen, which has been stained with a fluorescent dye. The laser beam is focused at a particular plane of the preparation. The image viewed is that of a cross section of the preparation in which only those cells that are in the plane of illumination appear on the display screen (see Figure 24.2).

Transmission Electron Microscope (TEM)

The light microscope has a useful magnification of about 1,000× to 2,000×. Although greater magnification can be achieved, as noted above, finer detail will not be visible, in part because of the properties of the illuminating source itself—light (see p. 62). It is not possible to observe objects well if they are smaller than the wavelength of the illuminating source. An analogy may help illustrate this point. Suppose you wish to make an impression of your hand and two materials are available: fine particles of wet clay (analogous to a short wavelength) or gravel (a long wavelength). You can make a much more detailed and accurate impression of your hand with the clay.

The **transmission electron microscope** can produce very short wavelengths by using a beam of electrons as the illuminating source. The wavelength is controlled by the voltage applied to an electron gun, which is the source of electrons. If the accelerating voltage is 60,000 volts, the wavelength of the electron beam is 0.005 nm, or 0.000005 µm. This is 100,000 times shorter than the wavelength of violet light (about 500 nm). As a result, the theoretical resolution of the electron microscope is about 2 Å (Å is an angstrom; 1 Å = 10^{-10} m), which is twice the diameter of the hydrogen atom.

In place of optical lenses, the TEM uses electromagnetic lenses to bend the electron beam for focusing. A vacuum is essential in permitting the flow of electrons through the lens system, so the entire electron microscope must have an enclosed chamber and accompanying vacuum pumps. This makes the TEM a much larger instrument than the ordinary light microscope (**Figure 4.9**). The real image is formed by electrons bombarding a phosphorescent screen, and the photograph, called an **electron micrograph**, is taken by a camera mounted below the screen, containing film sensitive to electron radiation.

The lenses of the TEM are equivalent to those of the compound light microscope—a condenser lens, an objective lens, and a projector (ocular) lens. The fundamental

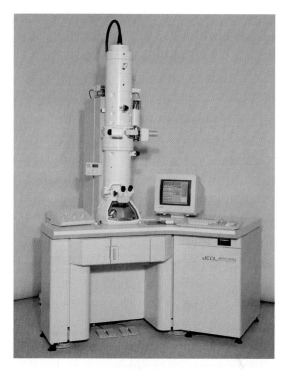

Figure 4.9 Transmission electron microscope (TEM) The TEM is very large because the entire device is contained in a vacuum and it uses electron magnets for lenses. The illuminating source is an electron gun located at the top of the microscope. The specimen is placed between the electron gun and the phosphorescent screen that is used to view the image. Courtesy of JEOL USA, Inc.

differences are that electrons are the illuminating source and magnets replace optical lenses (**Figure 4.10**).

Figure 4.11 compares the appearance of bacterial cells in a light microscope and in a TEM. Note the increased detail in the electron micrograph, even though the magnification is about the same for both preparations. Because the preparation to be observed in the electron microscope is placed in a vacuum chamber, it is not possible to observe living cells with the TEM. In place of a glass slide, cell preparations are placed on a small screen (4.0 mm in diameter, about the size of this **O**) called a **grid**. A thin plastic film is placed on the grid to hold the cell preparation. In the simplest preparation, cells are placed on the plastic-coated grid and allowed to dry. The whole cell preparation is then stained with a heavy metal stain such as phosphotungstic acid or uranyl acetate, allowed to dry, and then placed in the electron microscope.

The major application of the TEM in biology is to observe internal cell structures. This requires a more elaborate procedure for preparing the specimen for observation, a procedure called **thin sectioning**. Because the specimen will ultimately be observed in a vacuum, some procedure must be used to preserve the structure of the

(A) Light microscope

Light source

Condenser lens

Specimen

Objective lens

Intermediate image

Eyepiece

Projector lens

Retina of the eye or
photographic plate

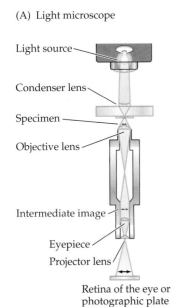

(B) Transmission electron
microscope (TEM)

Electron gun

Magnetic
condenser
lens

Specimen

Magnetic
objective
lens

Intermediate image

Magnetic projector

Observation screen or
photographic plate

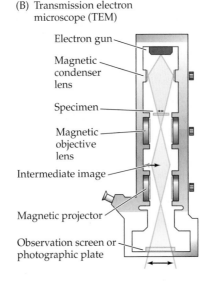

(C) Scanning electron
microscope (SEM)

Electron gun

Magnetic
condenser
lens

Scan coil

Magnetic
objective
lens

Specimen

Specimen
holder

Secondary
electrons

Cathode-ray
tube image

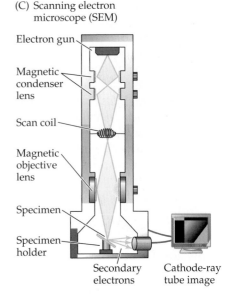

Figure 4.10 Illumination in light and electron microscopes
A comparison of the illuminating paths in the light micro-scope, the transmission electron microscope, and the scanning electron microscope. Note that all three have illuminating sources, condenser lenses, and objective lenses. Because a virtual image cannot be viewed, a projector lens is needed in the TEM and when cells are photographed in the light microscope. The image of the SEM specimen is viewed in a cathode ray tube.

(A)

(B)

This bacterium, like many prokaryotes, has fimbriae, short appendages made of proteinaceous material. In this species the fimbriae are too small to be seen by light microscopy.

2 μm

2 μm

Figure 4.11 Resolution in light and electron microscopy
A large unidentified colonial bacterium, obtained from a lake, is viewed under (A) the electron microscope and (B) the light microscope. Fimbriae are discussed later in the chapter (see also Figure 4.54). Courtesy of J. T. Staley and Joanne Tusov.

specimen in the absence of water. This is accomplished through a process called **dehydration and embedding**, in which the cells are taken from their aqueous environment and transferred into a plastic resin. Although more complex, this process is similar to embedding insects in plastic resins, as practiced by some hobbyists.

First, cells are harvested from their growth medium by centrifugation. They are gradually dehydrated by transferring them step by step from the aqueous medium into ethanol solutions of increasing concentration, until they are placed in 100% ethanol. The next step is to transfer the cells through an acetone-ethanol series until they are suspended in 100% acetone. Unlike ethanol and water, acetone is a plastic solvent and is miscible with plastic resins. At this point, the preparation is completely dehydrated. The next step is to embed the cells in a plastic mixture. Again, a series of transfers is made until the cells are in 100% plastic resin. Sufficient time is permitted to ensure that the resin completely displaces the acetone in the cells. The preparation is "cured" (polymerized to form a solid) by heating at 60°C in an oven. The organism is now embedded.

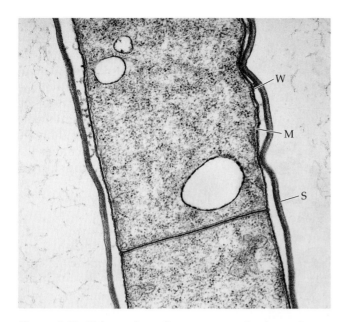

Figure 4.12 Thin section of a gram-negative bacterium
A thin section of *Thiothrix nivea* showing (from the outside)
its sheath (S), cell wall (W), and cell membrane (M).
Courtesy of Judith Bland and J. T. Staley.

The embedded cells are sliced into thin sections with
an **ultramicrotome**. The ultramicrotome is analogous to
a meat slicer, except that it has a diamond knife and cuts
extremely thin sections (about 60 nm thick). The thin
sections are stained with heavy metals (lead citrate and
uranyl acetate) to increase contrast. They are then placed
on TEM grids and examined. A typical thin section is
shown in **Figure 4.12**.

Scanning Electron Microscope (SEM)

As in the TEM, electrons are the illuminating source for
the **scanning electron microscope** (SEM), but the electrons are not transmitted through the specimen as in
transmission electron microscopy. In SEM the image is
formed by incident electrons scanned across the specimen
and back-scattered (reflected) from the specimen, as
described for the fluorescence microscope. The reflected
radiation is then observed with the microscope, in the
same manner that we observe objects illuminated by sunlight. Solid metal stubs are used to hold the preparations.
Before the organism is viewed, it is dried by **critical point
drying**, carefully controlled drying in which water is
removed as vapor so that structural damage to cells is
minimized. The specimen is then coated with an electron-
conducting noble metal such as gold or palladium.

The SEM has a larger depth of field than the TEM so
there is a greater depth through which the specimen
remains in focus. Thus, all parts of even a relatively
large specimen, such as a eukaryotic cell, remain in
focus when viewed by the SEM (**Figure 4.13A**). The
result is an image that looks three-dimensional. This is

the best procedure available to examine colonial forms
of microorganisms, such as the fruiting structures of the
myxobacteria (Figure 4.13B).

An SEM can be equipped with an X-ray analyzer,
enabling the researcher to determine the elemental
composition of microorganisms or their components.
Elements with atomic weights greater than 20 can be
assayed in a semiquantitative manner using this instrument (**Figure 4.14**).

The varieties of microscopes and associated procedures described above were instrumental in the developing field of microbiology. In the next section we discuss the various morphological attributes of prokaryotes

(A)

100 μm

(B)

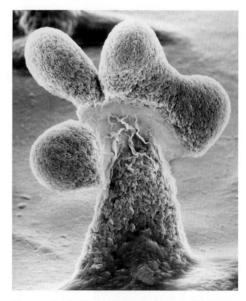

Figure 4.13 Scanning electron microscopy (SEM)
(A) A radiolarian, a eukaryotic protist with a siliceous (silica-
containing) shell. (B) The fruiting structure of the myxobacterium *Stigmatella aurantiaca*. The entire structure is about 1
mm in length. A, courtesy of Barbara Reine; B, ©K. Stephens
and D. White/Biological Photo Service.

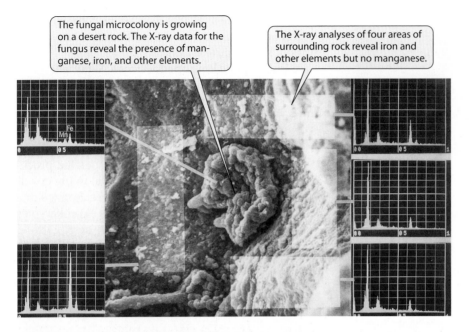

The fungal microcolony is growing on a desert rock. The X-ray data for the fungus reveal the presence of manganese, iron, and other elements.

The X-ray analyses of four areas of surrounding rock reveal iron and other elements but no manganese.

Figure 4.14 SEM-elemental analysis
SEM of a fungal microcolony growing on a desert rock. The X-ray findings suggest that the colony is accumulating manganese. Courtesy of F. Palmer and J. T. Staley.

at both the organismal and the subcellular level, their chemical composition, and their functional role in the organism. Keep in mind that most microbiologists study the behavior of microorganisms in the laboratory, although their ultimate interest lies in understanding the function of a given structure in the organism's natural environment.

MORPHOLOGY OF *BACTERIA* AND *ARCHAEA*

Bacteria and *Archaea* come in a variety of simple shapes. Most are single-celled but some are multicellular forms consisting of numerous cells living together.

Unicellular Organisms

The simplest shape for a single-celled prokaryote is the sphere (**Figure 4.15**). Unicellular spherical organisms are called cocci. Some cocci

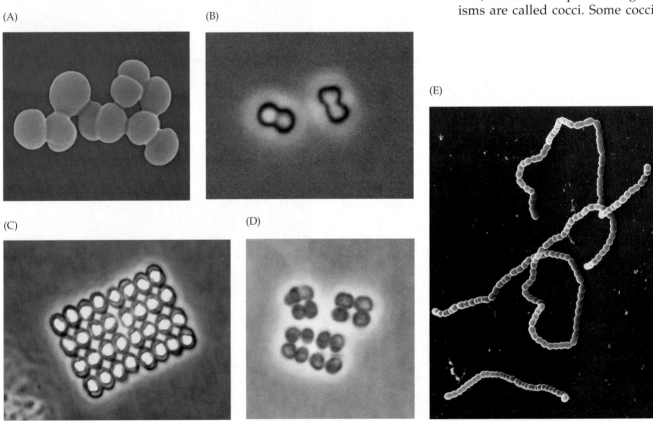

(A)   (B)   (C)   (D)   (E)

Figure 4.15 Cocci
Various formations of cocci, spherical cells, as shown by microscopy. (A) staphylococcus, a grapelike cluster. (B) diplococcus, a pair of cells; (C) sheet (internal bright areas of each cell are gas vacuoles); (D) eight-cell packet, or sarcina; (E) streptococcus, a chain of cells. A, ©Dennis Kunkel Microscopy, Inc.; B, ©M. Abbey/Visuals Unlimited; C,D, courtesy of J. T. Staley and J. Dalmasso; E, ©David M. Phillips/Visuals Unlimited.

grow and divide along only one axis. If the cells remain attached after cell division, this results in the formation of chains of cells of various lengths. A diplococcus is a "chain" of only two cells (Figure 4.15B); a streptococcus can contain many cells in its chain (Figure 4.15E). Some cocci divide along two perpendicular axes in a regular fashion to produce a sheet of cells (Figure 4.15C). Other cocci divide along three perpendicular axes, resulting in the formation of a packet or sarcina of cells (Figure 4.15D). Finally, random division of a coccus produces a grapelike cluster of cells referred to as a staphylococcus (Figure 4.15A).

The most common shape in the prokaryotic world is not a sphere, however. It is a cylinder with blunt ends, referred to as a **rod** or **bacillus** (**Figure 4.16**). Some rods remain attached to one another after division across the transverse (short) axis of the cell, forming a chain.

A less common shape for unicellular bacteria is a **helix**. A very short helix (less than one helical wavelength long) is called a **bent rod** or **vibrio** (**Figure 4.17A**). A longer helical cell is called a **spirillum** (Figure 4.17B) if the cell shape is rigid and unbending or a **spirochete** (Figure 4.17C) if the organism is flexible and changes its shape during movement.

Variations of these common shapes of unicellular bacteria also exist. For example, some bacteria produce appendages that are actually extensions of the cell, called **prosthecae**, which give the cell a star-shaped appearance (**Figure 4.18**). The morphological diversity of *Bacteria* and *Archaea* is described further in Chapters 18 to 22, where individual genera are discussed.

Multicellular Prokaryotes

Numerous prokaryotic organisms exist as multicellular forms. One such group is the **actinobacteria**. These rod-

(A)

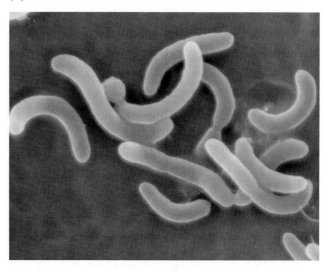

(B)

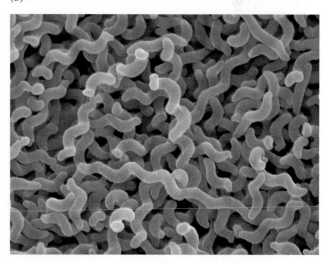

(C)

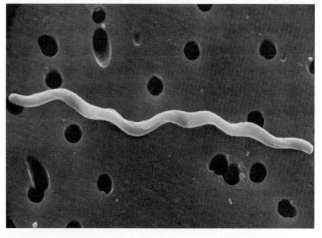

Figure 4.17 Curved and helical cells
(A) Bent rod, or vibrio; (B) spirillum; (C) large spirochete, from the style of an oyster. A, B, ©Dennis Kunkel Microscopy Inc.; C, ©Paul W. Johnson and John Sieburth/Biological Photo Service.

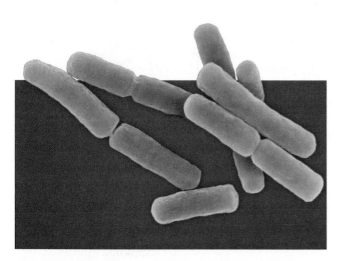

Figure 4.16 Bacilli, or rods
Rod-shaped cell of a unicellular *Bacillus anthracis,* shown by phase contrast microscopy. ©Dennis Kunkel Microscopy, Inc.

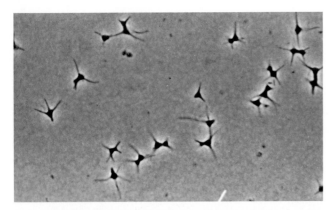

Figure 4.18 Prosthecate bacterium
A star-shaped bacterium, *Ancalomicrobium adetum*.
Courtesy of J. T. Staley.

(A)

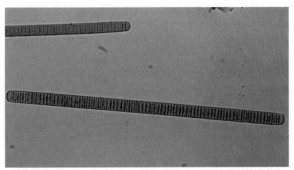

(B)

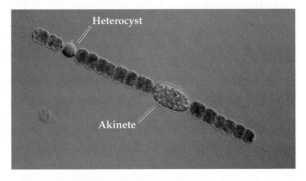

Heterocyst

Akinete

Figure 4.20 Filamentous bacteria
(A) *Oscillatoria*, a multicellular filamentous cyanobacterium, showing the close contact between cells in the trichome. (B) An *Anabaena* sp. filament with typical and specialized cells. Heterocysts are nonpigmented cells, the site of nitrogen fixation; the akinete is a resting stage. A, ©James W. Richardson/Visuals Unlimited; B, ©Paul W. Johnson/Biological Photo Service.

shaped organisms produce long filaments containing many cells. The filaments form branches, resulting in an extensive network comprising hundreds or thousands of cells. This network is referred to as a **mycelium** (**Figure 4.19**).

Another common multicellular shape is the **trichome**, which is frequently encountered in the cyanobacteria (**Figure 4.20A**). Although a trichome superficially resembles a chain, adjoining cells have a much closer spatial and physiological relationship than do the cells in a chain. Motility and other functions result from the concerted action of all cells of the trichome. And some cells in the trichome may have specialized functions that benefit the entire trichome. For example, the **heterocyst** (Figure 4.20B), seen in some filamentous cyanobacteria, is the site of nitrogen fixation (see Chapter 21).

CELL DIVISION OF *BACTERIA* AND *ARCHAEA*

Prokaryotes maintain their shapes during the process of **asexual reproduction**—a process in which a single organism divides to produce two progeny. For unicellular prokaryotes there are two ways in which this may be accomplished, either by binary transverse fission or by budding.

Binary Transverse Fission

The most common type of bacterial cell division is **binary transverse fission**. In this process, the cell (which may be a coccus, rod, spirillum, or other shape) elongates as growth occurs along its longitudinal axis (**Figure 4.21**). When a certain length is reached, a **septum** (wall structure) is produced along the transverse axis of the cell midway between the cell ends. When the septum has completely formed, the two resulting cells become separate entities. This process is called binary

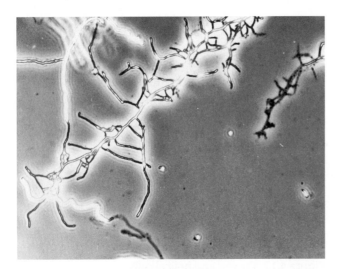

Figure 4.19 Mycelial bacterium
Streptomyces sp. illustrating the complex network of filaments called a mycelium. Courtesy of J. T. Staley and J. Dalmasso.

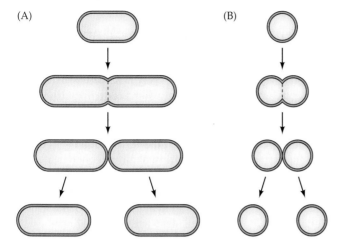

Figure 4.21 Binary transverse fission
Typical binary transverse fission in (A) a rod-shaped bacterium and (B) a coccus.

fission because two cells are produced by a division or "splitting" of one original cell. The process is described as transverse because the septum that separates the two new cells is formed along the transverse, or short, axis of the original cell.

In binary transverse fission, DNA replication precedes septum formation. The two resulting cells are mirror images of one another. Analyses of cell wall components of dividing cells indicate that the chemical constituents of the original "mother" cell wall are equally shared in the cell walls of the two "daughter" cells. In multicellular prokaryotes with trichomes, the organism divides by transverse fission and the trichome ultimately separates into two separate trichomes.

Budding

Budding, or bud formation, is a less common form of cell division among prokaryotic organisms. As in binary transverse fission, this is an asexual division process that results in the formation of two cells from the original cell. In the budding process, however, a small protuberance, a **bud**, is formed on the cell surface. The protuberance enlarges as growth proceeds and eventually becomes sufficiently large and mature to separate from the mother cell (**Figure 4.22**)

Binary transverse fission and budding differ in several ways. During binary transverse fission, the symmetry of the cell with respect to the longitudinal and transverse axes is maintained throughout the entire process (Figure 4.21). This results in the mother cell producing two daughter cells and losing its identity in the process. In the budding process, however, symmetry with respect to the transverse axis is not maintained during division (Figure 4.22). Also, in contrast to binary transverse fission, most of the new cell wall components are used in the synthesis of the bud instead of being divided equally between the two progeny cells. The result is that, during budding, the mother cell produces one daughter cell while retaining its identity generation after generation. Whether there is a limit to the number of buds a mother cell can produce during its existence is as yet unknown.

Fragmentation

Another type of cell division process occurs in the mycelial bacterial group, the actinobacteria or streptomycetes. These organisms have "multinucleate" filaments that lack septa between the cells; these filaments are referred to as **coenocytic** and are analogous to the filaments found in some fungi. Some bacteria of this type undergo a multiple fission process. Actinobacteria undergo a **fragmentation** in which the filament develops septations between the nuclear areas, resulting in the simultaneous formation of numerous unicellular rods. In an analogous fashion, some cyanobacteria produce numerous smaller daughter cells called **baeocytes** from a single large mother cell during cell division (see Chapter 21).

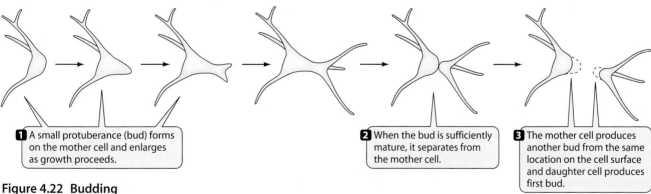

1 A small protuberance (bud) forms on the mother cell and enlarges as growth proceeds.

2 When the bud is sufficiently mature, it separates from the mother cell.

3 The mother cell produces another bud from the same location on the cell surface and daughter cell produces first bud.

Figure 4.22 Budding
Bud formation in *Ancalomicrobium adetum,* a prosthecate bacterium. The budding process can be repeated as long as nutrients are available for growth.

Cell Fractionation, Separation, and Biochemical Analyses of Cell Structures

In order to determine the composition of various components of the cell, scientists separate these components from the rest of the cell, purify them, and analyze them biochemically. The initial step is to break open the cells (see figure). Either chemical or physical procedures can be used to break open small, prokaryotic cells. For example, chemical procedures include lysis of the cells by enzymes or detergents. Physical methods include ultrasound (called **sonication** by biologists), in which high-frequency sound waves vibrate cells until they break. A sonicator probe is inserted into a cell suspension for this purpose, as shown in the illustration. Alternatively, cells can be broken by passing thick suspensions of frozen cells through a small orifice (French pressure cell) at high pressure.

Once the cells have been broken, the various structural fractions are separated, usually by centrifugation. Two types of centrifugation can be used. In **differential** or **velocity centrifugation**, fractions are separated by the length of time they are centrifuged at different gravitational forces. Denser structures such as unbroken cells or bacterial endospores, cell membranes, or cell walls sediment at low speeds ($15,000 \times g$ for 10 minutes). The supernatant is removed and centrifuged at higher speed to spin out less dense structures. For example, ribosomes sediment only after centrifugation at higher speeds ($100,000 \times g$ for 60 minutes). The remaining material that does not sediment in the centrifuge tube contains soluble constituents such as cytoplasmic enzymes.

Alternatively, **buoyant density** or **density gradient centrifugation** can be used to separate the various cell fractions. In this procedure a density gradient is set up in the centrifuge using different concentrations of a solute, such as sucrose. The sample is layered on the surface and centrifuged at moderate speed until the cellular fractions equilibrate with the layer in the gradient that has the same buoyant density. They can then be removed with a pipette and studied as purified fractions.

When the cell fraction that is of interest to the microbiologist is separated and purified by the procedures outlined above, it can be analyzed chemically. The electron microscope is used to check the identity and purity of the material at each step in the process.

FINE STRUCTURE, COMPOSITION, AND FUNCTION IN *BACTERIA* AND *ARCHAEA*

The remainder of this chapter is devoted to the description of various prokaryotic structures, their chemical composition, and their functions. The terms "fine structure" and "ultrastructure" refer to subcellular features that are best observed using the electron microscope. Studies of the fine structure of microbial cells began in the 1950s and 1960s, when electron microscopy procedures were perfected. Scientists used a combination of procedures to "break open" or **lyse** cells, followed by centrifugation to separate the various subcellular components. These components were purified and then analyzed biochemically. The electron microscope was used at various steps in the procedure to identify and assess the purity of the structures (**Box 4.2**).

We begin with internal structures found in the cytoplasm and then consider the outer layers of the cell. The discussion in this chapter is confined to structures commonly found in many prokaryotic phyla. Thus, we do not discuss structures such as spores and magnetite crystals, which are covered in descriptions of the particular phyla that have these structures in Chapters 18 through 22.

Internal Structures

Foremost among the intracellular materials of all cells is their DNA, the hereditary material of the cell. DNA and other intracellular components commonly found in many different *Bacteria* and *Archaea*, including ribosomes, gas vesicles, and various reserve materials, are discussed individually below.

DNA The deoxyribonucleic acid (DNA) of prokaryotes is a circular, or more rarely a linear, double-stranded helical molecule. The two strands are held together by hydrogen bonds, the nucleotide bases of one strand forming hydrogen bonds with the bases of the opposite strand: adenine with thymine and cytosine with guanine (see Chapter 3).

The DNA appears as a fibrous material in the cytoplasm when prokaryotic cells are viewed in thin sections (**Figure 4.23**). As noted in Chapter 1, the DNA of prokaryotes is not surrounded by a membrane and thus does not appear in a confined area within the cell; rather, it appears as a somewhat diffuse, dispersed fibrous material. For this reason the region is not called a nucleus but a **nucleoid** or **nuclear area**.

If gentle conditions are used to lyse the cells (like eggs, cells can be broken carefully; the cytoplasm can be freed

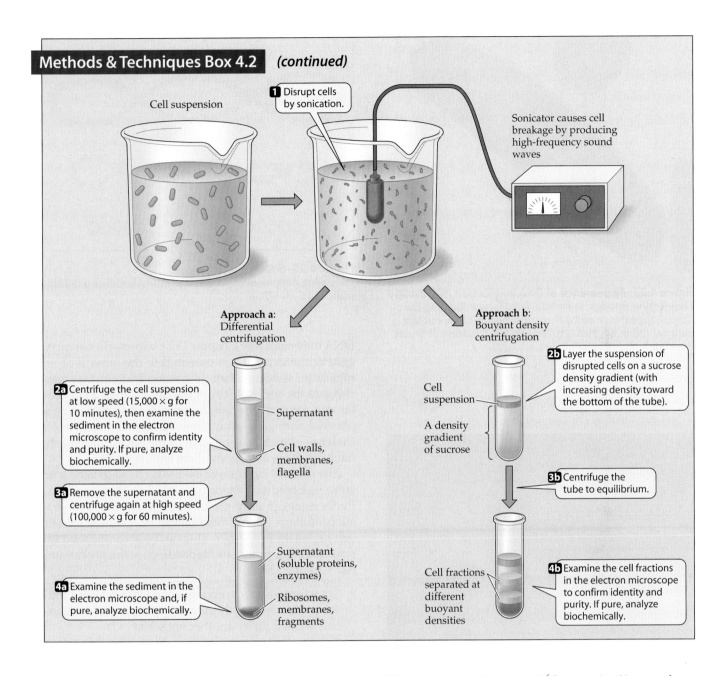

Methods & Techniques Box 4.2 *(continued)*

Cell suspension

1 Disrupt cells by sonication.

Sonicator causes cell breakage by producing high-frequency sound waves

Approach a: Differential centrifugation

2a Centrifuge the cell suspension at low speed (15,000 × g for 10 minutes), then examine the sediment in the electron microscope to confirm identity and purity. If pure, analyze biochemically.

Supernatant

Cell walls, membranes, flagella

3a Remove the supernatant and centrifuge again at high speed (100,000 × g for 60 minutes).

Supernatant (soluble proteins, enzymes)

4a Examine the sediment in the electron microscope and, if pure, analyze biochemically.

Ribosomes, membranes, fragments

Approach b: Bouyant density centrifugation

2b Layer the suspension of disrupted cells on a sucrose density gradient (with increasing density toward the bottom of the tube).

Cell suspension

A density gradient of sucrose

3b Centrifuge the tube to equilibrium.

Cell fractions separated at different buoyant densities

4b Examine the cell fractions in the electron microscope to confirm identity and purity. If pure, analyze biochemically.

from the cell membrane and wall), the DNA is released and appears as a coiled structure spilled from the cell (**Figure 4.24**). When stretched out, the length of the DNA molecule is about 1 mm, about a 1,000 times longer than the 1 to 3 μm length of the typical prokaryotic cell! In order to package all of this material within the cell, the DNA molecule is tightly wound in **supercoils** (**Figure 4.25**). Special enzymes are responsible for supercoiling and for controlling the unwinding of the DNA during replication (DNA synthesis) and transcription (production of RNA from the DNA template; see Chapter 13).

The molecular weight of the DNA molecule of prokaryotes ranges from about 10^9 to 10^{10} Da (Da is a dalton, a unit of mass approximately equal to the mass of the hydrogen atom, ^{1}H). The typical prokaryotic DNA contains about 4×10^6 base pairs (4 mega-base pairs, or 4 mgb). However, some intracellular symbiotic bacteria such as *Buchnera* species have smaller genomes (0.65 mgb), and some prokaryotic genomes are as large as 10 mgb. This is considerably smaller than the size of eukaryotic genomes, whose chromosomes may be ten times or more larger than prokaryotic chromosomes, but larger than those of viruses.

Prokaryotic cells may have more than one copy of the DNA molecule. For example, when the cell is growing and dividing rapidly, two or four or more copies, or partial copies, may be present.

In addition to the genomic DNA, the cell often contains other, extrachromosomal circular molecules of DNA called **plasmids**. These, too, are double-stranded

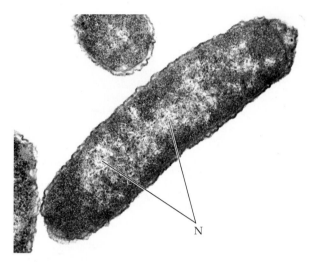

Figure 4.23 Appearance of DNA by electron microscopy
Thin section through *Salmonella typhimurium* showing the fibrous appearance and diffuse distribution of the nuclear material (N) in a typical prokaryotic cell. Courtesy of Stuart Pankratz.

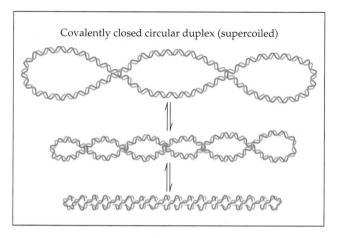

Figure 4.25 Supercoiled DNA
Increasing degrees of supercoiling of DNA produce a tightly compacted molecule.

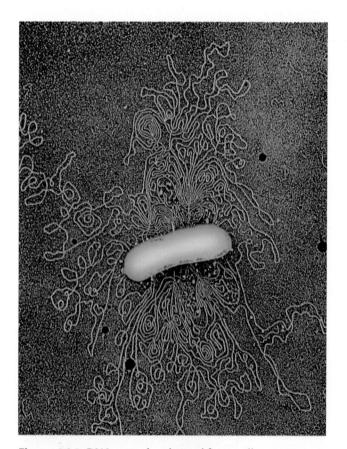

Figure 4.24 DNA strands released from cell
Photomicrograph showing DNA strands released from a lysed bacterial cell. ©Dr. Gopal Murti/SPL/Science Source/Photo Researchers Inc.

DNA molecules (see Chapter 15). Plasmids do not carry genetic material that is essential to the growth of an organism, although they do contain features that may enhance the survivability of the organism in a particular environment. For example, some bacteria carry a plasmid with genes that allow them to degrade naphthalene—which is not useful to the bacterium unless naphthalene is in its immediate environment.

The primary function of the prokaryotic genome is to store its hereditary information, carried in its genes. Furthermore, in some prokaryotes, under the appropriate conditions genetic material can be transferred from one organism to another. This can be accomplished by three different processes, depending on the prokaryote (see Chapter 15):

- **Transformation,** occurring when DNA released into the environment by lysis (cell breakage) of one organism is taken up by another organism
- **Conjugation,** in which transfer occurs during cell to cell contact between two closely related bacterial strains
- **Transduction**, in which prokaryotic viruses are involved in transferring DNA from one organism to another

Ribosomes As mentioned in Chapter 1, ribosomes are small structures that carry out protein synthesis (**Figure 4.26**), a process referred to as **translation** (see Chapter 13) in which messenger RNA (mRNA) carries the message in nucleotides from the genome to the ribosome, where amino acids are linked together by peptide bonds to form protein.

At high magnification the prokaryotic ribosome can be seen to consist of two subunits, the small 30S subunit and the larger 50S subunit (**Figure 4.27**). Note that the Svedberg units (S), sedimentation densities, are not

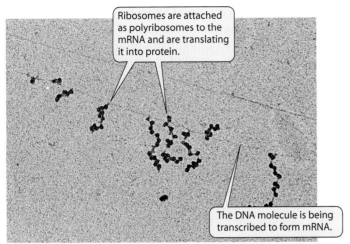

Figure 4.26 **Protein synthesis**
Ribosomes in action in *E. coli*. The DNA and RNA appear as filaments. Courtesy of Oscar L. Miller.

(A)

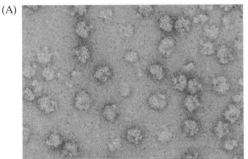

additive: the 30S and 50S ribosome subunits comprise a 70S ribosome. This is because the Svedberg unit is directly related not to molecular mass but to the density of particles in ultracentrifugation. (Likewise, the eukaryotic 80S ribosome consists of a small 40S subunit and a larger 60S subunit.) Ribosomes consist of both protein and a type of ribonucleic acid called ribosomal RNA (rRNA). Figure 4.27 (B and C) shows the RNA and protein components of each of these subunits from the 70S and 80S ribosomes.

Even though bacterial and archaeal ribosomes have the same sedimentation coefficients, they differ somewhat in structure and composition. As a result, these organisms respond differently to the same **antibiotic** (a substance produced by one organism that inhibits or kills other organisms). For example, certain antibiotics, including chloramphenicol and the aminoglycosides, disrupt ribosome activity (and therefore inhibit protein synthesis) in *Bacteria*, but have no adverse effect on *Archaea*. Likewise, eukaryotic ribosomes are not sensitive to some of the antibiotics that affect bacteria.

Gas Vesicles Among the most unusual structures found in prokaryotes are **gas vesicles**, produced by some aquatic species. These special protein-shelled structures provide buoyancy to many aquatic prokaryotes, and they are not found in any other life forms. When bacteria with gas vesicles are observed in the phase microscope they appear to contain bright, refractile areas, called **gas vacuoles**, with an irregular outline (**Figure 4.28A**). When gas vacuolate cells are viewed with the TEM, the vacuoles are found to consist of numerous subunits, called **gas vesicles** (Figure 4.28B).

Gas vacuoles are found in widely disparate prokaryotes. They are common in photosynthetic groups such as the cyanobacteria, proteobacteria, and green sulfur bacteria (*Chlorobi*). They are also found in heterotrophic bacteria such as the genus *Ancylobacter* and in the

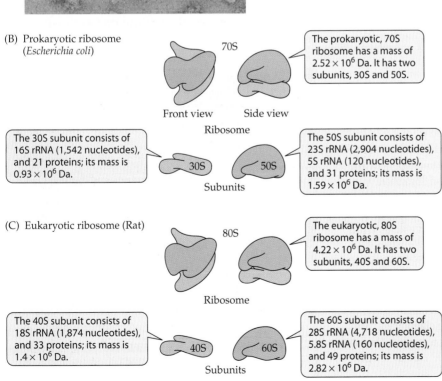

(B) Prokaryotic ribosome
(*Escherichia coli*)

70S

The prokaryotic, 70S ribosome has a mass of 2.52×10^6 Da. It has two subunits, 30S and 50S.

Front view   Side view
Ribosome

The 30S subunit consists of 16S rRNA (1,542 nucleotides), and 21 proteins; its mass is 0.93×10^6 Da.

30S   50S

The 50S subunit consists of 23S rRNA (2,904 nucleotides), 5S rRNA (120 nucleotides), and 31 proteins; its mass is 1.59×10^6 Da.

Subunits

(C) Eukaryotic ribosome (Rat)

80S

The eukaryotic, 80S ribosome has a mass of 4.22×10^6 Da. It has two subunits, 40S and 60S.

Ribosome

The 40S subunit consists of 18S rRNA (1,874 nucleotides), and 33 proteins; its mass is 1.4×10^6 Da.

40S   60S

The 60S subunit consists of 28S rRNA (4,718 nucleotides), 5.8S rRNA (160 nucleotides), and 49 proteins; its mass is 2.82×10^6 Da.

Subunits

Figure 4.27 **Ribosome structure**
(A) High-resolution electron micrograph of 70S ribosomes. (B) Composition of the *Escherichia coli* ribosome. (C) Composition of a eukaryotic (rat) ribosome. Photo courtesy of James Lake.

(A)

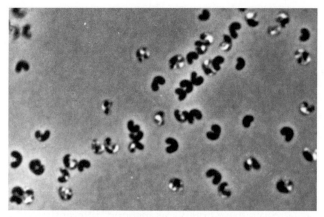

2.0 μm

(B)

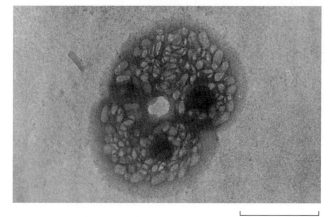

1.0 μm

Figure 4.28 Gas vacuoles and gas vesicles
Gas vacuoles and gas vesicles in *Ancylobacter aquaticus*, a vibrioid bacterium. (A) Phase photomicrograph showing gas vacuoles and (B) electron micrograph showing the numerous transparent gas vesicles that comprise a gas vacuole. Courtesy of M. van Ert and J. T. Staley.

prosthecate genera *Prosthecomicrobium* and *Ancalomicrobium*. The anaerobic gram-positive genus *Clostridium* also has gas vacuolate strains. Finally, some archaea produce gas vacuoles, including members of the genera *Methanosarcina* (methanogens) and *Halobacterium*.

Gas vesicles have been isolated from bacteria and studied biochemically. They are obtained by gently lysing cells to release the vesicles, then separating the vesicles from cellular material by low-speed differential centrifugation (Box 4.2). Cell material is relatively dense and is spun to the bottom of the centrifuge tube, while the buoyant gas vesicles float to the surface. The purified vesicles have a distinctive shape. They appear as cylindrical structures with conical end pieces (**Figure 4.29**). The size varies from about 30 nm in diameter in some species to about 300 nm in others. The vesicles can exceed 1,000 nm in length, depending upon the organism.

Each vesicle consists of a thin (about 2 nm thick) protein shell that surrounds a hollow space. The shell is composed of one predominant protein whose repeating subunits have a molecular mass of about 7,500 Da. Amino acid analyses of this protein from different prokaryotes indicate that its composition is highly uniform from one organism to another. About half of the protein consists of hydrophobic amino acids (such as alanine, valine, leucine, and isoleucine). It is thought that the hydrophobic amino acids are located on the inside of the shell and that their presence prevents water from entering the vesicle. Gases freely diffuse through the shell and are thus the sole constituents of the interior.

Gas vesicles do not store gases in the same way as a balloon. Gases freely diffuse through the vesicle shell, and the structure maintains its shape not because it is inflated but because of its water-impermeable, rigid protein framework. Because all gases freely diffuse in and out of the gas vesicle, the gases found in the vesicles are those present in the organism's environment.

The primary function of gas vesicles is to provide buoyancy for aquatic prokaryotes (**Box 4.3**). The density of the organism is reduced when the cell contains gas vesicles, thus permitting the organism to be buoyant in aquatic habitats. The mechanisms by which organisms regulate gas vacuole formation in nature are only poorly understood. Some cyanobacteria can descend in the environment by producing increased quantities of dense storage materials such as polysaccharides or by collapsing weaker gas vesicles. Depending upon their requirements for light, oxygen, and hydrogen sulfide, various bacterial groups are found in different strata in lakes during summer thermal stratification (see Chapter 24).

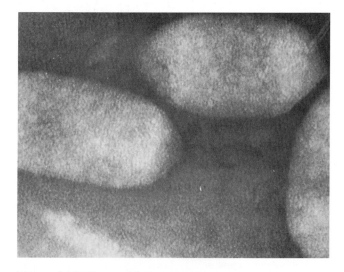

Figure 4.29 Gas vesicles
Gas vesicles isolated from *Ancylobacter aquaticus*. The vesicle diameter is about 0.1 μm. Courtesy of J. T. Staley and A. E. Konopka.

Milestones Box 4.3   The Hammer, Cork, and Bottle Experiment

In the early 1900s, C. Klebahn conducted an important but simple experiment called the "hammer, cork, and bottle" experiment, which provided the first evidence that the gas vacuoles of cyanobacteria actually contain gas. For the experiment, cyanobacteria containing the purported gas vacuoles were taken from a bloom in a lake. Microscopic examination showed that the organisms contained bright areas indicative of gas vacuoles. Samples were placed into two bottles, one to be used as a control for the experiment.

In both bottles, the cyanobacteria initially floated to the surface. A cork was then placed in one bottle and secured in such a way that no air space was left between the cork and the water. Then, a hammer was used to strike a sharp blow on the suspension of cyanobacteria in the corked bottle. The blow briefly increased the hydrostatic pressure of the water in the experimental bottle.

The cyanobacteria in the experimental bottle soon sank to the bottom, whereas those in the control bottle remained floating on top. Furthermore, an air space formed

(A)   (B)

Figure A Two bottles containing gas vacuolate cyanobacteria collected from a lake. The cell suspension with collapsed gas vesicles has a darker appearance. Gas vacuolate cells cause much greater refraction of light, so the bottle on the right appears more turbid. **Figure B** Appearance of the bottles after a few minutes. Courtesy of A. E. Walsby.

between the water and the cork in the experimental bottle. When the cells from this bottle were subsequently examined in the microscope, no bright areas remained in the cells, showing that the cyanobacteria had lost their buoyancy and their gas vacuoles at the same time. The air space that had

collected in the experimental bottle was due to the gas released from the broken vesicles and contained by the cork.

This simple experiment showed that, indeed, gas vacuoles do contain gas and that they are essential in providing buoyancy to the cyanobacteria.

The vertical position that gas vacuolate species occupy in the lake depends on their cell density, which is determined by the proportion of cell volume occupied by gas vesicles at a particular time.

Intracellular Reserve Materials Prokaryotes store a variety of organic and inorganic materials as nutrient reserves. Almost all these materials are stored as polymers, thereby maintaining the internal osmotic pressure at a low level (see the discussion of osmotic pressure in "Function of the Cell Wall" below).

The main organic compounds stored by bacteria are:

- Glycogen
- Starch
- Poly-β-hydroxybutyric acid
- Cyanophycin

Glycogen and **starch** are common storage materials in prokaryotic organisms. They are polymers of glucose units linked together primarily by α-1,4 linkages (see Chapter 11). These storage materials cannot be seen using a light microscope, but when observed with the electron microscope they appear as either small, uniform granules, as in some cyanobacteria (see Figure 21.16), or as larger spheroidal structures, as in heterotrophic bacteria. Glycogen and starch can be degraded as energy and carbon sources. Interestingly, glycogen is also formed by animal cells, and starch is formed by eukaryotic algae and higher plants.

Unlike glycogen and starch, **poly-β-hydroxybutyric acid** (PHB) and related polymeric acids appear as visible granules in bacteria when viewed with a light microscope. In phase microscopy, these lipid substances appear as bright, refractile, spherical granules (**Figure**

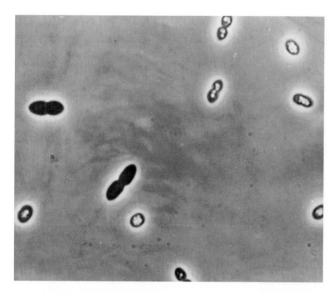

Figure 4.30 PHB granules
Poly-β-hydroxybutyrate (PHB) granules appear as bright refractile areas by phase microscopy, as seen in *Azotobacter* sp. Courtesy of J. T. Staley.

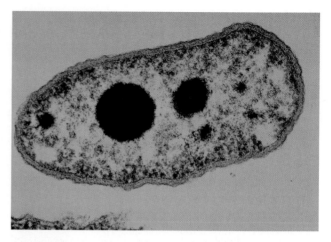

Figure 4.31 Polyphosphate granules
Thin section of a bacterial cell (*Pseudomonas aeroginesa*) shows polyphosphate granules as dark areas. ©T. J. Beveridge/Visuals Unlimited.

4.30). PHB granules are usually synthesized during periods of low nitrogen availability in environments that have excess utilizable organic carbon. In contrast to glycogen and starch, PHB is not found in eukaryotic organisms. PHB is a polymer of β-hydroxybutyric acid and is synthesized from acetyl-coenzyme A (see Chapter 11).The granules are enclosed within a protein membrane, which may be responsible for their synthesis or degradation, or both.

The only organic nitrogen polymer stored by prokaryotes is **cyanophycin**. As implied by the name, these granules occur only in cyanobacteria. They consist of a copolymer of aspartic acid and arginine and have a distinctive appearance when viewed in thin section with an electron microscope (see Figure 21.16).

Prokaryotes also store a variety of inorganic compounds, including polyphosphates and sulfur. **Volutin**, or **metachromatic granules**, consists of linear polyphosphate molecules, polymers of covalently linked phosphate units. Volutin appears as dark granules in phase microscopy. The granules can be stained with methylene blue, which results in a red color caused by the metachromatic effect of the dye-polyphosphate complex. They appear as dense areas when viewed by the electron microscope (**Figure 4.31**), and they may volatilize under the electron beam, leaving a "hole" in the specimen.

Volutin is synthesized by the stepwise addition of single phosphate units from ATP onto a growing chain of polyphosphate. Although its degradation has not been studied, volutin may serve as an energy source for the

synthesis of ATP from ADP, at least in some bacteria, as well as a reserve material for nucleic acid and phospholipid synthesis. Volutin is apparently stored by organisms when nutrients are limiting their growth. However, some bacteria are known to store volutin during active growth when there is no nutrient limitation, an effect called **luxurious phosphate uptake**.

Sulfur is stored as elemental sulfur by certain bacteria involved in the sulfur cycle. Sulfide-oxidizing pho-

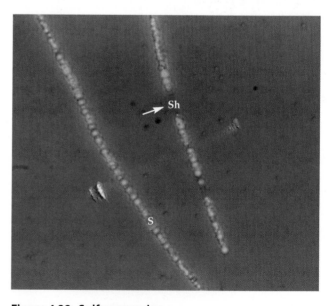

Figure 4.32 Sulfur granules
Phase photomicrograph of the sulfur bacterium *Thiothrix nivea* showing many bright yellow and orange sulfur granules (S), arrow pointing to sheath (Sh). Courtesy of Judith Bland.

tosynthetic proteobacteria and colorless filamentous sulfur bacteria, both of which produce sulfur by the oxidation of sulfide, store the sulfur as granules, which appear as bright, slightly yellow spherical areas in their cells (**Figure 4.32**). The sulfur can be further oxidized to sulfate by both groups of organisms. Thus, the granules are transitory and serve as a source of energy (when oxidized by filamentous sulfur bacteria; see Chapter 19) or a source of electrons (for carbon dioxide fixation in photosynthetic bacteria; see Chapter 21).

Cell Membranes

Bacterial Cell Membranes

The **cell membrane** or **cytoplasmic membrane** serves as the primary boundary of the cell's cytoplasm. In prokaryotes, it looks like a typical cell membrane—two thin lines separated by a transparent area—when viewed in thin section by the electron microscope (Figures 4.12 and 4.23). At the molecular level, a model known as the **fluid mosaic**

model best explains the structure and composition of the cell membrane (**Figure 4.33**).

The cell membranes of bacteria are composed of phospholipids and proteins. As many as seven different phospholipids and approximately 200 proteins have been identified in the typical membrane of bacteria such as *Escherichia coli*. One of the phospholipids is phosphatidyl serine (**Figure 4.34**). This complex lipid consists of a glycerol moiety covalently bonded to two fatty acids by an **ester linkage** and, at its third carbon, to a phosphoserine. The result is a bipolar molecule with a hydrophobic end (the hydrocarbon chains of the fatty acids) and a hydrophilic end (the glycerol phosphatidyl serine). In an aqueous environment, phospholipid molecules form a bilayer consisting of two membrane leaflets: the hydrophobic chains, or tails, of the two fatty acids extend into the center of the bilayer and the hydrophilic portions face away from the bilayer (Figure 4.33).

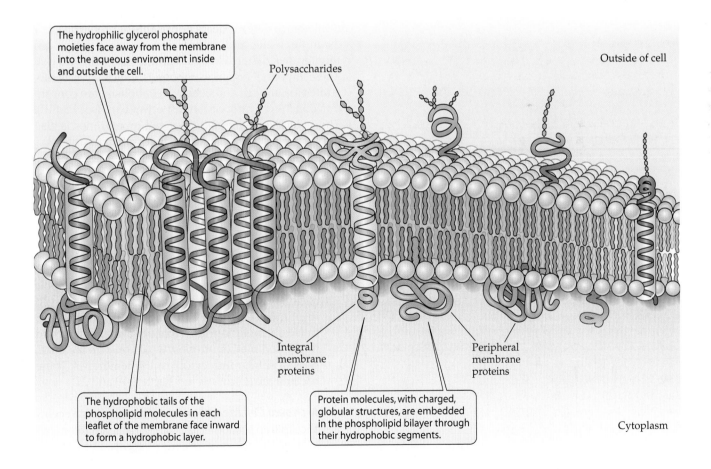

The hydrophilic glycerol phosphate moieties face away from the membrane into the aqueous environment inside and outside the cell.

Polysaccharides

Outside of cell

The hydrophobic tails of the phospholipid molecules in each leaflet of the membrane face inward to form a hydrophobic layer.

Integral membrane proteins

Protein molecules, with charged, globular structures, are embedded in the phospholipid bilayer through their hydrophobic segments.

Peripheral membrane proteins

Cytoplasm

Figure 4.33 Bacterial cell membrane structure
The fluid mosaic model of cell membrane structure. The bilayer structure consists of two phospholipid leaflets.

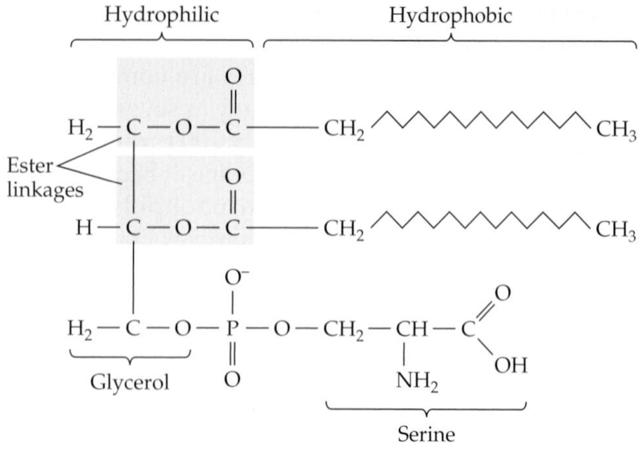

Figure 4.34 Phospholipid
Chemical structure of phosphatidyl serine, a typical phospholipid. The hydrocarbon side chains are fatty acids, typically containing 12 to 18, sometimes more, carbon atoms. Each is attached to glycerol by an ester linkage. Serine is attached to the phosphate moiety.

(A) Cholesterol

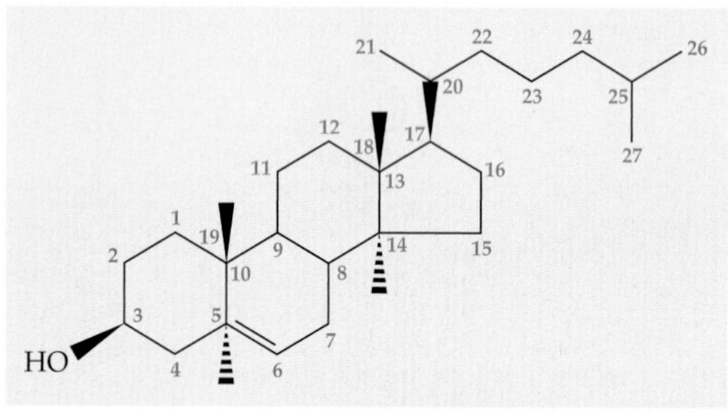

(B) A hopanoid from a cyanobacterium

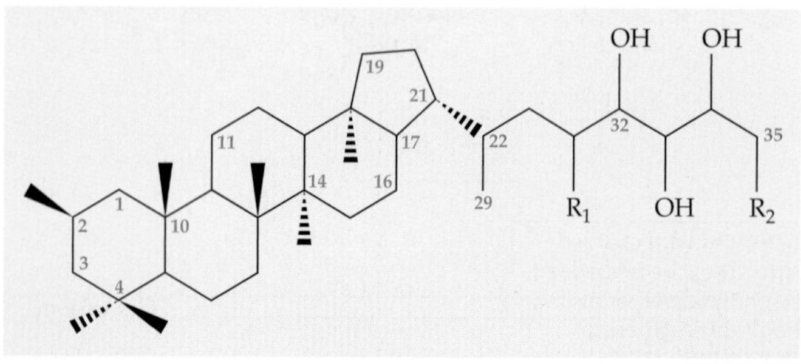

Figure 4.35 Sterols and hopanoids
(A) A sterol and (B) a hopanoid found in some bacterial cell membranes. The hopanoid shown here is from a cyanobacterium. R_1 and R_2 indicate two different hydrocarbon chains.

The proteins of the cell membrane are embedded in the phospholipid bilayer (Figure 4.33). Many of the proteins serve in the transport of substances into the cell. The membrane is stabilized by divalent cations including calcium and magnesium.

Sterols are known to have a stabilizing effect on cell membranes because of their planar chemical structures (**Figure 4.35**). However, most prokaryotic organisms do not have sterols in their cell membranes. The mycoplasmas (*Tenericutes*), which lack cell walls, are exceptions. They do not synthesize their membrane sterols but derive them from the environment in which they live. The only other group of prokaryotes that is known to contain sterols is the methanotrophic bacteria, which have special membranes involved in the oxidization of methane as an energy source. Some bacteria, such as the cyanobacteria, produce **hopanoids**, compounds that resemble sterols chemically and probably provide membrane stability (Figure 4.35).

The cell membrane is the ultimate physical barrier between the cytoplasm and the external environment of the cell. It is selectively permeable to chemicals; the molecular size, charge, polarity, and chemical structure of a substance determine its permeability properties. For example, water and gases diffuse freely through cell membranes, whereas sugars and amino acids do not. These important organic compounds, which serve as substrates for growth and energy metabolism, are concentrated in the cell by special transport systems (see Part III). This selective permeability of cell membranes is also important in cell energetics, as discussed in Chapter 8.

One additional function of the cell membrane is its role in the replication of DNA, which is attached to the membrane. Following replication, the two new DNA molecules are physically separated by new cell membrane synthesis. The eventual outcome is the separation of the two new nuclear structures prior to formation of the septum and cell division.

Some prokaryotes contain membranes that extend into the cell itself. For example, phototrophic bacteria have internal membranes that are involved in photosynthesis. Some chemoautotrophs (also called chemolithotrophs) that oxidize ammonia and nitrite and methanotrophs, which oxidize methane, possess similar intracytoplasmic membranes. These bacteria are discussed in Chapters 19 and 21.

Archaeal Cell Membranes Although their appearance in thin section is like that of bacterial cell membranes, the cell membranes of archaea have a different chemical composition. One of the major differences is that the lipid moieties of the membrane are not glycerol-linked esters but glycerol-linked *ethers*. Furthermore, rather than fatty acids, most archaea have **isoprenoid** side chains with repeating five-carbon units (Figure 4.36). Two patterns are found in archaeal cell membranes:

- A bilayer, with two leaflets of glycerol ether-linked isoprenoids held together by their hydrophobic isoprenoid side chains (**Figure 4.36A**)
- A monolayer, consisting of diglycerol tetraethers (Figure 4.36B)

In the remainder of the chapter we consider the layers external to the cell cytoplasm, beginning with those farthest from the cytoplasm and working toward the cell membrane.

Extracellular Layers

Some prokaryotes produce discrete layers external to the cell wall. These are of several types, varying in structure and function:

- Capsules, or glycocalyxes
- Sheaths
- Slime layers
- Protein jackets, or S-layers

If the external layer is removed from the cell, the organism does not lose its viability. Thus, they do not appear to be essential to the organisms that produce them, at least under some conditions of growth.

Capsules Capsules constitute barriers that protect the cell from the external environment, no doubt aiding the cell in unknown ways, such as preventing virus attachment or slowing the desiccation process when the cell encounters dry conditions.

In the laboratory, capsules can be removed without affecting the cells' viability, indicating they are not always essential to survival. Nevertheless, several functions have been attributed to capsules, and under certain conditions they may be crucial in determining the survival of the organism in its natural environment. For example, the capsules of some species are important virulence factors (factors that enhance the ability of the organism to cause disease). This can be illustrated by the species *Streptococcus pneumoniae*, one of the bacteria responsible for bacterial pneumonia. Unencapsulated strains of this bacterium are not pathogenic, that is, they are not capable of causing disease. The reason is that the unencapsulated cells are more readily killed by phagocytes (white blood cells) in the host organism's circulatory system. Apparently the white blood cells can ingest bacteria without capsules much more readily than those

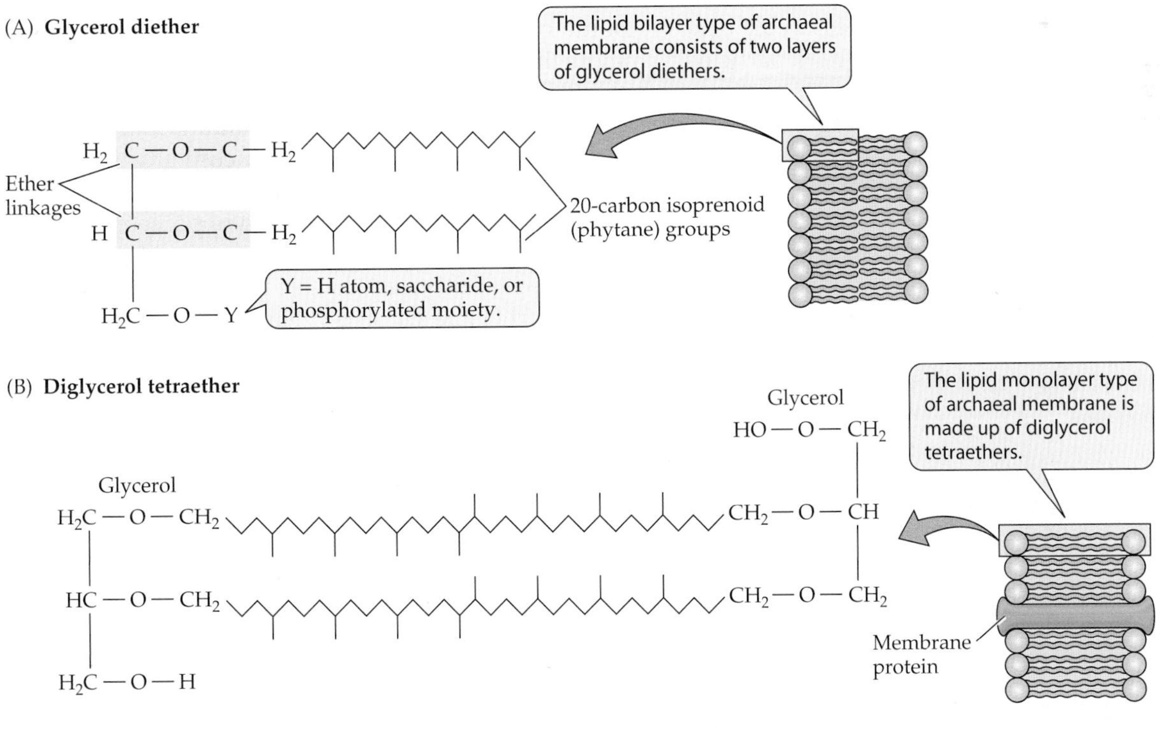

Figure 4.36 Archaeal cell membrane structure
Archaeal membranes are of two types: (A) a lipid bilayer, consisting of two leaflets of glycerol-linked isoprenoids, and (B) a lipid monolayer, consisting of a single layer of diglycerol tetraethers.

with capsules. Thus, capsules may provide a means by which the bacteria evade the host defense system in their natural environment.

Capsules are also important in mediating attachment of some bacteria. This is well illustrated by one of the bacteria responsible for dental cavities. In the presence of sucrose, *Streptococcus mutans* produces a polysaccharide capsule that permits the bacterium to attach to dental enamel. As *S. mutans* grows on the surface of the tooth, acid produced by fermentation of sucrose causes etching of the enamel, and a cavity may eventually develop.

Capsules are diffuse structures that are often difficult to visualize without special techniques. This is well exemplified by the capsule of *Streptococcus pneumoniae*. Its capsule is not visible by ordinary phase microscopy or by simple staining techniques, because of the diffuse consistency of the polysaccharide that forms the capsule. A special technique has been developed to permit visualization of the capsule, called the **Quellung** (German for "swelling") reaction. An antiserum (see Chapter 27) is prepared from capsular material of *S. pneumoniae* and is used to "stain" *S. pneumoniae* cells, complexing with their capsules. After staining, the bacterium appears to have grown much larger or to have swollen. In reality, the antiserum has complexed with the extensive but previously transparent capsule and made it thicker and more visible (**Figure 4.37**).

Negative stains can also be used to stain capsules for light microscopy. These stains consist of insoluble particulate materials such as India ink. The dark particles of the stain do not penetrate the capsule, so a large unstained halo appears around the cell (**Figure 4.38**).

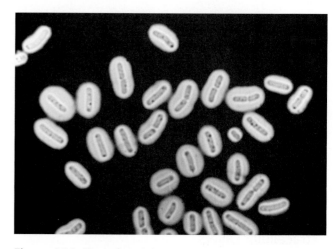

Figure 4.38 Negative stain
An India ink negative stain shows the capsule of a *Bacillus* sp. This is called a negative stain because the background around the capsule is stained, not the capsule itself. Courtesy of Carl Robinow.

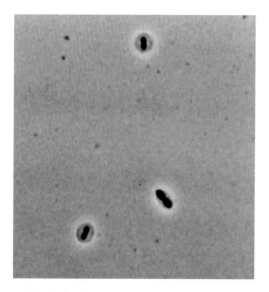

Figure 4.37 Capsule stain
Capsules of *Streptococcus pneumoniae* have been stained by the Quellung reaction. Courtesy of J. Dalmasso and J. T. Staley.

Most capsules are composed of polysaccharides (**Table 4.1**). Dextrans (polymers of glucose) and levans (polymers of fructose, also known as levulose) are common **homopolymers** (polymers consisting of identical subunits) found in capsules produced by lactic acid bacteria (*Streptococcus* and *Lactobacillus* spp.). However, some bacteria produce **heteropolymers**, containing more than one type of monomeric subunit. Hyaluronic acid is an example of a heteropolymer found in capsules; it consists of *N*-acetylglucosamine and glucuronic acid. Another example is the glucose-glucuronic acid polymer, also produced by some *Streptococcus* species.

Capsules can also be composed of polypeptides. For example, some members of the genus *Bacillus* have capsules made of polymers that are repeating units of D-glutamic acid (Table 4.1). It is interesting to note that the D stereoisomers of amino acids are found almost exclusively in the capsules and cell walls of some bacteria. Virtually all other biologically produced amino acids are of the L stereoconfiguration. Thus, other than capsule and cell wall peptides, all proteins in bacteria (as well as in other organisms) contain L-amino acids. The poly-D-glutamic acid of *Bacillus* is also noteworthy because the amino acids are linked together by their γ-carboxyl group, not the α-carboxyl group as is characteristic of proteins.

Sheaths Some prokaryotes, the **sheathed bacteria**, produce a much more dense and highly organized external layer, or sheath. Unlike a capsule, this structure is easily discerned with a light microscope (**Figure 4.39**). Sheathed bacteria form filamentous chains and grow in flowing aquatic habitats such as rivers and springs. The

Table 4.1	Chemical composition of capsules from various bacteria	
Bacterium	**Capsular Material**	**Structure of Repeating Units**
Leuconostoc mesenteroides	Dextran	α-1,6-poly D-glucose
Streptococcus pneumoniae	Polyglucose glucuronate	Glucuronic acid / Glucose
Streptococcus spp.	Hyaluronic acid	Glucuronic acid / N-acetyl glucosamine
Bacillus anthracis	γ-poly D-glutamic acid	γ D-glutamic acid

sheath protects the cells against disruption by turbulence in the water and ultimately against being removed from the area of the habitat that is most favorable for growth. The few chemical analyses of sheaths that have been made indicate that they are more complex than capsules, having, in addition to polysaccharides, amino sugars and amino acids.

Slime Layers Some prokaryotes, such as those of the *Cytophaga-Flavobacterium* group, move by a type of motility referred to as **gliding**. This type of movement requires that the organism remain in contact with a solid substrate. Gliding bacteria produce a chemical "slime," part of which is lost in the trail left by the organism during movement (**Figure 4.40**). Members of the *Cytophaga*

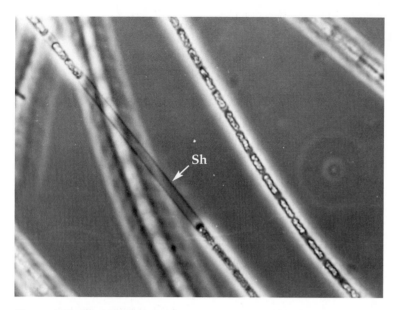

Figure 4.39 Sheathed bacteria
The filamentous gram-negative bacterium *Sphaerotilus natans* has a sheath (Sh). This species grows in flowing aquatic environments. Courtesy of J. T. Staley.

and *Flavobacterium* are noted for their ability to degrade substances such as cellulose and chitin and other high molecular weight polysaccharides found in plant material (cellulose) and insect or arthropod shells (chitin). It is advantageous for organisms that degrade these materials for nutrition to be in physical contact with them. As the bacterium glides on the surface of the material, the cellu-

Cells gliding together   Slime trail

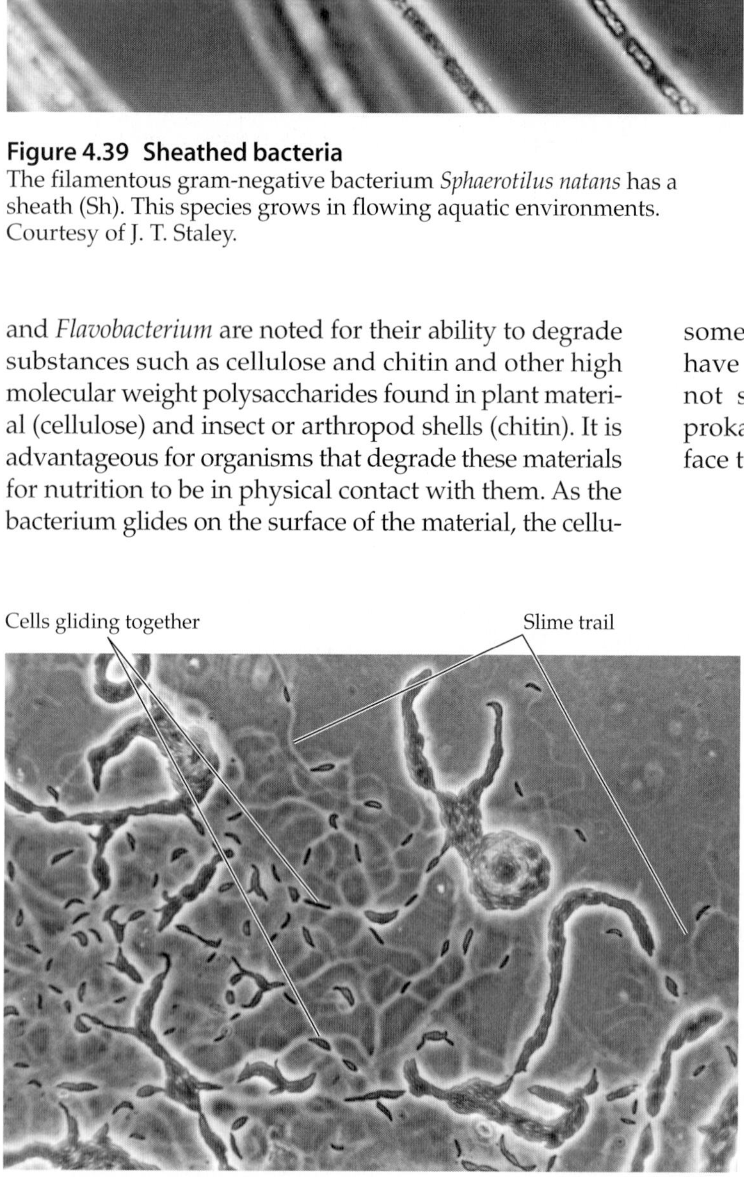

Figure 4.40 Gliding motility
The myxobacterium *Stigmatella aurantiaca* produces slime trails as the cells glide across an agar surface. Note that even the smallest units contain at least two cells gliding together. Courtesy of Hans Reichenbach.

lase and chitinase enzymes, bound to the organism's extracellular surface, are kept close to the substrate as well as the cell. In this manner the cell can derive nutrients more effectively.

The chemical nature of the slime varies. In some species it is a polysaccharide. For example, *Cytophaga hutchinsonii* produces a heteropolysaccharide of arabinose, glucose, mannose, xylose, and glucuronic acid. The mechanism of gliding motility is unknown, but recent interesting research on *Flavobacterium johnsoniae* indicates that a special sulfur-containing lipid (sulfonolipid) found on the surface of the cell is essential for gliding in this organism.

Protein Jackets External protein layers are produced by certain prokaryotes, such as the genus *Spirillum*. These **protein jackets**, also called **S-layers**, are often highly textured surfaces (Figure 4.41). Their function is unknown, although in some archaea they make up the only layer external to the cell membrane and therefore may serve as a cell wall. However, some *Spirillum* species and certain cyanobacteria that have these layers also have a cell wall, so S-layers cannot serve that role in these organisms. For some prokaryotes, the S-layer serves as a site on the cell surface to which bacterial viruses adhere.

Cell Walls

The cell wall and cell membrane are referred to as the **cell envelope.** The cell wall is the outermost cellular constituent of prokaryotic organisms. In contrast to extracellular layers, cell walls are essential to the microorganisms that produce them. Without this protection, the organisms could not survive in their normal habitats because the cells would lyse (break open and lose their cytoplasm; see below). Considerable variation in cell wall composition exists among prokaryotes. One special class of bacteria, the *Tenericutes* (commonly called the mycoplasmas) lack a cell wall entirely. Bacteria without cell walls survive because their cell membranes differ from those of typical bacteria. For example, their membranes contain sterols or other compounds (see above) that help stabilize membrane structure, and in this respect they are similar to animal cell membranes, which also lack cell walls.

As mentioned in Chapter 1, prokaryotes are classified as either *Archaea* or *Bacteria* based on evolutionary studies of 16S rRNA. In addition to this difference in their RNAs, the two

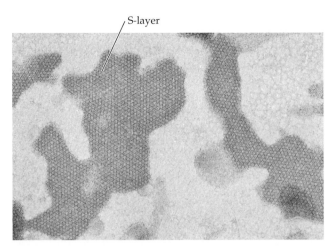

Figure 4.41 S-layers
Electron micrograph of negatively stained fragments of a highly textured protein jacket, or S-layer, from *Aquaspirillum serpens* strain VHA. The dark areas show the highly organized nature of the structure. Courtesy of S. F. Koval.

domains also differ in the chemistry of their cell walls. Almost all *Bacteria* have a chemical polymer in their cell walls called **peptidoglycan,** or **murein;** *Archaea* do not, although some have a pseudomurein (see Chapter 18). Peptidoglycan is a complex polymeric substance containing amino sugars and amino acids, as described in detail below.

Bacterial Cell Wall The peptidoglycan cell wall structure varies from one bacterial species to another. However, all have the same general chemical composition. Two amino sugars, glucosamine and muramic acid, are joined together by β-1,4 linkages to form a chain, or linear polymer (a glycan). The chains of amino sugars are cross-linked by peptides. **Figure 4.42** shows the peptidoglycan of *Staphylococcus aureus*. Note that the amino sugars are N-acetylglucosamine and N-acetylmuramic acid. Almost all bacteria have these N-substituted acetyl groups; some *Bacillus* species have unsubstituted amino groups on the sugars, and mycobacteria and some of the streptomycetes have N-glycolyl groups. The amino

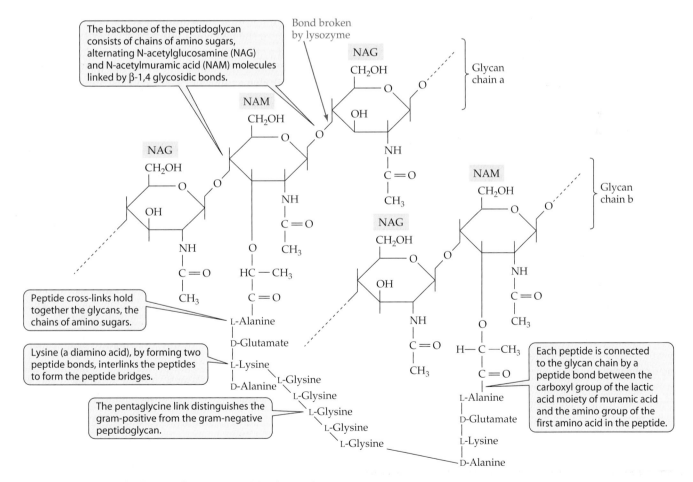

Figure 4.42 Peptidoglycan of a gram-positive bacterium
Chemical structure of the peptidoglycan layer of *Staphylococcus aureus*. Note that some amino acids in the peptide cross-links, alanine and glutamic acid, are in the D-stereoconfiguration. See also Figure 4.44.

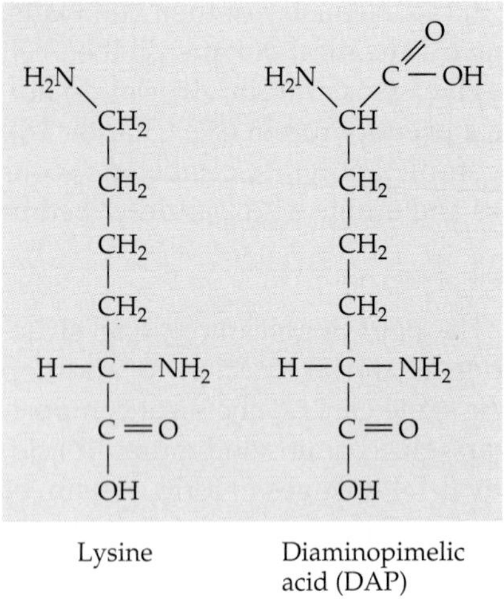

Lysine   Diaminopimelic
                          acid (DAP)

Figure 4.43 Diamino acids
Lysine and diaminopimelic acid are diamino acids found in peptidoglycans.

group of the first amino acid in the peptide cross-link is joined to the carboxyl group of the lactic acid moiety of N-acetylmuramic acid by a peptide bond. Some of the amino acids (e.g., alanine and glutamic acid) exist in the D stereoconfiguration.

Note that one of the amino acids in the peptide—in Figure 4.42, lysine—is a diamino acid, that is, it contains two amino groups (**Figure 4.43**). It is essential that one of the amino acids in the cross-linking peptide bridge is a diamino acid, because the single amino groups of the other amino acids are tied up in peptide linkages all the way back to the muramic acid. The diamino acid can link one of its amino groups to one peptide chain and its other amino to the other chain, thus cross-linking the peptides and their amino sugar chains. The result of this cross-linking is the formation of a two-dimensional network around the cell (**Figure 4.44**). It is this two-dimensional sac that provides the rigidity and strength of the cell wall. Lysine and diaminopimelic acid (Figure 4.43) are the most common diamino acids responsible for this cross-linking. Diaminopimelic acid is not found in proteins; it occurs uniquely in the peptidoglycan of virtually all gram-negative bacteria (Figure 4.44).

Considerable variation exists in the amino acids that form the cross-linking peptides of peptidoglycans. About 100 types of cell wall peptidoglycan structures are known. However, certain amino acids are never found in the peptide bridge, including the sulfur-containing amino acids, aromatic amino acids, and branched-chain amino acids, as well as arginine, proline, and histidine (see Chapter 3).

Gram-Positive and Gram-Negative Bacterial Cell Walls The domain *Bacteria* is divided into two groups based upon the cells' reaction to a staining procedure called the **Gram stain** (**Box 4.4**). The differences between gram-positive and gram-negative bacteria relate to differences in their cell wall structure and chemical composition. Thin sections of gram-positive bacteria reveal walls that are thick, nearly uniformly dense layers (Figure 4.45A). In contrast, the cell walls of gram-negative bacteria are more complex, because in addition to a peptidoglycan layer they have another layer, called an **outer membrane** (**Figure 4.45B**).

The structural differences between the cell walls of gram-positive and gram-negative bacteria reflect differences in biochemical composition. When techniques were developed to permit the separation of cell walls from cytoplasmic constituents, scientists could chemically analyze the cell wall. The major constituent found in the cell wall of gram-positive bacteria was peptidoglycan. It makes up about 40% to 80% of the dry weight of the wall, depending upon the species. Peptidoglycan has a tensile strength similar to that of reinforced concrete. Other constituents of gram-positive cell walls include teichoic acids and teichuronic acids. **Teichoic acids** are polyol phosphate polymers, such as polyglycerol phosphate and polyribitol phosphate (**Figure 4.46**). Sugars (such as glucose and galactose), amino sugars

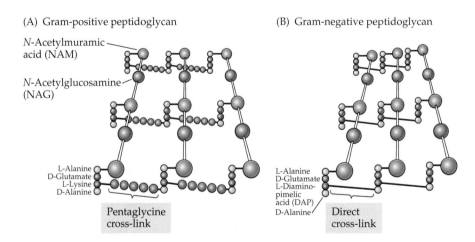

(A) Gram-positive peptidoglycan

N-Acetylmuramic acid (NAM)

N-Acetylglucosamine (NAG)

L-Alanine
D-Glutamate
L-Lysine
D-Alanine

Pentaglycine cross-link

(B) Gram-negative peptidoglycan

L-Alanine
D-Glutamate
L-Diamino-pimelic acid (DAP)
D-Alanine

Direct cross-link

Figure 4.44 Cell walls of gram-positive and gram-negative bacteria
The diagrams show the two-dimensional network of the peptidoglycan sac surrounding (A) a gram-positive and (B) a gram-negative cell. This layer is the major structural component of bacterial cell walls. The N-acetylglucosamine (NAG) and N-acetylmuramic acid (NAM) are linked to form the amino sugar backbone (glycan). The glycan chains are held together by peptide bridges.

Milestones Box 4.4   The Gram Stain

The Gram stain procedure was developed unwittingly in 1888 by Christian Gram, a Danish physician studying in Berlin. He was examining lung tissues during autopsies of individuals who had died of pneumonia. He noted that *Streptococcus pneumoniae* found in the lung tissues retained the primary stain known as Bismark Brown (crystal violet is now used), whereas the lung tissue did not. It was subsequently determined that certain bacteria, now termed gram-positive, like *S. pneumoniae* retain the purple dye when stained by this method, whereas others, the gram-negative bacteria, do not. The Gram stain is called a **differential stain** because it distinguishes between these two groups of bacteria.

Procedure

1 Primary stain: prepare a smear of the bacterium and stain with crystal violet.

2 Mordant: flood the preparation with a solution of potassium iodide and iodine, which complexes with both the crystal violet and the cellular material.

3 Decolorization: add ethanol dropwise to the stained smear.

4 Counterstain: stain the preparation with safranin.

Color of the cells

Gram +   Gram −

Both gram-negative and gram-positive bacteria are stained purple.

The dye complex is washed off the gram-negative bacteria. The gram-positive bacteria retain some of the dye and remain intensely purple in color.

The gram-negative bacteria are now stained red. The gram-positive bacteria retain their purple color.

The modern Gram stain procedure is illustrated here.

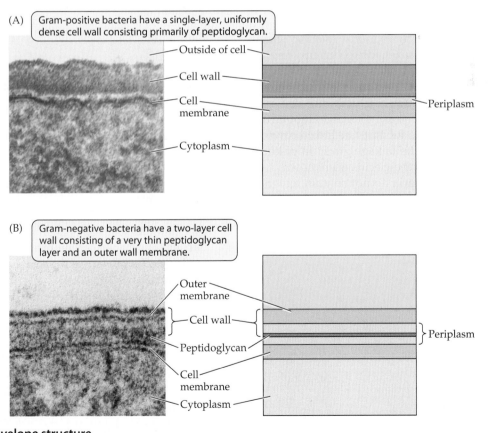

(A) Gram-positive bacteria have a single-layer, uniformly dense cell wall consisting primarily of peptidoglycan.

Outside of cell
Cell wall
Cell membrane
Cytoplasm
Periplasm

(B) Gram-negative bacteria have a two-layer cell wall consisting of a very thin peptidoglycan layer and an outer wall membrane.

Outer membrane
Cell wall
Peptidoglycan
Cell membrane
Cytoplasm
Periplasm

Figure 4.45 Cell envelope structure

Thin sections and diagrams showing the cell envelopes of (A) a gram-positive bacterium, *Bacillus megaterium*, and (B) the gram-negative bacterium *Escherichia coli*. Note the characteristic outer membrane of the gram-negative bacterium. The thin peptidoglycan layer of gram-negative bacteria is not always evident in electron micrographs. Note also the location of the periplasm in each type of organism. Photos courtesy of Peter Hirsch and Stuart Pankratz.

(A)

(B)

(C)

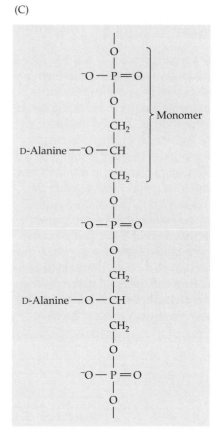

Figure 4.46 Teichoic acids
Chemical structures of two teichoic acids found in gram-positive bacteria. Monomers of (A) the polyribitol teichoic acid of *Bacillus subtilis* and (B) the polyglycerol teichoic acid of *Lactobacillus* sp. (C) The monomers are joined by phosphate linkages. Shown here is a polyglycerol teichoic acid.

(such as glucosamine), and the amino acid D-alanine are found in some of these compounds. Teichoic acids are attached covalently to the 6-hydroxyl group of muramic acid in the peptidoglycan. **Teichuronic acids** are polymers of two or more repeating subunits, one of which is always a uronic acid, such as glucuronic acid, or the uronic acid of an amino sugar, such as aminoglucuronic acid (**Figure 4.47**). The subunits are linked covalently to peptidoglycan, but the linkage group in unknown. The specific function of these acids is unknown.

Gram-negative bacteria do not contain teichoic or teichuronic acids. However, in other respects their cell wall structures are more complex than those of the gram-positive bacteria, as suggested by their multilay-

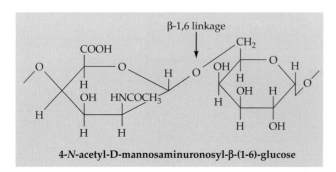

4-N-acetyl-D-mannosaminuronosyl-β-(1-6)-glucose

Figure 4.47 Teichuronic acids
A teichuronic acid found in the cell wall of *Micrococcus luteus*.

ered appearance (Figures 4.45B and **4.48**). Peptidoglycan is present, but it makes up a much smaller proportion of the cell wall material. In contrast to gram-positive bacteria, which have several layers of peptidoglycan encasing the cell, gram-negative organisms have only a single or a few macromolecular sheets. Indeed, the peptidoglycan comprises only about 5% by weight of the gram-negative cell wall. Despite the reduced amount of this material, it still serves an important role as a rigid barrier outside the cell membrane, because it is covalently attached to one of the major proteins in the outer membrane.

The outermost layer, or outer membrane, of gram-negative cells appears similar to a cell membrane when viewed by electron microscopy (Figure 4.45B). Lipids and proteins predominate; however, polysaccharides extend into the aqueous environment. As in the cell membrane, each of the two leaflets of the outer membrane consists of lipid molecules with their hydrophilic moieties (glycerol ester linkages) facing away from the membrane and toward the aqueous environment. Likewise, the hydrophobic portions (hydrocarbon chains) face inward and hold the two leaflets together.

The outer leaflet of the membrane serves as the site of attachment for lipid A (**Figure 4.49**). Bonded cova-

(A) Gram-positive cell envelope

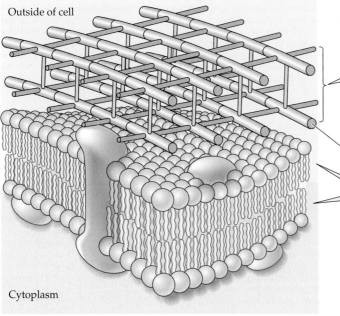

Outside of cell

The cell wall consists of thick peptidoglycan layers.

Peptidoglycan

The cell membrane consists of two phospholipid leaflets held together by hydrophobic forces.

Cytoplasm

(B) Gram-negative cell envelope

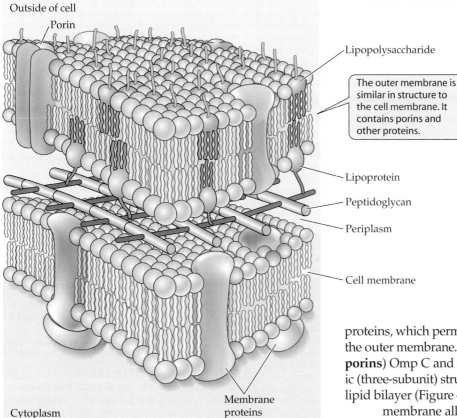

Outside of cell

Porin

Lipopolysaccharide

The outer membrane is similar in structure to the cell membrane. It contains porins and other proteins.

Lipoprotein

Peptidoglycan

Periplasm

Cell membrane

Membrane proteins

Cytoplasm

Figure 4.48 Cell envelopes of bacteria
Models of cell envelopes of (A) a gram-positive bacterium, showing the cell membrane and cell wall, and (B) a gram-negative bacterium, showing the cell membrane, peptidoglycan, and outer membrane.

lently to lipid A is a core polysaccharide (**Figure 4.50**) that has specific side-chain polysaccharides (called somatic polysaccharides or *O*-polysaccharides) that vary from one species of gram-negative bacterium to another. This lipid A-polysaccharide complex is referred to as a **lipopolysaccharide**, or simply **LPS**. The LPS that contains lipid A is called **endotoxin** because it is toxic to a variety of animals, including humans. Animals injected with endotoxin may experience fever, hemorrhage, shock, miscarriage, and a variety of other symptoms depending upon the dose and source. Many gram-negative bacteria, including *Salmonella* species (which cause enteric diseases, including typhoid fever) and *Yersinia pestis* (which causes bubonic plague), owe their toxic properties to this compound. Other gram-negative bacteria with LPS contain a version that is nontoxic because it lacks specific parts of the lipid A moiety.

The outer membrane of gram-negative bacteria also contains proteins, but the amounts are less than in the cell membrane. The most prevalent protein is a small polypeptide (which has a molecular weight of 7,200 Da) containing a lipid moiety. This lipoprotein is joined covalently to the diamino acid of the peptidoglycan to form a complex called **peptidoglycan-lipoprotein**. Thus, the lipoprotein serves to anchor the outer membrane to the cell wall peptidoglycan layer. Other important proteins serve as transport proteins, which permit the passage of molecules through the outer membrane. For example, the matrix proteins (or **porins**) Omp C and Omp F of *Escherichia coli* are trimeric (three-subunit) structures that form channels across the lipid bilayer (Figure 4.49). These water-filled pores in the membrane allow the free diffusion of water-soluble nutrients, such as sugars and amino acids, as well as inorganic ions. Finally, some proteins are responsible for the entry of specific compounds into the cell, such as vitamin B_{12}, iron chelates, disaccharides, or phosphorylated compounds.

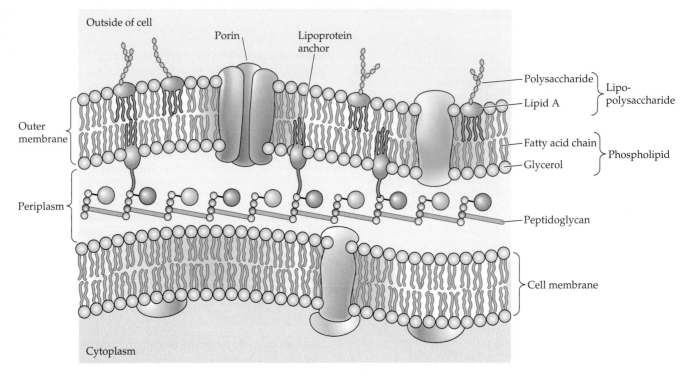

Figure 4.49 Outer membrane
Diagram of the outer membrane of a gram-negative cell. Note the location of the lipid A of LPS, the transmembrane porin proteins, and the lipoprotein anchor that attaches the outer membrane to the peptidoglycan layer. It also has a lipopolysaccharide complex consisting of lipid A attached to the outer leaflet of the outer membrane and covalently bonded to a polysaccharide. The polysaccharide tail extends away from the cell into the external environment.

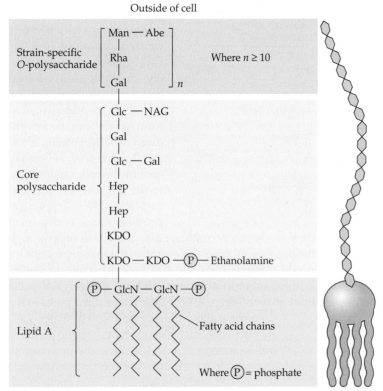

Archaeal Cell Wall The domain *Archaea* consists of three major phyla: *Crenarchaeota*, which contains the hyper-thermophilic organisms; *Euryarchaeota*, with two subgroups, the methane-producing group (methanogens) and the extreme halophilic group (salt-loving bacteria; see Chapter 18); and *Korarchaeota*, much less well known because isolates have not yet been obtained.

Some methanogens have proteinaceous cell walls. Others have cell walls with a material similar to peptidoglycan, called **pseudopeptidoglycan** or **pseudomurein (Figure 4.51)**. The amino sugar chain of the pseudomurein contains *N*-acetylglucosamine, but *N*-acetyltalos-

Figure 4.50 Polysaccharide of LPS
Chemical composition of the polysaccharide portion of LPS in a *Salmonella* strain. The polysaccharide portion extends outside the cell, where it acts as a strain-specific antigen, and may serve to help attach the cell in its environment. Abe, abequose; Gal, galactose; Glc, glucose; GlcN, glucosamine; Hep, heptose; KDO, ketodeoxyoctanoate; Man, mannose; NAG, *N*-acetylglucosamine; Rha, rhamnose.

```
— NAG — TAL — NAG — TAL —
                |
               Gal
                |
               Ala — Gal
                |
        Glu — Lys
         |     |
        Lys   Glu
         |     |
        Ala   Ala
         |
        Glu
         |
    — TAL — NAG — TAL — NAG —
                        |
```

Figure 4.51 Pseudomurein
The pseudomurein of *Methanobacterium thermautotrophicum.*
The amino sugar backbone consists of alternating NAG (*N*-acetylglucosamine) and TAL (*N*-acetyltalosaminuronic acid).
The amino acids are alanine (Ala), glutamate (Glu), and
lysine (Lys), which is the cross-linking amino acid.

aminouronic acid replaces the *N*-acetylmuramic acid
found in bacteria. Furthermore, the two amino sugars
are linked in a 1,3 rather than a 1,4 configuration, and
the amino acids of the peptide cross-links are not D
stereoisomers.

The extreme halophilic and hyper-thermophilic
archaea and some of the methanogens have glycopep-
tide cell walls containing a variety of sugars linked covalently to protein (see Chapter 18 for more detail). The
differences in cell wall composition may explain how
these species are able to survive in their special environments. For example, halophilic *Archaea*, which live
in high osmotic concentrations, may not need the rigid
peptidoglycan cell walls typical of *Bacteria*.

Function of the Cell Wall Prokaryotes generally grow
in hypotonic aquatic environments, that is, the concentration of particles (ions and molecules) is greater inside
the cell than outside the cell. The cell membrane, which
is freely permeable to water, permits uptake of water by
osmosis so that the concentration of particles on both
sides of the cell membrane tends to equalize. If this
osmotic equalization occurs, the cell swells and the
membrane distends to accommodate the increase in
volume. This osmotic pressure within the cell, called **turgor pressure**, is the normal consequence of living in
hypotonic environments. The cell wall prevents turgor
pressure from stretching the membrane too far.
Excessive distension of the membrane would cause it to
break and result in lysis (rupture) of the cell.

Thus, one function of the cell wall is to protect the cell
from **turgor cell lysis**, or **plasmoptysis**. Gram-positive

bacteria such as those of the genus *Bacillus* readily illustrate this function. The cell wall layer can be removed
by treating the cell with lysozyme, an enzyme found in
tears and other human body excretions that breaks the
covalent bonds between the amino sugars (Figure 4.42)
of the cell wall peptidoglycan. If this occurs in a normal
hypotonic growth environment, the cells simply lyse.
However, if the environment is rendered isotonic (i.e.,
the osmotic pressure on both sides of the cell membrane
is equalized)—for example, by suspending the cells in a
2 M sucrose solution—then the wall-less cells do not
lyse. Instead, the rod-shaped cells round into spherical
units called **protoplasts**, which lack a cell wall (**Figure
4.52**). A protoplast can persist in an isotonic environ-

(A)

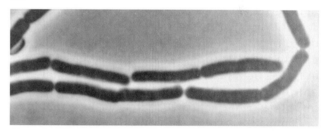

(B)

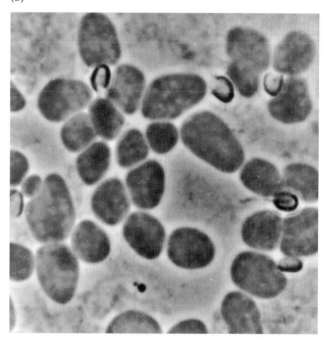

Figure 4.52 Protoplasts
(A) Normal cells of *Bacillus megaterium*, chains of rod-shaped
cells. (B) Protoplasts of the same *B. megaterium* preparation,
produced by lysozyme treatment and stabilized by sucrose.
Note the distension of protoplasts due to the low sucrose
concentration of the surrounding medium. Courtesy of Carl
Robinow.

ment because the turgor pressure of the cell is balanced by the osmotic pressure of (i.e., concentration of sucrose molecules in) the medium.

Protoplasts cannot be produced from gram-negative bacteria, because their cell walls cannot be completely removed in this way. Instead, removal of the peptidoglycan layer results in the formation of **spheroplasts**, which, although lacking in cell wall peptidoglycan, still possess the outer membrane.

Antibiotics and Cell Walls Some antibiotics exert their effect against bacteria by preventing the synthesis of normal cell walls. Antibiotics in the penicillin and cephalosporin groups (**Figure 4.53**), both of which are produced by certain fungi, act by inhibiting the enzymes responsible for biosynthesis of peptidoglycan (see Chapter 11). As a result, bacteria exposed to these antibiotics produce a defective cell wall, which results in the eventual lysis, and destruction, of the bacterial cell. It is noteworthy that penicillin causes lysis and death of bacteria only under conditions that permit growth and cell wall synthesis. Thus, when antibiotics are used therapeutically, bacterial cells that are not growing and dividing can survive and may begin growing following the antibiotic treatment. These cells must be eliminated by the host's own defense activities, such as by phagocytes.

The effect of antibiotics on the cell wall is different from that of lysozyme. Whereas lysozyme, as an enzyme, can continually degrade the peptidoglycan in cell walls, penicillin is active only on growing cells, because it affects only the *synthesis* of peptidoglycan. Nonetheless, treatment of growing bacteria with peni-

cillin in isotonic solution results in the formation of protoplasts and spheroplasts in much the same manner as lysozyme treatment.

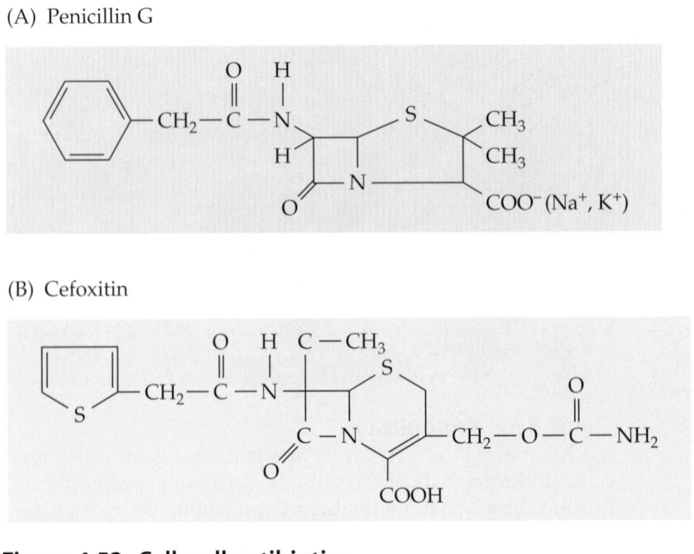

The antibiotics are selectively toxic to prokaryotic organisms because peptidoglycan is found only in *Bacteria*. Thus, humans treated with these antibiotics are not affected (unless they have an allergy to them), because human cells do not synthesize peptidoglycan. As one would predict, penicillin is generally much more effective against gram-positive bacteria than against gram-negative bacteria—first, because the major structural component of gram-positive cell walls is peptidoglycan, and second, because the outer membrane of gram-negative bacteria provides a permeability barrier to antibiotics such as penicillin G. For use against gram-negative bacteria, penicillin and other antibiotics can be modified to permit better solubility in the lipid outer membranes. Ampicillin, for example, is a penicillin with greater hydrophobicity that is used for gram-negative infections. Penicillin is effective in the treatment of certain gram-negative infections, including syphilis, caused by *Treponema pallidum*, and gonorrhea, caused by *Neisseria gonorrhoeae*, as well as in most infections by gram-positive *Bacteria*.

In summary, the cell walls of bacteria are rigid structures that not only give the cell its shape but also protect it from lysis. In addition, they serve a limited role in determining which molecules will pass through the cell membrane, especially for gram-negative bacteria.

Periplasm

The space between the cell wall and the cell membrane is called the **periplasm**. This area, though small (Figure 4.45), is important to the physiology of the cell. The periplasm is an area of considerable enzymatic activity. Several steps in the synthesis of the cell wall occur in this area. Chemoreceptors involved in the chemotactic response (discussed below) are also located here. Furthermore, the **binding proteins** produced by gram-negative bacteria are located in the periplasm. These binding proteins combine reversibly with substrate molecules, concentrate them, then release them to the membrane carrier proteins for transport into the cytoplasm (see Chapter 5).

Some controversy exists among microbiologists as to whether gram-positive bacteria have a periplasm. This is no doubt due to the extensive research that has focused on gram-negative bacteria, which have served as model organisms for the study of the periplasmic area. Furthermore, the periplasm of the two types of organisms undoubtedly differs, because gram-negative bacteria have the additional permeability layer, the outer membrane.

(A) Penicillin G

(B) Cefoxitin

Figure 4.53 Cell wall antibiotics
Chemical structures of (A) a penicillin and (B) a cephalosporin.

Cell Appendages

Many prokaryotes produce special appendages that extend from the surface of the cell. The two most common types, widely found among prokaryotes, are fimbriae and flagella.

Fimbriae, Pili, and Spinae Many prokaryotes have proteinaceous appendages, known as fimbriae, pili, or spinae, that are about 1 μm in length and range from about 3 to 5 nm in diameter in most bacteria (**Figure 4.54**) to about 100 to 200 nm in some aquatic varieties—sufficiently large to be seen with the light microscope. The three forms are structurally identical in that they all consist of repeating subunit protein molecules. The protein is helically wound, leaving a pore through the center. Some forms also have a glycoprotein tip.

The largest appendages of this type are **spinae** (Figure 4.54B and C). The spinae of aquatic bacteria serve to increase the cell's surface area and, hence, their drag, enabling the organisms to remain suspended for longer periods of time or to be carried farther by currents. Thus, the spinae assist the organism in maintaining its position in the plankton.

The terms "fimbriae" and "pili" are often used interchangeably, but **fimbriae** is usually considered a more general term for the thin fibrillar appendages so commonly found on prokaryotes. **Pili** is used to describe the fimbriae of many typical bacteria such as the enteric bacteria and various pathogenic species. A single bacterial cell may produce more than one type of fimbria or pilus. In fact, some bacterial viruses attach to one type of fimbria or pilus but not to the cell surface or to other types of fimbriae or pili produced by the same bacterial species.

Pili are known to be important for the attachment of some bacteria. For example, one type of pilus of *Escherichia coli* mediates the attachment of the two types of mating cells during sexual conjugation and is therefore specifically called the "sex pilus" (see Chapter 15). Likewise, pili produced by *Neisseria gonorrhoeae* are responsible for the attachment of pathogenic strains to endothelial tissue of the genito-urinary tract in humans. Only piliated strains cause gonorrhea.

In some *Pseudomonas* species, pili are involved in the extrusion of protein from the cell. Furthermore, a type of motility termed **twitching**, which occurs, for instance, in the gliding bacteria referred to as myxobacteria (slime bacteria), is also caused by fimbriae. Twitching is a poorly studied process whereby cells that seem to be sessile (immobile) suddenly jerk 1 μm or so in distance.

Flagella Prokaryotes that move by swimming produce **flagella**, long, flexible appendages resembling "tails." These proteinaceous appendages, generally about 10 to 20 nm in diameter and 5 to 20 μm in length, are found in some members of the gram-positive and gram-negative bacteria as well as in some archaea. They act like a propeller—the cell rotates the flagella and moves through the water. In motility, chemical energy generated from metabolism is converted into mechanical energy.

The number and location of flagella on the cell surface have been used as important features in the classification of

(A)

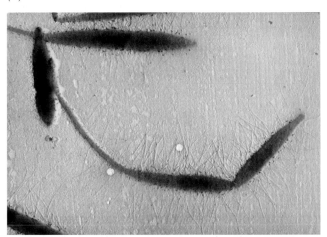

(B)

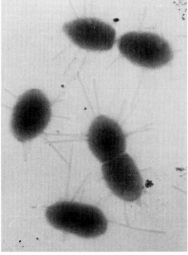

(C)　1.0 μm

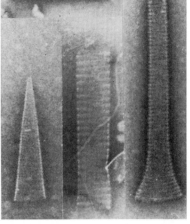

0.3 μm　　　0.1 μm

Figure 4.54　Fimbriae
(A) Electron micrograph of a cell of *Prosthecobacter fusiformis* showing many fimbriae. (B) A spinate bacterium with several large spinae. (C) Electron micrographs of isolated spinae. A, courtesy of J. T. Staley; B,C, courtesy of K. B. Easterbrook.

bacteria. Some bacteria have a single polar flagellum extending from one end of the cell and thus move by what is called **monotrichous polar flagellation**. These single flagellar filaments cannot be resolved by the light microscope but are readily discerned with the electron microscope (**Figure 4.55**). Other species produce polar tufts or bundles (**lophotrichous flagellation**; note that "flagellation" can refer both to the arrangement of flagella and the mode of locomotion), which can be seen with light microscopy (**Figure 4.56**). Still other prokaryotes produce flagella from all locations on the cell surface (**peritrichous flagellation; Figure 4.57**). Motile enteric bacteria, such as *Proteus vulgaris*, exhibit this latter type of flagellation. Finally, a few bacteria are known to move by **mixed flagellation**. Some *Vibrio* species, for example produce polar flagella for movement in water but produce lateral flagella when grown on an agar surface.

A flagellum consists of three major sections:

- A long **filament**, which extends into the surrounding environment
- A **hook**, a curved section connecting the filament to the cell surface
- A **basal structure**, which anchors the flagellum into the cell wall and membrane by special disk-shaped structures called **plates** or **rings**

The anchoring for a typical gram-negative bacterium involves L, P, S, and M rings associated with cell envelope components (**Figure 4.58**).

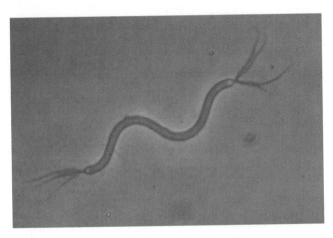

Figure 4.56 Polar flagellar tufts (lophotrichous flagellation)
Phase microscope image of a spirillum showing the polar tufts of flagella. ©Eric Grave/Science Source/Photo Researchers Inc.

The flagellar filament is composed of a protein called **flagellin**. With few exceptions, the bacteria studied thus far have flagellin with a single type of protein subunit (*Caulobacter* and some *Vibrio* spp. have two types). Repeating subunits of flagellin are joined together to form the filament, which is helically wound and has a hollow

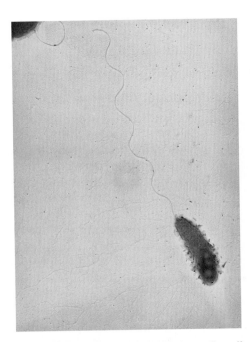

Figure 4.55 Polar flagellum (monotrichous flagellation)
Electron micrograph of a whole gram-negative bacterium showing its single polar flagellum. Courtesy of J. T. Staley.

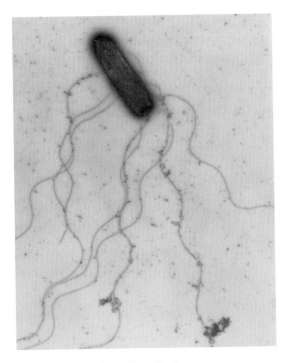

Figure 4.57 Peritrichous flagellation
Electron micrograph of a peritrichously flagellated bacterium showing many flagella about the cell surface. ©W. L. Dentler/Biological Photo Service.

(A)

20 nm

(B)

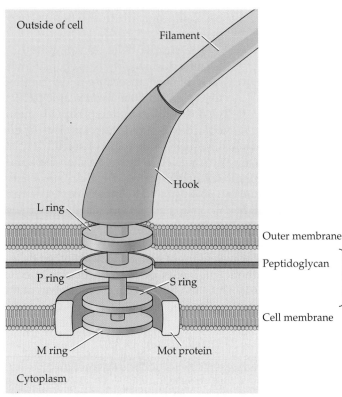

Outside of cell

Filament

Hook

L ring

Outer membrane

Peptidoglycan

P ring

S ring

Periplasm

Cell membrane

M ring

Mot protein

Cytoplasm

Figure 4.58 Flagellar structure
(A) Electron micrograph showing the basal ring structure of a gram-negative flagellum. (B) Model of a flagellum cross section showing the location of the rings and the Mot proteins in the cell envelope of a gram-negative bacterium. A, ©Jo Adler/Visuals Unlimited.

core. The filament (and thus the flagellum) is about 20 nm in diameter. The filaments are not straight. Each has a characteristic sinusoidal curvature that varies with each species.

The flagella of enteric bacteria have been studied most intensively. More than 40 genes have been identi-

fied in *Salmonella typhimurium* and *Escherichia coli* that are involved in motility. These include the genes responsible for producing flagellin subunits, flagellar motor proteins, and basal body structures, as well as those that regulate the synthesis of the flagellum.

Flagellar rotation, and hence cell movement, requires energy, which is generated by the proton motive force (see Chapter 8). This energy is imparted to the **Mot proteins**, which are embedded in the cell membrane and act as a motor that drives flagellar rotation. The rate of rotation varies among species but is in the range of 200 to more than 1,000 rpm. Motility requires a major expenditure of energy, but the result is immensely useful because it allows the cell to translocate to a place where conditions for growth are more favorable.

Tactic Responses: Chemotaxis and Phototaxis Heterotrophic bacteria use their flagella to propel themselves into areas where organic nutrients are concentrated. This phenomenon, termed **chemotaxis**, can be illustrated with a capillary technique. If a capillary containing an organic energy source such as glucose is placed in a suspension of heterotrophic bacteria, the bacteria will accumulate at the tip of the capillary from which the nutrient is diffusing (**Figure 4.59**).

Chemotaxis has been studied most thoroughly in enteric bacteria such as *E. coli*. These peritrichously flagellated bacteria orient their flagella in a polar bundle during motility. The bundle is rotated counterclockwise during forward movement called **running**. Running is disrupted when the flagella reverse to the clockwise direction: the cell stops, then somersaults or tumbles, a movement referred to as **twiddling**. After a twiddle, the flagella again rotate counterclockwise and the organism begins another run. However, the direction of cell movement is completely random. When the organism is in the gradient of a utilizable nutrient, less twiddling occurs, allowing the organism to spend more time in the favorable area (**Figure 4.60**). This phenomenon, in which an organism effectively becomes directed toward a utilizable nutrient, an attractant, is **positive chemotaxis**. **Negative chemotaxis,** movement away from toxic materials, called repellants, is also commonly exhibited by bacteria.

Scientists are just beginning to unravel the biochemical mechanism of chemotaxis. Many utilizable organic nutrients serve as attractant molecules in chemotaxis, causing positive chemotaxis. These molecules are bound to specific receptor proteins located either in the periplasmic space or in the cell membrane of gram-negative bacteria. For some attractants, such as the amino acid serine, the binding protein is a cell membrane protein that can be methylated in the presence of a methyltransferase, an enzyme that uses *S*-adenosylmethionine

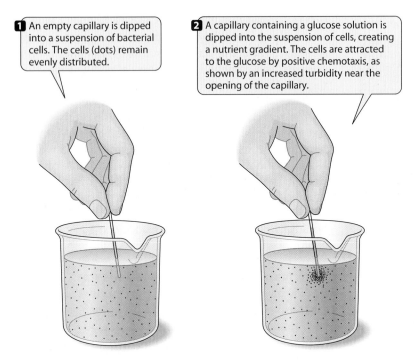

1 An empty capillary is dipped into a suspension of bacterial cells. The cells (dots) remain evenly distributed.

2 A capillary containing a glucose solution is dipped into the suspension of cells, creating a nutrient gradient. The cells are attracted to the glucose by positive chemotaxis, as shown by an increased turbidity near the opening of the capillary.

Figure 4.59 Chemotaxis
An illustration of the capillary chemotaxis effect.

as a substrate. The process of binding and subsequent methylation of the **methyl-accepting chemotaxis protein (MCP)** results in a signal that causes the flagellin subunits to form a bundle as the flagellum rotates in a

counterclockwise direction, thus resulting in a run. For some sugars that serve as attractant molecules, positive chemotaxis can occur at concentrations as low as 10^{-8} M.

It is noteworthy that in chemotaxis, both spatial gradients and temporal changes in the chemical environment have an effect on cell movement. Adaptation of the organism to its environment also occurs in chemotaxis. In the prolonged presence of the attractant substrate, the rate of twiddling is less than it would be in the absence of the attractant. This may be a mechanism whereby the organism can conserve energy that would be needlessly spent on changing direction when it is not needed.

The simple behavior exhibited by bacteria during motility is of great interest to biologists, because understanding this behavior at the molecular level may lead to a fuller understanding of more complex behavioral responses of eukaryotic organisms, such as the sense of smell.

Flagella are also involved in the phenomenon of **phototaxis** in photosynthetic bacteria. This light-associated phenomenon can be illustrated by projecting a spectrum of visible light onto a glass microscope slide. When a suspension of flagellated photosynthetic bacteria is placed on the illuminated slide, the organisms quickly concentrate at the wavelengths of light that are absorbed by the photosynthetic pigments of that particular species (**Figure 4.61**). Thus, the cells position themselves at the specific wavelengths used for photosynthesis.

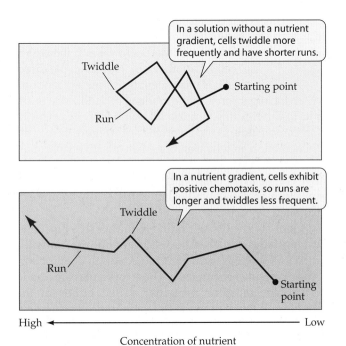

In a solution without a nutrient gradient, cells twiddle more frequently and have shorter runs.

Twiddle

Starting point

Run

In a nutrient gradient, cells exhibit positive chemotaxis, so runs are longer and twiddles less frequent.

Twiddle

Run

Starting point

High ◄─────────────────────► Low

Concentration of nutrient

Figure 4.60 Runs and twiddles in chemotaxis
The type of cell movement depends on the presence or absence of a nutrient gradient.

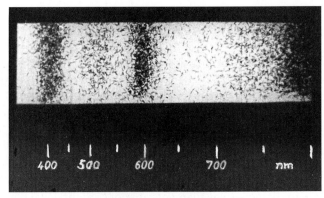

Figure 4.61 Phototaxis
A spectrum of visible light has been projected on a slide of *Thiospirillum jenense*, a purple sulfur photosynthetic bacterium. The bacteria have accumulated at those wavelengths at which their photosynthetic pigments best absorb light. Courtesy of Norbert Pfennig.

SUMMARY

▶ A **simple microscope** is analogous to a magnifying glass with a single lens; **compound microscopes** have two lens systems, including the eyepiece (or ocular) lens and the **objective lens**. In addition, a **condenser lens** system is used to concentrate light on the specimen.

▶ The **resolution** of the light microscope is determined by the **wavelength** of light, the **refractive index** of the suspending medium, and the closeness of the objective to the specimen, which determines the **angular aperture** of the lens.

▶ Under most conditions of growth, bacteria have a negative surface charge and are therefore stained with **basic dyes** (positively charged). In contrast, **negative stains**, such as India ink, do not stain cell material but stain the aqueous background.

▶ **Phase contrast microscopy** increases the contrast of bacteria by amplifying the difference in refractive index between the cells and the aqueous environment.

▶ **Transmission electron microscopes** (**TEMs**) have high resolution because they use electrons as their source of illumination. **Thin sections** to be studied for intracellular details are prepared by dehydrating and embedding cells in plastic resins, which are then sliced with ultramicrotomes.

▶ **Scanning electron microscopes** (**SEMs**) produce images of whole bacterial cells or colonies.

▶ Many prokaryotes are unicellular **rods**, **cocci**, **vibrios**, or **spirilla**. Some species form chains, filaments, or packets of cells. The **actinobacteria** are gram-positive bacteria that produce branching filamentous structures called **mycelia**.

▶ Most prokaryotes divide by **binary transverse fission**, in which a mother cell divides to produce two mirror-image daughter cells; a few prokaryotes divide by **budding** or **fragmentation**.

▶ The prokaryotic chromosome is a double-stranded circular molecule of DNA. Many prokaryotes harbor **plasmids**, double-stranded DNA molecules containing genes that are not essential for the livelihood of the host cell but may carry important genes for determining its physiological capabilities.

▶ **Gas vesicles** are protein-membrane structures that exclude water and therefore accumulate gases and reduce cell density, enabling cells to become buoyant in aquatic habitats.

▶ Organic storage compounds of prokaryotes include glycogen, starch, **poly-β-hydroxybutyrate (PHB)**, and **cyanophycin**. Inorganic storage compounds include **volutin** (also called polyphosphate) and elemental sulfur.

▶ The cell membranes of bacteria contain proteins embedded in lipid bilayer membranes, which contain fatty acids linked to glycerol through **ester linkages**. The cell membranes of archaea contain proteins and lipids with isoprenoid groups that are linked to glycerol by **ether linkages**.

▶ **Extracellular layers**, found in some prokaryotes are **capsules**, **slime layers**, **protein jackets**, or **sheaths**, depending on the species.

▶ Almost all members of the domain *Bacteria* produce **peptidoglycan** or **murein**, a unique cell wall structure that contains a backbone of N-acetylglucosamine and N-acetylmuramic acid and peptide bridges cross-linking the amino sugar backbones. Peptidoglycan contains some unique D-amino acids, such as D-alanine and D-glutamic acid, not found in protein. Peptidoglycan contains a diamino acid, such as **diaminopimelic acid** or **lysine**, that cross-links the peptide bridge. Peptidoglycan can be hydrolyzed by lysozyme. Normal peptidoglycan synthesis is inhibited by penicillin and certain other antibiotics.

▶ Gram-positive cell walls may also contain **teichoic acids** and **teichuronic acids**, which are not found in gram-negative bacteria. In contrast, gram-negative bacteria contain, in addition to peptidoglycan, an **outer membrane** composed of lipid, protein, and polysaccharides. The polysaccharides of the outer membranes are linked to **lipid A** to form a **lipopolysaccharide** (or **LPS**), called **endotoxin**, which is toxic to vertebrates.

▶ Archaeal cell walls lack peptidoglycan; most have polysaccharide or protein cell walls.

▶ The **periplasm** is an area between the cell membrane and cell wall; it is the site of a variety of enzymatic activities, including cell wall synthesis, chemoreception, and transport.

▶ **Fimbriae** and **pili** are proteinaceous extracellular fibrillar appendages of some bacteria used primarily in attachment.

▶ **Flagella** are protein fibrils that prokaryotes use for motility.

▶ **Positive chemotaxis** is movement of a cell toward a nutrient. **Negative chemotaxis** is movement away from a toxic substance. Photosynthetic bacteria can respond to light by either positive or negative **phototaxis**.

REVIEW QUESTIONS

1. Why does the TEM have better resolution than the light microscope?

2. What types of microscopy are best suited for observing motility of microorganisms?

3. What type of microscopy would you use to view (a) ribosomes, (b) gas vesicles, (c) gas vacuoles, (d) colonies of bacteria, (e) periplasm?

4. How does fluorescence microscopy work?

5. Which prokaryotes are not sensitive to penicillin? Why?

6. In what ways do the *Archaea* differ from the *Bacteria*?

7. Why is a diamino acid important in peptidoglycan?

8. What are the different types of extracellular structures of bacteria? What uses do they serve for the cell?

9. What are the various types of storage materials found in prokaryotes? Would you categorize them as energy sources or nutrient sources?

10. It has been hypothesized that the gas vacuole is a primitive organelle of prokaryotic motility. What evidence is consistent with this?

11. Distinguish between positive and negative chemotaxis. What is known about the molecular basis for chemotaxis?

12. Is the response of photosynthetic bacteria to light considered phototaxis? Explain your answer.

SUGGESTED READING

Beveridge, R. L. and L. L. Graham. "Surface Layers of Bacteria." *Microbiological Reviews* 55 (1991): 684–705.

Gerhardt, P., R. G. E. Murray, W. A. Wood and N. R. Krieg. 1994. *Methods for General and Molecular Bacteriology*. 4. Washington, DC: American Society for Microbiology.

Neidhardt, F. C., J. L. Ingraham and M. Schaechter. 1990. *Physiology of the Bacterial Cell: A Molecular Approach*. Sunderland, MA: Sinauer Associates, Inc.

Schatten, G. and J. B. Pawley. "Advances in Optical, Confocal, and Electron Microscopic Imaging for Biomedical Researchers." *Science* 239 (1988): 164.

Microbial Physiology: Nutrition and Growth

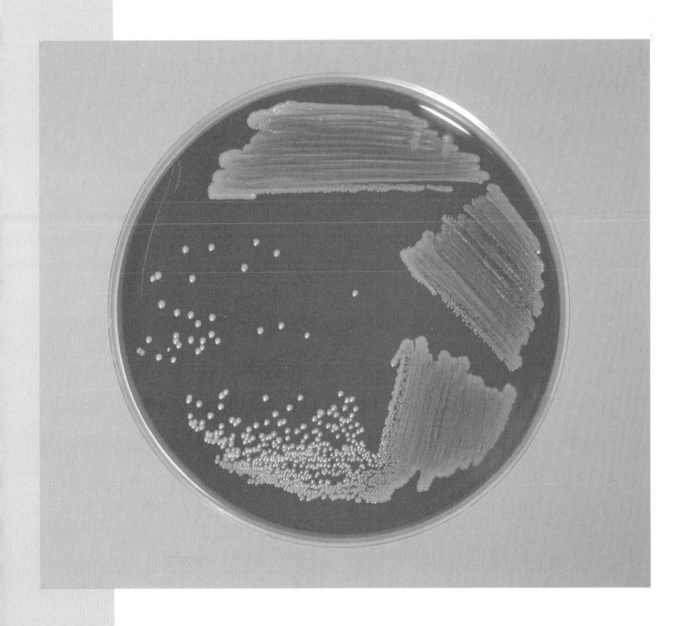

PART II Microbial Physiology: Nutrition and Growth

Previous page
This culture of *Corynebacterium diphtheriae* is growing on blood agar.
Courtesy of W. A. Clark/Centers for Disease Control.

Chapter 5

Isolation, Nutrition, and Cultivation of Microorganisms

It appears incongruous to delineate metabolism of bacteria from that of fungi, since the two overlap in so many respects and since each group displays, more than any other groups of microorganisms, a kaleidoscopic array of metabolic diversity.

— JACKSON W. FOSTER

Green plants are photosynthetic, gaining their energy from sunlight and using CO_2 as their carbon source. Animals, on the other hand, require preformed organic molecules for both energy and building material. These preformed molecules may be microbial, plant or animal in origin. A plant or an animal species is generally differentiated from any other plant or animal species by observable differences in size, shape, and component structures. Mites to elephants or moss to sequoias are examples of the extremes in size and structure in the animal and plant world.

In contrast to plants and animals, the various prokaryotic microorganisms are not markedly different from one another either in size or structure. Microorganisms are generally differentiated from one another by their nutrition, which is the sum of the processes whereby the organism takes in and utilizes food for energy and building material (carbon). Among *Bacteria* and *Archaea*, nutritional needs vary from species to species. It is axiomatic that a microorganism in nature can utilize, as nutrient source, any carbon-containing constituent that is or has been a component part of living cells. Many other compounds unrelated to living cells can also be utilized as substrate (source of food) for growth. Among these compounds that microorganisms can actually utilize as carbon/energy sources are carbon monoxide, methane, cyanide, acetic acid, or glucose. Others readily utilize aromatic compounds, such as benzene, that are toxic to most eukaryotes. These marked nutritional differences are referred to collectively as representing "metabolic diversity in the microbial world." This chapter examines the major nutritional differences that distinguish microorganisms, how different nutritional types can be isolated, and how they may be cultivated and preserved.

FUNDAMENTAL ASPECTS OF DIVERSITY

Microorganisms live in virtually every environmental niche on Earth. These microbial populations can utilize the naturally occurring organic compounds that fall into these niches as sources of carbon and/or energy. Many industrial chemical synthesis products, such as pesticides, are also metabolized by natural microbial populations. However, it should be noted that some chemically derived compounds are resistant to microbial degradation, and these tend to accumulate in the environment.

Evolution has resulted in a seemingly endless array of microbial types with broad metabolic capacities. These organisms differ from one another in their nutritional capabilities and their adaptation to the physical conditions in which they live. Selected *Bacteria* and *Archaea* are adapted to growth under wide-ranging physical conditions, including pH (<1 to >11), temperature (<0°C to >100°C), salinity (freshwater to saturated saltwater), or availability of O_2 (aerobic to anaerobic). These physical conditions are further imposed on the nature and availability of substrates for growth of a microorganism (form of carbon, nitrogen, or phosphorous).

Most bacterial species utilize simple monomeric substrates such as sugars or amino acids as growth substrate, whereas others digest complex proteins and polysaccharides. Although some bacteria, such as *Pseudomonas cepacia*, can utilize any one of over 100 different carbon-containing compounds for growth, most organisms can utilize only a limited number. Other microbes are restricted to growth on only one or a few chemically related substrates. For example, *Methylomonas methanooxidans* will grow only when methane or methanol is provided as growth substrate. Collectively, however, microbial populations are capable of growth on the unending variety of substrates encountered in nature. This broad metabolic diversity is essential in maintaining the level of CO_2 in the atmosphere. This CO_2 is the carbon source for the photosynthetic populations that are the basis for all viable systems on this planet.

The Evolution of Metabolic Diversity

The presence of diverse nutritional types, as the quote at the beginning of the chapter affirms, is a consequence of the evolutionary role that microbes have assumed. Much of the microbial community has evolved with the ability to mineralize constituents or end products of all biota. The word *mineralize* is used in this text to emphasize that microorganisms can break down organic molecules to basic constituents (NH_4, CO_2, SO_4^-, PO_4^-, etc.). In the course of microbial evolution, processes of mutation and selection gave rise to populations that are able to utilize all available carbon compounds as their source of carbon and/or energy source. Populations of microbes have also evolved that can obtain energy from light and from inorganic substrates: hydrogen (H_2), sulfides, reduced iron, and ammonia. Other organisms evolved that could utilize gaseous carbon sources, such as CO,

CO_2, methane, ethane, or propane. The ability to "fix" atmospheric nitrogen (N_2) lent a distinct advantage to selected *Bacteria* and *Archaea* occupying environments where little combined nitrogen (NH_4^+, NO_3^-, or organic) would be available. The ability to utilize a substrate better than others, or a substrate under physical conditions that others could not, gave a survival advantage. It also gave rise to a broad array of microbial species with small but significant differences in their nutrient requirements and tolerance to physical conditions. Basically, survival advantage has been the key during the four billion years of evolution and a major factor in generating microbial diversity (**Box 5.1**).

Nutritional Types of *Bacteria*

Bacteria can be divided into four major groups based on their source of carbon and their source of energy (the nutrition and characteristics of the *Archaea* is discussed in Chapter 18). All microorganisms must have a source of carbon to provide building blocks, and they require a source of energy to drive the reactions involved in the synthesis of cell components. The group distinctions are based on the source of carbon and the source of energy, that is, the use of CO_2 versus organic compounds as carbon source and light or chemicals as energy source.

Before discussing the characteristics of the individual groups, some features of bacterial metabolism should be considered.

The ratio of carbon/hydrogen/oxygen in a cell is 1-2-1, respectively, and an organism growing with CO_2 as carbon source would need a source of H^+ to *reduce* CO_2 to CH_2O. This source of H^+ is called the electron donor; for example, H_2S, NH_4^+, H_2O can all serve as a source of hydrogen (electron donor), as the electron would carry along the proton. An organism that utilizes reduced substrates such as methane (CH_4) as carbon source must contain an electron acceptor (O_2) to *oxidize* the substrate to the CH_2O level. Most microorganisms (and animals) utilize substrates such as glucose ($C_6H_{12}O_6$) that have a C/H/O ratio equivalent to that in a cell.

Although the categories presented here are not all inclusive, they do provide a framework for separating

---

Bacterial and archaeal species possess a significant number of unique capabilities not found anywhere among the eukaryotes. They:

- Fix atmospheric nitrogen (N_2)
- Synthesize vitamin B_{12}

- Use inorganic energy sources NH_4, H_2S, H_2, Fe^{2+}
- Photosynthesize without chlorophyll
- Utilize inorganic (CO_2, NO_3^-, SO_4^{2-}, $S°$, Fe^{3+}) terminal electron acceptor as an alternative to O_2

- Have extensive capacity for anaerobic growth
- Use H_2S, H_2, or organics as electron donor in photosynthesis
- Grow at temperatures in excess of 80°C

microbes into reasonable and concise nutritional units. These groups are:

- **Photoautotroph** (from *photo* meaning "light," *auto* meaning "self," and *troph* meaning "feeding"): Photoautotrophs use light as their source of energy and CO_2 as their source of carbon. An obligate photoautotroph is an organism that will grow only in the presence of both light and CO_2. Obligate photoautotrophs use inorganics such as H_2O, H_2, or H_2S as electron donor to reduce CO_2 to cellular carbon (CH_2O).

- **Photoheterotroph** (from *hetero* meaning "different"): Photoheterotrophs grow by photosynthesis if provided with an electron donor (H_2 or organic) for reductive assimilation of CO_2. Commonly, organisms in this group require selected growth factors, such as B-vitamins, and several species will grow on organic substrates if oxygen is available. They may also utilize light as an energy source while assimilating organic compounds from the environment as growth substrate.

- **Chemoautotroph** (from *chemo* meaning "chemical"): Chemoautotrophs utilize reduced inorganic substrates for both the reductive assimilation of CO_2 and as a source of energy. The oxidation of H_2, NH_3, NO_2^-, H_2S, or Fe^{2+} supplies energy for these microorganisms. Aerobic chemoautotrophs utilize O_2 as terminal electron acceptor, and some of the anaerobic *Archaea* can utilize inorganic sulfur as terminal electron acceptor (see Chapter 18).

- **Chemoheterotroph**: Chemoheterotrophs are microorganisms that assimilate organic substrates as source of both carbon and energy. This may be a single substrate such as glucose or succinate, or the sources of carbon and energy may be different. For example, the sulfate reducers use H_2 for energy but require an organic substrate for cellular biosynthesis. The vast majority of the microbes that have been studied are chemoheterotrophs.

This brief outline affirms that the various bacterial and archaeal species may have significant differences in their nutritional requirements. Bacteria representing each of these categories are presented in **Figure 5.1**. The ability of some organisms to fit into more than one of the preceding groupings is discussed in **Box 5.2**. Under laboratory conditions, microorganisms grow in an artificial environment generally not equivalent to that encountered in nature. A microorganism living in soil, water, or the intestinal tract of a warm-blooded animal survives there because its nutritional requirements are met. These requirements may be simple or they may be complex. If we wish to take a microorganism from an environmental niche and grow it in the laboratory, we must somehow mimic the conditions that permitted it to survive in the environment. These would include both physical conditions (pH, temperature, osmotic pressure, light) and substrate availability. A laboratory medium must provide nutrients that are essential for growth. The following section examines in some detail how to prepare a growth medium.

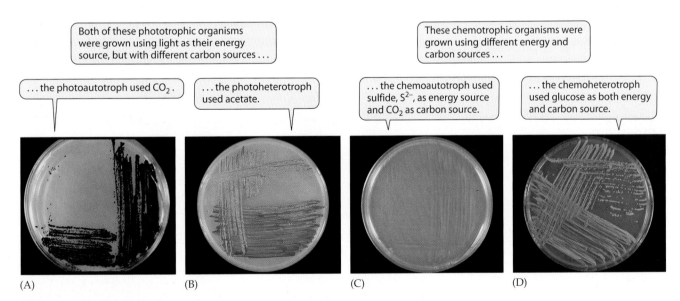

Both of these phototrophic organisms were grown using light as their energy source, but with different carbon sources …

… the photoautotroph used CO_2.

… the photoheterotroph used acetate.

These chemotrophic organisms were grown using different energy and carbon sources …

… the chemoautotroph used sulfide, S^{2-}, as energy source and CO_2 as carbon source.

… the chemoheterotroph used glucose as both energy and carbon source.

(A)   (B)   (C)   (D)

Figure 5.1 Nutritional types
Colonies of microorganisms representing the major nutritional groups: (A) *Anabaena viriabilis*, a photoautotroph; (B) *Rhodomicrobium* sp., a photoheterotroph; (C) *Thiobacillus perometabolis*, a chemoautotroph; and (D) *Micrococcus luteus*, a chemoheterotroph. A, courtesy of Robert Tabita; and B–D, courtesy of Jerome Perry.

Milestones Box 5.2   **Microbial Versatility: Obligate versus Facultative**

A microorganism that will grow only if provided with CO_2 for carbon and light as energy source is an **obligate photoautotroph.** One that grows only with CO_2 and NH_4^+ or S^{2-} as energy source is an **obligate chemoautotroph.** Restriction to an obligate lifestyle can be found in the microbial world, but versatility is the more common attribute. An obligate autotroph can actually assimilate a very limited amount of selected organic substrates while growing with CO_2 as bulk carbon source and light (for photo-) or inorganics (for chemo-) as energy source. However, many of the organisms that will grow as photo- or chemo- autotrophs can grow as chemoheterotrophs. Such organisms are termed **facultative.** The term *facultative* is also used to describe an organism that can grow aerobically if O_2 is available or anaerobically in the absence of O_2.

SUBSTRATE CONSTITUENTS NEEDED FOR GROWTH

Biochemical analysis of bacterial or archaeal cell mass affirms that their cellular composition is mostly equivalent regardless of the species or the composition of the medium on which they grew. They are composed of the same major components including the monomers that make up the major macromolecules (proteins, nucleic acids, carbohydrates, and lipids). A bacterium is 80% to 90% water. This elemental composition is likewise very similar, with little variance from the percentages of elements shown in **Table 5.1.** Consequently, a growth medium must provide a microorganism with these elements in one form or another. Requirement for the iron and the trace metals may vary. The macromolecules that are present in all microorganisms are composed of the first six elements listed in Table 5.1. These are discussed here as components of the growth medium.

Carbon

The most abundant element in any living cell is carbon. Carbon is the backbone of functional biological molecules. Microbial diversity reflects the ability of organisms to synthesize all of their component molecules from the carbon source that is available. Marked differences exist among microbes in this capacity. For example, a cyanobacterium can synthesize *de novo* all of its cellular components (protein, nucleic acids) from CO_2 if mineral salts are available (**Table 5.2**). Other organisms

Table 5.1	**A typical analysis of the elements that may be present in a bacterial or archaeal cell**

Element	Percentage of Dry Weight
Carbon	50
Oxygen	20
Nitrogen	14
Hydrogen	8
Phosphorus	3
Sulfur	1
Sodium	1
Potassium	1
Calcium	0.5
Magnesium	0.5
Iron	0.2
Cu, Zn, Mo, Bo, Se, Cl Ni, Cr, Co, and Wo	0.2
	100%

Table 5.2	**A typical mineral salts medium for the isolation and growth of free-living bacteria**

Constituents	Amount[a] (mg/l H_2O)
NH_4Cl	500
$NaNO_3$	500
Na_2HPO_4	210
NaH_2PO_4	90
$MgSO_4 \cdot 7\,H_2O$	200
KCl	40
$CaCl_2$	15
$FeSO_4$	1
Trace Elements	**µg/l H_2O**
$ZnSO_4 \cdot 7\,H_2O$	70
H_3BO_3	10
$MnSO_4 \cdot 5\,H_2O$	10
MoO_3	10
$CoSO_4$	10
$CuSO_4 \cdot 5\,H_2O$	5

[a]It would be necessary to add a carbon source such as glucose at 0.5 to 2.0%

may have a very limited synthetic capacity and require a complex growth medium, such as that shown in **Table 5.3**. A growth medium with all of the compounds listed in the table would be necessary for the growth of the

Table 5.3	A defined medium for the growth of the lactic acid bacterium *Streptococcus agalactiae*

Compound	Final Concentration (μg/mL)
L-alanine	250
L-arginine	320
L-aspartic acid	500
DL-asparagine	100
L-cystine	600
L-cysteine-HCl	600
L-glutamic acid	400
L-glycine	200
L-glutamine	100
L-histidine	320
L-leucine	200
L-lysine	320
L-isoleucine	200
DL-methionine	200
L-phenylalanine	200
L-proline	200
DL-serine	400
L-tryptophan	200
L-tyrosine	200
L-valine	200
Nicotinic acid	10
Ca pantothenate	10
Pyridoxal HCl	10
Thiamine HCl	0.6
Riboflavin	1
Biotin	0.2
Folic acid	0.02
K_2HPO_4	1,000
KH_2PO_4	1,000
$NaHCO_3$	500
$FeSO_4 \cdot 7 H_2O$	10
$MnCl_2$	25
NaCl	10
$ZnSO_4 \cdot 7 H_2O$	10
$MgSO_4 \cdot 7 H_2O$	80
Glucose	10,000
Adenine	10
Guanine	10
Xanthine	10
Uracil	10

From N. P. Willett and G. E. Morse. 1966. Long-chain fatty acid inhibition of the growth of *Streptococcus agalactiae* in a chemically defined medium. *J. Bact.* 91:2245.

fastidious lactic acid-producing bacterium *Streptococcus agalactiae*.

The carbon source for microbial growth can range from CO or CH_4 to any naturally occurring complex organic compound present in the biosphere (such as carbohydrates, peptides, and organic acids). A variety of microbes also exist that can, under appropriate conditions, grow with various synthetic organic compounds as substrate. This is the basis for "bioremediation," the removal of pollutant chemicals from an environment by use of selected microbes that can utilize these pollutants as a carbon/energy source.

Hydrogen

Hydrogen plays a number of roles in the life of *Bacteria* and *Archaea*—it is a structural atom in organic molecules and is a participant in the complex process of energy generation. A source of hydrogen is essential in autotrophic microorganisms to reduce CO_2 to cell material (CH_2O). Protons (H^+) are involved in the production of ATP via the ATP synthase system in the cell's membrane of most microorganisms (see Chapter 8). Microorganisms that respire anaerobically (CH_4 producers, denitrifiers, and sulfate reducers) gain their energy by transfer of electrons from a substrate to a selected acceptor (**Table 5.4**). Aerobic respiration involves the transfer of electrons to O_2, a factor in establishing a proton gradient for ATP synthesis (Chapter 8).

Nitrogen

Nitrogen is an integral constituent of amino acids (proteins), nucleotides (nucleic acids), phospholipids, and constituents of the cell wall. A unique property of some *Bacteria* and *Archaea* is their ability to obtain cellular nitrogen by fixation of N_2 from the atmosphere. These nitrogen fixers reduce N_2 to the level of NH_4^+, which is incorporated into carbon compounds by the synthetic machinery of the cell. The fixation of N_2 is not limited to a few species, as was long believed, but occurs in an array of bacterial and archaeal types, as outlined in **Table 5.5**. Most free-living microorganisms assimilate ammonia from their environment, or they can reduce nitrate. One or both of these are commonly included in a growth medium.

A requirement for an organic nitrogen source such as an amino acid is generally confined to microorganisms that evolved in richer environments where various amino acids, nucleic acids, and B-vitamins were readily available. For example, the medium presented in Table 5.3 is a growth medium for a microorganism that has a limited ability to synthesize nitrogen-containing intermediates. The lactic acid bacteria evolved in or on animals or plants where organic nitrogen compounds were available.

The growth of most heterotrophic bacteria is stimulated by adding rich nitrogenous material, such as yeast

Table 5.4	Types of respiration that occur in *Bacteria* and *Archaea*. Examples of organisms that perform this respiration are given

Aerobic respiration

Oxygen

$O_2 \rightarrow H_2O$ .... *Pseudomonas fluorescens*

Anaerobic respiration

Iron

$Fe^{3+} \rightarrow Fe^{2+}$ .... *Shewanella putrefaciens*

Nitrate

$NO_3^- \rightarrow NO_2^-, N_2O, N_2$ .... *Thiobacillus denitrificans*

Fumarate

Fumarate $\rightarrow$ Succinate .... *Proteus rettgeri*

Sulfate

$SO_4^{2-} \rightarrow HS^-$ .... *Desulfovibrio desulfuricans*

Sulfur

$S^o \rightarrow HS^-$ .... *Desulfurococcus mucosus*

Carbonate

$CO_2 \rightarrow CH_4$ .... *Methanosarcina barkeri*

$CO_2 \rightarrow CH_3COO^-$ .... *Acetobacter woodii*

extract (soluble portion of digested yeast cells) at 0.05%, to a basic mineral salts medium. Growth is stimulated because energy need not be expended to synthesize B-vitamins, amino acids, purines, pyrimidines, and other nitrogen-containing compounds.

Sulfur

Sulfur is a constituent part of two of the amino acids that make up proteins—cysteine and methionine. It is also present in certain B-vitamins (biotin and thiamine) and some other essential constituents of a cell. Sulfur is usually added to a growth medium as a sulfate salt. $MgSO_4$ added to a medium would serve as a source of both sulfur and magnesium. The capacity to reduce sulfate to the sulfide level (SH), the form present in cellular constituents, is a common attribute of bacteria. If an organism cannot reduce sulfate, the addition of the amino acid cysteine will permit growth. Addition of yeast extract or peptone (enzyme digest of protein) in a culture medium meets the need for reduced sulfur compounds in most microorganisms. Reduced inorganic sulfur compounds such as H_2S or pyrite (iron sulfide) can serve as the energy source for a group of bacteria called the *thiobacilli*. Oxidation of sulfides generates sulfate. Sulfur S (inorganic) can serve as a terminal electron acceptor in some bacteria, and many of the *Archaea* and sulfate serves this purpose in sulfate-reducing *Bacteria* and *Archaea*.

Phosphorus

This element has played a major role in the evolution of life systems on Earth. Phosphorus is a constituent of high-energy compounds, the phospholipids in cell membranes, and nucleic acids. Adenosine triphosphate (ATP) is the principle medium of energy exchange in cellular metabolism and is an indispensable contributor to biosynthetic reactions involved in reproduction and growth. For this reason, phosphates are an integral constituent of culture media. Inorganic phosphates are also an effective buffer at near neutral pHs and at concentrations that are not usually inhibitory to bacterial growth. Phosphate salts are commonly added to culture media to satisfy the phos-

Table 5.5	Some genera of *Bacteria* and *Archaea* that have nitrogen fixation ability

Bacteria

    Heterotrophs

        Aerobes

            Azotobacter

            Klebsiella

            Beijerinckia

        Anaerobes

            Clostridium

            Bacillus (facultative)

    Photosynthetics

        Cyanobacteria

            Anabaena

            Oscillatoria

            Gloeocapsa

        Purple and green bacteria

            Chromatium

            Chlorobium

            Rhodospirillum

    Symbiotic

        Legumes

            Clover + *Rhizobium*

            Soybeans + *Rhizobium* or *Bradyrhizobium*

            Bluebonnets + *Rhizobium*

        Nonlegumes

            Bayberry + *Actinomycete*

            Alder + *Frankia*

Archaea

    Methanogens

        Anaerobes

            Methanococcus

            Methanosarcina

            Methanobacterium

            Methanothermus

phate requirement and to provide a buffer to prevent significant changes in pH during growth.

Oxygen

The total number of oxygen atoms present as a cellular component is equivalent in aerobes and anaerobes. However, molecular oxygen (O_2) is toxic to most strictly anaerobic *Bacteria* and *Archaea*, so they obtain this element in a combined form from the substrate. As a general rule, anaerobes utilize growth substrates that are in an oxidation-reduction state equal to or more oxidized than cellular material (CH_2O). For example, anaerobes can utilize sugars, amino acids, or CO_2 as carbon source. Aerobic bacteria can grow on reduced substrates (such as methane and propane), when molecular oxygen is available. Aerobic microorganisms use oxygen as a terminal electron acceptor through the electron transport system. A number of aerobes use oxygen also as a reactant to incorporate it into their substrate.

Cations

Other elements are present in *Bacteria* and *Archaea* and are generally involved with enzymatic activity or in cell stability. These are mostly cations and include potassium, sodium, magnesium, iron, and cobalt (**Table 5.6**). Iron is a constituent of electron transport chains and therefore is an absolute requirement among aerobes. Copper, zinc, and molybdenum are required in small amounts and are called **trace** elements.

Growth Factors

Growth factors are specific relatively low molecular weight organic compounds that must be available in the growth medium of some microorganisms because they cannot synthesize them. The substances that generally serve as growth factors are selected amino acids, purines, pyrimidines, and B-vitamins. Microbes do not require fat-soluble vitamins such as A and D, as they are not constituents of microorganisms.

The three types of growth factor most often required in bacterial nutrition are as follows:

Table 5.6	The function of various elements in bacterial and archaeal nutrition
Element	**Function**
Potassium	Utilized in a number of enzymatic reactions as a cofactor and especially in protein synthesis
Sodium	Involved along with chloride in the regulation of osmotic pressure; affects activity of some enzymes; uptake of solutes in some species with Na^+ dependent transport systems
Magnesium	Integral part of chlorophyll; cation required in enzymatic reactions including those involved in ATP synthesis or hydrolysis
Iron	Reactive center of heme-containing proteins (cytochromes, catalase, etc.) and component of other proteins
Cobalt	Constituent of vitamin B_{12}, complexed to some enzymes
Copper, zinc, molybdenum, nickel, tungsten, and selenium	Essential components of some enzymes

- **Vitamins**: These organic compounds serve as the prosthetic group (nonprotein catalytic part) of a number of enzymes. Small catalytic amounts of vitamins are required, as they are present in cells in low quantity (varying from a nanogram of vitamin B_{12} to 250 micrograms/gram dry weight cell of nicotinic acid). The vitamins most frequently required are those less susceptible to destruction by light: thiamine, biotin, and nicotinic acid. The function of the B-vitamins in nutrition is outlined in **Table 5.7**.

Table 5.7	The function of the various B-vitamins in nutrition of *Bacteria* and *Archaea*
Compound	**Function**
p-Aminobenzoic acid	Precursor of folic acid, a coenzyme involved in one carbon unit transfer
Folic acid	A coenzyme involved in one carbon unit transfer
Biotin	A prosthetic group for enzymes that act in carboxylation reactions
Nicotinic acid	Precursor of NAD and NADP, which are coenzymes involved with hydrogen transfer
Riboflavin	A component of the flavin mononucleotide (FMN) and dinucleotide (FAD) involved in hydrogen transfer
Pyridoxine	A component of the coenzyme for transaminase and amino acid decarboxylase
Vitamin B_{12}	A coenzyme involved in molecular rearrangements
Thiamin	The prosthetic group for a number of decarboxylases, transaldolases, and transketolases
Pantothenic acid	A functional part of coenzyme A and the acyl carrier proteins
Coenzyme M	A coenzyme in methane-generating bacteria

- **Amino acids:** Proteins are made up from 21 α-amino acids (Chapter 3), and some *Bacteria* and *Archaea* cannot synthesize one or more of these. They would therefore require that they be present in the growth medium. For example, most strains of *Staphylococcus epidermidis*, a normal inhabitant of human skin, require six amino acids: proline, arginine, valine, tryptophan, histidine, and leucine. The lactic acid bacteria require a greater complement, as indicated in Table 5.3. The level required is proportional to the amount of that amino acid in the cellular protein. This can be calculated as follows: a cell is about 50% protein, and the aromatic amino acid phenylalanine present in protein averages about 5%. Therefore, to grow one gram of an organism that required this amino acid, one should add at least 25 mg phenylalanine to the growth medium (1 gram cell = 500 mg protein, $500 \times 0.05 = 25$ mg).
- **Purines and pyrimidines:** Requirements for the nucleic bases are most often observed in lactic acid bacteria (Table 5.3) and other fastidious organisms with many growth factor requirements. The need for added nucleic acid bases is rare in free-living soil microbes.

UPTAKE OF NUTRIENTS INTO CELLS

The cytoplasmic membrane of a bacterial or archaeal cell forms a highly selective barrier between the external environment and the cytoplasm. It permits or facilitates the entry of essential nutrients inward and rejects many of those that are not necessary. The hydrophobic nature of the lipid bilayer is responsible for the high degree of selective permeability inherent in cytoplasmic membranes. Polar solutes such as amino acids or sugars cannot traverse unaided across the cytoplasmic membrane. Water can pass freely across the membrane; alcohols, fatty acids, or other fat-soluble compounds may also pass through at varying rates.

Microorganisms in nature generally live in environments where virtually all nutrients are available at quite low concentrations. Effective growth can occur only if nutrients can be accumulated inside the cell at levels that far exceed those present externally. Microorganisms have evolved with mechanisms that permit them to concentrate nutrients such as sugars or required cations to levels inside the cell that are thousandfold or more higher than the level in the environment.

To accomplish this uptake of nutrients essential to growth and reproduction, a microorganism utilizes a number of distinct transport mechanisms. Among the transport mechanisms most utilized are **facilitated diffusion, active transport,** and **group translocation.** These carrier-mediated processes are necessary because simple diffusion would permit an internal concentration that is only equal to that present externally. Under most

environmental conditions, this would not support balanced growth. The following is a discussion of the mechanisms that are important in concentration of nutrient solutes inside a bacterial cell.

Diffusion

Two types of diffusion occur across cytoplasmic membranes: **passive diffusion** and **facilitated diffusion.** Diffusion, both passive and facilitated, requires a concentration gradient, and molecules will flow across the membrane from an area of high concentration to one of low concentration until equilibrium is established. Generally, passive diffusion occurs with fat-soluble compounds. Glycerol is a compound that may enter a cell by passive diffusion, and the rate of glycerol uptake is solely dependent on the external concentration.

Facilitated diffusion is a carrier-mediated transport process using transmembrane proteins termed **permeases.** One end of the transport protein protrudes to the outside of the membrane, and the other end extends to the interior. Some of these permeases are highly selective in that they transport only a single type of molecule but most are active with classes of substrates such as a group of similar amino acids or a series of related sugars.

Facilitated transport does not require energy, but depends on a conformational change in the transport protein. The solute to be transported binds to the external portion of the transport protein, then a change in conformation of the protein moves the solute inward, and it is released to the inside of the cell (**Figure 5.2**). Although facilitated diffusion speeds the entry of solutes into a cell, the total internal concentration will not exceed the level in the immediate external environment. Facilitated diffusion is effective because it "speeds up" the diffusion process, and generally the transported material is metabolized on entry. This maintains a constant internal concentration and promotes continued uptake.

Because facilitated diffusion will not function against a concentration gradient, a microbe in nature requires a better mechanism for concentrating a nutrient that is present at a low external concentration.

Active Transport

Active transport is the direct utilization of energy to move a solute from the side of a membrane where the concentration is low to establish a higher concentration on the other side. This accumulates a solute against a concentration gradient. As with facilitated diffusion, solute-specific transmembrane proteins are involved in active transport, and the solute transported is not chemically altered as it passes into the cytoplasm. For a single solute, an organism may have multiple transport systems that may differ in the energy source required and their affinity for the solute that is transported.

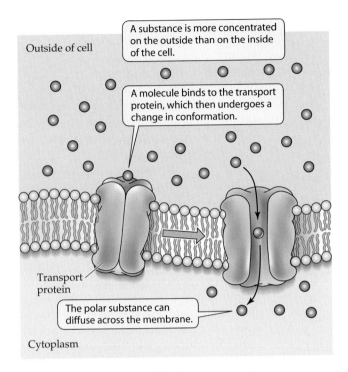

Outside of cell

A substance is more concentrated on the outside than on the inside of the cell.

A molecule binds to the transport protein, which then undergoes a change in conformation.

Transport protein

The polar substance can diffuse across the membrane.

Cytoplasm

Figure 5.2 Facilitated diffusion
In facilitated diffusion, the substance passes through a transporter protein, or permease. No energy source is required. If the internal concentration exceeds the external concentration, the process can be reversed.

The nutrients that are taken into the cell by active transport include some of the sugars, amino acids, organic acids, and inorganic ions. The energy for active transport generally comes from ATP or from a proton gradient established across the membrane. The proton gradient can result from the metabolism of inorganic or organic substrates, or it can be generated by light energy (in photosynthetics).

Transport of molecules by ATP-dependent systems involves **ATP-Binding Casette transporters** (ABC transporters). These transporters are present in bacterial, archaeal, and eukaryotic cells. An ABC transporter consists of hydrophobic transporter proteins that span and form a pore through the cytoplasmic membrane. On the cytoplasmic end of the transporter are nucleotide-binding sites where ATP is bound and hydrolyzed to promote uptake (**Figure 5.3**). The ABC transporter system depends on substrate binding proteins that are present in the periplasmic space in gram-negative bacteria and attached to the membrane near the outer end of the transporter in gram-positive bacteria. These binding proteins specifically bind the molecules to be transported and pass them on to the transporter protein. Conformational changes in the transporter protein promoted by energy from ATP-hydrolysis move the mole-

cule into the cell. In gram-positive microorganisms, the molecules to be transported would diffuse through the cell wall to the membrane surface where they would be available to the binding proteins. In gram-negatives, the molecules would pass through porins in the outer membrane (see Figure 4.48) and come in contact with the binding proteins in the periplasmic space.

Bacteria also transport solutes by utilizing the energy inherent in an electrical charge separation across the membrane. There are no periplasmic binding proteins involved in this transport. The charge separation, termed a **proton motive force,** is established by a proton gradient that is generated during electron transport. The proton force drives ATP synthesis (see Chapter 8) and can be used to drive the uptake of substances through the transport proteins. There are three types of these transport systems: uniporters, symporters, and antiporters.

Uniporters are transport proteins that take in cations such as K^+ through the electrochemical gradient established by a high positive charge on the outside of the membrane and negativity on the inside (**Figure 5.4**).

Symporters carry sugars, anions, or amino acids into the cell when accompanied by a proton. The binding of a proton to the transport protein and a change in conformation transports the solute inward.

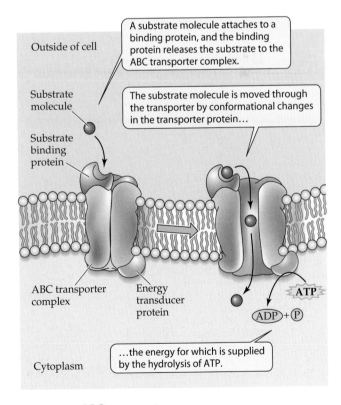

Outside of cell

A substrate molecule attaches to a binding protein, and the binding protein releases the substrate to the ABC transporter complex.

Substrate molecule

The substrate molecule is moved through the transporter by conformational changes in the transporter protein...

Substrate binding protein

ABC transporter complex

Energy transducer protein

ATP

ADP + P

...the energy for which is supplied by the hydrolysis of ATP.

Cytoplasm

Figure 5.3 ABC transporter
Active transport requires energy. In this transport system energy is supplied by hydrolysis of ATP.

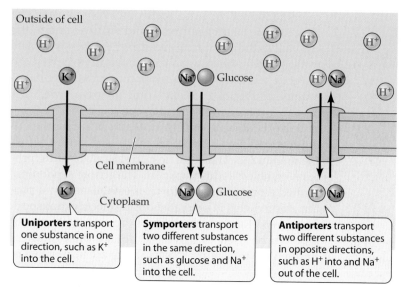

Figure 5.4 Proton-driven transport
Three types of active transport that use the proton gradient across the membrane as energy source. An electrochemical charge is generated across the membrane by electron transport, during which protons are "pumped" to the outside of the cell. The energy in this proton gradient is used to transport substances across the membrane.

Antiporters use the energy of a proton moving inward through the transport protein to drive a cation such as Na+ outward, creating a sodium concentration gradient. This sodium gradient can drive the uptake of an amino acid or other molecule. The sodium ion apparently effects a change in shape of the transport protein driving the solute inward.

All of these mechanisms depend on the generation of a proton gradient across the membrane to drive the transport processes.

Group Translocation

Group translocation is a transport process utilized by bacteria where the transported compound is chemically altered. The best definition of the group translocation is the **phosphotransferase system (PTS)** that is involved in transport of sugars into the cell including glucose, fructose, and α-glucosides. The PTS system is quite complex and involves a high-energy phosphate from phosphoenopyruvate and the direct participation of at least four enzymes to transport one sugar (**Figure 5.5**).

The first two enzymes, Enzyme I and a heat-stable protein (HPr), are soluble and present in the cytoplasm.

Enzymes IIa and b are peripheral proteins attached to the inner surface of the membrane at the site of Enzyme IIc.

The transmembrane protein Enzyme IIc is an integral part of the cytoplasmic membrane and the final recipient of the high-energy phosphate, which it passes on to phosphorylate the sugar in the transport process.

These enzymes are quite specific and are involved in the transport of a single sugar, such as glucose. HPr and Enzyme I are nonspecific and function in the PTS system for various substrates. The PTS transport system is energy conserving in that the high-energy phosphate from phosphoenolpyruvate ultimately becomes a part of the transported sugar.

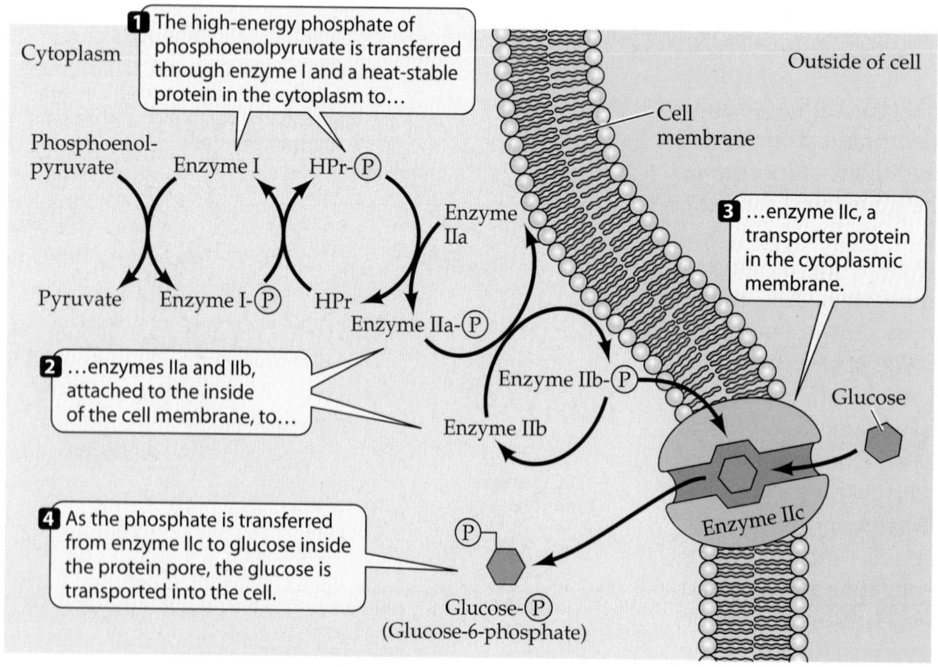

Figure 5.5 The phosphotransferase system
Active transport of glucose into the cell via the phosphotransferase system. HPr is a heat-stable protein; the other components of the system are enzymes.

The PTS system is present in obligately anaerobic bacteria such as *Clostridium* and *Fusobacterium*. It is also present in many genera of facultative anaerobes including *Escherichia*, *Staphylococcus*, *Vibrio*, and *Salmonella*. However, it is rare in obligately aerobic genera. Other substrates that may be transported by group translocation include purines, pyrimidines, and fatty acids.

Iron Uptake

Iron is a constituent part of the cytochromes and iron-sulfur proteins that are an integral part of electron transport in aerobic respiration. Most iron that is available in nature is in the insoluble oxidized state as the ferric ion (Fe^{3+}), which is rust and not freely available for transport into a cell. Iron may be provided in a culture medium in a complex with a **chelating agent.** A chelating agent is a compound such as ethylene diamine tetraacetic acid (EDTA) that binds to iron to prevent oxidation but can release it to the cell's uptake system.

Many microorganisms produce chelating agents that solubilize iron salts and make the iron available. These low molecular weight compounds are called **siderophores.** A typical siderophore produced by *Escherichia coli* is enterobactin. It is a derivative of catechol (**Figure 5.6**) and chelates iron through the oxygen molecules on the catechol ring. Microbes tend to secrete siderophores under growth conditions where little iron is available from the environment. Thus, even small amounts of environmental iron are made available to the cell. The siderophore-iron complex is bound by a specific siderophore receptor in the cell envelope, and the iron is transferred through the cytoplasmic membrane by specific transport proteins.

CULTIVATION OF MICROORGANISMS

Microbiologists have devised techniques and procedures that enable them to separate a single species of microorganism from a mixed culture. When successfully isolated, a strain can then be cultivated in **pure culture** (an **axenic culture**; *a*—without, *xenos*—strange). Some of the standard procedures for the isolation, growth, maintenance, and preservation of pure cultures are presented here.

A Culture Medium

A culture medium is composed of biological or chemically derived materials that provide the environment for the growth of a particular type of microorganism. There are hundreds of described media that can be used for the growth of bacteria. A variety of ingredients defined or undefined can be used in their preparation. Components of common media may include extracts of beef heart or brain, whole blood or serum, yeast extract (autolyzed yeast), peptone (protein digest), hydrolyzed

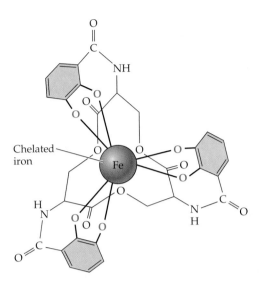

Figure 5.6 Iron transport
Enterobactin, a siderophore of *E. coli*, involved in transport of iron into the cell. This catechol derivative chelates iron through the oxygen groups on the catechol rings.

casein (milk protein), or soil extract. These ingredients are not chemically defined, and a medium composed of any of these would be termed a **complex medium.** They are not chemically defined because the various constituents have not been quantified and will vary from batch to batch. Sugars or organic acids might also be added to a complex medium. When a solid medium is desired, one would add the complex polysaccharide agar as solidifying agent. Silica gel can also be used for the solidifying agent when agar is undesirable. Agar contains small amounts of organic matter or sulfides that inhibit some species. Also marine bacteria may grow on agar.

A **defined** or **synthetic** medium can be employed for growth of many microorganisms. All of the ingredients in a defined medium are chemically characterized, and the amount of each added is known (Table 5.2). The defined medium may contain many components, for example the medium for culture of *Streptococcus agalactiae* (Table 5.3).

Isolation in Pure Culture

In nature, virtually all microorganisms coexist with countless other microbial species. For example, hundreds of different species live in the intestinal tracts of animals. A gram of fertile soil contains 10^7 to 10^9 bacteria and scores of species. This natural state where many different species coexist is referred to as a **microbial community** or a **mixed population.**

Enrichment procedures generally yield a mixture of microbial species. Therefore, it is necessary to apply techniques that will permit one to **isolate** individual species from this mixture. This is accomplished by use of **aseptic** isolation procedures. **Asepsis** means "in the absence of microorganisms" (its antonym is **sepsis**, which means nonsterile, or "in the presence of microorganisms"). In this context, asepsis mandates that one must handle materials in such a way that unwanted microorganisms are not introduced during isolation. Aseptic procedures include sterilization of medium, plate streaking, and pour plating.

Sterilization of Media

Sterilization is the process of killing or removing **all** living things from a particular environment. Heating with live steam under pressure (autoclave) is commonly used for sterilization.

Growth media are autoclaved in covered or cotton stoppered test tubes or flasks. Stainless steel or plastic caps are now used, as they are less cumbersome and permit an exchange of air into the culture vessel. Nutrient agar containing Petri dishes are prepared by autoclaving the medium in a flask and aseptically pouring the liquified agar/nutrient into a sterile Petri dish.

Incubation Conditions

Aerobic organisms require oxygen for growth and can be cultivated on solid medium in Petri plates or on agar slants (see later) by placing them in **incubators.** An incubator is an insulated chamber with a thermostatically controlled temperature that is maintained at 25°C or 30°C for common soil organisms or at 37°C (body temperature) for those from human or other mammalian sources.

Aerobic organisms grown in broth (liquid medium) culture tend to deplete the oxygen from the medium during active respiration, leading to curtailment of growth. Thus, mechanical **shakers** are sometimes used to agitate the cultures during growth so that more oxygen is stirred in at a constant rate (**Figure 5.7**). Obligately anaerobic organisms, of course, cannot be grown in the presence of oxygen, and special media and media preparation procedures are needed to cultivate these microbes (see Chapter 6).

Streak Plate Procedure

The classic method for isolation of bacteria is the **streak plate procedure**, which was developed in Robert Koch's laboratory (see Chapter 2). In this technique, a sample of a bacterial community is picked up with a metal wire that has a circular loop at its end (**Figure 5.8**). The wire is first sterilized by heating it directly in a burner flame until it is red hot. The sterile loop is then cooled by touching it on the sterile liquid or solid medium. By touching a colony or immersing the loop in a liquid medium, such as a sample of lake water or a liquid culture, a mixture of microbes would be collected on the loop. For example, lake water typically contains about

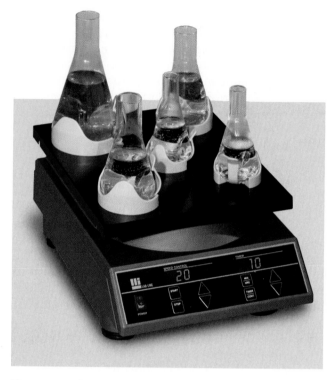

Figure 5.7 A bench-top shaker
The platform can gyrate at controlled rates, from 15 to 500 rpm. These shakers can be operated at room temperature or placed in a refrigerator or incubator at a desired temperature. The cotton stoppers ensure that only filtered, sterile air enters the flasks. Courtesy of Barnstead International/Lab-Line.

10,000 viable bacteria per milliliter (ml). The loop, which has an approximate volume of 0.01 ml, would thus pick up 100 or more bacterial cells. For identification, the bacteria in the droplet must be separated.

A common method of separation is by **streaking** the loop's contents on the solid surface of an **agar plate.** Streaking over the surface would be accomplished in three successive steps that separate individual cells (Figure 5.8A).

In the first streak, a loop bearing the sample is spread back and forth in nonoverlapping lines to cover about one-third of the agar surface, followed by flaming and cooling of the loop. In the second streak, the sterilized loop would be passed across the first streak to obtain some of the cells and streaked over another third of the plate. A third streak repeats on a third sector to further ensure the separation of bacteria.

In this streaking procedure, individual microorganisms from the original sample become spaced farther and farther apart, and the numbers are reduced in successive sectors. After streaking, the plate would then be incubated at the appropriate temperature to permit growth of the organisms.

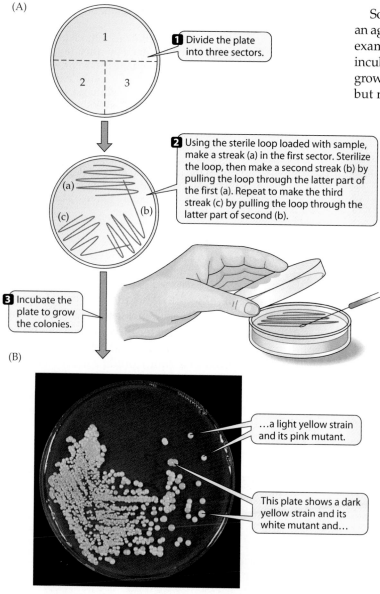

Figure 5.8 Streaking procedure
(A) The proper streaking procedure for obtaining isolated colonies.
(B) Results of a three-way streak of four strains of *Micrococcus luteus.*
Courtesy of Jerome Perry.

Some organisms may either not grow well or at all on an agar surface. There may be several reasons for this. For example, the composition or pH of the medium or the incubation conditions may not be satisfactory for their growth. Some bacteria cannot grow as separate colonies, but require closely associated bacteria to provide biochemicals that may not be provided in the medium. Organisms that require the presence of other organisms for growth are referred to as **syntrophic bacteria.** They grow very well in **consortia** (cultures containing more than one species), but not as pure cultures.

Spread Plate Procedures

Spread plates may also be used for isolation of bacteria. A small volume (typically about 0.1 ml) of the sample is placed on the sterile agar surface in a Petri plate. Then it is spread over the surface with a sterile bent glass rod, which resembles a "hockey stick" (Figure 5.9A). This process distributes the cells evenly on the surface, so if they are sufficiently diluted they grow as discrete colonies (Figure 5.9B). This procedure is less restrictive than pour plating because the cells are not exposed to the elevated temperature of the molten agar.

Pour Plate Procedure

Another method for separating mixtures, and one often used in counting the number of each colony type in mixtures, is the **pour plate procedure** in which a soil or water sample is mixed with a molten agar medium *before* it solidifies. After a thorough mixing, the molten agar (45°–48°C) is poured into a Petri dish and allowed to solidify. In this procedure, the colonies are distributed uniformly throughout the solid medium (Figure 5.9C). This procedure has one disadvantage in that the high temperature needed to keep the medium molten may actually kill some species. For example, bacteria from marine samples or temperate zone lakes rarely, if ever, encounter temperatures greater than 25°C to 30°C and may be killed by brief exposure to 45°C. Most species, however, are not harmed at this temperature.

Maintenance of Stock Cultures

After obtaining a pure culture, it can be transferred to a suitable agar medium and maintained as a **stock culture** by transfer at suitable intervals to a newly prepared medium. It is of utmost importance that a researcher retains such cultures in an axenic state, as faulty research will result from **contaminated cultures** (cultures that contain a mixture of two or more organisms).

Individual microorganisms that are spaced apart by the streaking process can multiply to produce a **colony** (or **clone**) made up of cells derived from a single parent. The colony represents an **isolated strain** of the microorganism (Figure 5.8B). To ensure purity of a desired species or colony type, however, it is necessary to repeat the streaking process one or more times on a freshly prepared plate. A pure culture is obtained when only one type of colony grows on the plate. This pure culture is referred to as a **bacterial strain.**

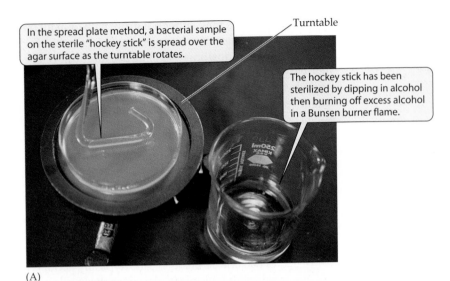

In the spread plate method, a bacterial sample on the sterile "hockey stick" is spread over the agar surface as the turntable rotates.

Turntable

The hockey stick has been sterilized by dipping in alcohol then burning off excess alcohol in a Bunsen burner flame.

(A)

Figure 5.9 Spread plate and pour plate procedures

(A) In the spread plate method, individual colonies are obtained by spreading a liquid sample of microorganisms over the agar surface. (B) Results of a spread plate inoculated with diluted nasal mucus, showing four *Staphylococcus* species: the small gray colonies are *S. epidermidis*; the small white colonies, *S. capitis*; the large white colonies, *S. hemolyticus*; and the single yellow colony, *S. aureus*. Contrast this with results from the pour plate method (C). The colonies "trapped" in the agar can be separated after incubation. Photos courtesy of Jerome Perry.

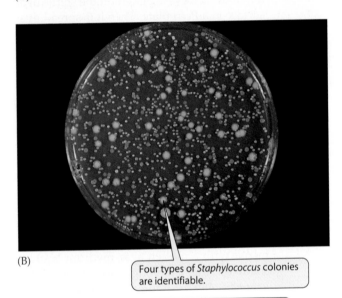

(B)

Four types of *Staphylococcus* colonies are identifiable.

It is often convenient to store stock cultures on an agar surface in a test tube. To increase the surface area for growth in the tube, the agar medium is allowed to solidify at an angle, resulting in a **slant** (**Figure 5.10**). Cultures transferred to a slant would be incubated and after growth placed in a refrigerator maintained at 4°C to 5°C. At this temperature, microbial cultures are generally viable for months, but this can vary from species to species.

This plate has been produce by the pour plate method: a culture of microorganisms in liquid was added to melted agar, and the mixture was poured onto the Petri plate, allowed to solidify, then incubated.

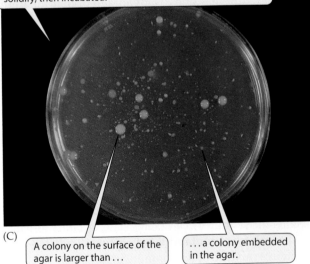

(C)

A colony on the surface of the agar is larger than . . .

. . . a colony embedded in the agar.

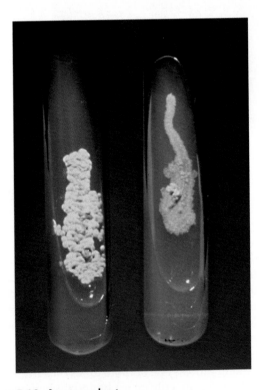

Figure 5.10 An agar slant

The organisms are two strains of the genus *Streptomyces*. Photo by CDC/Biological Photo Services.

Longer-term maintenance is accomplished either by **lyophilization** (freeze-drying) or by freezing bacterial or archaeal cell mass in glycerol at exceedingly low temperatures ($-70°C$ or lower in special freezers or in liquid nitrogen). Lyophilized cultures retain viability in freezers for extended periods of time—most cultures can be stored for 10 years or longer and therefore require minimum maintenance.

ISOLATION OF SELECTED MICROBES BY ENRICHMENT CULTURE

The basic requirements for the growth of bacteria have been described in the previous sections. The isolation of microorganisms that have selected nutritional requirements can be accomplished by preparing and using growth media that supply these needs. The physical conditions for growth can be adjusted to select for microorganisms that can grow under specific conditions such as temperature, pH, oxygen availability, and osmotic pressure. Isolation of specific types by a combination of nutrient and physical conditions is generally termed **enrichment culture**. Here we will focus on bacteria; the isolation of *Archaea* will be discussed in Chapter 18.

Origins of Enrichment Culture Methodology

The first scientists to apply enrichment culture on a broad scale were Martinus Beijerinck and Sergei Winogradsky (see Chapter 2). Their experimentation resulted in the isolation of a broad array of bacterial types and led to a rational approach to microbial ecology. Beijerinck's basic method was to prepare Petri dishes with a mineral salts agar medium; to this medium he added a sample of soil or water taken from diverse environments. Beijerinck then placed crystals of selected organic compounds (carbon source) on the surface of the inoculated agar. After a few days at room temperature, he noted that in virtually every case colonies developed on the agar surface where these organic compounds were added (**Figure 5.11**). Using this experimental approach, Beijerinck observed that equivalent conditions applied to soil or water samples obtained from widely separated habitats generally yielded similar bacterial types.

Thus, Beijerinck essentially demonstrated the Darwinian concept of natural selection in a

Petri dish. The organism or organisms most capable of development under the specific regimen of pH, temperature, and substrate were the ones that arose in greatest numbers from the soil or water inoculum. **Enrichment culture** is the selection of a specific organism(s) from all those in the inoculum (1 g of fertile soil generally contains 10^7 to 10^9 microbes).

If one were to set up enrichment cultures under equivalent selected conditions by using soil from North America, eastern Asia, and northern Europe, the organisms obtained from these three soils would be physiologically similar. The characteristics of the organisms from the separate continents would probably not differ any more than two cultures obtained from one acre of soil.

Applications of Enrichment of Culture

The use of bacteria obtained by enrichment was of considerable value in studies that led to the elucidation of both biodegradable and biosynthetic pathways. **Biodegradation,** in this sense, refers to the stepwise conversion of an organic compound to $CO_2 + H_2O$. Although notable exceptions exist, biosynthetic pathways generally occur through a reversal of the biodegradative route. **Biosynthetic pathways** are the chemical reactions that transform a substrate, whether glucose or CO_2, to the various macromolecular components of a cell.

It is generally easier to isolate sufficient quantities of pathway intermediate compounds for structural analysis by following biodegradation of a molecule than it is to accumulate identifiable quantities of biosynthetic intermediates. Identification of metabolic intermediates was particularly difficult when analytical methods were far less sophisticated than they are today. It should also

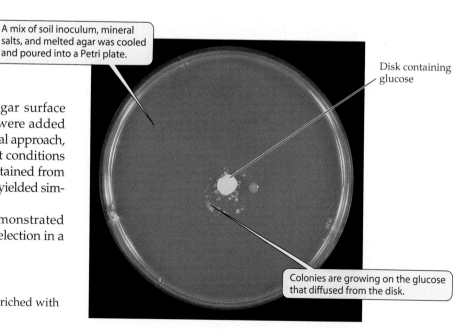

A mix of soil inoculum, mineral salts, and melted agar was cooled and poured into a Petri plate.

Disk containing glucose

Colonies are growing on the glucose that diffused from the disk.

Figure 5.11 Enrichment culture
Microorganisms grown in a culture enriched with glucose. Courtesy of Jerome Perry.

be noted that during the period when biosynthetic pathways were elucidated, the genetic manipulation of bacteria was also in its infancy. Genetic mutants that would accumulate intermediates were not available.

Often the goal in isolating microorganisms is to obtain a specific microbial type for practical purposes. For example, the isolation of thousands of actinomycetes has been accomplished over the past 50-plus years in the quest for those that yield useful antibacterial, antiviral, and antitumor agents. These organisms are commonly obtained from soil, compost, freshwater, and the atmosphere. A number of enrichment substrates have been devised for this purpose. One containing low levels (0.1%) of hydrolyzed casein (animal protein), soytone (plant protein), and yeast extract (microbial protein) has proven quite effective. The low level of sugars present in such a medium would curtail overgrowth by fast-growing motile genera, such as *Bacillus* and *Pseudomonas*, and permit the slow-growing actinomycetes to form colonies.

Colonies of microbes that have the capacity for extracellular digestion of insoluble higher molecular weight sugar polymers, lipids, and proteins can easily be selected. To do this, the polymer would be incorporated into the agar medium as the carbon source. The surface of the agar would then be inoculated with soil or water, and digestion of the polymer is evidenced by a clear area in the agar around the colony (**Figure 5.12**).

General Enrichment Methods

Both aerobic and anaerobic microorganisms can be isolated by enrichment methods. Samples should be obtained from environmental niches that favor the type of organism sought—soil for aerobes, mud for anaerobes. Basic conditions for enrichment would be as follows.

Aerobic Enrichment

Two general procedures that can be used to enrich for aerobic bacteria are liquid medium and solid medium (**Figure 5.13**). A streak from a liquid enrichment for an organism that might use a substrate such as sodium benzoate would appear as shown in Figure 5.13C. A pure and mixed culture can be distinguished by their colony appearance on plates (i.e., uniformly shaped, colored colonies of similar size versus diverse colony forms, shapes, and sizes). The direct application spreading of a sample on an agar surface would result in the growth of many different colonies (Figure 5.13D).

A shortcoming of liquid enrichment is a marked tendency for faster-growing organisms, such as pseudomon-

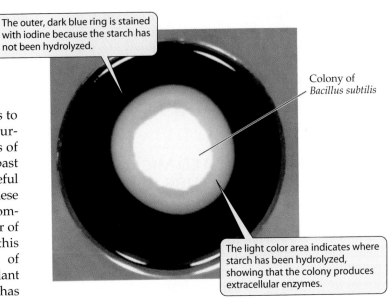

The outer, dark blue ring is stained with iodine because the starch has not been hydrolyzed.

Colony of *Bacillus subtilis*

The light color area indicates where starch has been hydrolyzed, showing that the colony produces extracellular enzymes.

Figure 5.12 Detection of extracellular enzymes
A plate containing starch (1%) was inoculated in the center with *Bacillus subtilis*. After 3 days, an alcohol solution of iodine (which stains starch) was flooded over the surface and the excess poured off. The technique of incorporating a polymer in the agar can also be effective in enriching cultures for microorganisms that utilize high molecular weight, water-insoluble compounds such as protein or lipids. Courtesy of Jerome Perry.

ads, to be favored, and these become dominant on continued transfer to a newly prepared medium. Enrichment on a solid medium will generally yield a greater variety of organisms, provided that one has the patience to wait for slower-growing colonies to appear. Unfortunately, organisms that are present as a small percentage of the total population may be overlooked by any enrichment procedure.

Anaerobic Enrichment

Enrichment for anaerobic bacteria is technically more difficult than enrichment for aerobes. This is particularly true for strict anaerobes that are killed by exposure to even minute levels of oxygen. Samples for enrichment must be handled carefully to preclude exposure to oxygen prior to isolation procedures. Elaborate methodologies have been developed for handling the obligately anaerobic bacteria, including the fabrication of entire rooms from which all traces of oxygen can be removed. A technique now in general use is an oxygen-free **glove box**, which permits one to apply techniques generally applicable to aerobes to the manipulation of strictly anaerobic bacteria.

The isolation of anaerobic organisms less sensitive to brief exposure to air can be accomplished in glass-stoppered bottles and in shake tubes (**Figure 5.14**). In this

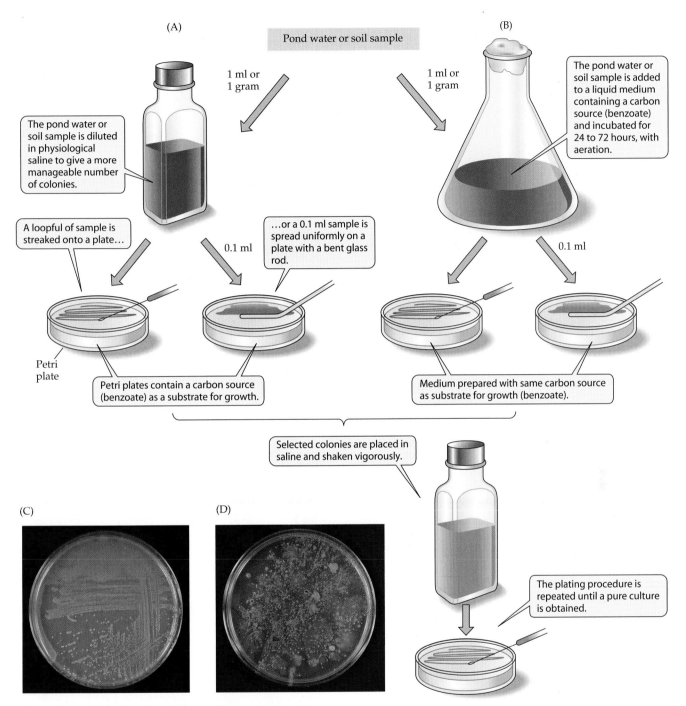

Figure 5.13 Isolation of microorganisms with specific nutritional requirements

Isolation of a bacterium that can utilize a specific compound as carbon and energy source. (A) Enrichment on a solid medium. (B) Enrichment in a liquid medium. (C) Plate streaked from a liquid enrichment. The liquid enrichment technique tends to favor rapidly growing organisms. Note that only one type of colony has developed. (D) Spreading samples directly onto the surface of an agar plate yields a variety of microorganisms. Photos courtesy of Jerome Perry.

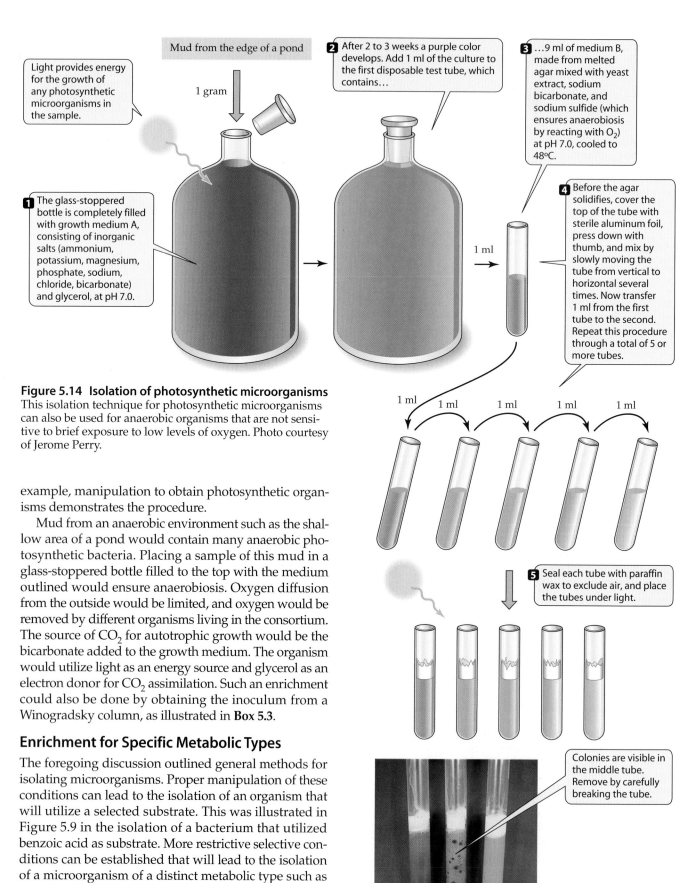

Figure 5.14 Isolation of photosynthetic microorganisms
This isolation technique for photosynthetic microorganisms can also be used for anaerobic organisms that are not sensitive to brief exposure to low levels of oxygen. Photo courtesy of Jerome Perry.

example, manipulation to obtain photosynthetic organisms demonstrates the procedure.

Mud from an anaerobic environment such as the shallow area of a pond would contain many anaerobic photosynthetic bacteria. Placing a sample of this mud in a glass-stoppered bottle filled to the top with the medium outlined would ensure anaerobiosis. Oxygen diffusion from the outside would be limited, and oxygen would be removed by different organisms living in the consortium. The source of CO_2 for autotrophic growth would be the bicarbonate added to the growth medium. The organism would utilize light as an energy source and glycerol as an electron donor for CO_2 assimilation. Such an enrichment could also be done by obtaining the inoculum from a Winogradsky column, as illustrated in **Box 5.3**.

Enrichment for Specific Metabolic Types

The foregoing discussion outlined general methods for isolating microorganisms. Proper manipulation of these conditions can lead to the isolation of an organism that will utilize a selected substrate. This was illustrated in Figure 5.9 in the isolation of a bacterium that utilized benzoic acid as substrate. More restrictive selective conditions can be established that will lead to the isolation of a microorganism of a distinct metabolic type such as chemoheterotrophs, chemoautotrophs, and phototrophs. Some of these selective methods are outlined here.

Methods & Techniques Box 5.3   Special Enrichment—The Winogradsky Column

The Winogradsky Column is an enrichment culture technique developed by Sergei Winogradsky in the latter part of the 19th century (see Chapter 2). A typical Winogradsky Column is a long glass tube ($1\frac{1}{2} \times 24$ inches), closed at one end, and with about two-thirds of the tube filled with rich mud. The column should be placed in a north window where it receives adequate but not intensive light. It is a microcosm where one can follow growth and succession in an anaerobic environment.

Organic-rich mud from a shallow area of a pond is a suitable source of mud for a column. Before placing the mud in the column, a few grams of calcium carbonate and calcium sulfate can be incorporated. In practice, it is preferable to mix the calcium sulfate in the mud that will occupy the lower fourth of the column along with some starch, cellulose powder, or shredded filter paper. These carbon sources can be varied depending on the imagination of the individual preparing the column. There should be no air pockets in the mud column, and any that form can be disrupted with a glass rod. The mud column is topped with a layer of pond water.

The appearance of the column after a few weeks would be as illustrated. Anaerobic bacteria ferment the cellulose or starch at the bottom to hydrogen, organic acids, and alcohols. These would be utilized as substrate by sulfate-reducing bacteria, and sulfate would serve as terminal electron acceptor. This would cause the formation of hydrogen sulfide resulting in an H_2S gradient from the bottom of the column upward. An O_2 gradient would occur from the top downward. The bottom area of the column becomes black due to formation of metal sulfides.

Restricting calcium sulfate to the lower area curtails an excessive production of black precipitate throughout. The green sulfur bacteria develop immediately above the dark area because of their tolerance for hydrogen sulfide. The purple sulfur bacteria are less tolerant and appear in the area above the green. The purple nonsulfur bacteria grow nearer the top in the absence of sulfide, and they may tolerate the low levels of oxygen present. Aerobic microorganisms, including cyanobacteria and algae, grow in the water layer. One would find anaerobic bacteria growing throughout and some sulfide oxidizers such as *Beggiatoa* and *Thiothrix* growing in the upper area. Samples can be removed periodically with a length of glass tubing narrowed at the end. Organisms can be isolated from these samples by aerobic or anaerobic techniques described in this chapter.

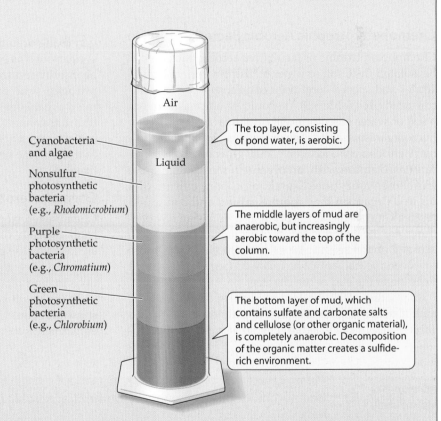

Cyanobacteria and algae

Nonsulfur photosynthetic bacteria (e.g., *Rhodomicrobium*)

Purple photosynthetic bacteria (e.g., *Chromatium*)

Green photosynthetic bacteria (e.g., *Chlorobium*)

Air

Liquid

The top layer, consisting of pond water, is aerobic.

The middle layers of mud are anaerobic, but increasingly aerobic toward the top of the column.

The bottom layer of mud, which contains sulfate and carbonate salts and cellulose (or other organic material), is completely anaerobic. Decomposition of the organic matter creates a sulfide-rich environment.

A Winogradsky column Anaerobic decomposition of organic matter in the bottom of the cylinder creates an anaerobic sulfide-rich environment. The green and purple sulfur bacteria are present immediately above the dark sulfide area. The purple nonsulfur bacteria would be above the purple sulfur bacteria. Cyanobacteria would grow at the top in the area exposed to air.

Table 5.8	Enrichment conditions for chemoheterotrophic aerobic or facultative aerobic organisms		
Substrates (Carbon and Energy Source)	**Inoculum**	**Special Conditions**	**Organism Favored**
Ethanol	Soil	N_2 as nitrogen source	*Azotobacter*
Uric acid	Pasteurized soil (80°C for 15 min.)		*Bacillus fastidiosus*
Glucose	Pasteurized soil		*Bacillus*
Casein + thiamine	Pasteurized soil	Add urea at pH 9.0	*Bacillus pasteurii*
Glucose + yeast extract	Soil	Incubate as 60°C	*Bacillus stearothermophilus*
Nutrient broth	Pasteurized soil	10% NaCl 0.5% $MgCl_2$	*Sporosarcina halophila*
Propane	Soil	Gas 50/50 in air	*Mycobacterium*
n-Hexadecane	Soil	Incubate at 60°C	*Bacillus thermolevorans*
Filter paper	Soil		Cytophaga
Chitin	Soil		Actinomycetes

Chemoheterotrophic Aerobic Bacteria

Chemoheterotrophic or facultative aerobic bacteria can be isolated from soil or water by using a variety of substrates and special conditions of atmosphere, temperature, and pH (**Table 5.8**). It should be emphasized that a soil or water inoculum will contain a great variety of microorganisms, and many of these can grow under primary enrichment conditions. They grow on products of, or in association with, an organism metabolizing the enrichment substrate. Obtaining a pure culture of a desired organism may require careful and continued restreaking on the appropriate medium. Such procedures are the general aspects of enrichment culture. To actually obtain specific types of microorganisms, more involved methods must be followed (see the Suggested Reading at the end of the chapter).

Many distinctly different bacterial species can fulfill their nitrogen requirement by fixing atmospheric N_2 (see Table 5.5). For the isolation of *Azotobacter* or any nitro-gen-fixing organisms, an inorganic or organic nitrogen source (such as NH_4^+ and NO_3^- or an amino acid) must be eliminated from the medium. Eliminating a fixed nitrogen source from the medium would select for microorganisms that can obtain this key element from the atmosphere. The metal molybdenum must be present in such enrichments because it is a constituent part of the enzyme nitrogenase that is involved in nitrogen fixation.

Chemoheterotrophic Anaerobic Bacteria

Chemoheterotrophic anaerobic bacteria can be isolated from a variety of sources (**Table 5.9**). Many anaerobes are killed by the presence of even very low levels of oxygen, and these microorganisms require special precautions in their handling.

To isolate nitrogen-fixing anaerobes, several different substrates could be used (Table 5.9). With starch as the enrichment substrate and pasteurized soil as inoculum,

Table 5.9	Enrichment conditions for chemoheterotrophic anaerobic *Bacteria* or *Archaea*		
Substrates (Carbon and Energy Source)	**Inoculum**	**Special Conditions**	**Type Organism**
Sugars + yeast extract	Plant material	pH 5–6	Lactic acid bacteria
Mixed amino acids	Pasteurized soil		*Clostridium*
Starch	Pasteurized soil	N_2 as nitrogen source	*Clostridium pasteurianum*
Uric acid + yeast extract	Pasteurized soil	pH 7.8	*Clostridium acidiurici*
Organic acids	Pond mud	Added SO_4^{2-}	Desulfovibrio
Organic acids	Soil	Added NO_3^-	Denitrifying bacilli and pseudomonads
Organic acids	Rumen fluid	Added CO_2	Methane producers
Lactate + yeast	Swiss cheese		*Propionibacterium*

Table 5.10	Enrichment for chemoautotrophic *Bacteria* or *Archaea* using CO_2 as carbon source[a]			
Energy Source	**Inoculum**	**Special Conditions**	**Organism Obtained**	
H_2	Soil or water	Aerobic	Hydrogen-utilizing bacteria	
	Rumen fluid	Anaerobic	methanogens	
NH_4^+	Soil or water	Aerobic	*Nitrosomonas*	
NO_2^-	Soil or water	Aerobic	*Nitrobacter*	
$Na_2S_2O_3$	Soil or water	Anaerobic + KNO_3	*Thiobacillus denitrificans*	
Fe^{2+}	Estuarine mud	Aerobic, pH 2.5	*Thiobacillus ferrooxidans*	

[a]$NaHCO_3$ may serve as source of CO_2.

the organism most likely isolated would be *Clostridium pasteurianum*. This anaerobe forms heat-resistant endospores that readily survive the pasteurization process and germinate and grow on the starch medium.

Chemoautotrophic Bacteria

To isolate chemoautotrophs, the carbon source is limited to carbon dioxide, but any of the following inorganics could be the source of energy: NH_4^+, NO_2^-, S^{2-}, Fe^{2+}, or H_2 (**Table 5.10**). Probably the most important factor in the isolation of the nitrifying bacteria (organisms that oxidize $NH_4^+ \rightarrow NO_2^- \rightarrow NO_3^-$) is patience. These are slow-growing organisms, and the appearance of visible colonies on Petri plates can take from one to four months. Enrichment and isolation of the nitrifiers is generally carried out by **serial dilution** of rich soil (for an example of serial dilution, see Figure 5.14). Growth of the ammonia (NH_4^+) oxidizers is measured by chemically analyzing for an increase in nitrite (NO_2^-) concentration in the medium. Growth of those that oxidize nitrite to nitrate (NO_3^-) is measured by adding measured amounts of nitrite and then determining the amount that disappears.

The sulfur-oxidizing autotrophic bacteria play a key role in the conversion of reduced sulfur compounds (S^{2-}) to sulfate (SO_4^{2-}). These bacteria are present in a wide range of habitats, from pH 1 to 9, and many species are thermophilic ("heat loving").

The hydrogen-oxidizing bacteria do not form a cohesive taxonomic unit, and the ability to grow

using hydrogen (H_2) as the energy source is an attribute of microorganisms in many of the bacterial phyla. Enrichment is rather straightforward because these organisms grow chemolithotrophically under an atmosphere of hydrogen, carbon dioxide, and oxygen. The hydrogen-oxidizing bacteria can be isolated by sprinkling soil on an agar surface and then incubating under the proper $CO_2/H_2/O_2$ atmosphere.

Phototrophic Bacteria

The major requirement for the isolation of phototrophic bacteria is a constant source of light. Other conditions of enrichment include an electron donor, minerals, and availability of CO_2 (**Table 5.11**). Provided light is available, photosynthetic bacteria can flourish under a broad range of conditions. Cyanobacteria, for example, grow in some very harsh natural habitats: hot springs, Antarctic lakes, deserts, and areas of high salinity. They are the photosynthetic symbiont with fungi in many lichen associations and are common inhabitants of terrestrial, marine, and freshwater habitats. As the cyanobacteria produce oxygen during photosynthesis, they are all aerobic, and isolation can be accomplished in vessels exposed to air.

The green and purple bacteria do not produce oxygen during photosynthesis, and they grow under anaerobic conditions. Nonsulfur purple bacteria can be isolated from mud or water samples from ponds, ditches, and shores of eutrophic lakes (lakes rich in nutrient but

Table 5.11	Enrichment for photosynthetic microorganisms that utilize light as energy source		
Special Conditions		**Source of Inoculum**	**Organism**
Aerobic			
	N_2 as nitrogen source	Surface water or soil	Nitrogen-fixing cyanobacteria
	NH_4^+ as nitrogen source	Surface water or soil	Cyanobacteria and eucaryotic algae
Anaerobic			
	Glycerol	Mud from pond edge	Purple or green nonsulfur bacteria
	H_2S (high levels)	Sulfide-rich mud	Green sulfur bacteria
	H_2S (low levels)	Mud from pond edge	Purple sulfur bacteria

low in oxygen). Nonsulfur purple bacteria utilize reduced electron donors other than sulfur compounds such as hydrogen (H_2). The mud from the bottom of a shallow eutrophic lake may contain as many as 1 million purple nonsulfur bacteria per milliliter.

The major component of an enrichment medium for these organisms would be hydrogen or a reduced organic compound to serve as electron donor in photosynthetic growth. Because the purple nonsulfur bacteria may require selected B-vitamins, it is essential that yeast extract (0.05%) be added to the enrichment.

The green sulfur bacteria and purple sulfur bacteria can also be obtained by enrichment culture, but are more difficult to isolate than nonsulfur bacteria. They utilize soluble sulfides (S^{2-}) as electron donor for photosynthetic growth. The sulfide concentration is critical for the enrichment of these bacteria. During enrichment, higher population density and species diversity can be obtained by repeated addition of sulfide during enrichment.

These sulfur-utilizing photosynthetic organisms are present in aquatic environments, especially where anaerobic conditions are stringent and constant. One can select for green sulfur bacteria by controlling the amount of light available to the enrichment. Low-intensity direct sunlight (north-facing window) or a tungsten light of low intensity (5 to 2000 lux) favors green sulfur bacteria, as these intensities are not sufficient to support growth of most other phototrophic bacteria. The purple sulfur bacteria can be enriched by higher light intensities.

SUMMARY

▸ *Bacteria* and *Archaea* **generally differ** from one another based on the compounds they utilize as a **food** and/or **energy source.**

▸ The **substrates** utilized by *Bacteria* and *Archaea* as sources of carbon include all constituents of living cells and many synthetic chemicals. This microbial biodegradation can occur over a broad range of pH, temperature, salinity, and other physical parameters.

▸ Microorganisms can be divided into four nutritional groupings: **photoautotrophs, photoheterotrophs, chemoautotrophs,** and **chemoheterotrophs**.

▸ Bacteria, regardless of species or growth substrate, are composed of the same **elements**. The monomers from which the macromolecules are synthesized are equivalent. A growth medium must, in one way or another, supply all elements that make up a microbial cell.

▸ Some microorganisms can synthesize all cellular components from CO_2, whereas others require a source of amino acids, vitamins, and nucleic acid bases.

▸ Many bacterial and archaeal species can **"fix"** atmospheric **N_2** into cellular material. N_2 fixation is not known among the eucaryotes.

▸ **Phosphate** is present in microorganisms in relatively low amounts (~ 3% dry weight), but it is an essential element in nucleic acids and in energy generation.

▸ The **growth factors** (needed in small amounts) most commonly required by microorganisms are B-vitamins, amino acids, and nucleic acid bases.

▸ A **trace element** is one required in very small amounts (traces). Among these are Ca^{++}, Mg^{++}, Na^+, and Zn^{++}.

▸ Microorganisms need specific nutrient uptake systems, as they generally live in environments where many nutrients are present at low levels. The three major systems for nutrient uptake are **diffusion (simple and facilitated), active transport,** and **group translocation**.

▸ **Facilitated diffusion** does not require energy input, but cannot concentrate nutrients internally to a greater concentration than the external concentration.

▸ **Active transport** requires energy to move a substance against a concentration gradient. The energy may come from ATP or a proton gradient.

▸ Proton gradients can affect uptake by three mechanisms— **uniport, symport,** and **antiport**.

▸ **Group translocation** results in chemical alteration of the compound transported. Sugars may be transported into the cell via this process. Group translocation is the most elaborate of the transport systems and involves a number of enzymes in uptake of a single solute.

▸ Iron is required by all organisms, and aerobic microorganisms may have specific chelators called **siderophores** for iron uptake.

▸ **Enrichment culture** is a method used for isolating microorganisms that grow on a specific nutrient and/or under selected physical conditions. The organism obtained from an environmental sample through this procedure will have the ability to grow under the specific enrichment conditions employed.

▸ Most microorganisms are maintained in collections as **pure (axenic) cultures**. All studies with pure cultures must be done under aseptic conditions.

▸ Pure cultures can be obtained by **streaking, pour plate,** or **spread plate** techniques.

▸ Microorganisms can be maintained for extended periods of time by **lyophilization** or by placing them in **liquid nitrogen**.

REVIEW QUESTIONS

1. What is metabolic diversity? How did it come about? How does this relate to the role of microbes in nature?

2. Microbes survive and actually thrive in virtually every environmental niche on earth. Describe how this relates to the perpetration of life on earth.

3. What are the basic requirements in the development of a growth medium? Why is each added? How does this relate to the composition of a bacterial cell?

4. Define "growth factor." Why do some organisms require them?

5. How would enrichment culture be used to isolate an organism capable of growth with *p*-aminobenzoic acid as sole source of carbon?

6. Why have bacteria been utilized in studying nutrition rather than using higher organisms such as rats or other animals? How does this relate to comparative biochemistry?

7. Where would one obtain samples to be used in isolating various photosynthetic bacteria? Would the source for cyanobacteria differ from that for green sulfur photosynthetics? Why?

8. What is a microbial community? Are microbial communities prevalent in nature?

9. What is the streak plate method, and how does one employ this to obtain a pure culture?

10. Define axenic culture, consortia, and a bacterial strain.

11. How does a pour plate differ from a streak plate? What are the advantages of each?

12. What are the two major methods for long-term storage of bacteria?

SUGGESTED READING

Gerhardt, P., ed. 1993. *Methods for General and Molecular Bacteriology*. 2nd ed. Washington, DC: American Society for Microbiology.

Gottschal, J. C., W. Harden and R. A. Prins. 1992. *Principles of Enrichment, Isolation, Cultivation, and Preservation of Bacteria*. In A. Balows, H. G. Truper, M. Dworkin, W. Harden and K. H. Schleifer, eds. *The Prokaryotes*. 2nd ed. New York: Springer-Verlag.

Neidhardt, F. C., J. L. Ingraham and M. Schaechter. 1990. *Physiology of the Bacterial Cell*. Sunderland, MA: Sinauer Associates Inc.

Norris, J. R. and D. W. Ribbons, eds. 1969. *Methods in Microbiology*. Vol. 3B. New York: Academic Press.

Staley, J. T. and A-L. Reysenbach. 2002. *Biodiversity of Microbial Life*. New York: Wiley-Liss.

Veldkamp, H. 1970. *Enrichment Cultures of Prokaryotic Organisms*. In J. R. Norris and D. W. Ribbons, eds. *Methods in Microbiology*. Vol. 3A. New York: Academic Press.

Williams, S. T. and T. Cross. 1971. *Actinomycetes*. In J. R. Norris and D. W. Ribbons, eds. *Methods in Microbiology*. Vol. 4 (pp. 295–334). New York: Academic Press.

Chapter 6

Microbial Growth

The paramount evolutionary accomplishment of bacteria as a group is rapid, efficient cell growth in many environments. Bacteria grow and divide as rapidly as the environment permits.

– J. L. INGRAHAM, O. MAALØE, F. C. NEIDHARDT

An overview of the nutritional requirements for the growth of representative bacterial and archaeal cultures was presented in Chapter 5. The indispensable need for carbon, nitrogen, sulfur, and various cations and anions was discussed. Methods were also outlined for the isolation of specific autotrophic and heterotrophic microorganisms. When the nutrients and physical conditions required by a specific organism are met, the organism will grow (see chapter opening quote).

In microbiology, the term *growth* generally refers to an increase in the number of cells in a population, or **population growth.** An individual bacterial or archaeal cell may increase in size, and growth in this sense would be called **cell growth.** During population growth, a bacterial or archaeal cell divides to generate two identical progeny. Each of the progeny receives a genome that is a precise copy of the genetic information in the parent cell (see Chapter 13).

A microorganism is generally considered viable only if it is able to reproduce. In fact, the test most often used for viability is to place the organism under the growth condition for that species, incubate, and examine for visible growth after a suitable time. It should be noted that growth, either as a process or a way of acquiring cells, is a major method by which microbes are studied. However, we must be aware, that most of the microorganisms in nature have not yet been cultivated in the laboratory. These organisms are most assuredly viable, and our inability to grow them is due to our lack of understanding of their growth requirements.

In this chapter, we will present the various parameters of microbial growth, including:

- Measurement of growth
- Effect of nutrient concentration
- Effect of environmental conditions

POPULATION GROWTH

Population growth is the culmination of a complex series of biochemical events that are driven by light or chemical energy. This energy is utilized to synthesize or assimilate monomers, and these are in turn assembled into macromolecules. During the course of growth and division, a bacterial or archaeal cell may synthesize an estimated 1,800 different proteins, more than 400 different RNA molecules, a complete copy of the genomic DNA, cytoplasmic membrane, and where necessary, sufficient cell wall to surround the newly formed cell.

Research Highlights Box 6.1   The Cell Cycle in *Bacteria*

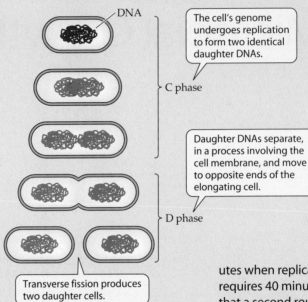

The cell cycle is the period of time in which a newly formed bacterium elongates, replicates its DNA, and divides to generate two cells. The replication of DNA and events involved in division to form two cells is under tight regulatory control.

In a newly formed cell there is a regulatory event that initiates the replication of the bacterial genome. The time required to replicate the 4.2×10^6 base pairs in the DNA of bacteria with a generation time between 20 and 60 minutes is 40 minutes. This is the C phase (see diagram). After completion of DNA replication (termination), there is a 20-minute period before division occurs. This is the D (delay) phase. During the D phase, the replicated DNA is separated into opposite ends of the elongated cell. The cytoplasmic membrane is involved in this separation. After separation of the DNA, the process of constructing cytoplasmic membrane and cell wall

DNA

The cell's genome undergoes replication to form two identical daughter DNAs.

C phase

Daughter DNAs separate, in a process involving the cell membrane, and move to opposite ends of the elongating cell.

D phase

Transverse fission produces two daughter cells.

begins at the midpoint of the cell. The division into two distinct daughter cells is called **transverse fission.**

You might ask how a species can have a generation time of 25 minutes when replication of DNA requires 40 minutes. The answer is that a second round of replication begins on the part of the DNA that has already replicated before the C phase of the first replication is complete. Thus, both daughter cells may receive a genome that has ongoing replication forks. This will be explained further in Chapter 13.

All the synthetic processes involved in population growth are regulated and integrated to produce a duplicate cell, and this occurs in a short period of time called the **generation time.** Generation time is defined as the time required for a population of cells to double in number.

A bacterial cell generally grows in size as a prelude to cellular division. Thus, this phase of the cell cycle (see **Box 6.1**) is considered cell growth, whereas cellular division results in an increase in the number of cells. When a bacterium is placed in a suitable growth medium, an initial adjustment period, termed the *log phase*, occurs, and during this time there is no increase in cell number. This phase is followed by an exponential increase in the bacterial population. This increase in numbers or cell mass per unit time occurs at a constant and reproducible rate for a given organism in a selected medium and is called the **growth rate.** Because growth of a population

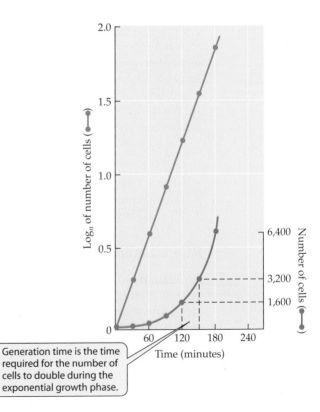

Generation time is the time required for the number of cells to double during the exponential growth phase.

Figure 6.1 Exponential growth
Exponential growth of a population of bacteria, plotted on logarithmic (left axis) and arithmetic (right axis) scales. The generation time for this microorganism is 30 minutes.

Table 6.1	Approximate generation times for several organisms growing in media optimal for growth.
Species	**Generation Time**
Escherichia coli	20 min
Bacillus subtilis	28 min
Staphylococcus aureus	30 min
Pseudomonas aeruginosa	35 min
Thermus aquaticus	50 min
Thermoproteus tenax	1 hr 40min
Rhodobacter sphaeroides	2 hr 20 min
Sulfolobus acidocaldarius	4 hr
Thermoleophilum album	6 hr
Thermofilum pendens	10 hr
Mycobacterium tuberculosis	13 hr 20 min

these values that some organisms divide much more rapidly than others. An organism such as *Escherichia coli* that can reproduce in 20 minutes could, starting with one cell and under unlimited growth conditions, yield 4.7×10^{21} progeny in 24 hours (**Box 6.3**). In 13 hours, a microbe with a generation time of 20 minutes can produce 1×10^{12} cells, approximately the number of cells in the entire human body! A single cell of a slower-growing soil organism such as *Bacillus subtilis* (generation time 30 minutes) would generate 1×10^{12} progeny in about 20 hours. These progeny numbers would apply only if an unlimited supply of all nutrients were available and if the culture medium were free of excreted inhibitory products.

is **exponential** (that is 1 cell → 2 cells → 4 → 8, etc.), it can be depicted on a logarithmic scale as a straight line or may be depicted on an arithmetic scale (**Figure 6.1**). The generation time for a species is the time required for the population to double during the exponential phase of growth. The generation time (also called doubling time) is usually determined under what would be considered optimal conditions for growth of that species (**Box 6.2**).

Typical generation times range from only 20 minutes to more than 10 hours (**Table 6.1**). It is apparent from

GROWTH CURVE CHARACTERISTICS

Adding a small population of bacteria (an inoculum) to a suitable volume of culture medium growth results in a predictable increase in cell numbers. If the number of organisms present is determined at intervals throughout population growth and plotted on semilogarithmic paper, a **growth curve** for that microorganism is obtained. A typical growth curve is shown in **Figure 6.2**. The numbers on the ordinate and abscissa of the graph vary quantitatively for different bacterial species, but the overall picture does not. The growth curve is divided into four distinct phases, each having a different slope. These phases are (1) lag, (2) exponential growth, (3) stationary, and (4) death phases. The cellular activities that are involved in each of these phases are discussed as follows.

Methods & Techniques Box 6.2   Calculating the Generation Time

The generation time for a microorganism that is growing under a determined set of conditions is constant. It can be calculated if three bits of information are available.

N_o The number of bacteria present at an early stage in exponential growth

N_t The number of bacteria present after a period of exponential growth

t The time interval between N_o and N_t

Exponential (logarithmic) growth must be understood as a geometric progression of the number 2. Expo-

nential growth proceeds:
2 cells → 4 → 8 → 16 → 32 ... or 2^n
where n = number of generations. Exponential growth can be described mathematically by the following equation:

$$N_t = N_o \times 2^n$$

To solve for n, one would take the logarithm of the two sides:

$$\log N_t = \log N_o + n \log 2$$

or solving for n

$$n = \frac{\log N_t - \log N_o}{\log 2}$$

Applying real numbers to the equation, one would proceed as follows:

Assume $N_O = 6{,}000$ cells

$N_t = 38{,}000{,}000$ cells

consulting a log table and filling in the equation:

$$n = \frac{7.5798 - 3.7782}{.301}$$

$n = 12.6$ generations

Assuming that the elapsed time between N_O and N_t was 5 hours (300 minutes), the generation time would be:

Generation Time $= \dfrac{300}{12.6} = 23.8$ minutes

A population increase where each cell divides during a unit of time is termed *exponential growth*. If the generation time of a microorganism is 20 minutes, then each cell will reproduce every 20 minutes. This rate of population growth will continue as long as the organism remains in the exponential phase. When growing in a culture medium, the exponential phase of a bacterium is relatively short. It can be limited by nutrient depletion, oxygen deprivation, the accumulation of inhibitory products, or other factors.

Imagine that we devise a means for sustaining a culture of *Escherichia* *coli* (generation time 20 minutes) in the exponential phase for 24 hours. What would the number of progeny be?

A single bacterium would yield 4,722,366,478,574,681,194,496 cells.

They would weigh 4,722,366,478 grams, which is 10,401,687 pounds or 5,200 tons.

Assuming the average weight of individuals in a crowd to be 160 pounds, this biomass would equal 65,010 people.

Assuming that each cell is 2 μm long, end to end they would stretch 9,444,732,957,149,362 meters or

9,444,732,957,149 kilometers or 5,855,734,433,432 miles.

Placed end to end, these bacteria would circle the earth 243,988,935 times or stretch to the moon and back 12,459,009 times.

Utilizing glucose as growth substrate, they would consume about 10,000 tons in the 24-hour growth period.

Assuming the biomass to be a satisfactory food source, it would feed the entire population of North Carolina for one day.

Never underestimate the power of a microbe.

Lag Phase

Transfer of a culture into a fresh medium results in a period of adjustment. During this initial **lag phase,** there is no increase in cell number in many cases, and generally there is actually a decrease. The cells, however, may grow in size, particularly if the inoculum is taken from a culture that is not actively growing. The lag phase is a period of unbalanced growth. During **unbalanced growth,** the various components of individual cells are synthesized to provide coenzymes and metabolites that are essential for **balanced** growth and division. During this phase, proteins necessary for the uptake and metabolism of available nutrients may also be synthesized. Balanced growth is a steady-state situation, a period where every component of a cell in culture increases by the same constant factor per unit time.

The length of the lag phase can vary considerably. It can be long for an inoculum obtained from cells in late stationary phase. It can be negligible or short when the inoculum is obtained from a culture in exponential growth, or it can be quite long if the inoculum is obtained from a rich complex medium, and the microorganisms are transferred to a mineral salts/glucose medium (Table 5.2). In the latter case, it would probably be necessary for the organism to synthesize many of the proteins that are involved in generating the various monomers that were readily available in the complex medium.

Exponential Growth Phase

During the exponential or logarithmic phase, each cell in the population doubles within a unit time (generation time; see Figure 6.1). If there is one organism present at the beginning of the exponential phase, the total number present after a defined period of time would be 2^n where n would be the number of generations. For example, starting with one organism with a generation of 20 minutes, after 4 hours or 12 generations there would be 2^{12}, or 4,096 progeny. This illustrates the explosive nature of exponential growth. The actual rate of expo-

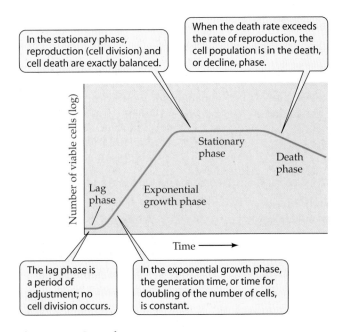

In the stationary phase, reproduction (cell division) and cell death are exactly balanced.

When the death rate exceeds the rate of reproduction, the cell population is in the death, or decline, phase.

Number of viable cells (log)

Stationary phase

Death phase

Lag phase

Exponential growth phase

Time

The lag phase is a period of adjustment; no cell division occurs.

In the exponential growth phase, the generation time, or time for doubling of the number of cells, is constant.

Figure 6.2 Growth curve
Growth curve of a typical bacterial population growing in batch culture, illustrating the four growth phases.

nential growth depends on the composition of the medium, with the most rapid growth occurring in a rich medium where microorganisms do not need to spend energy in synthesizing monomers.

Exponential growth is an important factor in the rapidity of food spoilage or the onset of an infectious disease. For example: consider that a human becomes infected with 32 disease-causing organisms (pathogens) with a generation time of 30 minutes. Assume further that the infecting organisms are in the exponential phase. During the next 30 minutes, each bacterial cell would divide once, adding 32 more organisms for a total of 64. However, should the disease causer continue growing exponentially, after 7.5 hours there would be 1,048,576 progeny, and these would double in the next 30 minutes, yielding 2,097,152 organisms after 8 hours. This rapid growth is obviously an important consideration in the early treatment of an infectious disease. The sooner therapy is initiated, the more effective it is.

Stationary Phase

Microbial populations, in general, will not maintain exponential growth for extended numbers of generations. Either the substrate or an essential nutrient will be exhausted or inhibitory products of microbial metabolism will accumulate in the medium. Other factors may also apply, including a form of population density sensing that limits the number of cells in the population. Many species placed under conditions that are considered to be ideal will grow to a low density, whereas other species will grow to heavy cell density before reproduction ceases. These various factors combined bring an end to exponential growth. The culture then enters the **stationary phase.**

During stationary phase, some of the cells die and lyse. These lytic products of cells can provide nutrients for other cells, and these divide and replace the dead ones. However, most of the population in the stationary phase survives but simply does not proliferate. The individual cells in this phase differ in certain biochemical components from cells in the exponential phase. Generally, stationary phase cells are more resistant than exponential phase cells to adverse physical conditions such as increased heat, radiation, or change in pH. They synthesize storage material such as β-hydroxybutyrate or glycogen. Members of the genera *Bacillus* and *Clostridium* that form endospores do so during this period.

The stationary phase is a period of survival and may mimic the conditions in nature when the microorganism is growing slowly or ceases to grow. There are a number of distinct genes that are expressed in microorganisms as they enter the stationary phase that are not expressed during exponential growth. Included among these are the genes for secondary metabolites such as

antibiotics (see Chapter 31). Other genes expressed during the stationary phase are known as the survival (*sur*) genes, and these are indispensable to the survival of an organism entering the stationary phase. *Escherichia coli* mutants have been obtained that lack or have defective *sur* genes, and these mutants die rapidly as they enter the stationary phase. Microbes living in nature often face conditions that are unfavorable for growth, and it appears that evolution has given them mechanisms for protecting themselves in times of stress.

Decline or Death Phase

If a culture is maintained in stationary phase beyond a certain length of time (depending on the species and the conditions), the microorganisms die. Estimation of cell mass by turbidimetry and direct microscopic counting of cells suggests that the total number of cells remains constant. Viability counts, however, suggest otherwise. Generally, the death phase is an exponential function during which a logarithmic decrease in the number of viable cells occurs with time. The actual death rate depends on the particular organism involved and conditions in the environment.

MEASURE OF GROWTH

A reasonably accurate estimation of total cell mass is essential in most studies involving the growth of microorganisms. The total biomass versus time is required to accurately measure growth rate, substrate utilization, effect of inhibitors, and other parameters. In this section, we will discuss methods that can be used to determine cell mass and cell numbers. Some methods are direct and others indirect. Methods generally employed to determine mass directly are wet or dry weight of cells. Indirect measurement of cell mass can be accomplished by chemical analysis of a specific cellular component (such as nitrogen). Total number of cells can be determined by direct counting of individual cells or by a viability count. Turbidimetry is an indirect method of obtaining a relative estimate of cell mass. Each of these methods is discussed briefly.

Total Weight of Cells

A liquid medium containing cultured cells can be centrifuged to concentrate the cells at the bottom of the centrifuge tube. The **wet weight** of packed centrifuged cells can be ascertained by placing them in a pan of known weight and determining the actual weight. This method is quite often employed to estimate a cell mass taken for enzyme isolation or lipid analysis.

The total **dry weight** is another reasonably accurate method whereby the cell mass is placed in a weighed pan and dried to constant weight. One can also filter cells onto a membrane filter with a pore size that will capture the desired organism. The filter and cells would then be dried to a constant weight. Drying is usually at 100°C to 105°C for 8 to 12 hours. Unfortunately, the drying process may render cells unsuitable for biochemical analyses.

Chemical Analysis of a Cellular Constituent

The total biomass can also be determined by analytical chemical procedures designed to quantitate the amount of a cellular constituent. The constituent most widely quantitated in laboratory research is total cell nitrogen. Because it has been established that cells are about 14% nitrogen, an analysis of the nitrogen present in a given sample will give a reasonable approximation of total biomass. In this procedure, the cells are washed free of residual medium and digested with sulfuric acid to release and convert the cellular nitrogen to ammonia, and the total NH_3^+ released is then determined by adding a reagent that combines with the ammonia to give a distinct color. The concentration can be determined colorimetrically.

Total cellular carbon can be determined by digesting a sample of cells with a strong oxidizing agent such as potassium dichromate in sulfuric acid. The amount of carbon present is proportional to the dichromate reduced.

Direct Count of Cells

The total number of microorganisms in a cellular suspension can be determined by a direct microscopic count. A sample of the suspended cells is placed in a **Petroff-Hausser counting chamber.** The Petroff-Hausser counting chamber is a thick glass slide with a grid etched onto its surface (**Figure 6.3**). Each square of the grid is of a known area (in square millimeters). After adding cells suspended in liquid to the chamber, it is placed under a microscope and the number of organisms in a large square (total volume of 0.02 mm³ in this example) determined by direct count. Several squares should be counted, and the average number present determined. This number can then be multiplied by a predetermined factor based on the volume of the liquid in a large square to give the total number of cells per milliliter in the original suspension. A concern with this technique is that both viable and dead cells can give the same appearance.

A direct counting of microorganisms in suspension can also be accomplished with an electronic instrument called a **Coulter counter.** The cells to be counted are suspended in a conducting fluid (saline solution) that flows

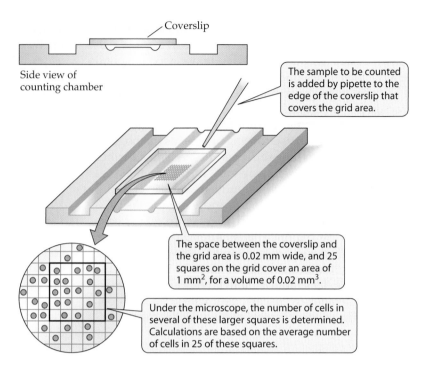

Figure 6.3 Cell counts using the Petroff-Hausser Counter
In this example, there are 16 cells in the 0.02 mm³ of the larger grid area, so the number of cells in a milliliter of original culture is 16 cells × 25 × 50 (number/mm²) × 10³ (number/mm³) = 2 × 10⁷ cells/ml.

through a minute aperture in the instrument. There is an electric current across this aperture. During passage of any nonconducting particle, such as a bacterium, the electrical resistance within the aperture is increased, causing a brief spurt of increased voltage across the aperture. The voltage pulse can be recorded electronically, and the number of pulses per unit volume is directly proportional to the number of particles flowing through the aperture.

Because the amplitude of the voltage pulse is proportional to the particle volume, the sizes of the cells can also be determined as they pass through the aperture. The aperture for bacteria is generally 10 to 30 μm in diameter, thus suspensions of cells to be counted must be diluted to minimize the probability of simultaneous passage of two or more cells.

Viability Count—Living Cells

A **viability count** is a determination of the number of living cells (able to reproduce) present in a suspension. A determination of the number of aerobic or facultatively aerobic bacteria present in a sample can be accomplished by adaptations of the spread or pour plate methods discussed in Chapter 5. Generally, viable counts are made on samples such as soil, ground meat, and milk that contain significant numbers of bacteria and therefore must be diluted. Dilutions are tenfold or hundred-fold and done in physiological saline (or seawater for marine organisms) to protect the microorganisms from the osmotic shock that might occur if diluted with tap water. To make a tenfold dilution, add 1 ml of sample to 9 ml of the diluent. If l ml of this 1/10 dilution is added to 9 ml diluent, it would produce a 1/100 dilution or 1×10^{-2}. The number of dilutions necessary would be determined by the nature of the original sample.

A dilution series is illustrated in **Figure 6.4**. The number of colonies that are countable on a plate and would be statistically valid is 30 to 300 per plate. Each of the dilutions should be plated in triplicate to increase statistical validity. The preferred volume of liquid that can be spread on the surface is 0.1 to 0.2 ml (more will flood the plate and result in indistinct colonies). Therefore, the dilution that would yield a countable plate should have 150 to 1,500 cells per ml. Spreading 0.2 ml of such a dilu-

tion would yield 30 to 300 colonies, a number that is readily countable. In the series illustrated in Figure 6.4, the number of cells in the original suspension would be:

$$\text{Spread } 0.2\text{ml} \rightarrow 124 \text{ colonies} = 620 \text{ cells/ml} \times 10^5$$
$$= 6.2 \times 10^7 \text{ cells/ml}$$

Concentrating the sample, rather than diluting it, can determine the number of organisms in samples that contain fewer than ten organisms per milliliter, such as some natural waters. Concentration can be accomplished by centrifugation of a measured volume of water. A known volume of water may also be filtered through a sterile membrane filter with a pore size that retains microorganisms. The filter may then be placed on the surface of suitable medium in an agar plate, and the number of colonies that develop can be determined after incubation (**Figure 6.5**).

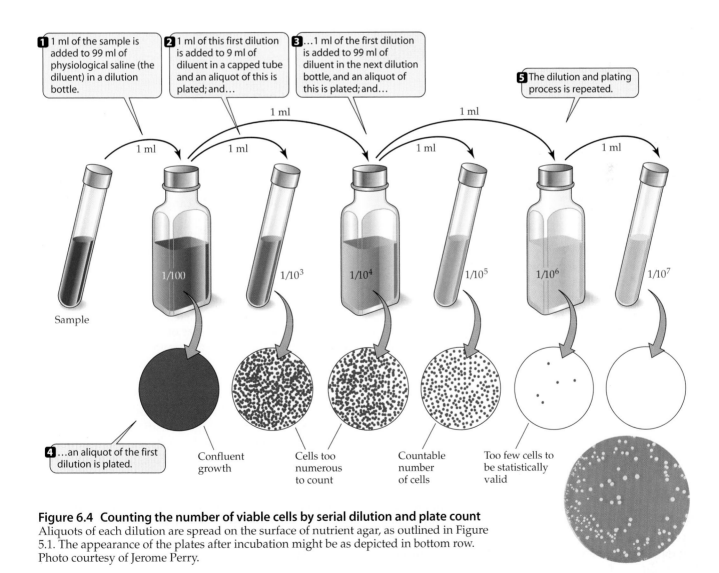

Figure 6.4 Counting the number of viable cells by serial dilution and plate count
Aliquots of each dilution are spread on the surface of nutrient agar, as outlined in Figure 5.1. The appearance of the plates after incubation might be as depicted in bottom row. Photo courtesy of Jerome Perry.

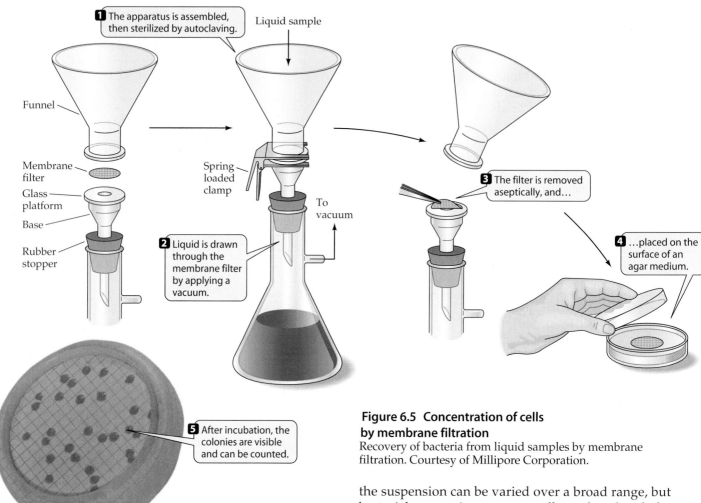

1 The apparatus is assembled, then sterilized by autoclaving.

Liquid sample

Funnel

Membrane filter

Glass platform

Base

Rubber stopper

Spring loaded clamp

To vacuum

2 Liquid is drawn through the membrane filter by applying a vacuum.

3 The filter is removed aseptically, and…

4 …placed on the surface of an agar medium.

5 After incubation, the colonies are visible and can be counted.

Figure 6.5 Concentration of cells by membrane filtration
Recovery of bacteria from liquid samples by membrane filtration. Courtesy of Millipore Corporation.

The problem with viable counts is in selecting a culture medium that will permit the growth of a reasonable number of the microorganisms present in the sample. A reasonable estimation of the viable microorganisms in a soil sample would not be possible because of the diverse nutritional requirements of a soil population. As mentioned previously, it is clear that we are unable to grow most of the microorganisms present in soil.

Turbidimetry: Scattered Light and Cell Density

The most convenient and rapid method for measuring cell mass, and consequently the one generally employed in the laboratory, is **turbidimetry,** which involves the use of a *spectrophotometer*. In a spectrophotometer, a beam of light is passed through a suspension of bacteria, and the light is scattered in proportion to the number of cells present. A bacterial suspension appears turbid because each cell scatters light. The more cells present, the more light scattered, and the greater the visible turbidity. The wavelength of light passing through

the suspension can be varied over a broad range, but bacterial suspensions are generally analyzed with the wavelength set between 400 and 600 nm.

Readings from the spectrophotometer are in absorbency (A), which is defined as the log of the ratio of incident light (I_O) to the light transmitted (I) through the suspension:

$$A = \log (I_O/I)$$

Because the shape and size of a bacterium affects the scattering of light, the absorbency is a relative figure. To estimate the actual cell mass for a particular species of microorganism, one must relate the absorbency to the actual cell dry weight or cell numbers that give a particular absorbency. This relationship is illustrated in **Figure 6.6**. Note that at high densities of cell mass, the absorbency deviates from linearity. As a consequence, one must dilute thicker suspensions to an absorbency that is in the linear range and multiply by the dilution factor to obtain a more accurate measure.

Turbidity is superior to other methods because it is quick, is readily repeatable, and does not adversely affect the cells. Turbidimetric measurements are not practical for cells that grow in clumps or for suspensions

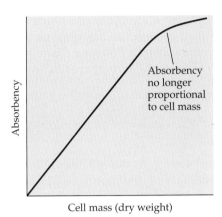

Figure 6.6 Relationship between light absorbency and cell mass
Measurements of absorbency in a spectrophotometer must be made using cell densities for which absorbency is proportional to cell mass (straight portion of the curve).

containing fewer than 10^7 cells per ml (lower limit of visible turbidity). Mycelial organisms such as actinomycetes or fungi do not form uniform suspensions. Consequently, one must resort to other methods for these organisms. Total nitrogen, total carbon, or wet/dry weight would be the methods of choice for any microbe that does not give uniform suspensions.

EFFECT OF NUTRIENT CONCENTRATION ON GROWTH RATE

Nutrient (substrate) is added to growth media to provide a source of building blocks and energy. The amount generally added is in excess of the level necessary to attain a maximum growth rate. The rate of growth is more rapid as the level of nutrient increases but levels off at a certain substrate concentration. When substrate is added below a threshold level, the growth rate is adversely affected. A decreased growth rate at low nutrient levels results from the inability of an organism to transport substrate at a rate sufficient to maintain balanced growth. The intracellular accumulation of essential intermediates is lower than at substrate saturation, and not all of the precursors are available simultaneously to provide for a maximal rate of cell synthesis and reproduction. The nutrient concentration also affects the total number of cells produced (**Figure 6.7**). As the nutrient level increases, the cell yield correspondingly increases. However, if the concentration of nutrient in a medium is increased beyond a certain threshold, the growth yield does not increase.

When the source of energy available to an organism is at a very limited level, growth will cease. Organisms require a fixed amount of energy, called **maintenance energy,** simply to stay alive and perform basic functions. Maintenance energy is used for DNA repair, accumulation of nutrient inside the cell, motility, turnover of macromolecules, maintenance of a potential across the cytoplasmic membrane, and other basic cellular needs. An increase in cell numbers can occur only when available energy exceeds these non-growth-related maintenance demands.

It should be emphasized that growth rates, total growth, and the concentration of substrate necessary to support growth all vary among species. As mentioned previously, some organisms will not grow to heavy suspension, even though they are presented with a favored nutrient and what are assumed to be optimal growth conditions.

Synchronous Growth of Cells in Culture

The sequential events that occur during the cell cycle (see Box 6.1) cannot be measured in a single bacterial cell. Techniques are just not available that are sensitive enough to follow DNA replication, protein synthesis, polymerization reactions, and the delay (D) period before division occurs in a single growing cell. The sequence of events during a cell cycle can be quantitated with reasonable accuracy only when we examine a **population** in which all of the cells are at the same stage of division. This means that an equal phasing of all growth processes exists in every cell in the population.

A culture of organisms where every cell is in the same stage of division is called a **synchronous culture**. Data obtained from experimentation on such a population may then be extrapolated to a single cell. A growth curve for a synchronous culture (**Figure 6.8**) is a periodicity in cell number versus time. This indicates that all cells in the culture divide at the same time (approximately 60 minutes).

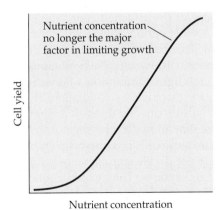

Figure 6.7 Relationship between nutrient concentration and total cell mass
Cell growth is directly proportional to nutrient availability, up to a certain point.

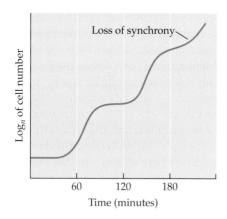

Figure 6.8 Growth curve for synchronous culture
A typical growth curve for a microorganism in synchronous culture. Cultures tend to lose synchrony after a few generations. Compare this curve with the straight line obtained in asynchronous growth (see Figure 6.1).

A number of techniques have been developed that permit one to establish a synchronous culture: filtration, light, stimulation, altered temperature, or nutrient limitation. A synchronous culture may be obtained by selecting organisms of equivalent size from a population. Young cells are generally the ones of choice in synchronous culture studies, and because they tend to be the smallest, physical means of obtaining this population of small cells have been devised.

One effective method is to filter a population of microorganisms through a cellulose nitrate filter with a pore size that retains most of the cells. Many bacterial species will adhere tightly to this type of filter. The filter can be turned over and liquid passed through to dislodge all microorganisms not tightly attached to the membrane. After removing these unattached cells, a nutrient is poured through to permit the attached cells to divide. The cells released by the nutrient wash are the small progeny of the attached cells. The effluent collected over a short period of time will contain bacteria of equivalent size and at the same stage of the cell cycle.

Synchronous cultures can also be obtained in certain species through light stimulation, alternating the temperature of growth, or nutrient limitation. The technique of passing a culture directly through a membrane filter of a pore size that allows the passage of small cells but retains mature large cells has also been employed.

Bacteria do not divide simultaneously for more than two or three generations. Thus, experimentation involving metabolic activities occurring during the cell cycle should be done in the first or second generation.

Growth Yield

The efficiency of an organism in converting a substrate to cell mass is termed the **efficiency of growth**. This is measured by comparing the grams biomass produced per mole substrate utilized. The ratio of the cell mass obtained to the substrate consumed can be defined as the **growth yield** and determined as follows:

$$y_{sub} = \frac{x_{max} - x_O}{\text{moles substrate utilized}}$$

In this case the x_{max} is the cell mass obtained, and x_O is the weight of cell in the inoculum. The growth yield can be expressed as y_{max} or y_{sub} (gram dry weight of cells/mole substrate). In organisms for which the catabolic pathway for utilization of the substrate is known, the energy yield (ATP formed) can also be calculated. Growth yield can then be expressed as y_{ATP}, which is the dry weight (in grams) of biomass per mole of ATP available.

Anaerobic bacteria, such as those that produce lactic acid, require a complex growth medium (see Chapter 5), and generally these anaerobic species utilize glucose solely as a source of energy according to the following reaction:

Glucose + 2 ADP + 2 Pi → 2 lactic acid + 2 ATP

One can readily determine the mass of cells produced per mole of ATP generated in organisms such as this, where the substrate is utilized solely as energy source.

The growth yield, y_{ATP}, is generally constant among bacteria. One mole of ATP will support the growth of about 10 grams of cell mass (dry weight). Consequently, the cell yield on a substrate utilized solely as the energy source is a direct function of the number of ATP molecules produced per mole of energy source utilized (**Table 6.2**).

This indicates that despite the diversity of microorganisms, the y_{ATP} value is constant, and the biosynthetic pathways that are fueled by ATP are similar in all species. The diversity of microbial life is therefore determined by the individual means through which microbes generate their ATP (i.e., diversity of catabolic pathways and substrates utilized).

The growth yield in a chemoheterotrophic bacterium utilizing a substrate as both energy and carbon source is more difficult to assess in terms of ATP generated. Aerobic organisms that utilize a sugar or other substrate as both carbon and energy source will vary in their efficiency.

Experimentation indicates that the efficiency of substrate conversion to biomass in chemoheterotrophs varies from 15% to 50%. The maximum conversion of substrate to biomass for a sugar is about 50% (one-half is assimilated into cells and one-half is released as CO_2 or waste product). An organism growing on methane

Table 6.2 **Growth yields of anaerobic bacteria utilizing glucose as the energy source**

	Mol ATP/Mol Glucose	y_{max} (g of cell/mol Glucose)	y_{ATP} (g of cell/mol ATP)
Lactobacillus delbrueckii[a]	2	21	10.5
Enterococcus faecalis[a]	2	20	10
Zymomonas mobilis[b]	1	9	9

[a]Homolactic fermentation, Embden–Meyerhof pathway (see Chapter 10).
[b]Alcoholic fermentation, Entner–Doudoroff pathway (see Chapter 10).

(CH_4) for which 1 mole of atmospheric oxygen is added per mole of methane assimilated can have a cell yield as high as 70%.

Continuous Culture of Bacteria

Bacteria, yeasts, or filamentous fungi are generally cultured in the laboratory by the batch method. In **batch culture**, a nutrient liquid medium is prepared in an Erlenmeyer flask, sterilized, and inoculated. The inoculated flask is then placed on a shaker (if aerobic) at the appropriate growth temperature for a suitable growth period.

Culturing bacteria by the batch method results in a population that is in a **balanced state** of growth for only a few generations. This balanced state occurs during the early to middle exponential growth phase (see Figure 6.2). Bacteria that grow in batch culture actually go through an "aging" process as they progress through the growth cycle. This means that cells in the latter stage of exponential growth can be physiologically different from cells present during the middle exponential growth phase. Differences can be detected in enzymatic composition, presence of storage material, capsules, and membrane lipid composition.

Techniques have been devised that permit maintenance of a bacterial culture in the exponential growth phase for an extended period of time. This is accomplished by establishing a **continuous culture** in an instrument termed a *chemostat* (**Figure 6.9**).

For continuous culture, sterile medium is fed into a culture vessel in a controlled manner. The culture vessel has a mixing system to ensure that the bacterial population is uniformly and maximally exposed to oxygen and substrate. An overflow line permits release of cell suspension at a rate equal to the rate of substrate addition. The dilution rate (D) in the culture vessel can be expressed as:

$$D = f/v$$

where f is the flow rate of the incoming medium and v is the volume of medium in the culture vessel.

The culture in a chemostat is controlled by the concentration of available substrate. Overall, the stability relies on this limitation of growth rate by the availability of a growth-limiting substrate such as an energy, nitrogen, or phosphate supply. As the concentration of the growth-limiting nutrient is increased, and the dilution rate remains constant, the cell concentration will increase, but growth rate will be unchanged. The growth rate increases only when the total energy available exceeds the maintenance energy. If the substrate level does not exceed the requirements for maintenance energy, the culture washes out. When the total cell concentration and the nutrient supply are balanced and in equilibrium, the system is in a **steady state**. The theo-

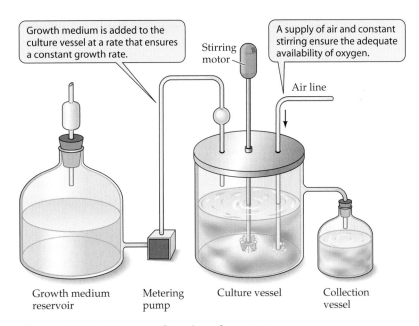

Growth medium is added to the culture vessel at a rate that ensures a constant growth rate.

Stirring motor

A supply of air and constant stirring ensure the adequate availability of oxygen.

Air line

Growth medium reservoir

Metering pump

Culture vessel

Collection vessel

Figure 6.9 Continuous culture in a chemostat
Model of a chemostat apparatus for continuous growth of a bacterial culture.

Figure 6.10 Steady-state relationship between substrate concentration and output of bacterial mass The relationship between substrate concentration, bacterial growth rate, and yield of bacterial biomass in a steady-state chemostat.

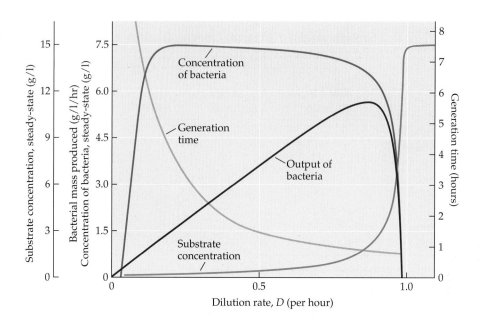

retical steady-state relationship between substrate concentration and output of bacterial mass is depicted in **Figure 6.10**. The maximum growth yield (y_{max}) is the ratio of the cell mass to the substrate consumed and is equal to about 0.5 for virtually all bacterial species.

A chemostat operates with a regulated flow rate such that the organism can grow at a rate equal to the dilution rate. Another instrument that functions solely by regulation of cell density is called a **turbidostat**. A turbidostat is equipped with a device that monitors the cell density in the growth chamber. If the cell density exceeds a preset level, the culture medium automatically enters the culture vessel, and the overflow is retained. If the cell density falls below the preset level, no culture medium flows into the vessel.

EFFECT OF ENVIRONMENTAL CONDITIONS OF GROWTH

A number of physical factors determine the type of microorganism that thrives in an environment. Among these are pH, temperature, presence of O_2, CO_2 availability, and availability of water. These also influence the growth of microbes in culture and will be discussed in this section.

Acidity and Alkalinity

The relative hydrogen ion concentration in a solution is expressed as the pH. It is based on the dissociation of water, in which there would be equal levels of H^+ and OH^-. This is defined as pH 7 (actually the negative log of the H^+ ion concentration in water). The pH scale is from 0 (the most acidic) to 14 (the most alkaline). Because pH is a log function, a difference of one pH unit represents a tenfold difference in hydrogen ion concentration.

Among the *Bacteria* and *Archaea* are species that grow at a pH of near 0. Other species can thrive at pH 11. Generally, microorganisms grow best at pH values between these extremes. Filamentous fungi grow over a pH range of 2 to 9, with a majority favoring a pH of 4 to 6. Yeasts cover the same range, but most species require a pH above 2. Most soil bacteria grow best at a pH near neutral, or a pH of 7.

The pH range for growth of any individual species is somewhat narrow. Most microorganisms, however, are quite insensitive to small variations in pH and can grow within a range of 2 to 3 pH units (which is actually a hundredfold to thousandfold difference in H^+ concentration). Growth of a microorganism in a culture medium requires that the pH be adjusted and maintained near that favored by the organism.

In the course of growth, an organism utilizing a fermentable substrate such as glucose may excrete acidic intermediates (acetic, lactic, and formic acids, among others) causing a lowering in the pH of the medium. Conversely, growth on proteins or amino acids results in ammonia production, causing the culture fluid to become alkaline. To prevent significant changes in pH, a chemical **buffer** is routinely incorporated into a culture medium. Buffers retard pH changes by removing excess H^+ or by donating H^+ to balance excess acidity or alkalinity, respectively. The most commonly used buffering system for organisms that grow near a neutral pH is a mixture of KH_2PO_4 and K_2HPO_4. This also provides the organism with phosphate for the synthesis of phosphorylated cellular constituents. Carbonates may be added to control pH when excess acid production is anticipated. For example:

$$CH_3COOH + CaCO_3 \rightarrow (CH_3COO)_2Ca \text{ (insoluble)} + CO_2$$

Microbes that grow at low pH are called **acidophiles.** The thiobacilli that are present in acid mine drainage are acidophiles and grow well at a pH of 1. *Thiobacillus thiooxidans* has a pH optimum of 2.5 and can survive at a pH of nearly 0. An interesting archaeal species is *Thermoplasma acidophilum*, which grows in coal refuse piles and can grow at 59°C and a pH of 1. This is remarkable because survivability of bacteria generally decreases at any given temperature as pH is lowered.

In contrast, **alkalophiles** are organisms that grow optimally under very alkaline conditions. The Wadi el Natrum, a lake in Egypt, for example, has a pH ranging from 9 to 11. The water contains a rich population of halophilic (salt-loving) and phototrophic bacteria. Many of the organisms in this environment have not yet been characterized.

Presence of Oxygen (O_2)

About 20% of the earth's atmosphere is molecular oxygen (O_2). As a consequence, most environmental niches are constantly exposed to this element. Swamps and subsurface areas may have little free oxygen present and may be completely anaerobic. Virtually all eukaryotic organisms are dependent on the presence of molecular oxygen for survival. Eukaryotes evolved on Earth in an aerobic environment, and most of the animal world generates energy utilizing CH_2O (biological material) in the presence of O_2 with the concomitant production of water and CO_2. Among the eukaryotes, only a few fungi and protozoa that inhabit anaerobic environments can survive without O_2.

Bacteria and *Archaea* exhibit a mixed response to oxygen. The atmosphere of the primordial earth was completely anaerobic, and for more than 1 billion years, bacterial and archaeal populations evolved without O_2 associated respiratory mechanisms for energy generation. Thus, throughout evolution, some bacterial and archaeal populations have occupied anaerobic niches and have retained lifestyles that are independent of molecular oxygen. Some of these organisms are unaffected by atmospheric levels of oxygen, whereas others are killed by a brief exposure. *Bacteria* and *Archaea* can be divided into three major groupings based on their response to molecular oxygen:

1. Aerobes: Obligate (strict) and facultative
2. Microaerophiles
3. Anaerobes: Obligate and tolerant

An **obligate aerobe** can grow only in the presence of molecular oxygen. It has no mechanism for energy generation other than a respiratory pathway that is geared to O_2 as a terminal electron acceptor. These microorganisms generally respond favorably to increased oxygen availability during growth. A **facultative aerobe** follows a respiratory pathway when oxygen is available, but it can employ alternate anaerobic energy-generating systems when oxygen is not available (see Chapters 8 and 9). In facultative organisms, growth is generally better in the presence of air.

Microaerophiles are organisms that occupy niches where the atmosphere has limited levels of oxygen. Although they must have oxygen for respiration, they grow best at reduced levels that range from 2% to 10% (dry atmospheric air is 20% oxygen).

An **obligate anaerobe** cannot grow in the presence of oxygen and may be killed by minute levels. Tolerant anaerobes are oblivious to the presence of oxygen and can survive both with or without oxygen, but they do not use molecular oxygen in energy generation.

Toxicity of Oxygen

Oxygen is a highly reactive molecule, and metabolic reactions involving molecular oxygen can generate toxic by-products. Most organisms that live in the presence of air synthesize enzymes that protect them from the toxic effects of oxygen by-products. The major toxic products are superoxide, hydrogen peroxide, hydroxyl radical, and singlet oxygen. These products, if not removed, would damage the cell by reacting with proteins, lipids, or nucleic acids. These toxic intermediates are generated as follows:

$O_2 + e^- \rightarrow O_2^-$	Superoxide
$O_2^- + e^- + 2H^+ \rightarrow H_2O_2$	Hydrogen peroxide
$H_2O_2 + e^- + H^+ \rightarrow H_2O + OH\cdot$	Hydroxyl radical
Chlorophyll + $h\nu \rightarrow$ chlorophyll*	
$Chl^* + O_2 \rightarrow {}^1O_2$	Singlet oxygen

Superoxide ($O_2 + e^- \rightarrow O_2^-$) is a free radical that has gained an extra unpaired electron. During respiration, the reduction of oxygen to water requires the addition of four electrons. This stepwise reduction occurs by single electron transfer. An intermediate product is O_2^-, and small amounts of this toxic compound may be released during respiration. Superoxide must be removed as it is formed to protect the cell. This is accomplished by the enzyme superoxide dismutase ($O_2^- + O_2^- + 2H^+ \rightarrow H_2O_2 + O_2$).

Hydrogen peroxide ($O_2^- + e^- + 2H^+ \rightarrow H_2O_2$) is the product of two electron additions to form O_2^{2-}. The anion is present as H_2O_2. It is commonly formed by a two-electron reduction of oxygen. Such reactions are mediated by flavoproteins, which are normal electron carriers. Once formed, hydrogen peroxide is destroyed by the enzymes catalase ($H_2O_2 + H_2O_2 \rightarrow 2H_2O + O_2$) or peroxidase ($H_2O_2 + NADH^+ \rightarrow 2H_2O + NAD$).

Hydroxyl radical ($H_2O_2 + e^- + H+ \rightarrow H_2O + OH\bullet$) is the most reactive of the toxic products of oxygen respiration and is a very strong oxidizing agent. It is readily formed by ionizing radiation. There is no specific defense against this toxic product.

Pigments such as chlorophyll are sensitive to light (photosensitizers). Exposure to light can convert these pigments to a triplet state (indicated by the asterisks), which can result in a secondary reaction that produces **singlet state oxygen** ($'O_2$). Singlet state oxygen is a very powerful oxidant and thus highly lethal to cells.

Aerobic organisms contain active superoxide dismutase and either catalase or peroxidase to break down hydrogen peroxide. These enzymes are an essential defense mechanism and a constituent part of all aerobic microorganisms.

Toxicity of O$_2$ and Growth of Anaerobes

Growing anaerobic *Bacteria* and *Archaea* in culture requires elaborate procedures and specialized appliances for handling because many anaerobes are destroyed by traces of oxygen. The major obligate anaerobic bacteria include the endospore-forming clostridia, the sulfate reducers, the methanogens, the bacteroids, and many microorganisms that inhabit the animal intestinal tract. Anaerobic protozoa are also present in the rumen of ruminants.

Anaerobic microorganisms can be grown in liquid media placed in glass tubes to which a chemical compound is added to remove any trace of oxygen. For example, the addition of thioglycollate is effective in removing oxygen from a culture medium. If a bacterial population is distributed in an agar culture medium prior to solidification, the growth patterns depicted in

Figure 6.11 occur after suitable incubation. This affirms the various response variations microorganisms have to O$_2$ availability.

Microorganisms can be grown in specialized chambers where all traces of oxygen can be removed from the atmosphere. An **anaerobic jar** is a convenient apparatus for growing anaerobes in the laboratory (**Figure 6.12**). After tubes or plates are placed in the jar and the jar is sealed, a catalyst is activated to remove all traces of oxygen, which is then replaced with hydrogen gas. These jars are suitable for growth of air-tolerant anaerobes that can be manipulated in the atmosphere but must have anaerobiosis for growth.

Strict anaerobes must be maintained in the total absence of O$_2$ at all times. This can be accomplished by manipulating them under a stream of nitrogen gas, an awkward procedure that is no longer in widespread use. It is much easier to carry out all manipulations of strict anaerobes in sealed chambers (**Figure 6.13**). The atmosphere in these boxes can be replaced with a nonreactive gas such as nitrogen. Oxygen scavenger chemicals are placed inside to remove all traces of O$_2$. These glove box anaerobic chambers permit one to perform all operations, including transfer and streaking, inside the box in the complete absence of molecular oxygen.

Effect of Temperature

Microorganisms are of widespread occurrence in environments in which the ambient temperature can range from less than 0°C to over 100°C. For example, saltwater beneath polar ice can reach −10°C, and volcanic areas under the sea can exceed 110°C. Microorganisms can thrive in both these environments, and those that live there have little competition from other life forms. It is clear that species diversity tends to decrease as the harshness of the environment increases.

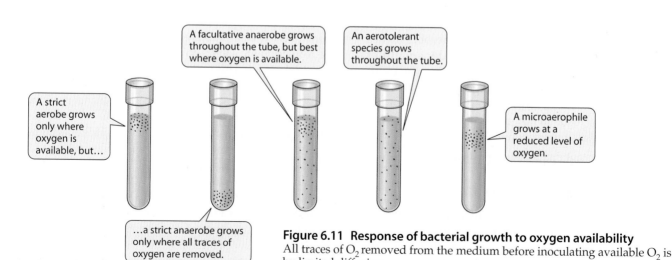

A facultative anaerobe grows throughout the tube, but best where oxygen is available.

An aerotolerant species grows throughout the tube.

A strict aerobe grows only where oxygen is available, but...

...a strict anaerobe grows only where all traces of oxygen are removed.

A microaerophile grows at a reduced level of oxygen.

Figure 6.11 Response of bacterial growth to oxygen availability
All traces of O$_2$ removed from the medium before inoculating available O$_2$ is by limited diffusion.

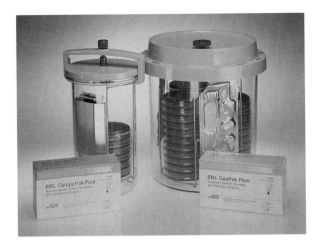

Figure 6.12 Anaerobic jars
Inoculated plates or tubes are placed in the jar along with a catalyst. After sealing the jar, the catalyst is activated to remove all atmospheric oxygen. Because transfer and manipulation of the culture occur outside the jar, this system cannot be used for strict anaerobes. Courtesy of Becton Dickinson Microbiology Systems.

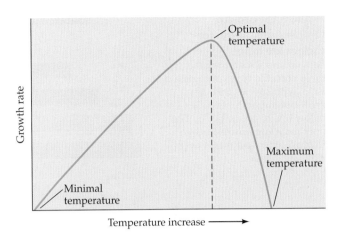

Figure 6.14 Response of bacterial growth to temperature
The cardinal temperatures for the growth of a bacterium.

The effect of temperature on the growth rate of a microorganism is depicted in **Figure 6.14**. As depicted in the figure, a given species has a minimum temperature, a maximum temperature, and an optimal temperature for growth. These three temperatures are referred to as **cardinal** temperatures. As the temperature rises above the minimum, the metabolic reactions in the cell accelerate, and growth becomes more rapid. This increase leads to a physiological state at which all cellular reactions are operating at their optimum level. A little above this temperature, the cytoplasmic membrane may lose function, proteins or nucleic acids may be denatured, and the cell loses function. The cellular reaction(s) that is least heat stable is primarily responsible for the demise of the microorganism as the temperature rises.

Temperature Range

Microbes are divided into four groups based on the range of temperature at which they can grow (**Figure 6.15**). These groups include the following: the *psychrophiles*, which grow at temperatures below 20°C; the *mesophiles*, which generally grow between 20°C and 44°C; the moderate *thermophiles*, which grow between 45°C and 70°C; and the *hyperthermophiles*, which are microorganisms that require growth temperatures above 70°C to over 110°C. The optimal growth temperature for microorganisms that fall into each category are also presented. **Table 6.3** presents a listing of organisms in each group and their ideal growth temperature range.

Psychrophiles A psychrophile (from *psychro* meaning "cool") is defined as a microorganism that grows at 0°C and has an optimal growth temperature at less than 15°C. An obligate psychrophile cannot grow at a temperature above 20°C. Truly psychrophilic types are present in seawater, where the average temperature is 5°C, and these microorganisms are particularly abundant in polar ice and water. The only requirement for survival is that liquid water be

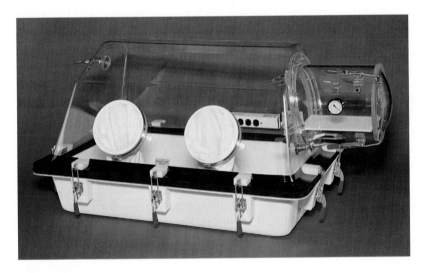

Figure 6.13 Anaerobic glove box
The atmosphere in a chamber of this type can be replaced by inert gases and maintained under anaerobiosis for the manipulation of cultures. This device may also be used in the purification of enzymes or other molecules that are sensitive to molecular oxygen. Courtesy of Plas Labs.

Figure 6.15 Growth temperature ranges for various life forms
Bacteria and *Archaea* are the most versatile types of organisms, with species that can grow at temperatures of below 0°C to about 115°C. The only microorganisms known to grow at temperatures over 92°C are *Archaea*.

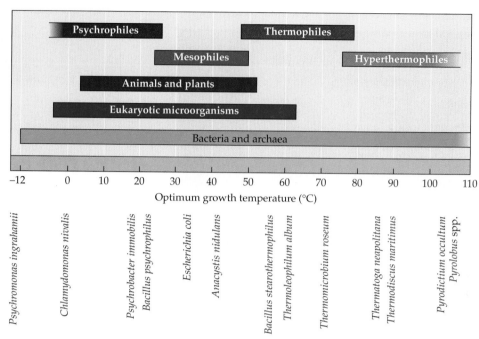

Table 6.3	Temperature ranges for growth of *Bacteria* and *Archaea*
Species	**Range (°C)**
Psychrophiles	
Cytophaga psychrophila	4–20
Bacillus insolitus	<0–25
Aquaspirillum psychrophilum	2–26
Mesophiles	
Escherichia coli	10–40
Lactobacillus lactis	18–42
Bacillus subtilis	22–40
Pseudomonas fluorescens	4–40
Thermophiles	
Bacillus thermoleovorans	42–75
Thermoleophilum album	45–70
Thermus aquaticus	40–79
Chloroflexus aurantiacus	45–70
Hyperthermophiles (*Archaea*)	
Hyperthermus butylicus	85–108
Methanothermus fervidus	65–97
Pyrodictium occultum	80–110
Thermococcus celer	70–95

available, and this does occur in microenvironments within sea ice.

The term *psychrophile* is somewhat less precise than the other terms used to describe the temperature ranges for growth. Many microorganisms that grow optimally at 25°C to 30°C can also grow very slowly at 0°C. The ability to grow at psychrophilic temperatures is not considered of major significance in classification.

Facultative psychrophiles are sometimes present in the soil and water from temperate climates. These microorganisms grow very slowly at 0°C but grow well at 22°C. They are often present in or on food and can cause spoilage in food that is refrigerated for extended periods of time. Cold-tolerant fungi also cause spoilage of refrigerated foods. Food is best preserved for extended periods of time by drying or freezing, because these processes eliminate liquid.

Mesophiles The mesophiles comprise the vast group of microorganisms that are ubiquitous in soil and water and inhabit nearly all animals and plants. They generally grow between 20°C and 45°C, although some will grow at temperatures as low as 10°C. The mesophiles are mostly responsible for the cycles of matter, disease, fermentation, deterioration of material, and other activities that are important to humans. Consequently, they have been the subject of the most research. Human pathogens such as *Streptococcus pneumoniae* grow optimally at 37°C, the temperature of the human body. Soil organisms generally grow best at 26°C to 30°C.

Thermophiles The thermophiles are organisms that grow optimally above 45°C and below 70°C. These temperatures commonly occur in hot springs, volcanic areas, and compost heaps. Soil exposed to direct sunlight can reach temperatures that are appropriate for moderate thermophiles. As a consequence, thermophiles are present in almost every environment, including the Antarctic. Thermophilic bacteria are present in significant quantities in heated industrial water, hot water heaters, water from cooling towers at nuclear energy plants, and other artificially created thermal environments.

Hyperthermophiles The hyperthermophiles are microorganisms that grow at temperatures above 70°C, and many grow at temperatures near or above the boiling point of water. The hyperthermophiles might appropriately be divided into two groups based on the cardinal growth temperature: those that grow from 70°C to 90°C in one group and those that grow from 80°C to 110°C in the other.

Liquid water is a necessity for all life forms, so these microorganisms that grow at temperatures near 100°C or higher are found in habitats that are under sufficient pressure to maintain water as a liquid. These conditions occur at volcanic areas beneath the ocean surface.

Several genera of the *Archaea* have been obtained in culture that grow at temperatures around 100°C. Not all *Archaea* are hyperthermophiles, but the microorganisms described to date that grow at temperatures in excess of 90°C are *Archaea*.

An organism has been reported that grows under pressure at considerably higher temperatures (>200°C), and reference to this still appears in the popular press. However, there is no creditable evidence for this. Present evidence suggests that the maximum temperature at which molecular constituents (ATP, protein) in bacteria can function is in the range of 120°C to 130°C.

Temperature Limitations

Proteins are one of the major cellular constituents that may limit the ability of a microorganism to grow at high temperatures. Thermophiles must have heat-stable proteins, and many of the proteins in mesophiles are heat labile (unstable at high temperatures). Generally, the cytoplasmic proteins in a mesophile begin to precipitate if heated at 60°C for 8 to 10 minutes. The equivalent protein in a thermophile would withstand this temperature.

Growth temperature can also influence the fatty acid composition of the membranes in bacterial cells. This is affirmed by comparison of the cytoplasmic membrane fatty acids in a microorganism grown at a minimum temperature with those present after growth at a maximum temperature. At the minimum temperature, the relative proportion of unsaturated fatty acids in the cellular lipids is high. Growth at the maximum temperature leads to a greater relative percentage of saturated fatty acids. The degree of saturation of fatty acids in the cytoplasmic membrane determines the fluidity of the membrane at a given temperature. Because the proper functioning of a cytoplasmic membrane depends on fluidity, it is not surprising that membrane composition reflects the growth temperature.

Carbon Dioxide Availability

Virtually all microorganisms require carbon dioxide (CO_2) for growth. However, the level necessary for optimal growth varies with the microbial species. The atmospheric level of carbon dioxide (0.03% by volume) satisfies the minimal requirement for virtually all heterotrophic and autotrophic microorganisms. Optimal growth of autotrophic microorganisms, however, is attained in culture when they are provided with carbonates or with air enriched in carbon dioxide. Heterotrophs cannot utilize carbon dioxide as the sole source of carbon, but they do require it for growth. These microorganisms assimilate carbon dioxide into a number of metabolic intermediates involved in synthesis and energy generation, and complete removal often leads to cessation of growth. Some animal and human pathogens, such as *Neisseria gonorrhoeae*, the agent of the venereal disease gonorrhea, will not grow in laboratory culture unless they are incubated in the presence of an atmosphere containing 5% to 10% carbon dioxide.

Water Availability

All microorganisms, whether bacterial or archaeal, require a source of water for survival. Availability of water is a major factor in determining whether a natural environment is colonized. It also markedly influences the type of microorganisms that can inhabit an environmental niche. The term **water availability** differs from **presence of water** because even if it is present, solids or surfaces in the environment may absorb molecules of water, rendering it unavailable to support life. For example, solutes such as sugar or salts that dissolve in water have a high affinity for water. If the concentration of a solute is high enough, it can withdraw water and make it unavailable to a cell. For this reason, microbes generally do not grow in solutions such as saturated salt or undiluted honey.

Water availability is expressed as **water activity** (a_w) and is defined as the vapor pressure of air over a substance or a solution divided by the vapor pressure over pure water. If a solute such as sodium chloride absorbs water, the vapor pressure in the atmosphere above that solution becomes lower. *Bacteria* and *Archaea* differ in their tolerance for decreased available water (**Table 6.4**).

If water is placed on one side of a semipermeable membrane and concentrated saltwater is placed on the other side, the water diffuses across the membrane into the salt solution. This process is called **osmosis**. A bacterium placed in a culture medium is faced with this physical problem. If the solute concentration is higher

Table 6.4	Tolerance of selected *Bacteria* and *Archaea* for decreased water activity a_w	
Type	**Organisms**	a_w
Nonhalophiles	*Aquaspirillum* and *Caulobacter*	1.00
Marine forms	Pseudomonads and *Alteromonas*	0.98
Moderate halophiles	*Vibrio* species and gram-positive cocci	0.91
Extreme halophiles	*Halobacterium* and *Halococcus*	0.75

in the medium than inside the cell, water leaves the cell, and the cytoplasmic membrane collapses inward. The organism may die of dehydration, a process called **plasmolysis**. Bacteria can prevent this from occurring by increasing the solute concentration inside the cell to one greater than in the suspending medium.

Microbes use a variety of solutes to establish this higher cytoplasmic solute concentration with normal cellular activities. Because they do not interfere with normal cellular activities, they are called **compatible solutes**. Compatible solutes may be ions pumped into the cell from the environment, or they may be organic solutes synthesized by the organism. Many organisms use potassium ions as the solute, including those as diverse as the halophilic organisms that live in highly saline environments and *Escherichia coli*, which lives in the gastrointestinal tract of humans. The cytoplasm of halophiles ("salt lovers") is virtually saturated with potassium ions. The gram-positive cocci synthesize the amino acid proline as solute. Yeasts that live in high salt or sugar concentrations synthesize polyalcohols such as sorbitol to serve as the compatible solute.

SUMMARY

▶ An increase in the number of cells in a culture medium is termed **population growth**. An increase in the size of a bacterial cell is **cell growth.**

▶ A **viable microorganism** is one that can reproduce in the laboratory. This definition is flawed, because conditions for the growth of most microorganisms present in nature have not yet been defined. Our inability to grow them results from our ignorance, not from their lack of viability.

▶ An increase in the number of microorganisms per unit time is the **growth rate**. The **generation time** is the time required for a population to double in number.

▶ The addition of a small number of microorganisms to a suitable growth medium and recording the number of microorganisms present versus time will produce a **growth curve**. The component parts of a growth curve

are: **lag phase, exponential growth phase, stationary phase,** and **decline** or **death phase.**

▶ During **balanced growth,** every component of a cell culture increases by the same constant factor per unit time.

▶ During the **exponential growth phase,** the number of organisms present increases logarithmically, that is $1 \rightarrow 2 \rightarrow 4 \rightarrow 8 \rightarrow 16 \rightarrow 32$, etc.

▶ The **stationary phase** is a period where a culture does not increase in numbers. It is a period of survival.

▶ The **wet weight** of packed cells is an effective method for determining biomass when survival of the microorganism is important for subsequent analysis, such as enzyme isolation or metabolic studies. **Dry weight** of cell mass is more accurate but tends to destroy the microorganism. Total **nitrogen** is an effective indicator of biomass, and it can be accomplished on a small amount of material.

▶ A **direct count** of bacterial or archaeal cells in a suspension can be performed microscopically with a **calibrated slide** or with a **Coulter counter.**

▶ When microorganisms grow in a clear liquid medium, they create a visible **turbidity**. This turbidity can be measured in a **spectophotometer** and is a reliable method for determining biomass. Turbidity is evident when there are at least 10^7 microorganisms present per milliliter.

▶ **Maintenance energy** is that amount necessary to stay alive.

▶ The events in the **cell cycle** can be followed in a population with all the cells in the same physiological state. A culture in this state is called a **synchronous culture.** A number of techniques are available for establishing synchronous cultures.

▶ **Growth yield** is a determination of biomass produced per mole substrate utilized.

▶ A microorganism growing on a liquid nutrient under favorable conditions for a set period of time is called a **batch culture**. A culture may be maintained in the exponential growth phase for extended periods of time in **continuous culture.** Continuous culture may be accomplished in a **chemostat apparatus** or a **turbidostat.**

▶ Selected species of *Bacteria* and *Archaea* can grow from a **pH** of near 0 to 11. Most soil bacteria favor a pH near neutrality (7); fungi generally grow better at an acidic pH (2 to 6).

▶ Virtually all eukaryotic organisms are **aerobic** and have a requirement for molecular oxygen in their energy generation. *Bacteria* and *Archaea* initially evolved in an **anaerobic** environment, and many species can thrive in the absence of O_2.

▶ Most microorganisms that live in the presence of air must have mechanisms for removal of toxic products of O_2 utilization, **superoxide, hydrogen peroxide,** and the **hydroxyl radical.**

▶ The most effective way to manipulate strict anaerobes in the laboratory is in a **glove box,** where all molecular O_2 can be excluded.

▶ The **temperature ranges** for growth of various microbial types: 0°C to 20°C, **psychrophiles;** 20°C to 44°C, **mesophiles;** 45°C to 70°C, **thermophiles;** and 70°C to over 110°C, **hyperthermophiles.**

▶ **Carbon dioxide** availability is necessary for the growth of virtually all microorganisms, and growth of autotrophs is stimulated by the presence of higher than atmospheric levels of CO_2.

▶ **Water activity** (a_w) or water availability is an important factor in the growth of microorganisms. Life cannot exist without available water.

REVIEW QUESTIONS

1. What is the difference between cell growth and population growth? How do these concepts fit into the cell cycle?

2. A growth curve is divided into phases designated lag, exponential growth, stationary, and decline or death. What is happening to individual cells during each of these phases? How does this relate to microorganisms in nature?

3. Total dry weight and viable count of cells are two methods employed in measuring cell mass. What are some of the pros and cons of each method? How can turbidity be employed in measuring microbial mass?

4. Why are some studies done with cells growing in synchronous culture? How would one obtain a synchronous culture?

5. Describe how a chemostat works. A turbidostat.

6. Define aerobe, anaerobe, and microaerophile. What do *strict* and *facultative* mean in reference to aerobes and anaerobes?

7. How have aerotolerant organisms adapted to tolerate the presence of toxic oxygen products?

8. What are some of the reasons why microorganisms have a fairly limited temperature range for growth?

9. Water availability is an important factor in microbial growth. How can an organism grow in high concentrations of sugar or salt?

SUGGESTED READING

American Waterworks Association. 1998. *Standard Methods for the Examination of Water and Waste Water*. 20th ed. Washington, DC: American Waterworks Association.

Cooper, S. 1991. *Bacterial Growth and Division: Biochemistry and Regulation of Prokaryotic and Eukaryotic Division Cycles*. New York: Academic Press.

Dworkin, M., S. Falkow, E. Rosenberg, K-H. Schliefer and E. Stackebrandt, eds. 2000. *The Prokaryotes: An Evolving Electronic Resource for the Microbiological Community*. 3rd ed. New York: Springer-Verlag. Release 3.7, latest update December 2001.

Gerhardt, P., ed. 1993. *Methods for General and Molecular Bacteriology*. 2nd ed. Washington, DC: American Society for Microbiology Press.

Neidhardt, F. C., J. L. Ingraham and M. Schaechter. 1990. *Physiology of the Bacterial Cell*. Sunderland, MA: Sinauer Associates, Inc.

Norris, J. R. and D. W. Ribbons. 1970. *Methods in Microbiology*. Vols. 2 and 3A. New York: Academic Press.

Chapter 7

Control of Microbial Growth

In order to use chemotherapy successfully, we must search for substances which have an affinity to the cells of the parasites and a power of killing them greater than the damage such substances cause to the organism itself, so that the destruction of the parasites will be possible without seriously hurting the organism.

— PAUL EHRLICH,
FATHER OF CHEMOTHERAPY, 1854–1915

The title of this chapter, "Control of Microbial Growth," is essentially a contradiction in terms. We live in a world where about one-half of the total biomass is microbial, and our bodies contain more microbial cells than human cells. In reality, we coexist with microorganisms. We attempt to keep the harmful ones at bay and foster the growth of those that are helpful. But regardless of our efforts, they are always around, and some can pose a serious threat. Their growth is never really under total "control."

Microbes do play an indispensable role in the cycles of matter, and many of these contributions are presented in Chapter 12. This positive role is a consequence of broad metabolic diversity among microorganisms. Microorganisms have evolved as recyclers and, as a consequence, can grow on and destroy our food, our clothing, our buildings, and even our bodies.

They have a virtually unlimited ability to metabolize organic compounds, thus returning carbon dioxide to the atmosphere. Some microorganisms are obligately parasitic, and these survive by feeding on plants, animals, or other microbes. Others are opportunists that, under appropriate conditions, cause disease.

To counteract a constant threat of destruction by microorganisms, humans have spent much money and effort in devising preservation methods and have met with moderate success. Human populations have historically suffered from the inability to understand or gain more than a limited amount of control over microorganisms. In early times before microorganisms were known and understood, whatever anyone did to thwart the destructive microorganism was the result of an empirical observation that something worked. Thus, Nicholas Appert discovered in 1810 that one could preserve food by heating it in a closed container (canning), American Indians dried meat, Norsemen salted fish, and many civilizations made cheese.

During the last 130 years, humans have come to understand the microbe, and our strategies have changed. We have developed procedures whereby, to a limited extent, microbes can be controlled. This chapter discusses many of the procedures and processes that are employed to control the microbe in areas where this is desirable. We sterilize with heat and chemicals, we filter air or liquids, and antibacterial chemicals are now available to cure disease. We have developed immunization methods and public health measures to survive in a world of the ubiquitous microbe.

HISTORICAL PERSPECTIVE

The early civilizations used many methods to protect food from microbial destruction. These methods included the salting of meat, fish, and vegetables. As discussed in the previous chapter, salting removes available water and prevents bacterial growth. Certain bacterial species either alone or combined with salt proved effective in preserving vegetables, such as with cabbage to produce sauerkraut. Long ago, people found that fermented milk products such as cheese, butter, and yogurt were stable for some length of time even without refrigeration. Acidification with vinegar (pickling) was also effective in preserving selected food products.

The development of effective control measures for microorganisms in general parallels the evolution of microbiology as a science (see Chapter 2). Several important early workers in medicine discovered microbial control techniques that significantly decreased the number of deaths during medical procedures. It is generally acknowledged that **Ignaz Semmelweis** (1816–1865) pioneered the use of disinfectants. He insisted that all individuals working in his hospital in Vienna wash their hands in chlorinated lime before working with patients. This brought about a marked decrease in disease and death in the obstetrics ward. **Joseph Lister** (1827–1912) was able to conduct aseptic surgery by sterilizing instruments with carbolic acid (phenol) and spraying phenol around the room during surgery. As discussed in Chapter 2, Robert Koch demonstrated that mercuric chloride would kill all bacteria tested, including the highly resistant endospore.

John Tyndall (see Chapter 2) developed an early and effective sterilization procedure, termed "tyndallization." In this procedure, a broth would be steamed for a few minutes on three or four successive occasions separated by 12- to 18-hour intervals of incubation at a favorable temperature. This would permit dormant spores to become vegetative cells and be killed by boiling.

The invention of the autoclave in Pasteur's laboratory revolutionized laboratory work, as it permitted one to sterilize media and to work under aseptic conditions.

HEAT STERILIZATION AND PASTEURIZATION

Sterilization is the destruction of all viable forms of life, including endospores. A properly packaged sterile material should be devoid of all life until opened and exposed to the outside environment. Sterilization can be attained by various procedures including fire, moist heat, dry heat, radiation, filtering, and toxic chemicals. Any procedure that disrupts an essential component of the cell—protein, nucleic acid, or cytoplasmic membrane—may lead to a loss of cell function. The more extensive the disruption, the greater the chance that the damaged microorganism will die. When you watch an egg fry, you are witnessing the effectiveness of heat in denaturing (breaking down) egg protein. Some proteins are more stable than others at high temperatures, but this is only a matter of degree. Given sufficient heat, the structural and intrinsic function of any protein will be lost. Heat, in some form, is the agent most often employed for routine sterilization.

Studies in a microbiology laboratory are dependent on the ability to sterilize media, inoculating loops, glassware, and all material involved in culturing microorganisms. Most studies are performed with pure cultures (only one strain present), and this can be accomplished only when all manipulations are done under sterile conditions. Sterility is also an important requirement for surgical instruments, bandages, and in most medical procedures.

Several factors can affect any sterilization procedure. First, the presence of organic matter can protect *Bacteria* or *Archaea* from the destructive effects of heat, chemicals, or radiation. Second, a material to be sterilized usually harbors a mixed population of microorganisms. Consequently, the regimen selected for sterilization must be designed to kill *all* of the most resistant organisms present and under the conditions that exist.

The pH of the suspending medium is also a significant factor. In an acidic medium, heat is considerably more effective in killing microorganisms than at neutrality. This has been a major factor in home canning, where acidic foods, such as fruits and tomatoes, rarely spoil. Foods that have a more neutral pH, such as corn, peas, and beans, spoil more often because on harvesting they harbor endospore-forming organisms that may survive the canning process. On rare occasions, the anaerobic spore-forming organism *Clostridium botulinum* may survive in improperly canned foods. This organism then may germinate and grow, producing botulinum toxin, which causes botulism. Ingestion of minute amounts of botulinum toxin can lead to death. This was a danger in the days of extensive home canning, when inefficient or improperly used pressure cookers were sometimes employed.

Technique of Heat Sterilization

Microbial death by heating occurs at an exponential rate (**Figure 7.1**); that, is a fraction of the cells present in a heated solution will be killed per unit of time. The effect of heat on the survival of a bacterium is illustrated in the figure and affirms that a higher temperature (95°C) kills more rapidly than the lower temperature (75°C). A survival curve can be obtained by suspending bacterial cells in water and heating to a selected temperature. At intervals, an aliquot can be removed and the number of viable cells present determined by a plate count as described in Chapter 6.

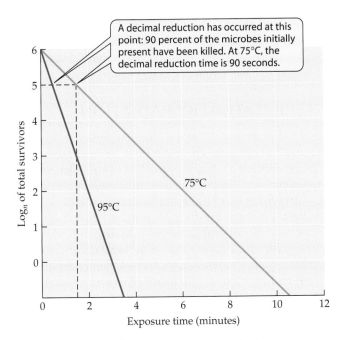

A decimal reduction has occurred at this point: 90 percent of the microbes initially present have been killed. At 75°C, the decimal reduction time is 90 seconds.

Figure 7.1 Effect of temperature on survival of bacterial cells
An exponential plot of number of surviving cells versus time exposed to heat, at two different temperatures, for the vegetative cells of a mesophilic bacterium.

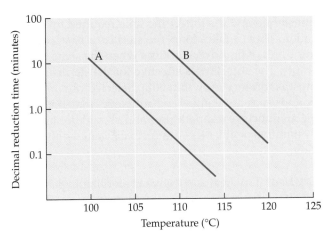

Figure 7.2 Decimal reduction time and heat sensitivity
The decimal reduction times for two organism: one relatively heat sensitive (A), the other more heat resistant (B). The rate of killing increases at higher temperatures.

Three general techniques are available for killing microorganisms by heat: boiling, steam under pressure, and dry heat. Boiling for 15 minutes will kill virtually all vegetative bacterial and archaeal cells, viruses, and fungi, but it does not generally kill bacterial endospores. The length of time required to reduce a population tenfold (for example, from 10^6 to 10^5) is the **decimal reduction time,** designated D. Endospores of various bacterial species have differing resistances to boiling (100°C), as indicated by the D values (**Table 7.1**).

A comparison of the decimal reduction times can distinguish between a heat-sensitive microorganism and one that is relatively resistant (**Figure 7.2**). The time required to sterilize a suspension of endospores varies over a wide range, with *Bacillus anthracis* requiring less than 5 minutes. In contrast, a suspension of *Clostridium botulinum* endospores can be heated for 5 hours or more at 100°C, and some of the endospores will survive. If placed in an appropriate medium, they can germinate and grow. It is clear that boiling water has its limitations for sterilization.

Moist Heat—Steam under Pressure The most effective sterilization system is steam under pressure. This method is widely used in the research laboratory and throughout the food

and pharmaceutical industries. The instrument available for accomplishing this is the **autoclave.** An autoclave is a closable metal vessel that can be filled with steam at greater than atmospheric pressure. The autoclave is designed so that the air present inside when the door is closed and sealed is driven out by steam. A valve closes automatically when the air passing outward reaches 100°C, an indication that steam now fills the entire inner space. Laboratory autoclaves operate at 15-psi (pounds per square inch) pressure, and at this pressure, the temperature rises to 121.6°C. The autoclave must be constructed to withstand the internal pressure produced when water boils. At this temperature, sterilization of routine material can be achieved within 10 to 15 minutes.

Large batches of material must be autoclaved for a longer time because heat transfer to the interior can be slow. The denser the material, the slower the heat is

Table 7.1	Heat resistance of bacterial endospores obtained from various genera and species in the genus *Bacillus*. Decimal reduction time at 100°C

Organism	D Value[a] (minutes)
Bacillus cereus	0.8
Bacillus megaterium	2.1
Bacillus cereus var. *mycoides*	10.0
Bacillus licheniformis	24.1
Bacillus coagulans	270.0
Bacillus stearothermophilus	459.0

[a]Time required to reduce the population tenfold.

transferred and the longer the time required for sterilization. Thus, it takes longer to sterilize a flask containing 1 liter of liquid than it does when the same flask is empty. The autoclave is effective for the sterilization of most laboratory growth media. Selected organic compounds such as vitamins and antibiotics may not be stable at autoclave temperatures and should be sterilized by filtration or other less-destructive means, as discussed later.

Dry Heat **Dry heat** is effective in sterilizing glassware, pipettes, and other heat-stable solid materials. Pipettes can be placed in metal canisters, glassware wrapped in paper, and flasks and other vessels capped with aluminum foil. This protects the objects from contamination when they are handled subsequent to sterilization. An oven that can be heated to 160°C to 180°C is used for dry heat sterilization, and the materials are retained at this temperature for 2 to 4 hours. Dry heat has been preferred for sterilizing pipettes, syringes, and other heat-stable objects, as autoclaving left them moist. However, most laboratory autoclaves now have a vacuum cycle that will remove moisture from such objects after sterilization.

Pasteurization

Pasteurization is a process that employs low heat and is often applied to milk products and other heat-sensitive foods. Pasteurization reduces the microbial population but does not sterilize the product. The method was devised by Louis Pasteur to control wine spoilage. Pasteurization was originally accomplished by heating the material to 63°C to 66°C and holding it at this temperature for 30 minutes. This method will kill 79% to 99% of the microorganisms present in milk. A short-term "flash method" is now generally employed, whereby milk is passed through coils where it is heated quickly to a temperature of 71.6°C for 15 seconds.

Pasteurization of milk was adopted as a common practice in the dairy industry because raw milk often contained *Mycobacterium tuberculosis*, the causative agent of tuberculosis (TB), and/or *Brucella abortus*, which causes brucellosis. These organisms have now been mostly eliminated from dairy herds in the United States through mandatory testing. However, pasteurization is still employed as a pre-

caution and to extend the shelf life of certain food products, including beer and milk. Due to the widespread consumption of dairy products, the potential spread of disease by this consumable is a constant threat.

STERILIZATION BY RADIATION

Selected wavelengths of the electromagnetic spectrum (**Figure 7.3**) can cause the death of living organisms, thus making **radiation** a practical method of sterilization. The two types of radiation that are effective microbe killers are **ultraviolet (UV) light** and **ionizing radiation.** Light in the UV range can disrupt cellular DNA and some other cell constituents, whereas **ionizing radiation** can cause harm directly and indirectly to all constituents of a cell. Gamma (γ) rays, X-rays, alpha (α) and beta (β) particles, and neutrons are all forms of ionizing radiation.

Ionizing Radiation

Gamma rays, as with other forms of ionizing radiation, kill cells by causing the formation of highly reactive free radicals (see Chapter 6). These free radicals, such as the hydroxyl radical (OH•), can react with and inactivate any of the various macromolecules present in a cell. It is probable that all macromolecules in a cell are equally susceptible to damage from ionizing radiation. Because there are multiple copies of most cellular macromolecules, damage to a limited number of the copies of a protein will not cause cell death. Generally, however, the cellular DNA will contain a single copy of a specific gene, and disruption of a gene can cause cell death. Ionizing radiation is destructive to genetic material

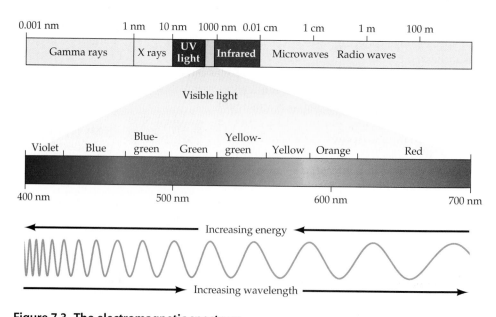

Figure 7.3 The electromagnetic spectrum
Humans can detect "visible" wavelengths, which range from about 400 nm (violet) to 750 nm (red).

because it causes single- and double-strand breaks in the DNA double helix.

The sensitivity of different organisms to ionizing radiation varies considerably. Endospores of *Clostridium* or *Bacillus* species are highly resistant, as are cells of *Deinococcus radiodurans*. *D. radiodurans* is exceedingly resistant to UV or gamma radiation and also to most mutagens. Materials that have been exposed to ionizing radiation often bear this organism as the sole survivor. The ability to repair DNA damage efficiently and rapidly is a major survival factor in radiation-resistant microorganisms.

Gamma rays are an effective sterilizing agent because they can penetrate deeply into objects such as containers. Consequently, sterilization can be accomplished after a product is packaged. Major sources of usable radiation for sterilization include cobalt-60 (gamma rays), X rays, and the neutron piles that are available at nuclear plant sites. The radiation source most often used commercially is cobalt-60. Generally, items that are heat sensitive and nonfilterable are sterilized by radiation. Among the items sterilized in this way are plastics, antibiotics, serums, various medicinals, and in some cases, food.

Ultraviolet Light

Wavelengths of electromagnetic radiation in the ultraviolet range (UV) between 220 and 320 nm are particularly important biologically because cellular constituents can absorb these wavelengths. More intense, shorter wavelengths of UV (less than 220 nm) are emitted by the Sun but are absorbed by the ozone layer in the Earth's upper atmosphere. This shorter wavelength UV radiation is lethal to microorganisms and harmful to many other organisms as well. This has caused the present concern over the damage to the ozone layer over the Arctic and Antarctic areas.

Germicidal lamps and lights can be constructed to emit the short, damaging UV wavelengths. UV lamps with this potential may be placed in sterile rooms and safety cabinets to sterilize the air, but cannot be operated when people are present. These wavelengths burn skin and may damage the eyes. UV radiation does not penetrate glass, water, or nongaseous substances to a significant extent. This limits the effective use of UV radiation to sterilization of surface areas and air.

Cellular components that strongly absorb UV light are the purines and pyrimidines. Both are present in DNA and RNA and absorb UV maximally at 260 nm. A UV lamp will produce levels of radiation that can effectively penetrate bacterial, archaeal, or other viable cells and cause alterations in cellular nucleic acids. The production of *thymine dimers* is probably the most important result of UV damage to DNA. A dimer is formed by the linkage of two adjacent thymine molecules on a single strand of DNA. Such linkages prevent the replication of DNA, and this can be lethal to a cell. Thymine-cytosine or cytosine-cytosine dimerization occurs less frequently and can disrupt DNA replication.

Microbes have DNA **repair systems** that can excise the dimers and replace them with unaltered thymine molecules. UV light is lethal if a sufficient number of dimers are created such that all of them cannot be repaired. In some cases, the enzyme repair system will make an error, which leads to mutations or death.

STERILIZATION BY FILTRATION

The ability of **filters** to retain bacterial and archaeal cells and allow viruses and other small particles to pass through has been recognized for over 100 years. In 1892, Dmitri Ivanowsky, a Russian scientist, demonstrated that tobacco mosaic virus would pass through a filter that would retain the smallest bacteria. This was confirmed and expanded by Martinus Beijerinck a few years later (see Chapter 2).

Filters have proven useful for the selective sterilization of liquids and gases. Heat-sensitive materials, such as enzymes, vitamins, and sugars, can be filtered to remove contaminating bacteria. Beer and wine are now filtered in bulk to remove bacteria that would oxidize the ethanol to acetic acid and render these drinks unacceptable (see Chapter 31). Virologists routinely filter blood serum and other heat-sensitive culture media to prevent unwanted bacterial growth. Media for plant or animal cell cultures are generally complex, and components break down at high temperatures. They, too, are sterilized by filtration.

Basically, a filter is constructed with a pore size that permits liquids to pass, but the pores are small enough to retain *Bacteria* and *Archaea*. These organisms, in general, are longer than 1 μm and are greater than 0.5 μm in diameter. Because some bacterial cells are smaller than this, effective filters must be able to retain cells that are quite small.

In practice, a filter can also be employed to separate or concentrate microbes based on their size. Bacteria can be separated from yeasts by employing a membrane with a pore diameter of about 3 μm. Most bacteria can pass through a filter such as this, but yeasts would be retained. Small bacteria such *Bdellovibrio*, an organism that parasitizes other bacteria, can be concentrated by passing water or dilutions of soil through membrane filters with a pore diameter of 0.45 μm. The *Bdellovibrio* are only 0.25 μm in width and would pass through, whereas most other microorganisms would be retained. The organisms in the filtrate can then be concentrated by centrifugation.

Most early filters were composed of stacks of asbestos, sintered glass (glass heated to fuse without melting), or diatomaceous earth. These materials formed circuitous passageways that retained microbial cells. Several different types of filters are used today. A **glass fiber depth filter** is a random stacking of glass fibers that forms a barrier impenetrable by bacteria and other particles (**Figure 7.4A**).

The filters that are now in common use are **membrane filters** made of cellulose acetate, polycarbonates, or cellulose nitrate (Figure 7.4B). These filters are quite thin (200 μm) and become relatively transparent when wet, allowing direct microscopic examination of microorganisms that are trapped on the filter. Membrane filters can be produced with pore sizes ranging from 0.1 to 0.5 μm, with 80% to 85% of the filter being open space. This allows a rather high rate of fluid passage. With this much open space available, a filter can be placed on the upper surface of an agar medium, and nutrients will diffuse to the cells through the filter by capillary action. Following incubation and growth, one can count the number of colonies that develop directly on the filter and observe their colonial morphology (see Figure 6.5).

The **Isopore** membrane filter has widespread application in research and industry. It contains clearly defined cylindrical holes that pass vertically through the membrane (Figure 7.4C). These filters are manufactured by treating thin sheets of polycarbonate with nuclear radiation. The radiation results in a random distribution of minute holes in the polycarbonate that can be enlarged by chemical etching. The length of time that the membrane is etched determines the size of the pores. A major advantage of these filters is that bacteria and other objects remain on the surface of the membrane and are thus readily observable with a scanning electron microscope (Figure 7.4C). With other types of membrane filter, the bacteria absorb to the sides of the pores and are not so readily observable. One problem with the Isopore filters, however, is the limited total space on the filter that is porous. This may result in low flow rates or clogging.

Presterilized disposable filter units available commercially are now routinely employed in the laboratory. In industry, much larger and more sophisticated filtration systems are used for large volumes of liquid. These larger volumes are sterilized by utilizing the filters placed in large stainless-steel cartridges. As an alternative to centrifugation, particularly for large volumes, bacteria can also be harvested from culture fluids by using membrane filters.

TOXIC CHEMICAL STERILIZATION OR CONTROL

We previously mentioned that some of the nineteenth-century scientists made use of chemical disinfectants. Included were Koch (mercuric chloride), Semmelweis (chlorinated lime), and Lister (phenol). Today we have a vast number of chemical agents available for use as disinfectants or to control the growth of undesirable microorganisms. These range from toxic gases to the medically important antibiotics. Following is a discussion of the chemicals now employed to kill or control microorganisms.

(A)   (B)   (C)

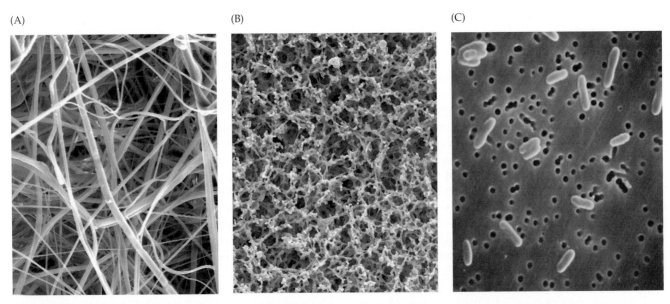

Figure 7.4 Scanning electron micrographs of filters
Scanning electron micrographs of (A) a glass fiber depth filter; (B) a membrane filter; (C) *Thermoleophilum album* trapped on an isopore filter. Courtesy of: A–B, Millipore Corporation; C, Jerome Perry.

Types of Antimicrobial Agents

An array of chemical agents are now available that can be used in the control of undesirable microorganisms. These chemicals are called **antimicrobial agents,** or antimicrobials (**Table 7.2**). Some of the agents are naturally occurring, some are synthetic, and others are a combination of the two.

An effective method for sterilizing surgical instruments and some heat-sensitive materials is to expose them to toxic compounds in the form of gases. Sheets and bedding in a hospital are sterilized in this manner. Toxic gases are also used to sterilize plastic Petri dishes and other plastic laboratory containers. Ethylene oxide and propylene oxide are generally the compounds of choice. Both function as alkylating agents in disrupting cellular DNA. Because these gases are exceedingly toxic to humans, their use in the sterilization process must be carefully controlled.

A much-desired property of an antimicrobial agent is **selective toxicity.** That is, the agent will eliminate a pest without doing significant harm to other species in the environment. Spraying plants with a chemical that kills invading bacteria or fungi but also harms the plant is of little value. Humans cannot be cured of disease by treating them with medicines that are overly harmful to the host (see the quote at beginning of this chapter).

In addition, humankind has become more and more concerned with the effects of chemical agents on the total environment. Experience has taught us the harm that can be done by careless application of antimicrobial substances. For example, for years, compounds containing mercury were used in the pulping industry and other industrial processes to control microbial contamination. These mercury compounds ultimately entered streams and rivers, resulting in considerable harm to fish and other wildlife. Stringent regulations are now in place for all potentially toxic compounds that are artificially introduced into the environment. They must be tested for safety, biodegradability, carcinogenicity, or other undesired effects on humans.

Another example of a carelessly applied agent is DDT, which was once applied widely as an insecticide. DDT was very effective against insects, but it accumulated in the natural environment and was very harmful to birds and other wildlife. Accumulation resulted from the inability of microbes to mineralize DDT at rates commensurate with application rates. DDT had a half-life of about 30 years in most environments, and its use has now been curtailed in the United States.

Table 7.2	Common antimicrobial agents and their uses
Compound	**Use**
Organic compounds	
Cresols	Disinfectant in laboratories
o-Phenylphenol	Disinfectant in laboratories
Phenol	Disinfectant in laboratories
Formaldehyde	Disinfect instruments
Quaternary ammonium compounds	Skin antiseptic
Ethanol and isopropanol	Disinfect instruments
Ethylene oxide	Sterilize instruments
Propylene oxide	Sterilize instruments
Halogens	
Chlorine gas	Disinfect water
Iodine solution	Skin antiseptic
Hexachlorophene	Skin antiseptic
Chlorine bleach	Domestic cleaning
Heavy metals	
Mercury chloride	Disinfectant
Silver nitrate	In infant eyes to prevent ophthalmic gonorrhea
Copper sulfate	Algicide
Other	
Hydrogen peroxide	Skin antiseptic
Mercurochrome	Skin antiseptic
Ozone	Drinking water and fish tanks

A compound that kills a target group is identified by combining the designated organism's name with the suffix -*cide*, from the Latin *cida,* meaning "having power to kill" (**Table 7.3**). Other agents are available that do not kill but rather prevent or inhibit growth. These are referred to as **bacteriostatic agents** (for bacteria) or **fungistatic agents** (for fungi). **Germicide** is a general term that describes substances that kill or inhibit microbes ("germs"). Some antibacterial substances bring about **lysis,** or disintegration, of the cell and are thus

Table 7.3	Categories of killing agents and their target organisms
Agent	**Target Organism**
Bactericide	*Bacteria*
Fungicide	Fungi
Algicide	Algae
Pesticide	Pests
Herbicide	Plants
Insecticide	Insects
Germicide	"Germs"

termed **bacteriolytic agents.** For example, lysozyme is a lytic enzyme that destroys the cell wall of gram-positive bacteria. Loss of the cell wall results in water uptake and subsequent bursting of the wall-less cell.

A **disinfectant** can destroy microorganisms on contact, including those that potentially cause disease. Phenol and mercuric chloride are disinfectants. Disinfectants are indiscriminate destroyers that, if applied to wounds, would also disrupt host tissue. An **antiseptic** is a substance, such as Merthiolate or Mercurochrome, that kills or prevents the growth of disease-causing pathogens. An antiseptic should not do significant harm to host tissue. As a rule, we use powerful disinfectants such as phenol on inanimate objects such as laboratory benches and less-potent antiseptics to treat wounds.

A **chemotherapeutic** agent is a chemical compound that can be applied topically, taken orally, or injected that will inhibit or kill microbes that cause infections. The term is also applied to compounds that are effective against viruses or tumor growth. **Antibiotics** are sometimes called chemotherapeutic agents and are widely used as antimicrobials, but the term "chemotherapeutic" is generally reserved for products of chemical synthesis. The antibiotics will be discussed later.

Tests for Measuring Antimicrobial Activity

Robert Koch (1843–1910) first devised a laboratory procedure for quantifying the effectiveness of antimicrobial compounds. Koch dried endospores of *Bacillus anthracis* on threads and placed these in solutions of potential antibacterial agents. Endospores were selected because they are more resistant to harsh physical conditions than other life forms.

Koch then estimated the effectiveness of a given compound by determining how long the endospores would survive exposure to the agent. At selected time intervals, threads were removed from the test solution and placed in a nutrient medium. Growth indicated survival. The shorter the time required to kill all of the endospores, the more effective the disinfectant. Koch discovered that mercuric chloride was one of the most effective of all available compounds. Mercuric chloride has been employed as a disinfectant up to the present day. Unfortunately, the use of large quantities of this mercury compound is discouraged because of the environmental problems it causes (see Chapter 24).

The relative strength of an antimicrobial can be determined by serial dilution of the parent compound. The more a compound can be diluted and still inhibit the growth of a test organism, the greater its relative strength. The lowest concentration that will inhibit growth

is designated as the **minimum inhibitory concentration,** or **MIC (Figure 7.5).** Many factors affect the MIC, including the test organism selected, the inoculum size, and the amount of organic material present. A strong germicide, such as ethylene oxide or formaldehyde (37% solution), will kill endospores and virtually all forms of life. Intermediate-strength antimicrobials such as phenolics can kill viruses and most vegetative cells but are less effective against endospores. Such agents are used for research and in hospital laboratories. Antimicrobials with less killing power are effective against vegetative cells (fungal and bacterial) but are not very effective against endospores, fungal spores, or some viruses. These weaker agents have less toxicity to humans and can be applied directly to the skin or can be included as a component in mouthwash. The general mode of action of various germicides is in the destruction of proteins (**Table 7.4**).

One standard that has been employed to compare the relative efficiency of germicides is the **phenol coefficient.** A phenol coefficient is a measure of the effectiveness of a test compound compared with the disinfecting power of phenol. For example, if a new antimicrobial at a dilution of 1:100 kills a standard population of *Salmonella typhi* that is killed by a 1:50 dilution of phenol, the phenol coefficient would be 100/50, or 2. The test antimicrobial would thus be twice as effective as phenol in destroying the test population.

However, the phenol coefficient has some limitations, as many factors can influence the effectiveness of disinfectants. Some are more affected by the physical envi-

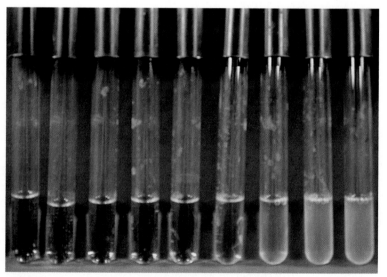

Figure 7.5 Dilution series for estimating antibiotic sensitivity
A tube dilution series for estimating the sensitivity of a microorganism (in this case *E. coli*) to antibiotic (here, chloramphenicol). The minimum inhibitory concentration (MIC) is the least amount of antibiotic that completely inhibits growth. Courtesy of Jerome Perry.

Table 7.4	Mode of action of some antimicrobial agents
Agent	**Mode of Action**
Alcohols	Denature proteins and dissolve membrane lipids.
Aldehydes	Combine with proteins and denature them.
Halogens	Iodine oxidizes cellular constituents and can iodinate cell proteins. Chlorine and hypochlorite oxidize cell constituents.
Heavy metals	Combine with proteins generally through sulfhydryl groups.
Phenolics	Denature proteins and disrupt cell membranes.
Quaternary ammonium compounds	Disrupt cellular membranes and can denature proteins.

ronment including light, oxygen, organic matter, and solid substrates. Gram-positive organisms are more readily killed than gram-negatives. Members of the genus *Pseudomonas* are not only resistant to, but may also grow on, selected disinfectants. The ability of pseudomonads to thrive in the presence of antiseptics is of particular concern for burn patients.

CHEMOTHERAPEUTIC AGENTS

The antimicrobial agents discussed in the previous section may be used to limit the growth of undesirable organisms in the environment. They may be employed as disinfectants on laboratory benches or other solid material but are not suitable for control of viral, bacterial, or fungal infections within the animal or human body. They would either be too toxic or be inactivated by the presence of organic matter.

From the time of **Paul Ehrlich** (1854–1915), scientists have sought chemicals that selectively inhibit the growth of infectious bacteria without harm to the human host. Ideally, such a **chemotherapeutic agent** would be **specific** for the microbe invading the body but with little or no harmful effect on normal body function. There is a constant search for better chemotherapeutics, including antibacterials, antivirals, and antitumor agents. Following is a discussion of some of the chemotherapeutics that have been synthesized and are effective against disease-causing microorganisms.

Chemically Synthesized Chemotherapeutics

Paul Ehrlich is considered the "father" of chemotherapy (see quote at the beginning of the chapter). He originated the concept that invading pathogenic organisms might be destroyed by selective chemicals that would not seriously harm the host (see Chapter 2). Others, aware of his contributions, have sought chemical compounds that might be employed as chemotherapeutic agents.

The **sulfa drugs** are chemotherapeutic agents that were discovered by Gerhard Domagk in the 1930s. Domagk was assessing the potential use of dye substances for antibacterial activity following the approach pioneered by Paul Ehrlich. Sulfanilamide was one of the compounds tested, and Domagk found that it possessed antibacterial activity both *in vitro* [Latin—in glass (outside the body)] and *in vivo* [Latin—living (inside the body)]. Sulfanilamide is a structural analog of the *p*-aminobenzoic acid, a component of the B vitamin folic acid (**Figure 7.6**). Folic acid is a coenzyme involved in nucleic acid synthesis. It is synthesized by enzymatically joining a

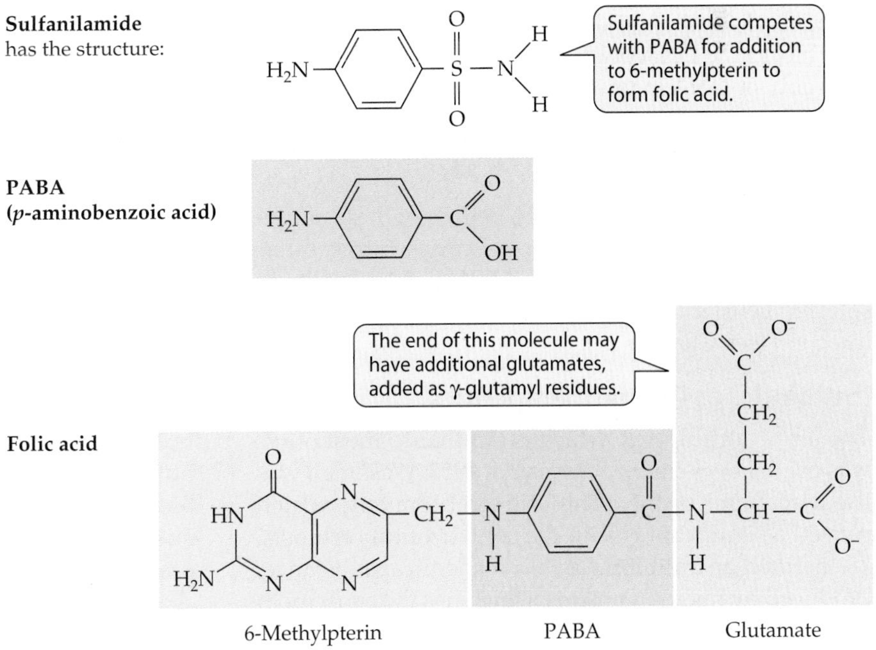

Sulfanilamide has the structure:

Sulfanilamide competes with PABA for addition to 6-methylpterin to form folic acid.

PABA (*p*-aminobenzoic acid)

The end of this molecule may have additional glutamates, added as γ-glutamyl residues.

Folic acid

6-Methylpterin   PABA   Glutamate

Figure 7.6 Sulfa drug as structural analog
Sulfanilamide is a competitive inhibitor of metabolic function in microorganisms. It is an analog of and competes with *p*-aminobenzoic acid in the synthesis of the vitamin folic acid.

molecule of 6-methyl pterin with *p*-aminobenzoic acid and a molecule of the amino acid glutamic acid. The presence of sulfanilamide results in the displacement of *p*-aminobenzoic acid in microorganisms that synthesize folic acid. The product thus generated cannot form a peptide bond with glutamic acid, as the inactive pteridine-sulfa conjugate predominates. Because folic acid is a mandatory catalyst in the synthesis of purine and pyrimidines, a microorganism without functional folic acid would not survive. Animals are unaffected by sulfanilamide because they obtain folic acid in their diet and do not synthesize it from the base constituents.

An understanding of the selective toxicity of sulfanilamide led to a concerted search for other chemotherapeutic agents. Structural analogs of amino acids, purines, pyrimidines, and vitamins have been synthesized. Emphasis has been placed on synthesizing analogs of the bases in DNA and RNA (adenine, thymine, guanine, uracil, and cytosine) in an effort to find chemical compounds that have antiviral or antitumor activity. Thousands of compounds have been synthesized, but a limited number have proven effective as chemotherapeutics.

Some important antiviral agents that have been developed in recent years, and these include acyclovir and azidothymidine (AZT). **Acyclovir** is a structural analog of deoxyguanosine (**Figure 7.7**) and can be used in the treatment of herpes virus infections. When a herpes virus (a DNA virus) multiplies in a cell, the enzyme thymidine kinase is activated. When acyclovir is present, this enzyme gratuitously phosphorylates, resulting in the formation of an unnatural triphosphate, which then blocks the DNA polymerase. As a consequence, the assembly of DNA in viral particles is curtailed.

AZT is an antiviral used in the treatment of AIDS (acquired immunodeficiency syndrome). This compound inhibits the multiplication of HIV, the retrovirus that causes AIDS. Unfortunately, neither of these antiviral agents can destroy the nonreplicating intracellular virus. Both the herpes virus and HIV persist for life in selected cells within the human host from the time of infection (see Chapter 29).

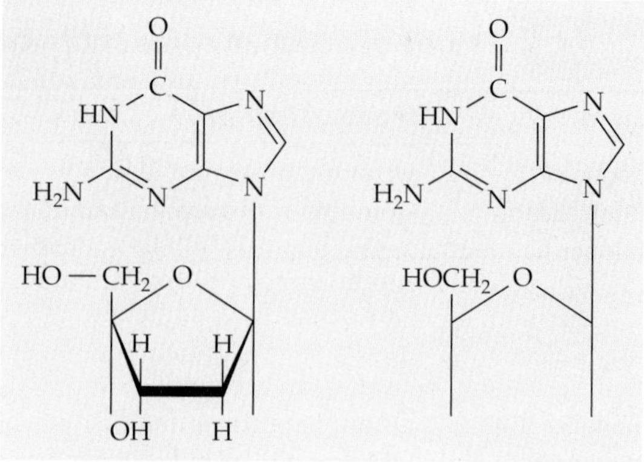

Antibiotics as Chemotherapeutics

The term "**antibiotic**," which is familiar to most of us, was coined by Selman Waksman in 1953. Waksman was the discoverer of the antibiotic streptomycin, which played a significant role in the control of tuberculosis. He defined an antibiotic as ". . . a chemical substance, produced by microorganisms, which has the capacity to inhibit the growth and even to destroy bacteria and other microorganisms, in dilute solutions." Antibiotics that are effective in controlling disease are produced during growth by certain bacteria, actinomycetes, and fungi. In fact, the ability to produce antibiotic substances

Figure 7.7   Acyclovir as structural analog
The antiviral agent acyclovir is structurally related to the nucleoside deoxyguanosine. Phosphorylation of acyclovir yields an analog of deoxyguanosine triphosphate, which inhibits the viral DNA polymerase.

is widespread in the microbial world, and the group of bacteria termed the *actinomycetes* is particularly adept at producing them. The antibiotics synthesized by actinomycetes have remarkable chemical diversity. However, the first commercially successful antibiotic used in chemotherapy was produced by a fungus. This antibiotic was **penicillin** and was produced by the fungus *Penicillium notatum* (**Box 7.1**). The search for antibiotic substances and their industrial production is presented in Chapter 31. The medical aspects of antibiotics are covered in Chapters 28 and 29. The present section will provide a brief discussion of their basic **mode of action** and the manner in which infectious organisms become resistant to them.

Antibiotics are selectively toxic to certain types of living cells and less so to others. This selectivity is due to distinct differences in fundamental physiological processes in affected cells and unaffected cells. Some antibiotics are effective against a limited range of bacteria and thus are considered to be **narrow-spectrum** antibiotics. Penicillin is an example of this type because it is mostly effective against gram-positive organisms. A **broad-spectrum** antibiotic, such as tetracycline, inhibits both gram-positive and gram-negative bacteria. Antibiotics that inhibit bacteria generally do not inhibit eukaryotes. Conversely, antibiotics that inhibit eukaryotes are generally not effective against bacteria. The *Archaea* are generally less sensitive to antibacterial antibiotics.

Antibiotic Actions: Sites and Modes

An antibiotic interferes in some manner with a normal physiological function in a susceptible cell (**Table 7.5**).

Table 7.5	The producing organisms and mode of action of several antibiotics	
Antibiotic	**Producer**	**Mode of Action**
Penicillin	*Penicillium* spp.	Blocks transpeptidation involved in cell wall synthesis
Erythromycin	*Streptomyces erythreus*	Binds to 50S ribosomal subunit and stops peptidyltransferase
Chloramphenicol	*S. venezuelae*	Binds to ribosomes blocking peptidyltransferase
Rifampicin	*S. mediterranei*	Blocks transcribing enzyme RNA polymerase
Novobiocin	*S. spheroides*	Binds to a subunit of DNA gyrase
Tyrocidine	*Bacillus brevis*	Ionophore disrupts cell membrane integrity and function
Polymyxins	*B. polymyxa*	Disrupts membrane transport and function
Streptomycin	*S. griseus*	Inhibits 30S ribosomes, blocks amino acid incorporation into peptides
Tetracycline	*S. aureofaciens*	Inhibits binding of aminoacyl tRNA to ribosomes

The major actions of antibiotics include:

- Disruption of cell wall synthesis
- Destruction of cell membranes
- Interference with some aspect of protein or nucleic acid synthesis

The most useful antibiotics employed to control human infections interfere with structural or physiological reactions that are unique to the invading pathogen. This is possible because the bacteria are physiologically quite different from the eukaryotes. Consequently, antibacterial antibiotics can exploit these differences and destroy the invader without significant harm to the human host.

Control of fungal, protozoan, and viral invaders is much more difficult because physiological reactions in these pathogens are quite similar to that in all eukaryotes including a human host. The physiological differences among the eukaryotes are simply not so distinct that these differences can readily be exploited. Thus, effective antifungal and antiprotozoan agents are generally toxic to humans. Because viral infections are intracellular, they are also difficult to control with chemotherapeutic agents. A virus is synthesized, for the most part, by the metabolic machinery of the host, and destruction of this machinery can lead to the destruction of the host.

Several antibiotics and their modes of action are listed in Table 7.5. Penicillin and chemically related antibiotics prevent the transpeptidation reaction, which is an important step in the assembly of the cell wall polymer, peptidoglycan (see Chapter 4). This results in a weakened cell wall, especially in gram-positive organisms. Because microorganisms generally live in osmotically unfavorable environments, one having a weakened cell wall will take up water and burst, or lyse. Gram-negative bacteria tend to be less sensitive to penicillin because their outer envelope can prevent the antibiotic from reaching the peptidoglycan layer of the cell. The

Archaea that do not have peptidoglycan in their cell wall are generally unaffected by antibiotics such as penicillin that interfere with the synthesis of this polymer. Some of the *Archaea* are sensitive to selected ionophores that disrupt membranes (see Table 7.5).

Tyrocidine and **polymyxin** are both polypeptide antibiotics that disrupt cell membranes. Both are produced by bacteria of the genus *Bacillus*. Tyrocidine is an **ionophore,** which destroys selective permeability by forming channels across the cell membrane, resulting in leakage of monovalent cations. As a consequence, the organism cannot establish a proton motive force, and transport into and out of the cell is impaired. Polymyxin causes similar cytoplasmic membrane damage. Peptide antibiotics are not taken internally but are applied topically to treat skin infections. Enzymes present in the intestinal tract would digest an ingested peptide antibiotic.

Certain antibiotics interfere directly with bacterial DNA replication. Among these is novobiocin, which binds directly to the beta subunit of the DNA gyrase responsible for unwinding supercoiled DNA during replication. Other antibiotics, such as rifampicin, block the RNA polymerase involved in transcribing the information on the DNA molecule to make messenger RNA for protein synthesis (see Chapter 13). Most of these antibiotics are ineffective in the *Archaea*.

Many of the clinically useful antibiotics inhibit protein synthesis in bacteria. Protein synthesis requires **ribosomes,** which are structures made up of subunits of unequal size. Ribosomes are involved in the synthesis of protein. Bacterial and eukaryotic ribosomes differ in size, protein content, and ability to bind antibiotics. On the basis of sedimentation velocity during ultracentrifugation, the smaller bacterial ribosome is designated **70S** and the larger eukaryotic ribosome size is **80S.**

Antibacterial antibiotics that inhibit protein synthesis do so by binding to the bacterial ribosome. This occurs with the antibiotics tetracycline, chlorampheni-

---

Milestones Box 7.1   The Antibiotic Age

Fortune favors the prepared mind.
— LOUIS PASTEUR

The more I practice, the luckier I get.
— LEE TREVINO

These quotes are a brief summation of the circumstances that led Sir Alexander Fleming (1881–1955) to the discovery of penicillin. All too often, scientific advances are attributed to chance or serendipitous good fortune, but usually they are not. More often they are the product of intuition and hard work. The discovery of penicillin resulted from Fleming's long-held and firm belief that naturally occurring substances existed that had useful antimicrobial properties.

    During World War I Fleming worked in a field hospital in France. There he treated casualties of battle and experimented with better ways of treating deep wounds. The accepted treatment at that time was copious quantities of antiseptic—

carbolic acid, boric acid, or peroxide. Unfortunately, in deep wounds these antiseptics were quickly neutralized by tissue, and serious infections often followed. Fleming and his colleague, Sir Almroth Wright (1861–1947; a noted developer of antityphoid vaccine), observed that wounds carefully cleansed and treated with minimal antiseptic healed faster and with fewer serious infections than wounds treated with massive amounts of antiseptic. They believed that stimulation of natural-defense mechanisms by encouraging the exudation of lymph into the wound site was particularly effective in promoting healing. They found that bathing wounds with a saline solution was effective in stimulating the exuding of lymph. The success of this experimentation convinced Alexander Fleming that naturally produced substances could be employed successfully in the treatment of infectious disease. Fleming explained to a physician visiting his laboratory at Boulogne, "What we are looking for is some chemical

Sir Alexander Fleming Photo by Sydney R. Bayne/NLM.

substance which can be injected without danger into the blood stream for the purpose of destroying the bacilli of infection, as salvarsan destroys the spirochaetes."

    Following the war, Fleming returned to his position at St. Mary's

---

col, streptomycin, and erythromycin. The antifungal antibiotic cycloheximide binds to eukaryotic ribosomes (80S) but not the 70S ribosome that is present in bacteria. Consequently, cycloheximide inhibits fungi but does not inhibit the growth of bacteria. As humans have 80S ribosomes, they too would be adversely affected by cycloheximide. A few antibiotics, including tetracycline, can bind to both 70S and 80S ribosomes. However, tetracycline can be used in humans because it does not inhibit the synthesis of protein in eukaryotic cells at the concentrations used chemotherapeutically. It does impede wound healing in humans under some circumstances.

RESISTANCE TO ANTIBIOTICS

Antibiotics have been effective in controlling many of the diseases that have been a scourge to humankind. Tuberculosis, bacterial pneumonia, syphilis, and many other human infectious diseases that were once fatal can now generally be treated with antibiotics. Antibiotics

have been called "wonder drugs" because they effect a dramatic cure for what had previously been incurable. But there is a problem with "wonder drugs" that became evident after widespread use.

    The extensive use of penicillin as a chemotherapeutic agent led to the evolution of pathogenic bacterial strains that were unaffected by the drug. These resistant organisms retained their potent pathogenicity but were no longer controlled by the administration of penicillin. For example, when penicillin G (see Figure 4.53) was first introduced, virtually all strains of *Staphylococcus aureus* were sensitive to the drug. Within a span of only 10 years, essentially all staph infections acquired in hospitals were caused by strains that were resistant to penicillin.

    Why do microorganisms become resistant to chemotherapeutics? Abundant microbial populations have existed throughout all of Earth's history because microbes are adaptable. For example, when oxygen appeared in the atmosphere or the climate became cool-

Milestones Box 7.1   *(continued)*

Hospital in London. As time permitted, he tested various naturally occurring substances for antibacterial activity. Among the materials he tested was nasal mucus, and he observed that addition of mucus to a suspension of *Staphylococcus aureus* led to a rapid clearing of the turbidity. He discovered that this clearing resulted from lysis of the bacterial cells. The factor involved was an enzyme termed *lysozyme* (see Chapter 4). This discovery, made in 1921, was important scientifically, but lysozyme was not useful as a chemotherapeutic agent. It did, however, offer confirmation to Fleming that his belief in natural antibacterials was warranted.

One of Fleming's research interests was *Staphylococcus aureus,* and he often had cultures growing on Petri dishes lying about the lab. One day while examining some of these old cultures, he observed that one was contaminated with a mold and that bacterial colonies did not develop in the area adjacent to the mold growth. Fleming was intrigued and

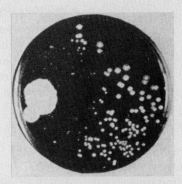

The original *Penicillium* plate From Maurois, Andre. 1959. *Life of Sir Alexander Fleming*, E. P. Dutton Co.

likened the phenomenon to the action of lysozyme on *S. aureus*. He then took a picture of the plate (shown in the drawing) and wrote up the observation in his research notebook. He showed the plate to a colleague with the remark—"Take a look at that, it's interesting—the kind of thing I like; it may well turn out to be important." Important, indeed; the antibacterial substance produced by the mold (*Penicillium notatum*) was penicillin. This ush-

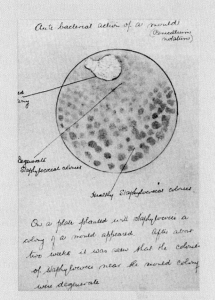

Fleming's drawings and notes

ered in the "antibiotic age." Fleming's monumental discovery led eventually to the conquest of such ancient scourges as pneumonia, tuberculosis, syphilis, staphylococcal infections, typhus, and a host of other human ills.

---

er, microbes evolved to survive these new environmental conditions. Environmental stresses and constraints became natural selection processes for microbes. Those that survived carried genes that better adapted them to serve in the altered conditions. Human control methods of all types create a challenge to bacteria that leads to natural selection for those organisms in the environment that can withstand the prevailing condition. Thus, the widespread use of penicillin resulted in the selection of strains that could survive in the presence of the drug. Some penicillin-resistant strains produce an enzyme, **penicillinase,** that destroys the antibiotic.

Microbial resistance to an antibiotic, a chemotherapeutic agent, or other chemicals such as heavy metals can occur for several of reasons:

Natural resistance

- The organism may lack the structure that the antibiotic inhibits, as occurs with mycoplasma, which lacks cell walls and is thus unaffected by penicillin.

- The cell wall structure or cytoplasmic membrane of an organism may be impermeable to an antibiotic or other chemical.

Acquired resistance

- A resistant microorganism may produce a substance that inactivates the antibiotic, as occurs in strains of *Staphylococcus aureus* in producing the enzyme penicillinase, which disrupts the penicillin molecule.
- A gradual accumulation of mutations in chromosomal DNA may result in cellular structures that will not bind the antibiotic or other chemical. For example, the gene for transpeptidase synthesis in staphylococci can mutate so that the enzyme does not bind penicillin.

Another significant problem in antibiotic and other forms of resistance is the acquisition of bits of extra chromosomal DNA that carry the information that renders a microorganism resistant. These bits of DNA are termed "plasmids" and can be transferred from cell to cell (see Chapter 15).

SUMMARY

▶ The role of microorganisms in the cycles of nature is essentially destructive; human survival often depends on counteracting their activities.

▶ An understanding of the nature of microorganisms led to more direct means for controlling them.

▶ **Sterilization** is the destruction of all viable life forms. It can be attained by fire, moist heat, dry heat, radiation, filtering, or toxic chemicals.

▶ Microbial death by heating occurs at an **exponential rate.** The length of time required to reduce a population tenfold is termed the **decimal reduction time**.

▶ The most effective means for routine sterilization of media and food is steam under pressure. The common laboratory apparatus employed for this is the **autoclave**.

▶ Dry heat at 160°C is effective in sterilizing glassware, pipettes, and other heat-stable material.

▶ **Pasteurization** is a method for reducing the number of microorganisms present in milk or other products. It can destroy selected disease or spoilage-causing microorganisms.

▶ The wavelengths of light that can cause the death of microorganisms are **ultraviolet light** (**UV**), which disrupts RNA/DNA, and **ionizing radiation**, which can potentially harm any constituent of a cell.

▶ Microorganisms have DNA repair systems that can correct damage caused by radiation.

▶ Filters with pores small enough to retain bacteria can be employed to sterilize liquids. A number of types are available, but membrane filters are the most widely used.

▶ Toxic gases such as propylene oxide is used to sterilize plastic Petri dishes, filters, and other disposable material.

▶ **Antimicrobial agents** may be products of chemical synthesis, natural products, or a combination of the two. The favored agent has selective toxicity in that it destroys the target microbe but produces limited activity toward other cells.

▶ A **disinfectant** is an agent such as phenol that destroys all living cells on contact. An **antiseptic** is less potent and prevents the growth of disease-causing organisms. Merthiolate that is applied to superficial wounds is an antiseptic.

▶ The **minimum inhibitory concentration** (**MIC**) is the lowest concentration of a **germicide** that will inhibit a test organism.

▶ The **phenol coefficient** is a measure of the effectiveness of a test compound as compared with phenol.

▶ The first effective chemically synthesized chemotherapeutic agent was **sulfanilamide**. An understanding of the role of this compound in blocking vitamin synthesis in bacteria encouraged the scientific world to search for other chemical agents that would be effective against viruses, protozoa, fungi, and cancer.

▶ By definition an **antibiotic** is a product of microorganisms. Antibiotics have proven to be effective in controlling human and animal diseases caused by bacteria. They generally interfere with metabolic activities or structures in disease-causing bacteria but do not interfere with them in a host eukaryote.

▶ The major **modes of action** of antibiotics include: disrupt cell wall synthesis, destroy cytoplasmic membranes, and disrupt nucleic acid or protein synthesis.

REVIEW QUESTIONS

1. What is the difference between sterilization and pasteurization? What agent is most often used in sterilization? Why are we so concerned with sterilizing things?

2. Dry heat is employed to sterilize selected materials. Name some of the materials best sterilized by dry heat.

3. What agent is most often used to sterilize air in a closed room? List what concerns there would be with other methods.

4. Give some advantages of sterilization by filtration. How is filtration applied in the laboratory? In industry?

5. How did Robert Koch determine the effectiveness of disinfectants? What compound did he find most effective with this method?

6. How would one calculate a phenol coefficient? Is this a useful number? List the pros and cons.

7. Sulfanilamide is a classical example of a competitive inhibitor. How does it function?

8. What are the basic shortcomings of antiviral agents such as AZT and acyclovir?

9. Why is it so difficult to find effective antiviral and antifungal agents when antibacterials are quite plentiful?

10. Discuss the mode of actions of some common antibacterials.

SUGGESTED READING

Block, S. S., ed. 1991. *Disinfection, Sterilization, and Preservation.* 4th ed. Philadelphia: Lea and Febiger.

Favero, M. S. and W. W. Bond. 1991. "Sterilization, Disinfection, and Antisepsis in the Hospital." In A. Ballows, W. J. Hausler, K. L. Herrmann, H. O. Isenberg and H. J. Shadomy, eds. *Manual of Clinical Microbiology.* 5th ed. (pp. 183–200). Washington, DC: American Society for Microbiology.

Hardman, J. G. and L. E. Limbird, eds. 1995. *The Pharmacological Basis of Therapeutics.* 9th ed. New York: Pergamon Press.

Harte, J., C. Holdren, R. Schneider and C. Shirley. 1991. *Toxics A to Z: A Guide to Everyday Pollution Hazards.* Los Angeles: University of California Press.

Levy, S. B. 1992. *The Antibiotic Paradox, How Miracle Drugs Are Destroying the Miracle.* New York: Plenum Press.

Microbial Physiology: Metabolism

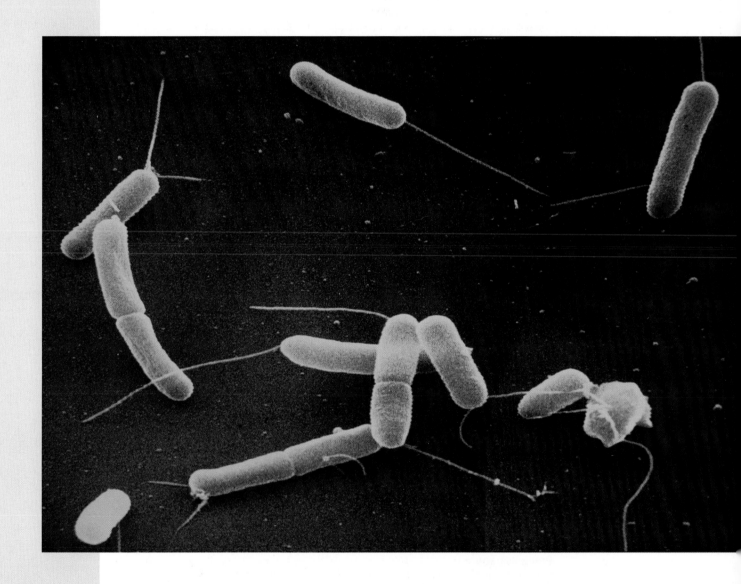

PART III Microbial Physiology: Metabolism

Previous page
The biochemical events that construct and power flagella are part of each
Escherichia coli's metabolism. ©David M. Phillips/Visuals Unlimited.

Chapter 8

Cellular Energy Derived from Chemicals

Life is driven by nothing else but electrons, by the energy given off by these electrons while cascading down from the high level to which they have been boosted up by photons. An electron going around is a little current. What drives life is thus a little electric current, kept up by the sunshine. All the complexities of intermediary metabolism are but lacework around this basic fact.

— *Albert Szent-Györgyi*

This chapter introduces the principal mechanisms whereby microorganisms can generate energy for biological processes. Bear in mind that virtually all energy for life on the earth's surface is furnished either directly or indirectly by sunlight, and this energy is captured during photosynthesis. Among the products of photosynthesis are ATP and potential energy-yielding compounds, such as sugars and starch, that provide sustenance for heterotrophic life. Animals survive by eating energy-yielding plant material and/or by devouring an animal that ate a plant. In this chapter, we will discuss the mechanisms whereby potential chemical energy trapped in the products of photosynthesis is converted to ATP. The following chapter will present the processes whereby light energy is converted to ATP.

BASIC PRINCIPLES OF ENERGY GENERATION

Among *Bacteria* and *Archaea*, one should be aware that the source of energy is, in many cases, different from the carbonaceous substrate used for synthesis of cell material. This is a fundamental difference between the microbial and the animal worlds. The one commonality in the evolution of all living systems is that **adenosine-5′-triphosphate (ATP),** formed from adenosine-5′-diphosphate (ADP) and inorganic phosphate (P_i), is a universal medium of biochemical energy exchange. The energy conserved in the "energy-rich" bonds of ATP can be employed by the cell to do work. Although ATP is common to all forms of life, the oxidation-reduction reactions that are involved in the synthesis of ATP are quite varied among bacteria.

Evidence suggests that a major event in the evolution of living organisms was the advent of the ability to establish a **charge separation** or **electrochemical potential** across the cytoplasmic membrane. This charge separation is a form of potential energy that may be equated to that of a car or flashlight battery that provides a current to activate a starter or produce light. The electrochemical potential established across a biological membrane can be converted to chemical energy in the form of ATP. A general representation of a charge separation across a cytoplasmic membrane is presented in **Figure 8.1**.

Solar energy is the ultimate source of energy, and photons from the sun can drive electrons across a membrane to establish an electrochemical potential. Light energy can also extract electrons from water. The electrons extracted from water ($HOH \rightarrow 2H^+ + 2e^- + \frac{1}{2} O_2$) may then be passed through

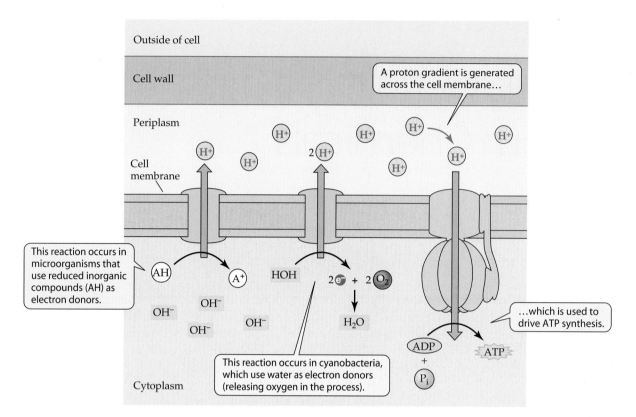

Figure 8.1 Establishing a charge separation across a cell membrane
The concentration of protons (H^+) on the outer surface and of negative ions (OH^-) inside the cell creates a charge difference across the membrane. This charge difference can be a source of energy for synthesis of adenosine triphosphate (ATP) and other energy-consuming activities of the cell.

a series of membrane-associated electron carriers to generate a charge separation. Energy generated in the form of ATP during photosynthesis is utilized for the assimilation of CO_2 from the atmosphere. Products of CO_2 fixation, such as glucose, store potential chemical energy yielding compounds.

Extracting electrons from the molecule and using these to establish a charge separation across the cytoplasmic membrane can recover the energy inherent in a glucose molecule. This charge separation can be the driving force in generating ATP for heterotrophic growth. In microorganisms, the source of the electrons that bring about the extrusion of protons can be either organic or inorganic substrates, for example:

$$Lactate \rightarrow pyruvate + 2\,H^+ + 2e^-$$

$$H_2 \rightarrow 2\,H^+ + 2e^-$$

Proton accumulation on the exterior of the membrane creates a positive charge (cathode), and there would be a negative charge (anode) on the interior; this charge separation is the electroachieved potential. The energy inher-

ent in the proton is captured as ATP when it passes through a specific membrane channel and returns to the interior. In an aerobe, the electrons involved in proton extrusion ultimately return to react with O_2 and protons to form H_2O (Figure 8.1). These processes will be explained in more detail later.

Bacteria and *Archaea* employ one or more of three general mechanisms to generate ATP: respiration, photophosphorylation, or substrate-level phosphorylation.

- **Respiration** is the production of ATP by oxidation of reduced organic or inorganic compounds (electron donors) coupled with the reduction of inorganic or organic electron acceptors.
- **Photophosphorylation** occurs in photoautotrophic microorganisms that generate ATP by using light energy to create a charge separation.
- **Substrate-level phosphorylation** is the synthesis of ATP by direct transfer of a high-energy phosphate from a phosphorylated organic compound to ADP to form ATP.

Aerobic oxidative phosphorylation utilizes O_2 as the terminal electron acceptor, whereas anaerobic respiration uses the electrons to reduce an oxidized electron acceptor. Some of the compounds that can serve as electron acceptors in anaerobic respiration are NO_3^-, SO_4^-, CO_2, ferric iron, or fumarate. The products generated from these compounds would be NH_4^+, H_2S, CH_4, ferrous iron (Fe^{3+}), or succinic acid, respectively.

OXIDATION-REDUCTION REACTIONS

Oxidation-reduction (redox) reactions occur because electrons flow from a component of higher potential (donor) to another of lower potential (acceptor). The donor is termed the **reductant** and the electron acceptor, the **oxidant.** One component in a redox reaction is oxidized and the other is reduced. The capacity for the occurrence of a redox reaction is defined as the **reduction potential,** or E_0 (see **Box 8.1**). The reduction potential of a compound is measured in volts and is arbitrarily based on the voltage required to remove an electron from H_2 under standard conditions. The hydrogen electrode has been assigned a standard value of 0.0 V (volts):

$$H_2 \rightarrow 2H^+ + 2e^-$$

When this reaction occurs under standard conditions, with reactants at a concentration of 1 M (molar) and at a pH of 7, the redox potential is actually –0.42 V. The reaction for H just presented would be termed a **half reaction** because electrons cannot exist alone in solution, and for the preceding reaction to occur, there must be another half reaction and a corresponding reduction. A half reaction (an oxidation) must be coupled to another half reaction (a reduction) such as $\frac{1}{2} O_2 + 2e^- + 2H^+ \rightarrow H_2O$. In a biological reaction, the amount of free energy released in a redox reaction is the difference in reduction potential between the electron donor and the electron acceptor. This energy difference is designated by $\Delta E_0'$. The term E_0' is the redox potential at pH 7 (neutrality). Thus, the free energy available from the reduction of oxygen by hydrogen is the product of two half reactions:

Methods & Techniques Box 8.1   Oxidation–Reduction

Redox reactions—Oxidation is the loss of electrons, and reduction is the gain of electrons. Because electrons cannot exist in solution, an oxidation reaction must be coupled with a reduction. Oxidation releases energy, and this can occur only if a substance is available that can be reduced. Biological reactions can involve the transfer of hydrogen atoms, which consist of an electron and a proton. When the electron is removed from an atom of hydrogen, one has a **proton** that will generally accompany the electron. For example:

$$2H_2 \rightarrow 4H^+ + 4e^-$$

$$O_2 + 4H^+ + 4e^- \rightarrow H_2O$$

Reduction potential—A quantitative measure of the tendency for a substance to give up electrons in biological systems. It is measured in volts and generally at pH 7, the pH of cytoplasm.

ATP—The energy generated by redox reactions is conserved in ATP, the major high-energy phosphate compound in living cells. ATP is an energy "carrier" and is formed during **exergonic reactions** (energy yielding), and it can drive **endergonic reactions** (energy requiring). In ATP and some other phosphorylated compounds, the outer two phosphate atoms are joined by an anhydride bond (see Figure 8.3). The free energy ($\Delta G^{0'}$) that can be released when the high-energy bonds are hydrolyzed is –31.8 kJ/mol. The low-energy bond of the AMP would yield only –14.2 kJ/mol. The phosphate bond in the glycolytic intermediate phosphoenolpyruvate is high energy ($\Delta G^{0'} = -51.6$ kJ/mol), and the phosphate bond in glucose-6-phosphate is low energy ($\Delta G^{0'} = -13.8$ kJ/mol). The free energy available in phosphate bonds is of value to cells only when it can be coupled to a second reaction that requires energy.

In redox reactions, electrons are transferred from one (a donor) component to another (an acceptor), which is reduced. Reductions may result in the gain of one or more protons.

Cytochrome c undergoes 1 e^- reduction:
$$Fe^{3+} - cyt\ c + 1e^- \rightarrow Fe^{2+} - cyt\ c$$

NAD^+ accepts 2e^- and gains 1 proton
$$NAD^+ + 2e^- + H^+ \rightarrow NADH$$

Ubiquinone accepts 2e^- and gains 2 protons
$$UQ + 2e + 2H^+ \rightarrow UQH_2$$

Respiration—Energy generation in which molecular oxygen or some other oxidant is the terminal electron acceptor. Among the latter are nitrate, sulfate, carbon dioxide, and fumarate.

Fermentation—Energy generation by anaerobic energy-yielding processes characterized by substrate-level phosphorylation and the absence of cytochrome-mediated electron transfer(s).

$$H_2 \rightarrow 2\,H^+ + 2\,e^- \ (E_o{}' = -0.42\ V)$$

$$\tfrac{1}{2}\,O_2 + 2\,H^+ + 2\,e^- \rightarrow H_2O \ (E_o{}' = +0.82\ V)$$

$$\text{Sum: } H_2 + \tfrac{1}{2}\,O_2 \rightarrow H_2$$

The total free energy change occurring in these two half reactions can be calculated by the following equation:

$$\Delta G^{o'} = (-nF)(\Delta E_h)$$

where ΔE_h is the free energy change under standard conditions, n is the number of electrons transferred, and F is the Faraday constant (-96.48 kJ/V).

$$\Delta E_h = E_o{}'\ (\text{oxidized}) - E_o{}'\ (\text{reduced})$$

Substituting in values for each factor gives the change in free energy as:

$$\Delta G^{o'} = (2)(96.48)(+0.82 - 0.42)$$
$$= 239.27\ kJ$$

A useful redox reaction occurs in a biological system when the electron and accompanying proton are enzymatically removed from the substrate and passed to an electron **carrier** such as nicotinamide adenine dinucleotide **NAD⁺** (**Figure 8.2**). An electron carrier is an intermediate that **carries** electrons from a **donor** to an **acceptor**. Reduction of NAD⁺ would yield NADH + H⁺, which for brevity is written NADH (see Box 8.1). This carrier (NADH) can pass the electrons derived from the

donor to a series of electron carriers in a cascade of increasing $E_o{}'$ (**Table 8.1**). The more negative $E_o{}'$ half reactions tend to donate electrons and become oxidized, and the more positive $E_o{}'$ half reactions accept electrons and become reduced. A typical biological reaction would proceed as follows:

$$NAD^+ + 2\,H^+ + 2\,e^- \rightarrow NADH \ (E_o{}' = -0.32) + H^+$$

$$\tfrac{1}{2}\,O_2 + 2\,H^+ + 2\,e^- \rightarrow H_2O \ (E_o{}' = +0.82)$$

$$\Delta G^{o'} = (2)(96.48)(1.14)$$
$$= 219.97\ kJ$$

The biological reaction releases considerable energy to drive cell reactions. Other redox reactions may yield considerably less energy. As an example, consider the inefficiency of growth with a donor such as iron [Fe^{3+}/Fe^{2+} (0.77)] as the energy source and oxygen [$\tfrac{1}{2}\,O_2/H_2O(0.82)$] as the terminal electron acceptor. Using values for $E_o{}'$ from Table 8.1, the calculation is:

$$\Delta G^{o'} = (1)(96.48)(0.05)$$
$$= 4.82\ kJ$$

This low value indicates that little energy is available from the oxidation of one iron atom as compared with the NADH $\rightarrow O_2$ reaction.

Electrons do not exist in the free state in biological systems but flow from donors to molecules that will accept them. Components of electron transport chains such as the cytochromes, in the cytoplasmic membranes of microorganisms, are efficiently organized to accept

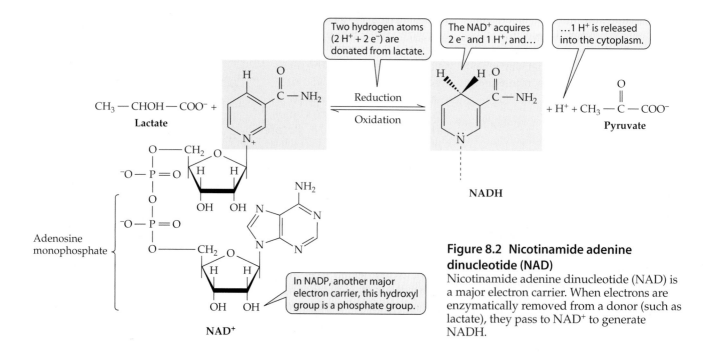

Figure 8.2 Nicotinamide adenine dinucleotide (NAD)
Nicotinamide adenine dinucleotide (NAD) is a major electron carrier. When electrons are enzymatically removed from a donor (such as lactate), they pass to NAD⁺ to generate NADH.

Table 8.1	Half reactions, the number of electrons transferred (n), and the electrode potential under standard conditions (E_o') compared to the hydrogen half cell		

Half Reaction		n	E_o' (V)
Ferredoxin (oxidized/reduced)		2	−0.43
$2 H^+/H_2$		2	−0.42
$NADP^+ + H^+/NADPH + H^+$		2	−0.32
1,3-di-P-glycerate + $2 H^+$/ glyceraldehyde-3-P + P_i		2	−0.29
Chlorophyll (P_{II})		1	−0.20
$FMN + 2H^+/FMNH_2$		2	−0.22
$FAD + 2 H^+/FADH_2$		2	−0.22
Standard half cell $2 H^+/H_2$		**2**	**0.00**
Methylene blue (oxidized/reduced)		2	+0.01
Fumarate + $2H^+$/succinate		2	+0.03
Ubiquinone (oxidized/reduced)		2	+0.06
Cytochrome b (Fe^{3+}/Fe^{2+})		1	+0.08
Cytochrome c (Fe^{3+}/Fe^{2+})		1	+0.25
Chlorophyll (P_I)		1	+0.40
$NO_3^- + 2 H^+/NO_2^- + H_2O$		2	+0.42
Fe^{3+}/Fe^{2+}		1	+0.77
$2 H^+ + \frac{1}{2} O_2/H_2O$		2	+0.82

electrons and then become donors as they pass the electrons along to an acceptor of lower redox potential. These are **fixed** electron carriers and are integrated into the membrane. There are electron carriers that move freely in the cytoplasm and function as links between metabolic pathways and the fixed-membrane bound electron transport chains. These are termed **diffusible** electron carriers. The major diffusible carriers are the aforementioned NAD^+, $NADP^+$ (NAD-phosphate), and the flavin nucleotides, which will be discussed later.

The reduction potential for $NAD^+/NADH$ and $NADP^+/NADPH$ is equivalent at −0.32V, and both are effective electron carriers. NAD^+ is principally involved in energy-generating (catabolic) reactions, and NADP is involved mostly in biosynthesis (anabolism). NAD^+ is a hydrogen atom carrier and can transfer two hydrogens in metabolic reactions, referred to as **dehydrogenation:**

$$CH_3-CHOH-COO + NAD^+ \xrightarrow[\text{dehydrogenase}]{\text{lactate}}$$

Lactate

$$CH_3-\overset{\overset{\displaystyle O}{\|}}{C}-COO^- + NADH + H^+$$

Pyruvate

The reduction potential in NADH can be delivered to the membrane-bound electron transport chain as electrons (e^-) from NADH move through the respiratory chain. The energy is used to translocate protons (H^+) to the *exterior* of the cytoplasmic membrane. This translocation creates a **charge separation** or an electrochemical potential, as mentioned earlier. The protons traverse back across the cytoplasmic membrane via a membrane-associated protein complex called the ATP synthase system, and ATP is generated. The ATP synthase system can undergo a proton-driven conformational change that covalently bonds a molecule of ADP and inorganic phosphate (P_i) to form ATP. The energy-rich chemical bond between ADP and P_i now bears the energy available from the substrate (lactate in the example) and carried by NADH to the electron transport chain.

ENERGY CONSERVATION IN THE MICROBIAL WORLD

According to the first law of thermodynamics, all forms of energy are convertible from one to another, but energy can neither be created nor destroyed. All living organisms are *dynamic systems* with steady inputs as well as outputs of both matter and energy. Every microbe has the metabolic capacity to transform chemical and/or light energy into biological energy in the form of ATP.

The nucleotide **adenosine triphosphate (ATP)** is involved in the biosynthesis of cell constituents, transport, motility, polymerization reactions, and other activities of the cell. Two of the bonds in ATP are high-energy bonds (**Figure 8.3**). Hydrolysis of the high-energy phosphate bond of ATP can raise the free energy level of metabolic intermediates so that they can participate in biosynthetic reactions.

In certain biosynthetic reactions, other high-energy **nucleotides** are involved. For example, uridine triphosphate (UTP) participates in the synthesis of polysaccharides, cytidine triphosphate (CTP) is involved in lipid synthesis, and guanosine triphosphate (GTP) may be an activator in protein synthesis. These nucleotides are thermodynamically equivalent to ATP.

The major mechanisms that have evolved for ATP formation in microbial cells are redox reactions and photon-driven reactions. Redox reactions fall into two categories: substrate-level phosphorylation and **electron transport phosphorylation**. The mechanism whereby ATP is generated by these redox reactions is discussed here in some detail. The photon-driven mechanisms are the subject of Chapter 9.

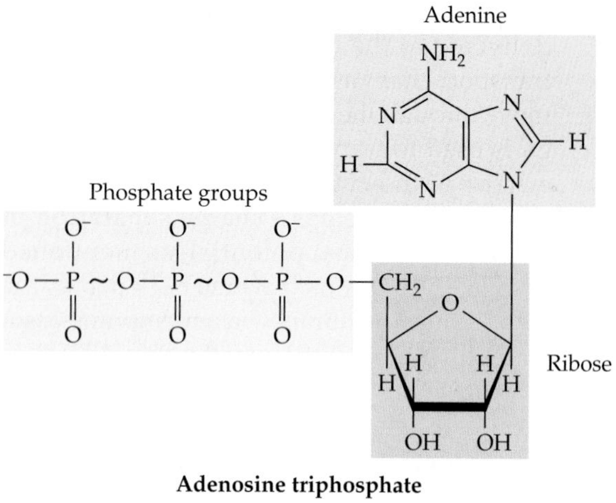

Adenine

NH$_2$

Phosphate groups

Adenosine triphosphate

Figure 8.3 Adenosine-5'-triphosphate (ATP)
ATP is a major carrier of chemical energy in living cells. The energy-rich phosphate bonds are designated by a squiggle (~).

Fermentation

Fermentations are the sum of anaerobic reactions that can provide energy for the growth of microorganisms in the absence of O$_2$. They can be defined as energy-yielding processes of substrate interconversions characterized by substrate-level phosphorylation and the absence of cytochrome-mediated electron transfer(s). During fermentation, one organic compound (the energy source or substrate) serves as a donor of electrons, and another organic compound is the electron acceptor. The principal substrates for fermentation include carbohydrates, amino acids, purines, and pyrimidines. The most studied fermentations include alcoholic fermentation by yeasts, lactic acid fermentation in bacteria, and anaerobic dissimilation of amino acids by some species of the clostridia. In amino acid **dissimilation,** one amino acid is oxidized and a product (or products) of the dissimilation of a second amino acid serves as electron acceptor.

Fermentation is essentially a **closed system** in which a substrate is oxidized and a product of the substrate is reduced.

Two examples of a balanced fermentation that occur under strictly anaerobic conditions are **glycolysis** and the **Stickland reaction**. In both cases, ATP is generated by substrate-level phosphorylation. The overall reactions in anaerobic glycolysis leads to the formation of two molecules of lactic acid:

$$C_6H_{12}O_6 \rightarrow 2\ CH_3\overset{\overset{\displaystyle O}{\|}}{C}-COOH \rightarrow 2CH_3-CHOH-COOH$$

2 NADH   2 NADH

Glucose   **Pyruvate acid**   **Lactic acid**

These reactions are balanced in that the oxidation level of the substrates and products are equivalent. There is no externally supplied electron acceptor such as O$_2$ involved in fermentation. The necessity for a balanced redox level in substrate and product limits the range of compounds that can serve as energy source for fermentation. Most such substrates are essentially neutral in the redox state (one atom of carbon to two of hydrogen or to one of oxygen). They cannot be highly oxidized or highly reduced and must provide intermediates that are sufficiently oxidized so that they can serve as electron acceptors. Another common intermediate that would be an effective electron acceptor is acetaldehyde. Reduction of this compound would produce ethanol. Because the carbons in the organic substrate in a fermentation are partially oxidized, only a limited amount of the potential energy available is released.

Glycolysis, also termed the Embden-Meyerhof pathway (**Figure 8.4**), is the energy-producing system in the lactic acid bacteria and many other microorganisms

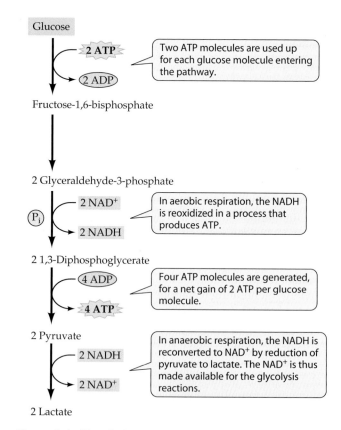

Glucose

2 ATP

2 ADP

Two ATP molecules are used up for each glucose molecule entering the pathway.

Fructose-1,6-bisphosphate

2 Glyceraldehyde-3-phosphate

P$_i$

2 NAD$^+$

2 NADH

In aerobic respiration, the NADH is reoxidized in a process that produces ATP.

2 1,3-Diphosphoglycerate

4 ADP

4 ATP

Four ATP molecules are generated, for a net gain of 2 ATP per glucose molecule.

2 Pyruvate

2 NADH

2 NAD$^+$

In anaerobic respiration, the NADH is reconverted to NAD$^+$ by reduction of pyruvate to lactate. The NAD$^+$ is thus made available for the glycolysis reactions.

2 Lactate

Figure 8.4 Glycolysis
The basic features of glycolysis, an anaerobic pathway in which a molecule of glucose is catabolized to yield two molecules of lactate. There is a net yield of 2 ATP per glucose and no net gain or loss of electrons. For more details of this pathway see Figure 10.2.

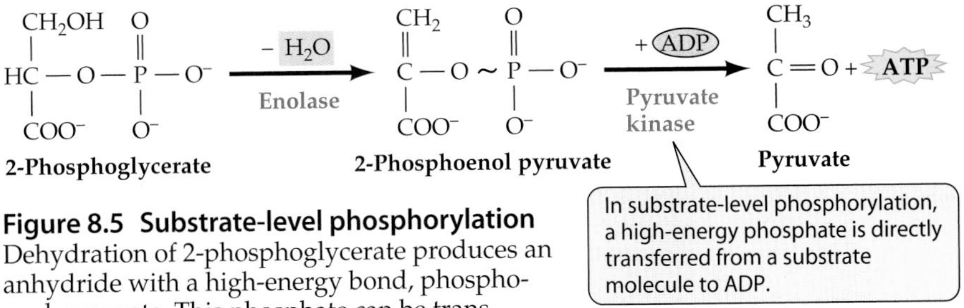

$$\underset{\substack{\text{2-Phosphoglycerate}}}{\overset{\substack{CH_2OH \quad O \\ | \qquad \| \\ HC-O-P-O^- \\ | \qquad | \\ COO^- \qquad O^-}}{}} \xrightarrow[\text{Enolase}]{-H_2O} \underset{\substack{\text{2-Phosphoenol pyruvate}}}{\overset{\substack{CH_2 \quad O \\ \| \qquad \| \\ C-O\sim P-O^- \\ | \qquad | \\ COO^- \qquad O^-}}{}} \xrightarrow[\substack{\text{Pyruvate} \\ \text{kinase}}]{+ADP} \underset{\substack{\text{Pyruvate}}}{\overset{\substack{CH_3 \\ | \\ C=O \\ | \\ COO^-}}{}} + \boxed{ATP}$$

Figure 8.5 Substrate-level phosphorylation
Dehydration of 2-phosphoglycerate produces an anhydride with a high-energy bond, phosphoenol pyruvate. This phosphate can be transferred to adenosine diphosphate to form ATP.

> In substrate-level phosphorylation, a high-energy phosphate is directly transferred from a substrate molecule to ADP.

that grow anaerobically. The pathway is also the major route of glucose catabolism in aerobic organisms. The lactic acid fermentation initially expends two molecules of ATP in going from glucose to fructose 1,6-diphosphate, which is cleaved to two molecules of glyceraldehyde-3-phosphate. In an oxidation carried out with NAD$^+$ and the enzyme glyceraldehyde-3-phosphate dehydrogenase, inorganic phosphate is incorporated into glyceraldehyde-3-phosphate to form energy-rich 1,3-diphosphoglyceric acid and NADH.

The energy-rich 1,3-diphosphoglyceric acid is catabolized to a molecule of 2-phosphoglyceric acid and generates a molecule of ATP via substrate-level phosphorylation. The 2-phosphoglyceric acid is dehydrated by the enzyme enolase to generate phosphoenolpyruvic acid, an anhydride with a high-energy bond. The enzyme pyruvate kinase can couple this high-energy phosphate bond to a molecule of ADP to generate a molecule of ATP and pyruvic acid (**Figure 8.5**). This is the second reaction in glycolysis that generates ATP.

The reactions in glycolysis result in the generation of four molecules of ATP from two molecules of 1,3-diphosphoglyceric acid (see Figure 8.4). The end product of the reaction is pyruvic acid. The reduction of pyruvic acid with NADH as electron donor (from the preceding oxidations) permits the regeneration of NAD$^+$. The net reaction is as follows:

> Glucose + 2 ADP + 2 P$_i$ →
> 2 lactic acid + 2 ATP

The total energy generated by glycolytic catabolism of glucose is two ATPs. This is meager compared with the 38 ATPs that can result from the complete aerobic oxidation of one molecule of glucose.

Amino Acid Fermentation by Clostridia

The strictly anaerobic clostridia (genus *Clostridium*) have evolved with an array of ATP-generating mechanisms that can employ many different substrates, including pyruvic acid, purines, pyrimidines, nicotinic acid, carbohydrates, and amino acids. Various species of clostridia utilize the Stickland reaction to anaerobically catabolize amino acids (**Figure 8.6**). In the classic Stickland reaction, one amino acid of a pair is oxidized and the other is reduced. As in glycolysis, an NAD$^+$ mediated oxidation generates the high-energy intermediate, acetyl-coenzyme A (**Figure 8.7**). Then substrate-level phosphorylation reaction results in the uptake of one inorganic phosphate to generate acetyl phosphate with the concomitant release of coenzyme A. The high-energy phosphoryl

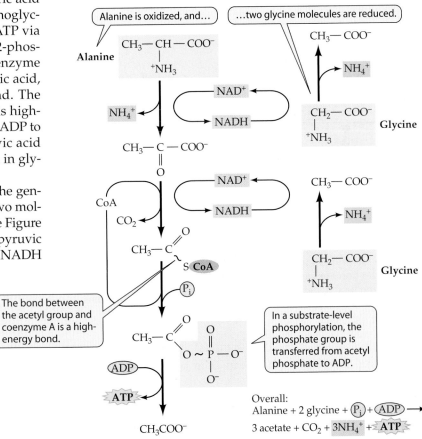

Figure 8.6 Stickland reaction
In the Stickland reaction, pairs of amino acids are used to generate ATP by substrate-level phosphorylation. This reaction occurs in anaerobic organisms such as *Clostridium sporogenes*.

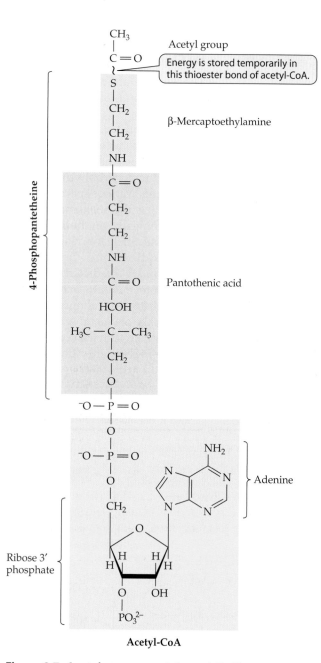

Figure 8.7 Acetyl-coenzyme A (acetyl-CoA)
The thioester bond between the β-mercaptoethylamine moiety of CoA and the acetyl groups (~) is an energy-rich bond.

Table 8.2 — Examples of products generated during fermentation of glucose and the microorganism involved

Type	Nongaseous Product	Micro-organism
Mixed acid	ethanol + acetate + lactate	*Escherichia coli*
Butanediol (neutral)	2,3-butanediol + ethanol	*Enterobacter aerogenes*
Alcoholic	ethanol	*Zymomonas mobilis*
Homolactic	lactate	*Lactobacillus acidophilus*
Heterolactic	lactate + ethanol	*Lactobacillus brevis*
Butanol/acetone	acetone + butanol	*Clostridium butyricum*

acids, CO_2, and molecular hydrogen (H_2). Other fermentations not involving sugars are listed in **Table 8.3**. One necessity is a **fermentation balance**, meaning that the oxidation/reduction state of the products is equivalent to that of the substrate.

Energy Generation Via Respiration

Aerobic and anaerobic **respiration** are both characterized by the transfer of reducing equivalents (electrons) from a donor, such as NADH, succinate, and methanol, to a terminal electron acceptor.

In microorganisms that generate ATP by respiration, this transfer of electrons occurs via sequential redox reactions involving highly organized electron transport chains in the cytoplasmic membrane. The terminal electron acceptor may be molecular oxygen or other oxidants such as sulfate or nitrate. The passage of electrons through an electron transport chain promotes the translocation of pro-

Table 8.3 — Bacterial fermentations that utilize substrates other than sugars

Type	Overall Reaction Involved	Organism
Homoacetic acid	$4H_2 + CO_2 \rightarrow CH_3COOH$	*Clostridium aceticum*
Propionic acid	Lactate → propionate + acetate	*Clostridium propionicum*
Acetylene	Acetylene + H_2O → ethanol + acetate	*Pelobacter acetylenicus*
Oxalate	Oxalate → formate + CO_2	*Oxalobacter formigenes*
Malonate	Malonate → acetate + CO_2	*Malomonas rubra*

group from acetyl phosphate can then be transferred to ADP to form ATP.

There are a number of distinct fermentations that bacterial species living under anaerobic conditions can utilize to generate energy. Among these are the alcoholic and homolactic fermentations (**Table 8.2**). Small amounts of other compounds may be produced during sugar fermentations including alcohols, organic

Energy Efficiency in Viable Systems

The efficiency of oxidative phosphorylation can be as follows:

For the complete oxidation of 1 mole of NADH via the electron transport chain with oxygen as terminal electron acceptor:

$$\Delta G^{O'} = -nF\Delta Eh$$

n = number of electrons

F = Faraday constant

ΔE_h = redox potential difference between two half reactions at a defined pH

$NAD/NADH + H^+ -0.32$

$2H^+ + \frac{1}{2} O_2 /H_2O + 0.82$

$\Delta G^{O'} = 2 (96.48)(0.82)(-0.32)$

$\quad = 219.97$ kJ/mol (theoretical)

The $\Delta G^{O'}$ for ATP hydrolysis

$\quad = 30.5$ kJ/mol

If 3 moles of ATP are synthesized/mol of NADH:

$$3 \times (30.5) = 91.5 \text{ kJ (actual)}$$

Efficiency would be:

$$\frac{91.5}{220} = 42\% \text{ of the energy is conserved as ATP}$$

tons to the exterior of the cytoplasmic membrane. The transfer of reducing potential from substrate to the membrane electron transport system is generally through NADH (see **Box 8.2**). Considerable variation exists in the constituents of the electron transport system used by various *Bacteria* and *Archaea*, and the total ATP yield will vary according to the efficiency of the system available. Facultatively anaerobic and obligately anaerobic chemoheterotrophs can utilize a variety of electron acceptors to derive energy from an electrochemical gradient generated by electron transport (**Table 8.4**). The following section will discuss some mechanisms for generating ATP via electron transport.

Energy-Transducing Membranes

The cytoplasmic membrane of respiring and photosynthetic prokaryotic cells, the inner membrane of the mitochondrion, and the thylakoid membrane of a chloroplast are all **energy-transducing membranes.** These membranes have electron carriers such as the flavins, nonheme iron, coenzyme Q, and cytochromes embedded in the bilayer (**Figure 8.8**). These carriers are organized to function in a sequential fashion (Figure 8.8). The orientation and carriers present in an electron transport chain may differ and are dependent on the nature of the microorganism, the primary energy source, and growth conditions. Carrier proteins that span the entire width of the cytoplasmic membrane mediate the chemical reactions that establish the proton gradient, or proton motive force. Among the common features of energy-transducing membranes are two distinct protein assemblies: (1) an ATP synthetase that catalyzes the "uphill" (energy-requiring) synthesis of ATP from ADP + P_i, and (2) a respiratory chain that catalyzes the "downhill" (energy-yielding) transfer of electrons to a terminal acceptor such as O_2 (Figure 8.8).

Generation of the Proton Motive Force

The proton motive force in respiring microorganisms is generated by the concerted actions of electron carriers. We will consider variations that may occur in the types of carriers and their orientation later. In the electron transport chain shown in Figure 8.8, NADH donates two hydrogen atoms and two electrons to the flavoprotein (see **Figure 8.9**). The reduced flavoprotein donates two electrons to

Table 8.4	Electron acceptors used by *Bacteria* or *Archaea* during aerobic and anaerobic respiration. Some organisms that utilize these acceptors are presented		
	Terminal Electron Acceptor	**Typical Product**	**Microorganism**
Aerobic			
	O_2	H_2O	*Micrococcus luteus*
Anaerobic			
	SO_4^{2-}	H_2S	*Desulfovibrio desulfuricans*
	Fe^{3+}	Fe^{2+}	*Geobacter metallireducens*
	Mn^{4+}	Mn^{2+}	*Desulfuromonas acetoxidans*
	NO_3^-	NO_2^-	*Escherichia coli*
	NO_3^-	N_2	*Thiobacillus denitrificans*
	CO_2	CH_4	*Methanosarcina barkeri*
	CO_2	CH_3COO^-	*Clostridium aceticum*
	S^0	H_2S	*Desulfuromonas acetoxidans*
	AsO_4^{2-}	AsO_3^{2-}	*Chrysiogenes arsenatis*
	Fumarate	Succinate	*Wolinella succinogens*

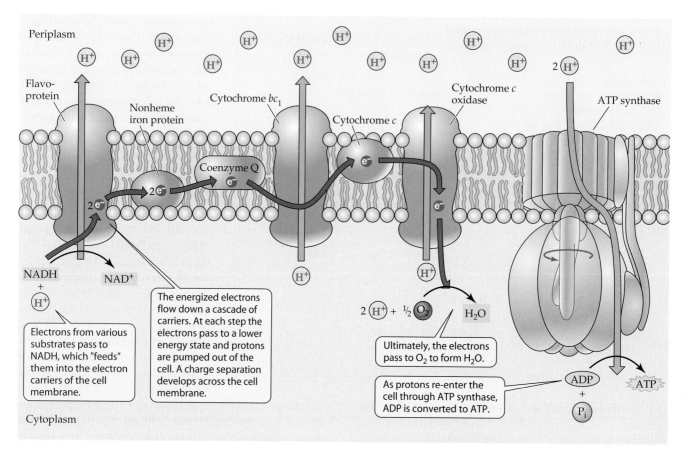

Figure 8.8 Electron transport and ATP synthesis
Model of an electron transport chain that translocates protons to the external surface of a cell. Re-entry of protons into the cell is accompanied by ATP synthesis. The overall reaction for the electron transport chain is NADH + H$^+$ + ½ O$_2$ → NAD$^+$ + H$_2$O. Each carrier in the chain alternates between its oxidized and reduced state.

the nonheme iron-sulfur protein complex (**Figure 8.10**), with two protons (H$^+$) moved across the cytoplasmic membrane and extruded to the periplasm. Two electrons are passed to the quinone coenzyme Q (**Figure 8.11**) and take up two protons from the cytoplasm in the reduction of CoQ. The "Q cycle" functions to transfer protons to the periplasmic (positive) side of the cytoplasmic membrane (**Figure 8.12**). Electrons flow from coenzyme Q to cytochrome bc, to cytochrome c and cytochrome o. The final step is the reduction of ½ O$_2$ to H$_2$O, and the electron transport reactions are completed. The overall results of this transfer of activated electrons "downhill" to the terminal acceptor is to generate a positive charge (H$^+$) on one side of the membrane and a negative charge on the other.

COMPONENTS OF ELECTRON TRANSPORT CHAINS

The several components of the electron transport chain participate in a series of reactions that carry the elec-

trons, from a high-energy state to a lower-energy state where the electrons react with the terminal acceptor. In the following discussion, we will focus on the links between electron transport and chemical structures of components of these chains in bacteria.

Flavoprotein

Flavoproteins are a family of molecules consisting of a protein of varied molecular mass bound to a nucleotide form of the B-vitamin riboflavin. The **flavins—flavin adenine dinucleotide** (FAD) and **flavin mononucleotide** (FMN)—are two electron (proton) carriers (Figure 8.9). FAD is generally covalently bound to the protein through a histidine group, whereas its negatively charged phosphate groups bind FMN. FAD and FMN have a relatively high redox potential (0.003 to 0.09) when bound to protein but have a redox potential of –0.22 V as coenzymes. The NAD-NADH couple (–0.32 V) can transfer electrons to either the FAD/FADH$_2$ or FMNH$_2$ couple as electrons flow freely from a more negative carrier to a more positive one.

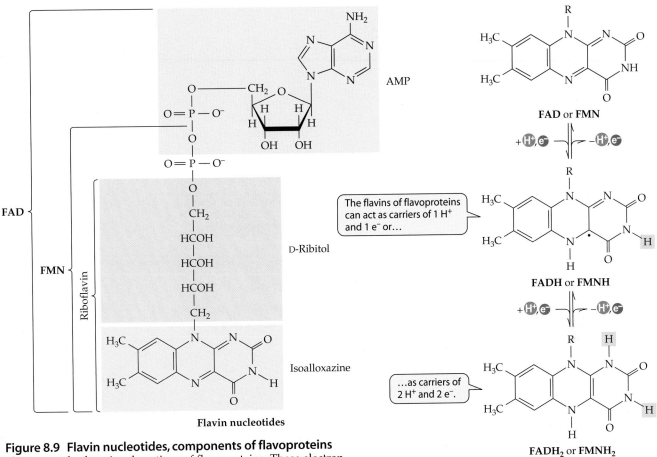

Figure 8.9 Flavin nucleotides, components of flavoproteins
Flavins are the functional portions of flavoproteins. These electron carriers are derivatives of the B-vitamin riboflavin.

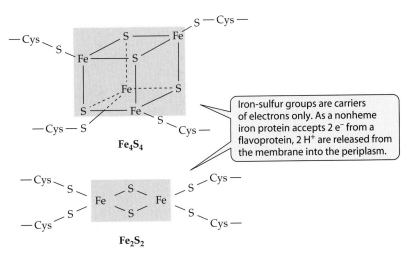

Figure 8.10 Iron-sulfur groups, components of nonheme iron proteins
The number and arrangement of iron and sulfur atoms in the nonheme iron proteins vary. The iron-sulfur groups are attached to the protein through the amino acid cysteine (Cys).

Iron-Sulfur Proteins

Iron-sulfur proteins are relatively small and contain equimolar amounts of iron and sulfur. There can be two, four, or eight atoms of each per molecule of protein (Figure 8.10), and the redox center in the 4Fe/4S complex forms a cube as shown. These centers are held in position by linkage through the sulfur of the amino acid cysteine. The redox potential for these membrane-bound carriers spans a broad range from –0.60 to +0.35 V. Consequently, they can function at different stages of electron transport depending on the microbial system involved. A common iron-sulfur protein in biological systems is **ferredoxin,** which has an Fe_2S_2 configuration.

Quinones

The terms **coenzyme Q, CoQ,** or **Q** denote a family of **quinones** that are of widespread occurrence in energy-transducing membranes of prokaryotes and eukaryotes.

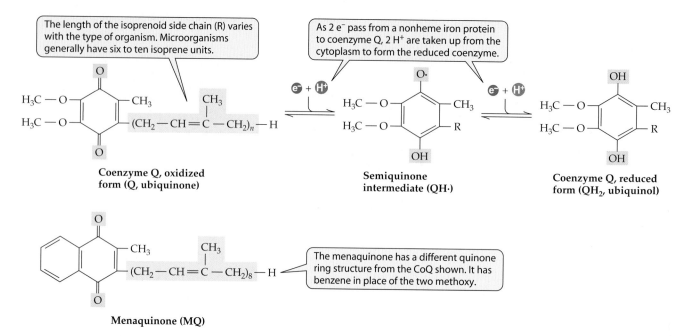

Figure 8.11 Coenzyme Q (ubiquinone)
Coenzyme Q is a quinone (also know as ubiquinone). Its fully reduced form is ubiquinol. Menaquinone (MQ) is the form present in some microorganisms, particularly gram-positive bacteria.

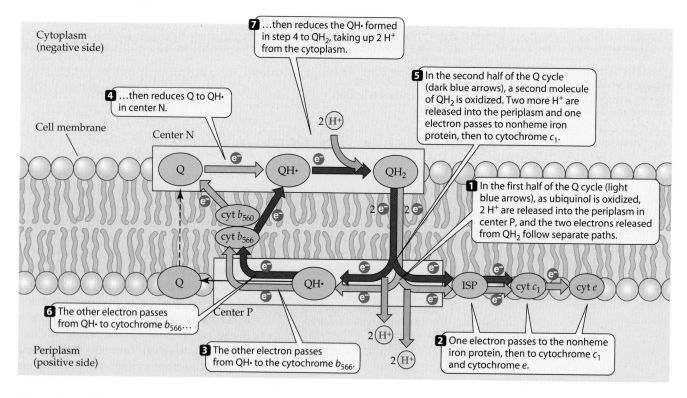

Figure 8.12 The "Q cycle"
In this scheme, Q (coenzyme Q) is ubiquinone, QH· is the semiquinone, and QH_2 is ubiquinol (see Figure 8.11). Cytochrome (cyt) b_{560} and cytochrome b_{566} are transmembrane proteins, part of the cytochrome bc_1 complex; the heme groups of these two cytochromes form an electrical circuit across the membrane. Fe_2S_2 is an iron-sulfur cluster in a nonheme iron protein. Center N is the site of the two-step reduction of Q on the negative side of the membrane. In one complete Q cycle, two QH_2 molecules are oxidized to Q and one Q is reduced to QH_2; cytochromes b_{566} and b_{560} are reduced and reoxidized twice. Cytochrome b_{560} reduces Q to QH· (semiquinone) in the first half of the cycle and reduces QH· to QH_2 in the second half. Overall, one QH_2 is oxidized to Q, two cytochrome c molecules are reduced, 2 H^+ are consumed on the cytoplasmic side of the membrane, and 4 H^+ are released into the periplasm.

Among these are the **ubiquinones,** which are the principle quinones in membranes of eukaryotes and gram-negative bacteria (Figure 8.11). Other bacteria (generally gram-positives) contain **menaquinone.** The structures of ubiquinone and menaquinone are presented in Figure 8.11. NADH generally funnels electrons into the respiratory chain via specific flavoprotein/NADH dehydrogenases. Other dehydrogenases can transfer electrons directly to alternate flavoproteins, and these electrons enter the respiratory chain through CoQ.

Cytochrome

Cytochromes are constituents of the respiratory chain in aerobic and facultatively anaerobic microorganisms. They are also present in some of the *Archaea*. Cytochromes are a heterogeneous group of compounds that are located primarily in the cytoplasmic membrane and are characterized on the basis of their spectrophotometric properties. Those cytochromes (cyt) with similar absorption properties are designated by lowercase letters as cyt *a*, cyt *b*, cyt *c*, and so forth. Within each of these groups there are measurable differences in spectral and biochemical properties. These are indicated by numerical subscripts, for example, cyt a_1, cyt a_2, and cyt a_3. In some cases, a cytochrome is identified by the wavelength of light at which it absorbs maximally. The functional portion of a cytochrome molecule is an iron-porphyrin electron transport component, called **heme** (**Figure 8.13**). The central iron atom in the porphyrin nucleus accepts a single electron on reduction.

The redox potential of the various cytochromes ranges from -100 to 400 mV, their long wavelength absorption maximums from 550 to 650 nm, and their molecular masses are between 12,000 and 350,000. In the electron transport chain, the cytochromes located between CoQ and O_2 transport a single electron per event, but the reduction of oxygen to two molecules of water requires four electrons ($O_2 + 4\,e^- + 4\,H^+ \rightarrow 2\,H_2O$). The cytochromes involved in the transfer of electrons to O_2 are called the cytochrome oxidases. The cytochrome oxidase in most cases is either cytochrome a_3 or cytochrome o.

Much of our understanding of the electron transport chain and respiratory ATP synthesis resulted from studies with chemical compounds that interfere with one or both of these processes. Early workers described two classes of chemicals that affect electron transport, and these were termed *uncouplers* or *inhibitors*. **Uncouplers** prevent the synthesis of ATP but do not interfere with electron transport. An **inhibitor** is a chemical that blocks both electron transport and ATP synthesis. Among the effective uncouplers are 2,4 dinitrophenol (DNP) and carbonyl cyanide-p-trifluoromethoxyphenylhydrazone (FCCP). Uncouplers permit electron transport to proceed unimpeded, but this flow cannot be coupled to ATP synthesis. This "uncoupling" occurs because DNP and FCCP are lipophilic and readily pass through a cytoplasmic membrane. They are also acidic and may bind protons on the outside of the membrane and carry them to the inside, thus dissipating the proton gradient. Inhibitors are compounds such as cyanide, carbon monoxide, or azide that can bind to the iron center of cytochromes and block electron transport. Without electron transport, there is no proton gradient and no ATP synthesis.

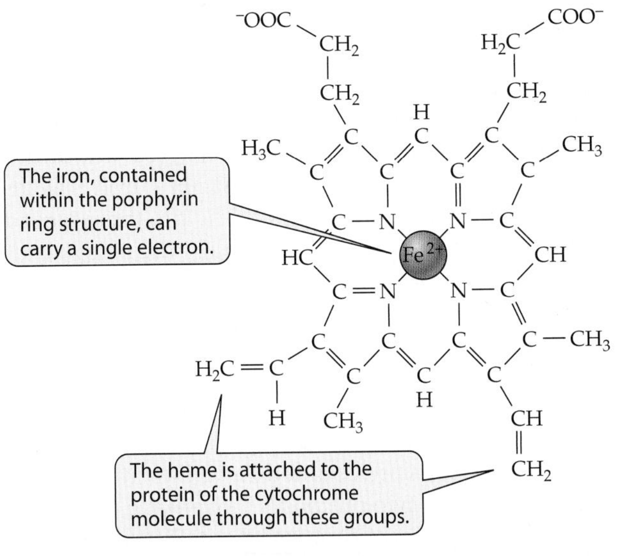

ORIENTATION IN ELECTRON TRANSPORT CHAINS

The orientation of carriers in the electron transport chains present in bacterial cytoplasmic membranes, the inner membrane of eukaryotic mitochondria, and the membranes of photosynthetic bacteria share a number of common properties. There is, however, much greater diversity in the components of the electron transport chains among the bacteria. Among the major differences between the transport chains in bacteria and eukaryotes:

- Bacteria have a greater variety of carriers
- Many bacterial species respond to changes in growth conditions by altering their respiratory chain
- The bacterial respiratory chains can be branched with alternate electron transfer pathways for different electron acceptors

The respiratory chain in *Paracoccus denitrificans* is typical for a bacterium growing aerobically (**Figure 8.14**) and is similar to the respiratory chain in mitochondria.

The iron, contained within the porphyrin ring structure, can carry a single electron.

The heme is attached to the protein of the cytochrome molecule through these groups.

Figure 8.13 Heme
The heme portion of a cytochrome molecule.

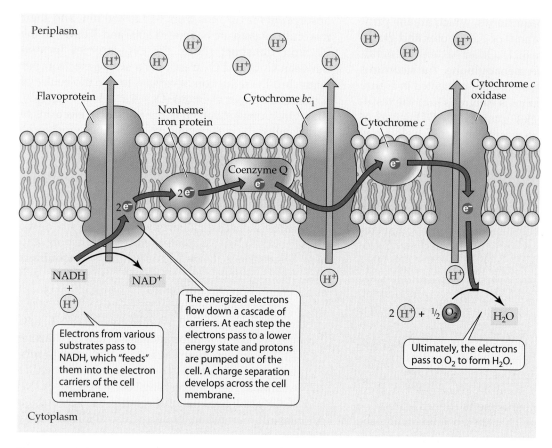

Figure 8.14 Electron transport chain in aerobic bacterium
The electron transport chain operating during aerobic growth in *Paracoccus denitrificans*.

When *P. denitrificans* is grown anaerobically with either NO_3^- or NO_2^- as the terminal electron acceptor, there is a distinctly different electron transport chain present in the cytoplasmic membrane. The electron carriers present in some bacterial species may be equivalent during aerobic or anaerobic respiration, although synthesis of some cytochromes, such as cytochrome oxidase, is oxygen dependent.

Diversity in electron carriers present in electron transport chains reflects the ability of bacteria to catabolize an array of different substrates. By using these various carriers, bacteria can extract available energy from substrates of markedly different redox potential.

Electrons may also flow into the respiratory chain at intermediate sites. The primary dehydrogenases such as succinate dehydrogenase can transfer electrons to CoQ at a different site than that utilized by electrons from NADH.

The orientation of the cytochrome system may be branched or linear or, in some cases, composed of a very limited number of cytochromes depending on substrate or O_2 availability. For example, *Escherichia coli* has a linear terminal pathway during growth at high oxygen levels and a branched system when oxygen availability is limited.

The aerobic iron-oxidizing bacterium *Thiobacillus ferrooxidans* has an abbreviated electron transport chain (**Figure 8.15**) and is an example of a bacterium that utilizes its environment to advantage in energy generation. *T. ferrooxidans* utilizes iron as an energy source in environments where the external pH is about 2. The pH in the cytoplasm of the microorganism growing under these conditions would be 6 to 6.5. At pH 2, iron is available as soluble Fe^{2+}, and at this acidic pH there would be an abundance of protons (H^+) in the area surrounding the outer membrane surface. At the outer surface of the cytoplasmic membrane in *T. ferrooxidans* is the copper-containing protein rusticyanin. Rusticyanin removes an electron from Fe^{2+} producing insoluble Fe^{3+}. These electrons flow from rusticyanin to cyt *c* and inward to cyt a_1 on the inner surface of the cytoplasmic membrane. There the electrons are coupled to $\frac{1}{2} O_2$ with the concomitant uptake of two protons (H^+) from the cytoplasm to form H_2O. Uptake of protons internally would result in a relatively high concentration of protons on the exte-

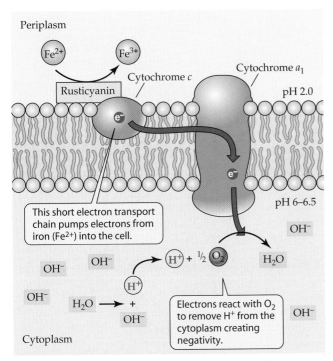

Figure 8.15 Abbreviated electron transport chain of a sulfur bacterium
The electron transport chain of *Thiobacillus ferrooxidans*, which uses Fe^{2+} as an energy source.

rior of the cytoplasmic membrane and negativity on the interior (H^+ outside, OH^- inside). Protons would be drawn from the acidic outer environment inward driving the ATP synthase reaction and generating ATP. The role of the abbreviated electron transport chain in *T. ferrooxidans* is to pump electrons inward, removing protons from the cytoplasm and creating a charge separation. This is a remarkable example of the versatility in the prokaryotic world.

CHEMIOSMOTIC THEORY OF ENERGY CONSERVATION

The mechanism by which the potential energy in cytoplasmic electron carriers such as NADH could be utilized in the synthesis of ATP was historically an intriguing question. For decades scientists exhaustively sought an "energy-transducing intermediate" that would be involved in the following reaction:

$$ADP + P_i \rightarrow (\text{"Activated Intermediate"}) \rightarrow ATP$$

The "intermediate" is no longer of much interest because evidence now available affirms that there is no such compound. About 40 years ago, Peter Mitchell, an English scientist, put forth the **chemiosmotic hypo-**

thesis, which proposed that the driving force for the "uphill" synthesis of ATP is a proton gradient that can be established across a cellular membrane. The chemiosmotic theory was the subject of much debate and experimentation, but is now widely recognized as the only feasible explanation for the data available. For his contributions, Mitchell received the 1978 Nobel Prize in chemistry (see Box 8.3).

The basic tenet of the chemiosmotic theory is that the electron transport chains in the respiratory membranes of mitochondria, chloroplasts, and bacteria are integrated with ATP synthesis through the establishment of a proton gradient across an intact membrane. There are two significant events involved:

1. Protons are translocated across the cellular energy-transducing membrane, thus creating a proton electrochemical potential.
2. The proton potential between the inner and outer surfaces drives the ATP-hydrolyzing proton pump (ATP synthase) backward toward ATP synthesis.

There are two distinct protein assemblies in the energy-transducing membrane:

1. An electron transport chain that catalyzes the transfer of electrons from substrates such as NADH or succinate to a terminal electron acceptor driving electrons in **one** direction to establish a proton potential across the membrane.
2. ATP synthase that catalyzes the synthesis of ATP from ADP and P_i.

The chemiosmotic theory is dependent on the concept of a **vectorial** (directional) **translocation.** The major role of the electron transport proteins in the electron transport chain is a passive one. The proteins hold the electron carrier prosthetic groups that define the vectorial pathways, which are spatially oriented to promote unidirectional processes.

The energy available from an electrochemical gradient is determined by the relative pH (concentration of H^+) on the opposing side of the membrane and is called the **electrochemical potential** or **proton motive force.** This is designated as $\Delta\mu H^+$, where μ represents the electrochemical potential. The cytoplasmic membrane of bacteria and the membrane of mitochondria maintain this potential because they are impermeable to both protons (H^+) and hydroxyl (OH^-) ions and have little electrical conductivity.

The proton motive force has two components: the pH gradient (ΔpH) and the electrical potential ($\Delta\psi$) between two phases separated by a membrane. This proton motive force drives the synthesis of ATP. The number of ATP molecules generated per transfer of two electrons from NADH through the respiratory chain is three in microbes having cytochrome *c* and two in those lacking this component.

The coupling of the $\Delta\mu H^+$ to the production of ATP uses a second major protein complex in energy-transducing membranes, and that is the ATPase system. The function of the ATPase system is presented in the following section.

ATP SYNTHASE SYSTEM

ATP synthase is present in the energy-transducing membranes of mitochondria, chloroplasts, aerobic, and photosynthetic bacteria (**Figure 8.16**). It is also present in *Archaea* and anaerobic bacteria that do not rely on substrate-level phosphorylation. The structure of the complex is remarkably similar regardless of the source. Staining of energy-transducing membranes with a negative stain (such as phosphotungstate) reveals that knobs or mushroomlike bodies protrude from one side of the membrane. In *Bacteria*, *Archaea*, and mitochondria, they protrude into the internal space (cytoplasm or matrix), and in isolated thylakoid or chromatophore membranes, they project outward. Regardless of direction, these projections are functionally equivalent because ATP is synthesized (or hydrolyzed) on the side from which the knobs project, whereas protons enter from the side lacking knobs.

The function of the ATP synthase is to utilize the $\Delta\mu H^+$ to maintain a mass-action ratio **away from equilibrium** and toward the synthesis of ATP. Remember that ATP synthase is actually an ATP-hydrolyzing enzyme. In fermentative bacteria, it utilizes ATP to maintain a $\Delta\mu H^+$ for active transport. Many organisms, including some of the photosynthetic bacteria, use electron donors that have a redox potential more positive than NAD^+ or $NADP^+$ (Table 8.1). These bacteria cannot produce NADH or NADPH directly and need a source of reduced pyridine nucleotide for synthetic reactions, and if they are CO_2-fixing autotrophs, they must contain NADH for the reduction of 3-phosphoglycerate (they reverse glycolysis to synthesize sugars). ATP can drive a reverse $\Delta\mu H^+$ generating system that generates reduced pyridine nucleotides.

There are in reality two forms of energy utilized by the bacterial cell; one is ATP and the other is the proton motive force. These energy forms are interconvertible by the membrane-bound F_1F_0 ATPase. The proton motive force is utilized for many purposes in the bacterial cell, including nutrient transport, membrane biogenesis, protein export, turning flagella, and maintaining a favorable intracellular pH.

The knob on energy-transducing membranes is a series of proteins designated $\mathbf{F_1}$. This is the site of ATP synthesis and can be readily separated from the mem-

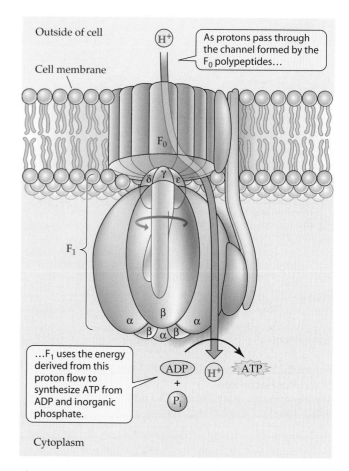

Figure 8.16 ATP synthase
The structure of ATP synthase, showing the F_0 and F_1 subunits.

brane (Figure 8.16). The F_1 is composed of at least five polypeptides, and isolated F_1 can catalyze an exceedingly rapid hydrolysis of ATP: One mole of F_1 can hydrolyze 10^4 moles of ATP per minute. The reverse reaction has not been observed in vitro (the proton motive force is required to drive the ADP + $P_i \rightarrow$ ATP reaction). The remainder of the ATP synthase complex, termed $\mathbf{F_0}$, is a hydrophobic protein that is buried in the membrane. This protein complex serves as a channel for passage of protons from the exterior of the membrane toward F_1.

The energetically favorable return of protons to the matrix is coupled to ATP synthesis by the F_1 subunit, which catalyzes the synthesis of ATP from ADP and phosphate ions (Pi). Structural studies have established the mechanism of ATP synthase action, which involves mechanical coupling between F_0 and F_1 subunits. In reality, the flow of protons through F_0 drives the rotation of F_1, which acts as a rotary motor to drive ATP synthesis.

Milestones Box 8.3   Evolution of a Concept—Chemiosmosis

Peter Mitchell was born in Surrey, England, in 1920. He attended Cambridge University, where he completed his Ph.D. in 1951. Mitchell remained at Cambridge until 1955, when he moved to the University of Edinburgh. There he directed the chemical biology unit for the next eight years. In 1961 Peter Mitchell, with his able assistant, Jennifer Moyle, published a landmark paper that revolutionized thinking on cell energetics. This was the *chemiosmotic hypothesis*, based on his belief that enzymatic reactions in solution could differ from those that might occur when enzymes are embedded in a membrane. Mitchell believed that one could have a chemical reaction that could result in translocation of a chemical group. This directional (vectorial) event could cause the movement of a proton from the inner surface of the membrane to the outer surface. He theorized that this would create an electrochemical proton gradient ($\Delta\mu H^+$) across the membrane. The relatively stable $\Delta\mu H^+$ was the central energy currency in respiration. In his hypothesis, the exterior of the membrane would be positively charged, and the interior would be rendered negative. The protons would be drawn toward the interior and pass through a proton-translocating ATP synthase. That would drive ATP synthesis. This hydrogen ion current would be the driving force for performance of chemical, osmotic,

and in the case of flagellar movement, mechanical work.

The scientific establishment working in the field of bioenergetics considered Mitchell's hypothesis bizarre and untenable. The central dogma of the time was that a high-energy intermediate composed of respiratory chain enzymes and an ATP-synthetase must exist. Intermediates were sought; many were considered, but all failed the ultimate test because no such factor existed. It was not until the mid 1960s, at which time many biochemists had worked long and hard and failed to find a bound high-energy phosphoryl intermediate associated with respiratory phosphorylation, that the world came around to Mitchell's ideas. There was no "coupling" factor between respiratory chain enzymes and the ATP synthase system.

In 1963 ill health led to Mitchell's retirement to a small laboratory in Cornwall. He rebuilt an eighteenth-century derelict Cornish mansion called Glynn House, and there he reestablished his research program. To the restful, calm environs of the Glynn Research Laboratory, he invited his detractors and gradually convinced all of them that the chemiosmotic hypothesis was indeed correct. In 1978 he received the Nobel Prize, and few doubted the wisdom of his theories.

Glynn House stands as an anomaly and a viable alternative to the

Peter Mitchell, winner of the Nobel Prize for chemistry in 1978, proposed the chemiosmotic theory for generation of ATP. Courtesy of *Times Union*.

huge collaborative efforts that receive funds from centralized funding sources. Is truly original research accomplished better in the hurly-burly of vast enterprises? Can original research always thrive best with centralized funding based on previous results? These are interesting questions in the age of big science, and Glynn House stands as mute testimony that bigger is not necessarily better.

SUMMARY

▶ The sun ultimately furnishes virtually all energy available for biological systems.

▶ The universal medium of biological energy exchange is **adenosine triphosphate (ATP)**. A major part of the ATP available to living organisms is generated by an **electrochemical proton gradient** across a membrane. **Light** from the sun can be employed to establish this

gradient in photosynthetic bacteria and eukaryotes. As products of photosynthesis are the ultimate source of food for heterotrophic life, this proton gradient provides energy for viable systems.

▶ There are **three major mechanisms for ATP generation:** (1) respiration, (2) photophosphorylation, and (3) substrate-level phosphorylation.

▶ **Redox** (oxidation-reduction) reactions require an **electron donor** and an **electron acceptor.** Oxidation is the loss of electrons, and reduction is a gain of electrons. In biological redox reactions, one molecule is oxidized and another reduced. The amount of **free energy** available in a redox reaction is the difference in **reduction potential (E_0')** between the donor and acceptor of electrons. The reduction potential is the inherent propensity of electrons to flow from a higher energy state to a lower state.

▶ In biological systems, **electron carriers** carry electrons enzymatically removed from a substrate to an acceptor. **Nicotinamide adenine dinucleotide** (NAD^+) is soluble and moves freely in cytoplasm. NAD^+ can accept two electrons and is a major electron carrier in cells as NADH.

▶ Electron carriers may also be **fixed** in membranes and form the electron transport chain. They carry electrons and accompanying protons across the cytoplasmic membrane to establish a proton gradient. **Cytochromes, iron-sulfur complexes,** and **coenzyme Q** are fixed electron carriers.

▶ A **fermentation** can be defined as an anaerobic reaction that provides energy when O_2 is unavailable. Fermentations are not involved with electron transport or proton gradients.

▶ **Substrate-level phosphorylation** is the direct formation of ATP from adenosine diphosphate and an activated phosphate group on a metabolic intermediate.

▶ **Glycolysis** is the major route of glucose catabolism in living cells. Anaerobic glycolysis is a **closed system** with products generated that are at the same redox level as glucose. For example, in the lactic acid fermentation:

$$C_6H_{12}O_6 \rightarrow 2\,C_3H_6O_3$$
glucose   lactic acid

▶ **Respiration** occurs when reducing equivalents (e^-) from donors are passed to electron acceptors. During this process, electrons move from substrate to fixed electron transport chains in the membrane. Components of these chains **translocate protons** to the exterior of the membrane. The ultimate acceptor of electrons that carry the protons outward would be oxidized molecules (O_2, NO_3^-, SO_4^{2-}, or fumarate).

▶ A cytoplasmic membrane that bears fixed electron carriers is called an **energy-transducing membrane.** The inner membrane of mitochondria, the thylakoid membranes in chloroplasts, and the cytoplasmic membrane of respiring and photosynthetic bacteria are energy-transducing membranes.

▶ Among the components of respiratory chains are **flavoproteins, iron-sulfur proteins, quinones,** and **cytochromes.** There is much greater diversity in the components of bacterial respiratory chains than in those of eukaryotes.

▶ According to the **chemiosmotic theory,** the proton gradient generated by electron transport is employed to drive ATPase, an assemblage of transmembrane proteins that synthesize ATP from ADP and P_i.

▶ The energy available from a proton gradient depends on the difference between the pH (concentration of H^+) on the outer surface and the pH on the inner surface of the membrane; this is termed the **electrochemical potential.** This separation of charge occurs because a cytoplasmic membrane is impermeable to protons and protons can flow to the cytoplasm only through the ATP synthase system.

▶ The ATP synthase complex can function in either **direction.** When protons flow inward, ATP is formed. ATP can be **hydrolyzed** to ADP + P_i and the energy utilized to establish a proton gradient.

REVIEW QUESTIONS

1. What is the universal medium of energy exchange in viable cells? There are three major mechanisms for formation of this compound. How do they differ? What do they have in common?

2. Define a charge separation. What are the terms we use to describe this charge separation?

3. A redox reaction occurs with a transfer of electrons. List some components that might function in a redox reaction. Think in terms of fermentation and respiration.

4. If ferredoxin served as a half reaction in a redox reaction and cytochrome c as the other half, what would be the $\Delta G^{o'}$? What is a half reaction? Do half reactions actually occur in biological systems?

5. Give examples of electron carriers. How does an electron carrier function? Explain how diffusible and fixed electron carriers differ.

6. How do substrate-level phosphorylation and electron transport phosphorylation differ?

7. Provide an example of a fermentation. What is the redox level of the substrate and products? Why? Cite examples of fermentation and how they are named.

8. What are some basic differences between fermentation and respiration? What do they have in common?

9. The electron transport chain in *Thiobacillus ferrooxidans* is unique. How does it differ functionally from that in *Escherichia coli*?

10. What is an energy-transducing membrane and why is it given this name?

11. How do electron transport chains in bacteria differ from those in eukaryotes? What do they have in common?

12. Outline the chemiosmotic theory. What is the role of the "Q cycle"? ATP synthase ?

13. Terminal electron acceptors in eukaryotes and bacteria differ in some respects. Explain what these differences are and what makes the prokaryotes unique.

SUGGESTED READING

Battley, E. H. 1987. *Energetics of Microbial Growth*. New York: John Wiley and Sons.

Harold, F. M. 1986. *The Vital Force: A Study of Bioenergetics*. New York: W. H. Freeman and Co.

Harris, D. A. 1995. *Bioenergetics at a Glance*. Cambridge, MA: Blackwell Science.

Shively, J. M. and L. L. Barton, eds. 1991. *Variations in Autotrophic Life*. New York: Academic Press.

White, D. 1995. *The Physiology and Biochemistry of Prokaryotes*. New York: Oxford University Press.

Youvan, D. C. and F. Daldal. 1986. *Microbial Energy Transduction: Genetics, Structure, and Function of Membrane Proteins*. New York: Cold Spring Harbor Laboratory Press.

Chapter 9

Cellular Energy Derived from Light

Life on planet Earth is completely dependent on the harvesting of the sun's light energy, and on the transmission and chemical storage of that energy by photosynthetic organisms and plants.

— PETER MITCHELL

P hotophosphorylation is the process whereby light energy is conserved as chemical energy ATP. This process has much in common with oxidative phosphorylation (respiration) that was covered in the previous chapter. The two processes, oxidative phosphorylation and photophosphorylation are, in a sense, mirror images of one another. In oxidative phosphorylation the hydrogens (H^+) on a reduced substrate (product of photosynthesis) such as glucose react with O_2 to generate energy, H_2O, and CO_2. In photophosphorylation, light energy is used to extract electrons (and H^+) from H_2O (or certain other reduced substrates). The light-activated electrons are employed to generate ATP and ultimately convert CO_2 to reduced plant or bacterial carbon compounds, the cellular components.

The reactions that generate ATP are termed **light reactions,** and those that convert CO_2 to organic matter are called **dark reactions.** Light is required to drive the former but not the latter. The chapter opening quote from Peter Mitchell affirms the importance of these processes. In this chapter we will discuss the diverse processes whereby light energy is conserved in photosynthetic bacteria. After a brief look at the evolution of photosynthesis and common features among diverse photosynthetic processes, we will examine the various mechanisms utilized by 191 diverse photosynthetic bacterial species.

PHOTOSYNTHETIC ORGANISMS

Early Evolution

The evolution of complex living organisms on Earth was dependent on the development of a capacity for **biosynthesis.** That is, the capacity of early life forms to synthesize the kinds and amounts of compounds needed to sustain life and reproduce. Simple accretion of molecules present in the "primordial soup" would have yielded beings of very restricted capabilities. Only when sources of chemical energy became available in the environment could the evolution of regenerative systems occur. The singular event that ultimately allowed for a systematic and orderly evolution of viable systems was the advent of the ability to capture the inexhaustible source of earthly energy— sunlight. Virtually all life on the surface of the Earth relies either directly or indirectly on photons from the sun for cellular energy. It is apparent that the orderly processes of life evolved from chaos when light-capturing chemicals evolved that coupled light energy of excitation to the production of relatively stable, "energy-rich" compounds.

The nature of primitive organisms on Earth is highly speculative, and many of their characteristics remain unknown. Examination of microorganisms now present in the environment and comparative analysis of their component parts and metabolic functions lead to questions of their origin and ancestry, but firm conclusions are elusive. Fermentative energy generation, akin to substrate-level phosphorylation used by present-day *Clostridium* species (discussed in Chapter 8), may have occurred in some of the earliest organisms. However, the abiotic synthesis (not produced by living organisms) of sugar under conditions that existed on the primeval earth is questionable. Consequently, an early metabolism based on the fermentation of sugars is improbable.

It is possible that an early life form may have utilized anaerobic respiration in energy generation. However, the absence of oxidized compounds to serve as electron acceptors would have placed limitations on the effectiveness of an anaerobic proton gradient for ATP generation. Evidence suggests that sulfate can be generated abiotically, and this compound may have served as a terminal electron acceptor for some early anaerobic microorganisms.

One attractive concept is that a simple photosystem, such as that present in *Halobacterium*, might have occurred early in evolution. The generation of a proton gradient across a membrane with a simple photosensitive carotene-type compound (see later discussion) might have been used to drive ATP synthesis phosphorylation.

A simplistic view of the energetics involved in life processes can be drawn as follows:

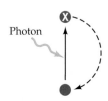

A quantum of energy from the sun, designated $h\nu$, drives an electron in a light-absorbing compound from the ground state (●) to a higher level of electronic excitation (X), illustrated by the vertical line. This excitation, *if untrapped*, would return to the ground state, as indicated by the dashed line. Energy for "life" would be available if excitation-capturing molecules were interposed between the two circles, thereby conserving the quantum of light energy (or **exciton**) inherent in X. This can be depicted as follows:

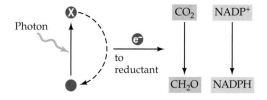

The driving force for all types of cellular processes, both autotrophic and heterotrophic, can be illustrated as components of this scheme:

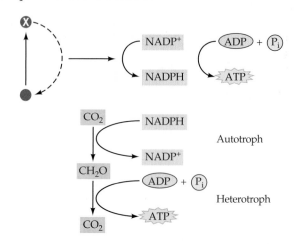

This scheme presents a simple and abbreviated illustration of the reactions that support most life forms. Photophosphorylation systematically captures light energy in a chemically stable intermediate (ATP) and generates the reductant (NADPH) to sustain biosynthetic reactions typically observed in autotrophs. The reduced products serve as source of carbon and energy for heterotrophs (**Box 9.1**). An understanding of the principles presented in this box is important in comprehending this chapter.

Origin and Types of Photosynthetic Organisms

As *Bacteria* and *Archaea* existed on Earth for more than 3 billion years prior to the advent of eukaryotes, it seems reasonable that light-energy generating systems originated with these microorganisms. It is certain that photosynthetic bacteria preceded the evolution of the eukaryotes (see Chapter 1). Three distinct photochemical energy-capturing systems exist in microorganisms.

- An **oxygenic** photochemical event
- An **anoxygenic** (does not produce oxygen)
- A **carotenoid**-based system that utilizes a pigment structurally similar to animal rhodopsin (an eye pigment)

The ability of green and purple photosynthetic bacteria to grow anaerobically (as well as other characteristics) suggests that microorganisms of this type preceded the oxygen generating cyanobacteria in evolution. The occurrence of chlorophyll-based photosynthesis in bacteria that are phylogenetically dissimilar supports the concept that photosynthesis may have been an early event in evolution.

The three photochemical systems in microorganisms involve either a chlorophyll or the carotenoid bacteriorhodopsin. These are the pigments that are responsible

Milestones Box 9.1 The Current of Life

A singular achievement in the evolution of viable microorganisms on Earth occurred when chemical systems arose that could capture light energy. The primordial soup contained compounds that could absorb light energy. Absorption of a quantum of light energy by a light-sensitive molecule could elevate a constituent electron to a higher energy level as shown in the diagram (M*). The energy of excitation now inherent in M* has four possible fates:

- The energy might be dissipated as **heat** as the energy is redistributed about the molecule.
- The energy of excitation might reappear as emitted light or **fluorescence.** A photon of light is emitted as the electron returns to a lower orbital.
- The energy of excitation may be transferred by **resonance energy transfer** to a neighboring molecule, thus raising an electron in the receptor molecule to a higher energy state as the photo-excited electron in the absorbing molecule returns to the ground state.
- The energy of excitation may change the reduction potential of the absorbing molecule to a level such that it can become an **electron donor.** Reacting this excited electron donor to a proper electron acceptor can lead to the transduction of light energy (photons) to chemical energy in the form of reducing power.

It is fate IV that permits the transduction of light energy into chemical energy, which is the essence of photosynthesis. Virtually all life on Earth depends on this simple equation:

light energy → chemical energy

The purposeful use of this chemical energy for synthesis and reproduction led to the evolution of the first viable microbe.

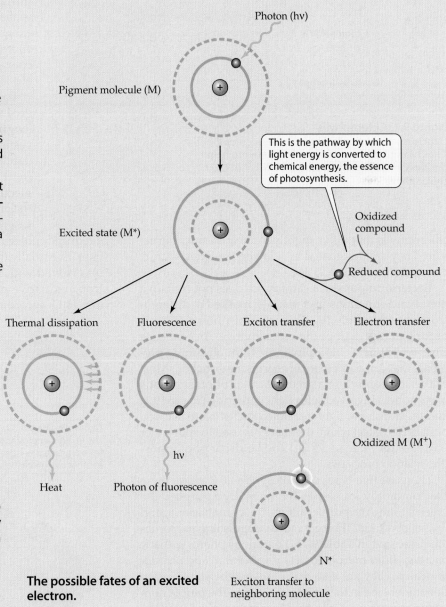

The possible fates of an excited electron.

for absorption of light energy and transforming this energy of excitation to stable intermediates. The **chlorophylls** are tetrapyrroles that are quite similar to the basic structure found in the heme portion of cytochromes (see Figure 8.13). The iron present in heme is replaced by an atom of magnesium in the chlorophyll molecule. Magnesium is crucial to light capture because it is a "close shell" divalent cation that changes the electron distribution in chlorophyll pigments and produces powerful excited states (**Figure 9.1**). There are a number of

CH₃
CH₂   H   CH₃

[Chlorophyll a structure diagram]

The phytyl side chain provides a hydrophobic tail that anchors the chlorophyll to membrane protein complexes, the site of the light reactions.

Chlorophyll *a*

CH₃ — C —
‖
O
Bacteriochlorophyll *a*

Figure 9.1 Chlorophylls

Structures of chlorophyll *a* and bacteriochlorophyll *a*. The chlorophylls are structurally related to heme, but the Fe^{2+} of heme is replaced by Mg^{2+} in the chlorophylls.

chlorophylls that differ slightly in their chemical structure. They are designated by a lowercase letter—*a*, *b*, *c*, *d*, *e*, and *g* (see Figure 21.1).

Bacteriorhodopsin, a carotenoid, does not contain a metal and functions in a manner markedly different from the chlorophylls. This type of photosynthesis occurs in the halophilic (salt-loving) *Archaea* (see Chapter 18). The chlorophylls are light-absorbing pigments that can participate in energy transduction. Most of the chlorophyll in a green plant or photosynthetic bacterium is involved with light gathering (**Figure 9.2**) and the transfer of light-generated excitation to reaction centers (see later) where the energy is used to split water ($H_2O \rightarrow 2H^+ + 2e^- + \frac{1}{2}O_2$) and/or drive photophosphorylation processes.

There are five types of photosynthetic *Bacteria*: these are commonly known as the nonsulfur purples, the green sulfurs, the purple sulfurs, the cyanobacteria, and the heliobacteria. Their general distinguishing properties are presented in **Table 9.1**. The various photosynthetic bacteria differ morphologically, in habitat, and physiologically. They are also distinguished by the electron donor involved in their photochemistry. The purple nonsulfur bacteria generally use H_2 or reduced organics, the green and purple sulfur bacteria use reduced sulfur compounds, the cyanobacteria use H_2O, and the heliobacteria use reduced organics as an electron donor.

The early anoxygenic photosynthetic bacterium, whether a purple or green sulfur bacterium, probably had a chlorophyll pigment similar to the chlorophyll *a* present in the cyanobacteria. Chlorophyll *a* absorbs light

of a wavelength (< 700 nm) that penetrates the deep anaerobic reaches of aquatic environments.

The absorption of an incident photon by a molecule of chlorophyll leads to a redistribution of electrons within the basic framework of the tetrapyrrole. This state of the molecule is known as an **excited state,** and it has an energy greater than the **initial** or **ground state.** Incident radiation is absorbed only if it provides the correct energy to achieve one of the allowed activated states. Photons with intermediate energies are transmitted by chlorophyll but leave it unaffected energetically.

Bacteriochlorophyll present in the green bacteria absorbs light at a wavelength of about 700 nm. This light-harvesting capacity permits growth at lower depths in water but does not have sufficient potential as an oxidant to extract electrons from water. Thus, evolution in the anaerobic environment yielded a photosynthetic population limited to the use of reduced molecules such as hydrogen (H_2) or sulfide (H_2S). Eventually (about 2.5 billion years ago) an organism evolved with a photosystem that had sufficiently strong oxidative power that it could extract electrons from H_2O, according to the following:

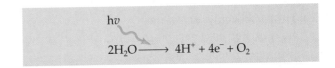

$$h\nu$$
$$2H_2O \longrightarrow 4H^+ + 4e^- + O_2$$

Light-harvesting antenna pigments absorb light energy and transfer it from one antenna molecule to another until it reaches…

…the specialized chlorophylls of the reaction center.

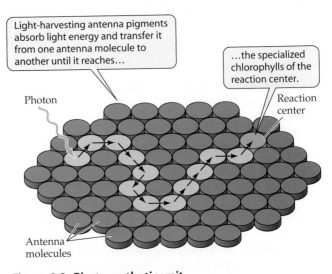

Photon

Reaction center

Antenna molecules

Figure 9.2 Photosynthetic unit

Diagram of a photosynthetic unit, showing the pathway of exciton transfer from antenna molecules to the reaction center (orange).

| Table 9.1 | Some general properties of the various photosynthetic bacteria |

	Nonsulfur Purple Bacteria	Purple Sulfur Bacteria	Green Sulfur Bacteria	Cyano-bacteria	Helio-bacteria
Source of reducing power (e⁻)	H_2, reduced organic	H_2S	H_2S	H_2O	Lactate, organic
Oxidized product	Oxidized organic	SO_4^{2-}	SO_4^{2-}	O_2	Oxidized organic
Source of carbon	CO_2 or organic	CO_2	CO_2	CO_2	Lactate pyruvate
Heterotrophic growth	Common	Limited[a]	Limited[a]	Limited[a]	Required

[a]Generally limited to assimilation of low molecular weight organics during autotrophic growth.

The development of this oxygen-generating system was the fundamental event that allowed eukaryotic organisms to evolve beyond the need for an aquatic habitat, thus opening up the endless possibilities of life on land (Chapter 1).

The primeval anaerobic environment on earth bore great quantities of ferrous iron (Fe^{2+}) that served as a trap for O_2 generated by the first cyanobacterial-type organisms. Bands of ferric iron Fe_2O_3, called banded-iron formations (BIFs), can be seen in geologic strata deposited during the era when oxygenic photosynthesis evolved, attesting to the probable role that ferrous iron played (**Box 9.2**). Free oxygen is a highly reactive molecule, and superoxide and singlet oxygen and peroxides would have damaged cells were it not for the presence of reduced iron to react with the O_2. The

Banded Iron Formations

During the Archean and Proterozoic ages (over 1.5 billion years ago) banded iron formations (BIFs) were laid down over extensive areas of planet Earth. These BIFs have been attributed to the chemical oxidation of reduced iron (Fe^{2+}) by molecular oxygen generated by the cyanobacteria. The cyanobacteria evolved with the ability to utilize water as an electron donor in photochemical energy generation. The resultant cleavage of water ($H_2O \rightarrow 2H^+ + \frac{1}{2} O_2$) was the original source of atmospheric oxygen. Recent evidence suggests that anaerobic oxidation of soluble ferrous iron could also have contributed to BIFs through the activities of the anoxygenic purple bacteria. These photosynthetic bacteria can utilize reduced iron that is present in anoxic iron-rich sediments as an electron donor for photophosphorylation. As the soluble ferrous iron is oxidized to insoluble ferric state, a brown precipitate is formed.

This is another example of the remarkable diversity that prevails in the microbial world.

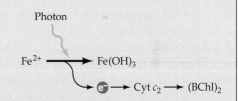

Banded Iron Formation Dales Gorge, Wittenoom, Western Australia. The formation is 2.47 billion years old. Courtesy of Eric Cheney.

reduced iron absorbed the toxic molecules as the biota adapted enzymatically to handle potential toxic oxygen products (see Chapter 6). The first line of defense within the primitive O_2-generating photosynthetic organisms may have been their **carotenoid pigments.** The carotenoids are long-chain hydrocarbons with extensive conjugated double bonds (see Figure 21.10). It is likely that the carotenoids evolved to provide light-harvesting capacity in the 400 nm to 550 nm range. Carotenoids also can intercept highly toxic single-state oxygen generated during photolysis of water and convert it to a less reactive state.

Ferrous iron was gradually lost through reactions with O_2, and over 2 billion years ago O_2 began accumulating in the atmosphere. This O_2 was subjected to the following reaction in the stratosphere: O_2 + sun (UV) → O_3. This formed an ozone (O_3) layer that screened out ultraviolet (UV) rays below a wavelength of 290 nm that would be harmful to any organism constantly exposed to unfiltered sunlight. Without this ozone layer, evolution would certainly have taken a different course because UV causes chemical changes in nucleic acids that absorb in the 260 nm range. Without this UV trap, most living creatures would have had to remain in aquatic environments because a thin layer of water absorbs UV.

With the ozone layer established, the lower wavelengths of UV no longer reached the surface of the Earth in great quantity, and life on land became feasible. The diversity of living forms that were possible under the ozone layer was virtually unlimited.

It is possible that UV-resistant microorganisms did evolve prior to the formation of the ozone layer. This resistance may have been lost after the ozone layer came into being, as it gave the microbe no selective advantage. Perhaps *Deinococcus radiodurans,* a microbe highly resistant to radiation, is a carryover from than era.

Common Features of Photosynthetic Prokaryotes

Photosynthetic microorganisms (excluding the halobacteria, which will be discussed in Chapter 18) have some common features, but differ in the way they achieve photosynthetic growth. These bacteria all employ light as an energy source, and carbon dioxide can generally be used as the source of carbon. The major energy-capturing pigment is chlorophyll (or bacteriochlorophyll) (**Table 9.2**), and all have auxiliary antenna pigments that aid them in capturing a broad range of light excitation. The sites of photochemical reactions are distinct areas called **reaction centers** and

are defined as a single photochemical unit because they transfer one electron at a time. Reaction centers are protein-chlorophyll complexes within the cytoplasmic membrane of photosynthetic bacteria where quanta of light energy are converted to stable chemical energy.

The next sections of the chapter focus on how the purple bacteria, green bacteria, and cyanobacteria capture energy from the sun.

PHOTOSYNTHESIS IN PURPLE NONSULFUR BACTERIA

The purple nonsulfur bacteria are prevalent on the surface of mud and in the water of lakes and ponds. Water that is rich in organic matter and has relatively low sulfide can have a population of these bacteria in numbers that give the water a distinct purple color.

The photosynthetic nonsulfur purple bacteria have been studied extensively because they grow readily in culture. The photosynthetic reaction centers have been isolated and crystallized from *Rhodobacter sphaeroides* and *Rhodopseudomonas viridis.* The reaction centers from *Rhodopseudomonas viridis* were obtained by breaking cells by sonication (ultrasound) and differential centrifugation (centrifuging at speeds that separate material by size or weight). The reaction centers were then purified by general protein separation techniques, crystallized, and studied. These reaction center crystals yielded much data on the structural relationships between the component parts and the mechanism whereby photosynthesis occurs. They have also been useful as models for studies on the photosystems in higher plants.

The reaction center in the nonsulfur purple bacterium *R. viridis* contains:

- Four molecules of bacteriochlorophyll and two molecules of **bacteriopheophytin** (a molecule of bacteriochlorophyll *a* that lacks magnesium)
- A molecule of **menaquinone**

| Table 9.2 | **The bacteriochlorophyll present in photosynthetic bacteria and primary acceptors involved in energy conserving reactions** |

	Electron Donor	Electron Acceptor
Purple nonsulfur bacteria	Bacteriochlorophyll *a* and *b*	Bacteriopheophytin *a*, Q_A, Q_B
Green sulfur bacteria	Bacteriochlorophyll *c*, *d*, and *e*	Bacteriopheophytin *a* and FeS-protein
Cyanobacteria photosystem I	Chlorophyll *a*	Chlorophyll *a* and FeS-protein
Cyanobacteria photosystem II	Chlorophyll *a*	Pheophytin *a*, Q_A, Q_B, and plastoquinones
Heliobacteria	Bacteriochlorophyll *g*	Bacteriochlorophyll *c*, FeS-protein

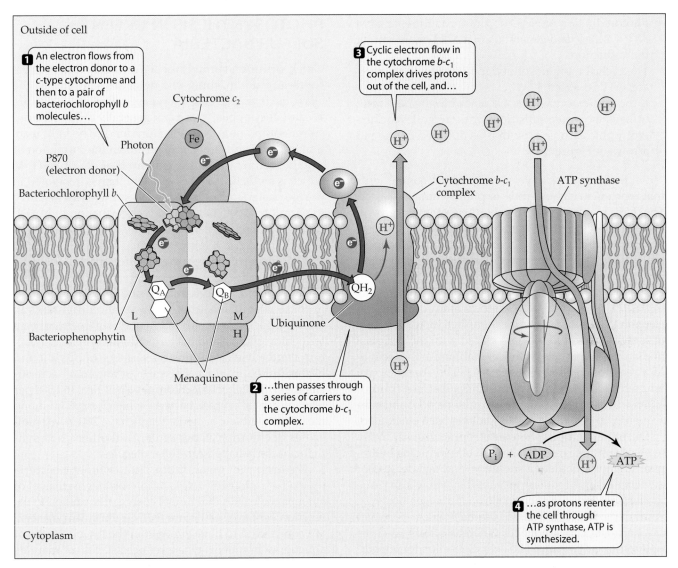

Figure 9.3 Reaction center of purple nonsulfur bacterium
Arrangement of electron carriers and electron flow in the reaction center of *Rhodopseudomonas viridis*. The subunits (polypeptides) of the reaction center are designated light (L), medium (M), and heavy (H), based on their relative size.

- A molecule of **ubiquinone**
- A nonheme iron that is present on the cytoplasmic side of the membrane

These components are covalently linked to two subunits (L and M) of the reaction center protein. Another subunit protein spans the membrane and is attached to the L-M protein complex. Two of the bacteriochlorophyll molecules are paired and share the function of a photochemical electron donor. One of the bacteriopheophytin molecules acts as the intermediary electron carrier between bacteriochlorophyll and menaquinone. The roles of the other two bacteriochlorophyll molecules and the one bacteriopheophytin molecule are not known at the present time.

Electrons from bacteriochlorophyll are passed to ubiquinone and then to the cyt b-c_1 complex where cyclic electron flow results in the extrusion of protons.

In reaction centers obtained from a strain of *R. sphaeroides*, one molecule of carotenoid is bound to each reaction center particle. The reaction center carotenoid is probably not involved in light gathering or excitation transport, but rather it functions in carrying stray energy away from the reaction center.

Cyclic and noncyclic electron transport in photosynthetic nonsulfur purple bacteria is illustrated in **Figure 9.3**. Many of the nonsulfur purple bacteria are facultative and grow aerobically on organic substrates. There are three major attributes of the electron transport system in these microorganisms, and these are:

- Photoreduction of NAD occurs through reversal of ATP synthase and generation of a $\Delta\mu H^+$ sufficient for this reduction.
- The cyclic photophosphorylation system that generates ATP is present.
- If these microorganisms are placed in the dark with utilizable organic substrates, the energetics for heterotrophic aerobic growth is as outlined by the red arrows in **Figure 9.4**.

All three systems have some common components that function in either aerobic or photosynthetic growth. The major requirement for aerobic (heterotrophic) growth is the synthesis of the terminal oxidase system.

Photophosphorylation occurs in the nonsulfur purple bacteria by a light-induced translocation of a proton across the energy-transducing membrane. The organization of a light-dependent transfer of electrons results in a $\Delta\mu H^+$ (see Figure 9.3). Cytochrome c_2 donates an electron to oxidized bacteriochlorophyll (Bchl 870). Two or more cytochrome c are associated with the periplasmic surface of the cytoplasmic membrane in the vicinity of a reaction center (Figure 9.4A). The specific path whereby electrons flow from the substrate (photoheterotrophic growth) to the ubiquinone pool and subsequently to cytochrome c_2 has not yet been elucidated.

During anaerobic photoautotrophic growth many reduced substrates can donate electrons, including hydrogen, ethanol, isobutanol, succinate, and malate. Recent evidence affirms that ferrous iron (Fe^{2+}) may also serve as electron donor in the purple bacteria (see Box 9.2). Some of the resultant oxidized products (such as acetate) can be further metabolized, whereas others (such as isobutyric acid) are not. The overall reaction for anaerobic photoautotrophic growth of *R. sphaeroides* can be depicted as:

$$H_2 + CO_2 \xrightarrow{h\nu} \text{cell material}$$

An increase in light intensity during growth curtails synthesis of photosynthetic pigments by *R. sphaeroides*, but the rate of growth is not altered. The synthesis of bacteriochlorophyll, however, is suspended temporarily, at higher light levels and resumed at a lower rate with the reestablishment of balanced growth.

The presence of oxygen also halts the synthesis of photosynthetic pigments. Cultures growing heterotrophically in the dark (facultative chemoheterotrophs) have little pigment per unit cell mass. A decrease in the oxygen supply results in pigment synthesis. The major mechanism for gaining energy anaerobically is through photophosphorylation, and a lowered oxygen supply is a signal to synthesize the light harvesting pigments.

PHOTOSYNTHESIS IN GREEN SULFUR BACTERIA

Green sulfur bacteria differ significantly from purple bacteria, and in some characteristics resemble the cyanobacteria. These bacteria are called green sulfurs because they appear green and generally utilize reduced sulfur compounds as an electron donor. The light reaction in green sulfur bacteria results in the formation of a stable reductant with a potential of about 540 mV, which is a much stronger reductant than is generated in purple bacteria. This reductant is sufficiently negative to permit a noncyclic, light-driven reduction of NADP (Figure 9.4B) without the reverse electron flow essential in purple bacteria (Figure 9.4A).

The reaction centers in green sulfur bacteria are more difficult to purify in a functional form, and thus, less is known about the mechanics of photophosphorylation in these organisms. The major antenna pigments in green bacteria are two different chemical forms of bacteriochlorophyll c and bacteriochlorophyll a. The reaction center itself contains bacteriochlorophyll a and bacteriopheophytin a. The ratio of antenna bacteriochlorophyll c to bacteriochlorophyll a is 1000 to 1500 per reaction center in green bacteria, which is unique among the photosynthetic bacteria. Plants have 200 to 400 molecules of chlorophyll per center, and other photosynthetic bacteria have 50 to 100 molecules.

The origins of green sulfur photosynthetic bacteria may offer a plausible explanation for this high level of antenna molecules that funnel energy of excitation to the reaction centers (see Figure 9.2). Green sulfur bacteria are considered to have evolved in the lower depths of aquatic environments where reduced sulfur compounds were available as electron donor. The green sulfur bacteria would therefore have existed under swarms of cyanobacteria and purple bacteria. To absorb the weak light that filtered down through the layers of cyanobacteria and purple bacteria, the green sulfurs evolved with an extensive network of antenna pigments to capture the excitation energy of the limited light available.

The way that the light-harvesting pigments are bound to the cytoplasmic membranes in green bacteria is similar to the manner in which these pigments are bound to thylakoids in the cyanobacteria. The antenna pigments in green sulfur bacteria are packed into vesicles called **chlorosomes,** which lie along the inside of the cytoplasmic membrane (**Figure 9.5**). The chlorosomes are bound by a proteinaceous (nonlipid) membrane. Each vesicle contains up to 10,000 bacteriochlorophyll c molecules, and each is attached to the cellular membrane by a **subantenna** of a bacteriochlorophyll a. Excitation energy, generated by light and absorbed by antenna pigments, passes through this subantenna to the reaction centers.

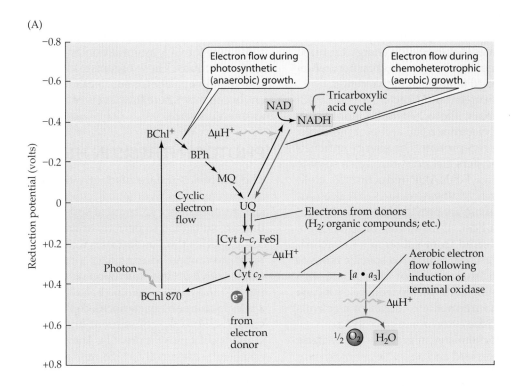

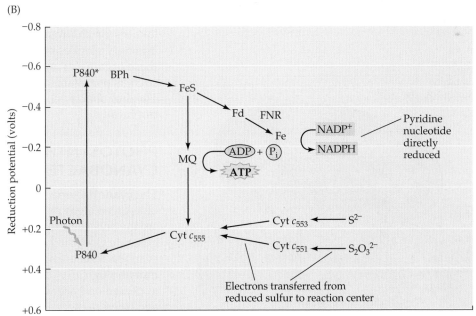

Figure 9.4 Cyclic and noncyclic electron flow in purple nonsulfur and green sulfur bacteria

(A) Electron flow in the purple nonsulfur bacterium *Rhodobacter sphaeroides* during photosynthetic (anaerobic) growth (black arrows) and chemoheterotrophic aerobic growth (red arrows). A number of components are common to both pathways. (B) Electron flow during photophosphorylation in a green sulfur bacterium. The P840* of these organisms has a sufficiently high reduction potential to directly reduce pyridine nucleotide. BChl, bacteriochlorophyll *b*; BPh, bacteriopheophytin; Fd, ferredoxin; FeS, iron-sulfur center; FNR, ferredoxin nucleotide reductase; MQ, menaquinone; UQ, ubiqinone.

Experimental evidence suggests that the reaction centers and probably some of the antenna bacteriochlorophyll *a* are built into a molecular structure, called a **base plate.** This base plate is embedded in the cytoplasmic membrane. The high efficiency of the photosystem in green bacteria indicates that excitation energy captured by the antenna pigments in a saturated reaction center can readily be passed to another receptive center.

Electron transfer in green sulfur bacteria (Figure 9.4B) begins with the primary photochemical donor (designated P840) passing electrons to an acceptor molecule, considered to be an FeS-protein. The energy enriched FeS-protein then transfers electrons through a soluble ferredoxin (Fd) to an Fd-NADP reductase to yield NADPH. This noncyclic pathway requires that electrons be replenished, and these are provided by reduced substrates such as sulfide, thiosulfate, or hydrogen. Electrons are transferred from substrates to bacteriochlorophyll in the reaction centers via specific types of cytochrome *c*; cytochrome c_{551} for thiosulfate and cytochrome c_{553} for sulfide. A specific reductase for each substrate acts as an intermediate carrier of electrons from the substrate to the cytochrome *c* involved.

The overall photochemistry in green sulfur bacteria is relatively straightforward and resembles photosystem I in cyanobacteria and green plants. ATP is generated through a cyclic photophosphorylation, and reduced pyridine nucleotides are produced by the noncyclic Fd-reductase system. Sulfides and other reduced sulfur compounds are of widespread occurrence in the environments in which green sulfur bacteria thrive. The primary requirements for survival of an autotrophic, anaerobic photosynthetic bacterium include a ready source of reductant for CO_2 assimilation and ATP to drive synthetic reactions. These organisms have evolved with a marvelous capacity for survival in sulfurous anaerobic environments where low levels of light are available as an energy source.

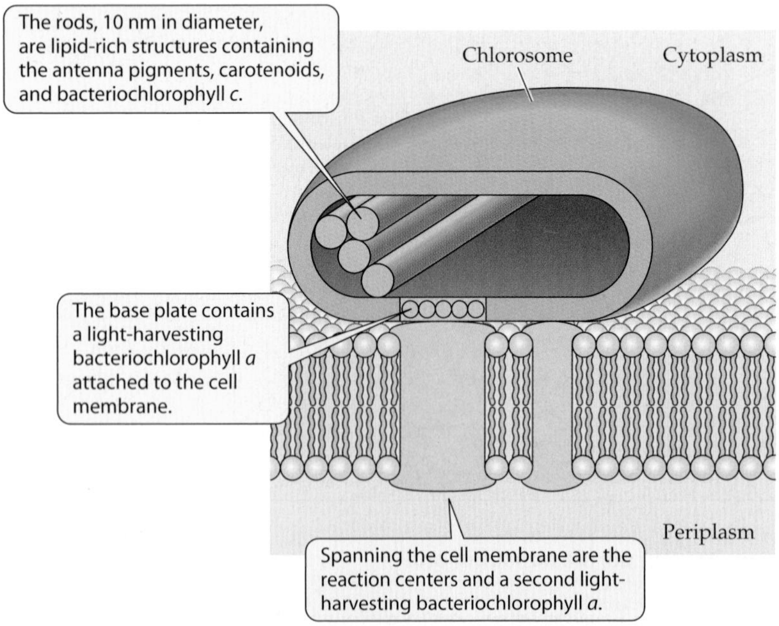

PHOTOSYNTHESIS IN HELIOBACTERIA

The heliobacteria are distinctly different from other anoxygenic photosynthetic bacteria. They are of widespread occurrence in tropical soils, particularly those that are subjected to periods of wetness and dryness. Some species form endospores that may provide a survival advantage during dry periods. The bacteriochlorophyll in these microorganisms is bacteriochlorophyll *g*, and they contain no bacteriochlorophyll *a*. Bchl *g* is unique among bacteriochlorophylls, as it contains a vinyl ($-CH=CH_2$) constituent on ring I, as is present in plant chl *a* (see Figure 9.1). The Bchl *g* in heliobacteria absorbs maximally at 788 nm, a wavelength not absorbed by other photosynthetic bacteria. The reaction centers of these microorganisms are embedded in the cytoplasmic membrane, and they do not form thylakoids or chlorosomelike structures. The only sources of carbon utilized by these organisms are simple monomers such as pyruvate, acetate, and lactate. The heliobacteria cannot grow autotrophically.

PHOTOSYNTHESIS IN CYANOBACTERIA

The **cyanobacteria,** formerly called the blue-green algae or blue-green bacteria, are the only photosynthetic bacteria that generate O_2 during photosynthesis. The cyanobacteria are a physiologically and morphologically diverse group that appeared on Earth over 2.5 billion years ago (see Chapter 21). During evolution, they developed two independent light reactions: photosystem I (PSI) and photosystem II (PSII), and these operate in series with a redox span from H_2O/O_2 to $NADP^+/NADPH$. The oxygen-generating photochemical systems of the cyanobacteria were instrumental in providing an O_2-containing atmosphere on

The rods, 10 nm in diameter, are lipid-rich structures containing the antenna pigments, carotenoids, and bacteriochlorophyll *c*.

The base plate contains a light-harvesting bacteriochlorophyll *a* attached to the cell membrane.

Spanning the cell membrane are the reaction centers and a second light-harvesting bacteriochlorophyll *a*.

Chlorosome   Cytoplasm

Periplasm

Figure 9.5 Organization of chlorosome of a green sulfur bacterium Chlorosomes are vesicular structures attached to the cell membrane at the site of a reaction center. The vesicular membrane is a polypeptide lipid matrix, not a unit membrane.

Earth that eventually permitted the commencement of aerobic respiration.

Photosynthetic reactions in cyanobacteria are similar to and a forerunner of the reactions in green plants. However the only relationship that the cyanobacteria have to eukaryotes is the similarity of the photosystem in these bacteria to the photosystem in the chloroplasts of green plants. The understanding of green plant photosynthesis has been enhanced by studies of the pho-

toapparatus in the more readily manipulated cyanobacteria. These rapidly growing microorganisms are easily grown under laboratory conditions and offer considerable functional diversity; thus, they have become a widely used model for study.

The cyanobacteria combine the salient features of photoreactions in both purple and green bacteria (see Figure 9.4 and **Figure 9.6**). The photosystem, designated PSI, is similar to the photosystem in the green sulfur

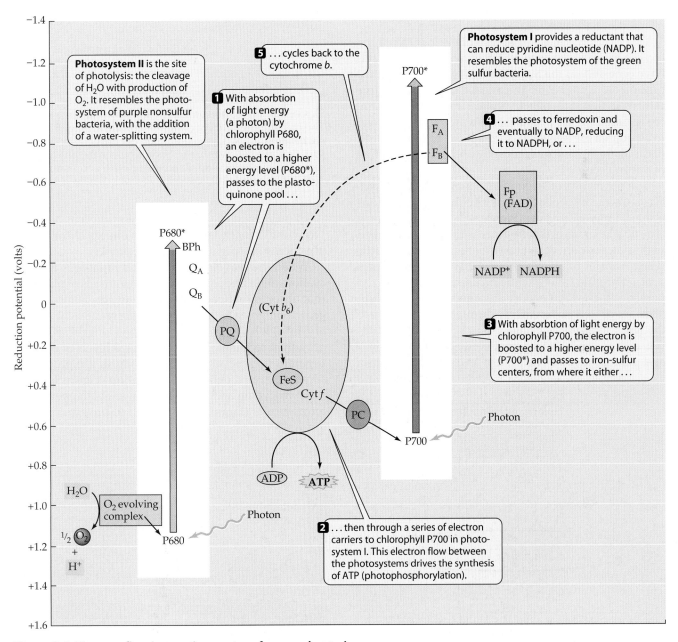

Figure 9.6 Electron flow in reaction center of a cyanobacterium
A cyanobacterium has two photosystems, designated photosystems I and II (PSI and PSII). BPh, bacteriopheophytin; Q_A, Q_B, PQ, plastoquinones; FeS, iron-sulfur center; PC, plastocyanin; F_A, F_B, iron-sulfur centers; Fd, ferredoxin; Fp(FAD), the flavoprotein enzyme ferredoxin-NADP$^+$ reductase.

bacteria. It provides a reductant of sufficient potential to reduce pyridine nucleotides. Photosystem II is similar to the photosystem in the purple nonsulfur bacteria except that the latter lacks the water-splitting system.

The light-activated system responsible for the splitting of water (photolysis) to generate oxygen and donate electrons to PSII requires four quanta of light energy. Photolysis occurs when two water molecules are bound to a light-sensitive enzyme, designated the S system, which absorbs four positive charges that extract four electrons from the water (**Figure 9.7**). The protons are released into the medium, and O_2 is produced.

One electron is released at each of the four steps, and oxygen is produced only after four separate photo acts have occurred on the S system. S_0 is the enzyme at the ground state, and after the release of O_2 the enzyme returns to this state of activation. Four quanta of light are required for the photolysis of water, and a total of 8 to 10 quanta are necessary for both the photosynthetic dissimilation of one molecule of water and assimilation of one molecule of CO_2.

The pigment-protein complex involved in photosynthesis in the cyanobacteria is present in projections termed a **phycobilisome** (**Figure 9.8**). A phycobilisome is a subcellular body that appears as tiny knobs attached to the outer surface of photosynthetic membranes. Little if any transfer of quanta occurs between these photosynthetic units, and each is a discrete miniaturized power cell. The light energy absorbed in the phycobilisomes is passed preferentially through chlorophyll *a* to the reaction center of photosystem II, but some excitation energy can be transferred between photosystem I and photosystem II. The system as depicted permits a

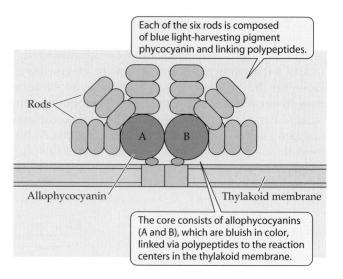

Figure 9.8 Phycobilisome of cyanobacteria
The antenna pigments of cyanobacteria are arranged in phycobilisomes. These knoblike structures project from the outer surface of the cell membrane. Shown here is the phycobilisome of *Synechococcus* sp. Reproduced with permission. Arthur Grossman, *Microbiological Reviews*, 57:725–749.

"downhill" cascade of excitation energy through the phycobilisome pigments to chlorophyll *a*.

When the wavelength of light available is altered, cyanobacteria undergo an induced synthesis of pigments that absorbs that particular wavelength. For example, growth in green light results in the synthesis of phycoerythrin, which absorbs green wavelengths, whereas growth in red light leads to increased levels of the blue pigment phycocyanin. This change in the phycobilisomes is termed **chromatic adaptation** (**Figure 9.9**).

Phycocyanin and phycoerythrin are both open-chain tetrapyrroles (**Figure 9.10**) coupled to proteins and col-

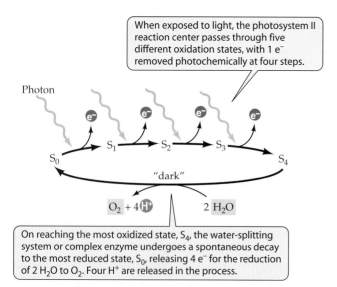

Figure 9.7 Photolysis reaction of photosystem II
Evolution of one molecule of oxygen requires the stepwise accumulation of four oxidizing equivalents in photosystem II.

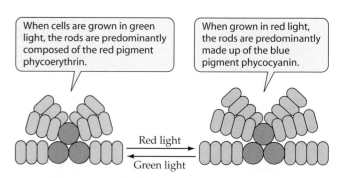

Figure 9.9 Chromatic adaptation of a phycobilisome
In some bacterial species, the composition of phycobilisomes can be altered by changing the wavelength of light provided for growth. This scheme shows chromatic adaptation in *Fremyella diplosiphon*. Reproduced with permission. Arthur Grossman, *Microbiological Reviews*, 57:725–749.

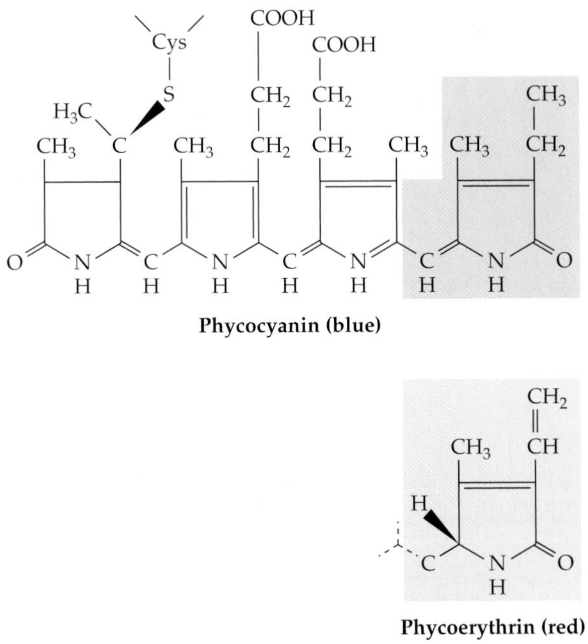

Phycocyanin (blue)

Phycoerythrin (red)

Figure 9.10 Chromophores of phycobilisomes
Structures of two of the chromophores (bilins) of the phyco-bilisomes of cyanobacteria. They differ only at the fourth pyrrole ring. Each bilin is joined to protein at a cysteine residue.

lectively are called phycobiliproteins. The phycobili-proteins are light-gathering antenna pigments. Phyco-cyanin absorbs light around 630 nm and phycoerythrin at around 550 nm. These chlorophylls absorb a rather narrow spectrum of light, and the accessory pigments allow the organism to capture more of the incident light that falls outside this narrow range.

The primary electron acceptors in photosystem I are soluble FeS proteins similar in function to those in green sulfur bacteria. Electrons flow cyclically back to chlorophyll P700 via cytochromes and the copper-containing protein plastocyanin. A theoretical scheme for electron transport, ATP synthesis, and NADPH production in cyanobacteria was presented in Figure 9.6.

PHOTOSYNTHESIS WITHOUT CHLOROPHYLL

The halophilic bacteria are a remarkable group of organisms that thrive in highly saline environments, as they require substantial concentrations of NaCl for growth (see Chapter 18). The halophilic bacteria are all in the *Archaea* domain. Of these organisms, *Halobacterium halobium* has been studied most extensively. *H. halobium* is obligately aerobic and operates a conventional membrane-bound respiratory pathway for ATP generation when sufficient O_2 is available.

Because the solubility of oxygen in water decreases as salinity increases, the availability of oxygen beneath the surface of water with a high salt concentration is limited. When the halobacteria encounter low oxygen concentrations, they synthesize purple patches that appear within the cytoplasmic membrane. These patches can cover over one-half the surface of the cellular membrane and provide a light-driven proton ejection system that is not inhibited by cyanide. Aerobic respiration in these organisms is inhibited by cyanide.

The purple patches are made up of flat sheets containing a crystalline array of a single protein—**bacteriorhodopsin**. This protein has a molecular mass of 26,000, and each protein assembly is attached to a molecule of retinal bacteriorhodopsin through a lysyl residue in the protein (**Figure 9.11A**). Bacteriorhodopsin (vitamin A aldehyde) is chemically similar to the visual pigments of animals. The bacteriorhodopsin absorbs light maximally at a wavelength of 570 nm and becomes bleached as it ejects a proton to the exterior of the cytoplasmic membrane. The deprotonated bacteriorhodopsin then takes up a proton from the cytoplasm of the cell (Figure 9.11). Bacteriorhodopsin absorbs at wavelengths at whichever maximum energy is available from sunlight, particularly when there is a water layer over the organism. No accessory pigments are necessary for the photochemical reaction to occur, and respiratory electron transport is not involved. The reaction is a simple light-driven proton pump.

Elucidation of this system has lent considerable credibility to the chemiosmotic theory of ATP generation. The isolated purple membrane has been reconstituted in closed vesicles, which when exposed to light, act as proton pumps. ATP synthase from beef heart has been incorporated into these vesicles, where they can perform a light-dependent synthesis of ATP (**Box 9.3**). It would be difficult to ascribe this phenomenon to any direct coupling reaction because no electron transfer is involved. Replacement of the halophile's ATP synthase with that from an animal confirms that no electron transfer-related phenomenon occurs in the native ATP synthase. Electron flow in respiration then serves only to transduce protons across the membrane.

The quantum requirement for the ejection of one proton by *H. halobium* is approximately 2, as compared with 0.5 for purple photosynthetic bacteria. Because light at a wavelength of 570 nm has a higher energy content than light at 870 nm, the overall efficiency of *H. halobium* is less than 25% of the efficiency of purple bacteria.

Recent studies, employing genomic analysis, indicate that there are many bacterial species in maritime habitats bearing genes that encode for a form of rhodopsin. These microorganisms have not yet been grown in culture. The rhodopsin present in these microorganisms differs from the bacteriorhodopsin

(A)

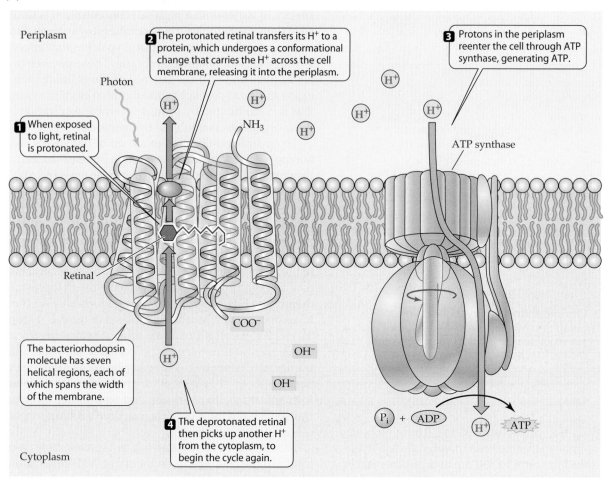

Periplasm

2 The protonated retinal transfers its H⁺ to a protein, which undergoes a conformational change that carries the H⁺ across the cell membrane, releasing it into the periplasm.

3 Protons in the periplasm reenter the cell through ATP synthase, generating ATP.

Photon

1 When exposed to light, retinal is protonated.

NH_3

ATP synthase

Retinal

COO^-

The bacteriorhodopsin molecule has seven helical regions, each of which spans the width of the membrane.

OH^-

OH^-

4 The deprotonated retinal then picks up another H⁺ from the cytoplasm, to begin the cycle again.

Cytoplasm

P_i + ADP

H^+

ATP

(B)

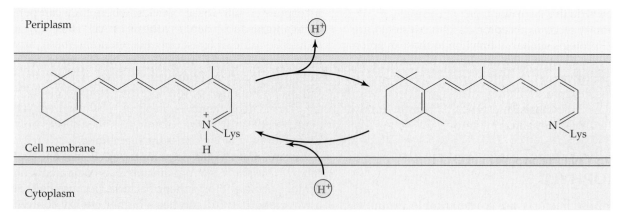

Periplasm

H^+

Cell membrane

N—Lys
H

N—Lys

Cytoplasm

H^+

Figure 9.11 Light-driven proton pump of halophilic bacteria
(A) The light-driven proton pump in the membranes of halophilic (salt-loving) bacteria consists of bacterorhodopsin, a pigmented protein molecule, to which is attached a retinal molecule. (B) The chemical reactions of retinal underlying the pumping mechanism. No electron transport is involved in this system.

Research Highlights Box 9.3 *Archaea*, Bovines, and Chemiosmosis

In 1974, Efraim Racker, a biochemist, and Walther Stockenius, a pioneer in elucidating the role of bacteriorhodopsin in halophilic bacteria, collaborated to perform an experiment that further confirmed Mitchell's chemiosmotic hypothesis. They reconstituted vesicles from the cytosplasmic membranes of *Halobacterium halobium* that were inverted so that the bacteriorhodopsin would pump protons inward when exposed to light. They also obtained bovine mitochondrial ATP synthase and incorporated it into the vesicle as shown. When illuminated, the bacteriorhodopsin pumped protons into the vesicles and generated a proton gradient inside that drove ATP synthesis by the ATP synthase. There were two distinct types of protein in the vesicles, one from an *Archaea* and the other from a mitochondrion. Together they synthesized ATP, as the Mitchell hypothesis predicted.

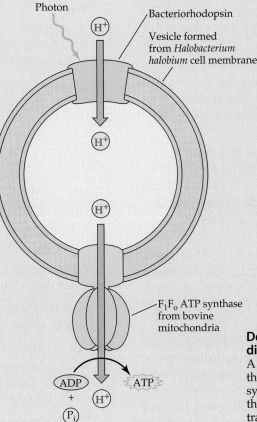

There were no electrons involved in the synthesis of this ATP, thus confirming that the role of respiratory electron transport chains is transduction of protons, not phosphorylation of ATP.

Demonstration that a proton gradient drives ATP synthesis.
A combination of archaeal photosynthetic pigments with bovine ATP synthase generated ATP, showing that a proton gradient, not electron transport, drives ATP synthesis.

---

present in the *Archaea* and is designated proteorhodopsin. The microorganisms involved are deep-sea dwellers and absorb blue light that penetrates to considerable depths. They are a potential source of biomass in these environments.

SUMMARY

▶ **Photophosphorylation** is the sum of the processes that generate ATP through the use of light energy.

▶ **All life** on the surface of the earth **relies** either directly or indirectly **on energy from the sun**. Photoautotrophs use photons directly, whereas heterotrophs exist by utilizing photosynthetic products.

▶ **Oxygenic** photosynthesis results in the release of O_2, as the source of electrons for this process is H_2O ($2H_2O \rightarrow 4H^+ + O_2 + 4e^-$). **Anoxygenic** photosynthesis does not result in O_2 production. The source of electrons in anoxygenic photosynthesis is reduced organic (ethanol $\rightarrow$ acetate) or inorganic compounds ($S_2O_3^{2-} \rightarrow SO_4^{2-}$).

▶ Much of the chlorophyll in photosynthetic organisms is involved in **light gathering**. The excitons are transferred to **reaction centers** where the **light energy** is converted to **chemical energy.**

▶ There are **several chlorophylls**—*a*, *b*, *c*, *d*, *e*, and *g*— and each has a **characteristic absorption spectrum**. This permits microorganisms with differing chlorophylls to grow at various depths in water because they absorb wavelengths of light that are not absorbed by those above them.

▶ The **ozone layer** above the Earth was probably formed when oxygenic photosynthesis evolved. The O_2 produced reacted with UV in the stratosphere to form O_3. The **ozone** layer **protected earthly creatures from harmful UV radiation** and was a major contributor to evolution of life on dry land. Aquatic life was not adversely affected by UV because this wavelength does not penetrate into water.

▶ **Reaction centers** have been crystallized from purple nonsulfur photosynthetics. Much has been learned about structural arrangements in reaction centers by studying these crystals.

▶ Most **purple nonsulfur bacteria** grow as either a **heterotroph or with light energy.** They use much of the same respiratory system when grown by either mode.

▶ The reaction centers in green **sulfur bacteria** are complex and have considerably more **antenna pigments** than are present in the purple nonsulfur reaction centers. This permits them to gather light of much lower intensity. These antenna pigments are arranged in **chlorosomes.**

▶ The **photosystem** in cyanobacteria and green sulfurs can **reduce NADP,** but that from the purple nonsulfur bacteria does not have sufficient energy to accomplish this. The latter must generate NADPH for CO_2 fixation by reversing ATP synthase to generate a change separation across the membrane.

▶ The antenna pigments in cyanobacteria are arranged in projections attached to the surface of membranes termed **phycobilisomes.** The pigments in these organelles can change as the wavelength of the light available changes.

▶ The **halophilic** *Archaea* are aerobic but can generate ATP anaerobically by utilizing **bacteriorhodopsin.** This compound is present in the cytoplasmic membrane and is similar to **retinal,** the eye pigment in animals. Bacteriorhodopsin can use light energy to move protons from the cytoplasm to the outer surface of the cytoplasmic membrane. The protons reenter the cytoplasm via ATP synthase, thus producing ATP.

REVIEW QUESTIONS

1. What is the basic difference between photophosphorylation and respiration? What are the common features?

2. There are basic differences in the electron donor for the three photosynthetic bacterial types—purple nonsulfurs, green sulfurs, and cyanobacteria. What are these differences? Is this related to habitat?

3. Which of the three bacterial types mentioned in question 2 would you consider the more ancient? Why? Consider the atmosphere of the primordial earth and the nature of compounds available.

4. Describe the structural make-up of chlorophyll and the relationship it has to heme in blood. Why does Mg^{2+} replace iron?

5. The evolution of the cyanobacteria with oxygenic photosynthesis was of major importance in the movement of life forms to terrestrial habitats. Explain.

6. Antenna pigments are important to photosynthetic organisms. Why? Consider all types of photosynthetic bacteria.

7. Why have the purple nonsulfur photosynthetic bacteria been a model system for clarifying the events that occur in a reaction center?

8. Compare the role of electron transport in purple bacteria grown photosynthetically to that present during aerobic/heterotrophic growth. How much actual difference is there?

9. How does a photosynthetic bacterium respond to increases in available light intensity? Consider chlorophyll and growth rate.

10. What is a chlorosome, and what role does it play in green sulfur bacteria? What are the major parts?

11. How have green sulfur bacteria adapted to life at the bottom of the pond?

12. List the differences in the photosynthetic apparatus among the three photosynthetic types: the cyanobacteria, the green sulfur bacteria, and the purple nonsulfur photosynthetic bacteria.

13. What is bacteriorhodopsin? How does photosynthesis with this pigment differ from that due to chlorophyll?

SUGGESTED READING

Blankenship, R. E., M. T. Madigan and C. E. Bauer, eds. 1995. *Anoxygenic Photosynthetic Bacteria.* Dordrecht, The Netherlands: Kluwer Academic Publishing.

Peschek, G. A., W. Loffelhardt and G. Schmetterer, eds. 1999. *The Phototrophic Prokaryotes.* New York: Kluwer Academic/Plenum Publishers.

Raghavendra, A. S., ed. 2000. *Photosynthesis: A Comprehensive Treatise.* Cambridge: Cambridge University Press.

Staley, J. T. and A-L. Reysenbach, eds. 2002. *Biodiversity of Microbial Life; Foundation of Earth's Biosphere.* New York: Wiley-Liss.

Chapter 10

Biosynthesis of Monomers

> *...a superficial survey of the biochemical field is apt to fill one with profound astonishment at the practically unlimited diversity of the chemical constituents of living organisms...We have to accept the undeniable fact that all these substances have ultimately been derived from carbon dioxide and inorganic salts.*
>
> *– A. J. KLUYVER*

*T*he previous chapter was concerned with the strategies employed by biological systems in the conservation of energy captured from the sun. The quanta of light that reach the Earth activate those systems that convert photons into stable energy-rich chemicals such as adenosine triphosphate (ATP) or reduced pyridine nucleotides (NADH and NADPH). These compounds then serve as key intermediaries in cell energetics.

If ATP is considered as the medium of exchange in energy transformation, then the C–H bond might be considered a "savings bank." Generally, ATP and the electron carriers act solely as transitory go-betweens, with the C–H bond present in sugars, amino acids, lipids, and other constituents of cells serving as a stable, life-giving energy source for heterotrophic growth. Thus, the formation of a C–H bond during the photosynthetic process provides a stable basic unit that can support the food chains on Earth. Annually, in excess of 10^{11} tons of CO_2 are fixed, principally through the mediation of roughly 40 million tons of the enzyme ribulose-1,5-bisphosphate carboxylase. This key enzyme in the Calvin cycle (see discussion later in this chapter), the major mechanism for CO_2 fixation in autotrophs, is among the most abundant proteins in the biosphere. Light is the ultimate source of energy on the surface of the Earth, and autotrophic CO_2 fixation provides sustenance for all life forms.

The combined mechanisms whereby the major constituents of the bacterial cell are synthesized from simple building blocks is the process that we term **biosynthesis** (anabolism). All cellular material ultimately is synthesized from carbon dioxide, as it is the primary substrate supporting life and is incorporated into cells via an assortment of mechanisms. Some of the other one-carbon compounds utilized by bacteria include carbon monoxide (CO), cyanide (CN^-), formic acid ($HCOO^-$), methanol (CH_3OH), and methane (CH_4). These substrates may be funneled through carbon dioxide and enter anabolic sequences through the classic Calvin cycle or may be assimilated through other reaction sequences.

Many organisms can satisfy their nitrogen requirements by fixation of gaseous N_2 available from the atmosphere; others utilize ammonia (NH_3) or nitrate (NO_3^-). Sulfur needs are met by inorganic sulfate (SO_4^{2-}). Phosphate for nucleic acid synthesis and generation of high-energy intermediates can come from inorganic sources.

This chapter considers the synthetic mechanisms that have evolved among the prokaryotes for the generation of the monomolecular units or **monomers** that can be polymerized to form the macromolecules of a cell. Synthesis by autotrophs is the basis for this discussion, as these microorganisms can construct all of their cellular components from CO_2, the most rudimentary of building blocks.

MACROMOLECULAR SYNTHESIS FROM ONE-CARBON SUBSTRATES

Evolution ultimately led to the selection of viable organisms that grew and reproduced by using the low molecular weight substrates that were available to them—primordial soup. It is possible that some of these early organisms may have utilized light as a source of energy. The use of low molecular compounds in biosynthesis permitted survival without relying on any tedious random process of abiotic (or biotic) generation of building blocks. It was essential that the development of counterpart biodegradative processes that continuously replenished the low molecular weight primary substrates occurred early in evolution. Ultimately, the central substrate for biosynthesis and also the end product of biodegradative activities came to be **carbon dioxide.** Thus, CO_2 was the starting material for anabolism (synthesis) and the end product of catabolism (breakdown).

A microorganism will grow and reproduce only if it can synthesize *all essential* cellular macromolecules from substrates available to it. Autotrophic organisms that use CO_2 as sole carbon source and N_2 as nitrogen source represent the ultimate in biosynthetic capacity. Research has shown clearly that the broad synthetic ability of autotrophs and other microorganisms that grow on low molecular weight substrates rests on the organisms' effective use of a limited number of biosynthetic pathways. In this section, emphasis will be placed on anabolic sequences as they function in *Bacteria* and *Archaea*. Eukaryotic organisms that can synthesize these cellular constituents do, in all likelihood, follow identical or very similar biosynthetic routes. Indeed, studies on the metabolic sequences in both bacteria and fungi have been of fundamental importance to our understanding of the biochemistry of animal systems (**Box 10.1**).

Milestones Box 10.1 **The Microbe as a Paradigm**

Back in 1930, the eminent Dutch microbiologist A. J. Kluyver gave a series of lectures on the metabolism of microorganisms at the University of London. His was a clear exposition on the status of the field at that time. These lectures were published in a classic treatise entitled *The Chemical Activities of Micro-Organisms* (1931, University of London Press, Ltd.). It was through the landmark studies of Kluyver and the Delft School of Microbiologists (M. Beijerinck, C. B. van Niel, and others) that bacteria became the paradigm for experimentation in biochemistry of the cell. Following is a quote from one of Kluyver's elegant lectures as it was presented to the audience in 1930.

"However, in so far as the biochemical achievements of higher animals are concerned, it is possible that we may be dealing mainly with a rearrangement of the component units of the food, since the necessity for a complex food for this class of organisms is a well-established fact. Also in the higher, chlorophyll-containing plants, the biochemical miracle still remains obscure, since attention is focused upon the essential function of light in the metabolism of these organisms. This external energy supply enables the green cell to carry out chemical processes and it can only be estimated with difficulty how far the influence of the light on metabolism extends.

All this becomes different when we consider the metabolism of the colourless microorganisms. For here we find the most remarkable fact—the significance of which is often underestimated, although known since the time of Pasteur—that a single organic compound suffices to ensure a perfectly normal development of these organisms, although they are cut off from any external energy supply. Here we find the biochemical miracle in its fullest sense, for we are bound to conclude that all the widely divergent chemical constituents of the cell have been built up from the only organic food constituent, and that without any intervention of external energy sources. The chemical conversions performed by these organisms rather resemble witchcraft than chemistry!"

In 1954, Kluyver and Van Niel presented the "John M. Prather Lectures" at Harvard University. These lectures were published in another classic, entitled *The Microbe's Contribution to Biology* (1956, Harvard University Press, Cambridge). This series of lectures confirmed that in the years between 1931 and 1954, studies on microorganisms did indeed teach us much about the biochemistry of the cell. The titles of the two books indicate the advances made in the 24 intervening years.

There are four major classes of macromolecules in a bacterial or archaeal cell: (1) proteins, (2) nucleic acids, (3) polysaccharides, and (4) lipids. These macromolecules are synthesized from a limited number of monomers. A generalized scheme of cellular synthesis is presented in **Figure 10.1**. According to this outline, CO_2 assimilation (fixation) reactions lead into glycolysis and the tricarboxylic acid (TCA) cycle. Other substrates lead into these two key cycles, as shown on the lower part of the diagram. All of the major monomers, such as amino acids, purines, and pyrimidines, that are polymerized to form macromolecules are readily synthesized from intermediates in glycolysis, TCA, and closely related pathways.

These metabolic reactions generate the **precursor metabolites** at a continuous and appropriate level to promote orderly cellular reproduction (**Box 10.2**). There are 12 key precursor metabolites that are involved in the biosynthesis of all building blocks, prosthetic groups, coenzymes, and the various cellular components. These key metabolites and their pathways of origin are listed in **Table 10.1**.

How are these key precursor metabolites generated in *Bacteria* and *Archaea*?

CORE REACTIONS THAT GENERATE BIOSYNTHETIC PRECURSORS

The precursor metabolites identified in Table 10.1 originate during glucose catabolism via four major metabolic pathways:

1. The Embden-Meyerhof (EM) pathway (see Figure 10.2)
2. The hexose monophosphate shunt (HMS) (see Figure 10.3)
3. Some metabolites are also intermediates in the Entner-Doudoroff (ED) pathway (see Figure 10.4)
4. The tricarboxylic acid (TCA) cycle (see Figure 10.5)

All free-living microorganisms obtain precursor metabolites by doing one or more of the following: (1) using these pathways, (2) operating variations of these pathways that yield equivalent products, or (3) having a nutritional requirement for the anabolic product generally synthesized from a given precursor metabolite.

The Embden-Myerhof (EM) pathway, also termed **glycolysis (Figure 10.2)**, is the most utilized pathway for glucose catabolism and yields a net of two molecules of ATP per molecule of glucose utilized. Two molecules of ATP are required for the conversion of glucose to fructose-1,6-bisphosphate. The aldolase cleavage of the fructose 1,6-bisphosphate generates two molecules of triose (glyceraldehyde-3-phosphate). Esterification with inorganic phosphate yields two molecules of 1,3-bisphosphoglycerate. Each phosphate group (two per molecule of 1,3-diphosphoglyceric acid) leads to the generation of a molecule of ATP, so four ATP are produced from the two 1,2-bisphosphoglycerates.

The **hexose monophosphate shunt (HMS)** is important in microorganisms during growth on hexoses and pentoses (**Figure 10.3**). The decarboxylation of 6-phosphogluconate provides the microorganism with pentoses that are essential for the biosynthesis of nucleotides and other cellular components. The hexose monophosphate shunt is also a ready source of NADPH, the favored nucleotide in biosynthetic reactions. Generally, the HMS is a secondary pathway operative in organisms that utilize the EM or ED as the major dissimilatory pathway for glucose.

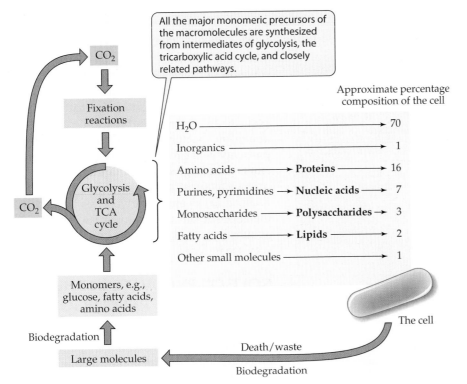

All the major monomeric precursors of the macromolecules are synthesized from intermediates of glycolysis, the tricarboxylic acid cycle, and closely related pathways.

Approximate percentage composition of the cell

H_2O	70
Inorganics	1
Amino acids → **Proteins**	16
Purines, pyrimidines → **Nucleic acids**	7
Monosaccharides → **Polysaccharides**	3
Fatty acids → **Lipids**	2
Other small molecules	1

The cell

Figure 10.1 Overview of the reactions of cellular synthesis and biodegradation

"Metabolic diversity" describes the seemingly endless capacity of microorganisms to utilize a vast array of naturally occurring chemicals (and products of chemical synthesis) as substrates for growth. They expeditiously convert these disparate substrates into the amino acids, purines, sugars, and fatty acids that make up the cell. How can a microorganism do this?

Evolution in microorganisms, as evident in the processes of anabolism, has proven to be exceedingly economical. Economical as defined by Webster— "marked by careful, efficient and prudent use of resources: operating with little waste." Once Mother Nature chanced on an effective mechanism for accomplishing a task, that mechanism became de rigueur. The heart and soul of anabolism begins with two

catabolic pathways that occur in one form or another in virtually all living cells—glycolysis and the tricarboxylic acid cycle. These are the main stream, and from this stream comes the precursors of all the macromolecules that make up the cell. A simple depiction might be as follows:

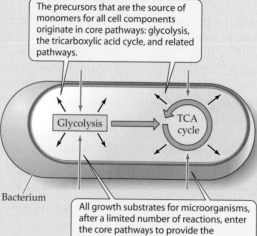

The precursors that are the source of monomers for all cell components originate in core pathways: glycolysis, the tricarboxylic acid cycle, and related pathways.

Glycolysis

TCA cycle

Bacterium

All growth substrates for microorganisms, after a limited number of reactions, enter the core pathways to provide the organism with the precursor metabolites.

What then of "metabolic diversity"? How does the ability to utilize so many different substrates fit into the economics of evolution? Neatly—because all of the processes of catabolism lead directly to the main stream. All substrates that can be utilized by microorganisms as carbon source are, by a limited number of enzymatic reactions, converted to acetate, sugars, keto acids, and other intermediates that can be funneled directly into the glycolytic or TCA main stream.

Together these catabolic and anabolic reactions generate a new cell economically and expeditiously.

Table 10.1   **Major precursor metabolites in biosynthetic reactions**

Metabolite	Predominant Pathway of Origin[a]			
	EM	TCA	HMS	ED
Glucose-6-phosphate	X		X	X
Fructose-6-phosphate	X			
Ribose-5-phosphate or ribulose-5-phosphate			X	
Erythrose-4-phosphate			X	
Glyceraldehyde-3-phosphate	X		X	X
3-Phosphoglycerate	X			
Phosphoenolpyruvate	X			
Pyruvate	X		X	X
Acetyl-CoA	X		X	X
Oxaloacetate		X		
α-Ketoglutarate		X		
Succinyl-CoA		X		

[a]EM = Embden-Meyerhof pathway; TCA = tricarboxylic acid cycle; HMS = hexose monophosphate shunt; ED = Entner-Doudoroff pathway.

The **Entner-Doudoroff (ED) pathway** yields one ATP with the formation of two molecules of pyruvate (**Figure 10.4**). This pathway is present in *Pseudomonas*, *Azotobacter*, and various gram-negative genera, but it does not occur to any great extent in gram-positive or anaerobic bacteria. The Entner-Doudoroff pathway is of greatest utility in microorganisms that use gluconate, mannonate, or hexuronate as substrate. Some organisms such as *Escherichia coli* employ glycolysis for glucose metabolism, but when gluconate is provided as substrate, the key enzymes of the ED pathway are induced.

ROLE OF PYRUVATE IN THE SYNTHESIS OF PRECURSOR METABOLITES

Pyruvate is a product of the three pathways of glucose catabolism—EM, HMS, and ED (see Table 10.1). Aerobically, a microorganism can use pyruvate for the production of reduced pyridine nucleotides, ATP, and precursor

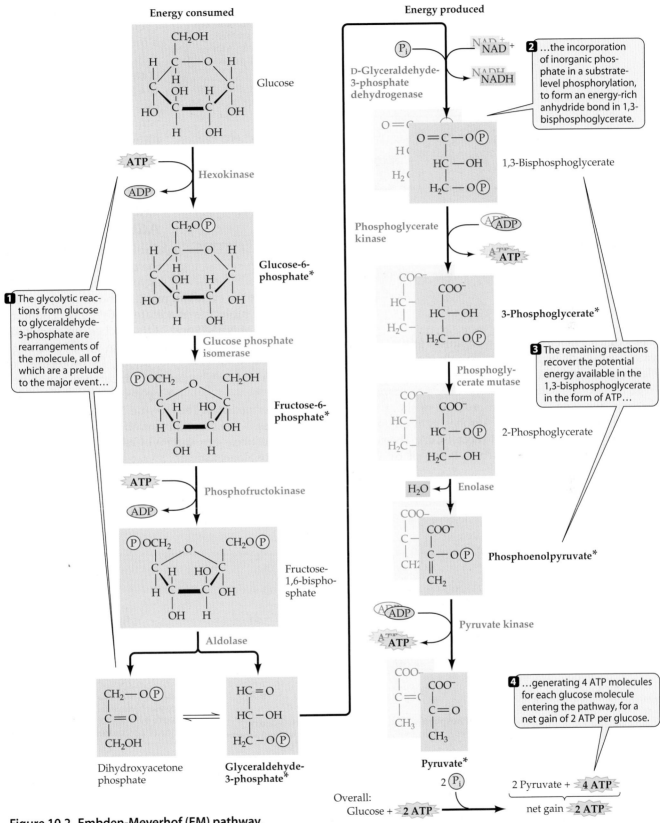

Figure 10.2 Embden-Meyerhof (EM) pathway, or glycolysis

Glycolysis is the major pathway of glucose metabolism and ATP synthesis in anaerobic organisms and the first stage of glucose metabolism in aerobic organisms. This pathway also produces a number of precursor metabolites (indicated by an asterisk*). Note that a single molecule of glucose yields two molecules each (shadow boxes) of the energy-producing three-carbon derivatives.

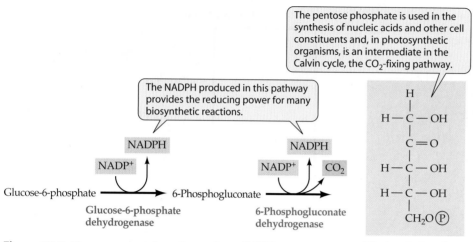

Figure 10.3 Hexose monophosphate shunt (HMS), or pentose phosphate pathway
The hexose monophosphate shunt is a major source of pentose sugars and NADPH.

metabolites. Anaerobically, an organism utilizes pyruvate or products of pyruvate catabolism as a terminal electron acceptor or a source of precursor metabolites. In an aerobe, the balance between CO_2/precursor production depends on the composition of the growth medium and other factors. Aerobes gain energy by decarboxylation of pyruvate to acetyl-CoA. The acetyl-CoA generated by the decarboxylation of pyruvate

enters the tricarboxylic acid cycle by condensing with a molecule of oxaloacetate. Aerobic organisms may completely oxidize acetate to CO_2 and water. The TCA cycle also supplies the organism with precursor metabolites—α-ketoglutarate, oxaloacetate, and succinyl-CoA.

THE TRICARBOXYLIC ACID CYCLE

The reactions involved in the tricarboxylic acid cycle are presented in **Figure 10.5**. One turn of the cycle can result in the complete oxidation of one molecule of acetate with the concomitant generation of NADH and CO_2. Aerobes can extract the maximum amount of energy from NADH through respiration and generate about 15 molecules of ATP/acetate (see Chapter 8). In anaerobes, a modified tricarboxylic acid cycle is employed to provide biosynthetic intermediates. The enzyme α-ketoglutarate dehydrogenase is not generally synthesized in anaerobes, and where necessary, succinyl-CoA is formed by reduction of oxaloacetate. In a microorganism growing

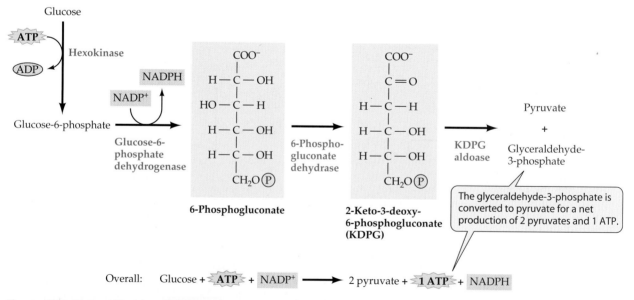

Figure 10.4 Entner-Doudoroff (ED) pathway
This pathway of glucose metabolism is common among the pseudomonads and some other gram-negative bacteria. The catabolism of one molecule of glucose to two molecules of pyruvate produces only 1 ATP (not 2 ATP as in glycolysis).

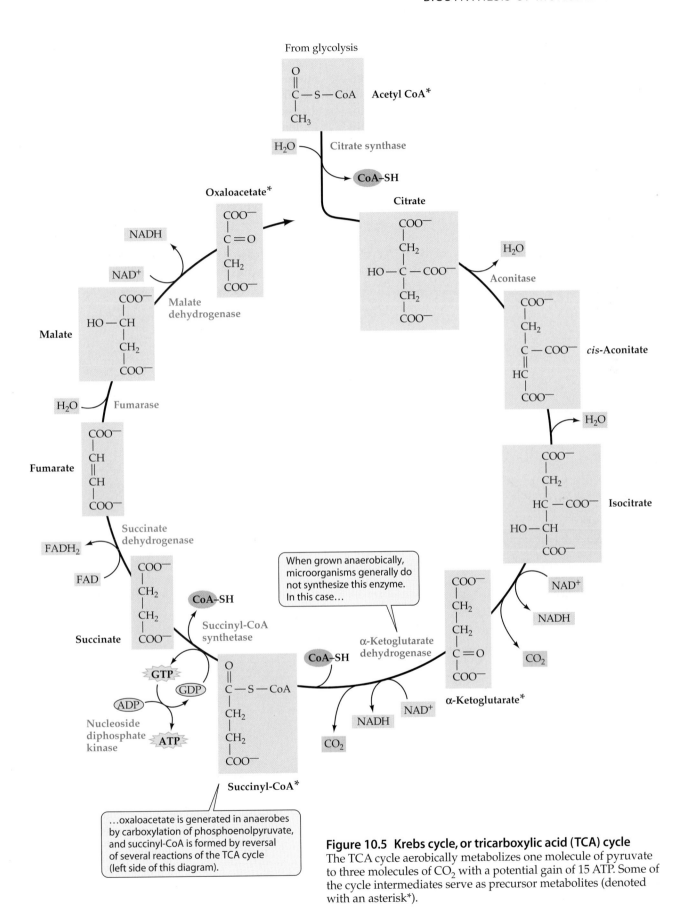

Figure 10.5 Krebs cycle, or tricarboxylic acid (TCA) cycle
The TCA cycle aerobically metabolizes one molecule of pyruvate to three molecules of CO_2 with a potential gain of 15 ATP. Some of the cycle intermediates serve as precursor metabolites (denoted with an asterisk*).

When grown anaerobically, microorganisms generally do not synthesize this enzyme. In this case…

…oxaloacetate is generated in anaerobes by carboxylation of phosphoenolpyruvate, and succinyl-CoA is formed by reversal of several reactions of the TCA cycle (left side of this diagram).

anaerobically, oxaloacetate can be synthesized by carboxylation of phosphoenolpyruvate.

Biosynthetic reactions withdraw intermediates from the TCA cycle: α-ketoglutarate to glutamic acid, oxaloacetate to aspartic acid, and succinyl-CoA to porphyrin. These reactions deplete the 4-carbon intermediate (oxaloacetate) that combines with acetate to form citrate. For the TCA cycle to continue, the oxaloacetate must be replenished. This can be achieved in most organisms by the carboxylation of pyruvate or phosphoenolpyruvate to form a molecule of oxaloacetate. Any reaction that replenishes intermediates in a central pathway such as this is called an **anaplerotic reaction**.

The **glyoxylate shunt** is a variation of the TCA cycle, and it is of considerable consequence to organisms growing on acetate, fatty acids, long-chain n-alkanes, or other substrates that are catabolized through 2-carbon intermediates. As anabolic reactions remove intermediates from the TCA cycle, it is essential that organisms growing on 2-carbon substrates have a mechanism for replenishing substantial quantities of oxaloacetate for continued operation of the cycle. Obviously while growing on acetate, the microorganism would not have available a 3-carbon intermediate such as phosphoenolpyruvate to carboxylate to form oxaloacetate. Organisms that utilize growth substrates such as acetate generally have the inducible glyoxylate shunt (**Figure 10.6**). In these shunt reactions, isocitrate is cleaved by the inducible isocitrate lyase, yielding succinate and glyoxylate. The glyoxylate thus formed can condense with a molecule

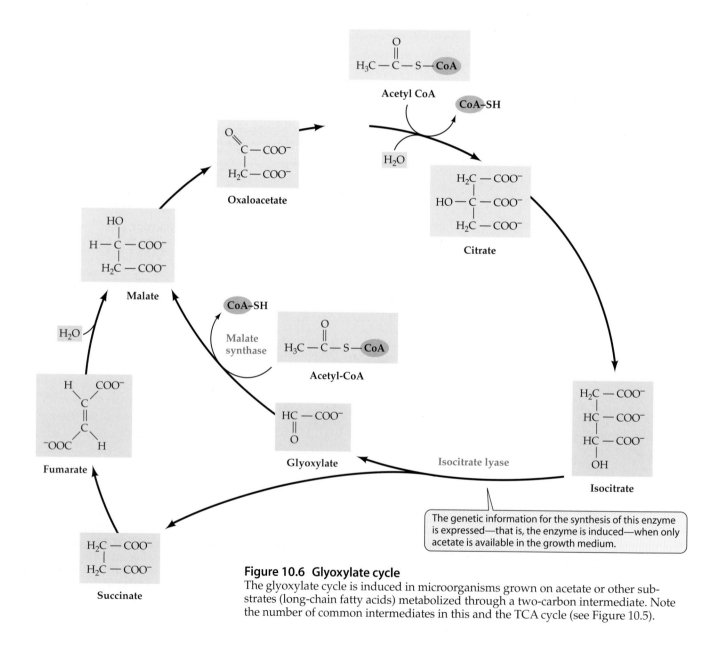

Figure 10.6 Glyoxylate cycle
The glyoxylate cycle is induced in microorganisms grown on acetate or other substrates (long-chain fatty acids) metabolized through a two-carbon intermediate. Note the number of common intermediates in this and the TCA cycle (see Figure 10.5).

of acetate to form a molecule of malate. Thus, an organism growing on acetate would have an unlimited supply of 4 carbon intermediates to continue the TCA cycle. Precursor TCA metabolites such as α-ketoglutarate and succinyl-CoA would be made available.

THE CALVIN CYCLE—CO$_2$ FIXATION

The major pathway for utilization of carbon dioxide as the sole source of carbon is designated the **Calvin cycle**. There are two major events that render an organism capable of autotrophic growth via this cycle (**Figure 10.7**).

- A carboxylation reaction that leads to the assimilation of a molecule of CO$_2$

- The molecular rearrangements that regenerate ribulose-1,5-bisphophate, which is the CO$_2$ acceptor molecule

A primary reaction in the Calvin cycle is the phosphorylation of ribulose-5-phosphate to produce ribulose-1,5-bisphosphate. The enzyme involved in assimilating a molecule of CO$_2$ is ribulose-1,5-bisphosphate carboxylase/oxygenase (RuBisCo). The RuBisCo carboxylase reaction incorporates a molecule of CO$_2$ into ribulose-1,5-bisphosphate with the simultaneous cleavage of the unstable intermediate formed to yield two molecules of 3-phosphoglycerate. This product, 3-phosphoglycerate, is an intermediate in glycolysis (see Figure 10.2), and glycolysis combined with the TCA cycle provides the organism with the precursor metabolites.

The regeneration of ribulose-5-phosphate would be of utmost importance to autotrophic organisms that utilize the Calvin cycle for CO$_2$ fixation. The first step in the regeneration process is the conversion of 3-phosphoglycerate to glyceraldehyde-3-phosphate, which can be coupled to form hexoses for biosynthetic reactions. The hexoses can be rearranged to generate 4-, 5-, and 7-carbon sugars (see Figure 12.3).

A key intermediate in regenerating ribulose-5-phosphate is the 7-carbon sugar, sedoheptulose-7-phosphate, as two carbons from this sugar can be added to one readily available glyceraldehyde-3-phosphate, resulting in the formation of two molecules of 5-carbon sugar (7 − 2 = 5; 3 + 2 = 5). The enzymes transaldolase and transketolase catalyze these interconversions of sugars (see Figure 12.3). These reactions also occur in heterotrophic organisms in conjunction with the hexose monophosphate shunt and supply an organism with ribulose-5-phosphate and erythrose-4-phosphate for biosynthetic purposes. The transaldolase and transketolase are also of importance in the biodegradation of various sugars (see Figure 12.3).

The Calvin cycle actually functions as a cycle because the products of CO$_2$ fixation, molecules of 3-bisphos-

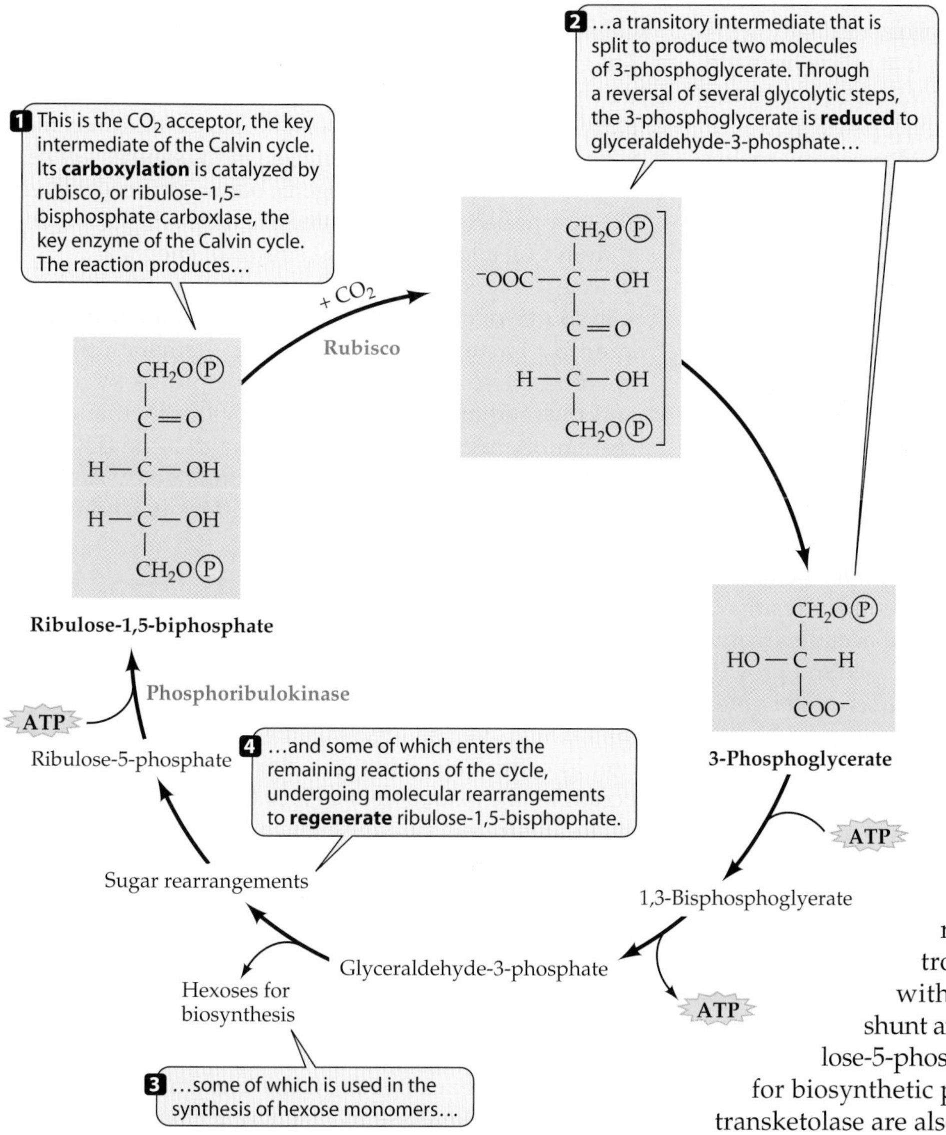

Figure 10.7 Calvin cycle
This CO$_2$-fixation pathway is predominant in plants, the cyanobacteria, and most other autotrophs.

phate, can be rearranged to generate ribulose-1,5-bis-phosphate. The enzyme RuBisCo incorporates a CO_2 into the ribulose-1,5-bisphosphate, and thus the cycle continues.

ALTERNATE PATHWAYS OF 1-CARBON ASSIMILATION

Elucidation of the Calvin cycle led to an assumption that this was the universal mechanism for autotrophic 1-carbon assimilation. However, experiments with green sulfur photosynthetic bacteria and some other autotrophs clearly indicated that some microorganisms utilize distinctly different pathways for CO_2 assimilation. When radiolabeled CO_2 was placed in the atmosphere supplied for growth of green sulfur photosynthetic bacteria, the primary products were not radiolabeled sugars associated with the Calvin cycle but were radiolabeled fatty acids. Similar experiments with organisms that utilize methane, methanol, or other 1-carbon compounds as substrates affirmed that they too utilized alternate pathways.

Anaerobic *Bacteria* and *Archaea* that occupy niches such as swamps in which electron donors more negative than NADH (see Table 7.1) are available can fix CO_2 by reductive carboxylation reactions. Among these organisms are the acetogenic anaerobes, green sulfur photosynthetic bacteria, and methanogens (methane-producing bacteria). Despite the fact that these organisms fix CO_2 via a common pathway (reductive carboxylations), they are not phylogenetically related.

Chlorobium thiosulphatophilum is a green sulfur photosynthetic organism that utilizes a reductive carboxylation mechanism for CO_2 fixation (**Figure 10.8**). This is essentially a reversal of the TCA cycle illustrated in Figure 10.5. Oxaloacetate is a principal intermediate in this cycle, and one complete turn essentially results in the net production of one molecule of acetate. Acetyl-CoA can enter into the synthetic TCA cycle and provide the organism with the essential precursor metabolites. Decarboxylation of oxaloacetate would provide the organism with pyruvate. Glycolysis leads from glucose to pyruvate and a reversal of glycolysis can generate a molecule of glucose (2 pyruvate → → → glucose).

Members of genus *Chloroflexus* have a unique pathway for CO_2 fixation, the hydroxypropionate pathway (Figure 10.8B). Hydroxypropionyl-CoA and propionyl-CoA are key intermediates, and one cycle leads to the formation of one molecule of glyoxylic acid. Glyoxylate can enter major cycles via malate (Figure 10.6).

A number of organisms generate acetate as a major product of anaerobic respiration. These organisms are called **acetogenic bacteria.** They produce acetate essentially by coupling two molecules of CO_2, utilizing H_2 as a reductant and their energy source. Among these organisms is *Acetogenium kivui*, a chemolithotrophic

anaerobe that obtains carbon and energy according to the following equation:

$$4H_2 + 2CO_2 \longrightarrow CH_3COO^- + 2H_2O$$

A general scheme for the energetics and acetate synthesis in acetogenic bacteria is presented in **Figure 10.9**. A key enzyme complex in the acetogenic pathway is carbon monoxide (CO) dehydrogenase. A molecule of carbon monoxide binds to this enzyme complex, and this is followed by the addition of a methyl group. The methyl group donated by the vitamin B_{12} corrinoid enzyme originates on a tetrahydrofolate-protein complex through the reduction of CO_2. The methyl group migrates and couples to the –CO attached to the CO dehydrogenase enzyme complex forming the acetyl derivative. A molecule of coenzyme A is attached to the complex, and the CO dehydrogenase catalyzes the release of acetyl-CoA. The three compounds—a molecule of coenzyme A (CoA), the methyl group, and carbon monoxide—are independently bound to the enzyme. The molecule of acetate formed in this series can be utilized in the synthesis of cellular components. The acetogenic bacteria are considered autotrophs because they synthesize their cellular material from CO_2 and utilize H_2 as source of energy.

Many microorganisms can grow with methane as the sole source of carbon and energy. Methanol is also utilized by bacteria and yeasts. The methane-utilizing microbes have the enzyme methane monooxygenase and the methanol utilizers do not. Both methane and methanol are oxidized to the formaldehyde (HCHO) level, and this molecule can be assimilated by two distinctly different pathways in different organisms (**Figure 10.10A,B**):

1. Glycine + HCHO → serine
2. Ribulose-5-phosphate + HCHO → fructose-6-phosphate

In the glycine-serine pathway(Figure 10.10A), the amino acid glycine serves as the acceptor of the 1-carbon intermediate. Energy is derived aerobically and occurs by the oxidation of CH_4 or CH_3OH to CO_2. In the cycle illustrated, one mole of formaldehyde and one mole of CO_2 would be assimilated in each turn.

The other distinct group of methane-utilizing microorganisms assimilates the formaldehyde via a **ribulose monophosphate pathway** (Figure 10.10B). This is somewhat similar to the ribulose-1,5-bisphosphate pathway operating in the autotrophs. Both employ a 5-carbon sugar and ultimately generate fructose-6-phosphate that can be rearranged by the enzymes transaldolase and transketolase to regenerate ribulose-5-phosphate. Fructose-6-phosphate and modifications of the tricarboxylic acid cycle present in anaerobes would be a source of precursor metabolites for cell synthesis.

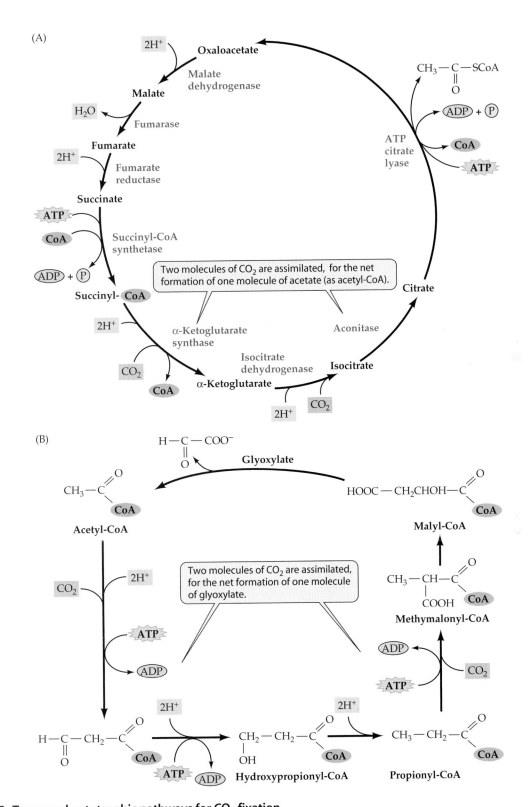

Figure 10.8 Two novel autotrophic pathways for CO_2 fixation
(A) The reductive citric acid pathway, present in green sulfur bacteria (*Chlorobium limicola*), thermophilic hydrogen-oxidizing bacteria (*Hydrogenobacter thermophilus*), and some of the sulfate-reducing bacteria (*Desulfobacter hydrogenophilus*). Note the similarity to the TCA cycle (see Figure 10.5). However, whereas the aerobic TCA cycle is a way of extracting energy from acetate by oxidizing it to CO_2, the reductive cycle shown here is an *anaerobic* pathway for *fixing* CO_2. (B) The hydroxypropionate pathway, present in *Chloroflexus*.

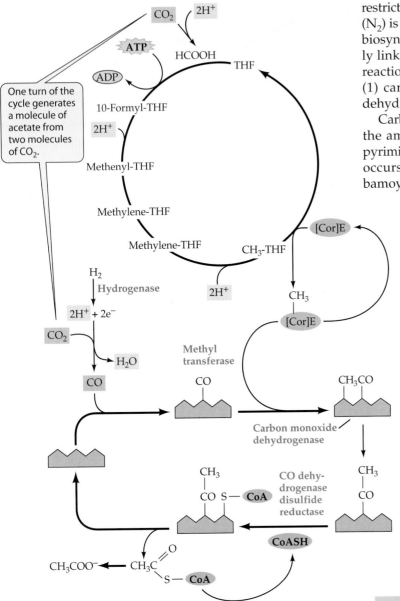

One turn of the cycle generates a molecule of acetate from two molecules of CO_2.

Figure 10.9 Pathway of CO_2 fixation in acetogenic bacteria
This pathway is present in homoacetogenic bacteria (*Clostridium thermoaceticum*), most of the sulfate-reducing bacteria (*Desulfobacterium autotrophicum*), and selected methanogenic archaea (*Methanosarcina barkeri*). THF, tetrahydrofolic acid; [Cor]E, vitamin B_{12} corrinoid enzyme.

NITROGEN ASSIMILATION

Nitrogen is a major component of all living cells and constitutes about 14% of the dry weight of a bacterium. The *Bacteria* and the *Archaea* generally use inorganic nitrogen in the form of the ammonium ion (NH_4^+) or nitrate (NO_3^-), whereas some species obtain cellular nitrogen by reduction of atmospheric dinitrogen (N_2). The ammonium ion (NH_4^+) is commonly utilized for bacterial growth; the ability to assimilate nitrate is more

restricted. The number of species that can fix dinitrogen (N_2) is even more limited. Both NO_3^- and N_2 enter into biosynthetic reactions through NH_4^+. Bacteria generally link NH_4^+ to organic intermediates via three major reactions. The enzymes mediating these reactions are (1) carbamoyl-phosphate synthetase, (2) glutamate dehydrogenase, and (3) glutamine synthetase.

Carbamoyl-phosphate is involved in the synthesis of the amino acid arginine and in the synthesis of the pyrimidine ring. The synthesis of carbamoyl-phosphate occurs via the following reaction catalyzed by carbamoyl-phosphate synthetase:

$$NH_4^+ + HCO_3^- + 2\,ATP \longrightarrow$$
$$H_2N-\overset{\overset{\displaystyle O}{\|}}{C}-O-PO_3^{2-} + 2\,ADP + P_i + 2H^+$$

The enzyme **glutamate dehydrogenase (GDH)** catalyzes the reductive amination of α-ketoglutarate to form glutamate (**Figure 10.11**). The reductant for this reaction is either NADH or NADPH. **Glutamine synthetase (GS)** catalyzes an ATP-dependent amidation of the γ-carboxyl of glutamate to form glutamine (Figure 10.11). Glutamine is a major amine donor in the biosynthesis of major monomers that make up a cell, including purines, pyrimidines, and a number of amino acids. Combined, the enzymes glutamic dehydrogenase and glutamine synthetase are responsible for most of the ammonium that is assimilated into cell material.

When a microorganism grows in an environment with an ample supply of NH_4^+ the sequence of reactions is:

$$NH_4^+ + \alpha\text{-ketoglutarate} + NADPH \xrightarrow{\text{GDH}}$$
$$\text{Glutamate} + NADP^+ + H_2O$$

$$\text{Glutamate} + NH_4^+ + ATP \xrightarrow{\text{GS}}$$
$$\text{Glutamine} + ADP + P_i + H_2O$$

When NH_4^+ availability is limited, the glutamate dehydrogenase (GDH) reaction is not effective, and glutamine synthetase (GS) becomes the major NH_4^+ assimilation reaction. This occurs because glutamate synthetase has a much higher affinity for ammonia. Because the glutamine synthetase reaction reduces the level of glutamate, there is a need for an alternate mechanism for glutamine production. The enzyme **glutamine synthase** (glutamate: oxoglutarate amino-tranferase or GOGAT) carries out the reductive amination of α-ketoglutarate with the amide-N of glutamine as N-donor according to the following reaction (page 214):

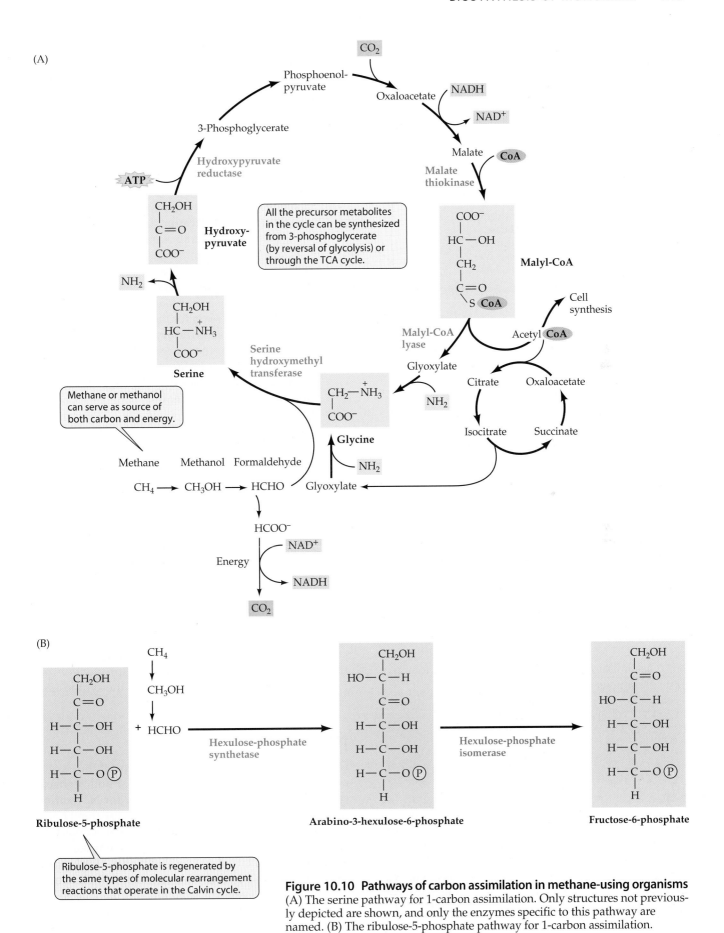

Figure 10.10 Pathways of carbon assimilation in methane-using organisms
(A) The serine pathway for 1-carbon assimilation. Only structures not previously depicted are shown, and only the enzymes specific to this pathway are named. (B) The ribulose-5-phosphate pathway for 1-carbon assimilation.

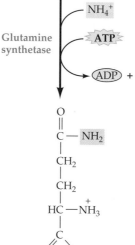

Figure 10.11 Assimilation of ammonia
Together, the glutamate dehydrogenase and glutamine synthetase reactions result in the assimilation of two ammonia molecules (shown as the ammonium ion, NH_4^+).

$$NADPH + \alpha\text{–ketoglutarate} + glutamine \longrightarrow$$
$$2\,glutamate + NADP^+$$

Two molecules of glutamate are generated—one from amination of α-ketoglutarate and the other from the deamidation of glutamine. Together the two enzymes GS and GOGAT constitute a significant pathway for NH_4^+ assimilation, and the role of GOGAT is to regenerate glutamate:

$$2\,NH_4 + 2\,ATP + 2\,glutamate \xrightarrow{\ GS\ }$$
$$2\,glutamine + 2\,P_i$$

$$NADPH + \alpha\text{–ketoglutarate} + glutamine \xrightarrow{\ GOGAT\ }$$
$$2\,glutamate + 2\,NADP^+$$

Sum:

$$2\,NH_4^+ + \alpha\text{-ketoglutarate} + NADPH + 2\,ATP \rightarrow$$
$$glutamine + NADP^+ + 2\,P_i$$

The amino group can be transferred from glutamic acid to other keto acids via enzymes termed **transaminases** (**Figure 10.12**).

Nitrate Reduction

Two enzymatic systems are principally responsible for nitrate reduction by bacteria: dissimilatory nitrate reductase and assimilatory nitrate reductase. The **dissimilatory pathway** is induced in selected facultatively anaerobic bacteria during anaerobic growth in the presence of NO_3^-. Nitrate serves as the terminal electron acceptor, and various gaseous-reduced nitrogen compounds are generated, including NH_2OH (hydroxylamine), N_2O (nitrous oxide), NO (nitric oxide), and dinitrogen (N_2). The process is termed **denitrification** and can occur in agricultural soil that becomes waterlogged and anaerobic. Denitrification is a dissimilatory reaction, as the reduced product is not incorporated into the cell. It is a serious problem, as it can deplete the soil of nitrogen necessary for the growth of crops.

The **assimilatory nitrate reductase** is induced in those organisms that can utilize NO_3^- as the nitrogen source and is present only in the absence of NH_4^+. The enzymatic reaction proceeds as follows:

$$\underbrace{NO_3^- \xrightarrow{\ 2e^-\ }}_{\text{nitrate reductase}}$$

$$\underbrace{NO_2^- \xrightarrow{\ 2e^-\ } HNO \xrightarrow{\ 2e^-\ } NH_2OH \xrightarrow{\ 2e^-\ } NH_3}_{\text{nitrite reductase}}$$

The molybdenum-containing nitrate reductase system reduces the nitrogen in NO_3^- with a valence of +5, to the ammonia level with a valence of –3. This results via the joint action of the two enzymes and requires NADPH and FAD, and there are no free intermediates involved. Considerable energy is expended in these reactions, and consequently, NH_4^+ is the preferred nitrogen source in the preparation of culture media. The product of this reduction is utilized by the microbe in the synthesis of cellular constituents (see Figure 10.11).

NITROGEN FIXATION

Dinitrogen (N_2) is the major constituent of the earth's atmosphere (80%), and some bacterial and archaeal species have evolved with the capacity for **N_2 fixation** (assimilation). This N_2-fixing capacity is apparently confined to the prokaryotic world, and these reactions ultimately provide a source of nitrogenous compounds for sustenance of eukaryotes. The mechanisms for N_2 fixation have received considerable attention in recent years. Dinitrogen is a very unreactive molecule due to the triple bond N≡N, and because of this unreactive nature,

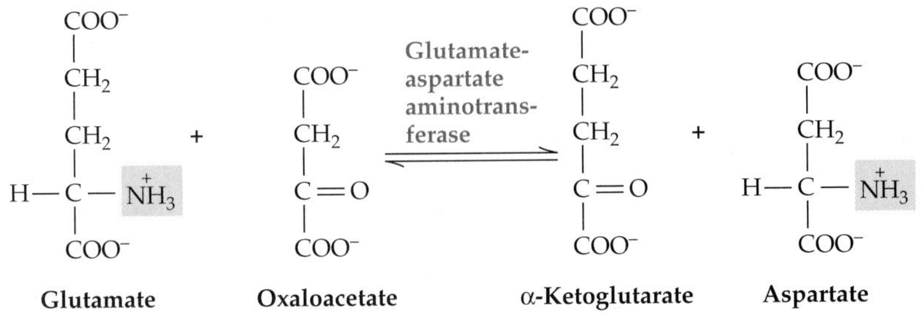

COO⁻ ... Glutamate-aspartate aminotransferase ... COO⁻

Figure 10.12 Transamination reaction
The glutamate-dependent transamination of an α-keto acid is a fundamental reaction of amino acid synthesis.

(Figure structures labeled: **Glutamate**, **Oxaloacetate**, **α-Ketoglutarate**, **Aspartate**)

considerable activation energy is required to reduce (fix) N_2 to NH_4^+.

Nitrogenase (**Figure 10.13**) is the enzyme complex involved in N_2 fixation, and it is inactivated by molecular oxygen (O_2). Nitrogenase is composed of two subunits: an iron-sulfur protein and a molybdenum-iron-sulfur protein. Nitrogen fixation occurs in anaerobic, aerobic, and photosynthetic prokaryotes. All N_2 fixers that grow in the presence of air have a mechanism for keeping O_2 away from their nitrogenase system. Many facultative anaerobes such as *Bacillus polymyxa* fix N_2 only when growing anaerobically. The cyanobacteria that fix N_2 develop specialized anaerobic cells (called **heterocysts**) that become the sites of nitrogen fixation (see Chapter 21).

The overall process of nitrogen fixation is a reductive one, and the electrons required for reduction are provided by photosynthesis, respiration, or fermentation. A considerable expenditure of ATP (20 to 24 per mole N_2 fixed) is also necessary.

SULFUR ASSIMILATION

Sulfur is an indispensable component of amino acids (cysteine, methionine), water-soluble vitamins (biotin, thiamine), and other key cellular constituents including coenzyme A. *Bacteria*, *Archaea*, and fungi can use sulfate as their source of sulfur, but it must be reduced to the sulfide S^{2-} level for incorporation into the constituents of cells. A few organisms, such as the methanogens, lack the capacity for sulfate reduction and require H_2S as source of sulfur. The key reaction in sulfur assimilation is as follows:

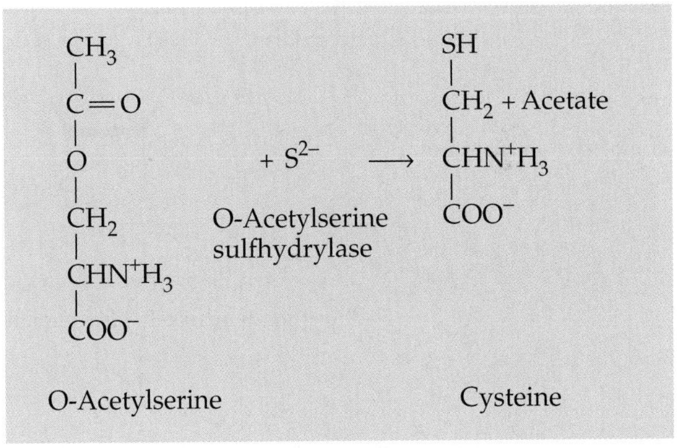

(Reaction structures: **O-Acetylserine** + S^{2-} → (O-Acetylserine sulfhydrylase) → **Cysteine** + Acetate)

The reduction of sulfate (SO_4^{2-}) occurs in a series of reactions involving ATP, NADPH, and the sulfur-containing protein thioredoxin. In the initial reaction (**Figure 10.14**), sulfate reacts with ATP to form adenosine-5′-phosphosulfate (APS). This APS reacts with another molecule of

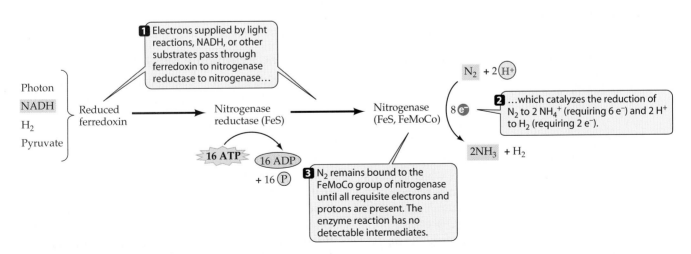

Figure 10.13 Nitrogen fixation
The primary electron donor in the nitrogenase reaction is reduced ferredoxin. When grown under iron-limited conditions, some bacteria can synthesize the flavoprotein flavodoxin, which is functionally equivalent to ferredoxin.

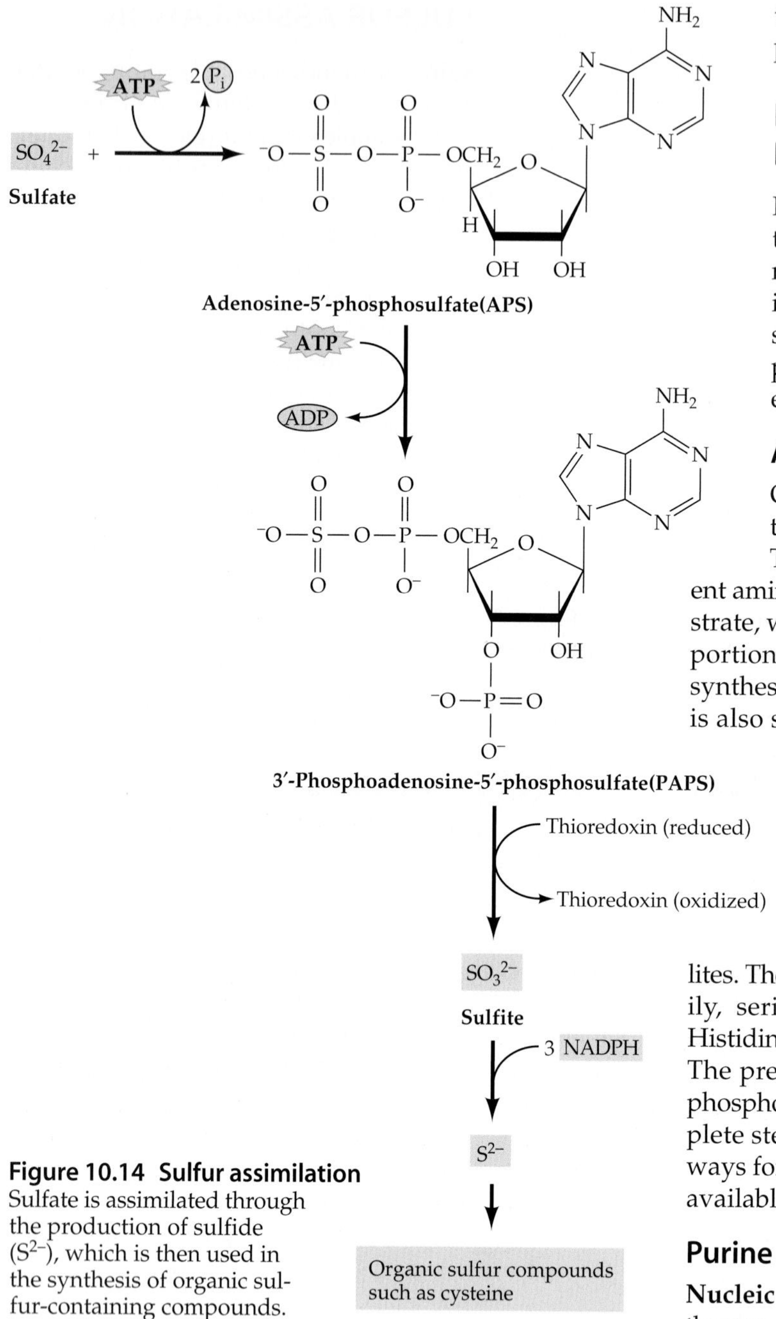

Figure 10.14 Sulfur assimilation
Sulfate is assimilated through the production of sulfide (S^{2-}), which is then used in the synthesis of organic sulfur-containing compounds.

taining amino acids, and other carbon-sulfur compounds.

BIOSYNTHESIS OF AMINO ACIDS, PURINES, AND PYRIMIDINES

Proteins are composed of amino acids, and proteins are the major constituents of a cell. Whereas nucleic acids (DNA and RNA) carry genetic information for a cell and are involved in transcribing this information, it is the proteins that provide the cell with structure, motility, and enzymatic capacity.

Amino Acid Biosynthesis

One gram of dry bacterial cell mass contains 500 to 600 mg of protein (50% to 60% of the cell). This protein is synthesized from about 20 different amino acids. An organism growing on a simple substrate, whether glucose or CO_2, expends a considerable portion of its energy and biosynthetic capacity on the synthesis of the monomeric amino acids. Much energy is also spent organizing and assembling these amino acids into functional protein units. The ability to synthesize many of the requisite amino acids is a trait shared by virtually all of the free-living bacteria. Obviously the strict autotrophs can synthesize them all.

Amino acids are synthesized as "families" from a limited number of precursor metabolites. The families are: the glutamate family, alanine family, serine family, aspartate family, and aromatics. Histidine is the sole product of its biosynthetic pathway. The precursors include α-ketoglutarate, pyruvate, 3-phosphoglycerate, and oxaloacetate (**Table 10.2**). A complete step-by-step depiction of the biosynthetic pathways for the amino acids from precursor metabolites is available in a biochemistry text.

Purine and Pyrimidine Biosynthesis

Nucleic acids are a constituent part of all living cells, as they are the genetic determinant in all creatures from the simple, acellular virus to the complex mammal. Before the advent of life, conditions on the primeval Earth included gaseous mixtures of hydrocarbon, hydrogen, and ammonia, as well as energy from ultraviolet radiation and lightning; these combined can lead to the formation of purines and pyrimidines. As the nucleic acid bases can be formed abiotically, they were apparently available early on in the evolution of viable organisms. Polymerization of these bases created a structure that evolved as the key inheritable element in all life forms. Nucleic acid bases are also a component part of ATP, NAD^+, $NADP^+$, and some of the B-vitamins.

ATP to yield 3'-phosphoadenosine-5'-phosphosulfate (PAPS). PAPS is reduced by thioredoxin, which releases a molecule of sulfite (SO_3^{2-}) that then is reduced by NADPH to the sulfide (S^{2-}) level. The sulfide is then incorporated into biomolecules through O-acetylserine.

There are actually limited amounts of sulfate in soil, and it is probable that microorganisms in the environment assimilate other sulfur compounds such as sulfur esters ($R-CH_2OSO_3$), sulfonates ($R-CH_2SO_3^-$), sulfur con-

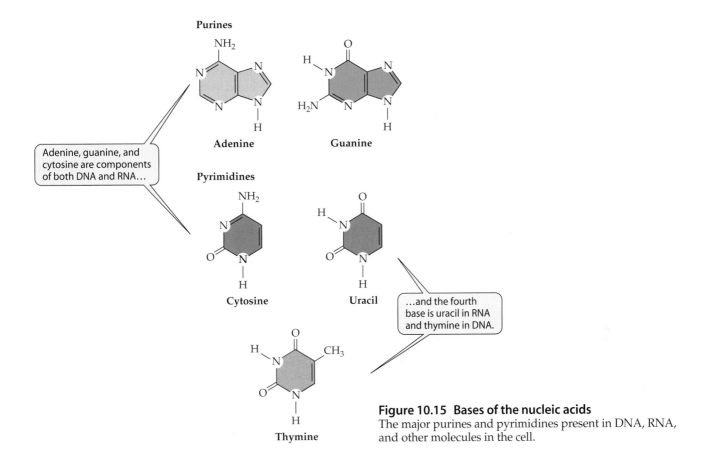

Figure 10.15 Bases of the nucleic acids
The major purines and pyrimidines present in DNA, RNA, and other molecules in the cell.

Table 10.2	Precursor of the major amino acids in protein and the family designations	
Family	**Precursor Metabolite**	**Amino Acid**
Glutamate	α-Ketoglutarate	Glutamate
		Glutamine
		Proline
Alanine	Pyruvate	Alanine
		Valine
		Leucine
Serine	3-Phosphoglycerate	Serine
		Glycine
		Cysteine
Aspartate	Oxaloacetate	Aspartate
		Asparagine
		Methionine
		Lysine
		Threonine
		Isoleucine
Aromatics	Phosphoenolpyruvate +erythrose-4-phosphate	Phenylalanine
		Tyrosine
		Tryptophan
Histidine	5-Phosphoribsyl-1-pyrophosphate	Histidine

The structures of the bases generally present in nucleic acids are presented in **Figure 10.15**. The purines adenine and guanine are constituents of both RNA and DNA. The two pyrimidines that are constituents of DNA are cytosine and thymine. Uracil replaces thymine in RNA. The origin of each component of the nine-member purine ring system is presented in **Figure 10.16**.

The biosynthesis of the pyrimidine molecule occurs by a more direct route than that for purines. Carbamoyl

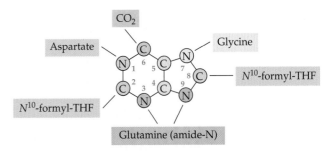

Figure 10.16 Origin of the nine atoms in the purine ring
The major contributor to the purine ring of adenine and guanine is the amino acid glycine.

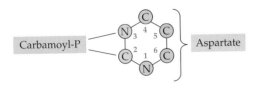

Figure 10.17 Origin of the six atoms in the pyrimidine ring

phosphate condenses with a molecule of aspartate to form the six-member ring (**Figure 10.17**). The detailed biosynthetic reactions involved in the synthesis are depicted in a biochemistry textbook.

BIOSYNTHESIS OF LIPID COMPOUNDS

The cytoplasmic membranes of bacterial cells are composed of phospholipids. The major component of these phospholipids is the long-chain fatty acids that form the hydrophobic core of the cytoplasmic membrane. The interior of the cytoplasmic membrane in archaeal species are compounds of ether-linked long-branched hydrocarbon chains. Both the long-chain fatty acids in bacterial cells and the branched chains of archaeal cells are synthesized by condensation of acetate via mechanisms that will be discussed.

Biosynthesis of Fatty Acids

Bacteria generally synthesize long-chain fatty acids by the condensation of a molecule of acetyl-CoA with one of malonyl-CoA and with NADPH as reductant (**Figure 10.18**). The malonyl-CoA is formed by carboxylation of acetyl-CoA by an ATP-driven reaction. The malonyl-CoA formed reacts with a protein, called the acyl carrier protein (ACP), to form the malonyl-ACP derivative.

A molecule of the acetyl-ACP reacts with β-keto-ACP synthase to form an acetyl Kase derivative. The acetyl Kase reacts with malonyl-ACP to form acetoacetyl-ACP. The decarboxylation of malonyl-ACP provides energy for the condensation of the acetyl-Kase and the methylene carbon of the malonyl-ACP molecule. Acetoacetyl is reduced stepwise to the saturated fatty acid. This fatty acid remains attached to ACP. Repetition of these reactions lengthens the chain two carbons at a time to the required length. There are eight independent enzyme reactions involved in fatty acid synthesis (each involves a different protein). **Acyl carrier protein (ACP)** is an important low molecular weight protein that is functionally equivalent to coenzyme A but activates and transfers fatty acids having chain lengths longer than the lower molecular weight groups involved with coenzyme A.

The unsaturated fatty acids in bacteria generally have one double bond. Those present in eukaryotic organisms commonly have more than one double bond. The cyanobacteria are the only *Bacteria* that have more than a single double bond in their cellular lipids. Many bacteria can incorporate exogenously supplied polyunsaturated fatty acids into their cellular lipids, and these apparently are not harmful.

Biosynthesis of the monounsaturated fatty acids can occur by either the aerobic or the anaerobic pathway. In the anaerobic pathway, the double bond is introduced between carbons 9 and 10 during fatty acid synthesis. This occurs in strict anaerobes and may occur in various aerobes, including *E. coli*, during aerobic growth. The aerobic pathway is utilized by micrococci, endospore-forming bacilli, some cyanobacteria, fungi, and animals. In the aerobic pathway, the double bond is introduced after synthesis between carbons 9 and 10 of the long-chain fatty acid.

Considerable variety exists in the fatty acids present in the major lipids of bacteria (**Figure 10.19**). The chain length varies from C_{12} to C_{20}, they are straight or branched, are saturated or unsaturated, they may have a cyclopropane or hydroxy constituent, and may (but rarely except in cyanobacteria) have more than one double bond. The hydroxy fatty acids that are constituents of the outer envelope of gram-negative bacteria are synthesized by the anaerobic pathway.

Biosynthesis of Branched Chains

The carotenoids and related long-chain branched hydrocarbons are common constituents of many bacterial and archaeal species. They are composed of isoprene subunits:

$$-CH_2-\overset{\overset{\displaystyle CH_3}{\|}}{C}=CH-CH_2-$$

These compounds, termed isoprenoids, are synthesized by condensation of molecules of acetyl-CoA (**Figure 10.20**). The yellow and red pigments of photosynthetic and other bacteria and the phytol chain that anchors chlorophyll to the photosynthetic membrane are composed of isoprene units. Isoprenoid molecules strung together are present in *Archaea* in place of fatty acids.

The phospholipids are assembled as outlined in **Figure 10.21**. The precursor metabolite, glycerol 3-phosphate, provides the backbone for polar lipid synthesis, and the addition of ACP conjugated fatty acids yields phosphatidic acid. Phosphatidic acid is a key intermediate in the synthesis of the various membrane lipids including phosphatidylserine, phosphatidylethanolamine, and cardiolipin. The synthesis of membrane lipids in the *Archaea* will be discussed in Chapter 18.

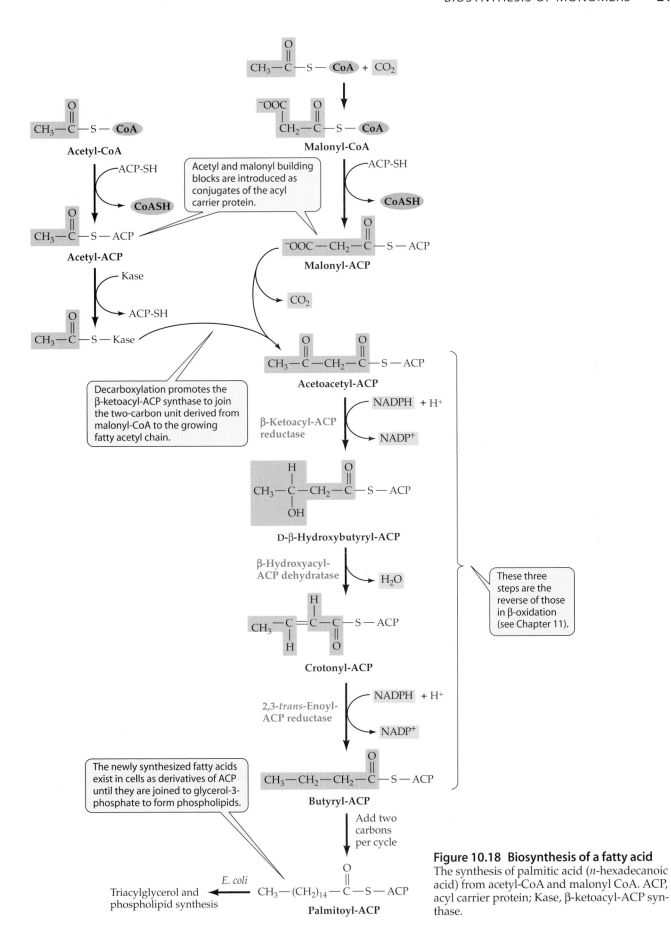

Figure 10.18 Biosynthesis of a fatty acid
The synthesis of palmitic acid (*n*-hexadecanoic acid) from acetyl-CoA and malonyl CoA. ACP, acyl carrier protein; Kase, β-ketoacyl-ACP synthase.

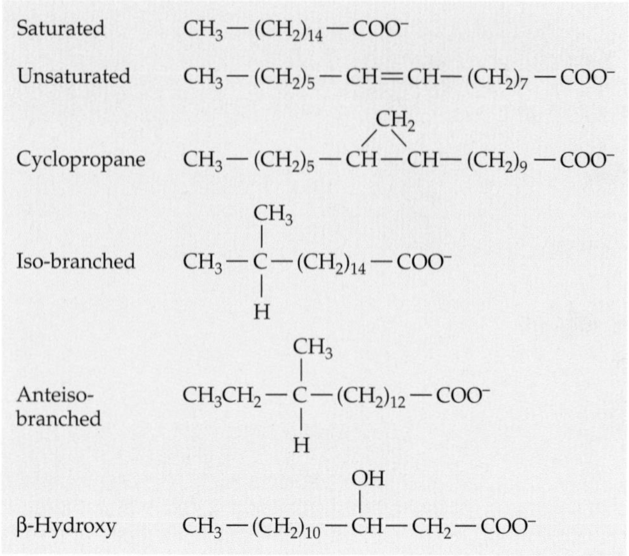

Saturated $CH_3 — (CH_2)_{14} — COO^-$

Unsaturated $CH_3 — (CH_2)_5 — CH=CH — (CH_2)_7 — COO^-$

Cyclopropane $CH_3 — (CH_2)_5 — \overset{\displaystyle CH_2}{\overset{\displaystyle /\backslash}{CH—CH}} — (CH_2)_9 — COO^-$

Iso-branched $CH_3 — \overset{\displaystyle CH_3}{\underset{\displaystyle H}{C}} — (CH_2)_{14} — COO^-$

Anteiso-branched $CH_3CH_2 — \overset{\displaystyle CH_3}{\underset{\displaystyle H}{C}} — (CH_2)_{12} — COO^-$

β-Hydroxy $CH_3 — (CH_2)_{10} — \overset{\displaystyle OH}{CH} — CH_2 — COO^-$

Figure 10.19 Fatty acids
Long-chain fatty acids commonly present in the phospholipids of bacteria.

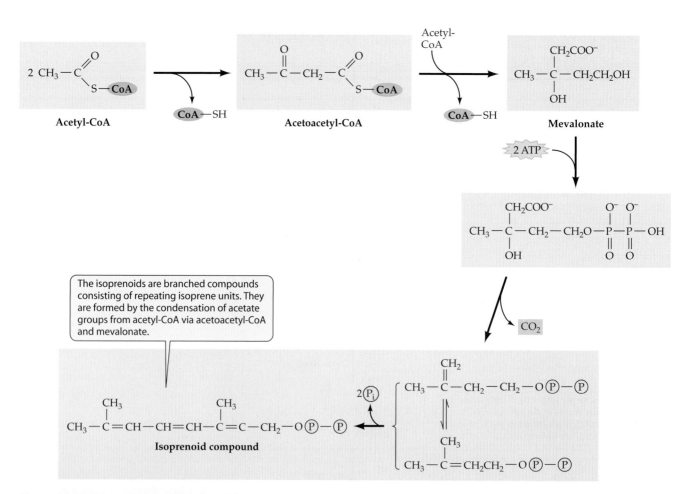

Figure 10.20 Biosynthesis of isoprenoids
The branched isoprenoids are commonly present in the membrane lipids of archaea and in the carotenoids of photosynthetic and other pigmented bacteria.

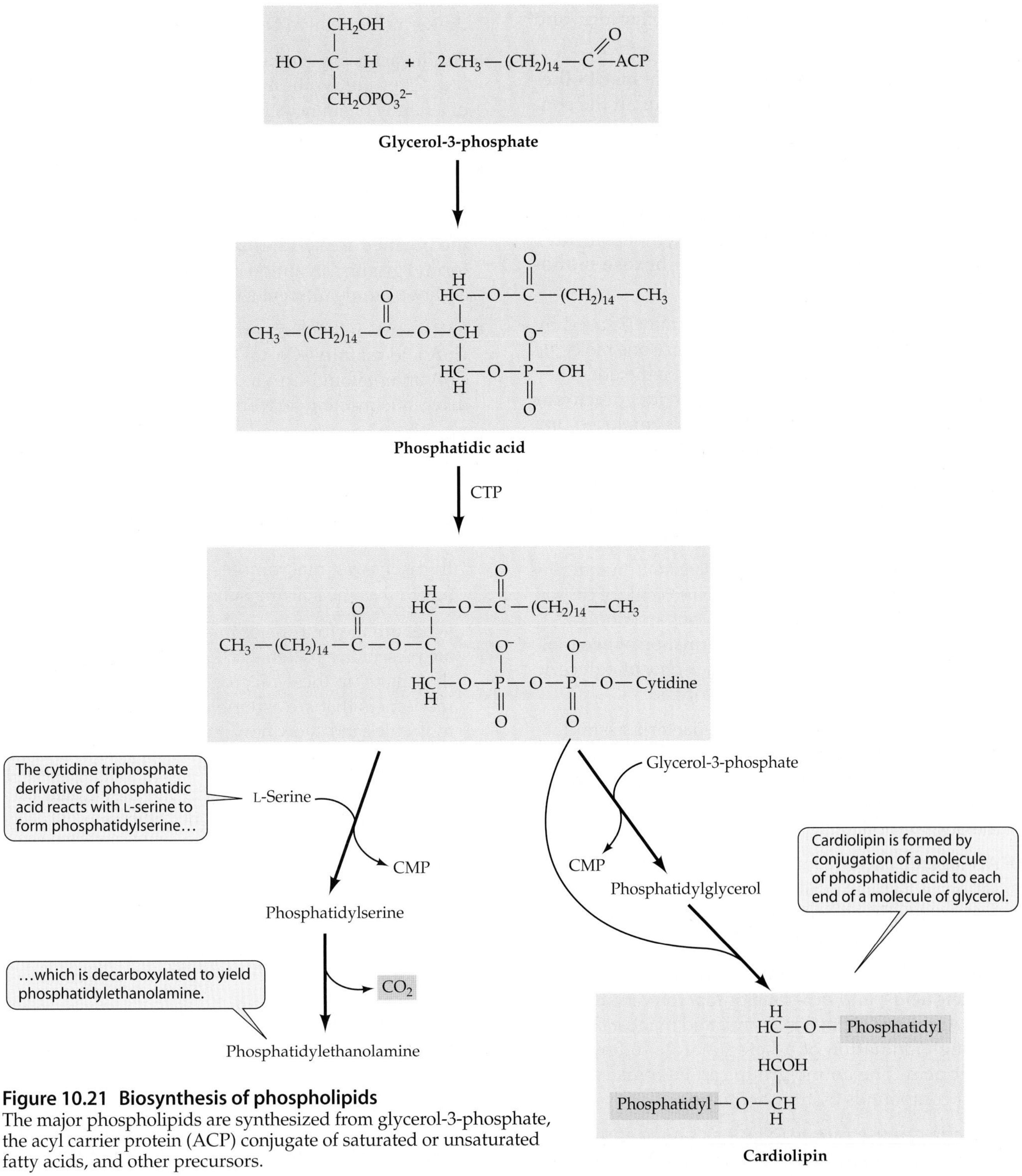

Figure 10.21 Biosynthesis of phospholipids
The major phospholipids are synthesized from glycerol-3-phosphate, the acyl carrier protein (ACP) conjugate of saturated or unsaturated fatty acids, and other precursors.

SUMMARY

▶ The **autotrophs** have the most thorough of biosynthetic capacities, as they are able to synthesize all constituents of a living cell from CO_2. Many can also obtain cellular nitrogen from atmospheric N_2.

▶ The **macromolecules** that make up a cell are of four types: proteins, nucleic acids, polysaccharides, and lipids.

▶ There are 12 monomers termed **precursor metabolites** that are involved in the biosynthesis of all the constituent parts of macromolecules in a cell. These metabolites are mostly products of **glycolysis** and the **tricarboxylic acid cycle.**

▶ Glucose is catabolized in bacteria by any of three pathways—**glycolysis** (Embden-Meyerhof pathway), **Entner-Doudoroff** pathway, or the **hexose monophosphate shunt.**

▶ The **tricarboxylic acid cycle (TCA)** may be used by **aerobic** organisms to completely oxidize one molecule of pyruvate to three molecules of carbon dioxide. They also utilize the TCA cycle to provide **precursor metabolites. Anaerobically,** the cycle is employed predominantly to generate **precursor metabolites.**

▶ An **anaplerotic reaction** is one that permits metabolic sequences such as the TCA cycle to continue by replacing intermediates removed for biosynthetic purposes.

▶ **Carbon dioxide** is generally assimilated into autotrophs via the **Calvin cycle.** In this cycle, a molecule of carbon dioxide is added to a 5-carbon sugar (ribulose-1, 5-bisphosphate) resulting in the formation of two molecules of 3-phosphoglyceric acid. 3-phosphoglyceric acid is also an intermediate in glycolysis.

▶ The green sulfur photosynthetic bacteria assimilate carbon dioxide via a mechanism that is essentially a reversal of the tricarboxylic acid cycle. This is termed a **reductive carboxylation mechanism** for carbon dioxide fixation.

▶ **Acetogenic bacteria** can effect a back-to-back condensation of two molecules of carbon dioxide to form **acetic acid.** This is another mechanism for autotrophic carbon dioxide assimilation.

▶ **Nitrogen** is an important constituent of proteins, nucleic acids, and other major macromolecules in a cell. Ammonia (NH_4^+) is assimilated in most bacteria through amination of α-ketoglutarate to form glutamic acid. The amino group can be transferred to other compounds by **transamination**.

▶ Bacteria can use **nitrate** (NO_3^-) as source of nitrogen and must reduce it to the amine level. The enzymes involved are **nitrate** and **nitrite reductase.**

▶ **Nitrogen fixation** (use of dinitrogen as nitrogen source) is confined to the bacterial and archaeal world. An array of free-living, commensalistic, and photosynthetic bacteria can supply their need for nitrogen by assimilating it from the air.

▶ The key reaction in the incorporation of **sulfur** into bacteria and fungi is the formation of **cysteine** from the serine derivative **O-acetylserine.**

▶ **Amino acids** are synthesized from precursor metabolites. They are synthesized as **"families."** An example of a family of amino acids would be those synthesized from 3-phosphoglycerate—serine, glycine, and cysteine.

▶ The nucleic acid bases are synthesized de novo from simple substrates. The **pyrimidines** are synthesized from aspartate and carbamoyl phosphate, whereas the **purines** are synthesized from CO_2, glycine, two formyl groups, an amine nitrogen from aspartic acid, and two amide nitrogens from glutamine.

▶ **Fatty acids** in bacteria are synthesized from **acetyl-CoA** and **malonyl-CoA.** During the joining of acetyl-CoA and malonyl-CoA a decarboxylation occurs that drives the reaction forward.

▶ The **isoprenoids** such as the **carotenoids** in photosynthetic bacteria are synthesized from acetyl-CoA.

REVIEW QUESTIONS

1. In discussing macromolecule synthesis, CO_2 was selected as the starting substrate. Why?

2. What are the four major macromolecular types that are present in a bacterial cell? What are the monomers that make up these macromolecules? Draw typical monomers that are assembled to form these macromolecules and show how are they linked.

3. A precursor metabolite is a molecule that is directly involved in biosynthesis. Why are they called precursors? What does the existence of these monomers tell us about biological systems? About metabolic diversity?

4. Review the three major pathways for glucose catabolism in microorganisms. What are some of the significant differences between them? What role might each play?

5. What is an anaplerotic reaction? What is the function of these reactions? How is CO_2 involved?

6. What are the key enzymes in the Calvin cycle? How different enzymatically is an organism utilizing CO_2 as substrate than one growing on glucose? How does the initial product of CO_2 fixation fit into the concept of the precursor metabolite?

7. The green sulfur bacteria are probably an ancient type of organism. How do they fix CO_2 and in what way is this related to aerobic metabolism?

8. Ammonia and sulfate are "funneled" into the metabolic machinery through two intermediates. Draw these reactions.

9. What is a family of amino acids? How many families are there?

10. What low molecular weight compounds are utilized in synthesis of a purine? A pyrimidine?

11. What is a fatty acid? List different types. Outline the mechanism of synthesis of phospholipids and isoprenoids.

SUGGESTED READING

Cooper, G. M. 2000. *The Cell: A Molecular Approach.* 2nd ed. Sunderland, MA: Sinauer Associates, Inc.

Garrett, R. H. and C. M. Grisham. 1999. *Biochemistry.* 2nd ed. Philadelphia: Saunders College Publishing.

Lawlor, D. W. 1993. *Photosynthesis: Molecular, Physiological, and Environmental Processes.* 2nd ed. Essex, England: Longman Scientific and Technical.

Neidhardt, F. C., R. Curtiss III, J. L. Ingraham, E. C. C. Lin, K. B. Low, B. Magasanik, W. S. Reznikoff, M. Riley, M. Schaechter and H. E. Umbarger. 1996. Escherichia coli *and* Salmonella: *Cellular and Molecular Biology.* Washington, DC: American Society for Microbiology.

Neidhardt, F. C., J. L. Ingraham and M. Schaechter. 1990. *Physiology of the Bacterial Cell: A Molecular Approach.* Sunderland, MA: Sinauer Associates, Inc.

Schlegel, H. G. and B. Bowien, eds. 1989. *Autotrophic Bacteria.* Madison, WI: Science Tech Publishers.

Chapter 11

Assembly of Bacterial Cell Structures

Usually, if nature invents something that is really good, it keeps applying the invention over and over again whenever this can help solve its problem.

— VLADIMIR P. SKULACHEV

The generation of **precursor metabolites** and the reactions that convert these monomers to the building blocks of a cell were discussed in Chapter 10. The precursor metabolites originate either from glycolysis, the tricarboxylic acid cycle, or allied pathways. Through a limited number of well-integrated reactions, the major monomers, including the amino acids, nucleotides, sugars, and fatty acids, are synthesized from these precursors. Rapid and orderly growth depends on the **polymerization** or assembly of monomeric building blocks to form macromolecules. Among the essential polymerization reactions are the formation of proteins from amino acids, polysaccharides from sugars, and the nucleic acids from nucleotides. Lipids are assembled from fatty acids. Once formed, macromolecules (DNA, RNA, proteins, and phospholipids) are assembled to generate a cell (**Figure 11.1**). Note that proteins and RNA make up the major part of a living cell. The actual number of molecules of each macromolecular component present in an *Escherichia coli* cell is listed in **Table 11.1**.

Replication of DNA, RNA synthesis, and the role of the nucleic acids in protein synthesis will be discussed in Chapter 13. The following is a brief discussion of the assembly of protein structures and the assembly of the cell wall and outer envelope (in gram-negatives).

PROTEIN ASSEMBLY, STRUCTURE, AND FUNCTION

The term **protein** broadly defines molecules that are composed of one or more polypeptide chains. A **polypeptide** is a polymer that generally exceeds several dozen amino acids in length. It is not a precise term. Some proteins are present in the cytoplasm; others are associated with the cytoplasmic membrane or associated with the outer envelope or periplasm of a bacterium (**Table 11.2**).

Microorganisms that can digest polysaccharides, fats, or other large molecules secrete enzymes (proteins) to the exterior of the cytoplasmic membrane that can hydrolyze large molecules to low molecular weight compounds. These low molecular weight compounds can then be transported into the cells. The outer envelopes of gram-negative bacteria contain the protein porins that are involved in passage of molecules into the periplasmic space, where binding proteins can retain them. Transport to the cytoplasm occurs through the action of proteins in the cytoplasmic membrane. The electron transport systems in the microorganisms capable of respiration are located in the cytoplasmic membranes as are proteins involved with the ATPase system. Proteins in the cytoplasm are involved with catabolic functions and the synthesis of DNA, RNA, and cellular components.

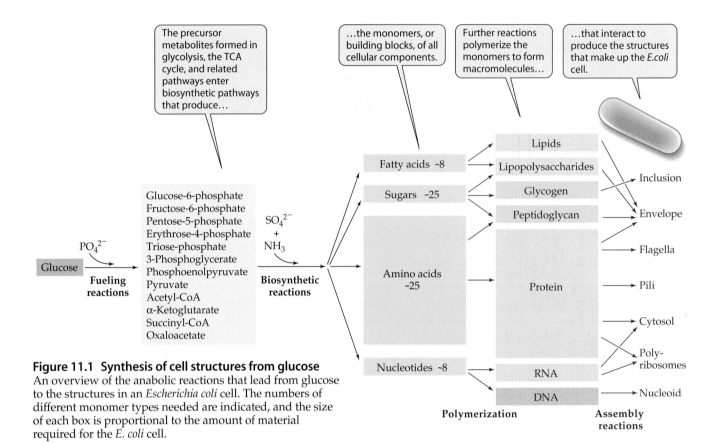

Figure 11.1 Synthesis of cell structures from glucose
An overview of the anabolic reactions that lead from glucose to the structures in an *Escherichia coli* cell. The numbers of different monomer types needed are indicated, and the size of each box is proportional to the amount of material required for the *E. coli* cell.

Protein Synthesis

Amino acids are joined together by **peptide bond** formation, and during this reaction a molecule of water is removed. A joining together of a series of amino acids results in a polypeptide chain referred to as the **primary structure** (**Figure 11.2**). The nature and character of a protein is determined to a considerable extent by the total number and sequence of amino acids in this chain. Functional proteins are generally composed of polypeptides that are folded or coiled into a three-dimensional structure, and this is the **conformation** that a functional polypeptide ultimately assumes.

The folding pattern is determined largely by the amino acid sequence, and the formation of a functional protein potentially can occur by unassisted self-assembly. However, there is evidence that suggests that random unassisted assembly can result in a structure that is nonfunctional. Consequently there are preexisting proteins in the growing cell termed **chaperones** that act to prevent the incorrect molecular interactions that would lead to nonfunctional secondary or tertiary structures. These chaperones assist in forming but are not a part of the final functional protein. Generally the folding of peptides to a functional secondary structure is energetically favorable; that is, it occurs without energy input.

Table 11.1	The overall macromolecular composition of an *Escherichia coli* cell	
Macromolecule	**Percentage Total Dry Weight of Cell**	**Number of Molecules per Cell**
Protein	55.0	2,360,000
RNA	20.5	
23S rRNA		18,700
16S rRNA		18,700
5S rRNA		18,700
transfer RNA		205,000
messenger RNA		1,380
DNA	3.1	2.1
Lipid	9.1	22,000,000
Lipopolysaccharide	3.4	1,200,000
Peptidoglycan	2.5	1
Glycogen	2.5	4,360
Soluble pool[a]	2.9	
Inorganic	1.0	

[a]Metabolites, vitamins
Source: Adapted from Ingraham, Maaløe, and Neidhardt, 1983.

Figure 11.2 Formation of a peptide

Formation of the primary, secondary, and tertiary structure of a peptide.

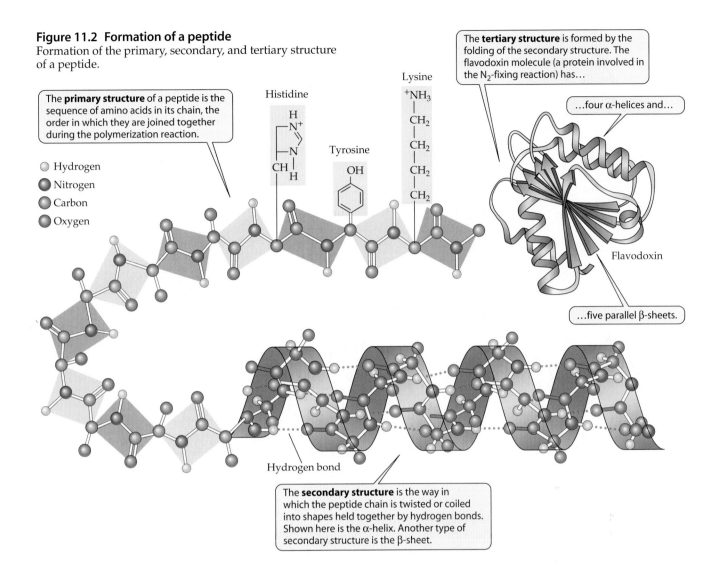

The **primary structure** of a peptide is the sequence of amino acids in its chain, the order in which they are joined together during the polymerization reaction.

- ○ Hydrogen
- ● Nitrogen
- ● Carbon
- ● Oxygen

Histidine

Tyrosine

Lysine

The **tertiary structure** is formed by the folding of the secondary structure. The flavodoxin molecule (a protein involved in the N_2-fixing reaction) has…

…four α-helices and…

…five parallel β-sheets.

Flavodoxin

Hydrogen bond

The **secondary structure** is the way in which the peptide chain is twisted or coiled into shapes held together by hydrogen bonds. Shown here is the α-helix. Another type of secondary structure is the β-sheet.

Structural Arrangements of Proteins

The types of bonding that occur between internal areas (domains) of a polypeptide giving the protein a secondary and tertiary structure are outlined in **Figure 11.3**. The variability in the side chains of the 21 major amino acids (see Figure 3.15), the total number incorporated, and their sequence in a polypeptide are all factors that contribute to the biochemical characteristics of the protein. Amino acids with nonpolar hydrocarbon side chains (valine, leucine, phenylalanine, and tryptophan) are hydrophobic (do not interact with water). These side chains point outward on the external surface of the proteins that are embedded in hydrophobic membrane lipids. In water-soluble proteins present in cytoplasm, these hydrophobic groups extend inward. The side chains of hydrophilic amino acids (aspartic acid, glutamic acid, lysine, arginine, and histidine) extend outward to the aqueous environment.

Typically, the twisting and coiling of polypeptide chains results in the formation of a **secondary structure** that is either helical in form, called an **α-helix** or a flat arrangement designated a **β-sheet**. The α-helix is a linear polypeptide that is wound like a spiral staircase (**Figure 11.4**). It is held together by hydrogen bonding between the amine hydrogen of one amino acid and an oxygen from another amino acid. The hydrogen bonding occurs as the polypeptide is formed and leads to a stable helical structure. Glutamic acid, methionine, and alanine are the strongest formers of the α-helix.

β-sheets (Figure 11.4) are formed when two or more extended polypeptide chains come together side by side so that regular hydrogen bonding can occur between the amide (NH) and the carbonyl (C=O) of an adjacent-chain peptide backbone. Addition of more polypeptides results in a multistranded structure. The amino acids valine, tyrosine, and isoleucine are common β-sheet formers.

In many proteins, the polypeptide chain is bent at specific sites and folded back and forth resulting in the tertiary structure (Figure 11.2). Although the α-helices

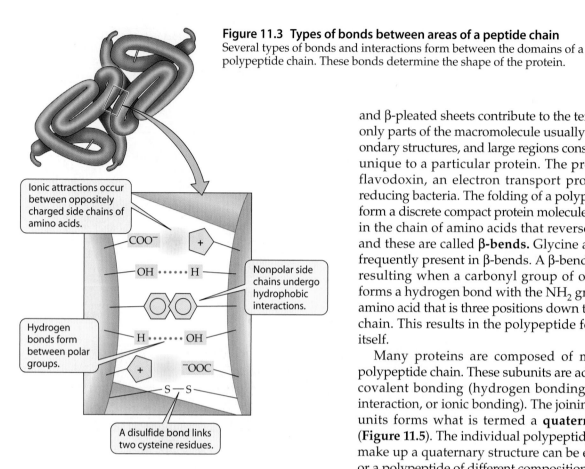

Figure 11.3 Types of bonds between areas of a peptide chain
Several types of bonds and interactions form between the domains of a polypeptide chain. These bonds determine the shape of the protein.

Ionic attractions occur between oppositely charged side chains of amino acids.

Nonpolar side chains undergo hydrophobic interactions.

Hydrogen bonds form between polar groups.

A disulfide bond links two cysteine residues.

Table 11.2	Location of proteins associated with a gram-negative bacterium
Location	**Proteins**
Extracellular	Carbohydrases
	Lipases
	Proteases
	Nucleases
Outer envelope	Receptors for bacteriophage
	Porin proteins
Periplasm	Binding proteins involved in transport and chemotaxis
	Phosphatases
	Esterases
Cytoplasmic membrane	Electron-transport chain
	Proton translocating ATP synthase
	Transport proteins
	Lipid biosynthesis enzymes
	Cell wall and outer-envelope biosynthetic enzymes
Cytoplasm	Enzymes involved in catabolism of soluble substrates
	Enzymes involved in DNA replication
Ribosomes	Structural proteins
External surface	Fimbriae
	Flagella
	Pili

and β-pleated sheets contribute to the tertiary structure, only parts of the macromolecule usually have these secondary structures, and large regions consist of structures unique to a particular protein. The protein shown is flavodoxin, an electron transport protein in sulfate reducing bacteria. The folding of a polypeptide chain to form a discrete compact protein molecule requires bends in the chain of amino acids that reverse the direction, and these are called **β-bends.** Glycine and proline are frequently present in β-bends. A β-bend is a tight loop resulting when a carbonyl group of one amino acid forms a hydrogen bond with the NH_2 group of another amino acid that is three positions down the polypeptide chain. This results in the polypeptide folding back on itself.

Many proteins are composed of more than one polypeptide chain. These subunits are adjoined by noncovalent bonding (hydrogen bonding, hydrophobic interaction, or ionic bonding). The joining of these subunits forms what is termed a **quaternary structure** (**Figure 11.5**). The individual polypeptide subunits that make up a quaternary structure can be either the same or a polypeptide of different composition or size. A classic example of a quaternary structure is the oxygen-carrying protein hemoglobin. This protein is composed of four polypeptides—of two different kinds—and is termed an $\alpha_2\beta_2$ tetramer. The individual subunits in a quaternary structure are proteins themselves having typical secondary and tertiary structures.

The **relative mass** (M_r) of a protein is the total molecular mass of the assembled subunits. Some representative proteins, their relative mass, and the number of subunits that form each protein are listed in **Table 11.3.**

ASSEMBLY OF THE CYTOPLASMIC MEMBRANE

One structure that is present in all *Bacteria* and *Archaea* is the **cytoplasmic membrane,** also called the cell membrane. The basic structure of a cytoplasmic membrane was described in Chapter 4 (see Figures 4.33 and 4.34). The bacterial cytoplasmic membrane is similar in its basic structure to other membranes—for example, those surrounding eukaryotic nuclei, mitochondria, and other organelles. The membranes of *Archaea* are quite different, as will be discussed in Chapter 18.

(A) α-helix

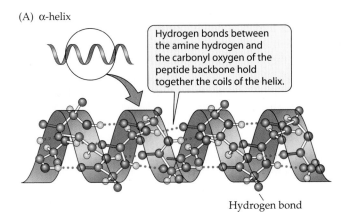

Hydrogen bonds between the amine hydrogen and the carbonyl oxygen of the peptide backbone hold together the coils of the helix.

Hydrogen bond

(B) β-sheet

Hydrogen bonds hold together neighboring parallel strands in the sheet.

Figure 11.4 Secondary structures
Hydrogen bonds hold a polypeptide chain in an α-helix or α-sheet configuration. These bonds are important determinants of secondary structure.

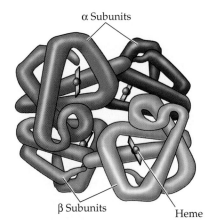

α Subunits

β Subunits   Heme

Figure 11.5 Quaternary structure
The quaternary structure of hemoglobin is composed of two α and two β polypeptide chains, or subunits. Note the compact symmetry of the structure.

Table 11.3	Molecular mass and number of subunits in selected proteins	
	Relative Mass (M_r)	**Subunits**
Ribonuclease	13,700	1
Lysozyme	14,388	1
Luciferase	80,000	2
Hexokinase	100,000	2
Lactic dehydrogenase	223,000	4
Urease	483,000	5

The average molecular mass of an amino acid residue in a protein is 120 daltons (average molecular weight less a molecule of H_2O), and an "average" protein molecule has about 300 amino acid residues, or an M_r 36,000. A dalton is defined as ¹⁄₁₂th of the mass of a carbon atom.

A cytoplasmic membrane in a growing cell must assimilate new membrane components (phospholipids and proteins) and integrate these into the membrane that is increasing in surface area. During integration of the newly synthesized phospholipids, the permeability and barrier functions of the membrane must be maintained.

The newly synthesized phospholipid is incorporated into the inner leaflet of the membrane bilayer and in a relatively short period of time appears in the outer leaflet. This movement from the inner to outer membrane leaflet occurs at a more rapid rate in actual cell membranes than it does in model bilayers. This suggests that rotation from the inner to outer leaflet in a viable cell may be controlled by proteins in the membrane. The process may require ATP, but this is uncertain at the present time.

Approximately 20% of the polypeptides synthesized by bacteria are located in the cytoplasmic membrane or are translocated across the membrane. These proteins pass into or through the membrane via the **general secretory pathway** (**GSP**). Proteins transported by the GSP are called **secretory proteins.** In some cases polypeptides secreted by the cell are then assembled outside the cytoplasmic membrane, and these polypeptides are termed presecretory proteins.

The distinguishing characteristic of secretory proteins is the presence of a **signal sequence** of at least 10 hydrophobic amino acids at a terminus of the protein. The function of the signal sequence is to direct the secretory proteins into the cytoplasmic membrane. There are three distinct regions to the signal sequence: the N, H, and C regions. The leading end of the signal peptide is termed "N" and is polar with a net positive charge (**Figure 11.6**). The middle region of a signal peptide, termed "H" has many hydrophobic amino acids and is inserted into the membrane. The "C" sequence is also

Classic experiments in the early 1960s by Christian B. Anfinsen led to the presumption that the sequence and nature of the amino acids in a polypeptide chain were the major contributing factors to the folding and final conformation of a protein. This was based on experimentation with bovine pancreatic ribonuclease, a small protein made up of 124 amino acids. The secondary structure of this protein is maintained by four disulfide bridges. Denaturation can be accomplished by cleaving these covalent disulfide linkages with that reduces the —s—s— linkage to —sh hs— as follows:

an inherent property of the primary structure of polypeptides. Thus folding to secondary and tertiary structures would be promoted by the proper spacing of amino acids whose side chains could interact with counterpart amino acids elsewhere in the polypeptide chain. The bonding that would give the functional structures would be those formed by the side chain interactions illustrated in Figure 11.3. This sort of in vitro study apparently does not completely reflect the conditions in the cytoplasm of a cell. The levels of polypeptide required for test tube refolding are much higher

associated with the proper folding of polypeptides in a cell. One class is the conventional enzymes that catalyze specific isomerization reactions that result in proper polypeptide conformations for proper folding.

The second class of proteins is the **chaperones** that stabilize unfolded or partially folded structures and preclude the formation of inappropriate intra or interchain interactions. A chaperone may also interact with individual proteins to promote protein-protein arrangements that yield quarternary functional structures. Among the roles that protein chaperones play are:

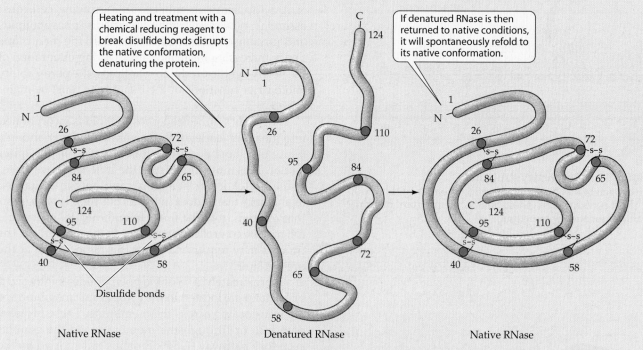

The secondary structure, and thus the activity, of bovine pancreatic ribonuclease is maintained by disulfide bonds between cysteine residues.

Disruption of the sulfide alters the conformation of the protein and loss of enzymatic activity. Removal of the denaturing agents and exposure to air results in a random reassembly and an inactive enzyme. However, exposure of the denatured protein to trace levels of α-mercaptoethanol promotes the proper rearrangement of the —s—s— cross-links results in a reformation of the active enzyme. This and other test tube in vitro experiments suggested that proper folding might be

than would occur in vivo and the physiochemical conditions of the in vitro experiments do not exist in the living cell. In reality, studies in vitro that mimic in vivo physiological conditions result in misfolding or aggregation of the polypeptide. This misfolding is uncommon in vivo except with mutant proteins or protein folding at elevated temperatures.

Studies in recent years affirm that there are two classes of proteins

- They prevent folding of secretory proteins before translocation.
- They are involved with assembly of bacterial viruses.
- They promote the assembly of pili and the assembly of the carboxysome in photosynthetic organisms.
- They prevent protein denaturation during environmental stress, such as elevated temperature.

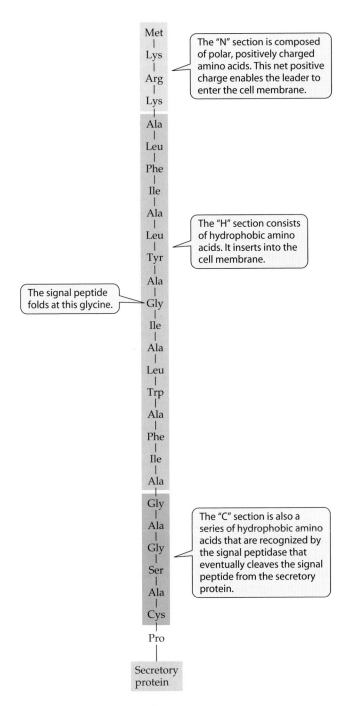

Met
|
Lys
|
Arg
|
Lys

> The "N" section is composed of polar, positively charged amino acids. This net positive charge enables the leader to enter the cell membrane.

Ala
|
Leu
|
Phe
|
Ile
|
Ala
|
Leu
|
Tyr
|
Ala
|
Gly
|
Ile
|
Ala
|
Leu
|
Trp
|
Ala
|
Phe
|
Ile
|
Ala

> The "H" section consists of hydrophobic amino acids. It inserts into the cell membrane.

> The signal peptide folds at this glycine.

Gly
|
Ala
|
Gly
|
Ser
|
Ala
|
Cys
|
Pro

> The "C" section is also a series of hydrophobic amino acids that are recognized by the signal peptidase that eventually cleaves the signal peptide from the secretory protein.

Secretory protein

Figure 11.6 Signal peptide
A typical signal peptide that may be present in a gram-negative bacterium. The signal peptide folds as shown in Figure 11.7.

hydrophobic and is recognized by a specific peptidase that cleaves the signal sequence from the secretory protein.

There are two general mechanisms for the passage of proteins or presecretory proteins into or through the cytoplasmic membrane. One of these is a receptor-independent insertion, and the other involves specific proteins that are embedded in the membrane. The **recep-**

tor-independent insertion occurs when a membrane protein is synthesized by polyribosomes in the cytoplasm and is then attached to and moves into the membrane with the assistance of the signal peptide (**Figure 11.7**). There are no other cytoplasmic or membrane proteins involved in this process. The number of proteins secreted by this mechanism is very limited.

Most secretory proteins are translocated with the mediation of specific **translocase complexes** composed of cytosolic and transmembrane proteins. The secretion of proteins occurs during protein synthesis and is generally chaperone-dependent. There are an estimated 500 translocase complexes in the cytoplasmic membrane of gram-negative bacteria, and these complexes span the membrane and are quite hydrophobic (**Figure 11.8**).

There are three general mechanisms for protein translocation via the translocase complex: cotranslational, translation-linked, and posttranslational.

In **cotranslational translocation** the signal sequence attaches to the secretory protein and carries it through the translocase complex as it is synthesized on the ribosome (Figure 11.8). This occurs in close proximity to the

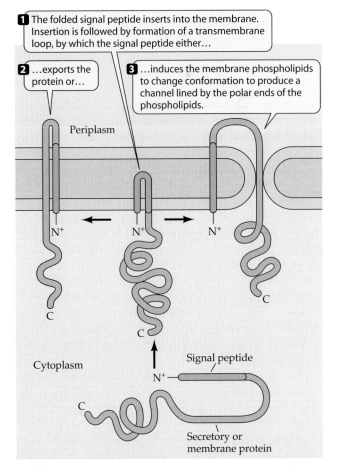

1 The folded signal peptide inserts into the membrane. Insertion is followed by formation of a transmembrane loop, by which the signal peptide either…

2 …exports the protein or…

3 …induces the membrane phospholipids to change conformation to produce a channel lined by the polar ends of the phospholipids.

Periplasm

Cytoplasm

Signal peptide

Secretory or membrane protein

Figure 11.7 Receptor-independent protein insertion
Model for receptor-independent insertion of a signal peptide into a cell membrane.

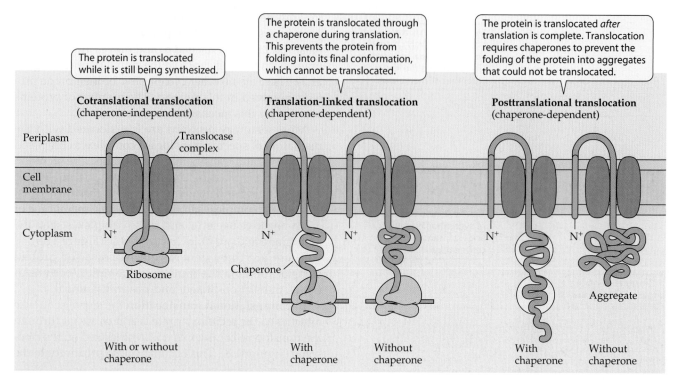

Figure 11.8 Translation-linked protein secretion
Model for three different modes of translation-translocation coupling for secretory proteins. After Anthony Pugsley. 1993. *Microbiological Reviews* 57: 50–108.

translocase. This form of translocation is chaperone independent. In chaperone-dependent **translation-linked translocation,** the protein is synthesized away from the translocase complex, and the protein is associated with a chaperone as it is synthesized. The chaperone protein then passes the secretory protein to the translocase complex. Proteins that are a part of the secretory (translocase) complex may act as chaperones. In the absence of a chaperone, these proteins might fold prior to translocation, and secretory function would cease. In **posttranslational translocation** the secretory protein must have chaperone molecules to prevent folding (Figure 11.8). An aggregate that would occur in the absence of chaperones would not pass through the translocase complex. After the secretory protein is in place, the signal peptide can be cleaved to a signal peptidase.

Membrane lipid synthesis and chemical modification of bacterial membranes can occur independently of the membrane proteins. A constant turnover of membrane proteins or phospholipids (or both) occurs as organisms adjust to environmental changes. Cells increase the proportion of membrane unsaturated fatty acids as a response to a decrease in growth temperature. This occurs because a functional membrane must be fluid,

and because of their double bonds unsaturated fatty acids are more fluid at lower temperatures. Newly formed transport proteins may also be inserted in the cytoplasmic membrane as a microorganism encounters a different substrate or other changes in growth conditions.

The production and secretion of fungal enzymes and other proteins is fundamentally different from that in bacteria. Fungi, of course, are eukaryotes, and fungal proteins are synthesized at the rough endoplasmic reticulum, transported in vesicles, and enter the Golgi apparatus. Vesicles then bud from the Golgi and are transported to the inner surface of the cell membrane. At the site of secretion, the membrane of the protein-containing vesicle protein fuses with the cytoplasmic membrane, and the protein is released to the outside.

BIOSYNTHESIS OF PEPTIDOGLYCAN

Peptidoglycan (murein) is the main structural component of the bacterial cell wall. It is the source of strength and provides shape to the organism. The composition, structure, and function of peptidoglycan in both gram-positive and gram-negative bacteria were discussed in Chapter 4.

The assembly of peptidoglycan units occurs in the cytoplasm and in the cytoplasmic membrane (**Figure 11.9**). At the inner surface of the cell membrane, both *N*-acetylglucosamine and *N*-acetylmuramic acid are synthesized and coupled to bactoprenol (undecaprenol

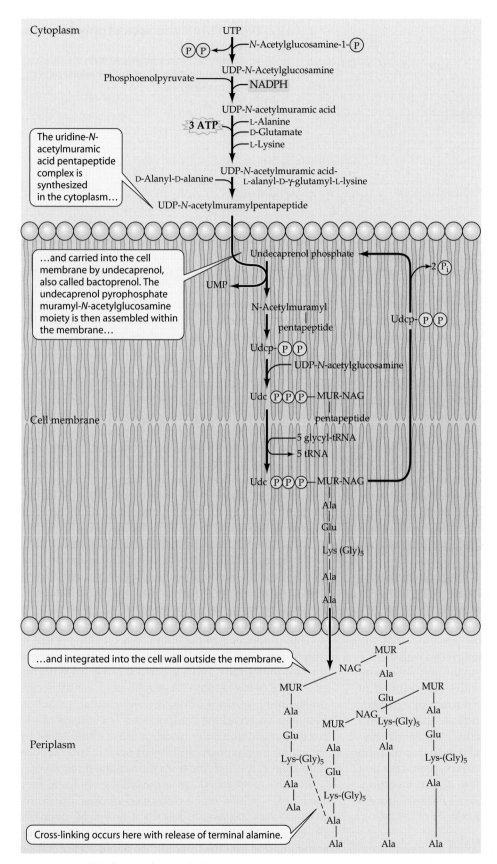

Figure 11.9 Synthesis of peptidoglycan
Steps in the synthesis of peptidoglycan. UTP, uridine triphosphate; NAG, N-acetylglucosamine; MUR, muramic acid; Udcp, undecaprenol phosphate (bactoprenol).

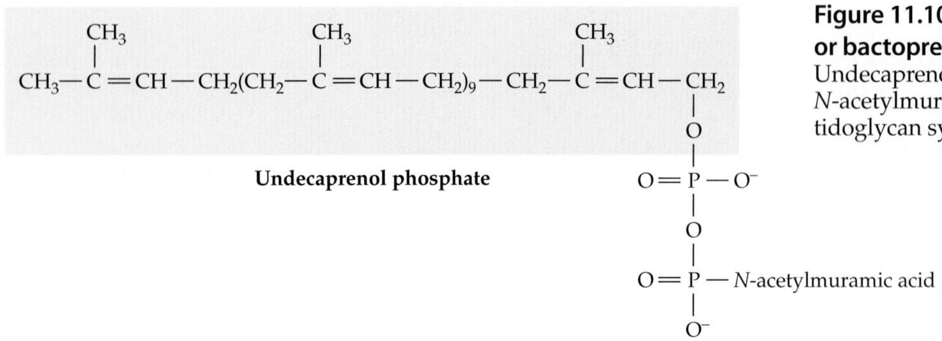

$$\text{CH}_3 - \overset{\overset{\displaystyle \text{CH}_3}{|}}{\text{C}} = \text{CH} - \text{CH}_2(\text{CH}_2 - \overset{\overset{\displaystyle \text{CH}_3}{|}}{\text{C}} = \text{CH} - \text{CH}_2)_9 - \text{CH}_2 - \overset{\overset{\displaystyle \text{CH}_3}{|}}{\text{C}} = \text{CH} - \text{CH}_2$$

Undecaprenol phosphate

$$\begin{array}{c} | \\ \text{O} \\ | \\ \text{O} = \text{P} - \text{O}^- \\ | \\ \text{O} \\ | \\ \text{O} = \text{P} - N\text{-acetylmuramic acid} \\ | \\ \text{O}^- \end{array}$$

Figure 11.10 Undecaprenol pyrophosphate, or bactoprenol
Undecaprenol pyrophosphate with an attached *N*-acetylmuramic acid, an intermediate in peptidoglycan synthesis.

phosphate, **Figure 11.10**) at the inner membrane surface. **Bactoprenol** (Figure 11.10) is a long-chain hydrocarbon that can enter the hydrophobic core of the cytoplasmic membrane and carry attached hydrophobic molecules into or through the lipid bilayer.

Bactoprenol displaces the uridine triphosphate (see Figure 11.9), and this renders the muramyl pentapeptide sufficiently hydrophobic to allow passage through the hydrophobic membrane. During this passage, a bridge peptide such as pentaglycine may be attached to the terminal amine of lysine via a peptide bond. This would occur in gram-positive bacteria. The peptidoglycan units then enter the periplasmic space and are inserted at a growing point in the cell wall.

The *N*-acetylmuramaic acid (MUR)–*N*-acetyglucosamine (NAG) units are added to the existing MUR-NAG backbone of the peptidoglycan. The peptide bridge, in this case pentaglycine, is joined to an alanine molecule on an adjacent MUR-NAG chain. The MUR-NAG backbone and the peptide cross-linking are responsible for the strength of the peptidoglycan cell wall.

A battery of enzymes is involved in the covalent reactions that result in extension and cross-linking between the peptidoglycan strands. The enzymes are also responsible for septation of the murein cell wall that occurs during cell division. The proteins that are involved in extension of the peptidoglycan wall during growth have a unique ability to bind the antibiotic penicillin and some related antibacterials. The number of penicillin-binding proteins on the surface of a bacterium varies with species. Studies indicate that the penicillin-binding proteins are involved with transglycosylation (elongation of glycan strands), transpeptidation (cross-linking), and the enzyme carboxypeptidase that cuts preexisting cross-links for new glycan insertion. Binding of penicillin to these proteins inhibits murein biosynthesis and can destroy the integrity of the cell.

Forming Gram-Positive Cell Walls

The structural relationship between the cytoplasmic membrane and the peptidoglycan in a gram-positive bacterium was described in Chapter 4. Peptidoglycan subunits are constantly synthesized and enter the space between cytoplasmic membrane and existing murein layer at points where the subunits are linked to the existing peptidoglycan layer. Thus, a growing cell is continually adding a murein layer in the area adjacent to the cytoplasmic membrane. About 40 layers of murein surround a gram-positive cell. The layers move outward as newly synthesized murein is added to the inner layer.

Gram-Negative Cell Walls

The cell wall structures of gram-negative bacteria are considerably more complex than those of gram-positives (Chapter 4). The gram-negative bacteria have a complex outer membrane (envelope) that surrounds the cell wall. Growth and expansion of the walls are consequently more elaborate. It is apparent that fatty acids, sugars, phospholipids, and proteins must move from the cytoplasmic side of the membrane to the periplasm during wall expansion in gram-negative bacteria (**Figure 11.11**).

Expansion occurs in minute openings in the cell wall, and these have been termed **zones of adhesion** because the inner and outer membranes actually make direct contact. The junctions occur in the peptidoglycan layer, and all components for the construction of the outer cell envelope passes through these gaps.

The components of the lipopolysaccharide present in the cell envelope are synthesized on the inner surface of the cell membrane and are carried outward with the assistance of bactoprenol. Assembly apparently occurs on the outer surface of the cell by mutual attraction between the envelope components. The phospholipid components of the outer membrane are translocated by the same mechanism, and hydrophobic interactions with lipid A result in the formation of the outer layer.

Proteins are secreted across the cell membrane as they are being synthesized on ribosomes in close proximity to the membrane, as was described earlier. Many major proteins of the outer membrane, such as porins, may self-assemble into the three-dimensional configuration shown (Figure 11.11). It is interesting to note that a mixture of lipopolysaccharide, porin proteins, and other components of the outer envelope can assemble under

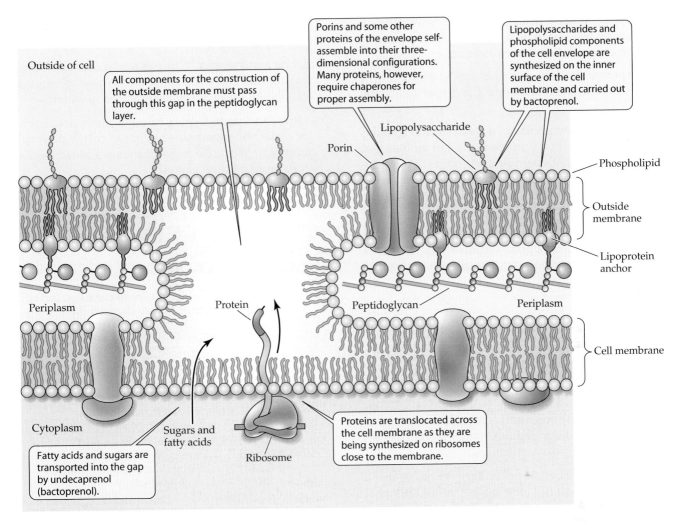

Outside of cell

All components for the construction of the outside membrane must pass through this gap in the peptidoglycan layer.

Porins and some other proteins of the envelope self-assemble into their three-dimensional configurations. Many proteins, however, require chaperones for proper assembly.

Lipopolysaccharides and phospholipid components of the cell envelope are synthesized on the inner surface of the cell membrane and carried out by bactoprenol.

Lipopolysaccharide

Porin

Phospholipid

Outside membrane

Lipoprotein anchor

Periplasm

Protein

Peptidoglycan

Periplasm

Cell membrane

Cytoplasm

Sugars and fatty acids

Ribosome

Fatty acids and sugars are transported into the gap by undecaprenol (bactoprenol).

Proteins are translocated across the cell membrane as they are being synthesized on ribosomes close to the membrane.

Figure 11.11 Synthesis of the outer envelope of gram-negative bacteria
A junction in the cell membrane–peptidoglycan–outer envelope complex of a gram-negative bacterium. This is an exaggerated view; in reality the junction is a minute pore. Junctions occur randomly at points around a growing cell.

laboratory conditions to form a structure similar to that present in a living cell. It is apparent, however, that many of the proteins that are assembled into structures outside the cytoplasmic membrane are complexed by chaperones that promote proper assembly. For example, the pilins were previously considered to self-assemble to form pili, but it is now known that pilin proteins are complexed by chaperone proteins until they arrive at the assembly site. Pilins are assembled and added at the base of the pilus. Pili are hairlike appendages on many gram-negative bacteria and are involved with attachment and genetic exchange.

The lipoprotein anchors that attach the cell envelope to the peptidoglycan are formed by the interaction of a protein with the lipids present on the outer membrane.

SUMMARY

▶ The **monomers** that are essential for the synthesis of cell matter are **polymerized** to form macromolecules, and these are the functional components of the cell. Major macromolecules in a cell are proteins, phospholipids, DNA, RNA, and polysaccharides.

▶ Over one-half of the dry weight of a bacterial cell is **protein.** About 20% is **RNA.**

▶ A **polypeptide** may become a functional **protein** by assuming a **tertiary configuration.** The tertiary configuration results from the joining of **secondary structures** such as the α-helix and β-sheets. Joining to two or more tertiary structures yields a **quarternary structure.**

▶ **Cytoplasmic membranes** are formed from bilayers of **phospholipid.**

▶ About 20% of the polypeptides synthesized by bacteria are **integrated** into the cytoplasmic membrane or **translocated** across the membrane.

▶ Most proteins are translocated across the membrane as they are **translated** on a ribosome. This process is termed **cotranslational translocation.**

▶ A **chaperone** is a protein that assists in the folding of a polypeptide chain to assume the proper functional configuration.

▶ The structural component of a eubacterial cell wall is **peptidoglycan.** It is composed of sugar amines that are complexed with peptides. Synthesis of peptidoglycan units occurs in the cytoplasm and within the cytoplasmic membrane. Assembly occurs outside the cytoplasmic membrane.

▶ **Gram-positive** microorganisms have a thick peptidoglycan layer whereas a **gram-negative** bacterium has a thinner peptidoglycan layer. The gram-negative bacteria have an elaborate cell envelope outside the peptidoglycan layer.

REVIEW QUESTIONS

1. Proteins make up over half of the dry weight of a cell. What are the diverse roles that proteins play and where are the various types located?

2. Draw a peptide bond and a peptide composed of several amino acids. Define primary, secondary, tertiary, and quaternary structures.

3. What are the two major secondary structures that a protein may assume? How do they differ, and what is our shorthand for depicting them? What is the major bonding that holds polypeptides in these configurations?

4. What are some structures (see question 3) that are involved in the functional configuration of proteins?

5. Draw a polar lipid. Which is the hydrophobic and hydrophilic end of the molecule? What happens if a number of phospholipids are added to water?

6. How can a peptide pass through the hydrophobic cytoplasmic membrane? What is cotranslational insertion? What role might chaperones play in secretory protein translocation?

7. Where are the major components of a gram-negative outer envelope assembled? Discuss the processes involved.

SUGGESTED READING

Garrett, R. H. and C. M. Grisham. 1999. *Biochemistry*. 2nd ed. Fort Worth, TX: Saunders College/Harcourt Brace Publishing.

Gerhardt, P. R., G. E. Murray, W. A. Wood and N. R. Krieg. 1994. *Methods for General and Molecular Bacteriology*. Washington, DC: American Society for Microbiology.

Neidhardt, F. C., J. L. Ingraham and M. Schaechter. 1990. *Physiology of the Bacterial Cell: A Molecular Approach*. Sunderland, MA: Sinauer Associates Inc.

Neidhardt, F. C., R. Curtiss III, J. L. Ingraham, C. C. Lin, K. B. Low, B. Magasanik, W. S. Reznikoff, M. Riley, M. Schaechter and H. E. Umbarger. 1996. Escherichia coli *and* Salmonella: *Cellular and Molecular Biology*. Washington, DC: American Society for Microbiology.

Purves, W. K., D. Sadava, G. H. Orians and H. C. Heller. 2001. *Life: The Science of Biology*. 6th ed. Sunderland, MA: Sinauer Associates Inc.

Chapter 12

Roles of Microbes in Biodegradation

All these tidal gatherings, growth and decay,
Shining and darkening, are forever
Renewed; and the whole cycle impenitently
Revolves; and all the past is future
— ROBINSON JEFFERS

The previous chapters in Part 3 dealt with the major processes microorganisms use to generate energy and the biosynthetic reactions involved in the construction of a living cell. It is axiomatic that survival, growth, and perpetuation of species depend on the continued generation of progeny. A source of building material and the assembly of this material into viable cells is an indispensable component of survival and growth. But building materials are limited. How long can nature generate new life without renewing the ultimate source of building blocks? Not long! The ultimate source of building blocks for new cell synthesis is through the destruction of dead and deficient cells. Such destruction replenishes the ultimate source of structural material—carbon—by generating CO_2. **Microbial biodegradation** is nature's way of recycling the organic and inorganic components of the inert products of life. A major step in the evolution of complex biological systems was the appearance, over 2.5 billion years ago, of viable organisms that could assimilate simple compounds as substrates during oxygen-generating (oxygenic) photosynthesis. These microorganisms were the cyanobacteria, a group that evolved with a synthetic system that utilized CO_2 as a carbon source and a photosystem that yielded molecular oxygen:

$$CO_2 + H_2O \longrightarrow CH_2O \text{ (cellular mass)} + O_2$$

Proliferation of the cyanobacteria, with the inexhaustible supply of solar energy, resulted in an atmosphere containing O_2, an oxidant suitable for the later evolution of aerobic heterotrophs. The broad diversity of microbes that evolved as a consequence of O_2 production was instrumental in the evolution of the multicelled eukaryotes.

Life on Earth can be sustained only if the cellular material produced either directly or indirectly by photosynthetic reactions is ultimately recycled back to CO_2, which constitutes only about 0.03% of the gases in the atmosphere. The success in meeting this requirement is apparent from the general balance and stability that is evident in nature. If it were not balanced by respiration, the massive biological productivity resulting from photosynthesis would markedly lower the CO_2 levels in the atmosphere within about 30 years. Various microorganisms (*Archaea*, *Bacteria*, and selected eukaryotes), through their metabolic activities, account for over 85% of the CO_2 released into the atmosphere. Without this constant renewal of atmospheric CO_2, plants would cease photosyn-

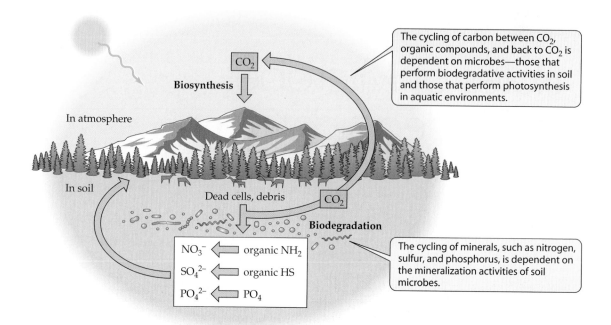

In atmosphere

In soil

Biosynthesis

CO_2

Dead cells, debris

CO_2

Biodegradation

The cycling of carbon between CO_2, organic compounds, and back to CO_2 is dependent on microbes—those that perform biodegradative activities in soil and those that perform photosynthesis in aquatic environments.

NO_3^-	⇐	organic NH_2
SO_4^{2-}	⇐	organic HS
PO_4^{2-}	⇐	PO_4

The cycling of minerals, such as nitrogen, sulfur, and phosphorus, is dependent on the mineralization activities of soil microbes.

Figure 12.1 Balance between biosynthesis and biodegradation
Biosynthesis is ultimately driven by light energy, and CO_2 is fixed into cells by photosynthesis. Biodegradation involves catabolic reactions that restore CO_2 to the atmosphere.

thesis, the pH of the ocean would increase, and much of life on Earth would be altered. Animals and plants would ultimately be doomed.

In a schematic depiction of the biosphere (**Figure 12.1**), biosynthetic reactions are those appearing at or above the soil's surface, and biodegradative reactions are those that occur below the surface of the soil. Biosynthesis in aquatic environments occurs mainly at or near the surface and to depths where light is available (about 100 meters). Biodegradative activities in oceans are mostly confined to the first 300 meters beneath the surface. Microbial activities are responsible for the cycling of carbon, nitrogen, and sulfur. The release of phosphate and other minerals present in living cells is also a result of microbial mineralization (breakdown of an organic compound to its component parts CO_2, NH_4^+, SO_4^{2-}).

The prominent role of microorganisms in perpetuating life can be summed up in two adages, termed **Van Niel's postulates**:

- A microorganism is present in the biosphere that can utilize every constituent part or product of the living cell as source of carbon and/or energy.

- Microbes with this capacity are present in every environmental niche on earth.

These postulates are based on an extensive body of knowledge available on the metabolism of microorganisms. The postulates are a clear statement of the role that microbes play in the balance that exists between biosynthesis and biodegradation. It is also evident that as eukaryotes evolved with complex molecular components, such as sterols or cerebrosides, a microbial population also evolved in due course that mineralized these unique molecules. This chapter covers some of the mechanisms involved in the biodegradation of biological material, selected xenobiotics (products of chemical synthesis), and other compounds. The examples presented are but a few representatives of the vast array of biodegradative pathways that occur in nature. Following an overview of the carbon cycle, we will examine the catabolism of individual carbonaceous compounds including starches, fatty acids, proteins, and aromatic hydrocarbons.

OVERVIEW OF THE CARBON CYCLE

The major pathways involved in the synthesis of cellular constituents from precursor metabolites were presented in Chapter 10. Recall that the precursor metabolites utilized in biosynthesis originate in glycolysis, the tricarboxylic acid (TCA) cycle, and the hexose monophosphate shunt. The precursor metabolites are converted to building blocks (amino acids, nucleotides, etc.), and these are then polymerized to form macromolecules. These macromolecules are then assembled to generate a cell. All of these processes constitute anabolism.

Archaea, Bacteria, and selected eukaryotes (mostly fungi) can reverse anabolism, and catabolic reactions can

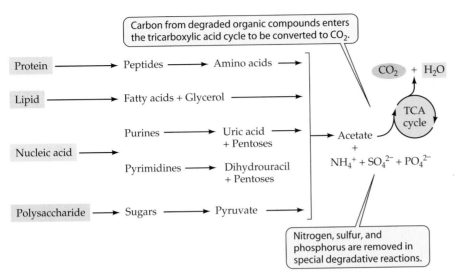

Figure 12.2 Fate of major biomolecules
Cell components and waste products are ultimately hydrolyzed by microorganisms to their constituent parts. All biodegradative pathways eventually lead to the tricarboxylic acid cycle or related cycles.

break down macromolecules (catabolic reactions) and release monomers that can be utilized to generate adenosine triphosphate (ATP). In many instances, such biodegradation reverses the pathways involved in synthesis. However, there are notable exceptions, and an example of this is fatty acid synthesis (see Figure 10.18) and biodegradation (see following) that follow somewhat different pathways and use different enzymes. Regardless of the route taken, macromolecules are disassembled to produce monomers quite similar to those from which the macromolecules were assembled. Thus, the reversal of anabolic sequences or other catabolic reactions can transform the constituents of the dead cell or the discarded products of living cells into low molecular weight compounds that can be completely mineralized (**Figure 12.2**).

The total mineralization of sugars and other metabolites is depicted in **Figure 12.3**. Note that each carbon in a sugar, from a three-carbon sugar to one with seven carbons, can become a molecule of CO_2. Acetate, pyruvate, or glyceraldehyde-3-P can enter into the TCA cycle or glycolysis and can be catabolized completely to CO_2 (see Figures 10.2 and 10.5). The microorganisms that are involved in these catabolic reactions gain energy and may also obtain precursor metabolites via these reactions. Approximately 35% to 45% of the organic carbon in a utilizable substrate can become a constituent of the growing bacterial, archaeal, or fungal cell involved in the biodegradative process.

Microorganisms can mineralize such compounds as plant polymers and alkaloids that are not constituent

parts of any bacterium. They have also evolved with the potential for biodegradation of a considerable number of aromatic hydrocarbons, alkanes, ketones, and xenobiotics.

CATABOLISM OF POLYSACCHARIDES

The number of compounds that can potentially serve as substrate for bacterial, archaeal, or fungal growth is virtually endless. Polysaccharides are abundant in nature and a ready food source for microorganisms. Polysaccharides are giant chains of monosaccharides (sugars) connected by glycosidic linkages. Starch, glycogen, and cellulose are huge polysaccharides composed of glucose, but they differ in the linkage between sugar units. Starch and glycogen are not chemically stable and are readily degraded by enzymes. They are therefore a good storage material. Cellulose is much more stable and is used by plants as a structural material. Xylans, too, are structural components of plants and are considered heterogeneous polymers because they contain small amounts of hexose and are quite complex. Chitin is a tough polymer of N-acetylglucosamine with hydrogen bonding between the long chains.

The following section presents pathways involved in the microbial catabolism of selected polysaccharide substrates.

Cellulose

Cellulose is considered to be the most abundant product of biosynthesis present on Earth. It is a major constituent of plants and is also synthesized by a limited number of bacterial species (*Acetobacter xylinum* and *Sarcina ventriculi*). Fungi and bacteria are the major decomposers of cellulose in nature. Fungi function mostly in acidic soils and in woody tissue where the cellulose present is protected from enzymatic attack by

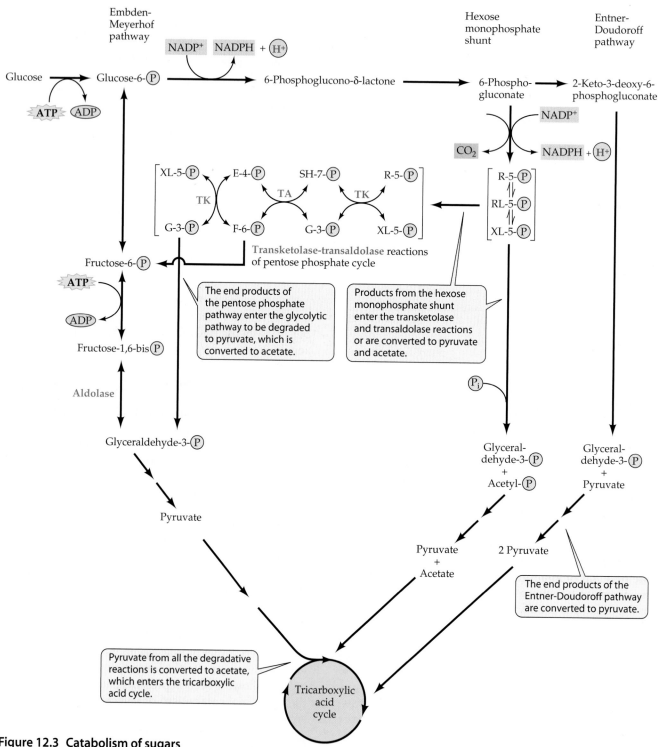

Figure 12.3 Catabolism of sugars
Transketolases (TK) and transaldolases (TA) are key enzymes in transforming sugars with three, four, five, or six carbons into intermediates that are further catabolized and enter the tricarboxylic acid cycle. Note that ATP and NADPH, both of which are used in biosynthesis, are generated during these catabolic reactions. In the pentose phosphate cycle, the sugars are: R, ribose; RL, ribulose; XL, xylulose; E, erythrose; SH, sedoheptulose; G, glyceraldehyde. Details of the other pathways are given in Figures 10.2 through 10.6.

lignin, whose composition and properties will be discussed later. Both wood-rotting and soil fungi are instrumental in the depolymerization and mineralization of cellulose.

The aerobic bacteria that decompose cellulose include myxobacteria, cytophaga, sporocytophaga, and some other species. Characteristics of these organisms will be

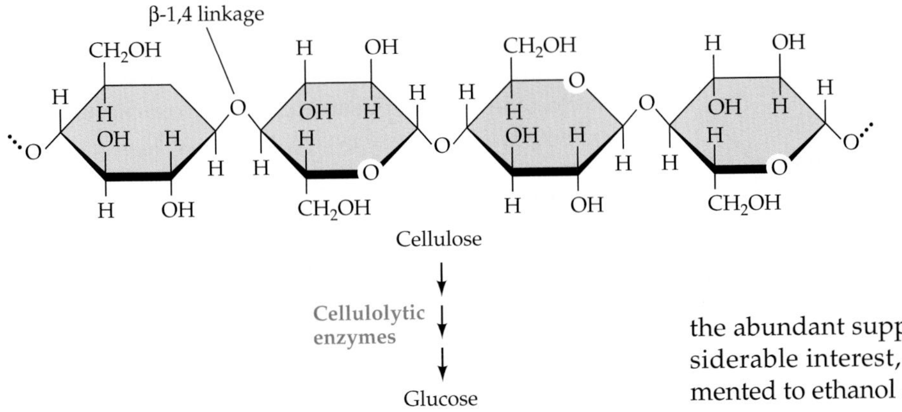

Figure 12.4 Cellulose
Cellulose consists of β-1,4-linked glucose units. The linkages in the polymer are twisted 180° (note the positions of the –CH₂OH groups in adjacent glucose units). This linkage permits the polymer to grow into long fibers, such as those in cotton.

Cellulose

↓ Cellulolytic enzymes ↓

Glucose

the abundant supplies of cellulose available is of considerable interest, as the glucose released can be fermented to ethanol and other industrial chemicals.

discussed in Chapter 20. Cellulose is fermented anaerobically in soil by clostridia and in the rumen of cattle and other ruminants by an array of anaerobic bacterial species.

The cellulose polymer is composed of glucose molecules that are joined by a β-1,4 linkage. An unbranched cellulose fibril can be composed of up to 14,000 glucose molecules (**Figure 12.4**). Decomposition of cellulose occurs by the concerted action of several enzymes:

- Endo-β-1,4-glucanase hydrolyzes bonds in the interior of chains releasing long chain fragments.
- Exo-β-1,4-glucanase removes two glucose units (called *cellobiose*) from the long fragments.
- α,β-Glucosidases hydrolyze cellobiose to glucose.

One problem in the biodegradation of cellulose is that the effectiveness of the endoglucanase is repressed (enzyme level is lowered) by the presence of cellobiose or by the glucose molecules as they are released. Industries have an interest in large-scale conversion of cellulose to glucose, and a successful operation must overcome this repression. Production of glucose from

STARCH

Starch is a major storage product in plants. There are two distinct polymers present in starch granules—**amylose** and **amylopectin**. Amylose consists of unbranched D-glucose chains in an α–1,4-glucosidic linkage with 200 to 500 sugar units per chain (**Figure 12.5**). The other polymer, amylopectin, is composed of the same glucosidic linkages but with branching in the 1,6-position every 25 to 30 sugar residues. Amylose is generally hydrolyzed by α-amylase or β-amylase. α-Amylase hydrolyzes the α-1,4 linkages to a mixture of glucose and maltose. **Maltose** is a disaccharide made up of two molecules of glucose in an α–1,4-linkage. β-Amylase hydrolyzes the α linkages of amylose at alternating positions to yield mostly maltose.

These two enzymes, α- and β-amylase, cleave amylopectin to a mixture of glucose, maltose, and a branched core. A different enzyme, amylo-1,6-glucosidase, cleaves the 1,6-linkage responsible for branching. This enzyme, in combination with α-amylase or β-amylase, can lead to the complete degradation of starch to maltose or glucose. The enzyme maltase is present in

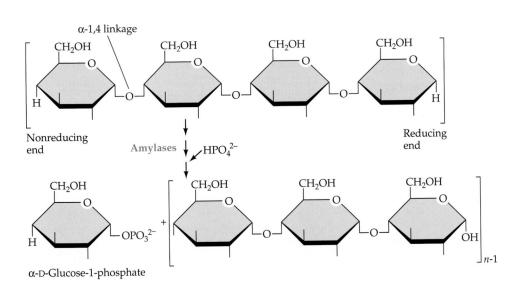

Figure 12.5 Starch
Starch consists of α-1,4-linked glucose units. Glycogen has a similar structure. The major difference between starch and glycogen is in the number of branches, formed through 1,6-linkages. The α-1,4 linkages result in a bending between glucose units and a helical structure to the starch molecule.

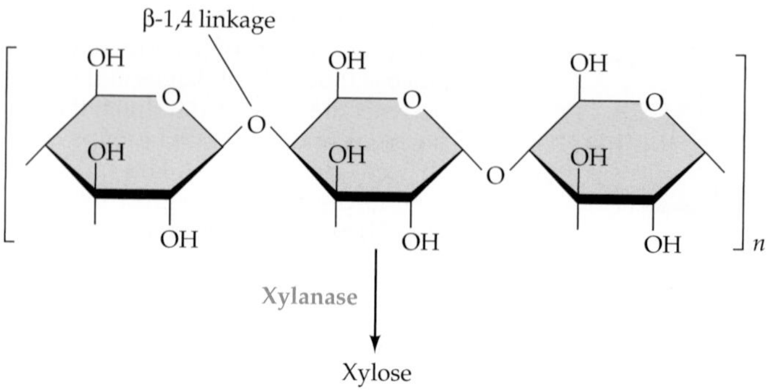

Figure 12.6 Xylan
Xylan is a polymer of the pentose D-xylan. The xylan chain consists of β-1,4-linked xylose units.

organisms that degrade starch and can hydrolyze one molecule of maltose to two molecules of glucose.

Amylases are present in many *Bacillus* species, including the obligate thermophiles *B. acidocaldarius* and *B. stearothermophilus*. The glucoamylases are produced by *Aspergillus niger, Rhizopus niveus,* and other fungi. Anaerobic thermophiles produce amylases that are stable at a range of temperatures and are of potential value in the production of commercial ethanol from starch. Starch digestion in the rumen of cattle and other ruminants is essential for the efficient utilization of plant material and grain. Rumen bacteria, such as *Streptococcus bovis* and *Bacteroides amylophilus*, produce α-amylase. *Entodinum caudatum*, a rumen ciliate, can hydrolyze starch to maltose and glucose.

Xylans

Xylans are heterogeneous polysaccharides composed mainly of xylose, a pentose sugar (**Figure 12.6**). They are second in abundance to cellulose among the sugar-based polymers that occur in nature. These polymers are present in the cell walls of virtually all land plants and are also a major constituent of mature woody tissue. Xylans are degraded by bacteria and fungi with two key enzymes:

- β-1,4-Xylanases that hydrolyze the internal bonds that link the xylose molecules yielding lower molecular weight xylooligosaccharides. An oligosaccharide contains a small number of monosaccharide units.

- β-Xylosidases release xylosyl residues through an endwise attack on the xylooligosaccharides.

When free xylose is produced, it is phosphorylated to xylose-5-phosphate rearranged to xylose-5-phosphate and enters the pentose cycle (see Figure 12.3).

Glycogen

Glycogen is a carbon and energy storage polysaccharide present in animals and bacteria. It is also called *animal starch* and is more branched than plant starch. Branching in glycogen occurs every 8 to 10 glucose residues, whereas branching in amylopectin occurs at every 25 to 30 residues.

Glycogen is hydrolyzed by a glycogen phosphorylase that removes one glucose and simultaneously incorporates an inorganic phosphate to yield a molecule of glucose-1-phosphate. The enzyme glycogen phosphorylase has been isolated from the yeast *Saccharomyces cerevisiae* and the cellular slime mold *Dictyostelium discoideum*. Because enteric bacteria, spore formers, and cyanobacteria store glycogen, it is apparent that these organisms also have the glycogen phosphorylase. The glucose-1-phosphate would be converted to glucose-6-phosphate by the enzyme phosphoglucomutase and enter the pathways outlined in Figure 12.3.

Chitin

Chitin is a major constituent in the cells of selected plants and animals. Many fungi possess chitin as a cell wall component, and it is present in crustaceans (lobsters, shrimp, crabs). Chitin is the tough, outer integument of grasshoppers, beetles, cockroaches, and other insects. It is a chemically stable polymer of N-acetylglucosamine units linked by β-1,4-glycosidic bonds. Its stability is derived from hydrogen bonding between adjacent chains through the N-acetyl moiety (**Figure 12.7**). The widespread occurrence of chitin in soil and aquatic environments has led to the evolution of a wide variety of microorganisms that can decompose this substance. The actinomycetes (filamentous soil bacteria, see Chapter 20) are a major utilizer of chitin. Addition of chitin as a substrate in an enrichment medium will consistently result in the isolation of diverse actinomycetes.

Chitin is cleaved by two enzymes:

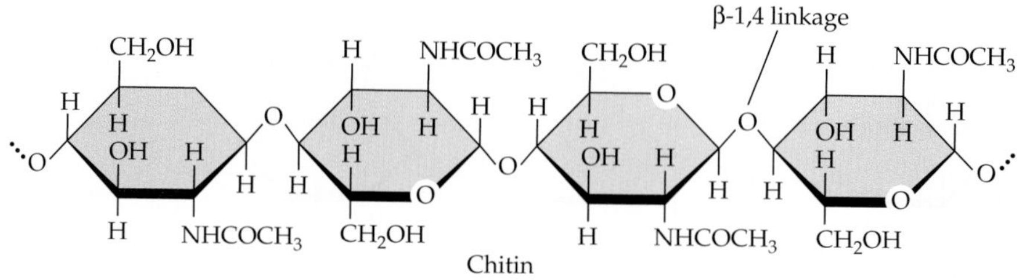

Figure 12.7 Chitin
Chitin consists of *N*-acetylglucosamine units. As in cellulose, the sugar units are joined by β-1,4 linkages. Chitin occurs as long ribbons in the cell walls of fungi and the exoskeletons of crustaceans, insects, and spiders.

1. Chitinase splits chitin at numerous points to mostly the disaccharide chitobiose and chitotriose.
2. Chitobiase hydrolyzes chitobiose and chitriose to monomers that are readily biodegraded.

CATABOLISM OF LIPIDS

An indispensable structure in all cells, prokaryotic and eukaryotic, is the cytoplasmic membrane. The membrane is composed mainly of polar phospholipids (Figure 4.34), with the exception of the ether-linked lipids in the *Archaea* (see Figure 4.36). The internal organelles of eukaryotes (mitrochondria, nucleus, etc.) are also enclosed in phospholipid membranes. The major components of the phospholipids are the long-chain fatty acids (or branched-chain alcohols in *Archaea*) that are attached to a glycerol moiety. Long-chain *n*-alkanes ($>C_{12}$) are also present in most living cells. The role of these hydrocarbons is not known.

For convenience in this section, we will first discuss the attack on *n*-alkanes as they are catabolized through the homologous long-chain fatty acid. This is followed next by the stepwise catabolism of the fatty acids and lastly by the dissimilation of the phospholipid itself.

Long-Chain Fatty Acids and *n*-Alkanes

Alkanes are saturated hydrocarbons of 12 to 20 carbons that are a minor constituent of most living cells, and *n*-alkanes of greater carbon chain length are present in many plant tissues. Paraffins in crude oil range from gaseous methane (CH_4) to solid paraffins of consider-able length ($>C_{20}$). Thus, significant amounts of these hydrocarbons are released into the environment on a regular basis. Microorganisms are abundant in soil and water that can utilize alkanes as growth substrate. The initial site of enzymatic attack on the *n*-alkane series from *n*-butane to eicosane (C_{20}) is at a terminal methyl group that ultimately yields the homologous carboxylic acid. There is substantial evidence that *n*-alkanes over 30 carbons long may be cleaved at sites in the interior of the chain to yield more manageable, smaller molecules.

Many bacteria, including pseudomonads, mycobacteria, actinomycetes, and corynebacteria, can utilize *n*-alkanes as substrate. Yeasts and filamentous fungi are also capable of growth on these compounds. A general scheme for the degradation of a molecule of *n*-alkane to the homologous alcohol and fatty acid is presented in **Figure 12.8**. Note that the carboxylic acid is coupled with CoASH to form the fatty acyl-CoA derivative.

The long-chain fatty acids commonly present in phospholipids, such as hexadecanoic (C_{16}) or octadecanoic acid (C_{18}), would be catabolized by coupling them with CoASH to form the fatty acyl-CoA. The fatty acyl-CoA would be catabolized by β-oxidation (**Figure 12.9**). This series of reactions is termed "β-oxidation" because the β-carbon is oxidized to a keto (C=O) prior to the cleavage of an acetyl-CoA from the chain.

Phospholipids

Phospholipids account for about 10% of the dry weight of a typical procaryotic or eukaryotic cell. They are constantly introduced into the environment by the decay of dead cells. The biodegradation of a phospholipid occurs by sequential removal of the fatty acid components, as illustrated in **Figure 12.10**.

The concerted action of the specific phospholipases involved in the disassembly of this phospholipid (Figure 12.10) would yield two long-chain fatty acids, one glycerol, and one phosphoserine. The fatty acid would be catabolized by β-oxidation (see Figure 12.9). Bacteria and fungi are commonly present in the environment that can disassemble phospholipids and utilize the acetyl-CoA released by β-oxidation.

An *n*-alkane undergoes a series of oxidation and reduction reactions to produce the corresponding fatty acid.

The fatty acid is linked to CoA to produce the fatty acyl-CoA, which can enter the β-oxidation pathway.

Figure 12.8 *n*-Alkane oxidation
The stepwise oxidation of a long-chain *n*-alkane to the homologous alcohol, aldehyde, and fatty acid. The fatty acid is converted to the fatty acyl-CoA.

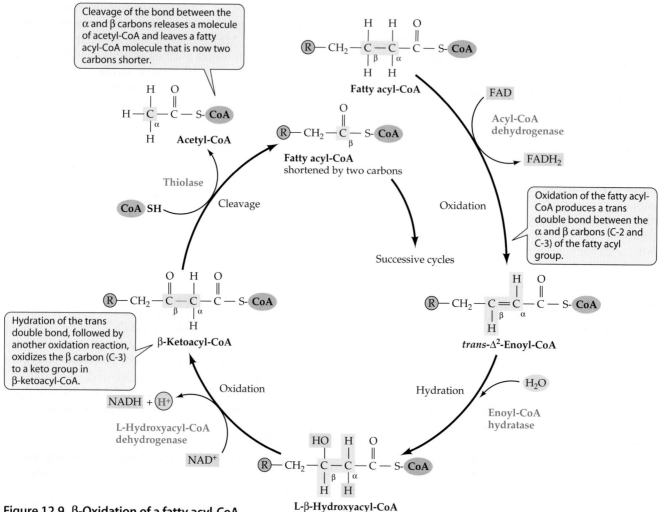

Figure 12.9 β-Oxidation of a fatty acyl-CoA
Each cycle of the β-oxidation pathway produces one molecule of FADH$_2$, one of NADH, and one of acetyl-CoA. The fatty acid is shortened by two carbons. Delta (Δ) indicates a double bond, and the superscript (2) indicates its position (between C-2 and C-3).

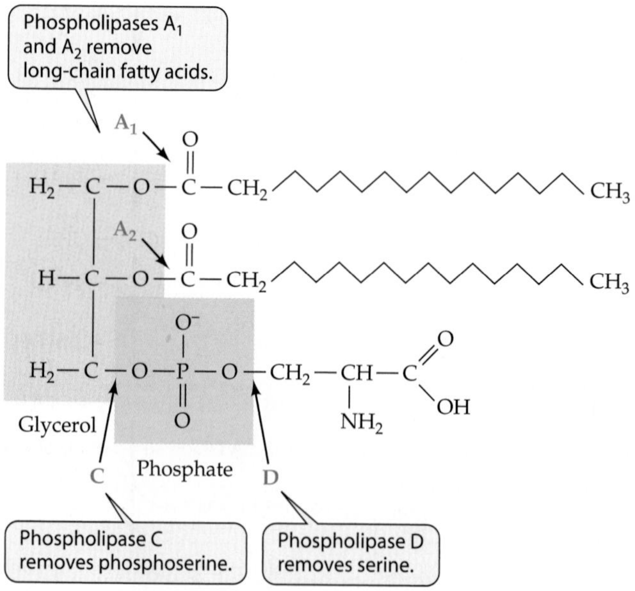

Figure 12.10 Phospholipase specificity
Phospholipases are the hydrolytic enzymes that cleave polar phospholipids. The different bonds are cleaved by different enzymes. Shown here are the effects of these enzymes on phosphatidylglycerol.

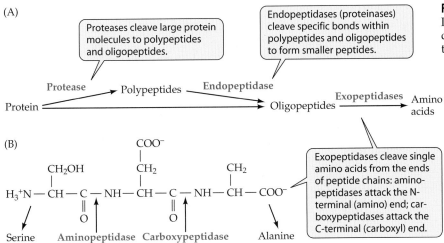

Figure 12.11 Protein biodegradation Proteases, endopeptidases, and exopeptidases act in concert to break down proteins to amino acids.

CATABOLISM OF PROTEINS AND URIC ACID

Proteins

Proteins are a major constituent of all cells and constantly enter into the environment from dead cells and waste materials. Microorganisms (*Bacteria*, *Archaea*, and fungi) that biodegrade proteins are widespread in nature. Microorganisms initiate digestion of these large molecules by transporting enzymes outward to the cell surface, which hydrolyze the protein to smaller molecules. These small molecules can be transported into the cell where they are utilized as substrates (**Figure 12.11A**). Extracellular **proteolytic enzymes** (**proteases**) hydrolyze proteins to polypeptides, and these in turn are cleaved by endopeptidases to oligopeptides. Endopeptidases (also called proteinases) hydrolyze peptide bonds within polypeptide chains. The endopeptidases are specific and hydrolyze the peptide bond only if specific amino acids are present at the cleavage site. For example, the endopeptidase **thermolysin** from *Bacillus thermoproteolyticus* cleaves peptide linkages where leucine, phenylalanine, tryptophan, or tyrosine is one of the amino acids present.

An **exopeptidase** is an enzyme that hydrolyzes amino acids from the end of a peptide chain (Figure 12.11B). A **carboxypeptidase** attacks the C-terminal end of the chain, and an **aminopeptidase** removes an amino acid from the N-terminal end of a peptide chain. The enzymes involved in proteolysis act in concert and with overlapping responsibility.

The amino acids released by the hydrolysis of a protein are transported into the fungal or bacterial cells by specific transport systems. The oligopeptides may also be transported into cells where they are hydrolyzed by peptidases. The free amino acids can be incorporated into the proteins of the cell or deaminated and catabolized to provide carbon and/or energy for growth.

Uric Acid

Birds, scaly reptiles, and carnivorous animals secrete uric acid instead of urea as a product of purine catabolism. Selected soil microorganisms catabolize uric acid as a source of carbon and energy (**Figure 12.12**). One of these organisms is the unique endospore-forming *Bacillus fastidiosus*, which cannot grow with any substrate except uric acid and its degradation products, allantoic acid and allantoin. Glyoxylate, CO_2, and urea are the products of uric acid degradation. Urea is degraded to ammonia and CO_2 by urease, an enzyme present in many species of bacteria.

CATABOLISM OF AROMATIC COMPOUNDS

Lignin

Lignin is the most abundant renewable aromatic molecule on Earth. Its natural abundance may be exceeded only by cellulose. Biodegradation of lignin, therefore, is an essential part of the carbon cycle. Lignin surrounds and protects the xylans and cellulose that are components of wood from destruction by microorganisms. Between 20% and 30% of wood and plant vascular tissue is composed of lignin. All higher plants, including ferns, contain lignin, but lower plants, such as mosses and liverworts, do not. Lignin is relatively indestructible, and it is probably not metabolized by any single microbial species as a sole source of carbon or energy.

The resistance of lignin to microbial destruction is founded in the reactions whereby it is synthesized. Lignin is synthesized from three precursor alcohols by free radical copolymerization (**Figure 12.13A**). The

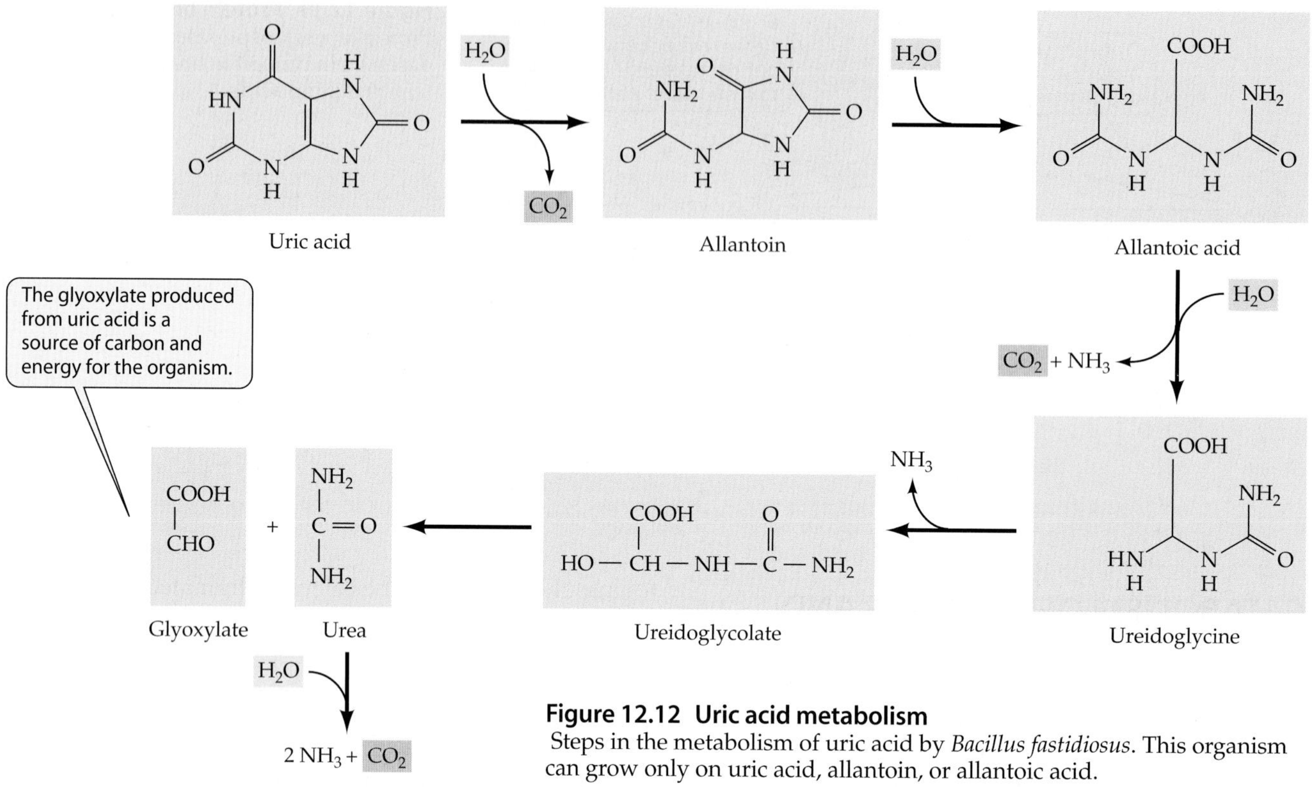

Figure 12.12　Uric acid metabolism
Steps in the metabolism of uric acid by *Bacillus fastidiosus*. This organism can grow only on uric acid, allantoin, or allantoic acid.

complex large polymer of lignin (Figure 12.13B) would present a microbial species with an insurmountable task in any attempt to break it down. The molecule is a heterogeneous polymer assembled in a haphazard way that would be attacked only by enzymes with very high levels of nonspecificity. Such enzyme specificity has apparently not evolved in nature.

Lignin is biodegraded in nature by white-rot basidiomycetes (fungi) via a mechanism befitting its synthesis, that is, a random oxidative attack. Lignin is not known to serve as growth substrate for this or any other fungus, so an alternate carbon or energy source, termed a cosubstrate, must be available. Xylans, cellulose, and carbohydrates can serve as this cosubstrate.

Experiments with a white-rot fungus, *Phanerochaete chrysosporium*, have affirmed that lignin is degraded by enzymatic "combustion." During growth on a cosubstrate, *P. chrysosporium* synthesizes enzymes that require H_2O_2 to function and catalyze the oxidative cleavage of both β-O-4 ether bonds and carbon-to-carbon bonds. The hydrogen peroxide is supplied by peroxidases in the fungus. Lignin is not degraded in the absence of O_2. The breakdown of lignin is initiated by the action of single electrons that produce aromatic cationic radicals in the lignin structure. These, then, function in nonenzymatic radical-promoted reactions that disrupt the lignin molecule. This process leads to a nonspecific oxidation of lignin to random, low molec-

ular weight products that are then available for mineralization by fungi or bacteria.

Aromatic Hydrocarbons

Aromatic and polyaromatic hydrocarbons such as phenol or pyridine are industrial chemicals that occur both as products of or intermediates in chemical syntheses. Polyaromatic hydrocarbons are also formed during combustion of fossil fuels and are natural constituents of unaltered fossil fuels. Benzene, toluene, xylenes, and other low molecular weight aromatics serve as commercial solvents and are constituents of paint. Various other aromatic compounds are intermediates in the synthesis of dyes, medicines, plastics, and other chemicals that have become a part of modern life. They are also a source of considerable environmental pollution. Aromatic compounds such as phenylalanine and tyrosine are constituents of living cells.

The biodegradation of aromatic hydrocarbons has been studied extensively. Benzene is one that has received much study, and research has affirmed that there are two major pathways for the biodegradation of this compound (**Figure 12.14**). The pathway followed depends on the bacterial strain involved. Both result in products (pyruvate, succinate, and acetate) that can readily be mineralized via the TCA cycle.

Derivatives of aromatic hydrocarbons, particularly chlorinated compounds, are widely used as pesticides, insulators, and preservatives. However, the accumula-

(A)

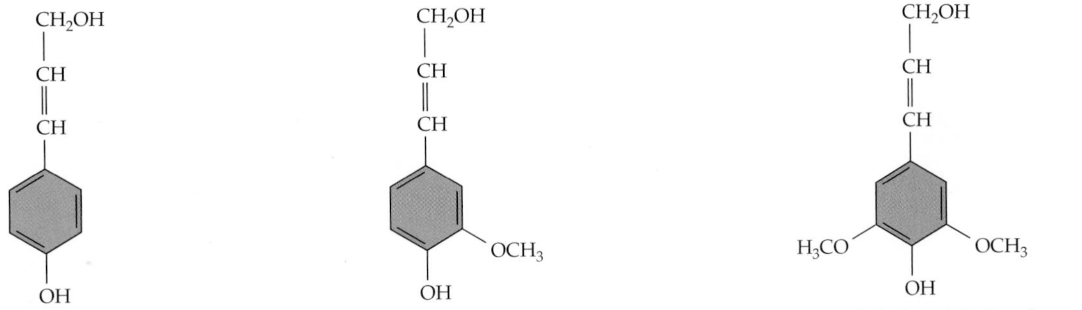

Coumaryl (*p*-hydroxy cinnamyl) alcohol

Coniferyl alcohol (4-hydroxy-3-methoxy cinnamyl) alcohol

Sinapyl alcohol (3,5-dimethoxy cinnamyl) alcohol

(B)

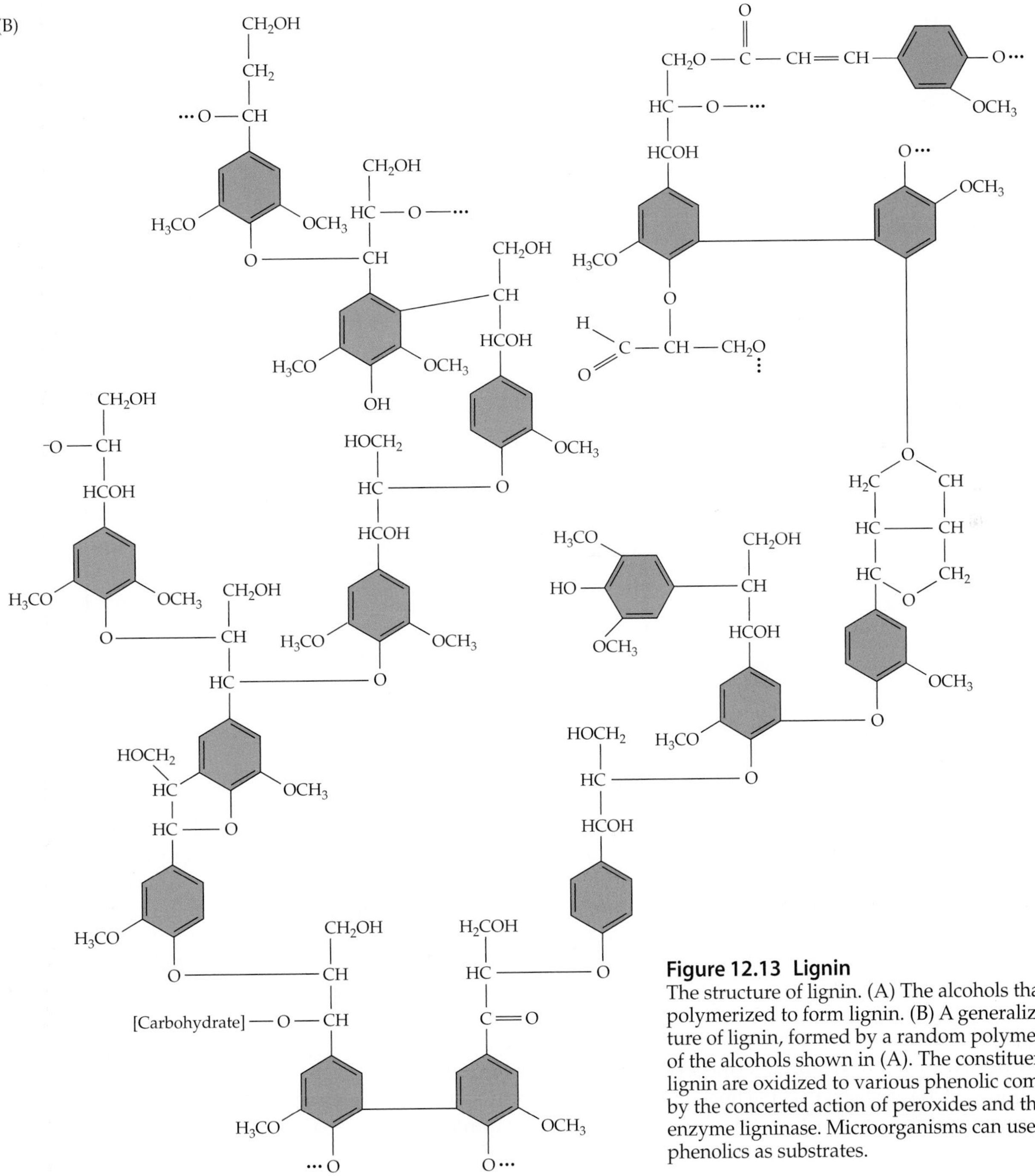

Figure 12.13 Lignin
The structure of lignin. (A) The alcohols that are polymerized to form lignin. (B) A generalized structure of lignin, formed by a random polymerization of the alcohols shown in (A). The constituents of lignin are oxidized to various phenolic compounds by the concerted action of peroxides and the enzyme ligninase. Microorganisms can use these phenolics as substrates.

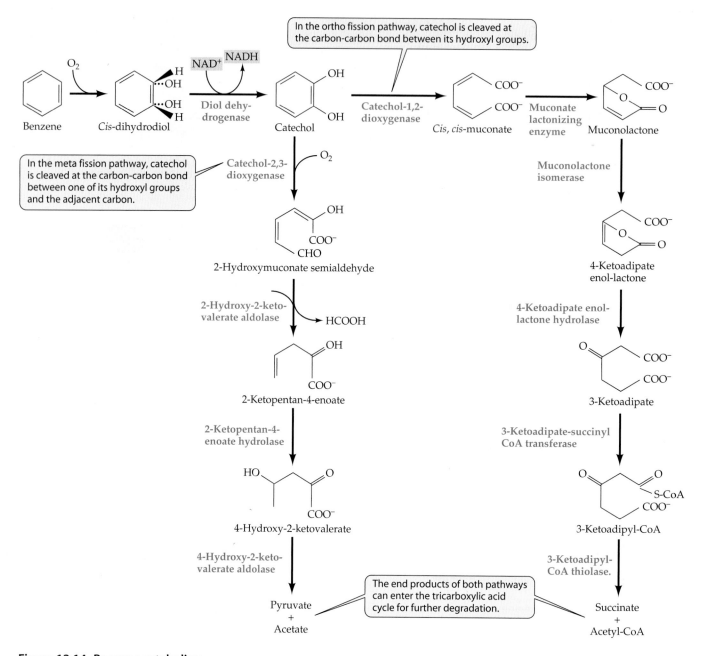

Figure 12.14 Benzene catabolism
The two pathways for benzene degradation in bacteria.

tion of these chloroaromatics has become a serious environmental problem. For example, the cumulative annual worldwide production of **polychlorinated biphenyls (PCBs)** (**Figure 12.15**) is presently estimated to exceed 750,000 tons. Ultimately, a considerable portion of this amount ends up in the environment. There are 210 possible PCBs, all having the same biphenyl carbon skeleton and differing only in the number and position of chlorine atoms. PCBs are used in capacitors, transformers, plasticizers, printing ink, paint, and a host of other materials and products. The use of PCBs has now been curtailed in the United States. **Pentachlorophenol** (Figure 12.15A) is another broadly applied compound that has become an environmental hazard. It is used in termite control, as a preharvest defoliant, and in wood preservation.

The biodegradation of chlorinated hydrocarbons occurs via dehalogenation reactions that yield compounds that are biodegradable (**Box 12.1**). The removal of chlorines from an organic compound may be a slow process, and cosubstrates often must be available to provide a source of energy and carbon for the dehalogenating microorganism. A pathway for mineralization of trichloroethylene is outlined in Figure 12.15B.

Recalcitrant—obstinately defiant of authority or restraint: difficult to handle or operate: not responsive to treatment.

Webster's New Collegiate Dictionary

Molecular recalcitrance measures the relative resistance a molecule has to microbial attack. Many recalcitrant compounds are halogenated aromatic or aliphatic compounds such as DDT and trichloroethylene. These are of considerable concern because they are harmful environmental pollutants. The persistence of many halogenated chemicals and their toxicity renders them hazardous to health and to the environment. Chlorinated compounds are widely applied as pesticides, solvents, and hydraulic fluids. and thus eventually enter the environment. For example, pentachlorophenol (Figure 12.15) is used as a wood preservative (telephone poles, fence posts) and is an effective long-term protector against rot.

Halogenated compounds have been considered impervious to microbial assault because they are the products of chemical synthesis, and microorganisms may not have evolved with enzyme systems capable of their mineralization. One may ask whether the inability of

microbes to dehalogenate is due to the presence of halogenated compounds in nature solely as products of chemical synthesis. Clearly this is not the case. There are over 700 naturally occurring halogenated compounds that have been identified thus far and at least seven distinct enzyme systems that can cleave the carbon-halogen bond.

Recalcitrance is not solely related to the presence of a —Cl, —Br, —F, or —I atom. The number, position, and electronegativity of the halogen-carbon(s) bonds are of fundamental importance in determining whether the dehalogenating enzymes can function. It is apparent that some halogenated compounds

are not recalcitrant, whereas others are impervious to microbial attack. The structures and half-life of some pesticides are presented in the figure below.

The half-life in soil would be for temperate regions such as the United States and would be somewhat shorter in tropical regions. Lindane and Mirex® are insecticides, Dalapon® is a herbicide, and DBCP is a nematocide. Dalapon would not accumulate in nature regardless of usage, whereas lindane or DBCP would gradually build up in soil with continued use. Mirex, which has been used to eradicate fire ants, is no longer applied in the United States.

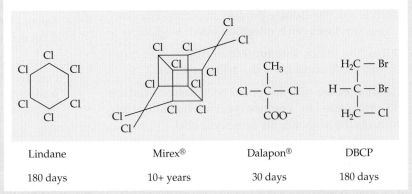

Lindane	Mirex®	Dalapon®	DBCP
180 days	10+ years	30 days	180 days

The half-life of some compounds that have been used as pesticides. DBCP is dibromochloropropane.

(A)

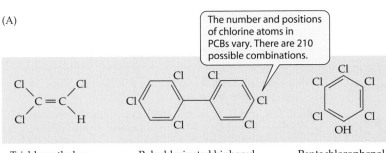

The number and positions of chlorine atoms in PCBs vary. There are 210 possible combinations.

Trichloroethylene   Polychlorinated biphenyl   Pentachlorophenol

Figure 12.15 Chlorinated hydrocarbons
Typical chlorinated hydrocarbons and their bacterial degradation. (A) Some significant environmental pollutants. Trichloroethylene (TCE) is a degreasing agent; polychlorinated biphenyls (PCBs) have been employed as insulators; and pentachlorophenol (PCP) is a wood preservative. (B) The biodegradation of trichloroethylene by *Mycobacterium vaccae*, following growth of this organism on propane.

(B)

This chlorine migrates to the other carbon to generate trichloroethanol.

Sequential removal of the chlorines as Cl⁻ ions produce ethanol.

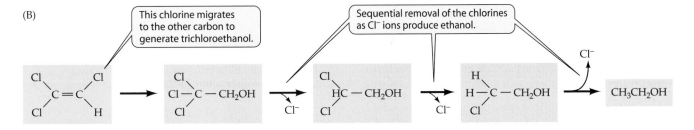

SUMMARY

▸ Sunlight supports biosynthetic reactions, and these reactions can continue only if photosynthate is **recycled** constantly by **biodegradation**.

▸ Without replenishment, the level of CO_2 in the atmosphere would support the present rate of photosynthesis on Earth for only **30 years**.

▸ **Microbial decomposition** is the source of 85% of the CO_2 in the atmosphere.

▸ Some **biodegradative pathways** are a **reverse** of **synthetic** reactions, whereas others follow distinctly **different** pathways.

▸ **All products** of living cells are biodegraded by some microorganism, **as are many products of chemical synthesis**.

▸ **Cellulose** is the most abundant product of **biosynthesis** present on earth. It is biodegraded by **both bacteria** and **fungi**.

▸ Starch is a major storage product in plants. Enzymes termed **amylases** can biodegrade starch to glucose.

▸ **Chitin**, the **tough integument** of insects and crustaceans, is mineralized by the **actinomycetes**.

▸ The **alkanes** are of widespread occurrence in nature and are oxidized by a **monooxygenase** that ultimately produces the homologous carboxylic acid.

▸ **Phospholipids** are biodegraded by the **sequential removal of the fatty acids** and polar group from the glycerol backbone by a series of distinctly different phospholipases.

▸ Initial attack on large molecules such as proteins occurs by the action of **extracellular** enzymes termed **proteases**. These enzymes release lower molecular weight polypeptides that are cleaved to **oligopeptides**. There are specific enzymes (**exopeptidases**) that hydrolyze amino acids from the oligopeptides.

▸ **Uric acid**, a product of purine catabolism, is utilized by *Bacillus fastidiosus* as sole source of carbon and energy.

▸ **Lignin** is an aromatic complex that makes up 20% to 30% of woody vascular tissue. It is **resistant** to **enzymatic attack** and is depolymerized by the peroxides generated by wood-rotting fungi. Both bacteria and fungi utilize the **products** of this **oxidative attack** as a carbon source.

▸ **Aromatic hydrocarbons** occur widely as **environmental pollutants**. Many are degraded by bacteria and fungi.

▸ The **chlorinated** hydrocarbons are generally quite resistant to enzymatic attack. Removal of the **chlorine atoms** through incidental attack while metabolizing **cosubstrates** is a key element in ridding the environment of these compounds.

REVIEW QUESTIONS

1. Why is biological materials recycling important in the perpetuation of life on Earth? What geochemical event was of significant importance in the evolution of heterotrophic microorganisms and multicelled eukaryotes?

2. What percentage of atmospheric CO_2 comes from microbial degradation? How long would the present supply last without replenishment? What would be the consequence of lowered levels of CO_2 in the atmosphere?

3. What are Van Niel's postulates? How do they relate to the role of microbes on earth?

4. Biodegradation is sometimes a reversal of biosynthesis. In some cases, biodegradation follows a different pathway from that involved in biosynthesis. Explain and consider examples of both.

5. How would a microbe attack a large molecule like cellulose? Like protein? What are the similarities and differences between cellulose and protein degradation?

6. Why are amylases important to the well-being of some animals? What types of organisms produce these enzymes?

7. Where does commonality occur in the biodegradation of long-chain *n*-alkanes and long-chain fatty acids?

8. What are the unique features of lignin biosynthesis? Biodegradation?

SUGGESTED READING

Alexander, M. 1999. *Biodegradation and Bioremediation*. New York: Academic Press.

Kirk, T. K. and R. L. Farrell. 1987. Enzymatic "Combustion": The Microbial Degradation of Lignin. *Annual Review of Microbiology* 41(1987): 465–505.

Young, L. and C. Cerniglia. 1995. *Microbial Transformation and Degradation of Toxic Organic Chemicals*. New York: John Wiley & Sons.

Zehnder, A. J. B., ed. 1988. *Biology of Anaerobic Bacteria*. New York: John Wiley & Sons.

Wackett, L. P. and C. D. Hershberger. 2001. *Biocatalysis and Biodegradation: Microbial Transformation of Organic Compounds*. Washington, DC: ASM Press.

Genetics and Basic Virology

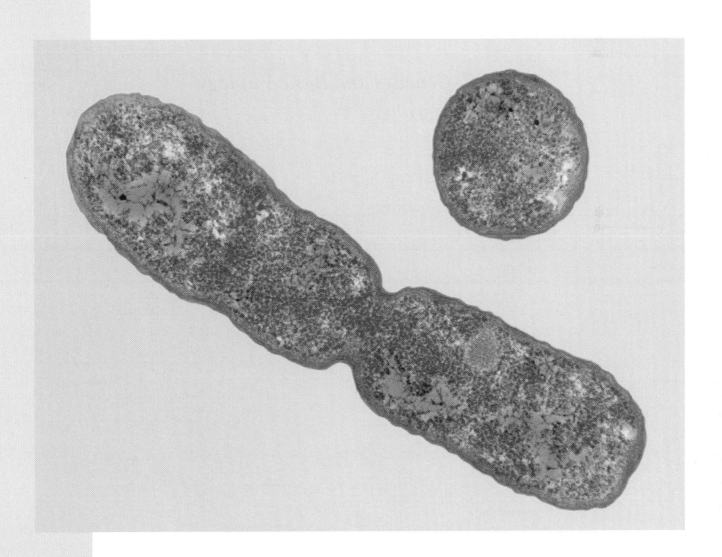

PART IV Genetics and Basic Virology

Previous page
DNA is visible as nucleoids in this lengthwise section through a dividing *Escherichia coli*. ©Dennis Kunkel Microscopy, Inc.

Chapter 13

Basic Genetics

A cell is DNA's way of making more DNA.
— *AUTHOR UNKNOWN*

All the members of a bacterial or archaeal species are remarkably similar in their structure and ability to survive in a given environment. The thousands of biochemical reactions carried out within the cell—synthesizing cellular constituents, responding to environmental stimuli—are nearly identical in all members of a species living under equivalent conditions. How is it that this species identity is maintained as each cell moves through a predetermined life cycle and divides, producing daughter cells that are identical to the parent? During replication, a copy of genetic instructions written in chemical code is transferred to each daughter cell. It is this genetic code, unique to each bacterial species, that safeguards the identity of the species from generation to generation.

This chapter will define the chemical nature of the genetic code, delineate the basic principles of informational transfer during cell division, and describe the cellular machinery responsible for translating the "words" of the code into the "actions" of the cell, such as enzymes and structures. We will also examine how the coded instructions carried by the genes can be altered and will investigate the consequences of such alterations on traits that bacteria possess and pass on to their progeny or, in some cases, to their siblings.

THE NATURE OF THE HEREDITARY INFORMATION IN *BACTERIA*

The hereditary information of all living cells is stored in ordered segments of deoxyribonucleic acid (DNA) called genes; genes are found strung together in large molecules called chromosomes. The complete set of an organism's genes is called its genome. Whereas higher organisms carry their genes in multiple, linear chromosomes, the genome in the vast majority of bacteria is contained on a single circular chromosome. Occasionally, bacteria contain additional, smaller, circular molecules of DNA that carry information necessary for survival in special environments. These molecules, called plasmids, are chemically identical to chromosomal DNA. The smallest genomes are found in viruses; viral genomes can be circular DNA (double- or single-stranded), linear DNA, or double- or single-stranded ribonucleic acid (RNA)

The flow of information within the cell may be viewed as ideas (DNA) transcribed to words (RNA) translated to actions (proteins), which necessarily include maintaining the store of DNA so that the flow can be repeated. These processes require energy and are therefore regulated at many levels. The rest of this chapter reviews the mechanics that make this regulated flow possible.

Structure of DNA

Of all the reactions in a cell, the consistently accurate copying of genetic information can be considered one of the most important. Species identity is written in DNA, and reliable replication is the guardian of this identity. The key to this accuracy is found in the double-stranded structure of DNA, the high fidelity of the machinery responsible for DNA replication, and the proofreading mechanism that corrects even the rarest of the mistakes.

Chemically, DNA is a relatively simple molecule: deoxyribonucleotides linked by phosphodiester bonds. All of these deoxyribonucleotides consist of the same sugar phosphate, 2-deoxyribose, but they differ with respect to their nitrogenous bases, as shown in **Figure 13.1**. Purines (adenine and guanine) and pyrimidines (cytosine and thymine) are attached to the sugar by specific *N*-glycosidic linkages. The long DNA chain is the result of links made by a phosphodiester bond between the 3' and 5'–carbon atoms of each sugar. It is the order of nucleotides in a gene that carries its unique meaning, just as the order of the letters in a word conveys its meaning. Again, the accurate copying of this sequence during bacterial division allows a bacterium to transmit its specific characteristics from the parent cell to the daughter cells.

Chromosomal and plasmid DNA are always present in a double-stranded form, with hydrogen bonds holding the two strands together. Base pairing takes place between adenine and thymine and between guanine and cytosine, with a purine bonding to a pyrimidine in each case (see Figure 13.1). The number of hydrogen bonds between the bases is not identical: two hydrogens participate in a bond between adenine and thymine, whereas guanine and cytosine form hydrogen bonds with three participating hydrogens. Nevertheless, this bonding maintains a distance of just under 3 Å (angstroms)

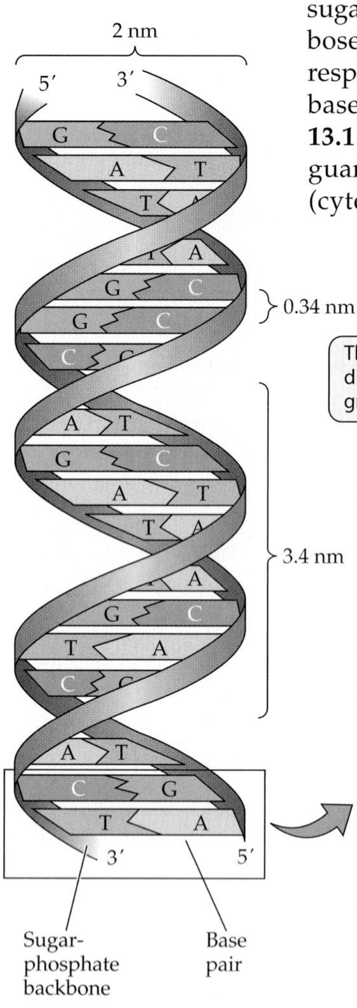

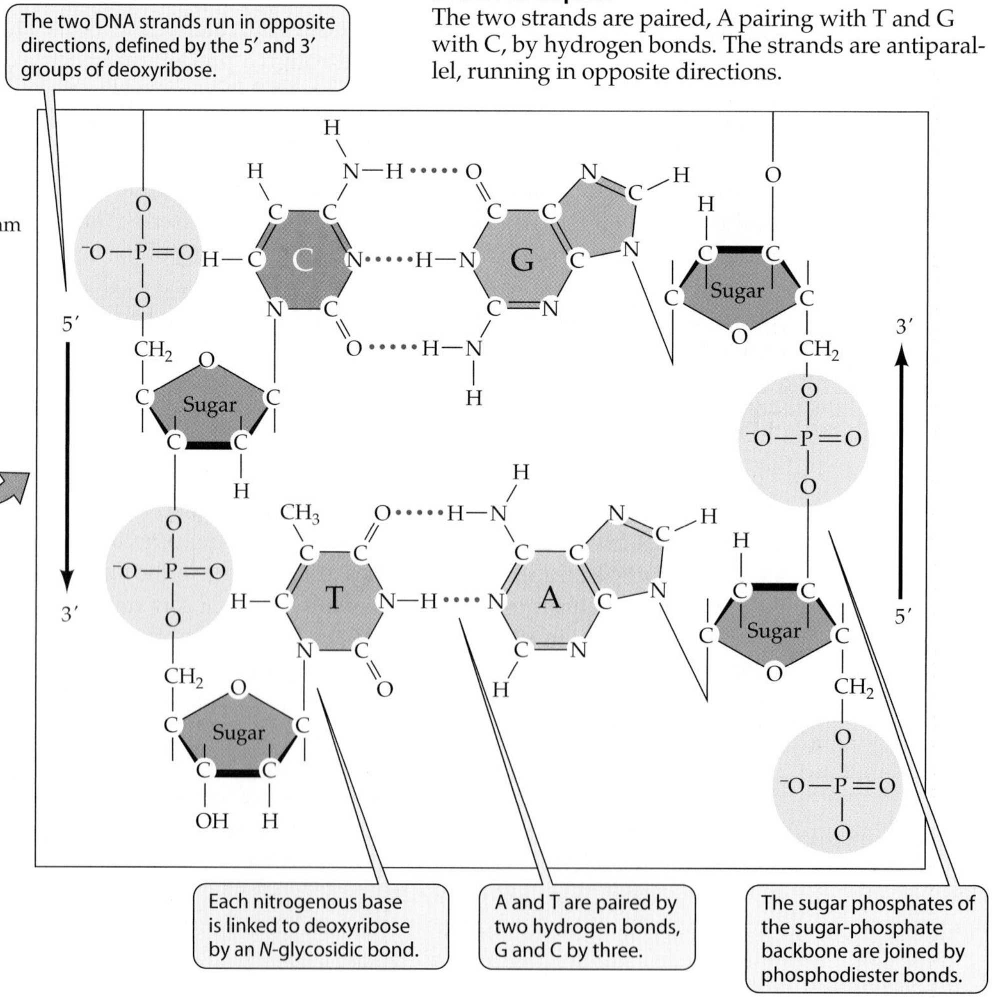

Figure 13.1 Structure of DNA strands in a DNA duplex
The two strands are paired, A pairing with T and G with C, by hydrogen bonds. The strands are antiparallel, running in opposite directions.

The two DNA strands run in opposite directions, defined by the 5' and 3' groups of deoxyribose.

Each nitrogenous base is linked to deoxyribose by an *N*-glycosidic bond.

A and T are paired by two hydrogen bonds, G and C by three.

The sugar phosphates of the sugar-phosphate backbone are joined by phosphodiester bonds.

between the bases of each strand, giving the DNA molecule a uniform width.

As a result of base pairing, the sequence in each strand is complementary; that is, a specific sequence in one strand determines the sequence in the other. Moreover, because the two polynucleotide chains run in opposite directions with respect to their 5′-to-3′ links, they are said to be antiparallel. This is depicted in Figure 13.1, where the phosphodiester linkages in one strand run in the 5′-to-3′ direction, and the complementary strand runs 3′ to 5′.

The linear arrangement of nucleotide bases, together with strict base pairing between specific purines and pyrimidines, puts additional constraints on the DNA molecule. Hydrophobic interactions between adjacent bases result in stacking, forcing the phosphate backbone into a helical configuration. The predominant form of double-stranded DNA is a right-handed helix, turning in a clockwise direction, with a regular periodicity of 10.6 base pairs, or 3.4 nm.

Typically, a bacterial cell contains about 5×10^6 base pairs of the four nucleotides, giving a total (unwound) bacterial chromosome length of approximately 1 mm. This single huge molecule fits into a 1- to 2-µm-long bacterial cell, a feat comparable to a 10-meter string folding into a 1-centimeter box. DNA accomplishes this by several additional higher-order structural features, most important of which is supercoiling. Like any circular object, if a circular DNA molecule is twisted, it will cross itself. When a right-handed, clockwise helix is twisted in a left-handed direction, it becomes underwound. DNA twisted in the same right-handed direction as the helix becomes overwound. The torsional stress is relieved by supercoiling, in which the double-stranded DNA twists onto itself. The condensed, supercoiled form of DNA brings together the negatively charged phosphate backbones, which are naturally repulsing. Neutralizing these charges is accomplished by a coating on the DNA that contains small, basic proteins and a variety of cationic molecules. These factors allow packaging of the genomic DNA into a compact space without disrupting its ability to provide genetic information to the cell.

DNA REPLICATION

Bacterial division is by an asexual process, in which a cell divides into more or less equal parts. This mode of reproduction is called binary fission and results in the formation of two daughter cells that are indistinguish-

Figure 13.2 Semiconservative DNA replication
Each new DNA molecule consists of a parent strand and a new, complementary daughter strand.

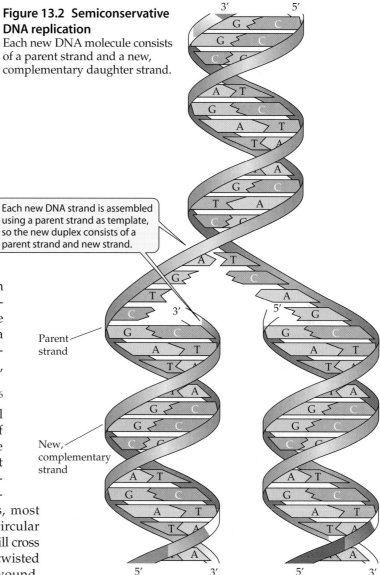

Each new DNA strand is assembled using a parent strand as template, so the new duplex consists of a parent strand and new strand.

Parent strand

New, complementary strand

able from the single parent cell. Constituents within the cell cytoplasm and cell envelope are divided approximately equally, and if necessary, minor differences in cytoplasmic content, such as enzymes and cofactors, can be made up by synthesis in the daughter cell immediately following cell division. The exact partition of the genetic material, however, is absolutely necessary. "Minor" errors in the genome can result in daughter cells permanently losing a genetic trait and, perhaps, in death.

The replication of the genome results in two double-helical molecules. One strand in each molecule is the parent DNA, whereas the other is a complementary, newly synthesized strand. The strand separation of a parent molecule, its use as a template, and its incorporation into the daughter genome is termed semiconservative replication and is shown in **Figure 13.2**.

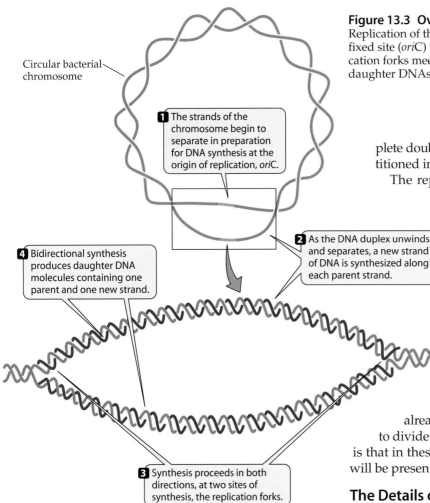

Circular bacterial chromosome

1 The strands of the chromosome begin to separate in preparation for DNA synthesis at the origin of replication, *ori*C.

2 As the DNA duplex unwinds and separates, a new strand of DNA is synthesized along each parent strand.

4 Bidirectional synthesis produces daughter DNA molecules containing one parent and one new strand.

3 Synthesis proceeds in both directions, at two sites of synthesis, the replication forks.

Figure 13.3 Overview of DNA replication
Replication of the circular bacterial chromosome begins at a fixed site (*ori*C) then proceeds bidirectionally until the replication forks meet, at which point two double-stranded daughter DNAs are formed.

plete double-stranded helix, which can then be partitioned into daughter cells.

The replication of this enormous molecule is very rapid; replication forks proceed at a rate of 1,000 base pairs (bp) per second, resulting in replication of the total 1-mm-long chromosome in about 40 minutes. This, however, does not explain how certain rapidly growing bacteria can divide in 20 minutes. To take advantage of optimal growing conditions, DNA replication is reinitiated before one round of replication is completed. The genomes partitioned into the new daughter cells will already be partially replicated, enabling them to divide again more quickly. A concomitant effect is that in these rapidly dividing cells, genes near *ori*C will be present in more than one copy.

The Details of Replication Cycle

Let's now examine in greater detail the events that establish the replication fork and subsequent replication. The bacterial replication apparatus is quite complex. It has been dissected with the aid of mutations in genes that specify its components and with meticulous biochemical reconstruction of the replicative events using purified enzymes. The major enzymes involved in DNA replication are presented in **Table 13.1**.

Replication of the chromosome in *Escherichia coli* is initiated at *ori*C, a 245-base-pair sequence that is highly conserved in gram-negative bacteria (**Figure 13.4**). There are four 9-base-pair repeats within *ori*C and another 3 repeats of a 13-base A:T rich sequence. The replication process is initiated by the binding of up to 40 molecules of the DnaA protein to the 9-base-pair repeats, causing the separation of the duplex DNA to single strands at the adjacent A:T rich 13-mers. A protein complex, designated DnaB hexamer, binds to the two replication forks with the assistance of a protein designated DnaC. The addition of the DnaB, which forms a ring around the DNA, completes the assembly of the prepriming complex. DnaB has helicase activity and assisted by DNA gyrase promotes unwinding of the DNA duplex. The single-strand binding protein (SSB) tetramers attach

Overview of DNA Replication

DNA replication in bacteria is rapid and continues throughout the cycle of cell division, in contrast to replication of DNA by eukaryotic cells, which takes place during one distinct stage of the cell cycle. The replication of DNA within the circular bacterial chromosome begins at a single specific site on the bacterial chromosome, termed *ori*C for the **origin of replication**. The DNA is unwound, and the double-stranded DNA is separated into two single-stranded templates, and synthesis proceeds from this point in opposing directions. This bidirectional replication mechanism requires the presence of two sites of synthesis of the new DNA. Each of these two sites, termed a replication fork (**Figure 13.3**), comprise a complex of proteins that incorporate complementary nucleotides, accurately copying the template of the parent cell. When the replication forks meet at approximately the opposite side of the circular DNA duplex, each strand has been copied into a com-

Table 13.1	Proteins required for replication of DNA in *Escherichia coli*
Protein	**Function**
DNA gyrase and helicase	Unwinding of DNA
SSB	Single-stranded DNA binding
DnaA	Initiation factor
DnaB	DNA unwinding
HU	Histone-like (DNA binding)
PriA	Primosome assembly, 3′ → 5′ helicase
PriB	Primosome assembly
PriC	DNA unwinding, 5′ → 3′ helicase
DnaC	DnaB chaperone
DnaT	Assists DnaC in the delivery of DnaB
Primase	Synthesis of RNA primers
DNA polymerase III	Strand elongation during DNA synthesis
DNA polymerase I	Excises RNA primers, fills in with DNA
DNA ligase	Covalently links Okazaki fragments
Ter	Termination of replication

to the single strands as they arise. Unwinding exposes the base sequences in the single strands to serve as templates in replication.

Further additions to the prepriming complex generate a replisome assembly that can replicate the genome directionally (**Figure 13.5**). The key addition is the enzyme primase, which synthesizes the RNA primer necessary for DNA synthesis and elongation of the DNA strand.

DNA polymerase is the enzyme that polymerizes deoxynucleoside triphosphate into DNA. Bacteria possess three distinct DNA polymerases, DNA pol I-III. The enzyme termed DNA pol III carries out the elongation stage of DNA synthesis as the replication fork moves forward (**Figure 13.6**). DNA pol I in concert with DNA ligase plays a crucial role in filling in gaps and removing RNA primers, annealing short gaps, and in DNA repair. The role of DNA pol II in replication of the chromosome is not known, and mutations in the gene encoding this enzyme have no effect on bacterial cell division.

DNA pol III can add only nucleotides to a preexisting 3′-hydroxyl; consequently the template must be "primed" with a nucleotide from which the polymerase can start. For this purpose, a short RNA primer, complementary to the exposed template, is synthesized by an RNA polymerase, called primase. This short RNA segment provides the necessary 3′-hydroxyl end to which single nucleotides are added. The polymerization reaction takes place in the 5′-to-3′ direction; that is, the incoming 5′-nucleoside triphosphate is linked by its α-phosphate to the 3′-hydroxyl group of the ribose on the primer. The accuracy in replication is dictated by the

same hydrogen-bonding rules that hold together the nucleoside triphosphates of the original duplex: a guanine (G) is paired with a cytosine (C), and a thymine (T) with an adenine (A).

The fact that DNA pol III can only polymerize in the 5′-to-3′ direction presents a special problem regarding one of the two strands of DNA. The strand depicted above the replication fork in Figure 13.5 contains constantly available primer, and synthesis is continuous in the 5′-to-3′ direction. This so-called leading strand is synthesized as a single molecule complementary to the template. The same 5′-to-3′ direction of synthesis on the opposite DNA template strand, called the lagging strand, however, requires discontinuous synthesis involving several steps. First, RNA primase synthesizes a short, 10-nucleotide RNA primer about 1,000 to 2,000 nucleotides upstream (that is, in the 5′ direction) from *ori*C. This process is analogous, although not identical, to the initiation of DNA synthesis at *ori*C. The 3′-hydroxyl group then serves as a primer for polymerization of DNA from the primer back toward the origin of replication. The DNA segments polymerized from such an RNA primer are called Okazaki fragments, named for the researcher who first characterized them. In leapfrog fashion, the primase continues laying primers on the lagging strand upstream of the last Okazaki fragment as DNA synthesis proceeds. The resulting complementary strand on the lagging strand is discontinuous, consisting of a series of RNA primers linked to Okazaki fragments. The primers are removed by DNA pol I via its 5′-to-3′ exonuclease activity (discussed in the section on DNA proofreading, p. 260), and the gaps formerly occupied by RNA are filled in by polymerization of deoxyribonucleoside triphosphates into DNA. The final stage of synthesis is sealing the nicks between fragments by another enzyme, DNA ligase.

The replication of the circular bacterial chromosome terminates when two replication forks meet. Being helical in nature, however, the DNA undergoes rotation to accommodate the moving replicative fork. This leads to excessive supercoiling and tension on the DNA molecule. As mentioned earlier, partial relief of this tight winding generated during synthesis, as well as the eventual untangling of the daughter strands, is directed by DNA topoisomerases. Two types of this enzyme have been isolated from bacteria. DNA topoisomerase I is an enzyme that creates a nick on one of the DNA strands, allowing free rotation around the phosphodiester bond opposite the nick, thus relieving the tension. This break persists only transiently and is rapidly closed. DNA

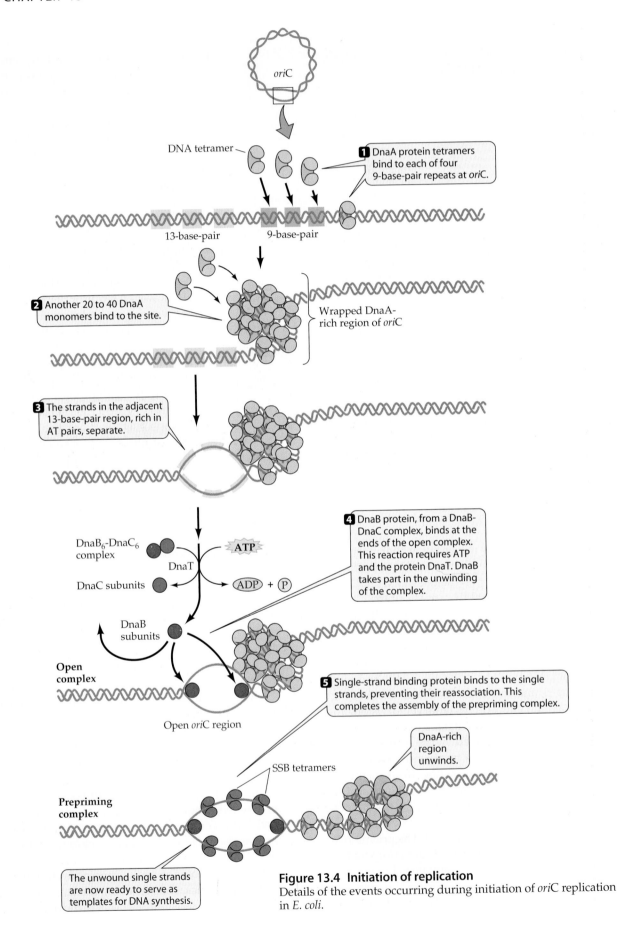

oriC

DNA tetramer

1 DnaA protein tetramers bind to each of four 9-base-pair repeats at *ori*C.

13-base-pair   9-base-pair

2 Another 20 to 40 DnaA monomers bind to the site.

Wrapped DnaA-rich region of *ori*C

3 The strands in the adjacent 13-base-pair region, rich in AT pairs, separate.

DnaB$_6$-DnaC$_6$ complex

DnaC subunits

ATP

DnaT

ADP + P

4 DnaB protein, from a DnaB-DnaC complex, binds at the ends of the open complex. This reaction requires ATP and the protein DnaT. DnaB takes part in the unwinding of the complex.

DnaB subunits

Open complex

Open *ori*C region

5 Single-strand binding protein binds to the single strands, preventing their reassociation. This completes the assembly of the prepriming complex.

DnaA-rich region unwinds.

SSB tetramers

Prepriming complex

The unwound single strands are now ready to serve as templates for DNA synthesis.

Figure 13.4 Initiation of replication
Details of the events occurring during initiation of *ori*C replication in *E. coli*.

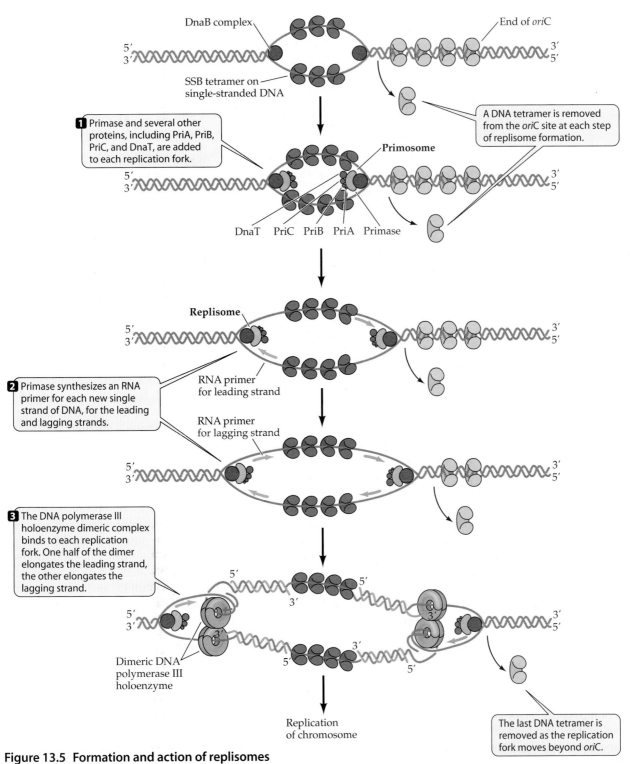

DnaB complex

End of *ori*C

SSB tetramer on single-stranded DNA

1 Primase and several other proteins, including PriA, PriB, PriC, and DnaT, are added to each replication fork.

A DNA tetramer is removed from the *ori*C site at each step of replisome formation.

Primosome

DnaT   PriC   PriB   PriA   Primase

Replisome

RNA primer for leading strand

2 Primase synthesizes an RNA primer for each new single strand of DNA, for the leading and lagging strands.

RNA primer for lagging strand

3 The DNA polymerase III holoenzyme dimeric complex binds to each replication fork. One half of the dimer elongates the leading strand, the other elongates the lagging strand.

Dimeric DNA polymerase III holoenzyme

Replication of chromosome

The last DNA tetramer is removed as the replication fork moves beyond *ori*C.

Figure 13.5 Formation and action of replisomes
A replisome assembly forms at each fork in the prepriming complex. Steps 1 through 3 occur at each replication fork. When the forks meet at the opposite side of the chromosome, replication is complete.

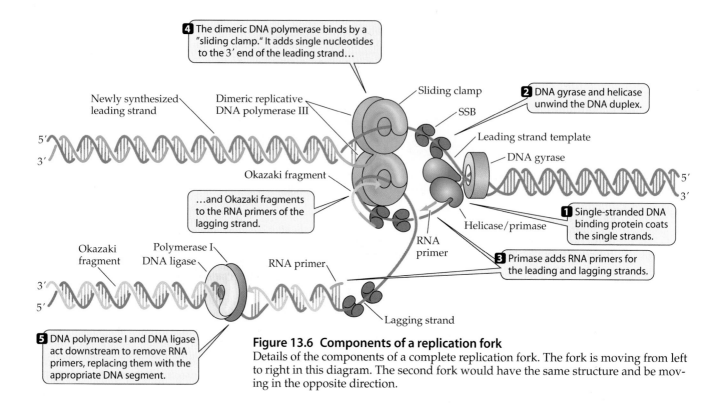

4 The dimeric DNA polymerase binds by a "sliding clamp." It adds single nucleotides to the 3′ end of the leading strand…

Newly synthesized leading strand

Dimeric replicative DNA polymerase III

Sliding clamp

2 DNA gyrase and helicase unwind the DNA duplex.

SSB

Leading strand template

DNA gyrase

5′
3′

Okazaki fragment

…and Okazaki fragments to the RNA primers of the lagging strand.

RNA primer

Helicase/primase

5′
3′

1 Single-stranded DNA binding protein coats the single strands.

3 Primase adds RNA primers for the leading and lagging strands.

Okazaki fragment

Polymerase I
DNA ligase

RNA primer

3′
5′

Lagging strand

5 DNA polymerase I and DNA ligase act downstream to remove RNA primers, replacing them with the appropriate DNA segment.

Figure 13.6 Components of a replication fork
Details of the components of a complete replication fork. The fork is moving from left to right in this diagram. The second fork would have the same structure and be moving in the opposite direction.

topoisomerase II (also called gyrase) causes double-stranded breaks, allowing two helices that cross each other to pass through the break on one of the strands before the break is resealed. The reaction catalyzed by DNA topoisomerase II is especially important in freeing the concatamer (two interlocked circles) on completion of replication, which allows free separation of each chromosome into each daughter cell during cell division.

The Mechanism of DNA Replication Proofreading

The specificity of base pairing (A:T, G:C) is the major mechanism by which reliable copying of the parent strand is guaranteed. Errors during replication take place when an incorrect base is added to the growing chain, creating a transient mismatch. When these strands separate and are partitioned into daughter cells, the chromosome of the daughter receiving the mismatched base will be different from the parent's. This change could be deleterious to the daughter cell if the change is in an important gene. The process of DNA replication is, however, remarkably efficient and has a very low error rate. It is estimated that in *E. coli*, the frequency of replication error is less than 10^{-9} per replicated base pair. This low frequency of mismatched bases is due to the proofreading and repair activities associated with DNA polymerases. One system for corrections is the mismatch repair system present in *E. coli* that monitors newly replicated DNA for mispaired bases, re-

moves the mismatched base, and replaces it with the correct one. Repair is by a DNA polymerase mediated local replication. To replace a mismatched base, the enzyme must identify which base in the pair should be replaced. A mismatch repair mechanism that can identify and remove an incorrect base is the methyl-directed mismatch repair system consisting of MutS, MutH, and MutL proteins. The activity of this repair system is shown in **Figure 13.7**. The parental strand in a duplex is methylated, whereas the newly formed strand is not. Methylation of bases occurs after replication. One of the components of this repair system, the MutS protein, binds to a mismatched base pair, whereas MutH continues along the DNA duplex until it encounters a methylated base. The third protein, MutL, binds simultaneously to both MutS and MutH. This bridging results in "nicking" of the unmethylated, newly synthesized strand by MutH. The portion of the DNA strand between MutS and MutH, including the wrong base that caused the mismatch, is removed. The final stage of repair process is the resynthesis of the repaired strand.

All three bacterial polymerases are capable of both 5′-to-3′ polymerizing and 3′-to-5′ exonucleolytic activity. The immediate repair of mismatched bases formed during replication is carried out by the 3′-to-5′ exonuclease activity of DNA polymerase III, the major polymerizing enzyme in the replicating fork. Filling in small gaps and repairing damaged DNA is accomplished by continuous degradation and synthesis in the 5′-to-3′ direction,

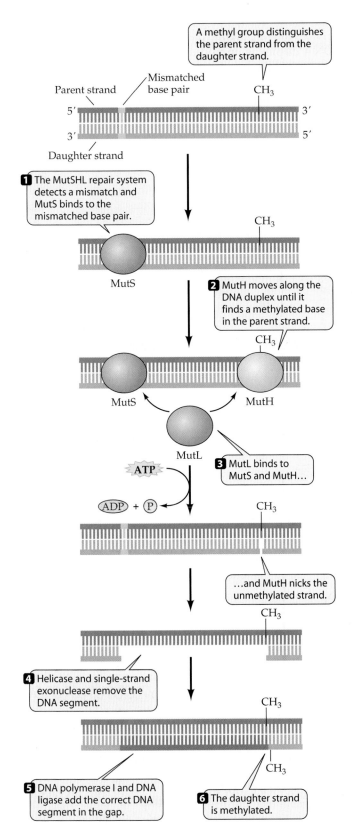

Figure 13.7 Mismatch repair system
Errors encountered during DNA synthesis are repaired by the methyl-directed mismatch repair system, MutSHL.

Image labels:

A methyl group distinguishes the parent strand from the daughter strand.

Parent strand · Mismatched base pair · CH₃
5′ · 3′
3′ · 5′
Daughter strand

1 The MutSHL repair system detects a mismatch and MutS binds to the mismatched base pair.

CH₃

MutS

2 MutH moves along the DNA duplex until it finds a methylated base in the parent strand.

CH₃

MutS · MutH

MutL

ATP

ADP + P

CH₃

3 MutL binds to MutS and MutH…

…and MutH nicks the unmethylated strand.

CH₃

CH₃

4 Helicase and single-strand exonuclease remove the DNA segment.

CH₃

5 DNA polymerase I and DNA ligase add the correct DNA segment in the gap.

CH₃

6 The daughter strand is methylated.

carried out by DNA polymerase I. Additional enzymes are involved in the repair of DNA damaged by physical or chemical agents; these types of repair are discussed later in this chapter.

TRANSCRIPTION AND ITS REGULATION

There are two occasions when the cell copies its genes. In the first, replication, the entire genome is copied from DNA to DNA for the purpose of continuing the species. In the second, transcription, genes are copied from DNA to RNA for the purpose of expressing those genes, that is, to make proteins and RNA needed for sustaining the life of the individual cell.

The genetic information encoded in the DNA almost always specifies proteins. The intermediary in the transfer of genetic information from DNA to proteins is a messenger molecule of ribonucleic acid (mRNA). Because the process of synthesizing RNA from DNA is called transcription, molecules of RNA are sometimes referred to as transcripts. Separate classes of RNA, not translated into protein, make up the protein synthesis machinery; these are ribosomal and transfer RNA (rRNA and tRNA). All types of RNA are synthesized by an identical enzymatic reaction. One significant difference between the types of transcripts is that whereas rRNA and tRNA are produced from larger RNA precursor molecules by enzymatic processing, bacterial mRNA is not processed after transcription. (This difference is not found in eukaryotes, where all types of RNAs are extensively processed, and the final form of the mRNA is the result of cutting and splicing from a large precursor molecule.)

Bacteria mRNA is rather labile, the average half-life being only 1.5 to 2 minutes, unlike eukaryotic mRNA, which can persist and function for hours. Short-lived mRNA enables bacteria not only to respond rapidly to environmental changes but also to conserve energy. Within minutes of a new signal from the environment, bacteria can stop putting their resources into transcription of genes that are no longer needed. Nucleotides from the degraded RNA become available for new transcripts. More important, the extremely energy-costly process of protein synthesis is not wasted on unnecessary proteins.

Steps in RNA Synthesis

The process of RNA synthesis is a chemical polymerization, which in principle resembles the copying of one strand of DNA during replication: initiation, elongation, and termination. Transcription, however, involves copying only a short segment of the DNA, and the product is built of ribonucleotides rather than deoxyribonucleotides. The steps in initiation of transcription are shown in **Figure 13.8**.

The enzyme RNA polymerase carries out the synthesis of RNA. Using a DNA template, it recognizes and binds to a specific sequence, which indicates the beginning of a gene. This site is termed the promoter; the transcription initiates 10 to 15 base pairs (bp) downstream from the middle of the promoter sequence. Because only one strand (called the sense strand) of the double-stranded helix is copied, it is necessary for the RNA polymerase to displace the complementary copy as it moves along incorporating free nucleoside triphosphates into a polymerized RNA strand. However, the DNA duplex is immediately re-formed following the passage of the transcriptional apparatus.

One method used most frequently to control gene expression is to regulate how often transcription is initiated. Regulation of RNA polymerase binding allows

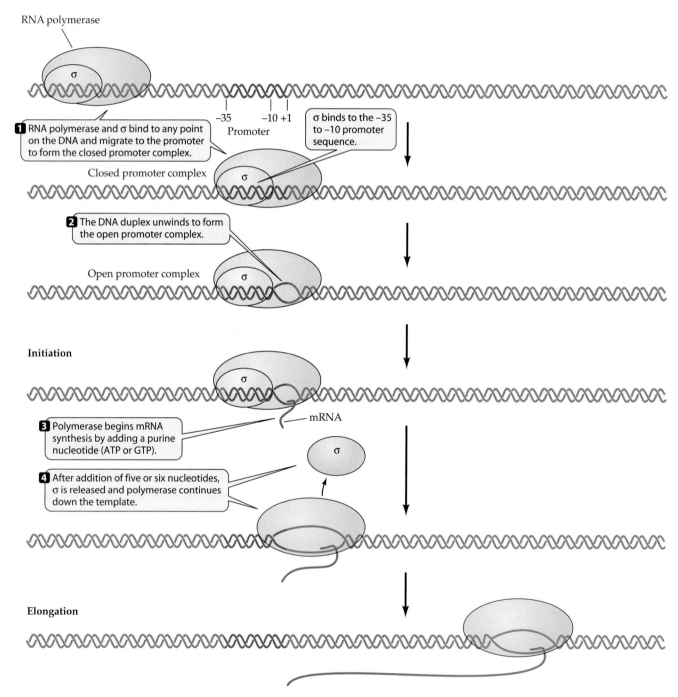

Figure 13.8 Initiation and elongation steps of transcription
The sequence of events in initiation and elongation during transcription.

bacteria to exert control over an entire cluster of genes specifying related functions, utilizing a common regulatory mechanism. A promoter serves as a start for transcription of a single gene, or that of an operon resulting in polycistronic mRNA. Because bacteria do not have a nucleus as eukaryotic cells do, the nascent transcript can be translated into proteins from its free end while it is yet being synthesized. In further contrast to eukaryotic transcription, bacteria frequently synthesize polycistronic mRNA; that is, mRNA that specifies more than one gene. Bacterial genes are frequently arranged in operons, sets of adjacent genes that are transcribed and translated together. Genes in an operon utilize a single promoter, preceding the first gene, and transcription terminates after the last gene. Therefore, the genes within an operon are expressed together, and control of their expression involves usually the regulatory sites near the single promoter (see following). Operons generally code for enzymes that function together in a common biosynthetic or catabolic pathway. This enables the bacterial cell to coordinately regulate expression of related genes, increasing its efficiency.

The bacterial RNA polymerase is a multisubunit enzyme, consisting of five subunits: two Alpha (α), one Beta (β), one Beta' (β'), and one sigma (σ). The subunits are assembled into a multimeric structure (called the holoenzyme), which binds to the promoter region. It is the sigma subunit that recognizes the promoter—the beginning of the region to be transcribed. The subsequent steps in the initiation of transcription are melting of the double-stranded DNA (i.e., separating the two strands) and polymerizing the first few nucleotides, after which the sigma factor is not needed. The sigma subunit then dissociates from the holoenzyme, leaving four subunits (2α, β, and β') as a complex termed the core enzyme that moves along the DNA strand and polymerizes nucleotides into RNA.

Signals built into the DNA allow the RNA polymerase to identify the end of the gene or genes it is transcribing. If RNA polymerase is transcribing an operon, the termination signal is present at the end of the operon. One type of termination sequence is identified as an inverted repeat followed by a stretch of polyuridines. As an RNA copy is made of this section, same-strand pairing of the inverted repeats forms a stem-loop structure that causes the RNA polymerase to be released from the DNA template, terminating transcription.

Another mechanism of termination of transcription involves both a signal on the template that causes RNA polymerase to pause and a specific termination protein called rho (ρ).

Rho factor is an ATP-dependent helicase that promotes unwinding of RNA:DNA hybrid duplexes. ρ binds to C-rich regions in the RNA transcript and advances in the 5'-3' direction until it reaches the RNA polymerase. At the site where the RNA polymerase stalls at a GC-rich termination site, the ρ factor unwinds the transcript from the template, and the mRNA is released.

The difference between the rho-dependent and rho-independent termination mechanism is the type of sequence information encoded at the termination site (i.e., whether it leads to pausing of the RNA polymerase, or whether it leads to an RNA stem-loop). Following release of the RNA polymerase, the core enzyme associates with a sigma subunit present in the cytoplasm, forming the holoenzyme capable of recognizing another promoter and initiating transcription of another gene.

Promoters and Sigma Factors

The key step in transcription is the recognition of the promoter by the sigma subunit of the holoenzyme. This interaction determines which gene is to be expressed. In bacteria there exist several types of sigma subunits that can form complexes with the RNA polymerase core enzyme. Each species of sigma subunit recognizes different promoter sequences, guiding the polymerase to specific genes.

σ^{70} (named for its molecular weight of about 70,000 daltons in E. coli) is sometimes referred to as σ^A, and it recognizes the majority of genes encoding essential functions in the bacterial cell. The promoter sequences of these genes are relatively conserved, although they are not identical. These promoters often have two regions of particular similarity: a sequence TATAAT approximately 10 base pairs upstream from the site of initiation of transcription and a sequence TTGACA, 35 base pairs upstream, termed the 35 region. (The first base pair copied into RNA is referred to as +1; hence, the TATAAT sequence is at position –10, TGACAT is –35.) The TATAAT is termed a Pribnow box and is named for David Pribnow, the first to recognize the importance of this sequence in transcription. These sequences were deduced by examination of a large number of different bacterial promoters. They represent what is called a consensus promoter for recognition by σ^{70}.

Bacteria contain alternative sigma factors, which also form complexes with the core RNA polymerase and lead to transcription of genes that contain their cognate promoter sequences. Some of these are summarized in **Table 13.2.** The genes requiring alternative sigma factors for transcription are involved in cellular responses to a variety of specialized conditions, including survival at elevated temperature, synthesis of genes for nitrogen metabolism, motility, and sporulation. The alternative sigma factors are a minority and do not effectively compete with σ^{70} under normal conditions.

One of the important features of all promoters, including those that are recognized by the alternative σ

Table 13.2 | **Some of the alternative sigma (σ) factors in _E. coli_ and their cognate promoter sequences**

σ Factor	Promoter Recognized	Genes Transcribed
σ^{70}	TTGACA-17bp-TATAAT	Many and diverse
σ^{32}	CNCTTGAA-14bp-CCCCATNT	Heat shock response
σ^{54}	CTGGNA-7bp-TTGCA	Many and diverse
σ^{28}	TAA-15bp-GCCGATAA	Chemotaxis, motility, flagellar components

factors, is that they contain specific sequences relative to +1. Therefore, promoters give RNA polymerase a direction for transcription; once RNA polymerase binds to a promoter, it will move unidirectionally, as determined by the relative position of the bases in the promoter (e.g., in the case of σ^{70} by the –10 and –35 sequences). Occasionally, two genes are transcribed from promoters that overlap, and the RNA is copied from opposite strands, but the position of the –10 and –35 sequence (or similar sequences for alternative σ factors) assure movement of the RNA polymerase in the correct direction for each gene.

The term "promoter strength" is used to describe the efficiency of transcriptional initiation from a given promoter. High frequency of holoenzyme binding to a promoter results in high levels of RNA synthesis, and consequently high levels of protein synthesized from a specific gene. A perfect match to the consensus sequence at both –10 and –35 would be expected to result in the highest level of gene expression. However, agreement of a specific promoter sequence is frequently not the sole determinant of the amount of a specific gene product made at any one time. In the next section, we will see that there are many ways of modulating gene expression.

REGULATION OF GENE EXPRESSION

Bacteria have the capacity to adapt to specific environmental conditions, and they do so by altering the levels of mRNA available for translation. From the point of view of energetics, this is most efficiently accomplished by controlling the initiation of transcription, to prevent the unnecessary use of nucleotide triphosphates and consequent energy-costly synthesis of protein. Although there are many genes expressed constitutively (i.e., without any control of the levels of mRNA other than the strength of the promoter), there are also a large number of genes for which the ability of RNA polymerase to initiate transcription is controlled by other proteins that respond to signals from the environment of the bacterial cell.

Control at the level of transcriptional initiation can be divided into two classes, based on whether the envi-

ronmental signals _facilitate_ transcriptional initiation (positive regulation) or _interfere_ with it (negative regulation). In certain instances, a gene or an operon can be regulated negatively under one set of conditions and positively under another. Moreover, both forms of regulation can involve small signaling molecules that regulate multiple operons scattered throughout the bacterial chromosome.

Negative Regulation

There are several modes employed to decrease the capacity of RNA polymerase to transcribe a specific gene or operon. The most common mechanism involves a regulatory protein called a repressor. Repressors usually bind to specific sites, termed operators, near the promoter region. The operator sequence is usually located between the promoter and +1 position on the DNA, and frequently overlaps the –10 region of the promoters. The binding of the repressor therefore prevents transcription by physically blocking the polymerase from either binding its operator or, if bound, from moving forward.

One of the best-studied examples of negatively regulated genes is the lactose utilization operon (_lac_) that controls catabolism of lactose by _E. coli_. The regulatory mechanism is shown in **Figure 13.9**. The genes of the _lac_ operon are expressed only when bacteria are growing in the presence of lactose as sole carbon source. The operon includes three genes: _lacZ_ (encoding the enzyme β-galactosidase), _lacY_ (lactose permease), and _lacA_ (transacetylase). The negative regulatory aspect of the _lac_ operon comes from the ability of a repressor protein, called the lactose repressor (product of _lacI_), to tightly bind to the operator site near the promoter from which the lactose operon is transcribed, preventing binding of RNA polymerase. In the absence of lactose, none of the three genes necessary for lactose catabolism are transcribed. When _E. coli_ finds itself in a medium where lactose is the only carbon and energy source, small amounts of lactose enter the cell and bind to the repressor protein. The interaction of lactose with the repressor prevents it from binding to the operator, and those repressor molecules that are already bound dissociate from DNA. The RNA polymerase now can bind to the promoter and initiate the transcription of the _lac_ operon. As long as there is lactose present in the bacterial environment (and hence inside the bacterial cell), the repressor is unable to bind to the operator and cannot interfere with transcription. It is precisely these conditions that necessitate the production of enzymes that are involved in lactose utilization. Interestingly, other synthetic molecules that resemble lactose, but are not metabolized by _E. coli_, can cause dissociation of the repressor from the operator. These small molecules and similar signaling molecules that interfere with the activ-

Lactose absent

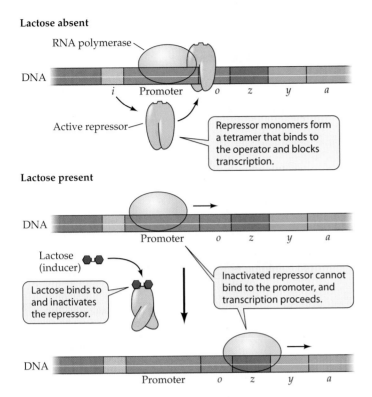

RNA polymerase

DNA

i   Promoter   *o*   *z*   *y*   *a*

Active repressor

Repressor monomers form a tetramer that binds to the operator and blocks transcription.

Lactose present

DNA

Promoter   *o*   *z*   *y*   *a*

Lactose (inducer)

Lactose binds to and inactivates the repressor.

Inactivated repressor cannot bind to the promoter, and transcription proceeds.

DNA

Promoter   *o*   *z*   *y*   *a*

Figure 13.9 Regulation of the *lac* operon
Regulation of gene expression in the *lac* operon. In the absence of lactose, repressor blocks the operator; when lactose is the sole substrate, it inactivates the repressor and the genes are transcribed.

ities of repressor proteins are called inducers. The common mechanism of all inducers is that they bind to the repressor, causing it to lose its ability to interact with the operator sequences.

Although all negatively regulated systems rely on the ability of repressor proteins to block transcriptional initiation, the precise interaction with small signaling molecules varies. The inducer (lactose or its structural analogs) causes dissociation of the repressor from the operator. Similarly, a regulatory protein (AraC), which controls genes of arabinose metabolism, acts as a true repressor, binding to a specific operator sequence and preventing the initiation of transcription in the absence of arabinose. In the presence of the inducer arabinose, the AraC-arabinose complex dissociates from the repressor, analogous to the behavior of the lactose repressor-lactose complex. The situation with the AraC protein is a little more complex, because the AraC-arabinose complex also participates in a positive regulatory mechanism (see following).

A slightly different mechanism of negative regulation involves interaction of a signaling molecule with a free, unbound repressor protein, in which this complex interacts with the operators. One of the most extensively studied systems of this type is the regulatory mecha-

nism involved in transcription of the tryptophan biosynthetic genes (**Figure 13.10**). The inactive "empty" repressor (called the aporepressor), which is not complexed with tryptophan, is unable to bind to the operator of the *trp* operon, thus tryptophan biosynthetic genes are transcribed, and bacteria synthesize their own tryptophan. In the presence of tryptophan in the growth medium, there is no need to synthesize the enzymes of the tryptophan biosynthetic pathway because external tryptophan is taken up and utilized; thus, the *trp* operon is not transcribed. The shutoff of the *trp* operon is due to formation of an active repressor, following its binding of tryptophan. In this system, the signal molecule is called a co-repressor, as it actively participates with the repressor in the negative regulation of transcriptional initiation.

Positive Regulation

Positive regulation involves direct or indirect action of signaling molecules with activator proteins, stimulating transcription. Although the mechanism of negative repression is obvious (i.e., interference with binding of RNA polymerase), the molecular basis of activation is less clear. The activator proteins interact with sequences usually upstream of the promoter, and these regulatory sites can be several hundred base pairs from the affected gene. The interaction of the activator protein with its cognate regulatory sequences results in a direct contact between the regulatory protein and RNA polymerase bound to the promoter. If the binding site for the regulatory protein is far from the promoter, this requires extensive looping of the DNA between the two sequences, to allow for the regulatory protein to contact RNA polymerase. Some activator proteins can recognize sequences that are found near or within promoters. In case of the promoters of positively regulated genes, which are recognized by the σ^{70} RNA polymerase, they have –10 regions that are well conserved with other promoters, but there is a considerable variation in the sequence at their –35 regions. The most plausible explanation for the stimulatory effect of the proteins involved in the positive regulation of gene expression is that they stabilize the RNA polymerase complex with the promoter sequence. Alternatively, they may help RNA polymerase in melting of the DNA duplex into single-stranded regions during the critical incorporation of the first few nucleotides at the initiation of transcription.

One of the most studied examples of positively regulated genes is the maltose regulon. The term "regulon" denotes a set of operons under the control of a common regulatory element. In the case of the maltose regulon, the positive regulatory protein MalT stimulates transcription of at least four different operons involved in uptake and metabolism of maltose or related sugars

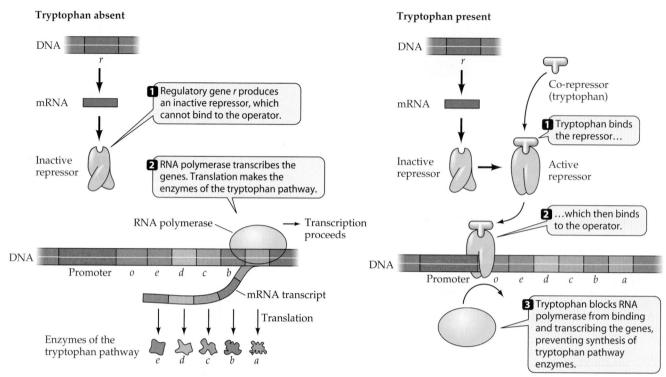

Figure 13.10 Regulation of the *trp* operon
In the absence of tryptophan in the growth medium, the genes are read constitutively and the enzymes required for formation of tryptophan are synthesized. When tryptophan is present in the medium, the promoter is blocked and transcription halted.

(**Figure 13.11**). The MalT gene product (Ti) becomes active (i.e., capable of binding to the DNA recognition sequences after complexing with maltose). All promoters of the maltose regulon contain a so-called maltose box, which consists of the sequence GGAG/TGA within the –35 region.

Several negatively regulated genes are positively controlled as well, whereas some genes are regulated only by activator proteins. The *lac* operon, for example, is regulated by the levels of cyclic AMP in the bacterial cell, through the activity of the catabolite activator protein (CAP, also called CRP), as shown in **Figure 13.12**. In this case, cAMP is a co-inducer, because it binds directly to the regulatory protein. In the presence of elevated levels of cyclic AMP (which takes place when preferred carbohydrate sources are depleted—another condition that

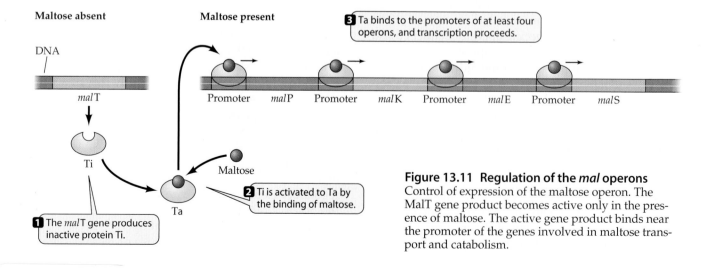

Figure 13.11 Regulation of the *mal* operons
Control of expression of the maltose operon. The MalT gene product becomes active only in the presence of maltose. The active gene product binds near the promoter of the genes involved in maltose transport and catabolism.

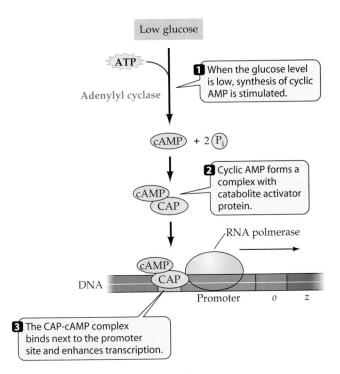

Figure 13.12 Catabolite activation
Catabolite activation occurs when products of catabolism regulate the rate at which genes are transcribed, as in the *lac* operon. Low glucose levels enhance transcription of the *lac* genes.

allows *E. coli* to take full advantage of lactose metabolism), the cAMP-CAP complex binds to a site adjacent to the promoter, and if the operator site is unoccupied by the repressor, the transcription of the *lac* operon is enhanced severalfold above that which would result in the presence of the inducer alone.

There are several cases of positive control of transcription, where the environmental signal is sensed by activator proteins through a covalent modification, which converts it into a form capable of binding to regulatory sequences and helps RNA polymerase in initiating transcription. The typical mechanism of conversion of an inactive regulatory protein into its active form involves a two-step process starting with autophosphorylation of a protein kinase, the sensor. The reaction involves a signal-induced conformational change in the sensor, which activates its latent self-modifying enzymatic activity (hence the term *autophosphorylation*), using ATP as a donor of the phosphate moiety. This is followed by transfer of the phosphate to the response regulator. The phosphorylated regulatory protein then binds to a specific regulatory sequence, which is located on the 5′ (upstream) of the gene and facilitates transcription of a promoter from an adjacent site. This mechanism of sensing the environmental signal and converting it to transcriptional activation of specific

genes always involves a pair of components, the protein autokinase and the regulatory protein, and is called the two-component regulatory system. The phosphorylation of the response regulator is reversible, and specific enzymes (phosphatases) can remove the phosphate from the regulatory proteins. In certain instances, the sensor kinase has the ability to carry out the removal of the phosphate, thus performing dual and opposite reactions, depending on the availability of the environmental signal.

The mechanism of sensing high osmolarity by the EnvZ (sensor) OmpR (response regulator) of *E. coli* is outlined in **Figure 13.13**. Similar phosphorylation mechanisms have also been shown to function in controlling selective expression of genes that respond to nitrogen limitation in enteric bacteria, nitrogen fixation in *Rhizobia*, signaling for transcription of the apparatus necessary for DNA transfer from *Agrobacterium* to plants, and biosynthesis of pili and flagella in an assortment of bacteria. Although the vast majority of the two-component regulators function in transcriptional activation, in some regulated systems, binding of phosphorylated response regulators leads to interference with RNA polymerase binding and a negative regulatory effect.

PROTEIN SYNTHESIS

Approximately 2,000 to 3,000 different polypeptides (the biochemical term for proteins) are present in a bacterial

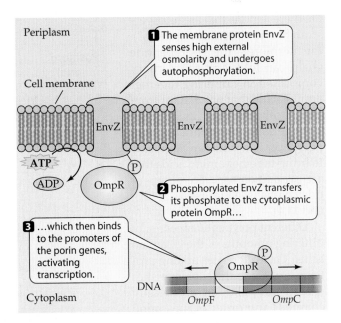

Figure 13.13 Osmoregulation of gene expression
Osmoregulation of porin gene expression in *E. coli* is controlled by a two-component regulatory system consisting of EnvZ, which senses external osmolarity, and OmpR, a transcriptional activator.

cell; the vast majority of these are the cellular structural components or enzymes. Chemically, proteins are relatively simple molecules, consisting of linearly linked amino acids. The order of the assembled amino acids—the total number and nature—distinguishes one protein from another. There are 20 different amino acids that can be used as building blocks. Amino acids are similar to one another, each consisting of a central carbon (called the α-carbon) with an amino group, a carboxyl group, a hydrogen, and a side group. Of the 20 amino acids, 19 differ only in the structure of the side groups. The twenty-first amino acid, proline, differs from the others by having its side group linked to the α-carbon, resulting in a ring structure. The structures of the 20 amino acids present in proteins were presented in Figure 3.15.

Each amino acid is linked to its adjacent amino acid by a peptide bond (see Figure 3.16), formed when the primary amino group of one amino acid condenses with the carboxyl group of the adjacent one. Short chains of bonded amino acids are therefore called peptides. Because of the directional bond, one end of the peptide is the amino terminus and the other is the carboxyl terminus. The sequence of linked amino acids makes up the primary structure of a polypeptide. Proteins, however, are rarely found in these extended configurations, but rather assume additional conformations. The secondary structure of a polypeptide refers to the shape of the backbone of each α-carbon and the positioning of the side groups (i.e., Does the backbone have a spiral shape with the side chains pointing outward, or are there regions that are relatively straight with intermittent turns?). Furthermore, the polypeptide chain can fold, and this tertiary conformation can bring together amino acids that are distant from one another in the linear polypeptide. Moreover, independent polypeptide chains can aggregate to bring together amino acids that are present on separate polypeptide chains (see Figures 11.2, 11.3, and 11.4).

Different polypeptides consist of differing amounts of the 20 amino acids. The differences among proteins are due to the characteristics (polar, nonpolar, or electrically charged) and order of the amino acids in their primary structure. The distribution of the specific amino acids along the polypeptide chain is determined by the specific order of the nucleotides in the gene

coding that polypeptide. The genetic information is first transcribed from DNA into linear RNA called messenger RNA (mRNA), as it serves as the messenger of genetic information between the genes and the protein synthesizing apparatus. Distinct triplet nucleotide sequences in mRNA, called codons, encode specific amino acids. The order of the codons, from a fixed point, determines the order of the amino acids in the polypeptide chain. Hence, the process of protein synthesis serves two main functions simultaneously: (1) it chemically catalyzes the linking of amino acids into a protein, and (2) it decodes the triplet codons on mRNA and converts this information to the primary structure of a protein. The process of protein synthesis is therefore called translation. One of the major breakthroughs in biological science took place in the early 1960s, when the genetic code was determined and the specific assignment of triplet nucleotides to each amino acid was made. There are 61 codons specifying all 20 amino acids that are components of proteins. Codon usage is relatively redundant. For example, any one of six different triplets (UUA, UUG, CUU, CUC, CUA, or CUG) in mRNA can specify leucine in the polypeptide chain. However, two amino acids have only one codon each: methionine (AUG), and tryptophan (UGG) (see **Table 13.3**).

In addition to codons specifying amino acids, several signaling codons determine the beginning and the end of polypeptide chains. Initiation codons signal the

| Table 13.3 | **The genetic code: the codons specify the amino acids incorporated into protein** |

First Position (5'-end)	Second Position				Third Position (3'-end)
	U	C	A	G	
U	UUU Phe	UCU Ser	UAU Tyr	UGU Cys	U
	UUC Phe	UCC Ser	UAC Tyr	UGC Cys	C
	UUA Leu	UCA Ser	UAA Stop	UGA Stop	A
	UUG Leu	UCG Ser	UAG Stop	UGG Trp	G
C	CUU Leu	CCU Pro	CAU His	CGU Arg	C
	CUC Leu	CCC Pro	CAC His	CGC Arg	C
	CUA Leu	CCA Pro	CAA Gln	CGA Arg	A
	CUG Leu	CCG Pro	CAG Gln	CGG Arg	G
A	AUU Ile	ACU Thr	AAU Asn	AGU Ser	U
	AUC Ile	ACC Thr	AAC Asn	AGC Ser	C
	AUA Ile	ACA Thr	AAA Lys	AGA Arg	A
	AUG Met[a]	ACG Thr	AAG Lys	AGG Arg	G
G	GUU Val	GCU Ala	GAU Asp	GGU Gly	U
	GUC Val	GCC Ala	GAC Asp	GGC Gly	C
	GUA Val	GCA Ala	GAA Glu	GGA Gly	A
	GUG Val	GCG Ala	GAG Glu	GGG Gly	G

[a]AUG signals translation initiation as well as coding for Met residues.

start (the amino terminus) of the polypeptide chain by coding for a methionine (AUG) or, rarely, valine (GUG). The end of the carboxyl terminus of the polypeptide is determined by termination codons, also called stop codons. Stop codons (UAA, UAG, or UGA) do not specify any amino acids; one or several of these on the mRNA will stop the RNA polymerase from continuing. The sequence of codons between the initiation and termination signals is the reading frame. Any one nucleotide sequence of mRNA contains three possible reading frames, depending on which of three positions is used as the beginning point, as illustrated by the following. The one translated is determined by the location of the initiation codon upstream of this sequence:

mRNA	...GUC	UUU	AUA	ACA	CAC	CCU	
Reading frame #1	Ala	Phe	Ile	Thr	His	Pro	
mRNA	...G	UCU	UUA	UAA	CAC	ACC	CU
Reading frame #2		Ser	Leu	Stop			
mRNA	...GU	CUU	UAU	AAC	ACA	CCC	U
Reading frame #3		Leu	Tyr	Asn	Thr	Pro	

If the desired protein contains the sequence Ala—Phe—Ile—Thr—His—Pro, the initiation codon must be located upstream, in multiples of three and in reading frame #1. If the initiation codon were in frame with either the second or third reading frames, proteins with totally different primary sequences would result. Note that reading frame #2 contains a stop codon, which would result in a short polypeptide chain having Leu as its carboxyl terminus.

Components of the Protein Synthesis Machinery

Protein synthesis takes place on complex particles called ribosomes. Ribosomes, together with a number of RNA molecules and polypeptides, recognize the coded information on mRNA and translate it into a polypeptide. Ribosomes in bacterial cells are relatively large, multisubunit complexes of RNA and proteins. In its functional stage, an individual ribosome consists of two subunits, one large and one small. As mentioned in Chapter 4, the physical properties of ribosomes were originally studied by their behavior during high-speed centrifugation. Hence, the size of a ribosome and

its subunits are in Svedberg units (S), the units of velocity of travel in a centrifugal field. The complete ribosome has a size of 70S, whereas the small subunit is 30S and the large one is 50S. A comparison of the size and the RNA/protein composition of individual subunits of the *E. coli* ribosome is presented in **Table 13.4.** Operons encoding ribosomal RNA genes are present in more than one copy in the bacterial chromosome. This redundancy of genes is rare in prokaryotes and enables bacteria to respond to sudden changes of growth conditions by synthesizing a large quantity of translational machinery.

The most critical components of the translational apparatus are small RNAs that carry specific amino acids and are responsible for the conversion of genetic information into protein sequences. These transfer RNAs (tRNA) are the "translators" of the translation process. The tRNA molecule is about 70 nucleotides long and folds into the configuration shown in **Figure 13.14.** Four extensive regions of complementarity are formed, and three open loops are present. One of these is the anticodon loop, containing a region precisely complementary to the codon sequence on mRNA. For example, the tRNA that is involved in adding tryptophan to the polypeptide chain contains an anticodon sequence ACC, which is complementary to the codon UGG in the mRNA. There are fewer tRNAs than there are possible codons, because incorrect base pairing in the third position of the codon can still result in incorporation of the correct amino acid into the polypeptide. This is referred to as third-base wobble. The tRNA genes, as with genes for rRNA, are redundant, and tRNA is often cut out from a larger precursor RNA molecules.

Table 13.4	**The structural components of *E. coli* ribosomes**		
	Ribosome	**Small Subunit**	**Large Subunit**
Sedimentation coefficient	70S	30S	50S
Mass (kD)	2520	930	1590
Major RNAs		16S = 1542 bases	23S = 2904 bases
Minor RNAs		5S = 120 bases	
RNA mass (kD)	1664	560	1104
RNA proportion	66%	60%	70%
Protein number		21 polypeptides	31 polypeptides
Protein mass (kD)	857	370	487
Protein proportion	34%	40%	30%

From: Garrett and Grisham, *Biochemistry*, 1995. Figure 32.1, page 1041.

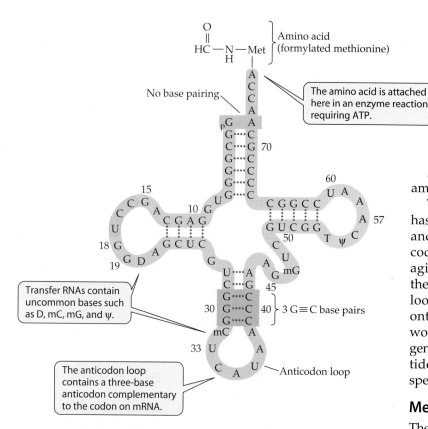

The attachment of the amino acid to tRNA is an enzymatic step involving recognition of a tRNA molecule by the enzyme aminoacyl-tRNA synthetase and attachment of the amino acid (aa) to the end of the tRNA:

$$aa + tRNA + ATP \rightarrow tRNAaa + ADP$$

A tRNA that is attached to its specific amino acid is said to be charged.

The enzyme aminoacyl-tRNA synthetase has a specific recognition for both the tRNA and the amino acid corresponding to the codon. For example, the amino acid asparagine is added only to the tRNA containing the sequence UUA or UUG in the anticodon loop. Errors (placing the wrong amino acid onto a tRNA) are exceedingly rare. They would result in an incorrect translation of the genetic code, thereby synthesizing a polypeptide with different amino acids than those specified by the genetic information.

Mechanism of Protein Synthesis

The process of assembling a polypeptide chain is an orderly reaction consisting of the stepwise transfer of amino acids directed by an mRNA sequence. The overall scheme of protein synthesis is outlined in **Figure 13.15**. The

Figure 13.14 The structure of *E. coli* N-formyl-methionyl-tRNA^fMet The base areas that differ from those in non-initiator tRNAs are highlighted. This is the initiatior of peptide synthesis. Several unusual bases can be also found in tRNAs; tRNA^fMet contains, D, mC, mG, and ψ, which are dihydrouridine, methyl cytidine, methyl guanosine and ribofuroanosyl uracil, respectively.

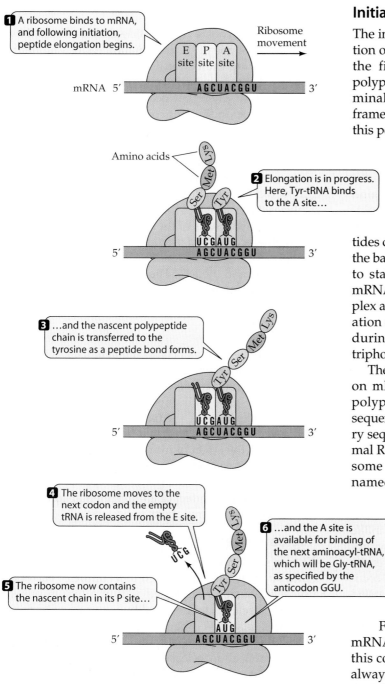

1 A ribosome binds to mRNA, and following initiation, peptide elongation begins.

Ribosome movement

mRNA 5′ E site P site A site AGCUACGGU 3′

Amino acids

2 Elongation is in progress. Here, Tyr-tRNA binds to the A site…

Ser Met Lys
Tyr

UCGAUG
5′ AGCUACGGU 3′

3 …and the nascent polypeptide chain is transferred to the tyrosine as a peptide bond forms.

Tyr Ser Met Lys

UCGAUG
5′ AGCUACGGU 3′

4 The ribosome moves to the next codon and the empty tRNA is released from the E site.

UCG

5 The ribosome now contains the nascent chain in its P site…

6 …and the A site is available for binding of the next aminoacyl-tRNA, which will be Gly-tRNA, as specified by the anticodon GGU.

Tyr Ser Met Lys

AUG
5′ AGCUACGGU 3′

Figure 13.15 Protein synthesis
The basic steps in protein synthesis. A ribosome has three sites that interact with tRNA during protein synthesis: the acceptor (A) site, the peptidyl (P) site, and the exit (E) site. During the continuous synthesis of a polypeptide, the P site is occupied by the tRNA with its nascent polypeptide chain, while the A site contains the next charged aminoacyl-tRNA. After addition of this amino acid, the ribosome moves along one codon and the empty tRNA is released from the E site.

process can be divided into three steps: **initiation**, **elongation**, and **termination**. Each of these steps is discussed here separately.

Initiation

The initiation of protein synthesis involves the interaction of ribosomes with the site on mRNA that encodes the first amino acid (the amino terminus) of the polypeptide chain. The precise placement of the N-terminal amino acid is important because the reading frame of the remaining polypeptide is determined from this point.

A series of soluble factors and the free 50S and 30S subunits are joined with mRNA and the aminoacyl-tRNA to form what is called the initiation complex. Initiation of protein synthesis involves binding the small ribosomal subunit to mRNA with the aid of several polypeptides called initiation factors (IFs). There are three IFs in the bacterial cytoplasm: IF1, IF2, and IF3. IF3 is required to stabilize the interaction of the 30S subunit with mRNA binding. IF1 is also part of the 30S-mRNA complex and assists IF3 although its precise role in the initiation of protein synthesis is not known. IF2 functions during binding of the initiator tRNA and guanine triphosphate GTP to the ribosome.

The association of the 30S subunit with a precise site on mRNA that encodes the first amino acid of the polypeptide is accomplished by pairing a specific sequence, GGAGGU, on the mRNA to a complementary sequence, ACCUCC, on the 3′ end of the 16S ribosomal RNA. This sequence on mRNA, the so-called ribosome binding site (or the Shine-Delgarno sequence, named after its discoverers), is approximately 4 to 6 nucleotides from the initiation codon, AUG. Thus, the interaction of the small subunits not only allows the ribosomes to find the beginning of the protein (even though there may be other AUG codons specifying methionine within a given polypeptide chain) but also determines the reading frame of the polypeptide.

Following binding of the 30S ribosomal subunit to mRNA, the first tRNA, carrying methionine, binds to this complex. In *E. coli*, the first amino acid is almost always N-formylated methionine (fMet), which is generated by first linking methionine to a specialized tRNA by aminoacyl-tRNA synthetase, and subsequently attaching a formyl group to the free amino group. The interaction of fMet-tRNA with ribosome involves interaction with initiation factor 2 (IF2) and formation of a hydrogen-paired region between the anticodon sequence UAC and the complementary codon AUG. Occasionally, initiation involves a mismatch, where the anticodon UAC pairs with GUG codon on mRNA. At the same time, a molecule of GTP binds to a specific site on the 30S subunit. The final stage in the assembly of the initiation complex involves association of the 50S subunit, hydrolysis of GTP to GDP, and release of all of the initiation factors. The ribosome is now ready to move along the mRNA and progressively synthesize a polypeptide chain.

Elongation

Following completion of the assembly of the initiation complex, the ribosome contains two sites that can bind aminoacyl-tRNA. The peptidyl site (or P-site) contains the fMet-tRNA, whereas the adjacent acceptor site (or A-site) is unoccupied, although it does contain the exposed triplet codon of the second amino acid in the reading frame. The formation of the first peptide bond between fMet and the second amino acid of the polypeptide chain takes place following the interaction of the aminoacyl-tRNA charged with the second amino acid and the A-site. This is directed by the correct hydrogen-bonded, double-stranded region formed between the codon and anticodon sequence of tRNA. The binding of aminoacyl-tRNA to the A-site is a stepwise process, involving several elongation factors (EFs) and GTP hydrolysis. The aminoacyl-tRNA first binds to a complex of elongation factor Tu (EF-Tu) and GTP. The EF-Tu places the aminoacyl-tRNA into the A-site, with the concomitant hydrolysis of GTP to GDP. The EF-Tu–GDP complex is released. The regeneration of fresh EF-Tu–GTP complex is necessary for the next step of the elongation process; another elongation factor, EF-Ts, releases GDP from EF-Tu and reloads it with fresh GTP.

The formation of the actual peptide bond between the two amino acids within the ribosomes is accomplished by the peptidyl transferase activity of the ribosome. This reaction is also unusual because it is catalyzed by the RNA component of the ribosome, unlike most biochemical reactions, which are catalyzed by proteins. The bond formed between the amino acids results in transfer of fMet to the second aminoacyl-tRNA occupying the A-site. The final step in elongation is called translocation, in which the ribosome moves three nucleotides and moves the free tRNA to a region of the ribosome called the exit (E)-site, from which it is immediately discharged. The translocation reaction is catalyzed by another elongation factor (EF-G) and requires hydrolysis of GTP. The translocation leaves the A-site open for the next incoming aminoacyl-tRNA, specified by the third codon of the mRNA. The subsequent elongation steps proceed in a manner entirely identical to the reaction between fMet and the second amino acid. In this process, all ribosomal elongation factors are reused, and the elongation of the polypeptide requires energy in the form of GTP, one to charge EF-Tu and one to fuel the EF-G–mediated translocation. Together with the requirement of ATP to charge a tRNA, the elongation stage of polypeptide chain formation is an energy-demanding process.

Termination

The termination of the polypeptide chain is directed by the location of one of the termination codons immediately following the codon for the carboxyl terminal amino acid. The termination codons do not code for any amino acid, so when they occupy the A-site, no tRNA has the corresponding anticodon in its loop structure. Termination codons are recognized by release factors (RFs) that bind to the A-site. The termination codons UAA and UAG are recognized by RF1, whereas RF2 recognizes UAA or UGA. The third release factor, RF3, complexed with GTP, then binds to the ribosome and catalyzes the cleavage of the peptide chain from the last tRNA. The hydrolysis of the GTP in RF3 ends the protein synthesis cycle and provides the energy for the dissociation of the ribosomes and the release of the polypeptide and the RFs. The polypeptide is now free to fold into its native tertiary structure.

MUTATIONS

A change in the base-pair composition of the DNA within a gene is an event called a mutation. One consequence of having a haploid chromosome is that change in a specific gene is immediately expressed by the daughter cells because the bacterial genome lacks the second copy of each gene that, in diploid organisms, often masks the effect of mutation in one of the chromosomes. This section discusses the types of mutations that occur in bacteria, the predicted consequences of mutations, and the effect of environmental conditions on the frequency of mutations in bacteria.

Fluctuation Test

When a small amount of streptomycin (an antibiotic made by *Streptomyces*) is added to a culture of bacteria, almost all bacteria are killed within a few minutes. Similarly, bacteria can be killed by exposure to bacterial viruses (also called bacteriophages) that adsorb to their cell walls, inject their genomes, and following a cycle of replication, release new viruses, with the concomitant death of the host bacterium. A small fraction, however—fewer than one in a million—can survive the exposure to potential killing agents. When these bacteria are propagated, they and their progeny are completely resistant to killing by the same agents that devastated the members of the previous sensitive population. The new, resistant population of bacteria owes its success to mutations. Changes in a gene that encodes a ribosomal protein results in the ability of the streptomycin-resistant bacterium to synthesize protein even in the presence of streptomycin. Similarly, resistant bacteria that are isolated from a population of cells killed by bacteriophages have altered genes for new cell wall components that do not allow bacterial viruses to attach or enter the cell.

Because these mutant bacteria appear to arise only following exposure to the appropriate agent, it had been debated for many years whether mutations are adaptive changes, wherein a specific agent would induce the required alteration in the genetic material. An alternative hypothesis, supported by the Darwinian theory of evolution, proposed that changes occur spontaneously and continuously in a culture, and it is this selective process that allows us merely to identify such bacteria that carry any one specific mutation.

The controversy was resolved in the early 1940s when M. Delbruck and S. Luria devised the fluctuation test, which demonstrated that mutations are spontaneous and independent of a selective environment. The original test was based on the ability of E. coli to mutate into a form that is resistant to the killing action of bacteriophage (phage) T1. When exposed to T1, a population of 10^8 bacteria gives rise to about 10 to 50 mutants that are T1 resistant; that is, they can grow in the presence of this phage.

The fluctuation test is outlined in **Figure 13.16**. A culture of E. coli propagated in a flask was divided into two tubes, each containing about 10^3 bacteria. The contents of one of the tubes were further subdivided into 50 smaller tubes. All of the cultures were grown to stationary phase. From the single large test tube, as well as from the 50 individual tubes, small aliquots were spread over an agar surface, densely seeded with T1 bacteriophage, which will kill most susceptible bacteria, unless the bacteria become resistant to T1 by virtue of a mutation. These resistant bacterial mutants can give rise to colonies even in the presence of large numbers of bacteriophage. After overnight incubation, the number of phage-resistant colonies was determined. The culture that was incubated in a single tube yielded about three to seven resistant colonies. The range of phage-resistant colonies obtained from the individually grown tubes was from none to 100 or more.

What was the reason for this variation in the number of resistant colonies? If mutations arise as a result of contact with the phage on the plate, aliquots from both large and small cultures should give rise to the same number of resistant colonies. If phage-resistant mutants are arising at all times, the individual cultures should have a

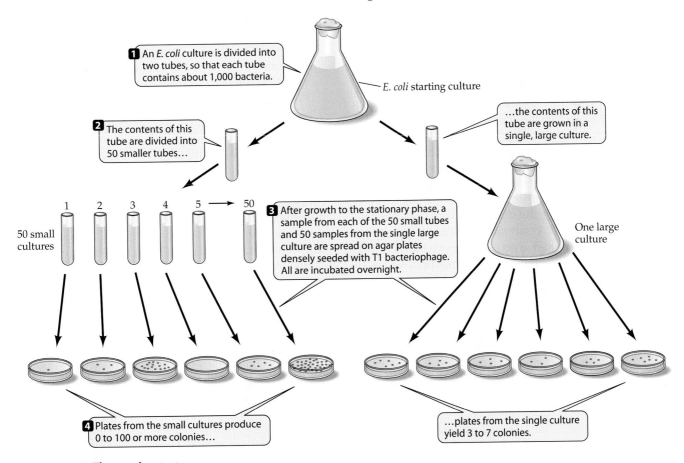

Figure 13.16 Fluctuation test
The fluctuation test provided conclusive proof that mutations occur in bacteria. In this experiment, T1-phage-resistant bacteria were produced by mutation.

great variation in number, depending on the time the mutation takes place during growth. That is, if mutation takes place immediately after subdivision, as many as 100 resistant bacteria will be identified in that tube following plating on a bacteriophage-containing agar plate. If mutation takes place long after subdivision or not at

sion of a specific gene. As an example, examine the effect of substitutions as well as deletions or insertions at the highlighted codon on the following polypeptide coding sequence: (In this example, DNA strand 1 is the template used by RNA polymerase to generate the mRNA for this protein.)

DNA strand 1	TAC	TTT	GCT	AAT	ACG	CTC	TAG	TTG	ACC	ATT
DNA strand 2	ATG	AAA	CGA	TTA	TGC	GAG	ATC	AAC	TGG	TAA
mRNA		AUG	AAA	CGA	UUA	UGC	GAG	AUC	AAC	UGG	UAA
Protein		Met	Lys	Arg	Leu	Cys	Glu	Ile	Asn	Trp	Stop

all, small numbers of phage-resistant bacteria will be identified. Clearly, results support the second hypothesis because of the great fluctuation in the number of resistant bacteria that arose from the small inocula.

Types of Mutation

A mutation involving a single-base-pair substitution, in which a purine (G or A) is replaced by another purine, or a pyrimidine (T or C) is replaced by another pyrimidine, is called a transition. A transversion is a type of mutation wherein a purine is replaced by a pyrimidine or a pyrimidine is replaced by a purine (Figure 10.15). These substitutions become part of a newly synthesized chain as a consequence of incorrect base pairing during the polymerization of the chain; for example, a G pairs with a T instead of a C. Deletions and insertions arise when the template or the newly synthesized strand loops out and the DNA polymerase makes a slip, resulting in duplication or elimination of the looped-out region in the newly synthesized strand. For any form of mutation to take effect, the change also has to escape the repair activity of DNA polymerase, namely the ability to back up in the 3'- to-5' direction, remove the incorrect base, and resynthesize the correct strand. Because base-pairing errors, loop-outs, and failure of the repair system are all very rare, mutations resulting from errors of this type arise infrequently.

Consequence of Mutations

Changes in the DNA sequence can have a number of effects in the bacterial cells that inherit the mutant ver-

Now, consider the following consequence of several changes in the DNA encoding this hypothetical nine amino acid peptide, all involving substitutions that change the mRNA from specifying the codon UUA, which codes for the amino acid, leucine.

1. A transversion of the AT base pair for GC results in the codon UGA in the mRNA. This is a termination codon, resulting in synthesis of a shortened three-amino-acid peptide.
2. A transition of the TA base pair for a GC pair changes the codon on mRNA into UUG. This mutation has no effect on the polypeptide because UUG, like UUA encodes leucine.
3. Another transversion of AT to TA alters the codon into one specifying isoleucine (AUA).

In addition to substitutions, insertions or deletions can take place. For example, insertion of one or two base pairs results in the alteration of the order of the triplet codons relative to the initiation codon. A new reading frame will originate from the site of insertion, encoding a completely different polypeptide. These so-called frameshift mutations often have the same result as a chain termination mutation at the same site because the new polypeptide, however long it is, usually does not specify a functional protein. Frameshift mutations terminate when ribosomes encounter a termination codon in the new reading frame. Deletion of one or two nucleotides also results in frameshift mutations.

A frameshift mutation due to insertion of a T paired with A would look like this:

DNA strand 1	TAC	TTT	GCT	TAA	TAC	GCT	CTA	GTT	GAC	CAT	T
DNA strand 2	ATG	AAA	CGA	ATT	ATG	CGA	GAT	CAA	CTG	GTA	A
mRNA		AUG	AAA	CGA	AUU	AUG	CGA	GAU	CAA	CUG	GUA	A
Protein		Met	Lys	Arg	Ile	Met	Arg	Asp	Gln	Leu	Val

Note that as a result of the insertion, the new reading frame continues beyond the stop codon of the wild type (that is, original) mRNA until a new termination codon is reached. The polypeptide sequence after the site of mutation is completely different from that of the original peptide. Insertions or deletions of three nucleotides (or multiples of three nucleotides) yield true insertion or deletions of amino acids, without altering the reading frame around the site of the mutation.

Changes in a specific base sequence can have variable effects on the polypeptide that is encoded by a specific gene. Alteration of the length of the protein by a mutation to a termination codon can abolish the function of that protein. Mutations that result in substitution of one amino acid for another may have only minor effects, especially if the amino acids are similar in size, such as the example in which leucine is substituted by isoleucine. However, if the substitution is in an important part of the molecule, it can also alter the functional area of the polypeptide. Mutations that have a more dramatic affect on a protein's function or stability are those that result in substitution of amino acids having opposite charges or very different side chains. Occasionally, the change will cause altered physical properties in the polypeptide. Certain substitutions render enzymes temperature sensitive, such that they can function only at lower temperatures. This is presumably due to incorrect folding of the polypeptide chain, making it more susceptible to thermal denaturation. The effect of the mutation would therefore be noticeable only under nonpermissive conditions, that is, when the protein is denatured. Under permissive conditions, however, less-than-perfect folding may still result in a protein with only slightly altered activity.

Because mutations often result in the alteration of cellular metabolism, such that the cell does not function optimally, there is sufficient selective pressure for secondary compensatory mutations that repair the original defect. Such restoration by a counteractive mutation is called suppression. Suppression of the mutation can occur by several mechanisms, the simplest of which is exact reversal of the mutation. However, other mechanisms of suppression involve genetic changes at adjacent or distal sites on the DNA. Suppression of a base substitution mutation can be accomplished by a mutation in a different amino acid, which would nullify the disruptive effect of the first mutation in the polypeptide chain.

A specialized category of suppression deals with nonsense mutations (**Figure 13.17**). Nonsense mutations occur when a base substitution changes a codon for an amino acid into one of the chain-termination codons. Nonsense suppressors are mutations in genes encoding the tRNA anticodon loop corresponding to an mRNA stop codon, such that the tRNA adds an amino acid to the growing peptide rather than

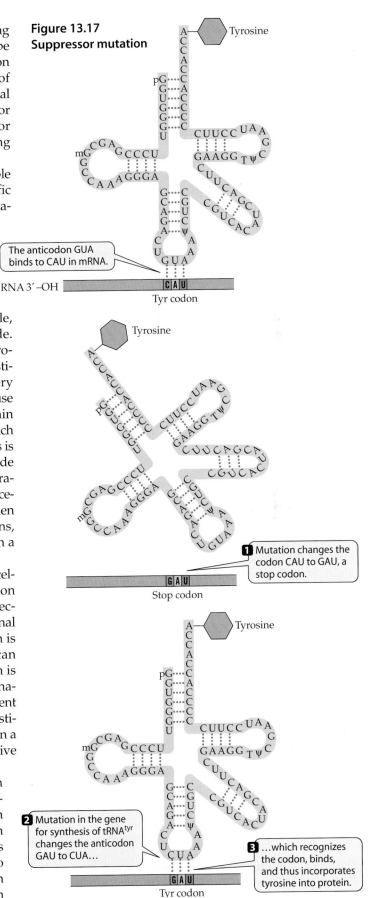

Figure 13.17
Suppressor mutation

The anticodon GUA binds to CAU in mRNA.

RNA 3′–OH

Tyr codon

1 Mutation changes the codon CAU to GAU, a stop codon.

Stop codon

2 Mutation in the gene for synthesis of tRNA^tyr changes the anticodon GAU to CUA…

3 …which recognizes the codon, binds, and thus incorporates tyrosine into protein.

Tyr codon

terminating it. Figure 13.17 presents an example of a chain-termination mutation suppressed by a mutant suppressor tRNA. Because cells still need the correct tRNA genes for normal protein synthesis, only those tRNAs that are present in more than one copy per cell can mutate to nonsense codon suppressors without killing the cell. Moreover, suppressor tRNA can interfere with normal termination of protein synthesis by a single termination codon, resulting in added carboxyl terminal tails extending through the adjacent gene to the next chain termination codon. This happens relatively infrequently, because most coding sequences of polypeptides terminate with more than one in-frame termination codon.

Often, two proteins associating with each other accomplish a cell function, and these complexes are formed by protein-protein interactions involving distinct regions of each polypeptide called interactive domains. A mutation in a gene encoding the interactive domain of one such protein can be suppressed by a compensatory mutation in the second protein's interactive domain. Some mutations are more difficult to suppress, such as frameshift mutations in which the reading frame has to be corrected near the site of the insertion or deletion by insertion or deletion of one or two base pairs to restore the original triplet reading frame.

MUTAGENS

The rate of spontaneous change in DNA is relatively low, owing to the effective repair mechanisms present in bacterial cells. Exposing cells to several physical or chemical agents can increase the rate of mutation. Physical mutagenesis almost exclusively involves radiation damage to DNA. A variety of chemicals can also cause mutations by interacting with DNA directly or by interference with the DNA replication and repair processes. Chemical compounds affecting mutational rates are termed mutagens and fall into three broad categories: structural analogs of nucleotides, chemicals that alter purine or pyrimidine bases, and agents that intercalate the DNA at the replication fork (**Box 13.2**).

Mutagenesis with Base Analogues

5-Bromouracil (BU) is an example of a base analog because of its similarity to thymine. During replication, it can substitute for thymine resulting in a BU:A pair. (The double colon indicates a hydrogen bond.) However, thymine can occasionally pair with guanine in a relatively unstable transient bond. When BU substitutes for thymine, the BU:G bond is relatively stable, and during subsequent replication the stabilized G pairs with C, resulting in a permanent transition of an AT pair for GC (**Figure 13.18**). Treatment with 5-bromouracil can also result in the opposite transition, in which an existing GC pair is replaced by an AT pair in daughter cells.

Mutagenesis with Agents That Modify DNA

Agents that modify DNA include a large number of compounds with nucleotide bases that readily incorporate into DNA.

1. *Nitrous acid* deaminates cytosine and adenine, converting them to uracil and hypoxanthine, respectively. During the subsequent rounds of replication, uracil pairs with adenine, whereas hypoxanthine pairs with cytosine. The net effect of the reaction of nitrous acid with DNA is the transition of AT to GC and GC to AT.
2. *Hydroxylamine* and *nitrous acid* specifically modify cytosine so it can pair with adenine, resulting in the transition of a GC pair to an AT.

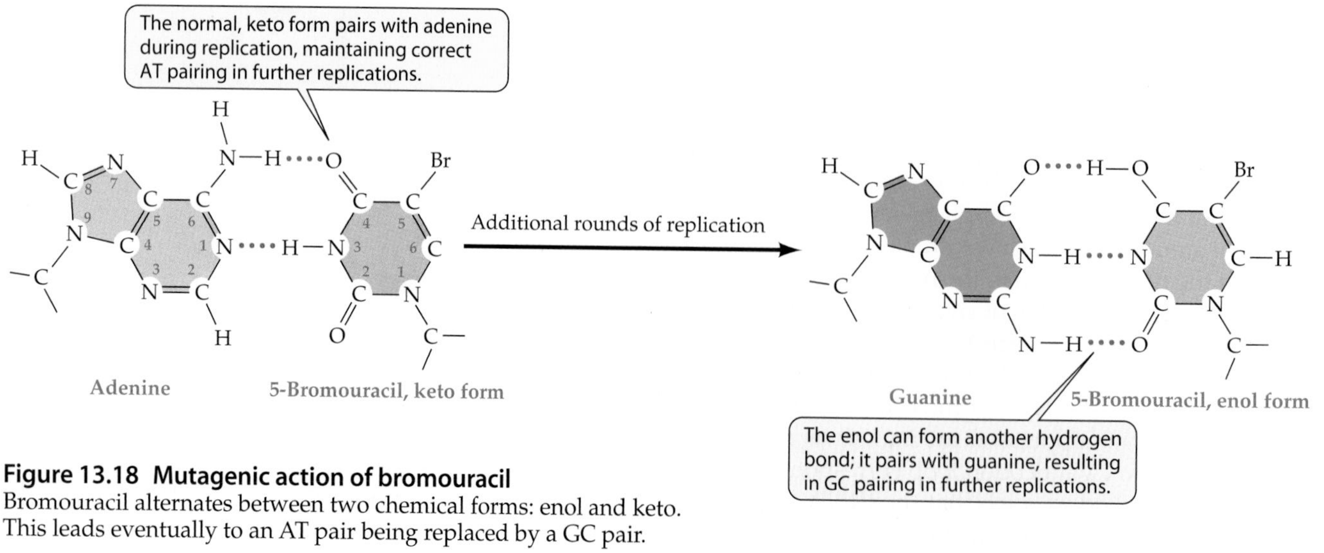

Figure 13.18 Mutagenic action of bromouracil
Bromouracil alternates between two chemical forms: enol and keto. This leads eventually to an AT pair being replaced by a GC pair.

Methods & Techniques Box 13.2   Ames Test

It is well known that many chemicals in our environment, including foods, are carcinogens; that is, they can induce cancers in animals and humans. Testing these chemicals is very complex, requiring the exposure of experimental animals to high concentrations of compounds over prolonged time periods and noting development of cancer. A simplified and sensitive test has been developed to examine carcinogens, based on the ability of the vast majority of these compounds to induce mutations. Mutagenic ability of a chemical is tested as outlined in the figure. The test compound is exposed to a strain of *Salmonella* that has a single mutation in an enzyme of histidine biosynthesis, designated His⁻ (unable to grow on media without histidine). If the compound is mutagenic, it can increase the frequency of mutations in the *Salmonella* genome and, at a certain frequency, will result in back mutations in the original gene responsible for the histidine requirement. Those bacteria can then synthesize their own histidine and grow on media that is not supplemented by histidine. In the Ames test, a pure culture of His⁻ *Salmonella* is mixed with the suspected carcinogen and plated on minimal media lacking histidine. As a control, the same culture is plated without the carcinogen. After the plates are incubated, the number of colonies is counted. A significant increase in the number of bacteria

on the test sample compared to the control, resulting from an increase in back mutations from His⁻ to His⁺, indicates mutagenic activity of the compound. This test can be adapted for those compounds that are not

mutagenic by themselves, but are metabolically converted to mutagens after ingestion. Usually, the compound is exposed to an extract from rat liver, to allow conversion to the mutagenic form.

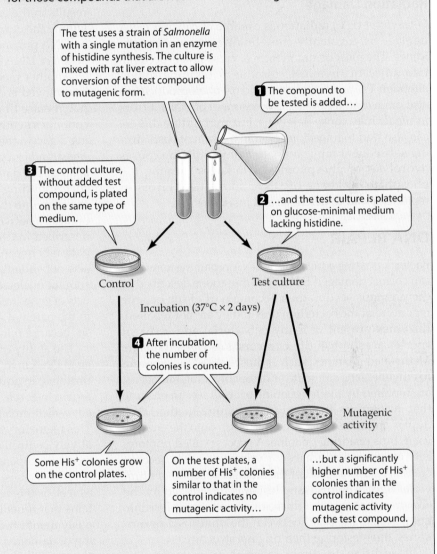

The test uses a strain of *Salmonella* with a single mutation in an enzyme of histidine synthesis. The culture is mixed with rat liver extract to allow conversion of the test compound to mutagenic form.

1 The compound to be tested is added…

3 The control culture, without added test compound, is plated on the same type of medium.

2 …and the test culture is plated on glucose-minimal medium lacking histidine.

Control

Test culture

Incubation (37°C × 2 days)

4 After incubation, the number of colonies is counted.

Mutagenic activity

Some His⁺ colonies grow on the control plates.

On the test plates, a number of His⁺ colonies similar to that in the control indicates no mutagenic activity…

…but a significantly higher number of His⁺ colonies than in the control indicates mutagenic activity of the test compound.

---

3. *Ethylmethane sulfonate* is an alkylating agent that modifies the oxygen at position 6 of guanine by adding an alkyl group (a methyl group), converting it to O⁶-methylguanine. This modified base can now weakly pair with thymine during subsequent replication. This modification leads to a GC to AT transition in one of the daughter cells. This mispairing can also lead to an AT to GC transition.

4. *Nitrosoguanidine* also alkylates guanines, giving a GC to AT transition. This often leads to small deletions.

Mutagens That Intercalate with DNA

Several complex organic molecules, such as acridine orange, proflavin, and ethidium bromide, interact with DNA and result in distortion of DNA. Distortion leads to DNA polymerase slippage, resulting in small insertions or deletions. Frameshift mutations are the most common types of lesions in DNA that occur as a result of interaction with intercalating compounds. The structures of all of these compounds are different, yet they all have conjugated rings of a size similar to a base pair in a DNA duplex, which they most likely insert between hydrogen-bonded adjacent bases in the duplex, resulting in the bending of DNA.

Radiation Damage

Ultraviolet (UV) radiation is capable of causing DNA damage by covalently cross-linking adjacent pyrimidines. The most common form of cross-link is between two adjacent thymines, forming a thymine dimer, although T-C and C-C dimers and 6,4 photoproducts are also often seen following UV treatment of DNA. Errors in repair may cause deletions, but pyrimidine dimers can also lead to transition and transversion of bases during subsequent replication. Ionizing radiation causes hydrolysis of phosphodiester linkages forming the phosphate backbone of DNA; subsequent errors in repair can lead to deletions and insertions.

DNA REPAIR

Bacteria possess a number of DNA repair systems that can correct damaged DNA. The mutations described in the previous section thus become part of the genetic repertoire of the bacterium only if they are not repaired. Enzymes present in many eubacterial and archaeal species accomplish direct reversal of DNA damage. Methylated guanines, such as those caused by exposure to ethylmethane sulfonate or other alkylating mutagens, are repaired by methylguanine transferase, an enzyme that recognizes the modified guanine in the DNA duplex and transfers the methyl group onto the enzyme itself, thus restoring guanine. A process called photoreactivation repairs adjacent pyrimidines that become cross-linked as a result of the mutagenic action of ultraviolet light. The resulting lesion is recognized by the enzyme photolyase, which, on absorption of visible light, breaks the bond between the dimerized pyrimidines, thus restoring their original structure.

In addition to direct reversal of mutations by enzymes, a common mechanism involves excision repair, in which the damaged DNA strand is enzymatically removed and the gap correctly resynthesized using the complementary strand as a template. For example, a mutation, created following deamination of cytosine to uracil, can be repaired by the enzyme DNA glycosylase, which removes uracil from the sugar-phosphate backbone. This region lacks bases and is called the AP-site. This defective lesion is removed by the combined action of two enzymes: AP endonuclease and deoxyribophosphodiesterase. DNA polymerase then repairs the nick by incorporating the correct base (dC) with U by DNA glycosylase.

Bulky distortions in DNA resulting from gross mispairing or ultraviolet light–induced pyrimidine dimers can be repaired by an excision mechanism (**Figure 13.19**) that relies on the abilities of a multisubunit endonuclease (in *E. coli*, encoded by genes *uvrA*, *uvrB*, and *uvrC*) to hydrolyze the phosphate backbone at two places flanking the damaged site. The action of DNA helicase results in the release of a 12-base oligonucleotide containing the damaged bases; the resulting gap is filled in by DNA polymerase I, and the backbone is closed by the action of DNA ligase.

Another mechanism by which thymine dimers can be removed is called recombination repair. When DNA polymerase III reaches a thymine dimer, it skips it and continues synthesis at the other side of the dimer, leaving a gap in the newly synthesized duplex. The opposite strand is replicated normally. The gap that is created around the damaged DNA is filled by excision of the corresponding fragment from the second strand and insertion into the gap by a mechanism similar to recombination between two related DNA molecules, as described in Chapter 15. This repair, just like recombination between DNA segments of similar sequence, is also dependent on the RecA protein, as well as a multisubunit nuclease (exonuclease V), which contains products of genes *recB*, *recC*, and *recD*. The newly created gap in the parental strand is then filled in by repair synthesis, because it involves copying an undamaged strand.

The ability of bacteria to sense the presence of damaged DNA involves synthesis of a number of specialized proteins, in addition to those that function in normal repair. In *E. coli*, a complex regulatory system is activated, which directs synthesis of many of these proteins. This regulatory network, called the SOS regulon, is negatively controlled by a repressor protein, LexA, which recognizes operator sequences for a number of genes for repair enzymes, including *umuC*, *uvrA*, *uvrB*, and *uvrC* as well as the *recA* gene. All of these genes express proteins that function at one stage or another in the previously mentioned repair systems. Clearly, it is the sensing of damaged DNA by the RecA protein, resulting in cleavage of the LexA repressor, that initiates the various DNA repair mechanisms by relieving the repression of the repair enzymes' genes. The SOS response is activated when the replication fork stalls and RecA protein binds to exposed ssDNA. The RecA-DNA complex binds LexA, and this binding induces a conformational change that causes LexA to cleave itself.

Figure 13.19 Excision repair ▶
Thymine dimers, created by exposure of DNA to ultraviolet light, are repaired by the UvrABC system.

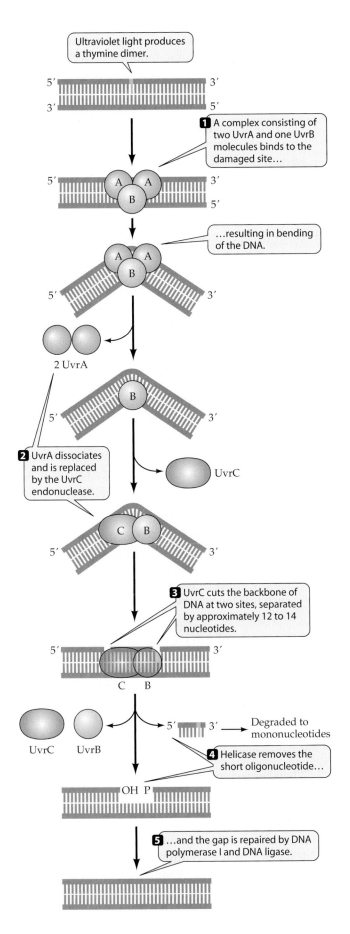

Ultraviolet light produces a thymine dimer.

1 A complex consisting of two UvrA and one UvrB molecules binds to the damaged site…

…resulting in bending of the DNA.

2 UvrA

2 UvrA dissociates and is replaced by the UvrC endonuclease.

UvrC

3 UvrC cuts the backbone of DNA at two sites, separated by approximately 12 to 14 nucleotides.

UvrC   UvrB

Degraded to mononucleotides

4 Helicase removes the short oligonucleotide…

OH P

5 …and the gap is repaired by DNA polymerase I and DNA ligase.

SUMMARY

▸ All living cells carry hereditary information in deoxyribonucleic acid (DNA), which is organized into **genes**. In bacteria, genes are usually found in a single circular **chromosome**.

▸ DNA is composed of four deoxyribonucleotides: deoxyadenosine monophosphate (**dAMP**), deoxythymidine monophosphate (**dTMP**), deoxyguanosine monophosphate (**dGMP**), and deoxycytidine monophosphate (**dCMP**).

▸ The deoxyribonucleotides are linked together via **phosphodiester linkages** between phosphoric acid and hydroxyls on neighboring deoxyribose sugars.

▸ Bacterial chromosomal DNA is **double stranded**; opposite, **antiparallel** strands are held together by hydrogen bonds formed between adenine and thymine and between guanine and cytosine bases of the deoxyribonucleotides.

▸ DNA replicates by a **semiconservative** mode of replication, where the strands are separated and each strand is copied, with the daughter genome receiving one parental strand and a newly synthesized complementary copy.

▸ DNA replication begins at a single, relatively well conserved site called the origin (*ori*). Two **replication forks** move away from the origin, as nucleotides are incorporated with the assistance of a number of proteins.

▸ Within each replication fork, one strand (the **leading strand**) is synthesized as single contiguous molecule, whereas the complementary strand (**lagging strand**) is synthesized in a discontinuous fashion, involving short RNA segments that serve as primers for DNA segments (**Okazaki fragments**). The newly synthesized DNA strand is completed by removal of RNA primers and filling in the gaps with DNA, followed by the joining of the fragment into a single molecule.

▸ The low frequency of mistakes made during DNA replication is due to the **proofreading** and **repair** activities associated with enzymes that replicate DNA.

▸ If genes specify proteins, this form of RNA is called messenger RNA (**mRNA**). If a single RNA can specify several genes, it is then called a polycistronic mRNA.

▸ Other forms of RNA specify components of ribosomes, the ribosomal RNA (**rRNA**) or the adapter molecules for protein polymerization, called transfer RNAs (**tRNA**).

▸ RNA is a copy of one of the DNA strands. It consists of polymerized ribonucleotides with the same base composition as DNA, except that thymine is replaced by uracil. The complementarity of the RNA copy is assured by hydrogen bonds base-pairing the nitrogenous bases during synthesis of the RNA strand.

▶ The beginning of the gene, a sequence called the **promoter**, where transcription initiates, is recognized by the sigma subunit. After binding, sigma dissociates from the RNA polymerase complex, leaving a form called the **core enzyme**, which continues elongating the RNA strand. RNA polymerase recognizes the end of a gene after formation of stem-loop structure or by termination assisted by the so-called **rho** protein.

▶ When the product of a gene is not needed, bacteria utilize **negative regulation** to prevent the gene's transcription. This often involves binding of a **repressor** protein to an **operator** site located adjacent to a promoter sequence. Tight binding of the repressor to a sequence that often overlaps the protein interferes with sigma-factor-directed binding of the RNA polymerase, and thus prevents transcription of the negatively regulated gene.

▶ Alternatively, protein factors (**activators**) stimulate transcription by binding to DNA sites upstream of promoters and facilitate the rate-limiting incorporation of the first bases into RNA.

▶ The information in mRNA is converted into proteins by a process of **translation**, which involves ribosome-assisted polymerization of 20 amino acids into a sequence specified by the triplet code of mRNA, copied from its complementary DNA strand. The specific order of such codons is called the **reading frame**.

▶ Amino acids are incorporated into polypeptides, after they are covalently attached to tRNA. Each tRNA recognizes a sequence of three bases on mRNA (called codons) via a complementary sequence.

▶ The ribosomes recognize and bind to a specific site at the beginning of the mRNA. This so-called ribosome binding site is adjacent to the codon for the first amino acid of the polypeptide, which is always methionine. Ribosomes terminate translation at specific codons (**termination**, or **stop codons**), for which there are no corresponding tRNAs.

▶ Changes in the base-pair composition of the DNA are called **mutations**.

▶ Various chemicals can affect the frequency of mutations in a specific gene, either by altering the DNA or by interfering with DNA repair during replication. These are called **mutagens**.

▶ There are several mechanisms through which cells repair DNA and eliminate the deleterious effects of mutations.

REVIEW QUESTIONS

1. Contrast the basic structural differences between deoxyribonucleotides and ribonucleotides.

2. Draw the structure of deoxycytidine-deoxyguanosine (dG-dC) and deoxyadenosine-deoxythymidine (dA-dT) base pairs as they appear in the antiparallel form in double-stranded DNA. Indicate the location of the hydrogen bonds holding the stands together.

3. Distinguish between DNA synthesis on a leading and on a lagging strand, addressing the different mechanisms of initiation and elongation of the DNA strand.

4. What is the mechanism for relieving the tension in DNA that is caused by supercoiling during replication?

5. What would be the consequences for *E. coli* in its ability to synthesize the enzyme β-galactosidase of the following mutations in the regulatory components of the lactose utilization operon? Explain your answer by considering bacteria that are present in medium with or without lactose.

   (a) A mutation in the operator sequence of the *lac* operon, such that it cannot bind the repressor protein.

   (b) Mutation in the DNA sequence of the *lacI* gene, such that the mutant repressor cannot bind an inducer (lactose or its structural analogue).

   (c) A two-base pair mutation in the *lacI* gene, changing, in mRNA, the AUG-initiating codon to UAG.

   (d) A double mutant containing both types of mutations described in (b) and (c).

   (e) Would a mutation in the gene coding for cyclic AMP-binding protein (CAP), resulting in the inability of bacteria to synthesize this protein, offset the effect of the mutation described in (b), such that the bacteria synthesize wild-type levels of β-galactosidase?

6. Consider a polycistronic operon consisting of genes X, Y, and Z. The promoter located upstream of the gene X undergoes a mutation, such that three of the bases in the –10 region are changed. How would this affect the synthesis of proteins encoded by genes X, Y, and Z? What would be the effect of a three-base-pair mutation in the ribosome-binding site of gene X on synthesis of proteins X, Y, and Z?

7. When bacteria are exposed to antibiotics, they are killed, except for a small population of mutants that are resistant, due to their ability to undergo mutations in the target of the antibiotic. Devise a set of experiments (analogous to the fluctuation test described in the chapter) that demonstrate that mutations arose spontaneously in the bacterial population and are not induced by exposure of the bacterial culture to antibiotics.

8. A bacterial culture was exposed to a mutagen, and some mutants were isolated that were unable to synthesize an essential biosynthetic enzyme. When these mutants were grown in pure culture and reexposed to the mutagen, a small fraction of bacteria regained their ability to synthesize the biosynthetic enzyme. Describe several possible mechanisms, explaining how the second exposure to a mutagen "corrected" the original mutation.

9. What are the mechanisms of removal of cross-linked thymine dimers in DNA?

10. What is the mechanism of induction of DNA repair processes, when the DNA replication fork stalls at the site of DNA damage, exposing single-stranded regions?

11. The following partial DNA sequence from the *E. coli* genome encodes a gene for a short polypeptide. The transcriptional start site is highlighted.

a. Identify the most likely promoter for this gene. Is it recognized by the major RNA polymerase?

b. Write out the sequence of the mRNA transcribed from this promoter.

c. Translate the sequence into the polypeptide according to the genetic code.

d. Identify the position in the DNA sequence, where, by a single-base-pair substitution or a frameshift mutation, the result would be synthesis of a peptide that would have the identical sequence but half the size of the wild-type peptide.

e. Identify the position in the nucleotide sequence where a mutation involving a single-base-pair change (insertion, deletion, or substitution) would result in synthesis of a peptide identical to the wild type throughout the sequence, but having four additional arginine residues at its carboxyl terminus.

+1

GGCTTGACACCCGGCTAGCGTAGTTGTATAATGGTCAGGCTTTGAGGAGGTCTAGGCATGAAAGCTCAACGTAAGGGGGATCCGGAGTAGGAGACGGAGGTAAG
CCGAACTGTGGGCCGATCGCATCAACATATTACCAGTCCGAAACTCCTCCAGATCCGTACTTTCGACTTGCATTCCCCCTAGGCCTCATCCTCTGCCTCCATTC

SUGGESTED READING

Birge, E. A. 1994. *Bacterial and Bacteriophage Genetics*. 3rd ed. New York: Springer-Verlag.

Joset, F. and J. Guespin-Michel. 1993. *Prokaryotic Genetics*. Cambridge, MA: Blackwell Science.

Kornberg, A. and T. A. Baker. 1992. *DNA Replication*. 2nd ed. New York: W. H. Freeman and Co.

Maloy, S. R., J. E. Cronan and D. Freifelder. 1994. *Microbial Genetics*. 2nd ed. Boston: Jones and Bartlett.

Primrose, S. B. 1995. *Principles of Genome Analysis*. Cambridge, MA: Blackwell Science.

Baunberg, S. 1999. *Prokaryotic Gene Expression*. Oxford: Oxford University Press.

Chapter 14

Viruses

It is, indeed, an RNA world, not a DNA world.
In present-day life, DNA acts entirely by way of RNA,
whereas RNA can act on its own. Viruses are a case in point.
— CHRISTIAN DE DUVE, BLUEPRINT FOR A CELL: THE NATURE AND ORIGIN OF LIFE

Self-perpetuation is the key to survival of a species, and as discussed earlier, the blueprint for "self" is present in nucleic acids (genome). Therefore, a bit of nucleic acid that can generate a like nucleic acid may be considered to have some properties of life. Thus we define a **virus**: a bit of nucleic acid—RNA or DNA—surrounded by a protective coat, and this small bit of nucleic acid can regenerate by commandeering the metabolic machinery of an appropriate prokaryotic or eukaryotic host. It is certain, however, that viruses are not alive nor do they have the characteristics associated with a viable cell.

Viral nucleic acid can gain entry into a suitable host where it mobilizes the essential cellular synthetic capacities and compels the host to generate viral particles. Viral RNA (message) can be synthesized by polymerases available in the host or furnished by the virus. However, viruses do not contain translational enzymes, and they depend largely on the translational machinery of a host to synthesize viral proteins.

An understanding of how viruses function begins with an overview of the basic structure of the virus (**Figure 14.1**). At the core of the virus is the viral genome, which can be composed of either RNA or DNA. Unlike the cells of animal, plant, or bacteria, a virus rarely contains both. The nucleic acid is bounded by a protective protein coat termed a **capsid**. The capsid and genome together form a **nucleocapsid**. In some viruses, a membrane envelope surrounds the nucleocapsid, and these are considered **enveloped viruses**. Host cellular membrane is the source of the viral envelope and surrounds the nucleocapsid as it exists from an infected cell. If virus-specific proteins are present in the envelope, they are embedded in the cellular membrane prior to virus envelopment.

All viruses exist in two states: extracellular and intracellular. In the **extracellular state,** the **virion** or **virus** particle is quite inert. The virion is the form in which a virus passes from one host cell to another host. In the **intracellular state**, the viral nucleic acids induce the host cell to synthesize the virus. Such viral reproduction is called infection. A viral infection can be productive, which leads to viral progeny, or nonproductive, whereby a virus may remain dormant in the cell, causing no apparent manifestations of infection. With some animal viral infections, multiplication is always accompanied by disease symptoms, as would occur with a case of measles. There are others where the inapparent infection rate may be 100 or more cases of viral multiplication for each clinical case. This chapter is concerned with the mechanisms whereby prokaryotic and eukaryotic viruses are propagated. The first half of the chapter presents a general overview of viruses: their size; their structure; how they are cultivated and quantitated; and how they reproduce. The second half considers selected viruses in some detail.

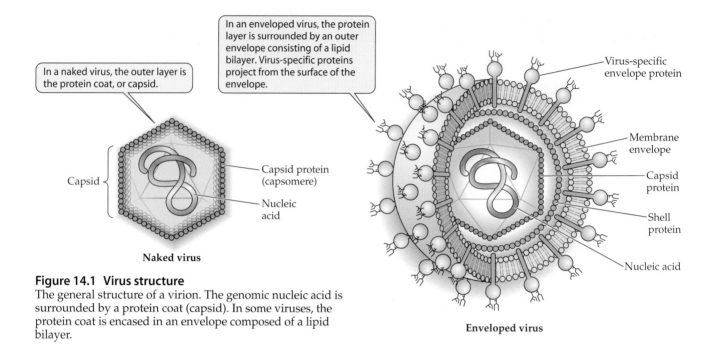

In a naked virus, the outer layer is the protein coat, or capsid.

In an enveloped virus, the protein layer is surrounded by an outer envelope consisting of a lipid bilayer. Virus-specific proteins project from the surface of the envelope.

Capsid

Capsid protein (capsomere)

Nucleic acid

Naked virus

Virus-specific envelope protein

Membrane envelope

Capsid protein

Shell protein

Nucleic acid

Enveloped virus

Figure 14.1 Virus structure
The general structure of a virion. The genomic nucleic acid is surrounded by a protein coat (capsid). In some viruses, the protein coat is encased in an envelope composed of a lipid bilayer.

ORIGIN OF VIRUSES

The origin of viruses is an intriguing question and has been the subject of considerable speculation in the scientific community. There is general agreement that viruses (bacteriophage) probably infected bacteria during those millennia when *Archaea* and *Bacteria* were the sole life form on Earth. As eukaryotes evolved, they too became subject to viral attack. There are three major theories that have been advanced for the origin of viruses:

1. Viruses originated in the primordial soup and co-evolved with the more complex life forms.
2. They evolved from free-living organisms that invaded other life forms and gradually lost functions.
3. Viruses are "escaped" pieces of nucleic acid no longer under control by the cell—also termed the escaped gene theory.

The great variety of viruses present in the living world affirms that they originated independently many, many times during evolution. Viruses also originated from other viral types through mutation. A brief discussion of theories of viral evolution follows.

Coevolution

We will consider first the theory of coevolution. The earliest self-replicating genetic system was probably composed of RNA. RNA can promote RNA polymerization, although this would be a slow process and proteins present in the primordial soup may have become promoters of RNA replication. The DNA template is much more effective and originated early in evolution. RNA

then became the messenger between the DNA template and protein synthesis. Thus the genetic code came into being and permitted orderly replication.

Gradually the early replicative forms gained in complexity and became encased in a lipid sac whose purpose was to separate its metabolic machinery from the environment. This may have been the ancestor of the progenote and later the *Archaea* and *Bacteria*. Another replicative form may have retained simplicity and may have been composed mainly of self-replicating nucleic acid surrounded by a protein coat. This entity may have been the forerunner of the virus and evolved to a dependence for duplication on its ability to invade and take over the genetic machinery of a host. Thus, there was coevolution of the bacterium and virus. This hypothesis has few supporters, but it does offer an explanation for the evolution of viruses with genomic RNA.

Retrograde Evolution

The theory that viruses originated as free-living or parasitic microorganisms is based on the concept that a microorganism became a predator and gradually lost unused genetic information. Genes for biosynthesis of intermediates supplied by the host could be lost by mutation without harm to the invader. Eventually this prokaryote may have evolved to nothing more than a group of genes now termed a virus. Prior to the advent of the eukaryote, bacterial and archaeal life forms were mostly free living but predators similar to *Bdellovibrio* may also have existed. An intracellular parasitic microorganism could have become more and more dependent on the host cell and would only need to retain the ability to replicate nucleic acids and a mechanism for traveling from cell to cell.

An equivalent occurrence in eukaryotic cells might account for the evolution of plant and animal viruses. An intracellular parasitic bacterial type such as the *Chlamydiae* would be an example of a bacterium that potentially could regress to the viral state. The *Chlamydiae* are bacteria that have no cell wall and cannot live outside of living cells.

Bacterial origin would be more likely for complex viruses such as the poxviruses, but the genetic information in these viruses differs so markedly from that in prokaryotes that the retrograde evolution theory has little support. The absence of any life form between intracellular bacterial pathogens and viruses is also cited as a concern, and it is difficult to explain how RNA viruses came into existence through loss of genetic information.

"Escaped" Gene Theory

This is the more plausible theory and has considerable support. It proposes that pieces of host-cell RNA or DNA "escaped" and gained independence from cellular control. Living organisms make duplicates of their genetic information by initiating replication at a specific site called the initiation site. The replication cycle ends when a full complement of the genome is synthesized. If initiation of replication begins elsewhere on the genome, this duplication would occur independently of host control. An entity that could recognize nucleotide sequences at sites other than the start site and carried the proper polymerase could have the capacity to produce RNA or DNA without interference from normal control mechanisms.

The origin of viruses may have been with episomes (plasmids) or transposons, circular DNA molecules that replicate in cytoplasm and can be integrated into or excised from various sites in the host chromosome. Plasmids can also move from cell to cell carrying information such as fertility or antibiotic resistance. Transposons are bits of DNA present in both prokaryotic and eukaryotic cells that can move from one site in a chromosome to another site carrying genetic information.

There is actually one type of transposon known that directs the assembly of an RNA copy of its own DNA. This transposon can use this messenger RNA to synthesize DNA via reverse transcriptase. This copy may then be inserted into the chromosome. The DNA of the transposon carries a gene for synthesis of the reverse transcriptase, and transposon elements with such properties have some similarity to retroviruses. Analysis of nucleotide sequences in viruses indicates that they are quite equivalent to specific sequences in the host cell. Evidence accumulated to date strongly supports the proposal that viruses originated from "escaped" host nucleic acid.

SIZE AND STRUCTURE OF VIRUSES

Electron microscopy has affirmed that viruses vary widely in size and structure. A very small virus, the poliovirus, is about 20 nm (a nanometer is 1/1,000 of a micrometer) in diameter. This virus actually approaches the size of a ribosome. The larger viruses, such as the poxviruses, are 400 nm long and 200 nm in width and are large enough to be visible by phase contrast microscopy. The size of the viral genome varies over a broad range from about 5,000 to 200,000 base pairs. Simian virus 40, a small virus, has a genome makeup of 5,224 base pairs, and vaccinia, a large virus, has 190,000 base pairs. A bacterial genome would range between one-and nine-million bases, depending on species. The smallest bacterial genomes are those of intracellular parasites and in the 600,000 base range.

Electron microscopy, X-ray diffraction, biochemical analysis, and immunological techniques have provided much information on the structure and composition of viruses.

Viruses are of several morphologic types including:

Icosahedral viruses. The capsids in these viruses are constructed from protein subunits termed **protomers**. In an icosahedral capsid, these protomers are arranged in capsomers composed of pentamers (five sides) and hexamers (six protein subunits). The bonding between capsomers is noncovalent and weaker than the bonding within a capsomer. The hexamers form the edges and faces of the icosahedron and pentamers are at the vertices. The 12 vertices in a polyhedron are shown in **Figure 14.2A**, a structure that has 20 equilateral triangles as faces. The capsid of a small virus such as poliovirus would be made up of 32 capsomers, and the capsid of a larger virus (adenovirus) would be composed of 252 capsomers. Generally the capsids follow the laws of crystallography. The pentamers and hexamers may be composed of one type protein subunit or the proteins in the pentamers may differ from those in hexamers (Figure 14.2B). The icosahedral capsid is assembled and then filled with nucleic acid. Viruses have specific mechanisms to assure that an entire genome is incorporated.

Helical viruses. These are long, thin, hollow cylinders composed of distinct protomers. The helical tobacco mosaic virus (TMV) is depicted in **Figure 14.3**. The protein subunits are arranged in a helical spiral with the genome fitted into a groove on the inner portion of the protein. The nucleic acid is enclosed in the capsid as the protomers are assembled. The helical viruses are narrow (15 to 20 nm) and may be quite long (300 to 400 nm). They may be rigid or flexible, depending on the nature of

(A) Icosahedron   (B)

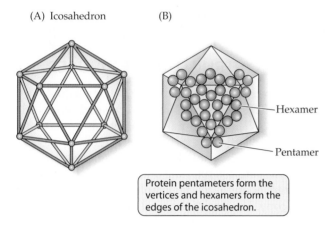

Protein pentameters form the
vertices and hexamers form the
edges of the icosahedron.

Figure 14.2 Icosahedral virus
Graphic representation of (A) an icosahedron and (B) the
arrangement of proteins in an icosahedral virus. (C) Electron micrograph of icosahedral viruses. Photo ©Hans
Gelderblom/Visuals Unlimited.

(C)

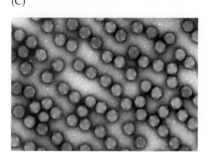

the constituent proteins. The helix of TMV has one type
of self-assembling protomer composed of 158 amino
acids. There are 2,130 copies of this one protein subunit
in the coat of a mature virus particle.

Enveloped viruses. These viruses have an outer
membranous structure that surrounds the nucleocapsid
(see Figure 14.1). The membranous envelope consists of

a lipid bilayer and may have specific glycoproteins
(spikes) embedded in the lipid. The glycoprotein spikes
and other coat proteins are incorporated into the host
membrane prior to exit of the virus and are picked up
by the virus as it leaves the infected cell (**Figure 14.4**).
The glycoproteins are encoded by the virus genome and
are not normal constituents of the host cell membrane.
These spikes are generally responsible for infection, as
they attach to specific sites on host cells.

Complex viruses. These have capsid symmetry, but
often the virus is assembled from parts synthesized separately (head, tail, capsomer). The *Escherichia coli* virus
T4 (**Figure 14.5**) is a complex virus. The virus is composed of a polyhedron head, a tail, whiskers, tail fibers,
and other parts. This virus will be discussed in some
detail later in the chapter.

Viral Propagation

There are four rather distinct phases in viral reproduction: (1) attachment, (2) penetration, (3) genome replication, and (4) assembly. These are as follows:

1. **Attachment or adsorption.** There are specific sites
   (virus receptors) on the surface of a bacterium where
   the virion attaches.
2. **Penetration or injection.** The nucleic acid of the virion gains access to the cytoplasm of the host. Bacterial
   and plant viruses initiating an infection must gain
   entry through the cell wall, whereas animal viruses
   enter through the cell membrane. This influences
   virus structure.
3. **Gene expression and genome replication.** Once
   inside, the viral genome is expressed and duplicated.
   This leads to the formation of gene products that control replication of viral nucleic acid and capsomer
   synthesis.
4. **Assembly and release.** The viral nucleic acid is packaged in the capsid, and the mature virus is released.
   A virus needs a host for replication, as ATP, ribosomes, and building blocks supplied by the host.
   Enzymes involved in viral replication are also supplied by the host with or without modification.
   Generally lytic enzymes cause host cell dissolution.

(A)

The protomers are arranged as a
hollow helix, with the genomic
RNA fitting into a groove formed
by the proteins.

Protomer   RNA

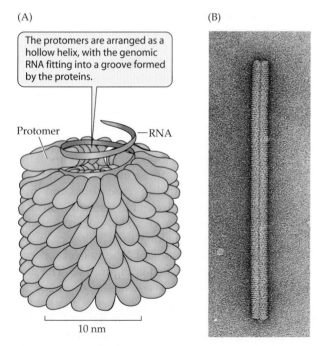

10 nm

(B)

Figure 14.3 Helical virus
(A) Graphic representation and (B) photomicrograph of
tobacco mosaic virus. Photo ©R. C. Williams/Biological
Photo Service.

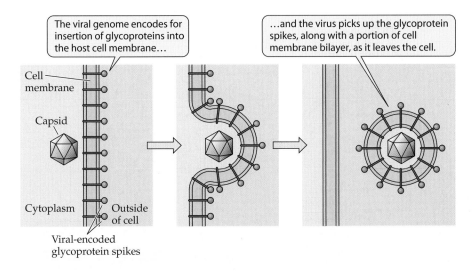

The viral genome encodes for insertion of glycoproteins into the host cell membrane…

…and the virus picks up the glycoprotein spikes, along with a portion of cell membrane bilayer, as it leaves the cell.

Cell membrane

Capsid

Cytoplasm   Outside of cell

Viral-encoded glycoprotein spikes

Figure 14.4 Enveloped virus
An enveloped virus acquires its lipid membrane coat as it exits through the cell membrane or nuclear membrane of the host cell.

These phases are illustrated in **Figure 14.6**. Entry of the viral genome generally results in an immediate cessation of normal host cell function and the replication of viral nucleic acid. Later in the infectious process capsid and other proteins are synthesized that are part of the mature viral particle.

CULTIVATION OF VIRUSES

Viruses can be grown only within living cells. This is because a virus has no biosynthetic capacity of its own, but can manipulate a cell genetically to synthesize viral particles. In early experimentation, viruses were culti-

vated by transfer from plant to plant, animal to animal, or bacterial culture to bacterial culture. The symptoms or manifestations of disease that resulted from inoculation were the principle means of identification. Animal viruses such as smallpox, polio, and mumps were routinely propagated in embryonated eggs. A fertile egg was incubated for 6 to 8 days at a temperature suitable for hatching and then inoculated with a viral suspension. Egg inoculation was often suitable for diagnostics, as the lesion produced in the embryo was characteristic of a given virus. More exacting cell-culturing techniques have been developed for use in research and diagnostic laboratories today.

Viruses that infect eukaryotic cells are now generally grown in monolayers of cells obtained from plant or animal tissue. Cells removed aseptically from a plant or animal can be placed in a sterile vessel made of glass or plastic. The cells may attach to the solid surface, and when fed properly, they can form a monolayer. This monolayer is called a **cell culture;** likewise, tissue can be maintained in culture and is termed **tissue culture.** A virus inoculum can be spread over a cell monolayer or introduced into tissue culture, and if the cells in the culture are susceptible, the virus will survive. After viral attachment to a monolayer, a layer of agar can be spread over the surface to limit the spread of virions to cells that

(A)   (B)

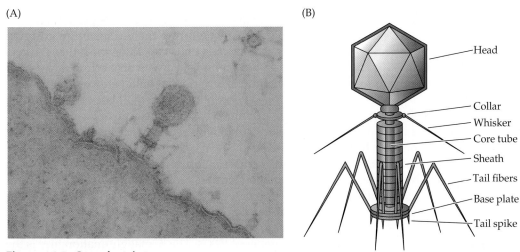

Head

Collar
Whisker
Core tube
Sheath
Tail fibers
Base plate
Tail spike

Figure 14.5 Complex virus
(A) T4 virus (phage) attacking an *Escherichia coli* cell. (B) Graphic representation of T4.
Photo ©J. Broek/SPL/Science Source/Photo Researchers, Inc.

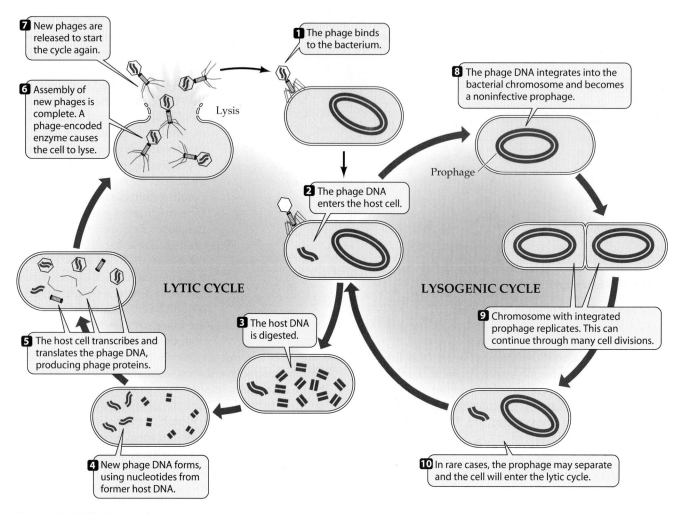

Figure 14.6 Viral reproduction
Generalized schematic for viral reproduction in a host bacterium, through the lytic and lysogenic cycles. In the lytic cycle, the virus (phage) multiplies in the host cell and the progeny viruses are released by lysis of cell. In the lysogenic cycle, viral DNA is integrated into the host genome and replicates as the chromosome replicates, producing lysogenic progeny cells.

are adjacent to the infected cell. In this way, areas of cell destruction can be observed.

Bacterial viruses can be cultivated by inoculating a suitable susceptible bacterial culture with a virus. Liquid cultures or cultures growing in an agar medium are suitable for this purpose.

Plant viruses can be cultivated in monolayers of plant tissue culture, in cultures of separated cells in suspension, in protoplast cultures, or in intact plants. The leaf may be abraded with a mixture of virus and an abrasive material, and localized necrotic lesions can be observed. Viral suspensions may be injected into a plant with a syringe, thus emulating transfer by insects in nature.

QUANTITATION OF VIRUSES

To understand the nature of viruses, their reproduction, and their properties, it is essential that methods be available to quantify the virus particles present in a suspension. Directly counting the virions or quantifying viruses as infectious units can accomplish this. A direct count can be accomplished with an electron microscope (EM). Dilutions of virions would be mixed with a predetermined concentration of small latex beads and the mixture examined by EM. The ratio of beads to viruses in a field would give a reasonable estimate of viral particles present.

Many viruses agglutinate red blood cells (RBCs), a phenomenon termed **hemagglutination**. This has been adapted as an effective method for quantitating selected viruses such as the influenza virus. Serially diluted suspensions containing virus are added to a standardized number of RBCs. The hemagglutination titer is the highest dilution of virus that causes visible aggregates.

Viruses are most often quantified as **virus infectious units**. A virus infectious unit is the smallest unit that causes a detectable effect on a specific host. The number of units present in a suspension can be determined by a plaque assay or by the direct effect on a host animal or plant.

Plaque assays are an effective means of measuring the number of infectious particles in a viral suspension and can be routinely applied to bacterial viruses. This can be accomplished by pouring a sterile agar medium that can support growth of the host into a Petri dish. Dilutions of the virus suspension would be added to a suspension of host bacteria in molten agar and, after mixing, the virus/bacterial suspension poured over the agar surface. A dilution of virus with significantly fewer virions than there are bacteria present will result in clear visible areas in the soft agar layer, and these are termed **plaques** (**Figure 14.7**). The plaque results from the destruction of a bacterium followed by destruction of those adjacent to it, and each plaque is assumed to have resulted from an infection with one virion. Counting the number of plaques on a plate permits one to estimate the number of **plaque forming units** (**PFUs**) in the original suspension.

The plaque assay may also be adapted to animal viruses. However, some animal viruses cause lethal cytopathic effects, but these do not result in discernible plaques. For these viruses, degenerative changes in host cells must be quantitated by microscopic examination.

Inoculating a suitable host with serially diluted virus and then determining the highest dilution that caused symptoms of disease or death can quantitate animal viruses that cannot be quantitated by plaque assays.

Leaf assays are useful in estimating the number of plant viruses present in a suspension. A dilution of virus particles would be combined with an abrasive and rubbed into a suitable host leaf. The number of necrotic lesions evident after incubation would be an estimate of the number of viruses present in the original suspension. Abrasion is necessary because plant viruses generally cannot penetrate intact cells.

PURIFICATION OF VIRAL PARTICLES

Virus particles must be separated from host cells to determine their chemical nature, for preparation of vaccines, and for other purposes. The procedure adapted depends on the nature of the virus and the host cell. No single technique is effective for all viral particles, but there are purification methods that are useful, including differential centrifugation, precipitation, denaturation, or enzymatic digestion.

Differential centrifugation. The infected host cells would be lysed, followed by differential centrifugation. The large host cell parts would precipitate at a relatively low speed, and the supernatant would contain the virus. The virus would then be precipitated by high-speed centrifugation.

Precipitation. Proteins from cell debris and virus would precipitate at different salt concentrations. Salts compete with protein for available water, and specific proteins tend to associate at a particular salt concentration. Each protein has a specific salting-out point.

Denaturation. Some viruses are tolerant of organic solvents such as butanol that denatures host cell debris. The virus would enter the aqueous phase, the cell debris would be present at the interface, and lipids would enter the solvent phase.

Enzymatic digestion. Selected proteases and nucleases would digest cell components, and a specific virus may be resistant to these enzymes. The cell components would be solubilized and the virus obtained by centrifugation.

VIRAL GENOME AND REPRODUCTION

The genomes of all cells are composed of double-stranded DNA. A virus genome differs in that it can be either DNA or RNA. Viruses have not been characterized that contain both RNA and DNA, although certain DNA viruses contain RNA primers. The genome of viruses that infects plants, animals, or bacteria can be composed of a single strand (ss) or double strand (ds) of RNA or DNA. An exception would be the hepadna viruses that are partially double stranded and partially single stranded. A viral genome may be a single strand of RNA that is mRNA (+ strand) or an RNA (– strand) that must be transcribed to produce mRNA that is translated to make proteins. The arena viruses and phleboviruses have an RNA genome that is part + and part – strand. There are no ssRNA (– strand) viruses yet characterized that infect bacteria.

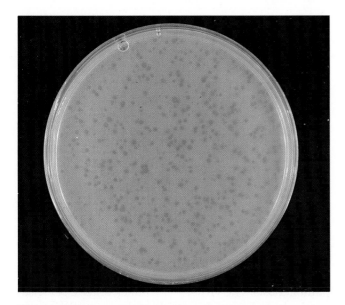

Figure 14.7 Bacteriophage plaques
Plaques formed in a lawn of *Escherichia coli* by phage T4.
Photo ©E. Chan/Visuals Unlimited.

Table 14.1	The pathway to mRNA followed in viruses that have differing types of nucleic acid

Virus Type

dsDNA ──────────┐
ssDNA ──→ dsDNA ──┴──→ mRNA
ssRNA (+) acts as mRNA
ssRNA (−)[a] ──────→ + strand RNA is mRNA
dsRNA ──────→ − strand transcribed ──────→ mRNA

Retroviruses

ssRNA (+) ──────→ ssDNA (−) ──────→ dsDNA ──────→ mRNA

[a]There are plant and animal viruses of this type, but bacterial ssRNA (−) have not been characterized. The minus strand of nucleic acid is the strand transcribed to mRNA. With dsDNA the minus strand is transcribed. By convention the mRNA translated into protein is considered the plus strand and is complementary to the minus strand of DNA.

The pathways followed in production of mRNA from the genomic nucleic acids of various viruses are presented in **Table 14.1**. By convention, the negative (−) strand of dsDNA is the one transcribed into mRNA. Consequently, the mRNA translated to protein is considered plus (+) and is a complementary copy of one strand of DNA. However, ssDNA geminiviruses that infect plants make a complementary double strand and transcribe mRNA from both strands. Capsid size limits the amount of nucleic acid a virus may contain. Thus, certain viruses have mechanisms for producing mRNAs that lead to genetic economy not known to occur in eukaryotic and prokaryotic life forms. For example, ϕ × 174, a ssDNA bacteriophage, contains enough DNA to code for four to five proteins (1795 base triplets; 400 amino acids/protein). The virus uses overlapping reading frames, different initiation sites, and frameshifts within a reading frame to produce 11 proteins. Parvoviruses, small ssDNA animal viruses, geminiviruses, ssDNA plant viruses, and Leviviridae, + ssRNA bacteriophage also employ overlapping reading frames.

Tobamoviruses, a group of plant (+) ssRNA viruses including tobacco mosaic virus, create long and a short polyproteins from the same genome sequences when the ribosome reads through the termination codon of the shorter. A (+) ssRNA strand viral genome would actually be mRNA. The RNA genome of retrovirus, such as the virus that causes AIDS, is transcribed to make dsDNA that is integrated into the host genome. From there it can direct the transcription to viral mRNA.

Some viruses produce early and late mRNA, and the early proteins resulting from translation of the early mRNA may direct the host cell to cease normal function. A series of reactions is then initiated that are directed toward production of viral messages. These result in the replication of viral nucleic acids and viral-specific enzymes. The late proteins synthesized from late mRNA are involved with the synthesis of capsid and other proteins needed for the assembly of the mature virus. A bacterial virus may direct the production of lytic enzymes that lyse the host cell and free the mature viral particles.

DNA viruses can transcribe and replicate their genomes using cell enzymes without modification, although some code for their own enzymes and/or modify host enzymes. Replication of viral DNA parallels host DNA replication by requiring a small RNA primer to provide the 3′–OH. The ssDNA parvoviruses self-prime by using palindromes to create a hairpin at the 3′ terminus (**Figure 14.8**). Pox viruses also self-prime by nicking one of their dsDNA covalently closed ends to give a free 3′–OH. Eukaryotic and prokaryotic cells, with the possible exception of some plants, do not contain the RNA-dependent RNA polymerase required to transcribe and replicate the genomes of RNA viruses. The (+) strand RNA viruses code for the polymerase, which is made early in the infection process while the polymerase is present in the virion of the (−) stranded RNA viruses and dsRNA viruses. Hepadna viruses make genomic DNA from an RNA template using a reverse transcriptase (RNA-dependent DNA polymerase) whereas retroviruses employ a reverse transcriptase to replicate their RNA using a DNA intermediate.

BACTERIAL VIRUSES

The first scientist to observe and report the manifestation of a viral attack on bacteria was Frederick W. Twort of England in 1915. Twort noted that an agent that caused dissolution of bacterial colonies could be transferred to intact colonies and that these also were lysed. Twort also realized that the agent was active at high dilution, was heat labile, and passed through a bacterial filter. He suggested that it might be a virus. In 1917, Felix D'Herelle of the Pasteur Institute in Paris observed the lytic phenomenon and called the agent a **bacteriophage**, which means "bacteria-eating." Today, the name *bacteriophage* can be used interchangeably with **bacterial virus**, as virions that attack bacteria have essentially the same attributes of animal or plant viruses—they simply have a different host specificity.

Bacterial viruses have been widely employed as a model system for host–parasite interactions in viral pathogenesis. The viruses that attack enteric bacteria, particularly *Escherichia coli*, have been studied most. There is a broad array of viral types that attack *E. coli*,

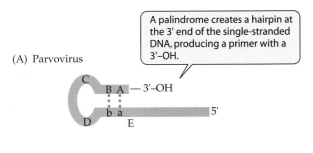

(A) Parvovirus

A palindrome creates a hairpin at the 3' end of the single-stranded DNA, producing a primer with a 3'–OH.

(B) Poxvirus

A nick in one strand of the double-stranded region of the closed DNA…

…produces a primer with a 3'–OH.

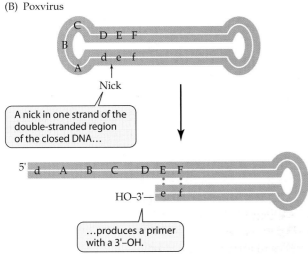

Figure 14.8 Priming of viral replication
Mechanisms used by single-stranded DNA viruses to prime replication of their genomes. (A) Parvovirus; (B) poxvirus.

Taxonomy

Bacterial viruses are generally identified by a Greek symbol or a combination of strain number and symbol. As viruses apparently are polyphyletic (arose multiple times), a grouping on phylogeny would not be possible. There are over 2,000 different bacterial viruses described in the literature. Unfortunately, the guidelines for describing a new virus were not established until recent years, and most of those described are inadequately characterized. Relatedness among viruses is determined by host range, virion morphology, type of nucleic acid present in the genome, and other factors. Immunological identification of viral components is now employed and is of value when characterizing a specific viral group and relating it to others. A classification scheme for bacterial viruses is presented in **Table 14.2**.

Lysogeny and Temperate Viruses

There are two potential outcomes of a viral attack on a susceptible bacterium. The virus may attach to gain

and *E. coli* is useful for study, as we know much about the genetics of the organism. The value of bacteria as a model is apparent when you consider that *E. coli* has a generation time of about 20 minutes, which means that considerable cell mass can be obtained overnight. A virus that attacks *E. coli* can produce mature progeny in around 22 minutes. Thus, one can follow the entire sequence of molecular events—from attachment of virus to the bacterium to lysis and release of virus—over a very short period of time.

There are viruses that attack virtually all microorganisms, including thermophiles and cyanobacteria. The number of viral types is broad, and the mechanism for viral reproduction is varied. Some of the viruses are complex, having a discrete head and tail.

Table 14.2		The families of some major bacterial viruses		
DNA ds – linear	**Size (nm)**	**Example**	**Organism Attacked**	
Siphoviridae	Head 90 Tail 200 × 15	λ	*E. coli*	
Myoviridae	Head 80 × 110 Tail 110 × 25	T4	*E. coli*	
ds – linear Corticoviridae	60	PM2	*Pseudomonas*	
Plasmaviridae	50 to 120	MV-L2	*Mycoplasma*	
ss – circular Microviridae	30	φ × 174	*E. coli*	
Inoviridae	9 × 890	M13	*E. coli*	
RNA ds – linear	**Size (nm)**	**Example**	**Organism Attacked**	
Cystoviridae	85	φ 6	*Pseudomonas phaseolicola*	
ss – linear Leviviridae	30	MS2	*E. coli*	

entry, reproduce, and destroy the host. A virus can also attach, gain entry, and establish a mutually beneficial interaction with a bacterium. This interaction allows both viral propagation and survival of the host in an association termed **lysogeny**. Those viruses that can establish this relationship are called **temperate** viruses. The nucleic acid of a temperate virus enters the host, becomes integrated into the host chromosome, and replicates along with the host genome. The progeny of a lysogenized cell are lysogenic (Figure 14.6). In nature, many bacteria may be lysogenic for one or more viruses. The lysogenic state can cause an alteration in the host outer envelope that can modify attachment sites for other bacterial viruses. Generally there are specific sites on a bacterium where the virus attaches, and infection cannot occur unless there is attachment. Lysogeny, then, may be beneficial in that it can protect the host bacterium from attack by more virulent viruses that are related to the temperate one. Lysis of all hosts could lead to extinction of a virus, whereas a lysogenized bacterium would retain the virus indefinitely.

In a lysogenic culture, 1/1,000 or less of the infected cells become lytic and produce virions. Proteins that repress the expression of genes for a lytic infection are part of the integrated prophage (provirus). It is these repressor proteins that maintain the lysogenic state. Under certain conditions the lysogenic state is lost, the virion is activated, and lysis occurs. This is called **induction of the lytic state**. Induction occurs at a low spontaneous rate but is accelerated in the presence of mutagens such as UV, X rays, and nitrogen mustards. Much research has been devoted to the lysogenic state because it has implications for gene regulatory mechanisms in all organisms and for dormancy in human viruses. The integration of viral nucleic acid into the host genome is also of considerable interest in diseases such as AIDS and cancer.

Lysogenic Conversion

As mentioned previously, lysogeny can alter surface components of the host bacterium and render it immune to attack by other bacteriophages. Lysogenic cells can undergo **lysogenic conversion**, thus gaining other properties that nonlysogenized counterparts do not have. These conversions result from expression of genes on the phage chromosome that are not repressed during lysogeny and will continue to be expressed as long as the bacterium retains the prophage. Lysogenic conversion changes the phenotype of a cell and can lead to an increase in pathogenicity in some disease-causing bacteria.

Lysogenic strains of *Corynebacterium diphtheriae* produce a potent exotoxin that is the cause of diphtheria. Scarlet fever results from a toxin released by lysogenized *Staphylococcus aureus,* and the botulinum toxin is also a product of prophage-bearing *Clostridium botulinum.* The

epsilon phage of *Salmonella* sp. alters surface lipopolysaccharides by altering the activity of enzymes involved in their synthesis. This changes the antigenic character of the surface and susceptibility of the pathogen to specific antibody. Elimination of the prophage from the aforementioned bacteria removes the virus-induced pathogenicity.

Lambda (λ) Phage

Lambda (λ) has long been a model system for understanding lysogeny and gene regulation mechanisms in its host, *Escherichia coli*. In the lysogenic state, λ DNA is integrated into the host genome at a specific site (*att*B) located between the *gal* operon (for catabolism of galactose) and the *bio* operon (for biotin biosynthesis). The integration of λ phage is effected by the protein integrase (*int*) a gene product that is encoded in the viral genome. The genome of λ has a region of DNA (*att*P) that is homologous to the *att*B site of the host genome. Immediately after λ infects the host, its linear dsDNA is circularized. The circularized virus integrates by a single reciprocal crossover between *att*P and *att*B, leading to the lysogenic state.

Whether lysis or lysogeny occurs with λ is determined by a number of host and phage factors, but ultimately depends on the concentration of two viral encoded proteins, Cro and cI (the λ repressor). The *cro* and *cI* genes are adjacent in the λ genome and are transcribed in opposing directions (**Figure 14.9**). The adjacent promoters are designated P_{RM} ("repressor maintenance") for *cI* transcription and P_R ("rightward") for *cro* transcription. cI and Cro proteins bind to the same transcriptional operators (O_{R1}, O_{R2}, and O_{R3}) adjacent to P_{RM} and P_R. When cI binds to O_{R1} and O_{R2}, it represses rightward transcription from P_R, but activates transcription from P_{RM}. Therefore, more cI repressor is made, and the lysogenic state is maintained. The cI repressor prevents expression of virtually all phage genes, including those transcribed from P_L (the leftward promoter). Thus, the cI protein is both a repressor and an activator of transcription. The level of the cII protein, an activator of λ *int* gene transcription, is also important for promoting lysogeny. The cII protein is unstable and affected by the physiological state of the bacterium. Induction of the lytic cycle depends on the relative amounts of the two repressor proteins cI and Cro. Although these repressors bind to the same operators (O_{R1}, O_{R2}, and O_{R3}) that regulate P_R and P_{RM}, they bind them in different order. The cI repressor has the highest affinity for O_{R1} and lowest for O_{R3}, whereas Cro has the highest affinity for O_{R3}. A high cI to Cro ratio will inhibit P_R and maintain lysogeny as described. If cI levels drop (by UV-induced cleavage, for example), Cro binds first to O_{R3}, repressor synthesis from P_{RM} is inhibited and lytic genes are transcribed from P_R and P_L. Expression of lytic genes

Figure 14.9 Lambda phage lysogeny
The genome of λ phage, with details of the critical area that determines whether the phage enters a lytic cycle or remains in the lysogenic state. The determining factor is the relative amounts of the two repressor proteins, cI and Cro.

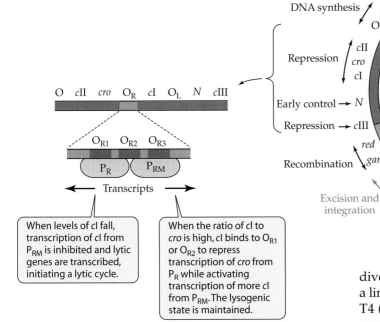

When levels of cI fall, transcription of cI from P_{RM} is inhibited and lytic genes are transcribed, initiating a lytic cycle.

When the ratio of cI to cro is high, cI binds to O_{R1} or O_{R2} to repress transcription of cro from P_R while activating transcription of more cI from P_{RM}. The lysogenic state is maintained.

results in excision of the prophage from the bacterial genome *att* site as a covalently closed circle.

The λ DNA replication cycle is presented in **Figure 14.10**. The excised, circularized genome becomes super-coiled, and early replication via "theta structures" yields additional copies of the circularized λ phage genome. Later, rolling circle replication leads to synthesis of linear concatemers (a chain composed of a number of repeated units of duplex genomic DNA). The concatemers are cleaved to form the individual λ phage genomes that are packaged in a separately synthesized capsid. Mature viruses then escape by lysing the host cell wall.

Mutagens and other agents that disrupt bacterial DNA and affect the integrity of the cI repressor are ultimately responsible for inducing the lytic cycle and virus release. Lysis can therefore be considered a survival mechanism for the phage.

We will now examine in some detail two other bacterial viruses that have been studied extensively and serve as examples of alternative mechanisms by which bacterial viruses propagate, M13 and T4.

M13

Bacterial viruses have proven to be excellent models for the study of viral infection and propagation. Some of the viral types and their size, structure, host, and nucleic acid components were presented in Table 14.2. The

diversity of viral types is too large to discuss more than a limited number. Consequently, only the complex virus T4 (Figure 14.5) and the filamentous M13 will be discussed in some detail.

The filamentous phage designated M13 has an unusual property in that it can invade a cell (*Escherichia coli*), reproduce, and exit without serious harm to the host. The M13 virion is a thin filament about 895 nm in length and 9 nm in diameter (**Figure 14.11**). The inner diameter of the protein capsid is 2.5 nm. This encloses a single-strand loop of DNA that extends the entire length of the filament. Stretched out, the viral genome would be nearly 2 μm meters in length. The capsid coat consists of about 2,700 individual protein subunits that overlap one another. At one end of the virion are four pilot proteins that guide the virus into and out of the host. In one generation, *E. coli* infected with M13 can produce 1,000 intact progeny virions. A considerable mass of virions can readily be collected (100 mg/liter growing culture) free of the host cell debris that would be associated with a lytic infection.

Genome The M13 genome is composed of 6,407 nucleotides, and there are ten defined genes. The genome has an intergenic space (about 8% of the genome) that is the origin of DNA replication. Single-stranded foreign DNA can be inserted into the intergenic space of the M13 genome to generate DNA fragments for sequencing and other studies.

Coat The M13 coat protein is composed of subunits that contain 50 amino acid residues. The capsid protein from gene VIII accumulates on the plasma membrane of the infected cell and is assembled on the genome as the

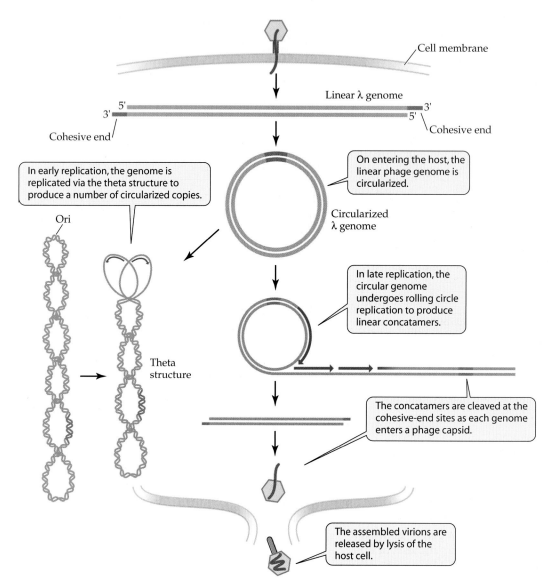

Figure 14.10 Replication of λ phage
Replication of λ phage. After entry into the host cell, the linear λ genome is circularized by joining of the cohesive ends.

Labels within figure:
- Cell membrane
- Linear λ genome
- 5' 3'
- 3' 5'
- Cohesive end
- Cohesive end
- On entering the host, the linear phage genome is circularized.
- In early replication, the genome is replicated via the theta structure to produce a number of circularized copies.
- Ori
- Circularized λ genome
- Theta structure
- In late replication, the circular genome undergoes rolling circle replication to produce linear concatamers.
- The concatamers are cleaved at the cohesive-end sites as each genome enters a phage capsid.
- The assembled virions are released by lysis of the host cell.

DNA exits from the cell. The binding proteins that protect the M13 genome are removed by endopeptidase as the pilot protein guides the genome outward.

Penetration and Infection The DNA of M13 is not injected into the host cell, as occurs with most bacterial viruses. The virion attaches at or near the F (sex factor) pilus of *E. coli.* (organisms lacking an F pilus are not susceptible to attack). Two theories have been advanced for viral penetration: (a) the virion follows a groove in the F pilus to a receptor site on the cell surface where it enters or (b) the virion stimulates a progressive depolymerization of the pilus that retracts the tip and brings the M13 to the surface.

Entry of the intact virion is evident, and decapsidation occurs within the cell cytoplasm. Replication of the (+) single strand of viral DNA occurs simultaneously with coat removal. Most of the capsid protein is conserved and deposited on the plasma membrane and reappears as capsid on progeny DNA. The gene III pilot protein present at the lead end of the virus remains bound to the genome, as it is initially replicated to the duplex form. The enzymes involved in initial replication are provided by the host and are normally associated with synthesis of extrachromosomal elements in the bacterium. M13 does not take over the genetic apparatus of the host. The growth rate of *E. coli* producing M13 virions is reduced by about one-third.

Replication The overall events involved in the replication of M13 viral DNA are presented in **Figure 14.12**. The process can be divided into three stages. In the first

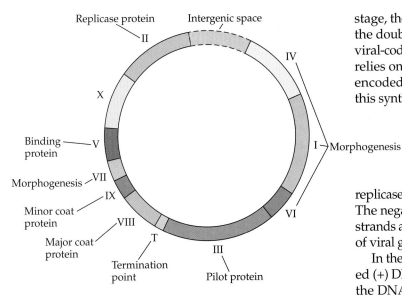

Figure 14.11 Genome of M13
A genome map of M13. The gene product or function of each gene is indicated (except for gene X, whose function is unknown). Figure 14.12 describes the role of the products of genes II, III, and V.

stage, the viral single-stranded (+) DNA is converted to the double-stranded replicative form (RF). There are no viral-coded proteins synthesized for this step. The phage relies on the host replicative system, and the gene (III) encoded A protein apparently is involved in directing this synthesis. Remember that in M13 infections the host synthetic machinery is not disrupted and the bacterial genome remains intact and functional.

In the second stage, a virus-encoded gene (II) is expressed and leads to synthesis of a replicase protein that promotes the multiplication of RFs. The negative strands of these RFs generate progeny (+) strands and serve as a template for continued production of viral genome.

In the third stage, the synthesized viral single-stranded (+) DNA is coated by binding proteins. This envelops the DNA in a form for assembly at the membrane. The function of the gene V binding protein is to facilitate the processes that generate (+) ssDNA. Binding also prevents the (+) strands from being employed as templates for synthesis of duplexes and protects against nucleases. As protein A guides the DNA through the cytoplasmic membrane, the capsid protein replaces the binding protein. The mature virus is extruded through the multilayered gram-negative bacterial cell wall without disrupting the host cell.

T4

The *Escherichia coli* phage T4 is a complex phage with several distinct functional parts including a head, tail, whiskers, and tail fibers (Figure 14.5). T4, along with T2 and T6, is one in a series of bacterial viruses called the T-even phages, which share about 85% of DNA homology. The most studied has been T4 because it is large, is complex, and has many capabilities, including the ability to propagate in a short period of time. As discussed previously, the host for T4, *E. coli*, has a doubling time of about 20 minutes under nutrient-rich conditions. The genome of *E. coli* has about 3,000 genes and can reproduce these in about 40 minutes. The organism can divide in 20 minutes because multiple rounds of DNA replication are ongoing in a rapidly growing cell. However, when *E. coli* becomes infected with T4, the virus assumes complete control of the cell's reproduc-

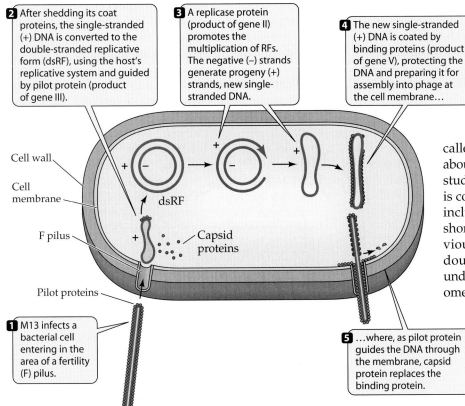

2 After shedding its coat proteins, the single-stranded (+) DNA is converted to the double-stranded replicative form (dsRF), using the host's replicative system and guided by pilot protein (product of gene III).

3 A replicase protein (product of gene II) promotes the multiplication of RFs. The negative (–) strands generate progeny (+) strands, new single-stranded DNA.

4 The new single-stranded (+) DNA is coated by binding proteins (product of gene V), protecting the DNA and preparing it for assembly into phage at the cell membrane…

1 M13 infects a bacterial cell entering in the area of a fertility (F) pilus.

5 …where, as pilot protein guides the DNA through the membrane, capsid protein replaces the binding protein.

Figure 14.12 Replication of M13
Summary of events occurring during the replication of M13.

tive machinery. Within about 22 minutes after infection, the *E. coli* host lyses and releases up to 300 viral particles. These viral particles each contain a genome bearing 200 genes. Essentially a normally dividing *E. coli* produces about 3,000 genes in 40 minutes whereas the machinery of an infected bacterium can generate 60,000 genes in an equivalent length of time. The ability to take over a cell and accomplish this feat in so short a time makes the T4 virion a prime object for study. The events that occur during a lytic infection of *E. coli* follow.

Genome   The T4 genome is among the largest known in viruses. It encompasses about 200 genes and is synthesized as a **concatamer**, a continuous series of separate genomes linked in a linear fashion. As the genome is packaged in the head of the virion, somewhat more than one genome (105%) enters. This ensures that the same gene will be present at or near the end of each genome (**Figure 14.13**), a phenomenon called **terminal redundancy**. The sequence of genes in each viral head is equivalent, but the genome may start and end with a different gene. Therefore, each viral particle has a complete genome plus two sets of at least one gene.

The genome of T4 is linear as packed in a phage head and enters a bacterial host as a linear strand. All markers or genes in the genome are linked. Mapping indicates that each marker identified is spatially linked to other known markers. Given that the gene at the end of each genome may differ, no single gene would consistently appear at the end of the linear chain. The genes as mapped would be linked on both sides by other known genes. Each gene would be in the middle of a linear genome with the same frequency. This phenomenon is called **circular permutation**.

The expression of T4 genes is tightly regulated. The organization of the genome is such that expressing the genes in the order they appear would follow the life cycle of the virus. Early genes are transcribed in a counterclockwise direction, and late genes tend to be read in the opposite or clockwise direction.

Absorption and Penetration   There are distinct sites on the outer envelope of *Escherichia coli* that act as receptors for the fiber tips of the T4 virion. The viral particle attaches at these recognition sites. This attachment may occur at a fusion point between the inner cytoplasmic membrane and the outer cell envelope (see Chapter 11). After attachment of the base plate, the sheath reorganizes, and the inner core of the tail penetrates through the cell envelope and cytoplasmic membrane. The sheath contains 24 protein rings but during penetration the proteins slide downward to form 12 rings. The tail sheath contains ATP, and the contraction has been compared with muscle activity. The core is unplugged, and the DNA genome is injected into the host.

Effect on Host   Within one minute after injection of viral DNA, the synthesis of host-specific macromolecules ceases. This stoppage occurs because the bacterial RNA polymerase has a strong affinity to promoters of T4 DNA, so T4 DNA transcription gets higher priority. Transcription of selected phage genes is initiated a few minutes after infection (**Figure 14.14**). In less than 2 minutes, the host RNA polymerase begins transcribing phage mRNA. This mRNA is translated to form early proteins from viral DNA and occurs before viral specific DNA is synthesized. These early proteins and enzymes take over various processes in the host cell and effect changes in the host genome. The synthesis of viral nucleic acid is initiated by these modifications. The host genome is broken up, and fragments of the DNA move to the cytoplasmic membrane. Degradation of the host nucleus to produce nucleotides that can be assembled to form viral DNA occurs within 5 minutes after infection. In less than 10 minutes the viral attack turns the host cell into a virus-synthesizing machine.

Replication   To induce the host to immediately express the T4 genome, the virus must take control of transcription and translation. Shortly after infection, proteins are synthesized, employing viral genome information that modifies the specificity of host RNA polymerase. Many of the enzymes actually involved in replication of viral DNA and encoded by the virus genome have catalytic functions that are equivalent to the enzymes synthesized by the host. The virus stimulates synthesis of enzymes leading to a more rapid synthesis of T4-specific DNA. The replication rate for phage DNA is ten times the rate of host DNA synthesis in an uninfected, normally dividing *bacterium*. Phage-induced enzymes are active in

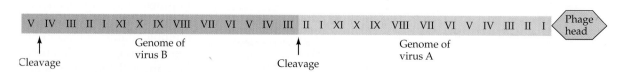

Figure 14.13 Terminal redundancy
A concatemer of T4, as it might be cleaved by endonuclease. Virus A would receive duplicates of genes I and II.

Figure 14.14 Lytic infection by T4

The order of events in a lytic infection of *E. coli* by virus T4.

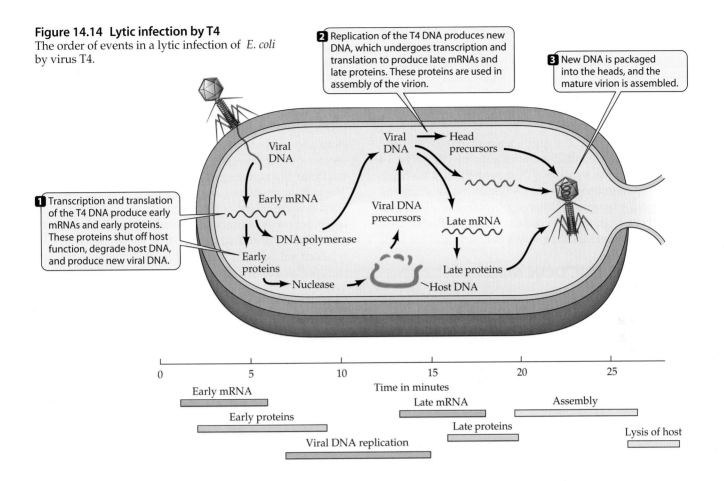

2 Replication of the T4 DNA produces new DNA, which undergoes transcription and translation to produce late mRNAs and late proteins. These proteins are used in assembly of the virion.

3 New DNA is packaged into the heads, and the mature virion is assembled.

1 Transcription and translation of the T4 DNA produce early mRNAs and early proteins. These proteins shut off host function, degrade host DNA, and produce new viral DNA.

promoting this increase. There are at least 20 viral-encoded proteins synthesized immediately after infection, and some of these effect the synthesis of new tRNAs. These tRNAs read the T4 mRNA more efficiently and rapidly than occurs in a normal growing cell. The phage also induces the synthesis of about 30 replication proteins. These proteins are involved in nucleotide biosynthesis, DNA synthesis, and modifications of DNA.

The pyrimidine cytosine is the nucleic acid base generally paired with guanine in cellular DNA. In T4 DNA, the cytosine is replaced by the unique base 5-hydroxymethyl cytosine (**Figure 14.15**). Endonucleases encoded by the viral genome hydrolyze the host DNA. To prevent the inclusion of unaltered deoxycytidine triphosphate (dCTP) as dCTP is formed, it is converted to dCMP. The dCMP reacts with hydroxymethylate to form 5-hydroxymethyl dCMP. The 5-hydroxymethyl dCMP is then phosphorylated to form 5-hydroxymethyl dCTP, which is incorporated into the viral genome. A glucosylation reaction adds a glucose molecule to the hydroxy group after the dCTP derivative is incorporated into viral DNA, thus protecting the DNA of the virion from attack by host-restriction enzymes in subsequent attacks on *E. coli*.

Assembly After the polyhedron head of T4 is completed, it is packed with viral DNA. A few low molecular weight core proteins and phage-induced enzymes are also incorporated into the T4 head. Magnesium and polyamines are also included to neutralize the genome DNA. It is remarkable that the 0.085 μm-wide capsid of T4 can incorporate a DNA genome that is 500 μm in length.

Translation of the late mRNA results in the synthesis of tail fibers and whiskers. The base plate is assembled

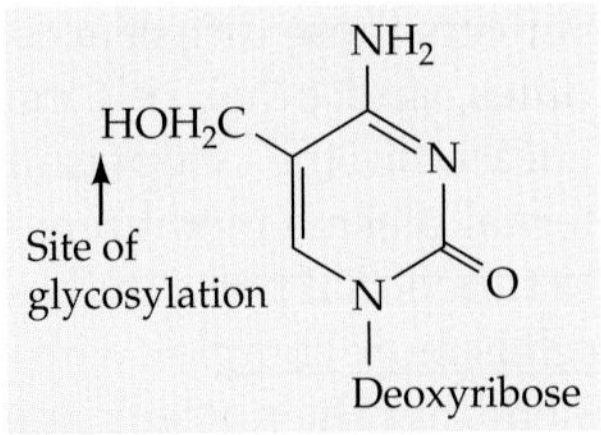

Figure 14.15 5-Hydroxymethylcytosine

The unique pyrimidine present in the DNA of T4, 5-hydroxymethylcytosine. This base is often glycosylated (has a glucose attached) at the hydroxyl, which renders it resistant to host restriction enzymes.

as a unit, and the core or tube through which the DNA is injected is assembled on this base plate. The core is capped at the distal end and encircled by the sheath proteins. The head and tail are then joined together. Whiskers are then put in place, and these probably direct the placement of the long tail fibers on the base plate.

Release Lysis of *E. coli* and release of 100 to 300 viral particles occurs about 25 minutes after infection. The T4 genes direct the synthesis of a protein that disrupts the cytoplasmic membrane. A lysozyme synthesized through T4 templates moves through the disrupted membrane to breach the peptidoglycan cell wall, and the cell disintegrates, releasing viral particles.

VIRAL RESTRICTION AND MODIFICATION

Bacteria have effective mechanisms for protection against viral invaders. The outer surface may be modified to prevent attachment or they produce enzymes, termed restriction enzymes, that cleave the DNA of an invader at one or more sites. Restriction enzymes are quite specific and attack only selected nucleotide sequences that are generally 5 to 6 base pairs in length. The host is protected against its own restriction enzymes by chemical alteration of its DNA at sites where the restriction enzyme might function. An example of chemical modification is methylation of bases at potentially susceptible sequences.

Viruses can counteract restriction enzymes by adding methyl or sugar groups (glycosylate) to the nucleic acid bases at susceptible sites. The T-even phages (T2, T4, T6) that infect *Escherichia coli* can glycosylate selected bases in DNA, rendering the genome resistant to restriction enzymes. The lambda phages methylate adenine and cytosine bases in their DNA. Methylation and glycosylation occur after replication of DNA.

EUKARYOTIC VIRUSES

Prokaryotic and eukaryotic cells differ significantly in structure. The strategy of viral attack on these two distinct cell types reflects these differences. The nuclear material, ribosomes, and other cell machinery of a prokaryote are encased in the cytoplasmic membrane. There are no internal plasma membrane-bound structures or separate functional parts in these organisms. The eukaryotic cell is more complex with discrete internal organelles including the nucleus, mitochondrion, and other structures that are membrane bound.

When a virus attacks a susceptible bacterial host, the genome of the virus must traverse the cell wall and cytoplasmic membrane. Once the viral genetic information enters the cytoplasm, it has immediate access to the transcription/translation machinery and energy-producing systems of the cell all in one compartment. In infecting a eukaryotic cell, the virus must pass through the cytoplasmic membrane, and in most DNA viruses and some RNA viruses, the viral genome must also gain access to the membrane-bound nucleus.

The receptor present on the surface of a virus attaches to a specific site on the surface of the host cell. This determines which host the virus can attack and, in multicellular hosts, the specific cells that will be infected. The cells in an animal are highly differentiated, and the surface components of these cells would differ. Potential receptors on the surface of the various organs in an animal would differ, and viruses that could attach to one type cell would not affect another. This is considered tissue specificity.

Classification, Size, and Structure

Some major animal and plant viruses are illustrated in **Tables 14.3** and **14.4**. These viruses, as with bacterial viral particles, are classified by the nature of the nucleic acid in the genome, although the presence or absence of an envelope is also considered. The animal viruses range from the tiny parvoviruses (20 nm) to very large poxviruses (250×350 nm). Many of the plant viruses are very long narrow filaments. For example, the potato γ virus is 10 nm across but 700 nm in length.

Viruses can differ in their effects on the host cell. An acute **infection** has a rapid onset and short duration but results in the destruction of the host cell. *Picornaviruses* such as the polio, influenza, and common cold viruses cause acute infections. In a **persistent infection** the virus replicates actively, there may be few symptoms, and the host cell retains viability and viral production continues for an extended period of time. The mature virus can leave the host cell by budding without disrupting the integrity of the cell. Measles and hepatitis B are examples of persistent or chronic infections. A **latent infection** is one in which the virus is generally not actively replicating and remains dormant within the host. During latency, there are no symptoms nor are antibodies produced against the virus. The herpesvirus (Herpes simplex type I), the cause of cold sores, would be an example of a latent viral infection. Most plant viruses cause persistent infections, and a few are involved in latent infections.

Some viral infections can cause **transformation** of the host cell. Transforming viruses can change a normal cell into a cancer cell with fewer growth factor requirements than for the normal cell. As a consequence, the transformed cell reproduces more rapidly, resulting in a mass of cells called a tumor. Some tumors are self-limiting, do not spread, and are called **benign**. Others are **malignant** tumors, and these spread and grow in other tissue and organs causing dysfunction or death of that tissue.

Malignant tumors initiated by viruses result from genetic changes either in expression or structure of the genes that lead to a loss of growth regulation. Abnormal masses of cells result, and these cancers or neoplasms can grow unchecked. Phosphorylations, transcription, or enzymes involved in DNA or RNA synthesis are among the altered functions caused by tumor viruses, and these altered functions can also result from other physical factors such as mutations. Chemicals, diet, and environmental factors can cause genetic alterations that ultimately lead to cancer.

ANIMAL VIRUSES

The genome of the viruses that infect animals is composed of either DNA or RNA (**Table 14.5**). The genomes of DNA viruses are mostly double stranded, but a few, such as the parvoviruses, are single stranded. The genome of RNA viruses is generally single-stranded RNA. The genome may be surrounded by a nucleocapsid, or an envelope of varying complexity may in turn surround a nucleocapsid. Attachment to the host cell may involve the nucleocapsid, the envelope, or the spikes that extend out from the surface of the virus. The unenveloped viruses enter a host cell by mechanisms that are not clearly understood. For example, the poliovirus attaches to a susceptible cell, the protein capsid loses structural integrity, and the RNA protein complex is translocated into the cytoplasm. It is not clear whether the capsid enters the cytoplasm in all cases with unenveloped viruses. The attachment and penetration of the genome of enveloped viruses into the host can occur by fusion of the viral particle with the cell membrane or by endocytosis (**Figure 14.16**). A limited number of representative viruses will be discussed in some detail.

Table 14.3	Some of the major animal virus genera		
ds – linear DNA	**Size (nm)**	**Example of Disease**	
Poxviruses	250 × 350	Smallpox, fever blisters	
Herpesviruses	180 – 120	Genital herpes	
Adenoviruses	75	Genital herpes	
s – linear Parvoviruses	20	Gastroenteritis	
ds – circular Papovaviruses	50	Warts (Human)	
Baculoviruses	40 × 400	Polyhedrosis (Lepiderm insects)	
ss – (+) RNA	**Size (nm)**	**Example of Disease**	
Picornaviruses	27	Polio, colds	
Togaviruses	50	Rubella, sindbis	
Flavivirus	40 × 50	Yellow fever	
Retroviruses	80	Sarcomas, leukemia	
Coronaviruses	25	Avian bronchitis	
ss (–) Orthomyxoviruses	110	Influenza	
Paramyxoviruses	200	Colds, measles, mumps	
Rhabdoviruses	70 × 170	Rabies	
Bunyaviruses	90	Encephalitis	
Filoviruses	50 × 1000	Ebola virus	
ds Retroviruses	65	Encephalitis, diarrhea	

Table 14.4   Some of the major plant virus genera

ds – linear DNA	Size (nm)	Example of Disease	
Caulimovirus (pararetrovirus)	50	Cauliflower mosaic	
ss – circular			
Geminivirus (pared genome segments)	15 × 30	Maize streak	
ss (+) linear RNA	**Size (nm)**	**Example of Disease**	
Tobamovirus	15 × 300	Tobacco mosaic	
Comovirus (2 genome segments)	30	Cowpea mosiac	
Carlavirus	15 × 165	Carnation latent	
Cucumovirus (3 genome segments)	30	Cucumber mosiac	
Potyvirus	10 × 70	Potato	
ds – linear			
Phytoreovirus (mutiple genome segments)	80	Wound tumor	
ss (–)			
Tospovirus	90	Tomato spotted wilt	
Rhabdovirus	70 × 170	Lettuce necrotic yellows	

Table 14.5   The basic mechanisms for production of viral proteins and genomic nucleic acid in animal viruses

Example of Disease Caused	Genome Type	Mechanism
Polio, cold	(+) ssRNA	Processing $\longrightarrow$ mRNA $\longrightarrow$ Proteins Replicase $\longrightarrow$ (±) RNA $\longrightarrow$ (+) RNA[a]
Influenza, enchephalitis, mumps	(–) ssRNA	RNA dependant polymerase $\longrightarrow$ mRNA $\longrightarrow$ Proteins Replicase $\longrightarrow$ (±) RNA $\longrightarrow$ (–) RNA[a]
Enchephalitis, diarrhea	(±) dsRNA	Transcriptase $\longrightarrow$ mRNA $\longrightarrow$ Proteins mRNA replicase $\longrightarrow$ (±) RNA[a]
Retrovirus		
AIDS, Rous sarcoma	(+) ssRNA	Reverse transcriptase—(–) DNA Reverse transcriptase DNA (±) $\longrightarrow$ (+) RNA[a] $\searrow$ mRNA

[a]Viral genome.

The glycoprotein spikes that protrude outward from the envelope are the means of attachment to specific host cell receptors. After fusion to the host cytoplasmic membrane, the nucleocapsid is extruded into the cytoplasm. The viral particle is uncoated, and the viral dsDNA traverses to the host cell nucleus. Initially the viral genome directs the synthesis of mRNA that is translated into regulatory proteins followed by mRNA that generates the proteins involved in DNA replication. A later mRNA is transcribed from the viral genome and codes for viral structural proteins and envelope proteins that are an integral part of the mature viral particle.

The herpesvirus DNA is synthesized in the host cell nucleus as long concatemers. These long concatemers are processed to viral genome length as they are incorporated into the nucleocapsid. The viral nucleocapsid capsomers are synthesized in the cytoplasm, and the nucleocapsid is assembled in the host nucleus. The amorphous fibrous coat that surrounds the nucleocapsid and the envelope are acquired as the virus buds through the host nuclear membrane. The virus particle passes to the endoplasmic reticulum of the host cell and exits the cell from this site. In many cases the accumulation of viruses may cause disruption of the cytoplasmic membrane.

Some of the herpesviruses, such as Herpes simplex I, can be latent for extended periods of time and become active when the host is under stress (fever, sunburn). The virus remains latent in neurons of sensory ganglia and is not integrated into the host genome.

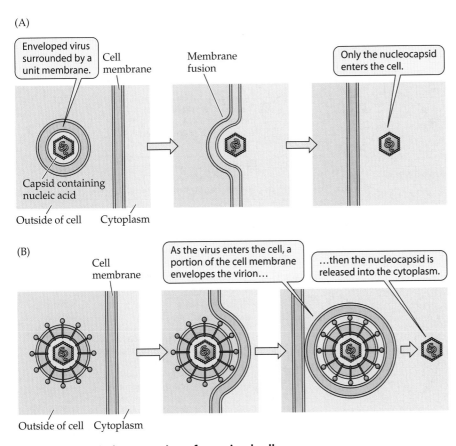

Figure 14.16 Viral penetration of an animal cell
Viruses attach to and enter a host cell by (A) membrane fusion between the host cell membrane and the viral envelope or (B) endocytosis, with invagination of the host cell membrane.

DNA Viruses

Most DNA viruses replicate in the nucleus of the host cell. Poxviruses are an exception, as they replicate in the cytoplasm. Herpesviruses are an example of a dsDNA virus that replicates in the nucleus of the host cell and are the causative agent of several human ailments including fever blisters, shingles, genital herpes, chicken pox, and mononucleosis (see Chapter 29). The Herpes simplex virus is presented in some detail as a typical DNA virus.

Herpesviruses

The herpesvirus particle is quite complex, with an icosahedral capsid enclosed by an envelope bearing elaborate glycoprotein spikes that protrude outward (**Figure 14.17**). The herpesvirus nucleocapsid is about 100 by 200 nm in size. Between the envelope and nucleocapsid is an electron-dense amorphous material of unknown function. The icosahedral nucleocapsid is composed of 162 capsomers made up of several distinctly different proteins. The genome is a dsDNA that carries genetic information for the synthesis of over 100 different polypeptides.

Pox Viruses

The pox viruses are complex viruses that have some features of a free-living cell. One disease caused by this group, smallpox, was important historically (see Chapter 2), but it has been eradicated by worldwide vaccination.

The pox viruses are large, and the vaccinia virus is a boxlike structure ($400 \times 240 \times 200$ nm) with an outer coat composed of protein filaments in an array that resembles a membrane (Figure 29.4). The vaccinia virus is taken up by host cells via a process that resembles phagocytosis. The plasma membrane of the host cell actually extends around the virus, resulting in viral

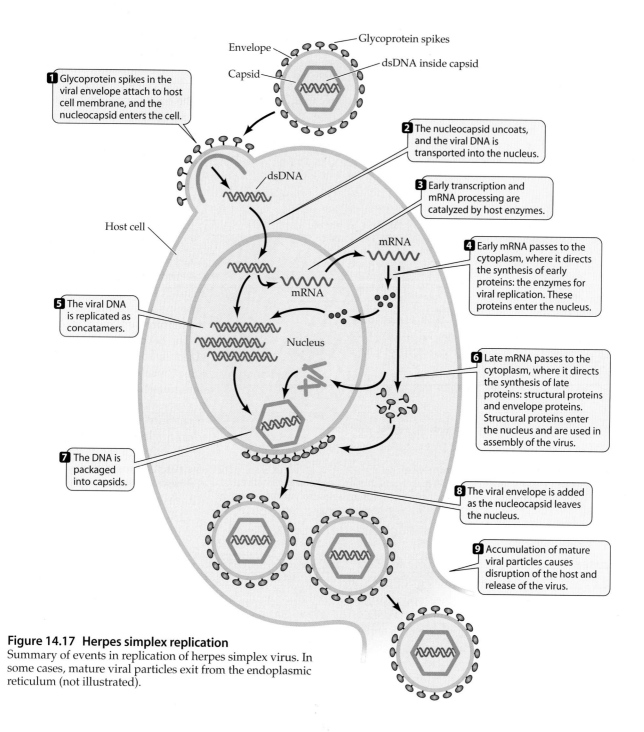

Figure 14.17 Herpes simplex replication
Summary of events in replication of herpes simplex virus. In some cases, mature viral particles exit from the endoplasmic reticulum (not illustrated).

entry. The core of vaccinia virus contains the DNA genome and several viral-encoded proteins. The pox virus replicates in the cytoplasm. As the virus does not penetrate the nucleus of the host, there are no proteins generated on infection that are transcribed from host DNA. Host genes are not expressed after infection, indicating that a molecular message does enter the host nucleus to turn off the host genetic machinery. All of the proteins expressed post infection are encoded in the vaccinia genome. Enzymes and DNA polymerase come from the virus. All of the enzymes necessary to tran-

scribe viral DNA are brought in by the virus upon infection. The mRNA transcripts of viral origin are capped and polyadenylated even before the viral coat is completely removed. The synthesis of pox virus DNA outside the nucleus is unusual, and such extranuclease DNA synthesis generally occurs only in specialized organelles such as mitochondria.

After a viral particle enters the host cytoplasm, an inclusion body is formed. If several viruses infect the host cell, there will be an inclusion body formed around each virus particle. It is in these inclusion bodies that transcription, replication, and encapsidation of mature viral progeny occurs. Disintegration of the host cell leads to release of virus particles.

RNA Viruses

The genome of most RNA viruses is a single strand of RNA (ssRNA), and this strand may be the equivalent of mRNA (plus strand). The ssRNA may be complementary to mRNA (minus strand) and in this case it must be replicated to generate mRNA. Major positive-strand RNA viruses that infect humans are polio and the cold (rhinovirus). Negative-strand viruses are the causative agents of rabies, HIV, influenza, and other human ailments. Two major diseases caused by RNA viruses—polio and HIV—will be discussed.

Poliovirus

The poliovirus has a genome composed of a single-strand of RNA. Polio was once a dreaded paralytic disease, but widespread use of a vaccine has led to its virtual elimination in the United States. The poliovirus multiplies in the intestinal tract of human beings, and infection is generally asymptomatic or causes mild symptoms. The paralytic form, poliomyelitis, results from passage of the virus to motor neurons in the spinal cord. Destruction of these nerve cells leads to paralysis. How a limited number of infections progress from the mild intestinal tract infection to a paralytic form is not known.

The poliovirus attaches to specific receptors on cells of the host intestine and enters host cells by endocytosis. Following entry of the virus, all protein synthesis in the infected cell ceases. The viral genome is mRNA (+ strand) and about 7,500 bases in length. The 5' end of the RNA codes for coat proteins, and replication proteins are coded at the 3' end. At the 5' terminus of the linear viral genome is a small protein attached covalently, and at the 3' end is a polyadenylate chain. The small protein (22 amino acids) is probably involved in formation of the mature viral particle. The polyadenylate chain is involved in infectivity. The (+) strand is copied by an RNA-dependent RNA polymerase. The (–) strand generated then serves as a template for the production of multiple copies of the (+) strand. The viral mRNA is a monogenic code that is translated into a single, very large protein molecule. This polyprotein bears all of the proteins involved in poliovirus synthesis. After formation, the large protein is cleaved into about 20 individual proteins. Some of these proteins are structural; others are the RNA polymerase involved with the synthesis of (–) strand templates and a protease that cleaves the polyprotein. Replication of viral RNA occurs a short time after infection by generating (+) strands from the (–) strand templates. An infected cell can contain hundreds of (–) strands that can generate up to a million viral (+) strands. The individual proteins cleaved from the polyprotein direct the synthesis of structural coat proteins and these are assembled to form the mature viral particle.

Retroviruses

The retroviruses are the causative agents of certain cancers, and Rous sarcoma, a form of chicken cancer, was the first virally induced cancer described. The avian leukemia virus and murine leukemia virus are also retroviruses. The human immuno deficiency virus (HIV) is another retrovirus and is the causative agent of AIDS (acquired immunodeficiency syndrome). HIV will be discussed in this section as a representative of these unique viruses.

The retroviruses are defined by their ability to reverse-transcribe a single-stranded RNA genome to form double-stranded (ds) DNA. The (+) single-stranded RNA genome is first copied into single-stranded (ss) DNA, and this ssDNA is replicated to form dsDNA. The viral dsDNA is an intermediate that is integrated into the host genome and resides there as a **provirus**. The dsDNA can be transcribed to generate mRNA for continued virus production.

Retroviruses are enveloped viruses with a diploid single-stranded RNA genome. They are considered diploid because they carry two copies of the single-stranded RNA genome (**Figure 14.18**). The inner core contains the ssRNA genomes and a number of enzymes that are responsible for the early replication events. Among the enzymes present are integrase, reverse transcriptase, protease, and ribonuclease H. This inner core is surrounded by a protein capsid. There is a protein matrix present on the inner surface of the lipid membrane, and this stabilizes the viral particle. The envelope is a lipid bilayer with spikes composed of glycoprotein encoded by the virus, plus numerous cellular proteins that protrude from the surface of the virion.

Infection occurs after direct exposure to the retrovirus HIV-1 through body fluids—blood, saliva, breast milk, semen, or vaginal fluids. The virus eventually encounters a cell with a high-affinity receptor for the virus surface glycoprotein. There is a highly specific binding between the glycoprotein protruding from the exterior of the virus and a surface molecule on these selected host cells. The host cell receptor is a glycoprotein present in considerable quantity on the outer surface of helper T-lymphocytes, and this glycoprotein is designated CD4. The glycoprotein is also present but in smaller amounts on the surface of monocytes and macrophages. Any cell bearing the specific glycoprotein is termed CD4+. Additional receptors, including CCR5 and CXCR4 plus minor coreceptors are also required for entry into a host cell. If the required receptors are present on a cell, HIV-1 attaches and enters the cell by fusion

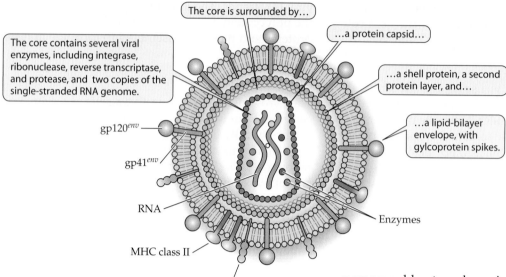

The core is surrounded by…

The core contains several viral enzymes, including integrase, ribonuclease, reverse transcriptase, and protease, and two copies of the single-stranded RNA genome.

…a protein capsid…

…a shell protein, a second protein layer, and…

…a lipid-bilayer envelope, with gylcoprotein spikes.

gp120^{env}

gp41^{env}

RNA

Enzymes

MHC class II

MHC class I

Figure 14.18 HIV-1
Cross-section of the human immunodeficiency virus, the causative agent of AIDS.

with the host cell cytoplasmic membrane (**Figure 14.19**).

After entry into a host cell, the viral RNA is transcribed to DNA by the action of a specific RNA-dependent DNA polymerase (reverse transcriptase). The enzymes responsible are contained in the virus particle. The order of the genetic information in the HIV-1 genome is outlined in **Table 14.6**. Synthesis of DNA occurs in the cytoplasm usually within the first 6 hours of infection. After transcription of the ssRNA viral genome to dsDNA, the RNA is destroyed by the viral ribonuclease. The viral reverse transcriptase that transcribes RNA into dsDNA is quite inaccurate, and there is no mechanism for correction of errors made. The error rate in HIV-1 may be more than ten times higher than in other reverse transcriptases. Therefore, variants occur continuously, and progeny have a DNA template that varies in base sequence from that of the parent virus. This variation generates progeny that may be unaffected by antibodies and the cellular arm of the immune system that would otherwise act against the progenitor virus. This would give a survival advantage.

Although the genome of HIV-1 is RNA, the ultimate genetic information that generates viral progeny is dsDNA. This dsDNA is integrated into the host genome by integrase, where it remains as a provirus. The viral integrase serves three major functions: (1) it trims the ends of the viral dsDNA, (2) it cleaves host DNA, and (3) the integrase covalently links the termini of the trimmed DNA to host DNA. The inserted viral DNA (provirus) has all of the characteristics of a cellular gene. A treatment for HIV-1 infection that might excise the provirus is improbable.

The synthesis of viral genome begins with an RNA transcript formed from the viral DNA template present in the host genome. Transcription rates are regulated in part by host proteins that are involved with the transcription of other host genes. Consequently, production of viral RNA transcripts can result from normal activities that occur in the host. An inactive T-cell that contains the viral genome would not produce viral RNA. Macrophages that are not dividing but contain the provirus do transcribe viral RNA and produce HIV-1.

The viral mRNA transcribed from the provirus DNA serves as the progenitor of capsid proteins and enzymes needed for viral assembly, and in a spliced form the RNA also encodes for the envelope proteins and other auxiliary proteins. The capsid proteins are synthesized as a polyprotein. The replicative enzymes and capsid proteins are fused. These fused proteins move to the inner surface of the host cytoplasmic membrane, and viral RNA binds to the capsid protein precursor. The precursor then surrounds the replicative enzymes and viral RNA, and the spherical particle buds through the host cell membrane. The glycoprotein surface complex present on HIV-1 is formed in the endoplasmic reticulum of the host and is added to the outer envelope of the virus as it buds through the host cell surface.

It is not clear how the HIV-1 virus curtails the immune response or causes the destruction of T cells. The CD4+ T-cells are an essential part of the immunological response. Despite this, an HIV-1 infection will ultimately result in loss of virtually all functional T-cells in the immune system. Destruction of T-cells may occur by apoptosis (programmed cell death) initiated by crosslinking of molecules by gp120 (envelope protein) of the HIV virus and binding of antigen to the T-cell receptor. It is probable that a number of factors combined are involved in the destruction of T cells by infection with HIV-1, and no single factor is responsible for AIDS pathogenesis.

INSECT VIRUSES

Some plant viruses are carried by and multiply in insect vectors, and these viruses may harm or kill the insect. The baculoviruses are one of the most studied of the insect viruses, as they are virulent and kill susceptible insect

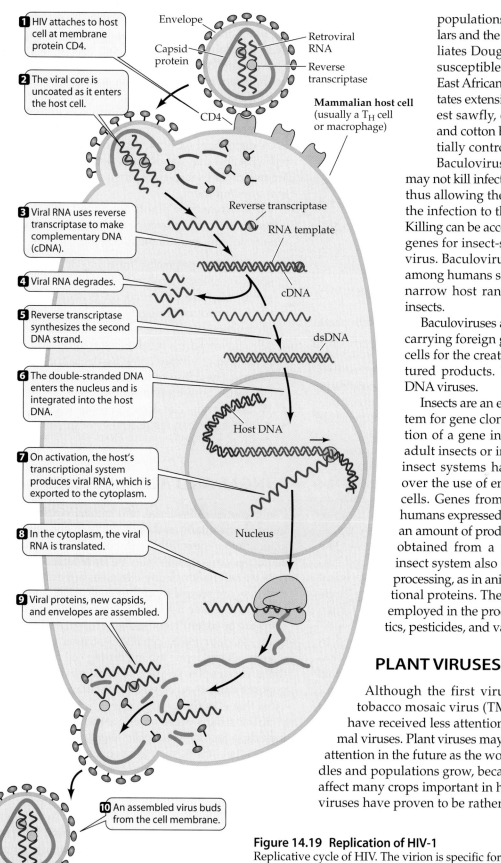

1 HIV attaches to host cell at membrane protein CD4.

2 The viral core is uncoated as it enters the host cell.

3 Viral RNA uses reverse transcriptase to make complementary DNA (cDNA).

4 Viral RNA degrades.

5 Reverse transcriptase synthesizes the second DNA strand.

6 The double-stranded DNA enters the nucleus and is integrated into the host DNA.

7 On activation, the host's transcriptional system produces viral RNA, which is exported to the cytoplasm.

8 In the cytoplasm, the viral RNA is translated.

9 Viral proteins, new capsids, and envelopes are assembled.

10 An assembled virus buds from the cell membrane.

Envelope

Capsid protein

Retroviral RNA

Reverse transcriptase

CD4

Mammalian host cell (usually a T_H cell or macrophage)

Reverse transcriptase

RNA template

cDNA

dsDNA

Host DNA

Nucleus

populations. Eastern tent caterpillars and the tussock moth that defoliates Douglas fir trees are highly susceptible to baculoviruses. The East African armyworm that devastates extensive grassland areas, forest sawfly, orchard codling moth, and cotton bollworm are all potentially controllable by baculovirus. Baculoviruses sprayed on crops may not kill infected insects immediately, thus allowing them to mate and spread the infection to the general population. Killing can be accelerated by introducing genes for insect-specific toxins into the virus. Baculoviruses can be distributed among humans safely, as the virus has a narrow host range and is confined to insects.

Baculoviruses are excellent vectors for carrying foreign genes into insect larval cells for the creation of insect-manufactured products. These are rod-shaped DNA viruses.

Insects are an effective eukaryotic system for gene cloning through introduction of a gene into cells cultured from adult insects or into their larvae. These insect systems have many advantages over the use of engineered mammalian cells. Genes from plants, animals, and humans expressed in insect cells generate an amount of product far in excess of that obtained from a mammalian cell. The insect system also does posttranslational processing, as in animal cells, to yield functional proteins. The insect systems can be employed in the production of pharmaceutics, pesticides, and various proteins.

PLANT VIRUSES

Although the first virus described was the tobacco mosaic virus (TMV), the plant viruses have received less attention than bacterial or animal viruses. Plant viruses may well receive increased attention in the future as the world food supply dwindles and populations grow, because viruses adversely affect many crops important in human nutrition. Plant viruses have proven to be rather difficult to study, but

Figure 14.19 Replication of HIV-1
Replicative cycle of HIV. The virion is specific for CD4$^+$ host cells. The timetable for the events depicted is highly variable, particularly the synthesis of viral RNA that initiates virus production (step 7).

Table 14.6	The known genes in the HIV-1 virion and their functions
gag	Nucleocapsid proteins
pol	Enzymes—integrase, reverse transcriptase, protease, and ribonuclease
vif	Involved in infectivity
vpr	Transcription
tat	Transcription
rev	Gene expression regulator
vpw	Budding
env	Coat glycoproteins, attachment, and fusion
nef	Unknown; signals reproduction

in recent years methods have been developed for growing many of them in plant cell culture.

The plant virus structures are essentially equivalent to the bacterial and animal viruses. Many are long, flexible, very narrow helices but of a length approaching that of a bacterium such as *E. coli*. The clostervirus, called beet yellow, is 1.3 μm in length but exceedingly narrow (<10 nanometers) and cannot be visualized in a light microscope. The capsid of plant viruses is generally composed of a single protein.

The genome of most plant viruses is composed of either single- or double-stranded RNA. There are exceptions, however, and the caulmovirus genome is composed of dsDNA and that in geminivirus is ssDNA. Caulmovirus, the causative agent of cauliflower mosaic, is a pararetrovirus, as it generates mRNA that is transcribed to dsDNA by a reverse transcriptase. The Geminivirus has a single (+) strand of DNA and employs host enzymes to form a dsDNA. The (–) strand produces multiple (+) strands via the rolling circle method. These (+) strands are encapsulated to form a mature virion.

A difficulty that a plant virus must overcome in nature is in passing through the plant cell surface barriers. Plant viruses are disseminated by wind or vectors such as insects or nematodes. The insects that transmit viruses include aphids, leafhoppers, whiteflies, and mealy bugs. Viruses infect the mouthparts of these insects during feeding and are transferred to uninfected hosts during normal feeding as the insects penetrate plant cells to withdraw sap. Some plant viruses are stored in the foregut of aphids and inoculated into the plant by regurgitation during feeding. Plant viruses also enter through breaks or abrasions on plant surfaces. Seeds or pollen transmits some plant viruses. Budding or grafting to nursery stock is also a concern, as parent stock must be virus free to ensure propagation of healthy progeny.

A select number of plant viruses propagate in both the plant and in the insects that transmit them, and in these the plant and insect are both host to the pathogen. The wound tumor viruses multiply in the tissue of leafhoppers before they move to the salivary glands. The virus is then transmitted by the bite of the insect. The RNA viruses that can reproduce in both plant and insect vectors produce viral genomes via a virus-carried RNA-dependent replicase. The tospovirus (tomato spotted wilt) and the rhabdovirus (lettuce necrotic yellow) are two other examples of viruses that infect the insects that transmit them.

Virus diseases of plants are classified according to apparent manifestations of disease symptoms. These manifestations include the following:

- Mosaic diseases cause mottling of leaves, yellow spots, blotches, and necrotic lesions. These symptoms sometimes are observed on flowers. The mosaic viruses produce variegations in leaves or flowers of ornamental plants. The variegated petals observed in tulip flowers are a result of viral infection.
- A number of viruses cause leaves to curl and/or turn yellow. Dwarfing of leaves or excessive branching can result from viral infections.
- Wound tumors can occur on roots or stems.
- Wilt diseases can result in the complete wilting and death of a plant.

FUNGAL AND ALGAL VIRUSES

Plant, animal, eubacterial, and archael viruses are transmitted horizontally as they multiply in one cell, exit, and enter another susceptible cell. Some animal viruses such as herpesvirus or HIV may be transferred to adjacent cells by cell fusion. Fungal viruses are quite unique in that they are spread by cell-to-cell fusion, and extracellular virions have not been observed. Latent viral infections are apparently common in fungi, and the majority of *Saccharomyces cerevisiae* strains carry dsRNA viruses. Cytoplasmic mixing during cell fusion transfers these viruses. There are also yeast viruses that have a genome composed of ssRNA circularized and a retrovirus that is ssRNA.

Killer strains occur among various genera of yeasts including *Saccharomyces*, *Hansenula*, and *Kluyveromyces*. These killer strains secrete a protein toxin that is encoded in specific regions of the genome of a latent dsRNA virus. Resistance to the toxin is also encoded in the viral genome rendering the host immune to the toxin. The killer toxin binds to specific glucans in the cell wall of nonimmune yeasts and causes cell death by creating cation-permeable pores in the cytoplasmic membrane. The killer trait (virus) can be introduced into yeast strains employed in the fermentation industry. The presence of the killer trait in a yeast that is immune to the toxin does not affect the fermentative ability of the strain.

Filamentous fungi, such as *Penicillium chrysogenum* are also susceptible to infection by viruses. A ssRNA virus (Barnavirus) can infect the cultivated mushroom *Agaricus bisporus* and is a concern in commercial production.

Algal viruses such as *Phycodnavirus* (dsDNA) are ubiquitous in fresh water. These viruses are host specific and attach to cell walls of susceptible unicellular eukaryotic algae. The virus causes dissolution of the cell wall at the site of attachment, and viral DNA enters the algal cell. Following virus multiplication, the viral particles are released by cell lysis.

It is apparent from this brief discussion that viruses have evolved that can infect all types of living cells. There is no specific chemotherapy for viral infections, and natural or induced resistance or vaccination is the only control measure now available.

VIROIDS

Viroids are the smallest nucleic acid-containing infectious agents known. They were first described by T. O. Diener, a plant physiologist, in 1971. Viroids have been identified only in plants and are composed solely of RNA. The RNA is made up of 270 to 380 nucleotides (**Figure 14.20**). This amount of RNA would be about one tenth the genetic material present in the smallest of the viruses that have been characterized. This number of nucleotides could theoretically produce a 100 amino acid protein if every base were used in a single reading frame, or two proteins if a double (frameshift) reading frame were used. However, much evidence suggests that viroids do not produce any proteins. The viroid is a circular single strand of RNA collapsed into a rod by intrastrand base pairing. Viroids are not surrounded by a protein capsid. Viroids appear in the nucleus of the plant cells and apparently do not function as mRNA. A viroid is synthesized by host RNA polymerases by employing the viroid particle as a template. The polymerase of the host cell recognizes the RNA as it would a piece of DNA.

The viroid most studied is called the potato spindle-tuber disease agent (PSTD). It is 359 nucleotides in length and may function by interfering with host gene regulation. Comparison of nucleotide sequences in viroids suggests that they are similar to the RNAs of introns. They may have evolved from introns and function by interfering with the normal splicing of introns in cells. A number of strains of PSTD have been isolated, and some cause mild symptoms in plants, whereas others cause lethal infections. The difference in virulence is due to alterations in the sequence of the nucleotides.

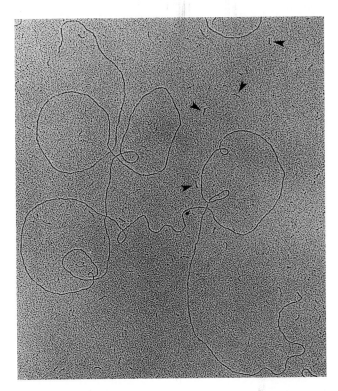

Figure 14.20 A viroid
Comparison of PSTD, a low molecular mass RNA viroid (arrows), with the DNA of T7, an *Escherichia coli* phage. Courtesy of T. O. Diener, U. S. Department of Agriculture.

Viroids are the causative agent of diseases in agricultural crops including cucumbers, potatoes, and citrus fruit, and these diseases cause losses in the millions of dollars each year. The spread of viroids by grafting in fruit trees and through the use of modern harvesting machinery may be involved in viroid transmission. Although viroids have not been identified in animal infections, they well may be. A viroid is not apparent in infected tissue without employing special techniques that involve nucleotide sequencing. Application of these techniques may lead to the identification of viroid-caused infections in humans.

SUMMARY

▶ Viruses vary in **size** from those that are close to the size of a ribosome (polio) to viruses that approach free-living bacteria (pox viruses). Smallpox is a large virus that is visible in phase-contrast microscopy.

▶ A viral genome is surrounded by protein **capsid** that protects the virus when outside a host cell. Many viruses also have a membranous **envelope** that surrounds the capsid, and there may be specific glycoproteins (spikes) embedded in this envelope. These spikes are the site of attachment to host cells.

▶ Viruses grow **only within living cells**, including the cells of plants, animals, *Bacteria*, and *Archaea*. Plant and animal viruses can be grown in cell or tissue culture.

▶ **Quantitation** of bacteriophage can be accomplished by placing dilutions of a viral suspension on a "lawn" of the susceptible host growing on an agar surface. The phage will lyse organisms in a confined area, forming clear areas called plaques. Plant or animal viruses can be quantitated in cell culture or by treating the host with dilutions of virus and determining the highest dilution that produces manifestation of disease.

▶ Viruses are generally **purified** by differential centrifugation, enzyme treatment, and other procedures generally applied to protein purification.

▶ The viral genome is composed of either **DNA** or **RNA**. This nucleic acid may be single stranded (ss) or a double strand (ds). A viral genome may be mRNA or transcribed to produce an early message for the proteins that initiate the infectious process.

▶ Some **tumor viruses** and the HIV-1 virus that causes **AIDS** are ssRNA. There are also DNA tumor viruses. The ssRNA viruses are transcribed in reverse to produce dsDNA. This dsDNA is integrated into the genome of the host.

▶ **All types of bacteria** are subject to viral attack. There are over 2,000 described bacterial viruses. A bacteriophage infection can result in lysis of the host, or in some the genome can be **integrated** into the host genome. The viral genome would be replicated along with host DNA and passed along to progeny without harm to the host. This is called **lysogeny.**

▶ A very elaborate bacterial virus is **T4**, which infects *Escherichia coli*. This virus can enter a host cell and produce 200 to 300 viral progeny in about 25 minutes. *E. coli* can be infected by a unique virus, **M13**, that can enter the cell, reproduce, and exit without serious harm to the host.

▶ Viruses can cause a **lytic** infection that destroys the host cell, a **persistent** infection where the host remains viable and sheds virus for an extended time, or a **latent** infection where the virus exists in a **dormant** state.

▶ Viruses that infect eukaryotes transfer genetic information into the cytoplasm, and, in most cases viral nucleic acid transcription occurs in the **nucleus**. RNA viruses and some DNA viruses (e.g., poxviruses) are transcribed in the cytoplasm. Some eukaryotic virus particles contain viral-encoded proteins that are involved in the infectious process. Enveloped animal viruses enter a cell by **fusing** to the cytoplasmic membrane or **endocytosis.**

▶ The **retrovirus** that is the causative agent of AIDS attaches to cells (macrophages, T-lymphocytes, monocytes) that have a specific glycoprotein on their surface. These cells are called **CD4⁺** cells. The **T-cells** are ultimately destroyed by the AIDS virus, resulting in a loss of immune function in the infected human.

▶ Herpesvirus is a **typical DNA virus** that can remain dormant in a host and cause an active infection when the host is under stress.

▶ Poliomyelitis is an infection caused by a **ssRNA virus**. The genome is actually mRNA that is copied to form an RNA template that is involved in production of viral genomes.

▶ Insects that suck sap from plants or enter through abrasions generally transmit plant viruses. Plant viruses can infect the insects that transmit them as well as plant hosts. Plants resistant to viruses can be obtained through selective plant breeding.

REVIEW QUESTIONS

1. What are the extremes in the sizes of viruses and how does this compare with sizes of bacteria?

2. What are the general morphologies of viruses? Can one predict the morphology of a virus based on the host it attacks—bacteria, animals, or plants?

3. Discuss cell culture and how it might be employed to propagate viruses.

4. Why are bacterial viruses employed as a model system for the study of viral replication?

5. What are the four phases involved in viral replication?

6. How can a potential host cell protect itself from viral invasion? How can the virus overcome these barriers?

7. From an evolutionary standpoint, why would the lysogenic state be favored? Why is lambda of interest?

8. Outline the steps involved in reproduction of M13. What is the unique attribute of M13? Why is M13 an attractive virus for recombinant DNA technologies?

9. Outline the steps involved in reproduction of the virus T4.

10. Compare the mechanism whereby an animal virus and a bacterial virus attach to and enter a host cell. What difference in structure of the two cell types is important in these processes?

11. Define *lytic*, *persistent*, *latent*, and *transforming* as these terms apply to animal viruses.

12. What is a retrovirus and why have these viruses been given this name?

13. Outline the life cycle of a virus like HIV. How does it attack, become integrated, and cause disease?

14. How do plant viruses spread? What difficulties do they encounter and how are these overcome?

SUGGESTED READING

Dimmock, N. J. and S. B. Primrose. 1994. *Introduction to Modern Virology.* Cambridge, MA: Blackwell Science, Inc.

Flint, S. J., L. W. Enquist, R. M. Krug, V. R. Rancaniells and A. M. Skalka. 1999. *Principles of Virology: Molecular Biology, Pathogenesis and Control.* Washington, DC: American Society for Microbiology.

Karam, J. D., ed. 1994. *Molecular Biology of Bacteriophage T4.* Washington, DC: American Society for Microbiology.

Kornberg, A. and T. Baker. 1992. *DNA Replication.* 2nd ed. New York: W. H. Freeman and Co.

Levy, J. A., H. Fraenkel-Conrat and R. A. Owens. 1994. *Virology.* 3rd ed. Englewood Cliffs, NJ: Prentice Hall.

Chapter 15

Genetic Exchange

If it is ultimately proved…that the transforming activity…is actually an inherent property of the nucleic acid, one must still account on a chemical basis for the biological specificity of its action…although the constituent units and general pattern of the nucleic acid molecule have been defined, there is as yet relatively little known of the possible effect that subtle differences in molecular configuration may exert on the biological specificity of these substances.

– O. T. AVERY, C. M. McLEOD, AND M. McCARTY

(1944, ORIGINAL PAPER ON DNA AS THE TRANSFORMING PRINCIPLE,

JOURNAL OF EXPERIMENTAL MEDICINE 79:137–158)

Tracing the effects of a mutation introduced into the genes of higher animals and plants would take decades, as their reproductive rates are measured in years and their numbers are comparatively sparse. Countering this inability to make quick changes in their genetic makeup, animals and plants have large and complex genomes that enable them to display an array of behaviors for coping with changes, opportunities, and threats in their environment.

By contrast, the effects of a new mutation in a bacterium can be seen in a matter of hours in the millions of cells that arise from the original. If the environment changes beyond their ability to respond by expression of genes already in their genomes, their only avenue of response is that one among them may happen to mutate in such a way that a new trait arises that meets or **adapts** to the new environment.

Random mutations in a bacterial population occur at a low but constant measurable rate. Some mutations can be deleterious and lead to the loss of a functional gene, and this leads to the death of the cell. Even so, mutations are one of the main mechanisms by which bacteria acquire new traits and adapt to life in changing environmental niches. However, the evolution of bacteria is greatly accelerated by their ability to acquire and express entire genes that they obtain from other bacteria. This chapter discusses the mechanisms of this genetic exchange. New traits that bacteria gain by genetic exchange allow them to adapt to new environments much more rapidly than would occur by mutation alone. This adaptability is important in circumstances where newly acquired genetic information permits bacteria to survive in what previously was a potentially lethal environment. Further, the new traits may augment the metabolic versatility of the bacterium, allowing survival in an environment in which the cell previously lacked the ability to utilize available nutrients.

PLASMIDS AND BACTERIOPHAGES

Bacterial plasmids and bacteriophages (bacterial viruses) are small genetic elements that are not part of the chromosome. They replicate in the bacterial cytoplasm and hence utilize the metabolic machinery and replicative apparatus of the host bacterial cell. Genes found on these extra-chromosomal elements can specify functions that may influence the life of the bacterial host. In general, however, the genes encoded by these elements are not essential; that is, they are useful only when the bacterium finds itself in a rather specialized environment.

Plasmids

Plasmids present in bacteria are generally circular, double-stranded DNA molecules. A single bacterial cell can be completely devoid of plasmids, or it can carry many different plasmids in its cytoplasm. Plasmids are variable in size, ranging from several hundred to many thousands of base pairs. In fact, some large plasmids that are present in certain species of the genus *Rhizobium* can be considered to be minichromosomes encoding hundreds of genes. Although many plasmids exist as in a single copy within a bacterial cell, some plasmids are present in several to as many as 20 or 30 copies. Many but not all plasmids can be transferred to other bacteria by a process called **conjugation** (see later discussion); thus, bacteria that are plasmid free can acquire plasmid-encoded genes. Usually, only a closely related member of the genus can receive plasmids, but some plasmids, called **promiscuous,** can be transferred to bacteria that are unrelated.

Plasmids Specifying Resistance to Antibiotics

Bacterial plasmids are classified on the basis of the information that is encoded in the genes of these plasmids. One very important group of plasmids is the **R factors** (or **R plasmids**) that encode antibiotic-resistance determinants. A bacterial species carrying one of these plasmids can be resistant to a specific antibiotic, whereas the same species lacking the plasmid can be killed readily by the antibiotic. A number of different mechanisms exist for plasmid-encoded antibiotic resistance, the most common of which is synthesis of enzymes that destroy the antibiotic. Another example is plasmid-encoded proteins that can alter the membrane structure such that the antibiotic cannot enter the cell and reach its target. Occasionally, a plasmid can encode a product that modifies the target of the antibiotic, rendering it resistant to the antibiotic. **Table 15.1** summarizes some common mechanisms of plasmid-encoded resistance.

The prevalence of R factors among pathogenic (disease-causing) bacteria is of concern to physicians treating infectious disease because it can limit the range of antibiotics that can be administered to treat a particular infection. The presence of a plasmid-encoding β-lactamase in *Neisseria gonorrhoeae* requires the use of costly

Table 15.1	Common mechanisms of plasmid-encoded antibiotic resistance
Antibiotic	**Mechanism of Resistance**
β-lactams	Synthesis of β-lactamases, enzymes that hydrolytically destroy the antibiotic.
Chloramphenicol	Synthesis of an enzyme that acylates chloramphenicol, rendering it inactive.
Aminoglycosides	Synthesis of one of several enzymes that inactivate the antibiotic by acetylation, phosphorylation, or adenylation.
Tetracycline	Synthesis of a membrane protein capable of pumping the antibiotic out of the cell before it can act on the ribosomes.
Erythromycin	Synthesis of an enzyme that methylates bacterial 23S ribosomal RNA; methylated ribosomes cannot bind the antibiotic.
Trimethoprim	Synthesis of a mutant, trimethoprim—insensitive form of dihydrofolate reductase.

and less-effective antibiotics for the treatment of gonorrhea, although previously susceptible strains had been killed by administration of a single large dose of penicillin to the infected patient. Many R plasmids can move from one bacterium to another by conjugation, thus allowing a pathogen that is normally sensitive to therapy with an antibiotic to become resistant. A plasmid can acquire additional resistance determinants from other R factors; therefore, plasmids carrying multiple antibiotic resistance determinants are common. **Figure 15.1A** is a schematic representation of RK2, a plasmid encoding resistance to tetracycline, kanamycin, and ampicillin. This plasmid is also transferrable to other bacterial hosts, including virtually all of the gram-negative bacteria. This promiscuous behavior of a plasmid implies that the plasmid or the bacterial host carrying the plasmid has the capability of transferring it to a recipient, and once in the new host, the plasmid can stably replicate.

Virulence Plasmids

Plasmids of pathogenic bacteria may encode genes that are required for virulence. Plasmids conferring pathogenic abilities on *Escherichia coli* have been well characterized. Certain strains of *E. coli* responsible for diarrheal diseases in humans and animals carry large plasmids containing genes for two types of toxins. The same plasmid can also carry genes for **adhesins,** surface molecules that are required for mucosal colonization. A diagram of plasmid pCG86 that encodes toxin production as well as resistance determinants to three different antimicrobial agents is presented in Figure 15.1B. Plasmids of *E. coli* that are important in virulence can encode both hemolytic factors (that is, proteins that destroy

(A)

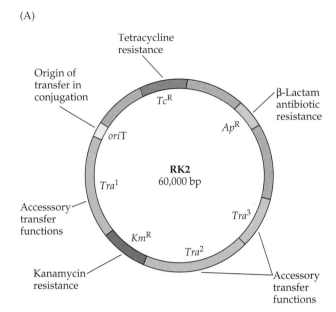

(B)

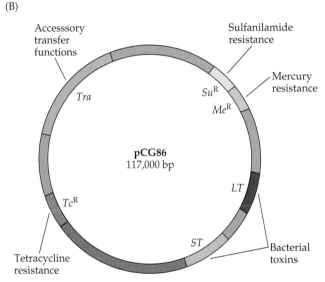

Figure 15.1 Plasmids of pathogenic bacteria
Genome maps of some plasmids found in pathogenic bacteria. (A) A widely distributed R plasmid, RK2. (B) Plasmid pCG86 of pathogenic *E. coli*. The gene product or function of the genes is indicated. β-Lactam antibiotics include ampicillin.

Infection of certain plants with the pathogen *Agrobacterium tumefaciens*, which carries specialized tumor-inducing (Ti) plasmids, results in formation of crown gall tumors. The plant tumor results from the expression in the plant of genes encoded on the bacterial plasmid, a portion of which is transferred into the nucleus of the host where it becomes incorporated into the plant genome. A schematic representation of a Ti plasmid is presented in **Figure 15.2**. Two regions of this plasmid are important for virulence, the T-DNA (so named because it carries the genes required for tumor-induction in the host plant) and the *vir* genes (for *virulence*). The T-DNA is about 30 kilobases (kb) in length and is flanked by two 25 base pair direct repeats. This is the region of the plasmid that becomes excised and transferred into the host genome. T-DNA genes encode enzymes responsible for the synthesis of opines, some of which are analogs of the amino acid arginine that can be used by *Agrobacteria* as a source of energy and nitrogen. In the plant cell, T-DNA genes also direct synthesis of plant hormones, resulting in uncontrolled growth of the plant tissue and formation of a tumor. The *vir* genes are not transferred into the plant, but nevertheless are responsible for transfer of T-DNA. They include regulatory genes that respond to signals from the plant, as well as genes involved in processing T-DNA prior to its transfer to the plant cell (**Box 15.1**).

Another group of plasmids can direct synthesis of proteins that are toxic to other bacteria. These bactericidal proteins, called **bacteriocins,** are encoded by specialized plasmids termed **bacteriocinogenic plasmids**. Commonly, the names of the plasmid and its toxic protein are derived from the genus or species name of the host bacterium. For example, *E. coli* can carry **colicinogenic plasmids,** which encode a variety of **colicins.**

membranes of a variety of cells, including red blood cells) as well as siderophores (polypeptides responsible for efficient iron uptake, a metal often limiting in infected tissues). Pathogenic gram-positive bacteria also carry plasmids that encode synthesis of toxins, such as plasmid-encoded toxin genes of *Bacillus anthracis*. A group of insecticidal toxins of *Bacillus thuringiensis* is plasmid encoded, as are *Clostridium* neurotoxins. Specific plasmid-encoded genes are required for cell invasion in the course of diseases caused by *Shigella*, *Yersinia*, and *Salmonella* species.

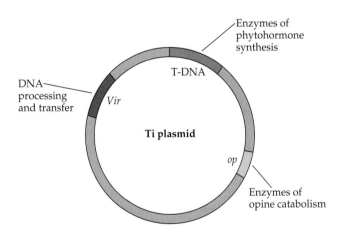

Figure 15.2 Ti plasmid
Genome map of the Ti plasmid of *Agrobacterium tumefaciens*, showing the gene product or function of the genes. Phytohormones are responsible for the induction of plant tumors.

Milestones Box 15.1   Sex among Bacteria

The discovery of transformation by Avery and his colleagues, as a means of bacterial acquisition of genetic information, raised the possibility that bacteria can acquire traits by more conventional (i.e., sexual) means. In 1946, Tatum and Ledeberg demonstrated the exchange of genetic information among bacteria. This simple experiment outlined in part A of the figure demonstrated transfer of genes between two strains of *E. coli*, each carrying two mutations in biosynthetic genes. One of the strains, which required both biotin and methionine for growth on minimal medium, whereas the second was a threonine and proline requiring double mutant of *E. coli*. When a culture of these two mutant strains was mixed and plated onto minimal medium, a few bacteria grew that did not require any of the supplements that characterized the mutations of the parental strains. When a comparable number of each of the *E. coli* mutants were plated onto minimal medium, no wild-type bacteria were ever recovered. This was the first demonstration that recombination can occur between two bacteria, even in the absence of apparent lysis and release of DNA, because DNase present during the coincubation of the two bacterial strains failed to block recombination.

The additional evidence that the observed recombination requires a cell-to-cell contact was

(A)

A strain requiring biotin and methionine for growth,…

…and a strain requiring threonine and proline for growth will not grow on minimal medium.

Incubation of a mixture of the two strains produces wild-type recombinants, which grow on minimal medium.

No colonies   No colonies   Wild-type colonies

(B)

When strains requiring methionine are incubated with strains requiring threonine, leucine, and thymine…

When the mutant strains are separated by a filter…

Glass filter

…wild-type recombinant colonies are produced.

…no wild-type bacteria are produced.

Wild-type colonies   No colonies   No colonies

Demonstration of (A) recombination among bacteria and (B) the need for intimate contact for genetic recombination to occur.

Colicins kill other bacteria through a variety of mechanisms, ranging from direct damage to the cell membrane to the destruction of ribosomes. Bacteria-carrying bacteriocinogenic plasmids protect themselves from the lethal bacteriocins by expressing **immunity determinants,** which are coded on the same bacteriocinogenic plasmids. The bacteriocin/immunity systems thus allow bacteria to survive while killing other bacteria that may compete with them in an environment of limited nutrients.

Plasmids Encoding Genes for Specialized Metabolism

Certain metabolic functions can also be encoded by plasmids, including the biodegradation of complex

Milestones Box 15.1   *(continued)*

provided by Davis in 1950. His experiment, shown in part B of the figure, utilized a U-shaped tube, which was separated by a fretted glass filter, allowing passage of liquid, but the pores in the disk were impermeable to the bacteria. One side of the tube was inoculated with an *E. coli* requiring threonine, leucine, and thymine, whereas the other side contained a methionine-requiring mutant of *E. coli*. In a separate experiment, the two bacterial mutants were mixed in one of the arms of the tube. The medium was flushed back and forth and incubated for several hours. Bacteria were recovered from the tube and plated onto minimal media.

The tube that contained a mixture of bacteria yielded several wild type recombinants, but the tube where bacteria were separated failed to give rise to any recombinants capable of growing on media lacking the appropriate nutritional supplements. Hence, genetic exchange by intimate contact was unambiguously demonstrated, which was subsequently shown to be mediated by plasmids mobilizing DNA from donor (male) to recipient (female) cells.

---

organic molecules by some species in the genus *Pseudomonas*. Large plasmids that encode catabolic pathways for aliphatic or aromatic compounds, such as toluene, naphthalene, chlorobenzoic acid, octanes, and decanes have been described in recent years. A map of one such degradative plasmid is presented in **Figure 15.3** together with the enzymes of naphthalene and salicylate oxidation. This plasmid codes for the enzymes that catabolize leading to naphthalene to pyruvate and acetylaldehyde. Such degradative plasmids allow the bacterium to utilize unusual organic compounds as their sole carbon and energy source. Because many environmental pollutants are complex organic molecules, bacteria that carry degradative plasmids are prime candidates for use in bioremediation (see Chapter 31).

Plasmid Replication

Plasmid replication is similar to that of replication of the host DNA. Starting from a specific origin of replication, the host replication machinery proceeds bidirectionally around the plasmid. In addition to the host replicative apparatus, most plasmids also require expression of plasmid-encoded genes. These genes code for various enzymes that assist DNA replication, such as unwinding and separating strands and ensuring correct partition of replicated plasmids into daughter cells.

Although a bacterial cell can carry several plasmids, not all plasmids can coexist in the same cell. Plasmids that use identical or closely related replication mechanisms are incompatible in the same cell and, therefore, belong to the same **incompatibility group**. If a cell receives two plasmids of the same incompatibility group by any one of the genetic exchange mechanisms, these two plasmids will segregate during cell division such that each daughter cell will have only one of the plasmids. If the incompatible plasmids can exist in more than one copy, it may take several cycles of cell division before a homogeneous population of daughter cells is reached, each containing plasmids from the same incompatibility group. This property allows yet another method of classifying plasmids—on the basis of plasmid compatibility. The mechanism of incompatibility is not well understood, but for a number of plasmids, it appears to be directed by a control system that coordinates replication with cell division.

Occasionally, plasmids can be "lost" by a bacterium, and this may occur naturally during cell division when both copies of a replicated plasmid end up in one daughter cell and the second daughter cell does not receive a plasmid. The daughter cell without a plasmid is said to be **cured**. Environmental stresses, such as exposure to extreme temperatures, nutritional limitation, or treatment with chemicals that interact with DNA, greatly stimulate curing. If the bacterium is in an environment where it depends on the expression of a plasmid-encoded gene, such as in the presence of antibiotics, the cured daughter cell will not survive.

Bacteriophages

Viruses are obligate intracellular parasites that can propagate only inside host cells. Virtually every type of living cell can be infected by one or several kinds of viruses, and bacteria are no exception. Although viruses can self-replicate, they do so only when inside a host cell, parasitizing virtually all of the host cell's metabolic machinery. Outside a host, they are inert molecules of nucleic acids surrounded by protein or lipid envelopes. For these reasons, viruses are not considered living organisms, although they are generally discussed in the context of their host. Bacterial viruses utilize the metabolic and biosynthetic machinery of their host, having replication and gene regulation patterns similar to those of bacteria or *Archaea*, whereas viruses that infect eukaryotic cells follow the patterns of gene expression that are seen in eukaryotic cells. The replication of viral nucleic acid and assembly of a virus was considered in Chapter 14.

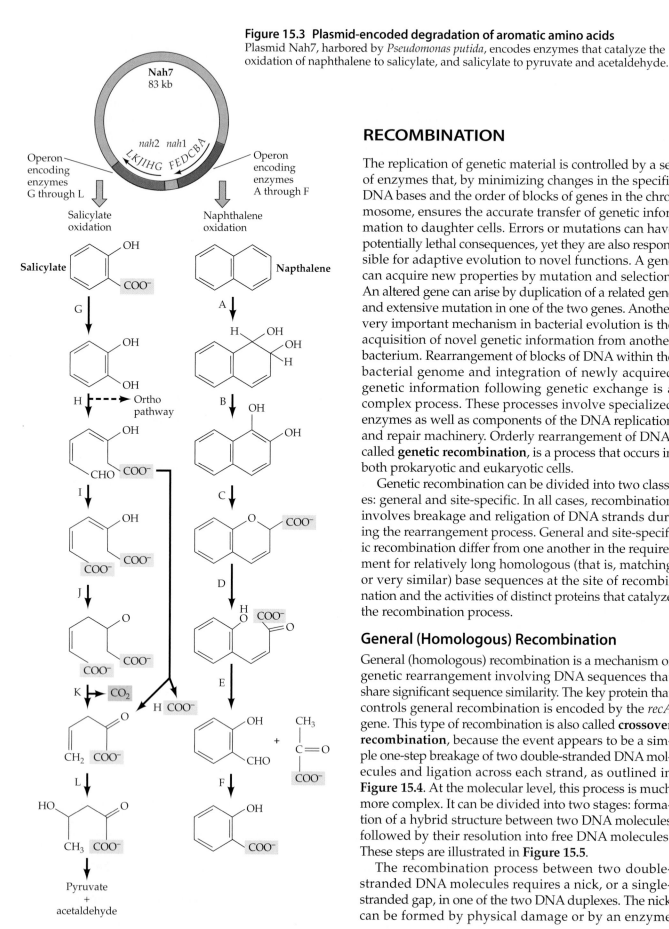

Figure 15.3 Plasmid-encoded degradation of aromatic amino acids
Plasmid Nah7, harbored by *Pseudomonas putida*, encodes enzymes that catalyze the oxidation of naphthalene to salicylate, and salicylate to pyruvate and acetaldehyde.

RECOMBINATION

The replication of genetic material is controlled by a set of enzymes that, by minimizing changes in the specific DNA bases and the order of blocks of genes in the chromosome, ensures the accurate transfer of genetic information to daughter cells. Errors or mutations can have potentially lethal consequences, yet they are also responsible for adaptive evolution to novel functions. A gene can acquire new properties by mutation and selection. An altered gene can arise by duplication of a related gene and extensive mutation in one of the two genes. Another very important mechanism in bacterial evolution is the acquisition of novel genetic information from another bacterium. Rearrangement of blocks of DNA within the bacterial genome and integration of newly acquired genetic information following genetic exchange is a complex process. These processes involve specialized enzymes as well as components of the DNA replication and repair machinery. Orderly rearrangement of DNA, called **genetic recombination**, is a process that occurs in both prokaryotic and eukaryotic cells.

Genetic recombination can be divided into two classes: general and site-specific. In all cases, recombination involves breakage and religation of DNA strands during the rearrangement process. General and site-specific recombination differ from one another in the requirement for relatively long homologous (that is, matching or very similar) base sequences at the site of recombination and the activities of distinct proteins that catalyze the recombination process.

General (Homologous) Recombination

General (homologous) recombination is a mechanism of genetic rearrangement involving DNA sequences that share significant sequence similarity. The key protein that controls general recombination is encoded by the *recA* gene. This type of recombination is also called **crossover recombination**, because the event appears to be a simple one-step breakage of two double-stranded DNA molecules and ligation across each strand, as outlined in **Figure 15.4**. At the molecular level, this process is much more complex. It can be divided into two stages: formation of a hybrid structure between two DNA molecules followed by their resolution into free DNA molecules. These steps are illustrated in **Figure 15.5**.

The recombination process between two double-stranded DNA molecules requires a nick, or a single-stranded gap, in one of the two DNA duplexes. The nick can be formed by physical damage or by an enzyme

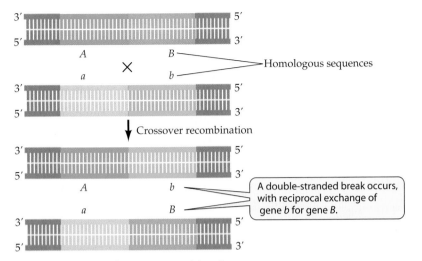

Figure 15.4 Homologous recombination
Exchange of two DNA segments by homologous recombination.

A double-stranded break occurs, with reciprocal exchange of gene *b* for gene *B*.

complex consisting of RecB, RecC, and RecD, which unwinds DNA and cuts one strand at a specific short sequence. In *E. coli*, it is the sequence GCTGGTCG, which is called the Chi site. A single-stranded tail is formed, which becomes coated by single-stranded DNA-binding proteins, including RecA. The bound form of RecA is now capable of aggregating with double-stranded DNA, searching for homology, and invading the duplex of the second DNA molecule, allowing formation of a stable duplex from the two homologous strands. The displaced "bubble" of the invaded strand is then nicked and forms a stable double-stranded structure with the

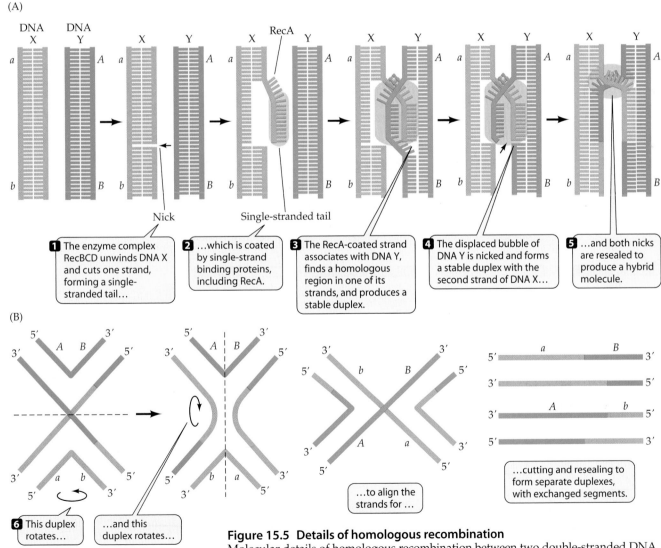

(A)

1 The enzyme complex RecBCD unwinds DNA X and cuts one strand, forming a single-stranded tail…

2 …which is coated by single-strand binding proteins, including RecA.

3 The RecA-coated strand associates with DNA Y, finds a homologous region in one of its strands, and produces a stable duplex.

4 The displaced bubble of DNA Y is nicked and forms a stable duplex with the second strand of DNA X…

5 …and both nicks are resealed to produce a hybrid molecule.

(B)

6 This duplex rotates…

…and this duplex rotates…

…to align the strands for …

…cutting and resealing to form separate duplexes, with exchanged segments.

Figure 15.5 Details of homologous recombination
Molecular details of homologous recombination between two double-stranded DNA molecules, here designated X and Y. (A) Formation of the hybrid molecule and (B) its resolution.

other strand, resulting in a transient hybrid molecule. The nicks in both of the strands are resealed. The final stage in recombination is **isomerization**, which leads to alignment of the two remaining outside strands across each other. Isomerization involves simultaneous rotation of the two strands, followed by cutting and ligation of the crossed strands. This process effectively exchanges a segment of two DNA strands.

Now let's consider some of the DNA rearrangements that can result from homologous recombination. When two plasmids, each containing similar DNA sequences, recombine, the result is a single, large, composite plasmid (**Figure 15.6A**). A similar process involving chromosomal sequences with sequences on a plasmid leads to incorporation of the plasmid into the bacterial genome (Figure 15.6B). The outcome of homologous recombination within a contiguous sequence (chromosome or plasmid) depends on the relative orientation of the homologous sequences Figure 15.6C). If such sequences are in an opposite orientation, the reciprocal

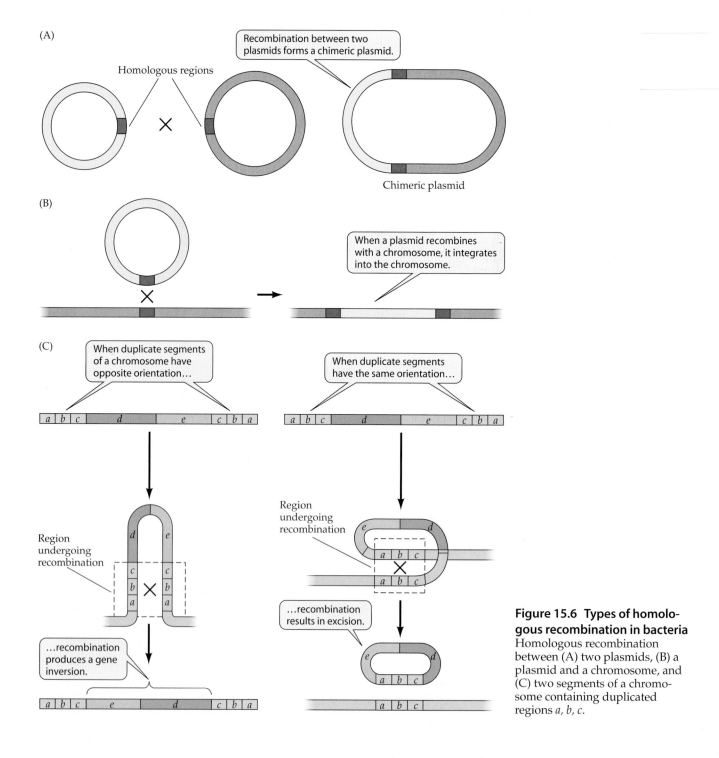

Figure 15.6 Types of homologous recombination in bacteria Homologous recombination between (A) two plasmids, (B) a plasmid and a chromosome, and (C) two segments of a chromosome containing duplicated regions *a*, *b*, *c*.

exchange results in inversion of the intervening region. If the sequences are in the same orientation (Figure 15.6C), the result of general recombination is excision of a circular DNA consisting of the intervening region and potential loss of the genetic information during subsequent cell division.

Site-Specific Recombination

Site-specific recombination between two DNA segments requires relatively short regions of homology between the base pairs involved in the exchange of DNA strands. The process is called site-specific because it takes place between specific sites on the DNA, as defined by their nucleotide sequence. Moreover, site-specific recombina-

tion requires a specialized set of enzymes that recognizes these sites and mediates the recombination process. This form of recombination mediates a number of important cellular processes, including integration of viral genomes into the bacterial chromosome, insertion of specific genes by a transposition mechanism (see following), and expression of genes by inversion of promoter or regulatory sequences.

The mechanism of inversion of DNA resulting in expression, or lack of expression, of specific genes has been extensively studied in a number of microorganisms. One of the best-understood examples is flagellar **phase variation** in the flagellin subunit proteins by *Salmonella typhimurium* (**Figure 15.7**). This bacterium

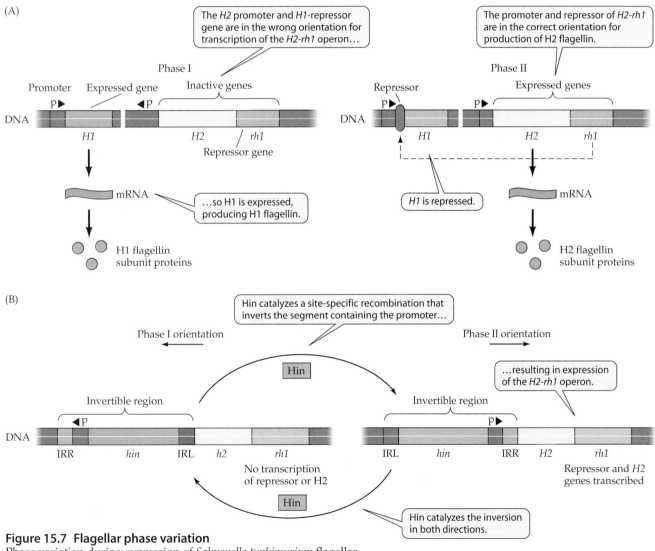

Figure 15.7 Flagellar phase variation
Phase variation during expression of *Salmonella typhimurium* flagellar proteins. (A) Overall scheme for expression of H1 and H2 flagellins and control of the H1 gene by the repressor. (B) Details of the inversion process. Hin mediates inversion in both directions, so cells alternate between H1 and H2 flagellin production.

alternates between expression of either one of the two possible immunologically distinct flagellin proteins (H1 and H2) encoded by two different genes. This cleverly designed molecular circuit not only assures that one of the flagellin genes is expressed, but also prevents expression of the other gene. The expression of gene for H2, and repression of the gene specifying H1, is accomplished because expression of H2 is linked to the expression of another protein, a repressor, which prevents expression of H1. The repressor gene *rh1* (repressor of H1) is a second gene of the two-gene operon, transcribed from a single promoter preceding the gene for H2. The expression of the H2-gene:*rh1* is controlled by an invertible 995 base pair (bp) DNA segment, flanked by short, 14 bp inverted repeats (IRL and IRR); the end nearest to the operon contains its promoter. At a regular frequency, the 995 bp segment is inverted, resulting in the promoter being oriented away from the genes for H2 and the repressor. The consequence of this rearrangement is concomitant loss of expression of H2 and lack of synthesis of the repressor. In the absence of the repressor protein, the gene for H1 is now expressed, and these new subunits are assembled into flagella. The inversion process can be reversed (going from H1 expression to H2) by the reversal of the same rearrangement, allowing for synthesis of H2 as well as the repressor of H1.

The molecular details of this type of site-specific rearrangement have been worked out (Figure 15.7). The key mediator of this process is Hin, a recombinase enzyme whose gene (*hin*) is encoded within the 995 bp invertible segment. Hin mediates this type of site-specific recombination by binding to the ends of the invertible segment, within the IRR and IRL, where with the aid of several ancillary proteins, the proteins, together with bound DNA, assemble. Within this protein complex, the DNA is cleaved and the ends that form the opposite ends of the segments are ligated together. The consequence of inversion of the segment is a reversal of the IRR and IRL sequences, thus moving the promoter away from its site adjacent to the H2 gene. Mechanistically similar reactions are responsible for expression of genes encoding receptors for certain bacterial viruses, and those specifying pili in a number of bacterial species.

GENETIC EXCHANGE AMONG BACTERIA

Bacterial recombination via a general or site-specific mechanism is most easily manifested following acquisition of foreign DNA by a bacterium. This is a normal process in higher eukaryotes in which sexual reproduction is initiated by fertilization of an egg that brings genetic material from two different sources. Bacteria also have the capacity to acquire DNA from other bacteria, but it is not tied to cell division. Genetic exchange in bacteria is an important vehicle for evolution and rapid adaptation to a new environment.

Bacteria can acquire novel genetic information, that is, new segments of DNA, by three different mechanisms: **transformation**, in which a bacterial cell takes up free DNA from its environment; **conjugation**, which is a plasmid-mediated transfer of DNA from one bacterium to another; and **transduction**, in which a bacteriophage carries bacterial DNA from one cell to another. The ability to detect microorganisms that have received new genetic information by any one of these methods depends on expression of the gene in the new host. Often the genes become part of the bacterial genetic repertoire, and the trait thus gained is inherited by daughter cells after cell division.

Transformation

The process of giving bacteria acquired heritable traits by exposure to a solution of DNA was a true landmark in the history of genetics, as it was the first demonstration that genes are located on DNA. In 1944, O. T. Avery, C. M. MacLeod, and M. McCarty reported that avirulent (non-disease-producing) *Streptococcus pneumoniae*, lacking a protective capsule, regained the ability to synthesize capsules and become virulent when exposed to crude DNA from virulent strains (**Figure 15.8**). They further showed that the ability of the DNA preparation to transform an avirulent *S. pneumoniae* into a virulent form was lost when the DNA preparation was exposed to DNAses (enzymes that destroy DNA), but not when exposed to proteinases or RNAses (enzymes that destroy proteins or RNA). This proved that the transforming genetic information was carried in cellular DNA, not in proteins or RNA.

Virtually all bacteria have an ability to take up DNA from the environment, provided they are **competent**. **Transformation competence** is a state of a bacterial cell during which the usually rigid cell wall can transport a relatively large DNA macromolecule. This is a highly unusual process for bacteria, for they normally lack the ability to transport large macromolecules across the rigid cell wall and through the cytoplasmic membrane. Competence in some bacteria, such as *Hemophilus*, *Streptococcus*, or *Neisseria* is expressed during a certain stage of cell division and most likely represents a stage when the reforming cell wall can allow passage of DNA. These bacterial genera possess **natural competence** because their cells do not require any special treatment to increase their ability to take up DNA. Not all bacteria, however, are naturally competent to take up DNA, but treatment of such cells with calcium or rubidium chloride alters their envelope, and they become competent. This form of competence, termed **artificial competence**, is one of the more important techniques of introducing recombinant DNA molecules into bacteria.

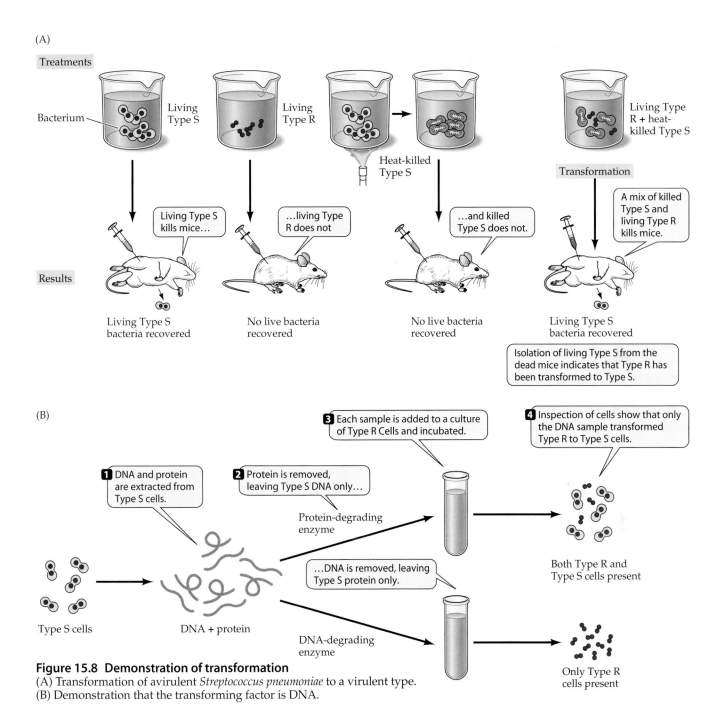

Figure 15.8 Demonstration of transformation
(A) Transformation of avirulent *Streptococcus pneumoniae* to a virulent type.
(B) Demonstration that the transforming factor is DNA.

What is the fate of a DNA molecule once it enters a competent cell? If the transformed DNA is a plasmid capable of replicating within that particular species of bacterium, the cell becomes plasmid-bearing, and plasmid DNA replicates as an extrachromosomal element. Genes encoded on the plasmid are expressed, and synthesis of plasmid-encoded proteins allows the bacterial host to take advantage of some additional traits that its siblings without plasmids lack. For example, if the bacterium took up a plasmid-encoding gene that specified aminoglycoside resistance, the transformant would then be resistant to aminoglycosides, a trait certainly useful if the bacterium is in an environment containing this class of antibiotic.

Transformation of bacteria with linear DNA fragments is more complex because the new DNA is a natural target for hydrolysis by intracellular enzymes that degrade noncircular DNA. One mechanism whereby foreign DNA can escape degradation is incorporation into the chromosome of the recipient cell via general recombination. Because incorporation of linear DNA into a bacterial chromosome by a single-crossover event

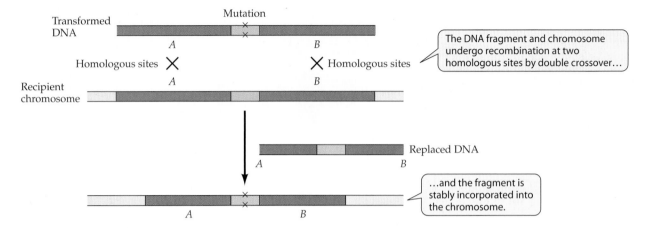

Figure 15.9 Double recombination
Introduction of a mutation into a bacterial chromosome from a piece of DNA acquired by transformation.

would disrupt the chromosome, stable incorporation of genes requires a **double-crossover** event. As shown diagrammatically in **Figure 15.9**, such double recombination effectively replaces the DNA sequence from the bacterial chromosome with that of the transforming DNA.

Because general recombination requires homologous sequences, such recombination can take place only with a nearly identical DNA sequence, thus having little or no effect on the bacterial genome. Note that homology is required only around the site of crossover and that the sequence between the two crossover sites need not be identical. Mutations can be introduced or eliminated by transforming DNA, provided the DNA participating in the crossovers flanks the DNA containing the mutation.

As illustrated in **Figure 15.10**, a newly acquired gene can be inserted into the chromosome by transformation, given that sufficient homology exists on either side of the acquired gene to allow for efficient general recombination. Analogous events can lead to loss of the gene. The double crossover, when applied to the introduction of new or modified genetic material, is sometimes referred to as **gene replacement**. The replaced DNA fragment is generally degraded by bacterial nucleases in the cytoplasm.

It is not clear to what extent transformation contributes to natural genetic exchange, although recent evidence suggests that for some bacteria that are naturally competent, individual genes can be acquired even from genomes of different species. However, transformation requires the presence of free DNA in the bacterial environment, and there it can be readily attacked by nucleases. The availability of DNA in the environment of a bacterium would be rather limiting. However, in certain environments where bacteria live in large numbers, some bacteria die and lyse, and it is not unreasonable to expect that some DNA escapes extracellular nucleases. As is the case with all mechanisms of DNA exchange, if the acquired trait gives the recipient bacterium significant survival advantage, the progeny of the rare transformant will outnumber the bacteria lacking such a trait within several generations, even if the DNA acquisition mechanism is inefficient.

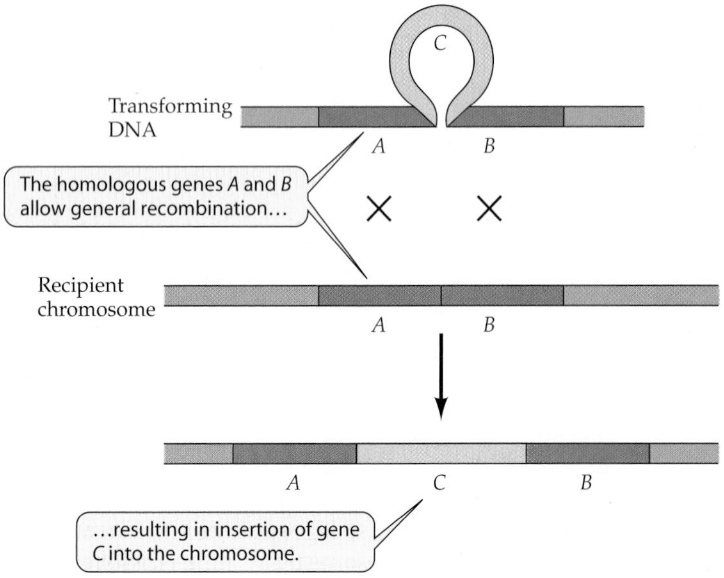

Figure 15.10 Gene replacement
Introduction of a new gene into a bacterial chromosome by transformation. General recombination occurs between homologous sequences that flank the gene.

Conjugation

Conjugation is a mechanism of DNA exchange mediated by plasmids. Some but not all plasmids are **transmissible,** which means they can promote their own transfer as well as mediate the transfer of other plasmids and even portions of chromosomal DNA. Bacteria that contain transmissible plasmids are called **donor** cells; those cells that receive plasmids are the **recipients.** Analogous to sexual transfer in higher eukaryotes, donor bacteria are often called male cells, whereas recipients are called female.

Plasmids, such as the F plasmid of *E. coli* or the large R plasmids of a variety of other bacteria, are circular molecules of DNA. These extrachromosomal genetic elements encode a number of genes responsible for their own replication and maintenance within the bacterial cell. A significant number of cellular genes are dedicated to the promotion of DNA transfer via conjugation. These transfer functions include several genes that encode the pilus components as well as proteins involved in the regulation of expression and biogenesis of the pilus. The pilus is an organelle formed on the surface of a plasmid-bearing donor cell that recognizes the recipient and makes the initial contact. A given plasmid can express a distinct pilus.

The process of conjugation can be divided into several steps, which are outlined in **Figure 15.11**. In addition to the initial contact mediated by pili, several other factors are necessary to stabilize the contact, resulting in a close association between the two mating cells and the formation of a channel in the bacterial cell wall. Plasmid DNA is then processed by nicking at a specific site called the **origin of transfer,** often abbreviated *ori*T. The transfer from this site proceeds by a rolling circle replication mechanism, in which the 5′-region of the DNA strand is transferred to the recipient cell. In the recipient, the transferred strand is converted to a circular double-stranded molecule. It is likely that the synthesis of the complementary strand in the recipient is initiated before the transfer of F plasmid has been completed.

The conjugative transfer described for F plasmids or large R plasmids requires that the plasmid contain all of the information for its own transfer. These plasmids are called **self-transmissible** because they can enter a recipient using entirely their own transfer machinery. The transfer of a self-transmissible plasmid into a recipient converts that recipient to a donor because, by inheriting a self-transmissible plasmid, it acquires the ability to conjugate with other recipients. Not all plasmids are self-transmissible. Some plasmids lack some of the genes that encode the transfer functions, rendering them incapable of transfer to another cell. These plasmids, however, need not be restricted to a permanent resi-

dence in a bacterial cell. Because a bacterium can contain more than one plasmid, these plasmids may provide various conjugal functions to other plasmids in that organism. For example, a plasmid lacking a certain transfer function, such as pili, can still transfer to a recipient cell, provided the donor cell has another plasmid

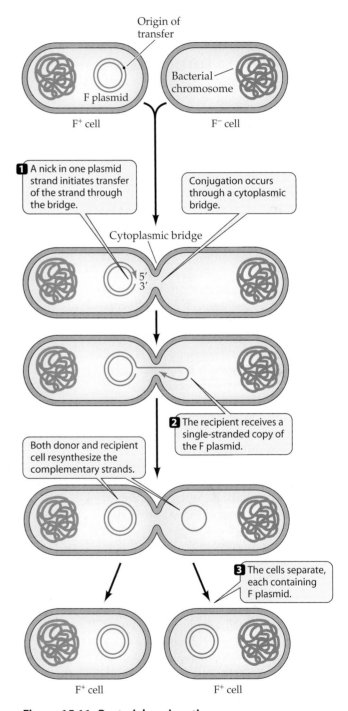

Figure 15.11 Bacterial conjugation
Transfer of F plasmid from donor to recipient cells by conjugation. Once transfer is complete, both cells have an intact copy of F plasmid and can act as donors.

that encodes pili. This "helper" plasmid thus is capable of mobilizing another plasmid for transfer. Note that, unlike the case of self-transmissible plasmids, the recipient receiving such an incomplete plasmid cannot be a donor, unless it already contains a helper plasmid.

Plasmids lacking *ori*T are not transmissible even with helper plasmids. However, these plasmids can encode transfer functions and can act as helper plasmids for other transfer-deficient plasmids. One way a plasmid lacking *ori*T or any additional transfer functions can transfer from one cell to another is by "hitchhiking" a ride on a transmissible element. This can be accomplished by having some homologous sequence on both plasmids such that a homologous recombination event fuses the two plasmids into a **chimeric** plasmid (that is, from two sources). The transfer of the chimeric plasmid is originated from *ori*T of the transmissible plasmid component.

Plasmid-Mediated Exchange of Chromosomal Genes

Transfer of genes among bacteria, mediated by self-transmissible plasmids, is not restricted to genes that are located on plasmids themselves. A number of conjugative plasmids have the ability to transfer chromosomal genes as well. This property was first discovered in the E. coli F plasmid. However, it is now apparent that similar mechanisms of chromosomal transfer exist among several R plasmids. The ability to mediate chromosomal genes' transfer is referred to as the **chromosome mobilizing ability**.

The F plasmid can mediate transfer of chromosomal genes by two related mechanisms. They both require formation of **high-frequency recombinant cells (Hfr)** in which the F plasmid has integrated into the chromosome (**Figure 15.12**). The integrated form of the F plasmid has no effect on the cell per se because it replicates along with the chromosome. The integration is not absolutely stable; in a population of cells, an equilibrium exists between bacteria with integrated F plasmid (Hfr) and cells where the F plasmid is not part of the bacterial chromosome. Integration is not random; a few preferred sites exist in the E. coli chromosome where most of the integration can take place. All of the sites of integration have a small DNA sequence in common.

Once the F plasmid has integrated, the cells can initiate a transfer process virtually identical to the one described for transfer of the F plasmid itself. Following contact of an Hfr cell with a recipient, a single-stranded nick is created in the integrated F plasmid, and transfer begins by the process analogous to the transfer of the free, nonintegrated F plasmid. However, unlike transfer of F plasmid, chromosomal genes are transferred in the

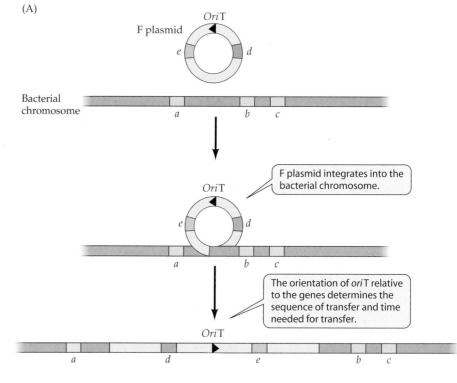

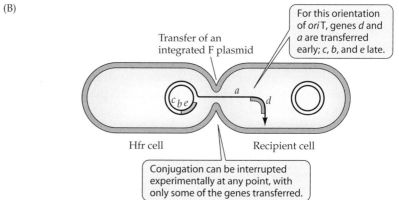

Figure 15.12 High-frequency recombinant cells
(A) Formation of an Hfr and (B) transfer of chromosomal genes into a recipient bacterium. Orientation of the inserted F plasmid in the opposite direction from that shown here would allow early transfer of genes *e*, *b*, and *c* and later transfer of *d* and *a*. Relative locations of genes in the bacterial chromosome can be mapped by mixing donor and recipient cells, interrupting the mating at various times, growing the cells on appropriate media, and identifying the transferred genes.

5′-to-3′ direction. The process continues as long as the cells remain in contact. The entire chromosome of a donor cell can be transferred into a recipient, resulting in a diploid cell containing a reformed Hfr chromosome. However, the association of donor and recipient is usually not very strong, and the amount of transferred donor chromosomal DNA is directly proportional to the duration of contact between the bacterial mating pair. Once DNA is transferred and the complementary strand is synthesized, the genes transferred by this process can undergo general recombination with the chromosome of the recipient, not unlike the fragments of DNA that a cell receives by transformation. Double-crossover events are again required for integration of blocks of DNA between the sites of crossover. The recipient cell usually receives only a small portion of the chromosome of a donor, containing only a fraction of the integrated F plasmid. Therefore, it is unable to donate DNA to another recipient. In complete matings in which the entire chromosome is transferred, however, the recipients are Hfr; that is, they are capable of further chromosomal transfer.

Bacterial conjugation via Hfr is a useful method of mapping genes in a bacterial cell. For each site of integration of F, one can mix donor and recipient cells and allow conjugation to take place in a controlled way. At various time intervals, conjugation is interrupted by shearing the suspension of bacteria, generally by disrupting contact by placing the bacterial suspension in a blender. Those recipients that received DNA from the donor and have undergone recombination are identified by their growth on specialized media. Thus, if a recipient is a mutant requiring an amino acid, recombination of the defective gene with DNA from a wild-type donor cell allows identification of such genes. The farther a gene is located from the site of integration of F, and hence, from the origin of transfer, the lower the frequency of transfer of that particular gene to a recipient.

F plasmids or similar R plasmids can mediate transfer of chromosomal genes by yet another mechanism. As alluded to earlier,

integration of the F plasmid into the bacterial chromosome is not permanent, and excision does take place occasionally. When the F plasmid excises from the chromosome by reversal of the reciprocal recombination, the site of integration is usually perfectly restored, yielding a complete F plasmid and uninterrupted chromosomal DNA. However, occasionally imperfect excision leads to removal of a small portion of the chromosome, such that genes adjacent to the site of recombination are now part of the excised plasmid. These plasmids are called **F prime** (**F′**), or if the excision involved an R plasmid, they are called **R′**. Transfer of DNA from an F′-donor (or an R′-donor) to a recipient proceeds by a mechanism identical to the transfer of the F plasmid (**Figure 15.13**). In the case of plasmids that contain most of the F plasmid's genome

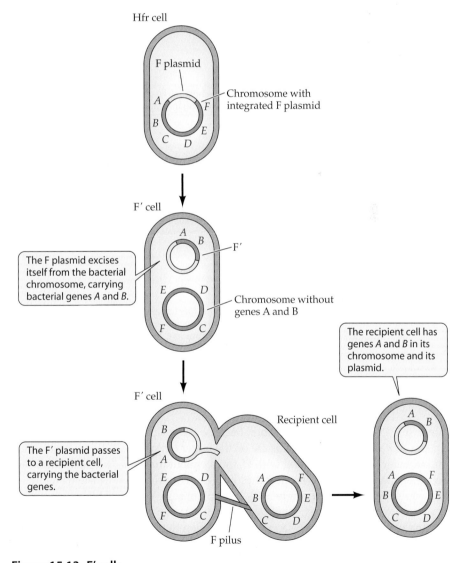

Figure 15.13 F′ cells
Formation of an F′ cell from an Hfr cell, and transfer of a bacterial chromosome segment to a recipient cell.

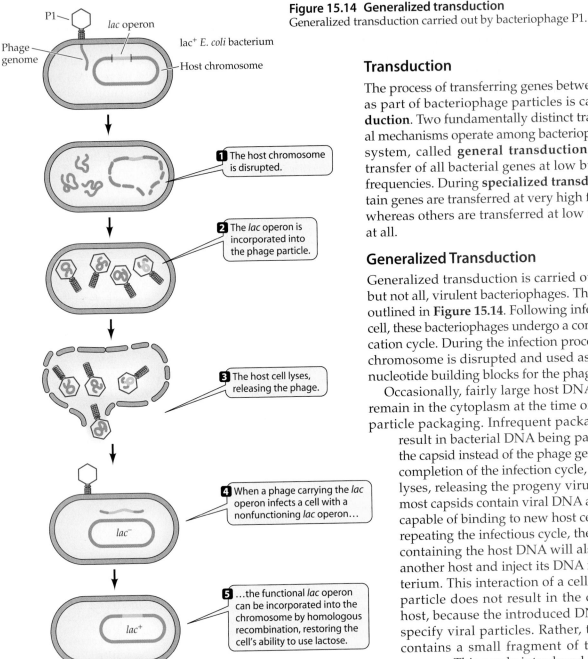

P1

lac operon

Phage genome

lac⁺ *E. coli* bacterium

Host chromosome

1 The host chromosome is disrupted.

2 The *lac* operon is incorporated into the phage particle.

3 The host cell lyses, releasing the phage.

4 When a phage carrying the *lac* operon infects a cell with a nonfunctioning *lac* operon…

lac⁻

5 …the functional *lac* operon can be incorporated into the chromosome by homologous recombination, restoring the cell's ability to use lactose.

lac⁺

Figure 15.14 Generalized transduction
Generalized transduction carried out by bacteriophage P1.

Transduction

The process of transferring genes between bacteria as part of bacteriophage particles is called **transduction**. Two fundamentally distinct transductional mechanisms operate among bacteriophages. One system, called **general transduction**, results in transfer of all bacterial genes at low but identical frequencies. During **specialized transduction**, certain genes are transferred at very high frequencies, whereas others are transferred at low rates or not at all.

Generalized Transduction

Generalized transduction is carried out by some, but not all, virulent bacteriophages. This process is outlined in **Figure 15.14**. Following infection of the cell, these bacteriophages undergo a complete replication cycle. During the infection process, the host chromosome is disrupted and used as a source of nucleotide building blocks for the phage genome.

Occasionally, fairly large host DNA fragments remain in the cytoplasm at the time of viral DNA particle packaging. Infrequent packaging errors result in bacterial DNA being packaged into the capsid instead of the phage genome. Upon completion of the infection cycle, the host cell lyses, releasing the progeny virus. Although most capsids contain viral DNA and are fully capable of binding to new host cells and thus repeating the infectious cycle, the rare capsid containing the host DNA will also adsorb to another host and inject its DNA into the bacterium. This interaction of a cell with a viral particle does not result in the death of the host, because the introduced DNA does not specify viral particles. Rather, the cell now contains a small fragment of the bacterial genome. This newly introduced DNA is now similar to any DNA fragment introduced to the cell by other gene exchange mechanisms, such as transformation. In the event that the transduced DNA shares homology with the bacterial chromosome, double reciprocal recombination can result in incorporation of the transduced DNA into the bacterial chromosome. This process, thus, results in transfer of DNA from one bacterium to another.

Specialized Transduction

Specialized transduction requires incorporation of the viral DNA into the bacterial chromosome, and it is therefore carried out only by lysogenic viruses.

intact, the recipient becomes a donor. Because the recipient can contain the same genes as are on the larger F plasmid transferred during mating with F′, the recipient may become a partial diploid (or merodiploid) (i.e., contains two copies of a gene in a single cell). Moreover, regions of extensive homology exist between the chromosome and the plasmid, allowing general recombination to take place along these sequences. Insertion of an F-like plasmid into the bacterial chromosome is one way new Hfr strains can be generated, thus allowing chromosomal gene transfer from a specific location.

Specialized transduction occurs following formation of prophage, whereby the viral DNA integrates at a specific site in the host genome. Although the prophage replicates for many generations with the bacterial chromosome, it can occasionally excise and initiate a virulent cycle, including viral DNA replication, packaging, and lysing of the host. Occasionally, excision of the prophage is imprecise, resulting in removal of some DNA on either left, right, or both sides of the site of integration, as shown schematically in **Figure 15.15**. This excised DNA now is part of the viral genome; it replicates and is packaged into every viral particle. Following lysis of the cell, the released virulent bacteriophage can infect other cells. Infection of a new host can lead to viral replication and continuation of the virulent cycle. However, as is the case with the original infecting phage, occasionally the transducing virus can enter into a lysogenic relationship with its new host. The

transduced genes would then be part of the genome of the second bacterial cell. Because the second recipient can also contain genes homologous to those carried by the transducing phage, reciprocal recombination can result in stable incorporation of such genes into the bacterial chromosome, without the necessity of establishing a bacterial lysogen.

Specialized transduction therefore results in phage-mediated transfer of genes that are near the attachment site of the lysogenic prophage. Although this process is limited to such genes, it is extremely efficient because, once the error in excision has occurred, such genes are part of the bacteriophage genome and are efficiently transferred.

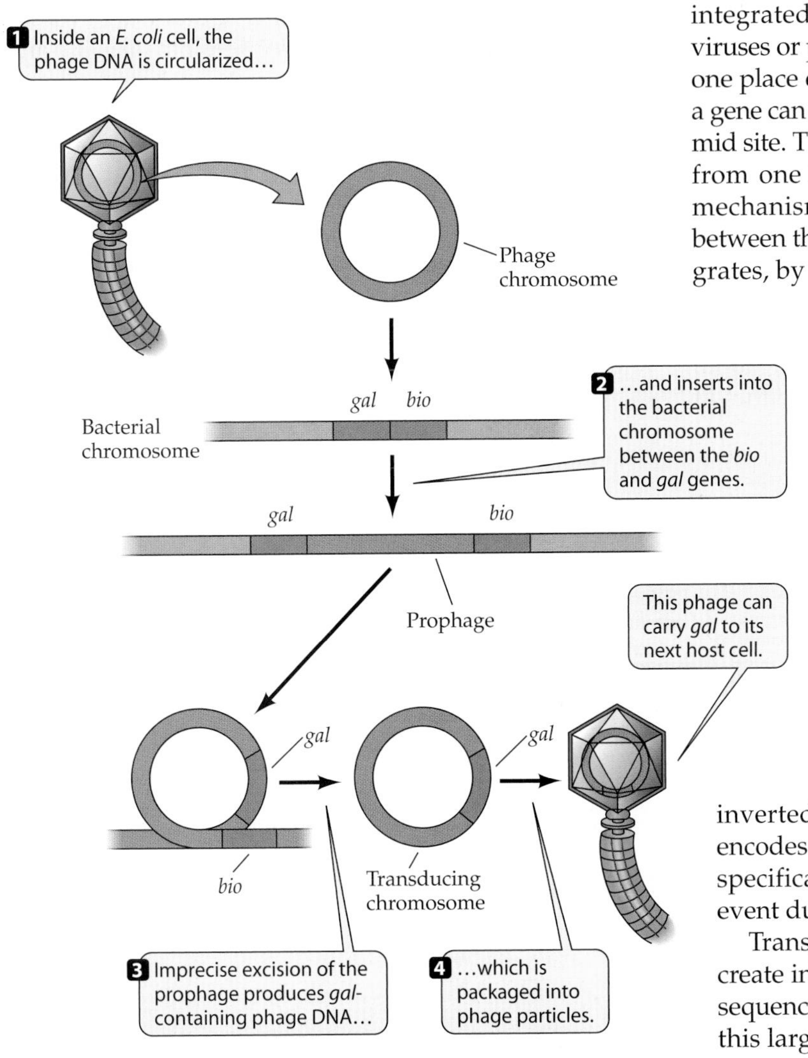

MOBILE GENETIC ELEMENTS AND TRANSPOSONS

This chapter has considered genetic exchange in terms of blocks of DNA that move from cell to cell by being integrated into defined, self-replicating units, such as viruses or plasmids. Occasionally, a gene can move from one place on the chromosome to another in that cell, or a gene can move between a chromosomal site and a plasmid site. This process is called **transposition**. Movement from one location to another by the transpositional mechanism requires recombination to take place between the transposing DNA and the site where it integrates, by distinct mechanisms that are similar to site-specific recombination because the DNA exchange leading to integration of a transposon does not require homologous sequences, as general recombination does. The site-specific recombination utilizes DNA sequences that define the site of transposition, and specialized enzymatic machinery that catalyzes the transpositional event.

The simplest form of such a mobile gene is an **insertion sequence**, abbreviated IS. A number of different insertion sequences have been characterized in bacteria, ranging in size from 700 to 5000 base pairs (bp). All IS elements have a common feature: they contain short (16 to 41 bp) inverted repeat sequences at their ends. The IS also encodes at least one enzyme, called **transposase**, that specifically mediates the site-specific recombination event during transposition.

Transposition of an IS element into a new site can create insertion mutations by physically disrupting a sequence that encodes a polypeptide. The presence of this large block of DNA in an **operon** (a set of genes transcribed from a single promoter) can cause polar mutations; that is, insertion in the first gene of an oper-

1 Inside an *E. coli* cell, the phage DNA is circularized…

Phage chromosome

Bacterial chromosome

gal   *bio*

2 …and inserts into the bacterial chromosome between the *bio* and *gal* genes.

gal   *bio*

Prophage

This phage can carry *gal* to its next host cell.

gal   *gal*

bio   Transducing chromosome

3 Imprecise excision of the prophage produces *gal*-containing phage DNA…

4 …which is packaged into phage particles.

Figure 15.15 Specialized transduction
Specialized transduction carried out by bacteriophage lambda.

on can block expression of genes downstream from the site of insertion. This is due to the inability of RNA polymerase to traverse through an IS element, presumably because of the presence of transcriptional termination signals within the element.

Transposons are more complex forms of mobile genetic elements and are characterized by the presence of genes in addition to the genes needed for transposition. Most common genes found within transposons are those specifying antibiotic-resistance determinants. However, other genes, such as those encoding toxins, have also been identified on these transposable elements.

Classification of Transposons

The simplest type of transposon is a **composite transposon**, which is a mobile genetic element in which two IS elements surround a gene. Transposition then involves translocation of a copy of the entire element (two ISs and the internal genes) into a new site, using the transposase gene of one of the IS elements. Composite transposons can serve as a source of IS sequence transposition because as long as the IS element contains a functional transposase, it can transpose independently from the transposon. Examples of such transposons are shown in **Figure 15.16**. As can be seen in the figure, some composite transposons, such as Tn5, contain multiple drug-resistance genes with the IS boundaries. Composite transposons represent a simple way for any bacterial gene to become mobile. Two identical IS sequences can surround a bacterial gene, and under some circumstances, that gene becomes a transposon. Every transposon requires expression of a functional transposase gene, which is generally found within at least one of the insertion sequences.

A distinct class of transposons is those modeled after transposon Tn3, shown in Figure 15.16B. This family of transposons does not contain repeated IS elements at its ends. However, the transposon is a single DNA segment that contains small (38 bp) inverted repeats at its ends. Within the element, there are antibiotic-resistance determinants (β-lactamase in Tn3), as well as two genes involved in transposition: transposase and resolvase. Resolvase is an endonuclease.

Mechanism of Transposition

Transposons are mobile and they "jump" from a **donor** site (phage, plasmid, or chromosome) into **target** sites, which can be another region of the bacterial chromosome, a site on a plasmid, or on a phage. The target sequence can vary, depending on the transposon. Some transposons require specific sites that occur only rarely within a chromosome or plasmid. Some transposons, such as Tn10, insert into a sequence NGCTNAGCN, where N can be any base. This sequence occurs many times in a bacterial chromosome. Other transposons, such as Tn5 and mu phage, can insert into virtually any target sequence.

It is also possible to make distinctions regarding the fate of different transposons when they move from the donor to the target sites. For some transposons, such as Tn10, this process is **conservative** and represents true translocation, where it is lost from the donor site after transposition to a new location. Other transposons (i.e., Tn3) move by a **replicative** mechanism: a copy of the transposon inserts into the target sequence, while a copy remains in its initial position.

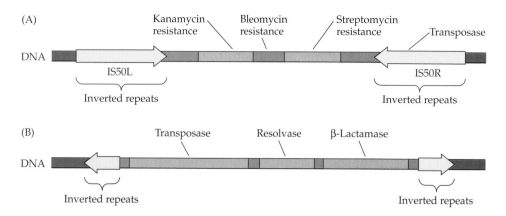

Figure 15.16 Composite transposons
Structures of some bacterial transposable elements. (A) A composite transposon contains antibiotic genes flanked by two insertion sequences as direct or inverted repeats Shown here is the Tn5 transposon, with inverted repeats. (B) The Tn3 transposon.

SUMMARY

▶ Although most of the essential genes are found on the bacterial chromosome, additional genes for a variety of cellular functions are encoded by extrachromosomal genetic elements called **plasmids**. A class of plasmids carrying antibiotic resistance genes is called **R factors** or **R plasmids**. A special class of plasmids, called virulence plasmids, carries genes that encode essential virulence determinants, such as **adhesins** and **toxins**.

▶ A plant pathogen *Agrobacterium tumefaciens* causes tumors in plants by transferring a portion of its plasmid (called tumor-inducing, or **Ti plasmid**) to the plant, where the Ti-encoded genes induce synthesis of plant hormones.

▶ **Homologous recombination** involves breakage and joining of two DNA strands along sequences of identity, or a high degree of similarity. Homologous recombination utilizes a DNA site that possesses a nick, generated by the **RecB**, **RecC**, and **RecD** nuclease complex, at a sequence called the *chi* site. Another key component, required for homologous recombination, is the **RecA** protein, which binds to single-stranded DNA and then identifies homologous regions in the opposite strand.

▶ There are three types of genetic exchange mechanisms among bacteria: (1) uptake of free DNA from the media, called **transformation**; (2) direct transfer of DNA between donor and recipient bacteria, called **conjugation**; and (3) bacteriophage-mediated transfer of DNA, called **transduction**.

▶ Uptake of DNA by transformation requires that the recipient bacterium be in a physiological state called **natural competence**.

▶ In some bacteria, chemical treatment can induce **artificial competence**, such that they can take up DNA fragments or plasmids.

▶ DNA exchange by conjugation requires that the donor bacterium carry a **transmissible plasmid**, which encodes a complete transfer machinery, including **conjugal (sex) pili**. Plasmid transfer by conjugation is **replicative**; that is, a copy of the plasmid is transferred from the donor to a recipient without the loss of the plasmid in the recipient bacterium.

▶ *E. coli* carrying a chromosomally integrated form of a well-studied F plasmid, called **high-frequency recombinant (Hfr)**, can transfer chromosomal sequences from this integrated form, starting from the *ori*T of the F plasmid. Upon entry into the recipient *E. coli*, the partial genome is incorporated into the chromosome. Occasionally, the copy of the entire genome can be transferred into a recipient, in which case the recipient can become a donor.

▶ **Specialized transduction** is the consequence of imprecise excision of the lysogenic bacteriophage from the chromosome, resulting in packaging into bacteriophage capsids, of portions of the chromosomal DNA from the region flanking the attachment site.

▶ **Generalized transduction** involves random packaging of chromosomal DNA fragments, instead of bacteriophage genomes, into bacteriophage capsids during the final stages of phage maturation.

▶ Movement of genes between various locations on the chromosomes or plasmids is referred to as **transposition**. Insertion of a transposon into a gene results in a mutation in the gene due to the disruption of its coding sequence.

▶ Simple mobile elements are **insertion sequences**, containing short **inverted repeats** at their ends; these encode one enzyme, **transposase**. Some genes can transpose when they are flanked by two insertion sequences. These mobile genetic elements are called **composite transposons**. Certain transposons, such as **Tn3**, lack flanking insertion sequences. However, they still contain inverted repeats as their ends; within the transposon, they encode two essential enzymes, **transposase** and **resolvase**. Depending on the type of transposon, the site of insertion—the **target site**—varies from a defined short sequence to any random DNA sequence.

REVIEW QUESTIONS

1. Does a bacterium bearing a self-transmissible plasmid require a functional DNA synthesizing machinery for it to transfer the plasmid into a recipient by conjugation? Does a recipient need a functional DNA synthesis for it to receive the plasmid from a donor?

2. How does the fate of a chromosomal DNA fragment differ from that of a plasmid, when it is introduced into bacteria by transformation?

3. When bacteria contain more than one compatible plasmid, some plasmids, which are not self-transmissible, can transfer to recipients. Describe the mechanism that makes it possible. If the same plasmid lacks an origin of transfer, can it ever transfer to a recipient by conjugation?

4. When a chromosomal gene is encoding an antibiotic-resistance determinant and is flanked by two identical DNA sequences, homologous recombination can lead to the loss of the resistance gene, such

that some bacteria become sensitive to the antibiotic. Explain.

5. Which method of DNA transfer between bacteria would not take place if the donor and recipient were separated by a filter with a pore size of 0.45 μm? Which method of transfer would be blocked by the presence of high concentrations of DNase (enzymes capable of degrading DNA)?

6. A transposon insertion into the *lacZ* gene of the lactose utilization operon results in inability of the bacterium to utilize lactose as a sole carbon and energy source. When this bacterium acquires an F' *lac* (an F plasmid with the intact *lacZ* gene), it is still unable to utilize lactose. Provide a reasonable explanation for this phenomenon.

7. Chromosome mobilization by the integrated F plasmid (Hfr) usually does not lead to the recipient being capable of acquiring donor functions. In contrast, F'-mediated transfer usually converts the recipient to a donor. Explain this difference in transfer mechanisms. What condition of DNA transfer results in the appearance of rare recipients that have acquired donor capabilities that are receiving DNA by the Hfr-mediated transfer?

8. How can the Hfr strain be used to map the relative order of genes?

9. A recipient *B. subtilis*, with mutations in the genes encoding enzymes of histidine and arginine biosynthesis, is transformed with DNA that came from *B. subtilis*, carrying a mutation in a ribosomal protein, rendering its resistant to streptomycin. The transformants were plated onto streptomycin, and subsequently replica plated onto minimal media containing either arginine or histidine. Growth (or lack of growth) on these supplemented minimal plates revealed that among the original streptomycin-resistant transformants, 12% did not require histidine anymore for growth, and 6% did not require arginine for growth, whereas the remaining 82% were streptomycin-resistant and retained the original requirement for both histidine and arginine. No streptomycin isolates that lost both of the arginine or histidine requirements were obtained. What is the location of the arginine and histidine biosynthetic genes, relative to the streptomycin-resistant determining ribosomal protein gene?

Compare and contrast specialized and generalized transduction.

10. A plasmid, carrying a kanamycin-resistance gene and containing a composite transposon (two insertion sequences flanking a tetracycline-resistance gene) was introduced into a recipient by conjugation. When the bacteria were passaged for many generations in an antibiotic-free medium, a fraction of these bacteria did not contain any plasmid. When these bacteria were individually analyzed for antibiotic resistance, it was found that some were resistant to tetracycline, because they contained a chromosomal copy of the transposon. However, some were resistant to kanamycin, and the kanamycin-resistance gene was inserted into the chromosome. No isolates that contained both of the resistance genes were recovered. Explain the events in the recipient bacterial cell following introduction of the original plasmid.

11. A wild-type strain of *E. coli* is transformed by DNA from *E. coli* that has a Tn5 transposon in a gene encoding an enzyme of the tryptophan biosynthetic pathway. After plating transformants on rich medium containing kanamycin (the resistance gene carried on Tn5), the colonies are individually transferred onto minimal medium. Of 100 colonies tested, 18 required tryptophan for growth, 70 did not require any nutrients, and 12 could not grow on minimal media, even when it was supplemented with tryptophan. Provide an explanation for the three types of transformants. If the recipient were a *recA* mutant of *E. coli*, would you expect a different distribution of the three classes of transformants based on their nutritional requirements?

SUGGESTED READING

Leach, D. R. F. 1995. *Genetic Recombination*. Cambridge, MA: Blackwell Science.

Old, R. W. and S. B. Primrose. 1994. *Principles of Gene Manipulation*. Cambridge, MA: Blackwell Science.

Schleif, R. 1993. *Genetics and Molecular Biology*. 2nd ed. Baltimore, MD: Johns Hopkins University Press.

Chapter 16

Microbial Genomics

If we should succeed in helping ourselves through applied genetics before vengefully or accidentally exterminating ourselves, then there will have to be a new definition of evolution, one that recognizes a process no longer directed by blind selection but by choice.
— M. A. EDEY AND D. C. JOHANSON, BLUEPRINTS, 1989

Previous chapters have dealt with the mechanisms for gene transfer and recombination that allow bacteria to acquire genes from other bacteria and express traits encoded by these genes. Our growing understanding of DNA replication, genetic variation, and transfer of genetic information among microorganisms, as well as of the molecular mechanisms that control gene expression, has led to the development of a new industry: biotechnology. Finally, the advanced technologies resulting in rapid determination of sequences of entire bacterial genomes have enormously enhanced our understanding of the potential of these organisms to carry out a multitude of known as well as unexpected biological processes.

In this chapter, we will discuss the rationale, methodology, and outcome of deliberate laboratory manipulations of genes, a process called **genetic engineering.** The aim of genetic engineering, generally, is to isolate genes from eukaryotic, or eubacterial, archaeal genomes and to introduce these genes into a different species where they will be propagated in a stable manner, resulting in a **transgenic** organism.

Moreover, the identification of the entire genetic repertoire of a bacteria or higher eukaryotes makes it possible to manipulate these organisms at a truly genomewide basis and, not surprisingly, provides unparalleled opportunities to engineer entirely new species with potentially beneficial, as well as harmful, effects on humankind.

BASIS OF GENETIC ENGINEERING

Early work by bacterial geneticists with transducing phages and conjugative plasmids (see Chapter 15) that carry fragments of the bacterial chromosome suggested a possible means of isolating and manipulating individual genes. This work has promoted the study of genetic organization and expression and has provided a mechanism for accumulating large amounts of protein for research and bulk amounts of protein for use in agriculture, industry, and medicine.

The isolation of individual genes by classic bacterial genetic techniques such as F' plasmids or transducing phages has been feasible for decades. However, it took the discovery of **restriction endonucleases** to launch the field of genetic engineering. By borrowing these specialized enzymes from bacteria, researchers gained a powerful tool for manipulating genetic information. Simultaneously, increased knowledge of the replication requirements for plasmids and phages provided vehicles for isolating individual genes and propagating them in pure form, that is, **cloning** them. Finally, development of readily applied methods for determining the precise nucleotide sequence of DNA allowed studies of genetic organization, genetic structure, and the function of gene products. Recombinant DNA technology has affected virtually every aspect of biomedical science, ranging from the production of human hormones in bacteria to the study of evolution. This chapter reviews some principles of genetic engineering and summarizes some of its more common applications.

Restriction and Modification

One of the most effective barriers to free genetic exchange of DNA between unrelated bacteria is the presence of **restriction-modification** systems. These are enzymes that bacteria employ in nature to maintain the integrity of their genomes and to degrade DNA that they may accidentally acquire from other bacteria by any of the genetic exchange mechanisms discussed in Chapter 15. **Restriction enzymes** are also useful in protecting bacteria from DNA viruses by degrading the viral genetic material after it enters the bacterial cytoplasm. Restriction enzymes (or **restriction endonucleases**) introduce double-stranded breaks in foreign DNA at specific base sequences. Because these sequences are short (generally four to six base pairs), they will undoubtedly be present in the bacterial chromosome as well. So the bacteria prevent destruction of self-DNA by using **modification enzymes** to modify the recognition sites of restriction enzymes such that they are no longer degraded by these enzymes. Each bacterial species has characteristic restriction-modification enzymes that recognize unique sequences. For example, certain strains of *Escherichia coli* contain an enzyme, *Eco*RI, which cleaves DNA whenever the sequence GAATTC occurs. The genome of *E. coli* itself has such sequences, and these are modified by methylation of the second adenine of this sequence on both strands, GAATTC, rendering the sequence resistant to *Eco*RI cleavage.

The best example of the advantage conferred by a restriction-modification system is observed in infections by bacterial viruses. The efficiency of infection of a particular host by bacterial viruses is inverse to the extent of damage to the viral DNA by the host's restriction endonucleases, following entry of the viral genome into the bacterial cell. An efficient restriction system can reduce the infection process to an exceedingly low level, such that most of the bacteria survive the infecting virus, and only 1 in 1,000 bacteria infected will ever yield viral progeny. However, when these rare infective virions that infect 1 in 1,000 are isolated and used to infect the same bacterial strain, the viral infection will be very success-

ful, with hundreds of virions obtained from each host. The reason for the high level of infectivity of the second generation virions is that they survived in the first host's cytoplasm long enough to have their DNA modified by the host's modification enzymes. Thus, during the subsequent infection, this modified DNA is not recognized as foreign by the host's restriction enzymes.

This same restriction barrier is effective against the uptake of foreign DNA by transformation or conjugation. Thus, the fate of a plasmid or a DNA fragment depends on whether it is a substrate for the host's particular restriction enzymes and, if so, whether it has been protected by modification. Overall, restriction-modification systems, along with other barriers to DNA exchange, favor recombination of DNA within identical or closely related species.

ANALYSIS OF DNA TREATED WITH RESTRICTION ENDONUCLEASES

In practice, restriction enzymes are powerful tools for genetic engineering because they permit hydrolysis of a DNA molecule into defined fragments containing individual genes or clusters of genes. Hundreds of restriction enzymes with unique recognition sequences have been purified for this purpose, providing an array of enzymes to fragment DNA to a desired size. Restriction enzymes can cleave at a sequence by hydrolyzing a phosphodiester bond in each strand that is staggered, or they can cleave across a double-stranded DNA at the same base pair, producing a blunt end, as illustrated in **Figure 16.1**. Some of the more commonly employed

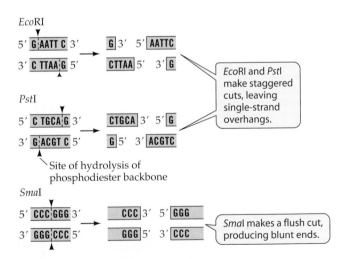

Figure 16.1 Restriction enzymes
Cleavage of DNA strands by restriction enzymes. *Eco*R1 creates a 5′ overhang; *Pst*I creates a 3′ overhang; *Sma*I produces blunt ends.

restriction endonucleases and their respective DNA recognition sequences are summarized in **Table 16.1**.

A simple method of DNA analysis is **agarose gel electrophoresis**, a procedure that separates DNA fragments by differential migration in an electrical field, applied across a hydrated porous matrix of agarose (a highly purified polysaccharide from seaweed). Among the critical parameters determining the rate of migration are the size and shape of the DNA. Agarose gel electrophoresis can thus separate plasmids based on size, and the plasmid content of independent isolates can be compared to see whether they contain identically sized plasmids. As illustrated in **Figure 16.2** (left box, A), different clinical isolates of the same organism can contain unique plasmids, and their respective electrophoretic mobilities can establish the size relationship among them.

The technique of comparing plasmids based on migration in an electrical field is rather unreliable, mainly because two plasmids can have similar sizes and yet be totally unrelated in sequence. Treatment of plasmids with restriction enzymes and subsequent separation by agarose gel electrophoresis allows a finer matching of sequences. If two plasmids are identical, the location of the restriction site recognition sequences is the same. Thus, following treatment with a particular restriction enzyme, an identical or similar pattern of fragments is obtained. Examples of such analysis are shown in Figure 16.2 (left box, B), which illustrates a case in which some of the fragments in strains A, B, C, and D are the same size and strains A and B may be identical.

To determine the relationship among any set of sequences, a method of DNA hybridization was developed by Edward Southern in 1975. This technique, called **Southern blotting** (Figure 16.2, center box), relies on the formation of stable duplexes (hybrids) between electrophoretically separated restriction fragments and radioactively labeled denatured DNA probes. The restriction fragments are denatured into single-stranded DNA by treating the gel with an alkali. The gel is then overlaid with a nitrocellulose sheet. Capillary action (blotting) moves the DNA onto the nitrocellulose sheet, where it firmly binds. In this way, a permanent copy of the gel is obtained, with the fragments immobilized in the exact position where they migrated on the gel. This blot can then be soaked in a solution of a denatured radiolabeled probe. If the probe is one of the plasmids (for example, plasmid A), it will anneal to any complementary sequences immobilized on the solid support of the nitrocellulose. By this method, similar sequences can be radioactively "tagged" and visualized following exposure of photographic film. Results presented in Figure 16.2 (left box, C) indicate that plasmid A is probably identical to plasmid B, but only partially similar to plasmid C. Plasmid D is completely different from plasmid A, despite having an identical size and sharing some common-sized restriction fragments. Extra sequences found in plasmid C are probably due to recombination or transposition from other plasmids or chromosomal sites having sequences similar to plasmid A.

This powerful technique can be applied to the analysis of sequences found on specific fragments of bacterial or even eukaryotic chromosomes. DNA is extracted from cells and treated with restriction

Table 16.1	Common restriction endonucleases (enzymes) and their DNA recognition sequences		
Microorganism	**Restriction Endonuclease**	**Target Sequences, Showing Axis of Symmetry (|) and DNA Cleavage Sites (▼)**	
Generates cohesive ends:			
Escherichia coli RY13	*Eco*RI	G ▼ A A | T T C	
		C T T | A A ▲ G	
Bacillus amyloliquefaciens H	*Bam*HI	G ▼ G A | T C C	
		C C T | A G ▲ G	
Bacillus globigii	*Bgl*II	A ▼ G A | T C T	
		T C A | A G ▲ A	
Haemophilus aegyptius	*Hae*II	Pu ▼ G C | G C Py	
		Py C G | C G ▲ Pu	
Haemophilus influenza R_d	*Hind*III	A ▼ A G | C T T	
		T T C | G A ▲ A	
Providencia stuartii	*Pst*I	C T G | C A ▼ G	
		G ▲ A C | G T C	
Streptomyces albus G	*Sal*I	G ▼ T C | G A C	
		C A G | C T ▲ G	
Xanthomonas badrii	*Xba*I	T ▼ C T | A G A	
		A G A | T C ▲ T	
Thermus aquaticus	*Taq*I	T ▼ | C G A	
		A G | C ▲ T	
Generates flush ends:			
Brevibacterium albidum	*Bal*I	T G G ▼ C C A	
		A C C ▲ G G T	
Haemophilus aegyptius	*Hae*III	G G ▼ C C	
		C C ▲ G G	
Serratia marcescens	*Sma*I	C C C ▼ G G G	
		G G G ▲ C C C	

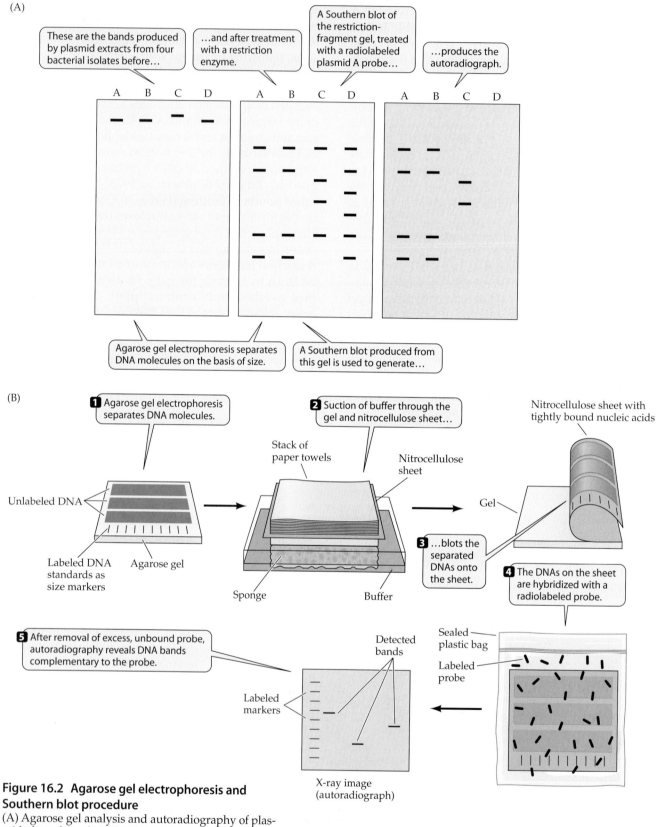

Figure 16.2 Agarose gel electrophoresis and Southern blot procedure
(A) Agarose gel analysis and autoradiography of plasmids from four clinical isolates of the same bacterium. The gels are stained with a fluorescent dye; the autoradiograph is an X-ray film exposure of a Southern blot. (B) Details of the Southern blot procedure.

enzymes. As with plasmid DNA, fragments are separated according to size by agarose gel electrophoresis. Unlike fractionation of plasmids, which yield only a handful of fragments, chromosomal DNA contains hundreds of recognition sites. Following gel electrophoresis, many bands appear, often preventing clear differentiation of one band from closely migrating bands of equivalent size. Nevertheless, following Southern blotting, a radiolabeled complementary probe can clearly identify a related sequence in one of the fragments, as shown in **Figure 16.3**. This method can detect proviruses in bacterial genomes (proviruses are viruses that have integrated their DNA into the host genome), as well as specific genes that encode determinants of antibiotic resistance or of bacterial pathogenicity, such as those that encode toxins.

Probes may be obtained from a variety of sources. Usually they are cloned pieces of DNA from different species that share some sequence similarity. Alternatively, a probe may be a short piece of chemically synthesized DNA or the DNA product from the polymerase chain reaction (**Box 16.1**). The synthetic oligonucleotide probe is made on the basis of a DNA sequence deduced from a known amino acid sequence of a protein. For example, an enzyme, X, has been purified and has an amino terminal sequence of Met-Trp-Asp-Trp. Based on the codons for these four amino acids (see Table 13.2), oligonucleotides ATGTGGGATTGG and ATGTGG-GACTGG can be synthesized chemically and can be used as probes in a Southern blot hybridization to bind to fragments containing the complementary sequence. (Note that codons GAT and GAC can both encode aspartic acid, so two probes must be made.) Labeled with radioactive phosphorus, these probes hybridize with their complementary sequence on one of the DNA strands within the gene that encodes enzyme X. This probe can also be used directly to identify recombinant plasmids containing the cloned gene for enzyme X.

CONSTRUCTION OF RECOMBINANT PLASMIDS

Extrachromosomal genetic elements, such as plasmids and bacteriophages, have become the logical choice for propagation of isolated genes because they can be readily isolated in pure form from their bacterial hosts. Many different plasmids and bacteriophages are used in genetic engineering as vectors, that is, vehicles of transmission for cloning isolated genes. Most of the useful plasmid cloning vectors are small (less than 10,000 base pairs) and contain antibiotic-resistance determinants. Bacteriophage vectors contain regions that are dispensable, and these can be replaced with foreign DNA. All plasmids contain an origin of replication, which is a piece of DNA that allows the plasmid to be autonomously maintained. Three commonly used plasmid vectors, pBR322, pACYC184, and pUC18, as well as a bacteriophage vector, lambda, are shown schematically in **Figure 16.4**.

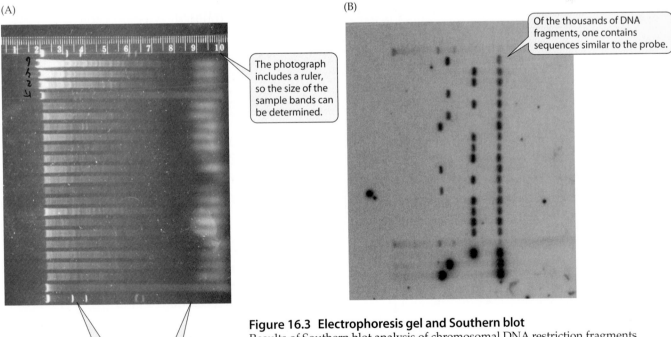

(A)

The photograph includes a ruler, so the size of the sample bands can be determined.

DNA fragments of known size (markers) are loaded on both sides of the gel.

(B)

Of the thousands of DNA fragments, one contains sequences similar to the probe.

Figure 16.3 Electrophoresis gel and Southern blot
Results of Southern blot analysis of chromosomal DNA restriction fragments from different clinical isolates of the same bacterium. (A) Agarose gel after electrophoresis, stained with fluorescent dye; (B) autoradiograph of the Southern blot. Courtesy of Steve Lory.

Methods & Techniques Box 16.1   The Polymerase Chain Reaction

One of the more powerful techniques for isolating genes that has been developed in recent years is called the **polymerase chain reaction** (PCR). The principle of PCR is the generation of copies of two strands of DNA by a repeated sequence of denaturation of double-stranded DNA, synthesis of the complementary strands, followed by the next round of denaturation and synthesis. Because synthesis of DNA strands by DNA polymers requires primers (see Chapter 13), this technique also necessitates knowledge of a limited sequence, such that the polymerization can be initiated from a short oligonucleotide that hybridizes to

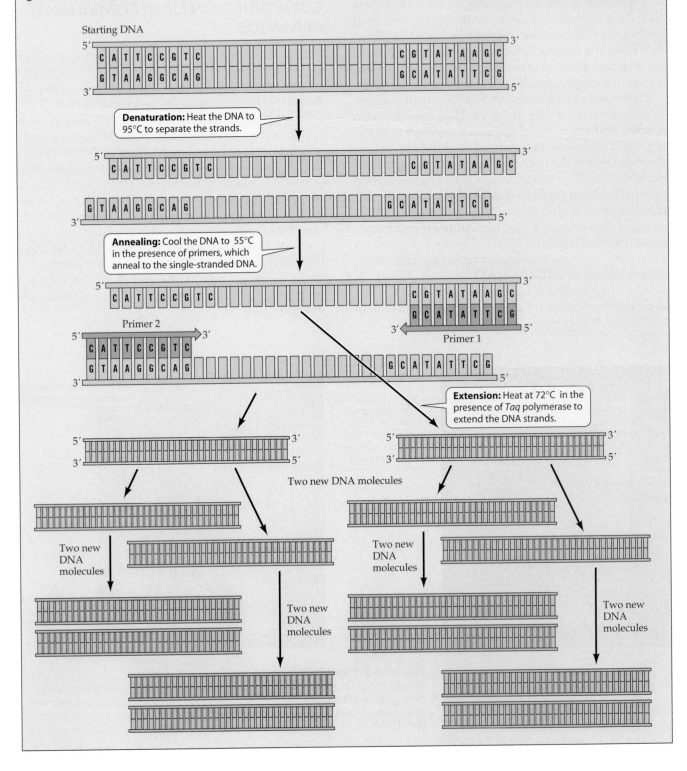

Methods & Techniques Box 16.1   *(continued)*

the ends. The PCR usually starts from a double-stranded DNA (called the template). However, the first DNA molecule can be generated by reverse transcription of mRNA to give a complementary DNA copy (cDNA), which is converted into a double-stranded template. The DNA strands are separated by heating, followed by hybridization of primers for the next round of polymerization. This modification of PCR is called RT-PCR. As illustrated in the figure, the reaction is divided into cycles, each cycle consisting of three steps. First, raising the temperature of the DNA solution denatures the double-

stranded DNA. Next, the temperature is lowered, and complementary oligonucleotides (the primers) are allowed to anneal to the DNA. Finally, DNA polymerase is added, together with dATP, dCTP, dTTP, and dGTP, and the complementary strand is synthesized. The denaturation-primer annealing-complementary strand synthesis is repeated 30 to 40 times. Because each newly synthesized complementary strand can serve as a template for the next round of PCR, the amount of DNA synthesis increases exponentially, and from a few starting DNA molecules, 20 cycles can generate over a million

copies. The PCR-generated DNA can be used for many manipulations, among them DNA sequencing and preparation of DNA probes for Southern hybridizations, or they can be directly cloned into plasmid or viral vectors. It is noteworthy, that the repeated denaturation exposes the DNA polymerase to temperatures that usually inactivate most enzymes. However, enzymes utilized in PCR are isolated from thermophylic bacteria (such as *Thermus aquaticus*), which usually replicate at high temperatures, and they possess thermo-stable DNS polymerases.

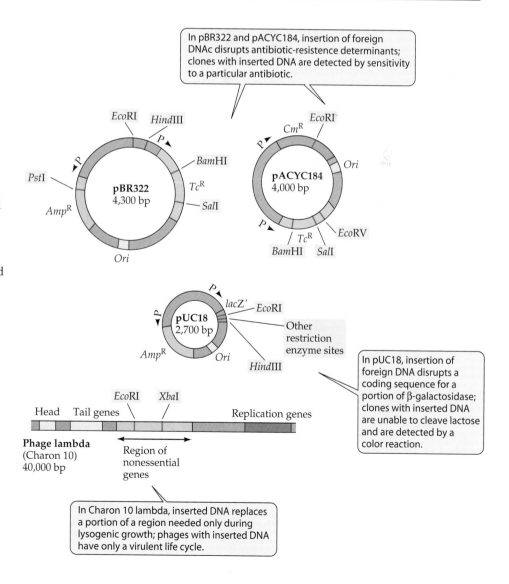

Figure 16.4 Vectors for construction of recombinant molecules

Plasmids pBR322, pACYC184, and pUC18 contain several unique restriction enzyme recognition sequences (shaded) as well as resistance determinants for ampicillin (Amp^R), tetracycline (Tc^R), and chloramphenicol (Cm^R). Plasmid pUC18 has a multiple cloning site with recognition sequence for 13 different restriction enzymes. In the case of pUC18, to screen for inserted DNA, the plasmids must be transformed into special *E. coli* strains that carry the gene for the non-*lacZ'* portion of β-galactosidase; the two portions of the protein form the active enzyme. Artificial formation of an active enzyme from two fragments (called complementation) allows screening for insertions by colony color in gel containing X-gal, the chromogenic indicator of β-galactosidase. Bacterial colonies with intact plasmid have a blue appearance; colonies with plasmids that carry inserts are white.

In pBR322 and pACYC184, insertion of foreign DNAc disrupts antibiotic-resistence determinants; clones with inserted DNA are detected by sensitivity to a particular antibiotic.

In pUC18, insertion of foreign DNA disrupts a coding sequence for a portion of β-galactosidase; clones with inserted DNA are unable to cleave lactose and are detected by a color reaction.

In Charon 10 lambda, inserted DNA replaces a portion of a region needed only during lysogenic growth; phages with inserted DNA have only a virulent life cycle.

Figure 16.5 Cloning of fragments from a large plasmid into vector pBR322
Cloning of DNA fragments from plasmid pXY1 using pBR322 as vector. pXY1 contains four sites recognized by the restriction enzyme *Pst*I.

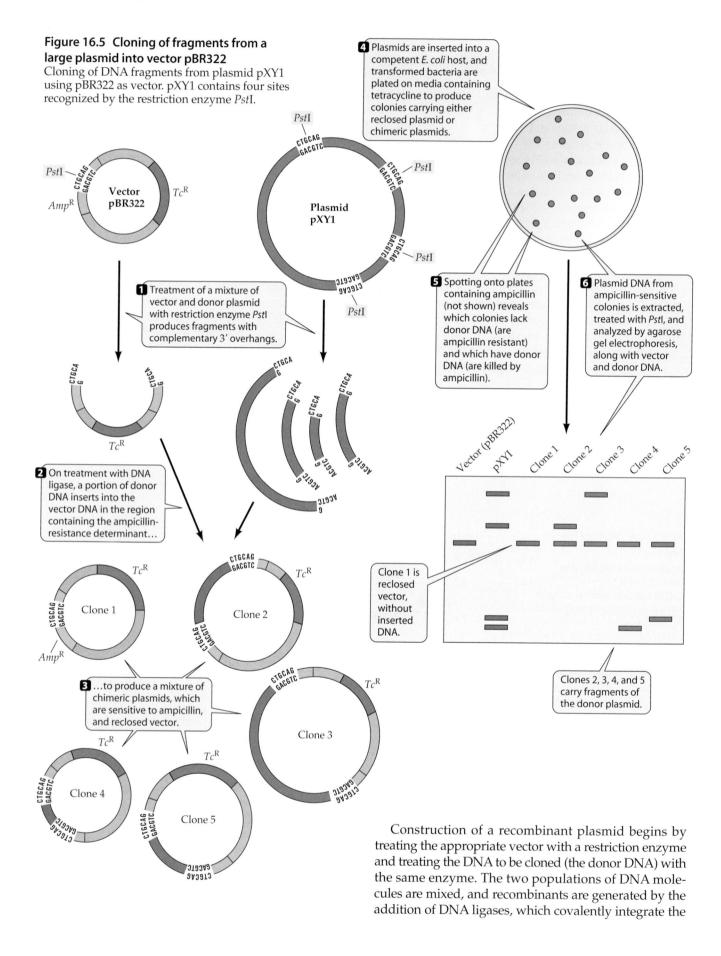

Construction of a recombinant plasmid begins by treating the appropriate vector with a restriction enzyme and treating the DNA to be cloned (the donor DNA) with the same enzyme. The two populations of DNA molecules are mixed, and recombinants are generated by the addition of DNA ligases, which covalently integrate the foreign DNA into the plasmid vector. DNA ligase also recloses the vector. The mixture of ligated DNA is introduced into bacteria (usually *E. coli*) by transformation of artificially competent cells. The bacteria are plated onto selective media containing an antibiotic to which the plasmid encodes a resistance determinant. Only the cells that have taken up a plasmid will grow on such a medium. Because the number of competent cells exceeds the number of reassembled plasmid molecules, each colony that arises from plating of transformants is expected to be a clone, containing a single plasmid with a DNA insert. Individual colonies can be isolated and plasmids extracted. After treatment with a restriction enzyme, they can be analyzed by agarose gel electrophoresis. Cloning of fragments from a large plasmid into vector pBR322 is outlined in **Figure 16.5**.

The mixture of bacteria containing the various inserts is called a **library** of DNA. A simple library, such as that outlined in Figure 16.5, is represented by four different clones plus one vector. Thus, the hundreds of possible transformants contain five populations of plasmids: four with inserts and one containing only the vector without an insert DNA. The identification of a specific clone, therefore, requires screening at least four randomly selected clones by agarose gel electrophoresis of the plasmid digests. A more complex library, prepared by restriction enzyme treatment of bacterial or eukaryotic chromosomal DNA ligated with a vector, contains thousands of bacteria, each carrying a recombinant plasmid with a single fragment of the chromosome. This is called a **genomic library** and is a collection of plasmids carrying all of the chromosomal DNA as fragments in recombinant plasmids. The task of isolating genes, then, is reduced to identification of a desired clone from a library of DNA fragments in a vector.

Identification of Recombinant Clones

In construction of recombinant plasmids, one often deals with the technically more difficult problem of identifying a specific clone from the collection of recombinant plasmids. A number of different techniques are available to investigators that allow them to identify a bacterium with a specific recombinant plasmid.

Expression of a Gene in a Heterologous Host

A library of DNA fragments containing thousands of bacteria can easily be screened for a desired cloned gene, provided that gene endows the host with a unique characteristic that distinguishes its particular host from others carrying different fragments of the donor chromosome. For example, to identify a clone of a hemolysin, a protein toxin made by *Pseudomonas aeruginosa*, a collection of *E. coli* carrying a library of *P. aeruginosa* DNA is plated on agar-based medium containing sheep blood. The clone containing the hemolysin gene can be identi-

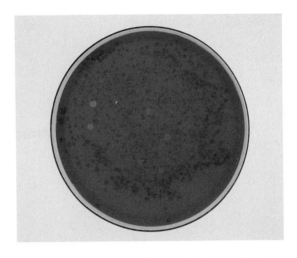

Figure 16.6 Identification of hemolysin-containing clones from a library of chromosomal DNA
Hemolysin is an enzyme that breaks down red blood cells. DNA from a hemolytic microorganism, *Pseudomonas aeruginosa*, was treated with restriction enzyme *Bam*HI, and a library of this DNA was prepared in pBR322. The ligation mixture was plated onto blood agar plates to detect clones containing the hemolysin gene. Courtesy of Steve Lory.

fied by a zone of clearing around the colony, as shown in **Figure 16.6**. Similarly, hydrolysis of starch or casein can be used as a test for expression of amylases or proteases, respectively. These assays rely on expression of the gene in the cloning host, and this is not always guaranteed, especially if the gene in its natural host is regulated. These regulatory signals may be absent in the host carrying the recombinant plasmid.

High-level expression of a cloned gene product can be ensured by using specialized cloning vectors containing regulatory sequences adjacent to the cloning sites. These sequences are usually promoters of highly expressed genes, and the vectors containing these promoters are called **expression vectors**. A modified strong promoter from the lactose operon and bacteriophage lambda, or T7, has been incorporated into some vectors (**Figure 16.7**). Expression vectors allow efficient transcription, provided the cloned gene is in the same orientation as the promoter.

Screening with DNA or Antibody Probes

Because many gene products do not have a simple assay such as screening on the agar surface of a Petri plate, clones carrying these genes must be identified by DNA hybridization techniques. This requires presence of the DNA sequence in the recombinant plasmid, but does not depend on expression of the cloned gene. The method for identifying such a clone is yet another application of a localized DNA hybridization technique, **colony hybridization**, which is analogous to Southern

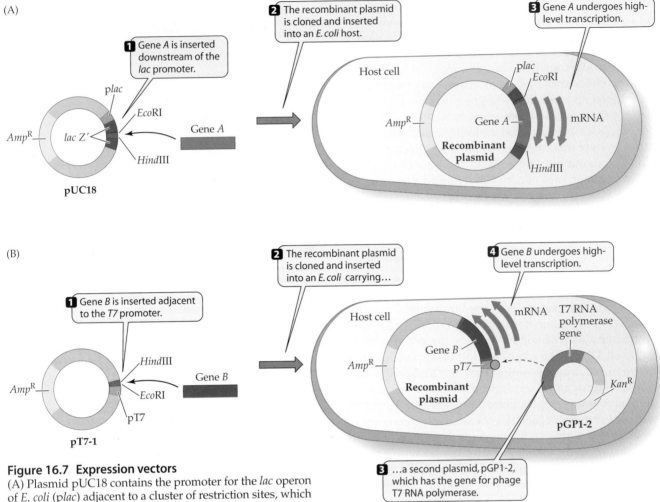

Figure 16.7 Expression vectors
(A) Plasmid pUC18 contains the promoter for the *lac* operon of *E. coli* (p*lac*) adjacent to a cluster of restriction sites, which can be used to insert foreign genes. Transcription from p*lac* results in high-level expression of the cloned gene. (B) An expression system based on components of a bacteriophage. The cloning vector pT7-1 contains a promoter from bacteriophage T7 (pT7), which is recognized by a specialized RNA polymerase from the same phage.

blotting (**Figure 16.8**). The oligonucleotide used in the previous example to identify enzyme X can be used to screen a gene bank of DNA fragments. Bacteria growing as colonies on a plate are allowed to stick to a nylon or nitrocellulose support, which is placed on the agar surface. Once the support is removed, these bacteria, which are bound to the support as an imprint of the original plate (hence called the **replica**), are lysed, and the DNA is denatured. The DNA becomes immobilized, and the radioactive probe hybridizes to the spot on the support that corresponds to the colony carrying a plasmid with the gene coding the enzyme.

A technique based on colony hybridization uses a radioactive antibody as a probe. This requires that the protein product of the cloned gene be obtained and injected into animals (usually rabbits or mice) whose immune system responds by producing an antibody that specifically binds the foreign protein; these are called primary antibodies. Lysed colonies on nitrocellulose filters, prepared as described for colony hybridization with a radiolabeled DNA probe, can be reacted with the radioactive antibody probe. To detect the binding of the antibody to proteins within the lysed colony, the filters are soaked in a solution of the primary antibody that has been "labeled" by chemically attaching to it radioactive compounds, such as the radioactive isotopes of iodine. Alternatively, a secondary antibody (which recognizes the primary antibodies) carrying a covalently attached enzyme that has a brightly colored substrate can be used to identify the colonies that now contain the primary antibody bound to the expressed recombinant protein. The probe will bind and identify only those colonies where protein is being expressed from the gene of interest.

Figure 16.8 Colony hybridization
Individual bacterial colonies are immobilized on nitrocellulose. The nitrocellulose replica is then exposed to a radioactive DNA probe, and positive colonies are identified.

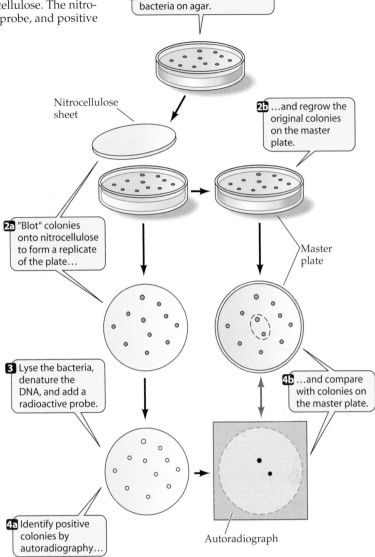

1 Grow colonies of transformed bacteria on agar.

Nitrocellulose sheet

2b ...and regrow the original colonies on the master plate.

2a "Blot" colonies onto nitrocellulose to form a replicate of the plate...

Master plate

3 Lyse the bacteria, denature the DNA, and add a radioactive probe.

4b ...and compare with colonies on the master plate.

4a Identify positive colonies by autoradiography...

Autoradiograph

CLONING OF EUKARYOTIC GENES IN BACTERIA

The relatively simple technique of isolating prokaryotic genes in recombinant plasmids or bacteriophages can be applied to isolating genes from eukaryotic microorganisms. However, if expression of a recombinant eukaryotic protein is desirable, it is likely that the gene cloned from the chromosome of a eukaryotic organism will not be expressed in bacteria. The majority of genes of higher eukaryotes contain intervening sequences (the final, or mature mRNA being assembled after extensive splicing of the primary transcript), and the genes may occupy long segments of DNA. Eukaryotic genes are often cloned as copies of mRNA, called complementary DNA (cDNA). The total mRNA is first isolated from a cell, and a copy of the RNA is made using the enzyme **reverse transcriptase** (see Chapter 14). This enzyme is an RNA-dependent DNA polymerase and can be isolated from selected animal viruses. The RNA-DNA hybrid is denatured, the RNA component hydrolyzed, and a second copy of the DNA is synthesized with DNA polymerase. The final double-stranded product can be cloned into phage and plasmid vectors (**Box 16.2**).

PRACTICAL APPLICATIONS OF RECOMBINANT DNA TECHNOLOGY

Recombinant DNA technology has revolutionized biological and medical research by providing scientists with the ability to obtain large quantities of products that were previously available from their natural sources in minute quantities only. For example, rare enzymes from both bacterial and eukaryotic sources can now be obtained in gram quantities by cloning the appropriate genes in expression vectors and isolating the recombinant products from bacterial extracts. Bacteria, as the source of recombinant products, offer an additional economic advantage because they can be propagated on relatively inexpensive media and grow rapidly.

Bacterial, and more recently yeast-based, expression systems have been extensively employed in industrial-scale production of various hormones, such as insulin and human growth hormone, as well as in vaccines against hepatitis B and herpes. An added benefit of recombinant products, recently realized, is that they are often safer than their natural counterparts, being free of potentially allergenic protein contaminants.

Currently, a great deal of research is devoted to development of genetically engineered vaccines for a variety of infectious diseases. This includes introducing specific mutations into the code for proteins that make up vaccine components, such that they retain their immunogenicity but do not have harmful effects on the vaccinated patients. Genetically attenuated bacterial and viral hosts are also being developed to provide a safe and efficient way to deliver such vaccines.

An integral part of the identification and study of genes is the knowledge of the precise nucleotide sequenced that allows deduction of the amino acid sequence of the protein product, using the genetic code table. Moreover, sequence information has been useful

Genetic Engineering of Plants Using *Agrobacterium tumefaciens* and Ti Plasmids

Crown gall disease in plants, described in Chapter 13, is caused by integration of genes from the bacterium *Agrobacterium tumefaciens* into a plant's chromosome. The genes that are transferred are T-DNA, found on the bacterium's Ti plasmid. *Agrobacterium's vir* genes encode proteins that promote the transfer of T-DNA not only into the plant cell, but also across the nuclear membrane and into the plant chromosome. Once integrated, the bacterial genes are expressed by the plant cell, giving the plant new properties (which in this case are beneficial to the bacterium, but harmful to the plant).

Genetic engineers have exploited this natural system by replacing the disease-causing genes with genes they believe will produce new, desirable traits in the plant. Genes "spliced" into the Ti plasmid will be carried into the plant during the process of infection and integrated into the plant chromosome under the direction of *Agrobacterium's vir* gene products. The steps involved are outlined in the accompanying figure.

A plasmid-carrying T-DNA is treated with restriction enzymes to allow cloning of a foreign gene such that it is flanked by T-DNA. This plasmid is introduced into a strain of *A. tumefaciens* that contains the *vir* region, but lacks its own T-DNA. The *vir* genes and the T-DNA need not be located on the same plasmid for infection to occur. These bacteria are incubated with plant cells, which incorporate the T-DNA into their genomes. When the plant cells car-

rying T-DNA divide, each daughter receives a copy of the cloned DNA. This can be verified by including a marker gene among the cloned genes. For example, genes for enzymes that cause color changes in the growth medium as the cell metabolizes will identify successfully transformed colonies by their color. These cells are hormonally induced to produce roots and shoots and grow into plants in which every cell

contains a copy of the cloned DNA. Such plants will now express the new trait introduced by the T-DNA-mediated gene transfer. The foreign genes used for genetically engineering plants include those that improve growth characteristics of the plant, speed up ripening of fruit, or render the plant resistant to killing by plant insects or disease-causing bacteria.

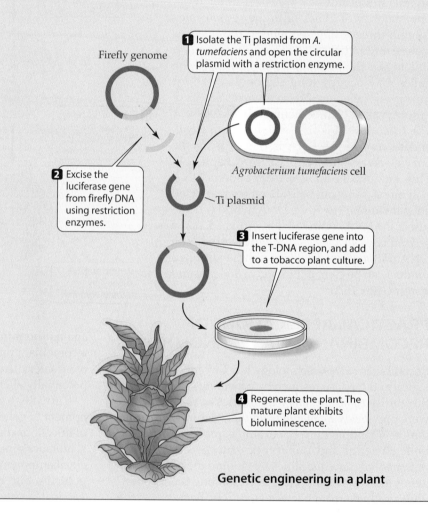

1 Isolate the Ti plasmid from *A. tumefaciens* and open the circular plasmid with a restriction enzyme.

Firefly genome

Agrobacterium tumefaciens cell

2 Excise the luciferase gene from firefly DNA using restriction enzymes.

Ti plasmid

3 Insert luciferase gene into the T-DNA region, and add to a tobacco plant culture.

4 Regenerate the plant. The mature plant exhibits bioluminescence.

Genetic engineering in a plant

in studies of molecular evolution, by comparing sequences between common genes and thus establishing relationships among genes and organisms (**Box 16.3**). The ability to detect the presence of certain genes in bacteria has been exploited in diagnostic microbiolo-

gy as well. Agarose gel electrophoresis of plasmid DNA, coupled with Southern blot analysis, often allows tracking of bacteria responsible for dissemination of antibiotic-resistance genes. This type of plasmid fingerprinting allows detection of specific size categories of plasmids, as well as specific DNA sequences that reside on such plasmids as a result of transposition from the chromosome or another plasmid.

MICROBIAL GENOMICS

One of the youngest disciplines of microbiology is based on increasing the efficiency of the simple method of DNA sequencing, which can then be applied to entire genomes of living cells. The era of microbial genomics refers to a flurry of sequencing activities during the last 5 years of the millennium, when the complete genomic sequences of over 50 different bacterial species were determined in a short time. The contribution of this field to the understanding of cellular processes will undoubtedly be felt for a long time. One of the most exciting results of the genomic sequencing projects is that a significant fraction (in some bacteria, nearly one-half) of all proteins deduced from the nucleotide sequence of corresponding genes cannot be assigned a biological function. This remarkable finding, in the face of the impressive body of knowledge in prokaryotic and eukaryotic genetics, biochemistry, and physiology, indicates that many biological processes remain undiscovered.

Rapid automation of DNA sequencing technology, described in Chapter 16, enabled the sequencing of large segments of DNA in relatively short periods of time. The most important accomplishment in this regard took place in 1995, when the entire genome of a living organism was sequenced. Although the genome of *Haemophilus influenzae* is relatively small (only 1,830 base pairs), it represented one of the major scientific accomplishments in biology. The strategy used to achieve this, called **whole-genome shotgun sequencing (Figure 16.9)**, became a model for all sequencing projects involving bacteria, and it has also been used in various parts of the project to sequence the entire human genome. This approach takes advantage of the high efficiency of automated DNA sequencing machines and computer-assisted tools, which are used to assemble the final sequence of the genome.

The process of generating the sequence of an entire bacterial genome begins with the isolation of chromosomal DNA from a culture of the organism and preparation of a DNA fragment library in a cloning vector. This is analogous to the procedure described earlier in this chapter (illustrated in Figure 16.5) for random cloning of DNA fragments into plasmids. Once a large collection of plasmids carrying chromosomal DNA inserts is generated, individual plasmids are isolated, and the ends of each insert are sequenced using dideoxy-termination methods. Approximately 500 to 800 bp of DNA sequence are routinely obtained from each end. These sequences are then entered into a computer, which identifies overlaps of these partial

sequences and aligns these overlaps. This stage is called sequence assembly. Once a sufficient number of short sequences has been generated (up to 60,000 sequences of 500 to 800 bp each for a genome of 4,000,000 base pairs) and assembled, the genome sequence is now represented as a series of long DNA segments, called contigs (short for contiguous), that are separated by regions where insufficient sequence information was generated. These assembly gaps often represent "toxic" sequences that are not clonable and that are therefore absent from the original insert library. Alternatively, it may represent repetitive sequence in the genome, which cannot be readily assembled. The gap closure and the final stage of assembly of the genome by contig linking involves PCR amplification across each gap, followed by sequencing the entire PCR product without cloning it into a vector.

The final stage of a sequencing project consists of the identification of sequences that code for proteins or ribosomal or tRNA. This process, called **annotation**, is carried out with the extensive use of computer-assisted tools. From the linear nucleotide sequence, the coding sequence for each gene is determined. Because there are three possible reading frames for a protein within each DNA strand (see Chapter 13), several criteria are used to assign the correct sequence for a gene. These include the recognition of initiation and termination codons, the presence of a ribosome-binding site (Shine-Delgarno sequence) preceding the initiation codons, and the codon usage pattern known to be employed by the protein synthesis machinery of the organism. Genes, as characterized by a contiguous coding sequence between an initiation and termination codons, are also called **open reading frames** (**ORFs**). Once these are identified, predictions of function for a product of the gene can be made. The process of annotation is perhaps the most important part of a genome-sequencing project, because it generates information about the possible function of all recognizable genes encoded in the genome, and it defines the genetic repertoire of the organism (see **Box 16.4** on p. 350).

Most bacteria have a single circular chromosome, but notable exceptions can be found, such as the two chromosomes for *V. cholerae* and a circular and linear chromosome in *Agrobacterium tumefaciences*. As shown in **Figure 16.10**, the DNA sequence is densely packed with genes, and only a small percentage of the DNA consists of intergenic regions, which are very likely binding sites for RNA polymerase and transcriptional regulators.

The annotation of the genome serves as a blueprint of the hundreds of interactive processes in the living cell. Even cursory examination of the various genes that can be tentatively assigned function can provide some interesting insights into the lifestyle of a particular

Methods & Techniques Box 16.3   Determination of a Specific DNA Sequence of a Gene

There are several methods available for the determination of a specific nucleotide sequence of a gene, primarily by sequencing of DNA fragments cloned in various cloning vectors. The technique most commonly employed for sequencing large fragments of DNA is based on a method devised by Frederick Sanger and colleagues in 1977. This method relies on synthesis of complementary DNA in a test tube that contains modified nucleoside triphosphates called 2′,3′-dideoxynucleotide triphosphates (ddNTP). These dideoxynucleotides can be incorporated into the growing chain; however, they lack the necessary 3′ hydroxyl group to form a phosphodiester bond with

the next deoxynucleotide triphosphate (dNTP). The consequence of incorporation of ddNTP bases is termination of DNA synthesis at a precise place. DNA sequencing by this method is carried out in four reactions, each containing a mixture with a single-stranded template, containing the DNA sequence of interest and a short radioactively labeled oligonucleotide primer, which anneals to its complementary sequence and one of four ddNTPs. The location of the primer determines the starting point of the sequence analysis. The reaction is started by adding DNA polymerase I, all four dNTPs in excess, and a small amount of one ddNTP. The chain

elongation reaction proceeds using dNTPs, but incorporation of a ddNTP terminates synthesis. For example, the reaction using ddATP will terminate at positions on the template containing a thymine (illustrated in the accompanying figure A). The other three ddNTPs will terminate reactions corresponding to the positions of their complementary bases on the DNA. These reactions give radioactive products of different length, which are size-fractionated by electrophoresis on acrylamide gels. The relative size of the fragments and the source of the ddNTP will allow determination of a corresponding sequence in the 5′ direction from the primer. This method

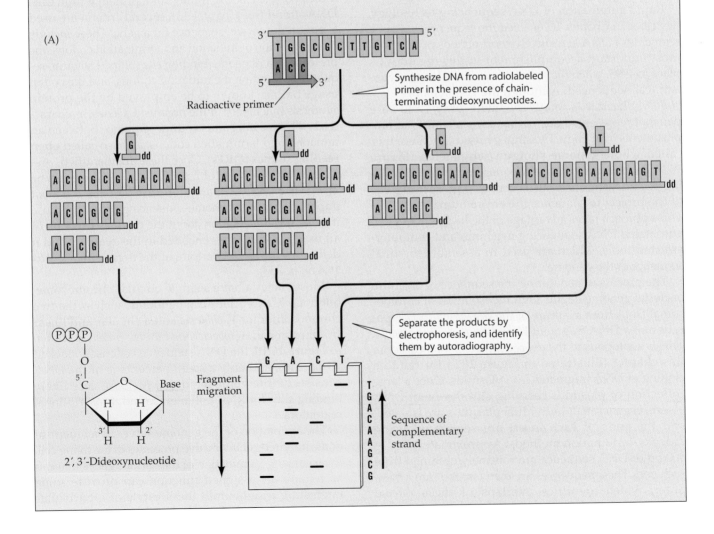

Methods & Techniques Box 16.3 *(continued)*

allows sequence analysis of a region of DNA, up to 600 to 700 base pairs from the site of the primer hybridization. Once this sequence is known, a new primer can be chemically synthesized, radioactively labeled, and annealed to the same DNA template or a new template carrying an additional portion of the gene; this will serve as a starting point for the next round of sequencing.

An innovation in DNA sequencing technology was made possible through the use of dideoxynucleotides that are labeled with fluorescent dyes (see accompanying figure, B). Each of the four dyes emits fluorescence at a different wavelength, and the bands migrating in the gel can be detected by a laser scanner. The four reactions are combined and analyzed in a single lane of a gel. This process can be automated, and a sequencing machine can analyze up to a hundred sequencing reactions at a time.

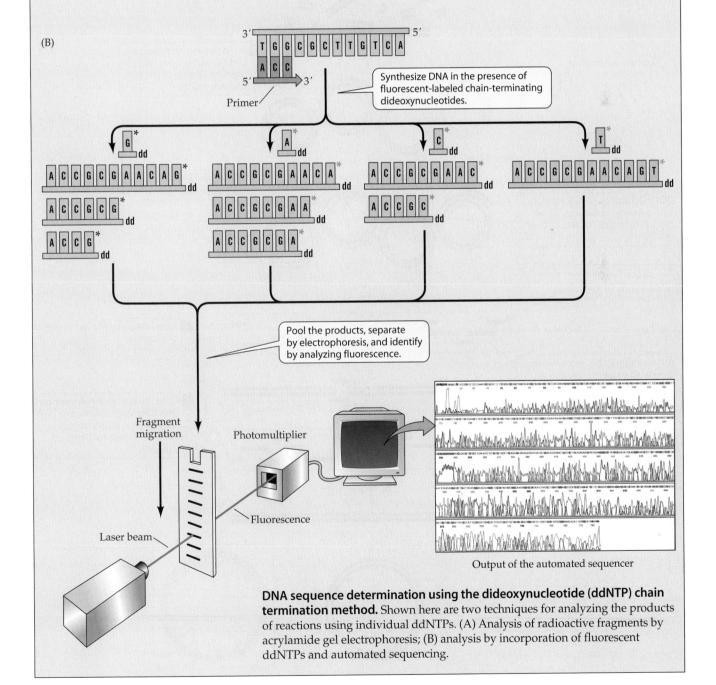

DNA sequence determination using the dideoxynucleotide (ddNTP) chain termination method. Shown here are two techniques for analyzing the products of reactions using individual ddNTPs. (A) Analysis of radioactive fragments by acrylamide gel electrophoresis; (B) analysis by incorporation of fluorescent ddNTPs and automated sequencing.

(A) **Construction of DNA library**

(B) **Random sequencing**

(C) **Assembly of sequences**

(D) **Annotation**

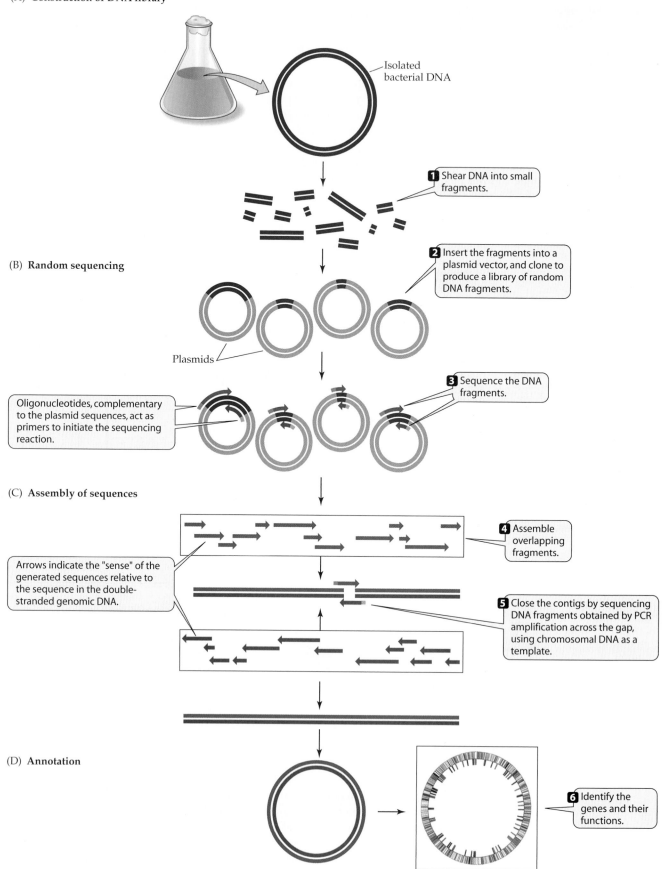

Isolated bacterial DNA

1 Shear DNA into small fragments.

2 Insert the fragments into a plasmid vector, and clone to produce a library of random DNA fragments.

Plasmids

Oligonucleotides, complementary to the plasmid sequences, act as primers to initiate the sequencing reaction.

3 Sequence the DNA fragments.

4 Assemble overlapping fragments.

Arrows indicate the "sense" of the generated sequences relative to the sequence in the double-stranded genomic DNA.

5 Close the contigs by sequencing DNA fragments obtained by PCR amplification across the gap, using chromosomal DNA as a template.

6 Identify the genes and their functions.

◀ **Figure 16.9 Whole-genome shotgun sequencing**
Determination of a microbial genome sequence. (A)
Construction of a random library of DNA fragments in a
cloning vector. (B) Random sequencing of clones. Short
sequences are obtained from each end of the cloned DNA,
and thousands of clones are sequenced (see Chapter 11 for a
detailed discussion of DNA sequencing methods). (C)
Assembly of the final sequence of the bacterial genome. The
thousands of short sequences are assembled into long segments called contigs, which are cloned. The assembled bacterial genome is usually a contiguous circle. (D) Completion
of the sequencing project. Annotation involves the identification of genes and, where possible, designation of a
biological function for the gene products.

bacterium. For example, the fraction of genes dedicated
to transcriptional regulation (**Table 16.2**) varies greatly
among the different bacteria. In *P. aeruginosa*, a free-living prototroph capable of colonizing a wide range of
environmental niches, 8.4% of its genome encodes puta-
tive transcriptional regulators. This may reflect the range

of environmental signals that this microorganism is
capable of responding to, and other prototrophic bacteria similarly dedicate a high percentage of their genomes
to transcriptional regulation. At the other extreme is *H.
pylori*, a human pathogen that thrived in a very specific
environment (the human stomach) with a small range
of environmental variables, and this is reflected in a very
limited regulatory circuitry. Only 1.1% of the *H. pylori*
genome encodes regulatory proteins.

One of the striking findings is the large number of
genes that encode proteins involved in unknown cellular processes (**Table 16.3**). These proteins of unknown
function can be further subdivided into two groups.
One category includes proteins of unknown function
that have similar proteins present in more than one bacterial genome. The other group contains proteins of
unknown function that are unique to a particular bacterium. Given that only a small fraction of bacterial
genomes have been sequenced, the complexity of life represented by bacteria is truly amazing. A survey of 24
genomes, encompassing 51,627 genes, shows that 21,248
of them encode products of unknown function, and
11,083 of these are unique.

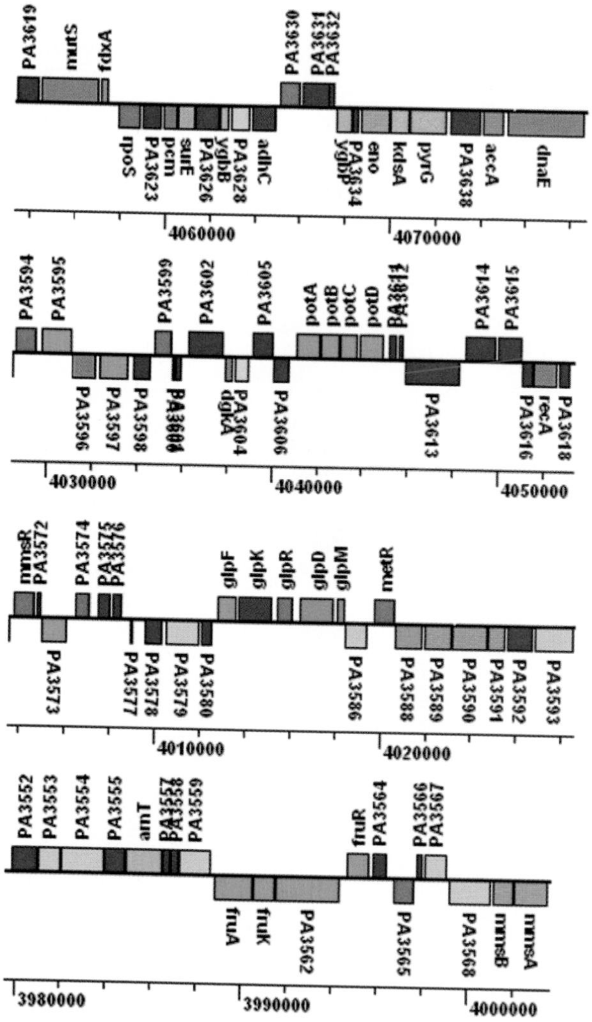

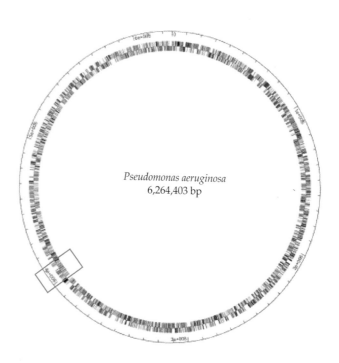

Figure 16.10 Genes in a portion of bacterial genome
Genome of *Pseudomonas aeruginosa*. The enlarged area shows
the dense packing of genes. Only a small percentage of the
DNA consists of intergenic regions, which are very likely
binding sites for RNA polymerase and transcriptional
regulators. Courtesy of Mathew Wolfgang.

Functional assignments can also provide a bird's-eye view of metabolic activities in the cell. Such reconstructions of the potential cellular processes are based on predicted functions deduced from the annotated genome sequence. They can provide comprehensive insights into the ability of a particular bacterium to function in specific environments. A transport and metabolic reconstruction, such as one shown for *Neisseria meningitidis* in **Figure 16.11**, can suggest various pathways of nutrient uptake, energy generation, biosynthesis, and survival in the environment of the infected host.

FUNCTIONAL GENOMICS

The major challenge faced by the research community, in light of the availability of complete genome sequences of a variety of organisms, is to understand the biological function of each gene product, specifically within the context of the parallel activities of the thousands of other components that make up a living cell. This postgenomic phase of research is called functional genomics. This field is

Table 16.2 Comparison of regulatory genes in selected bacterial genomes

Microorganism	# Genes in the Genome	# Regulatory Proteins	% of Total
Pseudomonas aeruginosa	5570	468	8.4
Escherichia coli	4289	250	5.8
Bacillus subtilis	4100	217	5.3
Mycobacterium tuberculosis	3918	117	3.0
Helicobacter pylori	1566	18	1.1

Table 16.3 Distribution of genes of unknown function among selected bacterial genomes

Organism	Genome Size (Mbp)	No. of ORFs (% coding)		Unknown Function		Unique ORFs	
Aeropyrum pernix K1	1.67	1,885	(89%)				
A. aeolicus VF5	1.50	1,749	(93%)	663	(44%)	407	(27%)
A. fulgidus	2.18	2,437	(92%)	1,315	(54%)	641	(26%)
B. subtilis	4.20	4,779	(87%)	1,722	(42%)	1,053	(26%)
B.burgdorferi	1.44	1,738	(88%)	1,132	(65%)	682	(39%)
Chlamydia pneumoniae AR39	1.23	1,134	(90%)	543	(48%)	262	(23%)
Chlamydia trachomatis MoP$_n$	1.07	936	(91%)	353	(38%)	77	(8%)
C. trachomatis serovar D	1.04	928	(92%)	290	(32%)	255	(29%)
Deinococcus radiodurans	3.28	3,187	(91%)	1,715	(54%)	1,001	(31%)
E. coli K-12-MG1655	4.60	5,295	(88%)	1,632	(38%)	1,114	(26%)
H. influenzae	1.83	1,738	(88%)	595	(35%)	237	(14%)
H. pylori 26695	1.66	1,589	(91%)	744	(45%)	539	(33%)
Methanobacterium thermotautotrophicum	1.75	2,008	(90%)	1,010	(54%)	496	(27%)
Methanococcus jannaschii	1.66	1,783	(87%)	1,076	(62%)	525	(30%)
M. tuberculosis CSU#93	4.41	4,275	(92%)	1,521	(39%)	606	(15%)
M. genitalium	0.58	483	(91%)	173	(37%)	7	(2%)
M. pneumoniae	0.81	680	(89%)	248	(37%)	67	(10%)
N. meningitidis MC58	2.24	2,155	(83%)	856	(40%)	517	(24%)
Pyrococcus horikoshii OT3	1.74	1,994	(91%)	589	(42%)	453	(22%)
Rickettsia prowazekii Madrid E	1.11	878	(75%)	311	(37%)	209	(25%)
Synechocystis sp.	3.57	4,003	(87%)	2,384	(75%)	1,426	(45%)
T. maritma MSB8	1.86	1,879	(95%)	863	(46%)	373	(26%)
T. pallidum	1.14	1,039	(93%)	461	(44%)	280	(27%)
Vibrio cholerae El Tor N1696	4.03	3,890	(88%)	1,806	(46%)	934	(24%)
	50.60	52,462	(89%)	22.358	(43%)	12,161	(23%)

From Fraser et al., *Nature* 2000, vol. 406. p. 800.

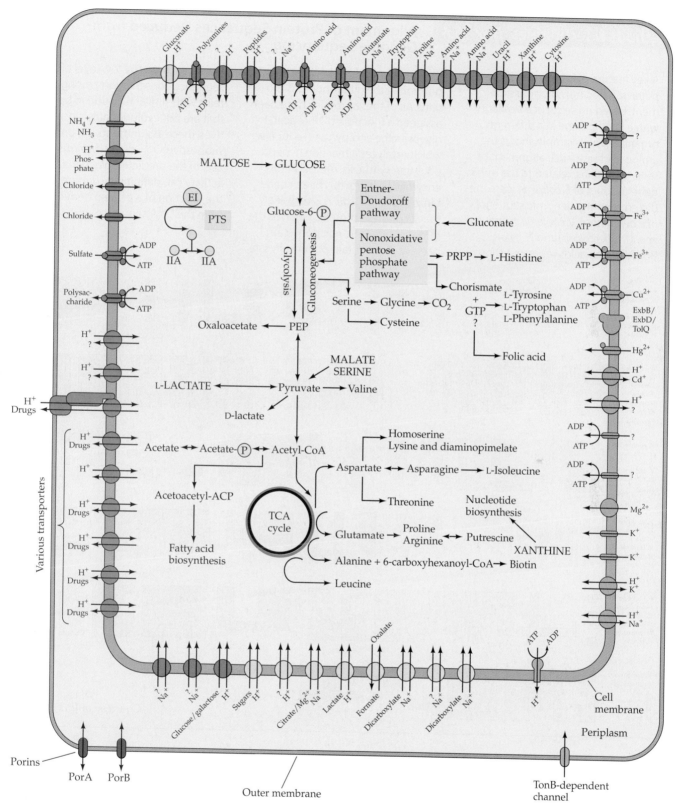

Figure 16.11 Cellular functions based on an annotated genome

Reconstruction of transport and metabolism of *Neisseria meningitidis*, based on the annotated genome. The reconstruction shows the potential pathways for the generation of energy and metabolism of organic compounds. Question marks indicate that a transporter's substrates are unknown, and functional assignment is based on its overall similarity to other transporters. After Nelson, K. T., I. T. Paulson and C. M. Fraser. 2001. *ASM News* 67: 310–317.

Methods & Techniques Box 16.4   Alignment of Protein Sequences Deduced from Genome Sequencing

A key feature of annotation is computer-assisted homology assignment, based on sequence similarity with a known gene in a different organism. This example shows the amino acid sequence alignment of several proteins, related to the *murB* gene product of *E. coli*, that were predicted from the genomic sequences of *P. aeruginosa*, *V. cholerae*, *H. influenzae*, *M. tuberculosis*, and *H. pylori*. Boxed amino acids represent those that are identical or similar at each position of the different MurB proteins. All of these proteins have been assigned a function: UDP-*N*-acetyl-glucosamine enolpyruvyl reductase, which catalyzes a reaction in peptidoglycan synthesis. Although this enzymatic activity has been characterized only in *E. coli*, it is very likely that all of these proteins catalyze the same biosynthetic reaction in all bacteria that contain peptidoglycan.

However, given the low level of sequence conservation among proteins with similar function in bacteria that are not evolutionarily close relatives, these assignments are only suggestive. Only direct demonstration of the predicted biochemical activity can definitively demonstrate the function of a protein deduced from the DNA sequence.

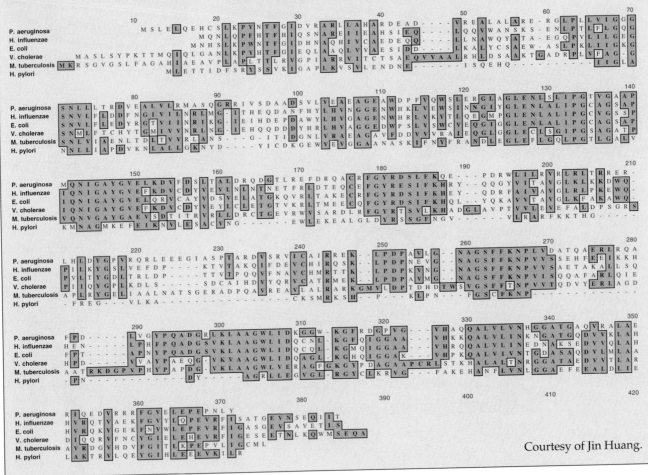

Courtesy of Jin Huang.

driven by technological developments in many different fields, including classical genetics, physics, engineering, molecular biology, and biochemistry. Most notable is the rise of computational biology as a central component of modern biomedical research. Computational tools played an important role in genome-sequencing efforts and will be continue to be the dominant technology supporting all disciplines of postgenomic research.

DNA Microarrays and Global Transcriptional Analysis

Of all the postgenomic technologies developed during the past five years, the most significant impact has been provided by the use of DNA arrays (also called microarrays or DNA chips) to study the transcriptional activities of cells. For any biological process, it is important to understand the flow of genetic information encoded in genes, and microarray technology provides a truly comprehensive view of transcriptional activity in the entire cell by simultaneously monitoring the levels of mRNA transcribed from every gene in the bacterial genome.

Microarrays are prepared in two stages. First, a complete collection of clones or PCR-amplified genes is generated. The guide for this process is the annotated genome. Usually, specific primers are designed for each gene, and the corresponding sequences are amplified and placed into multiwell dishes. The number of clones that are needed for a comprehensive, all-inclusive array depends on the size of the genome. For example, a complete gene array for *H. pylori* would require advance preparation of 1,566 unique clones or PCR products, whereas for *P. aeruginosa* this number would be considerably higher (5,570).

The spotting or "arraying" (sometimes referred to as printing) is carried out by robotic devices adapted from tools developed for industrial production of computer chips. These robots have movable arms that deposit very small (2 nl) aliquots at high density on a solid support, in a process called array printing. The arrays are often printed onto coated microscope slides. Once the droplets of DNA solution are spotted, the slides are exposed to ultraviolet light to covalently link the DNA to the slide. Each PCR product (representing a gene) is therefore immobilized at a specific location on the array.

The schematic diagram of a robotic arrayer is shown in **Figure 16.12**, and the overall procedure for microarray analysis is outlined in **Figure 16.13**.

The next step in the microarray analysis of RNA levels in bacteria is probe preparation. In most instances, gene expression levels are compared in bacteria grown under two different conditions, and the RNA for probe preparation is isolated from cultures grown under each of these conditions. For example, a comparison of gene expression levels in bacteria in exponential and stationary phases of growth requires isolation of RNA from cultures harvested at these two stages. The RNA is then labeled to provide a readout for hybridization to each gene on the microarray. One common method for generating probes is to use the eukaryotic enzyme reverse transcriptase to prepare a complementary copy of the RNA (this enzyme is also used for the RT-PCR procedure (see Box 16.2). During synthesis of the DNA strand complementary to mRNA, fluorescently labeled nucleotides are incorporated into the cDNA strand. These labeled nucleotides can be synthesized with tags that fluoresce at different wavelengths, so that different batches of probe will fluoresce at different wavelengths. For example, when comparing mRNA in bacteria at two different growth stages, RNA isolated from cells at the exponential phase probe can be labeled with modified deoxynucleotides that emit at a wavelength that is detectable in

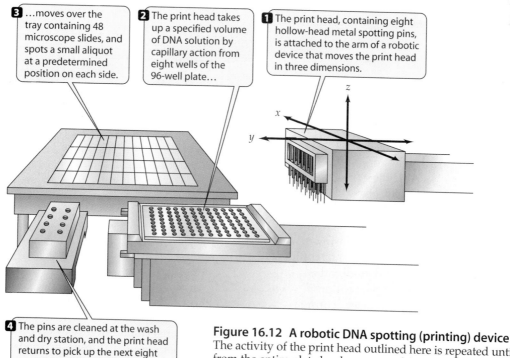

3 ...moves over the tray containing 48 microscope slides, and spots a small aliquot at a predetermined position on each side.

2 The print head takes up a specified volume of DNA solution by capillary action from eight wells of the 96-well plate...

1 The print head, containing eight hollow-head metal spotting pins, is attached to the arm of a robotic device that moves the print head in three dimensions.

4 The pins are cleaned at the wash and dry station, and the print head returns to pick up the next eight samples from the 96-well plate.

Figure 16.12   A robotic DNA spotting (printing) device
The activity of the print head outlined here is repeated until the DNA from the entire plate has been spotted (12 times in all for an eight-pin spotting arm). A fresh plate containing a new set of DNA samples is added, and the spotting process is repeated until an entire genome has been arrayed.

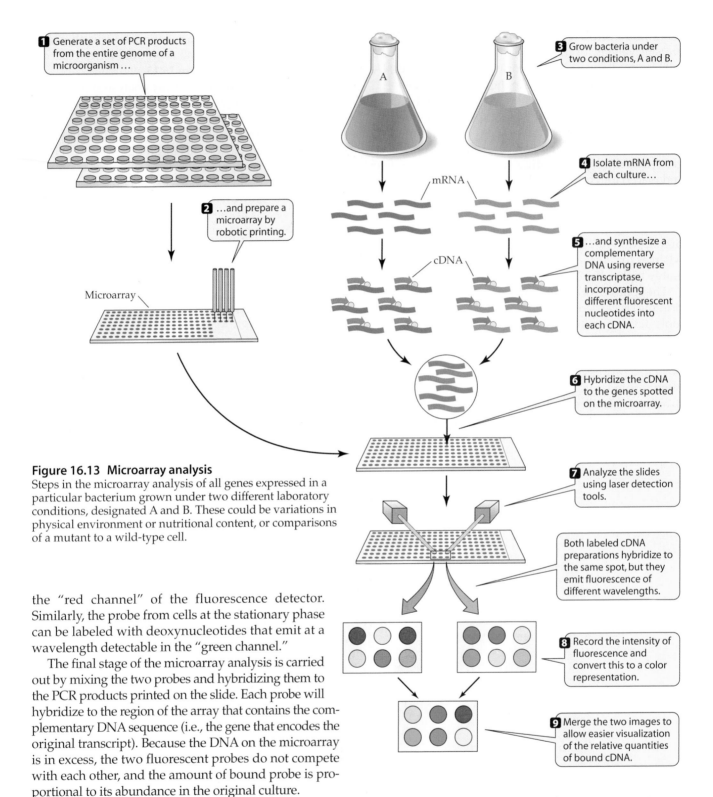

Figure 16.13 Microarray analysis
Steps in the microarray analysis of all genes expressed in a particular bacterium grown under two different laboratory conditions, designated A and B. These could be variations in physical environment or nutritional content, or comparisons of a mutant to a wild-type cell.

the "red channel" of the fluorescence detector. Similarly, the probe from cells at the stationary phase can be labeled with deoxynucleotides that emit at a wavelength detectable in the "green channel."

The final stage of the microarray analysis is carried out by mixing the two probes and hybridizing them to the PCR products printed on the slide. Each probe will hybridize to the region of the array that contains the complementary DNA sequence (i.e., the gene that encodes the original transcript). Because the DNA on the microarray is in excess, the two fluorescent probes do not compete with each other, and the amount of bound probe is proportional to its abundance in the original culture.

The images generated from the laser scanner can now be analyzed. As shown in **Figure 16.14**, the intensity of fluorescence emission from each spot on the array reflects the relative ratio of transcripts for the gene represented by that particular spot. In the example shown, gene A is highly expressed during the exponential phase of the growth, whereas gene B is pre-dominantly transcribed when the bacteria are in stationary phase. Gene C is not growth-phase regulated, and it is transcribed equally at all phases of growth. Other genes show differential levels and ratios of expression. Because microarrays allow comprehensive transcriptional analysis, quantification of the relative mRNA abundance in the cell can be carried out on a

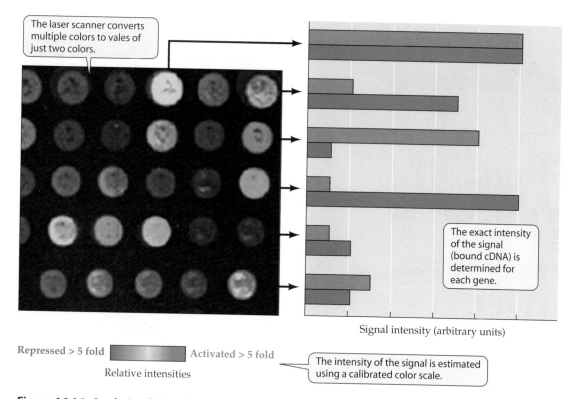

Figure 16.14 Analysis of data from a microarray experiment
The merged color representation (see Figure 16.13) reveals the relative quantities of bound cDNA.

genomewide basis. The total body of all RNA transcripts in a cell is called the **transcriptome.** Unlike a genome, which is a fixed and defined entity consisting of all genes, the transcriptome reflects the environmental conditions of the bacterium, and many different RNA profiles can be generated from a single genome. Moreover, perhaps only a fraction of all genes is expressed at any one time, reflecting the basic metabolic activities of the organism in that particular environment. **Figure 16.15** shows the *P. aeruginosa* transcriptome in media with varying amounts of calcium. The highlighted region shows the coordinate expression of genes under different calcium-limiting conditions, as they are displayed on the genetic map (see Figure 16.10). This provides an overview of genomewide transcriptional activity in a bacterial cell, and it serves as the starting point for dissection of the complex regulatory circuits that control the bacterial response to environmental changes.

The use of robotic devices to spot DNA solutions onto slides is only one of several methods currently available for producing DNA microarrays. An example of an alternative method is direct synthesis of short oligonucleotides at predetermined locations on solid supports. There are also at least a dozen different methods for preparing probes, involving covalent modifica-

tion of RNA, direct labeling of RNA, or other ways of preparing cDNA probes. New versions of fluorescent nucleotides are being developed that allow simultaneous detection of more than two conditions in a single experiment. Yet another exciting application, derived from DNA microarray technology, is the development of protein arrays, where all the individual proteins expressed by a microorganism are deposited on a support. Protein arrays have a large number of applications, ranging from detection of antibodies against pathogens, to studies of interactions between the proteins encoded in the genome.

SUMMARY

▸ **Functional genomics** is a term applied to the study of biological processes in the entire cell, using information generated from genome sequencing.

▸ The determination of the sequence of a microbial genome involves the process called the **whole-genome shotgun sequencing.**

▸ There are several distinct stages in the whole-genome sequencing process. The first step is the construction a library of fragments of DNA in cloning vectors. Thousands of randomly selected clones are sequenced,

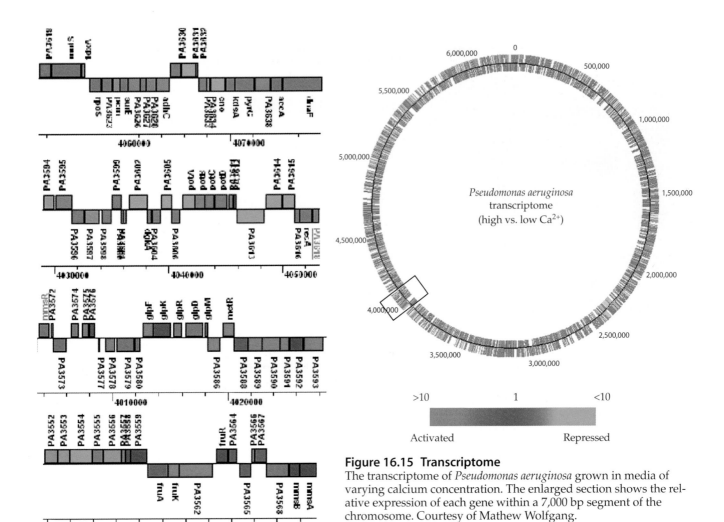

Figure 16.15 Transcriptome
The transcriptome of *Pseudomonas aeruginosa* grown in media of varying calcium concentration. The enlarged section shows the relative expression of each gene within a 7,000 bp segment of the chromosome. Courtesy of Mathew Wolfgang.

each providing a short 500 to 800 bp sequence from a portion of the genome. The short sequences are assembled into longer segments, called contigs, by aligning the short overlapping sequences. The genome sequence still contains gaps between the contigs, which are closed by specialized methods, such as PCR amplification across the gap and sequencing the PCR products.

▶ The final stage of work in genome sequencing is **annotation** of the genome. First the location of genes within the sequenced DNA is determined. This involves the identification of open reading frames, which define the most likely coding sequence of proteins.

▶ The function of an individual protein, whose sequence was deduced from the DNA sequence, can be sometimes determined by the extent of similarity (homology) to other proteins in different organisms that have a well-characterized biological function.

▶ A large number of genes in the genome encode proteins that have no known function, and a substantial fraction of these are unique to each organism.

▶ **DNA microarrays** are the tools to determine the levels of mRNA for each gene in the genome. Microarrays are immobilized DNA fragments corresponding to a complete or partial sequence of each gene in the genome. They are usually prepared using robots that apply aliquots of DNA to supports at every density. DNA is spotted on chemically treated microscope slides that allow its cross-linking and immobilization. An entire genome of a microorganism can be represented on a single slide.

▶ Microarrays are analyzed by hybridization of labeled RNA or a fluorescently labeled complementary DNA (cDNA).

▶ A list describing the abundance of RNA for each gene in a genome under a specified set of conditions is called a transcriptome.

▶ Classical bacterial genetics was the foundation for recombinant DNA research, which is based on the ability of scientists to **isolate individual genes** from more complex genomes.

▸ Restriction endonucleases, which introduce double-stranded breaks at specific sequences, are made by bacteria. These are purified and used to fragment DNA into pieces that contain intact genes or parts of genes. Restriction endonucleases, together with their cognate **modification enzymes**, serve as barriers to unrestricted DNA exchange among different bacterial species.

▸ DNA fragments, generated by treatment with restriction endonucleases, are usually analyzed by **agarose gel electrophoresis**, which separates DNA fragments according to their size.

▸ Specific DNA sequences found within a DNA fragment fractionated by agarose gel electrophoresis can be identified by a method of in situ **hybridization** of a single-stranded DNA probe with its complementary sequence in the gel. This procedure is known as **Southern blotting**.

▸ Construction of recombinants involves insertion of a DNA fragment into a **vector**, which can be either **plasmid** or **virus**, followed by introduction of the plasmid into the bacterial **host** by transformation. If the vector is a bacteriophage (usually bacteriophage lambda), the recombinant DNA is packaged into the bacteriophage head and is introduced into the bacteria by infection.

▸ Plasmid vectors contain sites where DNA is inserted by ligation of a fragment, generated by treatment of the DNA and vector by the same restriction endonuclease. The insert DNA is joined with the plasmid, to give a chimeric circular molecule, by the action of **DNA ligase**.

▸ Recombinant plasmids can be propagated in pure culture as **clones**. They are identified by expression of protein product using either immunological or activity screens.

▸ Specialized cloning vectors, called **expression vectors**, include specific sequences to maximize expression of the cloned genes.

▸ Cloned genes can be altered in the laboratory, and introduced into the bacterial genomes either on plasmids or bacteriophage vectors. Even plants can be genetically engineered, using **Agrobacterium**-mediated transfer of recombinant DNA cloned within its **Ti plasmid.**

▸ Eukaryotic genes encoding proteins are often cloned from mRNA sequences. First, a copy of mRNA, called **cDNA**, is made by the enzyme **reverse transcriptase**, and the second DNA strand, complementary to cDNA, is made by DNA polymerase. This double-stranded form is then inserted into the appropriate plasmid or bacteriophage vectors.

▸ Rare sequences from complex genomes, or from a mixture of different RNA species, can be isolated by repeated denaturation-complementary strand synthesis-denaturation, called the **polymerase chain reaction (PCR)**.

REVIEW QUESTIONS

1. What are the most common reasons for the failure to obtain a complete genome from the assemblies of thousands of short sequences in whole-genome shotgun sequencing?

2. Why do genomes of bacteria vary so much in size?

3. An annotation of a genome of a microorganism revealed the presence of 4,000 open reading frames. Does this imply that there are 4,000 different proteins present in the cell at any one time?

4. Describe the process of a genome-based microarray preparation.

5. Which genes bind the most to a probe in a genome microarray of a bacterium?

6. What is the relationship of a genome to a transcriptome?

7. Why is it important to name and classify bacteria?

8. What procedure(s) are necessary to identify a bacterial isolate as a species?

9. Differentiate between an artificial and a phylogenetic classification.

10. In what ways does the classification of *Bacteria* differ from that of eukaryotic organisms?

11. How do the *Archaea* differ from *Bacteria*? From eukaryotes?

12. How is DNA melted and reannealed, and why is this useful in bacterial taxonomy?

13. How would you go about identifying a bacterium that you isolated from a soil habitat?

14. Why is morphology of little use in bacterial classification? Is it of any use?

15. What is *weighting*, and should phenotypic features be weighted in a bacterial classification scheme?

16. Distinguish between lumpers and splitters.

17. If you were working in a clinical laboratory, outline the types of procedures you would use to identify isolates. Why do you recommend using the procedures you suggest?

18. Why is ribosomal RNA of use in bacterial classification?

19. Compare the information obtained from determining the DNA base composition (GC ratio) with that obtained by DNA reassociation experiments.

20. Why don't bacteria that produce restriction enzymes destroy their own DNA?

21. A portion of the *E. coli* chromosome containing the

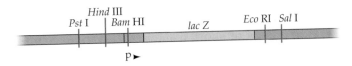

lacZ gene β-galactosidase) is shown.

a. Design a strategy to clone the minimal sequence, but still carry the full coding sequence for the enzyme, into plasmid pBR322.

b. Describe several methods to identify the recombinant *E. coli* carrying the cloned *lacZ* gene.

c. How would your cloning strategy change if you wanted to assure that the *lacZ* gene is expressed in a host strain other than *E. coli* ?

22. Some *E. coli* strains, used as cloning hosts, carry a mutation in their *rec*A gene. Why is it sometimes important to use such a mutant strain when cloning certain genes?

23. A solution of an oligonucleotide of the following sequence, 5' CGGCCCGGGATCCGAATTCCTAG-GCA 3', is heated and mixed with the following oligonucleotide: 5' TGGGATGGAATTCGGGATCC-CGGGCCG 3', and then allowed to cool.

a. Identify all the sites recognized by restriction enzymes (use Table 16.1).

b. Would a solution of the first oligonucleotide alone be recognized and cut by any restriction enzyme?

24. A patient was admitted to a hospital with a urinary-tract infection, and antibiotic-sensitive *Klebsiella pneumoniae* were recovered. During hospitalization he became nonresponsive to antibiotic therapy, and the new isolates of *K. pneumoniae* were now resistant to ampicillin and kanamycin. Stool cultures of all of the patients on the same floor revealed that several of them carried *E. coli* that was also resistant to ampicillin and kanamycin. You are in charge of determining if the *Klebsiella* somehow acquired an R plasmid from *E. coli*. What molecular tools would you use to prove the plasmid-transfer hypothesis?

25. You wish to isolate a gene from a chromosome, which is flanked by recognition sites for either *Bgl* II or *Sma*I restriction endonucleases. Is it possible to use either vector pBR322 or pACYC184, in spite of the fact that neither one of these plasmids has a recognition sequence for *Bgl* II or *Sma*I?

26. A procedure analogous to Southern blotting is called Northern blotting, where total RNA is extracted from bacteria and size-fractionated on agarose gels. Specific mRNA sequences are identified by in situ hybridization with radiolabeled probes, which can be deduced from protein sequence. If you wish to analyze for the presence of mRNA of the gene X (encoding the protein with the N-terminal sequence Met - Trp - Asp - Trp, described on pp. 333–335), could you use, in the Northern blot, the same oligonucleotide probes (5' ATGTGGGATTGG 3', or 5' ATGTGGGATTGG 3') used to detect the gene X in the Southern blot?

27. Describe the differences in colony hybridization aimed at detecting specific DNA sequences or the related immunological techniques for the identification of clones producing a specific protein.

28. Explain the difference in the initial steps of the polymerase chain reaction (PCR) when starting from mRNA, rather than DNA.

SUGGESTED READING

Ausubel, F. M., R. Brent, R. E. Kingston, D. D. Moore, F. G. Seidman, J. A. Smith and K. Struhl. 1989. *Current Protocols in Molecular Biology*. Vol. 2. New York: John Wiley & Sons.

Brown, P. O. and D. Botstein. 1999. Exploring the New World of the Genome with DNA Microarrays. *Nat Genet.* 21(1 Suppl): 33–37.

Duggan, D. J., M. Bittner, Y. Chen, P. Meltzer and J. M. Trent. 1999. Expression Profiling Using cDNA Microarrays. *Nat Genet.* 21(1 Suppl): 10–14.

Fleischmann, R. D., M. D. Adams, O. White, R. A. Clayton, E. F. Kirkness, A. R. Kerlavage, C. J. Bult, J. F. Tomb, B. A. Dougherty, J. M. Merrick et al. 1995. Nucleotide, Genome Whole-Genome Random Sequencing and Assembly of *Haemophilus influenzae* Rd. *Science* 269(5223): 496–512.

Fraser, C. M., J. A. Eisen and S. L. Salzberg. 2000. Microbial Genome Sequencing. *Nature* 406: 799–803.

Friedberg, E. C., G. C. Walker and W. Siede. 1994. *DNA Repair and Mutagenesis*. Washington, DC: American Society for Microbiology.

Hooykaas, P. J. and R. A. Schileperoot. 1992. Agrobacterium and Plant Genetic Engineering. *Plant Mol. Biology* 19: 15–38.

Leach, D. R. F. 1995. *Genetic Recombination*. Cambridge, MA: Blackwell Science.

MacBeath, G. and S. L. Schreiber. 2000. Printing Proteins as Microarrays for High-Throughput Function Determination, *Science* 289(5485): 1760–1763.

Old, R. W. and S. B. Primrose. 1994. *Principles of Gene Manipulation*. Cambridge, MA: Blackwell Science.

Microbial Evolution and Diversity

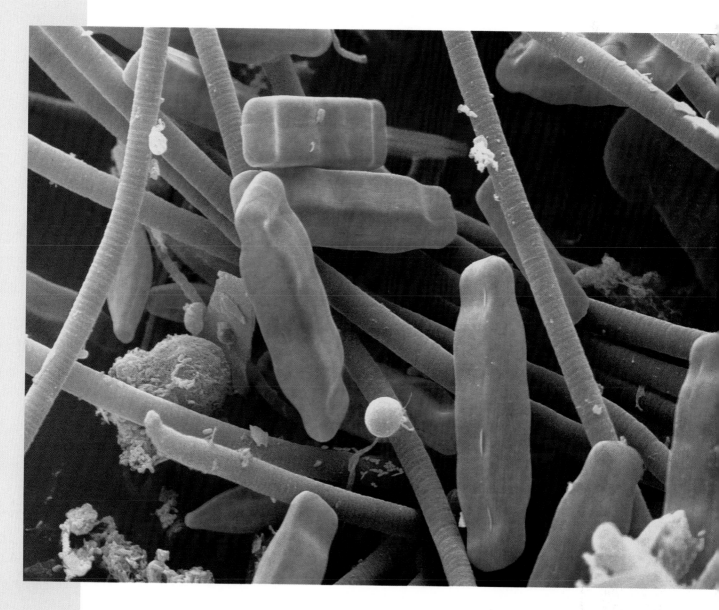

PART V Microbial Evolution and Diversity

Previous page
These filamentous cyanobacteria and the chloroplasts of these diatoms have evolved from a common oxygenic photosynthetic ancestor. ©Andrew Syred/SPL/Photo Researchers, Inc.

Chapter 17

Taxonomy of *Bacteria* and *Archaea*

It's just astounding to see how constant, how conserved, certain sequence motifs—proteins, genes—have been over enormous expanses of time. You can see sequence patterns that have persisted probably for over three billion years. That's far longer than mountain ranges last, than continents retain their shape.

– CARL WOESE, 1997 (IN PERRY AND STALEY, MICROBIOLOGY)

This part of the book discusses the variety of microorganisms that exist on Earth and what is known about their characteristics and evolution. Most of the material pertains to the prokaryotes; however, there is one chapter on eukaryotic microorganisms. Because greater emphasis is given to *Bacteria* and *Archaea* in this part, this first chapter discusses how they are named and classified. This chapter will be followed by several chapters (Chapters 18–22) concerned with their properties and diversity.

When scientists encounter a large number of related items, such as the chemical elements, plants, or animals, they characterize, name, and organize them into groups. Thousands of species of plants, animals, and bacteria have been named, and many more will be named in the future as more are discovered. Not even the most brilliant biologist knows all of the species. Organizing the species into groups of similar types aids the scientist not only in remembering them but also in comparing them to their closest relatives, some of which the scientist would know very well. In addition, microbiologists are interested in evolution. To unravel this puzzle, it is essential to understand how one species is related to another. For these reasons, about 5,000 bacterial species have been named and, based on their characteristics, placed within the existing framework of other known species. The branch of bacteriology that is responsible for characterizing and naming organisms and organizing them into groups is called **taxonomy** or **systematics.**

Taxonomy can be separated into three major areas of activity. One is **nomenclature,** which is the naming of bacteria. The second is **classification,** which entails the ordering of bacteria into groups based on common properties. In **identification,** the third area, an unknown bacterium, for example, from a clinical or soil sample, is characterized to determine what species it is. This chapter covers all three of these areas.

NOMENCLATURE

Bacteriologists throughout the world have agreed on a set of rules for naming *Bacteria* and *Archaea*. These rules, called the "International Code for the Nomenclature of Bacteria" (1992) state what a scientist must do to describe a new species or other **taxon** (-*a*, pl.), which is a unit of classification, such as a species, genus, or family. Each bacterium is placed in a genus and given a species name in the same manner as plants and animals. For example, humans are *Homo sapiens* (genus name first, followed by species), and a common intestinal bacterium is named *Escherichia coli*. This **binomial system** of names follows that proposed by the Swedish taxonomist Carl von Linné (Linnaeus; 1707–1778) for plants and animals.

By the rules of bacterial nomenclature, the root for the name of a species or other taxon can be derived from any language, but it must be given a Latin ending so that the genus and species names agree in gender. For example, consider the species name *Staphylococcus aureus*. The first letter in the genus name is capitalized, the species name is lowercase, and they are both italicized to indicate that they are Latinized. When writing species names in longhand, as for a laboratory notebook, they should be underlined to denote that they are italicized. The genus name *Staphylococcus* is derived from the Greek *Staphyl* from *staphyle*, which means a "bunch of grapes," and *coccus,* from the Greek, meaning "a berry." The *o* ("oh") between the two words is a joining vowel used to connect two Greek words together. The figurative meaning of the genus name is "a cluster of cocci," which describes the overall morphology of members of the genus. The species name *aureus* is from the Latin and means "golden," the pigmentation of members of this species. The *-us* ending of the genus and species names is the Latin masculine ending for a noun (*Staphylococcus* in this case) and its adjective (*aureus*). Successively higher taxonomic categories are family, order, class, phylum, and domain (**Table 17.1**)

The *International Journal of Systematic and Evolutionary Microbiology* (IJSEM) is a journal devoted to the taxonomy of bacteria that is published by the British Society for General Microbiology. IJSEM publishes papers that describe and name new bacterial taxa and contains an updated listing of all new bacteria whose names have been validly published. Thus, even though bacterial species may be described in other scientific journals, they are not considered validly published until they have been listed on a validation list in IJSEM.

Questions on a point of nomenclature are also published in the IJSEM. The question is then evaluated by the international Judicial Commission of the International Union of Microbiological Societies, which sub-

sequently publishes a ruling in the journal. One typical example of a problem considered by the Judicial Commission was the question about *Yersinia pestis*, the causative agent of bubonic plague. Scientific evidence indicates that it is really just a subspecies of *Yersinia pseudotuberculosis*, a species name that has precedence over *Y. pestis* because of its earlier publication. Because of the potential confusion and possible public health issues that could arise by renaming *Y. pestis*, *Y. pseudotuberculosis* subspecies *pestis*, the Judicial Commission ruled against renaming the bacterium despite its scientific justification.

Classification

Classification is that part of taxonomy concerned with the grouping of bacteria into taxa based on common characteristics. Classification systems can be either artificial or natural. **Artificial systems** of classification are based on *expressed characteristics* of the organisms, or the **phenotype** of the organism. In contrast, **natural** or **phylogenetic systems** are based on the purported *evolution* of the organism. Until recently, all bacterial classifications were artificial because there was no meaningful basis for determining their evolution. In contrast, plants and animals have a fairly extensive fossil record on which to base an evolutionary classification system. Although fossils of microorganisms do exist (see Chapter 1), the simple structure of microorganisms does not permit their identification to a taxonomic group by morphological criteria.

An important article published by Zuckerkandl and Pauling in 1962 suggested that the evolution of organisms might be recorded in the sequences of their macromolecules. Subsequent research in the late 1960s and 1970s has supported this concept. In particular, molecules such as ribosomal RNA (rRNA) and some proteins have changed at a very slow rate during evolution, and therefore their sequences provide important clues to the relatedness among the various bacterial taxa and their relatedness to higher organisms as well. The result is that a major breakthrough has occurred in the classification of prokaryotic organisms, which is quickly becoming phylogenetic based on sequencing of 16S rDNA. However, some controversies remain, so the phylogeny of the bacteria is still incomplete. This chapter first discusses the traditional system of classification and then covers what is being done to make it phylogenetic.

When considering bacterial classification, it is important to keep in mind that bacteria have been evolving on Earth for the past 3.5 to 4 billion years. Therefore, it should not seem surprising that two separate domains of prokaryotic organisms exist—the *Bacteria* and the *Archaea*—versus only one domain for eukaryotes, *Eucarya*. And furthermore, the *Eucarya* appear to have evolved more recently as the result of symbiotic events

Table 17.1	Hierarchical classification of the bacterium *Spirochaeta plicatilis*
Taxon	**Name**
Domain	*Bacteria*
Phylum	*Spirochaetes* (vernacular name: spirochetes)
Class	*Spirochaetes*
Order	*Spirochaetales*
Family	*Spirochaetaceae*
Genus	*Spirochaeta*
Species	*plicatilis*

between different early prokaryotic forms of life (see Chapter 1). Because of the long period of evolution of prokaryotes, the various groups display considerable diversity, particularly metabolic and physiological. In contrast, the metabolic diversity of the *Eucarya* is limited especially with respect to energy generation. The vast diversity of metabolic types of prokaryotes is discussed more fully in Chapter 5 and in Chapters 18 through 22.

The fact that two domains of prokaryotes exist was not at all appreciated until molecular phylogenetic studies were performed. Bacteriologists are now trying to sort out when the split occurred between the *Bacteria* and *Archaea* and what the nature was of their last common ancestor.

Artificial versus Phylogenetic Classifications

Conventional artificial taxonomy uses phenotypic tests to determine differences between strains and species. These tests are typically weighted so that characteristics that are considered to be more important are given higher priority. For example, in traditional taxonomy, the Gram stain has been given more weight in determining the classification of an organism than whether the organism uses glucose as a carbon source. Therefore, all gram-positive strains would be ascribed to one family or genus, and within that group, certain species or strains would use glucose and others would not.

Most bacteriologists favor a phylogenetic system for the classification of bacteria, and with the advent of molecular phylogeny, this hope is now being realized. The current accepted treatise that contains a complete listing of prokaryotic species and their classification is *Bergey's Manual of Systematic Bacteriology* (2001), published by Springer, and its more condensed edition, *Bergey's Manual of Determinative Bacteriology* (1994). In addition to containing a complete classification of prokaryotes, the more comprehensive version of *Bergey's Manual* contains a description of all known validly described bacterial species. Thus, it is the "encyclopedia" of the bacteria that is widely used by bacteriologists (**Box 17.1**).

Bergey's Manual of Systematic Bacteriology is now in its second edition. The first edition was based on an artificial classification because too little phylogenetic information was available. However, the second edition is phylogenetic and based on 16S rDNA sequencing, as discussed later.

Phenotypic Properties and Artificial Classifications

Phenotypic properties are those that are expressed by an organism. In classifications, it is important to select characteristics that clearly distinguish among organisms. Furthermore, it is important that the identifying characteristics selected be easy to determine. Two examples of simple phenotypic characteristics that have been widely used in artificial classification schemes are the Gram stain and cell shape. It turns out that each of these has some utility as an evolutionary marker, but they are rather crude indicators and, individually, yield only limited phylogenetic information. The Gram stain tells something about the nature of the cell wall (see Chapter 4). Furthermore, the Gram stain happens to be important phylogenetically because two of the phylogenetic groups of *Bacteria* are gram-positive (i.e., *Firmicutes* and *Actinobacteria*) and 21 are gram-negative. However, the mycoplasmas, which stain as gram-negative organisms, have been found to be members of the *Firmicutes* through 16S rRNA analyses (see Chapter 20). The reason the mycoplasmas stain as gram-negative is that they

---

Milestones Box 17.1　　*Bergey's Manual* Trust

David Bergey was a professor of bacteriology at the University of Pennsylvania in the early 1900s. As a taxonomist he was a member of a committee of the Society of American Bacteriologists (SAB—now called the American Society for Microbiology) that was interested in formulating a classification of the bacteria that could be used for identification of species. In 1923, he and four others published the first edition of *Bergey's Manual of Determinative Bacteriology*. This was followed by new editions every few years. Royalties collected by the publication activities of the committee were held in SAB. When David Bergey and his co-editor, Robert Breed, requested money from the account to be used for preparation of the fifth edition, the leadership in SAB refused. After a long and bitter fight, the SAB relented and turned the total proceeds (about $20,000) over to Bergey, who promptly put the money into a nonprofit trust with a board of trustees to oversee the publication of manuals on bacterial systematics. The Trust, now named in his honor as *Bergey's Manual* Trust, is responsible for the publication of *Bergey's Manual of Determinative Bacteriology,* which is now in its ninth edition, as well as other taxonomic books such as *Bergey's Manual of Systematic Bacteriology*. The trust is headquartered at Michigan State University and has a nine-member international board of trustees, as well as associate members from many countries.

lack cell walls altogether. Therefore, they have apparently evolved from a group of gram-positive bacteria that lost their peptidoglycan wall during evolution.

In contrast to the gram-positive bacteria, gram-negative organisms fall into many different phylogenetic groups, including peptidoglycan-containing types and non-peptidoglycan-containing types that are bacterial as well as archaeal. Therefore, the gram-negative bacteria are very diverse phylogenetically.

At one time, some bacteriologists had proposed that the simplest, and purportedly the most stable, cell shape—the sphere—must have been the shape of the earliest bacteria. They then developed an evolutionary scheme based on this theory, in which all of the coccus-shaped bacteria were included in the same phylogenetic group. The validity of this classification has not been borne out by research. For example, there are both gram-negative as well as gram-positive cocci. Some cocci are photosynthetic *Proteobacteria*, others are nonphotosynthetic, some are highly resistant to ultraviolet light (*Deinococcus*), some are *Archaea*, and others are *Bacteria*.

However, cell shape is important phylogenetically for one phylum, the spirochetes, which contains the helically shaped bacteria with axial filaments (see Table 17.1). Because of their morphology, these bacteria were correctly classified with one another in the order *Spirochaetales* and now in the phylum *Spirochaetes* long before phylogenetic data confirmed the grouping. Apart from this group, overall cell shape has little meaning at higher taxonomic levels. Nonetheless, it can still be significant at the species, genus, and even family levels.

Other phenotypic properties have also proven useful in both artificial and phylogenetic classifications. For example, because of their unique ability to produce methane gas, the methanogenic bacteria have always been classified together in artificial classifications. Likewise, from a phylogenetic standpoint, all the methanogenic bacteria are members of the *Euryarcheota* of the *Archaea*.

Of course, phenotypic properties have a special significance not found in molecular phylogeny in that these features provide information about what the organism is capable of doing. One cannot directly conclude from the 16S rRNA sequence that an organism is or is not a methanogen, for example, unless that particular feature has been tested for and determined. Thus, phenotypic tests provide valuable information about the capabilities of the organism that may help explain its role in the environment in which it lives.

Numerical Taxonomy When a large number of similar bacteria are being compared, computers are very useful in the analysis of the data. This aspect of taxonomy, which has been used in artificial classifications, is referred to as **numerical taxonomy.** Numerical taxonomy is most useful at the species and strain level where

phylogenetic relatedness has already been established by rRNA sequencing and DNA/DNA reassociation.

In numerical taxonomy *all characteristics are given equal weight.* Thus, metabolism of a particular carbon source is considered to be as important as the Gram stain or the presence of a flagellum. In characterizing strains in this manner, a large number of characteristics are determined, and the similarity between strains is then compared by a similarity coefficient. Each strain is compared with every other strain. The **similarity coefficient,** S_{AB} between two strains A and B, is defined as follows:

$$S_{AB} = \frac{a}{a+b+c}$$

where *a* represents the number of properties shared in common by strains A and B; *b* represents the number of properties positive for A and negative for B; and *c* represents the number of properties positive for B and negative for A.

Characteristics for which both strains A and B are negative are considered irrelevant because there would be many such features that would have no bearing on their similarity. For example, endospore formation is an uncommon characteristic for bacteria. It is not significant to incorporate this characteristic when comparing two species within a genus that do not produce endospores. It would, however, be of value in comparing endospore-forming organisms to closely related organisms. It should be noted that the similarity coefficient can be used to relate not only phenotypic features of one organism to another, but also to relate the sequence similarity of macromolecules of different organisms.

In numerical taxonomy, it is best to have as many tests of phenotypic characters as possible. Typically at least 50 *independent* characters are used, and many strains are usually compared simultaneously. Ideally, each characteristic should represent a single and separate gene. The same gene should not be assessed more than once, and therefore overlapping phenotypic tests must be avoided. S_{AB} values greater than 70% are expected within species and greater than 50% within a genus. An example of a numerical analysis is shown in **Box 17.2.**

In artificial classifications, bacteria are grouped into a hierarchy based on phenotypic properties. For example, the autotrophic bacteria that obtain energy from the oxidation of inorganic nitrogen compounds, such as ammonia and nitrite, would be classified in the group of nitrifying bacteria. This group would be regarded as an order. One subgroup, termed the family, would contain ammonia oxidizers, another nitrite oxidizers. Within each of these groups, features such as cell shape would be further used to define differences among the various genera and species. However, this artificial classification

In this example, eight strains, A through H, are compared to one another
by ten phenotypic tests. The results are shown in the first table:

Table 1 Results of phenotypic tests for eight strains, A–H

				Strains Tested				
Tests	A	B	C	D	E	F	G	H
1	+	–	+	+	+	+	–	+
2	–	+	+	–	+	–	+	+
3	+	+	–	+	–	+	+	+
4	+	–	–	–	+	+	–	+
5	+	–	–	+	+	+	–	–
6	+	+	–	+	–	–	+	–
7	–	+	+	+	+	–	–	+
8	+	–	–	+	–	+	+	–
9	+	–	+	+	+	+	–	+
10	–	+	+	–	+	–	+	+

Similarity coefficients are then determined by comparing the results of the tests for each of the strains against one another, using the
formula given in the text. The results are shown in Table 2.

Table 2 Similarity coefficients (×100 to give percent similarity) for the eight strains

Strains	A	B	C	D	E	F	G	H
A	100							
B	22	100						
C	20	38	100					
D	75	43	33	100				
E	40	33	71	40	100			
F	86	11	22	63	44	100		
G	33	67	25	33	50	22	100	
H	40	50	71	40	75	44	33	100

The information from this matrix is then used to group the strains into similar types, as shown in
the following matrix:

Table 3 Similarity matrix of grouped strains[a]

Strain	A	F	D	E	H	C	G	B
A	100							
F	86	100						
D	75	63	100					
E	40	40	40	100				
H	40	44	40	75	100			
C	20	22	33	71	71	100		
G	33	22	33	50	33	25	100	
B	22	11	43	33	50	38	67	100

[a]According to these tests, strains A, F, and D are very similar to one another and probably compose
a single species. Likewise, E, C, and H are very similar and appear to be a separate species. Strains
B and G may also be a different species, although more tests should probably be performed to sub-
stantiate this. As mentioned earlier in the chapter, phenotypic data such as this can provide an indication of relatedness at the species level, but if new species are being described, DNA/DNA reassociation tests should be performed.

does not take into account the evolutionary relatedness among the members of the nitrifying bacteria.

Phylogenetic Classification

During the 1970s a revolution occurred in bacterial taxonomy. By then, data had accumulated indicating that a true phylogenetic classification of bacteria was possible. What made it possible was, first of all, acceptance of the evidence that some of the macromolecules of bacteria were highly conserved, that is, changed very slowly during evolution, and that their sequences held the key to unlocking the relatedness of bacteria to one another and to higher organisms. Second, sequencing techniques were developed and improved so that it became easy to conduct sequencing analyses of ribosomal RNA and other macromolecules.

In this section, emphasis will be given to ribosomal RNA (rRNA) sequencing, in particular, the RNA of the small subunit of the ribosome, 16S rRNA (or 18S rRNA of *Eucarya*), as it is the most common conserved molecule used to study the phylogeny of microorganisms (**Figure 17.1**). In actual practice, the 16S rRNA gene, or 16S rDNA, is sequenced because the polymerase chain reaction (PCR) procedures are simple, and both strands can be used to confirm the actual sequence.

Several reasons justify the choice of ribosomal RNA as an evolutionary marker. First, the ribosome is a very complex structure that carries out a complicated function—protein synthesis (see Chapter 11). Keep in mind that the 16S rDNA is interacting in a three-dimensional structure with protein and other rRNA molecules as well as mRNA. Thus, the rate of evolutionary change in ribosome structure has been selected against during evolution. Mutant organisms with dysfunctional ribosomes would be unable to compete with existing types and have therefore not survived. In this manner, evolution has selected against *major* changes in the ribosome. Nonetheless, incremental modifications have occurred over the billions of years of biological evolution and these differences are used to construct evolutionary trees.

An additional advantage of using the ribosome in phylogeny is that *all* organisms, from bacteria to plants and animals, have ribosomes. Furthermore, the function of the ribosome as the structure responsible for protein synthesis holds true for all classes of life. Therefore, it is possible to compare the phylogeny of all organisms with one another by analysis of a single highly conserved structure with an important cellular function. Ribosomal RNAs are not the only macromolecules that have been considered in determining relatedness at higher taxonomic levels. The proteins cytochrome *c* and ribulose bisphosphate carboxylase are just two examples of other molecules that have been used. However, not all organisms synthesize these macromolecules. Also, not all cytochrome c–like molecules found in different organisms have the same physiological function. Because they are not universally distributed among all organisms, they cannot be used to compare distantly related taxa. In addition, proteins have not been used as much because it is often difficult to identify conserved primers as readily as for rRNA molecules due to the degeneracy of the genetic code. Nonetheless, protein sequences are very useful in constructing phylogenies of specific proteins such as nitrogenase.

Within the biological world, ribosomes share many similarities, indicating the conservative nature of the structure. Prokaryotic ribosomes contain three types of RNA: 5S, 16S, and 23S. Both 5S and 16S rRNA have been used to determine relatedness among organisms. Because the 16S molecule is larger (with about 1,500 bases), it contains more information (Figure 17.1) than the smaller 5S molecule with only about 120 bases (**Figure 17.2**). Less work has been done on the 23S molecule because it is longer (about 3,000 nucleotides) and therefore not as easy to study. Thus, scientists interested in the classification and evolution of bacteria have concentrated on 16S rRNA. The method of evaluation that provides the most information is sequence determination, especially for the complementary DNA of 16S rRNA, that is, 16S rDNA. It has been found that some regions of these molecules are more highly conserved than others. The more highly conserved regions permit one to compare distantly related organisms (**Figure 17.3**), and the more variable domains are used for comparing more closely related organisms. These regions that are unique to a given taxon are termed signature sequences. They can be used for the design of specific hybridization probes for identification (see section on Identification on pg. 380).

It appears that an analysis of 16S rDNA sequences provides important information on the evolution of prokaryotes. However, before concluding that the sequence of bases in rDNA accurately reflects the phylogeny of organisms, it is important to find totally separate and independent evolutionary markers to confirm the classification. Some work has been performed with sequencing of ATPsynthases, elongation factors, RNA polymerases, and other conserved macromolecules. The outcome of this research, which represents one of the most exciting areas of biology, is leading to the development of a complete phylogenetic classification of the *Bacteria, Archaea,* and other microorganisms. The 16S rDNA–based phylogeny will face its most stringent test as additional prokaryotic genomes are sequenced and compared.

Before we discuss the use of 16S rDNA sequences in arriving at the current phylogeny of *Bacteria* and *Archaea,* it is first necessary to consider phylogenetic trees.

Phylogenetic Trees Like the family tree, a phylogenetic tree contains the tree of descendants of a biological

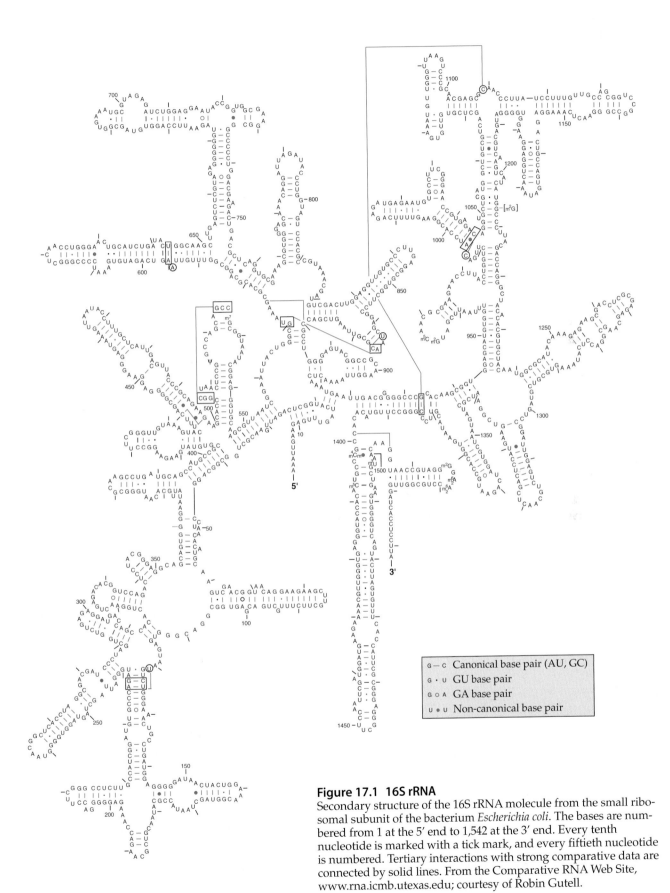

Figure 17.1 16S rRNA
Secondary structure of the 16S rRNA molecule from the small ribosomal subunit of the bacterium *Escherichia coli*. The bases are numbered from 1 at the 5' end to 1,542 at the 3' end. Every tenth nucleotide is marked with a tick mark, and every fiftieth nucleotide is numbered. Tertiary interactions with strong comparative data are connected by solid lines. From the Comparative RNA Web Site, www.rna.icmb.utexas.edu; courtesy of Robin Gutell.

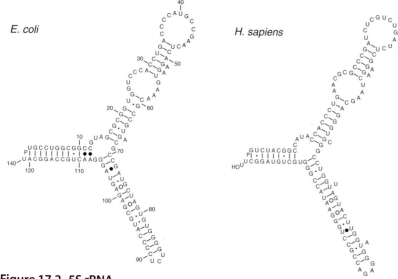

Figure 17.2 5S rRNA
Comparison of the secondary structures of 5S rRNA molecules from the bacterium *Escherichia coli* and the eukaryote *Homo sapiens*.

family or group. However, whereas the family tree traces the genealogy of a family of humans, phylogenetic trees trace the lineage of a variety of different species. Thus, **phylogenetic trees** reflect the purported evolutionary relationships among a group of species, usually through the use of some molecular attribute they possess, such as the sequence of their ribosomal RNA. In this particular section we will discuss molecular phylogeny based on a comparison of the 16S rRNA sequences of organisms.

Phylogenetic trees have two features—**branches** and **nodes** (Figure 17.4). Each node represents an individual species.

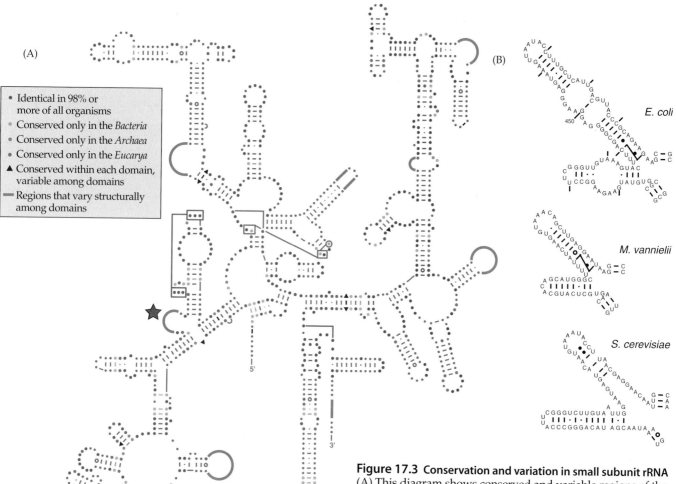

Figure 17.3 Conservation and variation in small subunit rRNA
(A) This diagram shows conserved and variable regions of the small subunit rRNA (16S in prokaryotes or 18S in eukaryotes). Each dot and triangle represents a position that holds a nucleotide in 95% of all organisms sequenced, though the actual nucleotide present (A, U, C, or G) varies among species. (B) The starred region from part A as it appears in a bacterium (*Escherichia coli*), an archaean (*Methanococcus vannielii*), and a eukaryote (*Saccharomyces cerevisiae*). This region includes important signature sequences for the *Bacteria* and *Archaea*. Figure by Jamie Cannone, courtesy of Robin Gutell; data from the Comparative RNA Web Site: www.rna.icmb.utexas.edu

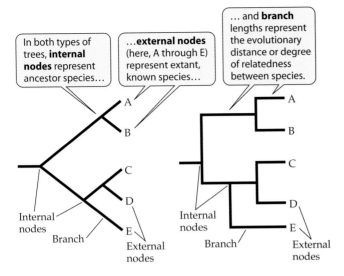

Figure 17.4 Phylogenetic trees
Two different formats of phylogenetic trees used to show relatedness among species.

External nodes (usually drawn to the extreme right of the tree) represent living species, and internal nodes represent ancestors. A branch is a length that represents the distance between or degree of separation of the species (nodes) from one another.

Trees may be rooted or unrooted. **Figure 17.5A** shows an unrooted tree containing three different species, A, B, and C. Unrooted trees compare one feature of a group of related organisms, such as the sequence of their 16S rDNA. There is only one shape to this particular tree with three species. In contrast, rooted trees need to have

either an outlying species that is distantly related to the species being compared or an additional gene with which to compare the species—this should be a different gene such as the sequence of a macromolecule that underwent a gene duplication event prior to when the taxa that are being studied diverged from one another. Thus, in this latter instance, the elongation factor gene, Tu, has been used to root the 16S rDNA Tree of Life. A rooted tree containing three species has three possible shapes (Figure 17.5B).

For bacterial phylogeny, trees are constructed from information based on the sequence of subunits in macromolecules. As mentioned earlier, ribosomal RNA, in particular the small subunit ribosomal RNA molecule, 16S rRNA, has been selected as the molecule of choice because of its conserved nature and length.

Sequencing 16S rDNA If one isolates a new bacterium and wishes to determine its phylogenetic position among the known bacteria, it is necessary to determine the sequence of its 16S rDNA. This can be accomplished in a number of ways. One of the most common ways is to use the polymerase chain reaction (PCR) to amplify the 16S rDNA from genomic DNA from the bacterium. The amplified 16S rDNA may be sequenced directly or ligated into a cloning vector and cloned into *E. coli*. This latter step allows a large quantity of 16S rDNA to be produced through growth prior to sequence analysis. The entire 16S rDNA can be sequenced using a standard set of oligonucleotide primers and standard sequencing techniques.

Alignment with Known Sequences The next step is to incorporate the determined linear DNA into an alignment with the sequences of other known organisms. An international database called the Ribosome Database Project, located at Michigan State University, contains the 16S rRNA sequences of those bacteria that have been sequenced. Sequences from this database can be retrieved electronically over the Internet (http:// rdp.cme.msu.edu/). Using a series of computer programs and careful manual examination, the sequence can be aligned with those retrieved from the database.

Phylogenetic Analysis Having the sequence is only half of the story. It is next necessary to compare the sequence of the unknown bacterium to that of other bacteria from the database. The determination of the evolutionary relatedness among organisms can be accomplished by one of a number of phylogenetic methods. There are several types of analysis that can be used. **Distance matrix** methods are one type of approach. In distance matrix methods, the evolutionary distances, based on the number of nucleic acid or amino acid monomers that differ in a sequence, are determined among the strains being compared. A second approach is to use

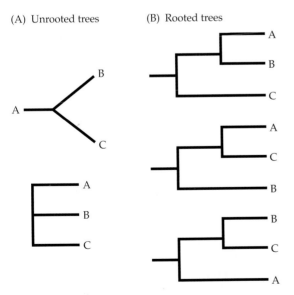

Figure 17.5 Unrooted and rooted trees
Representations of the possible relatedness between three species, A, B, and C. (A) A single unrooted tree (shown in both formats; see Figure 17.4). (B) Three possible rooted trees (in one format).

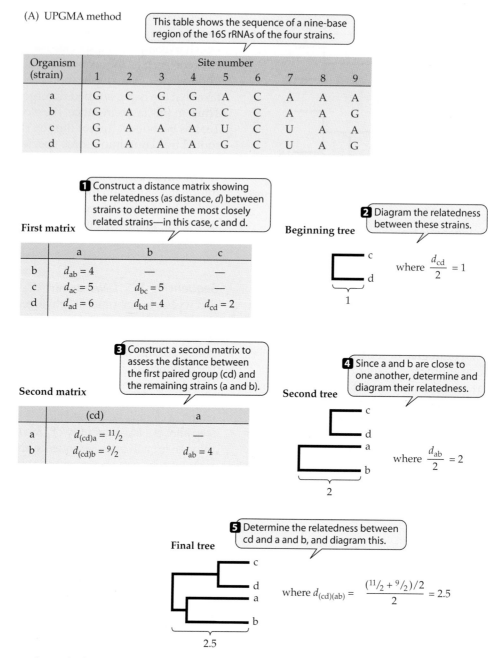

Figure 17.6 Phylogenetic analysis
Phylogenetic analysis of four different strains, a, b, c, and d, showing a hypothetical region of their 16S rRNA that contains nine bases. (A) The UPGMA method of determining a phylogenetic tree. (B) The maximum parsimony method (see text for details).

maximum parsimony methods. In maximum parsimony, the goal is to find the simplest or most parsimonious phylogenetic tree that could explain the relatedness between different sequences. In both approaches, the sequence of nucleic acid subunits among different strains is compared.

Figure 17.6A uses a distance matrix method called the "unweighted pair group method with arithmetic mean," or UPGMA, to analyze an aligned hypothetical

sequence region of ribosomal RNA from four different strains. This is one of the simplest analytical methods that can be used. Figure 17.6B using the same sequences shows an analysis of the data by maximum parsimony.

In the UPGMA method, a distance matrix is set up to compare the differences in sequence or "distance," d, between each of the four strains. There are four nucleotide differences between the sequence of organism a and the sequence of organism b, thus, d_{ab} is determined to be 4. Likewise, d_{ac} is 5, d_{bc} is 5, and so forth. In this instance, the shortest distance, 2, is between strains

(B) Maximum parsimony method

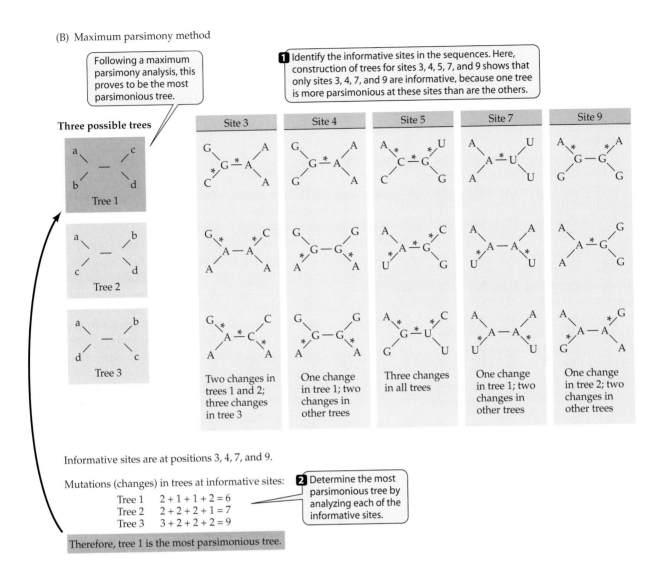

> Following a maximum parsimony analysis, this proves to be the most parsimonious tree.

1 Identify the informative sites in the sequences. Here, construction of trees for sites 3, 4, 5, 7, and 9 shows that only sites 3, 4, 7, and 9 are informative, because one tree is more parsimonious at these sites than are the others.

Three possible trees

Tree 1

Tree 2

Tree 3

Site 3	Site 4	Site 5	Site 7	Site 9
Two changes in trees 1 and 2; three changes in tree 3	One change in tree 1; two changes in other trees	Three changes in all trees	One change in tree 1; two changes in other trees	One change in tree 2; two changes in other trees

Informative sites are at positions 3, 4, 7, and 9.

Mutations (changes) in trees at informative sites:

Tree 1   2 + 1 + 1 + 2 = 6
Tree 2   2 + 2 + 2 + 1 = 7
Tree 3   3 + 2 + 2 + 2 = 9

2 Determine the most parsimonious tree by analyzing each of the informative sites.

Therefore, tree 1 is the most parsimonious tree.

c and d. From these two strains, which show the closest relationship to one another, a simple tree is constructed that shows c and d connected by a node that is half the distance between the two, that is, 1 unit. This is expressed in the actual tree as a horizontal branch length of one unit from each of the organisms to a common ancestral node (Figure 17.6A).

The next step is construct another matrix in which c and d are considered as a single composite unit (cd) and compared with a and b. From this matrix, the two most similar organisms are a and b, and the length of this branch is calculated as the distance, $d_{ab}/2$, which is equal to 2 units. From this an intermediate second tree is formed. Finally, a is different from (cd) by $(d_{ac} + d_{ad})/2$, or $(5 + 6)/2 = 5.5$. Likewise, b is different from (cd) by $(d_{bc} + d_{bd})/2 = 4.5$. To determine the connection between the composite branch cd and ab, the distance is calculated as the average of $d_{(cd)(ab)}$, or 5.5 + 4.5/2, which is 5.0. This is then divided by 2 to give the average distance between the two composites, cd and ab. Using this reasoning, a

final tree (Figure 17.6A) is produced showing the relationship among the four different strains.

The UPGMA method is the simplest of the distance methods used. More sophisticated distance methods include transformed distance and neighbor joining methods, which will not be discussed here.

As mentioned earlier, in maximum parsimony the goal is to identify the simplest tree that could explain the difference between two different sequences or species. This approach has its philosophical basis in **Occam's razor,** commonly used in the sciences, which states that *the likely solution to a problem is the simplest one.* In this case, to explain the evolutionary difference between two species, one looks at the tree that has the fewest changes (mutational events) that could explain their differences. This is accomplished on a computer that, at least in theory, considers all possible trees and then identifies the simplest one (the one with the fewest assumed mutational events).

In maximum parsimony it is important to recognize sites in the sequences that are useful for a comparison between organisms. These are termed **informative sites.** These sites are then used to determine the most parsimonious tree. For example, Site 1 in the example given is not informative because all the bases are identical. Site 2 is not informative either, because three of the strains have A and one has C, suggesting that a single mutational event has occurred. Site 3 is informative because trees 1 and 2, which have two changes, are more parsimonious than Tree #3, which has three. Site 5 is not informative because all trees constructed from the information at this site differ from one another by three mutations. In contrast, Site 4 is informative, because one tree (Tree #1) is more parsimonious than the other two. Sites 6 and 8 are not informative because all bases are identical, but both Sites 7 and 9 are informative. Site 7 favors Tree #1, whereas Site 9 favors Tree #2. Thus, for this set of data, Tree #1 is favored two out of four times, Tree #2 is favored one out of four times, and Tree #3 is not favored at any time. Adding the changes at those four sites gives the following data: Tree #1 is the most parsimonious because a total of only six changes (2 + 1 + 1 + 2 = 6) would explain its phylogeny, whereas in Tree #2, seven changes are required (2 + 2 + 2 + 1 = 7), and in Tree #3 nine changes are required (3 + 2 + 2 + 2 = 9).

Note that the two trees that were constructed from the distance matrix and maximum parsimony methods are identical in shape, indicating that two of the strains, a and b, are closely related to one another, but not to c and d. Likewise, c and d are closely related to one another. As you can imagine, some rather sophisticated computer programs have been developed to handle the immense amount of information inherent in longer sequences such as 16S rDNA, which contains about 1,500 base pairs. Moreover, when such large amounts of data are being analyzed, it is often not possible to determine that a proposed tree is, in fact, the true tree. Indeed, trees should be considered as hypotheses until additional information has been analyzed. To help support a given tree, other techniques are used. For example, in "bootstrap" analyses, random sequence positions are selected by the computer, and the trees formed from them are compared with the proposed tree to see if a simpler (more parsimonious) tree can be found. In this manner, some 100 or 1,000 different bootstrap comparisons might be made and provided as evidence that the proposed tree is indeed the most parsimonious one. Bootstrap analyses are applicable for all phylogenetic treatment procedures.

As mentioned previously, other analytical methods can be used to analyze sequence information and construct phylogenetic trees. A common one used by microbiologists is the **maximum likelihood** method, which involves selecting trees that have the greatest likelihood of accounting for the observed data. This is accomplished by assigning a probability to the mutation of any one base to any other base at each possible sequence position. From this, all possible topological trees are constructed. By integrating the probabilities for each mutation over each tree, a degree of improbability for a tree is assessed. The least improbable tree is chosen as the "true" tree.

Speciation

The process by which organisms evolve is termed **speciation.** As with plants and animals, bacteria evolve into habitats and ecological niches (i.e., physiological roles). However, unlike plants and animals, bacteria can evolve very quickly because of their rapid growth rates, high population sizes, and haploid genomes that allow for the rapid expression of favorable mutations through natural selection. Thus, a lineage of bacteria is determined in large part through **vertical inheritance,** the process by which the parental genotype is transferred to the progeny cells following DNA replication and asexual reproduction.

However, bacteria can also acquire genetic material from other organisms that are different from them through their various genetic exchange mechanisms, particularly transformation. This phenomenon is referred to as **horizontal (or lateral) gene transfer** to distinguish it from vertical inheritance. As a result, prokaryotic organisms may undergo dramatic changes in their population structure in a relatively short period of time. For example, we know that multiple-drug resistance can be rapidly acquired by a bacterial species that is sensitive to antibiotics if it is exposed to antibiotics in the presence of other antibiotic-resistant bacteria.

Consider the situation of a bacterium that is a member of the normal microbiota of the intestinal tract that is exposed to an antibiotic to which it is sensitive. The bacterium may either perish, or if a gene is available in the environment that confers resistance, it could acquire the resistance gene and survive. This example of a strong selective pressure likely explains how it is possible for sensitive bacteria to quickly become resistant to an antibiotic. This scenario applies equally to other environmental pressures that confront bacteria, such as exposure to potentially toxic hydrocarbons that are used as an energy source by other species in the environment.

Many of the known examples of rapid genetic change occur through the acquisition of plasmids from related organisms. Thus, in the preceding examples, some plasmids are known that carry multiple antibiotic resistant genes and others are known that carry hydrocarbon degrading genes.

Also, we do know that genes can be acquired from distantly related organisms. For example, it has been recently reported that a member of the *Proteobacteria* in the *Bacteria* has been found to contain a gene responsible for bacteriorhodopsin synthesis, which was only known previously from members of the *Archaea*. Thus, this represents a transfer of genetic material across two different domains of life.

Although the horizontal transfer of genes between widely disparate groups of microorganisms is known to occur, evidence points to this being a relatively rare event. If it occurred commonly, it could confuse phylogenetic classifications based on vertical inheritance to such an extent that it would render them utterly useless. From an analysis of genomes that have been sequenced, genes that have been derived from other organisms have been identified in some prokaryotic genomes (**Figure 17.7**).

TAXONOMIC UNITS

The basic taxonomic unit is the species, although as mentioned earlier, some species have subspecies categories as well. The categories above the species are (sequentially) genus, family, order, class, phylum, and domain (Table 17.1). It should be noted that uncertainties exist in bacteriology about the meaning of the higher taxonomic categories such as kingdom because oftentimes the phylogenetic markers that have been used (primarily 16S rDNA sequences) cannot definitively resolve the earliest branching points in the Tree of Life. Thus, although we know that each of these major branches is equivalent to the plant and animal "kingdoms," how the microbial "kingdoms" or phyla are related to one another is only poorly understood.

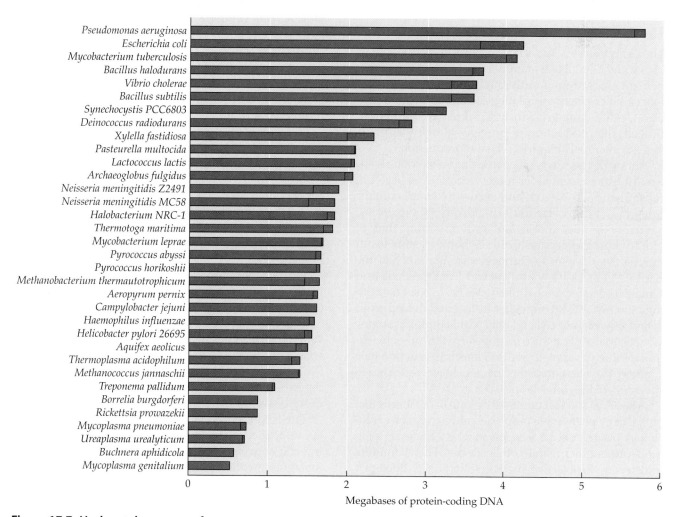

Figure 17.7 Horizontal gene transfer

Analyses of sequenced bacterial genomes indicate that a significant proportion of their genes can be traced to other phylogenetic groups, indicating the importance of horizontal gene transfer in bacterial speciation. This diagram shows the proportion of DNA that was acquired by horizontal gene transfer (in red) in some microbial genomes. Courtesy of Jeffrey Lawrence.

Each colony or culture of an organism represents an individual **strain** or **clone** in which all of the cells are descended from one single organism. In a somewhat different sense of meaning, a strain can also refer to a mutant of a species that has changed characteristics (for example, lacks a particular gene). The strain, however, is not considered a formal taxonomic unit, and Latin names are therefore not ascribed to strains; they have only informal designations, such as *E. coli* strain K12. There can also be **varieties** within species that exhibit differences. These are called **biovars**. For example, a serological variety such as *E. coli* O157:H7 is a pathogenic **serovar** that causes hemolytic uremic syndrome and can be lethal to children who become infected by eating contaminated food. Likewise, pathogenic varieties are termed **pathovars,** ecological types **ecovars,** and so forth. Now let us look at the individual taxa beginning with the species to see what features are typical at each taxonomic level.

The Species

The definition of a bacterial species differs from that of plants and animals. In mammals, the classical species is defined as a group of individuals (males and females) that exhibit evident morphological similarities and produce fertile progeny through sexual reproduction. Indeed, the production of progeny in many animals such as mammals requires sexual reproduction.

Although gene exchange occurs in prokaryotic organisms, it is not essential for reproduction. Most bacterial reproduction is asexual and occurs by simple binary transverse fission or budding. In prokaryotic organisms, sexuality is uncommon and different from that of eukaryotes. Eukaryotes produce haploid gametes in meiosis (see Chapter 1). During sexual reproduction, the haploid gametes (egg and sperm) from the male and female fuse to form the diploid zygote. In bacterial conjugation, DNA from one cell is transferred during replication to a receptor cell and only partial diploidy occurs. Genetic material can also be transferred by other mechanisms such as transformation and transduction (see Chapter 15). These transfers are not always restricted to members of the same species.

A bacterial species comprises a group of organisms that share many phenotypic properties and a common evolutionary history and are therefore much more closely related to one another than to other species. This definition, which is very subjective, has been interpreted differently by bacteriologists in describing species. For example, at one extreme some taxonomists are called **lumpers** because they group (or "lump") fairly diverse organisms into a single species or genus. An example of a lumper is F. Drouet, who has proposed reducing the number of cyanobacteria from 2,000 species to only 62! At the opposite pole are **splitters.** These are taxonomists who consider even the slightest differences sufficient for a new

species. For example, many years ago it was proposed that the genus *Salmonella* be "split" into hundreds of different species, a separate species for each of the hundreds of different serotypes (or serovars) that are recognized based on specific cell surface antigens of their lipopolysaccharides and flagella. However, the views of lumpers and splitters illustrated here are considered to be extreme and are not accepted by the majority of microbiologists.

In fact, more recently, a less-arbitrary, quantitative basis has been proposed to define a bacterial species. Agreement was reached by a group of prominent bacterial taxonomists to define a bacterial species based on genomic similarity between strains. Accordingly, a species is defined as follows: two strains of the same species must have a similar mole percent guanine plus cytosine content (mol % G + C) and must exhibit 70% or greater DNA/DNA reassociation. The procedures used to determine these features are described here.

Mole Percent Guanine Plus Cytosine (Mol % G + C) The **mol % G + C** refers to the proportion of guanine and cytosine to total bases (guanine, cytosine, adenine, and thymine) in the DNA. Recall that because G and C are paired in the double-stranded DNA molecule by hydrogen bonds, as are A and T, they occur in equal concentrations. The formula is given as:

$$\text{mol \% G} + \text{C} = \frac{\text{moles (G + C)}}{\text{moles (G + C + A + T)}} \times 100$$

Several methods can be used to determine the mol % G + C, sometimes also called the "GC ratio" of a bacterium. All of them require that the DNA be first isolated from a bacterium and purified. Thus, it is necessary to lyse the cells to release the cytoplasmic constituents including DNA, and the DNA must then be purified to remove proteins and other cellular material. Cell lysis is typically accomplished by treatment with lysozyme and detergents, and the DNA is precipitated with ethanol. When the DNA has been sufficiently purified, it can be analyzed chemically to determine the content of each of the bases. Several different procedures can be used to determine the GC ratio of the purified DNA. We will describe two of them here.

The first method is a chemical method and can be accomplished by hydrolyzing the DNA and determining the concentration of each of the bases using an instrument called a **high-pressure liquid chromatograph,** or **HPLC.**

Another common procedure to determine GC ratios is by **thermal denaturation.** The principle behind this method is that the hydrogen bonds between the double strands can be broken by heating dissolved DNA. As the hydrogen bonds are broken and the two strands separate, the absorbance of the DNA increases. This procedure, called **melting** the DNA, is conducted with a spectrophotometer set at 260 nm, a wavelength at which DNA absorbs strongly. The hydrogen bonding of the GC base pair is stronger than the AT pair in the double-stranded DNA molecule. Therefore, a higher temperature is required to melt DNA that has a high content of GC pairs, that is, a high GC ratio. **Figure 17.8** shows a graph of the melting of a double-stranded DNA molecule. This process is accomplished by gradually increasing the temperature of a solution of the DNA in an appropriate buffer (ionic strength is important). As the temperature is raised, the melting process begins and continues until the double-stranded DNA molecule is completely converted to the single-stranded form. The absorbance increases during this melting process. The midpoint temperature (T_m) is directly related to the GC ratio of the DNA. Thus, the GC ratio can be read from a chart showing the relationship between T_m and GC content (**Figure 17.9**). Once the DNA has been melted, it will reanneal if the temperature is slowly lowered. Thus, the process shown in Figure 17.8 is reversible. However, if the solution is cooled rapidly, hybrid formation does not recur, and the molecules of DNA are left in the single-stranded state.

Figure 17.10 shows the range of GC ratios in various groups of organisms. On this basis alone, one can see that bacteria, which have GC ratios ranging from about 20 to greater than 70, are truly a very diverse group. In contrast, higher organisms such as animals have a very restricted GC ratio range.

The GC ratio provides only the relative amount of guanine and cytosine compared to total bases in the DNA of an organism and says nothing about the inher-

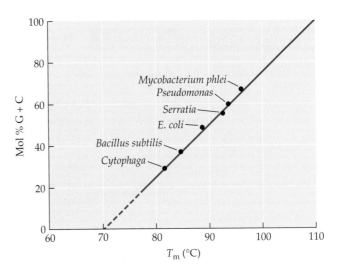

Figure 17.9 T_m and DNA base composition
Graph showing the direct relationship between mol % G + C and midpoint temperature (T_m) of purified DNA in thermal denaturation experiments.

ent characteristics of the organisms or what genes are present. Indeed, *two very different organisms can have similar or even identical GC ratios.* For example, the DNA of *Streptococcus pneumoniae* and humans have the same mol % G + C content.

DNA/DNA Reassociation or Hybridization
Although the determination of GC ratio is useful in bacterial taxonomy, it does not tell us anything about the linear arrangement of the bases in the DNA. It is the arrangement of the DNA subunits that codes for specific genes and proteins and therefore determines the features of an organism. DNA/DNA reassociation or hybridization is one method used to compare the linear order of bases

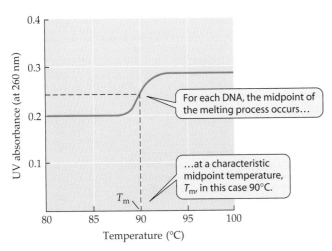

Figure 17.8 DNA melting curve
Melting curve for a double-stranded DNA molecule. As the temperature is raised during the experiment, the double-stranded DNA is converted to the single-stranded form and the UV absorbance of the solution increases. The midpoint temperature, T_m, can be calculated from the curve. This process is reversible if the temperature of the solution is slowly lowered to allow the single strands to reanneal.

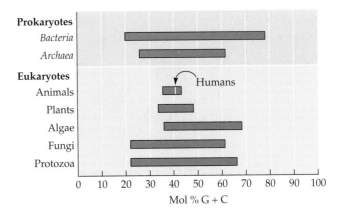

Figure 17.10 DNA base composition range
Range of mol % G + C content among various groups of organisms. Note the broad range of GC ratios for bacteria in comparison to plants and animals and other eukaryotes.

Methods & Techniques Box 17.3   DNA/DNA Reassociation

DNA/DNA reassociation can be performed by a variety of different methods. In all approaches, it is necessary to begin with purified DNA from the two organisms that are being compared. The DNA is then denatured by melting, and DNA from the two different strains are mixed and allowed to cool together to allow reannealing to occur. This reannealing will occur both between DNA strands of the same species and between strands of the comparison species. The degree of reannealing depends on how similar the DNAs are to one another. If two strains are very similar to one another, their DNAs will reanneal to a high degree. In contrast, if two strains are very different, then the extent of reannealing will be much less. One way to perform DNA/DNA reassociation is to radiolabel the DNA by growing the bacterium with tritiated thymidine or ^{14}C-labeled thymidine (other DNA bases or ^{32}P-phosphate labeling can be used as well). If the bacterium takes up this labeled substrate and incorporates it into DNA, then the DNA becomes labeled.

Alternatively, the DNA can be purified from the bacterium and labeled enzymatically in the laboratory.

After the DNA has been labeled and purified, it is sheared to an appropriate length by sonication. It is then ready for the hybridization experiments. First, single-stranded DNA is prepared. This is accomplished by heating the isolated DNA molecules to render them single-stranded and then cooling them rapidly to prevent reannealing.

First, let's look at the control assay for the DNA reassociation experi-

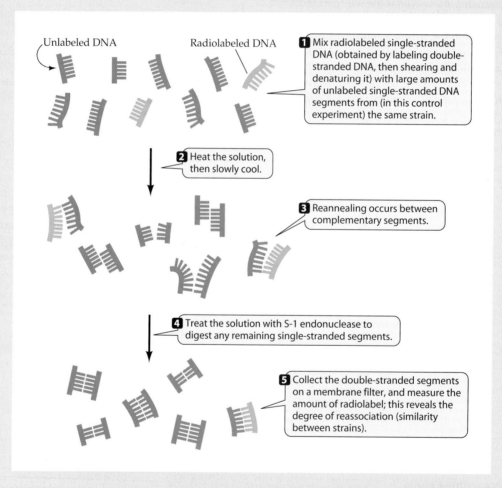

Unlabeled DNA   Radiolabeled DNA

1 Mix radiolabeled single-stranded DNA (obtained by labeling double-stranded DNA, then shearing and denaturing it) with large amounts of unlabeled single-stranded DNA segments from (in this control experiment) the same strain.

2 Heat the solution, then slowly cool.

3 Reannealing occurs between complementary segments.

4 Treat the solution with S-1 endonuclease to digest any remaining single-stranded segments.

5 Collect the double-stranded segments on a membrane filter, and measure the amount of radiolabel; this reveals the degree of reassociation (similarity between strains).

DNA/DNA reassociation   In this example, which is a control experiment (the radiolabeled sample is reannealed with unlabeled DNA from the same strain), the degree of reassociation is highest and treated as 100%. If a different strain is reannealed with the radiolabeled DNA, it will show a lower degree of reannealing (compared with the 100% attributed to the control), indicative of the similarity between the two strains being tested. Strains with reannealing values of 70% or greater are considered to be the same species.

ment. In this case, a small amount of sheared radiolabeled DNA is rendered single-stranded. This is then mixed with a much larger amount of unlabeled DNA obtained from the same bacterial strain. These are heated together and cooled slowly to allow the two single-stranded groups to reanneal to form hybrid double strands. Because the amount of labeled DNA relative to the unlabeled DNA is small, there is a very low probability that it will reanneal with other labeled strands. Most of the reassociations will occur between unlabeled strands, and most of the remainder will be between the labeled and unlabeled strands. The single-stranded fragments that did not reanneal are removed by enzyme digestion, using S-1 nuclease, which specifically degrades only single-stranded DNA, and the double-stranded frag-

ments are collected on a membrane filter or in a column. The amount of radioactivity remaining on the filter or on the column, after washing to remove low molecular weight material, represents the amount of hybrid formation between the labeled and unlabeled DNA for this identical strain. This is the *control reaction,* and the amount of radiolabel (the extent of hybridization) is considered to be 100%.

To determine the extent of reassociation between the strain described and an unknown strain, similar experiments need to be performed. In this instance, unlabeled single-stranded DNA from the unknown strain is prepared and mixed with the known strain for which we have labeled the DNA. As indicated earlier, those strains that show 70% or greater reassociation or hybrid formation with the labeled

strain (determined by the amount of hybrid DNA that is radiolabeled compared with the same strain control of 100% shown in the figure) are considered to be the same species. Anything less is considered a different species.

The temperature and salt concentration at which the DNA/DNA reannealing occurs will influence the degree of reassociation between single strands. Scientists conducting DNA/DNA reassociation experiments typically use a reannealing temperature that is 25°C lower than the average midpoint temperature (T_m) of the DNAs being compared.

This temperature is sufficiently high that only those sequences that are most complementary will reanneal. Thus, this is considered a **stringent** condition for reannealing.

in two different organisms (**Box 17.3**). It is important to recognize that the >70% level of reassociation used for the species definition does not indicate that the two DNAs are 70% homologous or identical.

In DNA/DNA reassociation, the actual order of bases in the DNA is not determined, but rather the extent of reannealing between the DNAs of two different strains is assessed. Ideally one would like to know the actual sequence of genes of a species. Indeed, sequencing entire bacterial genomes has become quite common (see Chapter 16). It is worthwhile noting that one could consider that the actual DNA base sequence of a strain is the ultimate definition of a strain—analogous to the chemical formula for a compound. However, it is important to recognize that, unlike chemical compounds, bacterial strains and species are not static; they continue to evolve.

Interestingly, the bacterial species definition appears to be much broader than that used for animals and plants when one considers DNA/DNA reassociation and other molecular criteria. For example, the bacterial species *E. coli* can be compared to its host mammalian species using a variety of molecular features, including the range in mol % G + C, 16S rDNA sequence (versus 18S rDNA sequence), and DNA/DNA reassociation (**Table 17.2**). Thus, although there is essentially no variation in the range in GC ratio for the human species, it

is about 4 mol % within the *E. coli* species. Indeed, there is less variation in GC ratio in the *Primate* order than there is in the species *E. coli*. Likewise, the 16S rDNA sequences of *E. coli* possess more than 15 substitutions, whereas that of 18S rDNA between the mouse (order *Rodentia*) and the human is less than 16. Finally, DNA/DNA reassociation data, which are used to define the bacterial species at 70%, indicate that humans are much more highly similar to one another in comparison

Table 17.2	Comparison of *E. coli* and its primate host species[a]		
Property	***E. coli***	*Homo sapiens*	**Primates**
Mol % G + C	48–52	42	42[b]
16S–18S rRNA variability	>15 bases	?	<16[c]
DNA/DNA reassociation	>70%	98.6%[d]	>70%[e]

[a]Adapted from J. T. Staley, *ASM News,* 1999.
[b]Value for all primates.
[c]Mouse 18S rRNA differs from humans by 16 bases.
[d]Comparison between *Homo sapiens* and chimpanzee.
[e]Comparison between *Homo sapiens* and lemurs.

with *E. coli*. Therefore, it is evident on this basis that this typical bacterial species is equivalent to a family or an order of mammals based on molecular divergence, indicating that a bacterial species is defined much differently from their eukaryotic counterparts.

The Genus

All species belong to a genus, the next higher taxonomic unit. When DNA/DNA reassociation is performed within a genus, some species of the genus may show little or no significant reassociation with other species. This does not indicate that they are unrelated to one another, only that this technique is too specific to identify outlying members of the same genus. Therefore, DNA/DNA hybridization has limited utility for determining whether a species is a member of a known bacterial genus.

The definition of genera is based on one or more prominent phenotypic characteristic that permits it to be distinguished from its closest relatives. Oftentimes some striking physiological or morphological feature is present that permits it to be distinguished from its most closely related taxa. For example, the genus *Nitrosomonas* is a group of rod-shaped bacteria that grow as chemoautotrophs gaining energy from the oxidation of ammonia. Other ammonia oxidizers with coccus-shaped and helical cells are placed in other genera. Of course, all strains of each of these genera need to be more closely related to one another phylogenetically than to strains of other genera in a phylogenetic classification. Ideally then, the genus makes up a **monophyletic** lineage (i.e., one in which all are members of the same phylogenetic cluster or clade).

Higher Taxa

Odd bedfellows are sometimes found in phylogenetic trees; thus, photosynthetic and nonphotosynthetic members of some closely related groups have been reported. Of course, loss of a key gene or two may result in converting a formerly photosynthetic organism to one that is not photosynthetic. Thus, although the plant and animal kingdoms are differentiated on the basis of whether or not they are photosynthetic, both features have been reported in two closely related bacterial genera. However, because phenotype is so important at the genus level, important features such as photosynthesis are sufficient to proclaim a separate genus, even though two groups are otherwise very closely related.

As mentioned earlier, the taxonomy of prokaryotes is undergoing major changes. Although by classical taxonomy each genus belongs to a family of similar genera, relatedness at the familial and higher level is often uncertain for bacteria. Bacteriologists have been reluctant to ascribe organisms to formal Latinized families and orders. However, as more becomes known about bacterial phylogeny, it is increasingly apparent that higher taxonomic levels do have meaning and can be distinguished from one another by comparing the sequences of certain macromolecules, as is reflected in the new edition of *Bergey's Manual of Systematic Bacteriology*.

MAJOR GROUPS OF *ARCHAEA* AND *BACTERIA*

Bacteriologists have begun to construct classifications using phylogenetic information from rRNA analyses. As mentioned in Chapter 1, some prokaryotes are very different from others, a revelation that came through an analysis of ribosomal RNA (**Box 17.4**). Ribosomal RNA data allow the division of all organisms on Earth, prokaryotic and eukaryotic, into three domains: *Bacteria, Archaea,* and *Eucarya* (eukaryotes) (see Figure 1.6). These domains can also be distinguished from one another by

---

Research Highlights Box 17.4 The Discovery of *Archaea*

Clearly, one of the most exciting developments in bacterial classification of the twentieth century was the discovery that there are two different groups of bacteria or prokaryotic organisms. The appreciation of the difference between the *Bacteria* and the *Archaea* was the culmination of years of research by microbiologists throughout the world. However, the final piece of evidence that convinced microbiologists of this dichotomy was the discovery that these organisms had very different 16S ribosomal RNAs. This research was performed in Carl Woese's laboratory at the University of Illinois. At that time, rRNA sequencing was not done routinely in laboratories. Instead, 16S rRNA was purified and digested by a ribonuclease. The oligonucleotide fragments produced were subjected to two-dimensional electrophoresis. The pattern of spots on a two-dimensional chromatogram represented the various rRNA oligonucleotides that were typical of each species. Studies from Woese's laboratory demonstrated that the patterns were very different for *Bacteria* and *Archaea*. Indeed, their studies of 18S rRNA from eukaryotic organisms indicated that the *Archaea* are as different from *Bacteria* as they are from eukaryotes. The seminal findings of this work have changed the way microbiologists view taxonomy and phylogeny.

phenotypic tests. For example, consider the cell enve-lope composition of the organisms. Peptidoglycan is found only in *Bacteria*, although two groups—the my-coplasmas and the *Planctomycetales*—lack it. Further-more, the lipids of the *Bacteria* and *Eucarya* are ester-linked, whereas they are ether-linked in the *Archaea* (see Chapters 1 and 18).

At this time, 26 different phylogenetic groups, referred to here as **phyla,** are known. Included are 23 phyla of *Bacteria* and three phyla of *Archaea*. Each of these phyla has specific signature sequences in their ribosomes that are distinctive to them. The prokaryot-ic phyla are listed here along with a brief description of their major features. They are treated in more detail in subsequent chapters (Chapters 18 through 22), and the groups in each chapter are indicated below. It is note-worthy that many new phyla of the *Bacteria*, in partic-ular, have been discovered in natural environments using clone library approaches that have not yet been isolated in pure culture (**Box 17.5**). Thus, at least ten to 15 additional major prokaryotic groups are very poor-ly understood.

The second edition of *Bergey's Manual of Systematic Bacteriology* has been largely followed in organizing the bacterial groups treated in this book. The *Archaea* are treated in Chapter 18 and the *Bacteria* are treated in Chapters 19 through 22 (**Table 17.3**).

Domain: *Archaea*

The *Archaea* are divided into the following three phylo-genetic groups or phyla.

Crenarchaeota   This phylum contains the most ther-mophilic organisms known. Some of these organisms grow at temperatures above the boiling point of water. Most rely on sulfur metabolism either as an energy source or as an electron sink. For example, some oxidize reduced sulfur compounds aerobically to produce sulfuric acid. Others reduce elemental sulfur and use it as an electron acceptor to form hydrogen sulfide. Some are iron and manganese reducers. Not all organisms are thermophilic. They are also significant in deep-sea environments as well as in polar seas.

Euryarcheota   The methanogens (methane producers) are noted for their ability to pro-duce methane gas from simple carbon sources. Some use carbon dioxide and hydrogen gas, whereas others use methanol or acetic acid. These bacteria are anaerobes, some of which grow at the lowest redox potentials of all prokaryotes. Some of these bacteria fix carbon dioxide, but they use neither the Calvin-Benson cycle nor the reductive TCA cycle.

Some of the *Euryarcheota* are hyperthermophilic. Extreme halophiles make up another phenotypic subgroup. These extremely halophilic bacteria grow only in saturated salt-brine solutions. They lyse if they are placed in distilled water.

It should be noted that there is an overlap between the extreme halophiles and methanogens. Thus, some species of methanogens grow in high-salt environments.

Korarcheota   These organisms have been found in hot springs, but no strains have yet been isolated in pure cul-ture, so little is known about their phenotypic properties.

Domain: *Bacteria*

Proteobacteria   The *Proteobacteria* comprises a very large and diverse group of organisms. All four of the major bacterial nutritional types are represented within this group. Some of these organisms are photosynthetic (treated in Chapter 21), whereas some are heterotrophic and others are chemolithotrophic. The chemolithotrophic bacteria include the nitrifiers, the thiobacilli, the fila-mentous sulfur oxidizers (*Beggiatoa* and related genera), and many species that grow as hydrogen autotrophs. Carbon dioxide fixation, when present, is via the Calvin-Benson cycle in all members of this phylum.

This phylogenetic group contains many of the well-known gram-negative heterotrophic bacteria such as *Pseudomonas*, the enteric bacteria including *E. coli*, *Vibrio* and luminescent bacteria, and the more morphologi-cally unusual bacteria such as the prosthecate bacteria. In addition, many symbiotic genera such as *Agro-bacterium*, *Rickettsia*, and *Rhizobium* are members of this group.

The gram-negative bacterial sulfate reducers such as *Desulfovibrio* are also found in this group. Also included in this phylum are the exotic myxobacteria that form fruiting structures as well as unicellular, nongliding forms such as the bacterial predator *Bdellovibrio*.

Finally, the mitochondrion present in almost all eukaryotes evolved from this group of bacteria.

Table 17.3	Overall organization for treatment of *Bacteria* and *Archaea*	
Bacterial Group	**This Book**	**Bergey's Manual of Systematic Bacteriology**[a]
Archaea	Chapter 18	Volume 1
Proteobacteria	Chapter 19	Volume 2
Gram-positive bacteria	Chapter 20	Volumes 3 and 4
Phototrophic bacteria	Chapter 21	Volume 1
Other bacterial phyla	Chapter 22	Volume 5

[a]For more complete treatment of organization of taxa see *Bergey's Manual of Systematic Bacteriology* (2nd ed.).

Research Highlights Box 17.5 **Novel Phyla Discovered by Molecular Analyses of Natural Habitats**

One of the major advances in exploring the diversity of microorganisms in natural environments has been the application of molecular approaches developed in Norman Pace's laboratory. In the most recent variation of this approach, DNA is extracted from the environment of interest. Then the PCR is used to amplify genes of interest from the DNA. For phylogenetic (i.e., diversity) information, 16S rDNA primers for the *Bacteria* or *Archaea* or universal primers that amplify both groups are commonly used. The segments retrieved, typically about 500 bp, can then be sequenced to identify the phylogenetic groups. Although PCR approaches are not quantitative, the various 16S rDNA types retrieved from a natural sample can provide important information about the diversity of prokaryotes that occur in that environment.

Using these approaches, it has recently been determined that at least 30 to 40 major phyla of *Bacteria* exist, yet isolates have been obtained of only 23 (Hugenholz et al., 1998).

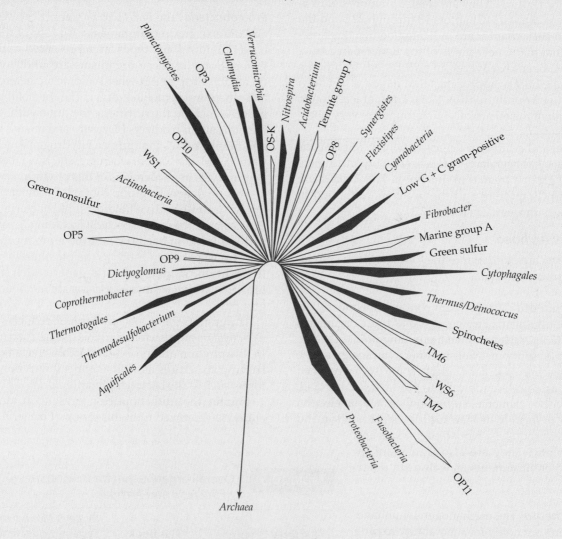

A phylogenetic tree of 16S rDNA sequences of *Bacteria,* based on pure cultures and clonal libraries from natural samples. Note the existence of many phyla (shown in outline rather than as solid black lines) that have not yet been cultivated. Courtesy of Phil Hugenholz and ASM Publications (Hugenholz, P., B. M. Goebel and N. R. Pace. 1998. *J. Bacteriol.* 180:4765-4774).

Firmicutes and *Actinobacteria*

Firmicutes The bacteria in this group are all gram-positive, although the *Mycoplasma* group lacks a cell wall altogether and therefore stains as gram-negative. All other members of the group contain large amounts of peptidoglycan in their cell wall structure.

The *Firmicutes* are unicellular organisms that have a low mol % G + C content. Most are cocci or rods, and some produce endospores. *Bacillus* are aerobic or facultative spore formers, whereas *Clostridium* species are anaerobic fermenters. Some are sulfate reducers. One group, the heliobacteria, are photosynthetic, and members produce a unique form of bacteriochlorophyll, Bchl *g*.

Actinobacteria These gram positive bacteria range in shape from unicellular organisms to branching, filamentous, mycelial organisms. Most are common soil organisms, some of which produce specialized dissemination stages called conidiospores that enable them to survive during dry periods.

Chloroflexi This group contains the genus *Chloroflexus*, a green gliding bacterium that is very metabolically versatile. Members can grow as heterotrophs or photosynthetically. Carbon dioxide is not fixed by either the Calvin-Benson cycle or the reductive TCA cycle but by a special pathway known only for this group of bacteria.

Chlorobi The green sulfur bacteria are anoxygenic photosynthetic bacteria. Some are unicellular forms, and others produce networks of cells. None are motile by flagella or gliding motility. Some have gas vacuoles. They use the reductive tricarboxylic acid (TCA) cycle rather than the Calvin-Benson cycle to fix carbon dioxide.

Cyanobacteria The cyanobacteria are the only bacteria that carry out oxygenic photosynthesis. This is a diverse group of bacteria ranging from unicellular to multicellular filamentous and colonial types. Some grow in association with higher plants and animals. All cyanobacteria use the Calvin cycle for carbon dioxide fixation. The chloroplast found in all eukaryotic photosynthetic organisms evolved from this group of bacteria.

Aquificae These bacteria are hydrogen autotrophs. This phylogenetic group contains the most thermophilic member of the *Bacteria* known and makes up one of the deepest branches of the *Bacteria*.

Thermotogae *Thermotoga* is a fermentative genus that contains some of the most thermophilic members of the *Bacteria* known. They grow at temperatures from 55°C to 90°C. Their cell lipids are unusual.

Thermomicrobia The genus *Thermomicrobium* contains small, rod-shaped thermophiles that grow as heterotrophs in hot springs with an optima temperature for growth of 70°C to 75°C. The cell wall contains very low amounts of diaminopimelic acid.

Thermodesulfobacteria This is a group of thermophilic sulfur-reducing bacteria.

Deinococcus-Thermus This is a very small group of organisms currently represented by very few genera. The genus *Deinococcus* contains gram-positive bacteria. However, they differ from other gram-positive bacteria in showing strong resistance to gamma radiation and ultraviolet light. *Thermus* contains thermophilic, rod-shaped bacteria. Ornithine is the diamino acid in the cell walls of both *Thermus* and *Deinococcus*.

Bacteroidetes This is a diverse group containing heterotrophic aerobes and anaerobes. Some are gliding heterotrophic bacteria that have a low DNA base composition (about 30 to 40 mol % G + C).

Planctomycetes The *Planctomycetes* group of *Bacteria* are budding, unicellular, or filamentous bacteria. These bacteria lack peptidoglycan.

Chlamydiae The *Chlamydiae* are a group of obligately intracellular parasites and pathogens whose closest relatives are the *Planctomycetales*. They also lack peptidoglycan.

Verrucomicrobia These bacteria are unusual in that some members have bacterial tubulin genes. Very few representatives of this phylum have been isolated in pure culture, although they comprise up to 3% of the microbiota from soils.

Spirochaetes The spirochetes are morphologically distinct from other bacteria. Their flexible cells are helical in shape. All are motile due to a special flagellum-like structure, the axial filament, not found in other bacteria.

Fibrobacteres These are anaerobic bacteria, some of which live in the gastrointestinal tracts of animals. Some are cellulose degraders.

Acidobacteria These bacteria are commonly found in soils and sediments, but few strains have been cultivated. In addition to the aerobic genus *Acidobacterium*, this phylum contains homoacetogenic bacteria, *Holophaga*, and iron-reducing bacteria in the genus *Geothrix*.

Fusobacteria These obligately anaerobic bacteria are commonly found in the oral cavities and intestinal tracts of animals.

Dictyoglomi Species of this group are thermophilic, obligately anaerobic fermentative bacteria.

IDENTIFICATION

The final area of taxonomy is identification. Bacteriologists are often confronted with determining the species of a newly isolated organism. Clinical microbiologists need to know whether a specific pathogenic bacterium is present so they can properly diagnose a disease. Food microbiologists need to determine whether *Salmonella*, *Listeria*, or other potentially pathogenic bacteria are present in foods. Dairy microbiologists need to keep their important lactic acid bacteria in culture in order to produce a uniform variety of cheese. Brewers and wine makers need to keep their cultures pure to inoculate the proper strains for quality control of their fermentations. Analysts at water treatment plants need to make sure that their chlorination treatment is effective in killing coliform bacteria in the treated water and distribution systems. Microbial ecologists need to identify bacteria that are responsible for important processes such as pesticide breakdown and nitrogen fixation.

The process of identification first assumes that the bacterium of interest is one that has already been described and named. This is the usual case for most clinical specimens or specimens from known fermentations. However, microbial ecologists often find that the organism they are interested in is new. It is estimated that less than 1% of prokaryotic species have been isolated, studied in the laboratory, and named. Therefore, it is not always possible to identify a bacterium that has been isolated from an environmental sample.

Phenotypic Tests

Phenotypic tests based on readily determined characteristics are often used to identify a species. Most are simple to perform and inexpensive. Furthermore, the amount of time and equipment required for conducting genotypic tests such as DNA/DNA reassociation preclude their use in routine diagnosis. By performing a battery of some 10 to 20 simple phenotypic tests, it is often possible to determine the genus, and perhaps even the species, of a clinically important bacterium, although some taxa are much more difficult to identify than others.

Traditional methods for identification require growing the organism in question in pure culture and performing a number of phenotypic tests. For example, if one wishes to identify a rod-shaped bacterium, the first test would be a Gram stain. If the organism is a gram-negative rod, the next questions to ask include the following: Is it motile? Is it an obligate aerobe? Is it fermentative? Does it have catalase? Can it grow using acetate as a sole carbon source? The answers to these questions will direct the investigator toward the next step in identification. On the other hand, if it is a gram-positive rod, then a completely different set of tests would need to be performed such as a test for endospore formation.

These tests take time to perform, and the appropriate tests for one genus of bacteria differ from that of others. Thus, the results of one set of tests will determine which tests will need to be performed for further clarification of the taxon. Sometimes several weeks might be required to conduct all the tests needed to identify a strain. Rapid tests are extremely helpful, especially in the medical, food, and water-testing areas. Fortunately, standardized, routine tests can be performed for most clinically important bacteria that allow for their rapid identification. Several companies have now produced commercial kits that are helpful in assisting in the identification of unknowns.

An increasingly popular approach to identification of bacterial unknowns involves characterization of their fatty acids. The fatty acids are found in membrane lipids and are readily extracted from the cells and analyzed. Different species of *Bacteria* produce different types and ratios of fatty acids. The *Archaea*, of course, do not produce fatty acids, so this procedure is not of value for their identification. However, they do produce characteristic lipids that are useful taxonomically, especially with the halobacteria.

The fatty acid analysis procedure involves hydrolyzing a small quantity of cell material (about 40 mg is all that is needed) and saponifying it in sodium hydroxide. This is acidified with HCl in methanol so that the fatty acids can be methylated to form methyl esters. The fatty acid methylated esters (FAME) are then extracted with an organic solvent and injected in a gas chromatograph. The resulting chromatograph (**Figure 17.11**) can be used to identify the fatty acids that are indicative of a species. Commercial firms have developed databases of fatty acid profiles that can be used for the identification of species. The advantage of this procedure is that many samples can be analyzed quickly and without great effort. However, all organisms must be grown under controlled conditions of temperature and length of incubation and on the same medium.

Nucleic Acid Probes and Fluorescent Antiserum

One exciting current area of research and commercial application involves the development of DNA or RNA

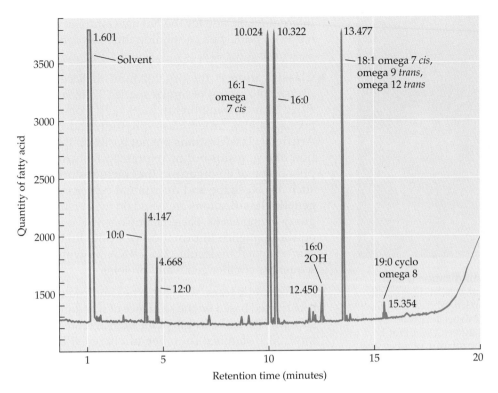

Figure 17.11 Fatty acid analysis
Fatty acid methyl ester (FAME) chromatogram of an unknown species, showing chromatographic column retention times and peak heights. Note: 10:0, 12:0, 16:0, and 19:0 indicate saturated fatty acids with 10, 12, 16, and 19 carbons; 16:1 and 18:1, monounsaturated 16-carbon and 18-carbon fatty acids; omega number, the position of the double bond relative to the omega end—that is the hydrocarbon end (not the carboxyl end)—of the fatty acid chain; *cis* and *trans,* the configuration of the double bond. For example, omega 7 *cis* indicates a *cis* double bond between the seventh and eighth carbons from the omega end of the fatty acid. Also, 2OH indicates a hydroxyl group at the second carbon from the omega end; cyclo omega 8, a cyclo-carbon at the eighth position from the omega end. The 18:1 omega 7 *cis*, omega 9 *trans*, and omega 12 *trans* peak results from either one fatty acid or a mixture of fatty acids with double bonds at the three positions indicated (the chromatographic column does not separate these three fatty acids). Courtesy of MIDI (Microbial Identification, Incorporated, Delaware).

"probes" that are specific for the signature sequences of rRNA or some other appropriate gene such as an enzyme that is characteristic of a species of interest. The probes are labeled in some manner so that the hybridization can be visualized. This is accomplished either by making them radioactive or by tagging them to a fluorescent dye or an enzyme that gives a colorimetric reaction.

An example of this process is shown in **Figure 17.12**. In this case, the procedure, which is called fluorescent in situ hybridization or FISH, entails the use of a fluorescent tag on a signature nucleotide sequence. This is then hybridized to a sample whose cells have been made permeable to the fluorescent tag. By the proper selection of probes, it is possible to identify an organism to a domain, genus, or species by demonstrating specific hybridization to the probe.

Potential applications for probe technology are considerable. A number of commercial firms are already marketing probes to identify pathogenic bacteria from clinical samples. A goal of this technology is to enable the rapid identification of organisms directly from clinical or environmental samples without actually growing them in culture first. Fluorescent antiserum tests are also useful. For example, *Legionella* spp., the causative agents of Legionnaire's disease, are very difficult to cultivate. However, good fluorescent antisera are available for the identification of these species directly from clinical samples or even from environmental samples.

Culture Collections

Unlike plants and animals, most bacteria can be easily grown in pure culture and preserved by **freeze-drying (lyophilization),** handled in small test tubes and vials, and readily sent anywhere in the world. As lyophils, many remain viable for 10 to 20 years or more and can be revived and studied by anyone anywhere. Cultures can also be frozen at −80°C in vials containing a suspension medium amended with 15% glycerol. These

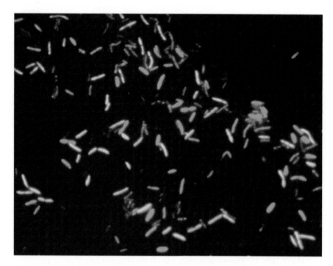

Figure 17.12 Fluorescent in situ hybridization (FISH)
This smear contains cells of *E. coli*, *Bordetella*, and *Legionella* hybridized with a bacterial probe, Eub338, labeled with fluoroscein (green) and a *Legionella micdadei* specific probe labeled with Cy3 (red). All bacteria are green except for *L. micdadei* which stains yellow due to the mixture of green and red probe dyes. Courtesy of Jinxin Hu and Thomas Fritsche.

remain viable for many years. Thus, unlike plants, for which an herbarium is used to preserve the specimens collected of an original species, bacteria are preserved as **type cultures** that are clones of the original viable **type strain** of a species. *The type strain of a species is the one on which the species definition has been based.* Through **culture collections** type strains are made available to professional microbiologists throughout the world. Many countries maintain national collections of microorganisms, such as the American Type Culture Collection (ATCC) in the United States (www.atcc.org). If a microbiologist from India wishes to determine if he or she has a new species, the original type strain can be obtained from a culture collection and used to conduct DNA/DNA reassociation assays and other tests to compare it with his or her own isolates.

Because of the importance of biological materials to science and industry, culture collections are rapidly becoming biological resource centers. Thus, strains that have been patented are also deposited in culture collections so that they are accessible. Clones of genomes that have been sequenced are also deposited as well as viruses and other biological materials.

SUMMARY

▶ Bacterial **taxonomy** or **systematics** consists of three areas: nomenclature, classification, and identification. **Nomenclature** is the naming of an organism. An International Code for the Nomenclature of Bacteria has been published containing the rules for naming *Bacteria* and *Archaea*.

▶ **Classification** is the organization of *Bacteria* and *Archaea* into groups of similar species. *Bacteria* are classified in increasing hierarchical rank from species, genus, family, order, class, phylum, and domain. **Artificial classifications** are not based on the evolution of organisms but on expressed features or the phenotype of an organism that includes properties such as cell shape and nutritional patterns. **Phylogenetic classifications** are based on the evolution of a group of organisms. Bacterial phylogeny is now based on the sequence information from the highly conserved macromolecule, 16S rRNA as well as the sequences of other genes and proteins.

▶ **Phylogenetic trees,** with branches and roots, can be constructed based on the sequence of macromolecules such as rRNA. The length of the branch represents the inferred difference (number of changes) between organisms. **External nodes** represent extant species, whereas **internal nodes** represent ancestor species. **Rooted** trees are based on a comparison of (a) related species and an outgroup species or (b) the sequence of one macromolecule such as 16S rRNA versus an entirely unrelated conserved macromolecule such as ATP synthase.

▶ **Speciation** is the process whereby organisms evolve. Typically genetic material is transferred from the parental bacterium to the progeny through **vertical inheritance. Horizontal gene transfer** occurs in which genetic material is transferred among bacteria that may or may not be closely related to one another.

▶ A **bacterial species** is a group of similar strains that show at least 70% DNA/DNA hybridization. Organisms of the same species will have similar if not identical DNA mol % G + C content. Organisms that have the same GC ratio are not necessarily similar. Based on molecular criteria, such as DNA/DNA reassociation, bacterial species are much more broadly defined than plant and animal species.

▶ **Identification** is the process whereby unknown cultures can be compared to existing species to determine if they are similar enough to be members of the same species.

▶ **Type strains** of all species must be deposited in at least two different types of **culture collections,** repositories where strains are preserved by lyophilization and deep freezing. The culture collections provide cultures to microbiologists from all over the world so they can compare unidentified strains to the official type strains.

REVIEW QUESTIONS

1. Why is it important to name and classify bacteria?

2. What procedure(s) are necessary to identify a bacterial isolate as a species?

3. Differentiate between an artificial and a phylogenetic classification.

4. In what ways does the classification of *Bacteria* differ from that of eukaryotic organisms?

5. How do the *Archaea* differ from *Bacteria*? From eukaryotes?

6. How is DNA melted and reannealed, and why is this useful in bacterial taxonomy?

7. How would you go about identifying a bacterium that you isolated from a soil habitat?

8. Why is morphology of little use in bacterial classification? Is it of any use?

9. What is *weighting* and should phenotypic features be weighted in a bacterial classification scheme?

10. Distinguish between lumpers and splitters.

11. If you were working in a clinical laboratory, outline the types of procedures you would use to identify isolates. Why do you recommend using the procedures you suggest?

12. Why is ribosomal RNA of use in bacterial classification?

13. Compare the information obtained from determining the DNA base composition (GC ratio) with that obtained by DNA reassociation experiments.

SUGGESTED READING

Garrity, G., D. Boone and R. Castenholz, eds. 2001. *Bergey's Manual of Systematic Bacteriology.* 2nd ed., Vol. 1. New York: Springer-Verlag.

Gerhardt, P., ed. 1993. *Methods for General and Molecular Microbiology.* Washington, DC: American Society for Microbiology.

Graur, D. and W. H. Li. 2000. *Fundamentals of Molecular Evolution.* 2nd ed. Sunderland, MA: Sinauer Associates, Inc.

Holt, J. G., editor-in-chief. 1994. *Bergey's Manual of Determinative Bacteriology.* Baltimore: Williams and Wilkins.

COMPUTER INTERNET RESOURCES

American Type Culture Collection. (http://www.atcc.org/)

This site has a listing of all the bacterial strains deposited in the American Type Culture Collection, as well as growth media and conditions.

Ribosome Database Project (RDP), Michigan State University. (http://rdp.cme.msu.edu/)

This site has information on the 16S rRNA sequences of over 10,000 bacterial strains. The database allows one to conduct phylogenetic analyses of unknown strains whose 16S rRNA sequence has been determined and to compare the sequence with those already reported.

Bergey's Manual Trust. Headquarters at Michigan State University. (http://www.cme.msu.edu/Bergeys/) This website has information on the current classification of *Bacteria* and *Archaea*.

Comparative RNA Web Site. (www.rna.icmb.utexas.edu)

A remarkable collection of RNA sequence information presented with secondary structure models, conservation diagrams, and more. Published as Cannone, J. J. et al., 2002. The Comparative RNA Web (CRW) Site: An Online Database of Comparative Sequence and Structure Information for Ribosomal, Intron, and other RNAs. BioMed Central Bioinformatics, 3:2.

Chapter 18

Archaea

Hence without parents, by spontaneous birth,
Rise the first specks of animated earth . . .
Organic life beneath the shoreless waves
Was born and nursed in ocean's pearly caves;
First, forms minute, unseen by spheric glass,
Move in the mud or pierce the watery mass;
These, as successive generations bloom,
New powers acquire, and large limbs assume;
Whence countless groups of vegetation spring,
And breathing realms of fin and feet and wing.

— ERASMUS DARWIN, FROM "THE TEMPLE OF NATURE," 1802

W hen Carl Woese and his collaborators at the University of Illinois began to study the phylogenetic relationships between prokaryotes during the mid-1970s, they discovered that one group was very different from all other prokaryotes. By sequence analysis, the 16S rRNAs of these prokaryotes were no more closely related to the bacterial 16S rRNAs than to the 18S rRNAs of eukaryotes. They called these different prokaryotes archaebacteria because they seemed to resemble the primitive organisms believed to exist in Archean times 3 to 4 billion years ago. This discovery piqued the interest of microbiologists around the world. After much debate, microbiologists now generally agree that these organisms are very different from other prokaryotes. In recognition of this fact, they currently are classified as *Archaea*, one of the three primary domains of organisms.

WHY STUDY *ARCHAEA*?

Microbiologists studying *Archaea* contribute to our basic knowledge in four areas. First, many *Archaea* are also extremophiles, and they flourish under conditions that are lethal to most *Bacteria* and eukaryotes. In fact, the most extreme thermophiles are *Archaea*. As such, they set one of the boundaries of the biosphere, especially in the deep Earth where the temperature increases with depth. The potential of microorganisms to transform the deep subsurface requires detailed knowledge of the upper temperature limit for life and the physiology of these extreme organisms. Studies of the extremely thermophilic and extremely halophilic *Archaea* also contribute to our understanding of growth mechanisms in these harsh environments. A question of special interest is how the macromolecules of extremophiles maintain their structure and activity under conditions that denature the macromolecules of most other organisms. This question is also of practical importance for the design of commercially valuable thermostable enzymes. The optimal growth temperature of many extremely thermophilic *Archaea* is also above melting temperature of their DNA, even at the salt concentrations of the cytoplasm. How these organisms maintain a functional chromosome is not fully understood.

Second, *Archaea* offer a valuable model system because some of their features closely resemble those of the eukaryotes. Their simpler cell structure and smaller size may prove to be a valuable research tool in understanding the function of the eukaryotic-type of RNA polymerase, DNA replication, and other processes common to both domains.

Third, as one of the most ancient lineages of living organisms, the *Archaea* represent one of the extremes of evolutionary diversity. Most of our knowledge of prokaryotes comes from studies of a relatively small number of model organisms, especially *Proteobacteria* such as *Escherichia coli*. Comparison with the *Archaea* has illustrated the enormous diversity of prokaryotes, even in mechanisms once thought to be universal. For instance, when the genomic sequence of *Methanocaldococcus jannaschii*, the first archaeon sequenced, was completed, only 16 of the 20 aminoacyl-tRNA synthetases believed to be essential for all living organisms were detected. Subsequent studies showed that many *Archaea* possess alternative synthetases and that these synthetases were common in the *Bacteria* as well. Previous studies limited to the *Proteobacteria* and a few eukaryotes had underestimated the extent of aminoacyl-tRNA synthetase diversity. Similarly, the first genomic sequence of an archaeon led to the discovery of a large number of genes that had never been encountered before. Although many of these genes have since been found to be widely distributed in living organisms, some have so far only been found in *Archaea*. Many of these uniquely archaeal genes encode enzymes and other proteins involved in processes unique to the *Archaea* such as methanogenesis. Presumably, those genes whose functions are currently unknown will also encode genes required for the unique aspects of the archaeal lifestyle.

Fourth, the *Archaea* offer valuable insights into the nature of the earliest organisms. Although most biochemical features of microorganisms are absent in the fossil record, these properties can be deduced from modern organisms. By studying modern organisms, we also know that form changes rapidly throughout evolution, but that even very remote ancestors inherit basic biochemical processes. Because the *Archaea* are an extreme in prokaryotic evolution, they offer the opportunity to obtain unique insights into the nature of the earliest prokaryotes. As extremes, they provide a unique picture of the full scope of modern life.

MODELS OF EARLY LIFE

Two general strategies have been used to study early life. In one strategy, abiotic reactions that might have formed precursors of cellular life are studied. One accomplishment of this strategy is the "RNA world" hypothesis (**Box 18.1**). In the second strategy, the phylogeny or ancestry of modern organisms is studied in an attempt to deduce the properties of the last common ancestor. The importance of the discovery of the *Archaea* to this second approach may be illustrated by the following example. First, consider that all modern organisms evolved from two lines of descent, a proposition that we know to be false (**Figure 18.1A**). Properties that are the same in both lines of descent, such as the genetic code, were probably inherited from the common ancestor; however, for different properties, such as the structure of the ribosome, very little may be concluded. For instance, did the 80S eukaryotic ribosome evolve by the addition of complexity to the simpler 70S bacterial ribosome, or did the 70S ribosome evolve by subtraction from and a more efficient construction of the 80S ribo-

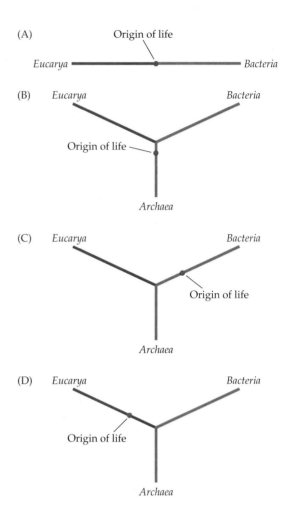

Figure 18.1 Possible models for ancient evolution
Of these four possible models for the origin of life, the one with two lines of descent (A) has been discounted. Only one of the other models (B, C, D), with three lines of descent, can be correct.

some? Either scenario is plausible, and this problem cannot be solved without additional evidence. Now, consider that all modern organisms evolved from three lines of descent that arose very early in life's history, a proposition that is probably true. The location of the origin of life is not known in this scenario, but it is probably close to the point where all three lines intersect and on one of the three branches (that is, not at the center). This scenario may be represented by the three models B, C, and D in Figure 18.1. Only one of these can be correct, but the correct one is not known. Again, properties that are common to all three lines of descent, such as the genetic code, were probably inherited from the common ancestor. But for properties that are different, it is now possible to draw firmer conclusions. For example, both the *Archaea* and the *Bacteria* possess 70S ribosomes, and the eukaryotes possess 80S ribosomes. (The chloroplastic and mitochondrial ribosomes are relatively modern and are not included in this discussion.) Did the com-

---

Research Highlights Box 18.1 **RNA World**

Proponents of an "RNA world" believe that RNA was the first macromolecule to evolve. A number of arguments favor an early form of "life" based on RNA. For example, RNA, like DNA, can store genetic information—as it does in RNA viruses. Furthermore, unlike DNA, some RNA molecules, called ribozymes, are catalytic and carry out biochemical reactions, as do enzymes. For example, the enzyme ribonuclease P is a protein whose active site has an RNA moiety that cleaves transfer RNA. In addition, in protein synthesis, it is the ribosomal RNA, not ribosomal protein, that is responsible for the formation of peptide bonds. Indeed, RNA is essential to protein synthesis and therefore probably existed before protein.

Proponents of an RNA world visualize a simple protoorganism that could carry out primitive metabolic activities. It contained RNA in its cytoplasm that was capable of catalytic and templating (reproductive) activities. A simple, probably inorganic barrier separated the cytoplasm and RNA from the environment. Later, proteins evolved, synthesized from the RNA template. Later proteins evolved, synthesized from the RNA template. Because RNA is chemically unstable, DNA itself may have evolved as a means of storing genetic information that was more stable and less prone to mutation. Increasingly, an RNA world is being accepted by biologists as a logical step toward the evolution of cellular organisms.

---

mon ancestor contain 70S or 80S ribosomes? If models B and C are correct, then it is very likely that the common ancestor contained a 70S ribosome. If model D is correct, then it is likely that the ancestor of the *Bacteria*, and the *Archaea* contained a 70S ribosome, but the properties of the common ancestor with the eukaryotes are still not determined. Thus, even without detailed knowledge of the early events, the presence of three lineages introduces new constraints on our models of early life. This example also illustrates the importance of the location of the origin of life on our interpretation of the properties of modern organisms. Although current evidence seems to suggest that the origin of life was on the bacterial lineage (model C), this evidence is by no means conclusive.

DIVERSITY OF *ARCHAEA*

In addition to being very different from other organisms, the *Archaea* are very different from each other. In fact, the variability within the *Archaea* is probably on the same order of magnitude as that found in the *Bacteria* even though many fewer *Archaea* have been identified. For instance, many of the morphological types common in *Bacteria* are also found in the *Archaea* (**Figure 18.2**). Two explanations for this current distribution seem plausible. Only a few investigators have systematically tried to isolate *Archaea*. Because little effort has been expended in this area, few have been found. In fact, surveys of natural populations that do not rely on culture techniques have discovered large populations of *Archaea* in the oceans, soil, and subsurface. Although it is not known at this time what fraction of the prokaryotes on earth are *Archaea*, it is clear that it is some significant proportion. The *Archaea* appear to be limited to extreme environments or to niches for which there is little direct competition with the *Bacteria*. Even the methanogens, which are abundant in temperate environments, possess a unique energy metabolism and do not compete with *Bacteria* and eukaryotes. It is possible that the *Archaea* are unable to compete successfully in the temperate environments found on most of the modern earth. Their numbers may be small because they may have been restricted to a few habitats. However, their diversity may be large because they are an ancient line of descent. Thus, there has been adequate time for their ancestors to specialize and form novel lineages.

BIOCHEMICAL DIFFERENCES AND SIMILARITIES TO OTHER ORGANISMS

The *Archaea* are composed of three major phenotypic groups. These include the methane-producing *Archaea*, the extreme halophiles, and the extreme thermophiles. Although the physiology of these *Archaea* is very different, some features are common to most if not all *Archaea* (**Table 18.1**). The cell membranes are composed of isoprenoid-based glycerol lipids. Murein is absent in their cell walls, and a protein envelope usually replaces it. The enzyme DNA-dependent RNA polymerase, which copies DNA to form messenger RNA, is different from the bacterial type. The proteins in DNA replication are more similar to the eukaryotic than bacterial homologs. Lastly, *Archaea* are insensitive to many common antibiotics that are potent inhibitors of *Bacteria* or eukaryotes.

Archaeal Membranes

Like the membrane lipids of the *Bacteria*, the archaeal lipids are composed of hydrophobic side chains linked to glycerol. However, many of the details are different

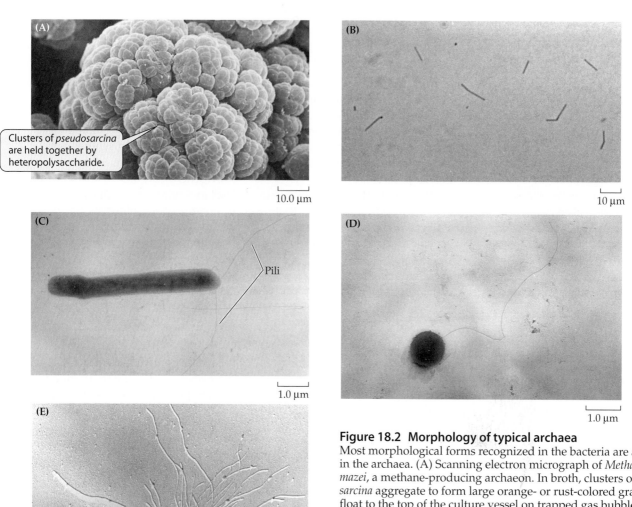

Clusters of *pseudosarcina* are held together by heteropolysaccharide.

10.0 μm

10 μm

Pili

1.0 μm

1.0 μm

Flagellar bundle

1.5 μm

Figure 18.2 Morphology of typical archaea
Most morphological forms recognized in the bacteria are also seen in the archaea. (A) Scanning electron micrograph of *Methanosarcina mazei*, a methane-producing archaeon. In broth, clusters of *pseudosarcina* aggregate to form large orange- or rust-colored grains that float to the top of the culture vessel on trapped gas bubbles. (B) Phase contrast micrograph of *Caldivirga maquilingensis*, a rod-shaped hyperthermophile isolated from a hot spring in the Philippines. This strict anaerobe grows best at 85°C. (C) Electron micrograph of *C. maquilingensis* shadowed with platinum/palladium. (D) Electron micrograph of a novel, extremely thermophilic archaeon of the family *Thermoproteaceae*, isolated from a hot spring in Japan. (E) Platinum-shadowed electron micrograph of *Pyrococcus furiosus*, a heterotrophic coccus with temperature optimum near 100°C. A, photo ©Ralph Robinson/Visuals Unlimited; B–D, courtesy of Dr. T. Itoh, Japanese Culture Collection; E, courtesy of K. O. Stetter and R. Rachel.

(**Figure 18.3**). Instead of fatty acids joined by an ester linkage of glycerol, most archaeal lipids are composed of isoprenoid side chains joined by an ether linkage to glycerol. Isoprenoids are branched-chain alkyl polymers based on a five-carbon unit synthesized from mevalonate. In *Bacteria* and eukaryotes, they are found in the side chains of quinones and chlorophyll, intermediates in the biosynthesis of sterols like cholesterol, and in rubber. They are never major components of glycerol lipids. Although branched-chain fatty acids are sometimes found in bacterial lipids, they are not synthesized from mevalonate. Instead, they are usually synthesized from the branched-chain amino acids or volatile fatty acids. In *Archaea*, the ether linkage of the isoprenoid side chains to glycerol is also distinctive. The substituted glycerol contains one asymmetrical carbon. Therefore, it has two stereoisomers (**Figure 18.4**). In the *Archaea*, only the stereoisomer called 2,3-*sn* glycerol is found. The glycerol lipids of the *Bacteria* and the eukaryotes contain only the other stereoisomer of glycerol, 1,2-*sn* glycerol. Although a few *Bacteria* contain ether-linked fatty acids, they have the typical bacterial stereochemistry. Therefore, the biosynthesis of these ether linkages may be very different in the *Bacteria* and the *Archaea*.

Table 18.1 Comparison of *Archaea* to *Bacteria* and *Eukaryotes*

	Archaea	Bacteria	Eucaryotes
Typical organisms	Methane-producing archaea, haloarchaea, extreme thermophiles	Enteric bacteria, cyanobacteria, *Bacillus*, etc.	Fungi, plants, animals, algae, protozoa
Typical size	1–4 μm	1–4 μm	>5 μm
Physiological features	Aerobic and anaerobic; mesophiles, extremely thermophilic and halophilic	Aerobic and anaerobic, chlorophyll-based photosynthesis	Largely aerobic, chlorophyll-based photosynthesis
Genetic material	Small circular chromosome, plasmids, and viruses; genome associated with histones	Small circular chromosome, plasmids, and viruses; no histones	Complex nucleus with more than one large linear chromosome, viruses, genome associated with histones
Differentiation	Frequently unicellular, cellular differentiation infrequent	Frequently unicellular, cellular differentiation infrequent	Unicellular and multicellular, cellular differentiation common
Cell wall	Protein, glycoprotein, pseudomurein, wall-less	Murein and LPS, protein and wall-less forms rare	Great variety, peptidoglycan absent
Cytoplasmic membrane	Glycerol ethers of isoprenoids, site of energy biosynthesis	Glycerol esters of fatty acids, site of energy biosynthesis	Glycerol esters of fatty acids, sterols common
Intracytoplasmic membranes	Generally absent	Generally absent; when present they frequently contain large amounts of protein	Common in organelles like mitochondria and chloroplasts, nucleus, Golgi apparatus, endoplasmic reticulum, and vacuoles; site of energy biosynthesis
Protein synthesis	70S ribosome, insensitive to chloramphenicol and cycloheximide, diphthamide present in elongation factor	70S ribosome sensitive to chloramphenicol, insensitive to cycloheximide, diphthamide absent in elongation factor	80S and 70S (organelle) ribosomes, insensitive to chloramphenicol (80S), sensitive to cycloheximide (80S), diphthamide present in elongation factor
Locomotion	Simple flagella	Simple flagella, gliding	Complex flagella, cilia, legs, fins, wings
RNA polymerase	Complex	Simple	Complex

Archaeal Cell Walls

The cell walls of the *Archaea* are different from the gram-negative and gram-positive cell wall types common in the *Bacteria*. Cell walls composed of murein are never found. Instead, the cell walls are usually composed of S-layers: protein subunits arranged in a regular array on the cell surface. Frequently, polysaccharides are also found associated with the envelope, although in many cases their structures are not known. The S-layers from many *Archaea* are very sensitive to detergents like SDS (sodium dodecyl sulfate), which solubilizes proteins.

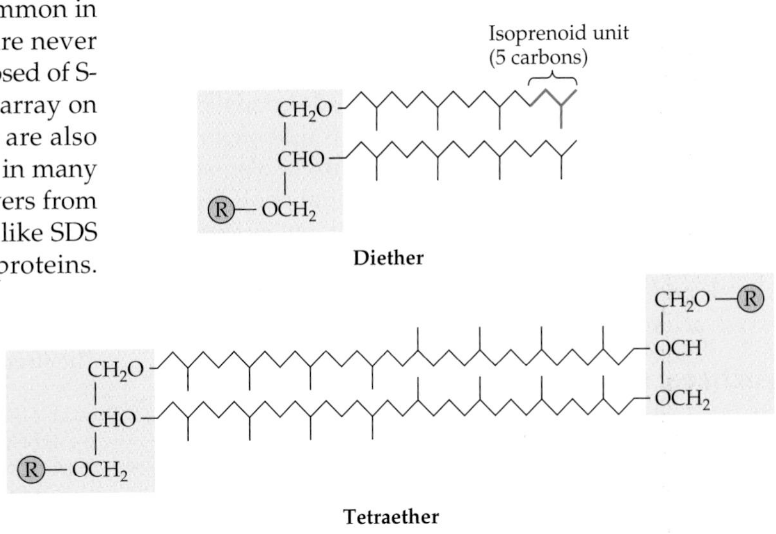

Figure 18.3 Archaeal glycerol lipids
Typical archaeal glycerol lipids contain isoprenoid side chains linked to glycerol by an ether bond, forming diethers or tetraethers. The tetraether is believed to span the membrane. (See Figure 4.36.)

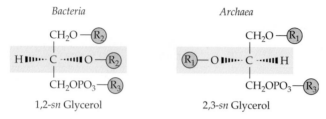

Figure 18.4 Stereochemistry of glycerol lipids in bacteria and archaea
The glycerol lipids of bacteria and archaea have different stereochemistry. Bacterial lipids contain 1,2-*sn* glycerol; archaeal lipids contain 2,3-*sn* glycerol. R_1, ether-linked isoprenoid side chain; R_2, ester-linked fatty acid; R_3, hydrophilic amino acid or sugar.

The cells of these *Archaea* also lyse rapidly in solutions containing low concentrations of detergents. Although S-layers are common in *Bacteria*, they are usually only one component of a complex envelope and are found outside the murein layer. Some exceptions to this generality are the budding *Bacteria* of the genera *Planctomyces* and *Pasteuria* and the thermophilic anaerobe *Thermomicrobium roseum*. In these *Bacteria*, the S-layer is the major cell wall component.

The cell envelopes of some *Archaea* are also similar to the cell envelopes of *Bacteria*. *Thermoplasma* is an archaeon that lacks a cell wall, as do the bacterial mycoplasmas. Some methanogens, such as *Methanobacterium*, contain a cell wall polymer called pseudomurein, which is strikingly similar to murein, the peptidoglycan in *Bacteria*. Pseudomurein is composed of polysaccharides cross-linked by amino acids much like murein (**Figure 18.5**). However, the polysaccharide is composed of *N*-acetylglucosamine (or *N*-acetylgalactosamine) and *N*-acetyltalosaminuronic acid. Muramic acid, the common saccharide component of murein, is never found. Moreover, D-amino acids, which are common in bacterial peptidoglycan, are not found. In other respects, many of the chemical and physical properties of pseudomurein and murein are similar. Both are resistant to proteases, or enzymes that hydrolyze peptide bonds. Both provide the cell with a rigid sacculus. In addition, many of the *Archaea* that contain pseudomurein also stain gram-positive. The archaeal and bacterial peptidoglycans appear to be an example of convergent evolution, where two prokaryotic lineages have independently developed similar solutions to the problem of how to make an enzyme-resistant sacculus.

Archaeal RNA Polymerase

The enzyme DNA-dependent RNA polymerase synthesizes messenger RNA, which is complementary to the DNA sequence of genes. Because this function is essential to all living cells, this enzyme is believed to be very ancient. The bacterial enzyme is relatively simple in structure, and it consists of a core enzyme with three subunits, β, β', and α. Binding of the core enzyme to DNA also requires an additional subunit, called the sigma factor (σ). In contrast, the eukaryotic polymerase is much more complex and contains 9 to 12 subunits. The archaeal enzyme more closely resembles the eu-

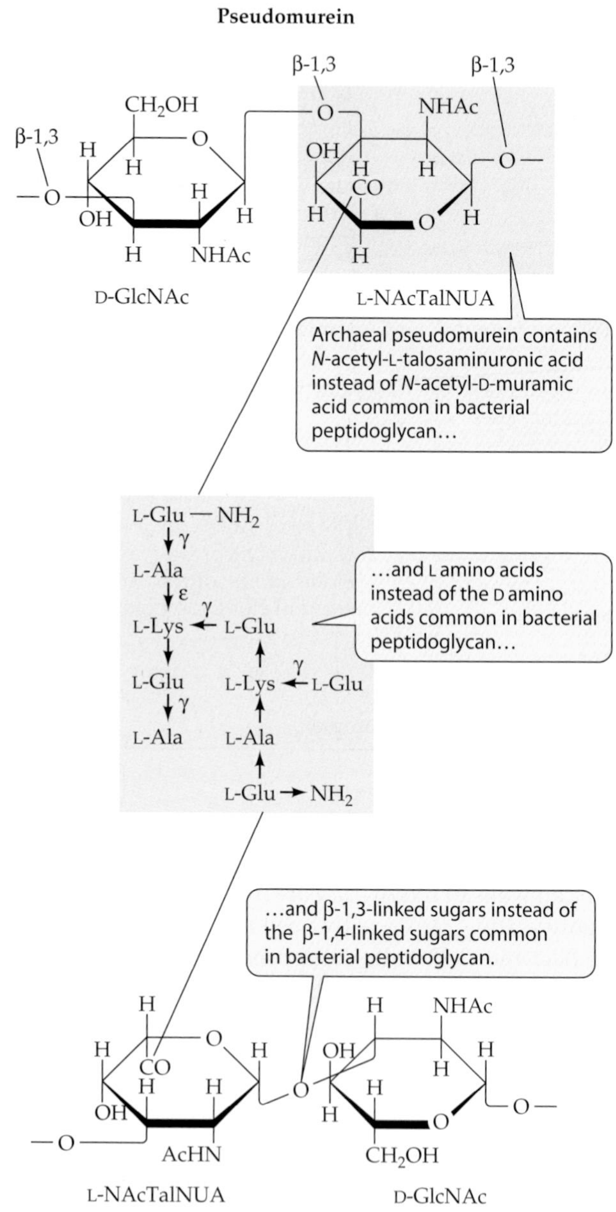

Pseudomurein

Archaeal pseudomurein contains *N*-acetyl-L-talosaminuronic acid instead of *N*-acetyl-D-muramic acid common in bacterial peptidoglycan…

…and L amino acids instead of the D amino acids common in bacterial peptidoglycan…

…and β-1,3-linked sugars instead of the β-1,4-linked sugars common in bacterial peptidoglycan.

Figure 18.5 Pseudomurein of *Archaea*
The structure of pseudomurein of the archaeon *Methanobacterium* compared with the peptidoglycan (murein) of bacteria. L-NAcTalNUA, *N*-acetyl-L-talosaminuronic acid; D-GlcNAc, *N*-acetyl-D-glucosamine. All Greek letters on the peptides indicate the type of peptide bond (other than a linkage). Modified from O. Kandler and H. Konig. 1985. In C. R. Woese and R. S. Wolfe, eds. *The Bacteria*, Vol. 8, pp. 413–457.

karyotic polymerase than the bacterial enzyme. It has a complex subunit structure, and the amino acid sequence of some of the subunits closely resembles that of the eukaryotic enzyme. As one might expect, the polymerases of *Archaea* as well as eukaryotes both initiate replication at promoters containing similar DNA sequences called TATA-boxes. The promoters of *Bacteria* are very different.

DNA Replication

In all living organisms, DNA replication is a complex process requiring binding of initiation proteins at the origin prior to the recruitment of DNA polymerase and other proteins required for elongation. *Archaea* appear to follow the same general pathway found in the eukaryotes and *Bacteria*. However, many of the proteins involved are clearly homologous to the eukaryotic and not the bacterial proteins.

Antibiotic Sensitivity of *Archaea*

Because the biochemistry of the *Archaea* is very different from other organisms (Table 18.1), it might be expected that the sensitivity to antibiotics would also be different. This is true. For instance, because the *Archaea* do not contain murein, they are not sensitive to most antibiotics that inhibit bacterial cell wall synthesis. Thus, most *Archaea* are not affected by very high concentrations of penicillin, cycloserine, vancomycin, and cephalosporin, all of which are inhibitors of murein synthesis. Because of this selectivity, these antibiotics are also useful additions to enrichment cultures for *Archaea*, where they prevent the growth of bacterial competitors. Likewise, the DNA-dependent RNA polymerase from *Archaea* is not inhibited by rifampicin, which inhibits the bacterial enzyme at low concentrations. Archaeal protein synthesis is also not affected by the common antibiotics chloramphenicol, cycloheximide, and streptomycin, although neomycin is inhibitory at high concentrations. Tetracycline is also a very poor inhibitor, even though it inhibits protein synthesis in both *Bacteria* and eukaryotes. These results suggest that the structure of the archaeal ribosome is very different from the bacterial and eukaryotic ribosome.

Similarities to Other Organisms

Equally interesting are the similarities between the *Archaea* and other organisms. For instance, even though the difference is more pronounced between the archaeal and bacterial 16S rRNAs than between any two bacterial 16S RNAs, archaeal and bacterial 16S rRNAs are still identical at 60% of their positions. Likewise, the genetic code is essentially the same in all organisms, many of the major anabolic pathways are the same, and many of the same coenzymes and vitamins are found in all organisms.

In conclusion, the differences found in the *Archaea* are a matter of degree, and a major challenge facing modern microbiologists is to understand the significance and evolution of this diversity. An illustration of the types of questions this understanding will enable us to address is the significance of "horizontal" evolution, or gene transfer between distantly related organisms, in bacterial evolution. In eukaryotes, the mitochondrion and chloroplast are two well-documented cases of horizontal evolution. Did similar events occur in prokaryotic evolution? Currently, the evidence is mixed. Certain properties such as methanogenesis and chlorophyll-based photosynthesis are found solely in the archaeal and the bacterial domains, respectively. For these properties, horizontal evolution probably did not occur between domains. However, with the availability of genomic sequences in prokaryotes, many examples of horizontal gene transfer of small numbers of genes have been observed between the *Archaea* and the *Bacteria* as well as within the *Bacteria*. Only one example of a potentially massive horizontal gene transfer, as massive as may have occurred in the formation of the mitochondrion or chloroplast, has been observed. In this case, the transfer seemed to be from a hyperthermophilic archaeon related to *Pyrococcus* to the ancestor of the hyperthermophilic bacterium *Thermotoga*. To fully understand the significance of these transfers, more information is needed about the function of the genes transferred and the frequency of genetic transfers.

MAJOR GROUPS OF *ARCHAEA*

As mentioned previously, the *Archaea* contain three major phenotypic groups: the methane-producing *Archaea* or the methanoarchaea, the extreme halophiles or haloarchaea, and the extreme thermophiles. The phylogenetic relationships between these *Archaea* are not simple (**Figure 18.6**). The cultivated *Archaea* are divided into two phyla on the basis of their rRNA structure. The cultivated members of one phylum, the *Crenarchaeota*, include only extreme thermophiles. The second phylum, the *Euryarchaeota*, contains all the methanoarchaea. The extreme halophiles are closely related to each other and to one branch of the methanogens within this group. Some of the extreme thermophiles also appear as deep branches of this second group. Thus, both the methanogens and the extreme halophiles are related phylogenetically as well as phenotypically. In contrast, the extreme thermophiles are not a closely knit phylogenetic group.

In addition, a number of deep phylogenetic lineages of *Archaea* probably exist that have never been cultivated. At least on the basis of rRNA sequences, some of these lineages, such as the *Korachaeota*, may represent novel phyla. Other lineages of special interest include a large group of *Crenarchaeota* that have only been detect-

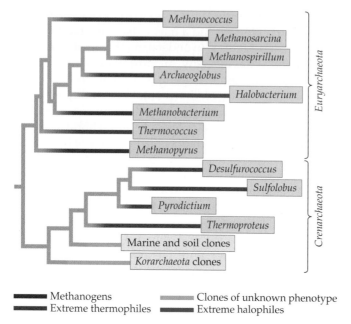

Methanococcus

Methanosarcina

Methanospirillum

Archaeoglobus

Halobacterium

Methanobacterium

Thermococcus

Methanopyrus

Euryarchaeota

Desulfurococcus

Sulfolobus

Pyrodictium

Thermoproteus

Marine and soil clones

Korarchaeota clones

Crenarchaeota

▬▬▬ Methanogens   ▬▬▬ Clones of unknown phenotype
▬▬▬ Extreme thermophiles   ▬▬▬ Extreme halophiles

Figure 18.6 Major groups of Archaea
Phylogeny of the major groups of archaea based on the sequences of 16S rRNA. The major groups of the methanogens are related. The extreme halophile *Halobacterium* and the sulfur reducer *Archaeoglobus* are also related to the methanogens in the orders *Methanosarcinales* and *Methanomicrobiales*. In contrast, the extreme thermophiles are found in both of the two major lines of descent. There is also evidence for many groups that have never been cultivated, including marine and soil clones and a possible phylum *Korarchaeota*.

ed in marine and soil environments. Because of their habitats, these types are expected to be mesophiles.

Methane-Producing *Archaea* or *Methanoarchaea*

These prokaryotes are the only *Archaea* currently cultivated that are truly cosmopolitan. By this we mean that these strict anaerobes are found in most anaerobic environments on earth, including waterlogged soils, rice paddies, lake sediments, marshes, marine sediments, and the gastrointestinal tracts of animals. They are also found over the full temperature range of life, from psychrophilic to mesophilic to hyperthermophilic. Usually they are found in association with anaerobic *Bacteria* and eukaryotes, and they participate in anaerobic food chains that degrade complex organic polymers to CH_4 and CO_2.

Methanogenesis is a significant component of the carbon cycle on earth. About half of the methane produced by these *Archaea* is oxidized before it reaches the atmosphere, and the total biological production on Earth is estimated to be about 0.78×10^{15} g of methane C per

year. Given that microbially produced methane also results in the production of an equal amount of CO_2 (see following discussion on p. 423), the *process* of methanogenesis results in the mineralization of about 1.6×10^{15} g of the 100×10^{15} g of C fixed every year, or about 1.6 % of the net primary productivity. Large amounts of methane are also released into the Earth's atmosphere each year. Because methane is a greenhouse gas, it contributes to the warming of the planet. The present atmospheric concentration of methane is about 1.7 ppm. However, this value is more than twice the concentration found in air frozen in ice cores before the industrial revolution. Currently, the concentration of atmospheric methane is increasing at a rate of about 1% per year, and methane's contribution to the greenhouse warming is expected to increase.

A summary of the properties of the major genera of methanoarchaea is given in **Table 18.2**. All methane-producing *Archaea* have the ability to obtain their energy for growth from the process of methane biosynthesis. So far, no methanogens have been identified that can grow using other energy sources. Thus, these *Archaea* are obligate methane producers. In the nomenclature used to describe prokaryotes, the names of genera of methanoarchaea contain the prefix "methano-." This prefix distinguishes them from an unrelated group of *Bacteria*, the methylotrophic *Bacteria*, which consume methane. The names of genera of methylotrophic *Bacteria* contain the prefix "methylo-."

The substrates for methane synthesis are limited to a few types of compounds (**Table 18.3**). The first type is used by the methanogens that reduce CO_2 to methane. The major electron donors for this reduction are H_2 and formate. In addition, some methanogens can use alcohols like 2-propanol, 2-butanol, c-pentanol, and ethanol as electron donors. In the oxidation of 2-propanol, acetone is the product. Because eight electrons are required to reduce CO_2 to methane, four molecules of H_2, formate, or 2-propanol are consumed. Even though formate is a reduced C-1 compound, it is oxidized to CO_2 before CO_2 is reduced to methane. The second type of substrate for methanogenesis includes C-1 compounds containing a methyl carbon bonded to O, N, or S. Compounds of this type include methanol, monomethylamine, dimethylamine, trimethylamine, and dimethylsulfide. The methyl group is reduced to methane. The electrons for this reduction are obtained from the oxidation of an additional methyl group to CO_2. Because six electrons can be obtained from this oxidation and only two are required to reduce a methyl group to methane, the stoichiometry of this reaction is three molecules of methane are formed for every molecule of CO_2 formed. An exception to this rule is found in *Methanosphaera stadtmaniae*, which can use H_2 only to reduce

Table 18.2 Summary of properties of the methane-producing *Archaea*[a]

		Morphology	Major Energy Substrates[b]	Temperature Optimum (°C)	Cell Wall[c]
Order *Methanobacteriales*					
Family *Methanobacteriaceae*					
Genus	*Methanobacterium*	Rod	H_2, (formate)	37–45	Pseudomurein
	Methanothermobacter	Rod	H_2, (formate)	55–65	Pseudomurein
	Methanobrevibacter	Short rod	H_2, formate	37–40	Pseudomurein
	Methanosphaera	Coccus	H_2 + methanol	37	Pseudomurein
Family *Methanothermaceae*					
Genus	*Methanothermus*	Rod	H_2	80–88	Pseudomurein + protein
Order *Methanococcales*					
Family *Methanococcaceae*					
Genus	*Methanococcus*	Coccus	H_2, formate	35–40	Protein
	Methanothermococcus	Coccus	H_2, formate	60–65	Protein
Family *Methanocaldococcaceae*					
Genus	*Methanocaldococcus*	Coccus	H_2	80–85	Protein
	Methanotorris	Coccus	H_2	88	Protein
Order *Methanomicrobiales*					
Family *Methanomicrobiaceae*					
Genus	*Methanomicrobium*	Rod	H_2, formate	40	Protein
	Methanogenium	Irregular coccus	H_2, formate	15–57	Protein
	Methanoplanus	Plate or disk	H_2, formate	32–40	Glycoprotein
	Methanolacinia	Rod	H_2	40	Glycoprotein
	Methanoculleus	Irregular coccus	H_2, formate	20–55	Glycoprotein
	Methanofollis	Irregular coccus	H_2, formate	37–40	Glycoprotein
Family *Methanospirillaceae*					
	Methanospirillum	Spirillum	H_2, formate	30–37	Protein + sheath
Family *Methanocorpusculaceae*					
Genus	*Methanocorpusculum*	Small coccus	H_2, formate	30–40	Glycoprotein
	Methanocalculus[d]	Irregular coccus	H_2, formate	30–40	ND[e]
Order *Methanosarcinales*					
Family *Methanosarcinaceae*					
Genus	*Methanosarcina*	Coccus, packets	(H_2), MeNH$_2$, Ac	35–60	Protein + HPS
	Methanococcoides	Coccus	MeNH$_2$	23–35	Protein
	Methanolobus	Irregular coccus	MeNH$_2$	37	Glycoprotein
	Methanohalophilus	Irregular coccus	MeNH$_2$	35–40	Protein
	Methanohalobium	Flat polygons	MeNH$_2$	40–55	ND
	Methanosalsus	Irregular coccus	MeNH$_2$	35–45	ND
Family *Methanosaetaceae*					
Genus	*Methanosaeta* (*Methanothrix*)	Rod	Ac	35–60	Protein + sheath
Order *Methanopyrales*					
Family *Methanopyraceae*					
Genus	*Methanopyrus*	Rod	H_2	98	Pseudomurein

[a]All of the methanoarchaea are members of the phylum *Euryarchaeota*.
[b]Major energy substrates for methane synthesis. MeNH$_2$ is methylamines, Ac is acetate. Parentheses means utilized by some but not all species or strains.
[c]Cell wall components include HPS for heteropolysaccharide.
[d]Placement in higher taxon is tentative.
[e]ND, not determined.

methanol to methane. Apparently, this archaeon lacks the ability to oxidize methanol. The third type of substrate is acetate. In this reaction, the methyl (C-2) carbon of acetate is reduced to methane using electrons obtained from the oxidation of the carboxyl (C-1) carbon of acetate. This reaction is called the aceticlastic reaction because it results in the splitting of acetate into methane and CO_2.

Table 18.3	Free energies for typical methanogenic reactions	
Reaction		$\Delta G^{o\prime}$ (kJ/mol of CH_4)
Type 1: CO_2-reducing		
$CO_2 + 4 H_2 \rightarrow CH_4 + 2 H_2O$		–130
$4 HCOOH \rightarrow CH_4 + 3 CO_2 + 2 H_2O$		–120
$CO_2 + 4$ (isopropanol) $\rightarrow CH_4 + 4$ (acetone) $+ 2 H_2O$		–37
Type 2: methyl-reducing		
$CH_3OH + H_2 \rightarrow CH_4 + H_2O$		–113
$4 CH_3OH \rightarrow 3CH_4 + CO_2 + 2 H_2O$		–103
$4 CH_3NH_3Cl + 2 H_2O \rightarrow 3 CH_4 + CO_2 + 4 NH_4Cl$		–74
$2 (CH_3)_2S + 2 H_2O \rightarrow 3 CH_4 + CO_2 + 2 H_2S$		–49
Type 3: aceticlastic		
$CH_3COOH \rightarrow CH_4 + CO_2$		–33

Biochemistry of Methanogenesis Elucidation of the pathway of methane synthesis has been a major challenge to microbial physiologists. Despite simple substrates, the complexity of the pathway is comparable to bacterial photosynthesis. In addition, many of the enzymes of methanogenesis are extremely sensitive to oxygen, and their study requires specialized equipment and techniques. Even after extensive research, many features of archaeal methanogenesis are still not well understood.

The reduction of CO_2 to methane requires five unusual coenzymes or vitamins (**Figure 18.7**). Elucidation of the structure and function of these coenzymes was achieved primarily in the laboratories of R. S. Wolfe at the University of Illinois, G. D. Vogels at the University of Nijmegen, and R. K. Thauer at the University of Marburg. The structures of these coenzymes represent a fascinating blend of novel and familiar themes in biochemistry. Methanofuran is an unusual molecule containing an aminomethylfuran moiety. During methanogenesis, the aminomethylfuran is a C-1 acceptor, forming formylmethanofuran. The use of a furan moiety as a reactive center is unique. In contrast, methanopterin is very similar to folate in structure and function. In both molecules, the reactive species is the tetrahydro-form and composed of a pterin and *p*-aminobenzoic acid (PABA). However, the side chains are strikingly different. Interestingly, unusual pterins are common in other *Archaea* as well, and methanopterin has also been found in the methylotrophic *Bacteria*. Coenzyme F_{430} is also familiar. Named because it absorbs light at 430 nm, it is a tetrapyrrole with Ni at the active center. Although the ring structure is more reduced than other common tetrapyrroles containing Fe (heme), Mg (chlorophyll),

and Co (cobamide), the biosynthesis and function of coenzyme F_{430} is probably similar. Coenzyme M and 7-mercaptothreonine phosphate are two thiol-containing coenzymes. Although thiols are common functional groups in other compounds in the cell such as coenzyme A and cysteine, the coenzymes from methanogens have some very unusual features. 7-Mercaptoheptanoylthreonine phosphate contains a threonyl phosphate residue, which has only been previously found in phosphorylated proteins. A phosphorylated amino acid in a coenzyme is extremely unusual. Likewise, coenzyme M is a sulfonate while most highly polar coenzymes are phosphate derivatives.

The requirement for a large number of unusual coenzymes in methanogenesis has profound implications. The ability to utilize methanogenesis as an energy source requires a large amount of genetic information for coenzyme biosynthesis in addition to methanogenesis. For this reason, methanogenesis probably evolved only once, and all methanogens are related.

Methanogenesis from CO_2 Reduction Occurs Stepwise CO_2 is bound to carriers and reduced successively through the formyl, methylene, and methyl oxidation-reduction levels to methane (**Figure 18.8**). Initially, CO_2 binds methanofuran (MFR) and is reduced to the formyl level. The mechanism whereby this initial binding occurs is not known. The formyl group is then transferred to tetrahydromethanopterin (H_4MPT), which is the C-1 carrier for the next two reductions. Thus, after dehydration of formyl-H_4MPT to methenyl-H_4MPT, the C-1 moiety is reduced to methylene-H_4MPT and then to methyl-H_4MPT. The methyl group is then transferred to coenzyme M. The last reaction forms methane from the reduction of methyl coenzyme M (CH_3-S-CoM). The methylreductase, the enzyme that catalyzes this reaction, requires two additional coenzymes, coenzyme F_{430} and 7-mercaptoheptanoylthreonine phosphate (HS-HTP). HS-HTP reduces CH_3-S-CoM to form methane plus the heterodisulfide of coenzyme M and HS-HTP (Figure 18.8). This mixed disulfide must then be reduced to regenerate the reductant.

The electrons for the reduction of CO_2 to methane are obtained from H_2, formate, or alcohols. For some of these reductions, the electron carriers are not known

Figure 18.7 Coenzymes of methanogenesis ▶
Although discovered in the methanoarchaea and closely related archaea, many of these coenzymes have also been found in some bacteria. The numbers in red refer to the position of the coenzyme structures.

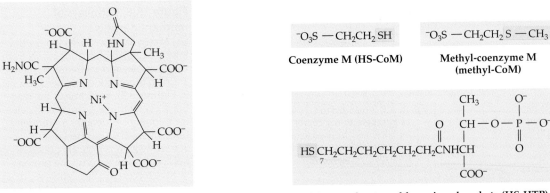

Methanofuran (MFR)

One-carbon acceptor

During methanogenesis, the aminomethylfuran moiety accepts a one-carbon unit to form formylmethanofuran.

N^5-Formylmethanofuran (N^5-Formyl-MFR)

Tetrahydromethanopterin is the biologically active form.

Methanopterin (MPT)

Tetrahydromethanopterin (H_4MPT)

5-Formyltetrahydro-methanopterin (N^5-Formyl-H_4MPT)

5,10-Methenyltetrahydro-methanopterin (Methenyl-H_4MPT)

5,10-Methylenetetrahydro-methanopterin (Methylene-H_4MPT)

5-Methyltetrahydro-methanopterin (Methyl-H_4MPT)

Coenzyme F_{420} (F_{420})

Coenzyme F_{430} (F_{430})

Coenzyme M (HS-CoM)

Methyl-coenzyme M (methyl-CoM)

7-Mercaptoheptanoylthreonine phosphate (HS-HTP)

(A)

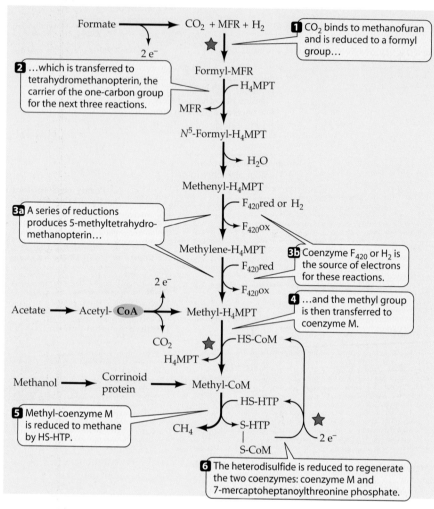

(B)

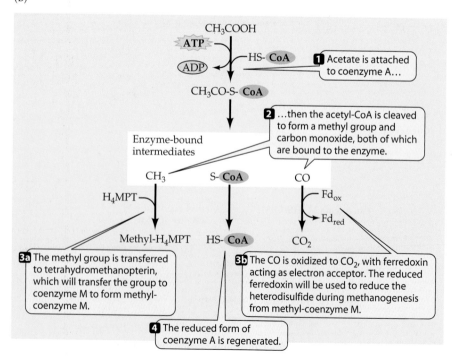

with certainty. However, for at least the reduction of methenyl-H_4MPT to methylene-H_4MPT and then methyl-H_4MPT, the electron carrier coenzyme F_{420} is utilized. Coenzyme F_{420} is named because of its absorption maximum at 420 nm. Although it was discovered in methanogens, it is also present in other organisms. In *Streptomyces*, it functions in antibiotic synthesis. In many *Bacteria* and eukaryotes, it is also essential for photoreactivation of UV-damaged DNA. In methanogens, coenzyme F_{420} is present in very high levels, and it is an electron carrier in many biosynthetic reactions in addition to methanogenesis.

Methanogenesis from Other Substrates Methyl coenzyme M is also a central intermediate in the pathway of methane synthesis from methyl compounds and acetate. For methyl compounds, the C-1 group is first transferred to a cobamide-containing protein and then to coenzyme M. The methyl group is then either reduced to methane or oxidize by the reverse of the pathway of CO_2 reduction. Acetate must first be activated with ATP to form acetyl-CoA (Figure 18.8B). Acetyl-CoA is then cleaved to form an enzyme-bound methyl group, an enzyme-bound carbon monoxide (CO), and HS-CoA. The methyl group is trans-

Figure 18.8 Pathways of methanogenesis
(A) Methyl-coenzyme M is a central intermediate in methanogenesis from CO_2, acetate, and methanol or methyl-amine (abbreviations of coenzymes as in Figure 18.7). For some of the reactions, coenzyme F_{420} or H_2 is the electron donor; 2 e⁻ denotes an unidentified immediate electron donor. Starred reactions indicate potential points of coupling with the proton motive force. (B) The aceticlastic reaction for biosynthesis of methyltetrahydromethanopterin from acetate.

means of energy generation in these archaea. Three coupling sites to the proton motive force have been identified: the reduction of CO_2 to formylmethanofuran, the methyl transfer from methyl-H4MPT to HS-CoM, and the reduction of the heterodisulfide to HS-HPT and HS-CoM (Figure 18.8). The first of these reactions is endergonic in the direction of methane synthesis, and the proton motive force is required to drive the formation of formylmethanofuran. At the remaining sites, the reactions are exergonic, and either a sodium or proton motive force is generated. These ion gradients may then be used for ATP biosynthesis, CO_2 reduction, or other energy requiring reactions in the cell.

Ecology of Methanogenesis

The methanoarchaea flourish in anaerobic environments where sulfate, oxidized metals, and nitrate are absent. In these environments, the substrates for methanogenesis are readily available as the fermentation products of *Bacteria* and eukaryotes. The methanogens catalyze the terminal step in the anaerobic food chain where complex polymers are converted to methane and CO_2. These principles are illustrated in a typical methanogenic food chain for a thermophilic bioreactor as shown in **Figure 18.9**. Polymers are first degraded by specialized microorganisms, such as the cellulolytic *Bacteria*. The major products are simple sugars such as glucose and the disaccharide cellobiose; common fermentation products such as lactate, ethanol, and volatile fatty acids (VFAs) such as propionate and butyrate; and acetate. The intermetabolic group further metabolizes these products. These microorganisms convert simple sugars to VFAs and alcohols. They also convert VFAs and alcohols to acetate and H_2, which are major substrates for the methanoarchaea.

Interspecies Hydrogen Transfer

Molecular hydrogen (H_2) is a key intermediate in this process. Under standard conditions, when the H_2 partial pressure is 1 atmosphere, the fermentation of ethanol and other alcohols and VFAs to acetate and H_2 are thermodynamically unfavorable (**Figure 18.10**). Therefore, the microorganisms that catalyze these reactions cannot grow. The VFAs then accumulate to very high concentrations that are toxic to the cellulolytic and intermetabolic groups of microorganisms, and the fermentation of cellulose ceases. However, if the methanoarchaea are present, H_2 is

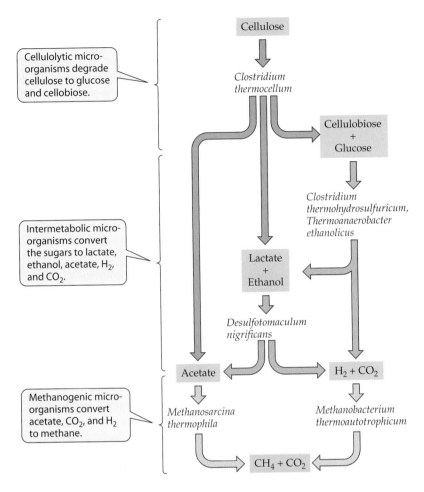

Figure 18.9 Typical anaerobic food chain in a thermophilic bioreactor Cellulose is converted to methane and CO_2 by the combined action of cellulolytic, intermetabolic, and methanogenic microorganisms. In this fermentation, about 95% of the combustion energy of the cellulose is released as methane. Modified from J. Wiegel and L. G. Ljungdahl. 1986. *CRC Critical Reviews in Biotechnology* 3:39–108.

ferred to coenzyme M via tetrahydromethanopterin. The enzyme-bound CO is oxidized to CO_2, and the electron carrier ferredoxin is reduced. The ATP for acetyl-CoA biosynthesis is recovered at this step. The ferredoxin is then used to reduce CH_3-S-CoM to methane. The enzyme complex that oxidizes acetyl-CoA is also called CO dehydrogenase because it oxidizes CO to CO_2. A similar enzyme is found in the homoacetogenic clostridia and many anaerobic autotrophs including the hydrogenotrophic methanogens. In the autotrophs, acetyl-CoA biosynthesis is the first step in CO_2 fixation, and all the organic molecules in the cell are derived from it.

Bioenergetics of Methanogenesis

Methanogenesis is a type of anaerobic respiration and serves as the primary

(A) 1 atmosphere of H_2

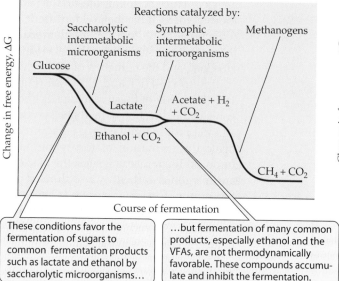

(B) 10^{-5} atmospheres of H_2

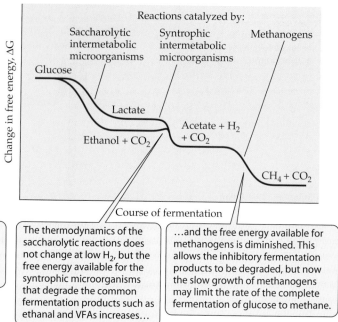

Figure 18.10 Free energy changes during fermentation of glucose to methane
Changes in free energy ($\Delta G'$) during a typical fermentation of glucose to methane under (A) standard conditions (1 atmosphere of H_2) and (B) low partial pressures of H_2 (10^{-5} atmospheres of H_2).

rapidly metabolized, and its partial pressure is maintained below 10^{-3}–10^{-4} atmospheres. Under these conditions, the fermentations of ethanol and other alcohols and VFAs are thermodynamically favorable (Figure 18.10). Because these compounds are rapidly metabolized, their concentrations are maintained below the toxic levels. This interaction between the H_2-producing intermetabolic organisms and the H_2-consuming methanogens is an example of interspecies hydrogen transfer. In many anaerobic environments, it is a key regulatory mechanism. Because the H_2-consuming methanogens play a critical role, they are said to "pull" the fermentation of complex organic polymers to methane and CO_2.

Interspecies hydrogen transfer is not limited to methanogenic food chains. In environments rich in sulfate or nitrate, anaerobic *Bacteria* oxidize H_2, VFAs, and alcohols. When sulfate is present, the sulfate-reducing *Bacteria* catalyze this activity. Because the oxidation of H_2 with sulfate as the electron acceptor is thermodynamically more favorable than when CO_2 is the electron acceptor (as in methanogenesis), the sulfate-reducing *Bacteria* outcompete the methanogens for H_2. For similar reasons, the sulfate-reducing *Bacteria* also outcompete the methanogens for other important substrates like acetate and formate. Therefore, methanogenesis is greatly limit-

ed in marine sediments that are rich in sulfate. The oxidation of H_2 with nitrate, Fe^{2+}, and Mn^{2+} as electron acceptors is also thermodynamically more favorable than methanogenesis. The denitrifying and iron- and magnesium-reducing *Bacteria* also outcompete the methanogens when these electron acceptors are present. In conclusion, with few exceptions, methanogenesis only dominates in habitats where CO_2 is the only abundant electron acceptor for anaerobic respiration.

Methanogenic Habitats Just as aerobic microorganisms rapidly deplete the O_2 in environments rich in organic matter to establish anaerobic conditions, sulfate-reducing *Bacteria*, iron- and magnesium-reducing *Bacteria*, and denitrifying *Bacteria* frequently consume all the sulfate, Fe^{3+}, Mn^{+4}, and nitrate in anaerobic environments and rapidly establish the conditions for methanogenesis. In these environments, CO_2 is seldom limiting because it is also a major fermentation product. Thus, methanogenesis is the dominant process in many anaerobic environments that contain large amounts of easily degradable organic matter. Especially important environments of this type include freshwater sediments found in lakes, ponds, marshes, and rice paddies. Here methane synthesis can be readily demonstrated by a simple experiment first performed by the Italian physicist Alessandro Volta in 1776. In a shallow pond or lake, find a place where there has been an accumulation of leaf litter or other organic debris. When stirred with a long stick, gas bubbles will escape from the sediment. Collect the bubbles in an inverted funnel that has been closed at the top with a short piece of tubing and a pinch clamp. Collect about four liters of gas. To demonstrate that the gas is methane, ignite it. First, the funnel is partially submerged to pressurize the gas. When the clamp is slowly opened, a match is touched to the escaping gas, which will burn with a blue flame. Care must be taken to keep your eyebrows clear of the opening and not to be burned.

Methanogenesis also occurs in other habitats. Methane is formed by the methanoarchaea in the anaerobic microflora of the large bowel. About one-third of healthy adult humans excrete methane gas. Some methane is also absorbed in the blood and excreted from the lungs. The most numerous methanogen in humans is *Methanobrevibacter smithii*, which is a gram-positive coccobacillus that utilizes H_2 or formate to reduce CO_2 to methane. In people who excrete methane, it is found in numbers of 10^7 to 10^{10} cells per gram dry weight of feces, or between 0.001 and 12% of the total number of viable anaerobic *Bacteria*. Why the numbers of *M. smithii* fluctuates so greatly in apparently healthy individuals remains a mystery. *Methanosphaera stadtmaniae* is also present in the human large bowel, but in much lower numbers. This interesting methanogen will grow only by the reduction of methanol with H_2. It cannot reduce CO_2 to methane or utilize acetate and methylamine. Like *M. smithii*, it is also gram-positive and contains pseudomurien in its cell walls.

The rumen is another major habitat for methanogens, and about 10% to 20% of the total methane emitted to the earth's atmosphere originates in the rumen of cows, sheep, and other mammals. In the rumen, complex polymers from grass and other forages are degraded to volatile fatty acids, H_2, and CO_2 by the cellulolytic and intermetabolic groups of *Bacteria*, fungi, and protozoans. The VFAs are absorbed by the animal and are a major energy source. Thus, little methane is produced from acetate. The H_2 is used to reduce CO_2 to methane, which is emitted. Methanogenesis represents a significant energy loss to the cow, and up to 10% of the caloric content of the feed may be lost as methane. *Methanobrevibacter ruminantium* is the predominant methanogen in the bovine rumen. It is a coccobacillus that utilizes H_2 and formate. Some strains require coenzyme M in addition to volatile fatty acids for growth. Because growth is proportional to the concentration of coenzyme M in the medium, these strains have been used in a bioassay for coenzyme M.

In the rumen, in freshwater sapropel or swamp slime, and in marine sediments, many of the anaerobic protozoans are associated with methanogenic symbionts. For rumen ciliates, methanoarchaea are attached to the cell surface. For some species, up to 100% of the individual ciliates are associated with methanogens. In aquatic sludge or sapropel, many protozoans contain endosymbiotic methanogens. These include both ciliates and amoebae. The methanogens are present in high numbers and distributed throughout the cytoplasm. The marine ciliate *Metopus contortus* also contains numerous endosymbiotic methanogens of the species *Methanoplanus endosymbiosus*. This disc-shaped methanogen has a protein cell wall and lyses rapidly in 0.001% sodium dodecyl sulfate. It is located in parallel rows on the cytoplasmic side of the pellicle and on the surface of the nuclear membrane, and it is present in very high numbers, greater than 10^{10} methanogens per ml of cytoplasm. *M. endosymbiosus* is probably associated with the hydrogenosomes, which are microbodies that convert pyruvate to acetate, H_2 and CO_2. The methanogen is then the electron sink for H_2. Conceivably, the ciliate may use this mechanism to divert reducing equivalents away from the sulfate-reducing *Bacteria* and limit the production of H_2S, which is very toxic.

Although methanogens are most frequently found at the bottom of anaerobic food chains associated with the intermetabolic microorganisms, in some ecosystems they are the primary consumers of geochemically produced H_2 and CO_2. The submarine hydrothermal vents found on the ocean floor expel large volumes of very hot water containing H_2 and H_2S. As the water cools from several hundred degrees Celsius to the temperature of the ocean, zones suitable for the growth of thermophilic methanogens are established. *Methanocaldococcus jannaschii* is found in such an environment. It has an optimal temperature of 85°C and a protein cell wall. Methane produced by this bacterium escapes into the surrounding water where it is utilized by symbiotic methylotrophic *Bacteria* living in the gills of mussels. In this ecosystem, the methanogen is at the top of the food chain.

Methanoarchaea are common in bioreactors that convert organic wastes to methane gas, which can then be used as fuel. *Methanospirillum hungatei* is a common methanoarchaea found in this and other freshwater habitats. A hydrogenotrophic methanogen, it has a distinctive spirillum morphology. It forms long filaments, where the individual cells are encased in a sheath (**Figure 18.11**). A complex spacer region composed of protein layers separates the cells. *Methanosaeta concilii* is another sheathed archaeon. This obligately aceticlastic methanogen has a very high affinity for acetate, and it is common in bioreactors where the acetate concentrations are low. Because *M. concilii* has a generation time of one day, it is frequently observed only after the bioreactor has been running for an extended period of time.

Extreme Halophiles

The extremely halophilic *Archaea* or haloarchaea are all closely related, and they comprise a single family of 14

Figure 18.11 Sheaths of methanoarchaea
Some methanoarchaea have complex cell walls. (A) Thin section of *Methanospirillum hungatei*, a species common in bioreactors and freshwater sediments. The rod-shaped cells are enclosed in a sheath and often form long filaments. (B) Thin section of *Methanosaeta concilii*, an aceticlastic methanogen. The spacer region appears less complex than that in (A). A, ©T. J. Beveridge/Visuals Unlimited; B, courtesy of T. J. Beveridge, U. of Guelph, Canada.

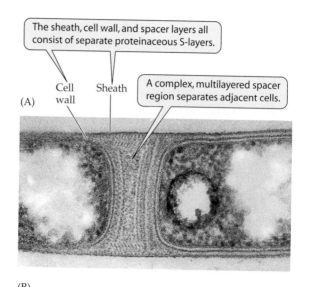

The sheath, cell wall, and spacer layers all consist of separate proteinaceous S-layers.

A complex, multilayered spacer region separates adjacent cells.

(A) Cell wall / Sheath

(B)

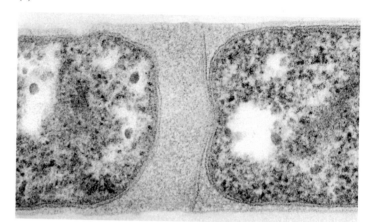

genera (**Table 18.4**). They are all halophilic, and most grow only at concentrations of NaCl above 1.8 M. Most species will also grow at concentrations greater than 4 M, which is close to the saturation point for NaCl in water. In addition, a number of genera are alkalinophilic and grow only at basic pHs. The genera of extremely halophilic methanoarchaea, such as *Methanohalophilus* and *Methanoalobium*, are in the *Methanosarcinales*, and the 16S rRNA phylogeny suggests that these two taxa are closely related.

In addition to the haloarchaea, some *Bacteria*, algae, and fungi can also grow in very high concentrations of NaCl. Examples of extremely halophilic *Bacteria* include the purple photosynthetic bacterium *Ectothiorhodospira* and the actinomycete *Actinopolyspora*. In addition, *Halomonas* is a halotolerant eubacterium that grows optimally near 1 M NaCl, but it also grows slowly at 4 M NaCl. Although these *Bacteria* share with the haloarchaea the ability to grow in high concentrations of salt, they are unrelated, and their biochemistry and physiology are quite different.

Table 18.4 Summary of properties of extremely halophilic *Archaea*[a]

		Morphology	Substrates[b]	pH Optimum	Optimal NaCl (M)
Family *Halobacteriaceae*					
Genus	*Halobacterium*	Long rod	Amino acids	5–8	3.5–4.5
	Haloarcula	Short pleomorphic rods	Amino acids, CH_2O	7.0–7.5	2.5–3.0
	Halobaculum	Rod	Organic	6–7	1.0–2.5
	Haloferax	Pleomorphic	Amino acids, CH_2O	7	2.5
	Halococcus	Coccus	Amino acids, CH_2O	6.8–9.5	3.5–4.5
	Halogeometricum	Pleomorphic	Amino acids, CH_2O	6–8	3.5–4.0
	Halorubrum	Rod	Amino acids, CH_2O	ND[c]	1.7–4.5
	Haloterrigena	Rod or oval	CH_2O	ND	2.5–4.3
	Natrialba	Rod	Organic, CH_2O	9.5	3.5
	Natrinema	Rod	Amino acids, CH_2O	7.2–7.8	3.4–4.3
	Natronobacterium	Rod	CH_2O, organic acids	9.5–10.0	3.0
	Natronococcus	Coccus	CH_2O	9.5–10.0	2.5–3.0
	Natronomonas	Short rod	Amino acids	8.5–9.5	3.5
	Natronorubrum	Pleomorphic	Organic	9.0–9.5	3.4–3.8

[a]All of the haloarchaea are members of the phylum *Euryarchaeota* and the order *Halobacteriales*.
[b]CH_2O is carbohydrate. "Organic" includes complex C sources, the exact nature of the components used are unknown. Oxygen is the electron acceptor for all the extremely halophilic *Archaea*. Some species may also grow anaerobically by fermentation or denitrification.
[c]ND, not determined.

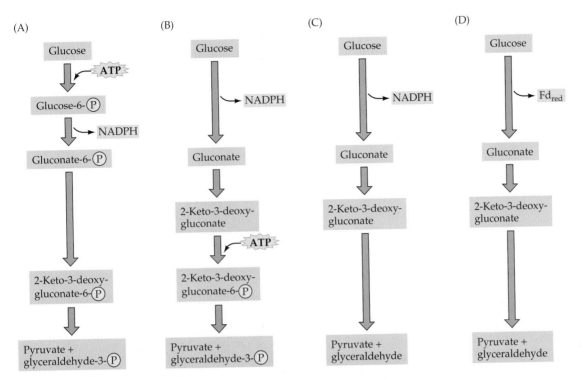

Figure 18.12 Modifications of the Entner-Doudoroff pathway in archaea
Shown here are common pathways in (A) *Pseudomonas* and other bacteria; (B) *Halobacterium* and *Clostridium aceticum*; (C) *Sulfolobus* and *Thermoplasma*, extremely thermophilic *Archaea*; and (D) *Pyrococcus*, another hyperthermophilic *Archaea*. Fd_{red} is reduced ferredoxin.

The haloarchaea are obligate or facultative aerobes. Most utilize amino acids, carbohydrates, or organic acids as their principle energy sources. *Halobacterium* oxidizes glucose by a modification of the Entner-Doudoroff pathway (**Figure 18.12**). In this modification, 2-keto-3-deoxygluconate-6-phosphate is the first phosphorylated intermediate, and glucose-6-phosphate is not found. A further modification in which none of the intermediates are phosphorylated is also found in the extremely thermophilic *Archaea* (Figure 18.12). Carbohydrates are further oxidized by pyruvate oxidoreductase and the complete tricarboxylic acid cycle. An electron transport chain very similar to that found in the *Bacteria* oxidizes NADH generated by the TCA cycle. In the absence of oxygen, *Haloferax denitrificans* is also capable of growth by the anaerobic respiration of nitrate. *Halobacterium salinarium* can ferment arginine in the absence of oxygen or nitrate.

Bacteriorhodopsin and Archaeal Photosynthesis

Many haloarchaea are also capable of an interesting type of photosynthesis. Under low partial pressures of oxygen, they insert large amounts of a protein called bacteriorhodopsin into their cytoplasmic membrane. This protein forms patches that are called purple membrane. The color is due to the presence of the pigment retinal, which is also the common visual pigment in eyes. The

retinal is covalently bonded to a lysyl residue of bacteriorhodopsin via a Schiff base (see Figure 9.11). In the dark, the retinal is in the all-*trans* configuration, and the Schiff base is protonated. Upon the absorption of light, the retinal isomerizes to the 13-*cis* configuration, and the Schiff base is deprotonated. A proton is also expelled outside the cell. The Schiff base then returns to the protonated form by removing a proton from the cytoplasm, and the retinal reverts to the all-*trans* isomer. In this fashion, bacteriorhodopsin functions as a light-driven proton pump. The proton motive force it generates is then used for ATP synthesis and transport. The haloarchaea have at least two additional retinal-based photosystems. One is a chloride pump called halorhodopsin. The second is important in phototaxis and is called "slow rhodopsin."

Although long believed to be unique to *Archaea*, bacteriorhodopsin has also been found in marine proteobacteria. First, a gene homologous to the archaeal bacteriorhodopsin gene was found in DNA cloned from an uncultured gamma-proteobacterium, SAR86. This gene was expressed in *E. coli*, and the recombinant protein formed a functional proton pump. Although currently uncultured, proteobacteria containing this bacteriorhodopsin gene are now known to be widely distributed in the ocean.

Retinal is also the visual pigment in the eye, and the animal rhodopsin functions very differently than bacteriorhodopsin. In rhodopsin, the retinal is not covalently bonded to the protein opsin. In the dark, retinal is in the 11-*cis* configuration. Exposure to light causes an isomerization to the all-*trans* configuration. Proton movement does not occur. Instead, the change in retinal causes a conformational change in the protein, which leads to the activation of a second messenger in the visual response. Because the mechanisms of the two retinal-based photosystems are fundamentally different, they may be an example of convergent evolution.

Photosynthesis in the haloarchaea is also very different from the chlorophyll-based system found in the *Bacteria* and chloroplasts. The electron transport chain does not participate in retinal-based photosynthesis, and it is not possible to directly reduce $NADP^+$ from electron donors that have unfavorable redox potentials, such as water or elemental sulfur. Moreover, extensive light-trapping pigments like the light-harvesting chlorophylls, which are elaborated by photosynthetic *Bacteria* in dim light, have not been described in the haloarchaea. Therefore, the retinal-based system is more suitable for growth at high light intensities.

Adaptations to High Concentrations of Salt

The haloarchaea are adapted specifically for growth in high concentrations of salt. The water activity (a_w) of the cytoplasm of all microorganisms is equal to or less than the a_w of their medium. To the first approximation, this means that the concentration of solutes in the cytoplasm equals or exceeds the concentration outside the cell. For halophilic microbes, the concentration of intracellular solutes must be very high. In the haloarchaea, the intracellular solute is KCl, which is often found at concentrations greater than 5 M. High concentrations of salts also tend to weaken the ionic bonds necessary to maintain the conformation of proteins and the activity of enzymes. To maintain their structure, the proteins from the haloarchaea have an increased number of acidic and hydrophobic residues, and their enzymes frequently require high concentrations of salts for activity. In the absence of salts, the acidic residues on the surface of their proteins are no longer shielded by cations, and many of their enzymes denature. Below NaCl concentrations of 1 M, the protein cell walls of many haloarchaea lose their structural integrity for the same reasons, and the cells lyse.

The haloarchaea are common in salt lakes and salterns, or ponds used to prepare solar salt by the evaporation of seawater. Frequently, they are abundant and impart a red color to the brine. The red color is due to the presence of C_{50} carotenoids called bacterioruberins in the cell envelope. Although nonpathogenic, the haloarchaea can spoil fish and animal hides treated with solar salt.

In salt lakes, the total concentration of salts is generally between 300 and 400 grams per liter, but the specific ions present may vary greatly. For instance, the Dead Sea in Palestine contains abundant chloride, magnesium, and sodium ions. In the Great Salt Lake in Utah, chloride, sodium, and sulfate ions predominate. The pH

of these lakes is near neutrality. In contrast, the Wadi Natrun in Egypt contains high concentrations of bicarbonate and carbonate ions, which maintain the pH near 11. The alkalinophilic haloarchaea, *Natronobacterium* and *Natronococcus*, have been found in these lakes. In the salt lakes, the haloarchaea mineralize organic carbon to CO_2. The organic carbon is produced by other halophilic microorganisms including the green algae *Dunaliella*, cyanobacteria, and *Ectothiorhodospira*.

Extreme Thermophiles

Although somewhat of a misnomer, the remaining *Archaea* are usually called the extreme thermophiles. This name refers to a disparate group of *Archaea* that are not particularly related, either phylogenetically or physiologically. Although most species are extremely thermophilic, some are only moderate thermophiles. Moreover, some methanogens are extremely thermophilic but are not included in this group.

The name is justified because extreme thermophily is almost exclusively a property of *Archaea*. Although many more *Bacteria* have been described, only two genera, *Thermotoga* and *Aquifex*, have optimal temperatures for growth equal to or greater than 80°C. In contrast, many *Archaea* have optimal temperatures greater than 80°C, and *Pyrolobus*, the most extreme thermophile yet isolated, has an optimal temperature of 106°C (**Figure 18.13**). Organisms such as *Pyrolobus* that can grow at 90°C or above are also called hyperthermophiles. Interestingly, extremely thermophilic *Archaea* become abundant just above the temperature where

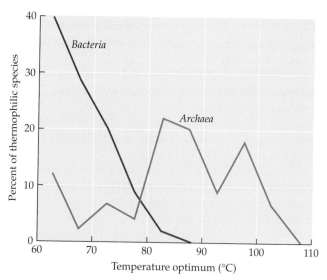

Figure 18.13 Optimum temperatures for growth of thermophilic organisms
The graphs show the optimum temperature for growth of thermophilic *Archaea* and *Bacteria*. The vertical axis indicates the percentage of 45 thermophilic species from each domain. Modified from T. D. Brock, 1985. *Science* 230:132–138.

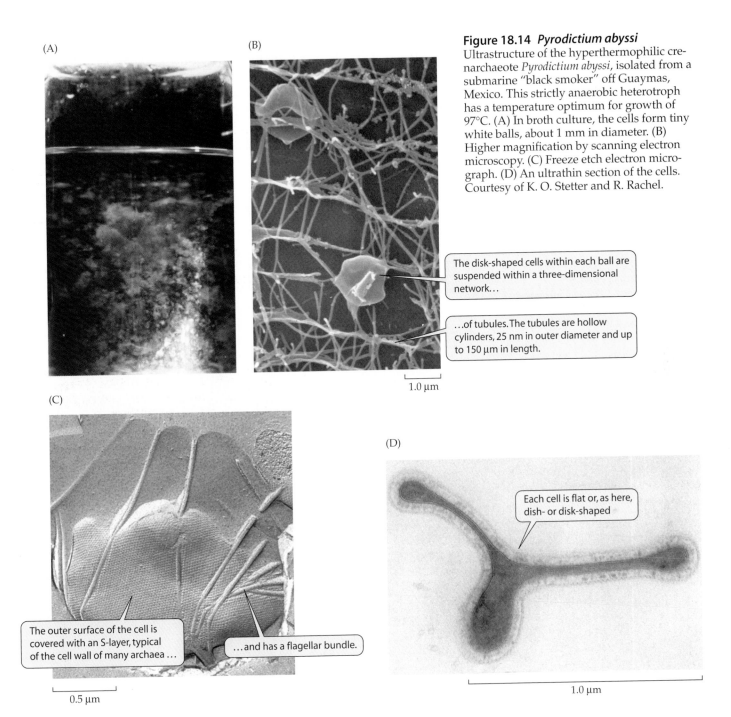

(A)

(B)

Figure 18.14 *Pyrodictium abyssi*
Ultrastructure of the hyperthermophilic cre-narchaeote *Pyrodictium abyssi*, isolated from a submarine "black smoker" off Guaymas, Mexico. This strictly anaerobic heterotroph has a temperature optimum for growth of 97°C. (A) In broth culture, the cells form tiny white balls, about 1 mm in diameter. (B) Higher magnification by scanning electron microscopy. (C) Freeze etch electron micrograph. (D) An ultrathin section of the cells. Courtesy of K. O. Stetter and R. Rachel.

The disk-shaped cells within each ball are suspended within a three-dimensional network…

…of tubules. The tubules are hollow cylinders, 25 nm in outer diameter and up to 150 µm in length.

1.0 µm

(C)

(D)

The outer surface of the cell is covered with an S-layer, typical of the cell wall of many archaea …

…and has a flagellar bundle.

Each cell is flat or, as here, dish- or disk-shaped

0.5 µm

1.0 µm

thermophilic *Bacteria* become rare. The *Archaea* that have optimal temperatures below 80°C are either methanogenic, extremely halophilic, or both moderately thermophilic and acidophilic. Thus, they occupy niches in which *Bacteria* are absent or rare. A possible explanation for this distribution is that *Bacteria* exclude the *Archaea* from temperate environments and that *Archaea* are only abundant either in habitats where *Bacteria* grow poorly such as at high temperatures or high salt or if they utilize a means of producing energy unavailable to the *Bacteria* such as methanogenesis.

The extremely thermophilic *Archaea* may be further subdivided into the obligate and facultative aerobes and

the obligate anaerobes (**Table 18.5**). The aerobes are usually acidophilic, and have pH optima close to 2.0. The obligate anaerobes are mostly neutrophilic and have pH optima greater than 5.0. In spite of the extreme conditions in which these organisms are usually found, the adaptations to their environment are of the same complexity as found in mesophiles. For instance, the hyperthermophile *Pyrodictium abyssi* forms tiny white balls in broth at 97°C composed of a complex three-dimensional network of tubules (**Figure 18.14**). Although the physiological rationale for this unique structure is not known, the cells obviously dedicate a major proportion of their resources in its construction.

Table 18.5 Summary of properties of the extremely thermophilic *Archaea*

	Morphology	Electron Donor[a]	Electron Acceptor	Temperature Optimum (°C)	pH Optimum
Phylum *Crenarchaeota*					
Order *Sulfolobales*					
Family *Sulfolobaceae*					
Genus *Sulfolobus*	Irregular coccus	Organic, S^0, H_2S, Fe^{2+}, H_2	O_2	65–85	2.0–4.5
Acidianus	Irregular coccus	H_2, S^0, organic	S^0, O_2	70–90	1.5–2.5
Metallosphaera	Coccus	S^0, FeS_2, organic	O_2	75	1.0–4.5
Stygiolobus	Irregular coccus	H_2	S^0	80	1.0–5.5
Sulfurisphaera	Irregular coccus	Peptone	S^0	84	2.0
Sulfurococcus	Coccus	S^0, FeS_2, organic	O_2	60–75	2.0–2.6
Order *Thermoproteales*					
Family *Thermoprotaceae*					
Genus *Thermoproteus*	Thin rod	Organic, H_2	S^0	85–88	5.0–6.8
Caldivirga	Thin rod	Organic	S^0	85	3.7–4.2
Pyrobaculum	Thin rod	H_2	S^0	100	6.0
Thermocladium	Thin rod	Organic, H_2	S^0	75	4.2
Family *Thermofilaceae*					
Thermofilum	Thin rod	Organic	S^0	85–90	5.0–6.0
Order *Desulfurococcales*					
Family *Desulfurococcaceae*					
Genus *Desulfurococcus*	Coccus	Organic	S^0	85–92	6.0
Aeropyrum	Coccus	Organic	S^0	90–95	7
Ignicoccus	Coccus	H_2	S^0	90	6
Staphylothermus	Coccus	Organic	S^0	92	ND[b]
Stetteria	Disk	Organic	S^0	95	6
Sulfophobococcus	Coccus	Organic		87	7.5
Thermodiscus	Disk	Organic	S^0	90	5.5
Thermosphaera	Disk, aggregates	Organic		85	6.5–7.2
Family *Pyrodictiaceae*					
Genus *Pyrodictium*	Disk	H_2, organic	S^0	97–105	5.5
Hyperthermus	Irregular coccus	Peptides, H_2	S^0	95–106	7.0
Pyrolobus	Irregular coccus	H_2	NO_3^-, $S_2O_3^{-2}$, O_2	106	5.5
Phylum *Euryarchaeota*					
Order *Archaeoglobales*					
Family *Archaeoglobaceae*					
Genus *Archaeoglobus*	Coccus	H_2, organic	SO_4^{-2}, $S_2O_3^{-2}$	80–85	6.0–7.0
Ferroglobus	Irregular coccus	Fe^{+2}, H_2S, H_2	NO_3^-, $S_2O_3^{-2}$	85	7.0
Order *Thermoplasmatales*					
Family *Thermoplamataceae*					
Genus *Thermoplasma*	Pleomorphic	Organic	O_2, S^0	55–59	1.0–2.0
Family *Picrophilaceae*					
Genus *Picrophilus*	Irregular coccus	Organic	O_2	60	0.7
Order *Thermococcales*					
Family *Thermococcaceae*					
Genus *Thermococcus*	Coccus	Organic	S^0	75–88	6.0–8.0
Pyrococcus	Coccus	Organic	S^0	100	7.0

[a]Organic substrates include amino acids, sugars, or yeast extract.
[b]ND, not determined.

Importantly, the extremely thermophilic *Archaea* are a recent discovery. More than 90% of the species currently known were isolated after 1980, largely due to the pioneering work of Karl Stetter at the University of Regensburg and Wilfram Zillig at the Max-Planck-Institut in Martinsried and their colleagues in Germany. Because of their recent discovery, many aspects of the species' biology, physiology, and biochemistry are not yet known. However, even at this early stage in their study, the extremely thermophilic *Archaea* have challenged many of the basic tenets of modern biology.

Habitats and Physiology   The extremely thermophilic *Archaea* grow above the temperature limit of photosynthesis. Therefore, they are generally restricted to environments where geothermal energy is available. These areas include hot springs, solfatara fields, geothermally heated marine sediments, and submarine hydrothermal vents. Here H_2, H_2S, and S^0 are abundant energy sources. Many of these *Archaea* are chemolithotrophic and are capable of some form of sulfur metabolism. *Acidianus infernus* is a particularly interesting example. It is an obligate autotroph and uses CO_2 as its sole carbon source. During aerobic growth, S^0 is oxidized to H_2SO_4 (**Figure 18.15**). Acid production maintains the pH at very low levels, and this organism is also an acidophile. Under anaerobic conditions, H_2 is oxidized, and S^0 is now reduced to H_2S. Thus, S^0 functions as either the electron donor in the presence of O_2 or the electron acceptor in the presence of H_2. Other extremely thermophilic *Archaea* are capable of either one of these two modes of sulfur metabolism. The aerobe *Sulfolobus* oxidizes S^0 and organic compounds. It is also an acidophile. The anaerobes in the order *Thermoproteales* oxidize H_2, sugars, or amino acids using S^0 as an electron acceptor. In addition, growth by fermentation or other types of anaerobic respiration may also be possible.

Thermoplasma is an unusual archaeon that was first isolated from burning coal refuse piles. It has since been found in hot springs as well, and two species have been identified, *Thermoplasma acidophilus* and *T. volcanium*. *Thermoplasma* is a facultative aerobe and grows on the peptide components of yeast extract. It has not been grown in defined medium. During aerobic growth, it utilizes an electron transport chain consisting of cytochrome *b* and a quinone. Nevertheless, its bioenergetics are poorly understood, and ATP is probably synthesized by substrate level phosphorylation. *Thermoplasma* lacks a cell wall, yet it is osmotically stable and contains glycoprotein and an unusual lipopolysaccharide in its cell membrane. Interestingly, *Thermoplasma* also contains a basic DNA-binding protein that is functionally very similar to the histones found in eukaryotes. However, a comparison of its amino acid sequence reveals that it is more closely related to the bacterial DNA-binding proteins than to the histones.

Archaeoglobus is another exceptional extreme thermophile. It is the only archaeon capable of sulfate reduction. Phylogenetically, it is a close relative of the methanogens and contains tetrahydromethanopterin, methanofuran, and coenzyme F_{420}. However, coenzyme F_{430} and coenzyme M are absent. During growth on lactate, *Archaeoglobus* utilizes an aceticlastic system similar to that found in the acetotrophic methanogens to oxidize acetyl-CoA (**Figure 18.16**). Methyltetrahydromethanopterin is then oxidized by the reverse of the pathway of CO_2 reduction.

Viruses   Like the *Bacteria* and eukaryotes, viruses specific for *Archaea* are common. The viruses of *Sulfolobus* have been studied extensively, and four morphological types are recognized (**Figure 18.17**). These viruses are ubiquitous in acidic hot springs, environments where *Sulfolobus* is common. So far all the *Sulfolobus* viruses found are temperate (i.e., they replicate and are shed from the host without killing it). This may be an adaptation to the extreme habitat of high temperature and low pH. Under these conditions, the free viruses are quickly inactivated.

Perspectives and Applications   The extremely thermophilic *Archaea* are especially interesting from at least four very different points of view. First, these microorganisms grow at the upper temperature limit for life. Although the upper limit is not known with certainty, its precise value is of importance because it sets one boundary on the biosphere. Thus, earth scientists studying geothermally heated aquifers, hot springs, deep vents, and similar geological formations need to know this boundary to know where biological transformations can be significant. Additionally, in the search for life on other planets, the upper temperature limit determines where life as we know it can exist. Currently, the upper temperature limit for life appears to be about 113°C, which is the temperature maximum for growth

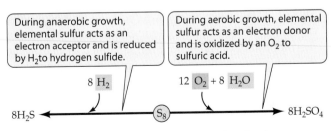

Figure 18.15 Sulfur metabolism of the facultative aerobe *Acidianus infernus*
The aerobic metabolism of *Acidianus* is typical of the obligately aerobic extreme thermophiles. The anaerobic metabolism is typical of the obligately anaerobic extreme thermophiles.

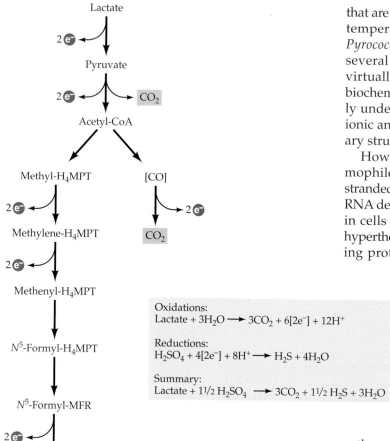

Oxidations:
Lactate + $3H_2O \longrightarrow 3CO_2 + 6[2e^-] + 12H^+$

Reductions:
$H_2SO_4 + 4[2e^-] + 8H^+ \longrightarrow H_2S + 4H_2O$

Summary:
Lactate + $1\frac{1}{2} H_2SO_4 \longrightarrow 3CO_2 + 1\frac{1}{2} H_2S + 3H_2O$

Figure 18.16 Lactate oxidation by the sulfate-reducing archaeon *Archaeoglobus*
Oxidation of lactate by *Archaeoglobus* proceeds through pyruvate to acetyl-CoA. Using an enzyme similar to that found in the aceticlastic methanogens, acetyl-CoA is oxidized to CO_2 and methyltetrahydromethanopterin (Methyl-H_4MPT; see Figures 18.7 and 18.8). Methyl-H_4MPT is then oxidized by the same pathway used by the methylotrophic methanogens (see Figure 18.8).

of *Pyrolobus fumarii*. By studying the growth conditions and physiology of these organisms as well as isolating new species from geothermal environments, it may be possible to increase the known temperature limit.

Thermoadaptation A second area of interest is how these bacteria cope with high temperatures. Most proteins and nucleic acids from mesophilic organisms denature at temperatures well below 100°C. Although a few mechanisms for surviving high temperatures are known, how organisms grow in this extreme condition is still poorly understood.

One important mechanism of thermoadaptation is the biosynthesis of specialized proteins and enzymes

that are both thermostable and maximally active at high temperatures. For example, many enzymes from *Pyrococcus* are most active above 95°C and stable for several days at that temperature. These enzymes are virtually inactive at room temperature. Although the biochemical basis for this thermoadaption is only partly understood, it probably requires an increase in the ionic and hydrophobic bonds that determine the tertiary structure.

However, not all biomolecules from extreme thermophiles are stable at high temperatures. Double-stranded DNA and the secondary structure of ribosomal RNA denature, even at the salt concentrations common in cells well below the temperature optima of many hyperthermophiles. In some *Archaea*, special DNA-binding proteins enhance the thermostability of double-stranded DNA in vitro. Presumably, these proteins protect the chromosome in the living cell. Similarly, the extremely thermophilic archaea possess an enzyme reverse gyrase that introduces positive supercoils into DNA. This enzyme may play a role in thermoadaptation. Some small molecules are also thermolabile. NADH and NADPH are very unstable above 90°C. During growth on sugars, *Pyrococcus* utilizes the pyroglycolytic pathway, where glucose is oxidized by a ferredoxin-dependent glucose oxidoreductase instead of the conventional $NADP^+$-dependent glucose dehydrogenase (Figure 18.12). *Pyrococcus* also contains a ferredoxin-dependent aldehyde oxidoreductase instead of the conventional glyceraldehyde dehydrogenase. Presumably, this ferredoxin-dependent pathway is less temperature sensitive than the $NADP^+$- dependent pathway. A similar rationale may explain why other extreme thermophiles replace some ATP-dependent reactions with ADP reactions.

Some extreme thermophiles also contain high concentrations of intracellular anions, which may contribute to thermostability. Two compounds, called "thermoprotectants," which may have this role are di-inositol-1,1'-phosphate from *Pyrococcus* and cyclic 2,3-diphosphoglycerate (or cDPG) from some methanogens. However, thermoprotectants have not been identified in all extreme thermophiles, so this mechanism may not be general.

Industrial Applications Because of their high temperature optimum and thermostability, enzymes from extreme thermophiles may have important commercial applications. Activity at high temperatures is important because most industrial processes operate between 50°C and 100°C. Enzymes with a temperature optimum in this range are cheaper to use because less protein is required for the same amount of activity. Thermo-

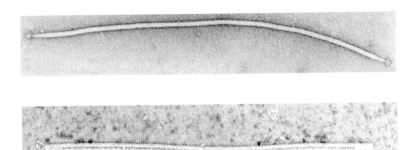

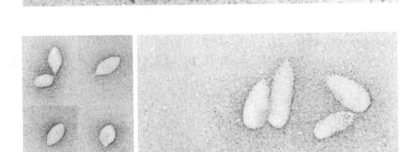

Figure 18.17 Viruses of the extremely thermophilic acidophile *Sulfolobus*

These representatives of each of the four families of viruses known to infect the *Sulfolobus* demonstrate a diversity of morphological types. Within the *Archaea*, viruses are known with either linear or circular DNA genomes. Some viruses are also known that possess envelopes. From D. Prangishvili, K. Stedman and W. Zillig. 2001. *Trends in Microbiology* 9: 39–43.

stability is important because enzymes are expensive. Thermostability is also correlated with greater stability at mesophilic temperatures and resistance to denaturing chemicals. One of the major uses of moderately thermophilic enzymes is alkaline proteases in laundry detergent. These enzymes have high activity at 50°C, the temperature of wash water. They are very stable at room temperature, which extends their shelf life in the supermarket. They are also resistant to the other components of laundry detergent that can denature enzymes, such as surfactants and metal chelators. Although proteases from extreme thermophiles have not yet been used in laundry detergents, many are being tested in this capacity.

An example of extremely thermophilic enzymes currently used industrially is the DNA polymerase from *Pyrococcus* and *Thermococcus*. The biotechnology industry uses these enzymes for DNA sequencing and the polymerase chain reaction. In addition to their high activity and thermostability, these enzymes are especially attractive because they have a very low error rate. For instance, the popular enzyme Pfu was obtained from *Pyrococcus furiosus*. Because of its proofreading activity, its error rate is much lower than that of the common Taq enzyme, which was discovered in the bacterium *Thermus aquaticus*.

Early Life The extreme thermophiles represent very deep branches of the *Archaea*. Similarly, the extreme thermophiles, *Thermotoga* and *Aquifex*, represent the deepest branches of the *Bacteria*. Based on these observations, it has been proposed that the earliest organisms were extreme thermophiles. If true, extant extreme thermophiles may have retained some ancient characteristics that were lost in the evolution of mesophiles. Thus, the unusual enzymes and pathways common in extreme thermophiles may be remnants of the earliest organism. These microorganisms have been described as "living fossils" for this reason. Even if this hypothesis is not true, it is a refreshing perspective on the origin of life.

Uncultured **Archaea** Molecular methods have been used to clone 16S rRNA genes directly from the environment. These studies have discovered an enormous diversity not represented among the cultured organisms. Of particular interest is a large group of uncultivated crenarchaeotes found in the ocean and soil (Figure 18.6). In the oceans, these organisms are widespread, representing 1% to 2% of the total number of prokaryotes in the surface layer and up to 30% in the deeper water. Phylogenetically similar organisms are also present in many soils, where they typically represent 1% to 2% of the total number of prokaryotes. Because they are found in temperate environments, it is unlikely that they are extreme thermophiles like the cultured crenarchaeotes. However, their physiology and growth properties are essentially unknown. If it is true that *Archaea* are limited to extreme habitats or physiological adaptations, these crenarchaeotes may possess some undiscovered physiological adaptation not found in the *Bacteria*.

Some of the *Archaea* yet to be cultivated probably represent novel phyla. In hot springs, one new phylum, the *Korarchaeota*, is abundantly represented in rRNA clones (Figure 18.6). Similarly, 16S rRNA genes with very low similarity to cultured *Archaea* have been detected in samples from the deep ocean.

SUMMARY

▶ The ***Archaea*** are a diverse group of prokaryotes distantly related to the familiar ***Bacteria***.

▶ Unusual biochemical features of the ***Archaea*** are the presence of **isoprenoid-based glycerol lipids**, protein cell walls, a eukaryotic type of RNA polymerase and DNA replication system, and insensitivity to many common antibiotics.

▸ The major physiological groups of the *Archaea* are the **methanogens**, the **extreme halophiles**, and the **extreme thermophiles**.

▸ The *Archaea* are widely distributed in aerobic and anaerobic, mesophilic and thermophilic, and freshwater and extremely halophilic environments.

▸ Methanogenesis is a form of anaerobic respiration and requires a complex biochemical pathway that contains unusual coenzymes: **methanofuran, methanopterin, coenzyme M, coenzyme F_{420}, coenzyme F_{430}, and 7-mercaptoheptanoylthreonine phosphate**.

▸ Methanogens catalyze the last step in the anaerobic food chain in which organic matter is degraded to CH_4 and CO_2. This food chain is the basis of many **symbiotic associations** of methanogens with eukaryotes and *Bacteria*.

▸ The extreme halophiles among the *Archaea* contain **bacteriorhodopsin**, a light-driven proton pump that contains **retinal**. This type of photosynthesis differs fundamentally from chlorophyll-based photosynthesis common in plants and *Bacteria* because it is not coupled to an electron transport chain.

▸ The enzymes from the extreme halophiles are unusually salt tolerant and many also require salt for activity and stability.

▸ Organisms with a growth optimum greater than 80°C are called **extreme thermophiles**. Organisms that grow above 90°C are called **hyperthermophiles**. Most extreme hyperthermophiles are *Archaea*. These organisms set the upper temperature limit for life.

▸ The extreme thermophiles are common in geothermal habitats where H_2 and reduced sulfur compounds are abundant.

▸ The extreme thermophiles are a rich source of thermostable enzymes, which may have valuable applications in biotechnology.

REVIEW QUESTIONS

1. What was the basis for establishing the *Archaea* as a domain? On what is the three-domain concept of life based?

2. Give the major reasons for studying *Archaea*. What can be learned from studying the *Archaea* that can not be learned from studying the *Bacteria*?

3. What are some distinguishing features of *Archaea*?

4. Where would one find the methanogens in nature?

5. In what ways might one consider the methane-producing archaea specialists?

6. Methanogens possess some unusual coenzymes: to what common bacterial coenzymes are the methanogen coenzymes related and what are their functions? Give examples of coenzymes that are unique to methanogens.

7. How does the presence of sulfate or nitrate affect methanogenesis? For respiration, why is CO_2 considered the electron acceptor of last resort?

8. Are the haloarchaea autotrophs or heterotrophs? Give reasons for your answer.

9. What is bacteriorhodopsin? How does photosynthesis with this compound differ from chlorophyll-based photosynthesis?

10. Define an extreme thermophile. What habitats do they occupy?

11. What can we learn from the hyperthermophiles? How might they contribute to industry?

12. Many of the *Archaea* have adapted to growth under extreme conditions of high salt or high temperature. What adaptations are necessary for growth under these conditions?

SUGGESTED READING

Balows, A., H. G. Trüper, M. Dworkin, W. Harder and K. H. Schleifer. 1992. *The Prokaryotes*. 2nd ed. New York: Springer-Verlag.

Dworkin, M., ed. 2001. *The Prokaryotes: An Evolving Electronic Resource for the Microbiological Community*. New York: Springer-Verlag.

Ferry, J. G., ed. 1993. *Methanogenesis: Ecology, Physiology, Biochemistry & Genetics*. New York: Chapman & Hall.

Woese, C. R. and R. S. Wolfe, eds. 1985. *Archaebacteria*. New York: Academic Press.

Chapter 19

Nonphotosynthetic *Proteobacteria*

The extraordinary diversity…among the bacteria is shown by a simple enumeration of the properties which they can possess. They can be unicellular, multicellular, or coenocytic; permanently immotile, or motile by any one of three distinct mechanisms; able to reproduce by binary fission, by budding, or by the formation of special reproductive cells, such as the conidia of actinomycetes; photosynthetic or nonphotosynthetic.

— R. Y. STANIER, IN THE BACTERIA, 1964

The *Proteobacteria*, also called the "purple bacteria" because of its photosynthetic species, makes up the largest and most heterogeneous phylogenetic group of the *Bacteria*. Photosynthetic, chemoautotrophic, and heterotrophic bacteria are all found within this single phylum, indicating a complex and extensive evolutionary history. Most are unicellular rods, vibrios, and cocci that divide by binary transverse fission. However, some have life cycles with two or more types of cells that divide by budding, and one group, the myxobacteria, have a complex developmental cycle culminating in the formation of an elaborate fruiting structure. Many are flagellated, but others move by gliding motility or gas vacuoles. Some produce stalks. Some are obligate aerobes, whereas others are facultative or obligate anaerobes. The current evolutionary chart, based on 16S rRNA analyses, indicates that there are five major subgroups of *Proteobacteria*, designated as the **alpha**, **beta**, **gamma**, **delta**, and **epsilon** subdivisions. Many species of common, well-known heterotrophic bacteria are included in these groups. Please note that all of the photosynthetic members of the *Proteobacteria* will be treated in Chapter 21. In the next sections, the major representatives of the heterotrophic and chemolithotrophic *Proteobacteria* (**Table 19.1**) are discussed.

FERMENTATIVE RODS AND VIBRIOS

Many of the important members of the *Proteobacteria* are fermentative. None of these is obligately anaerobic, but instead they grow as facultative aerobes. Several important groups are known, all of which are members of the gamma group of the *Proteobacteria*.

Enteric Bacteria

The **enteric bacteria** are small, nonsporeforming rods typically about 0.5 μm in width and from 1 to 5 μm in length. In general, they have simple nutritional requirements. Some, however, require vitamins and/or amino acids for growth. Most of the bacteria in this group are motile, and if so, they have peritrichous flagella. Enteric bacteria are facultative aerobes that, under anaerobic conditions, ferment glucose and certain other sugars to form an array of end products. They are catalase positive and oxidase negative, and most reduce nitrate to nitrite when oxygen availability is limited. The DNA from the various species exhibits considerable homology, based on hybridization tests (**Figure 19.1**). These organisms are metabolically and genetically similar but demonstrate

Domain *Bacteria*

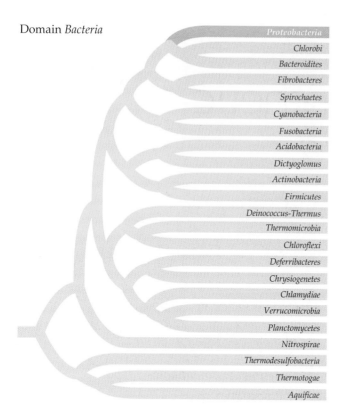

Proteobacteria
Chlorobi
Bacteroidites
Fibrobacteres
Spirochaetes
Cyanobacteria
Fusobacteria
Acidobacteria
Dictyoglomus
Actinobacteria
Firmicutes
Deinococcus-Thermus
Thermomicrobia
Chloroflexi
Deferribacteres
Chrysiogenetes
Chlamydiae
Verrucomicrobia
Planctomycetes
Nitrospirae
Thermodesulfobacteria
Thermotogae
Aquificae

Table 19.1	Major groups and subgroups of heterotrophic and chemolithotrophic *Proteobacteria*[a]	
Groups and Subgroups		**Page Number**
Fermentative rods and vibrios		435
Oxidative rods and cocci		442
Sheathed bacteria		450
Prosthecate bacteria		452
Bdellovibrio: The bacterial predator		459
Spirilla and *Magnetospirillum*		460
Myxobacteria: Fruiting, gliding bacteria		461
Sulfate and sulfur reducers		464
Nitrifiers		465
Sulfur and iron oxidizers		468
Hydrogen oxidizers		473
Methylotrophs		474

[a]See Chapter 20 for photosynthetic members.

considerable diversity in ecology and pathogenic potential for humans and other vertebrate animals (cold and warm blooded), insects, and plants. The intimate association of many enteric bacteria with higher eukaryotes suggests a long evolutionary relationship. Major genera in the enteric group and their habitats are shown in **Table 19.2**.

The enteric bacteria have had considerable influence on human history (see Chapter 2). Their medical importance is evident, as members of this group are the causative agents of diseases such as plague (*Yersinia pestis*), typhoid fever (*Salmonella typhi*), and bacillary dysentery (*Shigella dysenteriae*). Those that are not path-

ogens can cause disease under appropriate conditions and are therefore called **opportunistic** or **secondary pathogens.** Because they are so firmly associated with humans, a huge number of enterobacteria have been isolated and characterized. For example, 1,464 different serogroups have been defined as serovar subspecies in the genus *Salmonella*. The considerable medical and epidemiological importance of the enterics is discussed in Chapters 28 and 30.

The enteric bacteria are separated into two physiological groups, based on the products they generate during glucose fermentation. The **mixed-acid fermenters** (**Figure 19.2**) produce significant amounts of organ-

Figure 19.1 Enteric bacteria Relatedness among the enteric bacteria. Numbers indicate DNA/DNA reassociation values, a measure of the relatedness among the various genera.

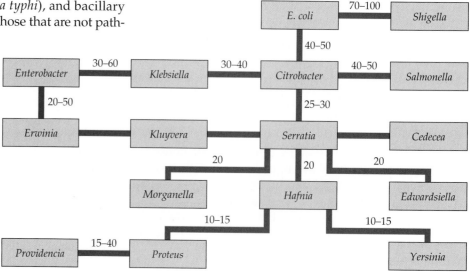

Table 19.2	Normal habitat of the enteric bacteria
Genus	**Habitat**
Escherichia	Normal microbiota of intestinal tract of warm-blooded animals
Salmonella, Shigella, and *Providencia*	Intestinal pathogens of humans and other primates
Hafnia	Feces of humans, other animals, and birds
Edwardsiella	Cold-blooded animals; may be pathogenic for eels and catfish
Proteus	Intestinal tracts of humans and other animals; also soil and polluted water
Morganella	Mammal and reptile feces
Yersinia	Humans, rats, and other animals
Klebsiella, Citrobacter, Enterobacter, and *Serratia*	Human intestine; also soil and water
Erwinia	Plants as pathogenic, saprophytic, or epiphytic microflora

Table 19.3	Relative amounts of product from a mixed-acid and a neutral (butanediol) fermentation	
	Moles of Product per 100 Moles of Glucose Fermented	
	Mixed-Acid Fermentation (*Escherichia coli*)	**Neutral Fermentation (*Enterobacter aerogenes*)**
Acetic acid	36	0.5
2,3-Butanediol	0	66
Ethanol	50	70
Lactic acid	79	3
Succinic acid	11	0
Formic acid	2.5	17
H_2	75	35
CO_2	88	172
Total moles of acid	128.5	20.5
Ratio CO_2/H_2 produced	1.2	4.9

ic acids. In contrast, the **2,3-butanediol fermenters** (**Figure 19.3**) produce mostly neutral compounds. *Escherichia coli* is a mixed-acid fermenter. It metabolizes glucose via the Embden-Meyerhof pathway to pyruvate. The reduced NADH generated by glyceraldehyde-3-phosphate dehydrogenase is reoxidized with the formation of three major acids: acetic, lactic, and succinic acids (Figure 19.3). Formic acid is catabolized by the enzyme formic hydrogen lyase to CO_2 and H_2 in a ratio of one to one (**Table 19.3**). Fermentation of 100 moles of glucose by *E. coli* leads to the formation of about 128 moles of organic acids.

Butanediol fermentation results in the formation of neutral products and considerable quantities of CO_2 (Table 19.3). *Enterobacter aerogenes* is a typical butanediol-fermenting enteric bacterium. Formic acid is cleaved to form $H_2 + CO_2$, but the significantly higher proportion of CO_2/H_2 occurs because other reactions produce CO_2 but not hydrogen. Only about 20 moles of acidic product are generated from 100 mol of glucose. A distinctive and therefore diagnostically

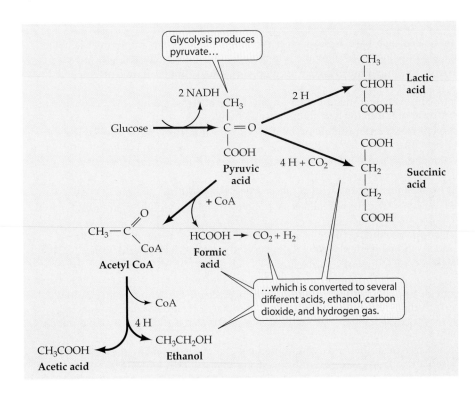

Figure 19.2 Mixed-acid fermentation
Mixed-acid fermentation of glucose by *Escherichia coli.* Pyruvate produced by glycolysis (the Embden-Meyerhof pathway) is catabolized to a mixture of products, high in acids. Acids are shown here and in Figure 19.3 in their nonionized form.

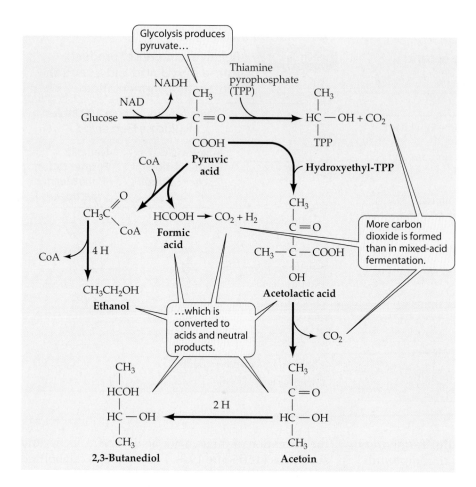

Figure 19.3 2,3-Butanediol fermentation
In the butanediol variation of mixed-acid fermentation, more neutral products are formed, as well as a higher ratio of CO_2 to H_2.

important intermediate called **acetoin** is formed in this fermentation (Figure 19.3).

The precise identification of enterics is of considerable importance in public health microbiology and epidemiology. Consequently, a variety of diagnostic tests (see **Table 19.4** for examples) have been devised to identify genera in this group (**Table 19.5**). Identification of a strain from a clinical specimen entails a large variety of diagnostic tests. Typically, the clinical microbiologist performs these tests routinely along with whole cell fatty acid methyl ester analysis (see Chapter 17) using prepackaged automated or semiautomated for-

Table 19.4	Types of tests used to differentiate enteric bacteria from one another
Test	**Description**
Urease	Streak the organism on urea agar base. Urea agar contains high levels (2.0%) of urea and phenol red indicator. Organism producing urease liberates ammonia that turns indicator red near streak.
Indole	Grow organism in a peptone medium with high tryptophan content. Test for indole production after growth.
Motility	Tubes contain tryptose in a soft agar medium (0.5% agar). Inoculate by stabbing down the center of the agar. Diffuse growth out from the inoculum indicates motility.
Methyl red	A buffered glucose-peptone medium containing methyl red indicator. Turns red if large amount of acid is produced, as in mixed-acid fermentation.
Acetoin (Voges-Proskauer)	A liquid medium containing glucose is inoculated and after growth, assayed for acetoin.
Citrate	Organism is tested for its ability to grow with citrate as sole carbon source.
H_2S production	Inoculate agar medium containing ferrous sulfate by stabbing. Blackening of the agar indicates sulfide production.
Phenylalanine	After growth on a medium containing phenylalanine deaminase (0.1%), assay for production of phenylpyruvic acid by adding ferric chloride. A green color indicates deaminase activity.
Ornithine decarboxylase	Measured manometrically by following CO_2 release.
Gas from glucose	Gas production from glucose or other sugars is measured by placing a Durham tube in a broth containing sugar. A Durham tube is a small test tube that is placed inverted in the broth before autoclaving. Autoclaving causes the tube to be filled. Gas production is indicated by a gas bubble within the Durham tube.
β-galactosidase	Grow organism on a lactose medium. Add a disc containing O-nitrophenyl-β-D-galactoside to the surface. After 15 to 20 minutes the yellow O-nitrophenol is released if the enzyme is present.

Table 19.5   **Characteristics used to distinguish the common genera of enteric bacteria**

Genera	Methyl-Red	Voges-Proskauer	Citrate Utilization	Urease	Indole Production	Motility	H₂S Production	Gas from Glucose	β-Galactosidase	Ornithane Decarboxylase	Phenylalanine Deaminase	Mol % G + C
Mixed-Acid Fermenters												
Eschericia	+[a]	–	–	–	+	+	–	+	+	+	–	48–52
Salmonella	+	–	V	–	–	+	–	+	V	+	–	50–53
Shigella	+	–	–	–	V	–	–	–	V	V	–	49–53
Edwardsiella	+	–	–	–	+	+	–	+	–	+	–	53–59
Citrobacter	+	–	+	+	V	+	–	+	+	+	–	50–52
Proteus	+	V	V	+	V	+	V	+	–	V	+	38–40
Morganella	+	–	–	+	+	+	–	+	–	+	+	50
Providencia	+	–	+	V	+	+	–	+	–	–	+	39–42
Yersinia	+	–	–	+	–	–	–	–	+	V	–	46–50
2, 3-Butanediol Producers												
Klebsiella	+	+	+	V	V	–	–	+	+	V	–	56–58
Enterobacter	V	+	+	–	–	+	–	+	+	+	–	52–60
Serratia	V	+	+	–	–	+	–	V	+	–	+	52–60
Erwinia	+	+	–	–	–	+	+	–	V	–	–	50–58
Hafnia	V	V	–	–	–	+	–	+	+	+	–	48–49

[a]Symbols: + = positive for most strains; – = negative for most strains; V = variable within genus.

mats that are available from commercial manufacturing firms. The results of these tests can be subjected to computer analysis to find the best-fit species identification for the unknown strain.

Escherichia coli, the best-known species of the enteric bacteria and arguably the most widely known of all the bacteria, has been more thoroughly studied than any other living organism. This common bacterium is a normal inhabitant of the intestinal tract of humans and most other warm-blooded animals. As a facultative aerobe of the intestinal tract, it is well equipped to survive anaerobically and consume oxygen that enters its habitat. This activity is essential for maintaining anaerobic conditions in the large intestine. Although it is an important species in the human intestinal tract, it is not the dominant one. Many other, but not all, bacteria in this group also live in the intestinal tracts of animals, hence accounting for the general name of "enteric bacteria" for these organisms.

Because *E. coli* is found in the intestinal tract of warm-blooded animals, it is an important test organism for fecal contamination of food and drinking water. Because *E. coli* does not grow or survive for long periods in food, water, or soil, its presence in these environments is indicative

that they have been contaminated by fairly recent fecal material and, therefore, should not be consumed. Thus, *E. coli* is termed an **indicator bacterium**, whose presence in the environment indicates fecal contamination from warm-blooded animals (see Chapter 32).

Shigella dysenteriae is the causative agent of **bacillary dysentery**, a severe type of gastroenteritis. The species of this genus are so closely related to *E. coli*, as determined by DNA/DNA reassociation, that they could be considered the same species. However, for historical reasons and because of their distinctive pathogenesis, a separate genus has been maintained.

Salmonella typhi causes **typhoid fever** and gastroenteritis. As mentioned earlier, there are hundreds of serovars in this genus. The antigens for the serovars are determined from three types of surface polymers, including the outer cell membrane lipopolysaccharides ("O" antigens), the flagella ("H" antigens), and the outer layer polysaccharides ("Vi" antigens). These serovars are helpful in tracing the source of organisms during an outbreak of typhoid fever.

Another important pathogenic genus is *Yersinia*, which contains the dreaded agent of **bubonic plague**, *Yersinia pestis*. During the fourteenth century, in a scourge called the Black Death, bubonic plague spread throughout Europe resulting in the death of over 25% of the population, a proportion much greater than caused by any war (see Chapter 2).

Klebsiella pneumoniae can cause a type of bacterial pneumonia as well as urinary tract infections, but it is not normally pathogenic. As a soil and freshwater group, this genus is noted for its lack of motility, its formation of a capsule, and the ability of many strains to fix nitrogen. In other respects, it is very similar to the common soil bacterium, *Enterobacter aerogenes*.

The genus *Proteus* is well known for its active **urease**, which breaks down urea to ammonia and carbon dioxide. It is also noted for its active motility, called **swarming**, which occurs when cells align themselves in the long axis and rapidly spread over the surface of an agar plate.

Serratia marcescens is a distinctive enteric bacterium because of its red pigment. The pigment of this bacterium is a bright red, pyrrole-based compound called **prodigiosin**. Prodigiosin is a tripyrrole (**Figure 19.4**) akin to the tetrapyrroles of chlorophyll, cytochrome, and heme. The function of prodigiosin in *S. marcescens* is still unknown.

One of the most recent isolates of the enteric group is *Xenorhabdus*, which grows in association with nematodes. Some strains can actually produce light, a phnomenon referred to as **bioluminescence.** This fascinating capability is more widely found in the *Vibrio-Photobacterium* group discussed next. Bioluminescent bacteria are usually associated with symbiotic associations, and they are therefore discussed in more detail in Chapter 25.

Vibrio and Related Genera

Vibrio, Photobacterium, Aeromonas, and *Plesiomonas* are gram-negative facultatively anaerobic straight or curved rods (**Table 19.6**). All are motile via polar flagella. All are capable of respiration or fermentation. None can use nitrate as the terminal electron acceptor in respiration. Most are oxidase positive and utilize glucose as the sole source of carbon and energy. These three genera are primarily aquatic organisms; those that live in seawater require 2% to 3% NaCl or a seawater-based medium

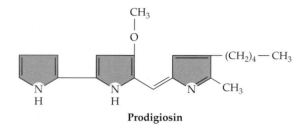

Figure 19.4 Red pigment
Chemical structure of the tripyrrole prodigiosin, the bright red pigment of *Serratia marcescens*.

for optimal growth. Some species are pathogenic to humans and other vertebrates. Some are pathogenic to fish, frogs, and certain invertebrates.

Certain species of *Photobacterium* and *Vibrio* are bioluminescent and emit a blue-green light. The bacteria live as symbionts in the animal light organs, special areas on the animals where these bacteria are maintained in high concentrations (see Chapter 25). The mechanism of bioluminescence has been extensively studied in *Vibrio harveyi* and *Vibrio fischeri*.

The polar flagellum of *Vibrio* species is enclosed in a sheath. The sheath is continuous with the outer envelope of this gram-negative bacterium. Some species also produce lateral flagella when they are grown on solid media. These species lack a sheath and have a shorter wavelength than the polar flagellum. Such flagellation, which is unique to this group, is termed **mixed flagellation**. Apparently, synthesis of the lateral flagella occurs when the microenvironment is particularly viscous, such as on an agar surface. In this manner, the additional flagella aid in the cell's movement.

Vibrio species live primarily in marine and estuarine environments (**Box 19.1**). Many live in the intestines or on the outer surfaces of marine animals. Almost all grow best with added sodium ions in the medium and therefore are typically grown in media with marine salts. Most species grow on a variety of organic compounds (sugars, amino acids, and organic acids) and do not require growth factors. Extracellular hydrolases are pro-

Table 19.6	Characteristics of the *Vibrio* group				
Genus	**Morphology (Size μm)**	**Luminescence**	**Mol % G + C**	**Extracellular Enzymes**	**Habitat**
Vibrio	Straight or curved rod (0.5–0.8 × 1.4–2.6)	+ or –	38–51	+	Aquatic, intestine of marine animals
Photobacterium	Rods (0.8–1.3 × 1.8–2.4)	+ or –	40–44	+	Marine environment
Aeromonas	Rods (0.3–1.0 × 1.0–3.5)	–	58–62	+	Freshwater and sewage
Plesiomonas	Rods (0.8–1.0 × 3.0)	–	51	–	Fish, aquatic animals, some mammals

duced by many species and include amylase, lipase, chitinase, and alginase. Also, some are agar digesters, which produce depressions or cavities on the surface of agar media where their colonies grow. One species, *Vibrio natriegens*, has the shortest-known doubling time, six minutes, of any organism.

Vibrio cholerae is the most thoroughly studied species in the group because it is the causative agent of **cholera.** *Cholera* is an epidemic human disease that occurs in India and more recently in Peru. It is transmitted by fecal contamination of water and food. Effective water treatment has eliminated the disease from most developed countries.

Photobacterium species, like *Vibrio*, are widespread in marine habitats and are also found associated with fish. The luminescent species, *Photobacterium phosphorum,* can be isolated by incubating marine fish, squid, or octopus partially submerged in seawater at 10°C to 15°C. After 15 to 20 hours, luminescent areas develop, and material from them can be streaked on plates for isolation of strains.

Aeromonas species live both in fresh and marine waters as well as in sewage. *Aeromonas salmonicida* is a strict parasite that causes severe diseases of salmon and trout and can also infect humans. It lives in the blood and kidneys of fish.

Plesiomonas shigelloides causes a disease similar to shigellosis, a type of bacterial dysentery. Although it has been isolated from fish and land mammals, it is not a constituent of the normal flora of humans. The disease is apparently spread through contaminated water.

Pasteurella and Hemophilus

Pasteurella and *Hemophilus* are parasites of invertebrates and are often pathogens for both mammals and birds. All

Table 19.7	Characteristics of the *Pasteurella* group	
Genus	**Morphology**	**Mol % G + C**
Pasteurella	Ovoid or rod (0.3–1.0 × 1.0–2.0 µm)	40–45
Hemophilus	Coccobacillus (0.3–0.5 × 0.5–3.0 µm)	37–44

are nonmotile rods that are oxidase positive. They are generally fastidious microbes requiring organic nitrogen sources, B-vitamins, amino acids, and hematin for growth. Characteristics of this group are shown in **Table 19.7**.

Pasteurella spp. are the causative agents of diseases in cattle and of a "fowl *cholera*" in poultry. They are parasites of the mucus membranes of the respiratory tracts of mammals and birds. Some species are found in the digestive tracts of animals, but they rarely cause human disease.

Hemophilus species are obligate parasites of the mucous membranes of humans and other animals. Blood or blood derivatives are required for growth. *Hemophilus influenza* was originally isolated from patients with viral influenza and was thought to be the causative agent of the disease, hence the name. The organism is present in the nasopharynx of healthy individuals and is probably a secondary invader in infections of the respiratory tract. This is the first organism whose entire genome was sequenced.

Zymomonas

Zymomonas is a gram-negative nonmotile rod that is facultatively anaerobic and oxidase negative. The mol % G

Research Highlights Box 19.1 — Are Marine Bacteria Unique?

The oceans began forming very early in the evolution of Earth. Their saltiness increased as land was eroded by rainfall and the rivers carried the dissolved salts into the sea. The total concentration of salts now equals a salinity of about 3.5% or, as biological oceanographers state, 35 °/oo (parts per thousand). Thus, marine bacteria must be able to grow in a saline environment with a lower water activity than their freshwater relatives.

Some microbiologists have defined marine bacteria as those that grow optimally when the salinity of the medium, or growth environment, is at about 3.5%. To test this, special media are prepared that are amended with the appropriate concentrations of sea salts—this is referred to as artificial seawater, or sterile seawater itself may be used.

In addition to the preference to grow best at 3.5% salinity, some microbiologists have proposed that true marine bacteria must also have an absolute requirement for sodium ions. This can be tested by preparing a medium in which the sodium ions are replaced entirely by potassium ions. Using this definition, if the organism cannot grow under these conditions, it is not regarded as a marine organism.

Recently it has been discovered that strains of *Caulobacter* isolated from marine environments differ phylogenetically from those isolated from freshwater. This is consistent with the notion that, millions of years ago, an independent evolution of at least some groups of marine bacteria occurred, so that they have now separated phylogenetically from their freshwater relatives.

+ C is 47 to 50. Strains are occasionally microaerophilic, and some are strictly anaerobic. *Zymomonas* is often a spoiler of beer and cider in which it produces a heavy turbidity and unpleasant odor due to formation of acetaldehyde and sulfide. The organism ferments glucose anaerobically via the Entner-Duodoroff pathway (see Chapter 8):

(1) Glucose → Ethanol + Lactate + CO_2

Small amounts of acetaldehyde and acetyl methyl carbinol are also produced during this fermentation. The palm wines of the Far East and Africa use *Zymomonas mobilis* as the fermenting agent. The organism is also involved in transforming the sugary sap of various agaves to pulque, a fermented beverage produced in Mexico. *Zymomonas* is found on honeybees and in ripening honey because it tolerates high concentrations of sugars.

Chromobacterium

Chromobacterium violaceum is one of the most striking of all bacteria because it produces deep-violet colonies on a solid medium. The pigment, called **violacein** (**Figure 19.5**), is produced in significant amounts during growth on tryptophan, and it bears a structural relationship to this amino acid. Members of the genus are motile by polar flagella and are rod shaped. *C. violaceum* occurs primarily in soil and water and is especially common in tropical soils.

OXIDATIVE RODS AND COCCI

Many of the *Proteobacteria* are obligately aerobic, nonfermentative rods and cocci that obtain their energy through aerobic respiration. They use a large variety of organic compounds as energy sources, depending on the genus. For example, some species of *Pseudomonas* can use more than 100 different organic monomers, including amino acids, sugars, and organic acids. Others in this group are nitrogen fixers. Some members of this group are widespread in aquatic and soil environments, whereas others are obligate parasites of humans and other animals.

Pseudomonads

The pseudomonads are a group of strict aerobes that utilize oxygen as a terminal electron acceptor. Some species also utilize nitrate as an alternate electron acceptor in anaerobic respiration, and therefore can grow anaerobically. The pseudomonads, however, do not ferment, which is a hallmark of the enteric and *Vibrio* groups. The pseudomonads are gram-negative rods (generally $0.5 \times 1–4$ µm), and virtually all species are motile by one or more polar flagella (**Figure 19.6**). The pseudomonads

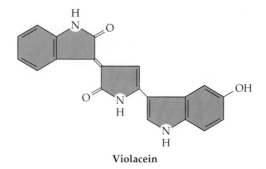

Figure 19.5 Violet pigment
Chemical structure of violacein, the purple pigment of *Chromobacterium violaceum*.

are oxidase and catalase positive. **Oxidase positive** organisms produce large amounts of cytochrome oxidase. The pseudomonads can be differentiated from the *Vibrio* and enteric groups by a few simple diagnostic tests (**Table 19.8**). The pseudomonad group consists of several genera in the gamma *Proteobacteria*, including *Pseudomonas*, *Xanthomonas*, and *Zoogloea*, as well as others in the alpha and beta groups. These are differentiated from one another by the tests outlined in **Table 19.9**.

The genus *Pseudomonas* is the most thoroughly studied and characterized of the group. It is of widespread occurrence in nature. Some species reside in soil, in water, and on surfaces such as human skin. Species are defined on the basis of physiological characteristics. Some species produce water-soluble fluorescent pigments. These are yellow-green pigments called **pyocyanin** and **pyoverdin** that diffuse into the medium and fluoresce under ultraviolet light. The pyocyanins are blue-colored phenazines (**Figure 19.7**). The chemical structure of the pyoverdins is not completely known because they are quite unstable. However, they are believed to be **siderophores**, which are

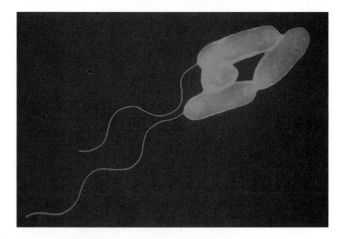

Figure 19.6 *Pseudomonas*
Electron micrograph of typical *Pseudomonas* cells with polar flagella. Photo ©Kwangshin Kim/Photo Researchers, Inc.

Table 19.8	Differentiation of the pseudomonads from the Vibrios and the enteric bacteria		
Characteristic	**Pseudo-monads**	**Vibrios**	**Enteric Bacteria**
Flagella	Polar	Polar or mixed	Peritrichous
Oxidase[a]	+[b]	+	–
Glucose fermentation	–	+	+

[a]The oxidase test is a measure of the enzyme, cytochrome oxidase. It is assayed by picking up a colony of the test organism with a platinum loop and placing a smear directly on filter paper containing two or three drops of a 1% solution of tetramethyl-*p*-phenylenediamine dihydrochloride. If the smear turns violet, the organism is considered oxidase positive.
[b]Some *Xanthomonas* strains are oxidase negative.

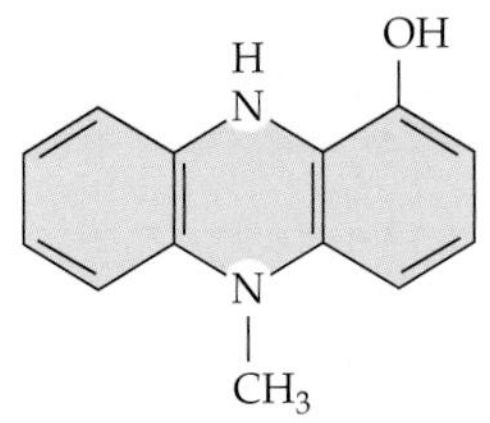

Figure 19.7 Fluorescent pigment
Chemical structure of pyocyanin, the blue phenazine pigment of *Pseudomonas aeruginosa*.

iron-binding compounds that bring ferric ion from the external environment into the cytoplasm.

Members of the genus *Pseudomonas* exhibit remarkable nutritional versatility. Most can grow on 50 or more different substrates, and some can use over 100 different organic compounds. Because of this metabolic versatility, *Pseudomonas* species are very important in the degradation of organic compounds in soil and aquatic environments. Among the substrates utilized are sugars, fatty acids, di- and tricarboxylic acids, alcohols, aliphatic hydrocarbons, aromatic hydrocarbons, amino acids, various amines, and other naturally occurring compounds as well as many manufactured chemicals such as chlorinated hydrocarbons. Glucose is catabolized by many of the pseudomonads via the Entner-Duodoroff pathway (see Chapter 8). *Pseudo-monas* species also have a limited capacity to hydrolyze polymeric compounds.

Pseudomonas species are used extensively as biochemical tools in the elucidation of catabolic pathways. They have been particularly useful in ascertaining pathways in the catabolism of aromatic compounds (see Figure 12.14). The genetics of many fluorescent strains has been clarified. Both conjugational and transduction systems have been established. The genetic information for synthesis of the enzymes involved in catabolism of many uncommon organics occurs on transferable plasmids. Among these are genes for catabolism of salicylate, camphor, octane, and naphthalene. It should be noted that the genus *Pseudomonas* has changed greatly since its taxonomy was evaluated by rRNA analysis. The original genus has now been split into several additional genera, including *Comamonas* (an alpha subdivision genus that does not produce fluorescent pigments); *Deleya* and *Halomonas*, which are marine; *Stenotrophomonas* (with its single species, *Stenotrophomonas maltophilia*); *Hydrogenophaga* (H$_2$-oxidizing facultative chemolithotrophs); and *Burkholderia*.

Table 19.9	Characteristics used to distinguish the common genera of pseudomonads[a]					
	Characteristics					
Subphylum/Genus	**Growth Factor Requirement**	**Hydrogen Autotroph**	**Flock Formation**	**Marine**	**Plant Pathogen**	
γ *Proteobacteria*						
Pseudomonas	–[a]	V	–	V	V	
Xanthomonas	+	–	–	–	+	
Syntrophomonas	+	–	–	–	–	
Zoogloea	+	–	+	–	–	
α and β *Proteobacteria*						
Comamonas	–	–	–	–	–	
Deleya/Halomonas	–	–	–	+	–	
Hydrogenophaga	–	+	–	–	–	

[a]Symbols: + = positive for most strains; – = negative for most strains; V = variable by strain within a genus.

Research Highlights Box 19.2   **Ice Nucleation**

Certain gram-negative plant pathogenic bacteria, including some species of *Pseudomonas*, *Xanthomonas*, and *Erwinia*, cause ice nucleation. These bacteria normally colonize plant leaves in the spring and summer. In the fall, when the temperature begins to dip toward freezing, they cause frost damage by forming ice crystals at temperatures somewhat higher than normal freezing temperatures. The ice crystals damage the plant leaf cells, which then exude nutrients that these bacteria utilize. Ice nucleation activity is caused by a protein located in the outer cell membrane of these bacteria. This protein has a distinctive repeating amino acid region, which is the site of ice nucleation.

Pseudomonas syringae, one of the ice-nucleating species, is also used to prevent ice nucleation damage to crops. To accomplish this, a mutant strain that has a deletion of the ice nucleation gene, an "ice-minus" strain, is sprayed on crop plants early in the spring. The concept is to enable the mutant bacterium to colonize the leaves before a wild-type, "ice-plus" strain reaches the leaves. Then, later on when a pathogenic, "ice-plus" strain arrives at the plant leaf, it will be prevented from colonizing because the ice-minus mutant is already there. This ecological concept is termed "preempting the niche," the result of which is that the ice-plus phytopathogen cannot colonize and subsequently cause damage to the plant.

The ice-nucleating bacterium, *Pseudomonas syringae*, is used commercially in snowmaking. The strain is added to water in snowmaking machines used in ski areas to raise the temperature at which the water freezes to make snow. In this manner, less energy is required to cool the water to sufficiently low temperatures.

Pseudomonas aeruginosa is a human pathogen. It is especially important in humans as an opportunistic pathogen of compromised hosts. For example, it causes serious skin infections of burn victims and grows in the lungs of patients with cystic fibrosis. However, it can also cause urinary tract and lung infections of normal individuals. Normally, it is found in soils and is an important denitrifying genus. Certain other *Pseudomonas* species are animal pathogens such as *Pseudomonas mallei*, which causes glanders in horses.

A number of species of *Pseudomonas*, including *Pseudomonas syringae*, are plant pathogens. *P. syringae* owes its invasiveness to its ability to form ice crystals, which damage the plant tissue (**Box 19.2**).

Xanthomonas is a genus of yellow-pigmented plant pathogens. The yellow pigment is a brominated aryl polyene of unknown function (**Figure 19.8**) called **xanthomonadin**. Disease symptoms caused by *Xanthomonas* vary with plant species and the strain involved. Stem wilt, leaf necrosis, and other manifestations of disease are observed. These are ascribed to toxins, enzymes, ice nucleation activity, and other metabolic products that accumulate during the growth of the bacterium in its host tissue.

Zoogloea is another pseudomonad genus. It forms masses of cells, called **zoogloea**—for which the genus was named—enclosed in a polysaccharide gelatinous matrix (**Figure 19.9**). *Zoogloea* occur in organically polluted freshwater and in wastewaters such as activated sludge in sewage treatment plants.

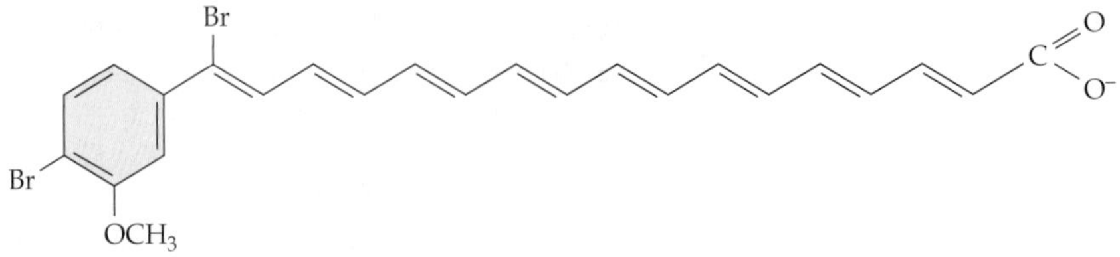

Azotobacter and Other Free-Living Nitrogen-Fixing Bacteria

The ability to utilize atmospheric N_2 as a sole source of cellular nitrogen is widespread among prokaryotic genera. Several otherwise unrelated aerobic gram-negative species are characterized in **Table 19.10**. Although these organisms are genetically distinct, they share a number of characteristics and are thus considered here as a group. All are free-living bacteria found in soil and waters. Most strains are motile. These aerobic organisms have an oxygen-sensitive, N_2-fixing nitrogenase system (see Chapter 10). Under favorable conditions, they fix about 10 mg of N_2 per gram of carbohydrate consumed. Because nitrogenase is sensitive to oxygen, it must be protected from excessive

Figure 19.8 Yellow pigment
Chemical structure of xanthomonadin, the brominated pigment from *Xanthomonas* spp.

O_2 in order for the fixation process to be induced and function. All of the organisms can utilize ammonia as their nitrogen source, and nitrogenase is induced as a response to the depletion of ammonia. All species produce copious quantities of a slimy, mucoid capsular material when growing with atmospheric N_2 as nitrogen source. In nature, this mucoid layer may limit the amount of oxygen that reaches the cell, or it might harbor aerobic bacteria that consume O_2 in the immediate vicinity.

Martinus Beijerinck at Delft University originally isolated *Azotobacter,* and it is the most thoroughly studied of the free-living nitrogen fixers. The cells are large, pleomorphic ovoid rods, 1.2 to 2.0 μm in diameter (**Figure 19.10**). *Azotobacter* has an exceedingly high respiration rate (O_2 uptake in the presence of substrate), which may aid in retaining low O_2 levels in the cell. As mentioned, nitrogenase is not functional in the presence of molecular oxygen.

Members of the genus form cysts that are resting bodies resistant to drying and radiation but not to heat. The cyst coat surrounds a cell that in other respects is quite similar to a typical vegetative cell. The cysts are not completely dormant because they can oxidize exogenous energy sources. The cyst-forming process can be triggered by supplying the organism with 1-butanol. Removal of the cyst coat by mechanical disintegration or metal chelators yields a viable dividing cell.

Azotobacter spp. are generally found in rich soil that is slightly acid to alkaline in pH and high in phospho-

rus. Sugars, organic acids, and fatty acids serve as substrates for their growth. The species most prevalent in soil is *Azotobacter chroococcum. Azotobacter vinelandii* is the most studied and characterized from a biochemical and genetic standpoint. The more efficient nitrogenase in *Azotobacter* species contains molybdenum, but recently a vanadium-containing nitrogenase has been characterized. This enzyme may serve as a backup in soils containing low levels of molybdenum.

Beijerinckia is the genus of nitrogen fixers found in acidic soils, particularly of tropical regions. *Beijerinckia* spp. form straight or curved rods with rounded ends. The cells are often misshapen with polar lipid bodies surrounded by a membrane. The lipid bodies are composed of poly-β-hydroxybutyric acid. Nitrogen-fixing colonies form copious quantities of tenacious elastic slime. Fixation is enhanced by reduced O_2 tension. Cysts

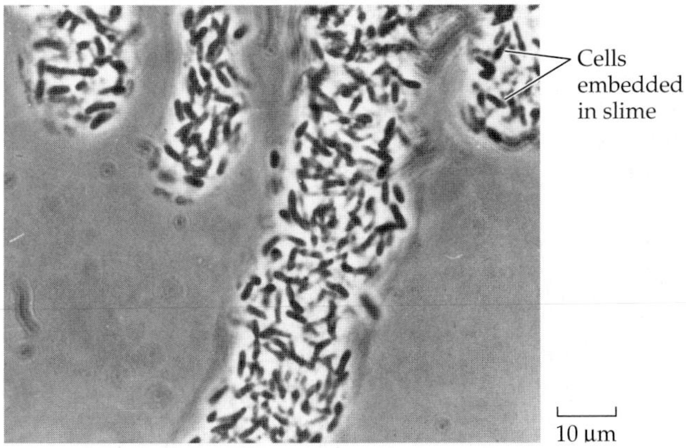

Table 19.10	Free-living nitrogen-fixing bacteria			
Genus	**Form Cysts or Resting Bodies**	**Autotrophic Growth**	**Mol % G + C**	**Habitat**
Azotobacter	+	–	63–67	Rich soil, high in phosphate, slight acid to alkaline pH
Beijerinckia	+	–	55–61	Tropical soil; acid pH
Azomonas	–	–	52–59	Soil and water
Azospirillum	+	+/–	69–71	Free-living in soil or with roots
Derxia	–	+	69–73	Tropical soil
Xanthobacter	–	+	65–70	Wet soil and water

Cells embedded in slime

10 μm

Figure 19.9 *Zoogloea*
Zoogloeal cell mass. Each finger-like projection contains numerous cells embedded in an extracellular slime matrix, which appears transparent. Courtesy of R. Unz.

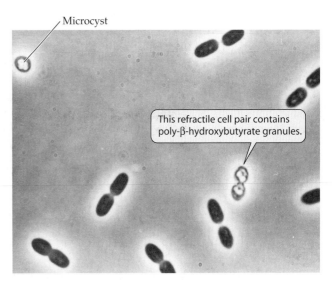

Microcyst

This refractile cell pair contains poly-β-hydroxybutyrate granules.

Figure 19.10 *Azotobacter*
Cells of an *Azotobacter* sp. Courtesy of J. T. Staley.

containing one cell may occur in *Beijerinckia*, however, during growth on a nitrogen-free medium, several individual cells will be enclosed in a common capsule. *Beijerinckia* spp. use glucose, fructose, and sucrose as the sole source of carbon, but grow poorly on glutamate.

Several other genera are also important nitrogen fixers, including *Azomonas*, which resembles *Azotobacter* and *Derxia*, which is common in acidic tropical soils. *Derxia* spp. can grow on sugars, alcohols, and organic acids. They can be distinguished from *Beijerinckia* by their ability to grow as a hydrogen-utilizing chemoautotroph. *Xanthobacter* species can also grow as facultative hydrogen-utilizing chemoautotrophs. *Xanthobacter* use a variety of carbon sources including l-butanol, l-propanol, and tricarboxylic acid intermediates. *Xanthobacter* occurs in wet soil and water containing decaying organic matter.

Azospirillum is unique in that it is a nitrogen fixer that grows in association with the roots of grains such as corn. These are plump curved rods that grow either as free-living bacteria in soils or may be associated with the roots of monocotyledonous plants including cereal grains, grasses, and tuberous plants. They do not produce root nodules. *Azospirillum* spp. form distinct capsules around the cell that give some resistance to desiccation. Sugars and organic acids such as malate are used as substrates; some strains are also hydrogen chemoautotrophs. Nitrogen is fixed only under microaerophilic conditions.

Rhizobium and *Bradyrhizobium*: Symbiotic Nitrogen-Fixing *Bacteria*

Rhizobium spp. may be free living or they may develop a symbiotic nitrogen-fixing association with leguminous plants (**Box 19.3**). Legumes are plants that bear seeds in pods and include soybeans, clover, alfalfa, peas, vetch, lupines, beans, and peanuts. The bacteria penetrate the roots or, in some cases, stems of the plant and form a nodule. The bacteria reside in this nodule and, under appropriate conditions, fix atmospheric nitrogen providing a source of nitrogen for plants growing in soil deficient in this nutrient. The symbiotic association of these bacteria and legumes is discussed more fully in Chapter 25.

Some of the characteristics of *Rhizobium* and *Bradyrhizobium* are given in **Table 19.11**. *Rhizobium* is generally involved in nodule formation in plants in the temperate zone. It is faster growing (with a generation time of 4 hours) and produces colonies 2 to 4 mm in diameter within 3 to 5 days on a yeast-mannitol-mineral salts agar medium. *Bradyrhizobium* grows slowly with a generation time of 8 hours. It nodulates some plants in the temperate zone but is also effective in nodulation of tropical leguminous plants. *Bradyrhizobium* produces colonies on the yeast-mannitol medium that do not exceed 1 mm in diameter in 5 to 7 days. There is a specificity in the species of legume nodulated by organisms within these genera. The leguminous plants infected by a particular strain are considered to be a cross-inoculation group (**Table 19.12**). The three described species of *Rhizobium* are *Rhizobium leguminosarum*, *Rhizobium meliloti*, and *Rhizobium loti*. The three major strains of *R. leguminosarum* are *trifolii*, *phaseoli*, and *viceae*.

Agrobacterium

Members of the genus *Agrobacterium* induce tumors in dicotyledonous, but not monocotyledonous plants. The disease is commonly called **crown gall**, as it frequently occurs at the soil-stem interface, the crown of the plant (**Figure 19.11**). Other varieties of the disease are hairy root and cane gall. The major species involved in gall

---

Milestones Box 19.3 | **History of Symbiotic Nitrogen Fixation**

Root nodules were observed on leguminous plants by early scientists. Marcello Malpighi (1628–1694), an Italian anatomist, drew elaborate pictures of leguminous plants depicting root nodules. During the early nineteenth century, scientists realized that the amount of combined nitrogen (nitrate and ammonium) in soils controlled the production of cereal grains. In contrast, the growth of legumes appeared to be independent of the nitrogen content. It was also noted that the total

combined nitrogen measured in soils after the growth of legumes was greater than what was present before. For this reason, it became common practice to grow legumes in nitrogen-poor soil and plow the resultant crop back into the soil. The soil was then fit for growth of cereal grains. This alternation of legumes and grains was the beginning of the routine practice of crop rotation.

In 1888 Martinus Beijerinck isolated and cultivated bacteria from root nodules. He demonstrated that

these bacteria could cause nodule formation in legumes grown in sterile soil. This confirmed that a relationship existed among the nodule, the bacterium, and nitrogen replenishment in soil. The potential utility in taking advantage of this for growing food crops without addition of nitrogen fertilizer was evident to agriculturalists. Since that time, research on the symbiosis between legumes and rhizobia has become an important and exciting area.

Table 19.11	Characteristics of free-living *Rhizobium* and *Bradyrhizobium* species

Gram-negative rods (0.5–0.9 × 1.2–3.0 μm)
Aerobic
Motile
Produce copious amounts of extracellular slime
Mol % G + C = 59–64
Metabolize glucose via Entner-Duodoroff pathway
Fix N_2 under microaerophilic conditions
Some species can grow as H_2 chemoautotrophs

Figure 19.11 Crown galls
Photograph of crown galls on a plant. This plant was artificially wounded by cutting with a knife then inoculated with *Agrobacterium tumefaciens*. Courtesy of E. W. Nester.

formation is *Agrobacterium tumefaciens*, whereas *Agrobacterium rhizogenes* causes hairy root disease. The genus *Agrobacterium* is closely related to *Rhizobium*. They are gram-negative rods (0.6–1.0 μm × 1.5–3 μm) that are motile by peritrichous flagellation. They are aerobes with a mol % G + C of 57 to 63. Agrobacteria utilize a variety of carbohydrates, organic acids, and amino acids as carbon sources. Some species use ammonia or nitrate as nitrogen sources, whereas others require amino acids and other growth factors. *Agrobacterium* spp. are common in soil and are often abundant in the rhizosphere, or root zone, of plants. In the past, *A. tumefaciens* received considerable attention as a model of tumor induction

with possible implications for tumor formation in humans. Now it is of commercial importance as a mechanism for genetic engineering of plants.

A. tumefaciens generally attacks plants at a wound site at the root stem interface. It is believed that the bacterium attaches at the wound site by one of two mechanisms: (1) specific components of the bacterial envelope, such as glucans, interact with plant cells, or (2) cellulose fibrils are formed by the *A. tumefaciens* strains that anchor the bacterium to the plant cells. The entire bacterium does not enter the plant. Instead, all oncogenic (tumor-producing) strains have a large conjugative plasmid, designated Ti for tumor inducing, that is, released into the site of infection. This plasmid carries several important genes:

1. Virulence genes that induce tumor formation

2. Genes for substances that regulate the production of plant growth hormones, auxins, and cytokinins

Table 19.12	Cross-inoculation groups nodulated by *Rhizobium* and *Bradyrhizobium* species[a]

Bacterial Species	Plant Genus	Plant Common Name
Rhizobium leguminosarum		
	Pisum	Field pea
	Lathyrus	Pea
	Vicia	Vetch
	Lens	Lentil
	Phaseolus	Bean
	Trifolium	Clover
Rhizobium meliloti		
	Melilotus	Sweet clover
	Medicago	Alfalfa
Rhizobium loti		
	Lotus	Trefoil
	Lupinus	Lupines
	Mimosa	Mimosa
Bradyrhizobium japonicum		
	Glycine	Soybean
	Arachis	Peanut

[a]Species listed normally cause root nodules on some, but not necessarily all, genera of legumes listed.

3. Genes that direct the plant to synthesize opines (Figure 19.12), special amino acids that serve as specific substrates for *Agrobacterium* species in their soil environment

4. Genes for enzymes that degrade opines

Only a specific part of the Ti plasmid enters the plant cell. This portion, designated T-DNA (where T stands for transforming) is a series of genes residing in the Ti plasmid between two 23-base-pair direct repeat sequences. The T-DNA carries the information listed earlier, is inserted into the plant chromosome at various sites, and is maintained in the plant cell nucleus. Once a plant cell is transformed, its progeny continue to exhibit the tumor-producing characteristic in the absence of bacteria. There is considerable evidence that T-DNA gene expression is controlled by plant regulatory elements; that is, the plant actually assists in the disease process.

The Ti plasmid is an excellent vector for introducing genetic information into a plant. For example, genes can be introduced into tobacco plants by genetic engineering (see Box 16.2). Some traits that are beneficial to establish in crop plants include herbicide resistance, resistance to plant pathogens, resistance to adverse conditions (drought, heat, and salinity), and improved nutritional value. Although dicotyledonous plants are normally infected by *Agrobacterium*, research is focused on introducing the Ti plasmid into monocotyledonous plants such as corn, wheat, rice, and other cereal grains that are not susceptible to the bacterium.

Vinegar Bacteria: *Acetobacter* and *Gluconobacter*

Vinegar has been in use as a preservative and flavoring agent throughout human history. As early as 6000 B.C., the Babylonians depicted methods for converting ethanol from beer to vinegar (acetic acid). Two genera of bacteria, *Acetobacter* and *Gluconobacter*, are the major microbes responsible for vinegar production (**Table 19.13**). Both of these genera are natural inhabitants of flowers, fruit, sake, grape wine, brewers yeast, honey, and cider. *Gluconobacter* is found on the beech wood shavings of vinegar generators (see Chapter 31). *Gluconobacter* and *Acetobacter* can be differentiated from other gram-negative genera by their ability to oxidize ethanol to acetic acid at low pH (4.0–5.0). Both genera also cause pink disease of pineapple and the rotting of apples and pears. They are pests in the brewing and wine industry because they convert ethanol to acetic acid. Besides acetification, they cause ropiness, turbidity, and off flavors in beverage products, including soft drinks.

Gluconobacter cells are also ellipsoidal to rod shaped and often form enlarged, irregular forms. Some strains are motile by polar flagel-

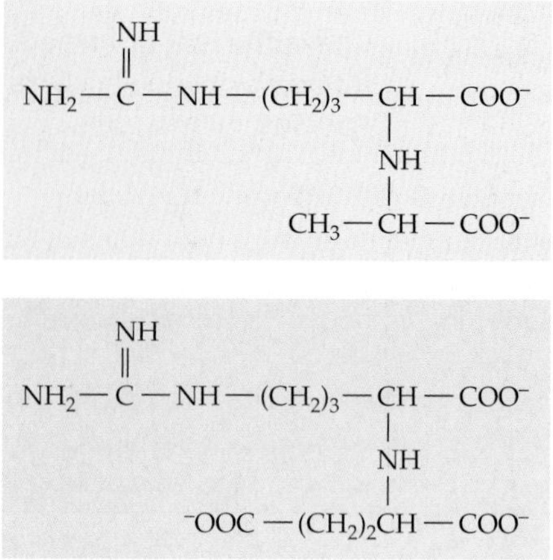

Figure 19.12 Opines
Chemical structures of two unusual amino acids, octopine and nopaline, opines produced by plants infected with *Agrobacterium tumefaciens*. The genes for producing these opines are transferred from the Ti plasmid to the plant genome, the plant excretes the opines, and they are degraded by the same *Agrobacterium* strain, living in the soil. This suggests one way in which *Agrobacterium* benefits from the crown gall association.

la. Extensive analysis of about 100 strains indicates there is only one species in this genus, designated *Gluconobacter oxydans*. *Gluconobacter* is a strict aerobe, but it lacks a complete tricarboxylic acid cycle and is therefore called an **underoxidizer.** As a result, it produces acetic acid as a terminal product. It is **ketogenic,** meaning that it oxidizes glucose to 1-ketogluconate, and some strains produce 5-ketogluconic acid. *Gluconobacter oxydans* also can produce intermediates such as dihydroxyacetone from glycerol.

Acetobacter cells are also ellipsoidal to rod-shaped gram-negative organisms, but they are motile by peritrichous flagellation. These are obligately aerobic bacteria that prefer an acid pH for growth. *Acetobacter* is called an **overoxidizer** because it can oxidize acetic acid all the way to carbon dioxide via its complete tricarboxylic acid cycle. *Acetobacter* is also ketogenic. It converts sorbitol to 2-keto sorbitol and glucose to 2-keto- and 5-ketogluconic acids. For this reason, this genus is used commercially for the production of ketonic acids such as ascorbic acid (vitamin C).

Table 19.13	Differences between *Acetobacter* and *Gluconobacter*	
Characteristic	***Acetobacter***	***Gluconobacter***
Cellulose production	+ or −	−
Complete tricarboxylic acid cycle	+	−
Flagellation	Peritrichous	Polar
Mol % G + C	56–60	56–64

Some strains of *Acetobacter* form extracellular cellulose microfibrils that surround the dividing cell mass. Therefore, the cells become embedded in a large mass of cellulose microfibrils. The cellulose is formed internally and secreted through pores in the lipopolysaccharide outer envelope. In static culture, cellulose-synthesizing cells are favored, and in shake culture, cellulose-free mutants predominate. Cellulose production is rare among bacteria, and its purpose is unknown. It has been postulated that cellulose production aids in maintaining the organism on the surface of liquids so they have oxygen available for growth. However, many aerobic organisms survive in equivalent environments without such an elaborate mechanism for flotation.

Legionella

DNA/DNA hybridization and 16S rRNA sequencing confirm that *Legionella* species and related organisms form a distinct group among the aerobic, gram-negative rods. *Legionella pneumophila*, which is the causative agent of Legionnaires' disease, is the type species. All species in the genus have been implicated in human respiratory disease. Morphologically, they are relatively small bacteria (0.5×2 µm). All strains are motile by polar or lateral flagella. Members of this genus have complex nutritional requirements, and L-cysteine and iron salts must be provided for growth. All are strict aerobes that utilize amino acids as their source of carbon and energy. They cannot oxidize or ferment carbohydrates.

Strains of *Legionella* and associated genera are commonly found in ponds, lakes, and wet soil. They also thrive in the warm water associated with evaporative cooling towers. It is probable that aerosols from this source caused an outbreak of the severe respiratory infection among the attendees at an American Legion Convention in Philadelphia in 1976. Several fatalities resulted from this outbreak. Difficulties in culturing the organism and assessing its nature pose a problem in prevention and treatment of the disease. It is not transferred from one human to another by contact. Sensitive immunofluorescence techniques indicate that *Legionella* strains are of widespread occurrence in the environment. Thus, they must either be of low pathogenicity or transferred to humans only rarely.

Neisseria and Related Genera

The *Neisseria* group are gram-negative, aerobic bacteria of varying morphology (**Table 19.14**). Most are nonmotile. *Neisseria* is the major genus of the group. The best-known species of *Neisseria* is *Neisseria gonorrhoeae*, the causative agent of the common venereal disease, **gonorrhea** (see Chapter 28). Another pathogenic member of this genus is *Neisseria meningitidis*, a causative agent of **cerebrospinal meningitis**. Other species are typically found in the nasopharynx and respiratory passages of warm-blooded animals. These rarely cause disease and are considered part of the normal microflora. *Neisseria* is best cultivated on chocolate-blood agar at 37°C in a 3% to 10% atmosphere of CO_2.

Moraxella is found in the eyes and upper respiratory tracts of humans and other warm-blooded animals. Some species cause conjunctivitis (eye inflammation) in humans and bovines. *M. catarrhalis* is a normal inhabitant of the nasal cavity of humans, but it is an opportunistic pathogen associated with disease in unhealthy patients.

Most strains of *Acinetobacter* have simple nutritional requirements, and as the genus name implies, are immotile. *Acinetobacter* is a group of nutritionally diverse organisms that can utilize a wide array of substrates, including selected hydrocarbons. They cannot use hexoses as the sole source of carbon and energy, although they can oxidize aldose sugars to the corresponding sugar acid. For example, glucose + O_2 → gluconic acid. *Acinetobacter calcoaceticus* is a prominent member of this genus and is readily isolated from soil and water.

Simonsiella is a unique filamentous, gliding bacterium. It resembles a watchband (**Figure 19.13**). One side of the filament, the ventral side, which is in contact with the surface on which it glides, is flattened. The other, or dorsal, side of the filament is rounded.

This genus has a very distinctive habitat as well. The organism is found in the oral cavity of animals including humans, dogs, cats, and sheep, each animal species harboring its own distinctive species. These organisms are aerobic and respire using a variety of sugars as sub-

Table 19.14	*Neisseria* group of gram-negative bacteria			
Genus	Cell Shape (Size)	Mol % G + C	Oxidase Reaction	Habitat
Neisseria	Cocci (0.6–1.0 µm)	47–54	+	Mucus membranes of mammals
Moraxella	Plump rods (1.0–1.5 × 1.5–2.5 µm)	40–48	+	Mucus membranes of warm-blooded animals
Simonsiella	Filamentous (2–8 × 10–50 µm)	41–55	+	Mucus membranes of mammals
Kingella	Rods (1 × 2–3 µm)	47–55	+	Mucus membranes of humans
Acinetobacter	Rods (0.9–1.6 × 1.5–2.5 µm)	38–47	−	Soil, water, sewage

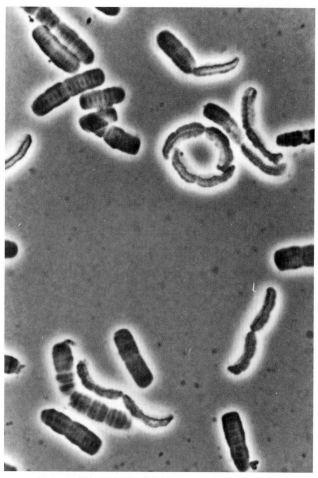

Figure 19.13 *Simonsiella*
Phase photomicrograph of *Simonsiella* cells. Note that the filaments appear as watchbands. The flat, ventral side is the surface used for gliding. The dorsal side is rounded and shows the cell segments. Courtesy of J. T. Staley.

strates for growth. They are not harmful to humans or other animals but are instead part of the normal flora.

Kingella, the final genus of this group, is a common organism in the mucus membrane of humans as part of the normal flora.

Rickettsia

Rickettsia species are unable to grow outside a living host cell. Instead, they live as intracellular parasites of animals and are therefore called **obligate intracellular parasites**. *Rickettsia* are parasitic in humans and other vertebrates. Members of this genus are the causative agents of typhus (*Rickettsia prowazeki*), Rocky Mountain spotted fever (*Rickettsia rickettsii*), and scrub typhus (*Rickettsia tsutsugamushi*). Rickettsias are often associated with arthropods, usually intracellularly. Arthropods such as ticks, lice, and fleas are vectors that introduce the *Rickettsia* spp. into humans and other mammals

through a bite (the medical aspects are considered in more detail in Chapter 28).

Rickettsia spp. are short rods (0.3–0.5 × 0.8–2.0 μm) that have not yet been cultivated away from host tissues (**Figure 19.14**). They multiply readily in embryonated eggs or metazoan cell culture, provided that the host cells are viable. They have a generation time in vivo of 8 to 9 hours. Their cell walls are typical of gram-negative bacteria with muramic acid and diaminopimelic acid in their peptidoglycan.

The rickettsias derive energy by oxidation of glutamic acid via the tricarboxylic acid cycle. The generation of NADH is coupled to an electron transport system for ATP synthesis. They require a source of AMP and other nucleoside monophosphates. *Rickettsia* spp. have limited synthetic capacity but can synthesize low molecular weight proteins and lipids. Monomers such as amino acids must be provided by the host cell.

Rickettsia prowazeki multiplies to a high density in the cytoplasm of a host causing disruption of host cells. The released bacteria move on and infect other cells. *R. rickettsii* produces small numbers of cells in the cytoplasm of infected cells that escape to extracellular spaces. These then go on to infect other host cells.

SHEATHED BACTERIA

Some bacteria produce **sheaths,** which are distinctive layers formed external to the cell wall (**Figure 19.15**). The three major genera in this group are *Sphaerotilus*, *Leptothrix*, and *Crenothrix*. They are differentiated from one another on the basis of their deposition of ferric hydroxide and manganese oxide as well as other features (**Table 19.15**).

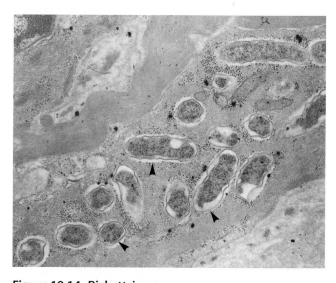

Figure 19.14 Rickettsia
Photomicrograph showing *Rickettsia* cells (arrowheads) infecting animal cells. Photo ©Science VU/Visuals Unlimited.

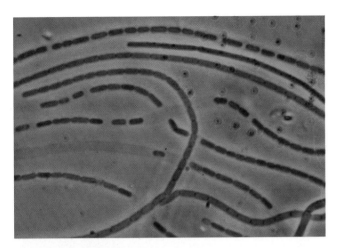

Figure 19.15 *Sphaerotilus*
Micrograph of *Sphaerotilus* filaments, showing the cells in chains within a sheath. The sheath appears as a transparent tube about 1.5 μm in diameter. Photo ©Michael Richard/Visuals Unlimited.

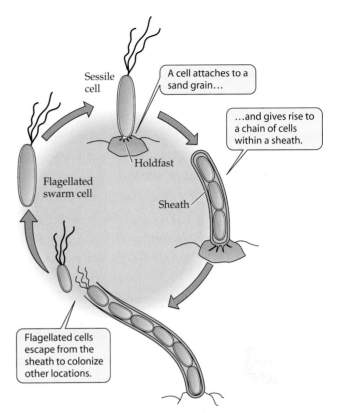

Figure 19.16 Life cycle of *Sphaerotilus natans*
Diagram of the life cycle of *Sphaerotilus natans*.

Sphaerotilus and *Leptothrix* are large rod-shaped bacteria (1–2 × ~10 μm) that are motile by polar flagella. They live in flowing aquatic habitats where they attach to inanimate materials such as rocks and sticks. They undergo a characteristic life cycle (**Figure 19.16**). Single motile cells have a holdfast structure that allows them to attach to inanimate substrata such as sand grains. When they are attached, they become sessile. These cells then reproduce by binary transverse fission to form a chain of cells enclosed in the sheath. As the chain grows and elongates, motile cells are produced and released from the unattached end. These motile swarmer cells repeat the life cycle and provide the organism with a means of dispersal to other habitats.

These bacteria are aerobic heterotrophs that use a variety of organic substrates for growth, including sugars, alcohols, and organic acids. All species known require vitamin B_{12} for growth. They are common in flowing aquatic habitats where they utilize the nutrients that pass by them in the water. They are particularly common in the Pacific Northwest, where they grow downstream of pulp mills and sewage treatment facilities. They can be a nuisance because when they grow profusely in enriched habitats, clumps can break off and clog fishermen's nets and water treatment inlets. Their growth has been controlled considerably since pulp mills have been required by the Environmental Protection Agency to treat pulping plant effluent prior to discharge into receiving streams.

The sheaths of *Sphaerotilus natans* have been analyzed chemically. They have a complex composition similar to that of the outer membrane of gram-negative bacteria consisting of protein, carbohydrate, and lipid.

Leptothrix species appear dark brown or black due to the deposition of iron and manganese oxides in their sheaths (**Figure 19.17**). These organisms are commonly found in iron springs. Samples from such springs are filled with encrusted sheaths that are devoid of cells. Conditions away from the mouth of the spring are not conducive to the growth of these bacteria, and the cells leave the sheaths or are lysed. However, samples taken close to the mouth of the spring where conditions are more reduced contain cells within the sheath.

Some controversy exists over the ability of these bacteria to obtain energy from the oxi-

Table 19.15	Differences among *Sphaerotilus, Leptothrix,* and *Crenothrix*		
Genus	**Mol % G + C**	**Flagellation**	**Distinguishing Features**
Sphaerotilus	70	Polar tuft	May deposit iron oxide
Leptothrix	69–71	Single polar	Deposits iron oxide; oxidizes Mn^{+2}
Crenothrix	Unknown	Unknown	Tapering sheath

Figure 19.17 *Leptothrix*
Phase photomicrograph of a filament of a *Leptothrix* sp. from a natural sample. The FeO(OH) and MnO_2 accumulated in its sheath have imparted a brownish color. The sheath is about 1.5 μm in diameter. Courtesy of J. T. Staley.

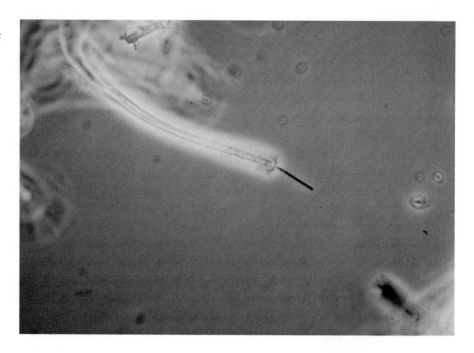

dation of reduced iron and manganese ions. As yet there has been no conclusive demonstration of chemolithotrophy using these inorganic substrates. The pH of the environment in which these organisms grow is near neutral, and virtually all iron would be already oxidized to the ferric state. However, manganese is available in part in the Mn^{2+} state and could be used as a source of energy, as suggested early on by Winogradsky. However, although enzymes have been found that enable the oxidation of Mn^{2+} to the Mn^{4+} state in *Leptothrix*, as yet there has been no demonstration of ATP production in this process. Many other bacteria are known to be able to oxidize and/or deposit manganese and iron, but with the exception of *Thiobacillus ferrooxidans* and *Gallionella ferruginea*, none are known to be able to derive energy from this process.

Crenothrix is a genus that has been known for over a century, but has not yet been successfully grown in pure culture. Although it resembles the other sheathed bacteria, it has a tapered sheath resembling a cornucopia

Figure 19.18 *Crenothrix*
Photomicrographs of *Crenothrix polyspora* from a freshwater source. Note the slightly tapered sheaths with the smallest cells near the tips in (A) and (B). Cells are being released from the sheath in (C). These are Nomarski interference photomicrographs. Courtesy of P. Hirsch.

(**Figure 19.18**) and produces small cells from the larger unattached end of the sheath.

THE PROSTHECATE BACTERIA

Prosthecate bacteria are unicellular bacteria that have appendages extending from their cells that give them remarkable and distinctive cell shapes (**Figure 19.19A**). The appendages are called **prosthecae,** which, by definition, are **cellular appendages,** *or extensions of the cell that contains cytoplasm* (Figure 19.19B). There are two important consequences of having these appendages. First, they increase the surface area of the cell, resulting in a higher surface area to volume ratio, thereby allowing the cells increased access to the nutrients in the environment. Because these organisms live in aquatic environments that have low nutrient concentrations, the prosthecae provide greater exposure to the nutrients

(A)

(B)

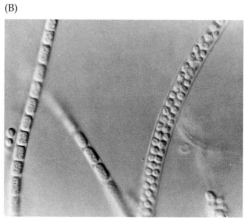

(C)

(A)

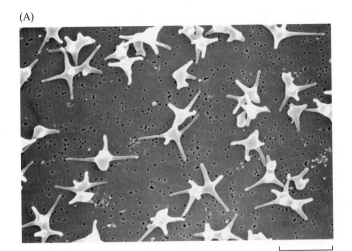

5.0 μm

(B)

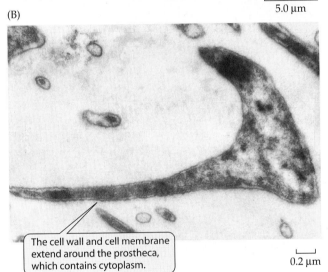

The cell wall and cell membrane extend around the prostheca, which contains cytoplasm.

0.2 μm

Figure 19.19 *Ancalomicrobium*
(A) Scanning electron micrograph of *Ancalomicrobium adetum*. Each cell has several long prosthecae extending from it. (B) Thin section through *A. adetum* cell and prostheca. Courtesy of J. T. Staley.

also important for these planktonic organisms because it provides greater friction or drag and therefore slows their settling out from the plankton.

There are three major groups of prosthecate *Proteobacteria*: the **caulobacters**, the **hyphomicrobia**, and the **polyprosthecate** bacteria (**Table 19.16**). Almost all heterotrophic prosthecate bacteria are members of the alpha group of the *Proteobacteria*, but not all alpha *Proteobacteria* have prosthecae. Next, each of the prosthecate groups is considered individually.

Caulobacters

There are several genera of caulobacters (**Table 19.17**). The best-studied genus is *Caulobacter* with a single polar prostheca, which in this genus is commonly called a "stalk" (**Figure 19.20**). *Caulobacter* spp., and certain other prosthecate and or budding bacteria, have a life cycle (**Figure 19.21**). The life cycle of *Caulobacter* spp. consists of two stages: (a) a motile stage and (b) a sessile, or prosthecate, stage. In the motile stage, the cell is called a **swarmer cell**, which has a single polar flagellum and, at the same site, bears a **holdfast**, a special adhesive organelle that mediates attachment. The holdfast allows the cell to attach to detritus or other particulate material in the environment. Before the swarmer cell divides, it develops a polar prostheca at the same site as the flagellum and loses motility. Normally by this time in its development, the cell would be attached to a detritus particle or a dead algal cell in the environment by virtue of its sticky holdfast. The *Caulobacter* cell then elongates and produces a daughter swarmer cell at the nonprosthecate pole. Next, the daughter cell becomes motile and separates from the mother cell to become free in the planktonic environment and ready to repeat the cycle. The mother cell continues to produce daughter cells from the

they require for growth. Studies have shown that these organisms have very high affinity nutrient uptake systems, which enables them to take up nutrients at very low concentrations. They are therefore examples of **oligotrophic** bacteria. Second, the increased surface area is

Table 19.16		**Groups of heterotrophic prosthecate bacteria**	
Group	**Budding**	**Prosthecae (No. per Cell; Location)**	**Reproductive Role of Prosthecae**
Caulobacters	–	One (rarely two); polar	–
Hyphomicrobia	+	One to several; polar[a]	+
Polyprosthecate bacteria	+	Several to many	–

[a]The genus *Pedomicrobium* can produce prosthecae at other locations on the cell surface.

Table 19.17	**Genera of caulobacters**	
Genus	**Prostheca Position**	**Habitat**
Caulobacter	Polar	Freshwater
Maricaulis	Polar	Marine
Asticcacaulis	Subpolar	Freshwater

(A)

(B)

These refractile cells contain poly-β-hydroxybutyrate.

5.0 μm

5.0 μm

(C)

(D)

5.0 μm

1.0 μm

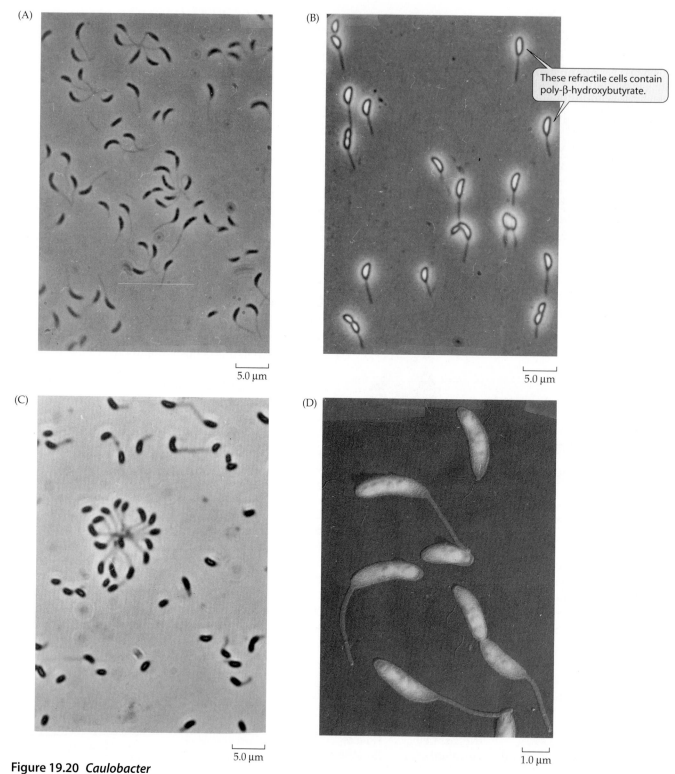

Figure 19.20 *Caulobacter*
(A–C) Phase photomicrographs of *Caulobacter crescentus*, showing the polar prostheca.
(D) Electron micrograph of a shadowed preparation of *C. crescentus*. Courtesy of J. Poindexter.

same position on its cell surface as long as conditions are favorable for its growth. A life cycle such as this, with two separate morphological forms (a flagelled nonprosthecate cell and a nonflagellated prosthecate cell), is termed a **dimorphic life cycle**. Because of their life cycle and their simple prokaryotic nature, *Caulobacter* spp. have become model organisms in the study of **cellular differentiation** or **morphogenesis (Box 19.4)**.

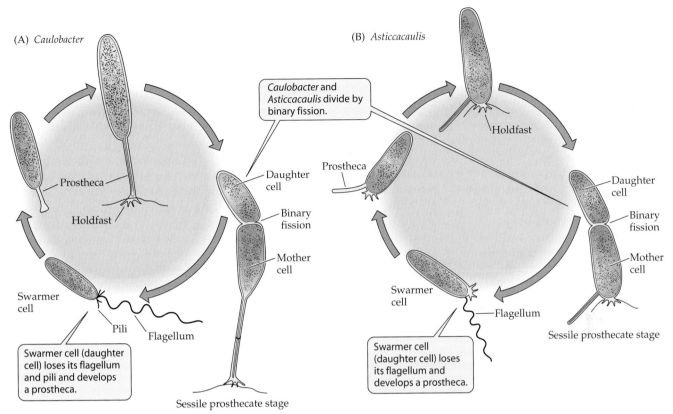

(A) *Caulobacter*

Prostheca

Holdfast

Swarmer cell

Pili — Flagellum

Caulobacter and *Asticcacaulis* divide by binary fission.

Daughter cell

Binary fission

Mother cell

Swarmer cell (daughter cell) loses its flagellum and pili and develops a prostheca.

Sessile prosthecate stage

(B) *Asticcacaulis*

Holdfast

Prostheca

Daughter cell

Binary fission

Mother cell

Swarmer cell

Flagellum

Swarmer cell (daughter cell) loses its flagellum and develops a prostheca.

Sessile prosthecate stage

Figure 19.21 Life cycles of prosthecate bacteria
Diagrams of life cycles of (A) *Caulobacter*, (B) *Asticcacaulis*, and (C) *Hyphomicrobium*. Note that all produce a flagellated swarmer cell.

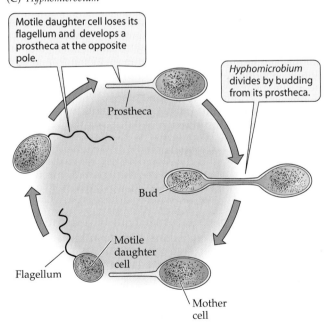

(C) *Hyphomicrobium*

Motile daughter cell loses its flagellum and develops a prostheca at the opposite pole.

Prostheca

Hyphomicrobium divides by budding from its prostheca.

Bud

Motile daughter cell

Flagellum

Mother cell

In pure cultures, the holdfasts of *Caulobacter* enable several cells to attach together to form a rosette structure (Figure 19.20C), something that is not found in the natural environment.

These organisms also produce structures called **crossbands** in their prosthecae (Figures 19.20D and **19.22**). Studies have shown that these are produced at the time of cell division. Therefore, it is possible to determine the "age" of a mother cell by counting the number of crossbands (hence, the number of times it has divided), much like counting tree rings.

These bacteria are metabolically similar to *Pseudomonas* species. They utilize a wide variety of soluble organic carbon sources, including sugars, amino acids, and organic acids, which are found in low concentrations in aquatic habitats (see Chapter 24). They use the Entner-Duodoroff pathway and TCA cycle and are active in aerobic respiration. They occur in both freshwater and marine habitats as well as in soils.

The genus *Asticcacaulis* is less frequently encountered in most habitats. This genus is very similar to *Caulobacter* in all respects, except that its flagellum and prostheca are in a subpolar position on the cell surface and one species produces two lateral prosthecae. However, the holdfast remains in a polar position. Thus, rosettes formed by pure cultures hold the cells together so that the prosthecae extend away from the center.

Caulobacter is widespread in aquatic habitats. Studies indicate they are found in lakes of all trophic states from oligotrophic to eutrophic. They comprise a significant proportion of the total heterotrophs in oligotrophic and mesotrophic habitats, making up as much as 50% of the

Research Highlights Box 19.4

Caulobacter crescentus, a Model Prokaryote for Studying Cell Morphogenesis or Differentiation

Several laboratories have been studying the dimorphic life cycle of *Caulobacter crescentus*. This organism is ideally suited for such analysis because it undergoes a transformation from a flagellated swarmer cell to an immotile, prosthecate cell during its life cycle (see figure). In the cell cycle, the cell must be programmed to "switch on" and "switch off" various biosynthetic processes at specific times. Furthermore, the cell has the capability of localizing these events at one pole of the cell or the other.

Let us consider initially the flagel-lated swarmer cell. This stage of the life cycle not only has a polar flagellum and is active chemotactically, but the flagellated pole of the cell also contains the holdfast, a specific pilus structure, and DNA phage receptor sites. Thus, there is strong polar orientation of functions. Then, at some time during the differentiation process, the swarmer cell loses its flagellum and pili and begins to develop a prostheca (called a "stalk") and a holdfast at the same pole of the cell.

Only after the stalk forms does the cell switch on DNA replication.

The stalked cell elongates, DNA replication occurs, and the polar differentiation process whereby the new swarmer cell is formed occurs.

Not surprisingly, the length of time it takes for a swarmer cell to undergo cell division is longer (typically an additional 30 min.) than it takes a cell with a stalk to divide (90 min.). This is due to the length of time required for the swarmer cell differentiation process. At this time, the molecular biological events during the life cycle of *C. crescentus* are being analyzed genetically and biochemically in several laboratories.

Life cycle of *Caulobacter crescentus*

Flagellum   Holdfast

Swarmer cell   30 minutes   90 minutes

Pili

viable heterotrophic count during their periods of greatest abundance (see Chapter 24). Although the natural function of the *Caulobacter* prostheca is not yet well understood, it is known that it lengthens during conditions of phosphate limitation. Because phosphate is a common limiting nutrient in aquatic habitats, especially during summer algal blooms, this appendage may enhance its uptake and improve their ability to survive when it becomes limiting during the summer periods.

Hyphomicrobia

Like the caulobacters, hyphomicrobia make up a group of several genera with similar morphological traits (**Table 19.18**). They all have polar prosthecae and divide by forming a bud at the tip of it (**Figure 19.23**). Buds are usually motile by a single flagellum; some species produce a holdfast for attachment. In this group, the holdfast is located on the surface of the cell, not at the tip of the prostheca.

These organisms also have a dimorphic life cycle consisting of a nonprosthecate motile daughter cell and a prosthecate mother cell (Figure 19.21C). The motile daughter cell must first form a prostheca. It then divides by a budding process in which a bud develops from the tip of the prostheca. The daughter cell enlarges, becomes motile, separates from the mother cell, and repeats the cycle. Mean-

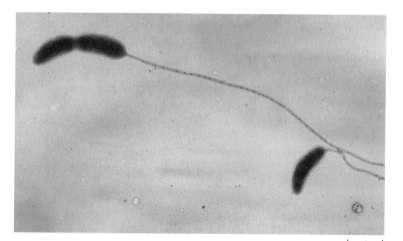

Figure 19.22 Crossbands in prosthecae
1.0 μm
Electron micrograph of a *Caulobacter* cell, showing 31 crossbands in its stalk. Courtesy of J. T. Staley and T. Jordan.

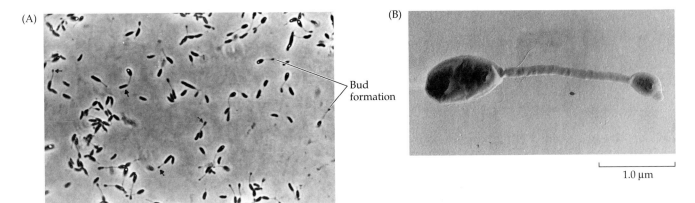

(A)

(B)

Bud formation

1.0 μm

5.0 μm

Figure 19.23 *Hyphomicrobium*
(A) Phase photomicrograph of cells of *Hyphomicrobium zavarzinii*, showing bud formation at the tips of the prosthecae. The bright area in some cells is poly-β-hydroxybutyrate. (B) Electron micrograph of a budding cell of *Hyphomicrobium facilis*. Courtesy of P. Hirsch.

while, the appendaged mother cell develops a new bud either at the same site as the earlier bud or on a newly formed polar appendage. Apparently, if conditions are favorable for growth, an unlimited number of buds can be produced from the mother cell, in the same manner as caulobacters.

Hyphomicrobium spp. are primarily methanol utilizers, although they can also use some other one-carbon compounds such as methylamine as their carbon source. They use the serine pathway for methanol degradation (see Chapter 8). Some carry out denitrification using nitrate as an electron acceptor in anaerobic respiration and oxidizing methanol as the energy source.

Hyphomonas and *Hirschia* look identical to *Hyphomicrobium*, but they do not use methanol as a carbon source. Instead, they use amino acids and organic acids as carbon sources. Amino acids are required. The life cycles of these bacteria are similar to that of *Hyphomicrobium*, despite their metabolic differences.

Pedomicrobium spp. are quite different from the other hyphomicrobia. They are soil and aquatic bacteria that grow on acetate and pyruvate as well as more complex organic compounds such as fulvic acid. They are best known for their ability to oxidize reduced iron and manganese compounds and form deposits of the oxides on the cell surface. Their life cycle is similar to that of the other hyphomicrobia. However, they commonly produce more than one prostheca per cell, and some of these are in nonpolar positions (**Figure 19.24**). The buds are more characteristically rod shaped rather than coccoidal such as the other hyphomicrobia.

The hyphomicrobia are widespread in soils and aquatic environments. *Hyphomicrobium* are commonly found in freshwater and soil habitats. Most *Hyphomonas*

Table 19.18	Genera of hyphomicrobia		
Genus	**Mol % G + C**	**Carbon Source**	**Special Features and Habitat**
Hyphomicrobium	59–65	Methanol	Polar prosthecae; soil, freshwater
Hyphomonas	57–62	Amino acids	Polar prosthecae; marine
Hirschia	45–47	Amino acids, organic acids	Polar prosthecae; marine
Pedomicrobium	62–67	Acetate, pyruvate	Lateral and polar prosthecae; soil, lakes

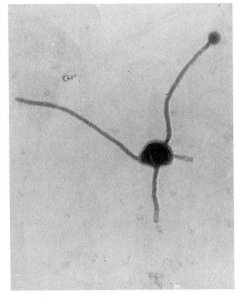

Figure 19.24 *Pedomicrobium*
Electron micrograph of a *Pedomicrobium* cell, showing a bud at the tip of one prostheca. Courtesy of R. Gebers.

and *Hirschia* spp. have been isolated from marine habitats. *Pedomicrobium* was first reported from soils, but they have also been obtained from freshwater lakes and water pipelines where they form a coating of manganic oxides if the water contains reduced manganese ions (Mn^{2+}).

Polyprosthecate Bacteria

Polyprosthecate bacteria are a diverse group of bacteria represented by several genera (**Table 19.19**). Unlike other prosthecate bacteria, they have several prosthecae per cell, which gives their cells starlike shapes. In fact, the genus *Stella* produces six prosthecae all in one plane, so they resemble a perfect six-pointed star (**Figure 19.25**).

The genus *Prosthecomicrobium* is the most common member of this group. They are found in freshwater, marine, and soil environments. Some species are motile by a single subpolar flagellum, whereas others are immotile or produce gas vacuoles (**Figure 19.26**).

Unlike all other heterotrophic prosthecate bacteria, the genus *Ancalomicrobium* (Figure 19.19) is a fermentative facultative aerobe. Indeed, it is a mixed-acid fer-

menter whose products are identical to that of *E. coli* (see Table 19.3). However, unlike *E. coli*, it produces prosthecae and gas vacuoles and is well adapted to live and compete successfully in aquatic habitats where nutrients are in much lower concentrations than in the intestinal tract of animals. *Ancalomicrobium adetum* uses a variety of sugars and other organic carbon sources for growth and requires one or more B-vitamins such as thiamine, biotin, riboflavin, or B_{12}.

Like the caulobacters, these bacteria live in aquatic habitats and soils, particularly those that are olig-

Table 19.19	Genera of polyprosthecate bacteria		
Genus		**Mol % G + C**	**Special Features**
Prosthecomicrobium		64–70	>10 Short prosthecae around cell; aerobic
Ancalomicrobium		70–71	<10 Long prosthecae; facultative aerobe; fermentative
Stella		69–74	Star-shaped; prosthecae in one plane

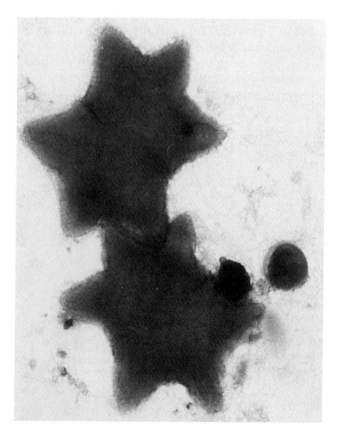

Figure 19.25 *Stella*
Two cells of *Stella humosa*, a six-pointed, star-shaped budding bacterium. Cells are about 1 μm in width. Courtesy of J. T. Staley.

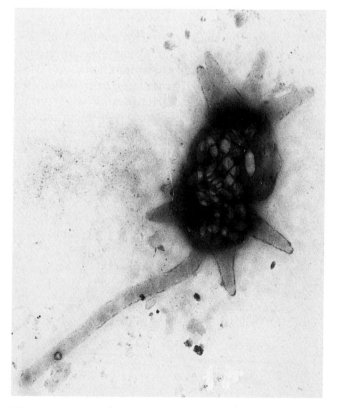

Figure 19.26 *Prosthecomicrobium*
Electron micrograph of a gas vacuolate cell of *Prosthecomicrobium pneumaticum*. Courtesy of J. T. Staley.

1.0 μm

otrophic to mesotrophic. Their appendages enhance the uptake of nutrients in habitats where nutrients occur in low concentrations. Each genus and species specializes in the uptake of a group of low molecular weight dissolved organic nutrients, such as sugars and organic acids, and are therefore responsible for keeping these nutrients in low concentrations in oligotrophic environments. In this manner, they maintain a competitive edge against eutrophic bacteria that require organic nutrients in much higher concentrations and are therefore not competitive in oligotrophic environments.

Conversely, the low maximum growth rates of the prosthecate bacteria (generation times are longer than 2 hours in laboratory culture) mean that they cannot outgrow fast-growing eutrophic bacteria such as the enteric bacteria in more enriched habitats, such as the intestinal tract, which has much higher concentrations of organic compounds.

BDELLOVIBRIO: THE BACTERIAL PREDATOR

The genus *Bdellovibrio* contains unique gram-negative bacteria that actually parasitize other gram-negative bacteria. They are small vibrios, only about 0.25 µm in diameter. The *Bdellovibrio* cell has a polar flagellum that rapidly propels the organism through the environment until it collides with an appropriate host cell. It hits the host cell with such force that the host is moved several micrometers. The *Bdellovibrio* cell then attaches to the gram-negative host at its nonflagellated pole and rotates rapidly at the site of attachment (**Figure 19.27**). After a few minutes during which it produces enzymes that are responsible for the breakdown of the cell wall, it enters the cell envelope of the host and moves into the periplasmic space. This process of penetration requires the production of enzymes such as proteases, lipases, and a type of lysozyme, all of which enable it to penetrate the outer lipopolysaccharide and peptidoglycan cell wall layers so that it can enter the cell. This entire process takes several minutes. The *Bdellovibrio* cell does not break the cell membrane and lyse the cell but instead resides within the periplasmic space during its parasitism. The host cell becomes rounded during this process, due to the destruction of its cell wall, and is called a **bdelloplast**. Once the *Bdellovibrio* cell enters the periplasm, it loses its flagellum and becomes immotile.

While in the periplasm, the *Bdellovibrio* cell produces enzymes that degrade the host cell macromolecules including its DNA, RNA, and protein. The degradation

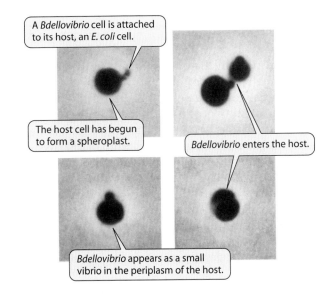

A *Bdellovibrio* cell is attached to its host, an *E. coli* cell.

The host cell has begun to form a spheroplast.

Bdellovibrio enters the host.

Bdellovibrio appears as a small vibrio in the periplasm of the host.

Figure 19.27 *Bdellovibrio* life cycle
Stages in the life cycle of *Bdellovibrio bacteriovirus*, shown in electron micrographs. Courtesy of R. J. Seidler.

products are then used by *Bdellovibrio* in making its own protein, DNA, and RNA. The *Bdellovibrio* cell enlarges as growth proceeds to form a long, helical cell during the next 4 hours or so (**Figure 19.28**). When mature, this cell divides by multiple fission to form up to six motile

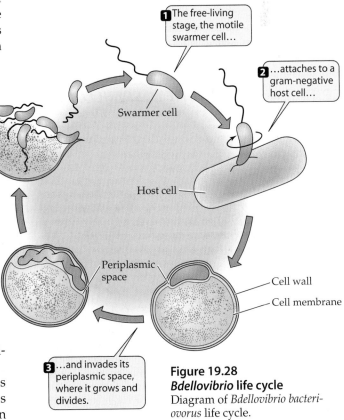

1 The free-living stage, the motile swarmer cell…

2 …attaches to a gram-negative host cell…

Swarmer cell

4 When the nutrients are exhausted, the four to six daughter cells lyse the host cell and leave.

Host cell

Periplasmic space

Cell wall

Cell membrane

3 …and invades its periplasmic space, where it grows and divides.

Figure 19.28 *Bdellovibrio* life cycle
Diagram of *Bdellovibrio bacteriovorus* life cycle.

daughter cells in a typical unicellular gram-negative bacterium such as *E. coli* or *Pseudomonas* spp. The *Bdellovibrio* progeny then swim away from the dead host cell and are free to repeat the life cycle.

One of the remarkable features of this organism is that it actually produces plaques such as phage plaques on plates of host bacteria. It was this feature that enabled the German microbiologist Heinz Stolp to discover the bacterium when he was isolating phage from soil and sewage samples. The plaques form after most phage plaques have already developed, and they keep on enlarging, making them unusual and suspicious. When wet mount preparations from such plaques were examined in the light microscope, the small bacterium was observed quickly darting about among the cells of the host species.

Bdellovibrio cells are obligately aerobic with a tricarboxylic acid cycle. They do not use sugars but rather degrade proteins and use the amino acids as their organic carbon source. Studies of the physiology and metabolism of this bacterium have been conducted with some mutant strains, the so-called host-independent strains that can be grown in the absence of host cells.

Bdellovibrio spp. are widespread in aquatic environments as well as in soils and sewage. Their major role appears to be analogous to that of virulent bacteriophages in controlling the numbers of gram-negative bacteria. However, they have a broader host range than typical phages and carry out a much more active type of parasitism. It is curious that they have not yet been found to be chemotactic, an ability that would seem to be of great significance in locating prey in the dilute environments where they reside.

SPIRILLA AND *MAGNETOSPIRILLUM*

The spirillum is one of the more common bacterial cell shapes. Like the rod and coccus, different phylogenetic groups contain spirilla. Many of the heterotrophic spirilla are members of the *Proteobacteria* (**Table 19.20**), but spirilla are also found in other phyla, too, such as the *Cyanobacteria* (genus *Spirulina*) and even in the *Archaea* (methanogen genus *Methanospirillum*).

Previously in this chapter we discussed *Azospirillum*, a nitrogen-fixing bacterium associated with monocotyledonous plants. It is a member of the alpha *Proteobacteria*, as is the unusual magnetite-containing genus, *Magnetospirillum*. The genera *Spirillum*, *Aquaspirillum*, and *Comamonas* also contain spirilla-shaped bacteria, but they are all in the beta *Proteobacteria*. Finally, the gamma *Proteobacteria* also contains spirilla such as the genus *Oceanospirillum* and the photosynthetic genus, *Thiospirillum,* which will be discussed in Chapter 21.

The heterotrophic proteobacterial spirilla are primarily microaerophilic and grow by the oxidation of organic acids. All are motile with either a single polar flagellum at each pole or with polar flagellar tufts (lophotrichous flagellation). The genus *Spirillum* contains the largest species, *S. volutans* (see Figure 4.57), which move by a flagellar tuft at each pole, and cells that are healthy are in constant motion. These bacteria live in natural aquatic habitats and are particularly successful at growing when nutrients are in low concentration.

Magnetospirillum magnetotacticum is one of the most fascinating bacteria known. Each cell contains small crystals of magnetite (Fe_3O_4) that are aligned along the long axis of the cell (**Figure 19.29**). The cells actually orient themselves and move in a magnetic gradient. Therefore, if a magnet is placed against a test tube containing the cells, they will move toward and accumulate where the magnet is located. The magnetite crystals are located inside a membrane called the **magnetosome**. It is unclear whether this membrane is in direct contact with the cell membrane.

These bacteria, which are microaerophilic, prefer to live in the sediment of freshwater and marine habitats. The magnetic forces of the Earth provide a direction for the cell to move down and toward the sediments where the concentration of oxygen is low and iron is readily available to the cell. This type of movement is termed **magnetotaxis**. The magnets allow the cell to orient themselves toward the North Pole if they are in the northern hemisphere or toward the South Pole if they are in the southern hemisphere. It should be noted that the magnetic force of Earth is not sufficient to "pull" the organism, but only to orient the cell. The flagella provide the force that permits cell movement.

Table 19.20	Common genera of spirilla		
Genus	**Mol % G + C**	**O₂ Requirement**	**Other Features**
Spirillum	36–38	Microaerophile	Large cells, freshwater
Aquaspirillum	49–66	Aerobic, micro-aerophilic	Freshwater
Magnetospirillum	65	Microaerophilic	Magnetotactic, freshwater
Oceanospirillum	42–51	Aerobic	Marine
Ancylobacter	66–69	Aerobic	Forms rings, freshwater
Campylobacter	30–38	Microaerophilic	Intestinal tract
Helicobacter	35–44	Microaerophilic	Upper digestive tract

(A)

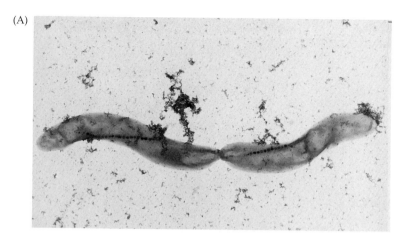

(B)

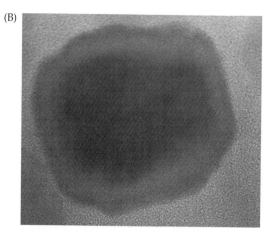

├──┤
0.5 μm

Figure 19.29 *Magnetospirillum magnetotacticum*
(A) *Magnetospirillum magnetotacticum* cell, showing its magnetosome and magnetite particles. (B) High-resolution electron micrograph. Courtesy of M. Sarikaya.

Materials engineers and scientists are very interested in knowing how bacteria can make such small and perfect magnets. For this reason, understanding the molecular biology of this process is one of the exciting areas of research for the future.

It is interesting to note that many other species of bacteria produce magnetite, but very few others have been isolated in pure culture. Some of them, such as *Magnetospirillum*, are members of the *Proteobacteria*, but recent studies indicate that some magnetotactic bacteria such as *Magnetobacterium bavaricum* are found in the phylum *Nitrospira* (see Chapter 23).

Ancylobacter (formerly *Microcyclus*) is unusual in that it forms ring shapes (see Figure 4.28). Most strains are gas vacuolate. It is quite versatile physiologically. In addition to growing as an ordinary heterotroph using sugars and amino acids as carbon sources, it is able to degrade methanol and also grows as a hydrogen autotroph. This is a common genus found in freshwater environments.

Campylobacter and *Helicobacter* are curved microaerophilic bacteria that can cause disease in humans and other animals. Both are members of the epsilon subphylum of the *Proteobacteria*. *Campylobacter fetus* causes enteritis in humans, and *Helicobacter pylori* is implicated as the causative agent of peptic ulcers.

MYXOBACTERIA: THE FRUITING, GLIDING BACTERIA

The **myxobacteria** (slime bacteria) are named for the production of extracellular polysaccharides and other material they excrete as they glide on surfaces. Myxobacteria are gram-negative, rod-shaped bacteria (**Figure**

19.30) that have complex life cycles. The cells live together in close association with one another as they grow. Most species require complex nutrients such as amino acids and protein for growth and are therefore commonly found on animal dung in soil environments. They produce extracellular hydrolytic enzymes to degrade protein, nucleic acids, and lipids. Using these enzymes, they are able to lyse and degrade bacteria and other microorganisms as their principal nutrient source. These bacteriolytic species do not use sugars as carbon sources for growth. Although they have an active tricarboxylic acid cycle, they do not have a complete Embden-Meyerhof pathway, which explains their inability to use sugars.

One genus of myxobacteria, *Polyangium*, lives on tree bark and degrades cellulose as its organic energy and carbon source and otherwise uses inorganic nutrients for growth. Therefore, it does not require the complex

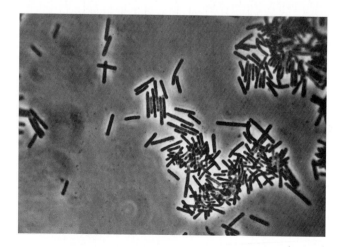

Figure 19.30 Myxobacteria
Typical rod-shaped vegetative cells of the myxobacterium *Chondromyces crocatus*. Courtesy of H. McCurdy.

├──┤
5.0 μm

organic nitrogen sources needed by others. All myxo-bacteria, including *Polyangium*, are aerobic respiratory bacteria.

The most remarkable aspect of the developmental process of these bacteria is their ability to aggregate together to form a **swarm** (**Figure 19.31**) and then, as nutrients or water become depleted, to develop a complex **fruiting structure** (**Figure 19.32**), a dormant stage in which reproductive structures called **microcysts** (**Figure 19.33**) are produced. Each cell of the swarm is converted into a **microcyst** (also called **myxospore**) in the fruiting structure. The microcysts are resistant to desiccation, thereby enabling the organism to survive until conditions are again favorable for growth. The microcysts can be dispersed much like the spores of fungi that form analogous fruiting structures for reproduction and dispersal. In some species, the microcysts are formed in larger structures called **sporangia**. These sporangia may be formed within the cellular mass or borne on **stalks**. Stalked varieties such as *Chondromyces crocatus* have an elaborate morphology (Figure 19.32). **Table 19.21** lists the important genera of these bacteria. Because they are simple, one-celled organisms with complex life cycles, these bacteria have become model organisms for the study of developmental processes. The species that has been most thoroughly studied is *Myxococcus xanthus*.

Gliding motility in these bacteria is poorly understood. As with other gliding bacteria, these organisms must be in contact with a surface to be able to move. They produce extracellular slime as they glide, leaving a trail of slime as well as an indentation or actual groove, called a slime track or trail, if they are on agar.

(A)

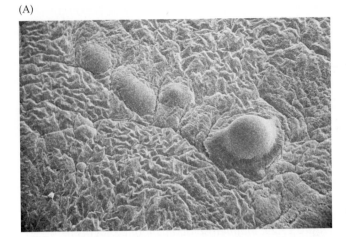

(B)

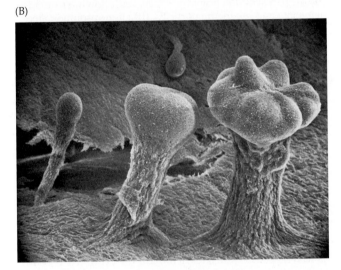

(C)

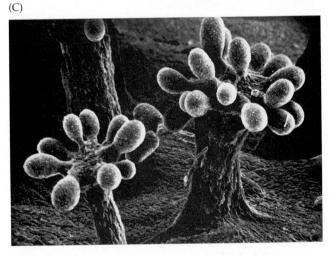

Figure 19.32 Fruiting structure of myxobacteria
Scanning electron micrographs showing development of a fruiting structure in *Chondromyces crocatus*. (A) Cells have aggregated and the structure has begun to rise. (B) Rising fruiting structures at various stages of differentiation. (C) The mature fruiting structure is about 1 mm in height. Each "fruit" is a sporangium filled with microcysts. Courtesy of Patricia Grilione and Jack Pangborn.

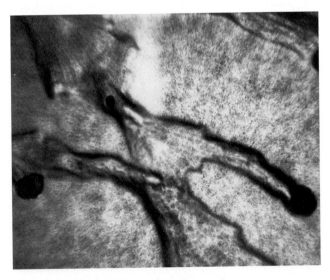

Figure 19.31 Swarming of myxobacteria
Swarm of myxobacteria moving across an agar surface. Each finger-like projection contains hundreds of cells. Courtesy of H. McCurdy.

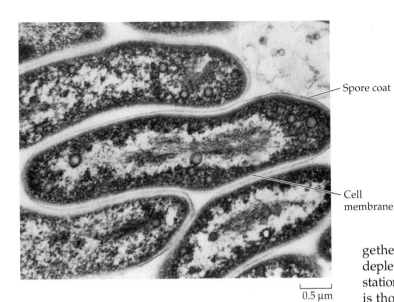

Spore coat

Cell membrane

0.5 μm

Figure 19.33 Microcysts of myxobacteria
Thin section through microcysts of *Cystobacter fuscus.*
Courtesy of H. McCurdy.

Once a track has been formed by one cell, others follow it in preference to making their own. The cells maintain a "herd" or "pack" instinct while gliding. Only rarely do they venture away from the pack except at the advancing edge of growth, and even then it is only for a

brief moment of time. In this period of growth, myxobacteria are actively involved in the degradation of particulate organic material that includes other microorganisms. They elaborate several lytic enzymes as they move, and if a host bacterium is approached by the pack, its cells are lysed even before physical contact is made. Therefore, although they resemble *Bdellovibrio* spp. in that they parasitize other bacteria, they do not actually enter individual host cells. Furthermore, they have a broader host range and can parasitize gram-positive as well as gram-negative bacteria.

Typically, the myxobacteria cells come together to aggregate at the onset of drying or nutrient depletion, although in some instances they merely enter stationary growth phase like other bacteria. Aggregation is thought to occur in response to the "A signal," the accumulation of a specific group of amino acids in the environment. This results in drawing the cells together. They pile on top of one another to form a mound, which eventually becomes the fruiting structure. The fruiting structure containing the microcysts develops into a distinctive shape characteristic for each species. These fruiting structures, as well as the vegetative cell masses, are pigmented in vivid yellow, orange, or red colors, depending on the species. A-signal mutants cannot aggregate or differentiate into microcysts.

In the simplest of the fruiting structures, such as the one formed by *Myxococcus xanthus*, the cells aggregate together to form a small, raised mound about 1 mm in diameter and height. Each of the cells in the fruiting structure is converted into a microcyst. The microcyst then, is a modified cell that shows greater resistance to drying and also ultraviolet light exposure than the vegetative cell. Although they also exhibit somewhat greater resistance to heating, they are not similar morphologically or physiologically to the more inert, heat-resistant endospores of the gram-positive bacteria.

The myxobacteria with the most elaborate fruiting structures form their microcysts in a special structure called a **sporangium.** Sporangia may contain hundreds of microcysts. In some species, the sporangium lies on the same surface as the vegetative growth. However, in the more exotic forms, the cells aggregate, climb on top of one another, and produce a stalk. The stalk consists of hardened, extracellular slime. The sporangia then develop on top of the stalk (Figure 19.32). Clearly, these are among the most complex prokaryotic organisms.

Table 19.21	Important genera of myxobacteria	
Genus	**Mol % G + C**	**Distinguishing Features**
Cells tapered		
Myxococcus	68–71	Spherical microcysts; no sporangia
Archangium	67–69	Rod-shaped microcysts; no sporangia
Cystobacter	68–69	Rod-shaped microcysts in in nonstalked sporangia
Mellitangium	Unknown	Rod-shaped microcysts in sporangium on single stalk
Stigmatella	68–69	Rod-shaped microcysts in sporangium borne singly or in clusters on stalk that may be branched
Cells have blunt, rounded ends, microcysts resemble vegetative cells		
Polyangium	69	Rod-shaped cells; form nonstalked sporangium; may degrade cellulose
Nannocystis	70–72	Coccoidal cells; form nonstalked sporangium
Chondromyces	69–70	Rod-shaped cells; sporangia formed singly or in clusters on stalks

Most myxobacteria are bacteriolytic and require complex organic materials for growth. They obtain these nutrients by killing and lysing other bacteria. To perform this activity, the myxobacteria produce lysozyme and some of their own antibiotics, most of which have not yet been characterized. Once they lyse the host bacterial cells, they degrade their proteins, lipids, and nucleic acids. Most myxobacteria live in animal dung, which contains other bacteria in high concentrations. However, *Polyangium* are cellulolytic and grow by degrading wood cellulose. Thus, they are found on logs and bark in forested areas.

SULFATE- AND SULFUR-REDUCING BACTERIA

The sulfate- and sulfur-reducing bacteria are a physiologically and ecologically related group of bacteria. They are strict anaerobes that generate energy by anaerobic respiration of a variety of organic compounds. Some can use hydrogen gas as an energy source. Elemental sulfur (S^0) or oxidized sulfur compounds such as sulfite and sulfate act as terminal electron acceptors that are converted to H_2S in the process of **dissimilatory sulfur reduction**. Some of the common genera of the sulfate and sulfur reducers and their characteristics are given in **Table 19.22**. It should be noted that many ordinary heterotrophic bacteria, such as some enteric bacteria and pseudomonads, can reduce sulfate to sulfide; however, these organisms cannot use sulfate as the sole terminal electron acceptor in anaerobic respiration, the hallmark of the dissimilatory sulfate reducers.

Not all members of this group are closely related to one another phylogenetically. Most bacterial sulfate reducers, however, such as the common genus *Desulfovibrio*, are members of the delta subphylum of the *Proteobacteria*. Although *Desulfonema* differs from the others in that it is filamentous, moves by gliding, and stains as a gram-positive bacterium, phylogenetic analysis indicates it is also a member of the delta *Proteobacteria*. Its closest relative is *Desulfosarcina variabilis*, a packet-forming (sarcina) coccus.

In contrast, *Desulfotomaculum* is not a *Proteobacterium*. It is a genus of endospore-forming bacteria with a broad range of DNA base composition (37–50 mol % G + C). 16S rRNA sequence analysis confirms that this endospore-forming genus is closely related to *Clostridium* and other gram-positive bacteria. Therefore it is treated in Chapter 20. Likewise, some *Archaea* (see Chapter 18) and other *Bacteria* (see Chapter 22) also carry out the dissimilatory reduction of sulfur compounds.

Sulfur- and sulfate-reducers are abundant in anaerobic soil as well as freshwater, estuarine, and marine sediments. They are especially common in estuarine and marine sediments, which are in constant contact with seawater that is rich in sulfate. Aquatic and terrestrial habitats that become anaerobic through decomposition of organic matter are rich in sulfate and sulfur reducers. Sulfate-reducing bacteria also live in the intestinal tracts of some animals.

The organic energy sources used by sulfur- and sulfate-reducing bacteria (such as acetate, lactate, ethanol, pyruvate, and butyrate) are end products of fermentation produced by other bacteria. As noted in Table 19.22, many sulfur reducers cannot catabolize acetate because they have an incomplete tricarboxylic acid cycle. Those with a complete tricarboxylic acid cycle oxidize acetate completely to CO_2. In addition to the respiratory generation of ATP with sulfate as the terminal electron acceptor, some genera carry out fermentations. Thus, *Desulfococcus*, *Desulfosarcina*, and *Desulfobulbus* ferment pyruvate or lac-

Table 19.22	Representative genera of sulfate- and sulfur-reducing bacteria[a]		
Genus	**Cell Shape**	**Carbon Source**	**Electron Acceptor**
Desulfovibrio	Vibrio	Lactate, ethanol	SO_4^{2-}; SO_3^{2-}; $S_2O_3^{2-}$
Desulfococcus	Coccus	Lactate, ethanol	SO_4^{2-}; SO_3^{2-}; $S_2O_3^{2-}$
Desulfosarcina	Ovoid; aggregates	Lactate, benzoate	SO_4^{2-}; SO_3^{2-}; $S_2O_3^{2-}$
Desulfobacter	Rod; vibrio	Acetate	SO_4^{2-}; SO_3^{2-}; $S_2O_3^{2-}$
Desulfobulbus	Oval	Lactate, acetate	SO_4^{2-}
Desulfomonile	Rod	Pyruvate, benzoate	SO_4^{2-}; SO_3^{2-}; $S_2O_3^{2-}$
Desulfomicrobium	Rod	Lactate	SO_4^{2-}; SO_3^{2-}; $S2O_3^{2-}$
Desulfobotulus	Vibrio	Longer-chain fatty acids	SO_4^{2-}; SO_3^{2-}; $S_2O_3^{2-}$
Desulfobacterium	Rod; vibrio	Lactate; acetate, longer chain fatty acids	SO_4^{2-}; SO_3^{2-}; $S_2O_3^{2-}$
Desulfurella	Rod	Acetate	S^0
Desulfuromonas	Oval, curved rod	Acetate, propionate	S^0

[a]Some gram-positive bacteria in the genus *Desulfotomaculum* and some thermophilic *Bacteria* and *Archaea* are also dissimilatory sulfur- and sulfate-reducing organisms (see Chapters 18, 20, and 22).

tate to propionate and acetate. This provides an electron flow that can be utilized in ATP generation.

CHEMOLITHOTROPHIC PROTEOBACTERIA

The **chemolithotrophic bacteria** obtain energy from the oxidation of reduced inorganic compounds. Several different groups exist and are differentiated from one another based on the inorganic compounds they oxidize as energy sources (**Table 19.23**). **Nitrifying bacteria** oxidize reduced nitrogen compounds, **sulfur oxidizers** oxidize reduced sulfur compounds, **iron bacteria** oxidize reduced iron, and **hydrogen bacteria** oxidize hydrogen gas. Another special group of hydrogen bacteria, the **carboxydobacteria**, oxidize carbon monoxide and hydrogen.

By and large, these chemolithotrophic activities are uniquely associated with bacteria. Only rarely have eukaryotic organisms been found that can carry out any of these processes, and when it has been found, as for example, nitrification by some fungi, their activity is not considered to be significant in the environment. In contrast, these processes carried out by bacteria are well known and considered to be extremely important in the biogeochemical cycles of nature (see Chapter 24). Each of the chemolithotrophic groups is discussed individually in greater detail. The chemolithotrophic *Proteobacteria* carry out carbon dioxide fixation using the Calvin cycle (Chapter 10). Ribulose-1,5 biphosphate carboxylase (RuBisCo) is the key enzyme in this pathway.

Nitrifying Bacteria

As their name implies, the nitrifying bacteria use reduced inorganic nitrogen compounds, ammonia and nitrite, as their sources of energy for growth. Ammonia is oxidized to nitrite by one group of nitrifiers, called the **ammonia oxidizers**:

$$(2)\ 2\,NH_3 + 3\tfrac{1}{2}O_2 \rightarrow 2\,NO_2^- + 3\,H_2O$$

Nitrite is oxidized to nitrate by another group called the **nitrite oxidizers**:

$$(3)\ NO_2^- + \tfrac{1}{2}O_2 \rightarrow NO_3^-$$

No single nitrifying bacterium is able to oxidize ammonia all the way to nitrate, although pure cultures of some fungi have been reported to carry out this process. These fungi, however, are not considered important in the environment. Normally the two groups of nitrifying bacteria grow in close association in the environment to carry out the two-step sequential oxi-

Table 19.23	Groups of chemolithotrophic proteobacteria	
Group	**Energy Source(s)**	**Product(s)**
Nitrifiers	NH_4^+ or NO_2^-	NO_2^- or NO_3^-
Sulfur oxidizers	H_2S, S^o, $S_2O_3^{2-}$	SO_4^{2-}
Iron oxidizers	Fe^{2+}	Fe^{3+}
Hydrogen bacteria	H_2	H_2O
Carboxydobacteria	$H_2 + CO$	$H_2O + CO_2$
Methane oxidizers[a]	CH_4	$CO_2 + H_2O$

[a]Methane is the simplest organic compound; however, the metabolism of the methane oxidizers is very similar to that of the nitrifiers, therefore they are considered here.

dation. The two processes appear to be tightly coupled so that nitrite, which can be toxic to higher organisms, does not accumulate in high concentration.

Nitrifying bacteria are found in all soil and aquatic habitats and are especially common in alkaline or neutral pH environments where ammonia rather than ammonium ion is abundant, consistent with ammonia being the form of nitrogen used as the energy source. Also, more carbonate is available for CO_2 fixation in moderately alkaline environments.

Nitrifiers are superb examples of chemolithotrophs. They grow well in pure culture on completely inorganic media. However, most species are able to use organic carbon sources such as acetate for growth and thus are facultative chemolithotrophs. They grow poorly in culture even under the best of conditions, having generation times of 24 hours or more. Rarely do they grow to high enough cell densities to produce turbidity in media, even though their activities in the medium can be readily demonstrated. The reason for their poor growth in pure culture is not well understood; however, it may be due in part to the toxicity of nitrite. We first discuss the ammonia oxidizers, which are sometimes also called the nitrosofying bacteria.

Ammonia Oxidizers The oxidation of ammonia to nitrous acid involves two major steps with hydroxyl amine, NH_2OH, as an intermediate (see also Chapter 8):

$$(4)\ NH_3 + O_2 + XH_2 \xrightarrow{\text{AMO}} NH_2OH + H_2O + X$$

$$(5)\ NH_2OH \xrightarrow{\text{HAO}} HNO + H_2O \xrightarrow{\text{HAO}} HNO_2 + 2\,H^+ + 2\,e^-$$

where AMO = ammonia monooxygenase
HAO = hydroxylamine oxidoreductase
X = a hydrogen carrier such as NADH
$\Delta G^{o'}$ (reactions 7 and 8 combined) = $-275\ kJ/mol$

Several enzymes are involved in the oxidation pathway to form nitrous acid. The first enzyme in this pathway is the key enzyme, ammonia monooxygenase (AMO), which carries out the oxidation of ammonia to hydroxyl amine. Ammonia monooxygenase shows rather broad specificity. Thus, it can also oxidize methane, making these bacteria resemble methane-oxidizing bacteria, which have a methane monooxygenase of similar broad specificity (see methanotrophic bacteria section that follows). A variety of inhibitors affect the ammonia monooxygenase, including acetylene, which is used as a blocking inhibitor in ecological studies, as well as 2-chloro-6-trichloromethylpyridine (nitrapyrin).

Energy is generated in the second reaction in which hydroxylamine is oxidized to HNO and subsequently to nitrous acid. This reaction is catalyzed by the enzyme hydroxylamine oxidoreductase (see reaction 4). HNO spontaneously degrades to produce N_2O, nitrous oxide (laughing gas). Electrons from the oxidations are passed through an extensive membrane-bound system of cytochromes (these give the cultures a red appearance). The free energy of this reaction is sufficiently high to generate ATP. Acid produced by oxidation of the inorganic nitrogen compounds results in the formation of protons that are passed through the membrane and ATP is formed by a cell membrane ATPase. Reducing power generated during the oxidation process is used for the carbon dioxide fixation reactions. Most ammonia oxidizers are members of the alpha group of *Proteobacteria*, but some are gamma *Proteobacteria*.

Nitrite Oxidizers

Nitrite oxidation is a one-step oxidation process carried out by the enzyme nitrite oxidoreductase. Water is the actual electron donor for the oxidation process:

$$(6)\ NO_2^- + H_2O \xrightarrow{\text{NOR}} NO_3^- + 2\,e^- + 2\,H^+$$

where NOR = nitrite oxidoreductase
$\Delta G^{o'} = -76$ kJ/mole

As shown in reaction 5, the protons are generated in the periplasmic space, thereby generating a proton motive force across the cell membrane. The electrons from reaction 5 are passed onto oxygen through the cytochrome system resulting in the formation of water in the cytoplasm (see Chapter 8). Under standard conditions, this reaction provides less energy per mole of substrate oxidized than ammonia oxidation. However, as with ammo-

Table 19.24   Important genera of nitrifying bacteria

Group/Genus	Mol % G+C	Special Characteristics
Ammonia oxidizers		
Nitrosomonas	45–54	Common rod
Nitrosospira	53–55	Tightly coiled helix
Nitrosococcus	48–51	Coccus
Nitrosolobus	53–56	Multilobed coccus
Nitrite oxidizers		
Nitrobacter	60–62	Common rod, divides by budding
Nitrospina	58	Thin rod
Nitrococcus	51	Marine genus
Nitrospira[a]	50	Separate phylum: tightly coiled helix; common marine genus

[a]See Chapter 22.

nia oxidizers, there is sufficient energy available in this process to generate ATP, and this is accomplished by chemiosmotic means. However, unlike the ammonia oxidizers, the nitrite oxidizers must produce NADPH by reverse electron transport (see Chapter 9).

Table 19.24 lists some of the genera of ammonia and nitrite-oxidizing bacteria, and representatives of selected genera are illustrated in **Figure 19.34**. Most nitrifying bacteria have a complex internal membrane system that contains cytochromes (Figure 19.34B, D, F, and H). These intracytoplasmic membranes are thought to contain the cytochromes systems that act as the site of ammonia or nitrite oxidation as well as the site of generation of NADH. *Nitrobacter* is a member of the alpha-*Proteobacteria*, whereas all others, with the exception of *Nitrospira*, which is in a separate phylum, are members of the beta-*Proteobacteria*.

Ecological Importance of Nitrifiers

Although nitrifiers occur in relatively low concentrations in habitats, they are very active metabolically. Nitrite does not accumulate in most environments because of the tight coupling between ammonia oxidation and nitrite oxidation. Therefore, nitrite, which is toxic (and mutagenic) to plants and animals, typically occurs in very low concentrations in environments because of its rapid oxidization to nitrate.

The activities of nitrifying bacteria are especially important in soil environments because of the role they play in nitrogen cycling (see Chapter 24). Ammonium nitrogen is a preferred source of nitrogen as a crop fertilizer. The principal reason for this is that ammonium is more readily retained in soils because it is positively charged. Nitrate, on the other hand, is highly soluble and readily leached from soils. Moreover, nitrate formed by nitrifiers can be converted to nitrogen gas by

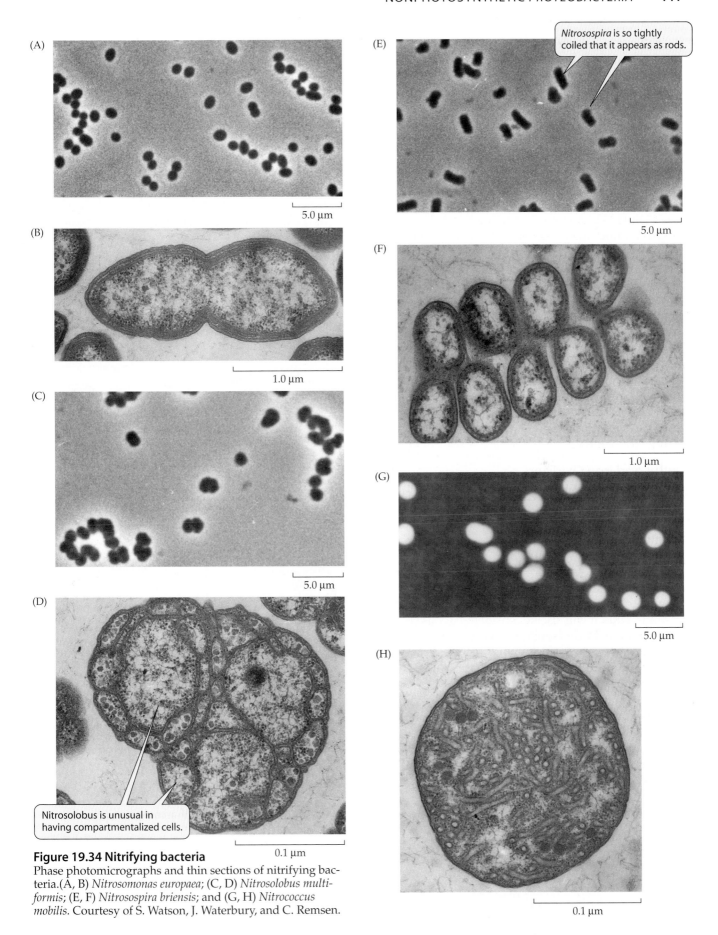

Nitrosospira is so tightly coiled that it appears as rods.

Nitrosolobus is unusual in having compartmentalized cells.

Figure 19.34 Nitrifying bacteria
Phase photomicrographs and thin sections of nitrifying bacteria.(A, B) *Nitrosomonas europaea*; (C, D) *Nitrosolobus multiformis*; (E, F) *Nitrosospira briensis*; and (G, H) *Nitrococcus mobilis.* Courtesy of S. Watson, J. Waterbury, and C. Remsen.

denitrifiers, and therefore lost from soils. For these reasons, the inhibitor nitrapyrin (see earlier) is produced commercially as a "fertilizer" to inhibit nitrification and thereby prevent nitrogen loss from soils.

It is also noteworthy that N_2O, an intermediate formed in nitrification as well as denitrification, is important environmentally in that it can react photochemically with ozone in the upper atmosphere (see Chapter 24) and destroy it.

Sewage contains high concentrations of ammonia due to the decomposition of proteins and amino acids (see Chapter 32). As a result, ammonia is discharged in large quantities in sewage outfalls. Like organic compounds, ammonia has a high biochemical oxygen demand (BOD, see Chapter 32), because the process of nitrification requires oxygen. Therefore, even treated sewage, which has low organic carbon concentrations, can cause low oxygen concentrations that reduce the quality of receiving waters because of nitrification. For this reason, modern wastewater treatment systems are designed to encourage the growth of nitrifying bacteria *within* the treatment plant so that downstream waters will be unaffected by ammonia (see Chapter 32). Otherwise, receiving waters could become anaerobic by the activity of these bacteria even though it contains little organic material.

Sulfur and Iron Oxidizers

Two different groups of lithotrophic *Proteobacteria* obtain energy by the oxidation of reduced inorganic sulfur compounds, the "unicellular sulfur oxidizers" and the "filamentous sulfur oxidizers." In addition, one proteobacterium that is a sulfur oxidizer, *Thiobacillus ferrooxidans*, can also carry out the oxidation of pyrite, FeS_2, with the formation of ferric oxide and sulfate and obtain energy both from iron oxidation as well as the oxidation of sulfur. Thus, it will also be considered in this section. Other sulfur oxidizers, such as the family *Sulfolobaceae*, are *Archaea* (see Chapter 18).

Some bacterial sulfur oxidizers use sulfide as an energy source. The sulfide can be supplied as hydrogen sulfide, or as a metal sulfide such as iron or copper sulfide. Some use elemental sulfur, S^0, whereas others use thiosulfate, $S_2O_3^{2-}$. Some can use all three forms of sulfur. The final product formed in all cases is sulfuric acid. The overall reactions for the complete oxidation of these various sulfur forms are shown following:

$$(7)\ S^{2-} + 4\,O_2 \rightarrow 2\,SO_4^{2-}$$

$$(8)\ 2\,S^0 + 3\,O_2 + 2\,H_2O \rightarrow 2\,H_2SO_4$$

$$(9)\ S_2O_3^{2-} + 2\,O_2 + H_2O \rightarrow 2\,SO_4^{2-} + 2\,H^+$$

The production of sulfuric acid (as sulfate or sulfuric acid) leads to the lowering of pH during the growth of these bacteria. Indeed, the acidophilic members of the sulfur oxidizers, as exemplified by *Thiobacillus thiooxidans*, can produce enough acid to lower the pH to 1.0, equivalent to 0.1 N H_2SO_4 and still remain viable. Clearly, these are examples of ultimate proton pumpers! These oxidations yield energy for the cells as shown in **Figure 19.35**.

The unicellular sulfur and iron oxidizers will be treated in the first section that follows, and this will then be followed by a discussion of the filamentous sulfur oxidizers.

Unicellular Sulfur Oxidizers There are three major genera of unicellular sulfur oxidizing bacteria: *Thiobacillus*, *Thiomicrospira*, and *Thermothrix*. *Thiobacillus* is a genus of

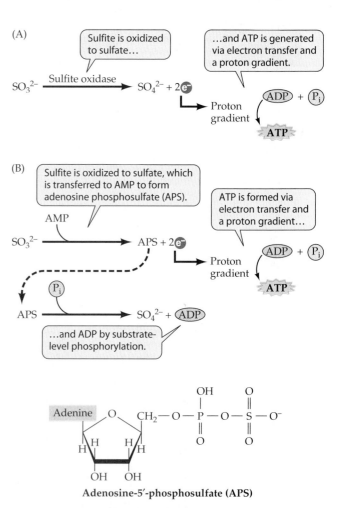

Figure 19.35 Oxidation of sulfur compounds by *Thiobacillus*
Pathways for oxidation of reduced sulfur compounds and generation of energy by *Thiobacillus* sp. Sulfide, elemental sulfur, and thiosulfate are oxidized by the enzyme sulfide reductase to form sulfite. The sulfite is metabolized by one of two different pathways, (A) and (B). (A) is the principal pathway.

Table 19.25	Characteristics of unicellular sulfur-oxidizing bacteria			
Genus/Species	**Nutrition Group**	**pH Optimum**	**Sulfur Source**	**Mol % G + C**
Thiobacillus thiooxidans	Chemolithoautotroph	2–4	$S^0, S_2O_3^{2-}$	51–53
T. ferrooxidans	Chemolithoautotroph	2–4	S^0, S^{2-} (incl. metal sulfides)	53–65
T. thioparus	Chemolithoautotroph[a]	6–8	$S^0, S^{2-}, S_2O_3^{2-}$	63–66
T. acidophilus	Mixotroph	2–4	$S^0, S_2O_3^{2-}$	61–64
T. novellus	Mixotroph	6–8	$S_2O_3^{2-}$	67–68
T. denitrificans	Chemolithoautotroph	6–8	$S^{2-}, S^0, S_2O_3^{2-}$	63–68
Thiomicrospira pelophila	Chemolithoautotroph	6.5–7.5	$S^{2-}, S^0, S_2O_3^{2-}$	44
Thermothrix thiopara	Mixotroph	6–8	$S^{2-}, S_2O_3^{2-}$	Unknown

[a]See Chapter 22.

polarly flagellated rods that is widespread in soil and aquatic habitats. *Thiomicrospira* is a genus of small, polarly flagellated vibrios that are common in marine habitats. The final genus, *Thermothrix*, resembles *Thiobacillus* spp. in being a polarly flagellated rod. However, it grows at temperatures from 40 to 80°C as a filamentous organism under the low oxygen conditions found in the thermophilic hot springs where it lives. **Table 19.25** lists important species from each of these genera and describes some of their salient features.

The genus *Thiobacillus* is especially diverse metabolically. At one extreme are obligate chemolithotrophs, such as *Thiobacillus thiooxidans*, which do not use organic compounds at all. These obligate chemolithotrophs are also called chemolithoautotrophs because they use (a) chemicals as energy source (hence, *chemo-*), (b) inorganic sources of electrons (that is, *litho-*, meaning rock), and (c) carbon dioxide as a sole carbon source (hence, *auto*trophic). Many autotrophic species produce carboxysomes in their cells, which contain high concentrations of RuBisCo (**Figure 19.36**). *Thiobacillus novellus* is an intermediate group of so-called facultative chemolithotrophs, or mixotrophs, that fix carbon dioxide but can also assimilate organic carbon sources for anabolic purposes. Finally, at the other extreme are some ordinary heterotrophs including some *Pseudomonas* species that oxidize reduced sulfur compounds but do not obtain energy from this process and cannot utilize carbon dioxide as a carbon source.

Filamentous Sulfur Oxidizers The filamentous sulfur oxidizers are found at the interface of sulfide-containing muds and the aerobic environment. For example, they are common on the surface of marine and freshwater sediments and in sulfur springs. These types of habitats, which have an active sulfur cycle, are called **sulfureta**. There are several important genera as presented in **Table 19.26**.

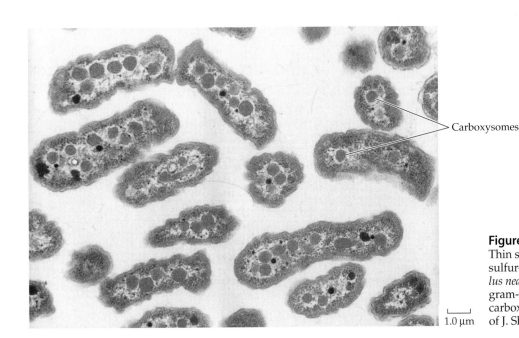

Carboxysomes

Figure 19.36 *Thiobacillus*
Thin section of rod-shaped cells of the sulfur-oxidizing bacterium *Thiobacillus neapolitanus*, showing the typical gram-negative cell structure. Note the carboxysomes in the cells. Courtesy of J. Shively.

1.0 µm

Table 19.26	Filamentous sulfur-oxidizing bacteria		
Genus	**Mol % G + C**	**Distinctive features**	
Beggiatoa	37–51	Long gliding filaments with sulfur granules	
Thiothrix	52	Rosette-forming; gliding gonidia	
Thioploca	Unknown	Gliding filaments in sheath	

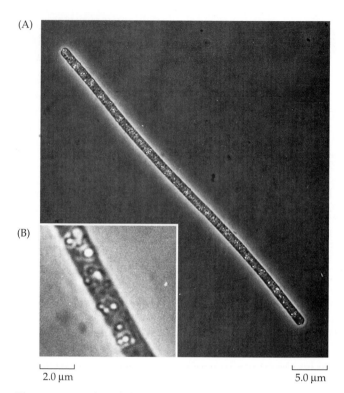

(A)

(B)

2.0 μm   5.0 μm

Figure 19.37 *Beggiatoa*
(A, B) Phase photomicrographs of the filamentous gliding sulfur-oxidizing bacterium *Beggiatoa*. Courtesy of J. T. Staley.

The genus *Beggiatoa* is by far the best-known member of the filamentous sulfur bacteria (**Figure 19.37**). Studies of this organism by Sergei Winogradksy led to his proposal of the concept of chemolithotrophy (**Box 19.5**). Although Winogradsky did not study *Beggiatoa* in pure culture, several strains of *Beggiatoa* have now been isolated. All isolated strains produce sulfur granules and are known to oxidize sulfides, but some are facultative chemoautotrophs that use acetate as a carbon source.

The classical metabolic type of *Beggiatoa*, as proposed by Winogradsky, is a strict chemolithotroph that uses sulfide as an energy source and oxidizes it completely to sulfuric acid. Furthermore, it is also grows chemo-

---

Milestones Box 19.5 ## Winogradsky and the Concept of Chemolithotrophy

Sergei Winogradsky, a scientist who was born and raised in Czarist Russia, became one of the most remarkable figures in microbiology (see Chapter 2). His most important discoveries relate to the phototrophic bacteria and the chemolithotrophic bacteria. For example, he described by direct microscopic observation several genera of the purple sulfur bacteria. During the past century, many microbiologists have worked on these bacteria, and his classification of them still holds largely unchanged today.

But we are most indebted to Sergei Winogradsky for his novel contributions to our understanding of bacterial chemolithotrophy. These studies began while he was a student in Switzerland. It was when he was in the Alps that Sergei Winogradsky first observed

Beggiatoa in sulfur springs. Based on his careful observations of their growth and sulfur chemistry, he concluded that these filamentous bacteria must obtain energy by oxidizing the sulfide emanating from the spring. He concluded this because he noted that the cells contained the more oxidized form of sulfur, elemental sulfur, in granules located inside their cells (see Figure 19.37). Therefore, he proposed that these bacteria were able to capture some of the chemical energy from the oxidation of sulfide to sulfur and to use this for their metabolic processes. Thus, the novel concept of chemolithotrophy was born.

Sergei Winogradsky was also the first to grow nitrifying bacteria successfully. He cultivated them on a silica gel medium because they do not form colonies on agar media. Most

isolates of nitrifiers are now obtained by dilution of enrichment samples to extinction using liquid inorganic media. To accomplish this, active enrichments containing the bacteria are diluted serially with many replicates. By chance, some of the highest dilution tubes with growth will receive only one cell of a nitrifier of interest, and from this a pure culture can be obtained. In this tedious manner representatives of several new genera have been isolated in pure culture. Many of these strains were first isolated in pure culture using this technique in the late Stanley Watson's laboratory at Wood's Hole Oceanographic Institute. Most pure cultures must be maintained in liquid media to keep them viable.

autotrophically using carbon dioxide as a sole source of carbon. Strains of this type have been isolated from the marine environment and they use the Calvin-Benson cycle for carbon dioxide fixation.

However, heterotrophic types are currently known and have been studied in pure culture. The heterotrophic strains are common in freshwater habitats and sulfur springs. This type oxidizes sulfide only to sulfur, not sulfate. They use acetate as a carbon source and obtain energy by its oxidation. These strains may also be able to fix carbon dioxide by the Calvin-Benson cycle, because low levels of ribulose bisphosphate carboxylase (RuBisCo), a key enzyme in this pathway, have been detected.

Thiothrix species are found not only in sulfur springs, but also in sewage treatment facilities as bacterium that causes bulking (see Chapter 32). The filaments contain sulfur granules and closely resemble *Beggiatoa* spp.; however, they produce holdfasts, form rosettes (**Figure 19.38**), and divide by generating gliding gonidia in a manner analogous to the genus *Leucothrix*. They also produce a thin sheath (**Figure 19.39**). Little is yet known about the physiology and metabolism of this group, although some strains have been cultivated in pure culture.

Another important filamentous sulfur bacterium is the genus *Thioploca*. It is found in lake sediments and in the sediments of certain marine intertidal zones, such as in the Gulf of Mexico and off the coast of Chile, where they serve as an important food source for

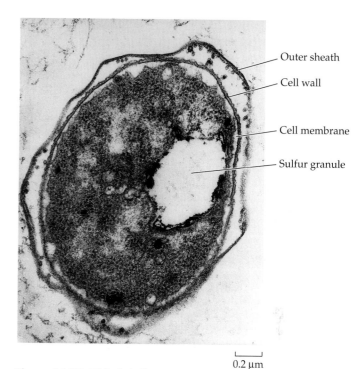

Outer sheath
Cell wall
Cell membrane
Sulfur granule

0.2 μm

Figure 19.39 *Thiothrix* fine structure
Thin section through a *Thiothrix* cell, showing the sheath, septum, and typical gram-negative structure. Note that the sulfur granules are bound by a membrane. Courtesy of J. Bland and J. T. Staley.

marine animals. Pure cultures have not yet been obtained of these bacteria. They resemble *Beggiatoa* spp. in that they produce filaments with sulfur granules; however, in addition, they produce a sheath in which one or more filaments reside (**Figure 19.40**). One of the unique attributes of this genus is that the cells appear "empty" when they have been examined by thin section in the electron microscope (**Figure 19.41**). Recent studies of these bacteria indicate that these vacuoles are used to store nitrate. These unusual bacteria have a unique capability. In the natural sediment habitat, the sheathed filaments are oriented vertically so that the trichomes can glide up and down through the aerobic/anoxic interface. They are able to migrate from anoxic zones in the sediments into aerobic zones, where they concentrate the nitrate from the environment. They can then glide down into the sediments, which are rich in sulfide, and carry out denitrification, thereby obtaining energy for growth. Thus, they have evolved to take advantage of both the oxidized and reduced zones in sediment habitats. *Thiomargarita*, which has the largest cells of any known bacterium, also carries out this process (**Figure 19.42**).

Iron Oxidizers Some bacteria are able to obtain energy by the oxidation of ferrous iron. Because iron oxidation at typical physiological pHs, that is, near neutrality (pH 4 to 8), occurs spontaneously in the presence of oxygen,

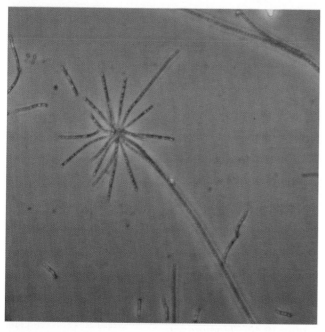

Figure 19.38 *Thiothrix*
A rosette of *Thiothrix* from a sulfur spring.
Courtesy of J. Bland and J. T. Staley.

(A)

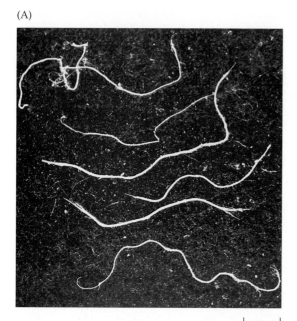

(B)

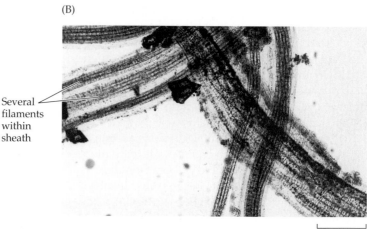

Several
filaments
within
sheath

0.2 mm

5.0 mm

Figure 19.40 *Thioploca*
(A) Sheathed filaments of *Thioploca* washed from marine intertidal sediments. These filaments are visible to the unaided eye. (B) Ensheathed filaments of *Thioploca* as they appear by brightfield light microscopy. Note that each sheath contains several filaments. Courtesy of S. Maier.

abundant, and these bacteria have been shown to be able to use it as an energy source (Figure 8.15).

In contrast, *Gallionella ferruginea* grows at neutral pH in an environment that is very low in oxygen. This bacterium lives in iron springs where it appears as masses of twisted filaments that appear reddish brown from the oxidized iron, FeO(OH). These are the stalks of the bacterium (**Figure 19.43A**). The cell appears as a small vib-

the bacteria that carry out this process either live in environments with very low pH or very low oxygen concentrations where ferrous ion remains reduced. Two different iron-oxidizers are discussed in more detail.

Iron oxidation by *Thiobacillus ferrooxidans* occurs at low pH. These bacteria grow in association with acidophilic sulfur oxidizers and are capable of acidophilic sulfur oxidation themselves. In this environment, ferrous ion is

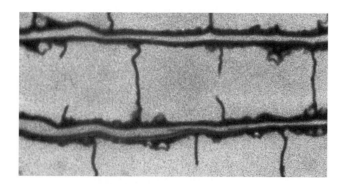

Figure 19.41 *Thioploca* fine structure
Thin section through *Thioploca* filaments, showing their empty appearance. Apparently the cytoplasm is pushed against the cell membrane by internal vacuoles. Courtesy of S. Maier.

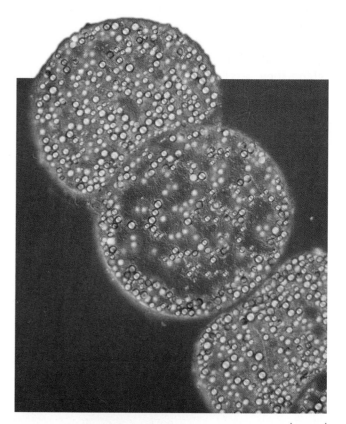

Figure 19.42 *Thiomargarita*
0.1 mm
This is the largest bacterium known. Bright yellow granules are sulfur. Photo by Ferran Garcia Pichel, from the cover of *Science*, Vol. 284, No. 5413. ©1999 AAAS.

(A)

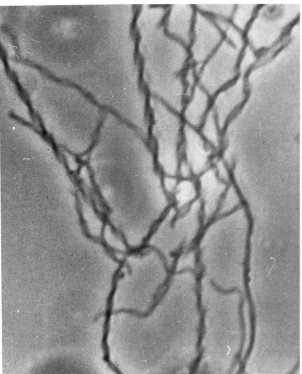

(B)

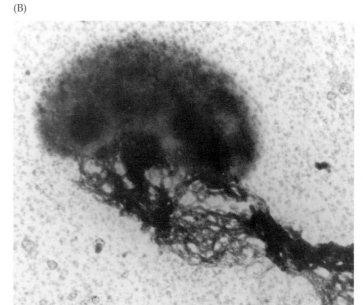

Figure 19.43 *Gallionella*
(A) Twisted stalks of *Gallionella ferruginea* as they appear when obtained from natural spring sources. (B) Electron micrograph of *G. ferruginea*, showing the vibrioid appearance of the cell and the stalk arising from the concave side of the cell. Courtesy of H. Hanert.

rio that produces the inorganic stalk of ferric hydroxide from the concave side of the cell (Figure 19.43B). As the organism grows, the stalk elongates and twists to form a helix. When the cells divide, the stalk bifurcates. Meanwhile, the cells remain small and almost indiscernible but continue to produce more stalk material as they grow. This iron-oxidizing bacterium is better known for its stalk than for itself, because when one observes it in the microscope, one primarily sees the massive amounts of iron oxide-encrusted stalks.

Gallionella ferruginea is a chemoautotrophic iron oxidizer that belongs to the alpha *Proteobacteria*. Although it has been difficult to obtain *Gallionella ferrugenea* in pure culture, some strains have been isolated. Results of physiological studies indicate that this bacterium is a true chemoautotroph because it has the enzymes of the Calvin-Benson cycle.

Hydrogen gas (H_2) is an excellent energy source for bacterial growth. Among the *Bacteria*, several types of hydrogen utilizers are found. One group is mesophilic, facultative hydrogen chemolithotrophs that can also use organic carbon sources for growth. Another group, the *Aquificales*, is thermophilic eubacteria. A third group also is involved in the oxidation of carbon monoxide. Still another eubacterial group is anaerobic bacteria, called acetogenic bacteria because they produce acetic acid from carbon dioxide. Each of these four groups is discussed individually next.

Mesophilic Hydrogen Bacteria

The ability to use hydrogen gas as an energy source is widespread among *Bacteria* and *Archaea*. At one time all hydrogen-oxidizing *Bacteria* were placed in the genus "*Hydrogenomonas*." However, many of these bacteria were found to be closely related to other heterotrophic bacteria from a variety of different genera (**Table 19.27**). Furthermore, with the exception of the thermophilic

Table 19.27	Genera in which hydrogen autotrophy has been reported
Gram-Positive Genera	**Gram-Negative Genera**
Arthrobacter	*Alcaligenes*
Bacillus	*Ancylobacter*
Mycobacterium	*Aquaspirillum*
Nocardia	*Derxia*
	Flavobacterium
	Hydrogenobacter
	Paracoccus
	Pseudomonas
	Rhizobium
	Spirillum

genera *Hydrogenobacter* and *Aquifex*, all aerobic hydrogen bacteria are facultative hydrogen bacteria that can use organic compounds as energy sources as well as hydrogen gas. Therefore, most are facultative chemolithotrophs or mixotrophs, not obligate chemolithotrophs. Typical hydrogen oxidizers obtain energy from the oxidation of hydrogen gas using a membrane-bound hydrogenase:

$$(10) \quad H_2 \rightarrow 2\,H^+ + 2\,e^-$$

The electrons generated are passed through an electron transport chain, and ATP is generated by proton pumping and membrane-bound ATPases. Only a few genes are needed for the hydrogen oxidation process. In some species, these have been found on plasmids. This suggests that the ability to generate energy from hydrogen can be genetically transferred from one species to another. We call this ability lateral gene transfer. However, in addition to having a hydrogenase and an electron transport system for energy generation, these bacteria also need to have the enzymes for carbon dioxide fixation in order to grow autotrophically. All of the hydrogen-oxidizing eubacteria studied so far that can grow on carbon dioxide use the Calvin-Benson cycle for carbon dioxide fixation. One of the most thoroughly studied members of this group is *Alcaligenes eutrophus.*

Carboxydobacteria   One special group of hydrogen bacteria obtains energy from the oxidation of hydrogen and carbon monoxide to form carbon dioxide. These are called **carboxydobacteria**. The biochemistry of the pathway has not been studied well, although it is thought that an oxidoreductase is responsible for the following overall reaction:

$$(11) \quad CO + H_2 + 1/2\,O_2 \rightarrow CO_2 + 2\,H^+ + 2\,e^-$$

Some hydrogen-oxidizing bacteria carry out this process, including *Alcaligenes eutrophus,* as well as members of the genera *Azotobacter* and *Hydrogenophaga.* However, not all hydrogen bacteria are capable of carbon monoxide oxidation. *Carboxydomonas* has been described as a distinct genus of carbon monoxide degraders. It is a thermophilic bacterium isolated from hot springs.

Carbon monoxide is produced through the incomplete combustion of wood, coal, and oil, and therefore it is common in the environment. CO is lethal to humans and other animals because it reacts with hemoglobin irreversibly. Fortunately for animals, this diverse group of bacteria appears to be widely distributed in soils and aquatic habitats and is very efficient in degrading CO in the environment. As a result, CO is found in extremely low concentrations in the atmosphere.

METHYLOTROPHIC BACTERIA

Methane and methanol are considered by some to be two of the simplest organic compounds. However, it could be argued that true organic compounds have carbon–carbon bonds. Of course, microorganisms are not mindful of human definitions. The methylotrophic bacteria treat methane and methanol from a purely physical and biochemical standpoint—as substrates that can provide energy for growth. **Methanotrophic bacteria** use methane as a carbon source and oxidize it to carbon dioxide. These bacteria can also use methanol and a variety of other one-carbon and a few two-carbon compounds as carbon sources. Another separate group of bacteria use methanol, but not methane gas, as a carbon source for growth. Therefore, these are not methanotrophic bacteria, but are instead called **methylotrophic bacteria**. It should be noted that methanotrophic bacteria are also methylotrophic because they can also use methanol and other one-carbon compounds for growth. For this reason, this section is entitled the methylotrophic bacteria.

Methanotrophs

There are two groups of methane-oxidizing bacteria, both of which have internal membrane systems similar to those of the nitrifying bacteria. The **type I methanotrophs** have an internal membrane system that lies perpendicular to the long axis of the cell (**Figure 19.44**), whereas the **type II methanotrophs** have membranes that lie parallel to the cell membrane (**Figure 19.45**). The genera that have been described are listed in **Table 19.28**. The type I methanotrophs are members of the gamma *Proteobacteria,* whereas the type II methanotrophs belong to the alpha *Proteobacteria.*

Some of the methanotrophs produce a resting stage, called cysts or exospores. The cysts are analogous to those produced by *Azotobacter* spp. in that they have complex cell wall layers and may store poly-ß-hydroxybutyrate. They are formed during conditions of nutrient depletion and can germinate during periods of nutrient sufficiency.

Energy Generation from Methane   The methanotrophs are all gram-negative obligate aerobes that derive energy from the oxidation of methane to carbon dioxide. The first reaction is unique to these bacteria and involves an enzyme called methane monooxygenase (MMO), which converts methane to methanol as shown following:

$$(12) \quad CH_4 + O_2 + XH_2 \rightarrow CH_3OH + H_2O + X$$

where X = hydrogen carrier (cytochrome-carrying electrons).

The MMO is a complex copper-containing enzyme that can be either membrane-bound or free in the cyto-

(A)

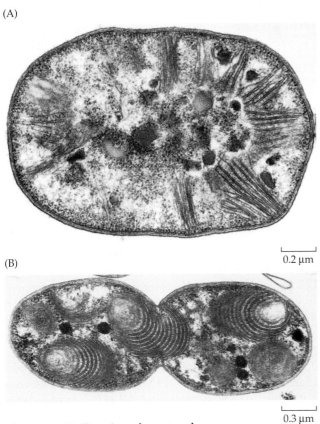

0.2 μm

(B)

0.3 μm

Figure 19.44 Type I methanotrophs
Thin sections through type I methanotrophs, (A) *Methylomonas agile* and (B) *Methylomonas methanica*, showing the typical transverse membrane systems. Courtesy of C. Murrell.

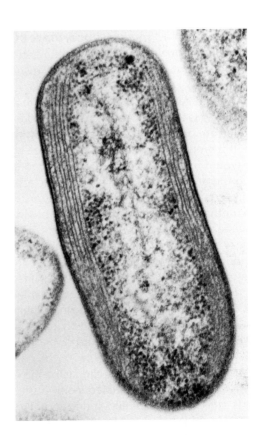

Figure 19.45 Type II methanotroph
Thin section of the type II methanotroph *Methylocystis parvus*, showing the membranes lying along the periphery of the cell. Courtesy of C. Murrell.

plasm. The second reaction involves a methanol dehydrogenase that converts the methanol to formaldehyde as follows:

$$(13)\ CH_3OH \rightarrow HCHO + 2e^- + 2\,H^+$$

The final reactions involved in carbon dissimilation for energy generation include either an oxidation via formate to carbon dioxide (see Chapter 10), or a more complex route to carbon dioxide called the ribulose monophosphate (RMP) pathway (Chapter 10). Many type I methanotrophs use the ribulose monophosphate pathway for energy generation, but some of these can also use the serine pathway. Some of the species using the serine pathway have a distinctive quinone, pyrrolo-quinoline quinone (PQQ), as a hydrogen acceptor.

Carbon Assimilation by Methanotrophic Bacteria To assimilate carbon, the type I methanotrophs use the RMP pathway (Figure 10.10B), whereas the type II methanotrophs use the serine pathway (Figure 10.10A). In both pathways, formaldehyde serves as the substrate for organic carbon synthesis. Thus, they are not autotrophs in that they cannot use carbon dioxide as their sole source of carbon.

Table 19.28	Genera of methanotrophic bacteria		
Group	**Genus**	**Mol % G + C**	**Special Features**
Type I methanotrophs			
	Methylobacter	50–54	Rod, polar flagellum, cyst
	Methylococcus	62–64	Coccus, nonmotile, cyst
	Methylomonas	50–54	Rod, polar flagellum, cyst
Type II methanotrophs			
	Methylocystis	62-63	Rod or vibrio, nonmotile
	Methylosinus	62–63	Curved rod, polar flagellum, exospore

Ecological and Environmental Importance of Methanotrophic Bacteria The methanotrophic bacteria are very important environmentally. They are responsible for the oxidation of methane produced by methanogenesis and natural gas seeps in the biosphere. Methane is a greenhouse gas that has been increasing in the atmosphere at a rate of about 1% per year during the past decade. If it were not for the activity of these bacteria, the gas would increase at much more rapid rates, with deleterious environmental effects. These bacteria reside in soil and aquatic habitats where oxygen is available as well as methane. Therefore, they live at the interface between the reduced zones of the environment in which methane is produced and oxygenated zones exposed to air. Although methane gas escapes in part from shallow ponds and marshes, in deeper water bodies it is largely oxidized by the activity of methanotrophs before it reaches the surface of the water column. Methane also reaches the atmosphere through the activities of termites as well as ruminant animals that harbor methanogenic bacteria.

The MMO of methanotrophs can also remove halogen ions from toxic halogenated compounds. This process is called dehalogenation or dechlorination if the compound is chlorinated. Brominated compounds can be similarly dehalogenated. The enzyme is most effective at dehalogenating low molecular weight, one- and two-carbon, aliphatic compounds such as chloroform and trichloroethene. The methanotrophs do not derive energy from these dechlorination reactions. This is an example of **cometabolism** (see Chapter 24) in that methane must be provided in order for them to carry out the dechlorination process. They cannot obtain energy from the dehalogenation reactions. These bacteria are important agents in dehalogenation of wastes containing low molecular weight halogenated compounds and are therefore of considerable importance in bioremediation (Chapter 32).

The methanotrophic bacteria share similarities with ammonia-oxidizing bacteria. Members of both groups have been shown to be able to carry out the oxidation of each others' substrates, ammonia, and methane, using their characteristic monooxygenases. In addition, both are members of the *Proteobacteria*, and therefore phylogenetically related to one another. However, neither group obtains energy from the oxidation of the other's substrate and therefore are not considered to be ecologically important in the other's niche.

Other Methylotrophs

As mentioned heretofore, the term **methylotroph** refers to a microorganism that can utilize one-carbon compounds such as methane, methanol, or methylamine as sole carbon and energy sources for growth. This is a general term that encompasses methanotrophs dis-

Table 19.29	Nonmethanotrophic methylotrophic bacterial genera	
Gram-Positive Genera	**Gram-Negative Genera**	
Arthrobacter	*Acinetobacter*	
Bacillus	*Alcaligenes*	
Mycobacterium	*Ancylobacter*	
Streptomyces	*Hyphomicrobium*	
	Klebsiella	
	Paracoccus	
	Pseudomonas	
	Rhodopseudomonas	

cussed previously as well as bacteria such as *Hyphomicrobium* spp. that can grow on methanol as sole carbon source. A list of heterotrophic genera that are known to be methanol or methylamine users is provided in **Table 19.29.** All of these bacteria that have been studied use the serine pathway for carbon assimilation and grow by aerobic or nitrate respiration. It should be emphasized that none of them are known to use methane gas, and none of them produce the enzyme methane monooxygenase. Also, all of them are facultative methylotrophs that can use other organic compounds for growth including sugars, amino acids, and organic acids, depending on the taxon.

One interesting newly discovered habitat for methylotrophic bacteria is on the surface of leaves. Methanol is derived from the degradation of methoxylated compounds produced by the plant. Strains especially adapted to growth on the plant leaf surface use the volatile methanol that is released. These strains are all pink in pigmentation.

SUMMARY

▸ The *Proteobacteria* is the most diverse phylum of eubacteria containing heterotrophic, chemoautotrophic, and photosynthetic bacteria. The *Proteobacteria* include the alpha, beta, gamma, delta, and epsilon subgroups.

▸ The **enteric bacteria** including *Escherichia coli, Enterobacter, Erwinia, Salmonella,* and *Shigella,* are all facultative aerobes that ferment sugars and belong to the gamma subgroup. Some enteric bacteria, such as *E. coli,* have a **mixed-acid fermentation** and produce acetic, succinic, and lactic acids as well as hydrogen and carbon dioxide. Some enteric bacteria, such as *Enterobacter aerogenes,* have a neutral or **butanediol fermentation** in which 2,3-butanediol, ethanol, carbon dioxide, and hydrogen are produced. Some enteric bacteria cause human disease, for example, **typhoid** fever (*Salmonella typhi*), **plague** (*Yersinia pestis*), and **bacterial dysentery** (*Shigella dysenteriae*).

▶ *Vibrio* species are typically marine or estuarine members of the gamma *Proteobacteria*—some are pathogenic to fish and humans. *Vibrio cholerae* is the causative agent of **human cholera**. The *Vibrio* group are facultative aerobic bacteria that ferment sugars. Some *Vibrio* and *Photobacterium* spp. produce light by **bioluminescence**, using bacterial luciferase—many luminescent bacteria live as symbionts in the light organs of fish.

▶ *Zymomonas* ferments sugars to ethanol, lactic acid, and carbon dioxide.

▶ Pseudomonads, also *Proteobacteria*, include the genera *Pseudomonas*, *Xanthomonas*, and *Zoogloea* and are obligately aerobic, oxidase positive, gram-negative rods that are motile by polar flagella. *Pseudomonas* species may be pathogens of humans (*P. aeruginosa*) or plants (*P. syringae*). *Pseudomonas* spp. are very versatile in their degradative abilities—many can use over 100 different sugars, organic acids, amino acids, and other monomers as sole carbon sources for growth. Some *Pseudomonas* spp. can degrade toxic compounds such as toluene and naphthalene, and the genes for this degradative ability are borne on plasmids. Some *Pseudomonas* and *Xanthomonas* spp. are plant pathogens that cause ice nucleation, which damages plant cells.

▶ *Azotobacter*, *Beijerinckia*, *Derxia*, and *Azomonas* are free-living soil bacteria capable of **nitrogen fixation**. *Rhizobium* and *Bradyrhizobium* are **symbiotic nitrogen-fixing** bacteria that are associated with legumes, where they form root nodules. *Rhizobium* cells multiply in the root nodule, and some become specialized cells, called **bacteroids**, which produce nitrogenase for nitrogen fixation. Nitrogen fixation in the legume is a truly symbiotic association—for example, **leghemoglobin**, an oxygen-scavenging compound is formed in part by the plant (globin portion) and in part by the bacterium (heme portion).

▶ *Agrobacterium tumefaciens* causes **crown gall** disease in dicotyledonous plants. *A. tumefaciens* carries a plasmid, called the **Ti (tumor-inducing) plasmid**, the DNA of which is incorporated into the plant host DNA upon infection—normally this plasmid contains genes that cause gall formation and plant disease. The Ti plasmid of *A. tumefaciens* carries genes for the production of unusual amino acids, called **opines**, which the plant then synthesizes. The opines synthesized by infected plants can be used by the same strain of *A. tumefasciens* that has caused the plant infection. The Ti plasmid can be **genetically engineered** to carry specific desirable genes into the plant.

▶ The *Proteobacteria* contain many other important genera. Included are *Acetobacter* and *Gluconobacter*, which are aerobes that produce acetic acid and are therefore called the **vinegar bacteria.** *Legionella pneumophila* is the pathogen that causes **Legionnaire's disease**. *Neisseria* is a genus of gram-negative cocci that cause **gonorrhea** (*N. gonorrhoeae*) and **bacterial meningitis** (*N. meningitidis*). *Rickettsia* and *Chlamydia* are two genera of obligate intracellular parasites. *Rickettsia* are members of the *Proteobacteria*, and *Chlamydia* have their own phylogenetic group. **Sheathed bacteria**, including *Sphaerotilus* and *Leptothrix*, live in flowing freshwater habitats—some deposit iron oxides and manganese oxides in their sheaths.

▶ **Prosthecae** are cellular appendages produced by some aquatic bacteria. Three groups of prosthecate bacteria exist: the caulobacters, the hyphomicrobia, and the polyprosthecate bacteria. The caulobacters, including the genus *Caulobacter*, live in freshwater and marine habitats. The hyphomicrobia produce **buds** from their appendages and include the genera *Hyphomicrobium*, which is a group of denitrifying bacteria that use methanol as a carbon source; *Hyphomonas* use amino acids and organic acids as carbon sources; *Pedomicrobium* are common in soils and freshwater habitats and are involved in the deposition of iron and manganese oxides. The polyprosthecate bacteria include *Prosthecomicrobium* and *Stella*, which are aerobic water and soil bacteria, and the genus *Ancalomicrobium*, which lives in aquatic habitats and ferments sugars by a mixed-acid fermentation identical to that of *E. coli*.

▶ *Bdellovibrio* is a bacterium that is a predator of other gram-negative bacteria. Spirilla are typically microaerophilic, heterotrophic bacteria that are members of the alpha, beta, and gamma *Proteobacteria*. *Magnetospirillum magnetotacticum* is a **magnetotactic bacterium** that has magnetite crystals that allow it to orient in a magnetic field.

▶ The **myxobacteria** are gliding bacteria that have a complex life cycle—cells move in "packs" when nutrients are plentiful; when nutrients are depleted, they aggregate and undergo a complex fruiting process in which the cells are converted to a more resistant stage called a **microcyst**, which are formed in a **fruiting structure**; microcysts can germinate when conditions for growth again become favorable.

▶ Several groups of chemolithotrophic bacteria exist, including the **nitrifiers**, the **sulfur oxidizers**, the **iron oxidizers**, and the **hydrogen bacteria**. The **nitrifying bacteria** comprise two separate groups, the **ammonia oxidizers** and the **nitrite oxidizers**. The **sulfur-oxidizing bacteria** also consist of two groups, the **uni-**

cellular species such as *Thiobacillus* and *Thiomicrospira* and the **filamentous** species, such as *Beggiatoa*. Many sulfur-oxidizing bacteria grow as **facultative chemolithotrophs** or **mixotrophs** by requiring or using various organic carbon sources.

▸ Only two genera of **hydrogen bacteria** are known to grow as chemolithotrophs. These are the thermophilic genera, *Hydrogenobacter* and *Aquifex*, which comprise a separate phylum, the Aquificae. **Carboxydobacteria** are a special group of hydrogen bacteria that can oxidize hydrogen and carbon monoxide.

▸ **Methylotrophic bacteria** use one-carbon compounds such as methanol or methane as a sole carbon and energy source for growth. **Methanotrophic bacteria** can use methane as a sole source of carbon; they comprise two groups, the type I and type II methanotrophs, which have internal membranes where the enzyme methane monooxygenase resides. Type I methanotrophs use the ribulose monophosphate pathway for carbon assimilation. Type II methanotrophs use the serine pathway for carbon assimilation. Methane oxidizing bacteria are all aerobic and oxidize methane produced by methanogens in natural environments.

REVIEW QUESTIONS

1. The *Proteobacteria* constitute an extremely diverse group of bacteria. Cite examples of their diversity. How could such great diversity occur in a single bacterial group?

2. Some bacteria move by gliding, some have flagella, and others are nonmotile. What are the advantages or disadvantages of each type of existence?

3. How is it that some gliding bacteria are found in the *Proteobacteria* and others in a completely different evolutionary group?

4. What is the role(s) of the prostheca? Why do some bacteria have more than one?

5. When you enter a winery you smell vinegar. Why?

6. What are the possible desirable features that could be engineered into plants by use of the *Agrobacterium* Ti plasmid? Could undesirable features be engineered, too?

7. What advantages does the sheath provide for an organism?

8. Compare the motility of spirochetes to that of flagellated and gliding bacteria.

9. Why is it that only prokaryotic organisms can carry out nitrogen fixation?

10. Why are *Legionella* species difficult to grow?

SUGGESTED READING

Balows, A., H. G. Trüper, M. Dworkin, W. Harder and K. H. Schleifer, eds. 1992. *The Prokaryotes*. 2nd ed. Berlin: Springer-Verlag.

Holt, J. G., editor-in-chief. 1994. *Bergey's Manual of Determinative Bacteriology*. 9th ed. Baltimore, MD: Williams and Wilkins.

Krieg, N., D. Brenner, J. T. Staley and G. Garrity, eds. 2002. *Bergey's Manual of Systematic Bacteriology*. 2nd ed., Vol II. New York: Springer-Verlag.

Chapter 20

Gram-Positive Bacteria: *Firmicutes* and *Actinobacteria*

The most important discoveries of the laws, methods and progress of Nature have nearly always sprung from the examination of the smallest objects which she contains.

– J. B. Lamarck, *Philosophie Zoologique*, 1809

The common gram-positive *Bacteria* comprise a spectrum of morphological types ranging from unicellular organisms to branching, filamentous, multicellular organisms. Most are heterotrophs that grow as saprophytes living off nonliving organic materials. Gram-positive bacteria are particularly abundant in soil and sediment environments. A number of them are important plant and animal pathogens. This phylogenetic group contains some of the simplest heterotrophic organisms from a metabolic standpoint, such as the lactic acid bacteria. A few gram-positive bacteria are hydrogen autotrophs that make acetic acid from carbon dioxide and H_2. One photosynthetic group, the heliobacteria, has also been described.

Gram-positive bacteria contain significant amounts of peptidoglycan in their cell walls (see Chapter 4). Their cell walls, however, lack lipopolysaccharide, a characteristic component of the cell walls of gram-negative bacteria. They also have a simpler cell wall architecture. However, one subgroup, the *Mollicutes* or **mycoplasmas,** lack a cell wall altogether. Nonetheless, they are members of the gram-positive bacteria based on 16S rDNA sequence analyses.

The gram-positive *Bacteria* are classified in two separate phylogenetic groups, the *Firmicutes* and the *Actinobacteria*, based on 16S rDNA sequences. The phylum *Firmicutes* contains unicellular gram-positive *Bacteria* that have a low mol % G + C content. In contrast, the phylum *Actinobacteria* contains many filamentous, multicellular species whose DNA has a high mol % G + C.

The phylogenetic relatedness of the gram-positive bacteria to other bacterial phyla and themselves is shown in **Figure 20.1**. Note that the low mol % G + C group branches off separately from the high mol % G + C group. Each phylum is discussed individually, beginning with the *Firmicutes*.

PHYLUM: *FIRMICUTES*

The *Firmicutes* contains a large number of important physiological groups and species that are discussed next.

Lactic Acid Bacteria

As their name implies, the **lactic acid bacteria** are noted primarily for their ability to ferment sugars to produce lactic acid. This family contains several key genera whose activities in human health and the environment are familiar to all of us. Most of us have had sore throats, called "strep" throat, caused by *Streptococcus* species. The genus *Streptococcus* contains several important pathogens, such as *Streptococcus pyogenes*, which causes scarlet fever and rheumatic fever.

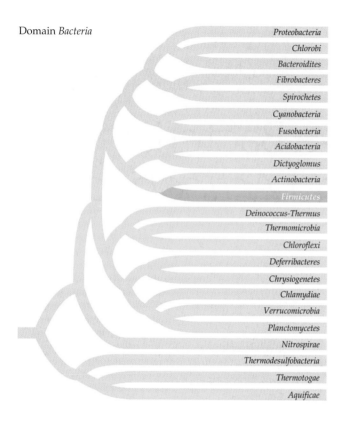

Domain *Bacteria*

Proteobacteria
Chlorobi
Bacteroidites
Fibrobacteres
Spirochetes
Cyanobacteria
Fusobacteria
Acidobacteria
Dictyoglomus
Actinobacteria
Firmicutes
Deinococcus-Thermus
Thermomicrobia
Chloroflexi
Deferribacteres
Chrysiogenetes
Chlamydiae
Verrucomicrobia
Planctomycetes
Nitrospirae
Thermodesulfobacteria
Thermotogae
Aquificae

The lactic acid bacteria are also significant in food and dairy microbiology as well as in agriculture. The fermentation of vegetable materials is used in the manufacture of pickles and sauerkraut. Milk fermentation by lactic acid bacteria leads to the manufacture of commercial products such as yogurt, buttermilk, and cheeses.

All lactic acid bacteria are **aerotolerant anaerobes;** that is, they are facultative aerobes that grow in the presence of oxygen but do not use it in respiration. Instead, they produce energy by fermentation of sugars. In addition, most of them have complex nutritional requirements. Many require amino acids and vitamins, and some even need purines and pyrimidines to grow. Not surprisingly, they are found in organic-rich environ-

Figure 20.1 Phylogeny of gram-positive bacteria
Diagram showing the two phyla of gram-positive bacteria based on 16S rRNA analyses. Note that the unicellular, low mol% G + C lactic acid bacteria and endospore-forming bacteria are found in the *Firmicutes* branch, whereas the mycelial, high mol% G + C *Actinobacteria* form a separate branch. Adapted from Ribosomal Database Project.

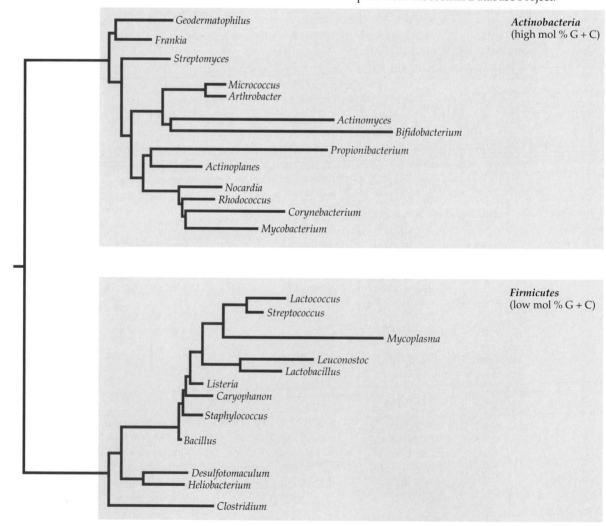

Table 20.1	Principal genera of lactic acid bacteria		
Genus	**Morphology**	**Fermentation Type**	**Lactic Acid Form**
Streptococcus	Cocci in chains	Homofermentative	L-
Leuconostoc	Cocci in chains	Heterofermentative	D-
Pediococcus	Cocci in tetrads	Homofermentative	DL-
Lactobacillus	Rods	Homo- or Heterofermentative	Varies with species

ments such as decaying plant and animal materials. **Table 20.1** lists the major genera of lactic acid bacteria as well as some of their common features.

Physiology and Metabolism Although all lactic acid bacteria carry out lactic acid fermentation, they can be separated into two different groups based on the type of fermentation process. The **homofermentative** bacteria carry out a simple fermentation in which lactic acid is the sole product from sugar fermentations:

Glucose → 2 Lactic acid

The Embden-Meyerhof or glycolytic pathway is used in this process (see Chapter 8). The pyruvic acid formed is reduced by the enzyme lactic acid dehydrogenase to produce the characteristic end product, lactic acid (**Figure 20.2**).

In contrast, the **heterofermentative** lactic acid bacteria produce ethanol and carbon dioxide as well as lactic acid:

Glucose → Lactic acid + Ethanol + CO_2

The heterofermentative pathway (**Figure 20.3**) lacks the key enzyme aldolase that is present in the glycolytic pathway of the homofermentative lactics. Aldolase cleaves fructose-1, 6-diphosphate to form the two phosphorylated trioses that ultimately lead to the production of two ATPs. Therefore, although homofermenters obtain 2 moles of ATP for each mole of glucose fermented, the heterofermenters obtain only 1 mole, as can be seen by examining their pathway. This pathway involves an initial oxidation of glucose to 6-phosphogluconic acid, which is decarboxylated to form CO_2 and ribulose 5-phosphate. The pentose formed is converted to (a) the three-carbon intermediate, glyceraldehyde-3-

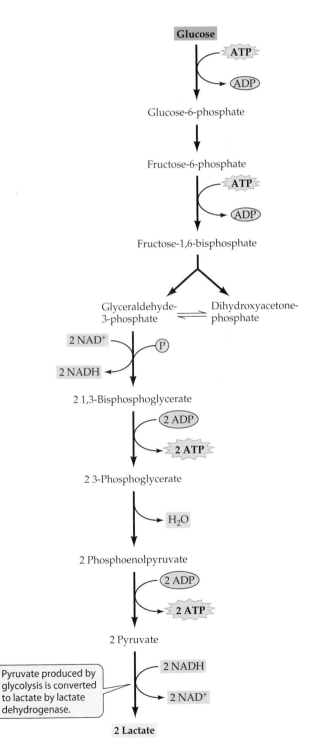

Figure 20.2 Homofermentation pathway
Pathway for dissimilation of glucose by homofermentative lactic acid bacteria. These bacteria use the glycolytic pathway for formation of pyruvate and lactate dehydrogenase to produce lactate as the final end product.

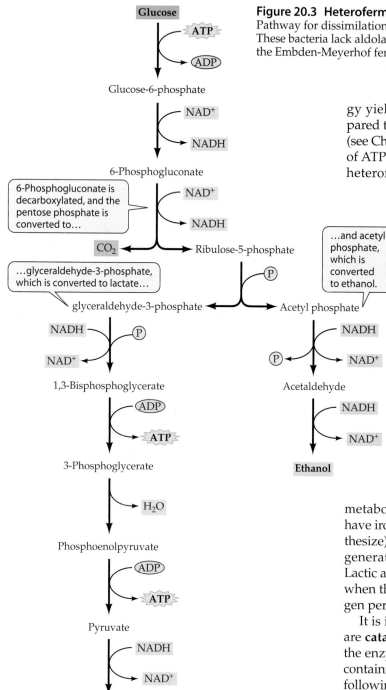

Figure 20.3 Heterofermentation pathway
Pathway for dissimilation of glucose by heterofermentative lactic acid bacteria. These bacteria lack aldolase, a key glycolytic enzyme, and therefore do not use the Embden-Meyerhof fermentation pathway.

gy yield per mole of glucose (Y_{ATP}) is very low compared to other bacteria, especially those that can respire (see Chapter 6). The homofermenters that produce 2 mol of ATP per mole of glucose are twice as efficient as the heterofermenters, and they capture about 23% of the total energy available in this process (**Table 20.2**). Although very little energy is available for these bacteria by such fermentations, they have been remarkably successful in establishing themselves in important niches in the environment, as discussed later.

The optical form of lactic acid produced by lactic acid bacteria varies in the different genera. Some produce one stereoisomer, the D-form, which rotates light toward the right (*dextro* rotary), whereas others produce the L-form (left or *levo* rotary). The reason for these differences lies in the stereospecificities of the lactic dehydrogenase enzyme itself. Some species produce enzymes that make both D- and L-forms, resulting in a racemic mixture of the two.

Lactic acid bacteria can grow in the presence of oxygen but are unable to use it metabolically. They lack cytochrome enzymes (which have iron-containing heme groups that they cannot synthesize) and an electron transport system with which to generate ATP by electron transport phosphorylation. Lactic acid bacteria do have flavoproteins, however, and when they are exposed to oxygen, they produce hydrogen peroxide that can be toxic to the cells.

It is interesting to note that most lactic acid bacteria are **catalase negative;** that is, they are unable to make the enzyme catalase, which like cytochromes, are iron-containing heme-proteins that degrade peroxides in the following manner:

$$H_2O_2 \rightarrow 2\ H_2O + O_2$$

A few lactic acid bacteria make a special manganese-containing catalase and are therefore phenotypically catalase positive. Those species lacking catalase have other enzymes, **peroxidases** that degrade hydrogen peroxide through organic compound-mediated reductions. Reactive oxygen compounds, such as hydrogen peroxide, can lead to damage of almost all cell components, including nucleic acids, lipid membranes, and proteins. The toxic effect caused to the cell is referred to as oxida-

phosphate, which gives rise to one molecule of ATP by substrate-level phosphorylation in the formation of lactic acid, and (b) acetyl phosphate, which is reduced to acetaldehyde, and then to ethanol.

The only energy available to lactic acid bacteria is through ATP generated by substrate-level phosphorylation. The metabolism of these bacteria is among the most simple of the various energy-yielding processes found in bacteria or other organisms. The resulting ener-

Table 20.2	Comparison of the efficiency of the homofermentative versus the heterofermentative lactic acid bacteria[a]	
Energy Sources and Efficiency		**Energy Available (kjoules)**
Glucose (1 mol)		2,870
Pyruvate (1,326 kjoules/mol × 2 mols)		− 2,652
Theoretical energy available in oxidation of glucose to pyruvate		218
Energy captured by ATP synthesis (29 kjoules/mol ATP):		
Homofermenters (2 ATP/mol glucose)		58
Heterofermenters (1 ATP/mol glucose)		29
Efficiency of process		
Homofermenters (58/218 × 100%)		27%
Heterofermenters (29/218 × 100%)		13%

[a]Because pyruvate is more oxidized than glucose, the two moles of pyruvate formed from glucose contain 218 kjoules less energy than the initial mole of glucose. Two molecules of ATP are synthesized in the homofermentative process, and each of these is equivalent to 29 kjoules of energy per mole of glucose. Thus, 58 kjoules of energy have been captured in the formation of ATP by homofermenters representing an efficiency of 27%. In contrast, heterofermentative lactic acid bacteria are only half as efficient because they produce only one molecule of ATP per molecule of glucose fermented. Also note that some 92% of the total energy available in the glucose molecule is still available in the two molecules of pyruvic acid formed (2652/2870 × 100 = 92%), indicating that much of the original energy still remains in pyruvate.

tive stress. Enzymes, such as peroxidases, degrade the toxic oxygen forms thereby providing a protective effect.

Another toxic form of oxygen is the superoxide anion, O_2^-, which is formed in various physical and biochemical ways by single electron reductions of O_2. The lactic acid bacteria as well as most aerobic and facultative aerobic bacteria protect themselves from this strong oxidizing agent by producing the enzyme **superoxide dismutase,** which decomposes superoxide as follows:

$$2\,O_2^- + 2\,H^+ \rightarrow O_2 + H_2O_2$$

Note that hydrogen peroxide is formed by the enzyme. The concerted action of the superoxide dismutase and either catalases or peroxidases results in the destruction of these toxic oxidizing agents. In contrast to the lactic acid bacteria, many obligate anaerobes do not produce these enzymes; therefore, these anaerobes can be exposed to lethal cellular oxidations if they encounter aerobic conditions.

Nutrition and Ecology The lactic acid bacteria are regarded as **fastidious** organisms because almost all species have complex nutritional requirements. With the exception of *Streptococcus bovis* from the intestine of cattle, virtually all require

preformed amino acids and vitamins, and some even require purines and pyrimidines. Thus, although they do not live as intracellular pathogens of other organisms, they live in habitats with other microorganisms, animal or plant tissues, or decaying organic substances that serve as sources of these materials. Because they require these compounds presynthesized, they conserve energy that would otherwise be required for their synthesis.

Despite their rather limited metabolic capabilities and their complex nutritional requirements, the lactic acid bacteria have survived well in selected environments due to their specialization in sugar fermentation. The principal habitat groups of the lactic acid bacteria are shown in **Table 20.3**.

One of the principal habitats of lactic acid bacteria is decomposing plant materials. These bacteria ferment the hexoses and pentoses that predominate as decomposition products. The lactic acid formed prevents the growth of many other organisms, especially if the pH is lowered to 5.0 or so. For example, in pickle fermentations, the cucumbers are placed in a vat with a small amount of salt and seasonings. They are covered with water and incubated at or about room temperature. Initially, aerobic and facultative aerobes grow and through aerobic respiration remove the free O_2 that is dissolved in the water in the vat. In a short time, the vat becomes anaerobic, and the plant sugars begin fer-

Table 20.3	Habitats of lactic acid bacteria	
Habitat	**Predominant Group**	**Activity or Product**
Decomposing plant material	*Streptococcus* spp. and *Lactobacillus plantarum*	Pickles, kimchee, silage, and sauerkraut
Dairy	*Streptococcus lactis,* *Lactobacillus casei,* *L. acidophilus,* *L. delbrueckii,* *Leuconostoc mesenteroides,* *L. lactis*	Cheeses, yogurt, etc.
Gastrointestinal tract of animals (oral) (intestinal)	*Streptococcus salivarius,* *S. mutans,* and *Lactobacillus salivarius* *Streptococcus faecalis*	Normal flora, dental caries Intestine; some urinary tract pathogens
Mammal vagina	*Streptococcus* spp. and *Lactobacillus* spp.	Normal flora

menting. The initial fermentation is accomplished by the genus *Streptococcus* because it grows at higher pH values than the genus *Lactobacillus*. The pH is lowered by this group until it reaches about 5.5 to 6.0, at which point *Lactobacillus* spp. begin to grow. They ferment the remaining sugars and lower the pH even further, to about 4.5. As long as conditions remain anaerobic, the plant materials are preserved very well due to the low pH of the lactic acid. Therefore, the pickles keep many weeks or, if refrigerated or canned, for many months.

The same general protocol is followed for the manufacture of sauerkraut and kimchee, the Korean bok choy product. The characteristic flavor of sourdough bread is also due to the fermentation of lactic acid bacteria. Likewise, in agriculture, silage is made by placing hay of the right moisture content in silos or, more recently, in large plastic bags to prevent the access of air and to allow the fermentation of plant materials. The resulting silage can be used during the winter months when fresh hay is not available.

Lactic acid bacteria are a critical component of the dairy industry as well. These bacteria are the natural souring agents of milk. They gain access to the milk through plant materials such as the feed or, in some cases, from the cattle themselves. If milk is permitted to sour naturally, these bacteria become established, with first the *Streptococcus* group followed by the *Lactobacillus* group, as discussed in pickle fermentation. They ferment milk sugar and lactose and form lactic acid in the process. When the pH drops sufficiently, the principal milk protein, casein, is precipitated. Thus, the **curd**, or the precipitated casein, is separated from the **whey**, the remaining fluid. It is this same process that is used for the commercial manufacture of cheeses. However, special strains called **starter cultures** of lactic acid bacteria are used by dairies for the manufacture of cottage cheese, yogurt, acidophilus milk, butter, buttermilk, and various cheeses. The distinctive flavor and aroma of butter is due to diacetyl formed spontaneously from acetoin produced by some lactic acid bacteria in butter production.

Another habitat of the lactic acid bacteria is the gastrointestinal tract of animals. They grow in the oral cavity where they ferment sugars such as sucrose with the formation of lactic acid. Both *Streptococcus* and *Lactobacillus* species occur in the mouth where the lactic acid they produce can cause decay of the enamel resulting in cavities of teeth. These bacteria also grow in the intestinal tract of humans and other animals where they are indigenous in the small intestine. The predominant intestinal species in humans is *Streptococcus fecalis*.

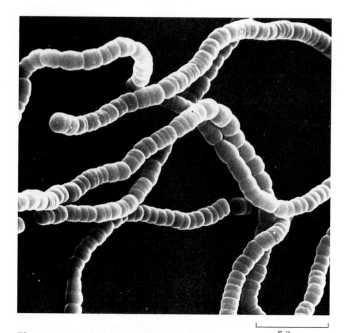

Figure 20.4 *Streptococcus*   5.0 µm
Typical appearance of a *Streptococcus* sp. in chains of cocci, as seen by scanning electron microscopy. Photo ©David M. Phillips/Visuals Unlimited.

A final habitat of significance is the vagina of female mammals. Both *Streptococcus* and *Lactobacillus* ferment glucose from the glycogen normally secreted by the vagina of adult females. This resulting lactic acid serves as a barrier to vaginal infection.

Some of the more common genera of lactic acid bacteria are discussed individually next.

Streptococcus This genus of cocci (**Figure 20.4**) contains several important species (**Table 20.4**). They all grow poorly on media and form only small, pinpoint colonies during growth. *Streptococcus lactis* is a common dairy organism found in milk and other dairy products. It does not cause disease but is of industrial significance. Note that it is incapable of causing the hemolysis of red blood cells (Table 20.4). This is because it lacks the ability to produce **hemolysins,** special enzymes that are

Table 20.4   **Differential characteristics of important species of the genus *Streptococcus***

Species	Group Name	Hemolysis of Red Blood Cells	
		alpha (greening)	beta (clearing)
S. lactis	Dairy group	–	–
S. salivarius	Viridans group	+	–
S. fecalis	Enterococcus	Variable	–
S. pyogenes	Pyogenic	–	+
S. pneumoniae	Pneumococcus	+	–

(A)

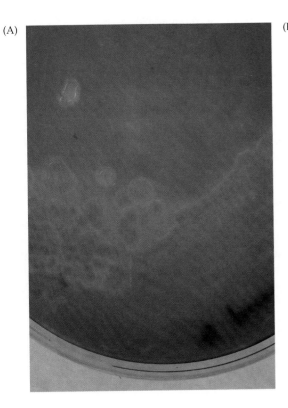

(B)

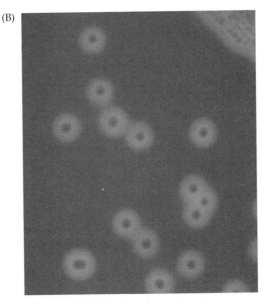

Figure 20.5 Blood cell hemolysis by viridans group
(A) Alpha-hemolysis, or "greening" hemolysis, on a plate of red blood cells. (B) Beta-hemolysis caused by *Streptococcus pyogenes*. A, ©L. M. Pope and D. R. Grote/Biological Photo Service; B, ©G. W. Willis/Visuals Unlimited.

responsible for this activity. Hemolysis is regarded as a virulence factor, a characteristic that enhances the pathogenicity of species.

Streptococcus salivarius is one of many bacteria that grow in the upper part of the gastrointestinal tract, the oral cavity of humans. This organism lives off the sugars supplied in the diet or produced from starches by amylases secreted by the salivary glands. A close relative of this species, *Streptococcus mutans,* produces a capsule in the presence of disaccharides that enables it to attach to enamel surfaces and cause tooth decay. Members of this genus, the so-called **viridans group** (from Greek meaning green) cause a partial clearing or greening effect on blood agar plates (**Figure 20.5A**). This phenomenon is referred to as **alpha-hemolysis**, but in fact red blood cells are not actually lysed in this process. In alpha-hemolysis hemoglobin is converted to methemoglobin in vivo—and this form of hemoglobin has a reduced capacity to carry oxygen. Most of the members of the viridans group are not pathogenic but are implicated in tooth decay.

Another group of the streptococci, appropriately informally named the **"enterococci,"** inhabit the lower part of the human digestive tract. *Streptococcus fecalis* is a normal and therefore nonpathogenic inhabitant of the human intestine—they are particularly abundant in the large intestine. This environment is anaerobic and contains ample nutrients, making it an ideal habitat for growth.

Some of the enterococci are pathogenic, causing infections of the urinary tract and sometimes endocarditis.

The **pyogenic** group of *Streptococcus* characteristically produces hemolysins that completely lyse red blood cells, resulting in a clearing on a blood agar plate prepared with mammalian blood. This type of hemolysis is termed **beta-hemolysis** (Figure 20.5B). The members of the pyogenes group of streptococci are serious human pathogens. In addition to hemolysins, they produce other virulence factors that are important for pathogenicity such as fibrinolysin, an enzyme that dissolves blood clots, an erythrogenic toxin, which causes the characteristic rash of scarlet fever, and leucocidin, which destroys white blood cells. Furthermore, they produce the enzyme hyaluronidase that attacks hyaluronic acid in connective tissue. These organisms are ideally adapted for growth in warm-blooded animals because their growth temperature range is between 10°C and 45°C. *Streptococcus pyogenes* is the causative agent of several diseases including scarlet fever, puerperal fever of postdelivery mothers, and endocarditis, an inflammation of heart tissue.

Streptococcus pneumoniae, one of the bacterial species that causes pneumonia, comprises its own group and is commonly referred to as the **"pneumococcus."** Unlike the other members of the genus, this species does not typically form long chains of cells under most conditions of cultivation, but rather grows as pairs of cocci, so it is also sometimes called the **"diplococcus."** Virulent members of this species have a capsule that is important for their pathogenicity (see Figure 4.34). If the capsule is removed, the cells are readily phagocytized. Only strains with capsules can cause pneumonia. Encapsulated strains form colonies that are described as "smooth" in texture and are distinguished from unencapsulated strains that form "rough" colonies. As discussed in Chapter 14, these bacteria served as important experimental organisms in Avery and McCleod's demonstration that genetic material can be passed from one organism to another, even if the donor organism is dead. This was the first demonstration of the phenomenon of genetic transformation.

Streptococcus is also known for the devastating tissue damage caused by some strains infecting humans. This rare "flesh-eating" disease has a rapid onset and, if not treated quickly, results in loss of limb or life.

Lactobacillus This genus of rod-shaped lactic acid bacteria is also widespread in the same habitats in which the genus *Streptococcus* resides, including plant materials and the oral and genital tracts of humans. However, unlike *Streptococcus*, the genus *Lactobacillus* does not harbor any pathogenic strains.

These bacteria can grow at lower pH values than the streptococci. As mentioned previously, they do not grow during the initial stages of fermentations, but only when the pH is lowered to 5.5 to 6.0. In turn, they continue to produce lactic acid and lower the pH of the environment to values below 5.0. This effectively eliminates the genus *Streptococcus*. This is a bacterial example of an ecological succession.

Other Important Genera Two other genera of importance are *Leuconostoc* and *Pediococcus*. *Leuconostoc* spp. are common in dairy products as well as in the oral cavity. They are especially well known as producers of **dextrans** (alpha-1,6-glucans), which they form as capsular materials in the fermentation of sucrose, a disaccharide consisting of glucose and fructose. *Leuconostoc mesenteroides* produces an extracellular enzyme called dextran sucrase that converts the glucose of the sucrose molecule to dextran and releases the fructose into the environment.

Sucrose → Dextran + Fructose

Therefore, the result is the formation of a polymer of glucose subunits linked in the alpha 1-6 position. Actually, the glucose sub-

units are added to the glucose on the initial sucrose molecule primer. Thus, each dextran molecule has one subunit of fructose as its terminal sugar. Dextran is used commercially as a blood plasma extender.

Pediococcus species also carry out important lactic acid fermentations. They can be troublesome in the brewing industry, as the name *Pediococcus damnosus* implies, in that they can interfere with the normal yeast alcoholic fermentation by producing undesirable products of the lactic acid fermentation.

Listeria

Listeria is a food-borne pathogen that causes **listeriosis,** a type of gastrointestinal disease. Although it is usually not fatal, it can cause fatalities in infants and debilitated individuals. Like *Lactobacillus*, *Listeria* is a gram-positive rod (**Table 20.5**) that is catalase positive and oxidase negative. It also ferments sugars to form L-lactic acid. However, unlike *Lactobacillus*, it is motile (peritrichous flagella), produces cytochromes, and grows aerobically by respiration. In addition, it grows at low temperatures and therefore multiplies in the refrigerator. For this reason it has been a problem with refrigerated foodstuffs such as cheese and fish.

Renibacterium

Renibacterium is a slow-growing, rod-shaped bacterium noted as the causative agent of bacterial kidney disease (BKD) in salmonid fish (Table 20.5). It has been cultivated in pure culture only recently. It is an obligate aerobe that uses sugars as carbon sources for growth and requires cysteine.

Staphylococcus

Like the lactic acid bacteria, members of the genus *Staphylococcus* grow anaerobically and produce lactic

Table 20.5	Other gram-positive cocci and rods		
Genus	Shape	Oxygen Requirement	Habitat (Mol % G + C)
Staphylococcus	Cocci in clusters	Facultative aerobe; fermentative	Warm-blooded animals; food (30–35)
Micrococcus	Cocci in clusters	Obligate aerobe	Soil; airborne dust (66–72)
Listeria	Rod	Facultative aerobe	Widely found; animal pathogen (36–38)
Renibacterium	Rod	Aerobic	Fish pathogen, salmonids (53)
Geodermatophilus	Filamentous	Aerobic	Soil (73–75)
Frankia	Filamentous	Aerobic	Alder—N$_2$ fixing symbiont (66–71)

acid from sugar fermentation. These bacteria are also gram-positive cocci, but they occur in grapelike clusters of cells (**Figure 20.6**), rather than chains (Table 20.5). The genus differs from the lactic acid bacteria in nutrition and physiology. For example, staphylococci do not have the complex nutritional requirements typical of lactic acid bacteria. Furthermore, they produce heme pigments and carry out aerobic respiration with an electron transport system containing cytochromes.

Staphylococci are associated with the skin and mucosal membranes of animals. Indeed, all humans carry *Staphylococcus epidermidis,* a nonpigmented coccus, as a normal inhabitant on the surface of their skin. *Staphylococcus aureus* is a human pathogen capable of causing a variety of health problems. Unlike *S. epidermidis, S. aureus* is pigmented a gold color. *S. aureus* also produces several virulence factors that enhance its ability to cause disease. One distinctive factor, called **coagulase,** is an enzyme that causes fibrin to clot. Another factor, leukocidin, attacks leukocytes. Several other factors can be produced by pathogenic strains including β-hemolysin, lipase, fibrinolysin, hyaluronidase, deoxyribonuclease, and ribonuclease.

S. aureus is a normal inhabitant of the nasopharynx region in humans. Some humans harbor a specific strain for many years and are therefore called **carriers.** Infants come into contact with the organism during their first week of life. The strain they receive may come from the mother or other close relatives or hospital personnel. Normally these strains do not result in any infection. However, if the infant is unhealthy, it is susceptible to the colonizing strain, and serious skin infections can result. Treatment of such infections can be difficult because many *S. aureus* strains are antibiotic resistant.

S. aureus can also cause a common type of food poisoning. Such strains produce an **enterotoxin,** a type of **exotoxin** (exotoxins are to be distinguished from endotoxin produced by gram-negative bacteria, which is their lipopolysaccharide—see Chapter 4). Foods that have been prepared and not refrigerated for extended periods of time allow for the growth of the toxin producer. Problem foods include cooked meats, potato salads, and cream desserts such as eclairs. During summer months when the weather is warm and families have picnics, conditions are ideal for potential problems. Foods that are not refrigerated properly and are served on subsequent days are especially hazardous. The symptoms of *Staphylococcus* food poisoning are acute and memorable. The toxin, which is produced while the bacterium

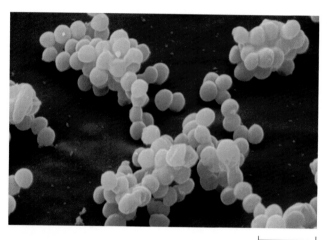

Figure 20.6 *Staphylococcus*   5.0 μm
Typical grapelike clusters of *Staphylococcus aureus.*
Photo ©Oliver Meckes/Photo Researchers, Inc.

grows in the food, takes effect within a few hours of ingestion and causes headache and dizziness, fever, diarrhea, and nausea that persists for 24 hours or so. During this period, the inflicted individual feels extremely uncomfortable and is totally incapacitated. Fortunately, it is self-limiting, and individuals recover fully after 24 to 48 hours.

Toxic shock syndrome (TSS) is a well-publicized malady caused by *S. aureus.* Women who use tampons during their menstrual periods are susceptible to this. *S. aureus* grows and produces toxins in the vagina under these conditions. If unchecked, TSS can be fatal.

Endospore Formers

The bacterial endospore is a very distinctive structure produced by relatively few genera (Table 20.6). The two principal genera are the aerobic to facultative anaerobic genus *Bacillus* (see Figure 4.8) and the obligately anaerobic genus *Clostridium.* Three additional genera of ecological importance include *Desulfotomaculum,* a group of anaerobic sulfate reducing bacteria discussed in Chapter

Table 20.6	Important genera of endospore-forming bacteria		
Genus	**Oxygen Requirement**	**Mol % G + C**	**Nutrition/ Physiology**
Bacillus	Aerobe or facultative aerobe	25–69	Aerobic respiration or fermentation
Clostridium	Obligate anaerobe	22–55	Fermentation
Desulfotomaculum[a]	Obligate anaerobe	37–50	Anaerobic respiration (sulfate reduction)
Thermoactinomyces	Obligate aerobe	52–55	Aerobic thermophile; compost
Sporosarcina	Obligate aerobe	40–42	Urea degrader

[a]See gram-negative dissimilatory sulfate reducers in Chapter 19.

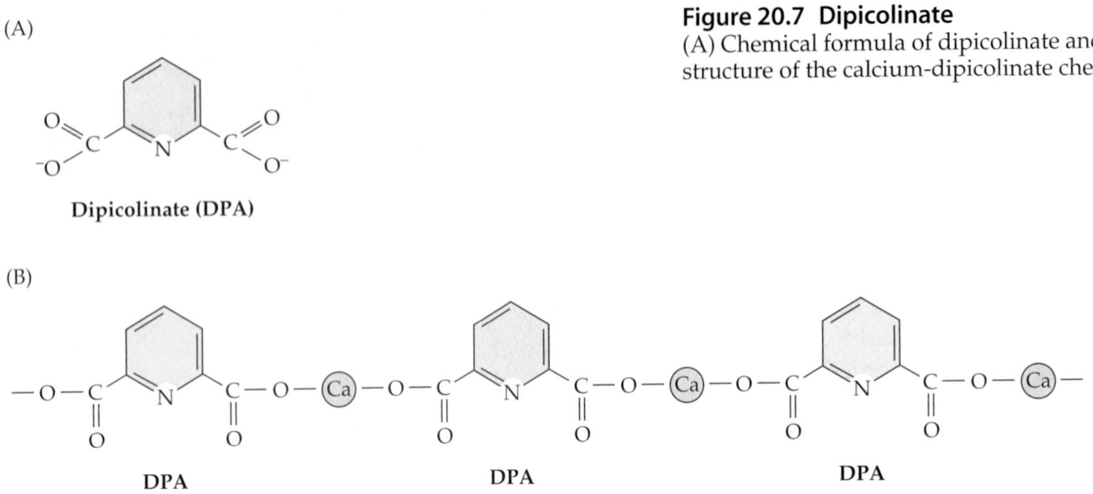

(A)

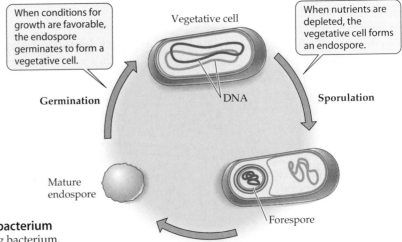

Dipicolinate (DPA)

(B)

DPA   DPA   DPA

Figure 20.7 Dipicolinate
(A) Chemical formula of dipicolinate and (B) proposed structure of the calcium-dipicolinate chelate.

19, *Sporosarcina,* a soil coccus, and *Thermoactinomyces,* a thermophile.

With the exception of *Sporosarcina,* which is a coccus, all endospore formers are rod-shaped bacteria that have the typical gram-positive cell wall ultrastructure and stain gram-positive to variable. Many species are motile with peritrichous flagella. The endospore has been the subject of many years of study, as it is an important example of cellular differentiation or morphogenesis.

Endospore Characteristics and Development The endospore, as the name implies, is a spore or resting stage unique to bacteria that is formed within the cell. The dormant spore allows the bacterium to survive extended periods of desiccation and high temperature. These two conditions are often encountered in soil and rock environments where these organisms are most commonly found. Not only are they more resistant to heat and desiccation than typical bacteria, but also to ultraviolet radiation and disinfection. Experiments have shown that the endospores of some species can survive storage in some environments for at least 50 years.

Endospores appear highly refractile when observed by phase contrast microscopy (Figure 4.8A). This high refractility is due to the low water content of the spore. The endospore is rich in calcium ions and a unique compound, dipicolinic acid (**Figure 20.7A**). These compounds are believed to form a chelate complex because of their equal molar concentrations and chemistry (Figure 20.7B), and are thought

to be responsible for the heat resistance of the endospore and other unique properties.

Endospores are very difficult to stain using ordinary staining procedures. For example, in the gram stain, the endospore does not take up either the primary stain or the counterstain and therefore appears colorless on completion of the staining process. Endospores, however, can be stained by special staining procedures. One common procedure involves steaming the smear with the dye malachite green. After this is completed, the smear is counterstained with safranin. In this procedure the cell appears red, whereas the endospore appears green.

Endospore-forming bacteria have two phases of growth, **vegetative growth,** which is the normal period of growth and reproduction, and **sporulation,** the period of spore formation (**Figure 20.8**). Sporulation is a highly regulated process. It appears to be specifically responsive to nutrient limitation when the density of the population is high. However, the first response of the cell to exhaustion of preferred nutrients is not sporulation. The bacteria induce a range of adaptive responses

When conditions for growth are favorable, the endospore germinates to form a vegetative cell.

Vegetative cell

When nutrients are depleted, the vegetative cell forms an endospore.

Germination

DNA

Sporulation

Mature endospore

Forespore

Figure 20.8 Life cycle of endospore-forming bacterium
Diagram of the life cycle of an endospore-forming bacterium.

(A)

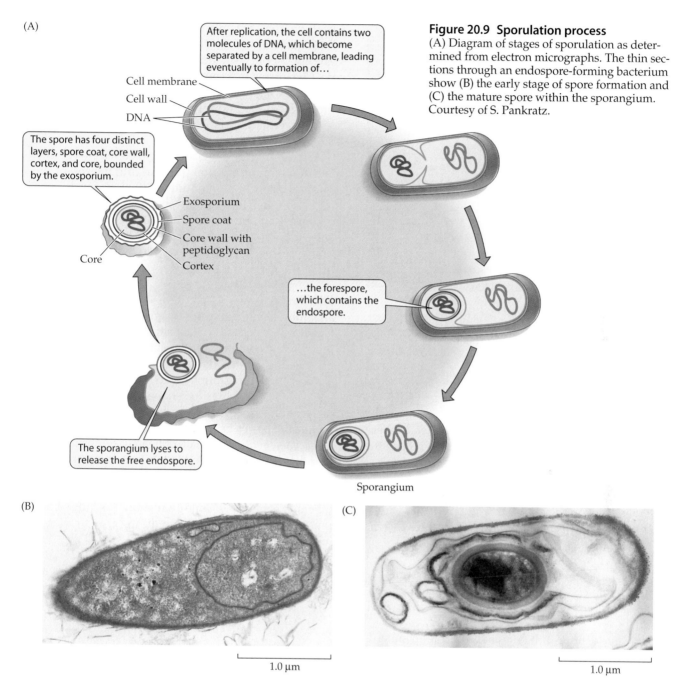

After replication, the cell contains two molecules of DNA, which become separated by a cell membrane, leading eventually to formation of...

Cell membrane
Cell wall
DNA

The spore has four distinct layers, spore coat, core wall, cortex, and core, bounded by the exosporium.

Exosporium
Spore coat
Core wall with peptidoglycan
Core
Cortex

...the forespore, which contains the endospore.

The sporangium lyses to release the free endospore.

Sporangium

Figure 20.9 Sporulation process
(A) Diagram of stages of sporulation as determined from electron micrographs. The thin sections through an endospore-forming bacterium show (B) the early stage of spore formation and (C) the mature spore within the sporangium. Courtesy of S. Pankratz.

(B)

1.0 μm

(C)

1.0 μm

such as chemotaxis, motility, and synthesis of extracellular degradative enzymes in order to procure secondary sources of nutrients. If these responses do not provide the cell with adequate nutrients for growth, the cells enter the sporulation pathway. Thus, the endospore is produced at the end of the exponential phase of growth.

A number of stages along with specific morphological changes are noted during sporulation (**Figure 20.9**). After the period of vegetative growth ceases, two resulting molecules of DNA are present in the cell. These first coalesce and are then separated from one another by a cell membrane formed during a special differentiation

process. An engulfment stage ensues in which the membrane from one incipient cell grows around the membrane of the other until it completely surrounds it. The engulfed incipient daughter cell, termed the **forespore,** eventually becomes the mature endospore. The other "cell," which contains the endospore, is called the sporangium. At this time the process of sporulation is irreversible, regardless of the presence of nutrients.

During sporulation the spore becomes increasingly refractile and develops into four distinct layers, a spore coat, the core wall with its modified peptidoglycan, the cortex, and the core wherein the calcium ions and dipicolinic acid are found along with the nuclear material.

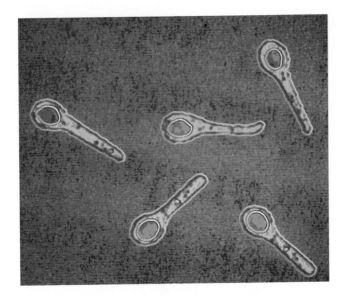

Figure 20.10 *Clostridium tetani*
Photomicrograph of *Clostridium tetani*. Note that the endospores are larger than the cell, giving the cells a club-shaped appearance. Photo ©Alfred Pasieka/Photo Researchers, Inc.

Thin sections through endospores show each of these features (Figures 20.9 and 20.14). A membrane called the exosporium binds these layers. Eventually, the original cell lyses to release the free endospore.

The endospore is a resting stage that can survive long periods of time in the environment. It remains viable, and under appropriate conditions it can undergo the process of **germination,** whereby it begins to develop into a vegetative cell again (Figure 20.8). In this process the endospore takes up water, swells, and breaks the spore coat and exosporium, releases its calcium dipicolinate, and begins multiplying as in normal vegetative growth.

Some endospores are known to be able to survive boiling temperatures for long periods of time. These were the heat-resistant structures that caused so many problems for the proponents and opponents of the theory of spontaneous generation discussed in Chapter 2.

The location of the endospore in the cell and its shape are important features in classification. Most species have oval-shaped endospores that are located in a central to subterminal location within the cell or sporangium. At the other extreme, some species, such as *Clostridium tetani*, produce a terminal endospore that is larger than the cell diameter, giving the sporangium a club-shaped appearance (**Figure 20.10**).

Habitats of Endospore-Forming Bacteria

The endospore-forming bacteria are found primarily in soils, aquatic sediments, and muds. Like other gram-positive bacteria, they are only occasionally reported in planktonic habitats such as oligotrophic lakes and the ocean. The endospore is an ideal survival device for soil bacteria. Organisms that can convert to resting stages during dry periods in soil environments can survive periods of prolonged desiccation. These bacteria are also found on rock weathering surfaces in deserts, where hot, dry periods are common.

Endospore-forming bacteria can be selectively isolated from habitats by first pasteurizing the soil or mud sample at 80°C for 10 minutes. Although this kills normal soil bacteria, the heat-resistant endospores of most of these bacteria will not be affected. In fact, some endospore formers do not germinate unless they are first heat shocked, that is, exposed to high temperature first.

Bacillus

This genus contains aerobic to facultative anaerobes. Unlike the strictly anaerobic genus *Clostridium*, *Bacillus* species produce catalase, which may explain the difference in the survival of these two genera in the presence of oxygen. The various species of this genus are classified into groups based on their cell morphology, using in particular the shape of the endospore and its location. **Table 20.7** provides a grouping of some of the major species based on cell morphology.

Bacillus subtilis is a small, obligately aerobic spore former that is a common soil inhabitant. It is noted for its ability to degrade plant polysaccharides and pectin, and some strains are even known to produce a rot in potato tubers. It can cause a condition known as "ropy bread" caused by polymer production during growth after baking contaminated bread.

Bacillus licheniformis resembles *B. subtilis*, but it is a facultative anaerobe that ferments sugars. It carries out a characteristic fermentation:

$$\text{3 Glucose} \rightarrow \text{2 Glycerol} + \text{2 2,3-butanediol} + \text{4 CO}_2$$

Most strains of *B. licheniformis* carry out denitrification, which is a relatively unusual attribute within this genus.

Bacillus megaterium (from the Greek *megaterium* meaning "big beast"), as the species name implies, is very large. Because it is large, bacteriologists use it as a teaching organism in introductory laboratories such as for gram staining as well as in research laboratories interested in morphology and ultrastructure. Like *B. subtilis*, this species in an obligate aerobe with very simple nutritional requirements (ammonium or nitrate can serve as the sole nitrogen source).

Bacillus cereus and *Bacillus mycoides* are also quite large and are among the most common soil bacteria. They grow anaerobically by sugar fermentation and require amino acids for growth. *B. mycoides* is a common

Table 20.7	Some major species of *Bacillus* and their properties	
Group	**Major Species**	**Distinctive Properties**
I. Oval spores		
A. Sporangium not swollen	*B. subtilis*	Common soil form; aerobic gramicidin-producer
	B. licheniformis	Denitrifier, fermentative; bacitracin producer
	B. megaterium	Large rod
	B. cereus	Common soil form, fermentative
	B. mycoides	Like *B. cereus* but distinctive hairlike colonies
	B. anthracis	Like *B. cereus*, but causes anthrax
	B. thuringiensis	Insect pathogen
B. Sporangium swollen		
	B. stearothermophilus	Thermophile
	B. circulans	Colony rotates on agar media
	B. polymyxa	Nitrogen fixer, polymyxin producer
	B. popilliae	Insect pathogen
II. Spherical spores, nonfermentative		
	B. pasteurii	Degrades urea, grows at high pH

air contaminant and is often found on plates in microbiology laboratories.

Closely related to *B. cereus* are two important pathogens. *Bacillus anthracis* causes anthrax of animals as well as humans. Unlike *B. cereus,* it is immotile. The source of infection of humans is largely through domestic farm animals, although many other animals including horses, mink, dogs, and even birds, fish, and reptiles can acquire the disease. In humans, the disease occurs most commonly in sheep and cattle tanners and meat workers. The organism normally invades through open cuts in the skin but can also infect through the lungs. The organism produces a gummy D-glutamic acid polypeptide capsule, an important virulence factor during invasion. The disease symptoms are produced by a protein exotoxin formed during growth in the host tissues. The toxin acts by causing physiological shock and ultimately kidney failure. The toxin is heat labile. The spores survive well in the environment, so animals that have died of the disease present a problem for disposal; they are usually buried. *B. anthracis* is the only member of the genus that is pathogenic to humans. This organism is a prospective agent in biological warfare (**Box 20.1**).

The best-known insect pathogen in this genus is *Bacillus thuringiensis.* This bacterium produces a large amount of a crystalline protein during the sporulation process. The protein crystal can be seen both with the light microscope and the electron microscope (**Figure 20.11**). It is formed in the sporangium along with the spore and is referred to as a parasporal body. This protein crystal is toxic to insects. Vegetative cells of *B. thuringiensis* and its spores are found on plant leaves

and are ingested by insects that feed on the leaves. The most common hosts are the larval stages of moths and butterflies (class *Lepidoptera*). These insects have an alkaline gut that dissolves the toxin. The toxin is a

Figure 20.11 *Bacillus thuringiensis* **crystal**
Thin-section electron micrograph of the protein crystal of *Bacillus thuringiensis.* Courtesy of S. Pankratz.

Milestones Box 20.1   Biological (or Germ) Warfare and Bioterrorism

Some pathogenic microorganisms have been studied as agents of germ warfare by many technologically advanced countries. *Bacillus anthracis* has been studied extensively because of the hardiness of the endospore, the ability to infect populations by aerial dispersal, and the deadliness of pulmonary anthrax.

At the international Biological Weapons Convention in 1972 many nations, including the United States and the Soviet Union, signed an agreement to ban research on the development and use of biological weapons. However, some countries continue to study biological warfare even today.

In 1979 a tragic incident occurred in Sverdlovsk, a city of 1.2 million inhabitants located 1,400 kilometers east of Moscow in the former Soviet Union. This city was the site of a biological weapons research facility. Apparently, due to an accidental release from the military research plant, 77 individuals were infected with anthrax, and 66 subsequently died. This epidemic occurred during the cold war between the Soviet Union and the United States after they were signatories to the 1972 Convention. Official information released by the Soviet Union regarding the outbreak stated that individuals developed gastrointestinal, not pulmonary, anthrax after eating contaminated meat.

Following the collapse of the Soviet Union, the real cause of the epidemic was determined in a study led by a U.S. scientist, Matthew Meselson (*Science*, 266: 1202–1208, 1994). This group of scientists discovered that, in fact, most individuals died of pulmonary anthrax, indicating that the outbreak was due to airborne dispersal and inhalation of spores, not from ingestion of meat as officially claimed. This man-made epidemic provides some insight into the deadliness of *B. anthracis* as a biological warfare agent.

During 2001, an incident occurred in the United States in which letters that had been heavily inoculated with anthrax spores were mailed through the U.S. Postal Service. The deadly letters were mailed to U.S. senators, congressional representatives, a private company, and the U.S. Department of Justice. Several postal workers and citizens died from pulmonary anthrax, even though they had not even opened the envelopes, indicating that the spores passed through the paper, contaminated other mail, and became aerosolized. A few postal workers and citizens died. These were some of the individuals who contracted pulmonary anthrax. Others contracted the less-deadly cutaneous form in which they developed skin lesions on their hands and arms. Once the disease is recognized, it can be readily treated with

antibiotics such as penicillins. However, those who contracted the pulmonary form died unless their illness was detected quickly. This incident indicates how vulnerable modern societies are to malevolent individuals and terrorist organizations who use weapons of mass destruction with the intent of killing innocent people.

In the past, the United States has also studied the offensive and defensive aspects of biological weapons. In the 1960s, airplanes released an aerosol of *Serratia marcescens* over the San Francisco metropolitan area with the intent of monitoring the effectiveness of its aerial dispersal, an important aspect of bacterial dissemination. Because of its distinctive red pigment, prodigiosin (see Chapter 19), it is easy to determine the concentration of this species when air samples are plated on agar media. At that time *S. marcescens* was not regarded as a pathogen. However, several individuals who were being treated in hospitals in the Bay area acquired previously unknown *S. marcescens* lung infections due to the aerial spraying, which resulted in one death.

Movement is now afoot to pass a revised treaty to ban biological weapons. It is hoped that in the near future all nations will agree to a complete ban accompanied by the necessary inspections by other countries.

neurotoxin that causes paralysis and death of the caterpillar. Presumably, the bacterium benefits from its host by growing on the dead carcass (**Box 20.2**).

Bacillus stearothermophilus is a thermophile with a temperature growth range between 45°C and 65°C. This particular thermophile has been known for many years and has been found in hot springs, deserts, and soils.

Bacillus polymyxa is a facultative anaerobe that ferments sugars with the production of 2,3-butanediol, ethanol, acetic acid, and CO_2. It is also distinctive in

being one of the few members of the genus to carry out nitrogen fixation, which it does when growing anaerobically.

Bacillus circulans is similar to *B. polymyxa* but is peculiar in its formation of motile colonies. The colonies of this organism actually rotate and can move across a Petri dish due to the concerted action of the flagella.

Another member of the genus that is particularly noteworthy is *Bacillus pasteurii*. This bacterium is especially suited for growth on urea. For example, it does not

Research Highlights Box 20.2   Biological Pesticides

Bacillus thuringiensis toxin, often referred to simply as **Bt,** is currently being used as a biological insecticide. Commercial preparations are available that are used to eradicate tussock moths and other troublesome insects. The advantage of using biological agents against such pests is that other animals and plants are not affected because of the high specificity of its action. Moreover, because the toxin is of biological origin, it is readily degraded in the environment. In contrast, DDT, which was used for control of these same pests, had devastating effects on other members of the food chain. Furthermore, it undergoes biomagnification (see Chapter 24) and persists in the environment for a long time. It was for this reason that DDT was banned in the United States and many other countries.

Some strains of *B. thuringiensis* as well as other species of *Bacillus*, *B. popillae*, and *B. lentimorbis*, produce toxins that are effective against other insects. The commercial development of the mosquito toxin promises to be useful in the control of malaria.

In a quite different approach aimed at controlling plant insect infestation, these toxin genes are now being incorporated into plants by genetic engineering using *Agrobacterium tumefaciens*. Thus, when insects eat plant leaves, they will ingest the toxin and be killed by the genetically engineered bioinsecticide.

grow on ordinary nutrient broth unless urea is added to it. As the urea is degraded, ammonia is produced and the pH rises:

$$CO(NH_2)_2 + H_2O \rightarrow CO_2 + 2 NH_3$$

The organism can tolerate pH values between 8.0 and 9.5, much higher than typical bacteria. Because of its ureolytic ability, it is commonly found in urinals and can be easily isolated from these sources or from soils by the addition of urea to nutrient broth. Although other bacteria can degrade urea, they cannot do so at the high pH values of this bacterium. Thus, other bacteria are eliminated from environments where this is a major activity because the resulting high pH is so selective. Some *Bacillus* species, such as *Bacillus alcalophilus*, grow at even higher pH values. These are typically found in alkaline soils such as in deserts. Some have been reported to grow at pH 10 and above.

Clostridium This genus of anaerobic spore formers is even more diverse than *Bacillus* and contains some important environmental species as well as human pathogens. *Clostridium* spp. are grouped according to their fermentative abilities into several major subgroups (**Table 20.8**).

These bacteria are obligate anaerobes that lack cytochrome and an electron transport system. Thus, they rely solely on the formation of ATP by substrate level phosphorylations during fermentation of various carbon sources. Although these organisms are obligately anaerobic, they are not as difficult to grow as methanogenic bacteria. Unlike methanogens, they can be transferred in the air and incubated easily in ordinary

Table 20.8   Major groups and representatives of *Clostridium*

Group or Substrate	Fermentation Type	Representative Species	Distinctive Products or Features
Cellulolytic	Acetate, lactate ethanol, H_2, CO_2	*C. cellobioparum* *C. thermocellum*	Rumen organism Soil and sewage
Saccharolytic	Butyrate (or butanol-acetone); proteolytic	*C. butyricum* *C. pasteurianum* *C. acetobutylicum* *C. perfringens*	Nitrogen fixer Wound infections (gas gangrene)
Amino acid pairs	Stickland reaction	*C. sporogenes* *C. tetani* *C. botulinum*	Causes tetanus Causes botulism
Purines	Uric acid and other purines fermented to acetic acid, NH_3, CO_2	*C. acidurici* *C. fastidiosus*	
H_2 and CO_2	Acetogen	*C. aceticum*	Acetic acid formed

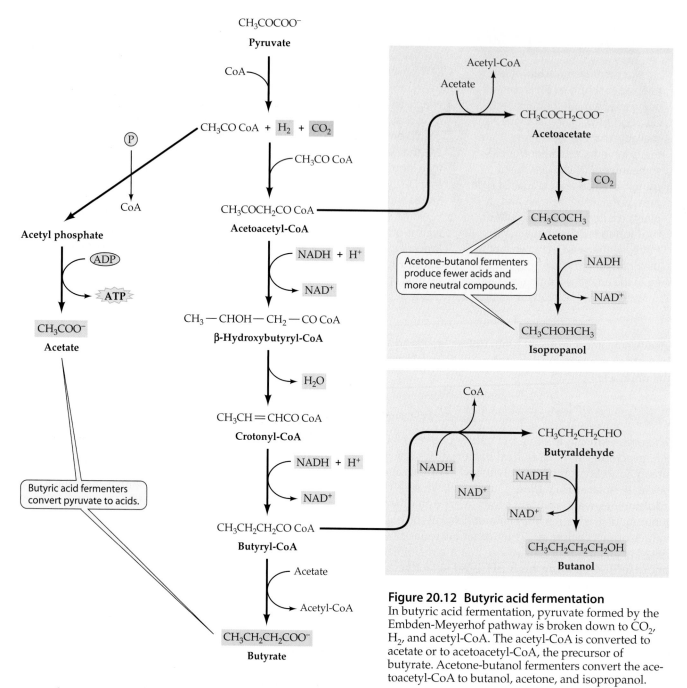

Figure 20.12 Butyric acid fermentation
In butyric acid fermentation, pyruvate formed by the Embden-Meyerhof pathway is broken down to CO_2, H_2, and acetyl-CoA. The acetyl-CoA is converted to acetate or to acetoacetyl-CoA, the precursor of butyrate. Acetone-butanol fermenters convert the acetoacetyl-CoA to butanol, acetone, and isopropanol.

anaerobe jars in which oxygen has been removed by reaction with H_2. Most species can be grown in liquid media in tubes that have been supplemented with sodium thioglycollate. Thus, they can be maintained in tubes on the lab bench without additional precautions.

The genus is metabolically diverse. Two species, *Clostridium cellobioparum* and *Clostriduim thermocellum,* carry out an anaerobic fermentation of cellulose to form the disaccharide cellobiose, which is ultimately fermented to produce acetic and lactic acids, ethanol, H_2, and CO_2 as the major end products. *C. cellobioparum* is found in the rumen of cattle and sheep and enables these higher animals to utilize cellulose in their diet (see

Chapter 25). Interestingly, this organism is inhibited when too much H_2 accumulates as a result of the fermentation. These bacteria live in close association with other ruminant microbes, the methanogenic bacteria, which remove the hydrogen by forming methane gas. The removal of the hydrogen gas lowers its concentration to a sufficiently low level that the formation of more hydrogen gas is favored. This interaction involving two different groups is referred to as an **interspecies hydrogen transfer** and is an example of synergism (see

Chapter 25). *Clostridium thermocellum* is common in decaying soils containing cellulose.

A major group of the clostridia ferment sugars and occasionally starch and pectin to form butyric acid, acetic acid, CO_2, and H_2 as the principal end products. The pathway for this fermentation is shown in **Figure 20.12**. In this **butyric acid fermentation,** glucose is fermented to pyruvic acid via the Embden-Meyerhof pathway. The pyruvate is split into carbon dioxide and hydrogen gas in the formation of acetyl-coenzyme A. Some of the acetyl-CoA is used for ATP generation in the formation of acetic acid via acetyl-phosphate. Also, the acetyl-CoA can be condensed with another molecule of acetyl CoA to form acetoacetyl-CoA, which is the precursor of butyric acid.

Some of the butyric acid bacteria produce fewer acids and more neutral products in prolonged fermentations. These are the so-called acetone-butanol fermenters. Butanol is formed from butyryl-CoA via butyrylaldehyde (see Figure 20.12). Acetone and isopropanol are formed from acetoacetyl-CoA by decarboxylation and subsequent reduction, respectively.

These so-called **butyric acid bacteria** include *Clostridium pasteurianum*, which can fix nitrogen, a property shared by several other species in this group. As the acids accumulate during butyric acid fermentation, some species, such as *Clostriduim acetobutylicum*, begin to produce more neutral compounds, including butanol and acetone. The acetone-butanol fermentation has been used commercially. It was especially important during World War I because of the need for acetone in munitions manufacture. In this regard, it is interesting to note that recently clostridia have been shown to be degraders of munitions, such as the explosive TNT (trinitrotoluene) (see Chapter 32).

Pectin-fermenting clostridia play a role in "retting," which is used in making Irish linen from flax. In this process, which was empirically determined centuries ago, natural plant stems of flax are bundled together and immersed in water. Conditions become anaerobic, and the pectin that cements the plant cells together is degraded by clostridia and other organisms, thereby freeing the fibers for linen manufacture.

Many butyric acid species of *Clostridium* are proteolytic; that is, they carry out the anaerobic hydrolysis of proteins resulting in amino acids. The amino acids can then be fermented with the production of ATP. **Figure 20.13** shows an example of a typical fermentation for glutamic acid. The end products of this particular fermentation are the same as the butyric acid fermentation, except that ammonia is also formed.

Some of the proteolytic clostridia carry out unique fermentation reactions called Stickland reactions in which two amino acids are catabolized. For example, L-alanine can be oxidized and L-glycine reduced by some

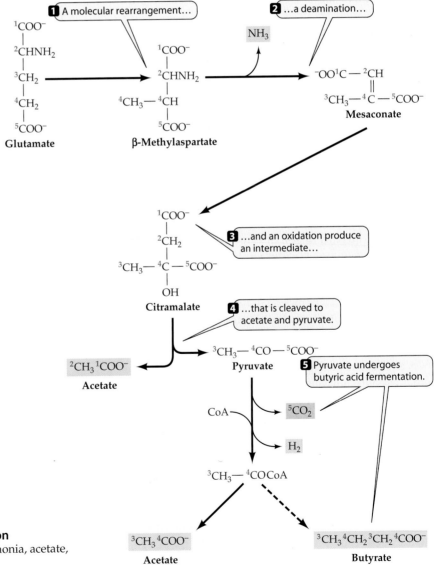

Figure 20.13 Glutamic acid fermentation
Many clostridia ferment glutamate to ammonia, acetate, and butyrate.

Milestones Box 20.3   Food to Die For!

The toxin produced by *Clostridium botulinum* is among the most deadly toxins known to humans. Most incidents of botulism occur from eating improperly sterilized home-canned foods, such as vegetables that are contaminated with soil containing the bacterium. If canning does not kill the bacteria, they can grow in the canned food and release the toxin into the contents of the can. As the toxin is readily denatured by boiling,

it is primarily a problem in foods such as canned peas that are often used cold in salads without prior cooking.

When botulism is detected in commercially canned food, it can have a disastrous financial impact on manufacturers. A few years ago, a handful of individuals from around the United States were diagnosed with botulism. By examining the consumption of infected individuals,

epidemiologists quickly identified the source of the problem as canned salmon. Ultimately, the cans were traced to a defective canning machine in one plant in Alaska. When the news broke, sales of canned salmon plummeted, and the canned salmon market took years to recover, despite assurances by canners that the problem had been corrected.

---

species, leading to the formation of the final products: acetic acid, CO_2, and NH_3 along with ATP (see Figure 8.6).

Some disease-producing amino acid fermenters include *Clostridium tetani,* the causative agent of tetanus, and *C. botulinum,* which is responsible for botulism. *C. tetani* can grow in wounds exposed to soil, which is the normal environment of the organism. The bacterium does not need to grow very much in the tissue because the toxin it produces is extremely potent. The disease is best handled by preventive medicine: children are given DPT (diphtheria, pertussis [whooping cough], and tetanus) immunizations to build up a natural antiserum immunity to the organism's toxin before any exposure to the bacterium through wound infections. However, if the skin of an immunized person is cut or punctured and exposed to dirt or soil, it is recommended that the individual receive a booster shot of DPT and receive antitoxin as well.

Clostridium botulinum is another common soil bacterium and is also a pathogen. It causes a fatal food poisoning called **botulism.** Botulinum toxin is a protein exotoxin that is excreted from the bacterium as it grows. It affects the normal release of acetylcholine from motor nerve junctions, thereby causing paralysis. It is one of the most potent toxins known (**Box 20.3**).

Clostridium perfringens is another important pathogen and food poisoning organism. Like *S. aureus,* this species produces an enterotoxin. The *C. perfringens* enterotoxin is formed during sporulation and can be seen as a protein crystal in the sporangium (**Figure 20.14**). This bacterium is widely distributed in soils and commonly contaminates foods. In addition, it is particularly troublesome to foot soldiers during times of war. Soldiers' wounds that are exposed to soil can become infected by the bacterium. It grows profusely in the tissues and produces gas that clogs blood vessels and can lead to

compromised blood circulation and eventually "gas gangrene," which may necessitate amputation.

Other *Clostridium* spp. are involved in the fermentation of purines such as uric acid (*C. fastidiosus*) with the resulting formation of acetic acid, ammonia, and CO_2 (see Figure 12.12). Uric acid is excreted as a nitrogenous waste product of birds, analogous to urine excretion by mammals. Therefore, these bacteria are widespread in soils and rookeries.

Desulfotomaculum Although this is also a genus of anaerobic spore formers, it differs from the genus *Clostridium* in two important ways. First, it carries out an anaerobic respiration involving sulfate and is therefore a

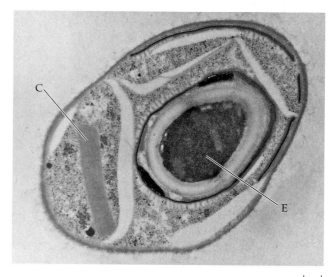

0.1 μm

Figure 20.14 *Clostridium perfringens* toxin
Thin section showing the endospore (E) and enterotoxin crystal (C) of *Clostridium perfringens.* Courtesy of Judith Bland.

sulfate-reducing bacterium (see Chapter 19 for more information on sulfate reducers). Second, it contains some cytochrome enzymes (cytochrome *b* but not *c3*), and therefore it has a limited electron transport system that is needed as a means of passing the electrons generated from the anaerobic oxidation of organic compounds onto sulfate, resulting in the formation of hydrogen sulfide. However, ATP is not generated by chemiosmotic processes, but only by substrate-level phosphorylation. Species using lactic or pyruvic acid as carbon sources oxidize these incompletely to produce acetic acid and carbon dioxide. Species using acetic acid oxidize it completely to carbon dioxide.

Desulfotomaculum is found in soils, geothermal regions, sediments, and anaerobic muds. It is also known to produce off-flavors in the canned food industry, referred to as "sulfur stinker." One species, *Desulfotomaculum nigrificans*, is thermophilic and grows at temperatures from 45°C to 70°C, whereas the others are mesophilic.

Sporosarcina This is the only group of endospore producing bacteria that have a coccus shape. They divide to form groups of cells in tetrads and packets of eight. They are obligately aerobic and motile. The two species are *Sporosarcina ureae*, from garden soils, and *Sporosarcina halophila*, from salt marshes. *S. ureae* is found in urban settings, particularly in soils frequented by dogs. Media supplemented with 3% urea select against most other bacteria due to the resulting high pH produced by the growth of this species.

Thermoactinomyces As the name implies *Thermoactinomyces* is a genus of thermophilic bacteria. Like *Bacillus*, this genus forms endospores and is aerobic. Likewise, studies of its 16S rRNA indicate it is most closely related to the other endospore-forming bacteria, although some taxonomists classify it with the actinomycetes. The organism is a septated, filamentous bacterium. Many endospores are formed within the same filament. The organism has been found in decaying compost piles where temperatures can become so hot from decomposition that they begin spontaneous burning.

Acetogenic Bacteria

Clostridium aceticum is a representative of a group of gram-positive bacteria that can grow as chemolithotrophs. This bacterium is a hydrogen autotroph that gains its energy by the oxidation of hydrogen. This is an example of an acetogenic bacterium in which carbon dioxide is reduced to form acetic acid in the following overall reaction:

$$2\ CO_2 + 4\ H_2 \rightarrow CH_3COOH + 2\ H_2O$$

Table 20.9	**Acetogenic bacteria**
Taxon	**Special Features**
Clostridium aceticum	Endospore-former
Acetobacterium spp.	Non-spore-forming rods; mesophilic; also ferments sugars to produce acetate
Acetogenium kivui[a]	Non-spore-forming rod; thermophilic (opt. temp. 66°C); also ferments sugars to form acetate

[a]Stain as gram-negative, but have a gram-positive cell wall type.

A number of non-spore-forming gram-positive bacteria are also acetogenic. These include the genera *Acetobacterium* and *Acetogenium* (**Table 20.9**). All of these bacteria fix carbon dioxide by the acetyl-coenzyme A pathway (see Chapter 10), a much different mechanism of carbon dioxide fixation from the Calvin-Benson cycle. These bacteria can also grow as ordinary heterotrophs by utilization of sugars in fermentations.

Mollicutes, Cell Wall-Less Bacteria

The *Mollicutes* are a major taxonomic group of cell wall-less bacteria that stain as gram-negative. However, studies of their 16S rRNA have clarified their relatedness to other bacteria and indicate that they are most closely allied with gram-positive bacteria in the genus *Clostridium*.

Because they completely lack a cell wall, the organisms have unusual shapes (**Figure 20.15**). They are also among the smallest of the bacteria and the simplest in structure. Some have a diameter of about 0.25 μm,

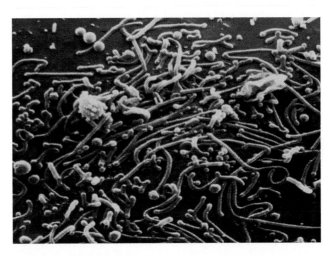

Figure 20.15 *Mycoplasma*
Electron micrograph of cells of *Mycoplasma pneumoniae*, a member of the *Mollicutes*. Note how the cells vary in shape. Photo ©David M.Phillips/Photo Researchers, Inc.

which approaches the theoretical minimum for the size of an organism, in other words, a structure large enough to contain the DNA, ribosomes, and necessary enzymes with which to perform the functions of life. Their genomes are also small, having an approximate molecular weight of 5×10^8, which is comparable to that of the obligately parasitic *Chlamydia* spp. (see Chapter 24).

Their lack of a cell wall makes the *Mollicutes* particularly fragile osmotically. Many species must be grown on complex media, and some require sterols to stabilize their plasma membranes (**Table 20.10**). Because they cannot synthesize sterols, they must obtain them from the medium in the form of cholesterol, often added as serum. It should be noted, however, that some species do not require sterols.

Mycoplasma One important genus of the *Mollicutes* is *Mycoplasma*. Colonies of *Mycoplasma* species appear as "fried eggs" with dense, granular centers and more transparent outer region. Some members of this genus obtain energy by fermentation of sugars through the Emden-Meyerhof pathway to form lactic, pyruvic, and acetic acids. Others degrade arginine as an energy source or derive energy from acetyl-coenzyme A by phosphoacetyl transferase and acetate kinase.

Many species of this genus are parasitic or pathogenic to animals. They are found particularly associated with mucoid epithelial tissues. *Mycoplasma pneumoniae* causes a type of bacterial pneumonia. Unlike pneumococcal pneumonia, however, it is not possible to use penicillin as an antibiotic in its treatment because *M. pneumoniae* lack peptidoglycan. However, mycoplasmas are inhibited by tetracycline and chloramphenicol.

Other Important Genera *Ureaplasma* spp. obtain energy through the degradation of urea to ammonium ions and the subsequent conversion of ammonium to ammonia plus protons, from which they produce ATP chemiosmotically. These are parasites in animal genitourinary tracts and respiratory tracts.

Acholeplasma spp. lack a requirement for sterol. They produce **lipoglycan,** which may help stabilize their cell

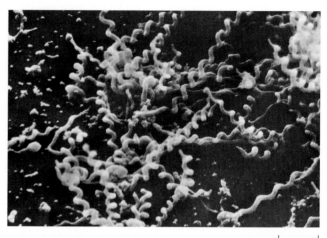

$\llcorner$————$\lrcorner$
2.0 μm

Figure 20.16 *Spiroplasma*
Cells of the cell wall-less genus *Spiroplasma* have a characteristic helical shape. Photo ©David M. Phillips/Visuals Unlimited.

membranes. Unlike the lipopolysaccharide of gram-negative bacteria, lipoglycan is not linked to a lipid A backbone. They resemble *Mycoplasma* species in that they obtain energy from sugar fermentation. None are known to be pathogens, but they are common parasites of animals.

Spiroplasma spp. have a characteristic helical shape (**Figure 20.16**). They obtain energy by sugar fermentation. They are found primarily as parasites of plants and insects. Insects feeding on infected plants can obtain the bacterium from plant phloem and carry it to another plant.

Anaeroplasma are strict anaerobic organisms found in the rumen of cattle and sheep. They ferment starch and other carbohydrates to produce acetic, formic, lactic, and propionic acids, as well as ethanol and carbon dioxide. They lyse other bacteria as well, but are not known to be pathogenic to their hosts.

PHYLUM: *ACTINOBACTERIA*

The "true" *Actinobacteria* (also referred to as the "actinomycetes") are mycelial organisms; that is, the cells produce branches as well as filaments. Therefore, they have a mycelial growth habit resembling that of the fungi. One of the first organisms studied in this group was the genus *Actinomyces,* after which the group is named. Some species produce only a substrate mycelium in which the growth is either within or on the surface of the agar or other growth medium. The sub-

Table 20.10	Important genera of *Mollicutes*			
Genus	**Mol % G + C**	**Sterol Requirement**	**Habitat**	**Distinctive Properties**
Mycoplasma	23–40	+	Animals	
Ureaplasma	27–30	+	Animals	Degrade urea
Acholeplasma	27–36	–	Animals	
Spiroplasma	25–31	+	Plants; insects	Helical shape
Anaeroplasma	29–34	+	Cattle rumen	Obligate anaerobes

Domain *Bacteria*

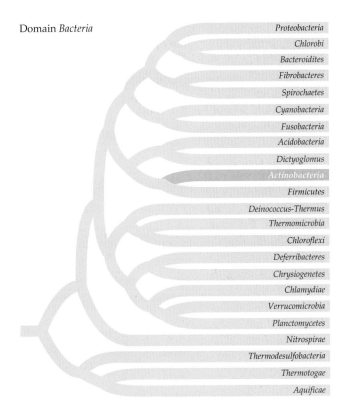

Proteobacteria
Chlorobi
Bacteroidites
Fibrobacteres
Spirochaetes
Cyanobacteria
Fusobacteria
Acidobacteria
Dictyoglomus
Actinobacteria
Firmicutes
Deinococcus-Thermus
Thermomicrobia
Chloroflexi
Deferribacteres
Chrysiogenetes
Chlamydiae
Verrucomicrobia
Planctomycetes
Nitrospirae
Thermodesulfobacteria
Thermotogae
Aquificae

strate mycelium is coenocytic, lacking cell septa. Some genera produce an aerial mycelium as well as a substrate mycelium. The aerial mycelium can form special reproductive spores called conidia or conidiospores (see Figure 20.27). Conidia are more resistant to ultraviolet light and can survive well under dry conditions. They are disseminated by the wind as a means of dispersing the organism from soil environments. However, unlike endospores, they are not particularly resistant to high temperatures. Almost all of the actinomycetes are non-motile; however, some types produce flagellated spores that permit dispersal in aquatic habitats.

Although many members of the *Actinobacteria* produce multicellular filaments or mycelia, some, such as *Micrococcus* are unicellular. We begin by discussing unicellular genera including *Micrococcus*, *Arthrobacter*, and *Geodermatophilus*, which are common soil genera, and *Frankia*, which is a plant symbiont (Table 20.5).

Micrococcus

Although *Staphylococcus* and *Micrococcus* resemble each another superficially in both being nonmotile, gram-positive cocci, they are considerably different. The genus *Micrococcus* is a group of obligately aerobic organisms with a high DNA base composition (66 to 72 mol % G + C), whereas the genus *Staphylococcus* is a group of facultative anaerobes having a low DNA base ratio (30 to 35 mol % G + C).

The genus *Micrococcus* is a common soil organism that is dispersed in air. Some species also reside on the skin of humans and other mammals. Most strains produce carotenoid pigments that give their colonies a yellow to red pigmentation. The pigments may protect these organisms from ultraviolet light during their dispersal in air by absorbing lethal ultraviolet radiation. Some of them produce packets of cells (**Figure 20.17**A). Some *Micrococcus* spp. produce capsules that give colonies a characteristic gummy appearance (Figure 20.17B).

Arthrobacter

Arthrobacter spp. are also common soil inhabitants. Sergei Winogradsky was the first to note that small coccoid cells

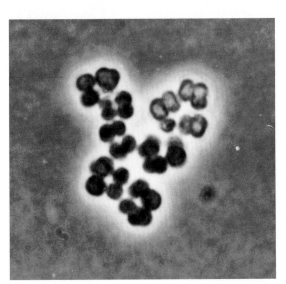

(A)

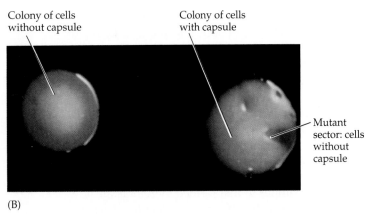

Colony of cells without capsule

Colony of cells with capsule

Mutant sector: cells without capsule

(B)

Figure 20.17 *Micrococcus*
(A) *Micrococcus luteus* as it appears in the phase microscope. Note how the cells occur in packets. (B) Colonies of *Micrococcus roseus*, without and with capsules. Note the larger, gummy appearance of the colony of encapsulated cells. A, courtesy of J. T. Staley; B, courtesy of Wesley Kloos.

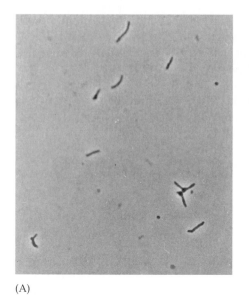

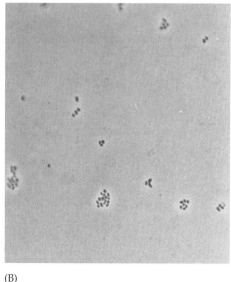

(A)   (B)

Figure 20.18 *Arthrobacter* **life cycle**
Cells of *Arthrobacter* sp. (A) during active growth and (B) during stationary-phase growth. Courtesy of Fred Palmer and J. T. Staley.

were abundant in soils. Actually, the coccoid cells are one of two cell types exhibited by this genus. When cells are actively growing, they grow as irregular rods (**Figure 20.18A**). However, as the cells enter stationary phase, they become shorter and rounded in appearance (Figure 20.18B). In fact, some studies have shown that the length of the rod is directly related to the growth rate of the bacterium. If they are growing slowly, they are shorter than if they are growing more rapidly. These bacteria are gram-positive, but some may stain as gram-negative cells. Some strains produce motile cells.

Arthrobacter spp. are obligate aerobes that use a variety of sugars as carbon sources. Many grow on a simple medium with ammonium as a nitrogen source, although some strains require biotin or other vitamins.

Geodermatophilus

This unusual genus, with the aptly named species, *Geodermatophilus obscurus*, is a common inhabitant of soils, particularly desert soils and rocks. Like *Micrococcus*, it is an obligate aerobe that utilizes a variety of sugars as carbon sources for growth. Colonies produce a greenish black pigment, probably a melanin. The organism grows as masses of cocci to form a gummy colony that rises above the agar surface. Motile cells are produced by some strains.

Frankia

This genus is an important plant symbiont. It produces root nodules similar to that formed by *Rhizobium*; however, it associates with a variety of nonleguminous plants such as *Alnus* (alder), *Ceanothus* (wild lilac), and *Casuarina* (Australian pine or she-wood). Like *Rhizobium*, *Frankia* carries out nitrogen fixation while growing in the plant as a symbiont, thereby benefiting the plant.

Frankia is very difficult to cultivate away from its host plant. The organism is microaerophilic and produces an

aerial as well as substrate mycelium. The aerial mycelium develops a sac or **sporangium** that is referred to as multilocular because it is compartmentalized into many individual spores (Figure 25.8).

Coryneform Bacteria

The coryneform bacteria are a group of organisms, primarily soil forms, that show characteristic branching cells with the formation of Y, V, or orthogonal shapes (**Figure 20.19**). The Y shapes are due to rudimentary branch formation, a common characteristic of the filamentous gram-positive bacteria, the actinomycetes (except for some of the cyanobacteria, gram-negative

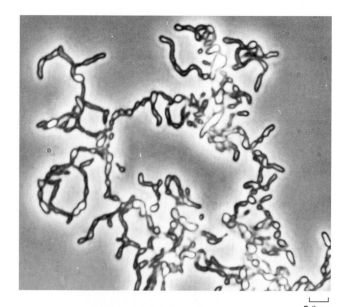

5.0 μm

Figure 20.19 *Corynebacterium diphtheriae*
Typical orthogonal arrangement of cells of the coryneform bacterium *Corynebacterium diphtheriae*. Courtesy of J. T. Staley.

Table 20.11	Important genera of coryneform bacteria				
Genus	**Shape**	**Mol % G + C**	**O₂ Requirement**	**Habitat**	
Corynebacterium	Irregular rods, V-shapes	51–60	Facultatively aerobic	Soil; animal pathogen	
Propionibacterium	Club-shaped rods	57–67	Facultatively aerobic	Animal intestine, cheese	
Mycobacterium	Rods, some branching	62–70	Aerobic	Animal pathogen	
Nocardia	Slight to extensive branching	64–69	Aerobic	Soil	
Rhodococcus	Coccus to filamentous mycelium	60–69	Aerobic	Soil	

bacteria do not produce true branches in their filaments). The characteristic V shapes occur after cell division. Cells have an outer cell wall layer not shared by other bacteria. When the cells have completed division, turgor pressure is exerted on the space between the dividing cells, and they separate incompletely. This is called a postfission snapping movement. After cell division, the two cells do not separate completely from one another and thereby produce a V-shaped configuration.

The coryneform bacteria appear to be intermediate morphological forms between lactobacilli and true mycelial actinomycetes. Five principal genera in this group include *Propionibacterium, Corynebacterium, Mycobacterium, Nocardia,* and *Rhodococcus* (**Table 20.11**).

Propionibacterium This genus is so named because they produce propionic acid as a principal product of their fermentation. They are gram-positive aerotolerant fermentative bacteria that are found in two different habitats. One habitat occupied by the classic type of *Propionibacterium* is the intestinal tract of animals. An allied habitat is cheese. Indeed, the characteristics of Swiss (Ementhaler) cheese are due to the growth of propionibacters. These organisms take the lactic acid formed by the lactic acid bacteria in cheese fermentation and further metabolize it to propionic and acetic acids as well as CO_2 (**Figure 20.20**). These reactions are more complex, comprising a series of steps beginning with the oxidation of lactic acid to pyruvate and then transformation of the pyruvate (**Figure 20.21**). The gas production

results in the characteristic holes, called "eyes" of Swiss cheese (if gas is not produced in adequate amounts, holes are not formed, and the cheese is said to be "blind"). Although cheese was the initial source of these bacteria, their true natural habitat has been traced to the rumen of cattle and other animals. Rennin, an enzyme used for the curdling of milk in the production of cheeses, is obtained from the stomachs of calves. Therefore, rennin contaminated with *propionibacteria* was

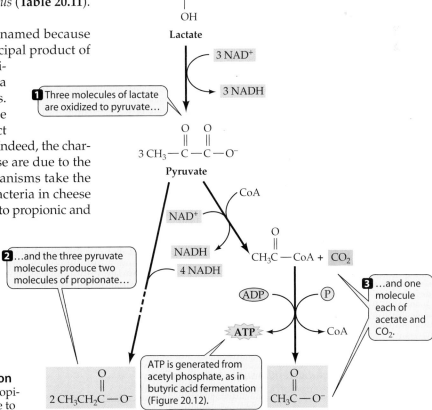

Figure 20.20 Propionic acid fermentation
Propionibacterium spp. ferment lactate to propionate and acetate. The route from pyruvate to propionate is shown in Figure 20.21.

the likely original source of these organisms in the cheese-making industry. Sugars are also fermented by propionibacteria. They use the Embden-Meyerhof pathway to produce pyruvate, which is then fermented as shown in Figure 20.21.

The other major habitat of propionibacteria is the skin of mammals. One species, *Propionibacterium acnes*, is found on the skin of all humans. It grows in the sebaceous gland (not sweat glands), where it produces propionic acid in abundance. It ferments the lactic acid produced by *Staphylococcus epidermidis* to form propionic and acetic acid as previously discussed. Humans fall into two groups based upon the numbers of *P. acnes* they harbor on their skin. Some have over 1 million per cm^2 of skin surface, whereas others have fewer than 10,000 per cm^2. Because propionic and acetic acids are volatile fatty acids with distinctive smells, they give animals, including humans, a characteristic natural scent.

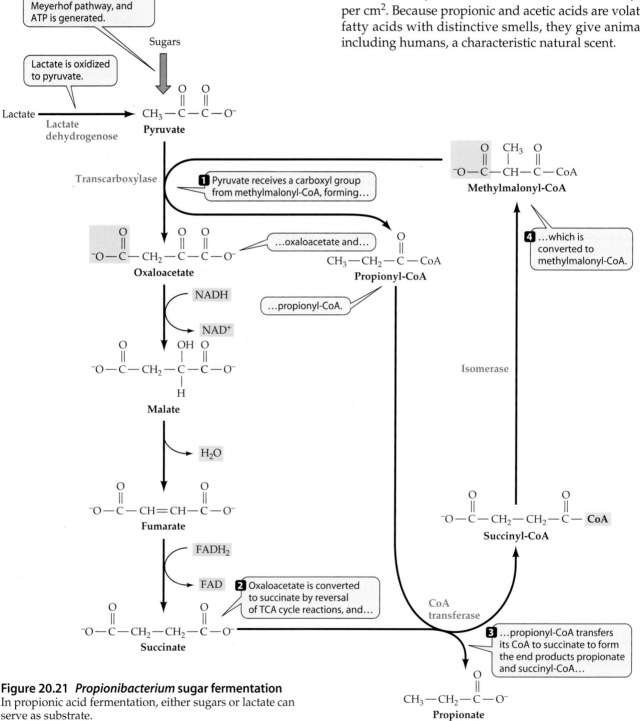

Figure 20.21 *Propionibacterium* **sugar fermentation**
In propionic acid fermentation, either sugars or lactate can serve as substrate.

Corynebacterium This genus is a group of common aerobic soil organisms. The first isolate of the genus was *Corynebacterium diphtheriae,* which is a normal inhabitant of the oral cavity of animals, where it lives as a parasite. Pathogenic strains of this species carry a piece of DNA that they have received from a phage by lysogenic conversion. This genetic material is responsible for production of the protein exotoxin, the principal virulence factor for diphtheria. The disease is rare now, having been controlled largely through vaccination.

Soil "diphtheroids," which appear as club-shaped rods (**Figure 20.22**), are similar nutritionally and metabolically to the animal parasites. All are immotile and most are facultative aerobes that produce propionic acid during fermentation of sugars. Some species cause diseases of plants.

Mycobacterium As mentioned earlier, there are many genera in the gram-positive bacteria that are transitional between unicellular forms and mycelial forms. One such group already discussed is the coryneform group, which shows evidences of true branching in their Y forms. Two other genera that appear to be intermediate are still largely unicellular, but are considered true actinobacteria because they commonly form true branches. These are the genera *Mycobacterium* and *Nocardia.* These will be discussed here along with *Bifidobacterium* and the more highly differentiated mycelial forms that include the genera *Actinomyces, Micromonospora, Streptomyces,* and *Actinoplanes.*

The mycobacteria are nonmotile rod-shaped bacteria that may show true branching and typically bundle

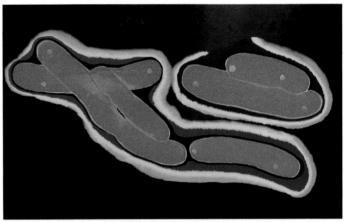

Figure 20.23 *Mycobacterium tuberculosis*
Cells of *Mycobacterium tuberculosis,* showing its tendency to aggregate into cordlike structures. Photo ©Dr. Linda Stannard/SPL/Photo Researchers, Inc.

together to form cordlike groups (**Figure 20.23**). They are aerobic rods that occur naturally as saprophytes in soils. Unlike other bacteria, they produce a distinctive group of waxy substances called **mycolic acids** (**Figure 20.24**) that are covalently linked to the peptidoglycan. In addition to glucosamine and muramic acid, they have a polymer of arabinose and galactose (called an arabinogalactan) bound to the peptidoglycan. The mycolic acids make the organisms difficult to stain using ordinary simple staining procedures; nonetheless, they are regarded as being gram-positive. These same mycolic acids make these bacteria **acid-alcohol fast,** which refers to the ability of these organisms, once stained with a solution of basic fuchsin in phenol, to withstand decol-

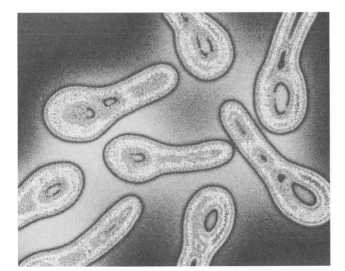

Figure 20.22 *Corynebacterium diphtheriae*
Photo micrograph showing club-shaped diphtheroids of *Corynebacterium diphtheriae.* Photo ©Alfred Pasieka/SPL/ Photo Researchers, Inc.

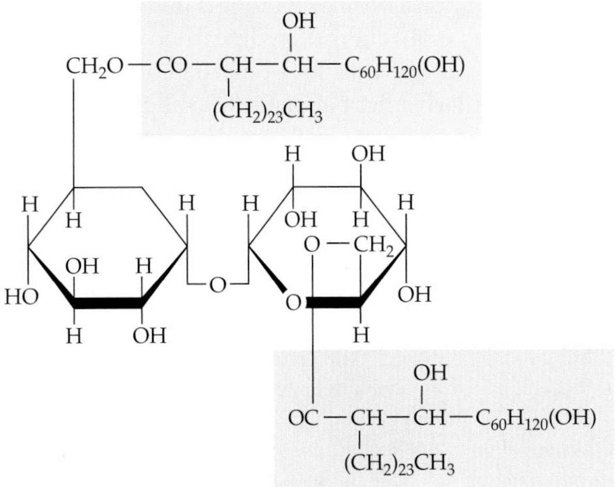

Figure 20.24 Mycolic acids
General formula for mycolic acids.

orization with acidified ethanol during a staining procedure called the Ziehl-Neelson stain. The basic fuchsin binds strongly to the mycolic acids. This acid-alcohol-fast property is not found in any other bacterial group, making it an excellent differential property for distinguishing these from all other bacteria.

Mycobacteria grow on simple inorganic media with ammonium as the sole nitrogen source and glycerol or acetate as a carbon source. Incorporating lipids into growth media tends to enhance growth of some species. Their slow growth is in part due to the hydrophobicity of the organisms, which prevents rapid uptake of dissolved nutrients.

Two species are important pathogens of humans. *Mycobacterium tuberculosis* is the causative agent of tuberculosis, and *Mycobacterium leprae* causes leprosy. See Chapter 28 for more detail on these diseases.

Nocardia The nocardia are separated into two groups. One group is acid-fast and therefore similar to the mycobacteria, whereas the other group is non-acid-fast. These nonmotile, rod-shaped to filamentous organisms show somewhat greater branching than the mycobacteria. Some strains even show extensive mycelial development (**Figure 20.25**) and may even produce an aerial mycelium. One feature characteristic of this group is that the mycelium, if formed, tends to fragment into small units during stationary growth phase.

Nocardia are common soil bacteria. They are obligately aerobic heterotrophs whose metabolism has not been well studied. As a group and individually, they utilize a wide variety of sugars and organic acids as carbon sources for growth.

Rhodococcus The genus *Rhodococcus* is widely distributed in soils and aquatic environments. It is particularly well known for its ability to degrade hydrocarbon compounds. It produces capsular materials that are excellent dispersants and emulsifiers of hydrocarbon compounds. The cells grow at the interface between oil and water and are particularly adept at degrading aliphatic hydrocarbons. Some species are pathogenic to animals.

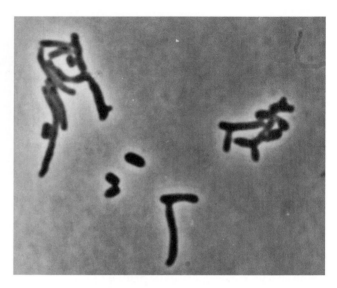

Figure 20.25 *Nocardia* sp.
Phase photomicrograph of a *Nocardia* sp., showing branching. Courtesy of J. T. Staley.

Highly Branched and Mycelial *Actinobacteria*

The more highly branched *Actinobacteria* are common soil organisms that are prokaryotic counterparts of the fungi (**Table 20.12**). These organisms are not acid-fast. Some of them produce an aerial mycelium that has a fruiting structure bearing spores called conidiospores. These are asexual spores, sometimes referred to as exospores to differentiate them from endospores.

Actinomyces This genus, for which the entire group was named, is atypical of other mycelial members, primarily because it is a group of anaerobic or facultative anaerobic bacteria. They ferment sugars such as glucose to produce formic, acetic, lactic, and succinic acids and do not carry out aerobic respiration. Organic nitrogen compounds are required for growth, and supplemental carbon dioxide greatly enhances it. Though they are mycelial, they do not produce an aerial mycelium. They

Table 20.12	Important genera of branching and mycelial bacteria				
Genus	**Shape**	**Mol % G + C**	**Habitat**	**Special Features**	
Actinomyces	Rods; some branching	58–63	Oral cavity; some pathogens	Only facultative aerobes in group	
Bifidobacterium	Rods	55–67	Animal intestine	Anaerobic	
Micromonospora	Substrate and aerial mycelium	71–73	Soil	Single conidium	
Streptomyces	Substrate and aerial mycelium	69–78	Soil	Many conidia	
Actinoplanes	Substrate and aerial mycelium	72–73	Soil	Motile sporangiospores	

Figure 20.26 *Bifidobacterium* fermentation
The bifidobacteria carry out an unusual lactic acid fermentation, as shown here.

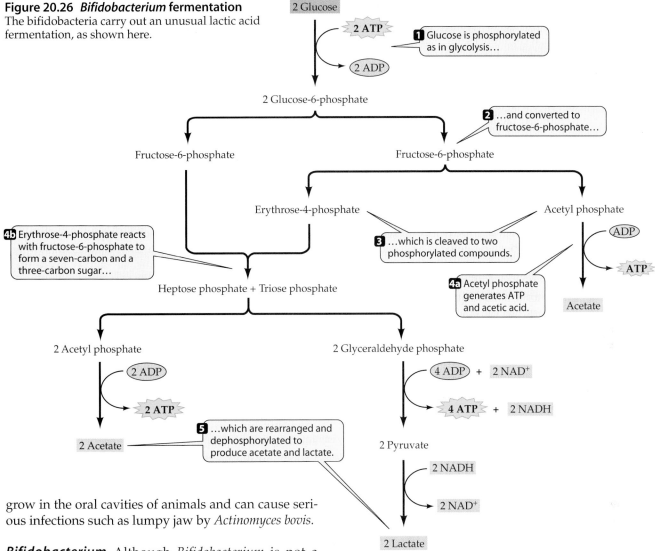

1 Glucose is phosphorylated as in glycolysis…

2 …and converted to fructose-6-phosphate…

3 …which is cleaved to two phosphorylated compounds.

4b Erythrose-4-phosphate reacts with fructose-6-phosphate to form a seven-carbon and a three-carbon sugar…

4a Acetyl phosphate generates ATP and acetic acid.

5 …which are rearranged and dephosphorylated to produce acetate and lactate.

grow in the oral cavities of animals and can cause serious infections such as lumpy jaw by *Actinomyces bovis*.

Bifidobacterium   Although *Bifidobacterium* is not a mycelial organism, its phylogeny places it as a close relative of *Actinomyces*. This is a genus of anaerobic irregular rod-shaped bacteria that ferment sugars to acetic and lactic acids (**Figure 20.26**). They are found primarily in the intestinal tracts of animals. One species, *B. bifidus*, is commonly found in the intestines of humans that are breast fed and is therefore a pioneer colonizer of the human intestinal tract. It is particularly well adapted to growing on human breast milk, which contains an amino sugar disaccharide not found in cow's milk. This species is unusual in that it requires amino sugars for growth.

Streptomyces   The foremost genus of the mycelial actinobacteria is *Streptomyces*. This is a diverse group of soil bacteria that produce aerial as well as substrate mycelia. The **hyphae,** or filaments, of the aerial mycelia differentiate to form asexual conidiospores. The spores of this genus are formed in chains at the tips of aerial hyphae (**Figure 20.27**). Species differences are based partially on

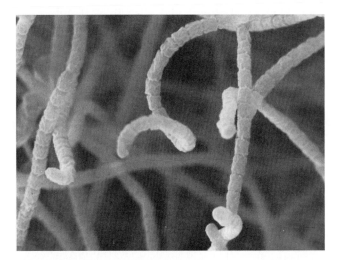

Figure 20.27 *Streptomyces*
The aerial mycelium of a *Streptomyces* sp., showing conidiospores. Photo ©Frederick P. Mertz/Visuals Unlimited.

the morphology of the spores. For example, some have a warty or spiny appearance, whereas others are smooth. Physiological differences, especially antibiotic production, are also important.

Members of this genus grow on simple inorganic media supplemented with a variety of organic carbon sources, including glucose or glycerol; vitamins are not required. They metabolize by aerobic respiration. In addition to simple organic carbon sources, some can use polysaccharides such as pectin, chitin, and even latex. If a culture is started with conidiospores, they first germinate to produce the vegetative or substrate mycelium. After this, the aerial mycelium is formed. If the medium has a sufficiently high carbon to nitrogen ratio, the aerial mycelium differentiates to produce conidiospores when the nutrients have been depleted. The conidia are actually formed within the outer wall of the conidiophore; however, they do not at all resemble endospores. The conidia become pigmented blue, gray, green, red, violet, or yellow colors, but the color can be influenced by medium composition. The substrate mycelium can also be pigmented.

Streptomyces occur in viable concentrations of 10^6 to 10^7 per gram in soil environments. They are responsible for imparting the characteristic "earthy" odor to soil. This is due to the **geosmins,** a group of volatile organic compounds that they produce during growth. Geosmins cause odors and flavors in drinking water supplies as well. However, cyanobacteria also produce geosmins and are more likely to be responsible for this problem in water supplies.

This group is noted for the commercially important antibiotics they produce including streptomycin, chloramphenicol, and tetracycline (**Table 20.13**). Approximately half of the commercially produced antibiotics are derived from this genus. See Chapters 7 and 31 for more detail. In addition, they are sources of anticancer drugs used in chemotherapy (see Table 31.9).

Although this is an extremely important genus of soil microorganisms, little is known of their ecological roles.

Intriguingly, it is unclear whether *Streptomyces* species produce the antibiotics while growing in their natural habitat. Certainly, it would appear to favor their activities because they could use them to inhibit competitors.

The proliferation of great numbers of species in this genus (over 500 species have been proposed in the past) largely reflects the need for pharmaceutical firms who study these bacteria to propose a new species when they wish to patent and manufacture a new antibiotic.

Other genera of mycelial actinomycetes are differentiated from this genus primarily by their cell wall composition and the morphology of their conidiospore-bearing structure.

Actinoplanes This is a genus of actinomycetes that produces a flagellated spore. Like *Streptomyces*, this organism has both a substrate and an aerial mycelium. However, the spores are produced within a sac or sporangium, a stage that can survive desiccation. When conditions in the environment are moist and favorable for germination and growth, the sporangium ruptures and releases the motile spores.

Heliobacteria

The **Heliobacteria** are one of the most recently discovered bacterial groups. These bacteria are the only phototrophic gram-positive bacteria known. They are photoheterotrophs that require organic compounds as carbon sources and use light for energy generation. They have a unique type of bacteriochlorophyll called bacteriochlorophyll *g* (see Chapter 21). Although they stain as gram-negative bacteria, an analysis of their 16S rRNA places them with the gram-positive bacteria. Furthermore, some members of the group produce heat-resistant bacterial endospores similar to those of *Bacillus* spp.

Two common genera in this group are *Heliobacterium* and *Heliobacillus*. *Heliobacterium* spp. are gliding bacteria, whereas members of *Heliobacillus* are motile by peritrichous flagella. The heliobacteria grow as obligate

Table 20.13	Important antibiotics produced by *Streptomyces* species		
Antibiotic Group	**Species**	**Common Name**	**Effective Against**
Chloramphenicol	*S. venezuelae*	Chloramphenicol	Broad spectrum
Tetracycline	*S. aureofaciens*	Tetracycline	Broad spectrum
Chlortetracycline	*S. aureofaciens*	Chlortetracycline	Broad spectrum
Polyenes	*S. noursei*	Nystatin	Fungi
Macrolides	*S. erythreus*	Erythromcin	Most gram-positives
Legionella	*S. lincolnensis*	Clindamycin	Obligate anaerobes
Aminoglycosides	*S. griseus*	Streptomycin	Most gram-negative
	S. fradiae	Neomycin	Broad spectrum

anaerobes carrying out anoxygenic photosynthesis using organic compounds such as pyruvate as carbon sources.

The finding of photosynthetic bacteria in the gram-positive phylum is consistent with photosynthesis being a primitive characteristic that is shared by many of the other bacterial phyla. This group is discussed further, along with the other photosynthetic bacteria, in Chapter 21.

SUMMARY

▸ Lactic acid bacteria ferment sugars to produce lactic acid either by the **homofermentative** or **heterofermentative** pathway. All energy generated by the lactic acid bacteria is via substrate-level phosphorylation. More energy is available to the homofermentative (2 ATP/mol glucose) compared to the heterofermentative (1 ATP/mol glucose) lactic acid bacteria.

▸ Lactic acid bacteria are **aerotolerant anaerobes** that are indifferent to the presence of oxygen in their growth environment. Most lactic acid bacteria are nutritionally **fastidious,** requiring complex nutrients for growth, including vitamins, amino acids, and purines or pyrimidines.

▸ Several groups of lactic acid bacteria exist, including those associated with plants, dairy products, and in animals in the mouth, intestines, and vagina. In natural fermentations of plant materials and dairy products, the fermentation undergoes a succession in which the *Streptococcus* group predominates first and lowers the pH to 5.5 or so, then the *Lactobacillus* group succeeds and lowers the pH even further, to about 4.5—the low pH acts as a preservative for the foodstuff.

▸ Some *Streptococcus* spp. cause dental caries, others cause scarlet fever, puerperal fever, pneumonia, tissue necrosis, and endocarditis. *Streptococcus* spp. may cause **alpha-** or **beta-hemolysis** of red blood cells.

▸ *Staphylococcus epidermidis* and *Propionibacterium acnes* are members of the normal microbiota of human skin—all humans are colonized by these two bacterial species. *Staphylococcus aureus* is a normal resident of the nasopharynx of humans, but it can also be a human pathogen—it is noted for causing skin infections, food poisoning, and toxic shock syndrome.

▸ *Bacillus* spp. are aerobic to facultative aerobic endospore-forming rods commonly found in soils. *Clostridium* spp. are obligately anaerobic spore-forming rods. **Endospores** are specially modified cells that are heat-resistant resting stages of various genera including *Bacillus, Clostridium, Sporosarcina, Desulfotomaculum,* and *Thermactinomyces.*

▸ Endospores are formed when the nutrients are depleted in the environment—in cultures this occurs at the beginning of stationary phase growth. Endospore formation involves producing a spore within a cell. After DNA replication and cell membrane separation, one of the resulting units engulfs the other, which will become the endospore; several stages are involved. Endospores have a low water content and high concentrations of calcium and **dipicolinic acid,** a unique compound not found elsewhere.

▸ Some *Bacillus* spp. such as *B. thuringiensis,* are insect pathogens that produce a characteristic protein crystal that is toxic to moths. *Bacillus anthracis* is the causative agent of anthrax in cattle and humans.

▸ *Clostridium* spp. ferment cellulose, sugars, amino acids, or purines. In the **Stickland fermentation,** two amino acids are catabolized: one amino acid is fermented, the other is oxidized. One species, *Clostridium aceticum,* is an acetogen that obtains energy from oxidation of H_2 and fixes CO_2 via the acetyl-coenzyme A pathway. *Clostridium botulinum* is the causative agent of botulism; *Clostridium tetani* is the causative agent of tetanus.

▸ The **mycoplasmas,** or *Mollicutes,* lack a cell wall, but are close relatives of *Clostridium*—many species, such as *Mycoplasma pneumoniae,* are pathogenic to humans.

▸ *Micrococcus* spp. differ from *Staphylococcus* in that they are obligate aerobes that are common soil inhabitants and have a higher mol % G + C content.

▸ *Propionibacterium* is noted for its production of propionic acid from lactic acid; it is responsible for Swiss cheese production. *Bifidobacterium bifidus,* which ferment sugars to acetic and lactic acids, are part of the normal microbiota of breast-fed infants.

▸ *Corynebacterium* is the formally named genus of the coryneform bacteria that are common soil organisms. *Corynebacterium diphtheriae* lives as a parasite of the oral cavities of humans—strains that cause diphtheria have received a piece of DNA from a phage that encodes for the diphtheria protein exotoxin.

▸ The **actinomycetes** are a group of branching filamentous organisms, some of which produce both a **substrate** and an **aerial mycelium.** *Mycobacterium* spp are acid-fast and produce mycolic acids; two species, *M. tuberculosis* and *M. leprae* (leprosy), are well-known pathogens.

▸ *Streptomyces* is a common soil actinomycete noted for its production of antibiotics such as tetracyclines, chloramphenicol, streptomycin, and erythromycin. *Streptomyces* impart an "earthy" smell to soil due to the production of geosmins.

▶ *Heliobacterium* and *Heliobacillus* are photoheterotrophic bacteria that use organic compounds such as pyruvate as carbon sources for growth and derive energy from sunlight.

REVIEW QUESTIONS

1. What evidence indicates that the Gram stain has phylogenetic significance? Name groups that are exceptions.

2. The primitive Earth was anaerobic and contained anaerobic bacteria. The modern world is aerobic (with anaerobic sediments), but it has highly evolved exotic aerobic organisms, such as humans. Do you believe there is a parallel to this in the evolution of gram-positive bacteria?

3. What bacterial groups do we ingest in large numbers with certain foods?

4. What is the significance of the bacterial endospore?

5. In what ways are gram-positive bacteria known to be commercially important?

6. List several genera of coccus-shaped gram-positive bacteria. How do they differ from one another?

7. What are the photosynthetic members of the gram-positive bacteria?

8. Compare the genus *Bacillus* to the genus *Clostridium.*

9. Identify pathogenic species of the gram-positive bacteria. What is the basis for their pathogenesis?

10. Identify agriculturally useful gram-positive bacteria and discuss their value.

11. What are some examples of fermentations carried out by gram-positive bacteria? Are they commercially important?

12. Why are some bacteria acid-fast?

SUGGESTED READING

Balows, A., H. G. Trüper, M. Dworkin, W. Harder and K. H. Schleifer, eds. 1992. *The Prokaryotes,* 2nd ed. Berlin: Springer-Verlag.

Garrity, G., editor-in-chief. 2001. *Bergey's Manual of Systematic Bacteriology.* 2nd ed., Vol. II. New York: Springer-Verlag.

Garrity, G., editor-in-chief. 2001. *Bergey's Manual of Systematic Bacteriology.* 2nd ed., Vol. IV. New York: Springer-Verlag.

Holt, J. G., editor-in-chief. 1993. *Bergey's Manual of Determinative Bacteriology.* Baltimore, MD: Williams and Wilkins.

Chapter 21

Phototrophic Bacteria

*The sulfur bacteria...life processes are played out according to a much simpler scheme;
all their activities are maintained by a purely inorganic chemical process,
that of sulfur oxidation.*

— S. WINOGRADSKY, 1887

The evolution of photosynthesis heralded an important event on Earth. Prior to that, organic materials were much less abundant. Also, the energy generated by photosynthesis allowed for the biological production of scarce oxidants such as sulfate, oxygen, and nitrate that are important now in sulfate reduction, aerobic respiration, and denitrification.

This chapter discusses *Bacteria* that obtain energy from light (**phototrophic metabolism**). The phyla we will discuss include the *Proteobacteria*, the *Chlorobi*, the *Chloroflexi*, the *Firmicutes*, and the *Cyanobacteria*. Some of these bacteria are **photoautotrophic;** that is, they are able to use carbon dioxide as their sole source of carbon for growth and to carry out photosynthesis. However, some of the phototrophic bacteria require organic compounds and are therefore termed **photoheterotrophic** (**Table 21.1**). Some *Archaea* can use bacteriorhodopsin in the generation of ATP from sunlight, but they do not produce the chlorophyll pigments that are the hallmark of all true photosynthetic organisms.

All photosynthetic organisms from phototrophic bacteria to higher plants obtain energy from sunlight. The reactions used to capture the energy from sunlight and transform it into chemical energy are referred to as the **light reactions** of photosynthesis (see Chapter 9). The energy generated from photosynthesis, as well as the reducing power produced as NADPH or NADH, are required for the carbon dioxide fixation reactions. Because the actual carbon dioxide fixation reactions do not directly involve sunlight, they are referred to as the **dark reactions.** Before discussing the various photosynthetic bacteria, the light and dark reactions that were treated in Chapter 9 are reviewed briefly.

LIGHT AND DARK REACTIONS

The light reactions initially involve absorption of radiant energy by pigments in the cells. The pigments fall into three groups. **Reaction center chlorophyll** pigments, located in the photosynthetic membranes, are chlorophyll molecules that play a direct role in generating ATP by photophosphorylation. The type of reaction center chlorophyll varies from one group of organisms to another (**Table 21.2**). Second, **antenna** or **light-harvesting chlorophyll** molecules are associated with the reaction center and are involved in harvesting light radiation for the reaction center chlorophyll. Finally, photosynthetic microorganisms also have **accessory pigments,** such as carotenoids, that collect light energy and transfer it to the chlorophyll molecules. The nature of these accessory pigments varies from one bacterial group to another (Table 21.2).

Domain *Bacteria*

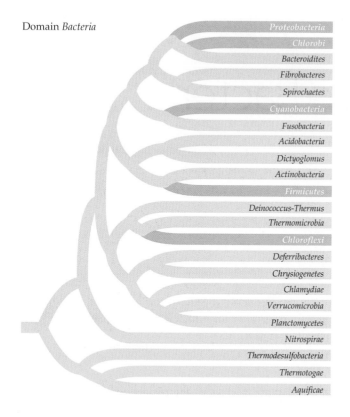

Proteobacteria
Chlorobi
Bacteroidites
Fibrobacteres
Spirochaetes
Cyanobacteria
Fusobacteria
Acidobacteria
Dictyoglomus
Actinobacteria
Firmicutes
Deinococcus-Thermus
Thermomicrobia
Chloroflexi
Deferribacteres
Chrysiogenetes
Chlamydiae
Verrucomicrobia
Planctomycetes
Nitrospirae
Thermodesulfobacteria
Thermotogae
Aquificae

Photosynthetic bacteria produce a variety of chlorophyll pigments (Table 21.2; **Figure 21.1**). Bacteriochlorophyll *a* (Bchl *a*) is a common reaction center chlorophyll used by both the purple and green bacteria. In contrast, Bchl *b* is produced only by purple bacteria, whereas Bchl *c*, Bchl *d*, and Bchl *e* are produced by various green bacteria, and Bchl *g* is produced by the heliobacteria. Unlike other prokaryotes, the *Cyanobacteria* produce chlorophyll *a* as their reaction center chlorophyll. It is interesting to note that the bacteriochlorophylls absorb very long wavelength light, some of which is in the infrared region (wavelength > 800 nm) not detected by human eyesight (Figure 21.1). In contrast, chlorophyll *a* and *b* absorb much shorter wavelength light in the visible range. Because each photo-

synthetic group has its own characteristic absorption spectrum for light, these groups can coexist in the same habitat without directly competing for available light.

ATP is synthesized by a proton gradient established across the photosynthetic membranes associated with the reaction center where the electron transfer reactions occur (see Chapter 9). Thus, like typical heterotrophic bacteria, photosynthetic bacteria generate ATP by use of proton pumps and their associated ATP synthases. The unique feature of photosynthesis is the generation of electron gradients using pigments that harvest sunlight.

A common biochemical pathway used by phototrophic bacteria for carbon dioxide fixation is the Calvin-Benson cycle (see Chapter 9). All photosynthetic bacteria, however, do not use this pathway. The green sulfur and green filamentous bacteria use different pathways for carbon dioxide fixation (see following discussion).

PHOTOTROPHIC PHYLA

First of all, it is noteworthy that several phyla of the *Bacteria* are either exclusively phototrophic or contain some phototrophic members (Table 21.1). In contrast, none of the *Archaea* is truly photosynthetic. Although it is true that some halophilic members of the *Archaea* produce bacteriorhodopsin and can generate ATP through a light-mediated reaction, their metabolism is quite different from that of the phototrophic bacteria, which have complex light-harvesting centers involving bacteriochlorophyll and accessory pigments such as carotenoids and phycobiliproteins, as discussed next.

The photosynthetic bacteria that can use light energy for growth have a distinct metabolic advantage over those that rely on organic or inorganic chemical energy sources, because of the much greater amount of energy available for their metabolism.

The phototrophic bacteria fall into five major groups: (1) the photosynthetic *Proteobacteria*, (2) the *Chlorobi* (i.e., green sulfur bacteria), (3) the *Chloroflexi* (green filamentous bacteria), (4) the photosynthetic *Firmicutes* (heliobacteria), and (5) the *Cyanobacteria*, which are differentiated from one another on the basis of phylogeny, carbon dioxide fixation pathways, and pigments (Table 21.2). Each of the photosynthetic bacterial groups is discussed in greater detail.

Proteobacteria

The photosynthetic *Proteobacteria* are commonly called the purple bacteria because of their reddish or purplish coloration. From an evolutionary perspective, the ancestors of this group may have been

Table 21.1	Phyla and properties of photosynthetic bacteria	
Phylum	**Carbon Source**	**Carbon Metabolism**
Proteobacteria		
Purple sulfur	CO_2	Calvin-Benson
Purple nonsulfur	CO_2 and organics	Photoheterotrophic
Chlorobi : Green sulfur	CO_2	Reductive TCA cycle
Chloroflexi: Green filamentous[a]	CO_2 or organic	Hydroxypropionate pathway
Cyanobacteria	CO_2	Calvin-Benson
Firmicutes: Heliobacteria	Organics	Photoheterotrophic

[a]The *Chloroflexi* are very versatile metabolically. See text for details.

Table 21.2	Characteristics of photosynthetic bacteria		
Group	**Reaction Center Chlorophyll**	**Antenna (Accessory pigments)**	
Proteobacteria	Bchl *a*	Bchl *b* (carotenoids)	
Chlorobi	Bchl *a*	Bchl *c*; Bchl *d*; Bchl *e* (carotenoids)	
Chloroflexi	Bchl *a*	Bchl *c* (carotenoids)	
Cyanobacteria	Chl *a*	Phycobilins or Chl *b* (in prochlophytes)	
Heliobacteria	Bchl *g*	Carotenoids	

the first photosynthetic organisms on Earth. The photosynthetic *Proteobacteria* comprise two separate subgroups, the **purple sulfur bacteria** and the **purple nonsulfur bacteria**. Like all *Proteobacteria* that fix carbon dioxide, both of these subgroups use the Calvin-Benson cycle, although they utilize different compounds as reducing power for the dark reactions. The purple sulfur bacteria use reduced sulfur compounds such as sulfide or elemental sulfur as their source of electrons. In contrast, the purple nonsulfur bacteria typically use organic compounds, especially organic acids and alcohols, as their source of reducing power. Photosynthesis in purple sulfur bacteria is represented by the following two reactions (reactions are not balanced):

$$(1) \quad H_2S + CO_2 \rightarrow (CH_2O)_n + S^0$$

$$(2) \quad S^0 + CO_2 + H_2O \rightarrow (CH_2O)_n + H_2SO_4$$

where $(CH_2O)_n$ represents organic carbohydrate carbon.

Photosynthesis in purple nonsulfur bacteria is represented by:

$$(3) \quad H_2A + CO_2 \rightarrow (CH_2O)_n + A$$

where H_2A = organic compound such as acetic acid.

As can be seen by the preceding three reactions for carbon dioxide fixation, oxygen is *not* formed during photosynthesis by these bacteria. This type of photosynthesis is therefore referred to as **anoxygenic photosynthesis.** As discussed in Chapter 10, the electron flow in anoxygenic photosynthesis is noncyclic, involving only one photosystem, Photosystem I, with Bchl *a*. Also, a characteristic of this type of photosynthesis is that it occurs at a relatively high redox potential, about –0.15 volts. Thus, NADPH cannot be directly formed from the electron flow; instead energy must be expended by reverse electron flow to obtain the NADPH needed for CO_2 reduction (see Chapter 10 for more detail).

The two purple photosynthetic groups are discussed individually.

Purple Sulfur Bacteria The purple sulfur bacteria are large unicellular bacteria (**Figure 21.2**) that may reach over 6.0 μm in diameter. Many are motile with polar flagella. Some have gas vacuoles, and all can utilize hydrogen sulfide in photosynthesis and form elemental sulfur granules inside their cells (**Figure 21.3A**).

The photosynthetic *Proteobacteria* have extensive intracellular membrane systems that contain the photopigments and photoreaction centers. These membranes are extensions of the cytoplasmic membrane and may be vesicular, lamellar, or even tubular (Figure 21.3). The photosynthetic pigments are all bacteriochlorophylls, with Bchl *a* found in all but one species, *Thiocapsa pfen-*

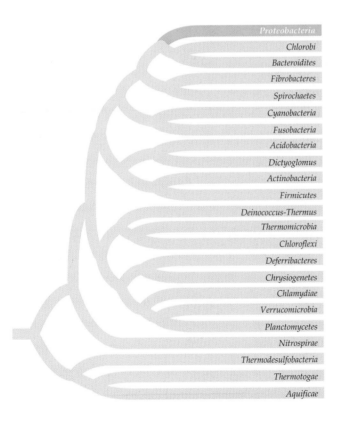

Proteobacteria
Chlorobi
Bacteroidites
Fibrobacteres
Spirochaetes
Cyanobacteria
Fusobacteria
Acidobacteria
Dictyoglomus
Actinobacteria
Firmicutes
Deinococcus-Thermus
Thermomicrobia
Chloroflexi
Deferribacteres
Chrysiogenetes
Chlamydiae
Verrucomicrobia
Planctomycetes
Nitrospirae
Thermodesulfobacteria
Thermotogae
Aquificae

(A)

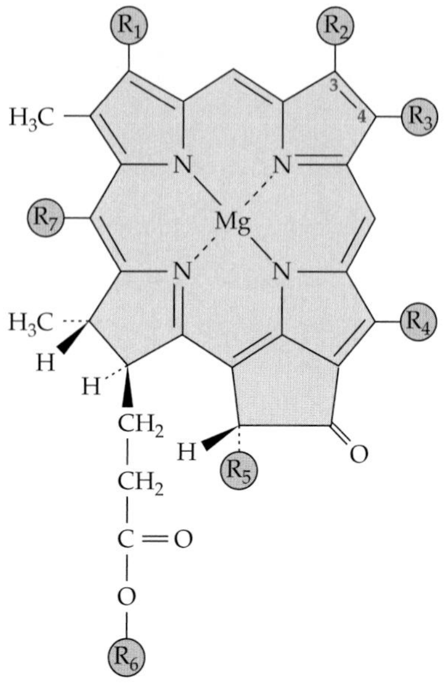

Figure 21.1 Bacterial chlorophylls
(A) General structure of the chlorophyll molecule. (B) Structures of the R groups and the absorption maximum for each chlorophyll type. P, phytyl ester ($C_{20}H_{39}$—); G, geranylgeraniol ester ($C_{10}H_{17}$—); F, farnesyl ester ($C_{15}H_{25}$—). Where two or more structures are shown for an R group, the native chlorophyll contains a mixture. Adapted from A. Gloe, N. Pfennig, H. Brockmann and W. Trowitsch, 1975, *Archives of Microbiology* 102:103-109, and A. Gloe and N. Risch, 1978, *Archives of Microbiology* 118:153-156.

(B)

Chlorophyll	R_1	R_2	R_3	R_4	R_5	R_6	R_7	Absorption maximum of cells (nm)
Bacterio-chlorophyll a	$-\overset{\displaystyle O}{\underset{\|\|}{C}}-CH_3$	$-CH_3$	$-CH_2-CH_3$	$-CH_3$	$-\overset{\displaystyle O}{\underset{\|\|}{C}}-O-CH_3$	P/G	$-H$	805 830–890
Bacterio-chlorophyll b	$-\overset{\displaystyle O}{\underset{\|\|}{C}}-CH_3$	$-CH_3$	$=\overset{\displaystyle}{\underset{H}{C}}-CH_3$	$-CH_3$	$-\overset{\displaystyle O}{\underset{\|\|}{C}}-O-CH_3$	P	$-H$	835–850 1020–1040
Bacterio-chlorophyll c	$-\overset{\displaystyle H}{\underset{OH}{C}}-CH_3$	$-CH_3$	$-C_2H_5$ $-C_3H_7$ $-C_4H_9$	$-C_2H_5$ $-CH_3$	$-H$	F	$-CH_3$	745–755
Bacterio-chlorophyll d	$-\overset{\displaystyle H}{\underset{OH}{C}}-CH_3$	$-CH_3$	$-C_2H_5$ $-C_3H_7$ $-C_4H_9$	$-C_2H_5$ $-CH_3$	$-H$	F	$-H$	705–740
Bacterio-chlorophyll e	$-\overset{\displaystyle H}{\underset{OH}{C}}-CH_3$	$-\overset{\displaystyle O}{\underset{\|\|}{C}}-H$	$-C_2H_5$ $-C_3H_7$ $-C_4H_9$	$-C_2H_5$	$-H$	F	$-CH_3$	719–726
Bacterio-chlorophyll g	$-\overset{\displaystyle H}{\underset{}{C}}=CH_2$	$-CH_3$	$-C_2H_5$	$-CH_3$	$-\overset{\displaystyle O}{\underset{\|\|}{C}}-O-CH_3$	F	$-H$	670, 788
Chlorophyll a	$-\overset{\displaystyle H}{\underset{}{C}}=CH_2$	$-CH_3$	$-C_2H_5$	$-CH_3$	$-\overset{\displaystyle O}{\underset{\|\|}{C}}-O-CH_3$	P	$-H$	440, 680
Chlorophyll b	$-\overset{\displaystyle H}{\underset{}{C}}=CH_2$	$-\overset{\displaystyle O}{\underset{\|\|}{C}}-H$	$-C_2H_5$	$-CH_3$	$-\overset{\displaystyle O}{\underset{\|\|}{C}}-O-CH_3$	P	$-H$	645

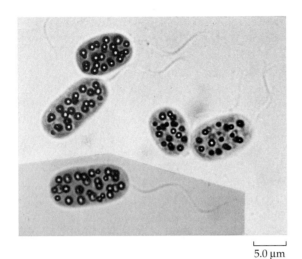

Figure 21.2 Purple sulfur bacterium
Phase photomicrograph of the purple sulfur bacterium *Chromatium okenii*. Note the polar flagellum (which is actually a tuft) and the internal sulfur granules. Courtesy of N. Pfennig.

5.0 μm

(A)

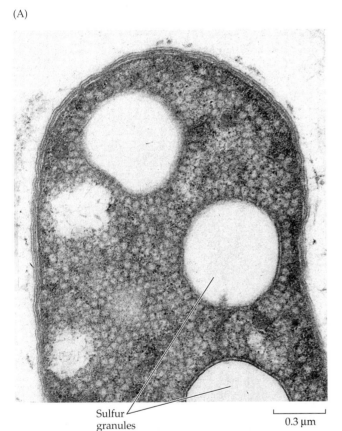

Sulfur granules

0.3 μm

(B)

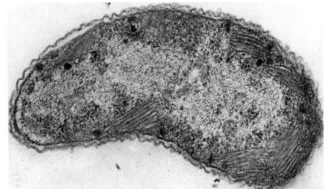

0.25 μm

(C)

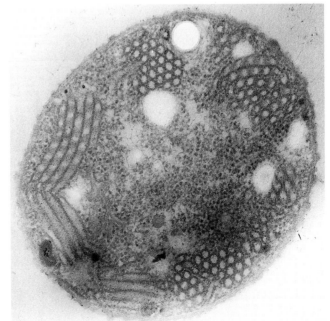

0.25 μm

Figure 21.3 Photosynthetic membranes of purple bacteria
Thin sections showing the various types of intracellular photosynthetic membranes of purple bacteria. (A) Vesicular membranes in *Chromatium vinosum*; (B) lamellar membranes in *Rhodospirillum molischianum*; (C) tubular membranes in *Thiocapsa pfennigii*. (A) and (C) courtesy of S. Watson, J. Waterbury, and C. Remsen, (B) courtesy of G. Drews.

nigii, which contains Bchl *b*. They also contain carotenoid pigments (**Figure 21.4**). These carotenoid pigments actually mask the green to blue bacterial chlorophyll pigments, thereby giving the organisms their characteristic red, orange, and purple colors (**Figure 21.5**). A variety of carotenoid pigments may occur, depending on the species. The pigmentation of an organism is related to the amount of the various carotenoid pigments produced. For example, lycopene, which is brown in color, is the biochemical precursor of spirilloxanthin, which is purple. The amounts of these two pigments and their intermediates and, hence, the color of an organism, will vary depending on the environment and growth state of the organism.

Table 21.3 lists important genera and their characteristics. *Chromatium* is one of the most common genera. This is a genus of large, polarly flagellated rods (Figure 21.2). The genus *Thiospirillum*, as the name implies, contains helical bacteria with polar flagellar tufts (Figure 21.5A). *Thiodictyon* is a genus of net-forming rods (Figure 21.5B), and *Thiopedia* species form flat plates of coccoid cells with gas vacuoles (Figure 21.5C). All purple sulfur bacteria are members of the gamma group of the *Proteobacteria*.

Under anaerobic conditions in light, purple sulfur bacteria grow as photolithoautotrophs using sulfide or elemental sulfur as an electron donor and carbon dioxide as carbon source. They are best known for this physiological activity. In addition, many species can also use organic compounds, such as acetate or pyruvate, as a reductant as shown in Reaction 3, which is typical of the purple nonsulfur bacteria. When organic carbon sources are used, these bacteria still require sulfide, however, because they cannot carry out assimilatory sulfate reduction.

Many species can also use hydrogen gas as a reducing agent for photosynthesis under anaerobic conditions. Virtually all species are nitrogen fixers. Although most are obligate anaerobes, some can grow under microaerobic to aerobic conditions in the dark. When doing this, they obtain energy from the oxidation of inorganic compounds such as hydrogen gas or sulfide or even organic carbon sources in the same manner as some chemolithotrophs and heterotrophs, respectively.

The purple sulfur bacteria are most commonly found in anaerobic environments where sulfide is abundant and light is available (see Chapter 24). Such conditions occur in the anaerobic hypolimnion of many eutrophic lakes, at the surface of marine and freshwater muds, or in sulfur springs (**Figure 21.6**). A characteristic vertical layering of phototrophic bacteria occurs in such environments. The *Cyanobacteria* are found nearest the surface, the purple sulfur bacteria are beneath them, and the green sulfur bacteria occur at the lowest depth. This pattern of vertical stratification is identical to that found in microbial mat communities as discussed earlier in Chapter 1.

The genus *Ectothiorhodospira* is closely related to the purple sulfur bacteria. It resembles the purple sulfur bacteria in its metabolism, pigmentation, and internal membrane systems. Like the purple sulfur bacteria, *Ectothiorhodospira* spp. are members of the gamma

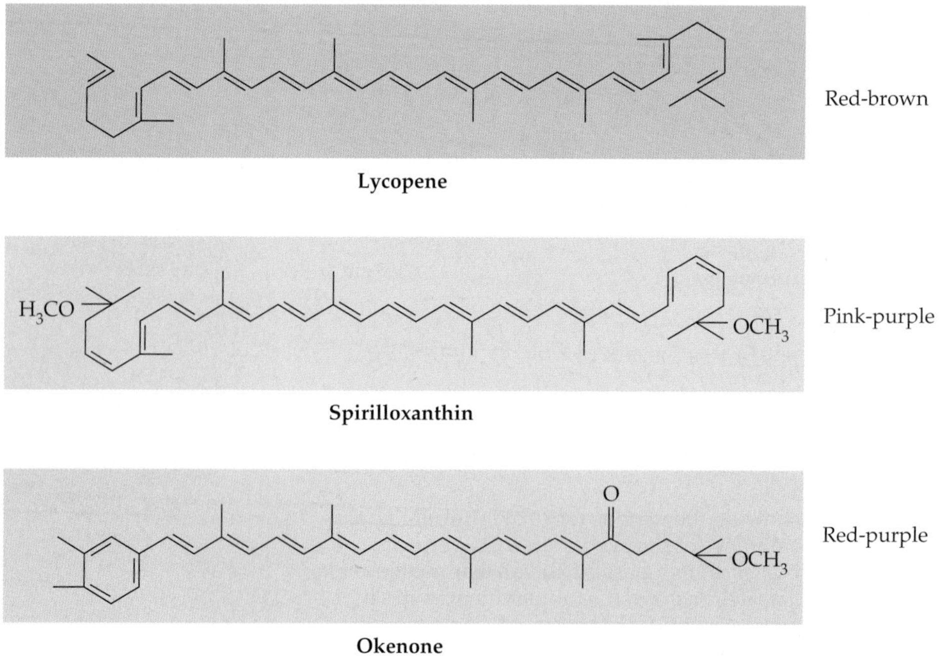

Figure 21.4 Carotenoids of purple bacteria
Structures of common carotenoid pigments of purple bacteria and the colors they impart to cells.

Lycopene — Red-brown

Spirilloxanthin — Pink-purple

Okenone — Red-purple

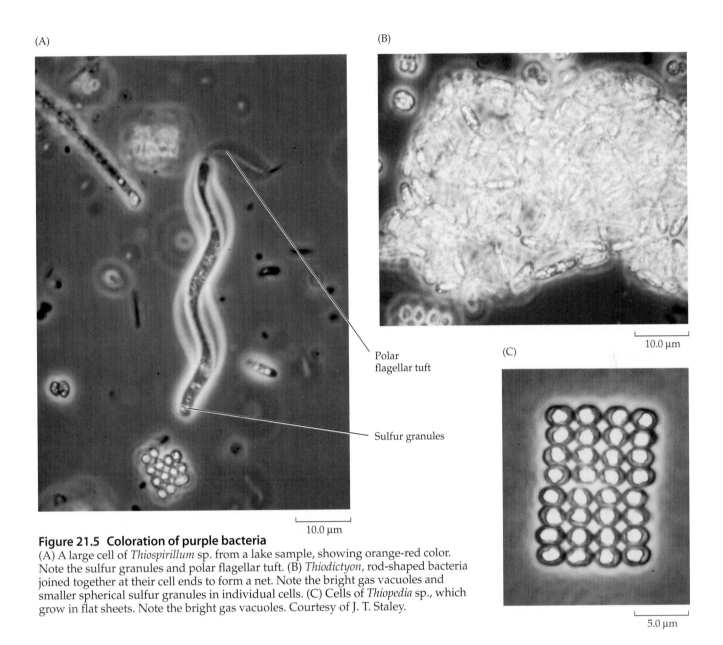

Figure 21.5 Coloration of purple bacteria
(A) A large cell of *Thiospirillum* sp. from a lake sample, showing orange-red color. Note the sulfur granules and polar flagellar tuft. (B) *Thiodictyon*, rod-shaped bacteria joined together at their cell ends to form a net. Note the bright gas vacuoles and smaller spherical sulfur granules in individual cells. (C) Cells of *Thiopedia* sp., which grow in flat sheets. Note the bright gas vacuoles. Courtesy of J. T. Staley.

Table 21.3	Characteristics of genera of purple sulfur bacteria			
Genus	**Flagella**	**Gas Vacuoles**	**Mol % G + C**	**Morphology**
Chromatium	+	−	48–70	Rod
Thiocystis	+	−	61–68	Coccus
Thiospirillum	+	−	45–46	Spirillum
Thiocapsa	−	−	63–70	Coccus
Lamprobacter	+	+	64	Ovoid cells
Lamprocystis	+	+	64	Coccus
Thiodictyon	−	+	65–67	Rod; forms network of cells
Amoebobacter	−	+	64–66	Coccus; single or clusters
Thiopedia	−	+	62–64	Cocci in one plane

(A)

(B)

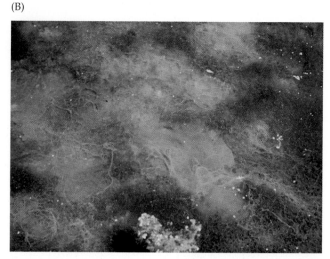

Figure 21.6 Sulfur springs
(A) Yankee Springs, a sulfur spring in central Michigan. The darker green growth on the surface in some areas is due to cyanobacteria. (B) Close-up of the "veils" in Yankee Springs, containing several different species of purple sulfur bacteria. Courtesy of Peter Hirsch.

Proteobacteria. Their cells are small, motile, and vibrioid to rod-shaped. Depending on the species, they live in marine habitats, alkaline soda lakes, or hypersaline lakes. Some species are gas vacuolate. Unlike the purple sulfur bacteria, however, the sulfur granules they produce are deposited *outside* of the cell (hence the name *ectothio-*, which means *extracellular sulfur*).

Purple Nonsulfur Bacteria The purple nonsulfur bacteria share characteristics of both the heterotrophic proteobacteria and the purple sulfur photosynthetic bacteria. Like many bacteria, they can grow aerobically (in the absence of light) as heterotrophs using certain organic substrates. However, if they are grown anaerobically in the light, they carry out photosynthesis, much like the purple sulfur bacteria. They differ from purple sulfur bacteria in that organic carbon sources, such as organic acids (for example, acetate, pyruvate, or lactate) or ethanol, rather than sulfide, are preferred as electron donors for carbon dioxide fixation. Some can also grow as photolithoautotrophs, using hydrogen gas or sulfide as reducing agents in the same manner as the purple sulfur bacteria. However, they do not tolerate high concentrations of sulfide and for this reason are called purple nonsulfur bacteria. When sulfide is available in low concentrations, those that use it as a reductant for photosynthesis do not form sulfur granules inside the cells. Elemental sulfur is either formed outside of the cell, or the sulfide is oxidized completely to sulfate.

The cells of most purple nonsulfur bacteria are smaller than those of the purple sulfur bacteria. Representatives of several common genera, *Rhodobacter, Rhodo-*

pseudomonas, and *Rhodospirillum,* are shown in **Figure 21.7**. They also differ from the purple sulfur bacteria in that individual cells do not appear pigmented when observed with the light microscope. This is because their smaller cells are not thick enough to contain sufficient pigments to reveal their true color.

Like the purple sulfur bacteria, they produce intracytoplasmic photosynthetic membranes when growing photosynthetically in the light under anaerobic conditions. These membranes are extensions of the cell membrane and may be either vesicles or lamellae. They produce more of these membranes and photosynthetic pigments under lower light intensity as a means of compensating for the decrease in available light. These organisms also contain the same carotenoid pigments as produced by the purple sulfur bacteria with characteristic types for each species. Many of the species in this group are motile by flagella. None is known to produce gas vacuoles. There are eight major genera (**Table 21.4**).

Most of the nonsulfur purple bacteria are members of the alpha purple group of *Proteobacteria* and are therefore phylogenetically related to the prosthecate, budding bacteria. Indeed, *Rhodomicrobium* spp. (Figure 21.7D) and some *Rhodopseudomonas* spp. produce prosthecae and divide by budding, confirming their close phylogenetic relatedness to the heterotrophic prosthecate *Proteobacteria*. Three genera, *Rhodocyclus, Rhodovivax,* and *Rhodoferax,* are members of the β-*Proteobacteria.* Unlike the purple sulfur bacteria, none are members of the γ-*Proteobacteria.*

When discussing the purple bacteria, it is important to note that several bacteriochlorophyll *a*–producing genera that superficially resemble purple nonsulfur bacteria have been discovered recently. However, unlike the purple nonsulfur bacteria, these bacteria produce Bchl *a*

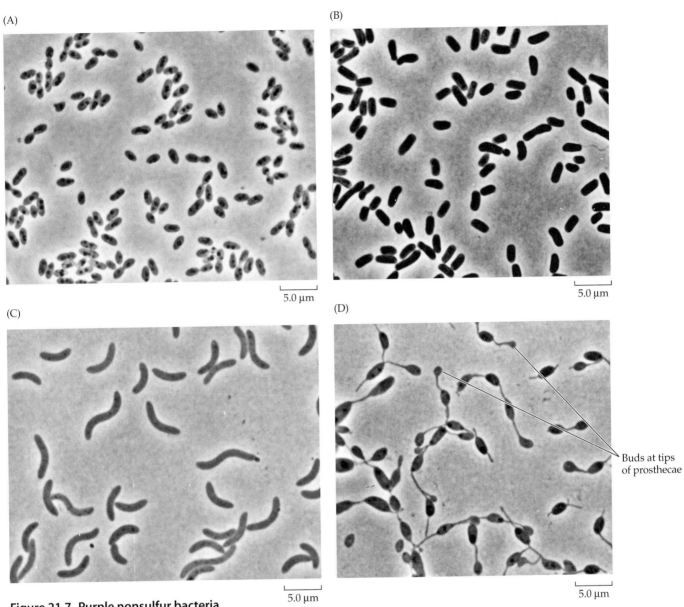

(A)

(B)

(C)

(D)

Buds at tips
of prosthecae

5.0 μm

5.0 μm

5.0 μm

5.0 μm

Figure 21.7 Purple nonsulfur bacteria
Phase photomicrographs of four representative purple nonsulfur bacteria. (A)
Rhodobacter spheroides, (B) *Rhodopseudomonas acidophila*, (C) *Rhodospirillum rubrum*,
and (D) *Rhodomicrobium vannielii*. Courtesy of N. Pfennig.

Table 21.4	Genera of purple nonsulfur bacteria	
Genus	**Mol % G + C**	**Cell Morphology and Features**
Alpha Proteobacteria		
Rhodospirillum	60-66	Helical cells
Rhodobacter	64–73	Ovoid to rod-shaped; neutrophilic
Rhodopila	66	Ovoid to rod-shaped; acidophilic
Rhodopseudomonas	61–72	Ovoid to rod-shaped; budding
Rhodomicrobium	61–64	Prosthecate, budding
Beta Proteobacteria		
Rhodocyclus	4–73	Curved rod to ring-shaped
Rhodoferax	59–61	Curved rod
Rhodovivax	70–72	Curved rod

Research Highlights Box 21.1   **Aerobic Bacteriochlorophyll *a*-Producing *Bacteria***

Phototrophic *Proteobacteria* and *Chlorobi* produce bacteriochlorophyll *a* only under anaerobic conditions. In contrast, several genera of aquatic bacteria have been recently discovered that produce Bchl *a* only under *aerobic* conditions. Many of these bacteria can make ATP by photophosphorylation, but all so far studied require organic compounds as carbon sources. Some of these bacteria have recently been isolated from marine habitats, including the genera *Erythrobacter* and *Roseobacter*, whereas others such as *Porphyrobacter* are from freshwater habitats. All are members of the alpha *Proteobacteria*, as determined by 16S rRNA sequence analyses.

Because of their novel physiology, the role of these bacteria in natural habitats has puzzled the microbiologists who study them. However, the common occurrence and widespread distribution of these bacteria suggest that they comprise an important part of the normal microbiota of aquatic habitats.

while growing under *aerobic* conditions. The role of these bacteria in nature is not yet understood (**Box 21.1**).

Chlorobi

The *Chlorobi*, or green sulfur bacteria, and the *Chloroflexi* (green filamentous bacteria) resemble one another in that they both have the **chlorosome** as their intracellular membrane system for photosynthesis. The chlorosome is a cigar-shaped structure situated next to the cell membrane (**Figure 21.8**). It is here and in the cytoplasmic membrane that the photopigments and photoreaction center of these bacteria are located. These two groups of bacteria are considered individually because they differ in their phylogeny, are found in different habitats, and have their own carbon dioxide fixation reactions.

The *Chlorobi*, or green sulfur bacteria, comprise a separate group of photosynthetic bacteria with an independent phylogeny. They are so named because most species are green due to the presence of characteristic bacteriochlorophylls (Figure 21.1), predominantly Bchl *c*, Bchl *d*, or Bchl *e*. Others appear brown in color, principally due to carotenoid pigments (**Figure 21.9**). In addition, all strains contain small amounts of Bchl *a* as their reaction center chlorophyll. They are termed "sulfur" bacteria because they carry out anoxygenic photosynthesis using H₂S as the reductant for carbon dioxide fixation in the same manner as the purple sulfur bacteria. However, they grow at a reduced redox potential and, because of their electron transport chains, are able to reduce NADPH directly without resorting to reverse electron transport, thereby saving energy.

The *Chlorobi* are small organisms, much smaller than the photosynthetic *Proteobacteria*. Like typical bacteria, their cell diameter is about 0.5 to 1.0 μm. None are motile by flagella, although many have gas vacuoles and use them to regulate their vertical position in stratified habitats (see Chapters 4 and 25). One genus, *Chloroherpeton*, moves by gliding motility.

Like the purple sulfur bacteria, the green sulfur bacteria use sulfide as an electron donor in photosynthesis and form elemental sulfur as an intermediate in this process. However, the sulfur granules are deposited *outside* the cells in the same manner as in *Ectothiorhodospira* (**Figure 21.10**). They can continue to oxidize the sulfur granules and ultimately produce sulfate. Therefore, the overall photosynthetic reactions are identical to that of the purple sulfur bacteria shown previously (Reactions 1 and 2). However, these bacteria do not use the Calvin-Benson cycle for carbon dioxide fixation. Instead, they use the reductive tricarboxylic acid cycle (Figure 10.8A). Also, unlike the purple bacteria, the green bacteria are strict anaerobes. Although most use carbon dioxide as their sole carbon source, some can also use simple

Proteobacteria
Chlorobi
Bacteroidites
Fibrobacteres
Spirochaetes
Cyanobacteria
Fusobacteria
Acidobacteria
Dictyoglomus
Actinobacteria
Firmicutes
Deinococcus-Thermus
Thermomicrobia
Chloroflexi
Deferribacteres
Chrysiogenetes
Chlamydiae
Verrucomicrobia
Planctomycetes
Nitrospirae
Thermodesulfobacteria
Thermotogae
Aquificae

Figure 21.8 *Chlorobium* sp., a green sulfur bacterium

Thin section through a green sulfur bacterium, showing the oval chlorosomes just inside the cell membrane. © T. J. Beveridge/Biological Photo Sevice.

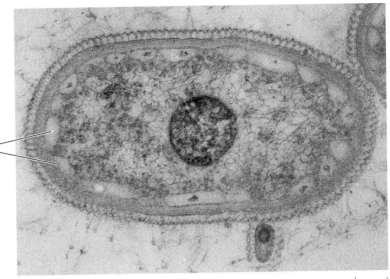

Chlorosomes

0.15 μm

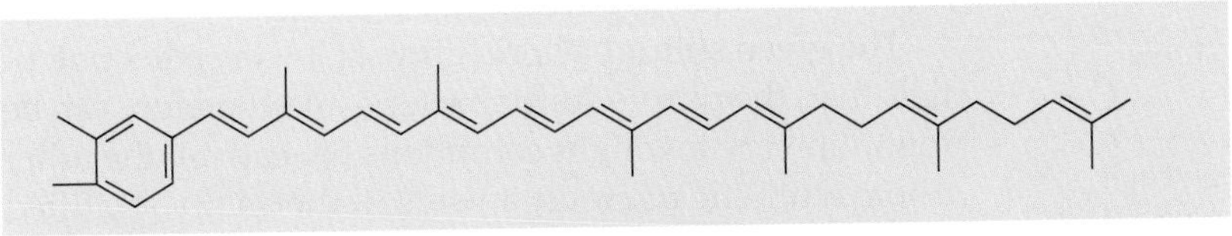

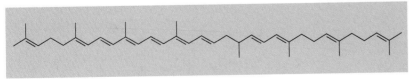

Green

Chlorobactene

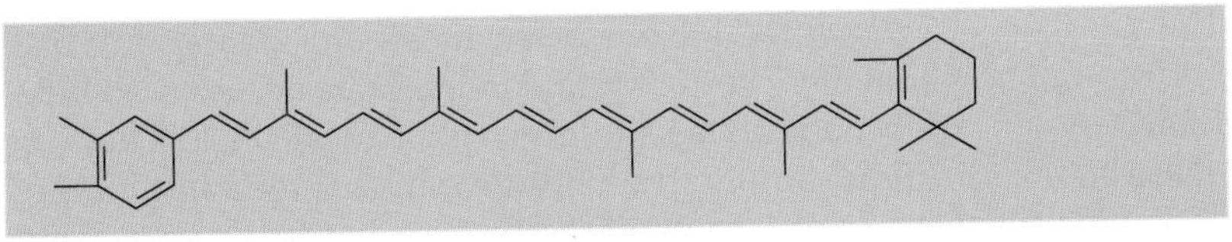

Brown

β-Isorenieratene

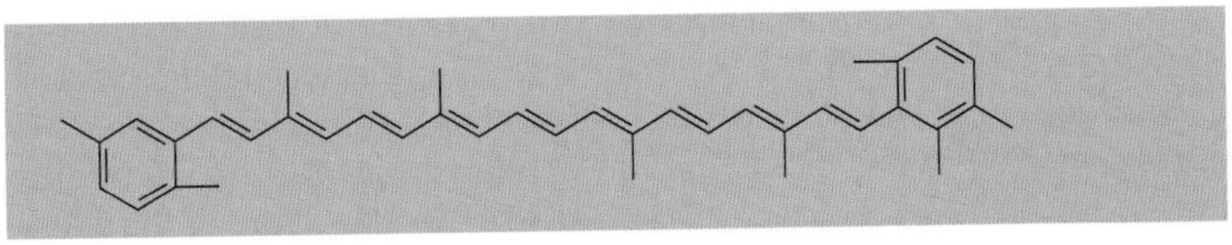

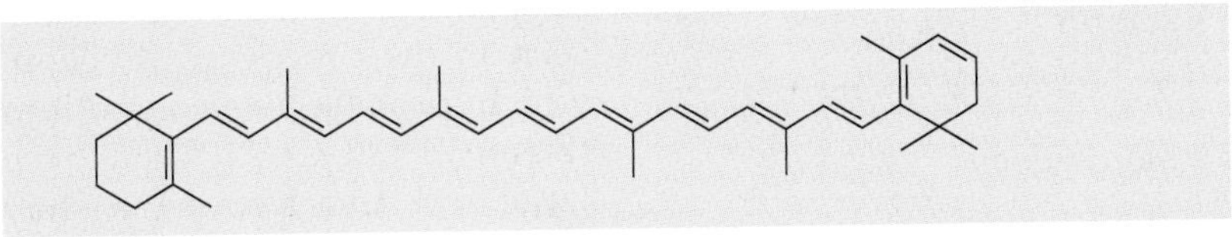

Brown

Isorenieratene

Orange

β-Carotene

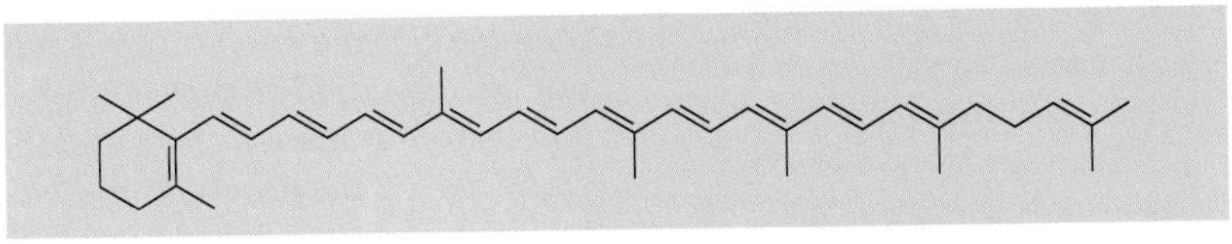

Green

γ-Carotene

Red-brown

Neurosporene

Figure 21.9 Carotenoids of green sulfur bacteria

Structures of carotenoid pigments of green bacteria and heliobacteria and the colors they impart to cells.

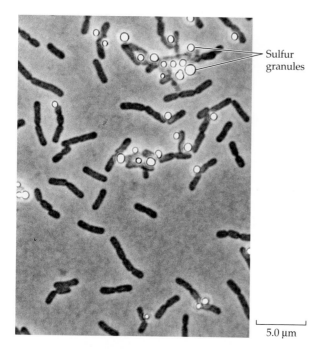

Sulfur granules

5.0 μm

Figure 21.10 *Chlorobium*
Phase contrast micrograph of a pure culture of *Chlorobium limicola*, showing external deposition of sulfur granules, which appear as bright spheres. Courtesy of N. Pfennig.

organic acids in the presence of sulfide and carbon dioxide. Again, as with the purple sulfur bacteria, these bacteria must have sulfide when using organic compounds, because they cannot carry out assimilatory sulfate reduction.

Several genera of *Chlorobi* are known, each with its own characteristic morphology (**Table 21.5**). *Chlorobium* is a common genus of unicellular rods whose species are

found in many anaerobic, sulfide-rich habitats (Figure 21.10). *Pelodictyon* is also very common. It is a genus of rod-shaped, gas vacuolate bacteria whose cells divide to produce a netlike structure similar to the purple sulfur genus, *Thiodictyon*. Some species appear green in color (**Figure 21.11A**), whereas others appear brown (Figure 21.11B) due to differences in pigment composition. *Ancalochloris* spp., which have not yet been isolated in pure culture, resemble *Pelodictyon* in that they form netlike microcolonies, but also produce prosthecate cells (Figures 21.11). These bacteria are found in the plankton of the anoxic hypolimnion of lakes, in freshwater and marine muds, and in marine or saline anoxic habitats. *Prosthecochloris*, as the name implies, is prosthecate and forms nets and is therefore similar to *Ancalochloris*, but it is found in saline habitats and does not produce gas vacuoles. The genus *Chloroherpeton* contains rod-shaped gliding bacteria with gas vacuoles and is found in marine habitats (**Figure 21.12**).

The green sulfur bacteria grow at lower redox potentials than the purple sulfur bacteria and therefore can be found in very low light conditions beneath all the other photosynthetic microorganisms in thermally stratified lakes and mat communities.

The consortium species, such as "*Chlorochromatium aggregatum*," are one of the most unusual forms of bacterial life. They are found in the same anaerobic sulfide-rich habitats as the green sulfur bacteria. They are unique among the phototrophic bacteria in that they consist of a symbiotic association between two separate organisms. For this reason, they cannot be officially named as a species, but instead quotation marks are used to designate this association between these two different species. They appear as multicellular aggregates in the anoxic, sulfide-rich lower depths of lakes (**Figure 21.13A**). At the center of the aggregate, or microcolony, is a large, nonpigmented motile rod, usually seen in the process of division, which propels the consortium (Figure 21.13B–C). Attached to the large rod and forming a shell around it are numerous rod-shaped green sulfur bacteria. Members of the consortium have not been studied in pure culture, so we do not have a clear understanding of the interactions between the two species. However, it appears that the large rod is an anaerobic heterotroph that obtains organic nutrients excreted from the green sulfur bacteria, and the green sulfur bacteria gain motility from the large rod.

Table 21.5	Important genera of green bacteria		
Group/Genus	**Mol % G + C**	**Gas Vacuoles**	**Morphology**
Chlorobi (Green Sulfur Genera)			
Chlorobium	49–58	–	Single cells, rods
Prosthecochloris	50–56	–	Prosthecate; marine or saline habitats
Pelodictyon	48–58	+	Rods in netlike clusters
Ancalochloris	Unknown	+	Prosthecate; freshwater
Chloroherpeton	45–48	+	Long rods; gliding
Consortium	Unknown	+/–	Microcolonial aggregates of two separate species
Chloroflexi (Green Filamentous Genera)			
Chloroflexus	55	–	Gliding; moderate thermophile
Heliothrix	Unknown	–	Gliding; moderate thermophile
Chloronema	Unknown	+	Gliding; mesophile

(A)

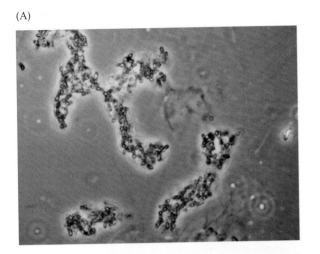

(B)

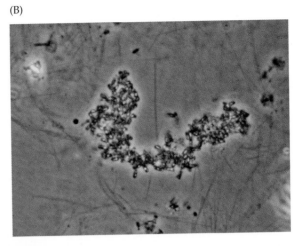

(C)

Prosthecae

Gas vacuoles

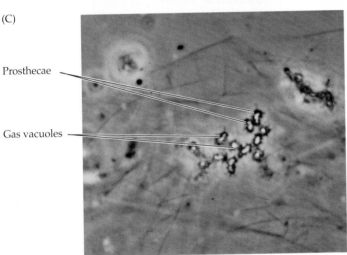

Figure 21.11 *Pelodictyon*
(A) Cells of *Pelodictyon clathratiforme*, showing net formation and small individual cells with gas vacuoles. Note the green color. (B) *Pelodictyon phaeum*, showing the brownish color of cells in the net. (C) *Ancalochloris*, a net-forming prosthecate bacterium found in lakes. Note the prosthecae and gas vacuoles. As yet, no *Ancalochloris* spp. have been isolated in pure culture. Courtesy of J. T. Staley.

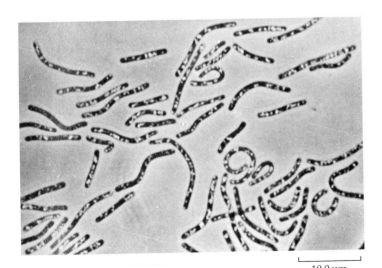

10.0 μm

Figure 21.12 *Chloroherpeton*
Phase photomicrograph of *Chloroherpeton* cells. Note the gas vacuoles. Courtesy of J. Gibson and J. Waterbury.

Chloroflexi

The *Chloroflexi* resemble the green sulfur bacteria only in that they contain chlorosomes and are phototrophic. They resemble the green sulfur genus *Chloroherpeton* most closely in that they move by gliding motility. However they are filamentous and multicellular organisms that evolved as a separate phylogenetic group. Three genera are currently known, including *Chloroflexus*, *Heliothrix*, and *Oscillochloris*. The latter two genera have not been isolated in pure culture. In addition, this phylum contains a genus of heterotrophic gliding bacteria, *Herpetosiphon*.

The most thoroughly studied genus is *Chloroflexus*. This long, thin (0.5 to 1.0 μm diameter) filamentous gliding bacterium (**Figure 21.14**) was first isolated from hot springs where it forms distinct layers in mats; however, some strains have also been found in mesophilic habitats. The temperature range for growth of the thermophilic strains is 45°C to 70°C.

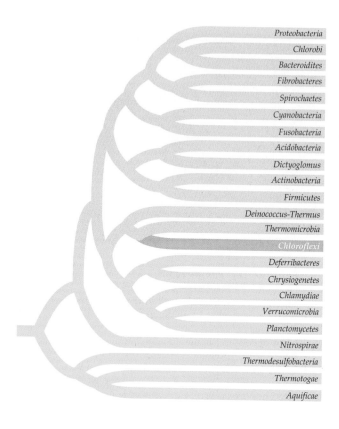

The nutrition of *Chloroflexus aurantiacus* is complex. It grows best as a photoheterotroph, anaerobically in the light. It utilizes a variety of organic carbon sources including sugars, amino acids, and organic acids during photoheterotrophic growth. *Chloroflexus aurantiacus* can also grow as a chemoheterotroph aerobically in the dark. Finally, it grows as a photoautotroph anaerobically in the light, using sulfide or hydrogen gas as an electron donor.

It is interesting to note that *C. aurantiacus* uses neither the Calvin-Benson cycle nor the reductive tricarboxylic acid cycle for carbon dioxide fixation. The best evidence available suggests that instead it uses a unique pathway, the hydroxypropionate pathway, for carbon dioxide fixation (see Figure 10.8B).

Like *Chloroflexus*, *Heliothrix* is a genus of filamentous gliding organisms that grow in alkaline hot springs at temperatures from 35°C to 56°C. It uses Bchl *a* as its primary photosynthetic pigment but is colored orange due to the large amount of carotenoid pigments produced. It has not been isolated in pure culture. *Chloronema* has not been isolated in pure culture either. It is a filamentous, gas vacuolate genus with chlorosomes and a sheath and is found in sulfide-rich lake habitats.

Herpetosiphon is a little-studied genus of filamentous gliding heterotrophic bacteria that produces a sheath and is mentioned here only because they are shown to be closely related to *Cloroflexus* by 16S rRNA sequence analysis. Thus, as in many bacterial phylogenetic groups,

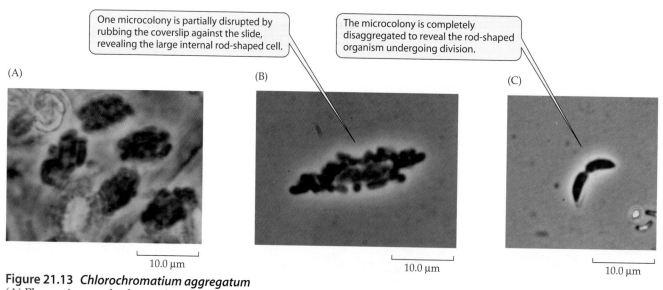

Figure 21.13 *Chlorochromatium aggregatum*
(A) Phase micrograph of several microcolonies of the consortium species *"Chlorochromatium aggregatum"* in a lake sample. (B) One microcolony has been partially disrupted by rubbing the coverslip against the slide. (C) After further treatment, the internal rod-shaped organism is revealed. Note the flagellar tuft. Courtesy of J. T. Staley.

(A)

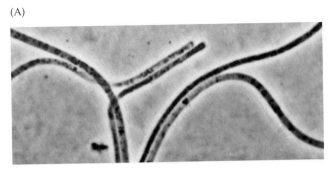

Figure 21.14 *Chloroflexus*
Filaments of *Chloroflexus aurantiacus* as seen by (A) light microscopy and (B) electron microscopy. Note the chlorosomes in
the electron micrograph. (A) courtesy of R. Castenholz and
(B) courtesy of M. Broch-Due.

(B)   Chlorosomes

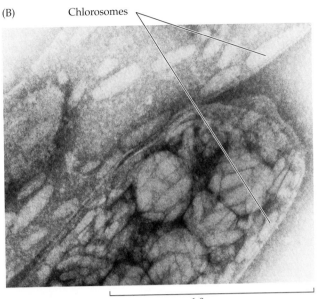

1.0 μm

both phototrophic and heterotrophic members occur.
Herpetosiphon spp. are aerobic heterotrophs that live on a
variety of organic carbon sources. Some species degrade
cellulose and chitin. Both marine and freshwater species
occur.

Firmicutes

The heliobacteria are the most recently discovered photosynthetic bacteria. They are unique because they comprise the first group of gram-positive phototrophic
prokaryotes and are therefore discussed in Chapter 20.

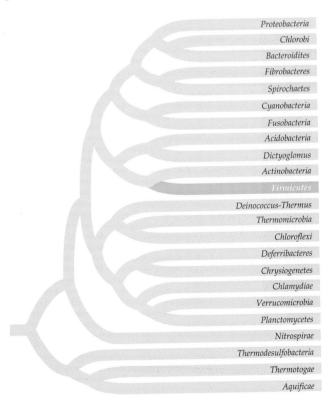

Proteobacteria
Chlorobi
Bacteroidites
Fibrobacteres
Spirochaetes
Cyanobacteria
Fusobacteria
Acidobacteria
Dictyoglomus
Actinobacteria
Firmicutes
Deinococcus-Thermus
Thermomicrobia
Chloroflexi
Deferribacteres
Chrysiogenetes
Chlamydiae
Verrucomicrobia
Planctomycetes
Nitrospirae
Thermodesulfobacteria
Thermotogae
Aquificae

The genera *Heliobacterium* and *Heliobacillus* appear to be
closely related to *Clostridium*, and some strains even produce true endospores that are rich in calcium and dipicolinic acid. They are all photoheterotrophic, requiring
an organic carbon source for growth, and in this respect
resemble the photoheterotrophic purple nonsulfur bacteria. Acceptable carbon sources include organic acids
such as acetate or pyruvate.

They are also unique in that their bacteriochlorophyll,
Bchl *g*, is not found in other organisms and most closely
resembles chlorophyll *a* (Figure 21.1B). Photosynthesis in the heliobacteria is reminiscent of the *Chlorobi*
in that the primary reduced electron acceptor is sufficiently reduced (–0.5 V) that NAD$^+$ can be reduced
directly without the use of reverse electron flow, as
required by the purple bacteria and green filamentous
bacteria. Cells are red-brown in color due to their characteristic carotenoid pigment, neurosporene (Figure
21.9).

The heliobacteria are found in alkaline soil environments. Their distribution and function in soil habitats is
not yet well understood.

Cyanobacteria

This group of photosynthetic bacteria differs from all
others in that they carry out **oxygenic photosynthesis**
in which oxygen is evolved:

$$(4)\ CO_2 + H_2O \rightarrow (CH_2O)_n + O_2$$

Oxygen is derived from water by "water splitting," a
process characteristic of all oxygenic photosynthetic
organisms including algae and plants. All oxygenic pho-

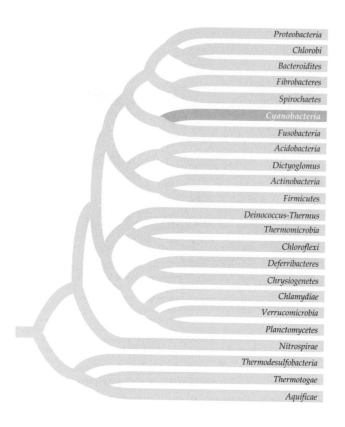

Proteobacteria
Chlorobi
Bacteroidites
Fibrobacteres
Spirochaetes
Cyanobacteria
Fusobacteria
Acidobacteria
Dictyoglomus
Actinobacteria
Firmicutes
Deinococcus-Thermus
Thermomicrobia
Chloroflexi
Deferribacteres
Chrysiogenetes
Chlamydiae
Verrucomicrobia
Planctomycetes
Nitrospirae
Thermodesulfobacteria
Thermotogae
Aquificae

tosynthetic organisms have photosystem II and chlorophyll *a* (Chapter 9). It is in photosystem II that the water-splitting reaction occurs, resulting in oxygen formation. Recent evidence suggests that the *Cyanobacteria* evolved from two anoxygenic photosynthetic groups, the *Firmicutes* and the *Chlorobi* (Chapter 9). Thus, their photosystems I and II are thought to be descended from the photosystems of each of these two groups.

The *Cyanobacteria* were traditionally classified with the algae as the blue-green "algae" because they have chlorophyll *a* and carry out oxygenic photosynthesis. However, because it was discovered by electron microscopy that their cell structure is truly prokaryotic (**Figure 21.15**), they have been claimed by bacteriologists and are now named by the bacteriological code and classified accordingly. Furthermore, phylogenetic analyses of their 16S rRNA indicate they that are a separate line of descent of the *Bacteria*. At this time, there is a dual system of classification of *Cyanobacteria*, because they are also named and classified by botanists using the botanical code.

Structure and Physiology The fine structure of *Cyanobacteria* is typical of other gram-negative bacteria in that they have a multilayered cell wall containing peptidoglycan and an outer membrane. However, not only do they have these layers, but their cell walls usually contain other layers as well (**Figure 21.16**).

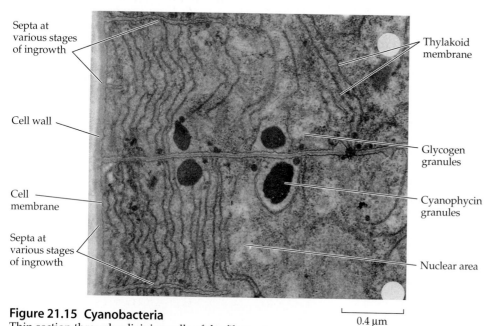

Septa at various stages of ingrowth

Cell wall

Cell membrane

Septa at various stages of ingrowth

Thylakoid membrane

Glycogen granules

Cyanophycin granules

Nuclear area

0.4 μm

Figure 21.15 Cyanobacteria
Thin section through adjoining cells of the filamentous cyanobacterium *Symploca muscorum*, an oscillatorian species, showing the prokaryotic nature of the cell (note the nuclear area) and the characteristic structures of cyanobacteria. The thylakoid lamellar membranes extend throughout the cell. Courtesy of H. S. Pankratz and C. C. Bowen.

The *Cyanobacteria* fix carbon dioxide via the Calvin-Benson cycle. One of the key enzymes in this cycle is ribulose-bisphosphate carboxylase (RuBisCo). Some *Cyanobacteria* store this enzyme in structures in the cell in carboxysomes, a feature shared by some of the thiobacilli discussed in Chapter 19. Most *Cyanobacteria* do not require vitamins and do not utilize organic compounds and are therefore excellent examples of photolithoautotrophs. Some species, however, can photoassimilate simple organic compounds such as acetate. They do not use these as energy sources, but simply as carbon sources for growth, thereby relying on photophosphorylation as their sole means of formation of ATP.

One of the recent exciting discoveries regarding *Cyanobacteria* is that some of them can perform anoxygenic photosynthesis as well as oxygenic photosynthesis. When they carry out anoxygenic photosynthesis, sulfide, but not sulfur, serves as the electron donor. Sulfur granules are, in fact, formed outside the cells. This discovery has major implications for evolution, suggesting that the *Cyanobacteria* may have evolved early on from anoxygenic photosynthetic bacteria, which is consistent with their purported descent from green sulfur bacteria.

One of the unique features of *Cyanobacteria* is that they can store a compound called **cyanophycin** (Figure 21.15). This is unique in that it a polymer of aspartic acid, with each residue of aspartic acid containing a side-group of arginine:

$$\begin{array}{c} \text{-asp-asp-asp-asp-} \\ | \quad | \quad | \quad | \\ \text{arg arg arg arg} \end{array}$$

Cyanophycin is unique, as it is the only nitrogen-containing storage granule of prokaryotic organisms. Because many environments contain low concentrations of nitrogen, it is often a limiting nutrient; thus, cyanophycin is a useful storage product. Also, cyanophycin can be used as an energy source during breakdown. Arginine dihydrolase leads to the synthesis of ATP:

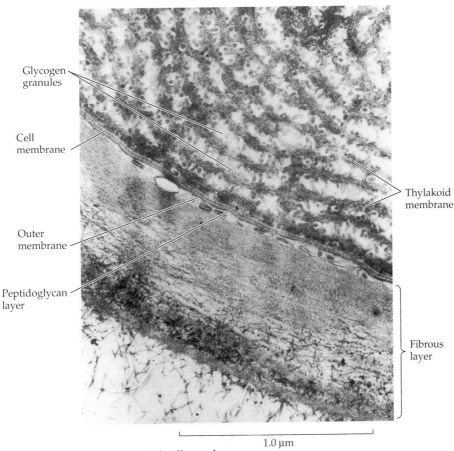

Glycogen granules

Cell membrane

Outer membrane

Peptidoglycan layer

Thylakoid membrane

Fibrous layer

1.0 μm

Figure 21.16 Cyanobacterial cell envelope
Thin section of *Dermocarpa* sp., showing the cell envelope layer. In addition to the cell membrane, peptidoglycan layer, and outer membrane typical of gram-negative bacteria, the organism has an additional outer layer of thick fibrous material. Courtesy of J. Waterbury and R. Stanier.

In addition to chlorophyll *a* and cytochromes, the *Cyanobacteria* have characteristic, unique pigments called **phycobilins,** including **allophycocyanin, phycocyanin,** and **phycoerythrin** (Figure 9.10). The phycocyanins give the *Cyanobacteria* their characteristic blue-green color. However, not all *Cyanobacteria* are blue-green in color. Some are red due to another phycobilin pigment called phycoerythrin. These pigments are covalently bonded to proteins in complexes that are called **phycobiliproteins**—phycocyanobilins for the blue pigments or phycoerthyrobilins for the red pigments.

The photosynthetic pigments of *Cyanobacteria* are located in lamellar membranes called **thylakoids** (Figures 21.15 and 21.16). These intracellular membranes are studded with small structures called **phycobilisomes,** which contain the phycobiliproteins (see Figures 9.8 and 9.9). The thylakoid complex serves as the photosynthetic reaction center for photosynthesis in *Cyanobacteria*. Thus, both the physical process of light harvesting as well as the chemical process of electron transfer occurs here.

$$\text{arginine} + \text{ADP} + P_i + H_2O \rightarrow$$
$$\text{ornithine} + \text{ATP} + NH_3 + CO_2$$

Ecological and Environmental Significance of Cyanobacteria The *Cyanobacteria* are very important ecologically. Perhaps their major significance is that they are important contributors to carbon dioxide fixation and, hence, primary production in aquatic environments. Whereas the anoxygenic purple and green photosynthetic bacteria contribute at best about 10% of the annual primary production of the habitats in which they

occur, the *Cyanobacteria* and prochlorophytes are responsible for up to half. Moreover, the *Cyanobacteria* are very common in all aquatic habitats, whereas the anoxygenic photosynthetic bacteria are restricted to special anaerobic habitats that are rich in sulfide. Thus, in the oceans where algae and *Cyanobacteria* are the dominant primary producers, up to almost half of the primary production is due to *Cyanobacteria* and prochlorophytes. Similarly high rates of primary production are attributable to them in freshwater habitats (**Box 21.2**).

Cyanobacteria are also important members of contemporary mat communities. They form the uppermost

---

Research Highlights Box 21.2   Cellular Absorption Spectra of Photosynthetic Bacterial Groups

As mentioned in the text, the various photosynthetic bacterial groups occupy the same aquatic habitats as mat communities in springs and intertidal zones or the water columns of lakes, ponds, and marine ecosystems. The pattern in the vertical distribution of these photosynthetic groups is always the same: the *Cyanobacteria* reside at the surface, nearest the oxygen, the purple bacteria lie beneath them, and the green bacteria occupy the deepest layer, which are farthest from the oxygen and closest to the sulfide. Each of these groups has its own characteristic bacteriochlorophyll or chlorophyll pigments and accessory pigments, such as the phycobiliproteins and carotenoids, and each of these pigments has its own characteristic absorption spectrum (see figure). Each group absorbs light of a different wavelength; thus, each is able to obtain some of the light radiating from the sun because the overlying groups filter out only some of the wavelengths.

It should be noted that the photosynthetic bacteria that reside below the *Cyanobacteria* in these vertically stratified communities receive less energetically favorable light for photosynthesis because of its longer wavelength. Furthermore, if these photosynthetic groups are

water column communities, little light remains available for the purple and green sulfur bacteria that

live in the deepest zones because much of the light is absorbed by the overlying water.

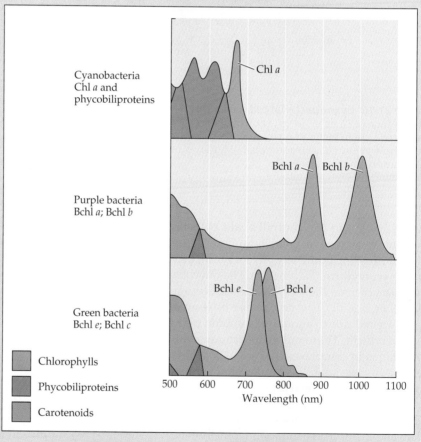

Absorption spectra of bacteriochlorophyll and chlorophyll from various photosynthetic bacterial groups shown in the vertical stratification in which they occur in natural communities. The cyanobacteria are shown at the surface, the purple bacteria underneath, and the green bacteria occupy the lowest layer.

layer in mats where they carry out oxygenic photosynthesis. The mat communities in which they dominate are common in thermal habitats, polar lakes, and hypersaline environments throughout Earth. Some species also grow well on the surface of soil.

Because *Cyanobacteria* require sunlight and oxygen for their metabolism, they grow at the surface of the aquatic habitats and mat communities where they reside. This position is especially high in solar radiation. In order to protect their cells from mutating ultraviolet radiation, many species produce pigments that absorb ultraviolet light. One such pigment is called scytonemin (**Box 21.3**).

Many *Cyanobacteria* are nitrogen fixers. Because nitrogen is a common limiting nutrient in many marine and some freshwater environments, their contribution to the nitrogen budget can be extremely important.

It should be noted, however, that some *Cyanobacteria* carry out undesirable activities in the environment. For example, certain gas vacuolate species are responsible for **nuisance blooms** in freshwater habitats (**Figure 21.17**). Nuisance blooms occur in the midsummer or fall. These gas-vacuolate *Cyanobacteria* float to the surface of the lake and are carried onshore by the wind. They accumulate and subsequently decay onshore, where they cause odors. Under normal conditions, the gas vacuoles allow the cells to float at or near the surface of lakes and ponds where they grow at a location where photosynthesis is most active.

Some species, such as *Microcystis aeruginosa*, may produce toxic compounds that can actually kill animals, including cattle and dogs, that eat them. Humans that ingest contaminated animals in the food chain have also died.

Cyanobacteria can cause odor and off-flavor tastes in drinking waters as well. This is due to the production of geosmin compounds, which are similar chemically to the geosmins produced by some *Actinobacteria*. This problem normally occurs in the late summer or fall

---

Research Highlights Box 21.3

Scytonemin, an Ancient Prokaryotic Sunscreen Compound

All organisms are susceptible to mutation caused by ultraviolet light. Humans can develop skin cancer if exposed to excess solar radiation. Animals produce melanin pigments to absorb ultraviolet radiation, and many plants produce flavonoid compounds. However, to further retard incoming radiation for humans, a variety of chemicals are commercially produced that are used as effective sunscreens to absorb damaging ultraviolet radiation. For example, *p*-aminobenzoic acid (PABA), a precursor in the synthesis of the vitamin folic acid (see Chapter 7), is commonly used in sunscreen and sunblocking lotions to absorb ultraviolet radiation. The reason for selecting these compounds is that PABA and other aromatic compounds are excellent absorbers of ultraviolet light.

Many *Cyanobacteria* produce their own remarkable sunscreen compound called **scytonemin.** Although this substance was named about 150 years ago, only recently has its chemical formula been determined by the *Cyanobacteria* expert Richard Castenholz and his colleagues at the University of Oregon.

Considering its complex aromatic structure, it is not surprising that this compound is an excellent UV blocking agent. Scytonemin is especially effective at absorbing UV-A (325 to 425 nm) but is also effective at absorbing UV-B (280 to 325 nm) and UV-C (maximum absorption at 250 nm).

Scytonemin is deposited in the extracellular sheath of the *Cyanobacteria* that produce it. There it can intercept the UV light before it reaches the cytoplasm and DNA of the cell. It has been estimated that at the normal concentrations of scytonemin found in a trichome, 85% to 90% of the incident UV will be absorbed by this biological sunscreen before it enters the cell.

Because sheathed *Cyanobacteria* are found in early fossil stromatolites, it is likely that scytonemin evolved early on through these *Cyanobacteria*. The UV radiation would have been much more intense earlier in Earth's history than it is now.

Scytonemin

The chemical structure of scytonemin.

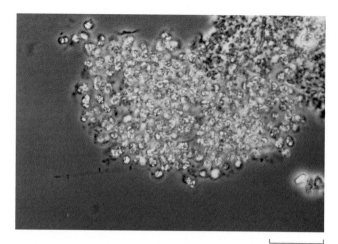

Figure 21.17 Gas vacuolate cyanobacteria
|————————| 10.0 µm

Phase photomicrograph of *Microcystis aeruginosa*. Bright areas are the gas vacuoles. Courtesy of R. W. Castenholz.

Table 21.6	Cyanobacterial Orders
Order	**Distinctive Features**
Chroococcales	Unicellular or multicellular, nonfilamentous
Pleurocapsales	Baeocytes formed
Oscillatoriales	Straight filaments without specialized cells
Nostocales	Straight filaments with heterocysts
Stigonematales	Branching filaments

when these *Cyanobacteria* are common in lakes and reservoirs that may be used for the water supply in municipalities.

Major Taxa of Cyanobacteria Several orders of *Cyanobacteria* have been named and classified on the basis of their morphological characteristics (**Table 21.6**). They range in morphology from unicellular rods and cocci to filamentous types, some of which show branching division. Many species have not yet been isolated in pure culture. This is due to the difficulty that microbiologists have in cultivating them. Part of the problem is that many *Cyanobacteria* produce an external capsule or sheath to which bacteria attach, making isolation especially difficult. Consequently, the cultivation of these bacteria in pure culture is a major area of research for bacteriologists. A brief description of each order and representative groups follows.

Chroococcales are common *Cyanobacteria* that are widely distributed in fresh and marine waters. **Table 21.7** provides a listing of the important genera of the *Chroococcales*, some of which are discussed next.

Synechococcus spp. (**Figure 21.18**), which are unicellular coccoid to rod-shaped *Cyanobacteria,* are among the most important photosynthetic organisms in marine habitats. It has been estimated that they account for approximately 25% of the primary production that occurs in typical marine habitats (see Chapter 24).

The genus *Gloeothece* is remarkable in being able to fix atmospheric nitrogen in the presence of oxygen. Perhaps the capsular material of this organism (**Figure 21.19**) is somehow involved in keeping the

oxygen concentration low within the cells. However, it is known also that some unicellular *Cyanobacteria* carry out nitrogen fixation during the dark period when oxygen is not formed by photosynthesis, and this may account for the ability of this genus to carry out this process.

Another notable genus in this order is *Microcystis*. Members of this genus are gas vacuolate and are one of the principal genera responsible for nuisance water blooms (Figure 21.17). Furthermore, these organisms may produce a toxin that can be deadly to animals that ingest them. *Chaemosiphon*, another genus in this order, is one of the few unicellular *Cyanobacteria* that divides by budding.

Pleurocapsales have an unusual morphology. They grow as unicellular organisms that enlarge in size and undergo internal cell divisions to form **baeocytes,** small spherical reproductive cells that are released when the division cycle is complete. The baeocytes then develop into larger cells and repeat the division cycle. Some members of this order form multicellular filaments, but baeocyte formation also occurs in these organisms.

Oscillatoriales are common filamentous gliding bacteria. Because the cells in the filaments are in very close contact to one another and act in a concerted fashion to effect motility, they are regarded as being truly multicellular. The name for this gliding multicellular filament, which may contain 50 to 100 or more cells, is the **trichome** (**Figure 21.20**A, B, and C). The trichome may divide to form groupings of 5 to 15 cells called **hormogonia,** which are short, gliding filaments used for

Table 21.7	Important genera of *Chroococcales*		
Genus	**Mol % G + C**	**Characteristics**	
Chaemosiphon	46–47	Divide by budding	
Synechococcus	47–56	Rods of freshwater, marine, and hot spring origin	
Gloeothece	40–43	Cocci in common sheath; nitrogen fixer	
Microcystis	45	Cocci; gas vacuolate; may cause toxic water blooms	

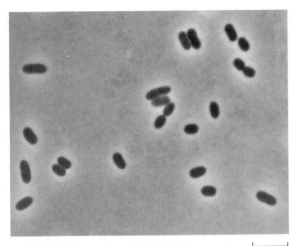

Figure 21.18 *Synechococcus*

5.0 μm

Cells of *Synechococcus*, a common marine
cyanobacterium. Courtesy of J. Waterbury and R. Rippka.

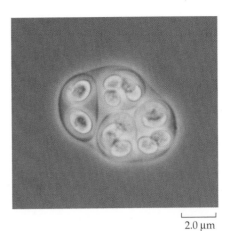

2.0 μm

Figure 21.19 *Gloeothece*
Phase photograph of *Gloeothece* sp., showing the cells within
a sheath. Courtesy of J. T. Staley.

(A)

(B)

10.0 μm

(C)

(D)

10.0 μm

10.0 μm

10.0 μm

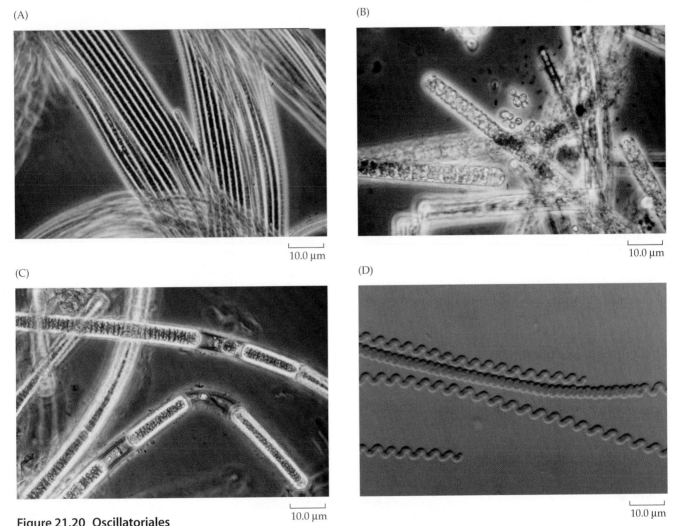

Figure 21.20 Oscillatoriales
(A) *Oscillatoria formosa*, a typical oscillatorian species, showing its filamentous growth habit.
Each multicellular filament is a trichome. (B) Gas vacuolate oscillatorian from a lake sample.
(C) *Lyngbya* sp., showing its dark sheath. (D) *Spirulina*, a helical, filamentous cyanobacterium.
A–C, courtesy of J. T. Staley; D, ©M. Abbey/Visuals Unlimited.

Table 21.8	Common genera of *Oscillatoriales*	
Genus	**Mol % G + C**	**Description**
Oscillatoria	40–50	Straight trichomes; some gas vacuolate; freshwater or hot springs
Trichodesmium	Unknown	Resemble *Oscillatoria*, but marine origin; fix nitrogen
Lyngbya	42–49	Sheathed; resembles *Oscillatoria*
Spirulina	54	Helical trichomes

dispersal and reproduction. These are commonly formed during asexual reproduction in all types of trichome-producing *Cyanobacteria*. A list of the common genera of this order is given in **Table 21.8**.

Oscillatoria spp. (Figure 21.20A and B) are common in freshwater habitats and hot springs and can be found growing on the surface of soils. They vary in pigmentation from light green to almost black. Some are even red due to phycoerythrin pigments. Lake forms may be gas vacuolate (Figure 21.20B). Their marine counterpart is the genus *Trichodesmium*. These bacteria are gas vacuolate and grow in clusters of filaments. *Trichodesmium* spp. are able to carry out nitrogen fixation even though they do not have specialized cells, that is, heterocysts (see *Nostocales*). *Lyngbya* spp. are similar to *Oscillatoria*. However, in addition, they also have a prominent sheath (Figure 21.20C).

Spirulina spp. are unusual in that they have a helical trichome (Figure 21.20D). They are used as a food source by Africans in Chad, and more recently *Spirulina* has been sold as a dietary supplement in health food stores.

Nostocales is an order comprising of an important group of nitrogen-fixing filamentous bacteria (**Table 21.9**). *Anabaena* spp. (**Figure 21.21**) are widely distributed in freshwater and saline habitats. They, like other members of the order *Nostocales*, have specialized cells called

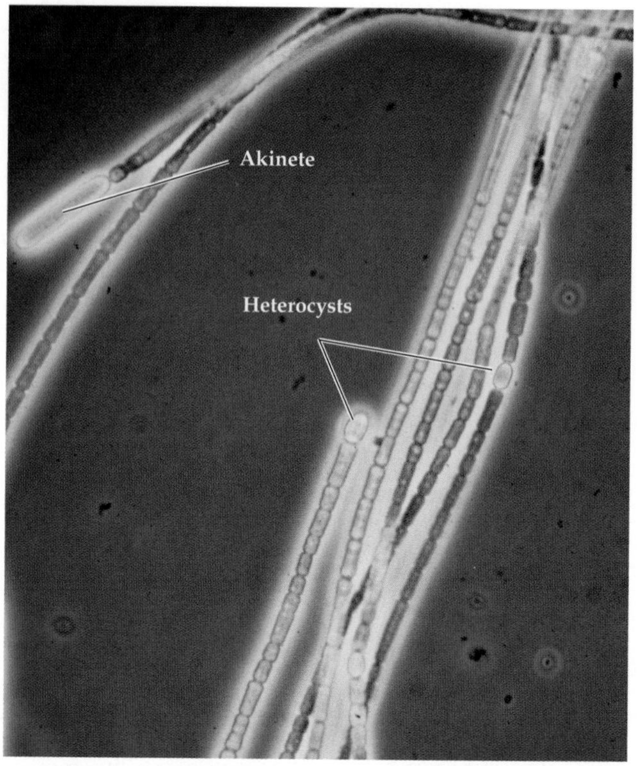

Figure 21.21 *Anabaena*   10.0 μm
Anabaena sp., showing typical filaments with heterocysts and akinetes. Courtesy of J. T. Staley.

heterocysts that are the sites of nitrogen fixation (Figure 21.21). Heterocysts are formed in the filaments from ordinary cells. Their location within the filament is distinctive for the species. Some are located at the end of the filament, whereas others are located within it. When ammonia or other forms of fixed nitrogen are depleted from the environment, heterocyst formation is induced. During the process in which the cell is converted to a heterocyst, the synthesis of phycobilins, the antennae pigments for photosynthesis, is stopped. Therefore, the heterocyst is no longer able to photosynthesize, so the production of oxygen ceases. Because oxygen is inhibitory to nitrogen fixation, this enables the heterocyst to carry out fixation when nitrogenase is synthesized. A considerable amount is known about heterocyst differentiation and the process of nitrogen fixation in *Anabaena* spp. (**Box 21.4**).

Some species produce **akinetes,** which are special cystlike resting cells. These are usually somewhat larger than the normal vegetative cells in filaments (Figure 21.21). They are usually formed after the heterocysts have been produced. Typically, they do not have gas vesicles, even though normal veg-

Table 21.9	Important genera of *Nostocales*	
Genus	**Mol % G + C**	**Distinctive Features**
Anabaena	35–47	Untapered trichomes, most gliding; many gas vacuolate
Aphanizomenon	Unknown	Like *Anabaena*; however, trichomes aggregate to form colonies; gas vacuolate
Nostoc	39–45	Like *Anabaena*, but sheathed
Rivularia	Unknown	Tapering filaments with polar heterocysts
Gleotrichia	Unknown	Tapering filaments with polar heterocysts in microcolonies

The Cyanobacterial Heterocyst

The heterocyst is a specialized cell produced by some filamentous *Cyanobacteria*. The heterocyst is formed from a typical vegetative cell through a developmental process that is induced when the concentration of fixed nitrogen forms such as ammonia and nitrate are depleted from the environment. Once the heterocyst has formed, it is unable to undergo cell division and therefore resembles a resting cell such as a cyst. A differentiation process occurs whereby the heterocyst develops a complex coat of three predominant layers, an outer fibrous layer, a homogeneous layer, and an inner laminated layer, which are thought to restrict the passage of oxygen and other gases, including nitrogen, into the heterocyst. Nitrogen, which is needed by nitrogenase, enters the heterocysts through the ends of the heterocysts, called the microplasmodesmata.

The preheterocyst cell loses its ability to produce phycobilins, the antennae pigments for photosynthesis. This shuts off photosystem II, and therefore, the production of oxygen as well as fixation of carbon dioxide cease. Photosystem I, however, remains intact, allowing the heterocyst to produce ATP by photophosphorylation. This ATP is needed by nitrogenase for nitrogen fixation. The reducing power needed for nitrogen fixation is produced through metabolism of carbohydrates, which are produced by adjacent photosynthesizing cells.

The decrease in oxygen concentration coupled with the decrease in fixed nitrogen available to the organism are signals for the induction of nitrogenase and the process of nitrogen fixation. The ammonia formed by nitrogenase (N_2ase) is carried to adjoining cells through the formation of glutamine as follows: The heterocyst has a high level of the enzyme glutamine synthetase, which carries out the following reaction:

$$\text{glutamate} + NH_3 \rightarrow \text{glutamine}$$

The glutamine (Gln) formed from the heterocyst is converted in the adjoining vegetative cells to glutamate (Glu) by an enzyme called glutamine oxoglutarate amido transferase (GOGAT), which couples glutamine to α-ketoglutarate to form two molecules of glutamate. Part of the glutamate is recycled to the heterocyst (see figure), and part is transferred to adjacent cells as a form of utilizable fixed nitrogen. Thus, organic, fixed nitrogen is produced that can be used by the vegetative cells in anabolic processes.

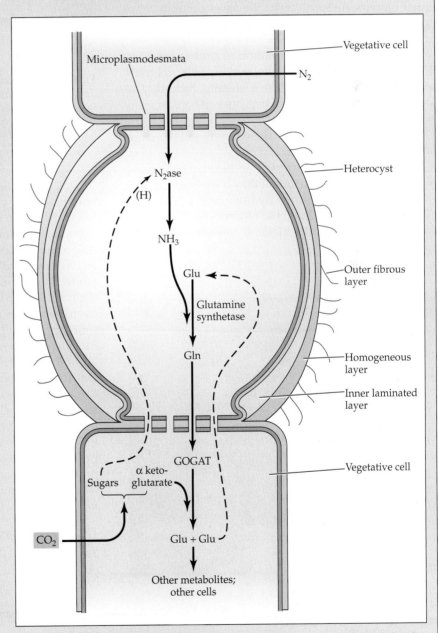

Diagram of the heterocyst showing its structure and the manner in which fixed nitrogen is formed and transferred from the heterocyst.

etative cells may have them. They can survive better under conditions of starvation and desiccation than the vegetative cells.

One species of *Anabaena, A. azollae*, grows in a symbiotic association with the small water fern *Azolla*. The bacterium is harbored in a special sac of the fern (Chapter 25). The water fern obtains fixed nitrogen from *A. azollae*, and in return, the bacterium has a protected place where it can grow. Water ferns are widely distributed in aquatic habitats such as rice paddies, and this association is important in ensuring the fertility of such habitats.

Aphanizomenon flos-aquae as well as *Anabaena flos-aquae* are common gas vacuolate organisms that cause water blooms. *Aphanizomenon flos-aquae* occur in characteristic colonies or "rafts" of trichomes that glide back and forth against one another.

Nostoc species are also common in aquatic habitats. They closely resemble *Anabaena* spp.; however, their filaments are enclosed within a sheath. Indeed, the sheathed structures can be as large as golf balls or even baseballs in habitats where they occur (**Figure 21.22**). Like *Anabaena* spp., they have heterocysts and therefore carry out nitrogen fixation.

The *Rivularia* group differs from the other heterocystous filamentous genera in that they have tapering filaments. However, as shown by the genus *Gloeotrichia* (**Figure 21.23**), they also have heterocysts, which are located at the base of the filament and are responsible for nitrogen fixation.

Figure 21.22 *Nostoc*
A large colony of *Nostoc pruniforme* from Mare's Egg Spring in Klamath County, Oregon. Note the new, smaller colonies forming on its sides. Courtesy of W. K. Dodds and R. W. Castenholz.

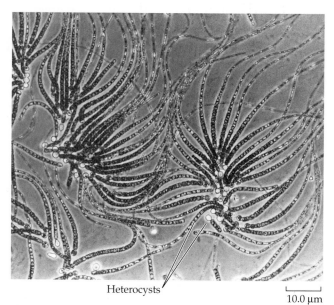

Heterocysts   10.0 μm

Figure 21.23 *Gloeotrichia*
Tapering filaments are typical of rivularian cyanobacteria such as *Gloeotrichia echinulata*. Note the heterocysts at the base of each tapering filament and the gas vacuoles. Courtesy of J. T. Staley.

Stigonematales is an order of *Cyanobacteria* that contains branching clusters of filamentous organisms. They have the most complex morphology of any of the *Cyanobacteria*. Some produce heterocysts as well as hormogonia and akinetes.

Prochloron and *Prochlorothrix* are *Cyanobacteria* that are referred to as prochlorophytes. They differ from typical *Cyanobacteria* in that they possess chlorophyll *b* as well as chlorophyll *a*, and they also lack phycobilins. At one time they were postulated as a possible evolutionary intermediate between *Cyanobacteria* and some of the algae. However, 16S rRNA phylogenetic analyses indicate they are members of the *Cyanobacteria*. The genus *Prochloron* grows in symbiotic association with marine animals called didemnids and has not been cultivated in pure culture yet. *Prochlorothrix* (**Figure 21.24**) is a free-living marine bacterium. Like the *Cyanobacteria*, prochlorophytes are significant contributors to the process of carbon dioxide fixation in the open oceans, where it is estimated they fix about 25% of the carbon in marine habitats.

The Chloroplast

Studies of 16S rRNA isolated from chloroplasts from plants and algae indicate they have a common origin. Furthermore, their ancestors all come from the *Cyanobacteria* branch of the *Bacteria,* and they would be placed with this group of the prokaryotes if they were free-living microorganisms. These results strongly support the endosymbiotic theory of evolution and argue

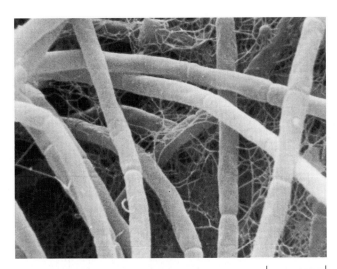

Figure 21.24 *Prochlorothrix*
Scanning electron micrograph of *Prochlorothrix*, a marine prochlorophyte. Courtesy of W. Krumbein.

3.0 μm

that the progenitor of the chloroplast was a cyanobacterium that evolved into a close association with eukaryotic cells early in evolution (see Chapters 1 and 17).

SUMMARY

▶ There are several major phylogenetic phyla that contain photosynthetic *Bacteria*, including the photosynthetic *Proteobacteria*, the *Chlorobi*, the *Chloroflexi*, the photosynthetic *Firmicutes* (heliobacteria), and the *Cyanobacteria*.

▶ The **phototrophic *Proteobacteria*,** which comprise two groups, the **purple sulfur** bacteria and the **purple nonsulfur** bacteria, carry out **anoxygenic photosynthesis** in which oxygen is not produced, and either reduced sulfur compounds or organic compounds are used, respectively, as electron donors for carbon dioxide fixation, which occurs anaerobically. The purple bacteria, like other *Proteobacteria*, fix carbon dioxide by the Calvin-Benson cycle. The coloration of the purple bacteria is determined by **carotenoid pigments** that mask the **bacteriochlorophyll *a*** that is found in the reaction center where **photophosphorylation** occurs.

▶ The *Chlorobi* and the *Chloroflexi* have a special structure called the **chlorosome,** which contains the photosynthetic pigments and reaction center. In addition to Bchl *a*, they produce other green or brown bacteriochlorophylls, Bchl *c*, Bchl *d*, and Bchl *e*. The *Chlorobi* use the **reductive TCA cycle** for carbon dioxide fixation. The **consortium species** consist of two separate organisms, a green sulfur bacterium and a heterotroph, living in close symbiotic association. *Chloroflexus aurantiaca*, a green filamentous member of the

Chloroflexi, has its own pathway for carbon dioxide fixation, termed the **hydroxypropionate pathway.**

▶ The **heliobacteria** are members of the *Firmicutes* and are photoheterotrophic. Some form endospores.

▶ The oxygenic photosynthetic bacteria are members of the *Cyanobacteria*. *Cyanobacteria* have photosystem I and photosystem II and carry out the water-splitting reaction to form oxygen from water. The accessory photopigments of the *Cyanobacteria* are phycobilins, **phycocyanin** and **phycoerythrin**. These are located in phycobilisomes, which are structures on the photosynthetic lamellae termed **thylakoids.** Some *Cyanobacteria* store an amino acid polymer called **cyanophycin,** which is composed of repeating units of aspartic acid and arginine. The **prochlorophytes** are *Cyanobacteria* that lack phycobilin pigments and contain chlorophyll *b*.

▶ The multicellular structure typical of many filamentous *Cyanobacteria* is termed a **trichome.** Most *Cyanobacteria* that are nitrogen fixing carry out the process in a specialized modified vegetative cell called the **heterocyst.**

REVIEW QUESTIONS

1. Distinguish between the terms chemoautotroph, chemolithoautotroph, photoautotroph, photolithoautotroph, photolithoheterotroph, and mixotroph.

2. List the various carbon dioxide fixation pathways used by photosynthetic *Bacteria*. Which organisms use which pathways?

3. What distinguishes anoxygenic from oxygenic photosynthetic bacteria? What groups fall into each category?

4. What is the evidence that prochlorophytes are or are not intermediates in the evolution of algae?

5. Compare the metabolisms of *Escherichia coli*, *Methylococcus* spp., *Beggiatoa* spp., and *Chromatium okenii*. How do you explain that they are all members of the γ-*Proteobacteria*?

6. Compare the morphological and metabolic diversity of the various cyanobacterial orders including the prochlorophytes.

7. Recent evidence indicates that the *Cyanobacteria* evolved from two separate photosynthetic phyla, the *Chlorobi* and the *Firmicutes*. How could oxygenic photosynthesis have evolved in this scenario?

8. Why are the *Cyanobacteria* so important in the evolution of life and on the biosphere of Earth?

SUGGESTED READING

Balows, A., H. G. Trüper, M. Dworkin, W. Harder and K. H. Schleifer, eds. 1992. *The Prokaryotes.* 2nd ed. Berlin: Springer-Verlag.

Holt, J. G., Editor-in-Chief. 1994. *Bergey's Manual of Determinative Bacteriology.* 9th ed. Baltimore, MD: Williams and Wilkins.

Staley, J. T., N. Pfennig and M. Bryant, eds. 1989. *Bergey's Manual of Systematic Bacteriology.* Volume III. Baltimore, MD: Williams and Wilkins.

Staley, J. T. and A-L. Reysenbach. *Biodiversity of Microbial Life: Foundation of Earth's Biosphere.* 2002. New York: John Wiley & Sons.

Chapter 22

Other Bacterial Phyla

Hence without parents, by spontaneous birth,
Rise the first specks of animated earth....
Organic life beneath the shoreless waves
Was born and nurs'd in ocean's pearly caves;
First forms minute, unseen by spheric glass,
Move on the mud, or pierce the watery mass;
These, as successive generations bloom,
New powers acquire, and large limbs assume;
Whence countless groups of vegetation spring,
And breathing realms of fin, and feet, and wing.
—*ERASMUS DARWIN,*[1] *FROM "THE TEMPLE OF NATURE," 1802*

The domain *Bacteria* contains many other phylogenetic groups than the six discussed in Chapters 19 through 21. Indeed, it has been estimated that 30 to 40 phyla of *Bacteria* exist altogether. As surprising as it may seem, bacteriologists do not even have representatives of some of these in pure culture. They are known to exist only from analyses of the diversity of 16S rDNA sequences extracted from the DNA from natural communities (Box 17.5). This chapter contains information about the remaining 17 phyla of *Bacteria,* based on knowledge of pure cultures. Those for which pure cultures are not yet available are not described, because virtually nothing is known about their phenotypic attributes. Thus, much about the vast diversity of the bacterial world still remains largely unknown.

Indeed, for some phyla in which representatives have been isolated, the few known strains represent only a small fraction of the diversity of the group. In fact, several phyla are represented by only a single species because they are so poorly understood. Thus, for those phyla, only brief thumbnail sketches of their descriptions are provided.

The chapter begins by discussing the various phyla that contain thermophilic members of the *Bacteria*. These phyla represent some of the deepest branches in the domain *Bacteria* of the 16S rRNA tree of life.

PHYLA CONTAINING THERMOPHILIC BACTERIA

Table 22.1 contains a listing of some of the genera of bacterial thermophiles that have been isolated from hot springs and other geothermal environments. All are gram-negative rods of various sizes with optimal growth temperatures in the range of 40° to 100°C. Most would be considered moderate thermophiles in contrast to some of the extreme thermophilic *Archaea* that grow optimally at 100°C or higher. However, some, such as *Aquifex pyrophilus,* are regarded as hyperther-

---

[1]Erasmus Darwin was the grandfather of Charles Darwin.

Table 22.1 Characteristics of thermophilic bacteria

Phylum/Genus	O$_2$ Requirement	Temperature Range (°C)	Mol % G + C	Habitat (Energy Source)
Aquificae				
Aquifex	Aerobe	60–95	40	Hot springs (H$_2$ autotroph)
Hydrogenobacter	Aerobe	60–80	38–44	Hot springs
Desulfurobacterium	Anaerobe	40–75	35	Hydrothermal vents
Thermodesulfobacteria				
Thermodesulfobacterium	Anaerobe	60–70	31–38	Hot springs (H$_2$; organic acids)
Thermotogae				
Thermotoga	Anaerobe	55–90	46–51	Marine thermal vents (sugars)
Fervidobacterium	Anaerobe	40–80	33–40	Hot springs (sugars)
Thermosipho	Anaerobe	35–77	60–63	Marine thermal vents (yeast extract; peptone) (H$_2$ autotroph)
Nitrospirae				
Thermodesulfovibrio	Anaerobe	40–70	30–38	Hot springs (H$_2$ + organic acids)
Deinococcus-Thermus				
Thermus	Aerobe	40–85	60–67	Hot springs (glucose; acetate)
Thermomicrobia				
Thermomicrobium	Aerobe	45–80	64	Hot springs (tryptone; yeast extract)
Dictyoglomus				
Dictyoglomus	Anaerobe	50–80	29	Hot springs (carbohydrates)

mophiles because they are capable of growth at very high temperatures, close to 100°C.

Some of the genera are aerobes, and others are anaerobes. It is interesting to consider that some of these phyla—*Thermodesulfobacteria, Thermotogae,* and *Aquificae*—represent rather deep branches in the bacterial lineage and are not closely related to any other known mesophilic bacterial groups.

Aquificae

Aquifex, Hydrogenobacter, and Desulfurobacterium

The *Aquificae* contains the most thermophilic species of the *Bacteria* known. The maximum growth temperatures of some species exceed 95°C, and therefore they can be regarded as hyperthermophiles. All cultured strains that do not grow on organic compounds are obligate hydrogen autotrophs.

Aquifex is the most thoroughly studied genus. The genome of *A. pyrophilus* has been sequenced. These bacteria are true hyperthermophiles that grow at a maximum temperature of 95°C. They fix carbon dioxide through the reductive citric acid cycle. In addition to using H$_2$ as an energy source, they can also use thiosulfate and sulfur, which they oxidize to sulfuric acid. They can also use nitrate as an electron acceptor and produce nitrite and N$_2$ gas. *Hydrogenobacter* is similar metaboli-

cally in that it also uses the reductive tricarboxylic acid cycle.

Desulfurobacterium grows chemolithotrophically by oxidizing hydrogen gas as an energy source and reducing thiosulfate, S^0, or sulfite to H$_2$S. This is an obligate anaerobe that has been isolated from deep-sea hydrothermal vents. Presumably it also uses the reductive citric acid cycle, but this has not yet been confirmed.

Thermodesulfobacteria

Thermodesulfobacterium is the sole genus in this phylum. These rod-shaped bacteria are heterotrophic sulfate-reducing bacteria that use lactate and pyruvate as energy sources and thiosulfate or sulfate as electron acceptors. H$_2$S is formed by their sulfate-reducing metabolism. The organic acids are incompletely oxidized to acetic acid and carbon dioxide. These bacteria live in hot springs and hot subterranean oil reservoirs.

Thermotogae

Thermotoga and Thermosipho

These genera are closely related anaerobic organisms isolated from submarine thermal environments. These bacteria represent one of the deepest branches of any known lineage of the *Bacteria*. The cells of both genera are surrounded by a proteinaceous sheathlike structure, a so-called **toga** that balloons over the ends of the rods. A number of rods in a chain may be surrounded by one sheath; the function of the sheath is unknown.

Thermotoga ferments sugars such as glucose to lactate, acetate, CO_2, and H_2. *Thermosipho* grows on richer media such as yeast extract and requires the amino acid cysteine. Some strains of *Thermotoga* are motile by monotrichous flagellation.

Nitrospirae

This phylum contains a variety of bacteria, most of which are mesophilic. However, in keeping with our theme of thermophiles, we begin by discussing the single known genus in the phylum that is a thermophile, *Thermodesulfovibrio*.

Thermodesulfovibrio *Thermodesulfovibrio* strains have been isolated from thermal hot springs in Yellowstone National Park and in Iceland. As their name implies, these are sulfate-reducing bacteria that use organic carbon sources as energy sources and reduce sulfate, thiosulfate, and sulfite to H_2S. Lactate and pyruvate are used as energy sources. Their optimal growth temperature is 65°C.

Nitrospira The namesake genus for this phylum is *Nitrospira*. These are spiral-shaped chemolithotrophic-nitrifying bacteria that oxidize nitrite to nitrate as an energy source. With the exception of some members of the planctomycetes, which are purportedly responsible for the anaerobic oxidation of ammonia to nitrate, the phylum *Nitrospirae* is the only other phylum apart from the *Proteobacteria* in which nitrification has been reported.

Nitrospira strains have been isolated from freshwater and marine sources as well as from soil and activated sludge. The pathway of carbon dioxide fixation has not yet been determined.

Magnetobacterium This is a genus of magnetotactic bacteria that has not yet been isolated in pure culture (**Figure 22.1**). However, the 16S rDNA sequence has been determined, and fluorescent probes have been used to identify the organism in the environment. This magnetic bacterium is a large rod that contains hook-shaped magnetite granules. It has been found at the oxic-anoxic transition zone in freshwater lake sediments.

Deferribacteres

Deferribacter and Geovibrio This phylum is characterized by having heterotrophic bacteria that respire anaerobically using reduced inorganic electron acceptors such as Fe^{+3}, Mn^{+4}, Co^{+3}, S^0, or nitrate. The genus *Deferribacter* contains anaerobic, rod-shaped bacteria that

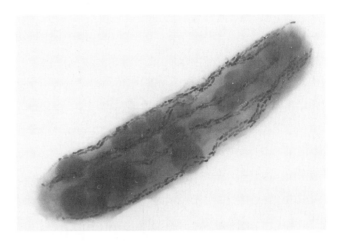

Figure 22.1 Magnetic bacterium
Electron micrograph of *Magnetobacterium bavaricum*, showing its intracellular hook-shaped magnetite particles. Courtesy of S. Spring.

oxidize organic acids and transfer the electrons to Fe^{+3}, Mn^{+4}, or nitrate. Complex organic compounds can also be used as substrates, none of which are fermented. This moderately thermophilic genus grows over a temperature range of 50° to 65°C. The original strain was isolated from a thermal petroleum reservoir in the North Sea.

Geovibrio is a genus of anaerobic vibrioid bacteria, whose metabolism is similar to that of *Deferribacter*. They are mesophiles that oxidize acetate using Fe^{+3}, S^0, or Co^{+3} as electron acceptors. Fe^{+3} can be used as an oxidant for oxidation of other energy sources, including H_2, proline, and a variety of organic acids and other amino acids. Strains have been isolated from soils.

Thermomicrobia

Thermomicrobium *Thermomicrobium* is the sole known genus. *Thermomicrobium roseum* is an aerobic, pink-pigmented, nonmotile species originally isolated from Toadstool Spring in Yellowstone National Park. The temperature at this site was 74°C. *Thermomicrobium* cells are short pleomorphic rods and dumbbell shapes. They grow best on a complex medium requiring low concentrations of nutrients.

Dictyoglomus

This phylum is named for the single thermophilic genus *Dictyoglomus*. These anaerobic, elongated, rod-shaped bacteria grow in alkaline hot springs. Cells are about 0.5 μm in diameter and 5 to 20 μm in length. They form bundles containing several cells in a distinctive spherical structure. The temperature range of growth is from 50° to 80°C. These are fermentative bacteria that use a variety of sugars as energy sources for growth.

(A)

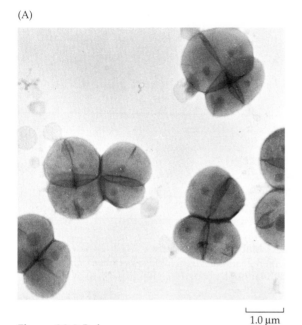

1.0 μm

(B)

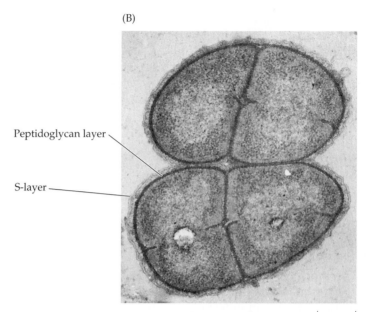

Peptidoglycan layer

S-layer

0.5 μm

Figure 22.2 *Deinococcus*
Deinococcus radiodurans. (A) Electron micrograph of actively growing cells, showing their tendency to grow as four-celled packets, or tetrads. Note that the septa are forming as "curtains" rather than as iris diaphragms. (B) Section showing the multilaminated cell wall structure, consisting of an outer single S-layer, outer membrane, and thick peptidoglycan layer. Courtesy of R. G. E. Murray.

Deinococcus and *Thermus*

Deinococci Members of the genus *Deinococcus* look superficially like *Micrococcus* species in their morphology, in that most species are nonmotile cocci (**Figure 22.2A**). However, they differ from the micrococci in their strong resistance to gamma radiation and ultraviolet light as well as in their complex cell wall structure (Figure 22.2B). The genus also differs because the diamino acid of its peptidoglycan is L-ornithine, and it lacks phosphatidylglycerol in its membrane. *Deinococcus radiodurans* was first isolated from foods that had been sterilized by gamma irradiation, indicating its importance in food microbiology. The genus is also noted for its pink to red pigments, which are carotenoids. The optimal temperature for growth is between 25°C and 35°C. Some members of the deinococci, such as *Deinobacter grandis*, are rod shaped (**Figure 22.3**).

Thermus The genus *Thermus* is quite different from *Deinococcus.* It contains aerobic nonmotile bacteria that form filaments from 5 to up to 200 μm in length. Unlike *Deinococcus*, it stains like a gram-negative bacterium, but like *Deinococcus*, it has ornithine as the diamino acid in its cell wall peptidoglycan. Colonies are often pigmented yellow, orange, or red due to the production of carotenoids. Many strains form rotund bodies that are

10 to 20 μm in diameter and are formed by the association of individual cells. The role of these bodies is unknown. *Thermus* is of widespread occurrence, and strains have been isolated from thermal environments throughout the world.

One species, *Thermus aquaticus,* is noted in molecular biology for its heat-stable DNA polymerase, the Taq polymerase, which is used in polymerase chain reaction technology. The enzyme withstands multiple heating and cooling cycles required in the polymerase chain reaction technology. The optimal growth temperature of *Thermus aquaticus* ranges from 70°C to 75°C. It is an obligate aerobe that grows as a heterotroph with ammonium salts as the nitrogen source and sugars as carbon source.

Outer membrane   Peptidoglycan layer

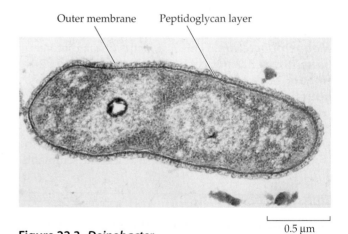

0.5 μm

Figure 22.3 *Deinobacter*
Thin section through *Deinobacter grandis*, a rod-shaped member of the deinococci group. The outer membrane appears as a looped layer surrounding the darker peptidoglycan layer. Courtesy of R. G. E. Murray.

Table 22.2	Principal genera of the *Cytophaga–Flavobacterium* group			
Genus	**Mol % G + C**	**Cell Length (μm)**	**Special Features**	
Cytophaga	30–45	1.5–15	Some degrade cellulose	
Capnocytophaga	33–41	2.5–6	Oral cavity	
Flavobacterium	30–45	1–6	Some degrade agar or chitin	
Flexibacter	37–47	15–50	Some degrade chitin or starch; freshwater	
Microscilla	37–44	10– >100	Like *Flexibacter*, but marine, and do not degrade chitin	
Saprospira	30–48	1.5–5.5	Helical filament; saprophyte	

Thermus aquaticus is one of the most common thermophilic bacteria known. It grows in most neutral pH hot springs, but it is also found in ordinary home water heaters. The usual way to isolate it is to inoculate a broth medium from hot tap water and incubate at 70°C.

REMAINING PHYLA

It is noteworthy that although we discussed those bacterial phyla that contain thermophilic genera that have been isolated in pure culture, it is known from DNA extracted from thermal environments that some of the remaining phyla discussed next also contain thermophilic members. However, apart from *Isosphaera*, which is a moderately thermophilic genus of the *Planctomyetes*, none of these have been grown in culture, so virtually nothing is known about them at this time.

Bacteroidetes

This diverse collection of bacteria contains heterotrophic, nonphotosynthetic organisms. Some are aerobic or facultative aerobic and move by gliding motility, as represented by the *Cytophaga-Flavobacterium* group, whereas others are anaerobic nonmotile fermenters represented by the *Bacteroides* group. These bacteria and their relatives are discussed here individually.

Cytophaga–Flavobacterium Group These are rod-shaped to filamentous bacteria (**Figure 22.4**), some of which are over 100 μm in length, depending on the genus (**Table 22.2**). They move by gliding motility or are immotile. Unlike the fruiting myxobacteria, which are also gliding heterotrophs, these bacteria have very low DNA base compositions (mol % G + C of about 30 to 48). Most are obligately aerobic, but some are fermentative. Because of their gliding motility, their colonies are thin and spreading, almost always with a yellow to orange (or rarely red) pigmentation.

The pigments are cell-bound carotenoids and flexirubins (**Figure 22.5**), some of which are unusual biologi-

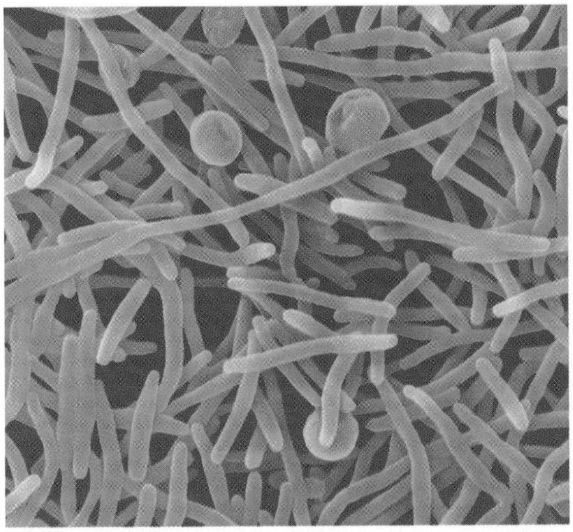

(A)

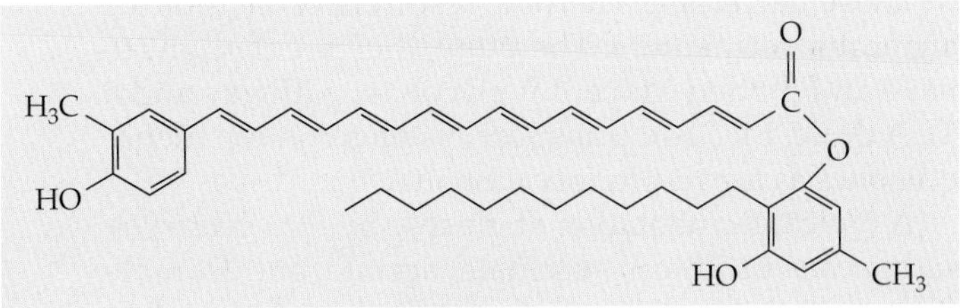

Flexirubin

(B)

Chlorinated flexirubin

Figure 22.4 *Cytophaga* group
Scanning electron micrograph of *Capnocytophaga ochracea* cells, showing morphology typical of the *Cytophaga* group. Photo ©Dennis Kunkel Microscopy, Inc.

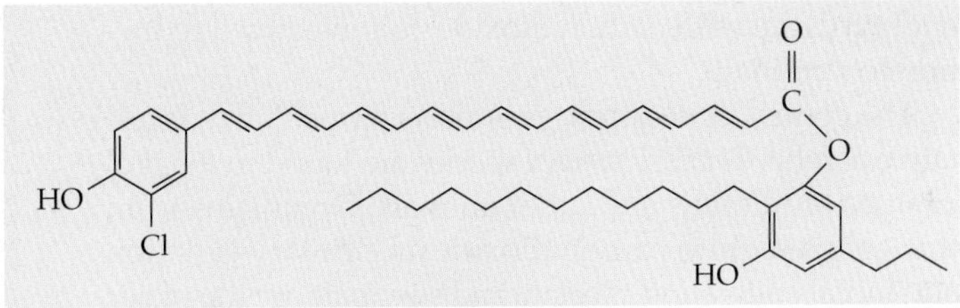

Figure 22.5 Flexirubin pigments
(A) The parent compound, flexirubin, from *Flexibacter filiformis*; (B) this chlorinated flexirubin compound is typical of those produced by *Flavobacterium* species.

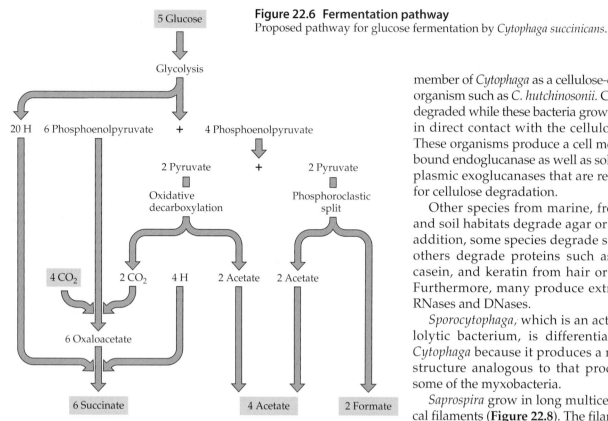

Figure 22.6 Fermentation pathway
Proposed pathway for glucose fermentation by *Cytophaga succinicans*.

Net reaction: 5 Glucose + 4 CO_2 ⟶ 6 succinate + 4 acetate + 2 formate

cally because they are chlorinated. One diagnostic feature of flexirubin is the "flexirubin reaction" in which the color of the colony changes from yellow to purple or red-brown when it is flooded with alkali (20% KOH).

These bacteria are aerobic respirers or facultative anaerobic fermenters. Therefore, they are thought to have an Embden-Meyerhof glycolytic pathway and a TCA cycle. Electron transport systems contain menaquinones as the respiratory quinone.

A typical fermentation of glucose by *Flavobacterium succinicans* produces succinate, acetate, and formate (**Figure 22.6**). *Flavobacterium succinicans* requires carbon dioxide for anaerobic growth, and it is thought that they produce succinate through condensation of phosphoenolpyruvate with carbon dioxide, followed by a reduction to succinate.

The genus *Capnocytophaga* is also a fermentative facultative aerobe. *Capnocytophaga* species are found in the oral cavities of humans and other animals. Organisms from this genus require carbon dioxide on primary isolation and initial cultivation. Acetate and succinate are the major acid by-products of the fermentation of carbohydrates.

Several genera in this group, including *Cytophaga*, *Flavobacterium*, *Flexibacter* (**Figure 22.7**), and *Microscilla*, are best known for their ability to degrade macromolecules. Indeed, Sergei Winogradsky described the first

member of *Cytophaga* as a cellulose-degrading organism such as *C. hutchinosonii*. Cellulose is degraded while these bacteria grow and glide in direct contact with the cellulose fibers. These organisms produce a cell membrane–bound endoglucanase as well as soluble periplasmic exoglucanases that are responsible for cellulose degradation.

Other species from marine, freshwater, and soil habitats degrade agar or chitin. In addition, some species degrade starch and others degrade proteins such as gelatin, casein, and keratin from hair or feathers. Furthermore, many produce extracellular RNases and DNases.

Sporocytophaga, which is an active cellulolytic bacterium, is differentiated from *Cytophaga* because it produces a microcyst structure analogous to that produced by some of the myxobacteria.

Saprospira grow in long multicelled, helical filaments (**Figure 22.8**). The filaments can be up to 500 µm in length and are composed of individual cells that are generally 0.8 to 1 × 1.5 to 5 µm. Most species produce pink, yellow, orange, or red carotenoid pigments. Some strains form coils, and others are uncoiled. Although the uncoiled filaments are nonmotile, the coiled strains glide via a screwlike motion along their long axis. *Saprospira* spp. are all strictly aerobic and are common inhabitants

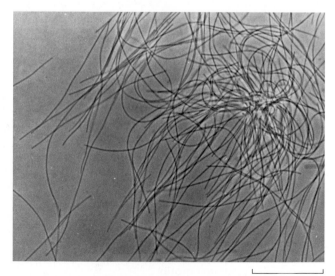

Figure 22.7 *Flexibacter*
12 µm
Flexibacter elegans grows as long, sinuous, gliding filaments. Courtesy of H. Reichenbach.

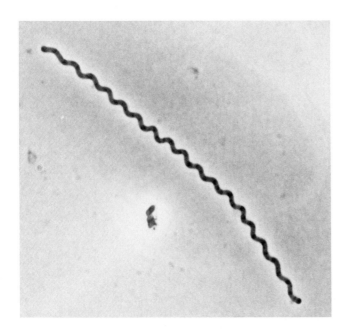

Figure 22.8 *Saprospira*
Filament of the gliding helical bacterium *Saprospira*.
Courtesy of J. T. Staley.

of sand, mud, and decaying organic materials that collect along seacoasts in temperate and warm climates around the world. Most strains are marine, but some live in freshwater.

The type species is *Saprospira grandis*, a marine organism with complex nutritional requirements. It grows in culture on peptones and amino acid mixtures. Unknown growth factors in yeast extract are also required. Glucose and other sugars stimulate growth of some strains. The generation time of *S. grandis* is 2 to 2.8 hours at 30°C. The mol % G + C is 30 to 37. The presence of these bacteria in decaying organic matter and as inhabitants of oxic areas in sewage treatment plants indicates that they play a role in the degradation of organic material in various environments.

Bacteroides Group   The genus *Bacteroides* members are all obligately anaerobic gram-negative heterotrophs. *Bacteroides* are rod shaped, often with terminal or central swelling of the cells. *Bacteroides* are the predominant microbe in the lower digestive tracts of mammals.

The *Bacteroides* group inhabits areas of the gastrointestinal tract where limited digestion occurs. They are predominant in the human caecum and colon. These areas are essentially fermentation chambers where food that is either not digested or is indigestible by the host is fermented by *Bacteroides*.

The *Bacteroides* group consists of oxygen-sensitive bacteria, but they can withstand exposure to air for 6 to 8 hours, particularly if blood and hemin are available. Many species grow best under increased CO_2 tension.

Bacteroides fragilis is a species found in the human alimentary tract. It ferments glucose to fumarate, malate, and lactate. The cells are 0.8 to 1.3×1.6 to 8 μm with rounded ends, occur singly or paired, and are usually encapsulated. Growth of *B. fragilis* is enhanced by 20% bile. Some species of *Bacteroides*, such as *B. fragilis*, chemically transform various bile acids. *Bacteroides fragilis* has a generation time of 8 hours without added hemin and 1 hour in a glucose-rich medium with added hemin.

As much as 30% of the fecal mass in humans consists of bacteria, and there are 10^{10} *Bacteroides* spp. and only 10^6 coliforms per gram. *Bacteroides* may cause disease in humans, generally in compromised individuals or those with other predisposing factors that make them susceptible to infection. Lowered oxygen levels in tissue caused by trauma, vascular constriction, necrosis, or concomitant infection by other bacteria all increase the risk of infection from these bacteria. Because *Bacteroides* are the most numerous bacterium in the human colon, they are found in significant numbers in fecal material and domestic sewage.

Fibrobacteres

The genus *Fibrobacter* is the only genus so far described in this entire phylum. *Fibrobacter* species are anaerobic gram-negative, rod-shaped bacteria. *Fibrobacter succinogenes*, which lives in the rumens of ruminant animals, ferments sugars to produce volatile fatty acids including acetic, propionic, and succinic acids that are taken up by the animal (see Chapter 25). The volatile fatty acids are absorbed in the rumen as a source of carbon and energy, and the protein-rich microorganisms pass into the small intestine, where they are degraded as a nitrogenous food source for the animal.

Spirochaetes   The spirochetes are noted for their distinct helical shape and unique mode of motility. Electron microscopy has revealed that the spirochetes have an unusual motility due to their **periplasmic flagella.** These flagella make up a structure called the **axial filament,** which extends around the helical body of these organisms within the outer membrane or outer envelope of these gram-negative bacteria (**Figure 22.9**). The axial filament is anchored in the cytoplasmic membrane in the same manner as are typical bacterial flagella with the characteristic flagellar plate. However, the hooks and filaments of the flagella remain in the periplasmic space (**Figure 22.10**). The axial filament is formed by the overlapping and apposition of two sets of flagella (called fibrils), which are inserted near the poles of the cell. The number of periplasmic flagella varies from 2 to more than 100 per cell, depending on the genus and species. Because of their helical shape and axial filament, the spirochetes are especially well adapted to movement through viscous liquids. Flagella movement is coordi-

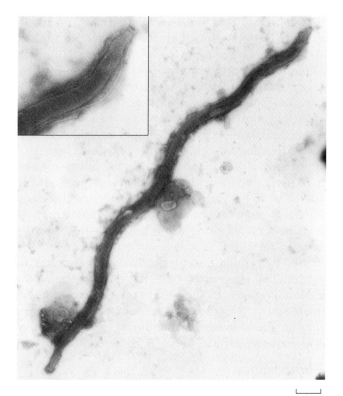

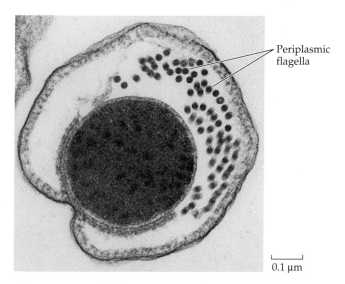

Periplasmic flagella

0.1 µm

Figure 22.10 Spirochete in cross-section
Electron micrograph of thin section through a termite spiro-chete. Note the location of the numerous periplasmic flagella of the axial filament. Courtesy of J. A. Breznak and H. S. Pankratz.

0.5 µm

Figure 22.9 Spirochete morphology
Electron micrograph of a spirochete from the hindgut of a termite, showing its characteristic helical morphology. Note the periplasmic flagella inserted at each pole (see inset). Courtesy of D. A. Odelson and J. A. Breznak.

nated at both poles. By their concerted action, the periplasmic flagella produce rotational movements that propel the organism forward in a corkscrew fashion, resulting in ordinary translocation. These are analogous to runs in typical bacteria undergoing positive chemo-taxis. Unusual flexing movements of the cell are analo-gous to twiddles and are caused by uncoordinated activ-ity of the two separate polar flagellar tufts, thereby interrupting translocation. They can also creep and crawl on solid media due to their unusual flexing move-ments. Many species are so slender that, despite their great length, the cells cannot be seen when observed unstained with the light microscope.

The cytoplasm and cell membrane of spirochetes are analogous to those of other gram-negative bacteria. However, apart from the genus *Leptospira*, their peptido-glycan contains ornithine as the diamino acid, not diaminopimelic acid. The flexible outer membrane of spirochetes is referred to as an outer sheath or outer enve-lope. Its composition varies from genus to genus; howev-er, all contain protein, lipid, and polysaccharide. The outer sheath is indispensable—its removal results in cell death.

Spirochetes are all heterotrophs, including obligate anaerobes, facultative aerobes, microaerophiles, and aer-obes. They divide by binary transverse fission and can

utilize carbohydrates, amino acids, long-chain alcohols, or fatty acids as carbon and energy sources. Many spiro-chetes placed in unfavorable conditions form spherical bodies, which appear as bulbous, swollen areas usually at the tips of the cells. These probably represent cells undergoing early stages of lysis.

The spirochetes range from free-living bacteria in soil and aquatic habitats to obligate parasites and pathogens of eukaryotes. Some grow in symbiotic associations with eukaryotes including the termite symbionts discussed later. There are several genera in the *Spirochaetes* (**Table 22.3**). Each of the five most important is discussed indi-vidually.

Spirochaeta *Spirochaeta* is a genus of anaerobes and facultative anaerobes that is present in freshwater and marine environments. As anaerobes they thrive in muds, ponds, and marshes. They are not pathogenic. They utilize carbohydrates as the source of carbon and energy, and their fermentation products are ethanol, acetate, CO_2, and H_2. The species *S. plicatilis* is found in H_2S-containing habitats. This species has not been obtained in pure culture, although Ehrenberg first described it over 150 years ago. Many other species in this genus have been isolated. During cell division, a new set of flagella appears at the middle of a cell, a sep-tum is laid down, and the two cells separate. Some strains are multicellular and grow to 250 µm or longer.

This free-living genus is the best understood of all the spirochetes. The most thorough studies on the metabol-ism, physiology, and motility have been conducted with members of this genus.

Table 22.3	Important genera of *Spirochaetes*			
Genus	Cell Length (µm)	Mol % G + C	Disease	Habitat
Spirochaeta	5–250	51–65	None	Sediments, mud, ponds, marshes
Cristispira	30–180	Unknown	None	Crystalline style; digestive tracts of molluscs
Treponema	5–20	25–53	Syphilis; yaws	Mouths, intestinal tracts, genital areas of animals
Borrelia	3–20	Unknown	Relapsing fever; Lyme disease	Mammals and arthropods
Leptospira	1–2	35–53	Leptospirosis	Free-living, warm-blooded animals

Cristispira   This genus was assigned the name *Cristispira* because members of the genus form a bundle of 100 or more periplasmic flagella. When these flagella are intertwined on the protoplasmic cylinder, they distend the outer sheath, forming a ridge or crest. *Crista* is the Latin word for *crest.* The genus is widely distributed in both marine and freshwater molluscs (clams, oysters, and mussels). They inhabit the crystalline style that is a part of the digestive tracts of these invertebrates. They can also be found in other organs in molluscs. The *Cristispira* are not pathogenic and occur in healthy univalve and bivalve molluscs. None, however, has ever been isolated in pure culture.

Treponema   Members of the genus *Treponema* live in the mouths, digestive tracts, and genital areas of humans and other animals. One of the ancient scourges of humans is due to a member of this genus, *Treponema pallidum,* the causative agent of syphilis. *Treponema pallidum,* subspecies *petunae,* causes yaws, a contagious disease in tropical countries (see Chapter 28). The treponemes are strictly anaerobic or microaerophilic. The human pathogens are probably microaerophiles, but have not been grown on artificial media. The treponemes that have been cultured are anaerobic and ferment carbohydrates or amino acids. Some require long-chain fatty acids for growth. Others require the short-chain volatile fatty acids present in rumen fluid.

Borrelia   Some species of *Borrelia* have been grown in culture. All are microaerophilic with complex nutritional requirements. Two noted pathogenic types occur in the genus. *Borrelia burgdorferi* is the causative agent of Lyme disease. This disease is transmitted to humans by the deer tick *Ixodes ricinus.* Other species are tick-borne pathogens of relapsing fever. Both diseases are discussed more fully in Chapter 28.

Leptospira   *Leptospira* is the simplest of the spirochetes morphologically. Species in this genus have only two flagella in their axial filament. The cells are tightly coiled, and the fibrils of the axial filament rarely overlap. Frequently, the cells end with a hook at both ends. Most species are free-living organisms that occur in soil, freshwater, and marine environments. Some are parasites of humans and other animals, whereas others are pathogenic. *Leptospira* spp. are obligate aerobes with a DNA base composition of 35 to 41 mol % G + C. The major cell wall diamino acid is diaminopimelic acid.

The source of carbon and energy for members of this genus is long-chain fatty acids (15 carbons or more) or long-chain fatty alcohols. They cannot synthesize fatty acids and directly incorporate substrates into cellular lipids. Based on DNA/DNA hybridization considerations, at least seven species of *Leptospira* are known, but only two are recognized. These are the free-living *Leptospira biflexa* and the pathogen, *L. interrogens.* The latter is the causative agent of leptospirosis, a disease known worldwide. The natural hosts for leptospirosis are rodents and other animals (see Chapter 28).

Spirochetes in Termites   Microscopic examination of the microbiota in the hindgut of termites and wood-eating cockroaches indicates that spirochetes are part of the normal flora. Some appear to be free-living in the gut fluids, whereas others are **ectosymbionts** that are attached to the surface of protozoa that also inhabit these areas. These bacteria were first observed in 1877 by J. Leidy and were regarded as vibrios or spirilla. Electron microscopic examination, however, confirmed that they are spirochetes. Many morphological types occur in the hindgut of each termite species. For example, one *Pterotermes occidentis* specimen examined had at least 15 types, based on size of the cell body, wavelength and amplitude of primary coils, and number of periplasmic flagella. The spirochetes vary in size from 0.2 × 3 µm to as large as 1 × 100 µm. The number of periplasmic flagella ranges from a few to over 100. Little is known of the physiology of these spirochetes, as they have not yet been grown in culture. They are apparently anaerobic because they lose motility when exposed to air. The presence of the spirochetes appears to be beneficial to the hosts, because their removal results in a shortening of the host's life span.

Only recently have studies shown some of the important physiological roles played by the spirochetes in termite ecology. Some of them carry out nitrogen fixation, and others are acetogenic. The acetogens compete with methanogenic bacteria for hydrogen gas released in the termite gut fermentation.

Planctomycetes

The planctomycetes comprise a group of unusual budding heterotrophic bacteria that lack peptidoglycan. They are most remarkable for their mode of cell division and for an unusual structural feature called **crateriform structures,** distinctive pits located on the cell surface (**Figure 22.11**). In addition, some species have unique noncellular appendages called **stalks** (Figure 22.11 and **Figure 22.12**). Several genera have been described and are listed in **Table 22.4**.

Table 22.4	Genera of *Planctomycetales*	
Genus	**Mol % G + C**	**Morphology**
Pirellula	54–57	Lack stalks; most have polar flagellum
Gemmata	64	Lacks stalks; polar bundle
Planctomyces	50–58	Stalks; polar flagellum
Isosphaera	62	Lack stalks; filamentous gliding bacterium

All of the described planctomycetes are heterotrophic, and many can grow by either fermentation or respiration of sugars, although they do not use a large variety of carbon sources. Recently, an uncultivated planctomycete has been implicated in a novel geochemical process, anaerobic nitrification, in which ammonia is oxidized to nitrate and nitrate is the electron acceptor. Although the organism has not been cultivated, good evidence links the organism to this unique transformation. Perhaps planctomycetes are involved in other

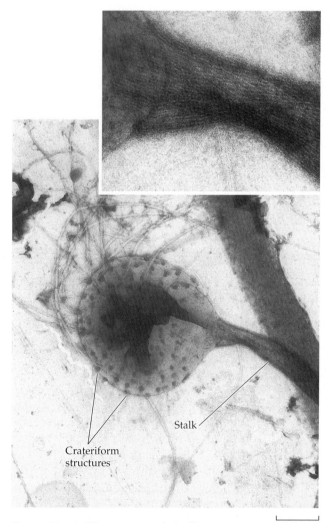

Figure 22.11 *Planctomyces bekefii*
The "pits," or crateriform structures, on the surface of *Planctomyces bekefii* are unique to these bacteria. Note also the fibrillar nature of the stalk, shown best in the inset. Courtesy of J. Fuerst and J. T. Staley.

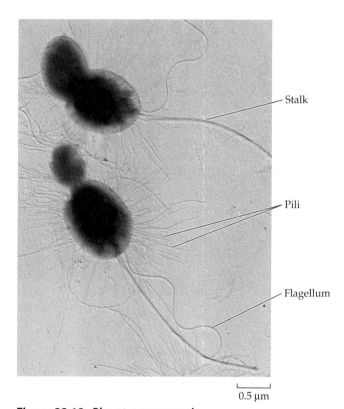

Figure 22.12 *Planctomyces maris*
Shadowed transmission electron micrograph of *Planctomyces maris*. This is a marine budding species. Note that in addition to the stalks, some cells have flagella and all cells have many pili. The stalk consists of several fibrils bundled together in a fascicle with a holdfast at its distal tip. Courtesy of J. Bauld and J. T. Staley.

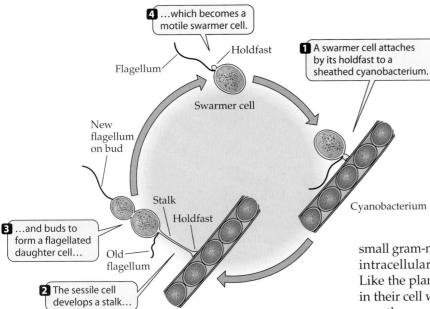

4 ...which becomes a motile swarmer cell.

Holdfast

Flagellum

1 A swarmer cell attaches by its holdfast to a sheathed cyanobacterium.

Swarmer cell

New flagellum on bud

Cyanobacterium

Stalk

Holdfast

3 ...and buds to form a flagellated daughter cell...

Old flagellum

2 The sessile cell develops a stalk...

Figure 22.13
Life cycle of *Planctomyces*
Diagram of life cycle of a *Planctomyces* species.

(**Figure 22.15**). It grows at temperatures between 35°C and 55°C. The organism is also gas vacuolate.

Chlamydiae

The genus *Chlamydia* is the sole known genus in the phylum. *Chlamydia* are small gram-negative rods (0.2 × 0.4 μm) that are obligate intracellular parasites of animals, including humans. Like the planctomyces group, they lack peptidoglycan in their cell walls. Nonetheless, for some unknown reason, they are susceptible to penicillin and other peptidoglycan-acting antibiotics, and their genomes contain the enzymes necessary for peptidoglycan synthesis. Perhaps peptidoglycan serves some other purpose in these bacteria. Analyses of 16S rRNA sequences confirm that the *Verrucomicrobia* and *Planctomycetes* phyla are their closest relatives.

There are currently three known species. *Chlamydia psittaci* causes diseases of parrots and other birds as well as psittacosis in humans who become exposed to fecal material. *Chlamydia pneumoniae* has not only been associated with pneumonia but has been implicated in heart disease as well. *Chlamydia trachomatis* causes a severe conjunctivitis called trachoma, an eye disease of humans

processes not yet discovered, as the group is very large and poorly studied.

Pirellula, Gemmata, and Planctomyces   *Pirellula, Gemmata,* and *Planctomyces* have life cycles of alternating motile and sessile cells (**Figure 22.13**). The sessile cells have a holdfast by which they can attach, much like *Caulobacter* species, to other organisms or detritus in the habitat. They are commonly found on sheathed organisms in aquatic habitats (Figure 22.13). The sessile mother cells divide by bud formation at the unattached pole to produce a motile daughter cell with either a single subpolar flagellum (in *Pirellula* and *Planctomyces*) or peritrichous flagella (in *Gemmata*). The motile daughter cell subsequently attaches to particulate matter and becomes immotile. It can then produce a bud and repeat the cycle. The mother cell remains sessile and periodically forms new buds from the budding or reproductive pole of the cell as growth proceeds.

Planctomyces species have a stalk (Figures 22.11, 22.12, and 22.14), which consists of a bundle of fibrils that resemble fimbriae, except that they have a noticeable holdfast at their tip. The holdfast allows them to attach to particulate material. Typically, unicellular species produce a stalk after the cell attaches to particulate material in the environment. In contrast, the microcolonial species aggregate to form rosettes in the environment (**Figure 22.14**). The rosettes are strikingly complex and beautiful, but none of these rosette formers has yet been isolated in pure culture. Some of the rosette formers, such as *Planctomyces bekefii*, deposit iron and manganese oxides in their stalks (**Box 22.1**).

Isosphaera   *Isosphaera pallida* is quite different from the other genera of this order. It occurs in hot springs, where it grows as a filamentous gliding bacterium

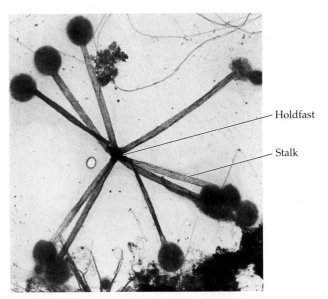

Holdfast

Stalk

Figure 22.14 *Planctomyces bekefii* colony
Electron micrograph of *Planctomyces bekefii*, showing the rosette structure as it appears when collected from lake waters. Courtesy of J. T. Staley.

Figure 22.15 *Isosphaera pallida*

Isosphaera pallida is a filamentous planctomycete found in hot springs. Note the sheath and the gas vacuoles (bright areas) in the cells. Courtesy of Steve Giovannoni.

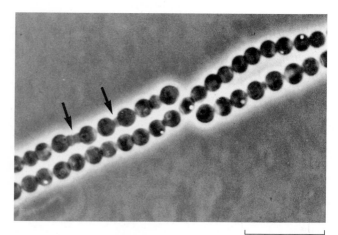

10 µm

and other animals; lymphogranuloma venereum, one of the most common venereal diseases of humans; and psittacosis.

The chlamydias closely resemble the rickettsias in size and dependence on hosts. Their genome size is also small, 4 to 6×10^8 daltons. They have been termed **energy parasites** because they cannot produce their own ATP but instead rely on the host tissues for this.

Research Highlights Box 22.1

Billions of Years of Evolution and Only 5,000 Species?

Although bacteria have lived on Earth for more than 3.5 billion years, fewer than 5,000 species have been described and named. In contrast, about 750,000 species of insects have been named, and they have lived on Earth less than one billion years. How can this be? There are a number of reasons. First of all, as discussed in Chapter 17, bacterial species are defined very differently from plants and animals. But there is also another major reason. For the most part, bacteria must be isolated in pure culture and studied in the laboratory before they can be officially named and described, and it is not always simple or inexpensive to do this. For example, consider the *Planctomycetes* phylum of *Bacteria*.

The Planctomycetes are among the most unusual members of the *Bacteria*. Although they were first observed in aquatic habitats in the early 1900s, none were reported in pure culture until 1971. Even now, some of the most morphologically striking species of this group have not been cultivated in pure culture, although they grow in ordinary habitats we have all visited. One example of this is *Planctomyces guttaeformis,* commonly found in freshwater lakes (see figures). This organism can be readily identified by its morphology using the light micro-

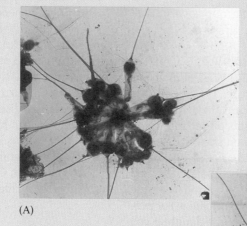

(A)

(B)

scope. Why then, hasn't it been grown in the laboratory? We simply do not know how to grow it, although microbiologists have tried.

One of the reasons that this organism is so difficult to grow is that we do not know what its requirements for growth are, and there is no obvious way to determine these requirements. One approach is to simply attempt to grow it under a variety of laboratory conditions using various carbon and nitrogen sources. But this approach is not particularly scientific and has not yet proven successful.

Strange as it may seem, the inability to cultivate bacteria is commonplace in the microbial realm. Thus, there are many bacteria in natural water and soil environments

Transmission electron micrographs of *Planctomyces guttaeformis* from a lake in North Carolina. (A) Note the long spikes, unique to this uncultivated species, extending from the pole of each ovoid cell in the rosette. (B) This smaller rosette contains some cells that are lysed. Note the buds at different stages of development. Courtesy of J. T. Staley.

that have never been grown in pure culture. As discussed in Chapter 24, it is estimated that less than 1% of all bacteria have been isolated and studied in pure culture in the laboratory. A similar situation exists for eukaryotic microorganisms. The good news is that microbiology will continue to be an exciting field of discovery for many years to come for microbiologists who have the imagination and patience to investigate these organisms.

However, their genome sequences indicate that they have the genetic capacity for ATP synthesis with proton pumping and ATP synthase. They are cultivated on the yolk sac of eggs or in tissue culture.

Chlamydia spp. have a complex life cycle (Figure 28.14). They are transmitted to the host epithelial tissues (in the eye or genitourinary tract) in a cyst stage (termed an **elementary body**). When they come into contact with host cells, they induce phagocytosis and are taken into the cell. They reside in the phagosome, where they are able to counter the host defense mechanism by interfering with the normal lytic activity of the lysozomes. While in the phagosome, they grow as vegetative cells, called **reticulate bodies,** and ultimately produce the elementary body cysts that are released into the environment when the cells are sloughed away.

Verrucomicrobia

The *Verrucomicrobia* are commonly found in soil and aquatic habitats. This phylogenetic group is quite diverse, as revealed by the clonal libraries of 16S rDNA sequences that have been obtained from environmental sources. However, only eight strains have been isolated in pure culture, including representatives of just two of the six major subphyla. Thus, one species of *Verrucomicrobium* and four species of *Prosthecobacter* have been isolated from one group and three strains of the anaerobic genus *Opitutus* from another group. Unfortunately, until strains are obtained of the other four groups, it will not be possible to understand the complete phenotypic diversity of this group because pure cultures are necessary for physiological studies.

Despite our poor understanding of this group, ecological studies indicate that they typically comprise as much as 3% of the total microbiota of soils. Therefore, they are significant members of soil communities in which they live, although their role(s) are poorly understood.

Prosthecobacter and *Verrucomicrobium* are heterotrophic, aquatic bacteria. *Prosthecobacter* attaches by a holdfast at the tip of its prostheca to algae and sheathed bacteria. Its life cycle resembles that of *Caulobacter* species except it has no motile stage.

Recent studies have indicated that at least some members of the *Verrucomicrobia*

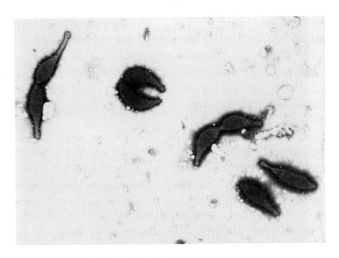

Figure 22.16 *Prosthecobacter*
Electron micrograph of *Prosthecobacter fusiformis*, showing several cells in the process of division. Courtesy of J. T. Staley.

have genes for tubulin synthesis. This is remarkable because until now, no prokaryotic organisms have been reported to contain tubulin genes. Alpha- and beta-tubulin gene homologs have been found in all four of the *Prosthecobacter* species (**Figures 22.16** and **22.17**). In

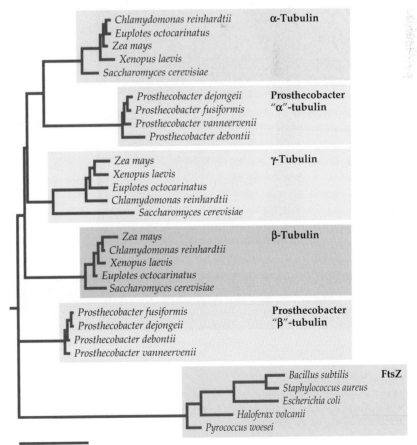

Figure 22.17 Tubulin tree
Phylogenetic tree showing relatedness between *Prosthecobacter* α- and β-like tubulins and eukaryotic tubulins. Also shown is FtsZ, a purported homologue of tubulin found in *Bacteria* and *Archaea*. Courtesy of C. Jenkins.

10% base change

eukaryotes, these two genes produce the subunits of microtubules that are responsible for movement of cell organelles, chromosomes, and flagella. Tubulin genes are found all eukaryotic organisms. Although evidence has been reported that they are being expressed in *Prosthecobacter* species, their role in these bacteria remains unclear.

One other interesting member of this phylum also appears to be able to make microtubules. This is a bacterium in the genus *"Xiphinematobacter,"* which is a protist symbiont. The bacterium, referred to as the epixenosome ("surface stranger"), grows attached to the outer layer of the protist (**Figure 22.18A**). When the protist is threatened by another protist, the bacterium produces a proteinaceous "harpoon" that is shot from its cells outward toward the invader (Figure 22.18B and **Figure 22.19**). This remarkable activity appears to be novel in the microbial world. As yet, tubulin genes for this bacterial symbiont have not been sequenced.

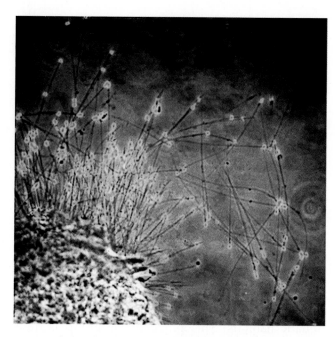

Figure 22.19 Extruded "harpoons"
Electron micrograph of host protist, showing extruded "harpoons" from *Xiphinematobacter brevicoli*. DNA from the bacterial cell is found at the apical tip of the harpoon. Courtesy of G. Petroni and G. Rosati.

(A)

Cells of *X. brevicoli*

(B)

Epixenosome remnant

Harpoon

Figure 22.18 Protist symbiont
(A) Electron micrograph showing cells of the epixenosome bacterium *Xiphinematobacter brevicolii* attached to the pole of its host protist, *Euplotidium arsenatus*. (B) Epixenosome remnant and the harpoon-like structure that has been extruded. Courtesy of G. Petroni and G. Rosati.

Chrysiogenetes

This phylum contains only a single genus and species at this time, *Chrysiogenes arsenatis*. *Chrysiogenes arsenatis* is an anaerobic, motile bent rod that grows at mesophilic temperatures. Acetate is used as an energy source and is oxidized to carbon dioxide using arsenate as the electron acceptor. Arsenate is reduced to arsenite in the process.

This bacterium was isolated from arsenic-contaminated muds of a wastewater purification system associated with a commercial gold-mining facility in Australia.

Acidobacteria

The *Acidobacteria* are gram-negative bacteria commonly found in 16S rDNA libraries from DNA extracted from natural samples. The original culture was obtained from an acidic mineral soil. However, clonal libraries indicate that these bacteria are found in many other soil habitats, including forests, deserts, grasslands, and agricultural soils. Yet, to date, only one species, *Acidobacterium capsulatus*, has been obtained in pure culture. This species of rod-shaped bacteria is orange pigmented. Cells contain menaquinones as their sole membrane quinones.

Fusobacteria

The genus *Fusobacterium* contains unicellular anaerobic organisms that are rod-shaped and sometimes spindle-shaped cells. They are nonmotile and fermentative. The major fermentation end product is butyrate along with acetate, lactate, and smaller amounts of propionate, succinate, and formate. These bacteria live in the mouths as well as in the intestinal and genital tracts of animals.

This phylum contains several other genera that are fermentative, such as *Propionigenium* and *Leptotrichia*, which produce propionic acid and lactic acid, respectively, as major fermentation end products. These bacteria have been found in the oral cavities of animals.

SUMMARY

▶ The deepest phylogenetic branches of the *Bacteria* contain thermophiles, including *Thermodesulfobacterium*, *Thermotoga*, and *Aquifex*—some of which grow at temperatures as high as 95°C. The *Thermotoga* group (*Thermotoga*, *Thermosipho*, and *Fervidobacterium*) produces a characteristic external sheath.

▶ *Aquifex* and *Hydrogenobacter* form the deepest bacterial branch and are **aerobic, obligate hydrogen autotrophs** that fix carbon dioxide by the reductive carboxylic acid cycle.

▶ The *Bacteroidetes* phylum contains **aerobic gliding bacteria** and **anaerobic fermenters**, respectively. *Cytophaga* and *Flavobacterium* are aerobic or facultative aerobic—most species are noted for **degradation** of **polymeric substances**. They are frequently pigmented yellow to orange in color, and many contain, in addition to carotenoids, special pigments called flexirubins—some of which may be chlorinated. *Bacteroides* species are common **fermentative bacteria** that are found in the digestive tracts of animals, where they occur in very high concentrations.

▶ The **spirochetes** comprise their own phylogenetic group of **helical bacteria** with unique **periplasmic flagella**, called an **axial filament**, that imparts a characteristic motility that is especially well adapted to highly viscous environments. Several spirochetes are **pathogenic**, including *Treponema pallidum* (**syphilis**), *Borrelia burgdorferi* (**Lyme disease**), and *Leptospira* (**leptospirosis**).

▶ The *Planctomycetales* comprise a separate phylogenetic group containing **budding bacteria** that are widespread in **aquatic habitats** and soils. *Planctomyces* species have a multifibrillar stalk with a holdfast at its tip. *Isosphaera* is a gliding planctomycete found in alkaline hot springs at 35°C to 55°C. *Chlamydia*, which are the closest relatives of the *Planctomycetes*, are obligate **intracellular parasites**. The *Verrucomicrobia* are unique among prokaryotes in that at least some species carry **alpha-** and **beta-tubulin genes**.

▶ Many additional phyla of the *Bacteria* are known from clonal libraries based on DNA extracted from natural samples. Some of these have cultured representatives, but much more remains to be learned about the diversity of the *Bacteria*.

REVIEW QUESTIONS

1. Why is it that microbiologists know so little about some of the major phyla of living organisms?

2. Do you believe that thermophilic *Bacteria* that branch deeply in the Tree of Life are among the earliest phyla of *Bacteria* to evolve?

3. In contrast to the *Archaea* in which none of the phyla contain chlorophyll-type pigments or are photosynthetic, many of the phyla of *Bacteria* contain photosynthetic members. One could argue that the *Bacteria* separated from the *Archaea* at the time photosynthesis evolved. If so, how do you explain that there are no known photosynthetic members in more than half of the bacterial phyla?

4. The phylum *Verrucomicrobia* contains some prosthecate bacteria. What other phyla contain them and how would you explain the evolution of this feature?

5. Why is it advantageous for bacteria that degrade particulate organic materials, like cellulose in plant tissues and chitin in the exoskeleton of insects and other animals, to have gliding motility?

6. Bioprospecting refers to the search for novel biological materials that might have commercial significance, such as antibiotics from *Actinobacteria*. What example(s) can you cite from the phyla in this chapter?

7. Explain why the discovery of tubulin genes in *Bacteria* might be important in understanding the evolution of eukaryotic organisms?

8. Compare the motility of spirochetes to that of flagellated and gliding bacteria.

9. How do you explain the occurrence of nitrifying bacteria in the *Nitrospirae* and in the *Proteobacteria*?

10. The antibiotic, ampicillin, is commonly added to media to selectively isolate *Planctomyetes* strains from enrichment cultures. Why is this so effective?

SUGGESTED READING

Balows, A., H. G. Trüper, M. Dworkin, W. Harder and K. H. Schleifer eds. *The Prokaryotes*. 2nd ed. 1992. Berlin: Springer-Verlag.

Garrity, G., editor-in-chief. 2001. *Bergey's Manual of Systematic Bacteriology*. 2nd ed., Vol. IV. New York: Springer-Verlag.

Holt, J. G., editor-in-chief. 1994. *Bergey's Manual of Determinative Bacteriology*, 9th ed. Baltimore, MD: Williams and Wilkins.

Chapter 23

Eukaryotic Micoorganisms

*And in the summer, when I feel disposed to look at all manner of little animals, I just take
the water that has been standing a few days in the leaden gutter up on my roof,
or the water out of stagnant shallow ditches: and in this I discover marvellous creatures.*
— FROM LETTER OF ANTONY VAN LEEUWENHOEK TO CONSTATIJN HUYGENS, DEC. 26, 1678

Comparisons of rRNA sequences suggest that there are three primary divisions in the living world. Members of two of these groups, the *Archaea* and *Eubacteria*, are prokaryotic. That is, they lack a membrane-bound nucleus that contains the cell's DNA. All known members of the third group are eukaryotic, which means that they contain a true nucleus. Eukaryotic cells are generally much more morphologically complex than prokaryotic cells, although presumably members of this group were not so structurally complex at the time they began diverging from the other two lineages.

The words *prokaryote* and *eukaryote* refer to the cardinal feature noticed by microscopists who first separated these two groups from each other in taxonomic schemes—the absence or presence of a nucleus. Although this is the feature most easily seen through a light microscope, it is neither the only nor, probably, the most important distinguishing characteristic. The cytoskeletal elements possessed by eukaryotes permitted the development of a novel way of separating pieces of DNA before cell division, allowed different mechanisms of motility to develop, and permitted the cell surface to be deformable. This last feature allowed cells to nourish themselves by engulfing and digesting other cells. This completely changed the nature of their trophic interactions with their surroundings and set the stage for the development of symbioses in which one or more types of cells were contained within another. The possession of their particular type of cytoskeleton, which freed them from living inside a little box as most prokaryotes do (although some have gone back to doing so), allowed eukaryotes to evolve along completely different lines than did the prokaryotes. The possession of cytoskeletal proteins like tubulin and actin is a less readily apparent feature of eukaryotes than compartmentalized DNA, but probably had a more profound effect on the cells possessing them.

One noticeable feature of eukaryotic cells is their knack for multicellularity, or their capacity to cooperate to form a single organism containing millions of cells. There are no prokaryotic equivalents of oak trees, whales, or mushrooms. All of the larger organisms in the world are eukaryotic, which makes it possible to overlook the existence of eukaryotic microorganisms. Nonetheless, such unicellular forms exist, and they form the bulk of the eukaryotic world. In age and genetic diversity, members of this group, the protists, dwarf the three multicellular eukaryotic kingdoms traditionally recognized—the green plants, animals, and fungi. Not all protists are unicellular; multicellularity has arisen independently many times among them.

After a discussion of general eukaryotic characteristics, the major protistan and fungal groups will be described in this chapter.

CHARACTERISTICS OF EUKARYOTIC CELLS

There are many major differences between eukaryotes and both groups of prokaryotes. Universal differences include the presence of a nucleus, mitosis, and cytoskeletal proteins. Frequently observed differences include the existence of meiosis, well-developed internal membrane systems, mitochondria, and chloroplasts in eukaryotes (**Table 23.1**). A discussion of these differences follows.

The Cytoskeleton and Movement

Prokaryotic and eukaryotic cells differ in the way that cell shape is maintained. Most prokaryotes have a stiff external envelope, the cell wall, but nothing inside the cell for maintaining its shape. Eukaryotes usually have an internal **cytoskeleton,** a system composed of fibrous elements. Some (algae, plants, and fungi) have also secondarily acquired an external cell wall of some type.

Three general types of fibrous elements having a wide distribution among eukaryotes are recognized. These are microtubules, microfilaments, and intermediate filaments. **Microtubules** are long, hollow fibers 24 nm in diameter. They are made up of molecules of the protein tubulin. **Microfilaments** are about 7 nm wide, and they consist of two strings of actin monomers wound around each other. **Intermediate filaments** are intermediate in size. Protists have all three of these ele-

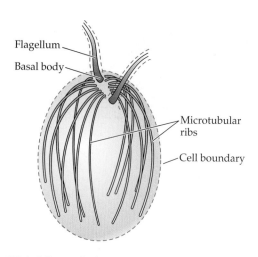

Figure 23.1 Microtubules in *Chlamydomonas*
In *Chlamydomonas*, microtubular ribs radiate from the flagellar basal bodies and run underneath the surface of the cell.

ments, as well as many others that don't fit into these categories. Microtubules may extend in parallel underneath the surface of the cell like so many microscopic ribs (**Figure 23.1**) or be grouped together into bundles to support extensions from the cell (**Figure 23.2**A and B). Microfilaments running underneath the cell surface in an ill-defined sheath give the surface of the cell some strength and rigidity. These structures do not necessarily have a fixed length. They can be made as long or short as necessary, and some are constructed temporarily for a particular task. This gives eukaryotic cells a plasticity in shape that is not ordinarily seen in prokaryotes.

By attaching accessory proteins to these stiff elements, they can be made to slide past each other. This provides a way of moving things around inside the cell. The spindle used in nuclear division is such a microtubular structure; the microtubules slide along each other to lengthen the spindle. Microtubules can also be used to move the cell itself or the water around the cell. The hairlike microtubule-containing entities that project from the cell and carry out these tasks are the cilia and flagella (see **Figure 23.3**). They are ultrastructurally alike and

Table 23.1	Major differences between eukaryotes and prokaryotes
Prokaryotes	**Eukaryotes**
No membrane around the cell's DNA	Nucleus present
No mitosis	Mitosis
No meiosis	Meiosis
No histones (few exceptions)	Most have histones
Few internal membrane systems	Many internal membrane systems
No tubulin	Microtubules composed of tubulin present
No actin	Actin present
No internal cytoskeleton	Internal cytoskeleton
Flagella with flagellin	Flagella with microtubules and dynein
No motility system based on actin or tubulin	Motility systems based on actin or tubulin
No ingestion of particles	Ingestion of particles (phagotrophy) widespread
No mitochondria	Mitochondria usually present
No chloroplasts	Chloroplasts sometimes present
Multicellularity rare	Strong tendency toward multicellularity; multicellular forms common
70S ribosomes	80S ribosomes with more proteins and larger ribosomal RNAs than in 70S ribosomes
No separate 5.8S rRNA	5.8S rRNA almost always present

(A)

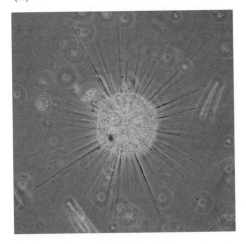

(B)

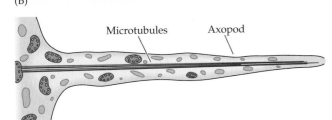

Microtubules · · · · · · · · Axopod

Figure 23.2 Microtubules in heliozoans
(A) Heliozoans use axopods, long thin projections, to capture small prey. Small organisms bump into the axopods, stick to them, and are drawn down into the body of the cell and digested. (B) Diagram showing the bundle of microtubules that supports an axopod. Photo ©J. Solliday/Biological Photo Service.

operate by the same basic mechanism. However, flagella tend to be longer, tend show a greater variety of movement patterns, and tend not to be used in large synchronously beating groups, as cilia often are.

A basal body, or kinetosome (two names for the same object), is present in the cytoplasm immediately under-

neath the cilium or flagellum. A cilium or flagellum is an outgrowth of a basal body. A basal body is ultrastructurally identical to a centriole, which also in some more indirect way organizes the growth of microtubules during mitosis. The basal body consists of nine groups of microtubules set near the perimeter of the organelle. Each group contains three microtubules. The cilium above it has two central microtubules and nine groups of two microtubules set around the periphery. A series of short arms extend along the length of one microtubule in each group. These are composed of the protein dynein. The dynein arms walk along the nearest microtubule of the neighboring pair. If the microtubules were free, they would slide along each other as the microtubules in a spindle do. But the microtubules are attached at the base of the cilium, so the walking action of the dynein arms causes the tubules to bend. This occurs rapidly and repeatedly, so that the cilia beat. A field of cilia beating this way generates sufficient force to propel the cell forward. Even a single flagellum can propel a small cell through the water. The bases of cilia and flagella are anchored in the cytoplasm by a complex array of microtubules and microfibrils that radiate away from the kinetosomes. The composition and arrangement of these accessory elements is characteristic for a given group of protists. Consequently, the ultrastructure of these elements (which together with the kinetosome and flagellum or cilium constitute the **kinetid**) has been used for taxonomy.

Actin is used in the muscles of animals, where actin and myosin molecules slide along each other in a manner similar to that seen in microtubular movement, although the details of the processes are very different. Actin is also part of the contractile system responsible for pinching a eukaryotic cell in two during division. However, actin is responsible for the movement of amoeboid organisms as well. Amoebae are highly deformable cells (**Figure 23.4**) that move through temporary extensions of the body called *pseudopods*. The body flows into an advancing pseudopod. Pseudopods can also be used to capture food particles. The way in which actin is used during amoeboid movement is by no means as clear as the way it is used in muscle contraction. Actin cross-linking and unlinking seems to be involved in the alternation between the fluid and more rigid states that allows pseudopod formation to occur.

The absence of a thick, stiff outer cell covering as a permanent feature of the eukaryotic cell together with the possibility of localized shape changes allow eukary-

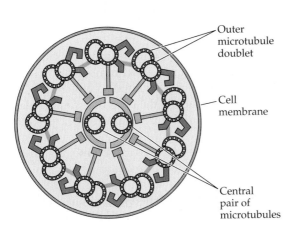

Outer microtubule doublet

Cell membrane

Central pair of microtubules

Figure 23.3 Eukaryotic flagellum
Diagram of a eukaryotic flagellum, showing the characteristics 9 + 2 microtubular arrangement. Bacterial flagella have an entirely different structure, consisting of a single fibril.

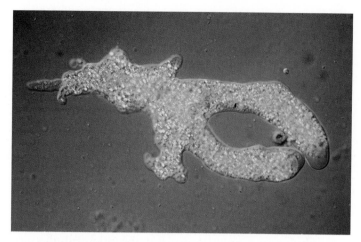

Figure 23.4 Amoeboid movement and phagocytosis
Amoebae use pseudopods, temporary extensions of the body, to move and to trap food by phagocytosis. The food vacuole containing the prey fuses with lysosomes, which contain digestive enzymes. Undigested remains are expelled. Photo ©M. Abbey/ Visuals Unlimited.

otes to engulf particles, a process that is not possible in prokaryotes. This process, called *phagocytosis* (Figure 23.4), is a basic method of feeding in eukaryotes. Large food particles are brought into the cell inside vacuoles that form at the cell surface where food is being trapped. Lysosomes then fuse with the vacuoles, releasing enzymes, which digest the food. In addition to phagocytosis, protists acquire food through **saprotrophy** (the absorption of dissolved organic material) and **phototrophy** (photosynthesis by chloroplasts).

Protists can be easily sorted by the way they move, a readily visible characteristic. Flagellates have flagella, ciliates have cilia, amoebae have pseudopods, and some parasitic forms don't move much and lack specialized structures for movement. These divisions were formerly incorporated into classification schemes at a high taxonomic level, because they were regarded as fundamental differences. Now the names are used only in a descriptive sense as a matter of convenience. The same organisms can have flagella and pseudopods at different stages in the life cycle. Conversely, rRNA comparisons have shown that organisms having the same motility system can be extremely distantly related to each other. Even humans illustrate the problem in using the type of motility system present to create the major divisions in a classification system. We have flagellated gametes, ciliated tracheal epithelium, and amoeboid macrophages. All are present simultaneously in the same organism, encoded by the same genome. These are merely common cell types, alternate states for the same cytoplasm, not markers of basic phylogenetic significance.

The Nucleus

The presence of a nucleus (**Figure 23.5**) in eukaryotic cells is only one manifestation of the differences that exist between prokaryotes and eukaryotes in the way that DNA is stored in cells and distributed between cells at the time of division. Nearly all eukaryotes have histones associated with their DNA. These are basic proteins around which the DNA is wrapped. Only a few prokaryotes, all of them *Archaea*, have histones.

Eukaryotes tend to have much more DNA than prokaryotes do. This may be due to the greater physical complexity of eukaryotes; it may take many more genes to make a eukaryotic cell. Whatever the reason, the difference in the amount of DNA can be dramatic. The length of DNA in an *Escherichia coli* genome is 1.3 millimeters. The length of DNA in a human cell nucleus is 1.8 meters! Eukaryotes are no longer able to divide their DNA in the simple way that bacteria employ during cell division. They have had to develop more elaborate mechanisms to ensure the proper segregation of chromosomes to daughter cells.

Eukaryotic nuclei contain many chromosomes instead of a single chromosome, as most prokaryotes do. Breaking the DNA into smaller pieces presumably makes it easier to move the DNA around in the cell at the time of division and easier to keep strands from being caught in the fission furrow that will constrict the parent cell into two new daughter cells. Humans have 46 chromosomes, and even the smallest of these contain more DNA than the entire *E. coli* chromosome does. Some eukaryotic microorganisms have hundreds of chromosomes.

Before division occurs, the chromosomes are duplicated. In prokaryotes, these replicated chromosomes are attached to the cell membrane. The points of attachment, which are initially next to each other, are gradually spread apart by the growth of the cell and the expansion of the membrane. A transverse extension of the cell wall forming between these sites divides the cell in two and segregates the chromosomes into the daughter cells.

Mitosis

Instead of splitting only two chromosomes apart, eukaryotes have a large number of chromosomes to separate. A complex three-dimensional scaffold called a spindle is required to place all the pairs of chromosomes into the proper orientation for separation and also to effect the separation (see Figure 1.8).

The chromosomes are spread apart by movement of the spindle fibers. After the chromosomes have moved toward the poles of the spindle and are divided into two groups, the spindle, which formed just before division, starts to break down again. As the spindle disappears,

Figure 23.5 Eukaryotic cell structure
Diagram showing the structure of a eukaryotic cell, in this case an animal cell.

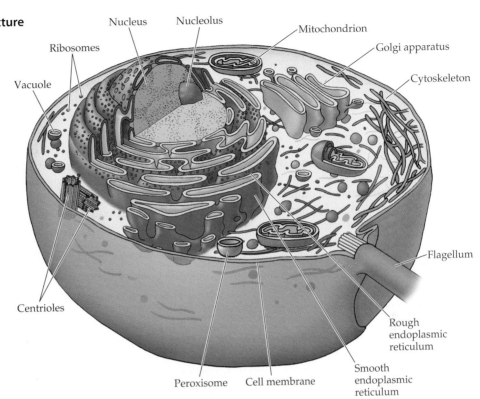

the two groups of chromosomes differentiate into two new daughter nuclei, and the cell divides between them.

In most eukaryotes, the individual chromosomes are visible only during nuclear division. During most of the life of a cell, the chromosomes exist as long extended strands of DNA and associated proteins too thin to be seen in a light microscope. Before division occurs, the chromosomes condense into tightly packed masses of DNA and protein that are thick and short enough to be seen with a light microscope. The actual distance that the spindle moves the chromosomes apart is usually quite small, so this condensation may be necessary for separation to occur. Without compaction, the chromosomes might not be adequately separated by the spindle, just as the strands in a plateful of spaghetti would not be separated from each other by sticking a fork into the mass and moving it one centimeter this way or that.

In most eukaryotic cells, nuclear division immediately precedes cell division; the two processes are coupled. However, this is not always true. In some eukaryotic microorganisms, growth is accompanied by nuclear division but not by cell division, so that a very large cell with many nuclei is formed. This shows that nuclear and cell division are separate processes, and because it is sometimes desirable to talk about only one or the other, these two types of division have their own names. Nuclear division is called mitosis, and cell division is called cytokinesis. Production of daughter cells through

mitosis and cytokinesis is called asexual reproduction.

The details of mitosis vary a great deal among the protists. In some, the spindle is formed inside the nucleus, and the nuclear envelope remains intact while the chromosomes move to opposite sides. The old nucleus is eventually pinched in half to form two new ones. This is called closed mitosis. In many protists, the spindle forms outside the nucleus, and chromosomes become attached to the spindle after the nuclear membrane breaks down (this also occurs in humans). Membranes form again at the end of mitosis around each of the two clumps of chromosomes to form two new nuclei. This is called open mitosis. Sometimes, the spindle fibers grow out of two specific cytoplasmic areas that have small bodies called centrioles in their centers. Centrioles are structurally like the kinetosomes found at the bases of cilia and flagella. The nucleating centers from which spindles grow in other protists do not all have centrioles. Other structures may be present, or there may be no visible structure at all. The chromosomes are condensed to varying degrees in different groups. In some cases, discrete chromosomes are never seen. In other cases, chromosomes are permanently condensed and may be seen whether the nucleus is dividing or not. All of these little details are useful in taxonomy and have sometimes been used to assign enigmatic organisms to the proper group. Some of these features will be mentioned in the discussion of individual groups. Even given the fact that eukaryotes have decided to fragment their genome and move the pieces around on a spindle, there is a great deal of variety possible in how this can be accomplished, and protists seem to have tried out many of the variations. This entire process, not seen in prokaryotes, depends on the cytoskeletal proteins found in eukaryotes.

Sexual Recombination

The process of cell division only allows mutations to build up slowly over evolutionary time. A set of mutations, which together may be quite advantageous, could appear in an evolutionary lineage only in the length of time it takes for all of them to independently appear. Evolution could proceed more rapidly if genes were shuffled between organisms in a manner that bypasses the limitations of mitosis, in which the genome is passed from generation to generation whole and uncontaminated with DNA from elsewhere.

There is such a process built into the life cycles of many eukaryotes, and it is called sexual recombination. It is frequently referred to as sexual reproduction, because in the familiar variant seen in multicellular organisms, this recombination accompanies the production of new individuals. Two gametic cells, one from each parent, fuse to form a zygote, which will grow into a new individual. However, in unicellular eukaryotes, in which a gamete is by necessity the whole organism, the fusion of cells actually results in a net loss of individuals (perhaps in this case, it would be more correct to speak of sexual deproduction). It is necessary to separate the concept of genetic recombination occurring through sex from the idea of multiplication of individuals in order to have an accurate impression of the role of sex in the lives of protists.

Genetic recombination also occurs in prokaryotes, but the way in which it is incorporated into the life cycle constitutes a major difference between the prokaryotes and eukaryotes. In the case of bacteria, recombination involving the direct passage of chromosomal DNA from one cell to another occurs only sporadically. It is effected by the temporary union of two bacterial cells and the passage of a strand of DNA from the donor cell to the recipient cell (see Chapter 15). The recipient cell is then left to integrate the new DNA into its own genome through the physical recombination of the two pieces of DNA by a process of snipping and gluing. Leftover pieces are discarded, so that at the end of the process, the recipient cell still has just a single chromosome. The donor cell does not lose any chromosomal genes, because the transferred segment is copied before being transferred. Chromosomal recombination in bacteria is a haphazard process in which variable amounts of DNA are passed from donor to the recipient cell. Bacteria may also pass plasmids to each other, and DNA may also be exchanged by viruses moving between cells.

Sexual recombination in eukaryotes is both more uniform and more complete. Instead of the temporary union of cells seen in bacteria, there is usually a complete fusion of cells. This is eventually followed by the fusion of the two nuclei, so that a cell is produced that contains the complete genomes of both parent cells. Both genomes are retained, and genes from both genomes are used in making RNA. Novel combinations of mutations are brought together in the same cell by combining genomes from different organisms in this way. Fusing together nuclei in each generation would result in organisms having enormous amounts of DNA and could not proceed indefinitely. Eukaryotes have invented a way of periodically halving the DNA in a cell. This is accomplished through a modified version of mitosis called meiosis (see Figure 1.12). In the life cycles of eukaryotes that have a sexual phase, zygote formation is balanced by meiosis. In mitosis, chromosomes are replicated once and then separated once by the action of the spindle into the two daughter cells. In meiosis, chromosomes are replicated once, but then separated twice on spindles into four daughter cells.

Although the most visible events of the sexual process in eukaryotes consist of this process of mixing and sorting of whole chromosomes, a process that has no counterpart in the prokaryotic world, the mechanical events seen in bacterial recombination at the DNA level are also occurring. During meiosis, homologous chromosomes (those chromosomes containing the same sets of genes) line up in physical contact with each other. Individual chromosomes sometimes break at this time. They may rejoin, reforming the same original chromosome, or they may recombine with homologues to form a new chromosome consisting of a mixture of pieces from the two different homologues. Thus, the processes occurring in prokaryotes remain in eukaryotes, although they have been overshadowed by the novel and more visible method of reshuffling entire chromosomes to produce new combinations of genes.

The **ploidy** level is the number of copies of a genome present in a cell. The gametes of organisms like us, with one copy of each chromosome, are **haploid.** The zygote they produce through their fusion, in which there are two copies of each different chromosome, is **diploid.** Shortly before its nucleus divides, it would be **tetraploid,** because the chromosomes would replicate, and there would be four copies of each different chromosome. The daughter cells produced after mitosis and cytokinesis would again be **diploid.**

The organisms most familiar to us—humans and other vertebrates—spend most of their lives as diploids. Haploid cells do not lead an independent existence in our life cycle. Eukaryotic microorganisms, on the other hand, vary tremendously in how meiosis and zygote formation are incorporated into the life cycle. Which ploidy state is dominant in the life cycle varies from group to group. The green algae like *Chlamydomonas* (**Figure 23.6**) are haploid organisms throughout most of their life. It is the haploid cells that swim around, photosynthesize, and divide to form more individuals asexually. Some of these cells occasionally fuse with each other to form a zygote. The zygote quickly undergoes meiosis to form more haploid cells. It is the diploid state that has a transient existence in the life cycle of a green alga. In other forms, both haploid and diploid stages are long-lived forms that lead an independent existence. Both cell types can multiply. This is true of some red algae, for example. Both the haploid and diploid stages undergo mitosis to produce multicellular "plants." These sometimes look so unlike each other that they were regarded as different species until the life cycle was determined (**Figure 23.7**). A few protists are polyploid and have nuclei containing multiple copies of the basic set of genes. These differences in ploidy level and life cycles are seen only in eukaryotes. They are not found in prokaryotes and exist only as a consequence of the type of genetic recombination that the eukaryotes have developed.

Internal Membrane Systems

The separation of chromosomes from the cytoplasm by a nuclear envelope is just one manifestation of the much greater use of internal membranes by eukaryotic cells than by prokaryotic cells. Most prokaryotes do not have well-developed internal membrane systems, whereas all eukaryotes have at least a nuclear envelope and usually much more than that.

The nucleus is actually bounded by two closely spaced membranes (Figure 23.5). These two membranes come together at a series of points around the surface of the nucleus. Channels, which connect the inside of the nucleus to the cytoplasm on the outside, are present at these junctions.

The space between the two membranes is continuous with the lumen of the endoplasmic reticulum, a system of channels and vesicles that extend throughout the cell and that are lined with an extension of the outer nuclear membrane. The endoplasmic reticulum is the site of fatty acid synthesis and metabolism. It is also where proteins that will be secreted from the cell are synthesized. They are made on ribosomes that line the endoplasmic

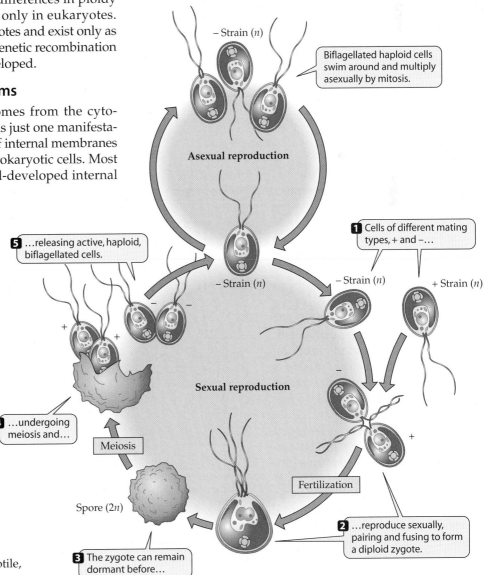

Figure 23.6 Life cycle of
Chlamydomonas
In *Chlamydomonas*, the active, motile,
multiplying cells are haploid.

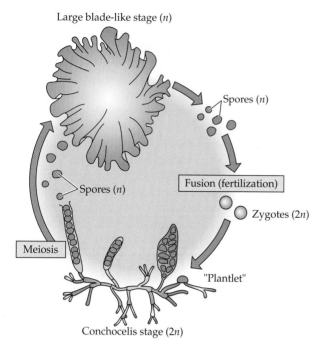

Large blade-like stage (*n*)

Spores (*n*)

Spores (*n*)

Fusion (fertilization)

Zygotes (2*n*)

Meiosis

"Plantlet"

Conchocelis stage (2*n*)

Figure 23.7 Life cycle of *Porphyra*
The haploid stage of *Porphyra* is a large leaflike structure; the diploid, conchocelis stage is an inconspicuous filamentous structure that grows in mussel shells. The two stages were originally described as different genera; the small filamentous form had the genus name *Conchocelis*.

reticulum, and they accumulate in the lumen of the endoplasmic reticulum as they are made. These secreted proteins are further processed in the Golgi apparatus.

The Golgi apparatus consists of a series of flattened, closely spaced vesicles that in cross section look somewhat like a stack of hollow plates or bowls. Proteins enter one side of the Golgi apparatus from the endoplasmic reticulum and leave the other side in small vesicles that bud off from the surface. The Golgi apparatus is a polarized organelle; material passes through in a one-way flow like a factory assembly line. In the Golgi, carbohydrates or phosphates can be added to proteins, or an inactive protein can be cleaved by enzymes to form a biologically active molecule.

Lysosomes also bud off the Golgi. These are small, membrane-bound organelles that contain digestive enzymes. They fuse with vacuoles in the cell and release their contents into it in order to digest the material within.

The vacuolar contents may be old organelles that need to be broken down and replaced with new organelles. They may also be deliberately brought into the cell from outside, such as an item of food, as in the case of a eukaryote feeding on bacteria.

Eukaryotic microorganisms also frequently contain contractile vacuoles, membranous structures that collect and expel water from the cell (**Figure 23.8**). Protists are permeable to water, and when the solute concentration

(A)

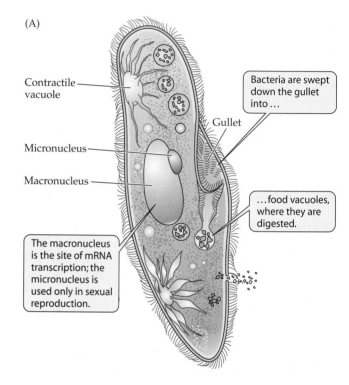

Contractile vacuole

Bacteria are swept down the gullet into …

Gullet

Micronucleus

Macronucleus

The macronucleus is the site of mRNA transcription; the micronucleus is used only in sexual reproduction.

…food vacuoles, where they are digested.

(B)

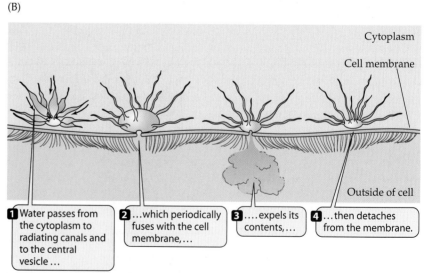

Cytoplasm

Cell membrane

Outside of cell

1 Water passes from the cytoplasm to radiating canals and to the central vesicle …

2 …which periodically fuses with the cell membrane, …

3 ….expels its contents, …

4 …then detaches from the membrane.

Figure 23.8 Contractile vacuole system of *Paramecium*
(A) Diagram of *Paramecium caudatum*, a ciliated protozoan. Not shown here are the extrusive organelles called trichocysts that line the cell surface. (B) The activity of the large contractile vacuole system is visible with the light microscope.

is higher inside the cell than outside, water is drawn into the cell by osmosis. Without a method to remove this water, the cell would swell and burst. Water collects inside the contractile vacuole, which periodically opens up to the outside of the cell and expels its contents.

These membranous systems are all connected with each other by an ongoing traffic of small vesicles that are constantly fusing with and budding from the larger membranous organelles. The membrane systems in eukaryotic cells are very active structures in a constant state of flux. The movements of these structures and other small intracellular bodies are often mediated by elements of the cytoskeletal system, organized somewhat like a complex system of conveyer belts to move things around in the cell.

Mitochondria and Chloroplasts

Many eukaryotes have two other membrane-bound organelles, quite unlike those just described. These are the mitochondria (see Figure 1.9) and chloroplasts (**Figure 23.9**). They are actually the remnants of *Eubacteria* taken into ancestral eukaryotic cells eons ago. Rather than being digested, they established a symbiotic relationship with their host cell. The host cells fed them and moved them around while the eubacterial metabolic pathways provided their hosts with much more efficient ways of generating ATP and with a way of using sunlight as an energy source. The pathways involved and the nucleotide sequences of genes in the organelles indicate that mitochondria are descendants of *Proteobacteria*, and that chloroplasts are highly modified *Cyanobacteria*. Mitochondria generate ATP through the oxidation of organic molecules and through the agency of a cytochrome electron transport chain. Photosynthesis takes place in chloroplasts in the same way that it occurs in *Cyanobacteria*. The bacteria have degenerated to such an extent that an independent existence is no longer possible for them. In most cases, all that remains of their bacterial past is a couple of biochemical pathways and a handful of genes. In a few cases, the endosymbiont genome is still quite large, and the "organelle" still has a very distinctly bacterial nature.

Mitochondria have both an outer and an inner membrane. The inner membrane runs parallel to the outer membrane in part, but is also drawn into a series of folds, which project inward into the mitochondrion. These projections of the inner membrane are called cristae (see Figure 1.9). Cristae may be tubular, lamellar, or discoid. The crista type is usually characteristic of a given group of eukaryotes, but what physiological significance these physical differences have, if any, isn't known. In some cases (e.g., among the trypanosomes), the type of crista varies between life cycle stages. Not all eukaryotes have mitochondria, but most do.

A smaller number of eukaryotes have chloroplasts (Figure 23.9). Protists that have chloroplasts are called

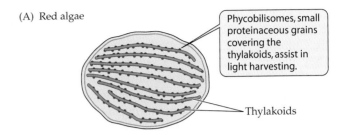

(A) Red algae

Phycobilisomes, small proteinaceous grains covering the thylakoids, assist in light harvesting.

Thylakoids

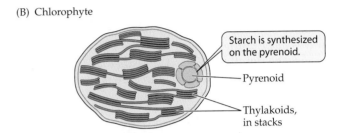

(B) Chlorophyte

Starch is synthesized on the pyrenoid.

Pyrenoid

Thylakoids, in stacks

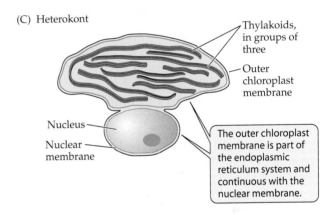

(C) Heterokont

Thylakoids, in groups of three

Outer chloroplast membrane

Nucleus

Nuclear membrane

The outer chloroplast membrane is part of the endoplasmic reticulum system and continuous with the nuclear membrane.

Figure 23.9 Chloroplasts of protists
Representations of chloroplasts found in (A) red algae, (B) chlorophytes, and (C) heterokonts.

algae. All eukaryotes that have chloroplasts also have mitochondria. Inasmuch as the acquisition of chloroplasts and mitochondria were independent events, there is no obvious reason why a eukaryote that lacked mitochondria could not have acquired chloroplasts, but this has apparently never happened. It may be that protists lacking mitochondria have a sensitivity to oxygen, which would prevent them from harboring an oxygen-generating organelle. Mitochondria-bearing protists require oxygen; mitochondria won't work without it. Any anaerobic protists originally sensitive to oxygen would have had to adapt to the presence of oxygen for mitochondria to become established in their cytoplasm. They would then be able to acquire chloroplasts without being damaged by them.

Chloroplasts also have internal membranes. The arrangement of membranes varies between algal groups and is of taxonomic significance. So are the types of pigments found in the chloroplasts. Pigments are the organic molecules that absorb the light used in photosynthesis. Chlorophyll is universally present in chloroplasts. They all contain chlorophyll *a* and may contain chlorophyll *b* or *c* as well. These chlorophylls differ from each other in only minor ways, involving the nature of the side groups attached to the main ring system.

Eukaryotes have acquired chloroplasts not only by consuming phototrophic prokaryotes but also by consuming other eukaryotes containing chloroplasts. The number of group-to-group transfers is still not known and will be difficult to determine. Even within one group, the dinoflagellates, it appears that chloroplasts have been independently acquired on several different occasions from separate algal taxa.

THE PROTISTAN GROUPS: ALGAE AND PROTOZOA

These two words are frequently seen in discussions of protistan classification, where they are used to divide the protists in half—into plantlike and animal-like groups. The word *algae* is used to denote photosynthetic protists, whereas the word *protozoan* is used to refer to most of the colorless, phagotrophic forms. Apart from the practical problem of protists that refuse to be either plant or animal and are both phagotrophic and photosynthetic, there is another problem. Both ultrastructural and molecular studies have shown that neither the algae nor the protozoa, no matter how they are defined, represents separate monophyletic units. Algae have arisen independently on several occasions through the independent acquisition of chloroplasts, and such groups are completely mixed together with colorless, or "protozoan" groups. This, in turn, means that the taxa "Algae" and "Protozoa" should not be used in a classification scheme that seeks to represent as accurately as possible the evolutionary history of the organisms being considered. The words can still be used as terms of convenience, to denote a set of organisms sharing some characteristics, in the same way that words like *predator* or *commensal* are used. The words *algae* and *protozoa* no longer have the taxonomic significance that they once did.

The arrangement of groups in this chapter follows a roughly phylogenetic order, insofar as the outlines of such an order can be apprehended. As in the case of prokaryotes, rRNA sequences are the most widely applied molecular indicator of evolutionary relationships among the protists. So many sequences have been determined that a phylogenetic tree containing representatives of most groups of protists, even quite obscure ones, could be created using rRNA sequence information alone. Unfortunately, it is clear that such a tree does not resolve some of the most interesting branching orders among the groups, and that some of branches strongly

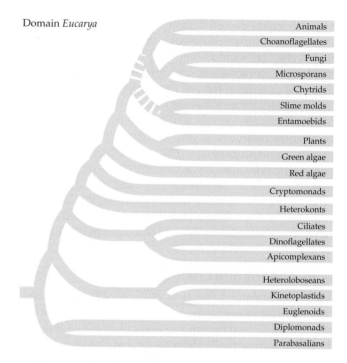

Domain *Eucarya*

Animals
Choanoflagellates
Fungi
Microsporans
Chytrids
Slime molds
Entamoebids
Plants
Green algae
Red algae
Cryptomonads
Heterokonts
Ciliates
Dinoflagellates
Apicomplexans
Heteroloboseans
Kinetoplastids
Euglenoids
Diplomonads
Parabasalians

Figure 23.10 Phylogeny of eukaryotes
Possible phylogenetic relationships among the eukaryotes. The phylogeny of protists discussed in this chapter is deduced from rRNA and protein sequences, morphology, and life cycles. The position of the slime molds and entamoebids is very uncertain; hence they are connected to the tree by a broken line.

supported by rRNA sequences are, in fact, resolved incorrectly. The phylogeny presented in this chapter is a "best guess" based on rRNA and protein sequences as well as on morphological and life cycle evidence (**Figure 23.10**).

Parabasalians

The **parabasalians** are predominantly symbiotic, although there are a few free-living forms. The symbiotic forms typically live in the digestive tracts of arthropods and vertebrates, where they consume the bacteria present in the same habitat. All the species living in the intestine are harmless, and those living in termites actually benefit their host. A few parabasalians, such as *Trichomonas vaginalis* of humans and *Tritrichomonas foetus* of cattle, live in the vertebrate reproductive system. These forms can cause disease.

In contrast to the diplomonads, the Golgi apparatus is highly developed in the parabasalians. It is large enough to be easily seen with a light microscope, although its structural details can't be made out, and it was given the name "parabasal body" before its identity was revealed by electron microscopy. The presence of this "parabasal body" (meaning "body beside the basal body") is one of the diagnostic features of the group.

The structurally simplest parabasalians, the trichomonads (**Figure 23.11**), are uninucleate. Each nucle-

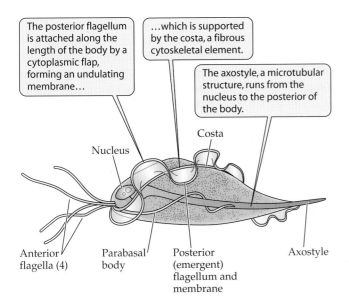

The posterior flagellum is attached along the length of the body by a cytoplasmic flap, forming an undulating membrane…

…which is supported by the costa, a fibrous cytoskeletal element.

The axostyle, a microtubular structure, runs from the nucleus to the posterior of the body.

Nucleus

Costa

Anterior flagella (4)

Parabasal body

Posterior (emergent) flagellum and membrane

Axostyle

Figure 23.11 Trichomonads
A trichomonad, the simplest parabasalian.

us is typically associated with four to six flagella. One of the flagella is recurrent, whereas the rest are directed anteriorly. The recurrent flagellum is usually attached for much of its length to the cell surface, forming an undulating membrane. A pipelike array of microtubules, called an axostyle, extends posteriorly from the basal bodies and nucleus at the anterior end of the cell and often passes all the way through the cell. The possession of an axostyle is another diagnostic feature of the group that is visible through light microscopy.

Parabasalians have a closed mitosis in which the nuclear membrane remains intact, but the spindle is extranuclear. Chromosomes and spindle fibers attached to them connect with each other at the nuclear surface. Meiosis also occurs in some species. It has been best studied in genera found in the hindgut of the wood-eating roach *Cryptocercus*. In a classic series of papers, L. R. Cleveland showed that the life cycles of these flagellates were controlled by the molting cycle of their insect host.

Parabasalians are also characterized by the possession of organelles called hydrogenosomes. They are double-membraned, mitochondrial-sized structures in which pyruvate is converted to acetate and H_2 under anaerobic conditions through the activity of iron-sulfur proteins. The processes carried out in hydrogenosomes are entirely unlike the oxidative processes that occur in normal mitochondria, but they are similar to processes found in some anaerobic bacteria. Hydrogenosomes occur in a few other protists occupying anaerobic habitats, and a series of transitional forms suggesting the derivation of hydrogenosomes from mitochondria has been found in ciliates. The source of their enzyme com-

plement is unknown; although gene sequences of some of the proteins have been obtained, it has not yet been possible to identify the bacterial source of the genes.

The largest and most impressive parabasalians live in the hindguts of termites and wood-eating roaches, where they are responsible for the digestion of cellulose. The protists ingest wood particles floating around with them in the hindgut and may become quite distorted by the large pieces they have taken in. They break down the cellulose anaerobically into small soluble molecules, mostly acetate, which they then excrete. The termite can absorb these and digest them further aerobically. This means that even though termites ingest wood, they actually live on dilute vinegar! This is a true mutualistic association; neither the flagellates nor the termites can live without each other. At first glance these large flagellates do not resemble their smaller trichomonad ancestors, because they may have hundreds of nuclei and thousands of flagella.

Diplomonads

The diplomonads are also flagellates. Many exist as symbionts in the intestines of both vertebrates and invertebrates, but a few are free-living and are found in sewage treatment plants and organically polluted water. *Giardia intestinalis* (= *G. lamblia*) (**Figure 23.12**), which Leeuwenhoek found in his own stools, was the first diplomonad discovered. It is the only member of the

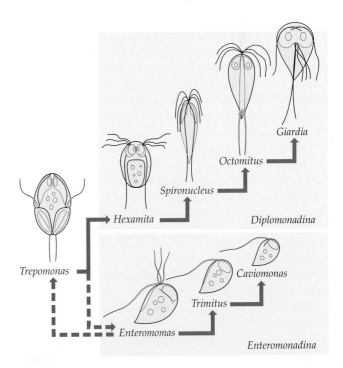

Hexamita

Spironucleus

Octomitus

Giardia

Diplomonadina

Trepomonas

Enteromomas

Trimitus

Caviomonas

Enteromonadina

Figure 23.12 Phylogeny of diplomonads
Diagram, based on cell morphology, showing the diversity of diplomonad genera and how they might be related. Dashed lines indicate possible alternatives for the origin of a genus.

group that has been intensively studied. The organism is somewhat pear shaped, as viewed from above, with the rounded end being anterior. It is dorsoventrally flattened, and the bottom surface bears a stiff bilobed disk. There are two nuclei, each of which is associated with four flagella. This appearance of being a double organism is the source of the group's name (*diplo,* meaning "double").

G. intestinalis lives and multiplies in the small intestine of many vertebrates, where it absorbs dissolved nutrients. It is able to adhere to the intestinal surface by using the ventral flagella and disk. The upper small intestine is completely coated with these parasites in heavy infections. This interferes with the absorption of fats and leads to some of the pathology associated with *Giardia* infections. Humans can experience severe diarrhea, flatulence, weight loss, and abdominal pain or remain asymptomatic. Campers and hikers in mountainous areas in the western United States frequently acquire *Giardia* infections by drinking directly from streams or contaminating their hands and utensils by washing them in streams. Wild animals are the source of *Giardia* in such situations.

Giardia cells dislodged from the small intestine and swept further back into the intestinal tract are induced to encyst when they enter the colon with feces and begin to dehydrate. These resistant cysts are then passed out in the feces. Cysts are not found in diarrheic stools, because dehydration, the stimulus for encystment, does not occur. Active flagellates are passed out, but these cannot survive outside of the body. This means that people with few or no symptoms spread *Giardia* most efficiently. Drinking water contaminated with cysts is the most common way of becoming infected with *Giardia,* but any means by which viable cysts can be passed from person to person serves to transmit the disease.

Ultrastructural studies have revealed the probable phylogeny of diplomonads (Figure 23.12) and shown that a series of uninucleate intestinal symbionts, the enteromonads, also belong to this group. Enteromonads double their cell organelles before undergoing division, indicating how the morphological state displayed by forms such as *Giardia* could have arisen from simpler forms through an arrest of cell division. Only asexual reproduction occurs; meiosis and cell fusion have never been observed in the diplomonads.

Although a Golgi apparatus is frequently said to be lacking, a membrane system possibly recognizable as a rudimentary Golgi system appears transiently during encystment while large quantities of material are being secreted from the cell to produce a cyst.

Diplomonads, like several other groups of protists, lack mitochondria. It was believed that some amitochondriate groups diverging near the base of eukaryotic phylogenies, such as diplomonads, may never have had mitochondria. However, genes that possibly have a mitochondrial origin have recently been found in the genomes of several of these species. It may be that these organisms once had mitochondria that were later lost or that the genes were transferred into them from some other source. It is currently an unsettled issue.

Discocristate Protists

The aerobic, mitochondria-possessing eukaryotes found in branches closest to the base of the phylogenetic tree are the protists that contain mitochondria with discoid or ping-pong paddle–shaped cristae at some stage in the life cycle (trypanosomes have discoid and tubular cristae at different stages). Two flagellate groups, the euglenoids and the kinetoplastids, belong here together with the more distantly related heteroloboseans.

Euglenoids The euglenoids are particularly common in organically enriched freshwater, but they may be found in virtually any aquatic environment. A few are parasitic. They usually have two flagella, but in some species, such as *Euglena gracilis* (**Figure 23.13**), one of the flagella is absent or reduced. Small hairs, called mastigonemes, project laterally along one side of emergent flagella. An unusual internal rod (paraflagellar rod) runs the length of the flagellum. Proteinaceous strips underlie the cell membrane (Figure 23.13, photo). Euglenoids may be phototrophic, phagotrophic, or saprophytic. Some combine saprotrophy with either phototrophy or phagotrophy. Species of *Euglena* can photosynthesize in the light or grow in the dark by taking up dissolved organic material. The trophic divisions that separate the animals (phagotrophy), land plants (phototrophy), and fungi (saprotrophy) are blurred together in the euglenoids.

Phototrophic species have chloroplasts containing chlorophylls *a* and *b*. They also have an orange carotenoid-containing eyespot in the cytoplasm near the base of one of the flagella, which is swollen. The eyespot and flagellar swelling interact in some way to produce the phototactic response characteristic of the photosynthetic species. They prefer subdued light, avoiding both bright light and darkness.

It is easy to produce permanently colorless strains of photosynthetic euglenoids by treatment with a variety of chemical agents, and it appears that chloroplast loss also occurs in nature. There are colorless euglenoids that, except for the lack of chloroplasts, are morphologically identical to *Euglena* species. Only about one-third of the euglenoids are phototrophic. The others are colorless phagotrophs or saprotrophs.

Kinetoplastids The flagellar apparatus of kinetoplastids is similar to that of euglenoids because it has

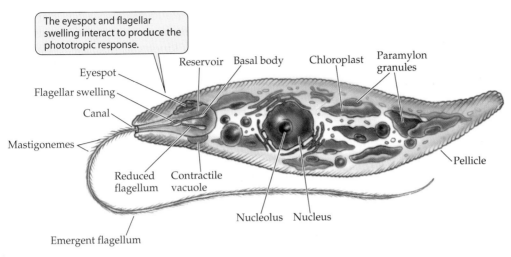

The eyespot and flagellar swelling interact to produce the phototropic response.

Eyespot

Flagellar swelling

Canal

Mastigonemes

Reservoir Basal body Chloroplast Paramylon granules

Reduced flagellum Contractile vacuole

Nucleolus Nucleus

Emergent flagellum

Pellicle

Figure 23.13 *Euglena*
Euglena gracilis, a phototrophic euglenoid. In this species, one of the two flagella is reduced. The micrograph shows the proteinaceous strips underlying the cell membrane. Photo ©Biophoto Associates/Photo Researchers, Inc.

(primitively) two flagella that emerge from a depression; paraxial rods on the flagella; and sometimes, mastigonemes on the flagella. The microtubular elements radiating from the flagellar basal bodies in the cytoplasm also have the same arrangement in the two groups. The euglenoids and kinetoplastids are sometimes classed together as the *Euglenozoa*; both ultrastructural and molecular evidence support such a union.

The kinetoplastids acquired their name from the kinetoplast, an object near the base of the flagellum that took up stains used by early protozoologists (**Figure 23.14**). It is now known to be a mass of

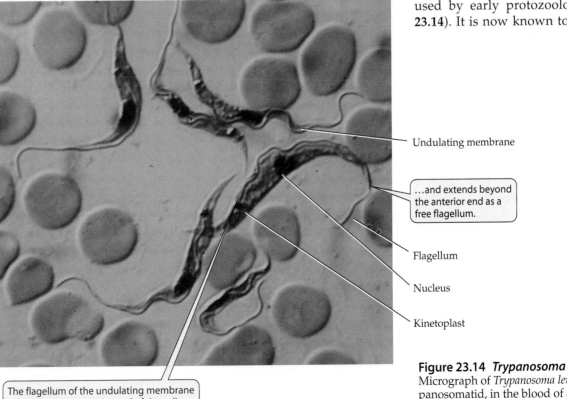

Undulating membrane

...and extends beyond the anterior end as a free flagellum.

Flagellum

Nucleus

Kinetoplast

The flagellum of the undulating membrane emerges at the posterior end of the cell...

Figure 23.14 *Trypanosoma*
Micrograph of *Trypanosoma lewisi,* a trypanosomatid, in the blood of a rat. Photo ©D. Snyder/Visuals Unlimited.

mitochondrial DNA, very different in organization from the mitochondrial DNA of other organisms. In most eukaryotes, the mitochondrial genome is contained in a small circular chromosome. There are several copies of this in each mitochondrion, but not so many that they can be stained and seen with a light microscope. In the few kinetoplastid species whose mitochondria have been well studied, the mitochondrial genome is a network of thousands of interlocked circular chromosomes. A small number of large chromosomes ("maxicircles") encode the genes commonly found in mitochondria, whereas thousands of much smaller "minicircles" exist that do not encode any complete genes. The coding regions are also unusual in that many of the transcripts produced from them do not encode functional proteins. These transcripts cannot be translated into normal mitochondrial proteins until U (uracil) residues are posttranscriptionally inserted into the proper places in mRNA. The reason for this type of organization is entirely unknown. Scientists who are interested in finding out how such a strange genetic system arose and how it works are intensively investigating the mitochondrial genomes of kinetoplastids.

There are two groups of kinetoplastids. One of these groups, the bodonids, contains many free-living forms that are very common in both soil and water. These ingest bacteria through a microtubule-lined cytopharynx (meaning "cell throat") (**Figure 23.15**). Traces of such a pharynx can be found in some euglenoids when they are examined ultrastructurally, which further supports the belief that euglenoids and kinetoplastids are related to each other. There are also a few parasitic species, which are found primarily in aquatic organisms.

The second group of kinetoplastids, the trypanosomatids (Figure 23.14), evolved from bodonids. They lack a cytopharynx and feed on dissolved organic material. All have only a single flagellum, but may have a second, barren, kinetosome. They are especially common in the intestines of insects, where they probably originated. They feed saprotrophically on the food present in the intestine of their insect hosts. Insects feeding on animal and plant juices have introduced them into other organisms, and they have gradually adapted to life in these secondary hosts. Members of the genera *Leishmania*, *Trypanosoma*, and *Schizotrypanum* include some important human parasites found in tropical and subtropical regions.

Leishmania species are transmitted to humans via the bites of sandflies. They cause a spectrum of diseases in which the pathology that develops is related to a great extent to a particular *Leishmania* strain's ability to spread through the body from the site of inoculation. Species in the genus *Trypanosoma*, transmitted by tsetse flies, are responsible for African sleeping sickness and serious livestock diseases. Trypanosomes are actually quite common in all classes of vertebrates. Leeches, rather than insects, transmit those found in fish. *Schizotrypanum cruzi*, transmitted by reduviid bugs ("kissing bugs") causes Chagas' disease in Central and South America. This is commonly a long, chronic disease. In the human body, *S. cruzi* cells preferentially multiply in nerve and muscle cells and gradually cause degenerative changes in the heart or intestinal tract. A high percentage of people in such areas who die of heart failure at an early age are actually succumbing to the effects of an *S. cruzi* infection.

Kinetoplastids have also been introduced into plants by their insect hosts. *Phytomonas* species are transmitted to a variety of plants by hemipterans. Most live in the latex of euphorbs and asclepiads, but they occur in other plants as well. Some species damage palm trees and coffee plants.

Heteroloboseans The third group of discocristate organisms is the heteroloboseans. They live in soil, freshwater, and seawater. Many have both an amoeboid and a flagellated stage in the life cycle (**Figure 23.16**). The production of these different stages is not connected with sexual recombination or ploidy level changes. Both amoebae and flagellates reproduce asexually, if they reproduce at all. Transformation of amoebae into flagellates is triggered by dilution of the medium. In nature, this could occur after a rain. It may be advantageous for a soil amoeba to detach from the substrate and rise above it so that it can be carried into a new habitat by the rainwater flowing over a particular patch of ground.

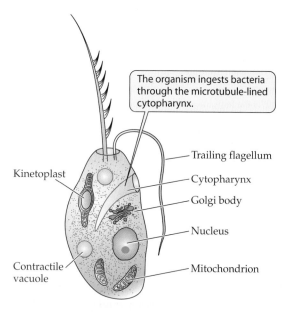

The organism ingests bacteria through the microtubule-lined cytopharynx.

Kinetoplast

Trailing flagellum

Cytopharynx

Golgi body

Nucleus

Mitochondrion

Contractile vacuole

Figure 23.15 *Bodonids*
A bodonid, showing details visible with an electron microscope.

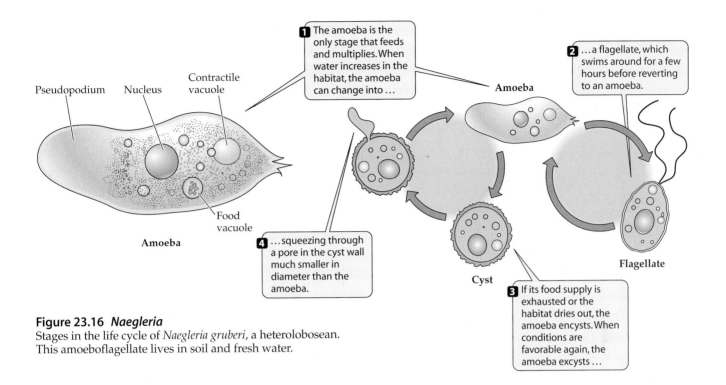

Figure 23.16 *Naegleria*
Stages in the life cycle of *Naegleria gruberi*, a heterolobosean. This amoeboflagellate lives in soil and fresh water.

The stability of the flagellate stage varies. In *Naegleria*, the flagellates are incapable of either feeding or dividing, and they revert to amoebae after several hours. In other genera, such as *Tetramitus* and *Percolomonas*, flagellates feed and divide indefinitely.

The manner of pseudopod formation by the amoebae is distinctive and characteristic of the group. In contrast to many other amoebae that send out exploratory pseudopodia in several directions at once, the amoeboflagellates tend to move through one rapidly advancing pseudopod at a time. This gives them a sluglike outline from above, which is responsible for their description in protistological literature as limax amoebae (*limax* means *slug* in Latin and is the genus name for the common garden slugs of Europe and North America). Pseudopod formation is eruptive, with hyaline cytoplasm flowing out ahead of the body in what looks like a series of spills. A cyst is also present in the life cycle and is formed by the amoebae whenever environmental conditions become unfavorable for further reproduction. Only the amoebae form cysts and emerge from cysts; the flagellates are not capable of cyst formation.

Heteroloboseans are bactivorous forms that multiply very rapidly under suitable conditions, some having a generation time of less than two hours. This might not be much compared to a prokaryote, but it is unusually fast for a eukaryote. Most species are completely innocuous free-living forms, but a thermophilic species, *Naegleria fowleri*, is capable of infecting humans, in whom it causes a fatal meningoencephalitis. Although it can live in any thermally elevated body of freshwater, people most commonly become infected in hot springs and warm, unclean swimming pools. The amoebae normally live on bacteria present in the water, but if they are accidentally drawn into the nose of a swimmer, they can invade the nasal mucosa and migrate up the olfactory nerves to the brain. There they cause the tissue destruction that results in death. There is no effective treatment.

Alveolates

The dinoflagellates, apicomplexans, and ciliates are the alveolates. These three groups share a unique type of cell surface consisting of a typical unit membrane underlain by a layer of vesicles developed to different extents in the three groups. In ciliates, the vesicles, called **alveoli**, are usually well developed and form a pellicle somewhat resembling bubble packing material. In dinoflagellates, these alveoli frequently contain cellulose plates, forming a cell wall just beneath the surface membrane. In apicomplexans, there is usually no space between the membranes, so three closely spaced unit membranes effectively surround the cell, one right on top of the next.

Dinoflagellates The dinoflagellates are an extremely diverse group of organisms, both morphologically and ecologically. Most have two flagella oriented at right angles to each other, which emerge halfway down the side of the cell (**Figure 23.17**). One flagellum is wrapped

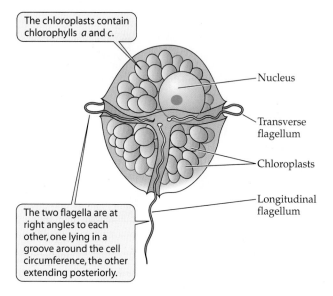

The chloroplasts contain chlorophylls *a* and *c*.

Nucleus

Transverse flagellum

Chloroplasts

Longitudinal flagellum

The two flagella are at right angles to each other, one lying in a groove around the cell circumference, the other extending posteriorly.

Figure 23.17 *Gyrodinium*
Gyrodinium, a dinoflagellate. Species of this genus are very common in coastal regions.

around the circumference of the cell in a groove, whereas the other extends posteriorly. Those species that have a covering of cellulose plates are called armored dinoflagellates. The most common dinoflagellates lack histones and have an extranuclear spindle during division, but this is not true of all species; these are secondary modifications of a more standard nuclear configuration. Most dinoflagellates whose life cycle is known are haploid organisms with a zygotic meiosis, such as the green alga *Chlamydomonas*.

Most are free-living photosynthetic forms whose chloroplasts contain chlorophylls *a* and *c*. Colorless saprotrophs and phagotrophs exist, and many photosynthetic forms are also phagotrophic. A number of dinoflagellates are parasites of invertebrates and fish. Symbiotic photosynthetic forms ("zooxanthellae") that benefit their hosts live in the tissues of a wide variety of marine invertebrates, especially in tropical water. They use carbon dioxide and ammonia produced by their hosts, and in turn they provide their hosts with sugars and amino acids. Zooxanthellae are abundant in reef-building corals and must be present in the coral for reef building to occur.

Some free-living photosynthetic forms produce toxic compounds that accumulate in the water during "red tides" (temporary blooms of dinoflagellates that occur when conditions for growth are optimal). These toxins are responsible for massive fish kills. Still others produce toxins that are passed up the food chain after the dinoflagellates are eaten, becoming more concentrated at each trophic level. Although the toxins are harmless to fish and shellfish, they can be lethal to humans. Paralytic shellfish poisoning and ciguatera poisoning both result from human ingestion of dinoflagellate toxins carried in the tissues of other organisms.

Apicomplexans Apicomplexans are parasites of animals. They probably arose from parasitic dinoflagellates living in the intestines of marine invertebrates. Most species still live in or on cells lining the intestine and are transmitted by resistant spores passed out with feces. A few have penetrated more deeply into the tissues of their hosts and must rely on other means of being passed from host to host. Apicomplexans are united by features of the life cycle and ultrastructural details rather than by readily visible gross morphological features; members of the different apicomplexan groups can look very much unlike one another at the light microscopic level. Most live intracellularly during at least a part of the life cycle. They all have a sexual phase in the life cycle, after which they all produce a distinctive stage known as a sporozoite. Sporozoites (**Figure 23.18**) con-

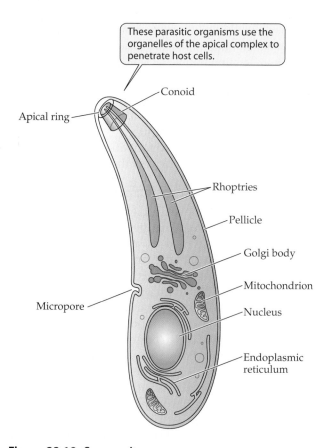

These parasitic organisms use the organelles of the apical complex to penetrate host cells.

Conoid

Apical ring

Rhoptries

Pellicle

Golgi body

Mitochondrion

Nucleus

Endoplasmic reticulum

Micropore

Figure 23.18 Sporozoites
Diagram of a sporozoite, showing features visible with an electron microscope. The conoid and apical ring are electron-dense bodies located at the tip of a sporozoite. Rhoptries and micronemes (not pictured) terminate within the conoid. Their contents are emptied as a sporozoite penetrates a host cell. These structures are used to invade host cells.

tain an apical complex, a group of organelles used in penetrating the cells of their hosts. Possession of this complex gives the group its name. All stages except the zygote are haploid, as in their dinoflagellate relatives.

Perhaps the best-known apicomplexans are the *Plasmodium* species, which are responsible for causing malaria in humans and other animals (**Figure 23.19**). This is one of the genera containing species that have migrated more deeply into host tissues. In this case, completion of the life cycle depends on a mosquito's picking up parasites from the blood of an infected vertebrate. Zygote formation occurs in the mosquito's intestine. The sporozoites formed afterward migrate to the salivary glands of the mosquito, to be injected when it takes its next blood meal. In humans, *Plasmodium* initially begins multiplying in liver cells, but eventually moves into blood cells. For unknown reasons, multiplication of all the individuals in the blood is synchronous, so that blood cells are lysed, and toxins are released in bursts. This causes the periodic chills and fever that a malaria victim experiences.

Ciliates The third alveolate group is the ciliates (**Figure 23.20**). They have a body with cilia in at least one stage

of the life cycle, two types of nuclei, and a peculiar method of sexual recombination called conjugation. Ciliates typically have two sets of cilia, somatic and oral cilia, as shown in Figure 23.20A. In *Tetrahymena* (**Box 23.1**), somatic cilia are organized into longitudinal rows called kineties, which run the length of the cell. The somatic cilia move the ciliate through the water. The oral cilia are set in groups on each side of the cytostome (meaning "cell mouth"). On one side, the cilia are aligned in a single closely set row, like the slats in a picket fence, to form an undulating membrane. This is structurally different from the undulating membrane of parabasalians, but they both have the same name because they looked alike to early microscopists. On the other side of the mouth, the cilia are arranged in three rectangular clumps, like three housepainter's brushes. These are called membranelles. The oral cilia are used in filtering bacteria from the water and sending them into the cytostome. This basic organization has been modified in the different lines of ciliate evolution, and differences in ciliary patterns served as the basis for early taxonomic schemes. Somatic cilia may be reduced or absent, and the oral cilia may be reduced or length-

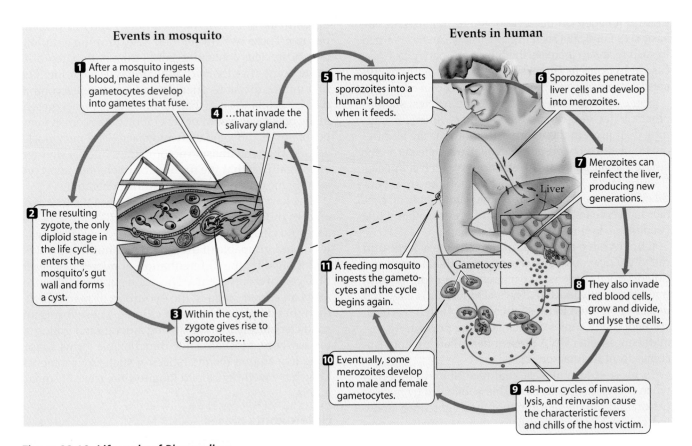

Figure 23.19 Life cycle of *Plasmodium*
Life cycle of *Plasmodium vivax*, one of the species that cause malaria in humans. The vector is an *Anopheles* mosquito. Merozoites have the same structure and functure as sporozoites, but are produced in the human host instead of the mosquito.

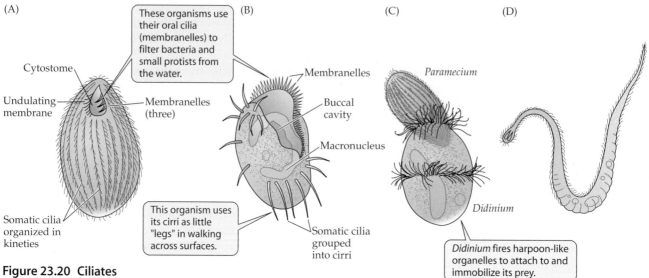

Figure 23.20 Ciliates
Some ciliates. (A) *Tetrahymena*. (B) *Euplotes*, a hypotrich; its oral cilia are arranged in the same way as those of *Tetrahymena* but cover a greater area. (C) *Didinium* feeding on a paramecium. (D) *Tracheloraphis*, a karyorelictid that lives in the sand of seashores.

ened in comparison to what is seen in *Tetrahymena*. The great majority of ciliates are phagotrophs, and many of these filter-feed on bacteria or small flagellates. The basic set of oral cilia seen in *Tetrahymena* can be enlarged in such forms to increase the filtering capacity, as is the case in hypotrichs (Figure 23.20B). A few types of ciliates lack mouths and the accompanying oral ciliature.

Some ciliates feed on very large particles, such as other ciliates, and cannot rely on ciliary filters for trapping their prey. In these forms, the oral cilia are also frequently reduced or absent. *Didinium*, which feeds on the ciliate *Paramecium*, is an example of such a species (Figure 23.20C). *Didinium* has harpoonlike organelles called toxicysts, which are fired into *Paramecium* on contact and serve to hold predator and prey together while *Paramecium* is slowly swallowed.

Ciliates have two types of nuclei. One is large and called the macronucleus, whereas the other is small and called a micronucleus. The micronucleus is diploid and transcriptionally inactive. The macronucleus is large and usually polyploid in those species in which it has been studied. Many copies of each gene are present. Almost all the messenger RNA present in a ciliate's cytoplasm comes from the macronucleus.

In most eukaryotes, sexual recombination occurs when two gametic cells fuse. This is followed by nuclear fusion and the development of the zygote into a new individual. However, in ciliates, there is usually no permanent cell fusion. Rather, two ciliates fuse temporarily and exchange haploid nuclei with each other. They then separate from each other. During conjugation, the micronucleus undergoes meiosis, but there is no cell

division. The consequence is that the ciliate contains a number of haploid nuclei. The macronucleus and all of the haploid micronuclei except two disintegrate. One of these nuclei is passed to the partner. Each partner winds up with one of its own haploid nuclei and one from its partner. These two nuclei then fuse, forming a diploid zygotic nucleus containing genes from both parents. At some point after the exchange, the cells separate. The first mitotic divisions after separation are not accompanied by cytokinesis, so that the number of nuclei present in the cells increases. After generation of the correct number of nuclei, differentiation into macro- and micronuclei occurs. Then cell division accompanies nuclear division once again, so that the number of nuclei remains constant in the cells through succeeding generations.

The karyorelictids (Figure 23.20D) are the group of ciliates thought to have nuclei most like those of the ancestral ciliates. The macronuclei are diploid (or nearly so) and incapable of division. Their number must be maintained by the differentiation of micronuclei each time cell division occurs. This state supposedly represents an earlier stage in the evolution of macronuclei than that displayed by other ciliates. These forms are characteristically found crawling among sand grains on the seashore. Most are long, thin, flexible forms adapted to live among the interstices of such environments. They have well-developed longitudinal rows of somatic cilia.

Heterokonts

The heterokonts (also referred to as the stramenopiles) and their relatives are a large group of diverse organisms that at first glance would not seem to be related to each other. However, most of them are grouped togeth-

Milestones Box 23.1   A Protistan *E. coli*

Experimental biology has always relied on a small number of organisms that are particularly amenable for use in certain types of investigations. These "model systems," as they are called, provide the means of elucidating phenomena that are believed to be widespread if not universal, but that cannot be studied conveniently in most organisms. As science advances and the questions that need answering change, so do the organisms that are used. Thus, the development of classical genetics depended on the fruit fly, but as genetics became molecular, fruit flies were displaced by *E. coli* and its phages (and ironically, as developmental biology became molecular, fruit flies replaced other metazoans!).

Earlier in this century, the development of biochemistry and cell biology required colorless, animal-like cells capable of jumping through the experimental hoops that scientists were holding up at the time. Growth factors were being identified, and it was desirable to see how universally these were needed, and if they played the same roles in all cell types. Metabolic pathways were being elucidated, and it was desirable to see how universal they were. It was becoming possible to study some basic eukaryotic processes, such as cell division and phagocytosis, and cells that could be induced to divide synchronously and phagocytize on command were needed. A eukaryotic cell as easy to

grow and as obedient as *E. coli* was required in order to study these things.

Andre Lwoff provided the necessary organism in 1923 by succeeding in getting *Tetrahymena* (then known as *Glaucoma*) (see Figure 23.20, A) into pure culture. This was the first ciliate to be grown **axenically** (meaning "in the absence of other organisms"). Getting it into pure culture was a remarkable achievement, as it was done in an age when antibiotics were not available to prevent the growth of bacteria in a rich organic medium, and nothing was known about the nutritional requirements of ciliates. Even more remarkable was the fact that Andre Lwoff was a 21-year-old student at the time. Both Andre Lwoff and *Tetrahymena* went on to greatness; Andre Lwoff gave up ciliates and received the Nobel Prize in 1965 for his work on lysogeny, and *Tetrahymena* became for a time *the* model eukaryotic cell.

The number of papers published on *Tetrahymena* increased exponentially in the years after it was first put into pure culture. Although the first studies were nutritional, it has now been experimentally poked and prodded in almost every way that a eukaryotic cell can be—with one exception. It is more difficult to use for modern genetic studies than some other eukaryotic cells are.

However, one very important genetic discovery was made by

using *Tetrahymena*. The Central Dogma (DNA makes RNA, which makes protein) contains an implicit evolutionary conundrum. If the formulation is true, how could it all have started? If proteins are all encoded for in DNA, where did the proteins required to make the first string of DNA come from, and how could they have been preserved in the coding system? Thomas Cech found the answer unexpectedly, while studying an intron in *Tetrahymena* rRNA. He found that the intron removed itself from the rRNA, without the involvement of any enzymes. rRNA itself displayed enzymatic activity. Subsequent studies on RNA showed that it is capable of carrying out a wide variety of activities. RNA is capable of producing larger strings of RNA on its own and is even capable of catalyzing peptide bond formation, the critical step in building up a protein. It is possible that the first genetic systems were RNA-based; RNA would have contained both the inherited sequence information and the catalytic capacities required for its own replication and inheritance. DNA was probably incorporated later, as a more stable repository of genetic information. Proteins were probably added in to increase catalytic versatility. *Tetrahymena* provided the first glimpse into a primeval RNA world, which may have preceded the genetic system on which all life is now based.

er by both morphological and molecular evidence. A common morphological type among members of this group is a small flagellate with two flagella. One flagellum is directed anteriorly and has small hairs projecting laterally on both sides. The second flagellum is naked and projects posteriorly. "Heterokont" refers to this difference in flagella. Many species also have a peculiar helical structure at the point where the basal body and flagellum come together.

Chrysophytes *Ochromonas danica* (**Figure 23.21**) is a well-studied chrysophyte, or golden-brown alga, and illustrates the basic chrysophyte features. The chloroplasts contain chlorophylls *a* and *c*. The outer chloroplast membrane is continuous with the outer nuclear membrane. There is a Golgi apparatus between the flagellar bases and the nucleus. Food vacuoles are frequently present; many chrysophytes are phagotrophic as well as photosynthetic. Chrysophyte chloroplasts fre-

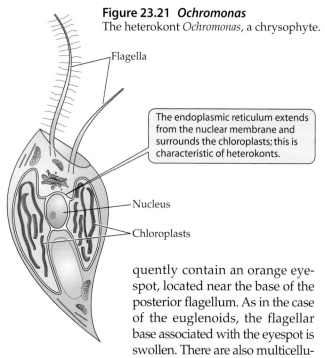

Figure 23.21 *Ochromonas*
The heterokont *Ochromonas*, a chrysophyte.

Flagella

The endoplasmic reticulum extends from the nuclear membrane and surrounds the chloroplasts; this is characteristic of heterokonts.

Nucleus

Chloroplasts

quently contain an orange eyespot, located near the base of the posterior flagellum. As in the case of the euglenoids, the flagellar base associated with the eyespot is swollen. There are also multicellular filamentous forms and a few amoeboid species as well. Chrysophytes tend to be freshwater species with a preference for slightly acidic water. Populations tend to reach maximum size in cool water, such as when ice is melting in the spring.

Xanthophytes These are the "grass-green algae," so called because of the color their pigment composition produces. Chlorophylls *a* and *c* are both present in the chloroplasts, but chlorophyll *c* is present only in very small amounts. Most live in freshwater. The majority of species are nonmotile coccoid or filamentous forms with cell walls. Many produce zoospores of a chrysophyte structure, which demonstrates their relationship to the golden-brown algae.

Diatoms The diatoms, or bacillariophytes (**Figure 23.22**), are the largest group of heterokonts; more than 10,000 species have been described. They are abundant in most aquatic habitats and are responsible for an estimated 20% of the primary productivity of the Earth. At first glance, they appear to have little in common with the chrysophytes and xanthophytes. The body is enclosed in two siliceous valves that fit together, and no flagella can be seen. However, the chloroplasts are ultrastructurally the same as those of chrysophytes, and the pigment composition is also the same. Furthermore, some species produce male gametes having flagella with lateral hairs. Other flagellated heterokont algae produce siliceous

scales and spines that adorn the body; diatoms merely carry this ability to an extreme.

Brown algae This group includes the kelps and wracks (**Figure 23.23**). Although the most obvious stages in the life cycle are large multicellular leaflike structures, they reproduce by means of chloroplast-bearing biflagellated unicells that have the appearance of chrysophytes. At all stages, the chloroplasts are chrysophytelike.

Oomycetes There are a number of colorless and non-phytosynthetic heterokonts, such as the oomycetes. Oomycetes are aquatic organisms similar in appearance to fungi; they form masses of white threads on decaying objects in the water such as dead fish or even minute objects such as pollen grains. They form motile zoospores with a typical chrysophyte structure, although without chloroplasts. Long regarded as a connecting link to the true fungi, oomycetes are shown by rRNA sequences to be quite unrelated to the true fungi. They have acquired their superficially similar life cycle and morphology through convergent evolution; they occupy a niche in aquatic environments comparable to that occupied by the higher fungi in terrestrial environments. rRNA sequences indicate that the first heterokonts were colorless, with chloroplasts being acquired later.

Cryptomonads

Cryptomonads (**Figure 23.24**) are small biflagellated unicells found in both freshwater and marine environ-

Figure 23.22 Diatoms
Some marine planktonic diatoms, or bacillariophytes. Photo ©R. Brons/Biological Photo Service.

Figure 23.23 Brown algae
Representative genera of brown algae. Not drawn to scale.

Macrocystis *Alaria* *Ectocarpus* *Chorda* *Fucus* *Nereocystis* *Laminaria*

remnants of ingested red algal cells. Instead of being digested, they remained in the cell and lost structures that were no longer needed. This process is parallel to the development of chloroplasts from *Cyanobacteria* in other algae.

Red Algae

The rhodophytes, or red algae (Figure 23.7), are predominantly photosynthetic forms, ranging from unicellular to large multicellular bladelike forms. Most are multicellular. In contrast with other major algal groups, there are no known free-swimming, flagellated unicellular forms. No such forms exist as independent species or even as a phase in the life history of a multicellular form. Red algae lack flagella, basal bodies, and centrioles at all stages of the life cycle. The life histories themselves can be quite complex, with different morphologically distinct stages associated with different ploidy

ments. They tend to be more common in cool water. Most of them are photosynthetic. Both flagella, which have mastigonemes, emerge subapically near a gullet. The gullet is lined with extrusive organelles called ejectosomes. In the body they look like very short rods, but unwind like a short roll of crepe paper when fired.

The chloroplasts contain chlorophylls *a* and *c*. The chloroplast structure is unusual in that a set of membranes outside the plastid itself contains eukaryotic-sized ribosomes and a nucleomorph, which appears to be a degenerate nucleus. This suggests that the "chloroplast" might represent the remnants of a eukaryotic alga ingested by some ancestral cryptomonad. rRNA sequencing supports this idea; small-subunit rRNA genes from the nucleomorph are eukaryotic in nature and group with the red algae in rRNA trees, whereas the nuclear cryptomonad sequences are quite different. The chloroplasts of cryptomonads are apparently the

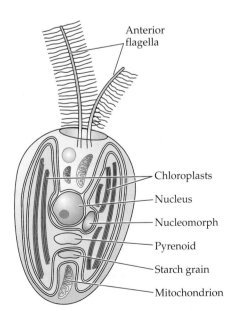

Figure 23.24 Cryptomonads
Diagram of a cryptomonad.

Anterior flagella

Chloroplasts
Nucleus
Nucleomorph
Pyrenoid
Starch grain
Mitochondrion

levels. Surprisingly, forms displaying simple asexual reproduction can also have life cycles composed of morphologically distinct generations.

Most red algae live in tropical marine environments. They are abundant in coral reefs. Some, referred to as corallines, produce an extracellular coating of calcium carbonate, as corals do. They can be the most abundant and productive members of a reef community and help to build the reef itself.

Red algae are economically important. They produce a series of unusual polysaccharides, two of which are of considerable economic value. Agar is used to make bacteriological media and also as a thickening agent in some foods. Carrageenan is also used as a thickening agent in foods such as ice cream and pudding and for stabilizing emulsions in paint. Red algae are also used directly as food. *Porphyra* is eaten in the Far East (it is the "nori" found in sushi) and is actually farmed in Japan and China.

Green Algae

The large and diverse group of green algae, the chlorophytes, branch off near the top of the tree, near the point at which the fungi, higher plants, and animals emerge. The land plants undoubtedly evolved from a chlorophyte ancestor. The morphology of the simplest forms is demonstrated by *Chlamydomonas* (Figure 23.6). Two naked flagella emerge from the anterior end of the cell. There is a single cup-shaped chloroplast containing chlorophylls *a* and *b*. It may also contain an eyespot. There is a cell wall, which is composed of glycoprotein. Many different lineages have produced multicellular forms, which may be filamentous, leaflike, or even spherical in shape.

Slime Molds

Slime molds were long regarded as being a type of fungus (hence the name "mold") because they produce spore-bearing, multicellular structures that extend up from the substrate into the air, as do true fungi. However, they are probably not closely related to fungi. They are quite unlike fungi in every way except in their possession of a multicellular spore-bearing structure. Their possession of similar structures is an example of **convergent evolution,** which is the development of similar characteristics by unrelated organisms because they have adapted in the same way to a similar environmental challenge. In this case, they have adapted in the same way to being small, relatively immobile, and terrestrial. The slime molds are terrestrial amoebae, living in soil, in leaf litter, or on the surfaces of decaying wood and plant leaves. They are not motile enough to transport themselves into new habitats. Aquatic protists are carried around by the movement of water. But stumps and forest soil don't move much; how are the amoebae that live there effectively dispersed? In the same way as are the fungi—by raising resistant spores above the soil, where they can be moved around by wind, rainwater,

or animals in whose fur they have been trapped. Such a strategy also encourages the development of multicellularity, because a single protistan cell cannot raise itself very far off the substrate. A multicellular structure can be much larger. There are a few unicellular slime molds, the protostelids, but most are multicellular. The two slime mold groups have very different life cycles, so it is likely that they are not closely related. The two kinds of slime molds and the fungi probably developed the same elevated spore-bearing structures independently.

The cellular slime molds (**Figure 23.25**) exist as uninucleate soil amoebae while feeding. They can be stimulated by starvation to aggregate and form multicellular masses called slugs, which rise from the substrate and differentiate into a stalk that supports a mass of spores. Under suitable conditions, single amoebae emerge from each spore to begin feeding and multiplying again.

The acellular slime molds (**Figure 23.26**) form similar resistant terrestrial structures, although their life cycles are different. Haploid amoebae emerge from the spores. They either fuse with one another or are transiently transformed into flagellates that fuse with one another before reverting to amoebae. The end product in either case is a diploid amoeba that grows tremendously in size while consuming bacteria on the substrate on which it crawls. Mitosis occurs repeatedly without cytokinesis so that the cytoplasm contains thousands of nuclei. This plasmodial stage is often encountered in forests as orange or yellow patches on decaying wood. Eventually the plasmodium is stimulated to form spore-bearing structures similar to those of the cellular slime molds. Meiosis occurs during spore formation.

Entamoebae

Many species of *Entamoeba* exist. Found in the intestines of vertebrate and invertebrate hosts, where most of them consume bacteria, they lack mitochondria. Several species live in humans. One of these, *Entamoeba histolytica*, causes amoebic dysentery. It alone, among the intestinal amoebae living in humans, is capable of secreting proteolytic enzymes that digest away cells lining the intestine, forming large ulcers. *Entamoeba histolytica* is transmitted by cysts in the same manner as *Giardia*. Several hundred million people worldwide are infected with this organism, which makes it an important human parasite. Fortunately, most people infected are asymptomatic. The factors that cause the amoebae to do very little harm sometimes and a great deal of harm at other times aren't known.

Other Amoebae

Amoeboid forms other than the entamoebae and slime molds are scattered throughout an rRNA tree, indicating that amoebae probably arose from flagellated cells on many occasions. Large amoeboid groups, containing many species, include the foraminifera (**Figure 23.27**),

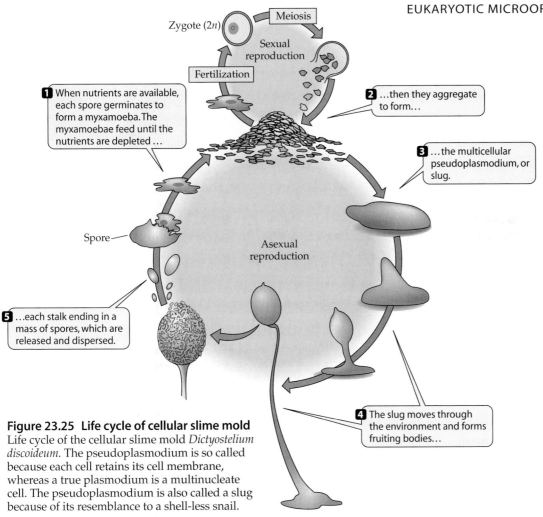

Zygote (2n)

Meiosis

Sexual reproduction

Fertilization

1 When nutrients are available, each spore germinates to form a myxamoeba. The myxamoebae feed until the nutrients are depleted …

2 …then they aggregate to form…

3 …the multicellular pseudoplasmodium, or slug.

Spore

Asexual reproduction

5 …each stalk ending in a mass of spores, which are released and dispersed.

4 The slug moves through the environment and forms fruiting bodies…

Figure 23.25 Life cycle of cellular slime mold
Life cycle of the cellular slime mold *Dictyostelium discoideum*. The pseudoplasmodium is so called because each cell retains its cell membrane, whereas a true plasmodium is a multinucleate cell. The pseudoplasmodium is also called a slug because of its resemblance to a shell-less snail.

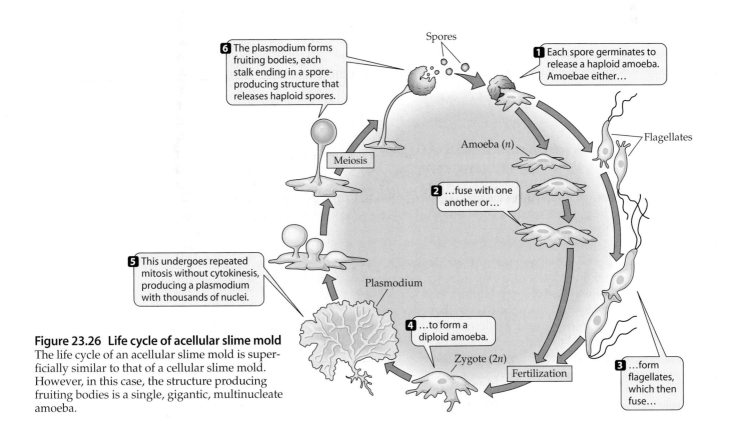

Spores

6 The plasmodium forms fruiting bodies, each stalk ending in a spore-producing structure that releases haploid spores.

1 Each spore germinates to release a haploid amoeba. Amoebae either…

Flagellates

Amoeba (n)

Meiosis

2 …fuse with one another or…

5 This undergoes repeated mitosis without cytokinesis, producing a plasmodium with thousands of nuclei.

Plasmodium

4 …to form a diploid amoeba.

Zygote (2n)

Fertilization

3 …form flagellates, which then fuse…

Figure 23.26 Life cycle of acellular slime mold
The life cycle of an acellular slime mold is superficially similar to that of a cellular slime mold. However, in this case, the structure producing fruiting bodies is a single, gigantic, multinucleate amoeba.

(A)

(B)

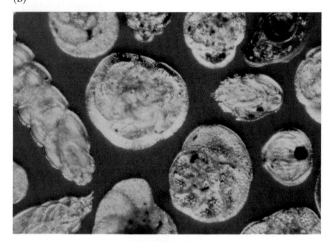

Figure 23.27 Foraminifera
A foraminiferan lives in a shell, or test, from which it extends a net of long pseudopods. (A) A living foraminiferan. Fine pseudopods can be seen radiating from the large test. (B) Foraminiferan tests. It is easy to see why foraminifera were regarded as minute mollusks when first discovered. A, ©Biophoto Associates/Photo Researchers, Inc.; B, ©Alfred Pasieka/SPL/Photo Researchers, Inc.

heliozoa (see Figure 23.2), and radiolaria (**Figure 23.28**). The foraminifera are marine amoebae with tests of calcium carbonate or an organic matrix in which sand grains are incorporated. Most are benthic forms that spread out an anastomosing net of pseudopods with which to trap food. The heliozoans and radiolarians extend long, thin rays from the body (*heliozoan* means "sun animal") composed of bundles of microtubules covered by a thin layer of cytoplasm. Small food organisms are trapped on these rays. Radiolarians are planktonic marine forms, whereas heliozoans are primarily freshwater species.

FUNGI

Fungi tend to show up as uninvited guests on Petri plates in microbiology labs, where they are immediately recognizable because of their fuzzy appearance. Such an uninvited guest landing on a Petri plate just over 70 years ago ushered in the age of antibiotics, a story recounted in Chapter 7.

Fungal spores are always in the air, in large-enough numbers to cause allergic reactions, and forever landing on things we would prefer they not. They spoil food, they consume wooden structures, they destroy fabrics and leather, and the long list of manufactured items they can attack includes such alimentary exotica as shoe polish and ink. This is done incidentally, as part of their broader role in recycling organic material. Complex macromolecules such as chitin, lignin, and cellulose are kept from accumulating in terrestrial habitats by the fungi. They are broken down into smaller molecules that can be used again by other organisms. Biomass production in forests, especially, is controlled by fungi, as their activities determine the rate at which nutrients from

Figure 23.28 Radiolaria
Artificially colored scanning electron micrograph of radiolarian tests. Photo ©Dennis Kunkel Microscopy, Inc.

dead plant material are released back into the environment. Fungi are relentless digesters of terrestrial macromolecules, and without their efforts, life on the dry parts of the planet would grind to a halt.

Fungi are also relentless producers of spores. A single mushroom can produce billions of spores. Such large numbers are necessary because higher fungi have given up moving themselves around and lack all traces of locomotor organelles. They are dependent on the passive movement of spores to move their protoplasm into places favorable for growth and reproduction. The constant rain of spores over the earth ensures that some small fraction will land on suitable food and survive. It also means that everything digestible by fungi will eventually be landed on and digested.

Fungi produce a large number of toxic substances, and people can be poisoned by eating misidentified mushrooms or unknowingly consuming fungi growing on food. For example, *Aspergillus flavus* likes to grow on grains and some nuts, including peanuts, and while growing there produces aflatoxins. They are among the most potent carcinogens known, causing cancer when present in diets in the parts per billion range. The role of most of these toxic compounds in the life of a fungus is unknown, and in many cases the toxicity is probably coincidental. In other cases, it is not. Fungi compete with each other and with bacteria for food, and their only effective means of staking a claim is chemical. If they can kill or inhibit the growth of another organism capable of using their food, they become sole proprietors of that resource. Penicillin is merely a fungus's way of saying, "This orange is mine."

Fungal contamination is not always bad. Millennia ago, when people were not as good at keeping bacteria and fungi out of their food as they are now, people learned to enjoy eating and drinking some foodstuffs contaminated with fungi and even learned how to deliberately inoculate fungi into the material, albeit without understanding either the process or organisms involved. The production of alcoholic drinks is as old as recorded history and is due to the anaerobic conversion of plant sugar to alcohol by yeasts. Yeasts are commonly associated with plants. They grow in the nectar of flowers and other places where they can obtain sugar, such as the surface of grapes. It is easy to imagine that wine was discovered almost as soon as grapes were first collected and stored. Some cheeses (blue cheeses and some soft cheeses like Camembert) are produced by the action of fungi on milk protein. Soy sauce is produced by the digestion of a mixture of wheat and soybeans by fungi and a *Lactobacillus*. We owe leavened bread to the bubbles of carbon dioxide produced by fungi consuming the dough.

Not all fungi are content to wait until an organism is dead before starting to consume its macromolecules. After all, cellulose is just as nutritious extending upward to the sky as lying flat on the forest floor. Fungi that are going to attack the tissues of living organisms have their prey's defenses to overcome, but the large number of fungal pathogens that exist indicates that overcoming those defenses is not an insurmountable problem. Fungi are the most important cause of plant diseases, with well over 5,000 species of plant pathogens known. The loss of chestnut trees from the forests of North America was due to a fungus. Wheat rust and Dutch elm disease are caused by fungi. A few fungi also cause disease in humans; ringworm and athlete's foot are both caused by fungi.

The association between phototrophs and fungi is not always destructive. The great majority of terrestrial plants have fungi growing in their roots in a mutually beneficial association ("mycorrhizae"). There is a two-way exchange of goods, with the fungus receiving photosynthate and the plant receiving inorganic nutrients, especially phosphorus, that the fungus has absorbed from the soil. For a modest nutritional investment the plant receives, in effect, an extended accessory root system. Lichens are an association between microbial phototrophs (both eukaryotic and prokaryotic) and fungi. Both partners benefit from the association and can live in places that neither partner could alone. Both mycorrhizae and lichens are described in more detail in Chapter 25.

Most fungi conduct their lives out of our sight in the form of microscopic threads, called hyphae, running over and through their food. They secrete enzymes to break the substrate into smaller organic molecules that the hyphae can then absorb. The collection of connected hyphae forming one organism is called a mycelium; a mycelium and its spores are called a thallus. During growth, division of the nuclei may be accompanied by the formation of septa, marking the boundaries of one cell. However, septa formation is usually incomplete, leaving a pore connecting compartments, and sometimes doesn't occur at all. This means that the contents of a mycelium are a single connected multinucleate mass of cytoplasm. If asked to imagine the body of a fungus, most people would conjure up mental images of mushrooms and puffballs rather than hyphae. However, these larger structures (in the fraction of fungi that actually possess them) are tiny and temporary appendages of a fungus's true body, formed only at the moment when spores are produced. A mushroom is just the tip of a mycelial iceberg; an acre of forest soil, with a mushroom popping up here and there, may contain several tons of hyphae.

The true fungi may be described as colorless, saprotrophic, predominantly terrestrial organisms that reproduce through the production of spores. Their cell walls are chitinous. Flagella do not occur at any stage of the life cycle. The different types of true fungi are clearly related to each other and form a monophyletic unit. How they arose from protists is a little clearer now than in the past. In the past, organisms such as slime molds and oomycetes were presumed to connect the fungi to the protists, because in addition to their protistan features, they produced fruiting bodies. Studies on ultrastructure and life cycles, as well as rRNA comparisons, show, in fact, that these forms are quite unrelated to fungi.

Chytrids

The Chytridiomycetes, or chytrids, which are the closest relatives of higher fungi in rRNA trees, are the one link remaining between the true fungi and the protists. They produce sporangia like fungi, but have flagellated zoospores. The zoospores usually have two basal bodies, only one of which has a flagellum. The flagellum emerges at the posterior end of the cell. The life cycle is extremely variable. Sexual reproduction may or may not occur. The entire organism may consist either of a single cell or a multinucleate branching mass resembling the mycelia of fungi, albeit much smaller. Most live in water or soil, but there are some multiflagellated Chytridiomycetes that live in the rumen of cattle. A few chytrids are parasitic. One of these, *Synchytrium endobio-*

ticum, causes black wart disease in potatoes. *Allomyces* species have some of the more complicated chytrid life cycles (**Figure 23.29**). These are aquatic species with an alternation of diploid and haploid phases in the life cycle. The diploid thallus (sporothallus) can produce diploid zoospores that swim off, settle on something, and produce another generation of sporothalli, or it can produce haploid zoospores through meiosis. These zoospores grow into haploid thalli (gametothallus) after settling onto the substrate. The zoospores produced by a gametothallus are gametes; they fuse to form diploid zygotes, which will grow into sporothalli.

The true fungi are divided into four major groups. These are the zygomycetes, ascomycetes, basidiomycetes, and the deuteromycetes (also called the Fungi Imperfecti). Three of these are natural groups whose members are united by a common evolutionary history. The deuteromycetes are an artificial group made up of members whose incompletely known life cycles prevent determination of their true taxonomic placement.

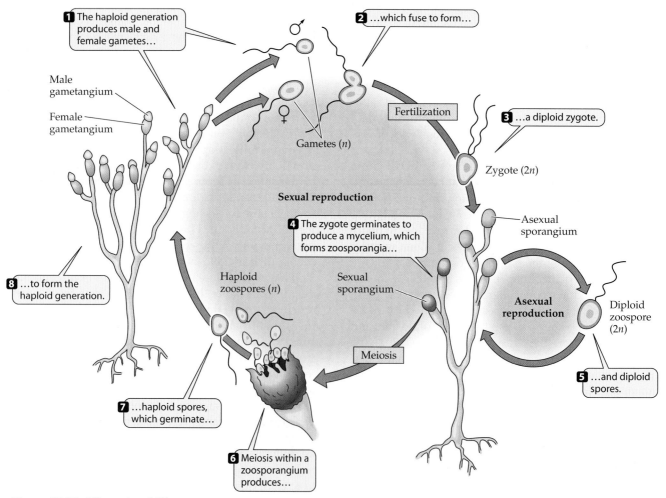

1 The haploid generation produces male and female gametes...

2 ...which fuse to form...

Male gametangium

Female gametangium

3 ...a diploid zygote.

Fertilization

Gametes (*n*)

Zygote (2*n*)

Asexual sporangium

Sexual reproduction

4 The zygote germinates to produce a mycelium, which forms zoosporangia...

Haploid zoospores (*n*)

Sexual sporangium

Asexual reproduction

8 ...to form the haploid generation.

Diploid zoospore (2*n*)

Meiosis

7 ...haploid spores, which germinate...

5 ...and diploid spores.

6 Meiosis within a zoosporangium produces...

Figure 23.29 Life cycle of *Allomyces*
Alternation of generations in the genus *Allomyces*, an aquatic fungus.

Zygomycetes *Rhizopus stolonifer* (**Figure 23.30**), a bread mold, is an example of the zygomycetes. The cycle begins when a spore lands on a piece of bread and germinates. Grayish hyphae grow rapidly across the bread.

Some grow down into it and absorb nutrients. Others grow up into the air. The hyphal tips swell, and nuclei collect in the swollen ends. The protoplasm here breaks up into spherical uninucleate masses, each of which is

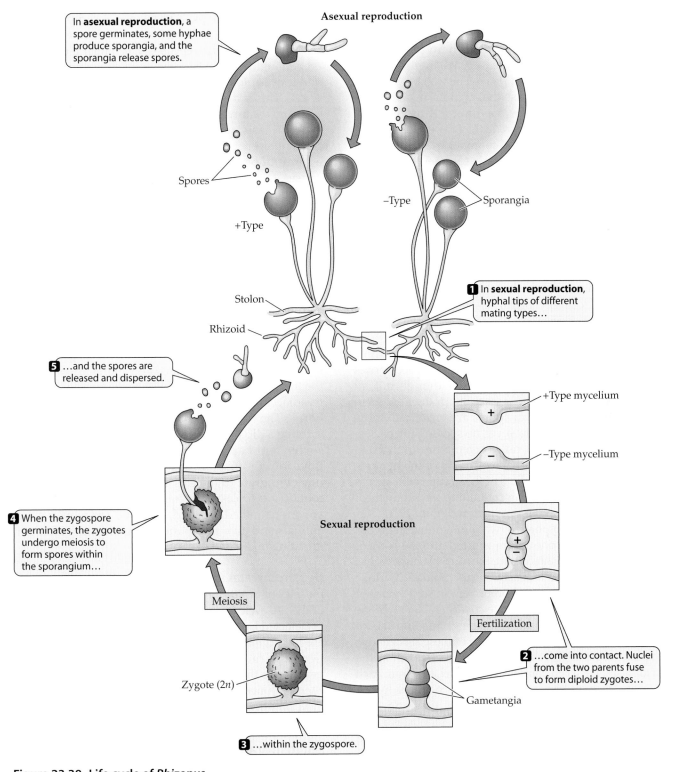

Asexual reproduction

In **asexual reproduction**, a spore germinates, some hyphae produce sporangia, and the sporangia release spores.

Spores

+Type

−Type

Sporangia

Stolon

Rhizoid

1 In **sexual reproduction**, hyphal tips of different mating types…

5 …and the spores are released and dispersed.

+Type mycelium

−Type mycelium

Sexual reproduction

4 When the zygospore germinates, the zygotes undergo meiosis to form spores within the sporangium…

Meiosis

Fertilization

Zygote (2*n*)

Gametangia

2 …come into contact. Nuclei from the two parents fuse to form diploid zygotes…

3 …within the zygospore.

Figure 23.30 Life cycle of *Rhizopus*
Life cycle of the bread mold *Rhizopus stolonifera*.

ultimately covered by a resistant cell wall. These are spores. The spherical hyphal tips, now called sporangia, eventually rupture and release the spores. If they settle on a new piece of bread, the spores germinate and the cycle is repeated. The asexual production of new generations can continue indefinitely in this manner. However, sexual reproduction can also occur. This necessitates the presence of two mycelia of compatible mating types on the bread.

When hyphal tips from two such mycelia contact each other, a septum is formed in each hypha at some distance behind the contact point. The wall between the tips is then broken down so that cytoplasm and nuclei from the two parental hyphae mix in this chamber. Eventually the two types of nuclei fuse in pairs, forming zygotes. The wall of the chamber becomes a thick black structure called a zygospore. After several months, the zygospore germinates, at which time the zygotes undergo meiosis, and a sporangium containing the division products emerges. The spores from this sporangium germinate and grow into hyphae if they land on a suitable piece of bread. They may reproduce either sexually or asexually. The organism is haploid throughout its entire life cycle except at the zygospore stage. Most zygometes are fairly inconspicuous terrestrial fungi. Some form mycorrhizae. A few are parasitic in animals. Some species have practical uses; one of these, *Rhizopus oryzae*, is used in making sake.

Ascomycetes
The ascomycetes may produce spores asexually as do the zygomycetes, although not all ascomycetes are capable of asexual reproduction. As in the zygomycetes, asexual spores are produced at the ends of raised hyphae. However, no sporangium encloses the spores. The spores, called conidia, are arranged in chains extending outward from the tip of the hyphae. The feature uniting members of the group is the formation of an ascus (a little bag containing the products of meiosis) during sexual reproduction (**Figure 23.31**).

When haploid hyphae of two different mating types meet, the nuclei of one are transferred into the body of the other. The hypha receiving the nuclei is regarded as female and is called an ascogonium. Somatic hyphae around the fertilized ascogonium are in some manner stimulated to form a protective covering. New hyphae grow from the ascogonium, containing a mixture of both parental nuclei. During the growth of these ascogenous hyphae, septa are produced that compartmentalize the cytoplasm. Each cell contains two nuclei, one from each parent. The penultimate cell of each ascogenous hypha grows into an ascus. The two nuclei fuse and then undergo meiosis to produce four daughter nuclei. These may undergo mitosis, in which case the developing ascus will contain eight nuclei. Eventually each nucleus and the cytoplasm around it differentiate into a spore. These will germinate when they reach a suitable food source and give rise to the next mycelial generation.

Superficially, the organisms in this group are very dissimilar, but all are united by the production of asci during sexual reproduction. Ascomycetes include cup fungi, powdery mildews, and even yeasts. Yeasts do not form mycelia; they are unicellular forms that reproduce by splitting in half or budding. Sexual reproduction also occurs when two haploid yeast cells of complementary mating types fuse. A diploid zygote is formed that can multiply by dividing repeatedly, producing many separate daughter cells. At some point, meiosis is induced, and four haploid cells come to lie within a single mother cell. This is the equivalent of an ascus.

Many ascomycetes are important to humans. In addition to yeasts, which are used in baking as well as in brewing beer and wine, the group includes truffles and a series of pathogens of crop plants. *Ceratocystis ulmi* causes Dutch elm disease. *Endothia parasitica* is the pathogen that has wiped out American chestnut trees. Powdery mildews on fruit can cause enormous losses.

Basidiomycetes
This group includes puffballs, smuts, mushrooms, and bracket fungi. Formation of the macroscopic spore-containing fruiting bodies requires the union of two mycelia of different mating types (see **Figure 23.32**). As in the ascomycetes, the two sets of haploid parental nuclei multiply together in growing hyphae for some time before fusing into diploid nuclei. At some point, the characteristic macroscopic fruiting body for the species is formed and spores are produced; the tip cells of the growing hyphae are transformed into basidia, club-shaped structures, on which spores will be borne. The two nuclei in a developing basidium fuse, forming a diploid zygote. This then undergoes meiosis, so that four haploid nuclei are produced. At the same time, four protuberances appear on the basidium. The nuclei migrate to the tips and, together with the surrounding cytoplasm, are transformed into spores. These are disseminated by wind, rain, and the movement of animals and give rise to the next generation after germinating on a suitable substrate.

Deuteromycetes (or Fungi Imperfecti)
These are ascomycetes or basidiomycetes that have lost the sexual part of the life cycle (or perhaps have one that has remained undetected by mycologists). Many species in this group are economically important. The fungi that produce penicillin (members of the genus *Penicillium*) belong here, as do the fungi used in the production of several cheeses (once again including members of the genus *Penicillium*). Among the pathogens are those that cause celery leaf spot, celery blight, and potato blight.

Microsporans

The **microsporans** (**Figure 23.33**) are a group of about 800 parasitic species that live intracellularly in the tissues of animals and in other protists. The infective stage

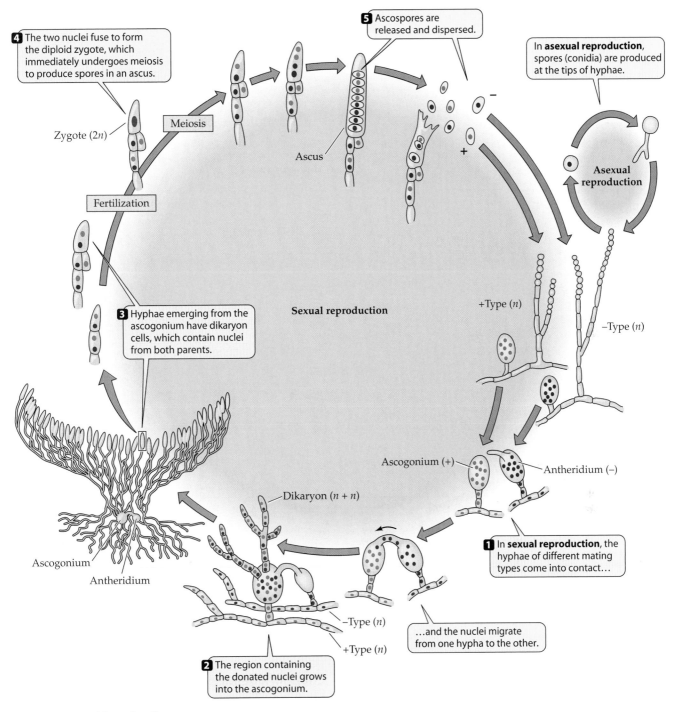

4 The two nuclei fuse to form the diploid zygote, which immediately undergoes meiosis to produce spores in an ascus.

Zygote (2*n*)

Meiosis

Fertilization

3 Hyphae emerging from the ascogonium have dikaryon cells, which contain nuclei from both parents.

Ascogonium

Antheridium

5 Ascospores are released and dispersed.

Ascus

In **asexual reproduction**, spores (conidia) are produced at the tips of hyphae.

Asexual reproduction

+Type (*n*)

–Type (*n*)

Sexual reproduction

Ascogonium (+)

Antheridium (–)

Dikaryon (*n* + *n*)

1 In **sexual reproduction**, the hyphae of different mating types come into contact…

–Type (*n*)

+Type (*n*)

…and the nuclei migrate from one hypha to the other.

2 The region containing the donated nuclei grows into the ascogonium.

Figure 23.31 Life cycle of an ascomycete
Life cycle of the cup fungus, an ascomycete. The cup (the fruiting body) consists of a mass of hyphae produced from two types of spores.

is transmitted from host to host in a resistant spore, which is typically about 4 μm to 5 μm in length (Figure 23.33, **1**). Most spores contain a coiled hollow filament that is everted and can penetrate a host cell after the spore is ingested. The microsporan cytoplasm then travels down the tube into the host cell where it begins growing and dividing (Figure 23.33, **2** through **4**). Ultimately, the microsporan goes through a division sequence in which a new generation of spores is produced. They may be passed from the body in feces or urine or retained until the death of the host, depending on the location in which sporulation occurs.

All microsporans undergo rounds of division leading to spore production (sporogony) (Figure 23.33, **6**). Most are also known to undergo at least one earlier round of multiplication (merogony) (Figure 23.33, **5**), leading to the production of more growth stages in the host.

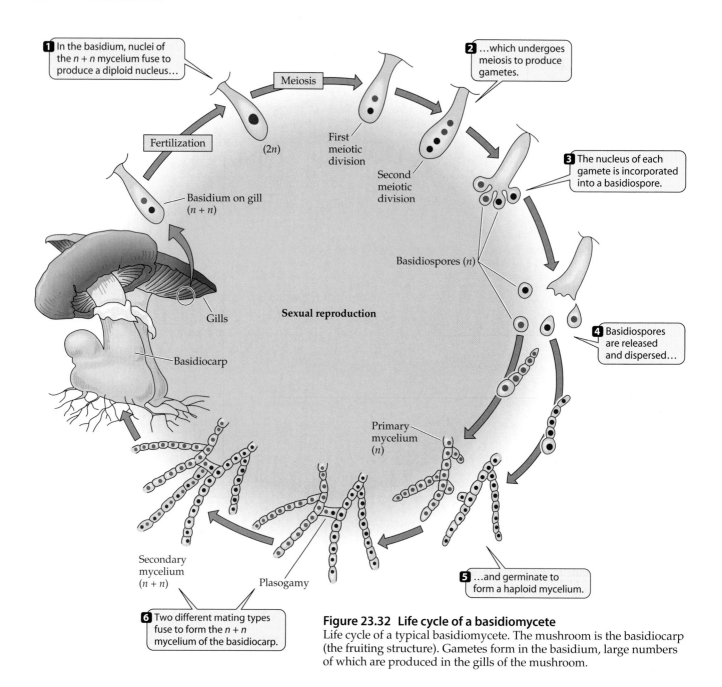

1 In the basidium, nuclei of the *n + n* mycelium fuse to produce a diploid nucleus...

2 ...which undergoes meiosis to produce gametes.

Meiosis

Fertilization

(2*n*)

First meiotic division

Second meiotic division

3 The nucleus of each gamete is incorporated into a basidiospore.

Basidium on gill (*n + n*)

Basidiospores (*n*)

Gills

Basidiocarp

Sexual reproduction

4 Basidiospores are released and dispersed...

Primary mycelium (*n*)

Secondary mycelium (*n + n*)

Plasogamy

5 ...and germinate to form a haploid mycelium.

6 Two different mating types fuse to form the *n + n* mycelium of the basidiocarp.

Figure 23.32 Life cycle of a basidiomycete
Life cycle of a typical basidiomycete. The mushroom is the basidiocarp (the fruiting structure). Gametes form in the basidium, large numbers of which are produced in the gills of the mushroom.

There are no flagella or centrioles at any stage of the life cycle. Dense plaques on the nuclear membrane serve as microtubular organizing centers in mitosis, which is closed. Meiosis has been seen, but how sexual recombination fits into the life cycle is mostly unknown.

Microsporans apparently lack lysosomes and may also lack a Golgi apparatus. Structures regarded as Golgi exist, but they do not have the morphology typically seen in eukaryotes. Microsporans also have unusual ribosomal RNAs. In addition to being prokaryotic in size, the 5′ end of the large RNA is not split off to form a 5.8S piece, as it is in other eukaryotes. Protein sequences indicate that microsporans are fungi of some type, but how they are related to other fungi is unknown.

Choanoflagellates

Choanoflagellates (**Figure 23.34**) are small, primarily marine, phagotrophic organisms. They have a single flagellum that beats in such a way as to draw a stream of water toward the cell body. As the water flows past the body, it is drawn through a filter consisting of a ring of tentacles running around the top part of the cell. Bacteria are trapped by these tentacles and consumed by the cell. Choanoflagellates frequently produce a delicate lorica, an external basket or shell, in which the flagellate sits.

In phylogenetic trees based on RNA sequences, choanoflagellates are the protistans that are closest to the animals, so animals probably arose from them or an unknown protistan group quite a bit like them. Cells

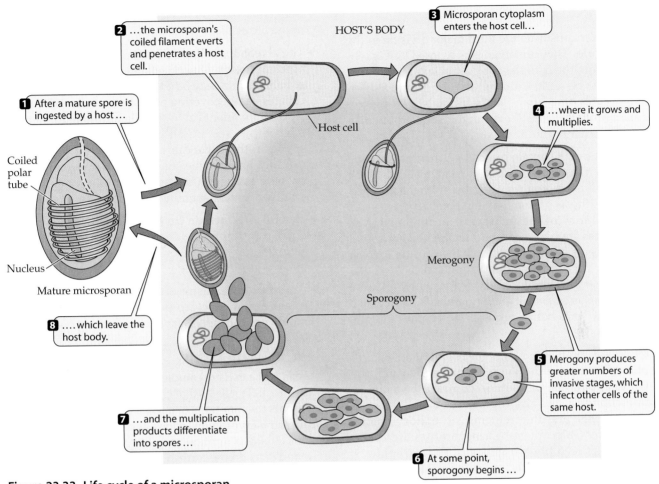

Figure 23.33 Life cycle of a microsporan
Life cycle typical of the microsporans, parasitic species.

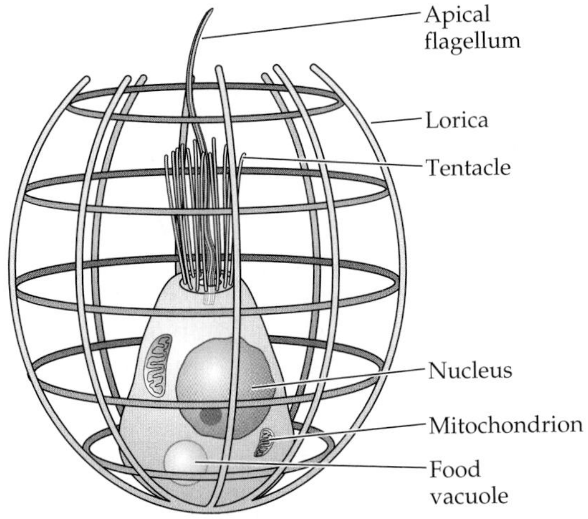

[Labels in figure:]

2 ...the microsporan's coiled filament everts and penetrates a host cell.

3 Microsporan cytoplasm enters the host cell...

HOST'S BODY

1 After a mature spore is ingested by a host ...

4 ...where it grows and multiplies.

Host cell

Coiled polar tube

Nucleus

Mature microsporan

Merogony

Sporogony

8 ....which leave the host body.

5 Merogony produces greater numbers of invasive stages, which infect other cells of the same host.

7 ...and the multiplication products differentiate into spores ...

6 At some point, sporogony begins ...

Figure 23.34 Choanoflagellates
Choanoflagellate with a lorica. The cell is contained within a delicate basket-like structure. The tentacles serve to catch food.

[Labels in figure:]
Apical flagellum
Lorica
Tentacle
Nucleus
Mitochondrion
Food vacuole

---

that have a similar collar of tentacles and that feed in the same way are found in sponges.

EUKARYOTES AND MULTICELLULARITY

All three of the primarily multicellular eukaryotic groups (plants, animals, and fungi) emerge from approximately the same place in an rRNA tree. There was no reason to believe that they would be so closely related, inasmuch as multicellularity has arisen in many protistan lineages. The red algae, heterokonts, chlorophytes, and ciliates (as well as many smaller groups) all show a series of forms ranging from uni- to multicellular. Genuine differentiation of cells occurs in these groups; the algae have cells specialized for reproduction, whereas in slime molds multicellularity and differentiation are necessary to produce the fruiting bodies. Plants, animals, and fungi, the eukaryotes most familiar to us, are small, recent offshoots of the protistan world that reflect the basic protistan evolutionary drive to escape the limitations of unicellularity.

SUMMARY

▸ **Eukaryotes** constitute the third major division in the living world. They are organisms whose DNA is separated from the cytoplasm by a **nuclear** membrane. The may be either unicellular or multicellular. Eukaryotes other than plants, animals, or fungi are called **protists**. Protists, even in the same group, may be unicellular or multicellular.

▸ Eukaryotic cells contain more DNA than prokaryotic cells, and transmission of DNA to daughter cells at the time of division requires a special mechanism called **mitosis**. Mitosis normally immediately precedes cell division, or **cytokinesis**. Genetic recombination in eukaryotes involves combining the chromosomal complements of two gametes by their fusion to form a zygote. This doubling of DNA must be balanced at some other point in the life cycle by a halving of the DNA. This is accomplished by **meiosis**, which also randomly sorts homologous chromosomes into the gametes.

▸ Other membrane-bound organelles found in eukaryotic cells include the **endoplasmic reticulum**, the **Golgi apparatus, lysosomes, contractile vacuoles, mitochondria,** and **chloroplasts**. Mitochondria and chloroplasts are the remains of endosymbiotic *Eubacteria.* Chloroplast-bearing protists are called algae. They feed **phototrophically. Phagotrophic,** or particle-eating, protists are called **protozoa.** Some members of both groups also feed **saprotrophically,** by absorbing dissolved nutrients.

▸ Protists contain structural elements collectively called the **cytoskeleton,** which are responsible for maintaining the cell shape. Three cytoskeletal components found among most protists are **microtubules, microfilaments,** and **intermediate filaments.** Locomotion is by means of **pseudopods, cilia,** or **flagella.** Cilia and flagella grow out of **basal bodies** or **kinetosomes,** which are ultrastructurally identical to the **centrioles** found at the poles of mitotic spindles.

▸ rRNA sequences are useful in revealing relationships among eukaryotes, just as they have been for prokaryotes. The deepest branches in rRNA trees lead to two groups of flagellated protists—**diplomonads** and **parabasalians.** Most live in the intestinal tracts of animals. A few are free living. Many parabasalians live in termites, where they are responsible for cellulose digestion.

▸ The aerobic protists found in the deepest branches of rRNA trees are those whose mitochondria contain discoid cristae—the **euglenoids, kinetoplastids,** and **heteroloboseans.** Euglenoids and kinetoplastids are flagellates. Several kinetoplastids cause serious diseases in humans and domestic animals. Heteroloboseans frequently have both flagellate and amoeboid stages in the life cycle.

▸ The alveolates include **dinoflagellates, ciliates,** and **apicomplexans.** Dinoflagellates live in a wide variety of habitats and may be either free-living or symbiotic. Ciliates have a covering of cilia, organelles of motility structurally identical to flagella but more numerous. Apicomplexans are all parasitic and include such species as those that cause malaria. All apicomplexans produce an invasive stage called a **sporozoite.** The sporozoite contains an apical complex of organelles involved in the penetration of host cells.

▸ Heterokonts include **chrysophytes, diatoms, brown algae, xanthophytes,** and **oomycetes.** Many species have chloroplasts. If chloroplasts are present, they contain chlorophylls *a* and *c.*

▸ Amoeboid organisms are scattered throughout the tree. **Entamoebae** are parasites of animals. **Slime molds** produce spore-containing fruiting bodies similar to those of fungi.

▸ The **red algae** are photosynthetic forms that lack flagella and can have complex life cycles. **Cryptomonads** are small flagellates that contain a chloroplast that represents the remnants of a red algal cell ingested by an ancestral cryptomonad long ago. **Green algae** may be unicellular or multicellular. They contain chlorophylls *a* and *b*, as do the terrestrial plants to which they are related. **Choanoflagellates** are small, colorless, filter-feeding organisms that trap food in a collar of tentacles surrounding their single flagellum. They are structurally similar to some cells found in sponges. **Chytrids** are related to fungi. They produce sporangia as do fungi, but have flagellated zoospores.

▸ The four groups of true fungi are very similar and are closely related. The basic structure of vegetative growth is the **hypha,** which is a multinucleated threadlike structure. Spores can be produced sexually or asexually. The groups are distinguished by differences in the sexual phase of the life cycle. In the **ascomycetes,** meiosis occurs in a sack called an **ascus.** In **basidiomycetes,** zygote formation and meiosis occur in a **basidium,** on which the spores are eventually borne. **Zygomycetes** produce neither of these. The **deuteromycetes** contain fungi that have lost the sexual phase of the life cycle. A group of entirely parasitic organisms, the microsporans, are related to fungi.

REVIEW QUESTIONS

1. What are some of the ways in which eukaryotes differ from the two prokaryotic groups?

2. List some of the different ways in which sexual recombination is incorporated into protistan life cycles.

3. Which protistans move by using flagella? Cilia? Pseudopods? More than one type of organelle?

4. Which protistan groups contain parasitic forms?

5. Which protists have chloroplasts?

6. List some organelles that are unique to a given group of protists and by which they may be identified.

7. How are the different groups of fungi distinguished?

8. Name some of the ways fungi affect human life.

SUGGESTED READING

Alexopoulos, C. J., C. W. Mims and M. Blackwell. 1996. *Introductory Mycology*. 4th ed. New York: John Wiley & Sons.

Graham, L. E. and L. W. Wilcox. 1999. *Algae.* Upper Saddle River, NJ: Prentice Hall.

Hausmann, K., N. Hulsmann, H. Machemer and M. Mulisch. 1996. *Protozoology*. 2nd ed. New York: Thieme Medical Publishers.

Kendrick, B. 1992. *The Fifth Kingdom*. 2nd ed. Waterloo, Ontario: Mycologue Publications.

Margulis, L. 1993. *Symbiosis in Cell Evolution*. 2nd ed. San Francisco: W. H. Freeman and Co.

Margulis, L., J. O. Corliss, M. Melkonian and D. J. Chapman. 1990. *Handbook of Protoctista*. Boston: Jones and Bartlett.

Patterson, D. J. and S. Hedley. 1992. *Free-Living Freshwater Protozoa: A Color Guide.* Boca Raton, FL: CRC Press, Inc.

Sze, P. 1997. *A Biology of the Algae.* 3rd ed. Dubuque, IA: William C. Brown Publishers.

Microbial Ecology

PART VI Microbial Ecology

Previous page
Microbial mat communities can flourish in habitats shunned by all other organisms, such as this thermal pool in Yellowstone National Park. ©Fritz Pölking/Visuals Unlimited.

Chapter 24

Microorganisms and Ecosystems

The ultimate aim of ecology is to understand the relationships of all organisms to their environment.

– R. E. HUNGATE, IN THE BACTERIA, 1962

Wkhat are the activities of microorganisms in the environment? How can these activities be measured? What microorganisms are responsible for which activities? Where are microorganisms located in the environment? Are they active at all times, or do they have dormant periods? What controls the rates of their activities and population sizes?

These are the some of the questions that the microbial ecologist is trying to answer. Microbial ecology is a young field of investigation, so there are tremendous gaps in our knowledge of the diversity of microorganisms, their distribution in natural environments, and their activities in natural communities. In fact, many important questions about microbial activities remain unanswered or have not yet been addressed due, at least in part, to the difficulties of studying such small organisms. Higher organisms such as birds and mammals can be readily observed in their natural environment and studied as they go about their daily activities. In contrast, microorganisms are too small to be seen without the use of a microscope, and because they are so simple structurally, most cannot be identified solely by microscopic observation.

Why is the study of microbial ecology so important? Microbial activities in ecosystems are important because:

- Microorganisms live in all ecosystems.
- Microorganisms, particularly *Bacteria* and *Archaea*, carry out unique activities not performed by other organisms.
- Microorganisms formed the first ecosystems and biosphere, and their activities are essential for all life on Earth.

As a group, microorganisms are more widely distributed on Earth than plants and animals. Microorganisms live in all **ecosystems,** which are defined as major geographic entities such as a lake that contain both biotic and the associated abiotic components of the environment. Indeed, microorganisms even occupy some of their own special ecosystems, **microbial ecosystems,** in areas that are too inhospitable for macroorganisms such as thermal, hypersaline, acidic, alkaline, and anoxic environments. Furthermore, microorganisms live in the intestinal tracts and on the surface of virtually all animals, form close associations with the plant rhizosphere, and are therefore found wherever animals and plants are located. For these reasons, the **microbial biosphere** is considerably more extensive than the biosphere of macroorganisms.

In addition, microorganisms are immensely important because, collectively, they are responsible for mediating many fundamental biologically mediated geochemical (Earth chemistry) transformations that are essential to all forms of life in the biosphere. Furthermore, many of the transformations of the elemental cycles that occur on Earth, called **biogeochemical processes**, are uniquely carried out by bacteria. Most of these reactions began billions of years ago, long before plants and animals evolved. Therefore, plants and animals are dependent on the transformations that microorganisms perform that are fundamental to the operation of the biosphere.

The chemical transformations that are carried out by microorganisms are often unapparent. For example, the process of nitrification and the bacteria that carry it out are not detectable by our unaided senses, so it is natural that most people are simply unaware that this important process even occurs. In point of fact, these and many other microbial processes that occur in nature cannot be readily differentiated from chemical processes, certainly not without careful scientific investigation.

Although it is true that the studies of the ecology of macroorganisms and microorganisms share common principles and ecosystems, the small size of microorganisms, their rapid growth rates, their large population sizes, the many unique geochemical transformations they perform, and the advantages of studying them in pure culture make microbial ecology a special area of investigation. Their small sizes mean that they create their own microscale environment as discussed next.

THE MICROENVIRONMENT

The typical bacterial cell is about one-millionth the length of a human being. Because microorganisms are so small, the natural environment, or **microenvironment**, in which they live and perform their activities is correspondingly small. The microorganism carries out important biochemical reactions that influence the physical and chemical conditions of the microenvironment immediately around its cells. As a result, the concentration of substrates and products is different in the vicinity of the cells as compared to the bulk environment that is measured with ordinary electrodes and by chemical analyses. Therefore, in order to study what is happening in the microenvironment, it is necessary to amplify these activities with appropriate instruments, a challenging aspect for scientists. For example, microscopes are needed to magnify the cells so they can be visualized. **Figure 24.1** shows typical microcolonies of bacteria that have grown on an electron microscope grid that was attached to a glass slide and immersed in a lake. Note that two of the cells can be identified as a *Caulobacter* sp. because of their distinctive prosthecae (see Chapter 19).

The small size of the microenvironment requires that special consideration be given to designing instruments to measure conditions in the vicinity of the organisms. **Microelectrodes** for the measurement of oxygen or pH are examples of such instrumentation. These fine probes come equipped with tips as small as about 5 to 10 μm in diameter. With these instruments, microbial ecologists

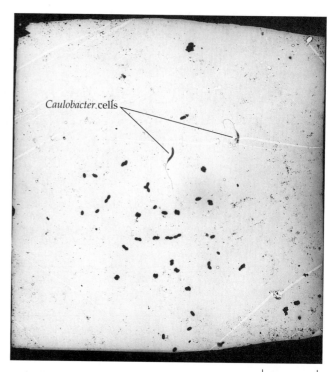

10.0 μm

Figure 24.1 Microcolonies
Microcolonies formed in situ. A microcolony of rod-shaped bacteria growing on an electron microscope grid attached to a glass slide that was immersed in a lake. Note the two attached *Caulobacter* cells with their prosthecae.

have demonstrated that the oxygen concentration and pH near colonized areas can be quite different in a span of a few micrometers (**Box 24.1**).

For these reasons, microbial ecologists need to be constantly aware that the growth and activities of microorganisms occur in the environment at a microscale. Nonetheless, the concerted action of enormous numbers of microorganisms are responsible for producing metabolic products and chemical gradients that are apparent at macroscopic scales in soils, aquatic environments, and the atmosphere.

Three basic issues that microbial ecologists need to address concern the identification, abundance, and location of microorganisms in the natural environment of interest. We begin by discussing autecology, the study of the individual microorganism in the microenvironment as well as population studies of the species in the macroenvironment.

AUTECOLOGY: ECOLOGY OF THE SPECIES

Three properties concerning the autecology of a species are of utmost importance:

- Identification of the species in the environment

Research Highlights Box 24.1

Thinking Small: A Microbial World Within a Marine Snow Particle

Marine snow is found in the water column of marine habitats. It is called marine snow because as these particles sediment in the water column, they are reminiscent of falling snowflakes to scuba divers. The marine snow particles range in size from 1 to about 10 mm in diameter and may contain photosynthetic algae and cyanobacteria as well as heterotrophic bacteria and protozoa.

Microprobes, with tips that are only 5 to 10 μm in diameter, were used by A. L. Alleridge and Y. Cohen to measure the pH and oxygen concentrations at the surface and at various points in the interior of marine snow particles. Microgradients of oxygen and pH were found from the surface to the interior of the particles. For example, the oxygen concentration in the interior of the particle can be half that of the bulk water in which the marine snow particle is found. It is lower inside the particle because of aerobic respiration that depletes the oxygen more quickly than it can be replenished by diffusion.

- Determination of the **habitat** (the *location* or "address") of the species
- Identification of the **niche** (the *activity* or "profession") for which the species is uniquely responsible

Identification of Species in the Environment

As mentioned previously, it is not usually possible to identify a typical unicellular bacterium at the level of species, genus, or perhaps even kingdom by observing it in the microscope. This is because most bacteria have very simple shapes, at least as they can be discerned by light microscopy. Thus, members of the genus *Nitrosomonas* appear very similar to *Escherichia coli*, although the former is a genus of chemolithotrophic ammonia-oxidizing bacteria, whereas the latter is an ordinary heterotroph. Two common approaches for identification are:

- Fluorescent antibody approach
- Fluorescent in situ hybridization (FISH)

Fluorescent Antibody Approach The problem of in situ identification of microorganisms can be solved by the use of **fluorescent antibodies** that are used to microscopically identify a species. A fluorescent dye, such as fluorescein, is covalently tagged to a whole-cell antiserum or to a monoclonal antiserum prepared from the organism of interest. An environmental sample that contains the organism can be stained with the fluorescent antiserum and examined microscopically using an ultraviolet or halogen light source. Cells of the organism of interest will fluoresce because they have been labeled by the specific antiserum. In fact, this is a preferred procedure for detection and identification of *Legionella* species in natural samples, because these bacteria are so difficult to grow.

Fluorescent in Situ Hybridization (FISH) More recently, considerable interest has been generated using nucleic acid probes (either labeled RNA or DNA) for the identification of microorganisms from the environment. In one application of this procedure, an oligonucleotide specific for a species or other taxonomic group (e.g., rDNA) can be tagged with a fluorescent dye. Then the hybridization reaction is prepared directly on a glass microscope slide with organisms from a natural sample and the fluorescent oligonucleotide. The preparation is then observed using a fluorescent microscope (Figure 17.12). The FISH label is usually targeted against rRNA because there are more ribosomes than DNA molecule sites, so the ribosome-labeled cell will fluoresce more brightly. Fluorescent rRNA probes require that the organism of interest have a sufficient number of ribosomes (about 1,000 per cell are needed) to serve as targets for the cell. Inasmuch as the concentration of ribosomes is directly related to the growth rate of an organism, only actively growing cells can be detected with these probes.

Counting Microorganisms in the Environment

In this section, we will concentrate on counting or enumerating bacteria, in particular heterotrophic bacteria, in natural habitats. When enumerating bacteria in natural environments, it is important to recognize that not all small organisms in the habitat are heterotrophic bacteria. Depending on the environment, large numbers of cyanobacteria, small algae, or autotrophic bacteria may also be found. As we will discuss, these other microorganisms can usually be readily distinguished from heterotrophic bacteria.

The fundamental problem of counting individual bacteria in natural environments is challenging for two primary reasons. First, because bacteria are so small, they must be visualized with a microscope. Second, even if we can see and count them, determining whether they are dead or alive is not a simple matter. Also, as we discussed previously, it is usually impossible to distinguish one species from another. Initially, we will address how bacteria are counted and then consider how it is possible to determine whether or not they are alive or metabolically active.

Total Microscopic Count One of the best ways to determine the total number of bacteria in a sample is to count

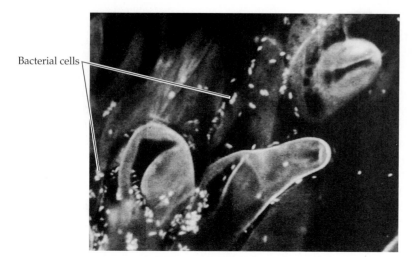

Bacterial cells

Figure 24.2 Fluorescing cells
Confocal laser photomicrograph of acridine orange–stained bacteria growing in the rhizosphere of a legume. Courtesy of Frank Dazzo.

them microscopically by **fluorescent dye staining** such as acridine orange or DAPI (4′,6-diamidino-2-phenylindole). These dyes bind to the RNA and DNA of a cell. In this staining procedure, the sample containing the bacteria is first fixed immediately after collection with formaldehyde or glutaraldehyde to preserve the organisms. They are then stained with acridine orange or DAPI. Cells that have taken up the dye appear green, blue, or orange when observed by fluorescence microscopy. This procedure is one of the best known for accurate counting of bacterial cells (**Figure 24.2**). Furthermore, because this is an epifluorescence procedure that uses incident light (Chapter 4), it is possible to observe cells attached to soil particles and other opaque materials. It should be noted that it is not possible to distinguish living from dead cells through this technique—*this is a total count of both living and dead cells.*

Cyanobacteria that might be present in the sample can be distinguished from other bacteria because the *Cyanobacteria* have chlorophyll *a*, which produces a natural red fluorescence without acridine orange staining. Thus, control samples are typically prepared to determine whether chlorophyll *a*–containing *Cyanobacteria* and *Algae* are present in a sample. It is not possible to distinguish lithotrophic bacteria from heterotrophic bacteria, although in most environments chemolithotrophs would be expected to occur in low numbers relative to the heterotrophs.

An alternative to microscopic counting is the use of **cell sorters** or **flow cytometers,** instruments developed for separation of blood cells in medicine. With minor modifications, these instruments can be used to separate microbial cells from one another, based on size. Furthermore, with special detectors it is possible to separate fluorescent particles (such as acridine orange–stained cells) from other particles and examine that group specifically. This automated procedure is being increasingly used in microbial ecology.

Viable Counts If one wishes to know the numbers of bacteria that are alive in a sample, it is necessary to use a viable counting procedure in which the bacteria are actually grown. The traditional method used for this is the spread plating procedure, in which a medium prepared to grow the heterotrophic bacteria is first poured into the Petri dish (Chapter 5), and a dilution of bacteria is spread over the surface of the plate with a sterile glass rod. This procedure is superior to pour plating because the high temperature required to maintain the agar in a molten state while mixing with the bacteria kills many psychrophilic and mesophilic bacteria.

An alternative to plate counts is the use of liquid media. For example, samples can be quantitatively diluted and inoculated into a liquid growth medium for viable enumeration such as the most probable number (MPN) procedure (see Chapter 32). This procedure has been found to provide higher counts of viable bacteria than plating.

The Bacterial Enumeration Anomaly

Major discrepancies are found between the total microscopic counts and viable counts of bacteria from many habitats. For example, in **oligotrophic** (low concentrations of nutrients) or **mesotrophic** (moderate concen-

Table 24.1	Bacterial enumeration anomaly illustrated for a mesotrophic lake, Lake Washington, compared with a eutrophic pulp mill aeration lagoon		
	Bacterial Cells/ml		
Trophic Status	**Total Count**	**Viable Count**	**% Recovery**[a]
Mesotrophic Lake Washington	3.0×10^6	2.0×10^3	0.067
Eutrophic Pulp mill oxidation lagoon	2.1×10^7	3.1×10^7	~100[b]

[a]This refers to the ability to cultivate the bacteria on a plating medium; it is the ratio of the viable count divided by the total microscopic count.
[b]In this particular instance, the viable count was actually somewhat higher than the total count, due to minor uncertainties in measurement. Therefore, the recovery is considered to be 100%.

trations of nutrients) aquatic environments, less than 1% of the total acridine orange–staining bacteria can be grown on the best of media. In contrast, in **eutrophic** environments (rich in nutrients) such as wastewaters, the recovery of bacteria can approach 100% of the total count (**Table 24.1**). The inability to recover viable bacteria from oligotrophic and mesotrophic environments is called the **bacterial enumeration anomaly,** or the "Great Plate Count" anomaly. There are three possible explanations for this result. First, there is no such thing as a "universal medium" that will permit the growth of all bacteria, so not all viable bacteria are capable of growth on the medium that is used. Second, it is possible that many of these bacteria do not grow well enough on artificial media under laboratory conditions to form colonies. And finally some, perhaps most, of these bacteria are dead. To address the enumeration anomaly issue, three alternative microscopic procedures have been developed to identify metabolically active ("living") bacteria from natural samples:

- Microautoradiography
- INT-reduction technique
- Nalidixic acid growth technique

Each of these procedures is described briefly next.

Microautoradiography

Microautoradiography entails the use of a radiolabeled substrate such as tritiated acetate or tritiated thymidine. If a microorganism can use acetate or thymidine as a substrate, it will incorporate the radiolabeled material into its cells as it would a normal nonradioactive substrate. When an organism takes up the radiolabeled substrate, the cells become radioactive and can then be identified by autoradiography (**Figures 24.3** and **24.4**). In using this approach the proportion of metabolically active organisms (that is, those that incorporate the label into their cells) to total cells can be readily determined.

INT Reduction

In the INT reduction procedure, a tetrazolium dye (INT) is used as the substrate. It is commonly metabolized by aerobic respiring bacteria if they are alive and metabolically active. They use it as an electron acceptor and reduce it to form a precipitate called formazan, which appears as a black deposit in the cell. When combined with acridine orange total counting, the organisms that are metabolically active can be distinguished from those that are not. More recently a fluorogenic dye, CTC, has been developed that produces a red fluorescent formazan. A fluorogenic compound, although not fluores-

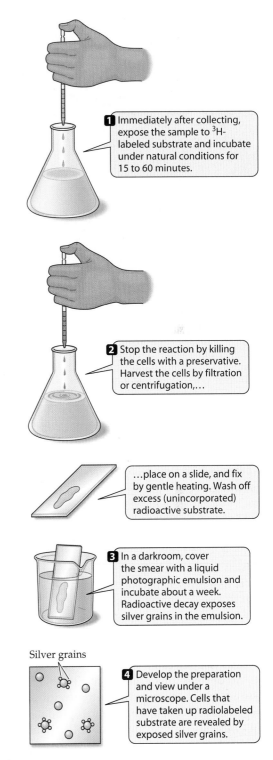

1 Immediately after collecting, expose the sample to ^{3}H-labeled substrate and incubate under natural conditions for 15 to 60 minutes.

2 Stop the reaction by killing the cells with a preservative. Harvest the cells by filtration or centrifugation,…

…place on a slide, and fix by gentle heating. Wash off excess (unincorporated) radioactive substrate.

3 In a darkroom, cover the smear with a liquid photographic emulsion and incubate about a week. Radioactive decay exposes silver grains in the emulsion.

Silver grains

4 Develop the preparation and view under a microscope. Cells that have taken up radiolabeled substrate are revealed by exposed silver grains.

Figure 24.3 Microautoradiography
Steps in microautoradiography. A radiolabeled substrate (or mixture of substrates) is added to a sample, and the sample is incubated for 15 to 60 minutes under the same conditions as in the natural habitat. The reaction is stopped, the sample fixed, and an autoradiogram prepared (see Figure 24.4) to determine what proportion of the organisms incorporated radiolabeled substrate, that is, what proportion were metabolically active. Radioactive cells are counted, and the percentage of active cells is determined by dividing the number of radioactive cells by the total cell count (determined by acridine orange counting).

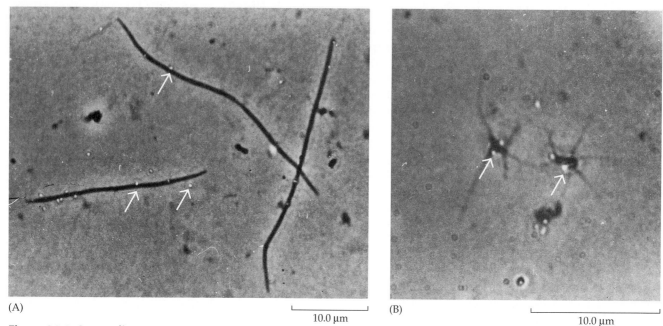

(A)   (B)

10.0 µm   10.0 µm

Figure 24.4 Autoradiogram

Photomicrographs showing (A) a filamentous bacterium and (B) *Ancalomicrobium adetum* from a pulp-mill treatment lagoon. The exposed silver grains (arrows) in the emulsion overlying the cells reveal which cells are metabolically active. The labeled substrate was tritiated acetate. Courtesy of P. Stanley and J. T. Staley.

cent itself, can be metabolized to produce a fluorescent compound. This dye provides a better assay inasmuch as fluorescent deposits are easier to detect in cells than nonfluorescent ones.

Nalidixic Acid Cell Enlargement In the **nalidixic acid cell enlargement procedure,** a small amount of yeast extract is added to a sample along with nalidixic acid, which inhibits DNA replication. Metabolically active cells that are sensitive to this antibiotic cannot divide, but their cells continue to enlarge in size as they grow. Therefore, the proportion of cells from a sample that become very large compared to those that have not enlarged provides a measure of the percentage of metabolically active cells.

All of these procedures have their own pitfalls and drawbacks; however, they all provide similar results when used to assess viability in typical mesotrophic and oligotrophic environments. Unlike viable plating in which less than 1% of the cells can be recovered, about half of the cells from most oligotrophic and mesotrophic habitats are metabolically active. Thus, either these organisms are different physiologically from known bacteria and therefore do not grow on typical plating media, or the cells are too debilitated to actually multiply on plating medium, or a combination of both.

Although there are drawbacks to using plating and other traditional viable counting procedures to deter-

mine total viable counts of heterotrophs, viable counting procedures work well if one is enumerating a specific group of organisms that are known to grow under certain laboratory conditions. For example, as mentioned previously, it is possible to determine the seasonal distribution of the genus *Caulobacter* in lakes using viable counting procedures (**Figure 24.5**).

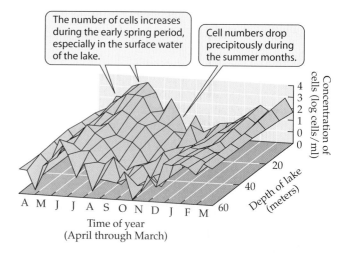

The number of cells increases during the early spring period, especially in the surface water of the lake.

Cell numbers drop precipitously during the summer months.

Concentration of cells (log cells/ml)

Depth of lake (meters)

Time of year (April through March)

Figure 24.5 Seasonal distribution study

The seasonal and spatial distribution of *Caulobacter* spp. in Lake Washington. Courtesy of J. T. Staley, A. E. Konopka, and J. Dalmasso.

Sampling Environments and Assessing the Habitat of a Species

By combining a sampling program of an environment with a means of identification of a bacterium, it is possible to conduct distributional studies of species to assist in determining its natural habitat. The first concern facing the microbial ecologist is, "How can I obtain samples from the environment?"

Sampling the Environment   Sampling environments is not a trivial problem. Imagine the difference between sampling for the bacteria that occur in the intestine of

an animal compared to sampling for marine planktonic bacteria from the Sargasso Sea. Completely different approaches are used to obtain samples of microorganisms from each habitat. For intestinal samples, the microbiologist has some difficulty in obtaining samples—stool samples could be used for lower intestinal bacteria, but more elaborate techniques would be required to sample the small intestine. Furthermore, the microbiologist needs to maintain anaerobic, aseptic conditions during the collection process. The marine microbiologist needs to find an appropriate location to obtain samples and uses entirely different collection gear (**Figure 24.6**). This text cannot describe all of the tech-

(A)

(B)

(C)

Figure 24.6 Sampling gear
Collection gear and instruments used by marine microbiologists. (A) Plankton net for sampling larger marine organisms. (B) A bathythermograph for measuring temperature and depth (hydrostatic pressure). (C) A Nansen bottle for collecting water samples for chemical analyses. A, ©G. Oliver/Visuals Unlimited; B, C, courtesy of J. T. Staley.

Table 24.2	Guidelines for microbial sample collection and processing[a]
Purpose of Sampling	**Preferred Handling Technique**
Total count (direct microscopic count)	Fix sample immediately after collection with formaldehyde or glutaraldehyde. Store at refrigerator temperatures in the dark. Aseptic collection is preferred, but is not always essential as long as vessel is clean and microbial counts are high.
Viable count	Collect aseptically. Enumerate as soon as possible after collection. Chill sample and keep under conditions of the habitat until processed. Incubate at or near the in situ conditions of the environment (i.e., pH, temperature, redox potential, etc.).
Activity measurements	Perform assay as soon as possible. Perform in situ if at all possible. May have to use a container for certain assays (e.g., radioisotopes), but these should be incubated under conditions of the environment, if not right in the environment.
Chemical analyses	Collection container should be chemically inert. Analyses should be performed soon after collection, especially for reactive compounds such as hydrogen sulfide, dissolved oxygen, etc. Special chemical fixation techniques are used for more reactive chemicals.

[a]As soon as a sample of microorganisms is taken from its natural habitat, conditions will change. Part of the change will be caused by the microbes themselves as they metabolize. However, physical and chemical changes also occur. It is important to minimize changes, as they can affect the results of a study. The actual sampling procedure will vary from one habitat to another.

niques used for sampling for microorganisms; however, some general guidelines for sampling are provided in **Table 24.2**.

In addition, it is desirable to measure the location and certain physical and chemical properties of the habitat at the time of sample collection. The location is determined from global positioning satellite (GPS) measurements. Also, in the intestinal and marine habitats described previously, it is important to measure the temperature at the time of collection as well as pH. In addition, in the marine habitat it is important to record the time of sample collection and the hydrostatic pressure and depth using special instruments such as the bathythermograph (Figure 24.6B). These parameters as well as others provide valuable information that is useful in interpreting what is happening in the environment.

Assessing the Habitat and Distribution of a Bacterial Species Autecological studies may be designed to assess seasonal changes in the distribution of a microbial species as well as its habitat. For example, nitrifying bacteria have been enumerated in marine environments and in soils by use of fluorescent antibody techniques. This procedure is quite species-specific at least for this group, so it provides a means of assessing the temporal and spatial distributions of particular nitrifying species.

Morphologically distinctive microorganisms such as many of the *Cyano-*

bacteria and anoxygenic photosynthetic bacteria can be enumerated microscopically. Thus, photosynthetic bacteria can be counted in lakes by identification using the light microscope (**Figure 24.7**). As with all microorganisms, the distribution of these groups is confined to certain strata in the lakes in which they reside. The lower depth of each photosynthetic bacterial group is determined by its need for sulfide obtained from the sediments of the lake, whereas its upper depth is determined by its need for light obtained from the surface of the lake.

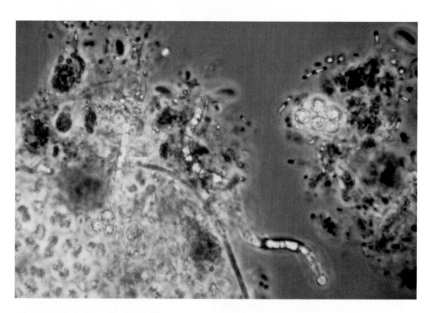

Figure 24.7 Anoxic hypolimnion sample
Sample from the anoxic hypolimnion of a lake, showing a variety of photosynthetic bacteria identifiable by their characteristic morphology. Bright areas in cells are gas vacuoles. Also, see book cover photomicrograph. Courtesy of J. T. Staley.

Viable counting procedures can also be used. For example, as mentioned previously, the seasonal and spatial distribution of *Caulobacter* spp. has been determined in freshwater lakes (Figure 24.5). Note that this bacterium is most abundant in the surface waters of the lake, and furthermore, that it is most numerous in the spring when diatom blooms occur in the lake. These results suggest that the *Caulobacters* may derive organic nutrients from the diatoms.

Evaluation of the Niche of a Species To assess the niche of a species in the environment, microbiologists may follow the "ecological" Koch's postulates approach. The approach can be summarized by paraphrasing the postulates used for studies of pathogenesis; however, in this instance the microbiologist is not attempting to verify that the bacterium causes a disease, but that it is responsible for a particular process or chemical transformation. These modified Koch's postulates for microbial ecology are as follows:

1. The species believed responsible for a transformation must be found in environments in which the process is occurring.
2. The species must be isolated from the environment where the process occurs.
3. The species must carry out the process in the laboratory in pure culture.
4. The species that carries out the process in the laboratory must be shown to carry out the process in the environment in which the process occurs.

Of course, unlike infections that are typically mediated by a single, specific pathogen or parasite, ecological processes in the environment are frequently mediated by groups of organisms growing in concert with one another. Such mixed groups containing two or more species are referred to as a **consortium.** An example of a consortium-mediated environmental process is the process of formation of stromatolites that are produced by a variety of organisms (see Chapter 1). This structure could not be formed in the laboratory using a single pure culture of an alga or cyanobacterium. However, it is formed in nature by the concerted action of several different species. Thus, it may not always be possible to use Koch's postulates with pure cultures to show that a specific microorganism causes a specific process. However, what has been said of the species would also be true of the consortium. So, the word *consortium* can substituted for *species* in the environmental Koch's postulates just stated. A microbial ecologist studying such a consortium should attempt to identify the members of the consortium and, ideally, also determine what each member of the consortium contributes to the overall activity.

Generally speaking, testing the first three postulates are straightforward: An organism suspected of causing a transformation can be found in various environments in which the process occurs and can be isolated in pure culture. Once in the laboratory, it can be shown to be able to carry out the process. The final step is much more challenging, however, because it is incumbent on the microbiologist to demonstrate that the species or consortium is carrying out this process in the environment. One approach in assessing this last postulate is to combine microautoradiography with an identification procedure such as the fluorescent antibody procedure. For example, if one wonders whether a specific heterotrophic bacterium is using acetate in the environment, then the species can be identified by fluorescent antibody, and the uptake of radiolabeled acetate associated with the chemical transformation is noted by the exposed silver grains on the same autoradiogram.

Microautoradiography can also be used with morphologically identifiable bacteria, such as *Ancalomicrobium adetum* (Figure 24.4B). Indeed, quantitative autoradiography can be achieved in some cases as with *A. adetum.* This species was found to comprise a minor component of the acetate-utilizing community in this study, in comparison with a filamentous organism (Figure 24.4A—data not shown). This does not mean that *A. adetum* is insignificant in this habitat. It is more likely that it has a different niche. For example, from pure culture studies in the laboratory, it is known that it preferentially uses sugars as carbon sources. Sugars are also abundant in pulp mill effluents and are probably being utilized as carbon sources by this species.

Molecular approaches are also possible. For example, it may be possible to detect the expression of a particular gene that codes for a specific enzyme of the organism of interest using RT-PCR.

Finally, it is very important that the organism that one has shown to be responsible for a process be kept in pure culture and deposited in a culture collection. If it comprises a new species, then it should also be described and named.

MICROBIAL COMMUNITIES

The community ecologist is not concerned with the activities or distribution of a single species but rather is interested in the processes of a group of species or all the species of the habitat. Thus, the community ecologist would be interested in all the sulfur-oxidizing species (i.e., the **guild** of sulfur oxidizers) in the lake and the rates at which this process is occurring, rather than the activity of a single species of *Thiobacillus.*

Microorganisms live in all ecosystems. These ecosystems contain a community of organisms comprising many different species that are interacting with one another and the abiotic environment. Typical ecosystems contain primary producers, consumers, and decomposers. Primary producers are organisms such as photosynthetic bacteria that fix carbon dioxide into organic material. Consumers are organisms that ingest organic matter (other organisms including primary producers) and use their organic material as an energy source. Protozoa are an example of a microbial consumer because they can ingest bacteria, algae, and other microorganisms. Decomposers are heterotrophic microorganisms that degrade organic material. When the complete degradation of an organic substance occurs to produce inorganic end products including CO_2 and H_2O, the process is called **mineralization**.

Some communities consist primarily or exclusively of microorganisms. Microbial mats and biofilms are examples of microbial communities.

Microbial Mat Communities and Biofilms

Chapter 1 describes microbial mats as the dominant manifestation of life on Earth from about 3.5 Ga ago until the rise of land plants about 0.5 Ga ago. Today's microbial mats are found in hot spring environments and intertidal zones of the marine environment (see Chapter 1).

Microbial mats are communities that consist of a vertical stratification of species in a layer that is typically about 1 to 4 cm in depth. Mats in hot spring environments have been studied most thoroughly. At higher temperatures near the orifice of hot springs, the primary producers are nonphotosynthetic *Archaea* such as *Pyrococcus*, or *Bacteria* such as *Aquifex*. These organisms obtain their energy from reduced sulfur compounds or H_2, not light. Downstream, at temperatures of about 70°C, *Cyanobacteria* become the major primary producers using light as their energy source. Eukaryotic consumers of the microbial mats cannot grow at the highest temperatures but begin to colonize the mats when the temperatures drop lower. Microbial decomposition occurs through heterotrophic bacteria and the sulfur- and sulfate-reducing bacteria (some of which are also autotrophic).

Not only is there a vertical layering of the species in hot spring communities, but also as the water flows away from the thermal source, the temperature drops. This gradient allows for the development of organisms with lower and lower temperature optima. Furthermore, day and night changes often occur in the community structure of mats. In hot springs with sulfide, *Cyanobacteria* dominate the surface layers of the mat during daylight hours. But at night, gliding sulfur oxidizers such as *Beggiatoa* migrate to the surface of the mat. *Beggiatoa* species do not require light, but gain greater access to oxygen by moving to the surface of the mat where they may oxidize their stored sulfur granules for energy.

A typical mat community in a hot spring may contain a guild of several different species of primary produc-

ers. For example, in Yellowstone National Park hot springs, several different cyanobacterial strains or species are known to occur in the community. Each type has its own optimal temperature, but several types may overlap one another at a particular location.

Biofilms are similar to mats in that they consist of a buildup of microorganisms on a surface but are not as thick as mats. Often, like microbial mats, they exist in flowing environments like on rocks in a stream. In such environments, the biofilm may not be visible to the eye, but the rock surface is slippery, indicating its presence. Another example of a biofilm is the layer of bacteria that develop on our teeth. Heterotrophic bacteria, such as *Streptococcus mutans*, grow on the teeth and obtain their organic nutrients from the food we eat. We attempt to remove this particular biofilm by brushing our teeth; however, it builds up again after every meal.

Unlike microbial mats, biofilm communities may not have all biotic components of a typical ecosystem. Thus, the biofilm on teeth or in a water pipeline do not have primary producers that are usually found in major ecosystems such as ponds and lakes. Nonetheless, biofilms contain many different species and, in that sense, comprise a microbial community.

Biofilms are very important economically. For example, biofilms that develop on the hulls of ships not only lead to corrosion but also increase the drag of the ship as it moves through the water. Likewise, microorganisms living in pipelines carrying water or oil products can corrode the pipes, impede the delivery of the fluid, and ultimately destroy the pipe itself.

Biomass and Biomarkers

How do microbial ecologists study microbial communities and assess their activities? We begin by discussing biomass and biomarkers.

Biomass In the same sense that autecologists are interested in knowing the numbers or quantities of a particular species in the habitat, the community ecologist is interested in knowing the quantity of all microbial cytoplasm. This is the **biomass,** or concentration of living microbial cell material in the environment. A variety of techniques have been used for the measurement of biomass, all of which have limitations. Because of its widespread use, the ATP biomass procedure is discussed here. ATP is found in the cytoplasm of all living organisms but not in dead organisms. The approach used in this procedure is to concentrate organisms from a specific area or volume of the environment and extract the ATP, the amount of which is indicative of the amount of living cytoplasm. Therefore, the microorganisms, or the bacteria in particular, are separated from other organisms in a natural sample by screening or filtration. The ATP concentration is determined, and the microbial or bacterial biomass is estimated. ATP is extracted from the harvested cells on the filter, and the ATP content is determined with an ATP photometer.

Although this is a straightforward analysis, note that it offers only an approximation of the quality of images. Different species have different ATP contents, and furthermore, cells of a given species will contain different concentrations of ATP, depending on how metabolically active the cells are. Thus, as with viable counting and other procedures, there are limitations. Nonetheless, this assay can provide important information to the microbial ecologist (**Box 24.2**).

Biomarkers Biomarkers are used to assess the presence and concentrations of microbial groups in natural communities. A **biomarker** is a chemical substance that is uniquely produced by a group of organisms. Ideally, the biomarker selected should be simple to analyze for, otherwise one might as well use other approaches such as the fluorescent antiserum procedure to identify specific groups.

A popular group of biomarkers currently used for bacteria is phospholipids. Different groups of bacteria have different membrane phospholipids and therefore can be differentiated from one another in natural communities by assessing the type of phospholipid present.

It should be noted that although biomarkers have also been used as biomass indicators for certain groups, biomarkers may survive in the environment even though the bacterium that produced them is no longer alive. Therefore, caution should be exercised in assessing the significance of finding a biomarker from a natural sample.

Biomarkers have useful applications. For example, some *Bacteroides* species that grow in the human intestine produce coprastanol, a product of cholesterol degradation. Because of its association with the human intestinal tract, coprostanol has been used as an indicator of human sewage contamination in natural environments. Therefore, if a water sample contains this biomarker, it is suggestive evidence that human fecal material has contaminated the water supply. However, a recent study has shown that this particular biomarker assay has some limitations. Coprastanol was found in marine sediments near Antarctica, far from human habitation and influence. Subsequent research determined that the source was from certain marine mammals including whales and some seals! So, coprastanol is not as much a marker for human feces as it is for the bacteria that live in the intestinal tracts of mammals that transform cholesterol to coprastanol.

Is It a Microbial or Chemical Process?

When examining a transformation in nature, the first question the microbiologist asks is, "Is this transforma-

tion a chemical or biological process?" It is not always clear whether a particular activity is performed by bacteria and other microorganisms or by a chemical mechanism. Consider for example, the weathering of rocks, buildings, and monuments. It is well known that a variety of microorganisms including algae, bacteria, fungi, and lichens grow on the surfaces of rocks and can be involved in weathering of the surfaces. They do this by producing organic acids and chelating agents that gradually dissolve away minerals. However, the chemical effects of acid rain and the physical effects of freezing and thawing are examples of nonbiological processes that cause weathering. For this reason, it is not always easy to determine if a specific rock is being eroded by microorganisms or by physical and chemical processes. Furthermore, both chemical and biological processes often work in concert with one another in effecting a transformation. And because these processes are so slow, they may be difficult to study in the field.

An early example in which a process was identified as a biological process is nitrification, which was studied by Schloesing and Müntz (Chapter 19). They added soil to a cylinder with a drain and trickled sewage effluent containing nitrogen in the form of ammonia through the column. In a few days, the effluent from the column contained nitrate, not ammonia. This indicated to them that the process of nitrification was occurring in the soil in the column. To determine whether this process was a biological process, they first added boiling water to the soil column. They noted that the process abruptly ceased. They reasoned that this must therefore be a biological process because, if it were chemical, the increased heat would speed up the process, not stop it. They then treated another nitrifying column with chloroform, which is toxic to most organisms. Again the process stopped. Because chloroform would not be expected to inhibit most chemical processes, they concluded that this result also supported the hypothesis that this was a biologically mediated process. Subsequently, nitrifying bacteria have been isolated from such columns, indicating that biological agents are indeed responsible for the transformation.

As discussed previously, Koch's postulates have been adapted for use in microbial ecology to verify that a microorganism is responsible for the process. This and similar approaches have been used for many years to establish whether natural processes are mediated by microorganisms or are chemical transformations. It is increasingly found that microorganisms are responsible for many reactions that were previously thought to occur only by chemical means. However, for processes such as some types of rock weathering that proceed very slowly in nature, it is not always easy to demonstrate whether the process is caused by biological or chemical and physical agents. Nonetheless, it has been well

Research Highlights Box 24.2

Effect of Mt. St. Helens' Eruption on Lakes in the Blast Zone

On May 18, 1980, Mt. St. Helens, a volcano in Washington State, erupted. The cataclysmic event devastated the forest in the blast zone area north of the mountain (Figures A and B). All of the trees and other vegetation in the blast zone were killed due to the heat and force of the blast, the ensuing mud slides, and the fall of ash and other pyroclastic materials.

Ecologists set out to determine the effect of the eruption on lakes in the blast zone on the north side of the volcano. Although it was difficult to obtain samples immediately following the eruption, after a short time helicopters were flown in, and small volume samples were collected that could be used for analysis of total and viable counts of bacteria, and analysis of chlorophyll *a* and ATP. Fortunately, some samples had been obtained from Spirit Lake prior to the eruption so that the chlorophyll *a* and ATP levels could be compared with those collected after the eruption (Table 1). Curiously those results indicated that, following the eruption, there was a decrease in chlorophyll *a*, but an increase in ATP. This was surprising because general-

ly the most significant biomass in a lake is due to algae and yet, in this instance, the concentration of chlorophyll *a* actually decreased. Thus, the increase in ATP could not be explained by an increase in phytoplankton growth.

The explanation for the increase in biomass, as indicated by the increase in ATP concentration, was that there was tremendous growth of bacteria. This result was first indicated because the initial plates used for viable counting of the bacteria showed there were much higher concentrations of bacteria in the blast zone lakes north of the volcano than were found in control lakes located south of the volcano (Merrill Lake and Blue Lake) that were not affected by the blast (Figure C). In fact, there were so many bacteria that higher dilutions were needed to accurately count them. The concentrations of viable heterotrophic bacteria in blast zone lakes north of the volcano (for example, Coldwater Lake), where the effects of the eruption were most severe, increased dramatically to over 10^6/ml in the first summer, and this pattern continued into the next

year after the eruption (Table 2). In contrast, the control lakes south of the mountain (Merrill Lake) were unaffected by the eruption, with a concentration of about 10^3 viable cells per ml (compare these numbers with Lake Washington and the pulp mill oxidation lagoon in Table 24.1). It was also interesting to note that not only did the viable counts of bacteria increase in the blast zone lakes, but they also approached the total microscopic cell count (acridine orange counts), thus indicating that the plating procedures for viable counts was able to recover most of the bacteria in the blast zone lakes. As mentioned in the text, the phenomenon of high recoveries of bacteria (>1% of total microscopic count) is found in lakes that are eutrophic (high in nutrient concentration) but not oligotrophic (low in nutrients). This suggested that the lakes in the Mt. St. Helens blast zone, which had previously been oligotrophic, became eutrophic due to the effects of the eruption. Indeed, the lakes were converted from oligotrophic mountain lakes to eutrophic lakes following the eruption (Table 2).

So, why were there such high concentrations of bacteria in the lakes? The dramatic increase in the concentration of heterotrophic bacteria in the blast zone lakes occurred because of the tremendous increase in nonliving organic matter (dead vegetation and animals) that became available when the plants and animals were killed following the eruption. The organic matter leached from the watersheds into the rivers and lakes where the bacteria thrived and made the lakes naturally eutrophic.

Table 1 Characteristics of Spirit Lake prior to and following 18 May, 1980, volcanic eruption

Feature	4 April 1980	15 July 1980	Enrichment
Chlorophyll *a* (µg/l)	2.5	0.3	
ATP (µg/l)	0.25	4.3	17-fold
DOC (µg/l)	0.83	39.9	48-fold
Dissolved oxygen (mg/l)	Unknown	2.35[a]	
Temperature	15°C	22°–24°C[b]	

[a]Although oxygen was not measured on 4 April, it would have been expected to be saturated at 4°C and therefore greater than 12 mg/l, close to the value normally expected on 15 July.
[b]A temperature of 35°C was actually measured in Spirit Lake on May 19, the day after the eruption. On July 15 the temperature was still much higher than it would have been normally at this time of year, due to hot water entering the lake from the volcano. (Data from R. C. Wissmar et al. 1982, *Science* 216:178–181.)

Research Highlights Box 24.2 *(continued)*

Table 2 Bacterial concentrations (per ml) and recovery from a blast zone lake compared to a control lake following Mt. St. Helens' eruption[a]

Date (mo/day)	Control Lake		%	Blast Zone Lake		%
1980–1981	Merrill			Coldwater		
	Viable	Total	Recovery	Viable	Total	Recovery
6/30/80	1.0×10^3	5.4×10^5	0.19	$>10^4$	7.4×10^6	—
9/11/80	2.0×10^2	1.4×10^6	0.01	2.5×10^6	6.4×10^6	39.1
4/30/81	ND[b]			2.2×10^6	2.7×10^6	81.5
6/29/81	ND			1.8×10^6	2.5×10^6	72.0

[a]Viable counts were on CPS (caseinate, peptone, starch) plates, and total counts were determined by acridine orange epifluorescence counting.
[b]Not determined.
Data from J. T. Staley, L. G. Lehmicke, F. E. Palmer, R. W. Peet, and R. C. Wissmar. 1982. *Appl. Environ. Microbiol.* 43:664–670.

(A)

(B)

(C)

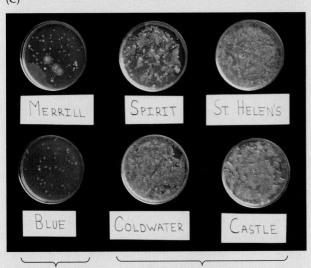

Spirit Lake, north of Mount St. Helens (A) before and (B) after eruption of the volcano. Note the dramatic impact of the eruption on the biological communities. (C) All plates contain 0.1 ml samples of water taken from lakes near Mount St. Helens shortly after the eruption. A, photo ©Pat and Tom Leeson/Photo Researchers, Inc.; B, C, courtesy of J. T. Staley.

Compared with the control lakes… …all lakes in the blast zone have an overgrowth of bacteria.

established that microorganisms play an important role in rock-weathering processes.

Stable isotopes can also provide important information about whether a process is biological because of isotope discrimination. For example, consider the sulfur cycle. There are nine isotopes of sulfur including four that are stable (nonradioactive): the normal isotope ^{32}S, and several others including ^{33}S, ^{34}S, and ^{36}S; the natural abundance of each in Earth's crust is, respectively, 95.0%, 0.76%, 4.22%, and 0.014%. Biological processes exhibit **isotope** discrimination, that is, organisms preferentially use lighter isotopes. For example, sulfate-reducing bacteria preferentially reduce $^{32}SO_4^{-2}$ rather than the heavier isotopic form of the ion, $^{34}SO_4^{-2}$. As a result, they produce a sulfide that is lighter (contains more ^{32}S than ^{34}S) than the substrate. From this information, it has been concluded that elemental S (S^o) deposits on Earth that have a lighter sulfur isotope content than that of Earth's crust are derived from biological sources. Presumably they are the consequence of an active sulfur cycle in which some sulfur-producing bacteria such as the photosynthetic sulfur bacteria used biogenic sulfide formed by sulfate reducers to produce a lighter sulfur deposit than would be expected if a chemical process were involved. Likewise, the sulfur cycle itself has been dated to about 3.5 Ga ago on the basis of observing lighter than expected sulfur compounds in sedimentary deposits.

Rate Measurements

One of the most important areas of microbial ecology is measuring the rates of microbial processes. Knowing this allows scientists to assess the significance of the contribution of the microorganisms to a process. The following section describes some of the various techniques for rate measurements and their applications.

Direct Chemical Analyses In some cases, rate measurements are rather straightforward and can be determined by simply analyzing for the changing concentrations of chemicals (substrates and products). An example of this is the measurement of primary productivity by measuring the oxygen evolved during photosynthesis. As discussed earlier, the overall reaction for primary production is given as:

$$(1)\ \ 6\,CO_2 + 6\,H_2O \rightarrow (C_6H_{12}O_6)_n + 6\,O_2$$

where $(C_6H_{12}O_6)_n$ = organic carbon formed.

Reaction 1 indicates that the amount of carbon dioxide fixed (that is, the primary production) is directly related to the oxygen evolved on a mole-to-mole basis. Therefore, the rate of primary production can be determined simply by measuring the amount of oxygen produced for a 24-hour period of time during photosynthesis. In aquatic habitats, this is accomplished in glass-stopper bottles containing water collected from a lake or marine habitat (**Box 24.3**).

Use of Chemical Isotopes Chemical isotopes can also be used for rate measurements. The major advantage of using them is the extreme sensitivity that they provide for rate analysis. Two types of isotopes are available, depending on the substrate. As mentioned previously, stable isotopes are not radioactive, but their atomic weight differs from that of the common form of the element. **Radioisotopes** exhibit radioactive decay. These isotopes occur normally in nature, but they are found in reduced amounts compared to the normal or common isotope. Both stable and radioactive isotopes are used to determine the rates of ecological transformations. Whether a stable or radioactive isotope is used is determined by several factors, such as the equipment available for analysis and the half-life (the length of time required for half of the radioactivity to be lost or to decay) of the radioisotope available. For example, for carbon isotope analysis, most scientists determining rates of processes use carbon-14 (^{14}C) because it can be simply quantified using a scintillation counter, and it has a long half-life (over 5,000 years). On the other hand, ^{13}C, a stable isotope, requires that a mass spectrometer be available for analysis, and this is much more expensive than a scintillation counter, and the analysis is much more labor intensive. However, with nitrogen, the choice is simple: ^{15}N is used even though a mass spectrometer is required, because the half-life of ^{13}N is only a few minutes, making it impracticable for virtually all ecological experiments.

The most common way to measure **primary productivity** is by the use of radioisotopic procedures, namely using ^{14}C-labeled sodium bicarbonate. The utility of this can be seen by referring to Reaction 1 for primary production. So, rather than measuring oxygen production, which is an indirect measure of primary production (Box 24.3), one measures carbon dioxide fixation into biomass using ^{14}C-labeled bicarbonate (**Box 24.4**). As mentioned earlier, this radioisotopic method is much more sensitive than ordinary chemical methods and is therefore the preferred technique.

The measurement of **heterotrophic activities** is more complicated than that of primary production. This is because in primary production, a single substrate, carbon dioxide, is being metabolized by primary producers, whereas for heterotrophic activities, there are hundreds of different organic substances that serve as substrates in the environment. Furthermore, some of these are soluble (DOC) and others are insoluble (POC). Nonetheless, radioisotopes are available for many DOC compounds, such as sugars, amino acids, and organic acids, and either tritium (^3H)-labeled or ^{14}C-labeled forms can be used.

Methods & Techniques Box 24.3   Chemical Measurement of Primary Production

The chemical measurement of primary production in an aquatic habitat is performed using freshly collected water from the depth(s) of interest. This is used to fill three bottles that are incubated at the depth (Figure A). One bottle is called the "dark bottle" because it is made opaque to prevent light from entering. Although photosynthesis cannot occur in this bottle, respiration, which is the reverse of the photosynthesis reaction, does occur. The second bottle, called the "fixed control," is fixed with formalin at zero time so that no biological activity can occur in it. The third bottle, called the "light bottle," or experimental bottle, is allowed to incubate and photosynthesize as it would in nature except that it is now occurring in an enclosed container. These bottles are incubated usually by submerging at the depth of collection preferably for an entire 24-hour period. The reaction is then stopped by addition of formalin to the experimental bottle and the dark bottle.

Primary production is determined from measurements of the concentration of oxygen in the different bottles (Figure B). At zero time the oxygen concentration would be expected to be the same in all bottles. After a 24-hour incubation period, the oxygen concentration would be expected to be highest in the experimental bottle in which photosynthesis occurs. The difference in oxygen concentration in this bottle during the time of incubation is a measure of the net primary production that occurred in the volume of water tested and is calculated as follows:

$$\text{Net primary production/day} = [O_2]_{L2} - [O_2]_{L1}$$

where $[O_2]_{L1}$ is the moles of oxygen per liter at zero time in the light bottle and $[O_2]_{L2}$ is the moles of oxygen per liter after the incubation period in the light bottle.

The gross primary production is calculated as the net primary production plus that amount of oxygen that was used in respiration (Figure B). This latter value can be calculated as the amount of oxygen used during incubation in the dark bottle where only respiration could occur. So, the gross photosynthesis is:

$$\text{Gross primary production/day} = \text{net primary production} - [O_2]_{D2}$$

The oxygen concentration in the fixed control should not have changed during the incubation. If it did, this would indicate either a leak in the bottle or some other problem with the assay that would indicate to the researchers that the experiment was flawed.

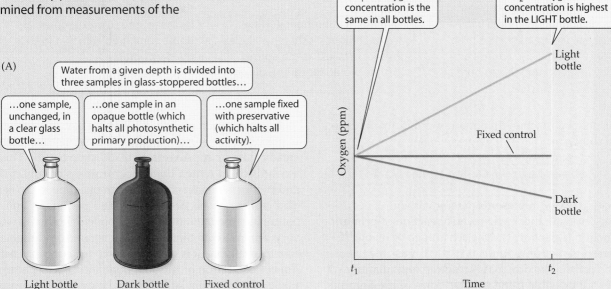

(A) Water from a given depth is divided into three samples in glass-stoppered bottles…

…one sample, unchanged, in a clear glass bottle…

…one sample in an opaque bottle (which halts all photosynthetic primary production)…

…one sample fixed with preservative (which halts all activity).

Light bottle   Dark bottle   Fixed control

(B) Oxygen is measured at the start (t_1), the samples are incubated for 24 hours at the depth under study, preservative is added to the light and dark bottles, then oxygen is measured again (t_2).

At t_1 the oxygen concentration is the same in all bottles.

At t_2 the oxygen concentration is highest in the LIGHT bottle.

Light bottle

Fixed control

Dark bottle

Oxygen (ppm)

t_1 Time t_2

Measurement of primary production in aquatic habitats by measuring oxygen production. (A) Experimental treatments of water samples from the depth at which photosynthesis is to be measured. (B) Graph showing the change in concentration of oxygen.

Methods & Techniques Box 24.4   **Radioisotopic Method of Measurement of Primary Production: An Example of the Tracer Technique**

The experiment is set up in the same manner as the chemical analysis experiment for primary production—that is, with both light and dark bottles as well as a fixed control (Box 24.3, Figure A). To initiate the experiment, a small amount (trace) of ^{14}C-labeled bicarbonate is added to each bottle. The amount of bicarbonate added is small relative to the total amount already in the water, which is measured chemically. Therefore, the concentration of the substrate, both unlabeled and labeled, in the sample is virtually unaffected by the addition.

This is called the tracer approach, because the pool size or natural substrate concentration is not affected by the amount of label added because so little is added relative to the total amount of natural substrate. By following what happens to the label, the biological transformation can be assessed.

In this procedure, the amount of carbon dioxide fixed is determined by the amount of label taken into the cells during photosynthesis. Thus, on completion of the incubation, the algae are harvested from the experimental and control bottles by filtration, and the amount of label incorporated is determined by scintillation counting of the filters. The net photosynthesis is the difference between the amount of label taken up in the experimental bottle less the amount taken up in the dark bottle. The fixed control is used to assess whether chemical effects such as adsorption might have caused artifacts in the experiment.

The photosynthetic rate is calculated as:

$$\text{Photosynthetic rate (mg c/l/h)} = \frac{cpm_L - cpm_D}{CPM_{added}} \times \frac{\text{Total } ^{12}C(mg/l) \times 1.06}{\text{Incubation time (hours)}}$$

where cpm_L and cpm_D equal the counts per minute found on filters in the light bottle and dark bottle, respectively, and CPM_{added} is the amount of label added to each bottle in the form of ^{14}C-bicarbonate. Total ^{12}C is the amount of carbon in the bicarbonate pool in the habitat; 1.06 is the isotope discrimination factor.

A major assumption of these tracer techniques is that the labeled substrate acts just as the normal substrate would for carbon dioxide fixation. In the case of carbon dioxide, the ^{14}C label is not only radioactive, but it is heavier than the naturally occurring [12]C. Therefore, the factor 1.06 is multiplied by the entire photosynthetic rate to correct for isotope discrimination (6% in this case) and determine the actual rate. As discussed previously, isotope discrimination occurs with both stable as well as radioactive isotopes whose masses are different from the normal isotope.

The measurement of primary production using the [14]CO_2 tracer procedure uses the same type of bottles as for the oxygen procedure. However, prior to incubation a trace amount of radiolabeled NaH[14]CO_3 is added to initiate the reaction. For the analysis of primary production, the amount of label taken into the cells as fixed organic matter is measured using a scintillation counter. Therefore, after incubation, the cells must be concentrated on a filter before placing them in the scintillation vial. Primary production is determined as the amount of carbon dioxide fixed by the algae in the experimental bottle compared with that fixed in the dark bottle.

The general reaction that occurs aerobically in respiration shown as:

$$(2)\ (C_6H_{12}O_6)_n + 6\,O_2 \rightarrow 6\,CO_2 + 6\,H_2O$$

where $(C_6H_{12}O_6)_n$ represents organic carbon.

Thus, overall, this is the reverse reaction of primary production (Reaction 1). One advantage of using ^{14}C-labelled organic compounds is that it is possible to collect the carbon dioxide that is respired and quantify it. This is not possible if the tritiated form of the organic substrate is used. Furthermore, the radioactive water formed with tritium is difficult to separate from the soluble organic substrate and products produced in the reaction. Therefore, ^{14}C-labeled substrates are most commonly used, except for microautoradiography.

Another major advantage of using isotopes is that it is possible to determine rates in systems in which a **steady state** or **approximate steady state** occurs. This is a condition in which the concentrations of substrates and products remain constant from one time to the next. This commonly occurs for extended periods of time in nature because the rate of formation of a product is often very similar to the rate at which it is removed or metabolized. Therefore, it is meaningless to measure the chemical concentration of the substrate and product at time zero and some time later, because they will be approximately the same, and if they are somewhat different, it is due to the combined activity of formation and utilization. Nonetheless, utilization is occurring in the natural habitat. By adding a labeled substrate to the natural substrate pool as a tracer (see Box 24.4), it is possible to then identify (label) that part of the substrate pool and follow it in the formation of the product.

This tracer approach is most commonly used in measuring, not only the rate of primary production but also the rate of the heterotrophic activity or metabolism of dissolved organic compounds in environments. This is usually accomplished with short-term incubations using sealed flasks (**Box 24.5**).

Radioisotopic methods can also be complemented with autoradiography as discussed earlier in this chapter. This permits the scientist to associate the uptake of a specific compound with a specific microorganism. Even though the organism might not be identifiable

morphologically, it can be rendered identifiable by labeling it with a fluorescent antiserum or by use of the FISH approach and then examining it by both fluorescence and by light microscopy.

Alternate Substrates For some activities, it is possible to use an alternate substrate in place of the natural substrate for determining the reaction rate. Perhaps the best example of this is the use of acetylene in place of nitrogen for the measurement of nitrogen fixation. The overall reaction for nitrogen fixation involves three reductive steps in which N_2 ($N \equiv N$) is reduced first to $HN = NH$, then to $H_2N - NH_2$, and then ultimately to $2\ NH_3$.

Methods & Techniques Box 24.5 — Radioisotopic Procedure for Assessing Heterotrophic Rates of Metabolism in Aquatic Habitats

Initially, the water sample is collected from the environment and distributed into flasks (bottom right). Both experimental and fixed control flasks are set up, and the reaction is initiated by adding the radiolabeled substrate. If the actual concentration of the substrate in the natural habitat can be measured, then the reaction can be quantified by labeling the pool with a tracer label of that organic compound.

By knowing the concentrations of the substrate, (C_o), the amount that is radiolabeled at zero time (cpm_s), and the amount of label in the product (or that which is taken up by the metabolizing bacteria), cpm_p, it is possible to calculate the total amount of substrate that has been incorporated into the product, X, during a certain period of time by a simple proportion:

$$C_o/cpm_s = X/cpm_p$$

where C_o is the concentration of the substrate; cpm_s is the counts per minute of substrate; cpm_p is the concentration of the product; and X is the amount of substrate taken up by metabolizing cells.

In these experiments, some of the substrate will be converted into carbon dioxide, which is collected in a separate vial in the experimental

vessel (below). Most of the remaining label in the product will have been incorporated into cells. This is determined by harvesting cell material on a membrane filter and counting its radioactivity along with that of the mineralized carbon dioxide using a scintillation counter. It is the sum of the $^{14}C\text{-}CO_2$ and ^{14}C-cellular material that is determined to provide the value for cpm_p.

In addition to the aspect of the decay of radioisotopes, there is the question of specific activity of the label. This refers to how "hot" the

substrate is. Because dissolved organic compounds occur at extremely low concentrations in natural environments, if we are to use the tracer approach, it is important to add small amounts of the radiolabeled substrate to the pool. This requires that the radiolabeled substrate received from the manufacturer contain as many labeled molecules as possible. Otherwise, the natural substrate will be diluted too much to provide a sufficiently accurate and sensitive assay.

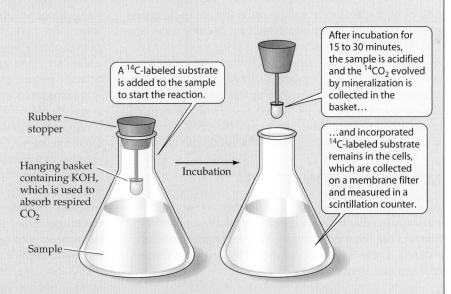

A ^{14}C-labeled substrate is added to the sample to start the reaction.

After incubation for 15 to 30 minutes, the sample is acidified and the $^{14}CO_2$ evolved by mineralization is collected in the basket…

…and incorporated ^{14}C-labeled substrate remains in the cells, which are collected on a membrane filter and measured in a scintillation counter.

Rubber stopper

Hanging basket containing KOH, which is used to absorb respired CO_2

Incubation

Sample

Apparatus for assessing rate of use of dissolved organic carbon by microorganisms in aquatic habitats using the tracer label approach.

Acetylene, which is also a triple-bonded compound, occupies the active site of nitrogenase and is reduced to ethylene as follows:

$$(3) \quad HC \equiv CH \rightarrow H_2C = CH_2$$

Remarkably, acetylene is actually preferred to N_2 by nitrogenase, so the reaction can be carried out even in the presence of the natural substrate, N_2. The main advantage of measuring nitrogen fixation this way is that both acetylene and ethylene can be easily quantified using a gas chromatograph, whereas the only stable nitrogen isotope that is useable is the stable $^{15}N_2$ isotope and its analysis is tedious and requires a mass spectrometer.

However, there are drawbacks to all of these types of assays. The disadvantage here is that the actual rate of the reaction is influenced by the substrate provided. Furthermore, the acetylene is reduced by only one reductive step rather than the three required for complete nitrogen fixation. For these reasons, it is not possible to directly determine in situ rates with these procedures. Nonetheless, they provide excellent relative measures of nitrogen fixation in systems where nitrogen fixation is known to occur. Furthermore, assays can be standardized using $^{15}N_2$.

Inhibitors Inhibitors are sometimes used to measure the rates of reactions. If the inhibitor specifically stops the process of interest and has no other effects, then the rate of the process can be determined by the rate at which the substrate accumulates. For example, molybdate ion is a known inhibitor of sulfate reduction. Therefore, it can be incorporated into assays to inhibit sulfate reducers. One example where this has been used is in the determination of how effectively sulfate reducers compete with methanogens for hydrogen in sediment communities. The sulfate reducers are inhibited with molybdate to assess for methanogen uptake of hydrogen in the environment, and the methanogens can be inhibited by the use of chloroform to assess the uptake of hydrogen by sulfate reducers. However, the use of such inhibitors may cause other effects on the community that may adversely affect the results, so they should be used with caution.

MAJOR ENVIRONMENTS

As previously mentioned, bacteria grow in some of the most unusual and extreme environments on Earth. In this section, the more common natural habitats are considered. These include freshwater, marine, and terrestrial environments.

Freshwater Habitats

Aquatic habitats comprise some of the most important habitats for microorganisms. Even waters of the most pris-

tine lake or areas in the open ocean contain hundreds of thousands or millions of bacteria per ml. Bacteria can be easily concentrated on filters of samples from aquatic habitats and therefore counted without interference from debris, as is found in the soil environment. Freshwater habitats include lakes, ponds, rivers, and streams. Most of these inland waters have very low concentrations of salt.

Lakes Many freshwater lakes of sufficient depth that are located in temperate zones become thermally stratified during the summer months. In such lakes, the surface water becomes warmer than the underlying water. This occurs because the surface is increasingly heated by the sun's radiation during the spring and summer months. Summer winds are not strong enough to mix the lighter warm surface layers with the more dense cold water that lies beneath. Therefore, the warm surface water layer becomes separated from the colder bottom water, resulting in **thermal stratification** of the lake. In contrast to deep temperate zone lakes, shallow lakes and ponds do not stratify because the winds keep them intermittently mixed.

In lakes where thermal stratification occurs, the surface layer of the lake is called the **epilimnion** and the lower body of water is called the **hypolimnion** (**Figure 24.8**). The zone of transition in temperature between the surface and the lower depths is called the **metalimnion**, or **thermocline.**

The period of thermal stratification persists throughout the summer season and into early fall. However, as the air temperature cools in the fall, the surface water temperature decreases, and the water becomes denser. At some point, the phenomenon of **fall turnover** occurs in which the density of the surface water and the underlying water are almost identical. Winds can then readily mix the lake thoroughly from top to bottom. This period of mixing continues throughout the wintertime, unless the temperatures become so cold that the lake actually freezes over. If freezing occurs, then the spring warming must first melt the ice and warm the melted surface water to 4°C (water is most dense at 4°C) before it will mix with the underlying water. In either case, at least one period of mixing occurs during the winter months in typical temperate zone lakes.

These physical features of lakes make them ideally suited for spring and summer blooms of algae and the resultant growth of bacteria. In the early spring months, the nutrients in the lake water are high in concentration because they have been brought up to the surface from the lower depths and sediments during the mixing period. At this time, the concentration of dissolved nutrients such as phosphate, ammonium, and nitrate is constant at all depths. Then, as the daylight period lengthens in the spring, it provides both increased light and warmer temperatures, conditions that are favorable for algal

(A)

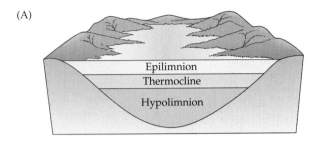

(B)

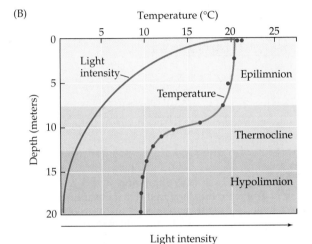

Figure 24.8 Thermal stratification
(A) Cross section through a thermally stratified lake, showing the epilimnion, thermocline, and hypolimnion. (B) A limnological graph is a plot of the depth of a lake versus the concentration or intensity of a parameter—in this case, temperature and light intensity in a typical thermally stratified lake during the summer months.

of dead algae and other organisms that inhabit the lake. These bacteria are called **mineralizers** or **decomposers** because they convert the organic material back into inorganic material so that it can be recycled. Most heterotrophic bacteria that are indigenous to planktonic aquatic habitats are referred to as **oligotrophs**, because they can grow using very low concentrations of organic nutrients. They are to be distinguished from **eutrophs**, which grow in organic-rich environments such as the intestinal tracts of animals.

Oligotrophs exhibit low growth rates and have high affinity uptake systems in their cell membranes that allow for the uptake of carbon sources that occur in low concentrations as found in planktonic habitats. By their activities, the concentrations of soluble organic substrates are kept at very low levels (µg concentrations) in mesotrophic and oligotrophic environments.

In contrast, eutrophic bacteria have rapid growth rates when they are provided with high concentrations of nutrients. In lake water, nutrients are available only in low concentrations, so the eutrophic bacteria are not competitive with oligotrophic species from these natural communities. However, the intestinal tracts of all of the consumer animals in aquatic food chains contain eutrophic heterotrophs, which play a major role in the decomposition of foodstuffs eaten by the consumers. These foods are converted into amino acids, vitamins, and organic acids that are absorbed in the intestinal tracts of the animals and used as their principal energy sources and sources of organic building materials. It is important to note that almost as much of the food consumed by the animal goes to produce microorganisms in the intestine, which are ultimately defecated, as goes to the animal's nutrition and energetics. Therefore, symbiotic microorganisms of the animal intestine play a major, often overlooked, role in all food chains and food webs.

The **microbial loop** model best explains what occurs in aquatic habitats (**Figure 24.9**) to account for the mineralization of organic material. In addition to the bacteria, protozoa and small invertebrate animals are involved in the degradation of organic materials. The heterotrophic bacteria utilize dissolved organic material excreted by algae and *Cyanobacteria*. Because they have high affinity uptake systems for organic compounds, bacteria grow principally on dissolved organic compounds (although through their extracellular cellulases, and chitinases they can also degrade particulate organic materials). The bacteria are then ingested by small protozoa, which also ingest small algae. The bacteria are commonly attached to organic **detritus** particles (nonliving organic particulate material), which they are degrading (**Table 24.3**). Unlike bacteria, protozoa and other invertebrates have mouths and therefore ingest the smaller bacterial cells and small detritus particles through **phagotrophic** nutrition. These protozoa, which

growth and other biological processes. Thus, spring and summer blooms of algae occur. At this same time, the increased warmth of the surface waters results in a decrease in surface water density and combined with wind, it enables the separation of the epilimnion from the hypolimnion.

The algae are **primary producers.** They serve as a source of food for **primary consumers,** such as protozoa, and small crustaceans, such as water fleas (*Daphnia* spp.). These primary consumers in turn serve as food for larger animals called **secondary consumers** (larger crustaceans and small fish), which feed on them, and they in turn provide food for **tertiary consumers** (larger fish, ducks, turtles, and so on) and higher members of the food chain called **quaternary consumers** (for example, raccoons and humans). Thus, a **food chain** or food web is established in which the carbon and energy flow from one group of organisms to another.

Heterotrophic bacteria degrade the excess organic material excreted by algae and decompose the remains

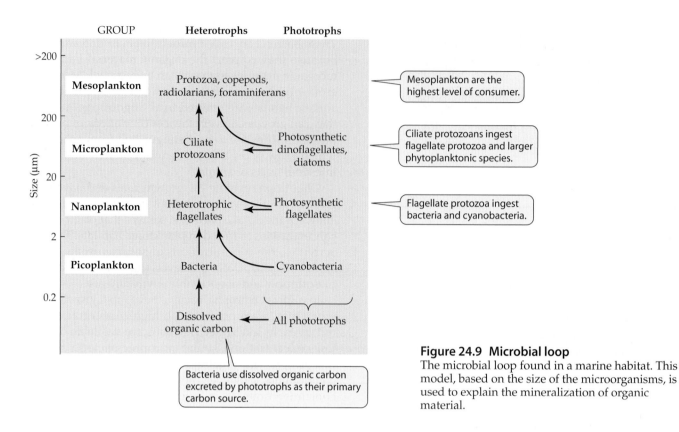

Figure 24.9 Microbial loop
The microbial loop found in a marine habitat. This model, based on the size of the microorganisms, is used to explain the mineralization of organic material.

are microbial **grazers** (consumers that ingest bacteria, algae, and bacterially coated particles), are in turn eaten by larger protozoa and so on.

Although grazers do not degrade all the particulate material during its passage through their digestive tract, their fecal pellets can be reingested again by another grazer after defecation, so the process is repeated until it is ultimately completely degraded. Therefore, the combined efforts of bacteria and these small grazing animals result in the ultimate decomposition of organic materials in aquatic ecosystems.

Eutrophic lakes that are sufficiently enriched with nutrients may develop an anaerobic hypolimnion in the summer months during the period of thermal stratification. This occurs because the hypolimnion is physically separated from the surface layers of the lake during stratification, so the amount of oxygen available in the hypolimnion is limited to what was initially there at the time of thermal stratification and the small amount that diffuses to it from the epilimnion. Anaerobic conditions develop in the hypolimnion in the following manner. The high concentration of organic nutrients in the eutrophic lake will be degraded in the hypolimnion initially by aerobes because aerobic respiration is such an energetically advantageous process. This results in oxygen depletion, and the hypolimnion becomes anaerobic.

Anaerobic conditions in the hypolimnion permit the growth of many other bacterial groups during the summer months. For example, if light penetrates to the hypolimnion and sufficient hydrogen sulfide is produced in the sediments, then photosynthetic bacteria grow and may produce hypolimnetic blooms. Common photosynthetic bacteria that may

Table 24.3	Composition of a natural water sample
Constituent	**Characteristics**
I. Dissolved material	Obtained by either centrifugation or filtration through a 0.2 μm filter to remove particulates
A. Organic components	
1. DOC	Dissolved organic carbon
2. DON	Dissolved organic nitrogen
B. Inorganic materials	
II. Particulate materials	Obtained as centrifuged material or collected on 0.2 μm pore size filter.
A. Living material	Living cellular material is called biomass
B. Nonliving materials	
1. Organic	Nonliving organic material is called detritus
2. Inorganic	

produce anaerobic blooms in this habitat include the purple sulfur bacteria and the green sulfur bacteria (Figure 24.7). Other anaerobic processes such as fermentation, denitrification, sulfate reduction, and methanogenesis occur at successively deeper depths in the hypolimnion or in the sediments of the lake.

Gas vacuolate organisms are especially common in thermally stratified lakes (Figure 24.7). These organisms can readily stratify in the lake at a light intensity or nutrient concentration level that is satisfactory for them. Thus, the gas vacuolate aerobic photosynthetic cyanobacteria are found at the surface of the lake where high light intensities occur, whereas the gas vacuolate purple and green sulfur bacteria are found at greater depths in the lake, where there is both sulfide and at least some light.

Rivers and Streams   Flowing water habitats are quite different from lakes. They are shallower and usually remain aerobic except in nonflowing backwaters and in their sediments. Frequently, microorganisms that inhabit these environments, such as *Sphaerotilus* spp., attach to rocks and sediments so that they can take advantage of the nutrients that flow by them. Streams in natural areas receive large amounts of leaf fall and other dead organic material in the autumn and winter and are therefore important habitats for leaf and litter decomposition.

Flowing habitats receive runoff from agricultural and urban areas and are greatly influenced by the nature of the material they receive. In urban areas, rivers frequently are used for sewage disposal and may receive varying loads of organic material and bacteria, depending on the degree of sewage treatment. Industrial efflu-

ents are also extremely important in larger cities and commercial waterways. In agricultural areas, excess fertilizers and pesticides reach the rivers through runoff and may pose serious pollution problems. These inputs provide nutrients for bacterial growth that may lead to anoxic conditions (see Chapter 32).

Marine Environments

Marine environments are extremely important because about two-thirds of the Earth's surface is covered by oceans. The predominant primary producers in the marine environment are planktonic algae (called phytoplankton), which serve as the base of the food chain. The consumers in the ocean that are highest in the food chain are larger fish, sharks, toothed whales and some other mammals. It is interesting to note that the krill-eating baleen whales, the largest mammals, occupy a rather low position in the food chain.

Because oceans cover so much of Earth, the role of marine microorganisms in primary production and other biogeochemical activities greatly surpasses that of microorganisms in inland lakes and rivers. Biological oceanographers study the biogeochemical processes that are occurring, some of which have important implications on the global climate.

Measuring primary productivity in the ocean is no small feat. Because of its vast size and seasonal and temporal variability, a few measurements here and there do not adequately assess activities. More recently, satellite-mapping procedures have been used to assess global productivity in the oceans (**Figure 24.10**). These procedures rely on analyzing photos that show chlorophylls

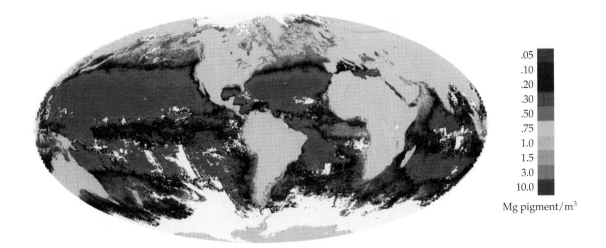

.05
.10
.20
.30
.50
.75
1.0
1.5
3.0
10.0

Mg pigment/m³

Figure 24.10 Global distribution of chlorophyll
Global satellite spectral image showing the absorption of chlorophyll-like pigments of the oceans during the summer months. Note the greater abundance (yellow and orange colors) of photosynthetic pigments off the coastal regions in the Northern Hemisphere. The Coastal Zone Color Scanner Project of the National Aeronautics and Space Administration's Goddard Space Flight Center produced and distributed the original data. This image was processed by David English at the School of Oceanography, University of Washington, using the University of Miami DSP system developed by Otis Brown and Robert Evans. Courtesy of David English.

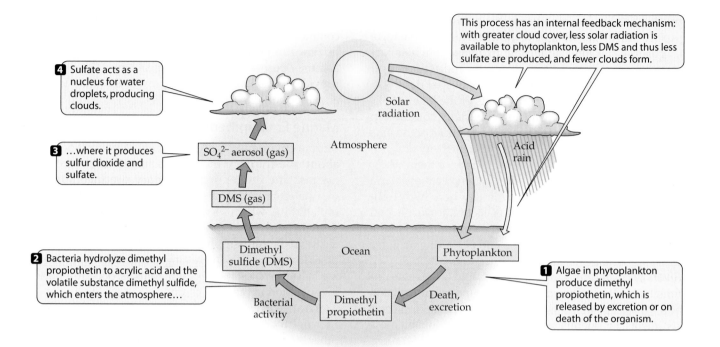

Figure 24.11 Microbial cloud formation
Proposed pathway for formation of clouds in the marine environment. It is interesting to note that the rainfall is acidic because of the sulfate, so this is an example of a process that causes a natural type of acid rain.

in the phytoplankton. Although the quantitative information of these photos is limited due to the inability to observe phytoplankton beneath the surface, they provide valuable information on the extent and intensity of blooms over large scales heretofore impossible to study.

Other marine microbial activities are also important. For example, it has recently been discovered that dimethyl sulfide (DMS) is produced as a product of metabolism by some phytoplankton. DMS is volatile and enters the atmosphere, where it reacts photochemically with oxygen

to form dimethyl sulfoxide (DMSO). This in turn serves as a raindrop nucleator, which results in cloud formation. Cloud formation affects the intensity of solar radiation on an area in the ocean, thereby allowing for a decrease in primary production and a decrease in DMS production. Thus, there is an internal feedback mechanism regulating primary production and temperature as well (**Figure 24.11**).

The anaerobic processes that occur in marine sediments are similar to those that occur in freshwater sediments. From the sediment surface going downward, the following are sequentially encountered: aerobic respiration, ferric oxide respiration, denitrification, sulfate reduction, and

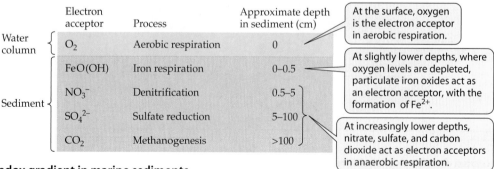

Figure 24.12 Redox gradient in marine sediments
Several types of electron acceptors are used in the biodegradation of organic substances in marine sediments, varying with depth. The actual depths at which these electron acceptors function will differ from one location in the sediment to another, depending on the amount of organic material, temperature, and other factors.

finally methanogenesis (**Figure 24.12**). However, there are some major differences. For example, in marine systems, there is much more sulfate than in freshwater habitats, so sulfate reduction is a much more important process in this habitat.

Typical marine bacteria are different from freshwater bacteria. The higher salt concentration (about 3.5%) makes marine bacteria moderate halophiles (see Box 19.1). Thus, one of the characteristics of a typical marine bacterium is that it will have an optimum salinity of about 3.5%, whereas a freshwater bacterium exhibits a lower optimal concentration. Also, most marine bacteria have an absolute requirement for Na^+, a feature not found in typical freshwater or terrestrial bacteria.

Terrestrial Environments

Terrestrial environments are extremely important for human civilization. Modern agriculture relies on good soils for both plant and animal productivity. Such soils contain a mineral component, which is derived from rock weathering, as well as an organic component, which is derived primarily from organisms that previously resided in the habitat.

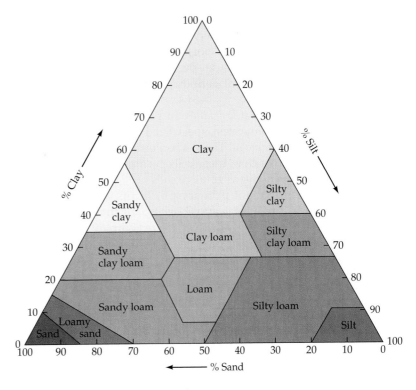

Figure 24.13 Soil textural triangle
Diagram illustrating how soils are named and described based on the proportion of sand, silt, and clay particles.

Soils The soil environment is very complex. It is made up of inorganic minerals, the weathered remains of rocks, and organic material, called **humus.** Humus comprises the partial decomposition products of plant organic material and as such contains many of the less-degradable or **refractory** organic plant substances such as lignin and humic acids. Soils vary greatly in composition and fertility depending on climate, geology, and vegetation. Depending on the composition of the soil, the climate, and the types of vegetation, soils are classified into various groups. Soil texture, one important feature of classifications, is determined by the relative proportion of clay, silt, and sand particles of the soil (**Figure 24.13**).

The soil profile is another important feature used in classification of soils (**Figure 24.14**). The profile of a soil is determined by making a vertical cross section or cut into the soil. Each layer, called a **horizon,** has a characteristic composition and color due to the activities that occur there. For example, the A horizon at the surface of the soil is dark in color due to the large amounts of organic matter (humus) that accumulate there. Below this is the E (for eluviation or "wash out") horizon, which is lighter in color and contains resistant minerals such as quartz. The deeper B horizon is a zone of accumulation (illuviation or "wash into"), which contains

materials such as Fe, Al, and silicate minerals. Each soil type has its own characteristic soil profile.

Unlike aquatic environments, water is not always available in soils. Nonetheless, water is absolutely

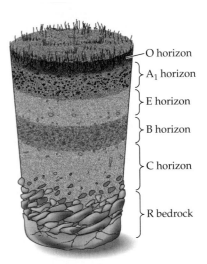

Figure 24.14 Soil profile
Each layer in this soil profile that is distinguishable by chemical composition, color, and microbial activities is referred to as a horizon.

required for the growth of all organisms. Thus, microorganisms that occur in soils grow intermittently, that is, only when there is sufficient soil moisture available for them. Microorganisms have made special adaptations to withstand these intermittent periods when water is unavailable. Some produce cysts (*Azotobacter* and myxobacteria); endospores (*Clostridium* and *Bacillus*); or conidiospores (the actinomycetes), specialized cells that survive periods of desiccation. Then, when moisture is available through rain or dew, these cells germinate to produce metabolically active vegetative cells.

In soils, the predominant primary producers are plants, not algae, although algae and *Cyanobacteria* may grow on the surface of soils. Plants synthesize complex organic constituents such as cellulose, hemicellulose, and lignin. These types of organic material are insoluble in water and generally more refractory to decomposition than algal polysaccharides and starch. Nonetheless, heterotrophic bacteria and fungi are known to degrade these substances. The rate and extent of degradation of these substances varies. For example, peat bogs are low pH environments that have complex organic constituents such as humic acids and lignin that degrade more slowly than they accumulate. Therefore, over many centuries the refractory organic substances may be converted to coal or petroleum deposits under the appropriate geological conditions. However, almost all of the organic material that is synthesized by plants is degraded biologically and goes back into the soil and air and thereby is made available for recycling.

The "health" of soils, called **soil tilth,** is extremely important for agriculture. Microorganisms play major roles in soil nutrient cycles, making nutrients available on the one hand and removing nutrients on the other hand, depending on the process. For example, nutrients are made available through the recycling of organic compounds as previously discussed. Likewise, important transformations in the nitrogen cycle occur in soils (see discussion later in the chapter on Nitrogen Cycle). Among the beneficial nitrogen transformations are nitrogen fixation, both symbiotically and asymbiotically, as well as ammonification, the removal of ammonia from organic sources. Less obviously beneficial is the process of nitrification that converts ammonia (normally a good source of nitrogen for plants) to nitrite and nitrate, which is an energetically less-favorable form of nitrogen for plants. The process of denitrification is actually harmful to soil fertility because it results in a conversion of nitrate nitrogen back into atmospheric nitrogen gas, in effect reversing the nitrogen fixation process.

Extreme Enivironments

As mentioned previously, some habitats have such extremes of temperature, salinity, low water activity, and pH that only microorganisms can grow in them (**Table 24.4**). Examples of some extreme environments are discussed here, and some are alluded to in Chapters 18 to 21, where bacterial diversity is treated, as well as in Chapter 25, which considers symbiotic associations. Acid mine environments are discussed in Chapter 32. Keep in mind that environments may have more than one parameter that makes them extreme. For example, some hot springs are also acidic. Nonetheless, certain bacteria such as *Sulfolobus* can grow in them.

Salt Lakes Some, such as the Great Salt Lake and the Dead Sea, are very salty because they do not have outlets to rivers or the sea. The saltiness occurs because the water evaporates from the lake, leaving behind the salt, which becomes more and more concentrated over time. After hundreds or thousands of years, the salt concentration increases as it has in the oceans. Sometimes these lakes become saltier than the sea. Table salt is obtained from salt lakes or from marine habitats, using special evaporation ponds (Figure 1.19C). The high salinity in

Table 24.4	Extreme conditions for microbial growth[a]		
Environmental Parameter		**Microbial Group**	**Habitat Source**
Temperature			
	High: > 110°C	Hyperthermophilic *Archaea*	Hot springs, marine hydrothermal vents
	Low: –12°C	Psychrophilic *Bacteria*	Polar sea ice
pH			
	High: > 12	Alkaliphilic *Bacteria, Archaea*	Desert soils, soda lakes
	Low: 0–1	Acidophilic *Bacteria, Archaea*	Acid mine drainages, sulfur springs
Saturated salts		Halophilic *Archaea*	Salt lakes, brines
High hydrostatic pressure		Barophilic *Bacteria*	Deep sea
High radioactivity		Radioduric *Bacteria*, e.g., *Deinococcus radiodurans*	Radioactive sites, soils
Low relative humidity (Low water activity)		Xerophilic fungi	Deserts, saps, brines

[a]Some environments have more than one condition that is considered extreme (e.g., acidic hot springs, which contain organisms such as *Sulfolobus*).

such habitats favors the growth of halophilic bacteria, including *Halobacterium* spp., and algae such as *Dunaliella*. These salt lakes will not be considered further here.

Sea Ice One special microbial environment found in oceanic polar regions of Earth is the sea ice microbial community (SIMCO). This habitat develops during the early spring when the sunlight reaches the polar region. The sea ice, which can be over 2 meters thick, is colonized by algae (mostly diatoms), bacteria, and protozoa (**Figure 24.15**A and B). They grow in the lower 10 to 20 cm of the ice just above the seawater (see Figure 1.19B). Thus, these organisms grow at temperatures below the freezing point of seawater (–1.8°C). They do this by growing in the brine pockets between ice crystals. One bacterial strain has been isolated that grows at –12°C.

The sea ice community is important on Earth because it is so extensive. Approximately 10% of the surface of the oceans is covered by ice on an annual basis. Furthermore, the SIMCO accounts for approximately one-third of the primary productivity of polar marine environments.

Rocks Although rocks are not normally thought of as an environment for living organisms, they are, in fact, common microbial habitats. Even the most inhospitable rock in the driest desert is likely to harbor microorganisms. The growth of microorganisms on rocks in conjunction with physical and chemical processes (such as freezing of water, wind action, and acid rainfall) causes rock weathering. As mentioned previously, weathering processes result in soil formation.

Algae and lichens are common primary producers of rocks (**Figure 24.16A**). They may grow on the surface of rocks (**epilithic**) (Figure 24.16B), in cracks (**chasmolithic**), or even on the underneath surface (**hypolithic**) if the rocks are translucent. Lichens are particularly well suited for growth on rocks because the fungal component that dominates in biomass can withstand severe desiccation and provide inorganic nutrients for the algal component. In turn, the algae produce organic nutrients for the growth of the fungus symbiont (see Chapter 25). Some fungi can also grow as microcolonies on rock surfaces in desert and arid areas in environments too severe for the growth of lichens (Figure 4.14). Apparently they obtain their nutrients from windblown dust.

One of the most remarkable findings has been that some lichens and algae can actually grow inside rocks (Figure 1.19D). Such growth is called endolithic. Indeed, endolithic microorganisms that grow in rocks in Antarctica are among the most extreme forms of terrestrial life on Earth (**Box 24.6**).

Subterranean Environments One of the major environments on Earth that is very poorly understood is the subterranean environment, which extends from the aphotic surface soils to 1 or more kilometer depths beneath the surface. It has been proposed that the primary production that occurs in this environment is derived from geochemical energy sources such as sulfide and hydrogen generated from thermal sources

(A)

(B)

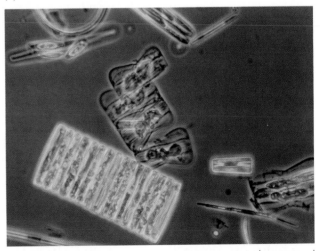

20.0 μm

Figure 24.15 Sea ice profile
(A) Diagram showing the sea ice overlying the water column at a North Pole location. The ice is approximately two meters thick, and the microbial community is shown as a brown layer at the interface between the ice and the water column. A core has been cut from the sea ice. (B) Diatoms predominate in the sea ice community. A, courtesy of John Gosink; B, courtesy of J. T. Staley.

(A)

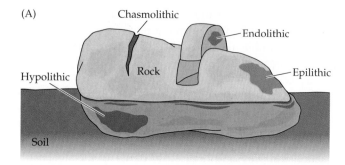

(B)

Figure 24.16 Microbial growth on rocks
(A) Diagram showing microbial colonization of rocks. Epilithic organisms grow on the surface, chasmolithics in cracks, endolithics inside, and hypolithics on the buried side of the rock. (B) Epilithic lichens colonizing rocks in Antarctica. B, photo ©R. Planck/Photo Researchers, Inc.

beneath the surface. These energy sources could be used by sulfate reducers and methanogens, respectively, which fix carbon dioxide to form a cellular biomass that serves as an energy source for fermenting microorganisms. Because this environment is so difficult to study, little is known about the primary production activities that occur. Nonetheless, its vast extent suggests that it could be a major and likely ancient environment dominated by primary producing *Bacteria* and *Archaea*.

Hydrothermal Vents Two major types of thermal environments occur: hot springs and hydrothermal vents. Microbial mats in hot springs were considered as an example of a microbial community earlier in this chapter.

Hydrothermal vents are among the most fascinating natural wonders of the world. They are a result of plate tectonic movements. Very briefly, the Earth's crust consists of large plates that are floating on the underlying hot mantle. The plate movements occur both in the Atlantic and Pacific Oceans. As the plates separate from one another, hot fluids from the mantle rise to produce the vents.

The most dramatic examples of the hydrothermal vents are the black smokers that rise from the ocean floor, spewing reduced gases and sulfide waters that contain iron and manganese sulfides, which impart the black appearance of a smoking chimney to the structures (**Figure 24.17**). These black fluids contain reduced inorganic nutrients that serve as substrates for the growth of chemolithotrophic *Bacteria* and *Archaea*.

What is especially fascinating about the hydrothermal vents is the unique biota that develops in association with them. Among the most fantastic are large tube worms that reach over 2 meters in height. In addition, clams and shrimp as well as other animals live in the vicinity of the vents.

The vents occur in many deep-sea environments that are 2,000 meters below the surface. Thus, they are too deep to receive sunlight. In these environments, photosynthesis cannot occur, and the amount of organic carbon raining on the ocean floor from surface photosynthesis is insignificant. Therefore, the biota in these vents is entirely dependent on the bacterial primary producers that obtain energy for their growth from the oxidation of the reduced chemicals such as hydrogen, hydrogen sulfide, iron, and manganese that are emitted from the vents. The result is that the animals have developed unique and special symbiotic associations with *Bacteria* and *Archaea*. Chapter 25 discusses some of these fascinating symbioses.

Both hydrothermal vents and subterranean environments are strong candidates for the evolution of life on Earth. The "warm little pond" model for evolution seems highly unlikely, considering that the period when life evolved on Earth closely followed the period of heavy bombardment by asteroids and comets (see Chapter 1). Although many of these impacts could have sterilized the surface of our planet, the hydrothermal vents and subterranean environments were much better protected from their destructive effects.

BIOGEOCHEMICAL CYCLES

Because microorganisms, particularly bacteria, evolved billions of years before higher organisms, they were largely responsible for the origin and evolution of the biogeochemical cycles and continue to play major roles in these transformations. Of course, the cycles have continued to change since the evolution of higher organisms and are also influenced by human industrial and military activities.

Atmospheric Effects of Nutrient Cycles

From a global perspective, the effects of the nutrient cycles are most keenly felt in the atmosphere. This is

Research Highlights Box 24.6   Endolithic Microorganisms of Antarctica

The Victoria Land dry valleys near the U.S. Antarctic base at McMurdo Sound are one of the most extreme environments on Earth. The two principal features that limit biological growth and activity are the low temperature and low water activity (humidity). As a result, "soils" taken from these environments appear sterile, because they have such extremely low counts of bacteria.

For this reason, the dry valleys have been compared to the surface of Mars.

However, some microorganisms do live in the terrestrial environments of Antarctica. They reside inside the rocks in the Dry Valleys. The rocks are porous sandstones that obtain some moisture from snowfall. During brief periods of midsummer, conditions allow for melting. During these periods the endolithic lichens obtain water and are able to photosynthesize and grow. For the remainder of the year, they exist in a natural freeze-dried or lyophilized state. Carbon dating by E. I. Friedmann, the microbial ecologist who discovered this remarkable community of microbial life, indicates that these lichens are thousands of years old.

(A)

(B)

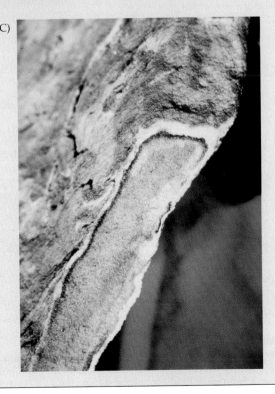

(C)

(A) View of the Victoria Land dry valleys. Rocks in foreground with a reddish color contain endolithic lichens. (B) Rocks being studied by microbial ecologists from the laboratory of the late J. Robie Vestal. (C) Broken sandstone rock showing the endolithic lichen community, which appears as a dark band beneath the surface. See also Figure 1.19D for a close-up of the algal and fungal layers of this community. Courtesy of J. R. Vestal.

because all higher terrestrial living forms exist bathed in the atmosphere. The animals are dependent on oxygen that is provided by primary producers and through respiration expel carbon dioxide, an essential nutrient for plants. Because of their reliance on air, animals and plants are profoundly affected by noxious airborne pollutants. Furthermore, the chemical composition of the atmosphere affects global temperatures and incident ultraviolet light radiation. Two major current concerns are the buildup of greenhouse gases that cause global warming and the release of anthropogenic chemicals such as the chlorofluorocarbons, which destroy the ozone layer that protects all living organisms from UV radiation.

Microorganisms are involved in many of the transformations that affect the atmosphere. For example, nitrous oxide, N_2O, an intermediate product formed in two different bacterial processes, nitrification and denitrification, reacts photochemically with ozone, O_3, in the atmosphere, thereby depleting it (see Nitrogen Cycle).

Both methane and carbon dioxide, other important constituents of the atmosphere, are greenhouse gases produced in part through microbial activities. Methane is produced largely through the activities of methanogenic bacteria, although it is also released from natural gas seeps and during oil drilling operations. Its concentration has been increasing in the atmosphere at a rate of about 1% per year for the past 15 to 20 years. The biological sources of methane include ruminant animals, termites, and bogs and swampy areas, all of which contain methanogenic bacteria. In the northern hemisphere an important nonbiological source comes from oil drilling operations, which release lighter hydrocarbons such as methane into the atmosphere.

Carbon Cycle

Microorganisms play major roles in the carbon cycle (**Figure 24.18**), an element that is a key constituent of liv-

(A)

(B)

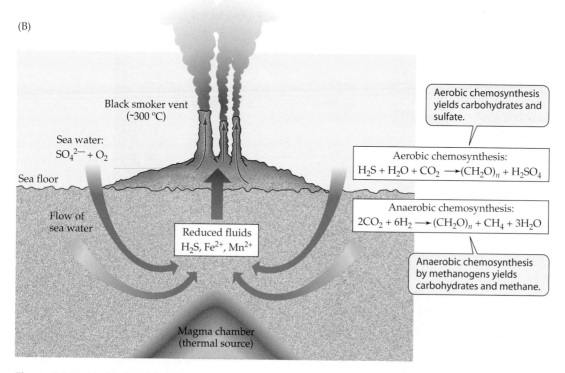

Aerobic chemosynthesis yields carbohydrates and sulfate.

Aerobic chemosynthesis:
$H_2S + H_2O + CO_2 \longrightarrow (CH_2O)_n + H_2SO_4$

Anaerobic chemosynthesis:
$2CO_2 + 6H_2 \longrightarrow (CH_2O)_n + CH_4 + 3H_2O$

Anaerobic chemosynthesis by methanogens yields carbohydrates and methane.

Black smoker vent (~300 °C)

Sea water:
$SO_4^{2-} + O_2$

Sea floor

Flow of sea water

Reduced fluids
H_2S, Fe^{2+}, Mn^{2+}

Magma chamber (thermal source)

Figure 24.17 Hydrothermal vents
(A) A "black smoker." (B) Diagram showing the flow of seawater from the seafloor into the sediments, where it is heated and chemicals are reduced to produce fluid containing sulfide, Fe^{2+}, and Mn^{2+}. Iron sulfides give the vent emissions their black color. These reduced substances serve as energy sources for chemosynthesis. A, photo ©D. Foster, WHOI Visuals Unlimited.

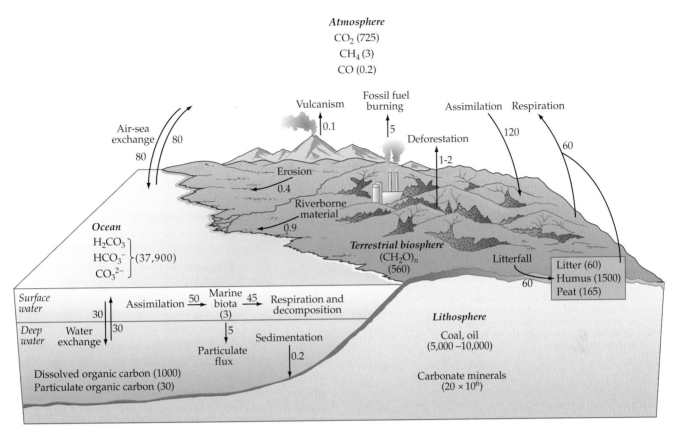

Figure 24.18 Global carbon cycle
Cycling of carbon through the atmosphere, biosphere, oceans, and lithosphere.
Reservoir amounts are in pentagrams of carbon (PgC) and fluxes (shown by arrows)
are in PgC/year. Freshwater environments are included in the terrestrial compartment.
Adapted from K. Holmén, *Global Biogeochemical Cycles*, Academic Press.

ing organisms. In the example shown, there are four separate **reservoirs** for carbon: the ocean, the atmosphere, the lithosphere, and the terrestrial biosphere. The amount of carbon in each reservoir, termed the *mass,* is given in PgC (petagrams of carbon) (where $1P = 10^{15}g$). Thus, the atmosphere of Earth contains 725 Pg carbon dioxide, 3 Pg methane, and 0.2 Pg carbon monoxide. Arrows designate the **flux,** or rate of movement, of carbon from one reservoir to another in terms of PgC/year. Therefore, in the ocean, a balance occurs between the flux of carbon to and from the atmosphere (80 PgC/year). The mass of the marine biota in the surface water of the ocean (3 PgC) is responsible for the processes of carbon **assimilation** (largely due to carbon dioxide fixation by primary producers), **respiration,** and **decomposition.**

Note that the fluxes to and from a box are balanced. In this example, 50 PgC/year are assimilated by the marine biota, 45 PgC/year are respired, and the final 5 PgC/year are particulate materials (largely nonliving phytoplankton and other organisms) that sediment into the deeper layers. The deeper water contains large

amounts of inorganic carbon in the form of carbonate, bicarbonate, and carbonic acid. It also contains substantial amounts of dissolved organic carbon (DOC) and particulate organic carbon (POC). Only a small portion of the carbon, 0.2 Pg/year, reaches the sediments.

This model also accounts for inputs of carbon to the atmosphere through volcanic eruptions (methane, carbon dioxide, and carbon monoxide) as well as fossil fuel burning and deforestation, all of which occur in the terrestrial compartment. Although this model is not very detailed, it provides important information about the major reservoirs of carbon and the most important fluxes that occur between the reservoirs. More and more complex models can be generated if more detailed experimental data are available.

Because this text is primarily interested in the microbial aspects of biogeochemical cycles, particular emphasis will be given to primary production, decomposition, and methane formation and oxidation.

Primary Production

Primary producers fix carbon dioxide and convert it to organic material (**Table 24.5**). One group of primary producers is the photosynthetic organisms that derive their

Table 24.5	Primary production (carbon dioxide fixation) among the *Bacteria* and *Archaea*[a]	
I. Photosynthetic Primary Producers[b]		
A. *Bacteria*		Cyanobacteria
		Proteobacteria
		Chlorobi
		Chloroflexi
B. *Archaea*		None
II. Chemosynthetic Primary Producers		
A. *Bacteria*		Nitrifiers
		Sulfur oxidizers
		Iron oxidizers
		(*T. ferrooxidans*)
		Aquificales
		(hydrogen oxidizers)
		Acetogens
B. *Archaea*		Methanogens
		Sulfur oxidizers

[a]Not all members of each group are primary producers.
[b]Some photosynthetic groups use light for photophosphorylation but do not fix carbon dioxide; these include the heliobacteria, the purple nonsulfur bacteria, and the halobacteria.

energy directly from sunlight. Another group is the chemolithotrophic bacteria. The chemolithotrophic bacteria require chemical sources of energy (reduced inorganic compounds such as hydrogen sulfide, ammonia, or hydrogen), and these are provided through geochemical activities or other biological processes, some of which may depend ultimately on sunlight. For example, chemolithotrophic sulfur oxidizers require reduced forms of sulfur derived either from geochemical sources or from reduced sulfur sources provided by sulfate reducers, who obtain their energy from organic materials ultimately produced by photosynthesis.

The principal terrestrial primary producers are plants that are the major producers of organic material. In freshwater and marine habitats, which account for about two-thirds of the surface area of Earth, the algal and cyanobacterial phytoplankton are the principal primary producers. In contrast, anoxygenic photosynthetic prokaryotes and chemolithotrophic bacteria play much more restricted roles in primary production except in specialized habitats like meromictic lakes (see following) and thermal habitats such as hydrothermal marine vents, respectively.

Decomposition of Organic Material

Organic material derived ultimately from primary producers resides in living organisms and the nonliving organic material derived from them. Plants and animals and most microorganisms carry out **respiration** to form carbon dioxide and water. For animals and aerobic, heterotrophic microorganisms, respiration is the mechanism by which they obtain ATP for their metabolism.

Bacteria and fungi are the ultimate recyclers of nonliving organic material. They live as saprophytes on organic material from dead plants and animals as well as other microorganisms. They are aided in this process by higher animals that ingest particulate organic materials (herbivores and carnivores) that contain bacteria associated with them or that are residing in their intestinal tracts. The process is chemically analogous to respiration, but involves the degradation of nonliving organic material to obtain energy for growth. This process is called organic decomposition or degradation. If the organic compound is degraded completely to inorganic products such as carbon dioxide, ammonia, and water, the process is called mineralization.

Bacteria and fungi are especially well suited for the degradation of polymeric organic compounds that are refractory to higher organisms. Cellulose, chitin, and lignin are examples of this. Cellulose, derived from plants, and chitin, derived largely from crustaceans, insects, and some fungi, are degraded by many bacteria and fungi that produce extracellular enzymes to attack these particulate substrates. The white rot fungi are the principal known degraders of lignin, a process that has only recently been chemically characterized. A wider variety of microorganisms can degrade soluble organic compounds including organic acids, amino acids, and sugars.

Almost all organic material produced on Earth is degraded by microbial activities. However, small amounts accumulate in sediments, as shown in Figure 24.18. Over time, these accumulations have resulted in the formation of coal and petroleum deposits.

Methanogenesis and Methane Oxidation

Major anaerobic processes of the carbon cycle result in the fermentation of organic compounds to organic acids and gases such as hydrogen and carbon dioxide. Additional degradation by methanogens results in the formation of methane gas from highly reduced sediments and the digestive tracts of ruminant animals and termites. Although many methanogens use carbon dioxide and hydrogen gas as substrates for methane formation, some use other fermentation products, including methanol and acetic acid to produce methane (see Chapter 18). Methane-oxidizing bacteria and certain yeasts degrade methane in the biosphere, but some escapes to the atmosphere, where it becomes a greenhouse gas.

Nitrogen Cycle

All living organisms require nitrogen because it is an essential element in protein and nucleic acids. Animals require organic nitrogen sources that they obtain through digestion of plant or animal tissues. Plants use inorganic nitrogen sources such as ammonia or nitrate. Most bacteria can use ammonia or nitrate as nitrogen for growth, but some, such as the lactic acid bacteria, may require one or more of the amino acids in their diet.

Microorganisms play several important roles in the nitrogen cycle (**Figure 24.19**). They are responsible for several processes not carried out by other organisms. A discussion of the major processes follows.

Nitrogen Fixation Only prokaryotic organisms carry out nitrogen fixation (see Chapter 10). Some members of the *Bacteria* and *Archaea* produce the enzyme nitrogenase, which is responsible for this process. Nitrogenase is found in both photosynthetic prokaryotes as well as in heterotrophic prokaryotes (**Table 24.6**). Because nitrogen is often a limiting nutrient in terrestrial and aquatic habitats, this process is important because it recycles nitrogen back into the biosphere.

Nitrogen is fixed into organic nitrogen (as the amino group in the amino acid glutamine). This can be transformed into other amino acids using various enzymes

available in the organism that fix the nitrogen or make it available to the organism's symbiotic partner (if it is a symbiotic fixer). In this manner, N_2, a rather inert gas, is made available to organisms directly from the atmosphere.

Ammonification Organisms release nitrogen from their cells or tissues in the form of ammonia, a common decomposition product. This process, called **ammonification,** is hastened by the activity of some bacteria that have deaminases that remove amino groups from organic nitrogenous compounds to form ammonia.

Nitrification Ammonia produced by ammonification can be used directly by many plants as a source of nitrogen for synthesis of amino acids and other nitrogen-containing organic compounds.

However, ammonia can be oxidized by a special group of bacteria called nitrifiers. These bacterial

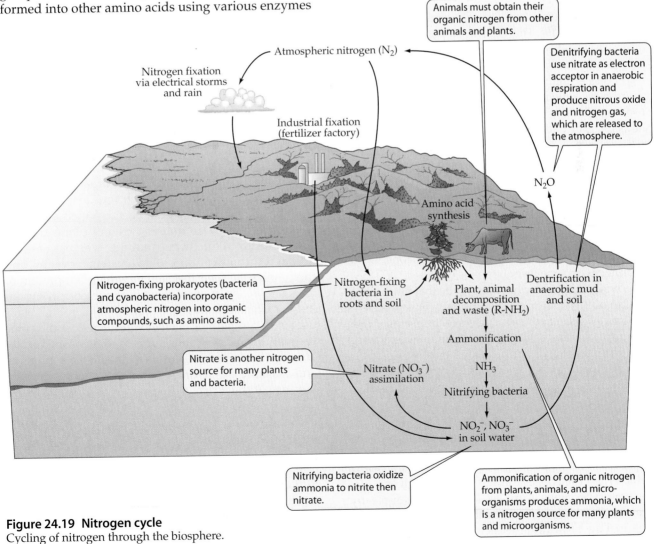

Figure 24.19 Nitrogen cycle
Cycling of nitrogen through the biosphere.

Table 24.6	Representative prokaryotic groups containing species that carry out nitrogen fixation[a]
I. *Bacteria*	
A. Nonsymbiotic	
1. Photosynthetic	*Proteobacteria* *Chlorobi* *Cyanobacteria*
2. Heterotrophic	*Azotobacter* *Clostridium* Some spirochetes
B. Bacterial Symbionts	*Rhizobium* (legumes) *Frankia* (alder trees) Cyanobacteria (lichens; *Azolla*)
II. *Archaea*	Methanogens

[a]This list is illustrative and does not contain all taxa that are capable of nitrogen fixation; not all members of each group are nitrogen fixers.

chemolithotrophs obtain energy from this oxidation process (Chapter 19). This is a two-step process in which ammonia is first oxidized to nitrite, and the nitrite is subsequently oxidized to nitrate. Like nitrogen fixation, this process is uniquely associated with bacteria. During this process, some nitrous oxide (laughing gas), N_2O, is produced. This is an important gas because it reacts photochemically with ozone according to the following reactions:

$$N_2O + h\nu \rightarrow N_2 + O$$
$$N_2O + O \rightarrow NO$$
$$NO + O_3 \rightarrow NO_2 + O_2$$

where $h\nu$ represents a photon.

The result of these reactions is that the ozone, O_3, in the protective ozone layer becomes depleted.

Recently, a discovery has been made that nitrification can also occur in anoxic environments. This process is purportedly carried out by members of the *Planctomycetes* phylum and is termed the anaerobic ammonia oxidation reaction. In this overall process, ammonia is oxidized to nitrate, and nitrate is the electron acceptor for the oxidation. The responsible organism has not yet been isolated in pure culture, suggesting that the activity may be due to a consortium of more than one species.

Nitrate is much more readily leached from soils than is ammonia. If excessive amounts of nitrate are leached from soils, it can accumulate in runoff water and in wells. When the concentrations become high enough, the water becomes unfit as a drinking source by humans. This happens because the nitrite formed in the intestinal tract by nitrate-reducing bacteria can have an

extremely adverse affect by interacting with hemoglobin in the bloodstream to produce methemoglobin. This causes **methemoglobinemia** and, if it is not properly diagnosed, may result in the death of infants because of its effect on respiration. The color of infants becomes blue (hence the vernacular name for the ailment, "blue babies"). This rare malady occurs primarily in agricultural areas that receive excess nitrogen fertilizer.

Denitrification A number of bacteria carry out nitrate respiration. In this process, which occurs preferentially in an anaerobic environment, nitrate is ultimately converted into nitrogen gas. This process is called denitrification. It occurs predominantly in waterlogged areas that have become anaerobic. From an agricultural perspective, this is an undesirable process because it results in a loss of fixed nitrogen back to the atmosphere. The intermediates of this process are similar to those of nitrification, including the formation of N_2O. Some of the bacteria that carry out this process include various *Pseudomonas* species, *Thiobacillus denitrificans*, and *Paracoccus denitrificans*.

Nitrate Ammonification In some anaerobic environments, such as cattle rumen, nitrate is not converted to nitrogen gas by denitrification, but is reduced to ammonia by the resident bacteria. This process is called nitrate ammonification.

Sulfur Cycle

The **sulfur cycle** is also an important biogeochemical cycle. Sulfur, like carbon and nitrogen, is needed by living organisms, where it is a constituent of protein. However, it is not required in such large amounts as carbon and nitrogen and has not been reported as a limiting nutrient in environments. Higher plants usually obtain sulfur in the form of sulfate, whereas animals obtain it through amino acids (cysteine, cystine, and methionine) in their diet, either from plant protein or from the protein of other animals or microorganisms (as in the ruminant animals). In contrast, microorganisms use sulfur compounds in other ways. For example, some use sulfur-containing substances as energy sources, others use it as electron acceptors in anaerobic respiration, and some use it as hydrogen donors for photosynthesis. Because of these diverse activities, the sulfur cycle is dominated by microbial processes and is one of the most fascinating of the elemental cycles (**Figure 24.20**). The oxidation state of inorganic sulfur ranges in its cycling from –2 in sulfide to +6 in sulfate.

Sulfur Oxidation Reduced inorganic forms of sulfur, including not only sulfide and elemental sulfur, but thiosulfate and other ions as well, can be oxidized by several different groups of organisms as has been discussed

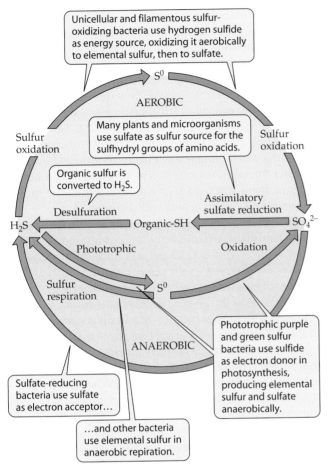

Figure 24.20 Sulfur cycle
Cycling of sulfur through the biosphere. S^0 indicates elemental sulfur.

eubacteria are involved in this process as well as some acidophilic, thermophilic archaeobacteria.

Sulfur Reduction The groups of sulfur-reducing bacteria are less well studied. The best-studied group is sulfate-reducing *Bacteria,* which obtain carbon from organic compounds and use sulfate as an electron acceptor in sulfate respiration. Some of these also use hydrogen gas as an energy source and grow autotrophically by carbon dioxide fixation. This process of sulfate reduction is referred to as **dissimilatory sulfate reduction** to distinguish it from **assimilatory sulfate reduction,** the process by which plants, algae, and many aerobic bacteria obtain sulfur for synthesis of amino acids. The dissimilatory sulfate reduction process (see Chapter 19) requires large amounts of sulfate and occurs in anaerobic muds. It is the sulfate-reducing bacteria that produce the sulfide smell typical of black muds from lake and marine sediments.

A number of thermophilic *Archaea* such as *Pyrodictium* spp. respire with sulfur. Elemental sulfur is used by most of them as an electron acceptor, although a few use sulfate.

Other Cycles

Microorganisms play important roles in many other elemental cycles. For example, all of the metallic elements have their own cycles, and bacteria are involved in all of them. More common microbial transformations occur with iron, manganese, calcium, mercury, zinc, cobalt, and the other metals. Ionized forms of the heavy metals precipitate protein and other macromolecules, explain-

previously (**Table 24.7**). This process is called **sulfur oxidation.** Photosynthetic bacteria use hydrogen sulfide produced by sulfate reducers in anaerobic environments as electron acceptors for the ultimate reduction of carbon dioxide for the synthesis of organic materials. Included in this group of organisms are the purple sulfur and green sulfur bacteria. These organisms oxidize the sulfide to elemental sulfur and ultimately to sulfate (see Chapter 21).

In addition, certain nonphotosynthetic bacteria also oxidize reduced sulfur compounds. Some of these are chemosynthetic and therefore use the reduced sulfur compounds as energy sources and inorganic carbon as a carbon source for growth. Others are heterotrophic. For the most part, these are obligately aerobic organisms, although a few can use nitrate as an electron acceptor in nitrate respiration. Both filamentous and unicellular

Table 24.7	Bacterial groups responsible for the oxidation of reduced sulfur compounds
Bacteria	**Oxidative Activity**
Photosynthetic Bacteria	
Purple Sulfur	$H_2S \rightarrow S^0 \rightarrow SO_4^{2-}$
Green Sulfur	$H_2S \rightarrow S^0 \rightarrow SO_4^{2-}$
Some Cyanobacteria	$H_2S \rightarrow S^0$
Chemosynthetic Bacteria	
Filamentous Sulfur Oxidizers (e.g., *Beggiatoa*)	$H_2S \rightarrow S^0 \rightarrow SO_4^{2-}$
Unicellular Sulfur Oxidizers (e.g., *Thiobacillus, Microspira*)	$H_2S \rightarrow S^0 \rightarrow SO_4^{2-}$
Heterotrophic Bacteria	
Filamentous Sulfur Oxidizers (e. g., *Beggiatoa*)	$H_2S \rightarrow S^0 \rightarrow SO_4^{2-}$
Unicellular Sulfur Oxidizers (e.g., some *Pseudomonas* spp.)	$H_2S \rightarrow S^0 \rightarrow SO_4^{2-}$
Archaea	
Acidianus, Sulfolobus	$H_2S \rightarrow S^0 \rightarrow SO_4^{2-}$

ing their toxicity. Many bacteria carry plasmids that have enzymes that carry out the oxidation/reduction or other transformations of these heavy metal ions, thereby detoxifying them. However, other bacteria can mobilize heavy metals. For example, some methanogens, in the presence of mercury, produce dimethyl mercury, which is volatile. In general, however, the metal cycles do not have an atmospheric stage, so they are not nearly as mobile as elements such as nitrogen and sulfur.

Iron and manganese oxides can serve as electron acceptors in anaerobic systems. These elements are accordingly reduced to their Fe^{2+} and Mn^{2+} states, which are soluble and therefore more mobile than their oxidized forms. These transformations are especially important in sediments. Because iron is needed by almost all organisms, its availability is restricted due to its insolubility under most aerobic conditions. Thus, bacteria have developed special iron transport compounds termed *siderophores* to bring Fe^{3+} into the cells (see Chapter 4).

DISPERSAL, COLONIZATION, AND SUCCESSION

In order to disseminate from one location to another, microorganisms need to be able to survive during transit under nongrowing conditions. Many bacteria and higher microorganisms have special dispersal and survival stages such as endospores, cysts, or conidiospores that allow them to maintain their viability for many days, weeks, or even years. Other organisms can survive for long periods of time in the dirt of birds' claws, in the digestive tract of birds, or on plant and animal materials eaten by birds. These birds can serve as vectors to carry a variety of microorganisms from one continent to another. Similar mechanisms occur with other animals, although their migration patterns are more limited in scale.

Perhaps the microorganisms that are best known for their abilities to traverse long distances are the fungi. Their airborne spores can be carried many miles under favorable conditions for dispersal. Some of these microorganisms cause plant diseases, and their long-distance dissemination is therefore a matter of concern to agriculture. Others are important allergens that affect the pulmonary systems of sensitive individuals.

A different illustration of dispersal and colonization is that of the ru-

minant animals (see Chapter 25). Ruminant animals maintain a dense community of anaerobic bacteria, including methanogens that are very sensitive to oxygen, in their forestomach. However, newborn calves have a sterile intestinal tract. After they are weaned and begin eating plant materials, they develop the typical anaerobic microflora of adult cattle. Because of the sensitivity of this community to oxygen, long-distance dissemination of the organisms is highly unlikely. The anaerobic microbiota is most likely transmitted from the mother through her saliva, which contains small numbers of these bacteria that are brought to the mouth through regurgitation and cud chewing.

Of course, newly exposed habitats, such as a new volcanic landmass, will not be colonized immediately. The process of colonization takes time. Furthermore, the colonizing or **pioneer** species may not necessarily be one of the species that survives in the new habitat. Colonization leads to the phenomenon of **succession,** in which the species composition of the habitat changes over time. An example of a microbial succession is one that occurs in the colonization the intestinal tracts of newborn mice (**Figure 24.21**). In this example, the initial colonizing organisms include a *Flavobacterium* species, a *Lactobacillus* species, and enterococci. Over a few days' time, the succession proceeds, and colonization by the obligately anaerobic genus *Bacteroides* occurs. Eventually, the *Flavobacterium,* one of the pioneer species and an obligate aerobe, disappears. Thus, changes have occurred in the intestinal tract as it matures, changes that are brought about in part by the colonizing organisms including the *Flavobacterium*. These changes have

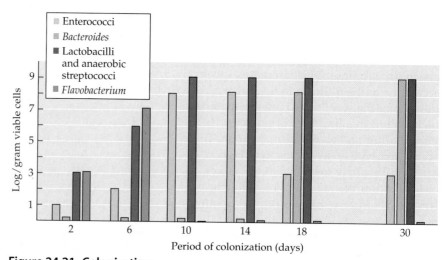

Figure 24.21 Colonization
Colonization of a newborn mouse's intestinal tract by various bacterial species. Not all early colonizers are successful. As the intestinal tract becomes anaerobic, obligate aerobes, such as *Flavobacterium* spp., can no longer survive. Adapted from R. W. Schaedler, R. Dubois and R. Costello. 1965. *The Journal of Experimental Medicine* 122: 59–66.

resulted in creating a habitat with new features (among them, anoxia) that no longer allow the *Flavobacterium* to survive. In contrast, the others, which are facultative anaerobes, remain, although the concentrations of all are affected by the changed conditions they helped bring about. Eventually a **climax state** is attained in which the types and concentrations of microorganisms attain an equilibrium in the intestine. This equilibrium can be disrupted by intestinal infections or antibiotic therapy.

BIOGEOGRAPHY OF BACTERIA

L. M. G. Baas-Becking, a Dutch microbiologist, stated that "Everything is everywhere, the environment selects." By this he meant, for example, that if a particular soil type in Europe is very similar to a soil type in North America with similar climatic conditions, the same species of bacteria, such as *Bacillus licheniformis,* is expected to live in both places. Thus, this hypothesis argues for most bacteria having a **cosmopolitan,** or worldwide, distribution. The basis for this hypothesis is that microorganisms are readily dispersed throughout Earth—they are carried by the wind across the ocean, as well as by animals, especially birds and humans, that travel long distances. When they find a favorable habitat, they will colonize it and grow. Therefore, over relatively short periods of time a bacterium from Europe could be brought to North America or vice versa. The rapid exchange of bacteria with other places on Earth, coupled with the ability of the best-fit strains to be selected and survive, is consistent with the cosmopolitan hypothesis.

Although this hypothesis is widely accepted by microbiologists, it has not been thoroughly tested. Recently, however, microbiologists are using molecular techniques to analyze samples from various habitats such as soil and sea ice to determine the validity of the hypothesis. Some recent data taken from diverse geographic regions suggest that perhaps some bacteria are endemic or indigenous to specific locales. However, it is difficult to preclude the possibility that the same organism might be found in much lower concentrations in another habitat. The answer to this question is important because it will help provide information on the total number of bacterial species on Earth. If there are no endemic bacteria, then fewer bacterial species would be predicted.

SUMMARY

▸ **Microbial ecology** is the study of the interactions between microorganisms and the **abiotic** and **biotic** components of their environment. The **habitat** of the organism is where it lives (its "address"), the **niche** of the organism is its role (its "profession") in the ecosys-

tem. **Ecosystems** are physical areas on Earth, such as lakes, where organisms live. The microbial biosphere extends to areas on Earth not occupied by other organisms.

▸ **Total counts** of bacteria, including dead and living organisms, are usually conducted by using fluorescent dyes and a fluorescence microscope. Viable counts are made using growth media. The **bacterial enumeration anomaly** refers to the fact that the recovery of culturable bacteria from typical mesotrophic and oligotrophic aquatic and soil habitats is typically less than 1% of the total count.

▸ **Autecology** is the study of an individual species in its environment. Species can be identified in natural samples by use of fluorescent antibody techniques or by nucleic acid probes. Using these procedures, it is possible to determine the spatial and temporal distribution of a species in its environment. The activity of an individual microorganism can be assessed in the environment by microautoradiography.

▸ **Biomarkers** are substances that are uniquely associated with a species or other group of organisms and can therefore be used for identification.

▸ The **biomass** of microorganisms is the sum total of all living microbial protoplasm in a habitat at a given time. ATP measurements are often used to assess biomass in microbial communities.

▸ The rates of microbial processes can be measured by direct chemical analyses or by use of radioisotopes. Radioisotopic measurements of rates typically use the **tracer approach** in which the substrate **pool** is labeled by small amounts of the radioisotope. After a short period of time, the amount of label that leaves the substrate pool or enters into the product pool is measured to provide an assessment of the rate of activity.

▸ Microorganisms inhabit all habitats known, including freshwaters, marine waters, soils, and rocks. These may be aerobic or anaerobic, acidic or alkaline, cold or hot. Indeed, microbial life is found in some of the most extreme environments known in terms of temperature, pH, and salt concentration.

▸ Microorganisms, particularly *Bacteria* and *Archaea,* play critical and unique roles in biogeochemical cycles. For example, they are solely responsible for such transformations as nitrogen **fixation, denitrification,** and **nitrification.** *Bacteria* and *Archaea* are also critical to sulfur cycling, where they are responsible for sulfate and **sulfur reduction** as well as **aerobic** and **anaerobic sulfide oxidation.** In the global carbon cycle, microorganisms are very important in decomposition of organic matter, **methanogenesis,** and **methane oxidation.** However, some microorganisms, such as the algae and *Cyanobacteria* are primary producers and others, primarily the protozoa, are consumers.

▶ Microorganisms are readily dispersed from one habitat on Earth to another. Therefore, it has been postulated that all species are **cosmopolitan** in their **biogeography** or distribution. However, this hypothesis has not been rigorously tested.

▶ Microorganisms live in all ecosystems. Microorganisms participate in all **biogeochemical cycles,** where they often play unique roles. For example, nitrogen fixation, nitrification, and aerobic and anaerobic sulfide oxidation are examples of unique microbial processes.

REVIEW QUESTIONS

1. What is meant by biomass and biomarkers, and how are they used in microbial ecology?

2. How would you go about determining the total and viable numbers of nitrifying bacteria in a lake? Do you believe probes would be of any use?

3. Discuss how the tracer approach is used to measure rates of microbial processes in nature.

4. Why does the hypolimnion of a eutrophic lake often become anaerobic during the summer months? Does this affect the temperature of the hypolimnion?

5. If you were counting only the total and viable heterotrophic bacteria in a mesotrophic lake and saw that the viable numbers increased 100-fold in a matter of two weeks, what would you believe was the most likely cause?

6. Define succession and provide an example. Some successions are seasonal and are repeated each year. Why?

7. In what ways are microorganisms important to the atmosphere? To the greenhouse gases? To the ozone layer?

8. How can microorganisms live on rocks?

9. Compare the sulfur cycle and the nitrogen cycle. Which do you believe is more important to the biosphere? Why?

10. What is the acetylene reduction technique and what is its use?

SUGGESTED READING

Hurst, C. J., G. R. Knudsen, M. J. McInerney, L. D. Stetzenbach and R. L. Crawford, eds. 2001. *Manual of Environmental Microbiology.* Washington, DC: American Society for Microbiology.

Jacobson, M. C., R. J. Charlson, H. Rodhe and G. H. Orians, eds. 2000. *Earth System Science.* New York: Academic Press.

Lynch, J. M. and J. E. Hobbie, eds. 1988. *Microorganisms in Action: Concepts and Applications in Microbial Ecology.* Oxford: Blackwell Scientific Publications.

Staley, J. T. and A-L. Reysenback, eds. 2002. *Biodiversity of Microbial Life: Foundation of Earth's Biosphere.* New York: John Wiley & Sons.

Chapter 25

Beneficial Symbiotic Associations

In symbiosis two dissimilar organisms become intimately associated. This type of vital relationship was apparently first observed by Reinke in 1872, who named the phenomenon consortism. In 1879, De Bary observed and recorded the same phenomenon, and introduced the term symbiosis, which has since been adopted into common usage.

— I. E. WALLIN, *SYMBIONTICISM AND THE ORIGIN OF THE SPECIES, 1927*

The term *symbiosis* refers to a more or less intimate association that occurs between two dissimilar organisms that live together. These interactions are outlined in **Table 25.1** and range from those that are beneficial to both species (mutualism) to those associations that are harmful to both partners (competition). Except for commensalism, which is discussed in Chapter 26, this chapter considers all the symbiotic associations involving microorganisms where the interaction is beneficial to at least one of the participants. Parasitism, antagonism (amensalism), and competition will not be discussed, and harmful interactions between humans, microbial species, and viruses will be discussed in Chapters 26, 28, and 29.

Keep in mind that prokaryotic life forms have lived on Earth for more than 3.0 Ga. This length of time is more than sufficient for the development of myriad, complex interspecies interactions or symbioses among prokaryotes. In addition, eukaryotes evolved in a prokaryotic world. As they evolved, eukaryotes also developed symbioses with prokaryotes. Some types, such as the endosymbiotic evolution of the mitochondrion, are so intimate that we no longer think of the prokaryote as a separate organism. Of course, most symbioses have not become that close or have had such a long time to evolve.

In this chapter, we cannot present all the types of important symbioses that occur between microorganisms and other organisms. However, we provide examples to illustrate the types and ranges of interactions that occur and how the two different organisms interact with one another (**Table 25.2**). Many of these symbiotic interactions are of critical importance to life on Earth. For example, the microbiota in the digestive system of humans and other animals are obligatory to the animal's survival.

FUNCTIONS OF SYMBIOTIC RELATIONSHIPS

Every symbiotic association is in itself rather unique, and the benefits or harm that may accrue to each of the partners is likely to be as varied as the number of such relationships. Symbionts often provide the partner with increased access to some physical or chemical element that a free-living organism might normally gain from its environment. To ascertain the role of the members in a symbiotic relationship, it is often helpful to separate the partners and determine the needs of each. This is impractical in many cases because some symbionts do not survive in the absence of a partner. Organisms in these associations are termed **obligate symbionts**. Those that can survive in pure

Table 25.1	Interactions that can occur between species in a symbiotic association[a]	
Interaction	**Species Interacting**	
	A	**B**
Mutualism	+	+
Syntrophy	+	+
Commensalism	+	0
Parasitism	+	–
Antagonism (amensalism)	–	0
Competition	–	–

[a]Symbols: + organism gains; 0 not affected; – organism harmed.

culture when separated may not perform as they would in their association.

Four functions are generally attributed to beneficial symbiotic associations. They are:

- Protection
- Access to new habitats
- Recognition aids
- Nutrition

Protection

Microbes in nature are quite often subject to adverse physical conditions. Availability of water, pH, and temperature extremes are among the physical conditions

Table 25.2	Examples of beneficial symbiotic relationships between microorganisms and other organisms	
Microbe–Microbe	**Symbiotic Partners**	
Bacteria–Bacteria	"*Chlorochromatium aggregatum*"	
*Archaea–*Protist	Methanogen–Protist	
Lichens	Cyanobacteria or algae–Fungus	
Plant–Microbe	**Symbiotic Partners**	
Rhizosphere	Bacteria–Plant	
Mycorrhizae	Fungus–Plant	
Actinorhizae	Actinomycetes–Plant	
Symbiotic N-fixers	*Rhizobium*–Leguminous plant	
	Cyanobacteria–*Azolla* (water fern)	
Animal–Microbe		
Marine invertebrates	Sulfur-oxidizing bacterium–Tube worm	
Prokaryotes–Insects	*Bacteria* and *Archaea*–Termites	
Prokaryotes–Birds	*Bacteria* and *Archaea*– Leaf eaters	
Microbes–Ruminants	Prokaryotes and protists–Cattle	

that microbes encounter. An endosymbiont that resides within its partner is generally protected from adverse environmental conditions. The partner can prevent desiccation and variances in osmotic pressure, predation, and in a warm-blooded animal, extremes of temperature. These symbionts in turn might protect the partner from invasion by pathogens (see Chapter 26).

Access to New Habitats

Photosynthetic organisms need access to light but cannot grow on rocks that are exposed to high light intensity, primarily because of the periodic lack of water and unavailability of nutrients. However, lichens, which grow on rocks, are a symbiotic association that consists of a photosynthetic microorganism, either a cyanobacterium or eukaryotic alga, and a fungus. The fungal partner adheres to a rock and provides water and nutrients to its photosynthetic partner. The photosynthetic partner is thereby enabled to photosynthesize and provide organic nutrients for the fungal partner.

Recognition Aids

Many marine invertebrates and fish have bioluminescent bacteria either on their surface or as endosymbionts in special organs. These bioluminescent bacteria emit light that is involved in schooling, mating, or attraction of prey.

Nutrition

The involvement of one symbiont or partner in providing nutrients for the other partner is common in favorable symbiotic associations. For example, the symbiotic nitrogen-fixing bacteria that infect alder, legumes, and other plants provide the eukaryote with fixed nitrogen in exchange for their own nutrients and a place to reproduce.

As will be evident from the examples presented in this chapter, a considerable amount of overlap exists among these functions in many associations.

ESTABLISHMENT OF SYMBIOSES

The evolution of symbiotic relationships occurred through a progressively greater interdependence of two different species. As the interdependence became more pronounced, it was essential that mechanisms evolved to ensure the continuity of the symbiosis from generation to generation. For example, in the lichen, special reproductive units containing both algae and fungal cells are dispersed through the air to allow for the colonization of new habitats. In some cases one of the partners transmits the other symbiont directly to its progeny, a common occurrence in insect/microbe symbioses. Other associations require that each generation establish the interaction within the environment.

Direct transmission occurs in many endosymbiotic associations. For example, the algal symbionts that are associated with protists would have no invasive ability and would have difficulty in gaining access to the protist partner. To ensure that each of the protist progeny receives its algal symbiont, a cell controls division of the symbiont. When two are present, the daughter cell may then proceed to division. This assures perpetuation of a favorable symbiotic interaction.

Sexually reproducing animals, generally insects, can transmit symbionts by infection of the egg cytoplasm. Some insects carry microbial symbionts in specialized cells called **bacteriocytes.** Symbionts released from bacteriocytes move to the reproductive tract of the insect and by various mechanisms are transferred to progeny.

Adult mammals have a distinct microbiota in their digestive tract that is essential to their well-being. At birth, the intestinal tract of the neonate is sterile, and an infant gains a normal population of microorganisms via their diet and by association with adults.

The remainder of this chapter is devoted to a discussion of several symbiotic associations that serve as examples of these interactions.

TYPES OF SYMBIOSES

Symbiotic interactions generally benefit at least one of the partners, and the terms *mutualism* and *symbiosis* are sometimes considered to be synonymous. This is not actually the case, as the interaction of species can be of several types, including **mutualism,** in which both partners benefit from the association, or **commensalism,** in which one partner benefits from the association and the other partner is neither harmed nor benefited. **Parasitism** is an association in which one partner benefits and the other is harmed; **antagonism,** where one partner is harmed and the other unaffected, and **competition,** which generally harms both partners, will not be discussed here. Usually, when one of the organisms is much larger than the other, the smaller organism in a symbiotic association is termed the **symbiont** and the larger one the **host.**

An **ectosymbiont** lives outside but in close proximity to the host, whereas an **endosymbiont** exists within the cells of the host. A cooperative interaction in which two or more organisms combine to synthesize a required growth factor or to catabolize a substrate or exchange nutrients is termed **syntrophy.**

We will begin by discussing microbe–microbe symbioses, then consider microbe–plant, and finally microbe–animal symbioses.

MICROBE–MICROBE SYMBIOSES
"Chlorochromatium aggregatum"

A remarkable association occurs between two prokaryotic partners, one of which is photosynthetic, the other of which is nonphotosynthetic. This mutualistic symbiosis is so intimate that it has been considered as one species and termed *"Chlorochromatium aggregatum."* The name is derived from the "aggregates" that are commonly present in anaerobic aquatic environments (see Figure 21.14). The aggregate or consortium consists of a central bacterium that is thought to be a sulfate reducer, which produces hydrogen sulfide. The sulfate reducer is surrounded by cells of an anoxygenic green sulfur photosynthetic bacterium that utilize sulfide as an electron donor in photosynthesis. The green bacterium resembles *Chlorobium limicola.* The photosynthetic partner may provide simple substrates such as acetate for the growth of the sulfate reducer. Like other green sulfur bacteria, the photosynthetic partner is immotile. However, its symbiotic partner is motile and has been shown to be chemotactic to sulfate. Interestingly, the consortium also exhibits tactic responses to light, suggesting that signal transduction occurs between the two partners. *"C. aggregatum"* is just one example of several similar related consortial species that have been described.

Bacteria–Archaea

The term *syntrophy* usually refers to a relationship where two or more species living together can utilize a substrate that neither can utilize alone. An example of this would be *Syntrophus aciditrophicus,* an anaerobic bacterium that grows in pure culture solely on crotonate. *Methanospirillum hungatei* uses hydrogen-formate for methanogenesis when it grows in pure culture. When the two species grow in a syntrophic relationship, the combined culture can utilize benzoate, butyrate, hexanoate, or heptanoate as a substrate. Thus, the coculture is able to utilize an array of substrates unavailable in axenic culture to either species. This interaction fits the classical definition of syntrophy.

Archaea–Protist

Sediments at the bottom of stagnant waters are rich in organic matter. These sediments are anaerobic but support the life of various ciliates, amoebae, and flagellates. These heterotrophic protists generate energy through oxidation of organic compounds and must have an electron sink to rid themselves of protons (H^+) generated during respiration. They accomplish this through the endosymbiotic methanogens that are spread throughout their cytoplasm. One amoeba can play host to more than 10,000 methanogens. Electron micrographs of protozoa from the surface of sediments indicate that many species that harbor methanogens lack mitochondria. An amoeba such as *Pelomyxa* excretes measurable quantities of CH_4. Much of the methane generated in the upper layer of anaerobic sediments originates from the methanogens that inhabit protozoa, and not from free-living methanogens.

Lichens

Lichens comprise a classical case of mutualistic symbiosis. Lichens are commonly observed as encrustations on rocks, tree bark, and the soil surface. A lichen is composed of a heterotrophic fungus and a photosynthetic cyanobacterium or a eukaryotic alga. Typical lichens are depicted in **Figure 25.1**. The fungus/cyanobacterium or the fungus/alga associations are so close that they are considered a unitary vegetative body and have been classified taxonomically as a distinct biological entity. The morphology and metabolic relationship of any particular lichen is so constant and reproducible that it can be assigned to both a genus and a species. More than 20,000 species of lichen have been described.

The fungus in most lichens is an ascomycete. Basidiomycetes are found in some lichens from tropical regions. One fungal species may associate with several different algae. Each resultant lichen is considered a separate species differing in morphology and metabolic interrelationship. Some of the fungal and phototrophic partners have been separated and grown in axenic culture. The phototrophic partner is generally one that lives as a free-living organism in nature, and these grow readily when separated. Those fungal components that can be grown in axenic culture grow poorly and generally require complex carbohydrates. Recombining the separated partners is most difficult.

In a lichen association, the fungus produces a mycelial structure that adheres to hard surfaces and allows the lichen to inhabit rocks and tree bark. The photosynthetic partner supplies the fungus with organic nutrients (photosynthate). The fungus scavenges inorganic nutrients such as phosphate for the association. In addition, it synthesizes sugar alcohols, such as mannitol or sorbitol, which, as compatible solutes, absorb moisture from the atmosphere. This water serves not only to prevent plasmolysis of cells, but also as the reductant for photosynthetic CO_2 assimilation. Many lichens grow at low temperatures in high altitudes or in polar environments. The so-called reindeer moss in the tundra of the arctic regions is actually a lichen. Lichens grow very slowly. For example, some of those from arctic tundra have been carbon-dated at more than 1,000 years of age.

Lichens can be used as indicators of air pollution, as they are highly sensitive to sulfur dioxide, ozone, and toxic metals. Apparently lichens absorb air pollutants, and the toxic components are harmful to the photosynthetic partner. Cities with air pollution problems are devoid of lichens.

The color of lichens can be gray, black, blue, yellow, green, or various shades of red or orange. The pigments have long been extracted and employed as textile dyes. The dyes that give the distinct color of Harris tweed, for example, have traditionally come from lichens. Other compounds produced by lichens are litmus, the acid-base indicator employed in chemistry, and some of the essential oils used in perfumes.

(A)

(B)

Figure 25.1 Lichen
Lichens (cyanobacterial-fungal symbiotic associations) vary in shape, color, and appearance. (A) Crustose lichen, commonly growing on rocks and tree trunks. (B) Foliose (leaflike) lichen. A, ©G. Meszaros/Visuals Unlimited; B, ©D. Sieren/ Visuals Unlimited.

MICROBE–PLANT SYMBIOSES

Various symbiotic relationships occur between plants and microbes. Symbionts may be present on a leaf surface (the phyllosphere) or in the soil surrounding the roots (rhizosphere). Microbes in such a relationship are considered ectosymbionts. Fungi may grow on the surface of or invade plant roots and form a relationship known as **mycorrhizae.** Bacteria and actinomycetes may also invade roots and form colonies with the formation of distinct nitrogen-fixing nodules. Microbes involved in these more intimate interactions are endosymbionts.

Rhizosphere

The **rhizosphere** is the thin layer of soil remaining on plant roots after taking the plant from its environment and shaking it. The microbial population in the rhizosphere is generally one to two orders of magnitude higher than in surrounding, root-free soil, and the number in the rhizosphere frequently reaches 10^9/gram of soil. The root system of plants can be quite extensive in area. For example, a typical cereal grain root system can be 160 to 225 meters in length with an average diameter of 0.1 mm. The total root surface area would be 62.8 cm^2, providing ample opportunity for microorganism–root interaction.

The rhizosphere has a higher proportion of gram-negative, nonsporulating, rod-shaped bacteria than would be present in adjacent soil. Gram-positive organisms are scarce in the rhizosphere. Ammonifying and denitrifying bacteria are present, and many of these require growth factors such as B-vitamins and amino acids, which are supplied by root exudates.

Ectorhizosphere organisms are those that grow in the soil immediately surrounding the root while endorhizosphere organisms penetrate the root itself, where they feed directly on the plant. The ectorhizosphere dwellers consume root exudates, lysates, and sloughed root cells. The total carbon released through the root system (including CO_2) can be as much as 40% of the total plant photosynthate. Root exudates include sugars, amino acids, vitamins, alkaloids, and phosphatides. The release of root exudates is frequently stimulated by the presence of selected microorganisms.

Here are some of the direct plant–microbe interactions that have been described:

- The organic fraction of wheat and barley exudate stimulates the growth of *Azotobacter chroococcum*. This bacterium is a nitrogen fixer that supplies nitrogen compounds for utilization by the plant. Many other species of the nitrogen-fixing genus *Azotobacter* are associated with root surfaces in the rhizosphere.
- Anaerobic clostridia occupy the *rhizosphere* of submerged seawater plants *Zostera* (eelgrass) and *Thallasia* (turtle grass). The root exudates released by these plants supply the clostridia with a carbon and energy source, and the bacteria are nitrogen fixers.
- *Azospirillum* species inhabit the cortical layer of tropical grasses and feed on plant-generated carbonaceous compounds. The bacteria in turn fix abundant quantities of nitrogen for the growth of both.
- *Desulfovibrio* are abundant around the roots of rice, cattails, and other swamp-dwelling plants. The sulfate-reducing members of the genus *Desulfovibrio* generate H_2S in quantities that would harm developing plants. The H_2S oxidizing bacterium, *Beggiatoa*, lives in the rhizosphere of these swamp-dwelling plants. Oxidation of sulfide by *Beggiatoa* under limited oxygen conditions reduces the sulfide concentration but results in the production of potentially toxic levels of H_2O_2. *Beggiatoa* does not produce catalase, the enzyme that decomposes H_2O_2. A catalaselike activity from the root tips breaks down the peroxide produced by the bacterium to water and oxygen (**Figure 25.2**). Thus, *Beggiatoa* removes sulfide that could limit plant growth, and the plant provides catalase to prevent peroxide autointoxication of the bacterium.
- *Pseudomonas fluorescens* living in the rhizosphere of some plant species produces antifungal agents, which ward off potential pathogens.

Mycorrhizae

Mychorrhizal plant root interactions are common in nature. The mycorrhizal relationship is beneficial to the plant in several ways:

- Provides longevity to feeder roots
- Improves nutrient absorption
- Enhances selective absorption of ions
- In some cases, increases resistance to plant pathogens

Plants growing in wet environments have mycorrhizal symbionts that increase availability of nutrients such as phosphorus. In arid environments, the fungi aid in water uptake and as a result, mycorrhizal plants often thrive in poor soil where nonmycorrhizal plants cannot. Most of the fungi involved in ectomycorrhizal associations are not obligate symbionts, and some of those that form endomycorrizal associations can grow independently. An estimated 80% of vascular plants have some type of mycorrhizal involvement, and over 5,000 species of fungi form mycorrhizal associations.

The dramatic effect of mycorrhizae on the growth of a citrus seedling is illustrated in **Figure 25.3**. The fungi obtain nutrients from the plant and, in turn, provide nutrients (especially phosphorus) and water for plant growth. In many cases, mycorrhizal fungi can be grown in culture and mixed with rooting soil resulting in better growth of the plant.

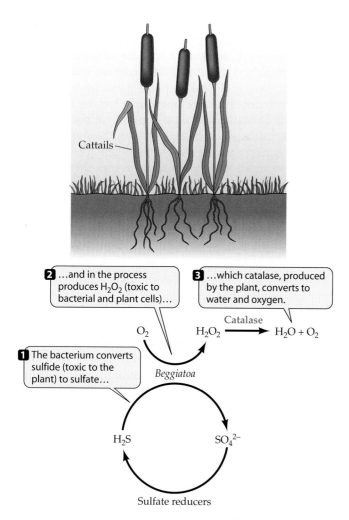

① The bacterium converts sulfide (toxic to the plant) to sulfate…

② …and in the process produces H_2O_2 (toxic to bacterial and plant cells)…

③ …which catalase, produced by the plant, converts to water and oxygen.

$$O_2 \qquad \underset{\text{Catalase}}{H_2O_2 \longrightarrow H_2O + O_2}$$

Beggiatoa

$$H_2S \qquad SO_4{}^{2-}$$

Sulfate reducers

Figure 25.2 Plant–bacterium symbiosis in rhizosphere
Growth of cattails (*Typha latifolia*) in an anaerobic swamp environment is sustained by a symbiotic association with the sulfur-oxidizing bacterium *Beggiatoa*. In the low-oxygen environment, the bacterium produces toxic H_2O_2 but lacks the enzyme (catalase) to remove it; the enzyme is supplied by the plant. Sulfide that would be toxic to the plant is converted to sulfate by the bacterium.

Two major types of mycorrhizal associations occur:

- **Ectomycorrhizae,** where fungi grow externally
- **Endomycorrhizae,** in which fungi grow inside root tissue

Ectomycorrhizae

These fungi grow as an external sheath about 40 nm thick around the root tip, as illustrated in **Figure 25.4**, and are present mainly on roots of forest trees, such as conifers and oaks. The fungi penetrate intercellular spaces of the epidermis and cortical regions but not into root cells. The relative growth of loblolly pine seedlings uninoculated and inoculated with the fungus *Pisolithus tinctoris* is illustrated in **Figure 25.5**A and B. Note the

Figure 25.3 Effect of mycorrhizal organisms on growth of citrus seedlings
Citrus seedlings grown with and without inoculation of mycorrhizal organisms. Note the significant difference in root and leaf structure. Courtesy of L. F. Grand.

superior and more uniform growth in the plants inoculated with the fungus (**Figure 25.6**). A marked contrast in the development of the uninoculated (Figure 25.6A) and inoculated (Figure 25.6B) roots is apparent. A close-up (Figure 25.6C and D) confirms the benefit of the fungus in root development.

Endomycorrhizae

In the endomycorrhizal association, there is significant invasion of cortical cells with some of the fungal hyphae extending outside the root (**Figure 25.7**). This fungal/plant relationship occurs in plants such as orchids and azaleas. In fact these plants cannot survive without these fungal associations. These symbiotic associations also occur in other plants, including wheat, corn, beans, and pasture and rangeland grasses. The fungi form intracellular structures termed vesicles and arbuscules, as depicted, and the association is designated a vesicular-arbuscular mycorrhiza. The fungi involved in these relationships are difficult or may not grow separately from the plant.

Frankia–Plant

Pioneer plants, such as alder (*Alnus*) and bayberry (*Myrica*), grow in moist, nitrogen-poor environments.

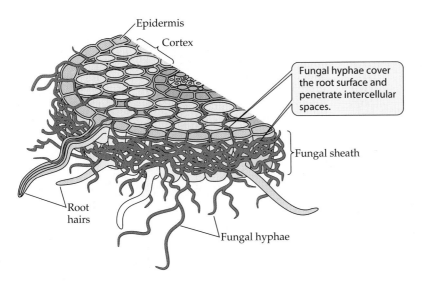

Figure 25.4 Ectomycorrhizae
Cross section of a root with ectomycorrhizae. The hyphae penetrate between, but not into, root cells.

These woody plants and shrubs are commonly seen in bogs, dredge spoil, and abandoned open pit mines. The ability to establish a symbiotic relationship with nitro-gen-fixing bacteria is, in part, the reason these plants survive in such an inhospitable environment. The nitrogen fixers in these plants are filamentous actinomycetes in the genus *Frankia*. The symbiotic relationship between plant and *Frankia* is functionally equivalent to that present in the legume/*Rhizobium* symbiosis (see later discussion). The striking difference is that *Rhizobium* sp. will only interact with one family, the legumes. *Frankia* can colonize the roots of plants that are phylogenetically distinct, and symbiotic relationships are known to occur between *Frankia* and 7 different plant orders, 8 families, and 14 genera. More will undoubtedly be discovered.

The endosymbiotic relationship formed between plant and *Frankia* apparently occurs through root hairs. *Frankia* detects a host and prepares for infection by responding to defense molecules secreted by the plant. Development of nodules leads to densely packed, coral-like branching roots that cease growth (**Figure 25.8**A), and the nodules can be quite large (Figure

Figure 25.5 Effect of mycorrhizal fungus on growth of loblolly pine seedlings
Growth of loblolly pine seedlings (A) without and (B) with the mycorrhizal organism *Pisolithus tinctoris*. Courtesy of L. F. Grand.

(A)

(B)

(C)

(D)

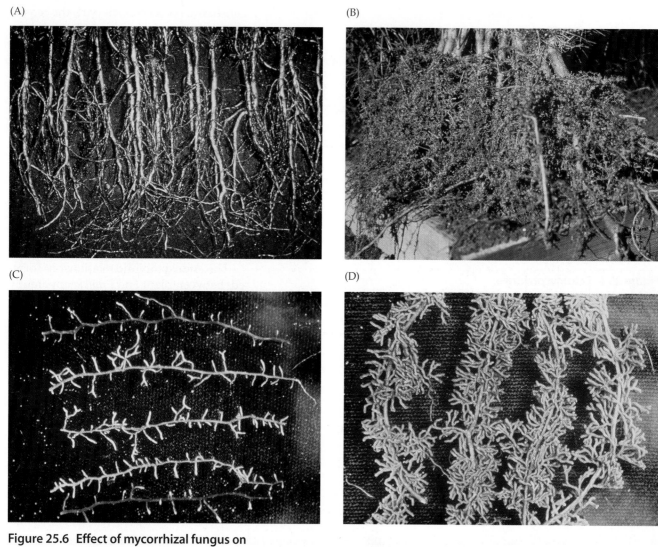

Figure 25.6 Effect of mycorrhizal fungus on development of loblolly pine roots
Development of loblolly pine roots (A) without and (B) with *Pisolithus tinctoris*. Close-up photographs show details of the root structure of seedlings grown (C) without and (D) with the mycorrhizal fungus. Courtesy of L. F. Grand.

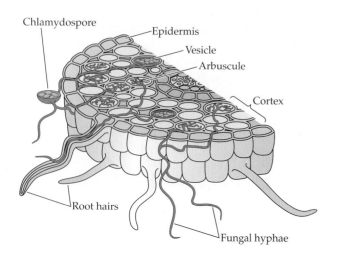

25.8B). The microorganism grows slowly in axenic culture, and intact cells can fix nitrogen under atmospheric levels of oxygen. Cell extracts from *Frankia*, however, are sensitive to oxygen. The development of nitrogenase activity coincides with a differentiation of terminal areas of the filaments as depicted in Figure 25.8C.

Rhizobium–Legume

Two major bacterial genera that establish a nitrogen-fixing endosymbiotic association with leguminous plants

Figure 25.7 Endomycorrhizae
Cross section of root with endomycorrhizae. Intracellular growth of the fungal hyphae is evident, along with "tree-like" structures (arbuscules) and "vesicle-like" structures (vesicles) inside root cells.

(A)

(B)

(C)

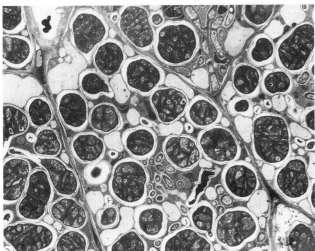

are *Rhizobium* and *Bradyrhizobium*. The general properties of these organisms were presented in Chapter 19. Nitrogen fixation by these microbes occurs within nodules that develop on the roots of leguminous plants. The development of the symbiotic association in legumes has been studied extensively, and the events leading to successful nodulation are now quite well understood. Both *Rhizobium* and *Bradyrhizobium* are free-living microbes that can move to, and specifically attach to, the root hairs of a specific legume. The rhizobia are quite selective in the plant species they infect, and the narrow range of plant species a given bacterial strain will infect is called a cross-inoculation group.

The establishment of the *Rhizobium*–plant symbiotic association requires a complex series of steps. The bacterium and the plant set up a "cross talk" using chemical signaling molecules. First the root releases flavinoids. These not only attract the *Rhizobium* to the vicinity of the root, but they also induce bacterial *Nod* genes, which encode Nod (nodulation) factors. These factors, secreted by the *Rhizobium*, stimulate cell division in the root cortex. This leads to the formation of a primary meristem (actively dividing cells).

The attachment of the bacterium to the root hairs is due to a specific adhesion protein, rhicadhesin, found on the root surface of *Bradyrhizobium* and *Rhizobium* species. *Rhicadhesin* is a calcium-binding protein and may function by binding calcium complexes on the root-hair surface. Lectins had been considered to be responsible for attachment but are less important than rhicadhesin.

Following binding to the root hair, the root curls back due to substances secreted, and the bacterium enters via an invagination process (**Figure 25.9**). The bacterium forms an **infection thread** as it grows and moves down the root hair. The root cells surrounding the infection thread become infected as growth of the bacterium continues. The bacterium divides rapidly resulting in the formation of a nodule. Most of the bacterial population is transformed to branched, club-shaped, or spherical, **bacteroids**. The **peribacteroid** membrane surrounds the bacteroids. The irregularly shaped bacteroids have a larger volume than the free-living bacteria and are incapable of cell division. A few dormant, normal rod-shaped bacteria live in nodules. They are the ones that survive to reproduce in the soil when the plant dies. They can infect legume roots in the vicinity or other plants at a later date.

Figure 25.8 Actinomycete–plant symbiosis
Growth of the nitrogen-fixing actinomycete *Frankia* on roots of alder (*Alnus*), a pioneering plant. (A) Large nodules on alder root caused by the symbiotic actinomycete. (B) Large nodule detached from an alder root. (C) Light micrograph of thick-walled terminal bodies of *Frankia*. These bodies are involved in nitrogen fixation. The vesicle is about 5μm in diameter. A, ©S. H. Wittwer/Visuals Unlimited; B, courtesy of J. M. Ligon; C, ©R. H. Berg/Visuals Unlimited.

Cortical cells   Root hair

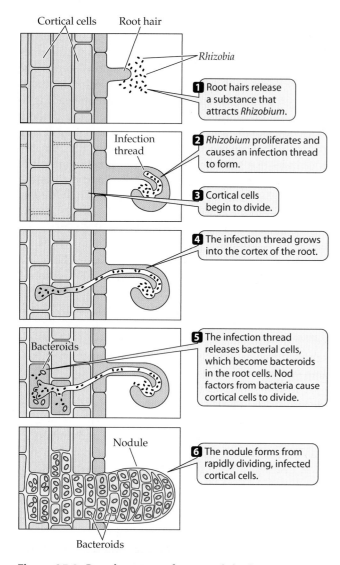

1 Root hairs release a substance that attracts *Rhizobium*.

2 *Rhizobium* proliferates and causes an infection thread to form.

3 Cortical cells begin to divide.

4 The infection thread grows into the cortex of the root.

5 The infection thread releases bacterial cells, which become bacteroids in the root cells. Nod factors from bacteria cause cortical cells to divide.

6 The nodule forms from rapidly dividing, infected cortical cells.

Figure 25.9 Development of root nodules in
***Rhizobium*–plant symbiosis**
Diagram showing attachment and invasion in development of a root nodule in a leguminous plant. *Rhizobium* attaches to the root hair of a susceptible plant, enters, and forms an infection thread as it moves into the root cells. Factors contributed by both the bacterium and the root result in nodule formation.

A number of biochemical events occur during the infection process. Over 20 polypeptides are synthesized either by the bacterium, plant, or both, which play a role in the development of the symbiotic association. The nitrogenase and associated genes are entirely contributed by the bacterium. Under tightly controlled microaerophilic conditions the free-living microorganism can fix nitrogen.

The Nod genes are highly conserved among *Rhizobium* species. These genes are on the large *Sym* plasmids. The Sym plasmids bear the **specificity genes** that are

(A)

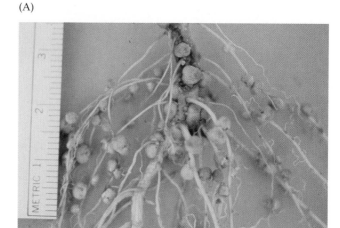

(B)

Figure 25.10 Root-nodulating bacteria in peanut plants
(A) Effective nodulation of a peanut plant (*Arachis hypogeal*) by *Bradyrhizobium* sp. (B) Peanut plants grown with and without inoculated root-nodulating rhizobia strains. Courtesy of T. J. Schneeweis.

responsible for the restricted host range of a *Rhizobium* strain. Host specificity can be transferred from one member of a cross-inoculation group to another by the Sym plasmid.

The consequence of a well-established association in the root of a peanut plant is illustrated in **Figure 25.10**, where nodulation of the roots is quite extensive (Figure 25.10A). Field-grown plants that have been inoculated have superior pigmentation, and total growth and development are better, as depicted in Figure 25.10B.

Leghemoglobin surrounds the bacteroids in the nodules. The synthesis of leghemoglobin is induced by genetic information from the plant and bacterium, as neither is able to synthesize it alone. Leghemoglobin is a red-pigmented iron-containing heme protein quite similar to animal hemoglobin. Leghemoglobin binds oxygen in the nodule and maintains an oxygen tension sufficient to allow the aerobic bacteroids to respire and

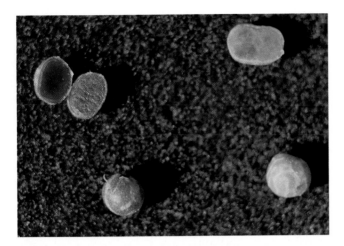

Figure 25.11 Nitrogen-fixing nodule
Effective nitrogen-fixing nodules (left) have a reddish coloration resulting from the presence of the oxygen-carrying protein leghemoglobin. Ineffective nodules (right) are unpigmented. Courtesy of T. J. Schneeweis.

generate ATP but restricts the oxygen available to a level that does not inactivate the oxygen-sensitive nitrogenase system. The ratio of oxygen bound to leghemoglobin to free oxygen in the nodule is in the order of 10,000/1. Nodules that are effective in nitrogen fixation are red (contain leghemoglobin) in the interior (**Figure 25.11**).

The bacteroids (**Figure 25.12**) receive photosynthate from the plant in the form of tricarboxylic acid cycle intermediates, mainly malate, succinate, and fumarate, which they oxidize to generate ATP.

Nitrogen fixation is a stepwise addition of three pairs of hydrogen atoms to nitrogen gas. This process requires considerable energy. This process requires 15 to 20 moles of ATP per mole of N_2 fixed. The product of fixation is ammonia, which is converted into organic form, glutamine, primarily by the plant. Glutamine is a major donor of amines in amino acid synthesis.

Azorhizobium–Legume

Rhizobia form stem nodules on leguminous plants that grow in tropical areas. These nodules are functionally equivalent to root nodules. The nodules are generally formed on the submerged portion of stems or near the water surface. The characteristics of the *Rhizobia* that form stem nodules are quite different from root nodulators and are placed in a separate genus, *Azorhizobium*.

Sesbania rostrata, a legume that grows in Senegal and Madagascar, forms stem nodules in association with *Azorhizobium caulin-*

odans. The establishment of this nitrogen-fixing system occurs through steps similar to those involved in root nodulation. The stem-nodulating *Rhizobia* apparently produce chlorophyll *a* and may carry out anoxygenic photosynthesis. Because N_2 fixation requires considerable energy, photophosphorylation would be an effective driving force for fixation reactions.

Cyanobacteria–Plant

Azolla is an aquatic fern that grows on the surface of still waters in temperate and tropical regions and also commonly grows on the surface of rice paddies after the harvesting of the rice crop. An ancient practice in Southeast Asia was to retain the water level after harvesting to allow the fern to grow, thereby ensuring a nitrogen supply for the subsequent growing season. The success of this practice resulted from ability of nitrogen-fixing cyanobacteria to grow within the tissue of the fern leaves (**Figure 25.13**A). The cyanobacterium involved is *Anabaena azollae*, and during symbiotic growth, 15% to 20% of the cells in a trichome are converted to heterocysts (Figures 25.13B), the specialized anaerobic nitrogen-fixing cell.

The leaf surface of *Azolla* contains cavities with mucilage. During development, these are opened to the outside, allowing free-living cyanobacterial cells that grow in the water to attach. The cavities then close as the *Azolla* develops, thereby entrapping filaments of *Anabaena* within the leaf structure.

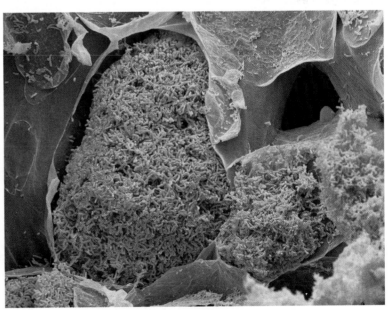

Figure 25.12 Bacteroids
Scanning electron micrograph of bacteroids inside a nodule. The cells lack a cell wall and are no longer able to reproduce. Photo ©A. Syred/SPL/Photo Researchers, Inc.

(A)

(B)

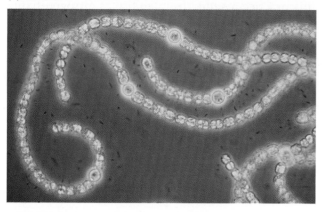

25.13 Cyanobacterium–plant symbiosis
(A) Fronds of the aquatic fern *Azolla*, which can harbor nitrogen-fixing cyanobacteria. (B) Trichomes of *Anabaena azollae*, showing the round heterocysts in which nitrogen fixation occurs. A, ©W. S. Ormerod/Visuals Unlimited; B, ©J. R. Waaland/Biological Photo Service.

Bacteria–Plant Leaf

Bacterial populations associated with leaf surfaces (phyllosphere) often reach 100 million per gram of fresh leaf material. Under the humid conditions of the tropics, the phyllosphere is constantly moist, and this provides a habitat for a diverse community of bacteria. The species present on leaves are significantly different from those that inhabit the soil in areas adjacent to the plant. Nutrients for the phyllosphere community come from the organic acids, sugars, and methoxyl compounds that leach from the leaf. Some of the organisms present on these leaves are nitrogen fixers of the following genera: *Klebsiella*, *Beijerinckia*, and *Azotobacter*. There is evidence that these nitrogen fixers provide the host plant with nitrogenous compounds, but the interactions involved are not yet clear. Also, methanol-oxidizing bacteria live in this environment.

MICROBE–ANIMAL SYMBIOSES

A better understanding of the microbiota of healthy animals has led to the realization that this resident microbial community plays more than a passive role. It is clear that the microbes living in the intestinal tracts of animals are indispensable to the health of the host. Some of the beneficial roles that microbial symbionts play in animal health have already been defined and will be discussed in this section.

Symbiotic Invertebrates

In 1977, during oceanographic studies on plate tectonics and related volcanic activities on the seafloor, dense and thriving populations of novel marine invertebrates were discovered at depths from 1,000 to 3,700 m with biomass orders of magnitude higher than could possibly be supported by a photosynthetic food supply. How was this possible? These complex animal communities were found near "hydrothermal vents," which emit hot water containing highly reduced molecules. Some of its major chemical constituents, H_2S, H_2, and CH_4, are known to serve as reductants for chemosynthetic growth of microorganisms. A variety of feeding mechanisms among the various planktonic or sessile animals make use of this unusual microbial base of the food chain at these ecosystems. However, the largest portion of the biomass appears to be produced by symbioses between chemolithoautotrophic bacteria and certain marine invertebrates.

Among the symbiotic invertebrates from deep-sea hydrothermal vent ecosystems in the Pacific Ocean are the vestimentiferan tube worms (*Riftia pachyptila*), **Figure 25.14**, mussels (*Bathymodiolus thermophilus*), **Figure 25.15**, and the "giant" white clams (*Calyptogena*

Figure 25.14 Symbiotic worms
A population of vestimentiferan tube worms (*Riftia pachyptila*) at the East Pacific Rise hydrothermal vent site. Courtesy of H. W. Jannasch.

Figure 25.15 Symbiotic mussels
Mussels (*Bathymodiolus thermophilus*) found at hydrothermal vent sites of the 9° N East Pacific Rise, at a depth of 2,520 meters (1.58 miles). Courtesy of H. W. Jannasch.

Figure 25.16 Clams associated with vents
Population of the "giant" white clams (*Calyptogena magnifica*), located within cracks between boulders of "pillow lava" at the East Pacific Rise hydrothermal vent site. Courtesy of H. W. Jannasch.

magnifica), specimens of which are up to 32 cm in length (**Figure 25.16**).

The vestimentiferan tube worms reach lengths of more than 2 meters and can occur at densities that appear to exceed the highest concentration of biomass per area known anywhere in the biosphere. They also have the highest known rates of growth of any animal. Detailed anatomical studies revealed the absence of a mouth, stomach, and intestinal tract, in short, the absence of any ingestive and digestive system. Instead, the animals contain a **trophosome** (**Figure 25.17**) in their body cavity, which accounts for about 50% of the weight of the animal. The tissue in the trophosome consists of large coccoid prokaryotic cells interspersed with the animal's blood vessels. There are about 4 billion bacterial cells per gram of tissue. The bacteria have not yet been characterized but resemble a large, marine sulfur-oxidizing bacterium.

A "chemoautotrophic potential" in these worms is evident from the presence of enzymes catalyzing the synthesis of ATP via sulfur oxidation (rhodanese, APSR, and ATP sulfurylase) as well as the Calvin cycle enzymes RuBisCo and ribulose 5′phosphate kinase (APK). ADP sulfurylase (ADPS) and phosphenolpyruvate carboxylase (PEPC) were also found. None of these enzymes were detected within the worm tissue proper. The necessary simultaneous transport of oxygen and hydrogen sulfide from the retractable plume of gill tissue

to the trophosome is carried out by an extracellular hemoglobin of the annelid-type blood system (**Figure 25.18**).

Electron microscopy and enzymatic studies of the bivalves have shown that the gill tissues contain prokaryotic endosymbionts that are involved in the production of ATP through the oxidation of reduced inorganic sulfur compounds (hydrogen sulfide and thiosulfate). Furthermore, they can reduce CO_2 to organic carbon. Specifically indicative of microbial metabolism (**Figure 25.19**) are the enzymes ribulose bisphosphate carboxylase (RuBisCo) and adenosine 5′phosphosulfate reductase (APSR). From the analyses of 5S and 16S

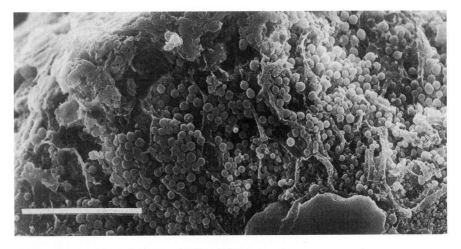

Figure 25.17 Internal tissue of tube worms
Trophosome tissue of *Riftia pachyptila* consisting of coccoid prokaryotic cells interspersed among blood vessels of the animal tissue. Courtesy of H. W. Jannasch.

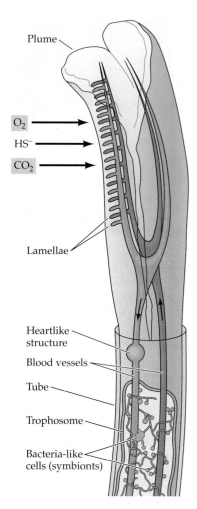

Figure 25.18 Oxygen transfer in tube worms
Scheme of the dissolved oxygen transport system in *Riftia pachyptila* from the gill "plume" to the symbiotic trophosome tissue.

Chemosynthetic symbioses based on chemoautotrophic microorganisms occur in environments that are dependent on an energy source other than sunlight. A number of these environments have been discovered in the deep sea. Light does not penetrate to deeper than 100 meters into the ocean, so most of the ocean is permanently dark. We know that more than 75% of our planet's biosphere lies below 1,000 m of seawater. At this depth, it is obvious that no photosynthesis can occur. The temperature ranges from 2°C to 4°C. The photosynthetic organic material produced by phytoplankton in the oceans' surface waters is largely recycled in the upper 300 m layer, and only a fraction, about 5%, reaches deeper waters in the form of sinking particulate matter. About 1% is estimated to reach the deep-sea floor at 3,000 to 4,000 m. This limited food supply controls the scarce animal populations and their life strategies in this desertlike deep-sea environment.

rRNA nucleotide sequences, it is apparent that the prokaryotic symbionts of all chemolithoautotrophic marine invertebrates so far investigated belong to different, albeit closely related, groups.

Therefore, the bivalves have chemosynthetic symbioses as well, although it should be noted that the mussels are also capable of filter feeding in environments that contain high concentrations of bacteria.

The discovery of chemosynthetic symbioses from the hydrothermal environments has led to the search for symbioses among invertebrates from other marine environments. Indeed, the phenomenon is much more widespread than possibly imagined. Thus, symbiotic methanotrophy has been observed in mussels similar to *Bathymodiolus* from anoxic, shallow hydrocarbon seepages in the Gulf of Mexico, as well as from so-called cold seeps at the bottom of the West Florida continental slope at a depth of 3,200 m.

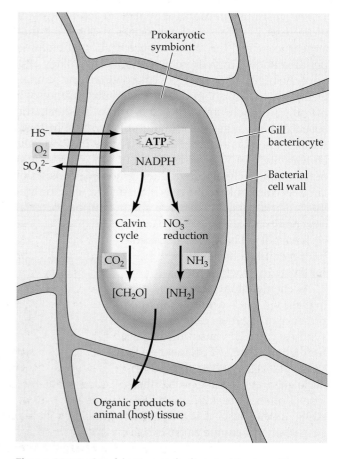

Figure 25.19 Symbiont metabolism in bivalve gills
Schematic of metabolic processes in prokaryotic symbionts within gill "bacteriocytes" of vent bivalves; the presence of enzymes catalyzing ATP production, the Calvin cycle, and nitrate reduction have been demonstrated. Adapted from Felbeck et al., Chapter 2 from *The Mollusca*, Academic Press 1983.

The most surprising characteristic of the chemo-lithotrophic-symbiotic sustenance of the copious deep-sea communities is its efficiency. Considering the point source of the geothermally provided energy for chemoautotrophic bacterial growth in the normal food chain, filter feeding on bacterial cells from the quickly dispersing vent plumes appears highly wasteful. This problem was overcome by transferring the chemosynthetic production of organic carbon to a site within the animal where the electron donor as well as the acceptor is made available with the aid of the respiratory system. This symbiotic association combined the metabolic versatility of the prokaryotes with the genetic and differentiative capabilities of the eukaryotes and takes advantage of a most direct and efficient transfer of the chemosynthate to the animal tissue. These transfer processes and the biochemical interactions between the microbial and animal metabolisms are presently the focus of research in this area.

Symbioses are also found in tectonic areas in the Atlantic Ocean. Although the large tube worms do not occur, smaller tube worms have been found at cold seeps. In addition, shrimp occur in these environments. The shrimp live in association with epsilon-proteobacteria, which have not been reported in such abundance in other environments. Clearly, much exciting research awaits microbiologists who study the microbial communities of tectonic areas in the oceans.

Termites–Microbes

The termite is an interesting social insect that relies on intestinal symbionts for its survival. The wood that termites eat cannot serve as a food for the insect unless it is predigested by its symbiotic partners. The enzymes essential for the degradation of the cellulosic compounds are supplied by microorganisms. In some cases, the cellulases are produced by microbes living in the termite gut; in other cases, the termite acquires digestive enzymes from externally grown fungi that they eat.

The "higher" termites are those that obtain digestive enzymes by ingesting fungi that are cultivated in the termite nest. The fungi are members of the genus *Termitomyces*, a basidiomycete. Termites moving from one habitat to another carry fungal spores to the site, where they establish a fungal garden anew. Growth of the fungi on the cellulosic material they are fed by termites ensures that cellulases will be present on the fungal mycelium. These fungi can digest lignin, which usually protects cellulose from microbial attack. How this occurs is not now known. The termites that cultivate and eat the fungi do not have internal cellulase-producing protists or bacteria.

The gut of "lower" termites has a population of protozoa that digest cellulose and release monomers and various nutrients for the termite. The gut protozoa are anaerobic and ferment cellulose to acetate, carbon dioxide, and hydrogen. The acetate is absorbed through the hindgut of the termite and is then metabolized aerobically to provide the termite with ATP and cell carbon. The termites may also harbor *Enterobacter agglomerans*, a nitrogen-fixing bacterium. Recently, nitrogen-fixing spirochetes have been reported in some termites, the first reported example of nitrogen fixation in this group of *Bacteria*. Thus, the microbial symbionts provide the termite with utilizable nitrogen as well as carbon sources.

The hindgut of lower termites also has significant populations of hydrogen-utilizing methanogens and acetogens. Recent studies indicate that the methanogens and acetogens compete for the hydrogen and carbon dioxide. The acetogens are located spatially closer to the hydrogen-producing bacteria in the hindgut. The methanogens are farther away near the outside wall of the gut where the concentration of hydrogen is lower. The acetate that the acetogens produce provides more than one-third of the energy requirements of the termite. The methane escapes to the air from the termite.

Aphid–Microbe

A fascinating mutualistic symbiosis occurs between the gamma-*Proteobacterium Buchnera* and the aphids that harbor them. These bacteria are endosymbionts that reside in a specialized cell termed a bacteriocyte inside the aphid. The bacterium provides essential amino acids (tryptophan, leucine, etc.) for the aphid, and the host aphid meets all of the nutritional needs of *Buchnera*. In order to enhance the production of amino acids, it is estimated that each *Buchnera* cell produces 50 to 200 copies of its chromosome. Neither host nor symbiont can survive without the other.

What is most remarkable about the *Buchnera*–aphid association is the length of time that it has persisted. The aphid lineage is about 150 to 250 million years old. Fossil evidence has been used to date when the different host species diverged from one another. By comparing the 16S rDNA sequence of the *Buchnera* with its host 18S rDNA sequences, phylogenetic trees have been constructed. These reveal (**Figure 25.20**) that the *Buchnera* species have evolved in concert with its aphid host species. This is seen by the mirror image of the two phylogenetic trees. This phenomenon indicates that as the aphid species diverged from one another, so did the *Buchnera* species, a process referred to as **coevolution**.

The long-term symbiotic association has led to a *Buchnera* chromosome now composed of 650 Kb where the ancestral chromosome was over 6× larger. Apparently most of the genes involved with survival of the ancestral *Buchnera* strain have been lost over the millennia because they are no longer needed. Furthermore, isolation of the *Buchnera* in the host organ has prevent-

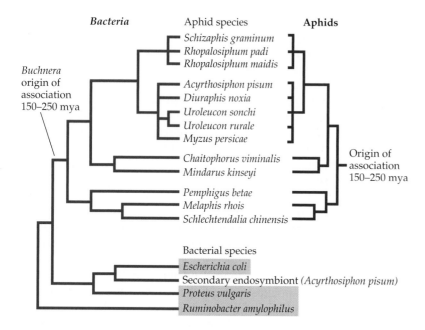

Figure 25.20 Phylogenetic trees of *Buchnera*–aphid associations
Congruence of the evolutionary relationships of *Buchnera* and its aphid hosts.
The endosymbiont tree is based on 16S rRNA; the aphid tree is based on 18S
rRNA and the fossil record (mya, million years ago). Shading indicates free-
living bacteria. Courtesy of Paul Baumann.

ed any acquisition of genetic information from other
bacterial species through horizontal gene transfer.

Microbe–Mealybug

Mealybugs are insects that feed on plant sap. Although
plant sap is available in abundance, it does not provide
these bugs with all of their required nutrients. These
essential nutrients are supplied by bacteria that are natural endosymbionts of the mealybugs and other sap-
sucking insects. This close association has evolved over
millions of years, and elaborate mechanisms are available to ensure that, in each generation, the endosymbionts are passed on to insect progeny. At an early stage
of the mealybug embryo development, specialized cells
are produced that reengulf symbionts that are released
by maternal cells.

Recent studies on the citrus mealybug (*Planococcus
citri*) affirm that this insect has two distinctly different
endosymbionts: one a gamma-type *Proteobacterium* and
the other a beta-type *Proteobacterium*. These endosymbionts are packaged in "symbiotic spheres." The symbiotic spheres are composed of insect cells that form an
oval organ that takes up about one-half of the insect's
abdomen. Within the symbiotic sphere of the citrus
mealybug are the beta-*Proteobacteria* and *within* the beta-
Proteobacteria are the alpha-*Proteobacteria*. There are 5 to
8 beta-*Proteobacteria* per specialized insect cell, and each

of these beta-proteobacterial cells contains
5 or more alpha-proteobacterial cells.
Analysis of RNA gene sequences indicates
that the beta-types are more ancient and
probably the original symbionts, whereas
the alpha-type are of more recent origin.
The beta-type endosymbionts apparently
contribute to the nutritional needs of the
mealybug, and the alpha-types serve
some needs of the beta-types. Thus, we
have within the mealybug a symbiont
within a symbiont, a fascinating interrelationship.

Wolbachia–Animals

Wolbachia, a bacterium discovered in 1924,
might be the most common infectious bacterium on Earth. For over 100 million
years, this alpha-*Proteobacterium* has
cospeciated with invertebrate hosts,
including, among others, fruit flies,
shrimp, spiders, and parasitic worms. The
organism is not known to infect vertebrates and has not yet been cultivated.
Wolbachia resides in the ovaries or testes of
insect hosts and manipulates the reproductive capacity of the host for its evolutionary benefit. When present in host
females, the infection is passed on to succeeding generations. In the male, as sperm develops, the microbes are
generally squeezed out and would not always be transferred to progeny. As a consequence, the female line is
favored by the symbiont. For example, if an uninfected
female mates with an infected male, some or all fertilized eggs die, but a female carrying *Wolbachia* can mate
with an infected or uninfected male and produce fertile
eggs. As a result, infected females outcompete those that
are parasite-free, and *Wolbachia* carriers increase in the
overall population. There is evidence that when present
in males, the bacterium produces a toxin that inactivates
sperm, whereas *Wolbachia* strains present in females produce an antidote that restores sperm to full viability.
Although *Wolbachia* controls the evolution of insects for
survival of the bacterial lineage, it does not cause apparent curtailment to survival of the insect species.

Bioluminescent Bacteria–Fish

The capacity for light emission (bioluminescence) is
quite widespread in the biological world. Bioluminescence has been observed in bacteria, fishes, insects,
mollusks, and annelids. In some instances, the animal
itself produces light, as in fireflies. However, in other
animals, the light is produced by symbiotic luminescent
bacteria living in or on the animal. Luminescence is
employed by the animal host as a recognition device

and may be involved in mating, schooling, for prey attraction, or in predator avoidance.

A number of bacterial species in the genus *Photobacterium* emit light, as do some members of the genus *Vibrio*. Both genera are abundant in marine environments and are often associated with indigenous fish. The microbes emit light only when grown to a dense suspension. The enzyme luciferase, a flavin (FMN); O₂; and a long-chain aldehyde (RCHO) are necessary for luminescence to occur. Luciferase is induced in *Vibrio fischeri* when a metabolic product, the autoinducer (N-β-ketocaproyl homoserine lactone), accumulates to a critical level in the growth medium. This is referred to as "quorum sensing," as induction occurs only when a dense bacterial population has accumulated that produces a sufficiently high concentration of the autoinducer. This critical level is attainable in a rich growth medium but does not occur among free-living organisms in the marine environment. The reaction is:

$$FMNH_2 + O_2 + RCHO \xrightarrow{\text{luciferase}}$$
$$FMN + RCOOH + H_2 + LIGHT$$

Some marine fish have luminous glands that are an integral part of the animal. These glands are actually pouches that contain cultures of bioluminescent bacteria. The Atlantic flashlight fish has special pouches under the eyes where *Photobacterium leiognathi* grows as an endosymbiont (**Figure 25.21A,B**). The fish can control the emission of light by drawing a fold of dark tissue over the gland. Some fish have luminescent bacteria in open glands, where they are nourished directly by the fish.

The squid (*Doryteuthis kensaki*) has luminous glands containing *P. leiognathi* embedded in the ink sac, which are partly enclosed in tissue.

In all of these animals, it is believed that the bacterial symbionts are a result of infection and not through egg passage or parental transfer.

Bacteria–Bird

Two fascinating cases of beneficial symbiotic associations occur in specialized species of birds. One of these is with the beeswax-eating honey guide and the other is the leaf-eating hoatzin. The honey guides (genus *Indicator*, **Figure 25.22**) are small, sparrow-sized birds that live in India and on the African continent. Honey guides seek out cracks and crevices in trees and cliffs where swarms of honeybees have established residence. The birds attract the attention of sweet-toothed honey badgers (or humans), who are aware that the presence of these birds affirms that honeycombs can be found nearby, and the honey guide instinctively knows that these animals will rip open the bees' habitation to obtain honey. The bird can then feed on the exposed honeycomb that remains.

The primary source of food for the honey guide is beeswax. The digestive tract of the bird does not produce esterases and other enzymes necessary for the utilization of the wax esters. However, the intestinal tract of the honey guide has two inhabitants that can digest beeswax: *Micrococcus cerolyticus* (cero-wax lyticus-splitting) and a fatty-acid-cleaving yeast, *Candida albicans*. The micrococcus requires a growth factor that is available in the digestive tract of the honey guide. Thus, we

(A)

(B)

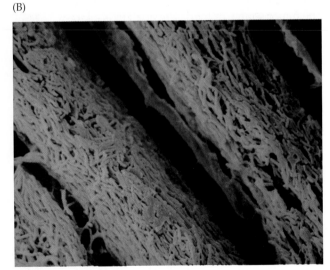

Figure 25.21 Bioluminescent bacteria in fish
Bioluminescence in marine fish. (A) The Atlantic flashlight fish *Kryptophanaron alfredi*, with a luminous organ under the eye. (B) Scanning electron micrograph of *Photobacterium leiognathi* inside the luminous organ. A, ©F. R. McConnaughey/Photo Researchers, Inc.; B, ©J. G. Morin/Visuals Unlimited.

Figure 25.22 Symbiosis in honey guides
A black-throated honey guide, perched on a honeycomb.
These birds (*Indicator* sp.) can use beeswax as a food source
because they are host to bacteria that break down wax to
yield long-chain fatty acids and a yeast that cleaves the long-
chain to short-chain fatty acids that the bird can metabolize.
Photo ©N. Dennis/Photo Researchers, Inc.

have a remarkable symbiotic association between a bird
species and microorganisms. The microbes have a home,
and the bird has a source of food.

The hoatzin (*Opisthocomos hoazin*) is a leaf-
eating (folivorous) bird for which leaves are
the major source of food. The hoatzin is a large
bird, about 750 grams, whose habitat ranges
from the Guianas to Brazil (**Figure 25.23**). It is
found mostly in riverine swamps, forests, and
oxbow lakes.

Cellulosic components of leaves are the
major food source for hoatzins. They have,
therefore, evolved with a rumenlike crop. The
crop and esophagus are the major digestive
organs in this species, and the pH and physical
environment in these organs support the
growth of bacteria at concentrations equivalent
to those in the ruminant (see following section).
Volatile fatty acids, including acetic, propionic,
butyric, and isobutyric acids, are produced by
the bacteria in the crop and esophagus. The
crop and esophagus make up about 75% of the
total digestive system. The volatile fatty acids
are absorbed by the small intestine.

The hoatzin is by far the smallest warm-
blooded animal that has a foregut that func-

tions as a fermentation vat (rumen). All other animals
that do this are the mammals, as discussed next. There
are some significant anatomical and behavioral modifications in the hoatzin to make room for the relatively
large crop. They have a reduced sternum area for flight-
muscle attachment, and the bird is a poor flyer. The
young take more than 60 days before they are able to fly,
and survival mandates that the juveniles have some
means for protection against predation. Hoatzins have
wing claws that are used in rapid crawling movements
to escape from predators. They dive into water when
threatened.

Ruminant Symbioses

Ruminants are herbivorous mammals that have a complex digestive system mainly composed of a large specialized fermentation chamber called the **rumen.**
Among the ruminants are domestic animals such as cattle and sheep, as well as giraffes, buffalo, and elk.

Ruminants feed on grasses and other plant materials
mainly composed of insoluble polysaccharides including cellulose, hemicellulose, and pectin. These animals
lack enzymes essential for the digestion of such refractory food sources, but their rumen harbors a complex
community of anaerobic bacteria and protozoa that use
and convert the polysaccharides to utilizable food for
the animal.

When a ruminant feeds on grass or other plant material, the food enters the mouth, where it is mixed thoroughly in saliva rich in bicarbonate. The material then
passes down the esophagus to the rumen (**Figure 25.24**).

Figure 25.23 Symbiosis in hoatzins
The foliage-eating hoatzin (*Opisthocomos hoazin*). The hoatzin is the smallest
animal—and the only bird species—with a rumenlike digestive system.
Photo ©F. Gohier/Photo Researchers, Inc.

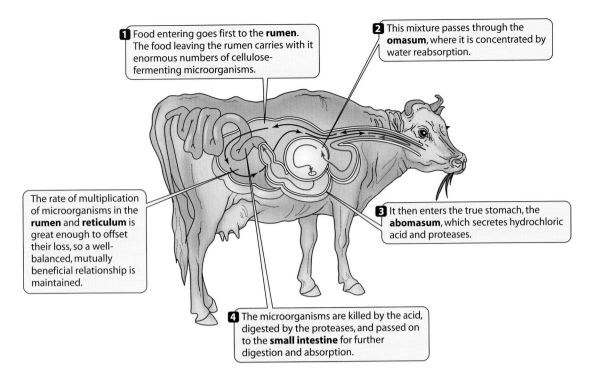

1 Food entering goes first to the **rumen**. The food leaving the rumen carries with it enormous numbers of cellulose-fermenting microorganisms.

2 This mixture passes through the **omasum**, where it is concentrated by water reabsorption.

The rate of multiplication of microorganisms in the **rumen** and **reticulum** is great enough to offset their loss, so a well-balanced, mutually beneficial relationship is maintained.

3 It then enters the true stomach, the **abomasum**, which secretes hydrochloric acid and proteases.

4 The microorganisms are killed by the acid, digested by the proteases, and passed on to the **small intestine** for further digestion and absorption.

Figure 25.24 Ruminant digestive system
Diagram of the rumen and gastrointestinal system of a ruminant. Food enters the rumen and is partially digested and fermented. It moves to the reticulum; cuds are formed, regurgitated, and move back through the digestive system via the omasum.

A rumen in a mature cow will hold 100 liters of material and has a microbial population approaching 10^{12} per ml. The temperature is 30°C, at a pH of 6.5, both highly favorable for a microbial fermentation. Because the rumen is anoxic, the microbial population is composed of anaerobes. The ingested food remains in the rumen for 10 to 12 hours, where the peristaltic action rotates the material breaking up the cellulosic mass to a fine suspension. Cellulolytic bacteria and protozoa attach to the pulpy mass and free up glucose and cellobiose, which are fermented to volatile fatty acids, mainly acetic, propionic, and butyric acids (**Figure 25.25**). These fatty acids pass through the epithelial of the rumen wall and are oxidized by the ruminant as energy source. The low redox potential ($^-0.30$ mV) in the rumen permits the growth of methanogenic bacteria that pro-

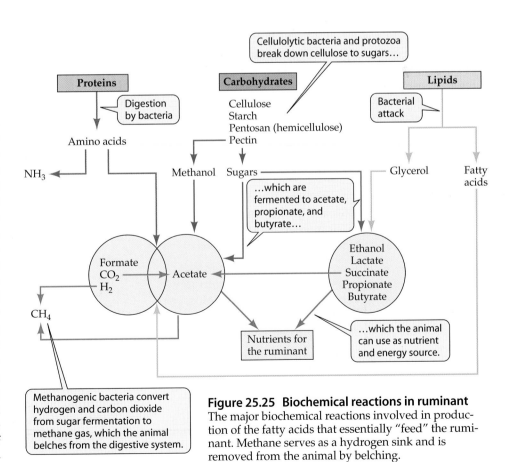

Cellulolytic bacteria and protozoa break down cellulose to sugars…

…which are fermented to acetate, propionate, and butyrate…

Methanogenic bacteria convert hydrogen and carbon dioxide from sugar fermentation to methane gas, which the animal belches from the digestive system.

…which the animal can use as nutrient and energy source.

Figure 25.25 Biochemical reactions in ruminant
The major biochemical reactions involved in production of the fatty acids that essentially "feed" the ruminant. Methane serves as a hydrogen sink and is removed from the animal by belching.

duce methane gas from the H_2 and CO_2 generated during the sugar fermentation, thus serving as a proton sink. A mature cow belches copious quantities of methane daily.

The pulpy mass then moves gradually from the rumen to the reticulum, where small clumps called cuds are formed. These cuds move up the esophagus to the mouth where they are rechewed to smaller particles and mixed with saliva. This mixture, which is quite liquid, is swallowed and moves to the omasum and thence to the abomasum, a more acidic area where digestive enzymes much like those of other mammals digest the material. The microbe-containing mass is digested to release amino acids and vitamins. Absorption of these nutrients occurs as the material moves through the small intestine.

From a microbial ecology perspective, the rumen fermentation has been a classic source of information on microbial interactions. For example, the first methanogenic bacteria that were cultivated were obtained from the rumen. Moreover, it was the study of ruminant metabolism that led to an understanding of the tight coupling in metabolism that can occur between different species. The concept of "interspecies hydrogen transfer," in which one species (a fermentative bacterium) produces hydrogen gas and another species (a methanogen) uses it quickly for energy with the formation of methane gas, originated from a study of rumen metabolism. The combined activity of these two species maintains an extremely low concentration of hydrogen gas. If hydrogen were to accumulate, it would inhibit the activity of the fermenter, and the fermentation could not proceed. This same tight coupling of interspecies hydrogen transfer also occurs in many anaerobic environments where methanogens proliferate.

SUMMARY

▸ Eukaryotes evolved in a prokaryotic world. Alliances between the eukaryotes and prokaryotes are of common occurrence. Two or more **dissimilar** organisms may form an alliance that is termed a **symbiosis.**

▸ Symbiotic associations can **benefit** the **symbionts,** among the advantages: **protection, access to new environments** in the biosphere, **recognition aids,** or **nutrition.** Microbe–microbe, microbe–plant, and microbe–animal symbioses are common in nature.

▸ Symbiotic relationships are often beneficial to at least one partner. **Mutualism** is an interrelationship where both partners benefit. Nutritional cooperation, in which the concerted activity of two species results in the metabolism of a substrate is termed **syntrophy,** a form of **mutualism.**

▸ **Commensalism** is an association in which one partner benefits and the other is neither harmed nor benefited.

▸ **Heterotrophic protists** survive in anaerobic sediments because they have methane-producing endosymbionts that serve as an electron sink. Many of these protozoa lack mitochondria, and their archaea serve the general function of this organelle.

▸ A **lichen** is a symbiotic association between a fungus and a cyanobacterium or eukaryotic alga.

▸ Fungi develop in or on roots, and these are termed **mycorrhizal relationships.** Bacteria may invade roots or stems of selected plants and form **nodules** in which nitrogen fixation occurs.

▸ Bacterial associations with plants that result in the formation of nitrogen-fixing nodules generally occur with **legumes** (peas, clover, etc.). The microorganisms involved are members of the genus *Rhizobium* or *Bradyrhizobium.* Formation of nitrogen-fixing nodules on roots of alder, a woody plant, is a result of invasion by *Frankia* sp., a filamentous actinomycete.

▸ Invertebrate animals survive at depths greater than 1,000 m in tectonically active areas of the ocean because they have developed beneficial symbiotic associations with **chemosynthetic** prokaryotes that live within their tissues. Among those animals are tube worms, mussels, shrimp, and clams.

▸ **Termites** depend on bacterial populations in their gut to **digest** the **wood** they consume.

▸ **Bioluminescence** is common in the biological world. Many **animals** in **marine environments** that emit light do so because they are hosts to **luminescent bacteria.**

▸ Symbiotic relationships in the **avian** world permit birds to utilize **beeswax** or **foliage** as foodstuff. The **honey guide** is a beeswax eater and is host to bacteria that cleave the wax to utilizable fatty acids. The **hoatzin** can survive on foliage because it has a **rumen,** much like that of mammals (cows and deer).

▸ **Herbivorous** animals survive on vegetation because they have a bacterial population in their digestive tract that converts **plant material** to utilizable foodstuff. This occurs in a large internal **fermentation vat** termed the **rumen.**

REVIEW QUESTIONS

1. What is symbiosis? What are the major functions of symbiotic associations? How does mutualism or commensalism differ from parasitism? Could many of these associations originate through parasitism?

2. How does a symbiotic association become established?

3. Define: ectosymbiont, endosymbiont, syntrophy, mutualism, and consortium. What are some examples of each?

4. Microbiologically, what is a rhizosphere and how does this environment differ from surrounding soil? What are benefits that the plant and microorganism receive in symbiotic associations of this type?

5. How do ectomycorrhizae and endomycorrhizae differ? Think of some of the ways that these relationships benefit a plant. What type of plant is involved in each of these?

6. What is *Frankia*? Where is it found? What does it do?

7. Outline the steps involved in the establishment of root nodules on a leguminous plant. What is the function of leghemoglobin?

8. How does a deep-sea chemosynthetic ecosystem differ from the ecosystems on the surface of the earth? Which animal systems on the earth's surface resemble those in the deep sea?

9. Outline the "feeding" of a tube worm. What are the unique features of a tube worm?

10. What is the chemical reaction involved in bacterial luminescence? Why have animals established relationships with luminescent bacteria?

11. Microbiologically, how do "higher" and "lower" termites differ?

12. What symbiotic associations occur in birds? Discuss the role of the bacterial symbionts in nutrition of the animal.

13. Outline the microbes' functions in a ruminant. What are the benefits to the animal? To the bacterium?

14. What function is served by methanogens present in the protozoa that reside in anaerobic environments? Is this similar to ruminants? How?

SUGGESTED READING

Harley, J. L. and S. E. Smith, eds. 1984. *Mycorrhizal Symbiosis*. San Diego, CA: Academic Press.

Margulis, L. and R. Fester, eds. 1991. *Symbiosis as a Source of Evolutionary Innovation: Specialization & Morphogenesis*. Cambridge, MA: MIT Press.

Margulis, L. and D. Sagan. 1986. *Microcosmos: Four Billion Years of Evolution from Our Microbial Ancestors*. New York: Summit Books.

Perry, N. 1990. *Symbiosis: Nature in Partnership*. New York: Sterling Pub. Co.

Staley, J. T. and A-L. Reysenbach. 2002. *Biodiversity of Microbial Life*. New York: Wiley-Liss.

Chapter 26

Human Host-Microbe Interaction

Healing is a matter of time, but it is sometimes also a matter of opportunity.
— HIPPOCRATES, 460–377 B.C.

Billions of bacterial cells reside in and on the human body. Most are harmless and play a beneficial role in our well-being. Every healthy human plays host to a distinct microbial community termed "**normal microbiota**." Bacteria are the major component of this microbiota along with a few fungi. Although the alliance between the human **host** and the normal microbiota is generally **mutualistic,** many bacteria are **commensals.** For example, most of the bacteria that live in our mouths benefit by growing on the food we eat. They are not disease agents and are not known to provide any particular benefit to us. This type of association in which one partner benefits and the other is unaffected is commensalism (see Table 25.1). Some of the normal microbiota are **opportunists** that can cause infection if the host defenses are lowered or if the natural defenses are breached. Humans also have a **transient** microbiota, composed of microbes that do not establish long-term residency.

Occasionally humans encounter **parasitic** organisms in daily life that can cause infectious disease. Most of these parasites do not colonize, but are transient and cause no apparent harm. However, in some instances parasites may colonize and injure the host. These microorganisms that cause harm are termed **pathogens**, and their presence leads to an **infection.** The level of injury resulting from an infection depends on the pathogenicity of the invader and the relative resistance of the host. **Virulence** is a quantitative measure of the capacity of a pathogen to inflict damage (**Box 26.1**). Some strains of a pathogenic species can be more virulent than others and cause a more damaging and/or persistent infection.

This chapter is concerned with three aspects of microbe–human interactions, first a description of the normal microbiota of the human host; second, an outline of host defenses against potential pathogens; and third, a discussion of the various mechanisms employed by pathogens to combat the normal host defenses.

THE NORMAL HUMAN MICROBIOTA

Prior to birth, the human fetus is virtually free of microorganisms. During birth and immediately thereafter, the newborn is exposed to and colonized by bacteria. The respiratory system and gastrointestinal tract of the neonate become colonized by microorganisms in the process of establishing the normal microbiota. These colonizers are picked up from the environment, and most are beneficial. Microorganisms gradually establish an "ongoing" residence in areas that are exposed to the external environment—the skin, the oral cavity, the respiratory tract, the gastrointestinal tract, and the genitourinary system. However, microorganisms are not normally present in organs, internal tissues, blood, or the lymphatic system. A **diseased state** is indicated if microorganisms populate these sites.

Milestones Box 26.1   Evolution of Virulence

English rabbits were introduced into Australia in 1859 by British sportsmen interested in game hunting. There were no carnivorous animals on the Australian continent that were natural enemies of the rabbit, so the population increased at an incredible rate. The agricultural interests in Australia became alarmed at the loss of grazing lands to the rapidly growing rabbit community, and they pressured the government to find a solution to this problem.

In 1950, a myxoma virus that was prevalent in South American rabbits was introduced into the Australian rabbit population. This virus had low virulence in the South American rabbit population but a much higher virulence for the European rabbit strains. The myxoma virus spread rapidly in Australia, and within a few years it had killed over 99% of the rabbit population. Death rates were highest in the summer, as mosquito vectors spread the virus. The scientific community in Australia watched this daring experiment in pest control closely.

Two interesting phenomena became evident over time—one in the rabbit population, the other in the virus itself. Controlled laboratory studies confirmed that over a 6-year span, the virus had become less virulent than the original strain. The rabbit became more resistant to the virus and less susceptible to ravages of the disease. Apparently, survivors of the initial epidemic were selected in some way for resistance to infection. The rabbit population in Australia now stays at about 20% of the previous high number. Experimental evidence indicated that the increased resistance in rabbits was not related to an improved immunological defense system but to a physiologically altered rabbit. This study indicates that both the susceptible host and the virulent virus evolved in a way that allowed both to survive. It is axiomatic that any pathogenic agent that kills all of its hosts will quickly become extinct. Evolution then favors survival of both the pathogen and the host.

Human health is dependent on the body's ability to establish and maintain this consistent nonthreatening indigenous microbiota. Yet certain factors, such as prolonged antibiotic therapy, can alter or suppress the normal microbiota and lead to serious health problems. Use of oral antibiotics can alter the normal gastrointestinal tract microbiota, increasing the risk of developing disease. Broad-spectrum antibiotics taken orally may also reduce the resident microbiota of the mucous membranes of the oral cavity and respiratory system, resulting in yeast or other fungal infections in the oral cavity or lungs. There are many examples of the adverse effects resulting from the destruction of the normal microbiota, confirming that the indigenous microbiota of humans and animals in general are a primary defense against invasion by harmful pathogens.

Some areas of the human body provide a more favorable environment for bacterial growth than others. The physical conditions such as pH, moisture, osmotic pressure, and temperature at a given site are constant, with some variability from site to site. The nutrients and other factors available result in the formation of a distinct microbiota associated in each area of the body. Many microorganisms can survive on human skin, but internal mucous membranes have a larger and more varied population. An estimated 10^{13} (100 trillion) bacteria live with an adult human, and this is far greater than the total number of eukaryotic cells of the human. The distribution of microbial groups that reside in humans is presented in **Figure 26.1**. The genera shown make up only a small part of the overall picture. For example, well over 350 named species have been identified in the large intestine, and countless others have not yet been characterized.

Skin

The intact skin or epidermidis is impermeable to microbes and is considered a first line of defense against invasion. Although much of the skin surface is dry and inhospitable, normal microbiota reside on human skin, and they require nutrients. Two key nutrient sources on human skin are the sweat glands and sebaceous glands (**Figure 26.2**).

The sweat glands are of two types:
- Eccrine glands that generate perspiration
- Apocrine glands that secret perspiration and nutrients

The apocrine glands are more active after puberty. They secrete lysozyme, an enzyme that hydrolyzes peptidoglycan in the cell wall of bacteria. The enzyme cleaves the β $(1 \rightarrow 4)$ glycosidic bond between N-acetylglucosamine and N-acetylmuramic acid in peptidoglycan (see Figure 4.42). These glands also secrete lactic acid, which lowers the local pH to 3 to 5, an acidity that discourages excessive colonization.

The sebaceous glands are associated with hair follicles. They secrete lipid materials (sebum) that prevent overwetting, overdrying, and abrupt changes in temperature of the epidermis layer. Catabolism of the lipids in sebum by the resident propionic acid bacteria releases fatty acids such as oleic acid that can be inhibitory to other species.

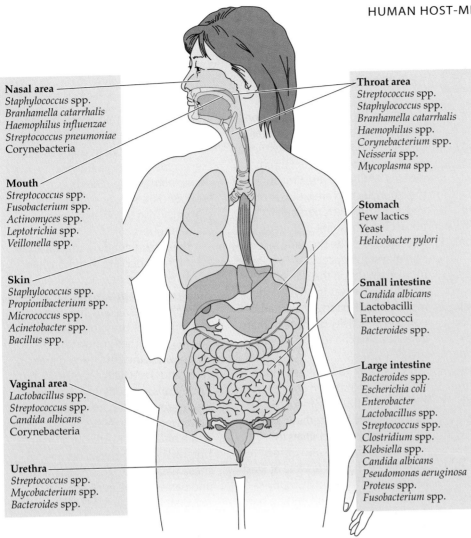

Nasal area
Staphylococcus spp.
Branhamella catarrhalis
Haemophilus influenzae
Streptococcus pneumoniae
Corynebacteria

Mouth
Streptococcus spp.
Fusobacterium spp.
Actinomyces spp.
Leptotrichia spp.
Veillonella spp.

Skin
Staphylococcus spp.
Propionibacterium spp.
Micrococcus spp.
Acinetobacter spp.
Bacillus spp.

Vaginal area
Lactobacillus spp.
Streptococcus spp.
Candida albicans
Corynebacteria

Urethra
Streptococcus spp.
Mycobacterium spp.
Bacteroides spp.

Throat area
Streptococcus spp.
Staphylococcus spp.
Branhamella catarrhalis
Haemophilus spp.
Corynebacterium spp.
Neisseria spp.
Mycoplasma spp.

Stomach
Few lactics
Yeast
Helicobacter pylori

Small intestine
Candida albicans
Lactobacilli
Enterococci
Bacteroides spp.

Large intestine
Bacteroides spp.
Escherichia coli
Enterobacter
Lactobacillus spp.
Streptococcus spp.
Clostridium spp.
Klebsiella spp.
Candida albicans
Pseudomonas aeruginosa
Proteus spp.
Fusobacterium spp.

Figure 26.1
Normal human microbiota
Microorganisms identified as "normal flora" on or in the human body.

The relatively high concentration of sebum, lipids, free fatty acids, wax alcohols, glycerol, and hydrocarbons produced by the sebaceous glands provides substrate for a microbiota of about 10^6 bacteria per square centimeter in these locales. The sweat glands also secrete urea, amino acids, lactic acid, and salts that favor bacterial sustenance. These areas have a constant population of mostly gram-positive bacteria including species of *Staphylococcus*, *Micrococcus*, *Corynebacterium*, and *Propionibacterium* (see **Box 26.2**).

In contrast to these moist areas on the skin such as the armpits, scalp, and feet, dry areas that have few or no secretory glands, such as the palms of the hands, have much lower microbial densities of 10^2 to 10^4 per square centimeter.

The bacterial community changes at puberty with increased concentrations of *Propionibacterium acnes* and the appearance of selected species of the staphylococci including *S. capitis* and *S. auricularis*. Diet has little measurable influence on the microbiota, but antibiotic therapy can alter the skin populations.

Respiratory Tract

The respiratory tract consists of the mouth, nasopharynx, throat, trachea, bronchia, and lungs. These moist locales are potential sites for microbial coloni-

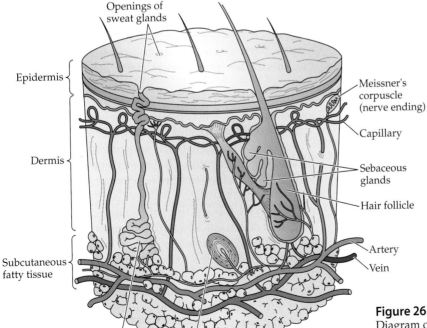

Openings of sweat glands

Epidermis

Dermis

Subcutaneous fatty tissue

Sweat gland

Pacinian corpuscle (nerve ending)

Meissner's corpuscle (nerve ending)

Capillary

Sebaceous glands

Hair follicle

Artery

Vein

Figure 26.2 Human skin
Diagram of a cross section of human skin. The sweat and sebaceous glands secrete materials that sustain bacterial populations on the skin.

Research Highlights Box 26.2   **Skin Staphylococci**

Staphylococci are one of the major bacterial colonizers of human skin. Of the bacteria present on the body surface and the anterior nares, over 50% are typically staphylococci. *Staphylococcus aureus* was the first member of this genus isolated and was described by F. J. Rosenbach in 1884. He isolated the organism from an infected wound. The second species described was *S. epidermidis*, which was isolated by C. E. A. Winslow in 1908. Both *S. aureus* and *S. epidermidis* are gram-positive cocci that are facultative anaerobes. They are generally differentiated from one another by the coagulase test. Most *S. aureus* strains produce coagulase (a blood-clotting enzyme), whereas strains of *S. epidermidis* do not. The common habitat for staphylococci is humans and other animals, and their occurrence in nature is generally attributed to human or animal sources. *S. aureus* is a serious, opportunistic pathogen in humans and the causative agent of boils, impetigo, pneumonia, osteomyelitis, endocarditis, meningitis, toxic shock syn-

drome, and several other infections. *S. epidermidis* strains are a clinical problem in patients with prosthetic heart valves or other implanted devices. Antibiotic-resistant strains of the staphylococci are a persistent and acute problem, particularly in nosocomial (hospital-acquired) infections. Despite the early recognition that staphylococci are a common organism on the skin of humans, there was relatively little research done on the characterization and infectivity of these human parasites. This changed markedly in the 1970s.

In the early 1970s, Wesley Kloos of the United States, in collaboration with Karl-Heinz Schleifer of Germany, initiated a systematic study on the staphylococci. Over the ensuring years, they and others determined that there were at least 12 species of the genus *Staphylococcus* that inhabit the human skin. *Staphylococcus* is the dominant organism on skin with members of the genus *Micrococcus* second to fifth in abundance. At least two to

three species of micrococci live on most humans, although some individuals may harbor as many as seven species. Coryneforms, *Acinetobacter* spp., *Enterobacteria* spp., *Bacillus* spp., and *Streptomyces* spp. may also be present in significant numbers.

These studies affirm that an array of staphylococci are common colonizers of the human skin. Other species are adapted to life on other animals: *S. intermedius*, the domestic dog and other carnivores; *S. hyicus*, the pig and other ungulates; *S. caseolyticus*, the cow, hoofed mammals, and the whale; *S. felis*, the cat; and *S. caprae*, humans and goats. There are now 32 named species of *Staphylococcus* specifically adapted to life on animal hosts. Despite the potential pathogenicity of many of these parasitic organisms, the host and parasite have evolved with a reasonable tolerance for one another. Unfortunately, this standoff is broken at times, and staphylococci can become a serious and deadly enemy.

---

zation. The **mouth** is a favorable environment for bacteria, as it is relatively nutrient rich and the temperature and pH are stable. About 10^8 bacteria live in each ml of saliva. Many of these microorganisms originate from colonies that adhere to surfaces of the tongue, gums, and teeth. Constant saliva production and swallowing tends to remove bacteria that do not stick to surfaces. Lysozyme and lactoperoxidase in saliva are inhibitory to bacterial growth. Saliva contains about 0.5% dissolved solids, consisting of proteins such as salivary enzymes and mucoproteins. *Streptococcus salivarius* proliferates by adhering to the surface of the tongue. **Teeth** are a calcium phosphate crystalline material that readily becomes colonized. In the newborn, before development of teeth, lactobacilli and streptococci predominate in the oral cavity. As teeth develop, oral microorganisms such as *Streptococcus mutans* invade and attach to the enamel surface. Oral streptococci produce glucans (polysaccharides) that bind bacteria together on the tooth surface. These aggregates of bacteria and organic matter are known as

plaques. Streptococcus sanguis and anaerobic actinomyces are among the microbes in these aggregations. The microbes in the plaque catabolize sucrose and other sugars to produce lactic and other organic acids that etch teeth, causing dental caries.

The **nasopharynx** and **nose** are areas with adequate moisture that are constantly inoculated with microorganisms as one breathes. The **nostrils** have a population similar to that of facial skin, with *Staphylococcus* species and other gram-positive bacteria predominating. The nasopharynx (above the soft palate) is colonized by avirulent strains of *Streptococcus pneumoniae, Neisseria meningitidis, Branhamella* sp., and *Haemophilus* sp. The **oropharynx** (between the soft palate and larynx) has a microbial population that is similar to the population of the nasopharynx but includes in addition some micrococci and anaerobes of the genera *Prevotella* and *Porphyromonas*.

In a healthy person, the **lower respiratory tract**, which includes the trachea, bronchi, and lungs, is typically devoid of bacteria. Most dust and other particles that are inhaled settle out on mucous membranes of the upper respiratory tract before reaching the lower regions. Particles bearing microorganisms that reach the lower respiratory tract are removed by the mucociliary escalator. The mucus lining the area traps particles, and the upward-beating cilia push the mucus layer upward, where the material enters the throat and is subconsciously swallowed.

Gastrointestinal Tract

The **stomach** is constantly exposed to bacteria through swallowing of saliva and mucus or by ingestion of food. Few survive—there are fewer than ten microbes per ml of stomach fluid. The low pH due to HCl and digestive enzymes destroys most of the organisms that reach the stomach. Some acid-tolerant lactobacilli and yeasts survive. The stomach of some humans is colonized by the microaerophilic bacterium *Helicobacter pylori* that is implicated in the development of peptic ulcers (see Chapter 28).

The **small intestine** is divided into three sections—the upper part is the duodenum, the middle the jejunum, and the lower part is the **ileum.** The number of bacteria present increases in concert with digestion of food as it moves downward through the small intestine. The duodenum has a limited bacterial population (fewer than 10^3 per ml) due to stomach acids, bile secreted by the gallbladder, and pancreatic secretions. The microbiota are mainly gram-positive cocci and bacilli. The jejunum population consists of enterococci, lactobacilli, and corynebacteria. The yeast *Candida albicans* is also commonly present in this area. The ileum microbiota resemble that of the large intestine, with large numbers of *Bacteroides* species and a limited number of facultative anaerobes such as *Escherichia coli*. The facultative organisms function to remove O_2.

The **large intestine** is also called the **colon.** This area is actually a fermentation vat populated by masses of anaerobic bacteria. Ingested food is the basic substrate for these organisms. About 10^{10} to 10^{11} bacterial cells per gram of mass live in the colon, and more than 350 different species have been characterized. Several hundred times as many strict anaerobes live in the colon in comparison to facultative aerobes. None are strict aerobes. An adult excretes 3×10^{13} bacteria per day—25% to 35% of fecal matter is bacterial mass.

The microbiota of the digestive tract supplies a human with several vitamins, including B_{12}, biotin, riboflavin, and vitamin K. The microbiota of the healthy adult colon consist of gram-negative bacteria, including bacteroides (*Bacteroides fragilis, B. melaninogenicus,* and *B. oralis*) and *Fusobacterium* species. Among the gram-positive bacteria are members of the following genera—*Bifidobacterium, Eubacterium, Lactobacillus,* and

Clostridium. Less than 1% of the microbiota consists of *Escherichia coli* and species of *Proteus, Klebsiella,* and *Enterobacter.* A few harmless protozoa such as *Trichomonas hominis* may also live in healthy human adults.

The newborn infant has a sterile gut that quickly becomes populated by bacteria. If breast-fed, the bacteria are mostly of the genus *Bifidobacterium*, obtained from the skin of the mother. Human milk contains an amino disaccharide that is required by this bacterium. Bottle-fed infants have a complex colon population with *Lactobacillus* species among the dominant organisms. As a child moves to a solid diet, the microbial population of the gut gradually changes to resemble that of the adult from whom they obtain their normal microbiota.

Genitourinary Tract

The kidney, ureter, and urinary bladder of the healthy adult are essentially free of microorganisms. The upper part of the urethra near the bladder is sterile because of mechanical flushing and possible antibacterial activity from urethral mucous membranes. Bacteria are present in the lower part of the urethra in both males and females. The adult female genital tract has a complex normal microbiota. The character of this microbiota changes during the menstrual cycle. The adult vagina is colonized by acid-tolerant lactobacilli that convert the glycogen produced by vaginal epithelia to lactic acid. This maintains the pH at 4.4 to 4.6. Microbes that can tolerate this pH—the lactic acid bacteria, enterobacteria, coryneforms, the yeast *Candida albicans*, and various other anaerobic bacteria—are commonly present.

INNATE HOST DEFENSES

The normal microbiota are important to our body's daily well-being because they prevent colonization by atypical and less-desirable species. Occasionally, some members of the normal microbiota may gain access to the bloodstream. This may occur through a scratch, an abrasion, a minor trauma that may occur when brushing one's teeth, or chewing food. Once in the tissue, the microorganisms can enter the bloodstream. Generally, these organisms are of little consequence, as they are removed quickly by specialized circulating cells, called **phagocytes,** that move to foreign material and engulf it. The phagocytes are also a part of the immune system that protects us against foreign invasion. (For a detailed discussion of the immune system, see Chapter 27.)

We begin this section with a discussion of the barriers that keep foreign invaders out of our system and specific host defenses that immediately attack any invader that passes through the skin or mucous membranes. A human has two types of defense against a potential harmful agent if it passes through the innate defense barriers of the host. These are the specific defenses mediated by the immune system (acquired immune defenses) and the nonspecific defenses that are continuously functioning in a healthy human.

Physical Barriers

We possess natural physical barriers that protect us from the potential invasive microorganisms that we encounter in our daily lives. These include skin and mucous membranes, which, under normal conditions, are quite impenetrable to microbes. Membranes in the eyes, lungs, intestines, and urinary tract are constantly washed by fluids that move over the surface and remove foreign matter that is not tightly attached. Mucus itself is very effective in trapping bacteria. Another barrier is the cilia of the respiratory passages that continually wave upward and push foreign material outward.

The acidity of the stomach (pH 2) destroys the majority of ingested microorganisms and leaves the natural microbiota of the intestines unchallenged. The moderately acidic pH of skin also inhibits microbial growth by all that are not adapted to grow there. In addition, the sebaceous glands in the skin release oils and waxes that are converted to organic acids, and these lower the skin pH. The acidity of the vagina precludes growth of organisms other than the resident lactic acid bacterial microbiota or other acid-tolerant species. Because pathogenic microorganisms are generally quite tissue specific and attach and colonize particular cell types, they are readily removed from nontarget areas. For example, washing your hands removes *Salmonella* and other intestinal pathogens. *Shigella* or *Vibrio cholerae* are also readily washed away, as they have no mechanism for attachment to skin.

Chemical Defenses

The body maintains an array of chemical defenses against invasion by microorganisms. Among the major chemical defenses is lysozyme, an enzyme that is present in blood, sweat, tears, saliva, nasal secretions, and other body fluids. Lysozyme is a particularly effective lytic agent for gram-positive bacteria. Gram-negative bacteria are more resistant to lysozyme because their peptidoglycan layer is located beneath the outer cell envelope.

Lactoperoxidase is an enzyme present in saliva and milk. Lactoperoxidase effects a reaction between chloride ions and H_2O_2, resulting in the generation of toxic singlet oxygen that can kill bacteria. In addition a basic polypeptide, termed b-lysin, is released from blood platelets that move to cuts and abrasions and is effective in killing gram-positive bacteria. The natural barriers, nonspecific enzymes, and other bactericidal agents keep us disease free for much of our lives. If a virulent pathogen breaches these defenses, then a diseased state

is possible. We become diseased at times because humans have little natural resistance to selected infectious agents. Bear in mind that many infectious agents survive only in a living host and have evolved with effective mechanisms for overcoming the natural defenses.

The Inflammatory Process

A cut, puncture, abrasion, or other wound to tissue initiates the **inflammatory response.** This nonspecific response consists of an ordered series of events that localizes and generally eliminates any microorganism(s) that may enter through the site of injury. The processes involved in the inflammatory response also remove damaged tissue and ultimately restore the area to its normal state. When an adverse condition, such as a puncture wound, initiates the inflammatory response, an immediate dilation of local capillaries, and an increased flow of blood occur at the site (**Figure 26.3**). The injured tissue releases chemical signals that activate the inner lining of adjacent capillaries including **selectins** (adhesion molecules) that are exposed on activated endothelial cells. The selectins attract neutrophils and cause them to tumble and slow down. Receptors on the neutrophils, termed **integrins,** bring the neutrophils to a stop as they adhere tightly to endothelial cells, change shape, and squeeze through the epithelial wall of the capillaries (**diapedesis**). After entering the tissue, the neutrophils migrate to the site of injury.

Disruption of tissue cells by the injury causes the release of mediators that lower the pH in the intracellular area. This activates an extracellular enzyme at the site of injury. Kallikrein forms **bradykinin** from a precursor molecule, and bradykinin attaches to sites on capillary walls and to the mast cells that surround local capillaries. The attachment causes openings to occur in the capillaries through which leukocytes and other phagocytic cells can then move into the infected area. The reaction between bradykinin and host cells leads to the release of histamine and serotonin. Histamine acts by dilating capillaries and makes the intercellular junctions in the cells of capillary walls wider, thereby releasing more blood into the damaged area.

The binding of bradykinin to capillaries stimulates the release of prostaglandins (PGE_2 and PGF_{2a}). These prostaglandins cause tissue swelling in the area of inflammation. Phospholipases release arachidonic acid, and this serves as a precursor for other prostaglandins (E_2 and F_{2a}), thromboxane A_2, and leukotrienes. Pain results when prostaglandins bind to nerve endings, causing them to produce a pain impulse. Physiologically, all of these factors together elicit the major symptoms of the inflammatory response—redness, pain, swelling, and increased localized temperature.

Phagocytes respond to inflammation and chemotactically move to the site of injury. The first phagocytic cells

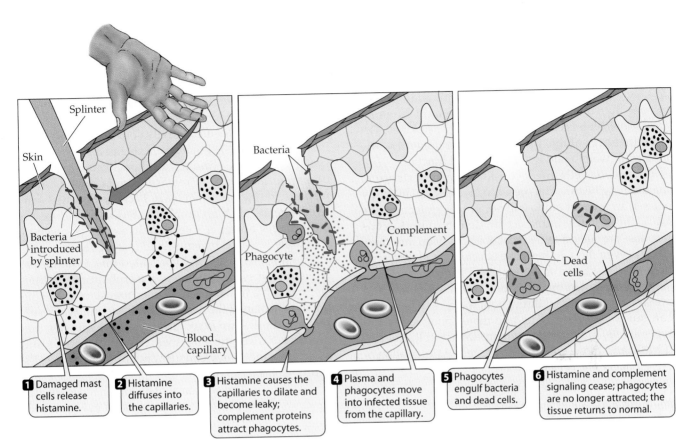

1 Damaged mast cells release histamine.	**2** Histamine diffuses into the capillaries.	**3** Histamine causes the capillaries to dilate and become leaky; complement proteins attract phagocytes.
4 Plasma and phagocytes move into infected tissue from the capillary.	**5** Phagocytes engulf bacteria and dead cells.	**6** Histamine and complement signaling cease; phagocytes are no longer attracted; the tissue returns to normal.

Figure 26.3 Inflammatory response
Nonspecific inflammatory response. The capillary is distended by an increased blood flow, and activation of endothelial cells (that line the capillaries) results in a slowing down of neutrophils carried in the bloodstream. The phagocytic cells change shape, squeeze through the capillary wall, and move toward invaders.

that enter the injured area are the **neutrophils,** which are closely followed by other macrophages. The local increase in temperature can stimulate the inflammatory response and accelerate the rate of phagocytosis.

The flow of fluids into the trauma area brings in large numbers of red blood cells, causing a restriction in blood flow. Greater numbers of leukocytes, polymorphonuclear neutrophils (PMNs), and monocytes enter the injured area and chemotactically move to any bacterial invaders. If the injured area is large, a fibrinous blood clot forms, which can prevent the spread of invaders further into the affected area.

After the phagocytic cells remove the invader, the wound is repaired by a proliferation of epithelial and endothelial cells. There is no single

event that results in the removal of an invading pathogen. The actions of several specific responses—dilation of capillaries, clot formation, release of chemotactic factors, and action of phagocytes—combine to eliminate the foreign invader.

Phagocytes and Phagocytosis

Human blood contains an array of distinct cell types (**Table 26.1**). Erythrocytes carry oxygen; platelets are the

Table 26.1	The cellular components of normal human blood	
Blood Cell Type	**Concentration**	**Function**
Red blood cells (Erythrocytes)	Male: 4.2–5.4 million/ml Female: 3.6–5.0 million/ml	Oxygen transport; carbon dioxide transport
Platelets	150,000–400,000/ml	Essential for clotting
White blood cells (Leukocytes)	5,000–10,000/ml	
Neutrophils	About 60% of WBCs	Phagocytosis
Eosinophils	1%–3% of WBCs	Role in allergic response; destroy parasites
Basophils	1% of WBCs	Mast cell progenitors
Lymphocytes	25%–35% of WBCs	Produce antibodies; destroy foreign cells
Monocytes	6% of WBCs	Differentiate to form macrophages

cellular component of the blood-clotting system; lymphocytes stimulate, regulate, and generate antibodies; and phagocytes are mostly mobile cells that ingest foreign invaders. All of the blood cell types listed in Table 26.1 are derived from a multipotent stem cell produced in bone marrow (**Figure 26.4**). The differentiation into the various blood cell types is effected by cytokines. The cytokines are soluble proteins that are generally low molecular weight and that regulate cell function.

Phagocytic cells are of considerable importance in maintaining health. The presence of a specific antibody against an invader alone will not prevent infection. We know this because humans with defects in phagocytic response are subject to frequent episodes of an infectious disease, even though they have effective circulating antibodies against that infectious agent. Phagocytes are constantly distributed by circulating blood into tissue and are strategically placed throughout the body where they can quickly respond. The two major phagocytic cell types are the granulated polymorphonuclear leukocytes (PMNs) and monocytes (macrophages). A description of these phagocytic cells follows.

Polymorphonuclear Leukocytes (PMNs) The **PMNs** are circulating leukocytes (white blood cells) that move about and remove dead cells and other debris, including bacteria, from the system. The PMNs move rapidly (40 meters/minute) and are the first phagocytic cells to arrive at a site of inflammation and enter tissue by diapedesis. The PMNs are actually short-lived cells (half-life of 6 to 8 hours) and are available as a response to a specific infection. This increased presence is the result of a stimulated rate of release from bone marrow, a response termed **leukocytosis.** Because they are short lived, a relatively high number of PMNs in the blood is indicative of an active infection. In a chronic infection, PMNs are replaced by long-lived macrophages.

The nucleus of a PMN is divided into a number of segments—hence, the name polymorphonuclear. PMNs also contain a number of granules that stain with neutral dyes (origin of the names *granulocyte* and *neutrophil*). These granules are **lysosomes,** small organelles that produce hydrolytic enzymes that can disrupt and hydrolyze the major constituents of dead host cells, debris, or bacterial cells.

The sequence of events that occurs during phagocytosis is depicted in **Figure 26.5**. The foreign object (here a bacterium) is engulfed by the phagocytic cell and is quickly

Figure 26.4 Origin of blood cells
Origin of the various cells of human blood. The multipotent stem cell is produced in bone marrow. Its differentiation is effected by cytokines.

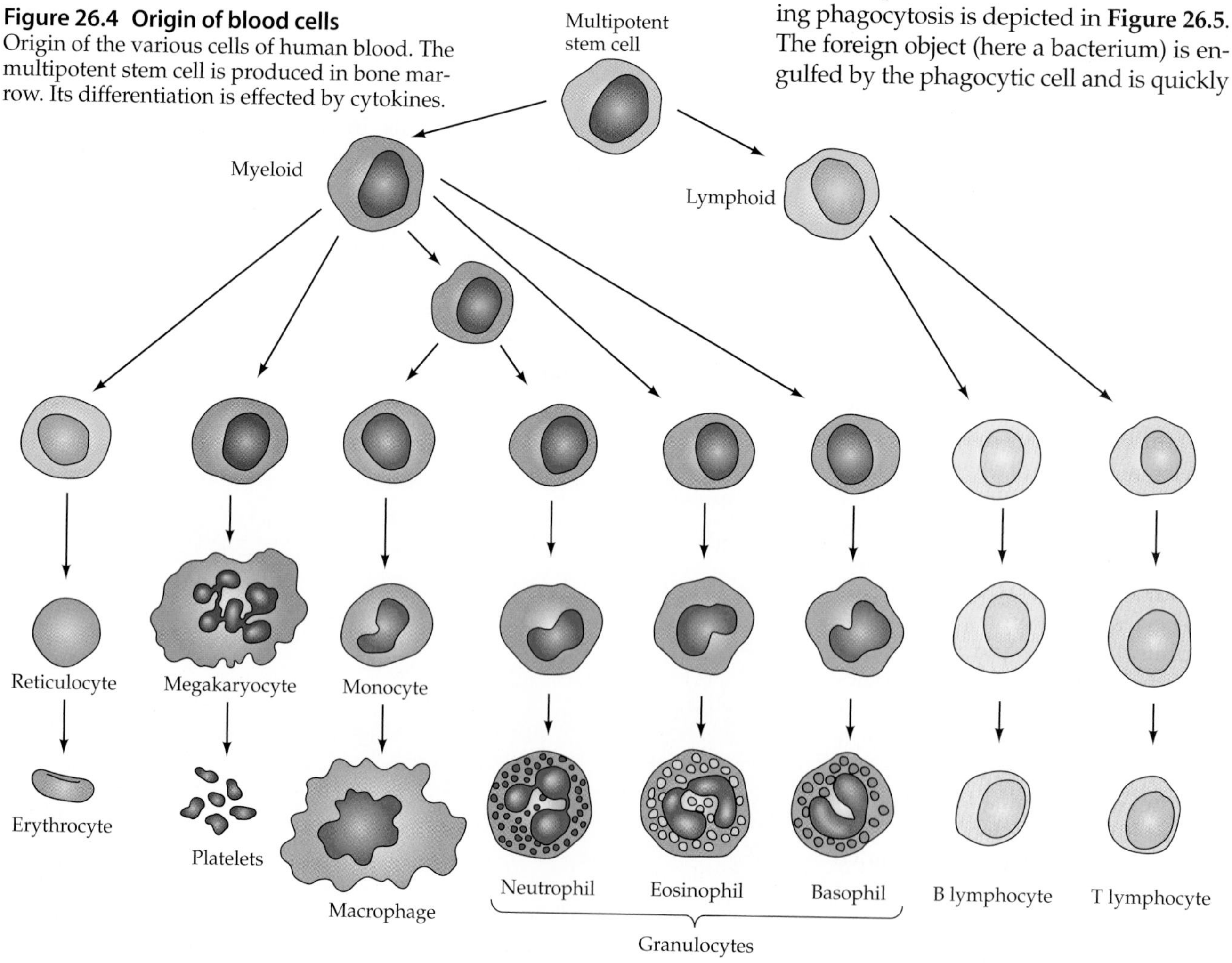

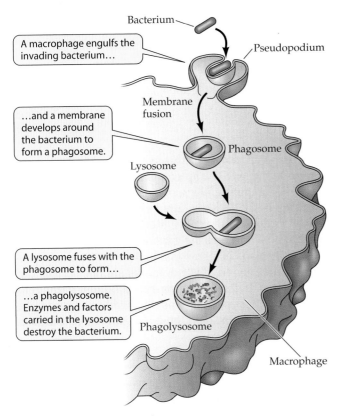

Figure 26.5 Phagocytosis
Events occurring in phagocytosis of a bacterial cell.

Bacterium

A macrophage engulfs the invading bacterium...

Pseudopodium

Membrane fusion

...and a membrane develops around the bacterium to form a phagosome.

Phagosome

Lysosome

A lysosome fuses with the phagosome to form...

...a phagolysosome. Enzymes and factors carried in the lysosome destroy the bacterium.

Phagolysosome

Macrophage

enclosed in a vacuole called a phagosome. The phagosome is formed by invagination of the phagocyte's plasma membrane. The phagosome separates from the plasma membrane and fuses with a lysosome, thus releasing a battery of hydrolytic enzymes into the fused cell. The fused cell is termed a **phagolysosome.** More than 60 distinct highly destructive enzyme systems are in the phagolysosome, including proteases, lipases, carbohydrases, RNAase, DNAase, acid phosphatase, peroxidase, and a myeloperoxidase system.

Following phagolysosome formation, a burst of metabolic activity occurs as stored glycogen from the lysosome is metabolized. Glucose released from glycogen is catabolized anaerobically via both glycolysis and the hexosemonophosphate shunt, resulting in the formation of reduced pyridine nucleotides (NADH and NADPH). The reduced pyridine nucleotides are utilized in production of peroxide and superoxide. These reactions are summarized as follows:

$$\text{Peroxide formation} \rightarrow H_2O_2$$
$$2\,NADPH + O_2 \rightarrow H_2O_2$$
$$\text{Superoxide formation} \rightarrow O_2^-$$
$$NADPH + 2O_2 \rightarrow 2O_2^- + H^+ + NADP^+$$
$$\text{Myeloperoxidase} - \text{halide} - \text{peroxide system}$$
$$\rightarrow \text{Hypochlorite}$$
$$Cl^- + H_2O_2 \rightarrow ClO^- + H_2O$$

The enzymes and oxidants (ClO^-, H_2O_2, O_2^-) kill and digest microbial cells, and the debris is excreted from the PMN by exocytosis. Normal body functions then remove the relatively harmless debris from the blood.

Monocytes (Macrophages)

Macrophages are large cells that play an immediate role in preventing invasion and are involved in the immune response (see Chapter 27). Monocytes that circulate in the blood pass through blood vessels into tissues where they are termed macrophages. Some of these become circulating or **wandering** macrophages and are carried by the lymph to various tissue locations. The wandering macrophages inhabit lung alveoli and are also found in the peritoneum, the membrane that lines the abdominal cavity. Other macrophages have a limited mobility because they are **fixed** to tissue. These macrophages occur on the epithelial surfaces that are exposed to the external environment. Fixed macrophages also line vessels through which blood or lymph flows and are found in the liver, the spleen, and lymph nodes. Macrophages sometimes are referred to as the mononuclear phagocyte system or **reticuloendothelial system.** The macrophages are essentially a second line of defense because they appear at a site of trauma after the PMNs respond in the inflammatory response.

The macrophages are long-lived cells that play a significant role in acute and chronic infections. They have fewer granules than PMNs, but the fixed macrophages continuously synthesize lysosomal enzymes. Macrophages engulf foreign invaders, form phagosomes, and fuse with lysosomes to generate a phagolysosome. They produce peroxide and contain the various hydrolytic enzymes. Macrophages also produce reactive nitrogen compounds including nitric oxide, nitrite, and nitrate that are generated from arginine when stimulated by cytokines. Macrophages apparently lack some of the activities of the PMN, including the hypochloride-generating myeloperoxidase system.

The Lymphatic System

The **lymphatic** system functions in conjunction with the cardiovascular system (**Figure 26.6**). The system consists of a network of lymph-carrying vessels, ducts, nodes, and other lymphatic tissues including the spleen (**Figure 26.7**). The lymphatic system is a significant part of the nonspecific host defense. The system carries lymph, a fluid that lacks red blood cells and most of the serum proteins, through tissue and back to the circulatory system. Lymph carries macrophages through tissue, picks up waste and debris, and transfers it to the lymphatic capillaries. The lymphatic capillaries are small tubes with a single layer of endothelial cells that are permeable to interstitial constituents. The interstitium is the fluid-filled space between tissue cells. Low levels of the interstitial fluid enter the

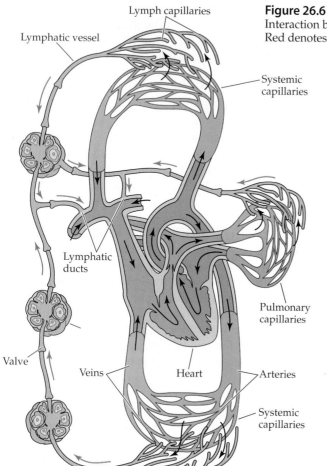

Lymph capillaries

Lymphatic vessel

Systemic capillaries

Lymphatic ducts

Pulmonary capillaries

Valve

Veins   Heart   Arteries

Systemic capillaries

Figure 26.6 Human cardiovascular and lymphatic systems
Interaction between the cardiovascular system and the lymphatic system. Red denotes oxygenated blood; blue, oxygen-depleted blood.

The spleen is part of this system and is the largest lymphoid organ. The spleen is made up of a vascular network containing macrophages. The macrophages of the spleen phagocytize aged erythrocytes and other waste particles and remove them from the system. The normal host defenses are generally successful in maintaining humans in a reasonable state of health. As previously mentioned, these defenses are not perfect and are at times ineffective against a pathogen. Pathogens have effective mechanisms that can overcome the defenses of the host, and these will now be considered.

PATHOGENESIS MECHANISMS

Despite natural barriers and various nonspecific host defenses, pathogens do invade and cause disease. Our natural defenses can be breached by abrasions or cuts, allowing entry to pathogens, which can initiate the infectious process. Disease can be caused by a pathogen

lymphatic capillaries and flow into the larger lymphatic vessels. The network flows through lymph nodes and eventually drains into the cardiovascular system. The lymphatic vessels have valves at certain points that permit only a forward flow.

The lymphatic system is nonspecific, monitors the health of tissue, and is also the site of the specific immunological (**humoral**) response. Thus it is the juncture where the cellular and specific responses to invasion come together. The lymph nodes are lined with macrophages that can engulf and digest foreign material carried by the lymph from interstitial spaces.

Figure 26.7 Human lymphatic system
In the human lymphatic system, lymph-carrying capillaries are widely distributed and lymph nodes cluster in specific regions.

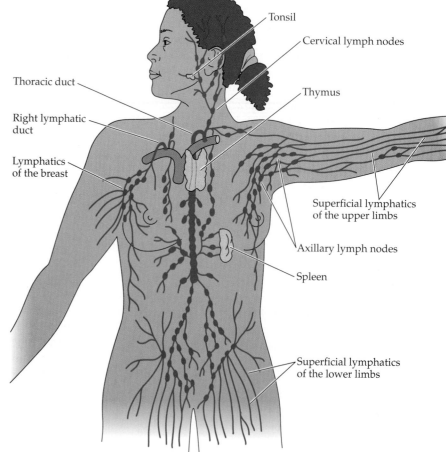

Tonsil

Cervical lymph nodes

Thoracic duct

Thymus

Right lymphatic duct

Lymphatics of the breast

Superficial lymphatics of the upper limbs

Axillary lymph nodes

Spleen

Superficial lymphatics of the lower limbs

that penetrates the skin, mucous membranes, or epithelial surfaces and successfully combats the natural defenses. Pathogens have the capacity to attach specifically to selected membranes or epithelia and produce toxins that damage host cells. Infection is inevitable if the pathogen successfully reaches a specific site and is able to colonize.

Disease-causing organisms can be categorized as **obligate, accidental,** or **opportunistic** pathogens. An **obligate pathogen** is an organism such as *Neisseria gonorrhoeae* or *Streptococcus pyogenes* that does not live outside a host. The pathogen's survival then depends on its ability to move from one host and to **adhere** and **colonize** another host. An **accidental pathogen** is an organism such as *Clostridium tetani* that is ubiquitous in nature and causes disease only under unusual circumstances. This organism causes tetanus, which can be fatal, if the microorganism accidentally enters the body through a deep wound. The ability to cause disease does not play a significant role in the survival of this species. An **opportunistic pathogen** is an organism that does not affect a healthy host but can cause disease of a compromised or unhealthy individual. For example, AIDS patients are more susceptible to tuberculosis and many other diseases than a normal individual. Also, some normal microbiota, such as *Staphylococcus* species on the skin, may cause infections if tissue is injured or when the resistance of the host is decreased.

Adherence and Penetration

As mentioned previously, infectious organisms rarely have the ability to penetrate intact skin, mucous membranes, or external epithelia. Rather, they depend on various **adherence factors** that permit them to attach to selected tissue where they colonize and initiate the infectious process. However, bacteria may cause disease without specific attachment. Minor breaks in the skin or mucous membrane can lead to infection by *Staphylococcus aureus* or *Streptococcus pyogenes,* as these organisms are commonly live at these sites. Organisms can also enter wounds or burn areas by passive transmission on airborne dust particles or water droplets. *Pseudomonas aeruginosa* is a constant threat to burn patients, as this opportunist can be carried passively on airborne dust particles.

Most acute infections originate on mucosal surfaces of the respiratory, gastrointestinal, or genitourinary tracts. The ability of the pathogen to initiate the infectious process is largely related to its innate ability to cling to specific cells. The organisms

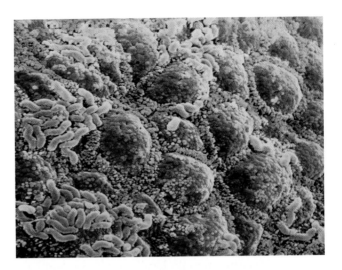

Figure 26.8 Adherence of microorganisms
Specific adherence of microorganisms to epithelium. Glycoproteins on the bacterial cell surface interact with specific sites on epithelial cells to cause the adherence. Photo ©P. Motta and F. Carpino/SPL/Photo Researchers, Inc.

that initiate infections on surfaces generally have factors that aid them in this process (**Figure 26.8**). These adherence factors are quite specific and do not interact with epithelial cells indiscriminately. Those that attach to throat mucosal cells would probably not attach to intestinal epithelia and vice versa. Some of the adherence factors that are involved in the adherence of pathogens to host cells are listed in **Table 26.2**.

An organism that can cause disease in a specific area of the body may have distinct fimbriae or other surface factors that are involved with attachment. Different types of fimbriae occur on strains that attach to different tissues. For example, *Escherichia coli* strains produce fimbriae with differing lectin specificities, and these affect adherence to different host tissues and cause distinctly different diseases. Human pathogenic *E. coli*

Table 26.2	Adherence factors involved in attachment of organisms to host cells
Adherence Factor	**Example**
Fimbriae (adhesion proteins)	*Proteus mirabilis*—urinary tract infections
	Neisseria gonorrhoeae—attach to urinary epithelia
	Salmonella—attach to intestinal epithelia
	Streptococcus pyogenes—M protein attaches to epithelia
Capsule (glycocalyx)	*Streptococcus mutans*—dextrans attach to teeth
	Streptococcus salivarius and *S. sanguis*—attach to tongue epithelia
Teichoic acids Lipoteichoic acids	*Staphylococcus aureus*—attach to nasal epithelia

strains that cause urinary tract infections have P (pyelo-nephritic) type fimbriae, whereas S-type fimbriae are produced by *E. coli* strains that cause intestinal infections in the neonate. Genetic alteration that results in the inability to synthesize the specific fimbriae leads to a loss of virulence in these organisms. *Neisseria gonorrhoeae* fimbriae attach to specific gangliosides on the surface of buccal mucosa or cervical epithelial cells in the genitourinary tract. Strains that do not have adherence factors (fimbriae) do not cause disease. These organisms are swept away by the action of cilia or flushing actions.

Diseases that affect one animal species generally do not affect animals of other species because of a difference in surface components on their epithelial cells. If the pathogen cannot adhere, it cannot infect and colonize.

Pathogens often remain localized on mucosal surfaces, where they reproduce and cause damage through the production of toxic substances. Diseases such as whooping cough (*Bordetella pertussis*), diphtheria (*Corynebacterium diphtheriae*), and cholera (*Vibrio cholerae*) are examples of such diseases. Other microbes cause infection because they penetrate the epithelia and grow in submucosa or the infectious agent may be carried by the lymphatic system or bloodstream to other areas of the body. *Neisseria meningitidis*, the causative agent of a type of meningitis, is one such disease. A localized infection caused by another species can disrupt epithelial cells of the throat, permitting the *N. meningitidis* to enter and be transported to the meninges, where it adheres and reproduces, resulting in meningitis.

The ability of microbes to adhere is not necessarily detrimental to the host. Our natural flora also possess adherence and survival mechanisms. The occurrence of these organisms at various sites precludes adherence or invasion by potential pathogens, which is referred to as "preempting" or "crowding out" an alien microorganism.

Colonization and Virulence Factors

To initiate the infectious process, pathogens must be able to grow in the host. If an organism can invade but can-not grow, it will not be able to surmount the natural defenses of the host. A pathogenic organism cannot colonize without a source of soluble nutrients such as sugars, amino acids, or fatty acids. Some minerals are also essential. As these are not available in extracellular areas, host-cell invasion is essential if the pathogen is to obtain nutrients and grow.

Host-cell disruption can occur through functional enzymes produced by the invading pathogenic bacterium. Some of these enzymes that are generated by pathogenic microorganisms are listed in **Table 26.3**. Collagenase, elastase, hyaluronidase, and lecithinase disrupt host cells by hydrolyzing collagen, the intercellular cement, or by disrupting phosphatidylcholine, a component of the host cytoplasmic membrane. These enzymes rupture the host cell releasing its soluble cellular contents to the pathogen. Coagulase, produced by *Staphylococcus aureus*, promotes the formation of blood clots that surround an infected site and prevent phagocytic cells from reaching and destroying the invader. *Staphylococcus aureus* and *Streptococcus pyogenes* can also synthesize streptokinase that dissolves clots, permitting the pathogen to move farther into tissue.

Virulence, as previously discussed, refers to the relative ability of a pathogen to cause disease. There are several virulence factors, and these include the ability to bind iron, phase variation, and plasmid-encoded factors. A highly virulent organism is one that, given the opportunity, will cause an acute infection. Most virulent microorganisms are highly invasive. However, some pathogens have little invasive ability but have the ability to synthesize potent toxins. *Streptococcus pneumoniae* has considerable invasive ability but does not produce a significant toxin. Yet, before the advent of antibiotics, this organism was a leading cause of death because it physically clogged the lungs of victims. Organisms such as *Clostridium tetani* and *Clostridium botulinum* have virtually no invasive ability but produce highly potent toxins. Introduction of *C. tetani* into a puncture wound can be fatal. Ingestion of botulinum toxin, even in small amounts, can also be fatal.

Table 26.3	Some enzymes produced by pathogenic bacteria that promote invasion of the host	
Enzyme	**Organism**	**Function**
Collagenase	*Clostridia*	Breaks down collagen in connective tissue
Coagulase	*Staphylococcus aureus*	Clot formation around point of entry protects from host defenses
Elastase	*Pseudomonas aeruginosa*	Disrupts membranes
Hyaluronidase	*Streptococcus* *Staphylococcus* *Clostridium*	Hydrolyzes hyaluronic acid–intercellular cement
Lecithinase	*Clostridia*	Disrupts phosphatidylcholine in membranes
Streptokinase	*Staphylococcus* *Streptococcus*	Digests fibrin clots

The virulence of a pathogenic organism has been measured by determining the number of microorganisms of a particular strain that must be injected or given to an animal to kill 50% of the population within a fixed period of time. This is the classical LD_{50} (lethal dose—50% killing).

The LD_{50} test has considerable inherent variability, and federal agencies, including the Food and Drug Administration and the Environmental Protection Agency, no longer consider the classical LD_{50} essential to toxicity testing. Alternatives are being developed.

Iron is essential for the synthesis of cytochromes, the electron carriers that are involved in energy generation. Vertebrates retain iron in their system in a soluble state by binding reduced iron to high-affinity glycoproteins: lactoferrin in milk, tears, saliva, mucus, and intestinal fluids and transferrin in plasma. The level of free iron in body fluids is less than 10^{-8} M. Unless an invading bacterium, all of which require iron, can take the iron from the lactoferrin or transferrin of the host, it cannot be an effective colonizer. For this purpose, many pathogens synthesize low molecular weight siderophores (Figure 5.6) that have a much higher affinity for iron than the host lactoferrin or transferrin. These siderophores appear as receptors on the cytoplasmic membrane of the pathogen and transfer the iron pulled away from the host glycoprotein to the colonizing bacterium. This role of iron can be confirmed by feeding soluble iron to animals infected with selected bacteria and observing a marked increase in the severity of the disease.

Another virulence factor is **phase variation.** Bacterial fimbriae are effective antigens, and host antibodies are often directed at proteins in the fimbriae of invasive pathogens. Fimbriae are prime inducers of antibody, as they are on the surface of the invading microbe. We previously discussed the role of fimbriae in adherence to host mucosa. The chromosomes of some pathogens have a constitutive gene for fimbriae synthesis that is routinely expressed. Other sites on the bacterial chromosome have unexpressed fimbriae genes that code for fimbriae of a different amino acid sequence. Recombination between fimbriae-coding genes can change either a part or the entire gene that directs fimbriae synthesis. This generates redesigned fimbriae that are not affected by antibodies that have been elicited by protein in the original fimbriae. A virulent bacterium with altered fimbriae would therefore have an advantage against host defenses.

Virulence factors may be encoded in the plasmids of pathogens. This permits a rapid passage of genetic information from a limited number of virulent pathogenic microorganisms to a population with lower virulence. The organisms in an infection that have **plasmid-encoded virulence factors**, such as antibiotic resistance, will survive and pass the information to others in the colony, increasing the level of antibiotic resistance in a large proportion of the population. Some plasmid-mediated virulence factors are listed in **Table 26.4.** An infection with a population that had a limited number of organisms with any of these virulence factors would rapidly become a population where almost all would be more highly virulent. Plasmid transfer enables them to generate toxins, additional enzymes, or antibiotic resistance.

Interaction of Pathogens and Phagocytic Cells

A pathogenic organism can gain entrance into a host and establish residence only by overcoming the natural defenses of the host. As mentioned earlier, a major defensive system that invading pathogens encounter are the phagocytes. Pathogens that are effective in invading the host have evolved mechanisms for combating phagocytic cells by producing antiphagocytic factors. Some of these antiphagocytic factors are listed in **Table 26.5. Leukocidins,** produced by pneumococci, streptococci, and staphylococci, attach to the membrane or leukocytes and promote disruption of the lysosomes. This releases the hydrolytic enzymes from the lysosomes that destroy the phagocytic cell. **Hemolysins** act by destroying the plasma membrane of host cells, including leukocytes. A primary function of hemolysins in invasion is to release nutrients from the cell for use by the pathogen.

Capsular material surrounding pneumococci and meningococci protect these organisms from phagocytosis. A pneumococcus without a capsule is not very effective in invasion, as it will be readily captured and destroyed by phagocytes. *Borrelia recurrentis*, the causative agent of relapsing fever, has multiple genes for cell surface proteins. Alterations in cell surface proteins negate the actions of the macrophage/antibody defense enabling *B. recurrentis* to survive in infected individuals for long periods of time with periodic relapses to the diseased state.

Table 26.4	Virulence factors that are generally encoded in plasmids		
Organism		**Factor**	**Disease**
Escherichia coli		Enterotoxin	Diarrhea
Clostridium tetani		Neurotoxin	Tetanus
Staphylococcus aureus		Coagulase enterotoxin	Boils/skin infections, food poisoning
Streptococcus mutans		Dextransucrase	Tooth decay
Agrobacterium tumefaciens		Tumor	Crown gall
Staphylococcus spp.		Antibiotic resistance	Various

Table 26.5	Antiphagocytic factors produced by bacteria and their mode of action
Factor	**Action**
Leukocidins	Specific lytic agent for leukocytes including phagocytes
Hemolysins	Form pores in host cells including macrophages. Streptolysin O affects sterols in membranes. Streptolysin S is a phospholipase
Capsules (glycocalyx)	Long polymers of carbohydrate—physically prevents engulfment
Fimbriae	(1) Bind to surface components of phagocytes, prevent close contact, and phagocytosis may not occur (2) Phase variation—a change in the antigenic composition

Brucella abortus, the causative agent of undulant fever in humans, has cell envelope components that inhibit phagocytic digestion after the bacterium is engulfed. By an unknown mechanism, the organism prevents degranulation (phagolysosome formation). *B. abortus* can multiply in phagosomes, where it is protected from host defenses, and survive in the human body for extended periods.

Legionella pneumophila forms microcolonies inside monocytes and is thereby resistant to normal host defenses. Also, virulent strains of *Mycobacterium tuberculosis* can grow in macrophages, where the intracellular bacteria accumulate to form a nodule. In an active infection, the nodules are surrounded by connective tissue, forming the **tubercle** that is characteristic of tuberculosis. Chest X rays are employed to detect the presence of these tubercles in lungs. *Salmonella typhi,* the causative agent of typhoid fever, survives and multiplies in macrophages. When growing in macrophages, their cells are protected from circulating antibodies.

The rickettsia that cause diseases such as Rocky Mountain spotted fever and typhus are intracellular pathogens. These rickettsia can grow in the cytoplasm or in the nucleus of the host cell. Apparently, the organism growing in cytoplasm can pass through the cytoplasmic membrane without disrupting it, thereby infecting adjacent cells. When growing in the nucleus, the rickettsia are protected from antibodies generated by the host.

TOXINS

Adherence, virulence factors, and a pathogen's defenses against phagocytosis are critical to a microorganism in gaining entry into a human host. The role of these factors has been discussed previously. The presence of bacteria is important in the infectious process, but their mere presence rarely is the cause of disease symptoms. The symptoms of disease are the result of **toxic materials** generated by the pathogen and released at the site of infection. Circulating blood can carry toxins to other areas of the body.

The toxins produced by bacteria are of two basic types:
- Endotoxins
- Exotoxins

The major differences between an endotoxin and an exotoxin are outlined in **Table 26.6**. A discussion of these toxins follows.

Endotoxins

An endotoxin is actually part of the lipopolysaccharide complex that is a component part of the outer envelope of a gram-negative bacterium. Endotoxins are released on lysis of the organism or, in some instances, during cell division. The lipopolysaccharide complex is composed of Lipid A, core polysaccharides, and the O-polysaccharide side chain (see Figure 4.51). The polysaccharide component has little apparent toxicity but is important because it tends to solubilize the complex. Lipid A exhibits virtually all of the toxic properties of the intact lipopolysaccharide complex. Endotoxins all have similar toxic effects on the human host regardless of the microorganism that produces them. They have little toxicity at low levels and are not highly site specific. Endotoxins are rather stable to heat and are weak inducers of an antibody response. Because they do not elicit a protective antibody, it is difficult to establish an effective immunity to endotoxins.

The release of endotoxins into the human body at high levels can induce general acute inflammatory responses that can lead to serious illness and/or death. Endotoxins damage the lining of blood vessels (endo-

Table 26.6	Characteristics of exotoxins and endotoxins
Exotoxins	**Endotoxins**
Heat labile 60°C to 80°C	Heat stable
Immunogenic	Weakly immunogenic
Cause no fever	Cause fever
Can be lethal at low concentrations	Toxic at high doses
Different genera produce different toxins	Similar regardless of source
Released by live bacterium	Released on lysis of bacterium
Inactivated by chemicals that affect proteins	Not generally harmed by chemicals that affect proteins

thelium) leading to the release of interleukin 1 and the Hageman factor. Interleukin 1 is associated with events in the immunological response and will be discussed in Chapter 27. The Hageman factor is a blood-clotting factor (Factor XII) that initiates a blood-clotting cascade through a series of reactions that are "triggered" by a single event. The clotting cascade leads to the activation of thromboplastin, a cross linking of fibrin and the development of blood clots. In the absence of adequate control mechanisms, thrombosis (blockage of blood vessels) can occur, leading to widespread coagulation of blood in the vascular system. The clotting removes platelets from the blood system at rates that exceed their replacement and may cause hemorrhaging elsewhere in the body. This hemorrhaging can lead to the failure of essential organs.

The Hageman factor activation initiates the cascade that leads to fibrinolysis (digestion of fibrin deposits). This may cause further hemorrhaging, with a decrease in blood pressure, depressed respiration, and loss of consciousness. High levels of endotoxin can bring about uncontrolled clotting/fibrinolytic cascades that cause circulatory collapse and potentially death. The Hageman factor also causes activation of the kallikrein system and the full inflammatory response discussed earlier. This response promotes vasodilation and drainage of blood from vessels, further decreasing blood pressure.

An exceedingly sensitive test for trace amounts of endotoxin is the *Limulus* amoebocyte lysate assay. This test employs amoebocytes of the horseshoe crab, *Limulus polyphemus*. An endotoxin, even in trace amounts, reacts with the clot protein from these circulating amoebocytes of *Limulus* and causes clotting. This clotting can be measured spectrophotometrically and is an effective and very sensitive specific assay for endotoxins.

Exotoxins

An exotoxin is a heat-labile toxic protein released into the surrounding medium by a growing microorganism. The exotoxin produced by a species is generally unique, differing in both structure and function from other exotoxins. If the bacterium releases an exotoxin within the human body, it may travel from the focus of infection to other areas of the body. The effects of exotoxins on tissues can be classified by site of action as follows:

- **Enterotoxins**—cause dysentery and other intestinal distress. An example would be the cholera toxin.
- **Neurotoxins**—affect nerve impulse transmission. Tetanus and botulinum toxins are neurotoxins.
- **Cytotoxins**—destroy cells by inhibiting the synthesis of proteins. They may also disrupt or disorder membranes. A cytotoxin can harm the heart, liver, and other organs. Diphtheria is caused by a cytotoxin.
- **Pyrogenic toxins**—stimulate the release of cytokines leading to fever and shock. An example is the exotoxin produced by *Staphylococcus aureus*.

The exotoxins are effective in eliciting an antibody response, and antigens are available to routinely inoculate infants (and adults) against many of the diseases caused by exotoxin-producing microorganisms. Among the diseases that can be prevented by vaccination are diphtheria, tetanus, and whooping cough. A number of exotoxins, the producing organism, and the disease symptoms are outlined in **Table 26.7**, which contains a

Table 26.7	Some exotoxins produced by bacteria		
Exotoxin	**Producing Organism**	**Disease**	**Effect**
Diphtheria toxin	*Corynebacterium diphtheriae*	Diphtheria	Inhibits protein synthesis; affects heart, nerve tissue, liver
Botulism toxin	*Clostridium botulinum*	Botulism	Neurotoxin; flaccid paralysis
Perfringens	*Clostridium perfringens*	Gas gangrene	Hemolysin, collagenase, phospholipase
Erythrogenic toxin	*Streptococcus pyogenes*	Scarlet fever	Capillary destruction
Pyrogenic toxin	*Staphylococcus aureus*	Toxic shock syndrome	Fever, shock
Exfoliative toxin	*Staphylococcus aureus*	Scalded skin	Massive skin peeling
Exotoxin A	*Pseudomonas aeruginosa*	—	Inhibits protein synthesis
Pertussis toxin	*Bordetella pertussis*	Whooping cough	Stimulates adenyl cyclase
Anthrax toxin	*Bacillus anthracis*	Anthrax	Pustules; blood poisoning
Enterotoxin	*Escherichia coli*	Diarrhea	Water and electrolyte loss
Enterotoxin	*Vibrio cholerae*	Cholera	Water and electrolyte loss
Enterotoxin	*Staphylococcus aureus*	"Staph" food poisoning	Diarrhea, nausea
Enterotoxin	*Clostridium perfringens*	Food poisoning	Permeability of intestinal epithelia
Neurotoxin	*Clostridium tetani*	Tetanus	Rigid paralysis

number of diseases commonly encountered among humans. It is apparent from this list that the effects of exotoxins vary widely, from diarrheal diseases to paralysis. The mode of action of specific exotoxins will be discussed in Chapter 27.

Toxins and Fever

Pyrogens are substances that cause an increase in human body temperature, a condition referred to as a **fever,** that we have all experienced. The normal body temperature of humans, 37°C (98.6°F) is controlled by the hypothalamus, a part of the brain. Body temperature generally increases with strenuous exercise and decreases during sleep. A major factor responsible for an increased body temperature during an infection is endotoxin. Some viruses and microorganisms in the body, certain immunological responses, and a limited number of exotoxins can also cause fever. Localized heat may also accompany an inflammatory response.

A temperature increase engendered by gram-negative endotoxins results from the interaction of the toxin with phagocytic cells. Phagocytes release **endogenous pyrogens** during phagocytosis, and these travel to the hypothalamus and interfere with the thermostat that controls body temperature. The factor released from phagocytes that acts as an endogenous pyrogen is the cytokine, interleukin-1. A rise in temperature can occur within 20 minutes of this pyrogen/hypothalamus interaction.

An increase in temperature may actually be a beneficial response, as it can stimulate T cells, increase the rate of phagocytosis, and promote immunological processes in general. There is some evidence that an increase in temperature may also decrease the amount of iron available in the system, thereby interfering with microbial growth. As many pathogens grow optimally at 37°C, an increase of 2°C to 3°C could also curtail microbial growth. "Chills" that occur with a fever are a response by the body to increase muscular activity to drive the temperature higher. However, a temperature much above 41°C (105.8°F) can lead to systemic damage and be fatal.

SUMMARY

▶ Most microorganisms normally present in or on humans have an alliance that is either **mutualistic** or **commensalistic.** In some cases, the microorganism is a **parasite.**

▶ A **parasite** that causes measurable injury to the host is a **pathogen. Virulence** measures the ability of a pathogen to inflict damage. An **infection** occurs when the pathogen grows in the host tissue.

▶ A healthy human has beneficial **natural microbiota** that consist of a myriad of different species.

▶ The **skin** has a natural bacterial population that is mostly **gram-positive bacteria.** The concentration on drier areas is 10^2 to 10^4 per square centimeter and about 10^6 per square centimeter in moist areas. **Natural** skin secretions include amino acids, fatty acids, urea, lactic acids, lipids, and mineral salts. These provide nutrients for the normal microbiota.

▶ Saliva in the mouth contains about 10^8 bacteria per ml. The bacteria **adhere** to the teeth, tongue, and other surfaces and are not dislodged by swallowing.

▶ The **nasopharynx** has adequate moisture and nutrients for growth of bacteria.

▶ The **lower respiratory tract** is mostly devoid of bacteria. Dust and other particles that enter this region are moved upward to the throat by action of **cilia** and swallowed.

▶ The concentration of organisms in the **stomach** is limited to some acid-tolerant lactobacilli and yeasts. There are fewer than **ten microbes** per ml of stomach fluid. Colonization of the stomach by *Helicobacter pylori* can cause peptic ulcers.

▶ The **small intestine** has an increasing number of bacteria as it proceeds downward.

▶ The **large intestine** is a fermentation vat, with 10^{10} to 10^{11} bacterial cells per gram mass and high **species diversity.**

▶ The upper **genitourinary** tract of a healthy human is relatively free of bacteria, but the lower part of the urethra may harbor numerous species of bacteria and even fungi.

▶ The most important **defense barriers** against infection are **skin, membranes,** and **cilia** that ward off invaders. **Acidity** in the stomach and vagina are also effective in protection against pathogens. **Colonization** by nonpathogens can **preempt** potential invaders.

▶ The **inflammatory response** is an important nonspecific defense against entry of pathogens. This response brings **phagocytes** and other protective cells to the site of injury.

▶ **Phagocytes** are body defense cells that **engulf and destroy invaders.** Phagocytic cells have **lysosomes,** compartments containing an array of destructive enzymes. Polymorphonuclear leukocytes and macrophages are important phagocytic cells.

▶ The **lymphocytes** are a nonspecific system that constantly monitors the health of tissue.

▶ **Adherence** to specific tissue is a major aspect of **pathogenesis.** A **pathogen** generally has to adhere and produce **toxic** substances to cause disease.

▶ **Pathogens** can be divided into three general categories: (1) those that live only in and infect a host are **obligate** pathogens, (2) those that can cause infection if given access by trauma are **accidental** pathogens, and (3) those that infect debilitated individuals are **opportunistic** pathogens.

▶ **Virulence** is measured by the number of microorganisms that must be administered to a host population to **kill 50%** in a fixed period of time. This is called an LD_{50} (lethal dose).

▶ Some **pathogens** can survive in **phagocytes,** and this disseminates the pathogen within the host.

▶ **Disease** is generally caused by **toxic substances** produced and not solely by the presence of microorganisms.

▶ The **toxins** produced by microorganisms are **endotoxins** that are part of the bacterial cell and **exotoxins** that are released into the surrounding medium.

REVIEW QUESTIONS

1. How do virulence and pathogenicity differ? How would one measure virulence?

2. Define: host, commensalism, infection, and normal microbiota.

3. What are some factors on human skin that promote growth of bacterial inhabitants? Some that inhibit or limit bacterial colonization?

4. Why is brushing one's teeth important in preventing dental caries? How do microorganisms contribute to caries?

5. How do the bacterial populations of the stomach and large intestine differ qualitatively and quantitatively? What role do the facultative aerobes play in the human gastrointestinal tract?

6. What is the source of the acidic pH in the genitourinary tract of females? How does the bacterial flora in this area prevent invasion by pathogens?

7. Is inflammation a positive or a negative in defending against infection? Give reasons for your answer.

8. What are phagocytes? How do PMNs differ from macrophages? What does a high PMN count in human blood signify?

9. The lymphatic system is important in removing bacteria and other debris from tissue. How does this occur? What is the role of the lymphatic system? How does a humoral response differ from a cellular response? How do these relate to the lymphatic system?

10. How does an obligate pathogen differ from an accidental pathogen? What is an opportunist? Cite examples.

11. Adherence is important in infection. Why? Cite diseases where adherence is not essential.

12. What are some of the virulence factors, and how do they function in promoting the survival of pathogens? What are some antiphagocytic factors?

13. How do endotoxin and exotoxins differ chemically, in toxicity, heat stability, and origin?

14. How do endotoxins and exotoxins differ in their mode of action? What are the reasons for this? How would you determine whether an intravenous fluid might contain endotoxin?

SUGGESTED READING

Cossart, P., P. Boquet, S. Normark and R. R. Appuoli. 2000. *Cellular Microbiology*. Washington, DC: American Society for Microbiology.

Groisman, E. A. 2000. *Principles of Bacterial Pathogenesis*. San Diego: Academic Press.

Kreier, J. P. and R. F. Mortensen. 1990. *Infection, Resistance and Immunity*. New York: Harper & Row.

Minion, F. C. and M. J. Wannemuehler. 2000. *Virulence Mechanisms of Bacterial Pathogens*. 3rd ed. Washington, DC: American Society for Microbiology.

Salyers, A. A. and D. D. Whitt. 1994. *Bacterial Pathogenesis*. Washington, DC: American Society for Microbiology Press.

Immunology and Medical Microbiology

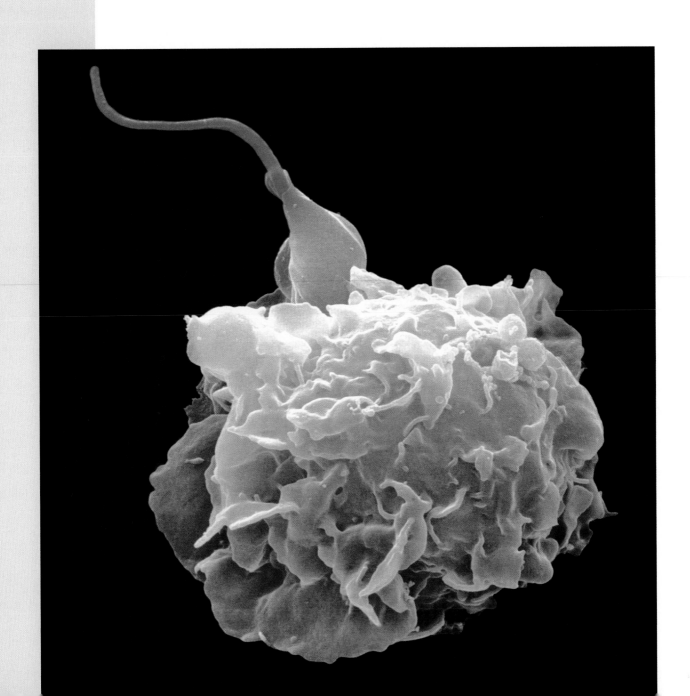

Previous page
A macrophage engulfs the protist that causes leishmaniasis.
©Jürgen Berger, Max-Planck Institute/SPL/Photo Researchers, Inc.

Chapter 27

Immunology

"Complex...Of course the immune system is complex. If it was simple, every Tom, Dick and Harry of a microbe would overwhelm it."

—ANONYMOUS PROFESSOR RESPONDING TO A DISGRUNTLED STUDENT

The living body of mammals is warm, moist, and full of nutrients. Thus, animal cells and tissues provide an excellent living environment for many viruses, bacteria, fungi, protozoa, and parasitic worms. Consequently, the survival of an animal depends on the successful defense of the body against invasion by these agents. Healthy animals resist invasion using an array of defense mechanisms that are encompassed within the discipline of immunology, the subject of this chapter. The immune system is described here in the context of responses to infections, with cells and molecules of the immune system introduced as they would be induced during an infection and encountered by an infectious disease agent. It is the goal of this chapter to provide a comprehensive introduction to the immune system and immune responses to infection.

The essential nature of defense of the body is exclusion of disease-causing agents (pathogens) from host tissues or their control and eventual elimination. Several layers of sequential, overlapping defense mechanisms accomplish this. In the order encountered by incoming microbes, these are:

- Barrier defense or immunity (excludes infectious agents)
- Inflammation and innate immunity (limits the replication of breakthrough microorganisms by relatively nonspecific responses)
- Adaptive immunity (eliminates invaders by highly pathogen-specific mechanisms that are more efficient on second exposure to the same pathogen and hence are considered to have immunological memory)

Each of these systems has an important role to play in host defense, because in the absence of any one of them, the person or animal will likely be more susceptible to infection. This is illustrated by the susceptibility of burn patients to infection as a result of a breach of barrier immunity. The susceptibility of people and animals with a defect in the ability of cells of the innate immune system to adhere to blood vessels and enter sites of inflammation and infection develope recurrent infections. People and animals with genetic defects in the adaptive immune system cannot survive for prolonged periods outside of sterile environments.

BARRIER IMMUNITY

Infectious agents are first encountered at body surfaces, where they are destroyed or repelled (**Figure 27.1**) by a combination of barrier immune mechanisms that can be grouped as physical, mechanical, chemical, and microbial. These are the first layer of defense against infection by bacteria, viruses, fungi, protozoa, and parasitic worms, since the protective systems at body surfaces

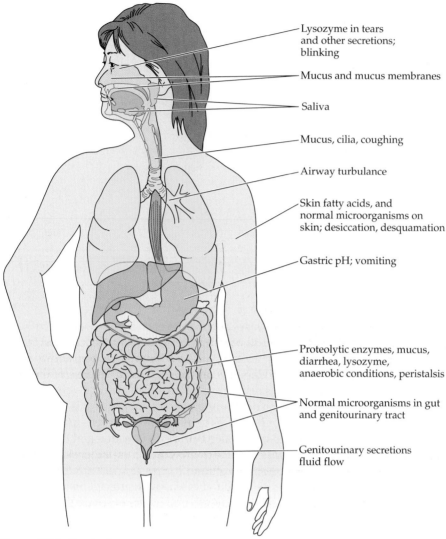

Lysozyme in tears
and other secretions;
blinking

Mucus and mucus membranes

Saliva

Mucus, cilia, coughing

Airway turbulance

Skin fatty acids, and
normal microorganisms on
skin; desiccation, desquamation

Gastric pH; vomiting

Proteolytic enzymes, mucus,
diarrhea, lysozyme,
anaerobic conditions, peristalsis

Normal microorganisms in gut
and genitourinary tract

Genitourinary secretions
fluid flow

Figure 27.1 Barrier immune system
The barrier immune system has physical (e.g., skin and mucus membranes), chemical (e.g., low pH, fatty acids, lysozyme), mechanical (e.g., peristalsis, coughing, blinking), and microbial (e.g., normal biota of skin and gut) components.

establish local environmental conditions suitable for only the most adapted microorganisms.

Physical Barriers

The skin and mucus membranes that line the mouth, nasal passages and upper respiratory tract, and gut are the obvious physical barriers to infectious diseases. When these barriers are burned, lacerated, or punctured, the risk of infection increases significantly. A familiar example is a cut on the skin, but infection can also occur when the intestine is punctured or torn, in which case bacteria can leak into the peritoneal cavity, causing peritonitis. Other physical barriers include saliva and mucus, which are present on several epithelial surfaces, including those lining the respiratory tract and the gas-

trointestinal tract. The viscous nature of mucus makes it more difficult for microbes to attach to host cells.

Mechanical Barriers

Particles that enter the respiratory tract are first removed by turbulence that directs them onto its mucus-covered walls, where they adhere. The turbulence is created by the shape of the turbinate bones, the trachea, and the bronchi. This turbulence filter removes particles before they reach the alveoli. The trapped particles are moved back toward the mouth by ciliated cells on what is called the **mucociliary escalator**. Swallowing clears microbes that have entered by the mouth or nasal passages and sends them to the more hostile environment of the stomach. Once swallowed, peristalsis in the esophagus and intestinal tract continues the journey for the ill-fated microbes. Similarly, microbes that enter the body through the urethra can be removed by the cleansing action of urination, and blinking and tears are mechanical mechanisms that clear the eyes. Many of these actions may occur without the notice of the individual while other mechanical mechanisms are more energetic or violent, such as coughing and sneezing, vomiting, and diarrhea.

Chemical Barriers

Low pH is a familiar manifestation of chemical barrier. Microbes that are swallowed with food or routed to the stomach by the mucociliary escalator following inhalation are killed by the low pH. The pH in the intestine is kept low and anaerobic by the resident bacterial flora and the pH of urine is also low. Resident microbes on the skin produce lactic acid, and sebaceous glands release oils and waxes that become organic acids, which together lower the pH of the skin. Lactic acid-producing bacteria in the vagina keep the pH low, at around 5.0. In addition to low pH, the antibacterial enzyme **lysozyme** is found in tears and saliva and is synthesized by the gastric mucosa. Its role is to break down the cell envelope of bacteria by disrupting the peptidoglycan (muramyl dipeptide) layer through cleavage of the *N*-acetylglucosamine and *N*-acetylmuramic acid bonds.

Microbial Barriers

The skin and mucosal membranes are populated by bacteria that are well adapted to survival there. These resident bacteria are unlikely to cause disease when confined to these primary niches and prevent colonization by other organisms that are not adapted to the environment and are potentially pathogenic. Resident flora also play an important role in defense of the digestive tract by taking up space and nutrients. Their importance is illustrated by the overgrowth of pathogens when normal microorganisms are either eliminated or altered by aggressive antibiotic treatment, after which pathogenic organisms can overgrow. The normal microbes also decrease the pH of some barriers as discussed above.

How Infectious Agents Overcome Barriers

Successful pathogens use sophisticated mechanisms to colonize individuals that do not have a preexisting breach to barrier immunity. Some infectious microbes are spread by insects and arthropods when they bite the host or defecate on the host's skin. For example, the protozoan that causes malaria (*Plasmodium*) is introduced in saliva from an infected mosquito when it takes a blood meal, whereas the protozoan that causes Chagas' disease, *Trypanosoma cruzi*, is spread via the feces of kissing bugs and enters the host through scratches and when the host rubs the feces into the mucosa of the eye. West Nile virus is also spread by mosquitos, whereas the rickettsial bacteria that causes Rocky Mountain spotted fever and the bacteria that causes Lyme disease (*Borrelia burgdorferi*) are spread by ticks. Parasitic worms hook into the musculature of the intestinal tract so they are not swept away by peristalsis. Other parasitic worms tunnel through the unbroken skin. For example, the infective larvae of *Schistosoma* secrete enzymes to digest the skin. Once through the skin, they can migrate via the blood to the lungs.

INFLAMMATION AND INNATE IMMUNITY

Any process that breaches barrier immunity and causes tissue damage (a burn, traumatic injury, surgery, the bite of a blood-sucking fly, local infection, or the passage of a parasitic worm through tissues) evokes **inflammation**, which is the focus of this section. Inflammation is a general term that describes the biochemical and cellular responses that result from tissue damage. The inflammatory response functions to:

- Seal lacerations and plug broken blood vessels

- Remove damaged cells from sites of tissue injury and repair damaged tissue
- Restrain the population growth rate of pathogens that may have invaded or been introduced at the site of tissue damage by killing them or creating conditions that limit their replication

These diverse functions are mediated by a combination of molecules that are present in normal **plasma** and to a lesser extent in tissue fluids, and by white blood cells (leukocytes) that circulate in the blood and move to the site of inflammation or that reside in the tissues and are activated by inflammatory events. Two important leukocytes involved in the inflammatory response are **neutrophils** and **monocytes** (also called **macrophages** when in tissues). These cells are part of the **innate immune system**. Both cells are **phagocytes**, which means that they can engulf cells and other particles, such as bacteria and viral particles, and subsequently destroy them. Because the molecules and cells that participate in the inflammatory and innate immune response are typically present in sufficient concentration or number (albeit in an inactive state) in the blood and tissues, the inflammatory response can be rapidly deployed through a series of interdependent steps following tissue injury.

Early Events in Inflammation

Inflammation of the skin causes redness, heat, swelling, and pain (**Figure 27.2**) and is detected within minutes after injuring that tissue. Redness and heat result from increased local blood flow caused by an increase in the diameter of blood vessels (**vasodilation**). Swelling results from local **edema** brought on by the leakage of plasma from the bloodstream into tissues at the site of injury and subsequently the infiltration of cells into this region. Pain results from tissue destruction by enzymes released by infiltrating phagocytes and from a reduced threshold of activation of nerve endings that results from local production of a fatty acid breakdown product called **prostaglandin E$_2$**. This is made by damaged cells and by infiltrating phagocytes. Pain focuses the attention of the afflicted individual on the site of inflammation. Wounded animals lick and cleanse damaged areas. Humans use additional strategies to cleanse wounds and expedite healing. Pain, therefore, plays a significant role in the control of infection.

Vasodilation results from relaxation of smooth muscle of the blood vessel walls. This is brought on by the interaction of the muscle cells with a number of molecules produced at the site of inflammation. These are called bradykinin, fibrinopeptides, complement factor C5a, and histamine. Generation of these vasoactive agents is initiated by the shearing of cells from their extracellular matrix at the site of tissue damage and the concomitant rupture of local capillary beds, which

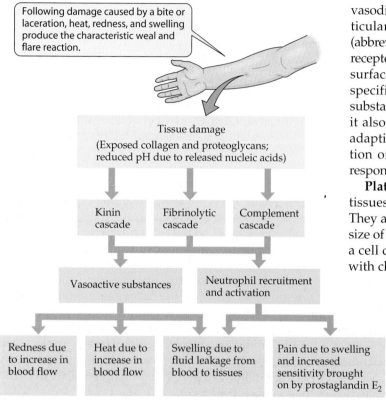

Following damage caused by a bite or laceration, heat, redness, and swelling produce the characteristic weal and flare reaction.

Tissue damage
(Exposed collagen and proteoglycans; reduced pH due to released nucleic acids)

Kinin cascade

Fibrinolytic cascade

Complement cascade

Vasoactive substances

Neutrophil recruitment and activation

Redness due to increase in blood flow

Heat due to increase in blood flow

Swelling due to fluid leakage from blood to tissues

Pain due to swelling and increased sensitivity brought on by prostaglandin E_2

Figure 27.2 Inflammation
The characteristics of the inflammatory response are heat, swelling, and redness resulting from several biochemical cascades that occur following tissue damage.

results in the influx of plasma and cells. Bradykinin is generated by the kinin system, an enzyme cascade that begins with a plasma clotting factor called Hageman factor. Hageman factor is activated upon contact with collagen exposed at sites of tissue damage. In its active form it is called Factor XIIa. Activation of the blood clotting cascade by Factor XIIa results in the generation of thrombin, which converts plasma fibrinogen to fibrin. This creates a mesh that promotes clot formation and seals breaches in the barrier, allowing repair to begin. Immediately following, some of the clot is dissolved by plasmin, which is a potent enzyme generated by the fibrinolytic system. This generates the fibrinopeptides.

Mast cells, another type of leukocyte in the innate immune system, also play an important role in inflammation. Plasmin acts on a plasma protein called complement factor 5 (C5), breaking it to yield complement factors C5a and C5b. C5a is one of a group of complement fragments called **anaphylatoxins**, small molecules that promote inflammation. (Other anaphylatoxins are complement fragments C3a and C4a.) The C5a fragment binds to receptors on mast cells. The mast cell then releases histamine, which it stored in its granules (i.e., it degranulates). The released histamine causes local

vasodilation. Mast cells also have a receptor for a particular type of antibody called immunoglobulin E (abbreviated IgE) (discussed on pg. 665). Because of the receptor, the mast cells become coated with IgE. When surface-bound IgE on mast cells is cross-linked by a specific **antigen** (a molecule derived from a foreign substance that elicits an adaptive immune response), it also causes release of histamine. In this way the adaptive immune response (which includes production of antibodies) is tied into the innate immune response.

Platelets have a central role in repair of damaged tissues. These small particles circulate in the blood. They are saucer-shaped disks about one-fifteenth the size of a red blood cell, generated by fragmentation of a cell called a megakaryocyte. Platelets are riddled with channels and filled with granules. They adhere to thrombin that forms at the site of damaged blood vessels and spread to form a plug that rapidly seals torn blood vessels. When activated, they contract vigorously and release these granules, which contain many biologically active molecules, notably:

- Platelet activating factor, which recruits more platelets
- Fibroblast growth factor, which facilitates repair of damaged ground substance by inducing fibroblasts to replicate and to produce collagen
- Platelet-derived growth factor, which helps recruit neutrophils and macrophages
- Serotonin, which can cause vasodilation or vasoconstriction, depending on the receptors expressed on the target blood vessels, and also activates neutrophils, macrophages, and fibroblasts
- Decay accelerating factor (DAF) and factor H, both of which limit activation of complement (a major source of pro-inflammatory agents)

In addition to causing vasodilation, bradykinin, fibrinolytic peptides, C5a, and histamine react with receptors on endothelial cells that line blood vessels, causing these cells to swell and contract. This opens gaps between the endothelial cells, allowing plasma fluids, including large molecules, to leak into surrounding tissues and to pool there, causing edema. So long as collagen is exposed at the site of plasma leakage, vasodilation, edema, clot formation, and tissue repair will proceed. To arrest inflammation, damaged tissues must be repaired. This requires both removal of the dead cells and their replacement. Phagocytes remove damaged tissue in addition to infectious agents. Recruitment and activation of the phagocytic cells for both purposes involves their interaction with the complement fragments C3a and C5a, discussed next.

Activating Complement

The key player in the inflammatory process is the **alternative pathway** of **complement** activation. The complement system consists of some 20 interacting plasma proteins that:

- Kill foreign cells by rupturing their plasma membranes
- Stimulate phagocytosis (act as an opsonin)
- Cause inflammation that isolates infectious agents and repairs tissue damage
- Attracts phagocytes (act as an anaphylotoxin)

The complement proteins are activated by two pathways, the classical pathway, which is discussed later in the chapter, and the alternative pathway discussed here. The classical pathway depends on the formation of antibody-antigen complexes during an adaptive immune response, while the alternative pathway is independent of antibodies and therefore occurs during inflammation and infection without involvement of the adaptive immune system. **Table 27.1** provides a partial list of complement fragments and their main biological activities.

Complement factor 3 (C3) is a key player in the alternative pathway. C3 is a protein that is produced by macrophages and constitutively present in plasma. Plasma C3 is subject to slow breakdown, yielding the fragments C3a and C3b. This goes on whether or not there is inflammation, and the concentrations of C3a and C3b generated are low and cause no adverse physiological response. However, the ongoing supply of C3b allows for the massive amplification of C3b and C3a, as well as other biologically active complement fragments, in the presence of many pathogens.

C3b that is generated in plasma has two fates. It can be broken down in a series of steps by factor H, or complement receptor 1 (CR1) together with factor I, or it can be converted to an enzyme complex, called the **alternative pathway C3 convertase**, which converts C3 to C3a

and C3b in the presence of pathogens that facilitate this reaction (**Figure 27.3A**). These include bacteria, yeast, some protozoa, and parasitic worms. Generation of the alternative pathway C3 convertase occurs at a site on pathogens that favors the binding of factor B to C3b over binding of factor H to C3b. In a sporting gesture of evolution, the co-selection of hosts and pathogens has equipped pathogens with such sites and masked them on host cells. In addition, host cells are endowed with a molecule that both inhibits formation of, and promotes disassembly of the alternative pathway C3 convertase. Selection through evolution has also equipped some pathogens with this capability. The human pathogen *Trypanosoma cruzi*, a protozoan parasite that causes South American trypanosomiasis (Chagas' disease), has a developmentally regulated complement inhibitor that has genetic and functional similarities to human DAF.

Although some pathogens inactivate complement, many do not and instead promote formation of the alternative pathway C3 convertase. The pathogens therefore increase the generation of products of the alternative pathway of complement activation, namely, C3b, C3a, C5a, and the **membrane attack complex**, $C5b678(9)_n$. The membrane attack complex is a hole-forming donut-shaped molecule that inserts into lipid membranes, lyses affected organisms and cells, and damages viral particles. Because the membrane attack complex forms where the alternative C3 convertase is formed, it damages pathogens and not host cells. The soluble complement fragments that are formed, namely C3a and C5a, have several pro-inflammatory properties. For example, by binding to receptors on endothelial cells, C5a increases vascular permeability and leakage of plasma proteins into the inflamed tissue as discussed above.

Recruiting Phagocytes to Sites of Infection

As long as pathogens that activate the alternative pathway of complement continue to replicate unchecked at

Table 27.1	Complement fragments and their biological activities					
	Activity					
Fragment	**Vasodilation**	**Recruitment: Neutrophils, Monocytes**	**Activation: Neutrophils, Monocytes**	**Opsonization**	**B Cell Activation**	**Membrane Attack Complex**
C3a	+	+	+	−	−	−
C4a	−	+	+	−	−	−
C5a	+	+	+	−	−	−
C3b	−	−	−	+	−	−
iC3b	−	−	−	+	−	−
C3dg	−	−	−	−	+	−
C3d	−	−	−	−	+	−
$C5b678(9)_n$	−	−	−	−	−	+

(A) **Alternative complement cascade**

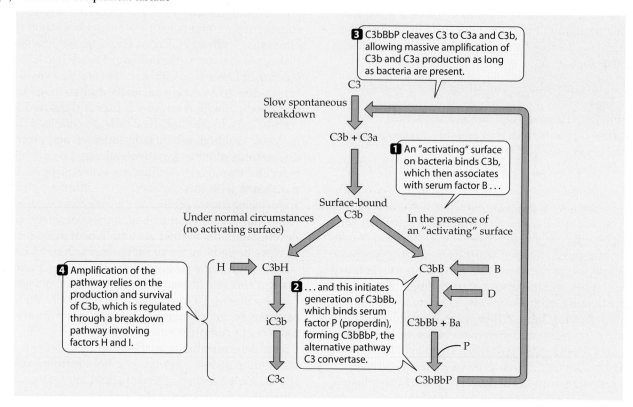

(B) **Classical complement cascade**

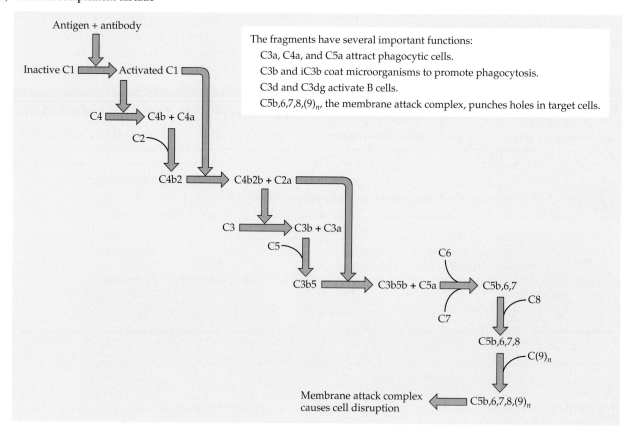

◄ Figure 27.3 Complement pathways
There are two pathways to activate the complement cascade: the classical and the alternative. (A) The alternative pathway is initiated because complement component C3 breaks down at a slow rate in plasma to C3a and C3b and bacteria have a conserved site that can bind C3b. This initiates a pathway to C3bBbP, the alternative pathway C3 convertase, which amplifies C3b and C3a production. The alternative C3 convertase also drives cleavage of C5 leading to formation of the membrane attack complex. (B) In the classical cascade once an antibody has bound antigen the antibody molecule undergoes a conformational change that allows it to bind to C1q, the first component of the complement cascade. This recruits C1r and C1s and initiates a series of enzymatic reactions involving several complement components. At each step, an activated component cleaves and activates the next. Fragments formed by these cleavages have several important functions.

the site of tissue injury, products of the alternative pathway of complement activation will increase exponentially. However, the alternative pathway of complement activation also creates conditions that limit the capacity of pathogens to replicate through the recruitment of phagocytes cells that engulf, kill, and degrade the pathogens (Table 27.1).

Recruitment of the phagocytes involves two distinct processes. First, circulating phagocytes adhere to the walls of inflamed blood vessels as a result of binding to a group of molecule, that become expressed on vascular endothelial cells within minutes after exposure to C5a and histamine. Second, the phagocytes migrate between endothelial cells, due to attraction by complement factors.

The neutrophils are normally swept along in the center of the stream of blood flowing through blood vessels but **marginate** (move to the outside of the stream) in dilated blood vessels, where blood flow becomes sluggish because of the increased internal diameter of the affected vessel. Selectins induced by the action of vasoactive agents on endothelial cells at the site of inflammation bind specific carbohydrate groups on blood neutrophils. At first they do not make a firm attachment to the endothelial cells and are said to "roll" along the vessel membrane. As the phagocytes roll slowly along the region of the affected vessel wall, they are exposed to platelet activating factor, which causes them to express molecules called integrins. Expression of integrins on neutrophils causes these cells to bind tightly to the endothelium and cease rolling. Neutrophils that bind to altered vascular endothelium on blood vessels at inflammatory sites have receptors for C3a and C5a. Binding of these complement fragments to their receptors on the phagocytes induces changes in the cells that

allow them to traverse the blood vessel wall. This process is called **diapedesis** and is diagrammed in **Figure 27.4**. Once the phagocytes have squeezed between endothelial cells, they release protein-digesting enzymes that break down the extracellular matrix/basement membrane and escape into the tissue fluids. Neutrophils have a life span of only about 24 hours once they have moved from the bone marrow into the blood, whereas monocytes live longer. The dying neutrophils give rise to much of the composition of **pus**.

Monocytes are recruited to the inflammatory site about 24 hours later than neutrophils. Their attachment to inflamed blood vessels is by different molecules than those used by neutrophils. Monocytes (also known as macrophages) that are recruited to the inflammatory site produce molecules called cytokines that increase inflammation and thus are referred to as pro-inflammatory cytokines. They also produce two cytokines called **interferon** (IFN)-α and IFN-β. These cytokines have the ability to inhibit virus replication. When these cytokines bind to receptors on the virus-infected cell, a series of

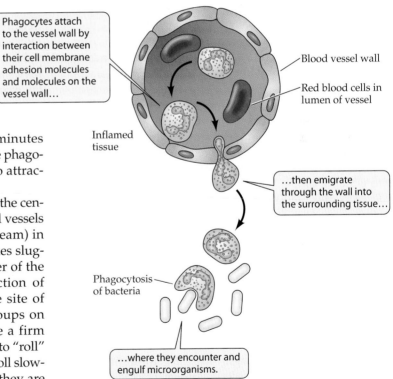

Phagocytes attach to the vessel wall by interaction between their cell membrane adhesion molecules and molecules on the vessel wall…

Blood vessel wall

Red blood cells in lumen of vessel

Inflamed tissue

…then emigrate through the wall into the surrounding tissue…

Phagocytosis of bacteria

…where they encounter and engulf microorganisms.

Figure 27.4 Recruitment of phagocytes from the blood
Phagocytic cells—neutrophils and macrophages—are attracted to the site of inflammation, where they may encounter and kill microorganisms. The inflammatory response results in an increased expression on local blood vessels of adhesion molecules that bind phagocytes, allowing these cells to pass into the surrounding tissue in a process known as diapedesis. At the site of the inflammatory response, the phagocytes are likely to encounter microorganisms.

events are initiated that ultimately result in activation of a ribonuclease that degrades viral RNA.

Neutrophils and macrophages engulf and destroy pathogens, but they can also release microbicidal compounds. As neutrophils and macrophages migrate toward areas of increasingly high concentration of soluble complement fragments, the binding of C5a and C3a activates these cells for enhanced phagocytic activity and enhanced microbicidal activity. At high concentration, phagocytes release microbicidal and tissue-destructive materials into the extracellular fluids. Although this process limits pathogen replication, it does so at the expense of collateral damage to host tissue, which, if unresolved, will lead to ulceration.

Molecular Tags Facilitate Phagocytosis

Complement factor C3b tags cells for **phagocytosis** by neutrophils and macrophages, both of which express a receptor, **complement receptor 1 (CR1)**, for this fragment. The product of C3b cleavage is iC3b, which remains attached covalently to the surface on which it forms. The iC3b binds to **complement receptor 3 (CR3)** on neutrophils and macrophages. Engagement of CR1 or CR3 is one of the signals needed for the phagocytic process to proceed. Molecules that tag cells and other molecules for phagocytosis are called **opsonins**. Other opsonins are C reactive protein and mannose binding protein, produced by liver cells during acute inflammation.

Antibodies of the immunoglobulin G (IgG) class (discussed later) are also opsonins. These antibodies bind to macrophages and neutrophils when complexed with their target antigen. The receptor they bind to is called an **Fcγ receptor**, after the region of the antibody to which it binds. This binding of the antibody to the Fcγ receptor facilitates phagocytosis. Other antibody classes, for example, IgM, initiate the classical pathway of complement activation (discussed on pg. 664) and cause C3b and iC3b to be deposited on pathogens and hence also facilitate phagocytosis of the pathogens. However, IgM is not referred to as an opsonin because it does not directly opsonize target cells and molecules.

Cytokine Signals Cause Wasting, Fever, and the Acute Phase Response

When of adequate magnitude, the inflammatory response has systemic effects, all of which act to control infectious agents, but which may also have an adverse effect on the host. Many of these effects are caused by cytokines, which are soluble proteins released by activated leukocytes. Cytokines direct the function of other cells bearing specific receptors for them. The cytokines released by activated macrophages include interleukin 1 (IL-1), IL-6, and tumor necrosis factor-α (TNF-α). TNF-α is responsible for the dramatic loss in body fat

that occurs during chronic infections. It binds to receptors on fat cells, preventing production of an enzyme, lipoprotein lipase, that is required for the restoration of fat reserves from dietary lipoproteins. In the absence of lipoprotein lipase, the fat cells lose their lipid reserves. This form of wasting is called cachexia.

IL-1, IL-6, and TNF-α induce hepatocytes in the liver to mount an **acute phase response**, which results in the production of C reactive protein, which binds to phosphorylcholine on bacteria and promotes their phagocytosis. The acute phase response also results in the secretion of haptoglobin into plasma. Haptoglobin is a hemoglobin-binding protein that limits loss of iron from the body in the event of hemolytic anemia and also limits its availability for invading microorganisms. The acute phase response also results in production of mannan-binding protein, which attaches to mannose residues on the surface of many bacteria facilitating their phagocytosis. Like C3b, iC3b, C reactive protein, and IgG, mannan-binding protein is an opsonin.

Mannan-binding protein adheres to mannose and N-acetylglucosamine on microorganisms. It does not react with normal host cells because on these cells mannose and N-acetylglucosamine are covered by other carbohydrates. Mannan-binding protein that has bound to microorganisms acquires the capacity to bind and activate mannan-binding protein-associated serine proteases and activates the complement cascade in a similar manner to the classical pathway of complement activation discussed on pg. 664. Anaphylatoxins and the membrane attack complex are generated, and target cells are coated with C3b and iC3b, which facilitates their phagocytosis. Thus, generation of mannan-binding protein during the acute phase response amplifies complement activation and the clearance of pathogens with exposed target sugars from the blood. Mannan-binding protein is implicated in the efficient killing of several microorganisms, including *Neisseria meningitides*. Deficiencies in mannan-binding protein are associated with chronic diarrhea in children and may enhance susceptibility to cryptosporidiosis in patients with acquired immunodeficiency syndrome.

Interleukin 1 and prostaglandin E_2 react with receptors expressed on cells in the brain's thermoregulatory center, resulting in vasoconstriction and shivering, which conserves and generates heat, leading to an elevation in body temperature (i.e., fever). The fever response is detrimental to growth of pathogens, as it creates conditions that are not optimal for function of cellular enzymes that regulate the cell division cycle. The goal here is to kill the pathogen before killing the host, although this is not always achieved.

When the site of tissue injury is not infected, the inflammatory response is self-limiting. In the event that a pathogen is introduced at the site of tissue injury, the inflammatory response will typically continue to amplify until a specific immune response develops that clears the pathogen, allowing conditions that provoke the inflammatory response to subside. When the specific immune response that develops is adequate to limit pathogen growth but not to cause elimination, chronic inflammation can ensue. This can also occur when a pathogen induces a local self-sustaining autoimmune reaction and when inflammation is induced by a substance that is not biodegradable (e.g., asbestos fibers). When infections are not controlled, inflammatory responses continue to amplify until, racked with fever and wasted by cachexia, the afflicted individual dies.

Phagocytosis and Killing of Infectious Agents

Macrophages and monocytes are phagocytes, meaning they can engulf particulate matter, including infectious agents (e.g., bacteria, protozoa, viruses, and yeast). They do this by adhering to the particle by any of a number of different cell surface receptors on the phagocyte, and counter-receptors, or ligands, on the infectious agent, some of which are mentioned in preceding sections. These include components of the complement system that have been deposited in the surface of microbes during the inflammatory response. The plasma membrane of the phagocytic cell then wraps around the particle by increasing the numbers of receptor-ligand interactions, thereby engulfing it. Alternatively, the phagocyte sends out pseudopodia that engulf the particle after the initial interaction of phagocyte and particle. The membrane-enclosed vesicle formed during phagocytosis is known as a **phagosome (Figure 27.5)**.

Phagocytosis is accompanied by several events that create a micobicidal environment. The first of these is the assembly of the enzyme NADPH oxidase that catalyzes the **respiratory burst** coincident with phagocytosis. The respiratory burst refers to the generation of superoxide anion and hydrogen peroxide. If iron or copper is present in the phagosome, they will mediate the subsequent generation of hydroxyl radicals from the superoxide anion and hydrogen peroxide. These reactive oxygen intermediates are antimicrobial, particularly the hydroxyl radicals, and may damage or kill microorganisms. Following formation of the phagosome, the pH of the phagosome is slowly decreased from neutral to pH 3 to 4. This low pH is detrimental to the survival of many microorganisms. Intracellular bags of enzymes, known as **lysosomes** and granules, also fuse with the phagosome, depositing enzymes that degrade lipids, proteins, and carbohydrates, as well as lysozyme that damages the cell wall of bacteria, as discussed in Barrier Defense. In some phagocytes, particularly in neutrophils and newly recruited macrophages, the lysosomes contain the enzyme myeloperoxidase that catalyzes the generation of hypohalides such as hypochloride (commonly known

as chlorine bleach, which is a very effective disinfectant) from hydrogen peroxide. Other components of lysosomes and granules of neutrophils include defensins and lytic peptides, which circularize and insert themselves into bacterial cell walls, resulting in bacterial killing.

Tissue Macrophages Originate from Monocytes

Monocytes may undergo an additional round of replication in the blood once they have left the bone marrow where they arise from stem cells. They then migrate by diapedesis into tissues where they mature. These resident macrophages of tissues, called by various names such as Kupffer cell and microglial cell, depending upon the tissue (**Table 27.2**), may be there for many years. They play a similar role as the monocytes that are recruited from blood as a result of inflammation, except that they are permanently available in organs and tissues, usually in positions at entry point for pathogens.

Surviving Phagocytosis—Intracellular Microbes

Some protozoa, bacteria, and fungi can survive inside monocytes or macrophages following phagocytosis. They are referred to as **intracellular organisms** because of this ability. This does not occur in neutrophils because of the very short life span of those cells. The survival mechanisms of intracellular organisms include expression of genes that allow resistance to the low pH, oxidative conditions, and the proteolytic environment of the phagolysosome. Protozoan parasites of the *Leishmania* sp. have a surface lipophophoglycan that has all of these properties, and to add insult to injury, they use C3b and iC3b to enter their new home. Other pathogens remodel the phagosome to suit their needs, prevent phagolysosomal fusion, or escape from the phagosome into the cytoplasm of the cell. Some of these are listed in **Table 27.3**. To combat these microbes, the immune system must rely on additional cells that produce a cytokine called interferon-γ (IFN-γ), which acts on macrophages to elevate their killing mechanisms. IFN-γ is secreted by many T cell subpopulations of the adap-

Table 27.2	Macrophage names
Name of Macrophage	**Tissue or Organ Where Found**
Monocyte	Blood
Kupffer cell	Liver
Microglial cell	Brain
Splenic macrophages	Spleen
Alveolar macrophage	Lung
Peritoneal macrophage	Peritoneum
Bone marrow macrophage	Bone marrow

(A) Cells of innate immune system

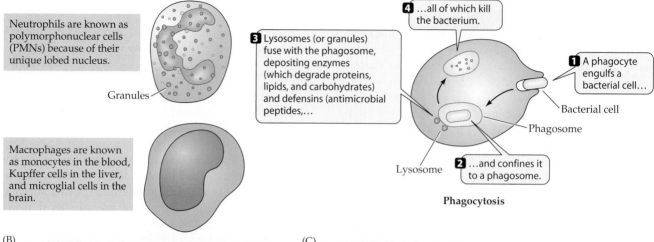

Neutrophils are known as polymorphonuclear cells (PMNs) because of their unique lobed nucleus.

Granules

Macrophages are known as monocytes in the blood, Kupffer cells in the liver, and microglial cells in the brain.

3 Lysosomes (or granules) fuse with the phagosome, depositing enzymes (which degrade proteins, lipids, and carbohydrates) and defensins (antimicrobial peptides,…

4 …all of which kill the bacterium.

1 A phagocyte engulfs a bacterial cell…

Bacterial cell

Phagosome

Lysosome

2 …and confines it to a phagosome.

Phagocytosis

(B)

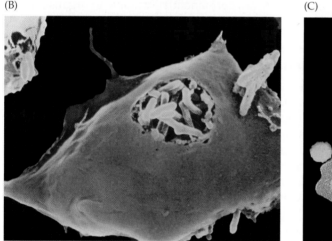

(C)

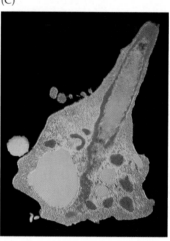

Figure 27.5 Phagocytosis
Neutrophils and macrophages are phagocytic cells that engulf microorganisms and confine them to an intracellular compartment known as a phagosome. The antimicrobial reactive oxygen intermediates produced in the phagosome include hydrogen peroxide, hypohalides, hydroxyl radicals, and nitric oxide. The process is diagrammed in (A) and shown by a scanning electron micrograph in (B) and a transmission electron micrograph in (C). The scanning electron micrograph depicts a macrophage (red) phagocytosing a bacterium, *Mycobacteria tuberculosis* (yellow). The transmission electron micrograph depicts a bacterium (green) that has been phagocytosed by a neutrophil (nuclear lobes in blue). B, Photo ©S. H. E. Kaufmann and J. R. Golecki/SPL/Photo Researchers, Inc.; C, photo ©K. Lounatmaa/SPL/Photo Researchers, Inc.

tive and bridging immune systems and is described in those sections. Although IFN-γ is by far the most powerful macrophage-activating cytokine, other cytokines also cause macrophage-activation. These include tumor necrosis factor α (TNF-α), which is produced by the macrophages themselves after they contact microbes, and a related molecule lymphotoxin (TNF-β) that is produced by subpopulations of lymphocytes.

IFN-γ, and TNF-activated macrophages have increased antimicrobial properties as a result of increased generation of reactive oxygen intermediates and production of

antimicrobial nitric oxide. Nitric oxide is particularly effective against protozoan parasites and diffuses from the macrophage, as well as acting inside it. It is, however, very short lived in blood, where it rapidly reacts with hemoglobin. In addition, activated macrophages decrease availability of intracellular tryptophan by increasing the enzyme indoleamine 2,3-dioxygenase. Tryptophan is required by many pathogens. This mechanism is particularly effective against intracellular *Chla-mydia*, a bacterium that causes infertility in women as well as blindness. Activated macrophages also have decreased expression of transferrin receptors on the cell surface. Transferrin is an iron transporting plasma protein. Reduced expression of transferrin receptors on macrophages results in decreased intracellular iron. Iron is needed for bacterial replication. As indicated previously, iron is also needed for generation of antimicrobial hydroxyl radicals, so a tenuous balance needs to be struck regarding iron levels to prevent growth of intracellular organisms.

Macrophage Pattern Receptors and Non-Phagocytic Defenses

Macrophages have molecules on their cell membranes that interact with conserved molecules on microbes.

Table 27.3	Mechanisms of survival in macrophages		
Type of Organism	Name of Organisms	Disease Caused	Mechanism by Which Pathogen Survives Intracellularly
Bacteria	*Shigella* spp.	Food poisoning	Escapes into cytoplasm
	Mycobacterium tuberculosis	Tuberculosis	Remodels phagosome with proton pump to neutralize the pH
	Brucella spp.	Brucellosis (Bang's disease, undulant fever)	Survives in phagolysosome by activation of specific acid-response genes
	Listeria monocytogenes	Listeriosis	Escapes to cytoplasm
	Legionella pneumophila	Legionnaire's disease	Transits to ribosome-studded autophagosomes
Protozoa	*Leishmania major*	Leishmaniasis	Survives in phagolysosome
	Leishmania donovani	Visceral leishmaniasis	Survives in phagolysosome
	Toxoplasma gondii	Toxoplasmosis	Prevents phagolysosomal fusion
	Trypanosoma cruzi	Chagas disease or South American trypanosomiasis	Escapes into cytoplasm
Fungi	*Candida albicans*	Candidiasis	Survives in phagosome

This interaction sends an alarm to the macrophage and activates it to increase transcription of genes that encode proinflammatory molecules, that is, that enhance inflammation as described in the preceding section. The receptors on the macrophages that react with conserved molecules on microbe surfaces are referred to as **pattern receptors** and include a receptor that is known as the lipopolysaccharide receptor and several receptors known as **toll receptors**. The bacterial molecules that interact with pattern receptors on macrophages include lipopolysaccharide, lipoteichoic acid, muramyl dipeptide (that are part of the bacterial cell envelope), bacterial heat shock proteins, and the dinucleotide CpG motif present on DNA released by dying bacteria. The interactions of ligands and pattern receptors on macrophages are summarized in **Table 27.4**.

While there are differences in expression of these molecules among bacteria, conserved portions that are common among bacteria are responsible for interaction with the pattern receptor. A recent study showed that gram-positive, gram-negative, and acid-fast mycobac-teria induce a similar response through pattern receptor stimulation, even though they differ in structures that are used to stimulate receptors. For example LPS is common to gram-negative bacteria but not to gram-positive or acid-fast bacteria.

ADAPTIVE IMMUNITY

Inflammation and the phagocytic cells of the innate immune system slow pathogen population growth rates but typically do not eliminate the pathogens. For this to occur, highly specific adaptive immune responses are required. The leukocytes involved in the **adaptive immune system** are subpopulations of **lymphocytes** known as **B cells** and **T cells**. B cells that respond to pathogens make and secrete water soluble proteins called **antibodies** that neutralize toxins, coat pathogens so that they are phagocytosed, or activate complement causing the membrane attack complex to form on the surface of pathogens. T cells are specialized to kill cells that are infected by pathogens, to enhance the capacity of macrophages to kill pathogens and to enhance responses by B cells. The hallmarks of responses by lymphocytes are specificity and immunological memory.

Lymphocytes have receptors on their plasma membranes that allow them to interact with specific portions (**epitopes**) of molecules referred to as **antigens**. These antigens are typically foreign (non-self) molecules of infectious disease agents. There are millions of different T cell and B cell receptors expressed which gives the specificity to the response even though all of the antigen-binding receptors on a single lymphocyte have the same specificity. The genetic basis of receptor diversity is discussed later in the text. Antigens recognized by the antigen-specific receptors of B cells (**B cell receptor**) are

Table 27.4	Pattern receptors	
Receptor Designation	Target Molecules	
CD14	LPS and yeast	
Toll 2	Lipoproteins, lipoteichoic acid of gram-positive bacteria, muramyl dipeptide, bacterial heat-shock proteins	
Toll 4	LPS, bacterial heat-shock proteins	
Toll 9	Bacterial CpG DNA	

present on the intact pathogens or their products. In the case of T cells, the antigenic epitopes are peptides (a small piece of a protein) derived from pathogen molecules that are presented on the surface of host cells that have taken up pathogens or their products, for example, macrophages that have phagocytosed the pathogen, or host cells that are infected by pathogens. The host cell molecules with which the antigenic peptides are displayed to the T cells are known as **major histocompatibility complex** (MHC) molecules and this process of display is known as **antigen presentation**.

The second hallmark of the adaptive immune response is memory, which enables more rapid and heightened responses of antigen-specific lymphocytes when they encounter a particular antigen the second time. Previously, memory was thought to result mainly from the replication of lymphocytes that had reacted with a particular antigenic epitope, consequently resulting in an expanded population of antigen-specific cells. Indeed there is an increase in the number of antigen-specific cells during an infection that can react with particular antigens. However, most of these cells subsequently die. Mammals would be unable to accommodate the large number of lymphocytes that would be acquired after each antigen encounter if this did not occur. However some of the cells that reacted with the antigen do not die and have a less stringent requirement for activation when they encounter those antigens the next time, meaning they respond more rapidly. These are referred to as memory cells. Immunological memory is the principle behind vaccination.

Initiation of the Adaptive Immune Response

To initiate an adaptive immune response, antigens from infectious agents must come in contact with the lymphocytes bearing receptors for that antigen. An antigen that is introduced into the blood is filtered out in the liver, spleen, and lungs, where phagocytes are found close to the organ blood vessels or sinuses and lymphocytes enter these organs via the blood vessels. Antigen introduced into the digestive tract comes in contact with lymphocytes in special areas of intestines known as Peyer's patches that are concentrated centers of lymphocytes. Antigen that is in peripheral tissues is delivered by the **lymphatic system** to lymphocyte-containing organs known as lymph nodes. The lymphatic system is a specialized one-way system of vessels that drains fluid that accumulates in tissues in spaces between cells. This drainage occurs continually from all peripheral tissues as a normal process, but pathogens and fragments of these, which are tagged by complement components C3b and iC3b at the site of inflammation, are also transported to lymph nodes, where they bind to receptors on macrophages and another specialized cell called a **dendritic cell**. The macrophages and dendritic cells present the antigen to T cells.

Blind-ended lymphatic capillaries originate in the tissues and have flap cells that open when local fluid pressure builds up, admitting fluid, antigen, and cells into the vessels, which is now known as **lymph**. Once the pressure has been relieved, the flap cells close. The lymph is pumped through the lymphatic vessels toward lymph nodes by the contractile action of muscles surrounding the vessels. This does not occur when one is inactive, which accounts for the swelling (edema) that occurs around the feet and ankles when one is immobile for a long time, such as on an airplane. Lymph nodes are found clustered at the base of the skull, where they drain the tissues of the head; in the armpit, where they drain the lymphatic vessels of the arm; on the flank, where they drain the body wall; and in the groin and behind the knees, where they drain the legs.

The lymph passes through a series of lymph nodes connected by lymphatic vessels. These vessels finally coalesce into one large lymphatic vessel known as the thoracic duct that empties into the blood system. Lymphocytes also circulate in lymph and blood. They reenter the lymph nodes from the blood, as a result of their interaction with a specialized type of cell that lines the blood vessel, as it passes through the lymph node. In this way lymphocytes continue to circulate around the body. This enhances the likelihood that they will encounter antigens that they can react with.

Lymphocytes that participate in adaptive immune responses arise from a stem cell in the bone marrow that derives from a hemopoietic stem cell. Hemopoietic stem cells give rise to stem cells of the myeloid lineage, which form neutrophils, eosinophils, basophils, monocytes, platelets, and red blood cells. The lymphoid stem cells differentiate into B cells and T cells (**Figure 27.6**). The final phase of maturation into a cell that is ready to respond to antigen occurs for T cells in the thymus, whereas for B cells it occurs in different organs, depending on the species of animal. In birds, B cell maturation occurs in the bursa of Fabricius. In humans and mice it occurs in the bone marrow.

T Cell Development

T cells develop in the thymus and their development is associated with the assembly of an antigen-specific receptor and selection of cells whose antigen-specific receptor has an appropriate specificity. The antigen-specific receptors of T cells are designed to recognize fragment of proteins, called peptides, that are expressed on the surface of cells. A group of host molecules has evolved to transport peptide fragments of pathogens onto the cell surface. These are called major histocompatibility complex (MHC) molecules. Antigen-specific receptors on T cells actually bind to the peptides and some amino acids residues of MHC close to where the peptide is located. There are two classes of MHC mole-

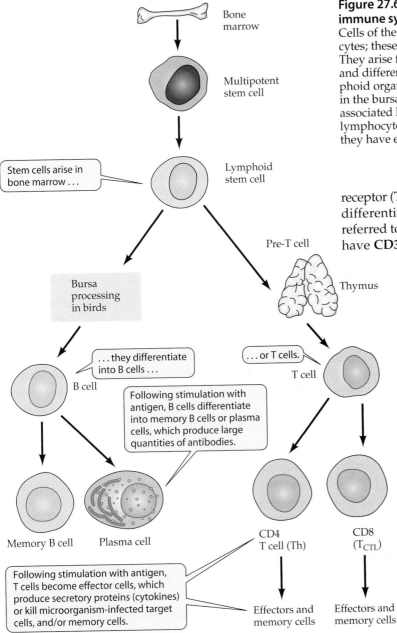

Figure 27.6 Development of cells of the adaptive immune system

Cells of the adaptive immune system are known as lymphocytes; these include B cells, CD4 T cells, and CD8 T cells. They arise from a common precursor in the bone marrow and differentiate into major subpopulations in primary lymphoid organs: T cells differentiate in the thymus, and B cells in the bursa in birds and in the bone marrow, spleen, or gut-associated lymphoid tissue in mammals. For each type of lymphocyte, they mature into effector or memory cells once they have encountered antigen.

- Bone marrow
- Multipotent stem cell
- Lymphoid stem cell
- Stem cells arise in bone marrow . . .
- Pre-T cell
- Bursa processing in birds
- Thymus
- . . . they differentiate into B cells . . .
- . . . or T cells.
- B cell
- T cell
- Following stimulation with antigen, B cells differentiate into memory B cells or plasma cells, which produce large quantities of antibodies.
- Memory B cell
- Plasma cell
- CD4 T cell (Th)
- CD8 (T$_{CTL}$)
- Following stimulation with antigen, T cells become effector cells, which produce secretory proteins (cytokines) or kill microorganism-infected target cells, and/or memory cells.
- Effectors and memory cells
- Effectors and memory cells

cules, called class I and class II. These different types of MHC molecule direct responses of two functionally different T cell subpopulations called **CD4 T cells** and **CD8 T cells**, which are discussed later in the text.

The MHC molecules do not discriminate between peptides of pathogen origin and of host origin and consequently can display a comprehensive sampling of these on the cell surface. This is important for educating the developing immune system to selectively respond to largely non-self-antigens, that is, to antigens from pathogens and not to antigens from normal self-molecules.

Immature T cells that enter the thymus after leaving the bone marrow express an antigen-specific T cell receptor (**T cell receptor**) and both the CD4 and the CD8 differentiation antigens on their surface and thus are referred to as double-positive T cells. These cells do not have **CD3**, which is the name given to a set of plasma membrane glycoproteins associated with the T cell receptor in mature T cells. The CD3 molecules are responsible for signal transduction after the T cell receptor binds antigen (**Figure 27.7**). T cells mature in the cortex of the thymus and then transit to the medulla. By the time they enter the medulla, they are single positive, express only CD4 or CD8 molecules. They also express CD3 and are ready to encounter antigen. While in the medulla of the thymus, T cells undergo programmed cell death unless spared from this fate by a threshold interaction with self MHC molecules. Only those that react strongly enough receive the appropriate signal to survive, that is, they undergo positive selection. Subsequently, if their T cell receptor reacts too strongly with MHC molecules that are presenting self-peptides, they are depleted or negatively selected to avoid immune responses to self-antigens or autoimmunity. This would be very detrimental to the host.

CD4 T cells react with peptide and class II MHC molecules, whereas CD8 T cells react with peptide and class I MHC molecules. The interaction of these cells with peptide MHC complexes is directed both by the antigen-specific receptors and by the accessory molecules CD4 and CD8. Antigen-specific receptors on CD4 T cells bind peptide and amino acid residues of the peptide-binding groove of the class II MHC molecule, while CD4 itself binds a conserved component of the class II MHC molecule. Antigen-specific receptors on CD8 T cells bind peptide and amino acid residues of the peptide-binding groove of the class I MHC molecule, while CD8 itself binds a conserved component of the class I MHC molecule. CD4 T cells produce cytokines that assist or help both B cells and CD8 T cells to proliferate and mature into fully func-

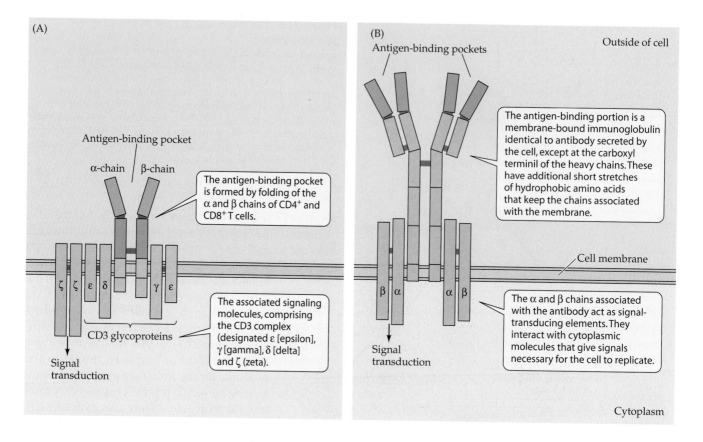

Figure 27.7 Structure of lymphocyte receptors for antigen
(A) The antigen-binding portion of the antigen-specific T cell receptor is composed of an α and a β chain for both CD4 and CD8 T cells. It is associated in the cell membrane with a complex of molecules called CD3 and zeta chains that signal the cell when the receptor has engaged antigen. (B) The B cell receptor interacts with antigen through membrane-bound antibody molecules and then signals the cell that it has found the appropriate antigen through associated molecules called α and β chains.

tional effector cells, and hence CD4 T cells are referred to as helper T cells, or Th. CD8 T cells bind to and kill other host cells that express the appropriate MHC and antigenic peptide. The CD8 cells are therefore referred to as cytotoxic T cells, or Tc.

B Cell Development

B cells first arise from hemopoietic stem cells in the yolk sac and liver of the embryo. After birth, they arise from hemopoietic stem cells that are mainly present in the bone marrow. Development of B cells is associated with development of their antigen-specific receptors. The B cell receptor is composed of an immunoglobulin (Ig) molecule that has two identical immunoglobulin heavy (IgH) and two identical immunoglobulin light (IgL) chains joined together by disulfide bridges (Figure

27.7B). The B cell receptor is held in the B cell plasma membrane by a short stretch of hydrophobic (fat-loving) amino acids at the opposite ends of the heavy chains to the antigen-binding sites.

The antigen-binding sites of the B cell receptor are created by the folding together of parts of the heavy and light chains called variable regions. This creates a groove lined with the reactive groups (R groups) of amino acid residues in segments of these domains, called hypervariable or complementarity-determining regions because they vary in amino acid sequence among different immunoglobulins. Each mature B cell has on its surface between 100,000 and 150,000 such receptors, and for any given B cell all hundred thousand or so receptors have identical antigen-binding sites. The capability of the B cell system to respond to several million different antigenic epitopes results from the presence of several million different antigen-binding pockets on several million different clonal populations of B cells. Development of this legion of antigen-specific B cells results from events that take place during early B cell development.

There are two families of IgL genes called the kappa (κ) and lambda (λ) families. Each of the families is on a different chromosome. The IgL genes are also on different chromosomes to the IgH genes. It is easiest to think of all of the germ line Ig genes as gene segments and to think of B cell development as a process through which individual Ig gene segments are moved together in such a way as to generate genes encoding a complete Ig κ or λ L-chain gene and a complete IgM H chain. This movement of gene segments together to form one mature gene is called **gene rearrangement**.

Ig gene segment rearrangements are a property of B cells only, and do not occur in other cells of the body. Rearrangement of Ig gene segments occurs independently in all precursor B cells (pre-B cells). The first Ig gene to be assembled during B cell development is the μ heavy chain gene. This is made by the moving together of a heavy chain variable gene segment (VH), a diversity segment (DH), and a joining segment (JH) so that they can be transcribed as a single messenger RNA, together with a gene segment encoding a heavy chain constant region, specifically that for μ.

Expression of an μ heavy chain by a developing B cell starts rearrangements that leads to assembly of a light chain gene from variable light (VL) and joining light (JL) segments. This is first attempted with segments encoding a κ light chain, and successful rearrangement (meaning it give rises to a gene that can successfully code for a protein) stops further light-chain gene-segment rearrangements. If rearrangement is unsuccessful for the κ gene (on either of the parental chromosomes), rearrangement continues for the λ gene. If L-chain gene rearrangement is unsuccessful for both κ and λ light-chain genes, the developing B cell dies. Successful rearrangement of a μ H-chain and an L-chain gene allows assembly of an IgM molecule and heralds the appearance of the immature B cell. This process of Ig gene assembly is irreversible in any B cell and generates millions of different B cell antigen binding receptors.

Some B cell antigen receptors that are generated are capable of binding self-antigens with high efficiency. Fortunately, immature B cells are negatively regulated by antigen. When surface immunoglobulin is cross-linked by binding antigen, the cell dies or undergoes a cycle of receptor editing in which further rearrangements occur in light-chain genes that can alter the antigen-binding specificity of the expressed receptor. Nevertheless, about 90% of B cells generated each day die without leaving the bone marrow. The joint processes of negative selection and receptor editing eliminate a significant portion of the self-reactive B cells that are generated. However, many self-antigens are present at only very low concentration, are absent from the bone marrow, or are developmentally regulated, hence, not all self-reactive B cells are eliminated. Any immature B cell that is not eliminated matures to a resting or virgin B cell, which can be stimulated to proliferate by cross-linking of the Ig receptor as discussed on pg. 660. How self-reactive B cells that escape elimination and reach this stage of development are regulated is still under investigation.

Receptors of αβ T Cells

The T cells discussed in this section on adaptive immune responses have a T cell receptor composed of two chains called α and β. Therefore these T cells are known as **αβ T cells**. The other major subpopulation of T cells is known as γδ T cells because their T cell receptor is composed of two chains called γ and δ and coded for by a different set of genes than those which code for the αβ T cell receptor molecules. The γδ T cells are discussed on pg. 701.

Although there are millions of different possible T cell receptors that an αβ T cell can express, and millions of T cells with those different receptors, an individual T cell and its progeny, known as a **clonotypic population** or clone, express only one type of T cell receptor. The T cell receptor can interact with only one type of antigen presented on one type of MHC molecule. The goal for an individual T cell is to constantly survey the body for the antigen/MHC combination with which its T cell receptor reacts. This is the method by which an individual T cell is notified that it is needed to counteract an infection. Once a T cell encounters the correct combination of antigen and MHC, it is stimulated or activated to undergo cell replication or proliferation so that there is an expanded population of that clone of T cells. The expanded T cells will express their **effector functions** to attempt to limit infection by a variety of mechanisms. Some portion of these antigen-stimulated T cells will also become long-lived **memory cells** that will be preserved to respond more rapidly in a secondary encounter with the same antigen.

The T cell receptor is constructed by combining gene segments selected from a large array of possibilities, very similar to the way in which Ig diversity of B cells is created. For example, the β chain is a protein coded for by recombination of a variable gene segment (called a V gene), a diversity segment (called a D gene), and a joining gene segment (called a J gene) with a constant region segment (called a C-β gene). The α chain is a protein coded for by recombination of a V gene, a D gene, and a J gene with an α constant region segment (C-α gene). The V, D, and J genes that code for the α chain are different from those that code for the β chain. The DNA coding for V, D, J, or C genes that were not selected is looped out and permanently removed so that the chosen gene segments can be fused together, or **recombined**, to make a single gene. The V, D, and J gene segments give rise to the portion of the protein that is the antigen-binding site (Figure 27.7A) that interacts with the antigen and MHC molecules. The many millions of different T cell receptors arise because of the high number of gene segments from which to choose. It is this diversity of T cell receptors that allows the adaptive immune system to be so specific for a particular antigen and therefore the adaptive immune system carefully regulated.

The second important component of the T cell receptor complex is the set of associated glycoproteins called CD3 (Figure 27.7A). The CD3 molecules are identical on all T cells and among all members of the same species. The CD3 molecules are named γ, δ, and ε and are associated with two other molecules called ζ. Together with the ζ chains, CD3 makes up the signaling portion of the T cell receptor. That is, when the T cell receptor encounters appropriate antigen, the CD3 and ζ chains undergo changes on the portions of the molecules that are inside the cell. This results in a series of chain reactions that eventually result in cell activation.

αβ T cells can be divided into two major subpopulations, designated CD4 and CD8. CD4 and CD8 are glycoproteins on the membranes of the T cells. The T cell receptor for the CD4 T cells and CD8 T cells may be the same, but the presence of either CD4 or CD8 on the cell surface determines whether the T cell responds to antigen with MHC class II molecules or with class I molecules. This is because CD4 molecules interact with a conserved region of MHC class II molecules, whereas CD8 molecules interact with a conserved region of MHC class I molecules. These interactions of CD4 and CD8 with MHC molecules are necessary for stabilizing the interaction of the T cell receptor with the MHC/antigen complex. CD4 and CD8 are also involved in sending signals that result in activation of the cell.

Antigen Presentation to CD4 T Cells

Presentation of antigen with MHC class II molecules occurs by a mechanism known as the **exogenous pathway** (**Figure 27.8**). This is because the presented antigen may be synthesized outside the presenting cells (produced exogenously). For example, the antigen presented may be from microbes that were phagocytosed by the macrophage. The phagolysosome containing the killed microbe that has been degraded into fragments fuses with other compartments in the macrophage that contain membrane-bound MHC class II molecules. The class II molecule has an antigen-binding groove and thus transports the antigen to the cell surface with it, where they are displayed to T cells.

Only select populations of host cells have MHC class II molecules and thus are equipped to present antigen to CD4 T cells. This is in contrast to the MHC class I molecules that are on all nucleated host cells. The cells with class II MHC include macrophages, B cells, and dendritic cells. However, because B cells and dendritic cells are not phagocytes as are macrophages, the antigens they present must be soluble antigens that have been endocytosed.

The stimulation of the T cell that occurs as a result of interaction of the T cell receptor with the antigen/MHC complex is sometimes referred to as signal 1. For a CD4 T cell to become activated to proliferate (clonally expand) and express its effector functions, it needs to receive a second signal that is referred to as costimulation. This second signal is received by the T cell either through its receptor for the cytokine IL-1, that is produced by macrophages, or a T cell membrane molecule called CD28 that interacts with a molecule on antigen presenting cells. In the absence of appropriate costimulation, T cells that have not previously encountered antigen may undergo programmed cell death as a result of signal 1 alone.

CD4 T Cell Effector Molecules: Cytokines

When a CD4 T cell encounters the antigen/MHC complex that interacts with its T cell receptor, it is known as a T-helper zero (Th0) cell and may produce a variety of cytokines. Th0 cells subsequently mature into one of two functional subsets called **type 1** (**Th1**) or **type 2** (**Th2**), distinguished from one another by the types of cytokines they produce (**Figure 27.9**). This depends not on their T cell receptor but on the cytokines they receive when encountering antigen. For example, IL-12 produced by dendritic cells and macrophages after they are stimulated through pattern receptors biases T cells toward a Th1 response. In contrast, IL-4, perhaps produced by mast cells during an inflammatory response and by other Th2 T cells, biases T cells toward a Th2 response.

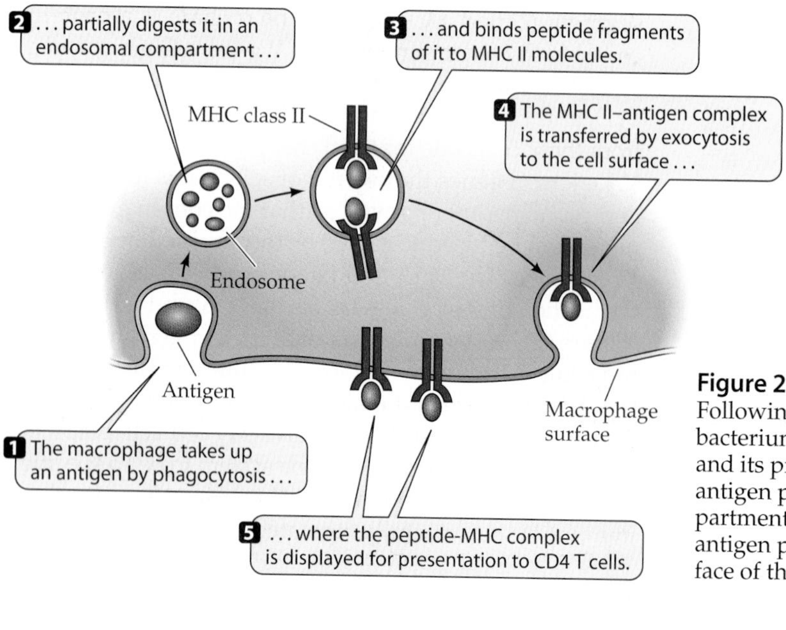

Figure 27.8 Presentation of antigen with class II MHC Following internalization of a foreign antigen (for example, a bacterium or virus) by the macrophage, the microbe is killed and its proteins degraded into small antigen peptides. These antigen peptides associate in specialized intracellular compartments with MHC class II and together the MHC and antigen peptide are transported and displayed on the surface of the antigen presenting cell.

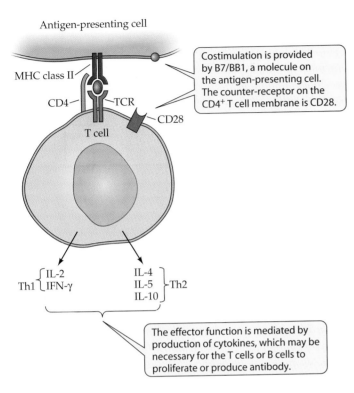

Antigen-presenting cell

MHC class II

CD4 — TCR

T cell

CD28

Costimulation is provided by B7/BB1, a molecule on the antigen-presenting cell. The counter-receptor on the CD4⁺ T cell membrane is CD28.

Th1 { IL-2 IFN-γ

IL-4 IL-5 IL-10 } Th2

The effector function is mediated by production of cytokines, which may be necessary for the T cells or B cells to proliferate or produce antibody.

Figure 27.9 Interaction of CD4 T cell with antigen presenting cell

The T cell receptor of CD4 T cells recognize pieces of antigen presented in conjunction with MHC class II molecules on the surface of antigen presenting cells. Thus, the antigen-binding pocket interacts not only with the foreign antigen fragment (or protein peptide) but with the MHC molecule as well. In addition, antigen-presenting cells such as macrophages have costimulatory molecules that react with counterreceptors on the T cells. The CD4 molecule interacts with MHC class II molecules.

The main Th1 cytokines produced are IFN-γ and IL-2. The general function of a Th1 cytokines is to promote **cell-mediated immunity** (also called cellular immunity). Cell-mediated immunity means production of IFN-γ that inhibits viral replication and activates macrophages for increased control of intracellular microbes, and generation of cytotoxic CD8 T cells. IFN-γ inhibits viral replication through stimulation of a cellular enzyme that results ultimately in generation of a cellular endoribonuclease that cleaves viral mRNA. As discussed on pg. 648, activated macrophages have increased production of reactive oxygen intermediates and nitric oxide, that are toxic for intracellular microbes, and decreased iron availability. IL-2 produced by Th1 CD4 T cells helps CD8 T cells to replicate. Subsequently, CD8 T cells become **cytotoxic**. This means they kill or lyse host cells that are infected with viruses or with intracellular pathogens. Thus, Th1 cells are particularly effective against infectious agents that live within host cells. However, IFN-γ also promotes B cells to develop into cells that produce a particular IgG type of antibody. IFN-γ produced by Th1 T cells also influences development of other T cells into Th1 cells.

Th1 T cells mediate two other responses: delayed type hypersensitivity and granuloma formation. These involve recruitment of macrophages to sites of infection. Delayed type hypersensitivity is a response to antigens that are presented in the skin. It takes approximately 48 hours to develop and involves the recruitment of CD4 T cells and macrophages to the site. This response is often called a skin test and is an in vivo determinant of whether a person or animal is sensitized to a particular antigen, such as tuberculosis. A granuloma is similar to a delayed type hypersensitivity response, but it is a long-term response that occurs when the influx of macrophages is unable to clear the antigenic stimulus. The macrophages adhere to one another, fusing to form multinucleate giant cells. Examples of this are the walling off of mycobacterium in the lungs, the reason chest X rays are given to people potentially infected with tuberculosis, and the walling off of eggs of the parasitic worm *Schistosomula mansoni* in the liver, resulting in enlarged livers.

The main Th2 cytokines are IL-4 and IL-10. These cytokines promote **humoral immunity**. Promotion of humoral immunity by Th2 cytokines means they promote B cell growth and differentiation into cells producing antibodies of the IgA, IgE, and the other IgG types. These antibody types are particularly effective against extracellular pathogens (i.e., those that do not live within host cells), including bacteria, protozoa, and parasitic worms. In addition, the Th2 cytokines activate mast cells and eosinophils, myeloid cells that are involved in immunity to parasitic worms (discussed on pg. 665) and influence the development of other T cells into Th2 cells (**Figure 27.10**).

Antigen Presentation to CD8 T Cells

Antigen that is presented with MHC class I molecules occurs by a mechanism known as the **endogenous pathway**. This is because the presented antigen is synthesized within the cells or produced endogenously. The association of antigens of infectious agents with MHC class I molecules first requires that these molecules be free in the cell cytoplasm. For example, viruses produce viral proteins only inside host cells because they do not actually contain the needed cellular machinery to do this independently. Thus, to make new viral particles, a virus must attach to a host cell, enter, and uncoat to expose its genetic code (DNA or RNA) to serve as a template for producing new viral proteins. Similarly, intracellular bacteria and protozoa that replicate either in a phagosome, phagolysosomes, or within the cytoplasm of host cells can have their protein peptides presented with class I MHC molecules. In addition to those intracellular microbes mentioned previously in this chapter that live within macrophages (see Table 27.3), protozoa and

(A)

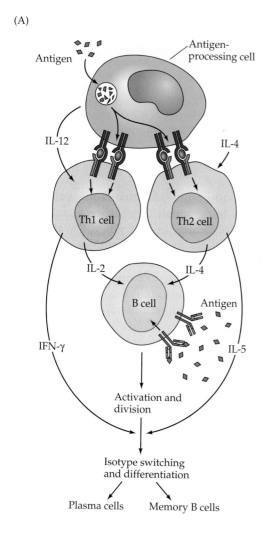

(B)

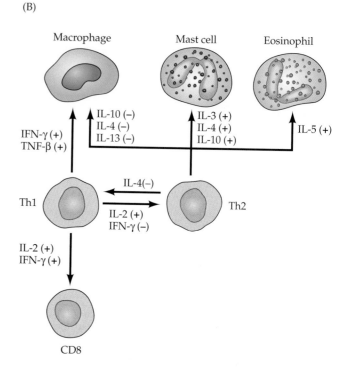

Figure 27.10 Cytokines produced by CD4 type 1 and type 2 T cells
(A) Following activation, CD4 T cells mature into type 1 or type 2 subpopulations, distinguished by the types of cytokines they produce. This depends not on their TCR but on the cytokines they receive when encountering antigen such as IL-12 and IL-4. CD4 T cells that produce type 1 cytokines are known as Th1 cells; those that produce type 2, as Th2 cells. Both type 1 and type 2 cytokines act on B cells. Type 2 cytokines such as IL-4 and IL-5 promote B cell differentiation and growth and the generation of specific antibody isotypes. The type 1 cytokine IFN-γ promotes generation of specific IgG subclasses not promoted by type 2 cytokines. (B) Cytokines produced by CD4 T cells also act on macrophages to activate (+) or deactivate (–) them for antimicrobial activity and presentation of antigen. The cytokines IL-4 and IFN-γ affect development of other T cells into type 1 or type 2 cells. IL-2 helps other T cells proliferate. Finally the type 2 cytokines activate mast cells and eosinophils.

bacteria may also live within host cells that are not professional phagocytes. Examples include the bacteria *Brucella abortus* that lives within trophoblast cells and the protozoan *Plasmodium falciparum* that initially lives in hepatocytes (liver cells). How the proteins from the infectious agents that are contained within intracellular endosomal compartments gain access to the cytoplasm is not yet entirely clear. There may be specific transport mechanisms to remove pathogen components from the compartment. It is also likely that some intracellular organisms die within the compartments, and it may be proteins from these that gain access to the cytoplasm.

Once the proteins are in the cytoplasm, they are degraded by an enzyme complex, called the **proteosome**, into peptides. Resulting peptides are moved by chaperonins to the endoplasmic reticulum and moved into its cysterni by the TAP transporter. Within the endoplasmic reticulum, the antigen peptides come in contact with MHC class I molecules, where they bind into the groove. Together, the MHC class I and the antigen peptide are transported to the surface of the cell, where they are displayed to CD8 T cells (**Figure 27.11**). Because all nucle-

ated cells of the host express class I MHC, the CD8 T cells can receive signal 1 for activation from any of them.

Some viruses have developed methods to foil the immune system by decreasing the level of class I MHC expressed on the cell surface or causing retrograde transport of viral peptides back into the cytoplasm of the infected cell. In this way, the CD8 T cells do not recognize the virally infected cells. However, the natural killer (NK) cells in the bridging immune system are alerted as a result of decreased MHC class I expression, which will be discussed on pg. 668.

Like CD4 T cells, CD8 T cells express CD28. However, CD8 T cells can respond to antigen in the absence of costimulation, although when antigen load is low, costimu-

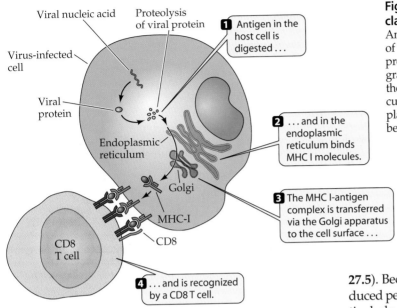

Figure 27.11 Antigen presentation with MHC class I molecules
Antigens produced in a host cell (whether as a result of a viral infection or of parasitism by a bacterium or protozoan that survives within phagocytes) are degraded into small antigen peptides that associate in the endoplasmic reticulum with MHC class I molecules. Together, the MHC and antigen peptide are displayed on the cell surface in such a way that they can be recognized by the T cell receptor of CD8 T cells.

lation through CD28 can enhance the CD8 T cell response. It is logical that CD8 T cells may not require costimulation, because CD8 T cells need to be able to respond to antigen on any infected cell of the host, and not all cells express costimulatory ligands for CD28. CD8 T cells also require IL-2 from CD4 Th cells, although occasionally a CD8 T cell clone has been reported to produce its own IL-2.

CD8 T Cell Effector Function: Cytotoxicity and Cytokines

The main effector functions of CD8 T cells are cytotoxicity and production of IFN-γ (**Table 27.5**). Because of the presentation of endogenously produced peptides with class I MHC molecules, they are particularly important in immunity to viral infections. There is evidence that in the control phase of a viral infection, their main role in protection is through production of IFN-γ, whereas in viral clearance phase it may be cytotoxicity. These same mechanisms have been shown to be important in immunity to infections with intracellular bacteria and protozoa, but in that instance the IFN-γ acts by activating macrophages.

The mechanism by which CD8 T cells kill target cells (i.e., those expressing the appropriate antigen peptide

---

Table 27.5   Summary of αβ T cell effector functions for control of infectious disease

T Cell Cytokine or Action	T Cell Type Involved	Effector Function	Infectious Agents Targeted
IFN-γ	Type 1 CD4 T cells mainly but some by type 1 CD8 T cells	Activates macrophages to kill or control replication of intracellular protozoa, bacteria, and fungi	Intracellular *bacteria* Intracellular *protozoa* Intracellular *fungi*
		Inhibits virus replication in host cells	All viruses after they have entered host cells
		Promotes class switching of B cells to specific IgG subclasses	Extracellular infectious agents in internal tissues and organs
TNF-α	Type 1 CD4 T cells	Assists in macrophage activation	All intracellular infectious agents (see above in this column)
IL-2	Type 1 CD4 T cells	Assists in T cell replication	
IL-4, IL-5, IL-6	Type 2 CD4 T cells	Promotes class switching of B cells to IgA, IgE, and other IgG subclasses	Immunity at body surfaces by IgA against entry of viruses and bacteria; engagement of mast and eosinophils by IgE to kill worms; IgG subclasses to promote immunity to all extracellular viral particles, bacteria, and protozoa by complement activation, agglutination and opsonization
Cytotoxicity	CD8 T cells	Kills infected cells	Virus-infected cells; cells infected with intracellular bacteria, protozoa, and fungi

Figure 27.12 CD8 T cell killing target cell
One of the main functions of CD8 T cells is to kill other host cells that are displaying foreign antigenic peptides with class I MHC molecules that are recognized by their T cell receptor. This is done through the production of perforins and granzymes and in some cases a protein called Fas ligand.

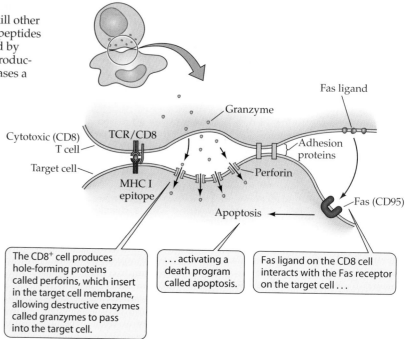

Granzyme

Fas ligand

Cytotoxic (CD8)
T cell

TCR/CD8

Adhesion
proteins

Target cell

MHC I
epitope

Perforin

Fas (CD95)

Apoptosis

The CD8⁺ cell produces hole-forming proteins called perforins, which insert in the target cell membrane, allowing destructive enzymes called granzymes to pass into the target cell.

... activating a death program called apoptosis.

Fas ligand on the CD8 cell interacts with the Fas receptor on the target cell ...

with MHC class I molecules) is diagrammed in **Figure 27.12**. At the apposition point of the cell membranes, the CD8 T cells insert perforin molecules into the target cell, thus forming pores. The perforin is stored in granules within the CD8 T cells. Granzyme is then deposited into the target cells through the pores, which results in the target cell undergoing programmed cell death or **apoptosis**. An additional mechanism not unique to CD8 T cells is the production of Fas ligand that interacts with its receptor called Fas or CD95 on the target cell (Figure 27.12), causing the cells to undergo programmed cell death or apoptosis.

Memory Responses by T Cells

The basic definition of immunological memory is a long-lived response following priming in vivo that can be demonstrated as a recall response in vivo and in vitro. Most importantly, it results in more rapid control of a particular infection on secondary challenge with the infectious agent. The requirements for activation of memory cells are less than those for activating naïve cells. That is, less antigen is required and no costimulation is necessary. These features, in conjunction with the clonal expansion in the first round of response to an antigen, result in the more rapid control of infection.

B cells and Antibodies

Antibodies belong to a family of proteins called immunoglobulins. They are water-soluble immunoglobulins that are secreted by plasma cells. **Plasma cells** are the end differentiative stage of antigen-activated B-lymphocytes. An antigen is any substance capable of stimulating an immune response and can be a protein or other macromolecule. Molecules present in one's own body are called self-antigens and typically stimulate no or only low-titer and harmless antibody responses in one's own body. When self-antigens do stimulate a high-titer antibody response, or harmful immune response, this is called an autoimmune response. All antigenic material that is not self is called foreign antigens. Some material is not capable of stimulating an immune response, and this is referred to as nonantigenic.

The purpose of antibodies is to bind and neutralize (render harmless) antigens. Binding occurs in an antigen-binding pocket present at the N-terminus of the polypeptides that make up an antibody molecule. The part of an antigen that binds in the pocket is called an antigenic epitope (**Figure 27.13**). Neutralization may be achieved by binding the antigen, as is the case with toxins; however, it is generally facilitated by additional properties of antibodies that are latent until antibody has bound to antigen. These are discussed later. There are several classes and subclasses of antibodies. Not all antibody classes and subclasses are present in all species of mammals, although the main classes are represented (i.e., IgM, IgG, IgA, and IgE). The different Ig classes have different molecular weights and different biological properties dictated by how the non-antigen binding parts of the antibody molecule (**Fc region**) interact with cells and molecules of the immune system. The genetic processes that lead to generation of the different Ig classes and their biological functions are discussed on pp. 661 and 663.

All secreted antibodies irrespective of class have the same subunit structure (**Figure 27.14**). IgG and IgE are composed of a single subunit. IgM is composed of five subunits, with each disulphide bridged to another, a process initiated by a small peptide called the J-chain. IgA is composed of two subunits held together by a J-chain. Antibody resembles cell surface Ig (Figure 27.7B), except that the membrane-spanning domain of the heavy chains is missing. The genetic processes that generate membrane and secreted Ig from the same gene are discussed on pg. 661.

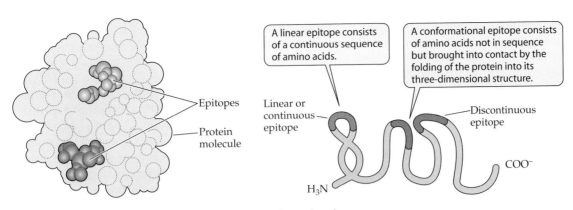

Figure 27.13 Antigenic determinants on a protein molecule
Specific portions of antigens, called the antigenic determinants or epitopes, are exposed on the intact antigen molecule and recognized directly by antibodies or B cell receptors in their native form (that is, they are not degraded by antigen-presenting cells and presented).

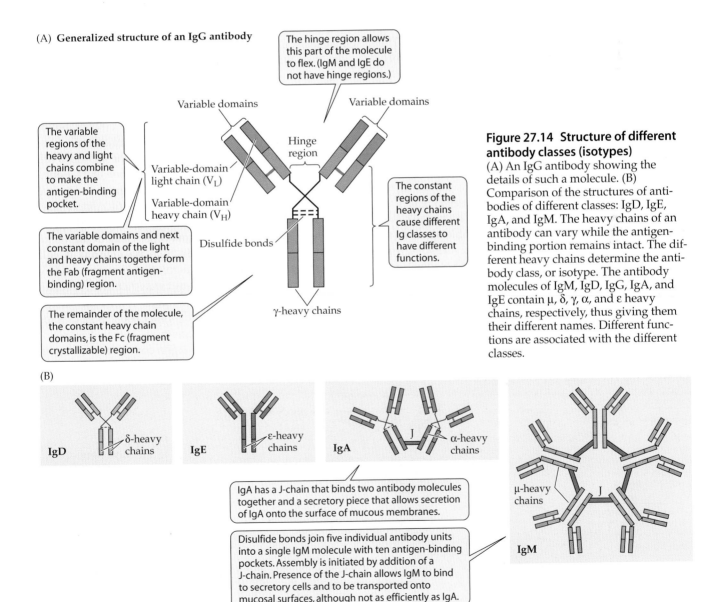

(A) **Generalized structure of an IgG antibody**

The hinge region allows this part of the molecule to flex. (IgM and IgE do not have hinge regions.)

Variable domains

Variable domains

Hinge region

The variable regions of the heavy and light chains combine to make the antigen-binding pocket.

Variable-domain light chain (V_L)

Variable-domain heavy chain (V_H)

The variable domains and next constant domain of the light and heavy chains together form the Fab (fragment antigen-binding) region.

Disulfide bonds

The constant regions of the heavy chains cause different Ig classes to have different functions.

The remainder of the molecule, the constant heavy chain domains, is the Fc (fragment crystallizable) region.

γ-heavy chains

Figure 27.14 Structure of different antibody classes (isotypes)
(A) An IgG antibody showing the details of such a molecule. (B) Comparison of the structures of antibodies of different classes: IgD, IgE, IgA, and IgM. The heavy chains of an antibody can vary while the antigen-binding portion remains intact. The different heavy chains determine the antibody class, or isotype. The antibody molecules of IgM, IgD, IgG, IgA, and IgE contain μ, δ, γ, α, and ε heavy chains, respectively, thus giving them their different names. Different functions are associated with the different classes.

(B)

IgD   δ-heavy chains

IgE   ε-heavy chains

IgA   α-heavy chains   J

IgA has a J-chain that binds two antibody molecules together and a secretory piece that allows secretion of IgA onto the surface of mucous membranes.

Disulfide bonds join five individual antibody units into a single IgM molecule with ten antigen-binding pockets. Assembly is initiated by addition of a J-chain. Presence of the J-chain allows IgM to bind to secretory cells and to be transported onto mucosal surfaces, although not as efficiently as IgA.

μ-heavy chains   J

IgM

Each Ig subunit is composed of two identical polypeptide chains, called heavy or H-chains, and two identical smaller polypeptide chains, called light or L-chains. The heavy chains are covalently attached to each other by a disulphide bridge. Each light chain is also covalently bonded to a heavy chain by a disulphide.

The variable domains of a heavy and a light chain (**Figure 27.15**) fold to create a small pocket capable of accommodating a structure that is equivalent in size to a small peptide, for example, one of 5 to 8 amino acid residues. This is called the antigen-binding pocket (**Fab region**). The amino acid residues of the antigen-binding pocket that make contact with the antigenic epitope lie within hypervariable regions of the variable domain of H- and L-chains. There are three such hypervariable regions, also called complementarity-determining regions (Figure 27.15), on the H- and L-chain V region, in each case representing about 15% of the variable regions. The remaining stretches of amino acids in the variable region are called the framework regions. The wide range of specificities of antibody molecules reflects variation in the amino acid sequences and lengths of the six hypervariable loops.

Material that binds in the antigen-binding pocket of an antibody or a B cell receptor is called an antigenic epitope. When size, shape, charge, and phobicity permit, an antigenic epitope that enters an Ig antigen-binding pocket can form a noncovalent association with amino acid residues that line the pocket, in which case the antigen and antibody now form a complex (Figure 27.15). This complex is held together by four forces, each weak, but in combination adequate to sustain a stable association that can be broken only by proteolytic enzymes or chemical conditions that break the noncovalent associations. The forces are: (1) hydrogen bonds in which a hydrogen atom is shared between two electronegative atoms; (2) ionic bonds between oppositely charged residues; (3) hydrophobic interactions in which the absence of polar groups prevents interaction with water, which consequently forces hydrophobic groups together; and (4) van der Waals interactions, which form between the outer clouds of two or more atoms due to transitory dipole formation.

An antigenic epitope could be a linear stretch of amino acid residues on a large polypeptide (called a linear or continuous epitope), a group of nonlinear amino acids brought together by folding of a protein (called a discontinuous or conformational epitope) (see Figure 27.13) or it might be a carbohydrate, a stretch of nucleic acid bases, a lipid, a glycolipid, a glycoprotein, or even a chemical group called a hapten. A hapten does not induce an immune response on its own, but when bound to a larger molecule, it can act as a B cell epitope.

B cell Activation

B cells can be activated in two ways, termed **T cell independent (TI)** and **T cell dependent (TD)**. TI antigens belong to two classes, TI-1 and TI-2, which activate B cells in different ways. TI-1 antigens include bacterial cell wall lipopolysaccharide and have B cell mitogenic (causes activation of most B cells) properties, for example, they directly stimulate B cell proliferation by reacting with molecules other than the antigen-specific Ig receptors. At low concentrations of TI-1 antigens, only the B cells specific for that antigen are activated. This is thought to result from the binding of the TI-1 antigen to the Ig receptor in an antigen-specific manner and focusing the remainder of the molecule onto the second B cell mitogenic receptor. At high concentrations of TI-1 antigens, the B cells specific for that antigen are silenced, possibly through overexposure to the mitogen, whereas most other B cells are induced to replicate and differentiate to IgM-secreting plasma cells. Thus, at high concentrations TI-1 antigens induce **polyclonal B cell activation**. Polyclonal B cell activation is often a feature of chronic parasitic diseases but may not always result from pathogen-specific TI-1 antigens. B cells stimulated by TI-1 antigens typically do not produce Ig classes other than IgM, and, hence, generate elevated levels of serum IgM at high antigen concentration. These antigens do not elicit immunological memory.

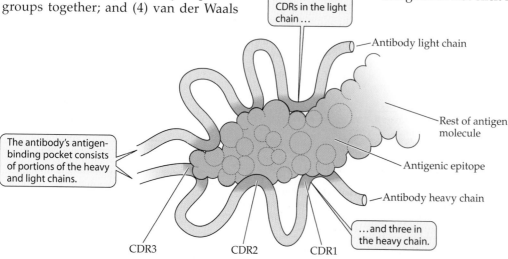

There are three CDRs in the light chain ...

Antibody light chain

The antibody's antigen-binding pocket consists of portions of the heavy and light chains.

Rest of antigen molecule

Antigenic epitope

Antibody heavy chain

... and three in the heavy chain.

CDR3   CDR2   CDR1

**Figure 27.15
Interaction between antigen and antibody**
Interaction of the antigen-binding pocket of an antibody molecule with epitopes of an antigen. The portions of the antibody molecule that interact with the antigen epitopes are hypervariable regions called complementary-determining regions (CDRs).

TI-2 antigens are highly repetitive molecules, such as polymeric proteins, bacterial cell wall material with repeating polysaccharide units, and bacterial flagellin. They activate B cells by extensive cross-linking of the Ig receptors. Development of B cell responses to TI-2 antigens requires low concentrations of cytokines supplied by T cells and does not result in polyclonal B cell activation at any concentration. TI-2 antigens mainly induce IgM antibodies, but a low amount of IgG may also be made. The TI-2 antigens do not elicit immunological memory.

Antigen cross-linking of the B cell receptor is also required for activation by T cell dependent antigens, which are typically soluble proteins with few repeating antigenic epitopes. In the case of TD antigens, B cell antigen-specific receptor cross-linking on its own is inadequate to stimulate cell division. Additional signals from T cells are required to elicit this response and interaction of the B cells and T cells is facilitated by presentation of antigenic peptides on B cell class II MHC molecules, as discussed below. A coreceptor complex on B cells modifies their activation. This coreceptor complex includes complement receptor 2 (CR2, which binds C3dg and C3d as discussed in the section on Inflammation). Binding of C3d or C3dg to this receptor is required for activation of T cell dependent B cells.

Antigen that binds to the B cell receptor is endocytosed by the B cell, it is processed, and peptides are presented on the B cell surface in association with class II MHC antigens leading to interactions with CD4 helper T cells whose T cell receptor is specific for the antigen. Activation of the B cell leads to their elevated expression of class II MHC antigens and the expression of the T cell costimulator molecule called B7. Recognition of peptide-MHC II by CD4 T cells together with costimulation through B7 causes the T cells to express CD40 ligand. This binds CD40 on B cells, providing the second signal required for their activation. At the same time, the associated T cells release IL-2 and IL-4, which support progression of the B cells to DNA synthesis and eventually to differentiate to memory cells and to high-rate antibody-secreting cells called plasma cells (see Figure 27.10).

Thus, stimulation with TD antigens in conjunction with signals provided by CD4 T cells causes B cells to replicate and give rise to plasma cells that secrete antibodies. The antibodies have the specificity of the B cell receptor because they result from transcription of the same IgH- and IgL-chain genes. Whether the IgH chain is made as a secreted (antibody) versus a surface form (B cell receptor) is determined by differential processing of the same primary mRNA transcript. In the case of secreted Ig, two exons of the gene segment encoding the C-terminal of the heavy chain are spliced out of the mRNA.

Antigen binding and stimulation of B cells with T cell cytokines can result in a process called **Ig class switching,** in which the VH-DH-JH gene segment assembly, which

Table 27.6	Cytokine-induced Ig class switching
Cytokine	**Ig Class Switch**
IFN-γ	IgG2a or IgG3
TGF-β	IgA or IgG2b
IL-4	IgE or IgG1
IL-2, IL-4, IL-5	IgM

encodes the variable region of the IgH chain, is spliced onto a constant region gene segment lying downstream of the IgH δ gene segment. In this way, the same variable region gene sequence can be joined to an IgH γ gene segment to generate an IgH γ chain, or to an IgH α gene segment to generate an IgH α chain, or to an ε segment to generate an IgH ε chain. In this way, members of the same B cell clone can produce either IgM, IgG, IgA, or IgE with the same antigen-binding specificity.

Class switching among IgH constant region gene segments is not reversible. It involves flanking sequences of DNA, called switch regions, that are located upstream of each Ig H chain constant region gene segment. It is hypothesized that various cytokines make these switch regions accessible to enzymes that loop out and excise intervening sequences between the chosen switch and variable regions. This may account for the ability of different cytokines to bias B cells to make antibodies of different Ig classes (**Table 27.6**). The different IgH constant region gene sequences imbue antibodies with different secondary functions. The functions of different antibodies are discussed on pg. 663.

B Cell Maturation, Memory, and Secondary Immune Responses

T cell dependent antigens on infectious agents, toxins, and venoms stimulate B cell responses that expedite clearance of these agents from the body, protect the body from immediate re-exposure to the same agents, and prepare the body for the rapid elimination of the same agents, should they be reintroduced at a more distant time. The B cell responses have three components:

- The development of a primary immune response that involves production of antibodies by short-lived plasma cells

- Ig class switching, somatic mutation of the variable region of the expressed Ig genes, and selection of cells with an improved capacity to bind the specific antigen, a process called affinity maturation, and the generation of memory B cells, for example, B cells that rapidly give rise to antibody secreting cells on re-exposure to antigen, and the development of long-lived antibody-secreting plasma cells

The development of these responses occurs in the spleen or in lymph nodes to which antigen that is coated with the complement components C3b and iC3b (and therefore C3dg and C3d) is drained. Purified proteins typically do not provoke inflammation and also do not invoke primary immune responses unless administered with material that elicits the inflammatory response. Primary antibody responses take several days to arise and result in an increase in serum IgM that is specific for the stimulatory antigens (**Figure 27.16**). IgM has a half-life of about 7 days in plasma and disappears fairly rapidly from circulation, indicating that it is produced for a short time only. Infectious agents that take some time to be eliminated from the body may continue to stimulate IgM antibody responses after the initial burst of IgM production, which reflects continuous recruitment of resting B cells into T cell independent (TI) responses, or expression of new antigens as a result of antigenic variation, or both.

Within the secondary lymphoid organs (i.e., the spleen, lymph nodes, and mucosa-associated lymphoid tissue), lymphocytes segregate into T cell and B cell areas, the latter in ovoid groups called follicles, which are regions where B cells develop into memory cells. Follicles also contain **follicular dendritic cells**, which are phagocytes that are specialized for antigen capture and retention and that form the scaffolding of an antigen-induced body called a germinal center. In addition, follicles contain a type of macrophage, called a tingible body macrophage, which endocytoses apoptotic cells that arise in germinal centers. Follicles also contain a few CD4 T cells that, together with TD antigens, are required for germinal center formation.

Follicles are divided into two types, called primary and secondary follicles. Primary follicles mainly contain small resting B cells and a loose network of follicular dendritic cells and are antigen independent, whereas secondary follicles contain an inner region of activated B cells associated with a denser network of follicular dendritic cells, called a germinal center, and are formed in response to antigenic stimulation. T cell independent antigens do not induce germinal center formation.

When C3b/iC3b-coated antigen is transported into a secondary lymphoid organ, a portion encounters dendritic cells, B cells, and T cells in the interfollicular areas, causing both T cell and B cell activation, antibody formation, and Ig class switching, as discussed earlier. A portion of the complement-tagged antigen also encounters follicular dendritic cells, is endocytosed, and is subsequently presented on the surface of these cells, as are antibody-antigen complexes. Activated B cells migrate from the interfollicular areas into follicles, in regions called germinal centers where they are driven through several divisions and at the same time subjected to modifications in the Ig variable region gene segments that result in modification of the antigen-binding sites of expressed antigen binding receptors.

This process of somatic mutation has two possible consequences with respect to antigen binding. It can be improved or diminished. This is critical to the fate of the germinal center B cells, because continued cross-linking of their antigen binding receptors is required for the expression of anti-apoptosis molecules by the B cells. Pro-grammed cell death is called apoptosis. In the ensuing competition among antibodies and antigen binding receptors on germinal center B cells, only those B cells whose receptors bind to available antigen survive. This process is called **affinity maturation** and results in B cells whose antigen-specific receptors form very strong interactions, called high affinity interaction, with the selecting antigen. B cells that cannot compete die and are consumed and their components broken down and recycled by tingible body macrophages.

B cell activation and selection in secondary follicles also results in the development of long-lived plasma cells producing IgG with high affinity for the selecting antigen, many of which relocate to the bone marrow. These cells continue to produce antibody for peri-

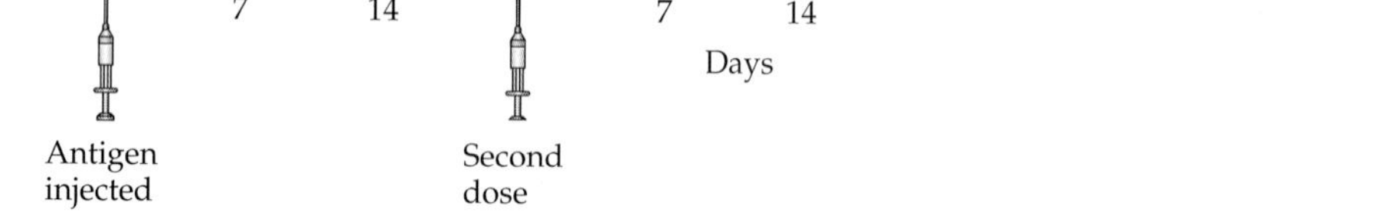

Figure 27.16 Time course of antibody response
A hallmark of the adaptive immune system is memory: On second exposure to an antigen, the immune response develops more rapidly and reaches a higher magnitude than at the first exposure. This results from clonal expansion during primary exposure to the antigen and less-stringent requirements for memory lymphocytes to respond; as a consequence, more antibody is made. Antibody circulates in the fluid component of blood and is usually measured in serum, obtained by collecting the cell-free components from clotted blood. The amount of antibody present for a specific antigen is measured and is typically expressed as the titer, the reciprocal of the greatest dilution of serum that gives a measurable reaction with the antigen.

ods in excess of a year, even when elicited by dead antigens. The conditions that ensure survival of the long-lived plasma cells are still under study.

In addition, B cell activation and selection in secondary follicles gives rise to memory B cells, which are long lived and responsible for secondary immune responses that arise on re-exposure to the priming antigen (Figure 27.16). Secondary immune responses develop more rapidly, are of higher magnitude and affinity compared to primary responses, and involve B cells that have Ig class switched. Immunological memory can last for many years. The extent to which this reflects the life span of individual memory B cells and T cells, the retention of pockets of antigen that could support replication and renewal of the memory pool, and interactions with antibodies and T cells with receptors that react specifically with the antigen-binding portion of receptors of memory B cells (called anti-idiotypic antibodies and lymphocytes) are still under study and are of particular importance to developing more efficient vaccines.

Orderly development of primary antibody responses, Ig class switching, germinal center formation, B cell affinity maturation, and memory cell formation are features of response to dead antigens and some pathogens that are rapidly cleared from the body. These processes can be disrupted, or not occur at all, during acute and chronic infections. For example, in mice that are infected with African trypanosomes, the spleen becomes massively enlarged, the distinctions between areas containing red blood cells (called red pulp) and leukocytes (called white pulp) blur, follicles do not develop, there is massive activation of splenic phagocytes, plasma cells are spread throughout the organ, memory development is poor, and necrotic areas develop. Furthermore, with time, the capacity of the spleen and lymph nodes to develop immune responses is diminished or lost as a result of immunosuppressive products released by activated phagocytes. These gross manifestations of immunopathology are common in hosts infected with disease causing organisms and are the subject of ongoing investigations aimed at elucidating mechanisms of resistance and susceptibility to disease.

Functions of Antibody Classes: IgG, IgA, IgM, and IgE

IgG is primarily viewed as an opsonin. IgG that has bound antigen is in turn bound by IgG-Fc-receptors on macrophages, and the complex is phagocytosed/endocytosed and degraded; the site on IgG that engages the macrophage receptor is exposed only after the IgG has bound antigen. Sites at the C-terminal domains of the IgG molecule also facilitate its transport across the placenta in some species to help protect the fetus and across the gut of the neonate to help protect the newborn before full development of its immune system.

IgG is one of the most abundant serum proteins. There are about 80 mg of protein/ml of serum in an adult cow, and about 10% of this protein is IgG. There is about 10 times less serum IgM than IgG and little or no free IgE (most IgE is bound to mast cells). Serum IgA is usually present at an even lower concentration than serum IgM because of its efficient transport out of the plasma and onto mucosal surfaces (**Figure 27.17**).

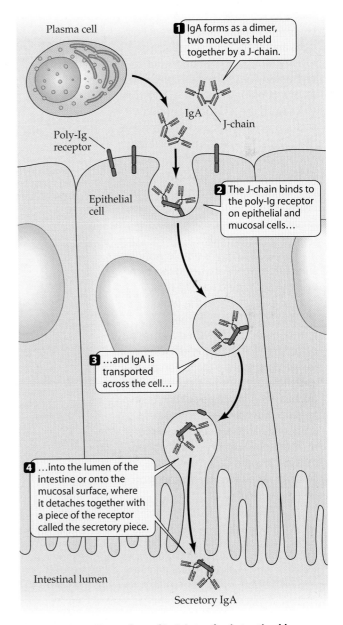

Figure 27.17 Secretion of IgA into the intestinal lumen Plasma cells that produce IgA are enriched in the lymphoid tissues associated with the digestive system. The specialized transport system allows the IgA to get into the lumen of the intestine where it can act against infectious agents that are swallowed and survive the barrier defense systems.

Research Highlights Box 27.1 — Monoclonal Antibody Production

Antigens typically have several different epitopes, each of which stimulates B cells with a distinct B cell receptor. The resulting immune sera recognize many different epitopes even when a single purified polypeptide is used as antigen. These polyvalent sera are not suitable to be used for diagnostic reagents because they often show cross-reactions among different antigens and are difficult to reproduce exactly. Monoclonal antibody (mAb) technology has solved this problem. mAb technology has made it possible to immortalize a clone of antibody secreting B cells and thus to generate an unlimited amount of its secreted antibody. Because all of the progeny cells make the same antibody, there is only one type of antigen-binding site represented. Thus, the immunoglobulin product of a clone of B cells is a monoclonal antibody.

Mab technology is simple. Antigen-stimulated B cells are fused to a tumor cell, and the hybrid has the transformed phenotype of the tumor but still makes and secretes the immunoglobulin specified by the stimulated B cell. A B cell tumor called a myeloma is used as the fusion partner, and the most commonly used myeloma has been modified so that its own Ig genes are no longer functional. Köhler and Milstein developed the method to make and select hybridomas in 1975. It is based on the way that cells obtain purines for synthesis of DNA and RNA.

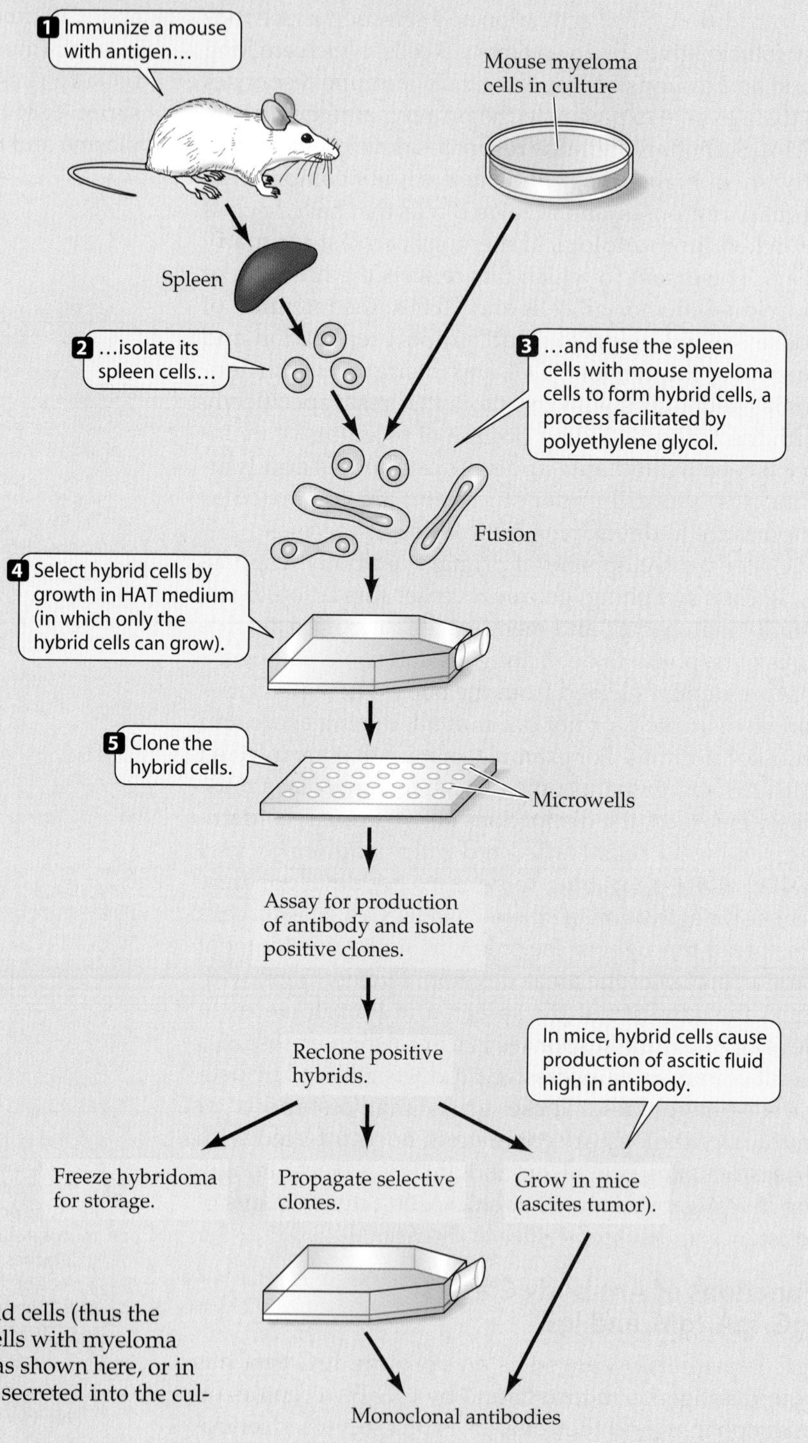

Monoclonal antibodies are produced by hybrid cells (thus the name *hybridoma*) by fusing activated spleen cells with myeloma cells. Clones can be expanded either in mice, as shown here, or in tissue culture flasks, where the antibodies are secreted into the culture medium.

1. Immunize a mouse with antigen…
2. …isolate its spleen cells…
3. …and fuse the spleen cells with mouse myeloma cells to form hybrid cells, a process facilitated by polyethylene glycol.
4. Select hybrid cells by growth in HAT medium (in which only the hybrid cells can grow).
5. Clone the hybrid cells.

Mouse myeloma cells in culture / Spleen / Fusion / Microwells

Assay for production of antibody and isolate positive clones.

Reclone positive hybrids.

In mice, hybrid cells cause production of ascitic fluid high in antibody.

Freeze hybridoma for storage. / Propagate selective clones. / Grow in mice (ascites tumor).

Monoclonal antibodies

IgM is renowned for its capacity to activate complement. IgM that has bound antigen activates complement by the classical pathway (see Figure 27.3A) leading to the generation of pro-inflammatory complement fragments (C3a, C4a, C5a), fragments that promote phagocytosis of materials to which they are attached (C3b, iC3b), fragments that are required for activation of T cell dependent B cells (C3d, C3dg), and assembly of the

The genetic material of cells is made up of purines and pyrimidines. Each time a cell replicates, it must make or obtain these essential components. Cells can make purines by the de novo synthetic pathway or obtain preformed purines from medium or extracellular fluids by the salvage pathway. The de novo synthetic pathway is sensitive to the drug aminopterin, a folic acid antagonist that blocks de novo purine synthesis. Tumor cells that are incubated in vitro with aminopterin continue to replicate in medium-containing purines because they can obtain the purines by the salvage pathway.

In the salvage pathway, preformed purines are imported into the cell and remodeled. Hypoxanthine guanine phosphoribosyl transferase (HGPRT) is a key enzyme in the purine salvage pathway. This enzyme adds a phosphoribosyl group to the purine base, thus converting it to a purine nucleotide, which is the form of purine present in DNA. Köhler and Milstein developed a myeloma cell line that lacked HGPRT by culturing mutagenized cells with a toxic purine analogue called 8-azaguanine. Cells with an active purine salvage pathway incorporated 8-azaguanine and died. 8-azaguanine resistant cells lack a purine salvage pathway and rely solely on their de novo purine synthetic pathway for purines required for DNA and RNA synthesis. 8-azaguanine resistant cells are sensitive to aminopterin, which blocks de novo purine synthesis.

Aminopterin-sensitive myeloma cells were fused to antigen-stimulated B cells to generate hybrids. Fusion was encouraged using polyethylene glycol (PEG), which causes the membranes of closely juxtaposed cells to meld into a single lipid bilayer incorporating the content of both cells including their nuclei. Fused cells randomly assort chromosomes at the next metaphase. The resulting progeny cells that arise at cell division contain a mix of chromosomes. In a medium containing aminopterin and hypoxanthine, only those cells that have obtained an HGPRT gene from the B cell, the transforming gene from the myeloma, and a full complement of relevant housekeeping genes multiply, and all other cells die. The growing hybridomas are cloned by placing single cells in a liquid medium or by growing single cells as colonies on a medium that contains gelling agents, and antibodies produced by the hybridoma cells are screened for antigen-binding activity.

---

membrane attack complex. The site on IgM that activates complement is exposed only after IgM has bound antigen. IgM is much more efficient at initiating the classical pathway of complement activation than IgG, whereas IgA and IgE do not activate complement at all. Activation of complement by antibody requires the binding of a plasma complement component called C1q, a large protein with six globular heads, two of which must be bound for a stable C1q antibody interaction. Each globular head group of C1q engages a site that is exposed on the Fc region of IgM or IgG when it is in complex with antigen. Because IgM is a pentamer, it potentially has five C1q binding sites, whereas IgG, being a single subunit, has only one. Thus, the concentration of IgG required to elicit complement activation is very high because two IgG molecules must attach to antigen within 30 to 40 nm of each other to provide two attachment sites for C1q. Once two sites of C1q are engaged, C1r and C1s are recruited into the complex, C1 is activated, and the pathway proceeds, as shown in Figure 27.3B.

IgA is specialized for transport across mucosa, where its presence on the lumenal surface helps prevent pathogens with target antigenic epitopes from crossing the mucosa as well as to bind and inactivate toxins.

Transport of IgA across the mucosal epithelial cells involves binding to a receptor on these cells called the poly-Ig receptor. This receptor binds to the J-chain of dimeric IgA. It can also bind to the J-chain of IgM but less efficiently than to that of IgA, even though these have the same J-chain. IgA is transported through the mucosal epithelial cell in association with the poly-Ig receptor, the piece of the receptor that binds J-chain is enzymatically cleaved in the cell. The dimeric IgA is released on the lumenal side of the cell with its associated section of the poly-Ig receptor, now called the secretory piece. IgA is commonly detected in tears, saliva, feces, and the fluids bathing the urogenital tract.

IgE binds to IgE-Fc-receptors on mast cells, and when the mast cell-bound IgE is engaged by antigen, a signal is propagated that causes mast cell degranulation; histamine released in this process is responsible for many of the symptoms of hay fever, which is an allergic response to pollen. IgE also binds directly to antigen, resulting in changes in its Fc region that causes it to bind to Fc ε receptors on eosinophils. Cross-linking of these receptors causes eosinophils to degranulate, releasing materials that are toxic for parasitic organisms (**Figure 27.18**). This is an important component of defense against parasitic worms.

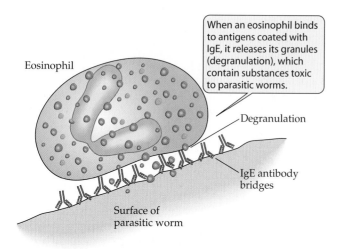

Figure 27.18 Eosinophil interaction with parasitic worm
An eosinophil interacts with a parasitic worm through antibody bridging. Eosinophil granules contain several substances that are toxic to parasitic worms, including eosinophil peroxidase, cationic proteins, neurotoxin, and lysophospholipase.

Figure 27.19 presents a summary of the roles played by antibodies of different classes in host defense. To get a pure population of antibodies of one particular isotype reactive with a single antigen epitope, hybridomas are made that secrete monoclonal antibodies (Box 27.1).

BRIDGING IMMUNE RESPONSES

The inflammatory/innate response is rapid because participating molecules and cells are already present in the blood, and cell replication is not required for this response to occur. The adaptive immune response is slow because participating lymphocytes have to replicate to achieve a critical number. The bridging immune response falls between these and involves lymphocytes that can respond rapidly to pathogens. The functions of cells that make up the bridging immune system are the same as those of cells in the antigen-specific adaptive immune response (**Table 27.7**). However, there are significant differences in how the cells of the bridging immune system are triggered to display their functional activities, even when they have an antigen-specific receptor that is structurally very similar to that of classical T and B cells. These include a class of B cells that arises early in ontogeny and responds to TI-2 antigens called B-1 B cells, γδ T cells, and CD1-restricted T cells. In most instances, these cells do not seem to display memory responses, constitute only a few percentage of the total leukocyte population of the blood, and have limited diversity of their antigen receptors. This limited diversity may allow a significant response by an otherwise small population of cells and in this way precludes the need for development into memory cells.

Primitive T Cells: NK Cells

Natural killer (NK) cells are in the lymphoid lineage and are generally considered to be primitive T cells because they have a common progenitor cell with other T cells. However, NK cells do not mature in the thymus as T cells involved in adaptive immune responses, nor do they express an antigen-specific T cell receptor. Nevertheless they share two functions in common with T cells of the adaptive immune system, that is, production of the cytokine IFN-γ and cytotoxicity.

NK cells are important during the first few days of an infection before the adaptive immune response has developed. For example, NK cell production of IFN-γ is important in protection against the intracellular bacteria *Listeria monocytogenes*. In the absence of NK cells, some infections have been shown to be lethal in mouse

Table 27.7	Cells of the bridging immune system		
Type of Cells	**Functions**	**Types of Antigen Interaction**	**Receptors Involved in Activation**
NK cells	IFN-γ production; cytotoxicity; ADCC	Microbial antigens via antibody; altered self-cells	NCR NKG2A NKG2D KIR Fc
γδ T cells	IFN-γ production; cytotoxicity	Soluble antigens; non-proteinaceous; self-antigens	γδ T cell receptor NKG2D
NKT cells	IL-4 production IFN-γ production	Lipid-containing bacterial proteins	Invariant αβ T cell receptor
CD5 B cells (B-1 B cells)	Make IgM antibodies	Common carbohydrates of bacteria, self-antigens	B cell receptor

	Effect	Types of infectious agent targeted	Examples of infectious pathogen component or structure targeted	Isotype of Ab involved	Where antibody is found
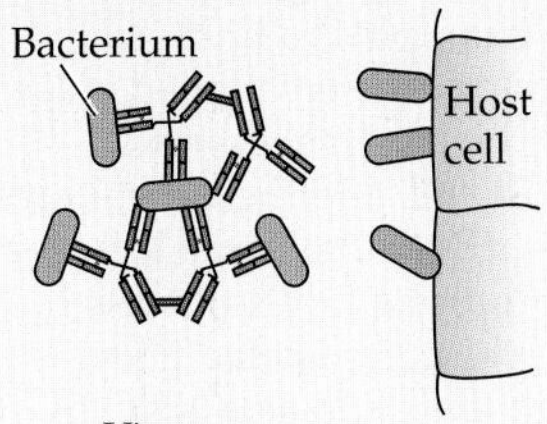 Bacterium / Host cell	Prevents attachment to host cells for colonization of the body	Bacteria	Fimbriae, pili, flagella, LPS, outer membrane proteins of bacteria	IgA	Mucus membranes in upper respiratory tract, intestinal tract, vagina
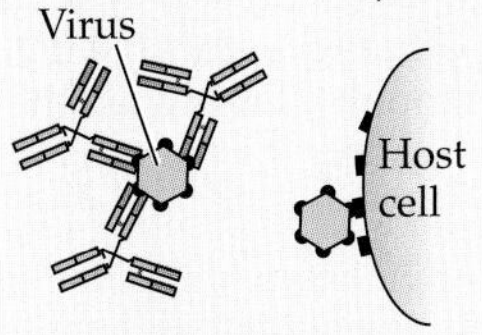 Virus / Host cell	Prevents attachment and entry into host cells for infection	Viruses	Influenza virus hemagglutinin for attachment and neuraminidase for entry into host cell	IgA IgG	Depends on the target cells for virus infection: IgA for cells associated closely with mucus membranes and IgG for internal tissues and organs
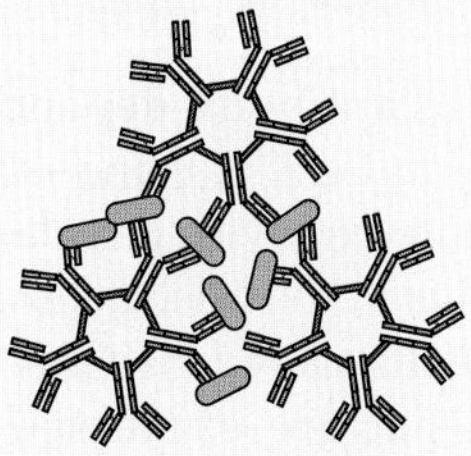	Agglutination	Bacteria, viruses, fungi, protozoa	Any external structures with antigenic epitopes	IgM IgG	Body tissue
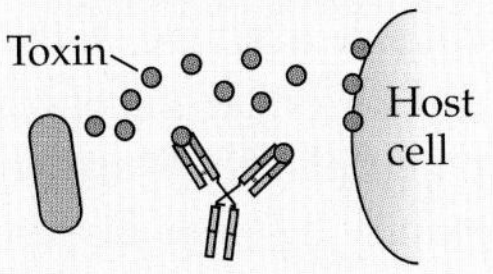 Toxin / Host cell	Neutralizes toxins	Produced by bacteria	Secreted toxins, such as cholera toxin, tetanus toxin	IgA IgG	Gut lumen and tissues
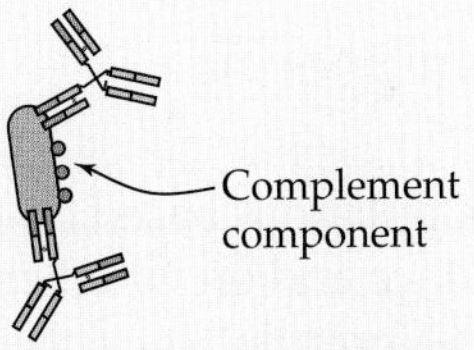 Complement component	Kills by activating complement onto microbial surface	Bacteria, viruses, protozoa	Any external surface	IgM IgG	Internal tissues and organs
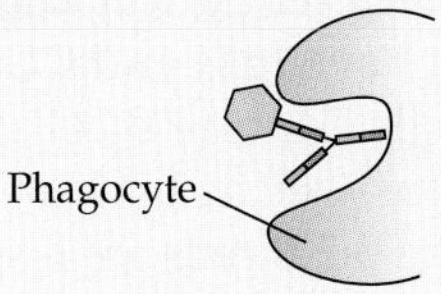 Phagocyte	Opsonization for phagocytosis	Bacteria, viruses, protozoa	Any external surface	IgG	Anywhere phagocytic cells found
See Figure 27.18	Tag for recognition by eosinophils	Helminths	Tegument	IgE	Intestinal tract
See Figure 27.20	Tag for ADCC	Cells infected with enveloped virus	Influenza virus	IgG	Where it comes in contact with NK cells in internal tissues and organs
	Kills by activating complement on surface of host cells infected with enveloped virus	Cells infected with enveloped virus	Influenza virus	IgG IgM	Internal tissues and organs

Figure 27.19 Antimicrobial activities mediated by antibodies
This diagram summarizes the various activities of antibodies to prevent infection.

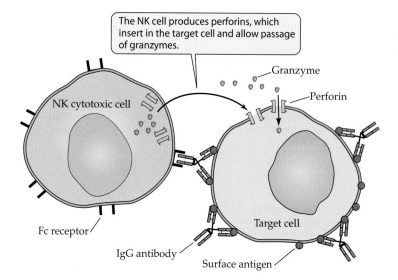

The NK cell produces perforins, which insert in the target cell and allow passage of granzymes.

NK cytotoxic cell

Granzyme

Perforin

Fc receptor

IgG antibody

Surface antigen

Target cell

(B)

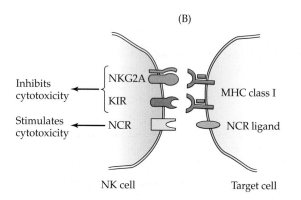

Inhibits cytotoxicity

NKG2A

KIR

MHC class I

Stimulates cytotoxicity

NCR

NCR ligand

NK cell

Target cell

Figure 27.20 ADCC by NK cells

(A) Cytotoxic NK cells have Fc receptors for IgG antibodies, which allow the cells to bind antibody-coated cells on which the antibody Fc piece is accessible, and kill them. This is known as antibody-dependent cellular cytotoxicity (ADCC). An example is a host cell infected with an enveloped virus; the enveloped virus inserts viral proteins into the host cell membrane before budding out of the host cell, and these viral proteins bind specific antibodies. (B) NK cells can kill host cells that do not express appropriate levels of class I MHC molecules, such as virus-infected cells. They evaluate MHC levels through their KIR and NKG2A receptors. They also express other receptors (NCR), which results in activation through CD94 when the receptors engage the correct ligand.

models. NK cells are also important in controlling virus infections by IFN-γ secretion.

NK cell cytotoxicity (**Figure 27.20**) is mediated by a mechanism quite similar, if not identical, to that used by CD8 T cells. When activated, the NK cells have granules containing perforin in their cytoplasm. These are visible under the light microscope and the NK cells are therefore known as large granular lymphocytes. NK cells kill target cells that do not express normal levels of MHC class I molecules. They have receptors for class I MHC molecules that, when engaged, signal the NK cells not to kill the target cell. The reader may recall that one way in which viruses try to thwart the adaptive immune system is by decreasing MHC class I expression, thereby decreasing the presentation of antigen on class I. However this backfires to an extent since reduced levels of class I MHC on these cells makes them vulnerable to killing by NK cells. In this way the NK cells act as a backup for the adaptive immune system.

NK cells also have other receptors. Fcγ receptors allow them to bind to IgG and antigen complexes. These receptors can link NK cells to target cells bearing antigen with IgG bound, allowing the NK cell to kill the tar-

get cell. This process is known as **antibody-dependent cellular cytotoxicity (ADCC)** (Figure 27.20) and might be effective against host cells infected with enveloped viruses that display viral proteins in the host cell membrane. NK cells also have a receptor called NKG2A, for MHC-class I-like molecules that are expressed by mycobacterial-infected cells. When this receptor is stimulated, the NK cells display cytotoxic activity. A third type of receptor, called the natural killer cell receptors, or NCR, also provides activation signals. NCR are only on NK cells and mediate cytotoxicity against Epstein-Barr virus-infected cells and against tumor cells.

Lymphoid Cells with Antigen-Specific Receptors: B-1 B Cells

B cells can be divided into subpopulations based on the Ig classes expressed on their surface and on whether or not they have an additional differentiation antigen called CD5. Mature B cells that have IgM and CD5 on their surface are called B1-B cells. Mature B cells that lack CD5 and have surface IgD as well as IgM, at least in mice, are called B-2 B cells. The B-2 B cells are responsible for adaptive immune responses against T cell dependent antigens.

The antigen binding receptors of B-1 B cells, and the antibodies they produce, are skewed toward reactivity with common pathogen-associated carbohydrate antigens and toward weak (and harmless) reactivity with self-antigens. B-1 B cells produce **natural antibodies**, which are antibodies of the IgM class that are sustained in plasma without obvious antigenic stimulation. B-1 B cells spontaneously produce these antibodies when placed in tissue culture or moved to an Ig-free host animal.

Natural antibodies produced by B-1 B cells are reactive with bacterial lipopolysaccharide (LPS), the archetype TI-2 antigen discussed earlier, and consequently limit nonspecific B cell activation and endotoxin shock induced by LPS. They also contribute to resistance to influenza infection and resistance to infection with *Streptococcus pneumoniae*. In addition, there is good evidence that a significant portion of IgA-secreting plasma cells in the lamina propria of the gut derive from B-1 B cells, suggesting an important role in controlling enteric infections.

Lymphoid Cells with Antigen-Specific Receptors: γδ T Cells

The antigen-binding receptors on γδ T cells (**Figure 27.21**) are similar to those on αβ T cells, and the generation of diversity occurs by the same mechanism as used by αβ T cells (i.e., rearrangement of specific gene segments). However, the genes that code for the γδ T cell receptor are largely distinct from those coding for the αβ T cell receptor, and overall, the T cell receptor diversity γδ T cells is quite limited. Within a tissue or organ there may be only a few γδ T cell receptor gene combinations expressed. The CD3 signaling complex of molecules

associated with the T cell receptor is identical for the two T cell lineages. However, γδ T cells do not mature in the thymus in the same manner as αβ T cells because they do not undergo positive and negative selection, and the γδ T cells in the gut do not go to the thymus at all.

Although very few molecules have been defined that stimulate γδ T cells through the T cell receptor, it is clear that they differ significantly from those that stimulate αβ T cells. The molecules that stimulate γδ T cells do not require presentation on MHC molecules, nor are they necessarily derived from proteins nor even foreign. For example, they include non-proteinaceous phospholigands, which are produced by species of *Mycobacteria* and the malaria-causing parasite *Plasmodium falciparum* and alkyl amines from bacteria. A number of self molecules that stimulate γδ T cells have been identified, including MHC class I-like molecules called MICA and MICB and heat-shock proteins that cross-react with mycobacteria heat-shock proteins.

The limited T cell receptor diversity means that within an organ, relatively large numbers of γδ T cells have identical T cell receptors. The only time that the frequency of αβ T cells with identical T cell receptors is as high is at the height of an infection after they have undergone tremendous clonal expansion. This peculiarity may enable γδ T cells to have an immediate effective response to conserved molecular components of microbial pathogens, abnormal host cells such as tumor cells, and even normal self-components under special circumstances, such as during an inflammatory response, without needing to expand to establish sufficient numbers. In support of this, an influx or increase in the number of γδ T cells has been reported for a large number of infectious diseases, including those mediated by viruses, bacteria, and protozoa.

γδ T cells are most often reported to produce IFN-γ, and thus it has been suggested that they may generally make a type 1 cytokine response. This is consistent with their role in limiting virus infections and playing a crucial role in bacterial infections, particularly the development of granulomas in mice with tuberculosis. However, it

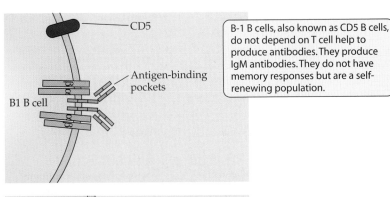

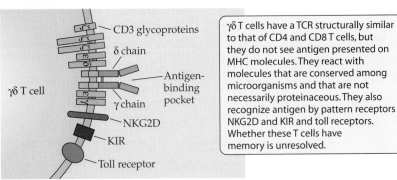

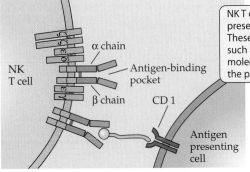

Figure 27.21 Bridging lymphoid cells
Stimulation of bridging lymphoid cells that have antigen-specific receptors.

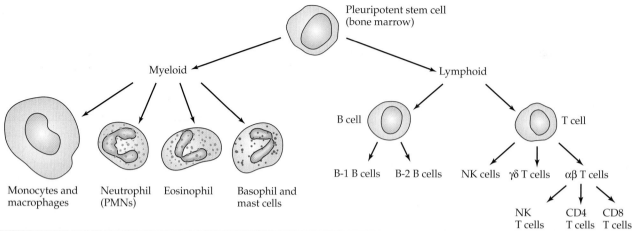

Cytokine	Produced by (cell type)	Cytokine function	Acts on (cell type)
IL-1	Macrophages, monocytes	Costimulatory molecule; fever inducer	CD4 T cells
IL-2	Th1 CD4 T cells	T cell growth factor	CD4, CD8 and γδ T cells; NK cells
IL-3	Th1 CD4 T cells	Colony stimulating action	Myeloid stem cells
IL-4	Th2 CD4 T cells, NKT cells, dendritic cells, macrophages	B cell growth factor; drives T cells toward Type 2 and induces proliferation	B cells; CD4 T cells
IL-5	Th2 CD4 T cells	Promotes growth and differentiation	B cells; eosinophils
IL-6	Th2 CD4 T cells, macrophages	Promotes terminal differentiation into plasma cells	B cells
IL-8	Activated macrophages	Recruits phagocytes to site of inflammation	Macrophages and neutrophils
IL-10	Th2 CD4 T cells, B cells	Decreases antigen presentation; decreases macrophage activation; decreases IFN-γ production by T cells	Macrophages
IL-12	Macrophages and dendritic cells	Induces IFN-γ production	T cells, NK cells
IFN-γ	Th1 T cells, γδ T cells, NK cells	Activates macrophages; inhibits virus replication; directs T cells toward Type 1	Macrophages, T cells
IFN-α and β	Macrophages, fibroblasts, virus-infected cells	Inhibits virus replication; activates NK cells for cytolytic activity	Virus-infected cells; NK cells
TNF-α	Th1 T cells, macrophages	Activates macrophages; fever inducer; cachexia	Macrophages
TGF-β	Macrophages, lymphocytes, mast cells, platelets	Antibody class switching	B cells

Figure 27.22 Relationship of all cell types involved in immunity and cytokines they produce
All the immune system cells discussed in this chapter arise from a common leukocyte precursor in the bone marrow and differentiate into two main lineages: myeloid and lymphoid. They interact through a variety of soluble molecules called cytokines that act on other cells of the immune system.

has also been shown that they may make the type 2 cytokine IL-4 in animals with a parasitic worm infection. γδ T cells also have been shown to have cytolytic activity against foreign cells, although cytotoxicity is not necessarily a common feature of all γδ T cells. Also, because of their potential ability to produce type 1 and type 2 cytokines and to respond immediately, it has been suggested that they may direct development of adaptive immune responses by αβ T cells and B cells by virtue of the cytokines they secrete.

Based on current evidence, γδ T cells may respond as part of the adaptive immune response framework as well as to respond similarly to innate/nonadaptive immune responses, bridging the two systems. The latter is supported by the fact that they share elements with natural killer (NK) cells, clearly cells of the nonadaptive system. That is, they express the receptor molecule NKG2D, which stimulates cytotoxic responses by γδ T cells. A role in adaptive immune responses for γδ T cells is supported by studies showing that they may have a memory response to antigens of the bacteria *Leptospira*, *Mycobacteria* and *Listeria*. In *Listeria* infections, γδ T cells play a perceptible role in resistance to secondary challenge, although it is less than that of αβ T cells and only marginally greater than in primary infections.

Lymphoid Cells with Antigen-Specific Receptors: CD1-Reactive T Cells

CD1 is a nonclassical MHC molecule that does not vary among individuals of a species and that presents certain microbe antigens to specific populations of T cells. In humans there is a population of αβ T cells that recognize CD1-presented antigens, appear to have memory responses, and produce IFN-γ and kill mycobacteria-infected cells. They respond to lipid-based antigens presented in the deep groove of the CD1 molecule, including those from mycobacteria, and thus are thought to have a role in protection of bacteria. The lipid-containing antigens can be obtained from various endosomal compartments in macrophages.

A second type of MHC CD1-reactive T cells, called NK T cells, expresses the natural killer cell marker NK1.1. These cells have one type of T cell receptor gene expressed only. They respond quickly to infection, and make either IFN-γ or IL-4 in response to glycolipids presented on CD1 molecules. The NK T cells have been shown to be important in preventing growth of the malaria-causing protozoan *Plasmodium* in liver cells and participate in granuloma formation to *M. tuberculosis*.

The relationship of the cells of the innate, adaptive, and bridging immune systems and their derivation from stem cells as well as the cytokines produced by these cells are summarized in **Figure 27.22**.

VACCINES

Immunological memory is a characteristic of the adaptive immune response by both αβ T cells and B-2 B cells and is the keystone of vaccination against infectious diseases.

Passive and Active Immunity

Immune protection may be induced by immunization (called active imunity) or passively conferred on an individual by transfer of antibodies (called passive immunity) (**Figure 27.23**). Passive immunity occurs naturally in humans where antibodies may cross the placenta. Antibodies are also delivered to newborns in the colostrum of mother's milk. Before birth, the antibodies become concentrated in the mother's breast ready to impart an instant immunity to the newborn, which will require considerable time to build up active adaptive immune responses to infectious disease agents.

Passive immunity is also sometimes given to people and animals in the form of antibodies called γ globulin. Antibodies produced against a particular infectious agent in one individual are purified and given to another individual. This is used in situations where the infection is acute and may kill the individual before a protective immune response can develop, for example, to treat exposure to rabies and tetanus. In some cases, it has been used as a temporary prophylactic, for example, against hepatitis. The disadvantage of passive immunity is that it is short lived because IgG antibodies have a half-life of less than 3 weeks.

Active immunity is induced as a result of survival of a natural infection or through vaccination. Vaccines may be live organisms that have been attenuated so that they do not cause full-blown disease, dead organisms, or portions of organisms that contain the antigenic determinants needed to induce a protective immune response (Figure 27.23). The use of vaccines to control infectious diseases such as smallpox, rabies, tetanus, anthrax, cholera, and diphtheria has been one of the great success stories of modern medicine. In the United States, the administration of effective vaccines has reduced the number of reported cases of diphtheria, measles, mumps, pertussis, poliomyelitis, rubella, and tetanus by at least 97%. No other form of disease control has had such an effect on the reduction of mortality. When vaccinated correctly, the immunity induced is long lasting and may be extended by booster vaccines given at intervals, such as for the tetanus vaccine.

The advantage of living vaccines is that the organisms that make up the vaccine will increase in number following immunization, thus increasing and prolonging the exposure of the host to antigen and minimizing the need for multiple doses or boosters. In addition, if engagement of a CD8 T cell response is necessary for

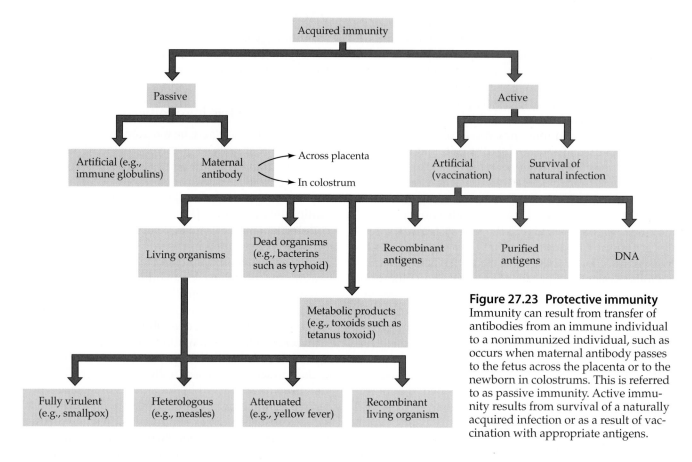

Figure 27.23 Protective immunity
Immunity can result from transfer of antibodies from an immune individual to a nonimmunized individual, such as occurs when maternal antibody passes to the fetus across the placenta or to the newborn in colostrums. This is referred to as passive immunity. Active immunity results from survival of a naturally acquired infection or as a result of vaccination with appropriate antigens.

protective immunity, for example, to a virus or intracellular bacteria or protozoa, a living vaccine is needed. However, if living vaccines are to be used, the organism must be an **avirulent** strain that will induce cross-protective immunity to the **virulent** (or disease-causing) strain. This can be achieved by passaging the organism in unusual culture conditions. For example, the intracellular bacteria *Mycobacteria bovis* was attenuated by growing it for 13 years on a bile-saturated medium and is now used in many countries as a vaccine against tuberculosis, caused by the heterologous bacteria *Mycobacterium tuberculosis*. The living poliovirus vaccine is obtained by growing the virus in monkey cells, whereas the rubella virus vaccine is grown in duck-embryo cells.

A more modern way of attenuating organisms for vaccine development is to disrupt genes responsible for virulence, while retaining those that encode host-protective antigens. Although these types of vaccines are theoretically possible, they are less easy to achieve than was predicted at the time of conception. For example, deleting genes to attenuate *Brucella* spp. has revealed that the loss of a single gene product can usually be compensated for and thus the virulence not reduced significantly. Alternatively, depletion of some genes may result in such severe attenuation that the organism does not survive long enough in the host to induce a protective immune response. Another option is to transfer genes that code for host-protective antigens of a disease-causing infectious agent into an unrelated nondisease-causing organism. The recipient organism will then express the gene from the virulent organism and thereby produce the antigens of the pathogen needed to stimulate an immune response. This is known as a recombinant vaccine.

The use of dead organisms (for example, intact killed bacteria, viruses, protozoa) or portions of those infectious agents has an advantage over using living organisms as vaccines. That is, the chance for causing disease is minimal. For example, the vaccine against tetanus consists only of the toxin of tetanus that has been inactivated (referred to as a **toxoid**). Antibodies to the toxin are sufficient to protect the recipient of the vaccine from disease. Specific antigens to be used as a vaccine can either be purified from virulent organisms using biochemical techniques or made artificially as recombinant proteins. Recombinant proteins are proteins made from genes of one organism in another harmless organism. The disadvantage of killed vaccines is that they often require more doses or boosters to achieve sufficient antigen load, and they do not usually engage responses by CD8 T cells. However, if the protective immune response is one mediated by antibodies, these vaccines may be appropriate.

One of the latest innovations in vaccinology is the **DNA vaccine**. Here DNA from an infectious agent is injected into muscle with a needle or stuck onto a gold particle and shot into a tissue by a gene gun. Either way, the DNA enters a host cell, allowing the encoded proteins to be made. The vaccinated host then makes an immune response to these foreign molecules even though the host's own cells are making the antigens. The usefulness of this type of vaccine is that only DNA that is needed to code for antigens needs to be given. Moreover, because those antigens are produced within the host's cells, the antigen can be presented by the endogenous pathway to CD8 T cells even though the host never received the living infectious agent. In addition, DNA is less affected by temperature and by long-term storage than proteins and its use as material for vaccination may remove a need for a cold-chain (i.e., refrigeration units) for transport and storage. This has obvious advantages for vaccination to be used in developing countries.

DYSFUNCTIONS OF THE IMMUNE SYSTEM

There are several situations when immune responses cause more harm to the host than benefit. Sometimes the immune system responds to antigens that are not actually a threat to the health of the individual. This type of immune response is commonly called **allergy**. In some cases, an overzealous immune response can actually result in death of an individual. This occurs with allergies, as well as in response to bacterial infections. The later are called **toxic shock** and **endotoxemia**. Finally, the immune system may also respond to self-molecules, recognizing these as appropriate antigens. This is known as **autoimmunity** and can result in a chronic debilitating state or even death. Current ideas about autoimmune responses suggest that they may actually be initiated as a response to antigens of infectious agents and that subsequently these same lymphocytes cross-react with self- or autologous molecules.

The other side of the coin is an inability to mount a sufficient immune response to control infections. Some individuals or animals are born with a defect in the immune system, known as an **immunodeficiency**, which interrupts one layer of the immune system and leaves the individual vulnerable to infection. In other cases, immunodeficiencies may be transient, resulting from poor health, poor nutrition, or stress. Infections themselves can also result in immunodeficiencies, such as human immunodeficiency virus, which causes acquired immunodeficiency disease syndrome (AIDS) primarily by infecting CD4 T cells. Because of this immunodeficiency, AIDS patients often die as a result of an uncontrolled secondary infection with other infectious agents.

These aberrations of an effective immune response are briefly discussed.

Allergy

Immune responses that are against antigens that do not pose a threat to the host but are persistent and annoying are often referred to as allergies. Allergies are actually immune responses that serve little useful purpose for the host. They fall within two categories of immune responses known as **hypersensitivities**: type I and type IV. Type I hypersensitivity is known as **immediate hypersensitivity**, because it takes minutes to occur, whereas type IV hypersensitivities are known as **delayed type hypersensitivities**, because this type of response takes 2 days to become apparent after contact with the allergy-causing agent.

Type I hypersensitivity responses are mediated by the IgE class of antibodies and mast cells. The reader will recall that in preceding sections, the IgE/mast cell collaborative response was discussed with respect to vasodilation, and the development of inflammation and edema. The mechanism is the same for an allergic response that falls under the type I hypersensitivity category. However, in this case, the antigen that cross-links the IgE molecules on armed mast cells is pollen, mold, insect or snake venom, food, antibiotics, or insect saliva. The mast cells degranulate, and the symptoms are those common among hay fever sufferers, that is, itchy eyes, running nose, and congestion, which are annoying but rarely life threatening. Treatment usually involves taking antihistamines to counteract histamine released by mast cells.

The response may progress to **anaphylactic shock** if enough mast cells are armed with the appropriate IgE and sufficient quantities of antigen are introduced to cause a generalized degranulation of mast cells throughout the body. Anaphylactic shock consists of itching, redness, hives, swelling of the throat, pulmonary edema, and heart failure. The hives represent areas of local mast cell degranulation, whereas the swelling and edema are the result of mast cell release of histamine that causes vascular permeability. The heart failure results from the difficulty presented to the heart of pumping blood through lungs filled with fluid. Death due to full-blown anaphylactic shock can occur in less than one-half hour.

People prone to these severe allergies may carry adrenaline (norepinephrine) syringes to counteract the anaphylactic shock and seek a treatment of desensitization. The principle behind desensitization is that injection of small, incrementally increasing doses of the allergen (i.e., the molecule to which the person is allergic) intradermally will stimulate production of IgG antibodies that will mop up the antigen when it is introduced, thus preventing it from cross-linking IgE on mast cells. The norepinephrine works by stimulating the β-receptor on mast cells that prevents their degranulation.

Allergies associated with type IV hypersensitivities are referred to as **allergic contact dermatitis** and are commonly associated with responses to poison ivy, nickel, formaldehyde, and neomycin. In this type of allergy the molecule that induces the response does so by complexing onto the surface of dendritic cells in the skin. CD4 T cells respond to this complex by producing cytokines that recruit macrophages. This can result in symptoms such as extreme itchiness and is treated with steroids that inhibit T cell responses. The type IV hypersensitivity reaction is used as a diagnostic test for tuberculosis. Injection of tuberculin (killed preparation of mycobacterial antigens) into the skin results in development of a small white bump at the site if the individual has T cells that have been sensitized to these antigens.

Toxic Shock Syndrome and Endotoxemia

Some microbial proteins are unique in that they can stimulate as many as 20% of the total T cell population. These proteins are called **superantigens** and result in a life-threatening condition called toxic shock syndrome. Normally, a limited number of clonal populations of T cells respond to any foreign antigen, which results in a controlled and coordinated immune response (**Figure 27.24**). Generally the proportion of T cells that would respond to an individual antigen would be much less than 0.01% of those in the body. Superantigens stimulate T cells by binding to those with a particular type of β chain in their T cell receptor and linking it to MHC class II molecules on other cells by binding outside of the antigen-binding grooves of both the T cell receptor and the MHC molecule (Figure 27.24). Because of the large number of T cells activated, a powerful immune response is

elicited that results in overproduction of cytokines such as IFN-γ by the T cells and thus overproduction of IL-1 and TNF-α by activated macrophages. This results in toxic shock syndrome due to fever, widespread blood clotting, and shock. Superantigens made by streptococci and staphylococci bacteria include the toxic shock syndrome toxin from *Staphylococcus aureus* and enterotoxins from *Streptococcus pyogenes*.

A second type of life-threatening immune response to bacteria is known as endotoxemia. Gram-negative bacteria have endotoxin, also called lipopolysaccharide, as a major component of their outer cell envelope. Because of the ability of bacterial endotoxin to stimulate macrophages through pattern receptors (see pg. 649), there may be an overproduction of macrophage cytokines including IL-1 and TNF-α resulting in endotoxemia. As for toxic shock, this results in development of a fever, drop in blood pressure, and widespread blood clotting, often resulting in death. Bacteria with endotoxin molecules that mediate this violent response include *E. coli*, *Klebsiella pneumoniae*, *Pseudomonas aeruginosa*, and *Neisseria meningitidis*.

Autoimmunity

Three processes contribute to the presence of lymphocytes that react with our own antigens (self-reactive lymphocytes) in the peripheral circulation, which is the cause of **autoimmunity**. These processes are: (1) failure of clonal deletion of thymocytes and pre-B cells to eliminate all self-reactive cells, because not all self-antigens are accessible in the thymus or in the circulation; (2) positive selection of weakly self-reactive B-1 B cells; and (3) development of B cells with a self-reactive antigen-specific receptor as a result of somatic mutation in the germinal center and positive selection by endogenous or acquired antigens present on follicular dendritic cells. Although all of us have self-reactive lymphocytes; only 5% to 7% of us will develop debilitating autoimmune disease. The rest of us are spared by good fortune (i.e., we do not encounter infections or other environmental conditions that stimulate these self-reactive lymphocytes).

While some self-reactive peripheral CD4 T cells encounter their target antigens in the absence of costimulator activation and enter an inactive state, other self-reactive lymphocytes encounter their target antigen under activating conditions. These are usually associated with infections and inflammation. Infectious agents with molecules that are cross-reactive with self, called **molecular mimicry**, induce costimulator activity on antigen-presenting cells and can stimulate autoimmune responses. For example, the immunodominant antigen epitope of *Borrelia burgdorferi* outer surface protein A (amino acids 165–173) is both homologous to a sequence in human lymphocyte function associated antigen-1 and predicted to bind in the groove of the human MHC class

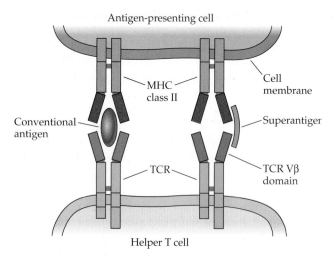

Figure 27.24 Bacterial superantigens trigger T cell responses
Bacterial superantigens directly link the TCR with MHC class II molecules. They bind to a region of the TCR V β domain outside the antigen-binding groove and tie this to a region of MHC class II molecules that lies outside the antigen-binding groove.

II allele HLA-DRB1*0401 and related alleles. Thus, in people with the appropriate MHC molecule, this self-like bacterial antigenic epitope can be presented to CD4 T lymphocytes. That this happens is shown by a high frequency of treatment-resistant Lyme arthritis in people with the HLA-DRB1*0401 allele.

Many other instances of molecular mimicry have been noted, including between poliovirus VP2 and the receptor for acetylcholine, between papilloma virus E2 and insulin receptor, between HIV-1 envelope proteins and HLA-DR4 and DR2, variable regions of T cell receptors, Fas protein and IgG, and between an antigen of *Onchocerca volvulus* and a 44 kDa protein of ocular tissue. Immune responses against the self-like epitopes on pathogens may contribute to autoimmune pathology. In this regard, cross-reacting B cell epitopes may account for infection-associated pathology, e.g., antibodies against *Mycobacterium leprae* react with human skin.

Autoantibodies that arise in infections may not result from cross-reacting pathogen and self-antigens. A variety of different autoantibodies are detected in people infected with African and South American trypanosomes and arise as a result of nonspecific B cell activation due to the infection. Other situations where pathogens elicit autoimmune responses by processes other than molecular mimicry include T cell responses to superantigens. These can lead to the activation of T cells with self-reactive T cell receptors (for example, staphylococcal enterotoxin A superantigen stimulates T cells that cause thyroiditis), and human endogenous retroviral superantigen is induced by IFN-α, an inflammatory cytokine and activates T cells that cause Type 1 diabetes. Infection can also break self-tolerance by a process called epitope spreading. Here, an ongoing cytolytic response results in release of self-antigens that stimulate an autoimmune response. This is implicated in a model of multiple sclerosis in mice, brought on by Theiler's murine encephalomyelitis virus.

Whether or not a pathogen induces an autoimmune response may reflect host regulatory processes. For example, the development of autoimmunity may require strong Th1 T cell help and not occur if a strong Th2 response is stimulated, or the response may be suppressed by natural regulatory T cells either through antigen binding receptor (idiotypic) recognition pathways, or through natural CD4+ CD25+ regulatory T cells.

Immunodeficiencies

While there are many types of inherited immunodeficiencies, an example of one that affects the innate immune system and one that affects the adaptive immune response are described here. **Leukocyte adherence deficiency (LAD)** is a defect of the innate immune system that occurs in humans and animal species, including cattle and dogs. The defect is in expression of the molecule required for neutrophils to adhere to the wall of blood vessels in inflamed tissue. As a result, neutrophils cannot be recruited from the blood to the site of infection. Animals and people with such a deficiency have recurrent bacterial infections because the ability to control them at the level of the innate immune response is severely impaired.

Severe combined immunodeficiency (SCID) affects the adaptive immune response and occurs in humans and animal species such as horses. Neither the T cells nor the B cells function, and therefore there is no adaptive immune response. Infants with SCID generally have recurrent infections during the first weeks of life, such as oral candidiasis, and pneumonias caused by organisms that are not normally significant pathogens, such as *Pneumocystis carinii*, and diarrhea. Without isolation from infectious microbes or treatment, these children will not survive. The cause of the immunodeficiency in humans is traced to a deficiency in an enzyme. Lack of the enzyme eventually results in the death of T cells and thus loss of any cell-mediated functions by these cells as well as the provision of help to B cells. As a consequence, there is a defect in antibody production as well.

SUMMARY

▶ The primary role of the **immune system** is to combat invasion by infectious agents by excluding disease-causing agents (pathogens) from host tissues or controlling and eliminating them.

▶ Several layers of sequential, overlapping defense mechanisms accomplish this. **Barrier defense** or immunity excludes infectious agents by physical, chemical, mechanical, and microbial mechanisms.

▶ **Inflammation** and **innate immunity** limit the replication of breakthrough microorganisms by relatively nonspecific responses.

▶ **Adaptive immunity** eliminates invaders by highly pathogen-specific mechanisms that are more efficient on second exposure to the same pathogen and hence is considered to have memory.

▶ Each of the overlapping systems has an important role to play in host defense, because in the absence of any one of them, the person or animal will be susceptible to infection.

▶ The molecules and cells that mediate inflammation and innate immunity are present in sufficient quantity or number constitutively, and thus can be mobilized immediately.

▶ Monocyte (or macrophages) and neutrophils are **phagocytic cells** that are called into areas of inflammation. They phagocytose and kill infectious agents present.

▶ Phagocytes use a variety of mechanisms to kill infectious agents. These include lowering the pH in the phagosome, generating toxic oxygen radicals and nitric oxide, and exposing the organisms to lysosomal enzymes and defensins.

▶ The adaptive immune system mediates its protective responses through soluble molecules called **antibodies** produced by B cells and **cytokines** produced by T cells. T cells are able to kill microbe-infected host cells by a process called **cytotoxicity**.

▶ B cells and T cells respond to foreign molecules called **antigens** that may be parts of infectious agents.

▶ The adaptive immune system has **specificity** and **memory**, and therefore it is a carefully regulated response to infection that is more effective upon the second encounter with the same antigen.

▶ **Cell-mediated immunity** refers to activation of macrophages by **interferon-γ** for more efficient killing of phagocytosed microbes and generation of cytotoxic **CD8 T cells**. This type of immune response is directed by Type 1 or **Th1 CD4 T cells.**

▶ Cytotoxic CD8 T cells kill cells infected with viruses and also cells infected with **intracellular microbial pathogens**.

▶ **Humoral immunity** refers to antibody-mediated control of pathogens, especially antibodies of the IgA, IgE, and several IgG subclasses. This is directed by Type 2 or **Th2 CD4 T cells**.

▶ Extracellular bacteria and protozoa are controlled by antibody-mediated humoral immunity. These include blocking entry into the host, **opsonization** for phagocytosis, **complement**-mediated killing, and neutralization of their toxins.

▶ **Antibodies** are also important in the control of viral infections by blocking adhesion to and entry into host cells and mediating **antibody dependent cellular cytotoxicity** against cells infected with enveloped viruses.

▶ Antibodies are important for control of parasitic worms through IgE mediated mechanisms that result in degranulation of **mast cells** and **eosinophils**.

▶ Other cells in the immune system are part of the **bridging immune system** and include natural killer cells, γ δ T cells, B-1 B cells, and NK T cells. These cells share function with the T and B cells involved in adaptive immunity.

▶ **Allergy, toxic shock syndrome, endotoxemia,** and **autoimmunity** are dysfunctions of the immune system since these types of responses are detrimental to the host and provide no protective value.

▶ **Vaccination** is a way to exploit the properties of the adaptive immune system since it has immunological memory.

REVIEW QUESTIONS

1. What are the principal components of the barrier defense system?

2. How does breach of the barrier defense and exposure of collagen initiate an inflammatory response?

3. How is it possible that individuals with an inherited deficiency in C1q synthesis are still able to generate the opsonins C3b and iC3b, and the anaphylatoxins C3a, C4a and C5a during inflammation?

4. What are the mechanisms by which phagocytic cells control infectious disease?

5. How does interferon-γ contribute to host control of bacteria, protozoa, and viruses?

6. What are the hallmarks of the adaptive immune response?

7. What are the endogenous and exogenous pathways of antigen presentation; what role do class I and class II MHC antigens play in this?

8. Th1 and Th2 CD4 T cells make distinct contributions to the control of pathogens. What are they?

9. How does diversity of B cell receptors and antibodies occur?

10. Why are live vaccines needed to engage responses by CD8 T cells?

11. What are the distinct functions of antibodies of the IgM, IgG, IgA and IgE subclasses?

SUGGESTED READING

Goldsby, R. A., T. J. Kindt and B. A. Osborne. 2000. *Kuby Immunology.* 4th ed. New York: W. H. Freeman and Co.

Kreier, J. 2002. *Infection, Resistance and Immunity.* 2nd ed. New York: Taylor and Francis.

Paul, W. E. 1999. *Fundamental immunology.* 4th ed. Philadelphia, PA: Lippincott-Raven.

Janeway, C. A., Jr., P. Travers, S. Hunt and M. Walport. 2001. *Immunobiology.* 5th ed. New York: Taylor & Francis, Inc.

Chapter 28

Microbial Diseases of Humans

Throughout nature infection without disease
is the rule rather than the exception.
— *RENE DuBos, IN MAN ADAPTING*

The symptoms and ravages of infectious disease have been prominently recorded in the annals of human history. Biblical and ancient writings describe the crippling and suffering caused by various scourges, and the words "plague" and "disease" have 17 citations in the concordance of the King James Bible. Ancient drawings show humans scarred by smallpox or crippled by polio. The epidemic of syphilis that spread through Europe in the latter part of the fifteenth century killed thousands. The devastations of "Black Death," the plagues that periodically ravaged Europe, are well recorded (see Box 2.1 and discussion in Chapter 2). Typhus was often the final arbiter in war; the disease destroyed Napoleon at Moscow and killed 3 million Russian soldiers during World War 1.

An understanding of the relationship between a pathogenic microbe and disease symptoms has increased markedly since the pioneering studies of Robert Koch (Chapter 2). It was Koch who clearly defined the causal relation between infectious disease and a specific bacterium, and he and his coworkers suggested definitive procedures for confirming this relationship. The minor limitation imposed by Koch's methodology (a microbe must be culturable outside the animal) has now been overcome by improved identification techniques.

This chapter will discuss some of the significant human infectious diseases and their manifestations. A limited number of fungal and protozoan diseases will also be described. The presentation will be by route of entry or site of infection such as respiratory, gastrointestinal tract, and so on. Presentation in this order is both convenient and appropriate because most infectious microbes are selective in the portal of entry. Major exceptions are *Staphylococcus aureus* and *Streptococcus pyogenes*, organisms that can cause disease at various sites in or on the human body.

SKIN INFECTIONS

Skin and external infections that are often encountered by humans include impetigo, wound infections, acne, ringworm, and athlete's foot (**Table 28.1**).

Impetigo is a skin infection that occurs frequently in child-care centers and schools where youngsters are exposed to one another under confined conditions. It is caused by *Streptococcus pyogenes* or *Staphylococcus aureus* and is transmitted by scratching or other interpersonal contacts. It generally occurs around the nose and spreads about the face. Streptococcal impetigo is characterized by the appearance on the skin of vesicles that are eventually covered by a reddish crust (**Figure 28.1**). The infection caused by *S. pyogenes* can be more serious, as it spreads rapidly and can cause scarring. Impetigo should be treated without delay and can be controlled by topical application of antibiotic ointments.

Table 28.1	Major diseases of the skin[a]	
Disease	**Organism**	**Symptoms**
Impetigo	*Staphylococcus aureus* *Streptococcus pyogenes*	Skin lesions
Wound infections (burns)	*Pseudomonas aeruginosa*	Infection followed by systemic invasion: produces toxins
Acne	*Propionibacterium acnes*	Redness and swelling of sebaceous glands
Ringworm (Tinea corporis)	*Trichophyton mentagrophytes*	Fungal invasion of skin; ring of inflamation
Scalp ringworm	*Trichophyton tonsurans*	Hair follicle infection and loss of hair
Athlete's foot (Tinea pedis)	*Trichophyton rubrum* *Trichophyton mentagrophytes* *Epidermophyton floccosum*	Peeling and cracking of skin between toes
Groin ringworm	*Epidermophyton floccosum*	Fungal infection of the groin

[a]The list is not exhaustive and other diseases occur. References to some of these are presented at the end of the chapter.

Another common skin infection is acne. **Acne** occurs during adolescence when the endocrine system (hormone-secreting glands) is most active. Hormonal activity stimulates sebaceous glands in the skin to overproduce sebum. The sebum can accumulate within the gland and become infected by a normal skin inhabitant, *Propionibacterium acnes*. The consequence is an inflammatory response with swelling and reddening in the area. Tetracycline is commonly given to patients suffering from chronic acne.

Another microorganism that is a concern is *Pseudomonas aeruginosa*. This gram-negative organism is commonly present in air and in water and is an opportunistic pathogen in large open wounds such as extensive burns. *P. aeruginosa* infects over 25% of burn patients and can produce toxins and proteases that contribute to pathogenicity. It is naturally resistant to many antibiotics and is rather difficult to control.

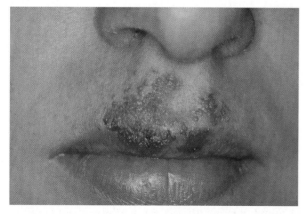

Figure 28.1 Impetigo
Impetigo on the face of a child. Note the inflammatory response (redness and swelling) around the lesions. Impetigo is caused by *Staphylococcus aureus, Streptococcus pyogenes*, or a combination of these microorganisms. Photo ©Biophoto Associates/Photo Researchers, Inc.

Staphylococcus aureus is probably the most versatile pathogen that afflicts human beings (**Table 28.2**). Infections by *S. aureus* are difficult to prevent, as virulent strains of the organism are carried in the nasopharynx by up to 50% of the human population. These pathogens also normally inhabit the skin, intestine, and vagina. The organism is readily spread from asymptomatic carriers by touching, sneezing, coughing, or passage on inanimate materials. Overall, there are at least 14 species of *Staphylococcus* that colonize humans, with coagulase positive *S. aureus* being most often associated with disease. *S. epidermidis* and *S. saprophyticus* can be infectious if resistance of the host is compromised by immunosuppressants.

S. epidermidis is the most prevalent of the staphylococci that infect in the vicinity of devices such as prosthetic joints, artificial heart valves, and catheters. The organism produces a polysaccharide capsule that adheres tightly to artificial materials. *S. saprophyticus* can cause cystitis, a urinary tract infection in women. Cystitis is an inflammation of the urinary bladder resulting from microorganisms traveling from the urethral opening. It occurs less frequently in males, as their urethra is longer.

Infections caused by *S. aureus* are listed in Table 28.2. Generally, these infections result from a breakdown in the natural defense system. Serious staphylococcal infections occur most often in individuals with a decreased ability to phagocytize and destroy the organism. This is frequently the case in the neonate, in surgical and burn patients, in individuals on immunosuppressants, or in individuals lacking a normal immune response. Scratches and blocked pores can lead to pimples, impetigo, and various superficial or deep-seated skin infections. Deep *S. aureus* infections, such as carbuncles, can lead to infections of the lymph nodes.

Most *S. aureus* strains have a cell wall component, termed protein A, that binds to immunoglobulins and interferes with the normal phagocytic response. Infected lymph nodes and localized internal infections can result

Table 28.2	Some infections caused by *Staphyloccus aureus*
Disease	**Site**
Pimples	Anywhere on the body
Impetigo	Generally on face
Boils	Anywhere
Carbuncles	Anywhere
Lymph nodes	Near site of infection
Septicemia (blood)	Carried from site of infection
Osteomyelitis	Bones
Endocarditis	Heart
Meningitis	Brain
Enteritis (food poisoning)	Gastrointestinal tract
Nephritis	Kidneys
Toxic shock syndrome	Hypotension
Staphylococcal scalded skin syndrome	Skin peeling
Wound infections	Skin
Inner ear infections	Ear
Respiratory infections	Larynx, bronchi, lungs

in rapid multiplication and dissemination of the organism into the bloodstream. The organism produces a number of toxins and enzymes that aid in the invasive process (see Chapter 26). One enzyme, **coagulase**, is particularly effective at inducing the host to produce a fibrin clot that surrounds the bacterial cell, preventing the normal host defenses from phagocytizing the invader. *S. aureus* also produces **leukocidin**, an antiphagocytic factor that destroys leukocytes. Production of leukocidin and coagulase permits the organism to overcome host defenses.

Viral diseases can also predispose an individual to invasion by staphylococci. Influenza can be followed by a staphylococcal pneumonia that may be fatal. It is probable that many deaths ascribed to viral influenza can be ascribed to staphylococci. Other viral infections of the respiratory system can cause tissue trauma that creates an opening for invasion by *S. aureus*, resulting in meningitis or laryngitis (see later discussion). Staphylococcal infections can be treated with cephalosporin, penicillin, cloxacillin, and other antibiotics.

Toxic shock syndrome occurs in surgical patients of both sexes where there is an instance of internal colonization by *S. aureus*. In women, toxic shock syndrome results from the overcolonization of the vagina by *S. aureus*. This disease, first described in 1978, is a syndrome that occurs most frequently in menstruating women who use tampons. The disease symptoms are a sudden fever, diarrhea, vomiting, red skin rash, and low blood pressure. The drop in blood pressure can lead to an irreversible state of shock and death. The death rate in confirmed cases is between 5% and 12%. Newer superabsorbent brands of tampons that bind magnesium and a lowered level of magnesium in the environment apparently stimulate the *S. aureus* strains to produce the toxin associated with the syndrome. Women can reduce the risk of contracting the disease by not using tampons, by not using the superabsorbent tampons, or by not using them continuously during the menstrual period.

Fungal diseases of the skin are a widespread occurrence in humans. Fungal infections such as ringworm and athlete's foot are familiar to most of us through personal experience and products advertised on TV. The level of advertising is indicative of the incidence of these

(A)

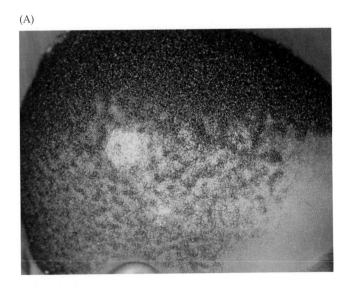

(B)

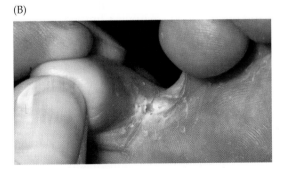

Figure 28.2 Fungal infection of the skin
(A) Ringworm on the head (tinea capitis) is caused by *Microsporum audouinii*. (B) Ringworm (athlete's foot) in the toe area (tinea pedis) is caused by *Trichophyton rubrum*, *Trichophyton mentagrophytes*, or *Epidermophyton floccosum*. A, photo ©K. Greer/Visuals Unlimited; B, photo ©J. Shemilt/SPL/ Photo Researchers, Inc.

infections in human populations. These fungi invade the keratinized part of skin and collectively are called **dermatomycoses** (**Figure 28.2**).

They most often affect the toe area of the feet but can cause infections around hair follicles, including on the scalp. Ringworm can also occur on the torso or limbs. It is estimated that up to 70% of the population in the United States is subclinically infected with athlete's foot. The causative agents of these diseases are presented in Table 28.1. The infections are mostly caused by fungi in three different genera—*Epidermophyton, Microsporum,* and *Trichophyton.* Ath-lete's foot causes a cracking between toes with itching. Ringworm is characterized by circular, red, clearly demarked lesions and also with itching. These infections can be treated successfully with topical applications of imidazole drugs. Severe fungal infections of the skin are best treated by a physician, as there is a danger of superinfection by bacteria.

RESPIRATORY DISEASES

Infectious diseases of the mouth and respiratory system are a concern because they are readily communicated from human to human via air droplets created by sneezing, coughing, or face-to-face talking and by kissing or other human contacts (**Table 28.3**). These microorganisms are in the air we breathe, and there is no possible way to completely avoid exposure. Sneezing and coughing are significant mechanisms for expelling infectious droplets into the air. A sneeze droplet can move at 100 meters per second for short distances, and a single sneeze can expel 10,000 to 100,000 or more bacteria into the air. The upper respiratory tract is more susceptible to infection, as microorganisms that enter tend to settle in this area (**Figure 28.3**).

Although microorganisms do not survive well in outdoor air, the atmosphere is constantly inoculated by organisms from soil, plants, and animals but mostly by our fellow humans. Indoor air has higher numbers of microbes, and the residents occupying the space generally disseminate these. If large numbers of humans are crowded into a room, the air will be populated by microbes common in the respiratory tract. The gram-positive organisms such as *Micrococcus, Staphylococcus,* and *Streptococcus* can survive in air, as they are quite resistant to drying and relatively resistant to ultraviolet light.

Immunization has reduced the scourge of many respiratorily transmitted diseases, particularly diphtheria and whooping cough. Antibiotics are effective against most of the others, including tuberculosis, but strains resistant to antibiotics are a concern. A discussion of some of these respiratory infections follows.

Diphtheria

Diphtheria is caused by *Corynebacterium diphtheriae,* a gram-positive, nonmotile pleomorphic bacillus that may

Table 28.3	Diseases transmitted predominately via the respiratory tract and the organisms involved
Disease	**Organism**
Diphtheria	*Corynebacterium diphtheriae*
Whooping cough	*Bordetella pertussis*
Pneumonia	*Streptococcus pneumoniae*
	Staphylococcus aureus
	Legionella pneumophila
	Mycoplasma pneumoniae
	Klebsiella pneumoniae
	Pneumocystis carinii
Streptoccocal infections	*Streptococcus pyogenes*
Meningitidis	*Haemophilus influenzae*
	Neisseria meningitidis
	Streptococcus pneumoniae
Tuberculosis	*Mycobacterium tuberculosis*
Psittacosis	*Chlamydia psittaci*
Leprosy	*Mycobacterium leprae*
Coccidioidomycosis	*Coccidioides immitis*
Aspergillosis	*Aspergillus fumigatus*
	Aspergillus flavus

appear to be club shaped. Diphtheria was once a leading cause of death in children, but immunization has reduced the number of cases in the United States to fewer than five per year. Diphtheria is still a concern in economically deprived urban areas of the world.

C. diphtheriae enters the body through the respiratory route and adheres to tissue in the throat. Lysogenic strains (see Chapter 14) excrete an exotoxin (cytotoxin), and inflammation is the immediate response. The cytotoxin kills host cells, and these are intermixed with leukocytes and bacteria, forming a dull gray pseudomembrane on the throat mucosa (**Figure 28.4**). The disease is diagnosed by the presence of this membranous material and by culturing of the microorganism. The pseudomembrane can extend over the tracheal opening, blocking the passage of air into the lungs, resulting in suffocation and death. Other symptoms of diphtheria are caused by transport of the cytotoxin elsewhere, causing lesions in the kidney and heart.

The diphtheria toxin has been studied extensively, and its mode of action is quite well understood. It was the first extracellular exotoxin described. Two French scientists, Pierre Roux and Alexandre Yersin, discovered that culture filtrates of the bacterium *Corynebacterium diphtheriae* were lethal to test animals, affirming that lethality in diphtheria infections was due to toxic material released by the bacterium.

The toxin is produced by strains of the bacterium lysogenized by bacteriophage B. Toxin is produced at low iron concentrations, as the repressor that prevents

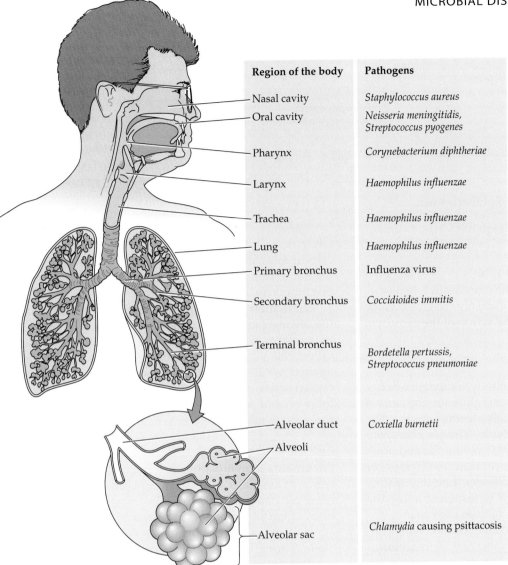

Region of the body	Pathogens
Nasal cavity	*Staphylococcus aureus*
Oral cavity	*Neisseria meningitidis,* *Streptococcus pyogenes*
Pharynx	*Corynebacterium diphtheriae*
Larynx	*Haemophilus influenzae*
Trachea	*Haemophilus influenzae*
Lung	*Haemophilus influenzae*
Primary bronchus	Influenza virus
Secondary bronchus	*Coccidioides immitis*
Terminal bronchus	*Bordetella pertussis,* *Streptococcus pneumoniae*
Alveolar duct	*Coxiella burnetii*
Alveoli	
Alveolar sac	*Chlamydia* causing psittacosis

Figure 28.3 Respiratory infections
The human respiratory system, showing the sites where various pathogens may localize and cause infections.

expression of the *tox*⁺ gene in phage is iron containing. If the iron concentration is low, the repressor is not formed and the *tox*⁺ gene is transcribed.

Diphtheria toxin has enzymatic activity that cleaves nicotinamide from NAD and catalyzes the attachment of the resultant ADP-ribose to elongation factor-2 (EF-2) necessary for the growth of the polypeptide chain during protein synthesis (**Figure 28.5**). The addition of the ADP-ribose to EF-2 completely inactivates the factor and halts protein synthesis. Diphtheria toxin is relatively potent because a few molecules of the enzyme will kill a single host cell. It is interesting to note that *Pseudomonas aeruginosa,* a common organism in water, soil, and food, secretes a functionally similar toxin. *P. aeruginosa* is not a major health problem, as it does not have the virulence factors necessary to sustain an infection.

Figure 28.4 Diphtheria
Diphtheria results from colonization of the oropharynx by *Corynebacterium diphtheriae.* In an active case, as here, a pseudomembrane develops in the throat. Photo ©Science VU/Visuals Unlimited.

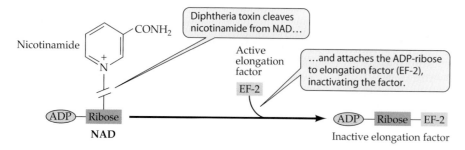

Figure 28.5 Mode of action of diphtheria toxin
By inactivating an elongation factor (EF-2), diphtheria toxin brings the host's protein synthesis to a halt.

Antibiotics, such as penicillin, are effective against *C. diphtheriae,* but the antitoxin should also be given, as the antibiotic does not prevent the action of the cytotoxin once it is produced.

Whooping Cough

Prior to the introduction of an effective vaccine, **whooping cough** afflicted over 95% of children in the United States and resulted in about 4,000 deaths per year. The etiologic agent is *Bordetella pertussis,* a fragile gram-negative aerobic coccobacillus that is nonmotile and often encapsulated. There are now fewer than 10,000 cases per year in the United States and fewer than ten deaths. Worldwide, whooping cough is responsible for the deaths of over 500,000 young children each year.

Whooping cough is highly contagious in young children and is communicated by respiratory discharges from infected individuals. The incubation period is 7 to 14 days. The organism binds to ciliated epithelial cells of the bronchi and trachea. The adherence factor that binds to cilia is **hemagglutin** because the purified factor will agglutinate erythrocytes. After binding to cilia, *B. pertussis* produces several virulence factors, including a pertussis toxin, an extracytoplasmic adenyl cyclase, the hemagglutin, and a tracheal cytotoxin. The pertussis toxin increases tissue sensitivity to histamine and serotonin that are related to increased levels of adenyl cyclase. An increased level of cyclic AMP induced by the adenyl cyclase inhibits the action of phagocytic cells. The tracheal cytotoxin is a *B. pertussis* cell wall peptidoglycan precursor. It destroys the host ciliated respiratory epithelial cells, and the destruction of these cilia in bronchi brings about an accumulation of mucus, bacteria, and host-cell debris in the lungs.

Infection with *B. pertussis* begins with a catarrhal stage, which causes the inflammation of the mucous membrane, sneezing, and relatively mild coughing. After one to two weeks, the paroxysmal stage is reached and is marked by violent coughing. These violent coughs are the body's effort to rid the lungs of accumulated debris. The violent cough is followed by a "whoop," the sound of incoming air; hence, the name. Coughing can be so violent that the victim suffers **cyanosis** (turns blue from O_2 insufficiency), vomiting, and convulsions. The catarrhal and coughing stage can last six weeks with the convalescent period lasting even longer.

Whooping cough is diagnosed by plating throat swabs on Bordet-Gengou agar, where they have a typical identifiable appearance. Smears may also be stained directly with the fluorescent antibody. The disease can be treated with erythromycin and other antibiotics.

Vaccination of infants is at two months of age, since the disease can occur in infants and has a high mortality in infants less than one year old. Until 1991, a whole killed cell vaccine was employed in immunization against pertussis. In rare instances, the vaccine itself caused severe systemic toxic reactions. The current acellular vaccine should be free of side effects, and toxic reactions would be exceedingly rare.

Pneumonia

Pneumonia is an inflammatory reaction in the alveolar region of the lungs. Viruses, mycoplasma, eubacteria, or fungi can cause the infection. The most frequent infectious agent is *Streptococcus pneumoniae,* the etiologic agent of around 70% of pneumonia cases in the United States. Formerly called *Diplococcus pneumoniae, S. pneumoniae* is a gram-positive coccus that grows in pairs. Cases of pneumonia generally result from strains that are present as normal host microbiota. Most often, pneumonia is a secondary infection that follows a viral infection, a respiratory tract injury, debilitating injuries, or chronic illness. There are over 300,000 pneumonia cases per year in the United States with a death rate between 10% and 20%.

S. pneumoniae has low invasive ability and does produce a cytolytic toxin that binds to cell membranes. The actual function of this toxin in pathogenesis is not clearly understood, and the capsule is considered to be more important in the development of the disease. The capsule protects the organism from phagocytosis. Phagocytes readily remove organisms without capsules. The bacterium grows rapidly in alveolar spaces, and the alveoli become filled with blood, bacteria, and phagocytic cells. Acute lung inflammation is a consequence of this fluid buildup.

Symptoms of pneumonia are a sudden onset of chills, labored breathing, and pleural pain. If untreated, the bacterium may escape from the lungs and cause a bacteremia (presence in blood) and is carried to the middle ear and sinuses, and acute infections can occur at these sites. The organism can travel to the heart and cause endocarditis (inflammation of the heart valves) or pericarditis (inflammation of the pericardium, the membrane surrounding the heart), both of which can have long-lasting effects on health. Penicillin has been effective against *S. pneumoniae*, but resistant strains have evolved, and these cases are treated with other antibiotics such as erythromycin. Diagnosis of pneumonia is by chest X ray or the culturing of sputum (expectorated matter). A vaccine composed of purified *S. pneumoniae* polysaccharide from over 20 pneumonococcal types is available and is recommended for adults over 50. A more limited pneumococcal vaccine was approved for children in the year 2000.

S. aureus pneumonia often occurs as a secondary infection following an initial infection with the influenza virus. Staphylococcal pneumonia is a serious infection, as this organism has considerable invasive ability. The symptoms are similar to those of *S. pneumoniae*. Individuals with cystic fibrosis are quite susceptible to staphylococcal pneumonia. Treatment would be with penicillin or other antibiotics that are effective against gram-positive organisms.

Another pneumonia that is of relatively recent description is **Legionellosis'** (**Legionnaires disease**), so named because fatal cases of the disease occurred in attendees at a 1976 American Legion Convention in Philadelphia. At that time, the disease could not be assigned to any known organism. Since then, *Legionella pneumophila* has been identified as the causative organism.

Isolation and characterization have confirmed that *L. pneumophila* is unrelated to any of the other organisms that cause respiratory disease. It is a gram-negative pleomorphic rod that stains poorly and does not grow on media employed in diagnosing pneumonia caused by staphylococci and streptococci. Hence, earlier isolation protocols did not recover *L. pneumophila*, and the disease may have erroneously been attributed to a virus.

Legionellosis is spread by inhaling mist from air-conditioning cooling towers or being in contact with water from areas where epidemics have occurred. There is no evidence that the organism is spread by human contact. Despite the fact that the organism has complex nutritional requirements, including a need for high iron availability, it apparently survives in cooling water towers, on vegetable misters, and on showerheads. *L. pneumophila* can infect amoebae such as *Hartmannella vermiformis*, and this may be a source of virulent organisms in nature.

Legionellosis occurs most often in elderly or debilitated individuals. Symptoms include fever, headache, chest pains, muscle aches, and bronchopneumonia. The organism produces a hemolysin, an endotoxin, and a cytotoxin, but the exact role of these invasive agents is unknown. The organism can survive and multiply in the phagosomes of macrophages and cause tissue destruction through the production of a cytotoxic exoprotease. *L. pneumophila* has also been implicated in skin abscesses and inflammation of internal organs such as the heart and kidneys. Diagnosis is generally by fluorescent antibody or by the presence of antibody in the serum of the host. Treatment is with erythromycin or rifampin.

Mycoplasma pneumoniae is the etiological agent for a disease called primary atypical pneumonia. It spreads from person to person among schoolchildren, men and women in the military service, and others living in close quarters. The symptoms include chills, fever, and a general malaise, but these symptoms are often mild and virtually unnoticed. The mild nature of the disease and its insidious (slow) onset is the reason for calling it atypical. Diagnosis is by chest X ray. *M. pneumoniae* can be isolated from the sputum of patients and identified by staining with fluorescent antibodies. However, the organism grows slowly, and isolation is rarely of clinical significance. Treatment is with tetracycline, erythromycin, and rifampin.

Pneumonia can be caused by *Klebsiella pneumoniae*, a bacterium that is present as part of the respiratory tract microbiota in about 5% of the population. It is a gram-negative nonmotile rod. The organism can cause severe pneumonia that may result in chronic ulcerative lesions in the lungs. Fewer than 5% of pneumonia cases can be ascribed to *K. pneumoniae*. The infection can be treated with gentamycin.

A fungus that was previously considered a protozoan causes another form of pneumonia. The organism is *Pneumocystis carinii*, and it occurs in many animal species, including humans. Serological examination suggests that all children have been infected with this organism. The organism remains dormant unless the host becomes immunocompromised, at which point it can cause a fatal pneumonia. In recent years, *P. carinii* has been identified frequently in cases of fatal pneumonia in immunocompromised individuals. *P. carinii* occurs in about 50% of AIDS patients and results in a fatal infection in about one-half of those infected. It is diagnosed by fluorescent antibody staining. Sulfamethoxazole and trimethoprim are given as an oral prophylactic.

Streptococcal Infections

Streptococcus pyogenes is the etiologic agent for diseases commonly known as **streptococcal infections**. It is a gram-positive coccus that divides in one plane, and the individual cocci tend to remain together and form long chains. The organism is the causative agent of skin infections (impetigo) as described previously.

The streptococci are **aerotolerant fermenters,** so called because most grow in the presence of air but obtain energy by a lactic acid fermentation of sugar. Generally, systemic infections by *S. pyogenes* are the result of a weakened host defense. Pathogenic strains of *S. pyogenes* secrete an active hemolysin that can lyse erythrocytes. The streaking of some strains of streptococci on blood-agar plates results in a greenish discoloration around colonies, indicating that there has been an incomplete destruction of red blood cells (β-hemolysis)—these strains are termed **β-hemolytic.** The **β-hemolytic** streptococci effect a complete lysis of red blood cells with a clear zone around the colony. The role of the hemolysin in pathogenesis is the destruction of host phagocytes.

The major human pathogens are β-hemolytic and have a virulence factor that promotes invasion. This factor is called the M protein and is located at the cell surface. The M protein is antiphagocytic and can maintain an infection in the absence of adequate treatment or specific antibody.

Many strains of *S. pyogenes* are surrounded by a capsule composed of hyaluronic acid. Because hyaluronic acid is a normal host cell constituent, this gives the bacterium protection from phagocytosis. Other invasive molecules produced by streptococci were discussed in Chapter 26.

S. pyogenes is the causative agent of several familiar acute infections, including streptococcal pharyngitis, scarlet fever, rheumatic fever, and glomerulonephritis. **Streptococcal pharyngitis** is the most common infection caused by *S. pyogenes,* an acute infection with reddened tonsils, a swollen pharynx, and the production of purulent exudate. Often the local lymph nodes become enlarged, and there is a high fever. The infection is commonly known as "strep throat" and may spread to the middle ear, and an acute infection there can cause a loss of hearing. On rare occasions, the infection breaks through the throat epithelia and is carried by the bloodstream to the meninges, causing a form of meningitis. Epidemics of strep throat are the result of personal contact with infected persons or carriers. Control with antibiotics is essential, as streptococcal sore throat can be followed by scarlet or rheumatic fever and glomerulonephritis.

Scarlet fever is caused by lysogenic (phage bearing) strains of *S. pyogenes.* The phage carries the information for the production of a **pyogenic exotoxin or erythrogenic toxin.** The erythrogenic toxin can diffuse throughout the body and induce the skin rash that is typical of scarlet fever (**Figure 28.6**). The pyogenic toxin is a low molecular protein with two functional parts. One part is a heat labile portion that induces fever and suppresses the immune system. The other part is heat stable and increases the pyrogenicity (heat and fever) and lethali-

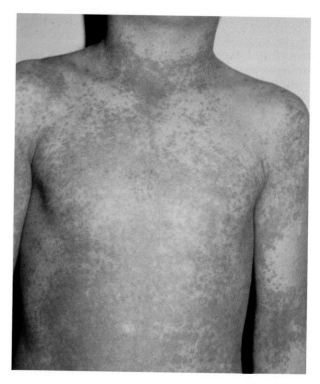

Figure 28.6 Scarlet fever
The rash associated with scarlet fever. The erythrogenic toxin produced by *Streptococcus pyogenes* causes a hypersensitivity that results in the pink-red rash. Photo ©Biophoto Associates/Photo Researchers, Inc.

ty of the toxin. The rash is a hypersensitive reaction (see Chapter 27) to the heat stable part of the molecule. Scarlet fever can be controlled with antibiotics. It is very important to halt the infection at an early stage for, untreated, the infection can lead to rheumatic fever.

Rheumatic fever can follow streptococcal sore throat caused by certain strains of *S. pyogenes.* This disease occurs in 3% to 5% of pharyngitis patients that fail to receive proper treatment. Rheumatic fever can result in permanent damage to the heart. There are two major theories that have been advanced to explain this heart damage. One theory is that antigens, possibly the M protein, are disseminated about the body and deposited in joints and in heart tissue. When antibodies produced by the host interact with the antigens adhering to heart tissue, some damage occurs. Another theory holds that some strains of *S. pyogenes* have an antigen that induces antibodies that cross-react with a similar antigen in the heart, resulting in heart damage. For unknown reasons, strains of *S. pyogenes* that induce rheumatic fever are now very rare.

Glomerulonephritis, often a consequence of streptococcal infection, is an inflammation of glomeruli in the kidney. It may be an autoimmune response, as the streptococcal toxin and kidney cells share a common antigen.

Antibodies to the toxin cross-react and harm kidney tissue. The incidence of this disease is low, and most patients undergo a slow healing of the damaged glomeruli. Others (under 10%) may have a chronic kidney disease.

Meningitis is an inflammation of the meninges, which are any of three membranous layers that envelope the brain and spinal cord. These layers are made up of the **dura matter**, located next to the bone; the middle **arachnoid** layer, and the **pia** matter, located next to the nerve tissue. The space between the arachnoid and the pia is filled with cerebrospinal fluid. When meningitis is suspected, the cerebrospinal fluid is tapped and examined for the presence of microorganisms.

A number of organisms involved in respiratory infections may be responsible for meningitis. These include *Haemophilus influenzae, Neisseria meningitidis,* or *Streptococcus pneumoniae,* all of which can be transmitted through the bloodstream to the meninges. It is estimated that 43% of the cases of meningitis are caused by *H. influenzae,* 27% by *N. meningitidis,* and 11% by *S. pyogenes.*

Meningitis symptoms are sudden fever, stiffness in the neck, headaches, and often delirium, followed by convulsions and coma. *N. meningitidis* is called the meningococcus and is a gram-negative diplococcus. The initial infection resulting from *N. meningitidis* invasion is a nasopharyngeal infection. In a limited number of cases, however, the organism enters the bloodstream and causes a systemic infection. This can be a rapidly fatal stage of the disease, as it may cross the blood–brain barrier and infect the meninges. The disease is more prevalent in younger individuals. Treatment of meningococcal infections is with penicillin. Erythromycin and chloramphenicol are also effective, and treatment must be initiated early, as the disease can be fatal otherwise. Meningitis is such a severe, fatal disease that anyone in contact with a victim is generally treated with antibiotics as a precautionary measure.

H. influenzae meningitis is also an exceedingly dangerous disease, and survivors can suffer neurologic disorders. Ampicillin has been an effective therapy, but ampicillin-resistant strains of *H. influenzae* are increasing in number, requiring the use of chloramphenicol as an alternative. Vaccination for *H. influenzae* is now available. Meningitis caused by *S. pneumoniae* can also be treated with chloramphenicol.

The etiologic agent of **tuberculosis** is *Mycobacterium tuberculosis,* commonly referred to as the tubercle bacillus or TB. It was isolated in 1882 by Robert Koch, who confirmed the relationship between the organism and the disease. Historically and currently, tuberculosis is a devastating disease that continues as a leading cause of human suffering and death. Worldwide, there are 10 million active new cases of tuberculosis each year and 3 million deaths. Nearly 2 billion people have latent tuber-

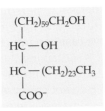

Figure 28.7 Mycolic acid
The structure of a typical mycolic acid, a long-chain waxy fatty acid present in acid-fast microorganisms such as *Mycobacterium tuberculosis.*

Mycolic acid

culosis. In the United States alone, there are over 10,000 new or reactivated cases per year. Prior to the availability of effective antibiotics against *M. tuberculosis,* victims of the disease were practically bedridden, as rest was the only known treatment, and patients were isolated from the general population.

M. tuberculosis is a thin rod, sometimes bent and club shaped. The organism is acid-fast, a staining property where cells stained with hot carbofuchsin are not decolorized by acid alcohol, and this can be used as a diagnostic test. *M. tuberculosis* is aerobic with relatively simple nutritional requirements. The organism has a high lipid content (over 40% cell dry weight), and the lipid contains unique fatty acids called mycolic acids (**Figure 28.7**). Mycolic acids vary somewhat in size and in amount of hydroxylation and are frequently linked to sugars to form glycolipids called mycosides. A mycoside associated with virulent mycobacterial strains is depicted in **Figure 28.8**. Here two molecules of mycolic acid are attached to the disaccharide trehalose forming cord factor. Cord factor is considered to be responsible for the severe weight loss (cachexia) observed in tuberculosis patients.

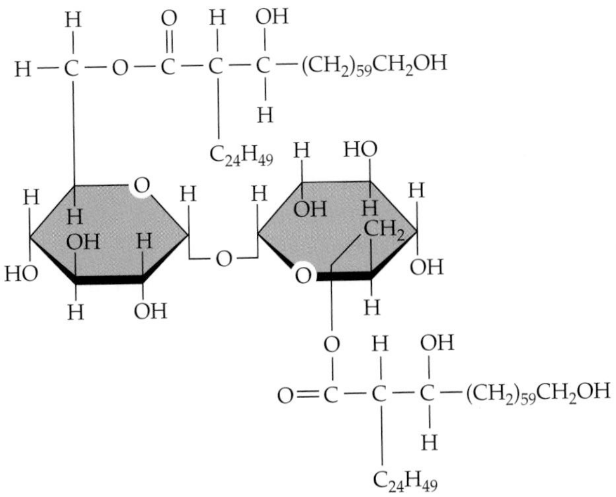

Figure 28.8 Mycoside
Structure of a cord factor, a mycoside present in virulent strains of *Mycobacterium tuberculosis.* Colonies of bacteria that synthesize this mycoside appear as pieces of cord sitting side by side.

Tuberculosis is transmitted from person to person respiratorily via small droplets introduced into the air by infected individuals. A single active case can infect many bystanders. Inhalation of droplets containing the tubercle bacillus can lead to propagation within the lungs. Ninety percent of infected people do not progress to clinical disease, but the bacillus remains latent. Should the patient become immunocompromised, for example, due to AIDS, the clinical disease will appear.

Growth of *M. tuberculosis* in the lungs results in inflammation and the development of lesions. The bacteria can reproduce within phagocytic cells and be disseminated by lymph and blood to other body organs and tissue. As the bacteria propagate in lesions, they produce a mass consisting of the waxy bacterium, lymphocytes, and macrophages. The lesion becomes enmeshed in fibrous connective tissue forming a nodule called a **tubercle.** A tubercle contains viable bacteria that can be dormant indefinitely. During dormancy, the infected individual is asymptomatic, and this would be considered a primary infection. The dormant primary infection may be activated by malnutrition, stress, hormonal imbalance, or decrease in immune function. Symptoms of the disease—fatigue, weight loss, and fever—are only obvious after extensive lesions are formed.

The formation of tubercles is mostly a delayed-type hypersensitivity reaction (see Chapter 27). Consequently, individuals in the primary stages of the disease are hypersensitive to protein fractions from *M. tuberculosis.* The **tuberculin test** given to children and to others during physical examinations is based on this hypersensitivity. A small amount of protein from *M. tuberculosis* is placed intradermally (beneath the skin) and observed for redness and swelling. Reddening after a few days (tuberculin positive) indicates that the individual has had an inapparent infection.

Diagnosis of tuberculosis is by the presence of *M. tuberculosis* in sputum. Presence of lesions results in the expectoration of bloody sputum laden with *M. tuberculosis.* X-ray examination of the chest is employed to reveal lung damage and tubercles. Treatment formerly was with streptomycin, but this has been supplanted with various combinations of rifampin, isoniazid (INH), pyrazinamide (PZA), and ethambutol (EMB) or streptomycin. The regimen depends on the susceptibility of the infecting organisms to the drugs. Treatment must last at least 6 months to ensure that all intracellular and intranodular organisms are destroyed. Unfortunately, these drugs all have nasty side effects, and patients are reluctant to continue to take them while they are feeling well. This is not only bad for the patient but has serious public health consequences. The solution is DOT (directly observed therapy), when a public-health nurse observes the patient take their medication three times a week.

In recent years, strains of *M. tuberculosis* have evolved that are resistant to antibiotics commonly employed to treat tuberculosis. These resistant strains are prevalent among AIDS patients and will ultimately spread to the general population. Poverty and inadequate efforts at eradication worldwide also contribute to the potential pool of resistant strains.

Leprosy

Leprosy is a dreaded disease that afflicts about 14 million people worldwide, mostly in tropical countries. About 100 new cases of leprosy are diagnosed in the United States each year. The organism responsible for leprosy is *Mycobacterium leprae,* a rod-shaped, acid-fast organism. *M. leprae* can be obtained from lesions but has not been grown in axenic culture. The organism will grow in nude mice that have an impaired immune function. The armadillo is a natural host where a low body temperature permits a systemic multiorgan infection. Death of the animal occurs within 2 years. Attempts to infect human volunteers have not been successful.

The epidemiology of leprosy is somewhat of a mystery. Most humans apparently are not susceptible to infection, although it afflicts twice as many men as women. Children are more susceptible than adults. Transfer by human contact is possible, as individuals with lepromatous leprosy have over 10^8 *M. leprae* in their nasal discharge. This may be the mechanism whereby the organism is transmitted to susceptible individuals, but epidemics do not occur. There are two major forms of leprosy in humans: tuberculoid leprosy or lepromatous leprosy.

Tuberculoid leprosy is often a mild self-limiting disease. It is nonprogressive and occurs as a delayed hypersensitivity reaction to proteins in *M. leprae* or a normal cellular immune response. Damaged nerves and loss of sensation occur in areas where lesions exist. Rarely are intact organisms isolated from material taken from these lesions. Spontaneous recovery from tuberculoid leprosy frequently occurs.

Lepromatous leprosy, on the other hand, is a progressive, nasty disease that ultimately leads to death. *M. leprae* is found in virtually every organ of the body. Skin lesions are common, often pigmented, and organisms can be obtained for examination by skin scraping. Skin nodules infected with *M. leprae* appear about the body. Nerves are damaged, and mucosal lesions are abundant. Lesions in the mucous membranes of the nose lead to the destruction of cartilage and nasal deformities. The eyes also can be infected, and this can lead to blindness. Loss of fingers or toes is common (**Figure 28.9**).

Diagnosis of leprosy in the early stages is difficult. The occurrence of numbness concurrent and the isolation of the acid-fast bacterium are the major diagnostic features. It has been suggested that a material (lepromin) purified from infected armadillo tissue might be employed as a diagnostic tool. This would be analogous

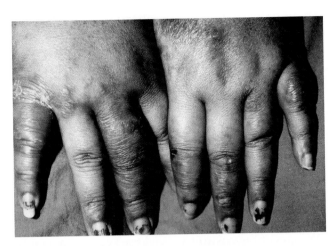

Figure 28.9 Leprosy
Hands of an individual with leprosy. The infection progresses from a loss of sensation in the skin to deformity and lesions and can result in the loss of fingers and toes. Photo ©K. Greer/Visuals Unlimited.

to the use of the tuberculin test for tuberculosis. In practice it has not been very effective.

Treatment of leprosy is difficult. The incubation period for tuberculoid leprosy is 3 to 6 years and for lepromatous leprosy 3 to 10 years. Without a reliable diagnostic tool, treatment is delayed until the full onset of visible symptoms. Dapsone (4,4'-sulfonylbisbenzamine) has been given to patients but is quite toxic. A variant of the drug, diacetyl dapsone, is less toxic and appears to be more effective. Unfortunately, *M. leprae* resistant to the dapsones have emerged. Rifampin is an alternative drug, but is much more expensive. Vaccines that might be employed in tropical areas where leprosy is a concern are under development.

Psittacosis (Ornithosis)

Psittacosis is an avian disease that afflicts many species of birds and can infect humans. The causative agent is *Chlamydia psittaci*, an obligate intracellular prokaryotic parasite related to gram-negative bacteria. The disease is generally latent and asymptomatic in birds. Crowding and unsanitary conditions, such as those encountered in transporting birds, can lead to active infections. The organism is present in virtually every organ of an infected bird, and considerable numbers are excreted in the feces. The consequent inhalation of dried feces by bird handlers can cause human infections. The inhaled organism travels to the liver, spleen, and lungs. Lungs become inflamed, hemorrhagic, and with symptoms of pneumonia. The pneumonia can be fatal.

A case of fatal meningitis can also occur in a limited number of psittacosis patients. The disease can be diagnosed by isolation of *C. psittaci* from blood or sputum and serologically. Treatment with tetracycline is effective.

FUNGAL DISEASES

There are a number of human respiratory diseases caused by fungi. Fungal spores are inhaled in normal respiration, and in some cases these fungi can propagate in the lungs. Among these diseases, we will consider histoplasmosis, coccidioidomycosis, and aspergillosis.

Histoplasmosis

Histoplasma capsulatum is the etiologic agent of **histoplasmosis**, the most common fungal disease in humans. An estimated 40 million Americans have been infected with histoplasmosis, and 250,000 new cases occur each year. *H. capsulatum* is widely distributed in soil and is generally associated with birds and bats. Abundant numbers are present in chicken coop litter, bat caves, and bird roosts. The disease is endemic in the Ohio and Mississippi River valleys. *H. capsulatum* is dimorphic, producing hyphae in soil and budding yeasts in the disease state. Infection occurs on inhalation of the reproductive spores (microcondia) that are present on hyphal filaments. In most cases, the normal host immune response is sufficient to rid the body of the infectious organism. A small number of infected individuals may develop a systemic histoplasmosis of variable severity. The symptoms vary from none to coughing, fever, and pain in joints. Lesions may appear in the lungs and become calcified. Generally the disease is resolved by host defenses but may become chronic, with bouts of lung infection occurring periodically. Diagnosis is by examination of sputum or by a test similar to the tuberculin test described earlier employing a fungal protein. The disease can be treated with amphotericin B or ketoconazole, but relapses may occur.

Coccidioidomycosis

The etiologic agent of **coccidioidomycosis** is *Coccidioides immitis*, a filamentous fungus present in arid soils. Most cases in the United States occur in the Southwest. Infection of the lungs occurs by inhalation of the reproductive arthrospores. Over half of primary infections are asymptomatic, and most of the remainder produce a mild to a severe pulmonary disease. Chronic or acute disseminated infections occur in less than 0.5% of those infected. Symptoms of the disease are a mild cough, fever, chest pains, and headache. Chronic cases occur if lung tissue develops localized cavities filled with the spherules (cylindrical bodies) of *C. immitis*. Amphotericin B is the treatment of choice for coccidioidomycosis.

Aspergillosis

Aspergillosis occurs in individuals frequently exposed to conidiospores of filamentous fungi in the genus *Aspergillus*. The disease occurs more often in immunosuppressed individuals. Conidiospores inhaled into the lungs cause a delayed-type hypersensitivity reaction that appears as a typical asthma attack. Inhaled conidiospores can enter the lungs where they form colonies. These colonies may remain small and confined or can spread to other organs. In patients with severely limited immune function, the mycelia can completely fill the lungs. Treatment of aspergillosis is difficult, as the fungus is resistant to most antifungal agents.

Candidiasis

Candida albicans is a part of the normal microbiota of the gastrointestinal tract, mouth, and vaginal area. The organism does not usually cause disease because the bacterial populations in these areas suppress the growth of *C. albicans*. In the absence of the normal microbiota, the fungus can proliferate and cause disease involving the skin or mucous membranes. It has become an important nosocomial pathogen (develops in hospital patients).

Oral candidiasis (thrush) infects neonates born to mothers with *Candida* infections of the vagina. In cases of candidal vaginitis, the infection can be transmitted to the male during sexual intercourse. The resultant infection of the penis is termed balanitis. Thrush also occurs in immune-compromised individuals. Systemic candidiasis can occur in AIDS patients. Individuals whose hands are immersed in water for long periods, such as cannery workers, often develop candidiasis infections of the hands or around the nails. Candidiasis is best treated by relieving the underlying conditions that result in infection. But several effective drugs are now available for treatment including clotrimazole and fluconazole.

Cryptococcosis

Cryptococcosis is a yeast infection caused by *Cryptococcus neoformans*, a common soil inhabitant. Most human infections result from inhalation of the organism on dried pigeon feces. *C. neoformans* does not infect pigeons but grows prolifically on the droppings. The primary infection is in the lungs, and most cases are asymptomatic or undiagnosed and cure is spontaneous in healthy individuals. However, cryptococcosis is a serious problem in AIDS patients, where it causes infections in the brain and meninges. Cryptococci are present in the spinal fluid of 10% to 15% of AIDS patients. Cryptococcosis is fatal in immunocompromised individuals if untreated. Cryptococcosis can be diagnosed with latex beads coated with rabbit antibodies to the polysaccharide capsule of the yeast. Presence of *C. neoformans* is indicated by aggregation of the beads. Amphotericin and 5-fluorocytosine are the drugs of choice for treatment.

BACTERIAL DISEASES OF THE GASTROINTESTINAL TRACT

Diseases of the gastrointestinal tract occur frequently and range from a mild upset stomach to fatal cases of botulism. Proper sanitation, clean water, refrigeration, and public health measures have been effective in lowering the incidence of these diseases in the Western world. Because many gastrointestinal diseases are spread by fecal contamination of food and water, they are a serious concern in developing countries where sewage treatment is often inadequate. Typhoid fever and cholera are endemic in many parts of Southeast Asia and South America due to a lack of proper sewage treatment.

The intestinal tract diseases are of two major types:

- Food poisoning results from ingestion of food on which bacteria have grown. In such cases, bacteria release toxins that cause the disease symptoms. These toxins are then ingested along with the food.
- Food-borne infections occur if one ingests food or drinks liquids that are contaminated with disease-causing bacteria. The bacteria propagate in the intestinal tract. In some cases, the infection occurs directly from hand to mouth after touching or handling fecal-contaminated objects.

Food poisoning symptoms generally occur within a few hours after ingestion of contaminated food (**Table 28.4**). Duration of the poisoning by *Staphylococcus,*

Table 28.4	Gastrointestinal diseases caused by food poisoning bacteria[a]		
Organism	**Time to Onset of Symptoms in Hours**	**Symptoms**	
Staphylococcus aureus	1–6	Nausea, vomiting, diarrhea	
Clostridium botulinum	18–36	Dizziness, double vision, swallowing, and breathing problems	
Bacillus cereus	1–6	Vomiting	
	10–12	Abdominal pain, diarrhea, nausea	

[a]These organisms grow and produce toxins in the food prior to ingestion, except *B. cereus* may also grow for a short time and produce enterotoxin in vivo.

Table 28.5	Food-borne and water-borne infections of the gastrointestional tract		

Organism[a]	Infection	Source of Infection
Salmonella typhimurium	Salmonellosis	Poultry, eggs, animals
Salmonella typhi	Typhoid fever	Water, food
Vibrio cholera	Cholera	Water, food
Shigella dysenteriae	Shigellosis	Water, food
Campylobacter jejuni	Campylobacteriosis	Poultry, shelfish
Escherichia coli	Travelers' diarrhea	Uncooked vegetables, salads, water
Yersinia enterocolitica	Enterocolitis	Milk
Listeria monocytogenes	Listeriosis	Milk, cheese
Vibrio parahaemolyticus	Hypersecretion	Shellfish
Clostridium perfringens	Hypersecretion	Meats, gravy, stews
Clostridium difficile	Pseudomembranous colitis	Natural inhabitant of GI tract

[a]These organisms survive in the intestinal tract and produce toxins.

Bacillus, or *Clostridium perfringens* is relatively short, and recovery is generally rapid. The exception, of course, is botulism, which, though rare, can be a fatal disease. The incubation time for food-borne infections is generally longer than for food poisoning cases, and the duration of symptoms can also be considerably longer. The major route of transmission for food infections is directly or indirectly related to fecal contamination. Most of the organisms listed in **Table 28.5** are inhabitants of the intestinal tract of various animals and do not survive in nature for any extended period of time. Consequently, they survive by passing from one animal to another through fecal contamination of food or water.

In this section we will consider some of the food- and water-borne infections and the causal agents of food poisoning.

Ulcers

Peptic ulcers occur in the stomach and duodenum of humans. These ulcers result from erosion of the highly alkaline mucus that forms a protective layer over the gastric epithelia. This mucus protects the epithelial cells from the acid and pepsin (proteolytic enzyme) secretions that regularly enter the stomach to aid in digestion of food. Medically, ulcers have been ascribed to several factors including genetic predisposition, excess secretion of acids and pepsin, and diet. It has long been the prevailing dogma that acidity and other factors rendered the stomach an inhospitable environment and that it was free of bacterial colonization. However, in 1977, a microaerophilic bacterium, later named *Helicobacter pylori*, was discovered that could colonize the human stomach. Studies conducted since 1985 have confirmed that this bacterium is a causative agent of peptic ulcers and may cause gastric cancer.

Infections with *H. pylori* occur throughout the world and are more prevalent in developing countries than developed. The organism synthesizes an adhesion that is involved in attachment to stomach epithelial cells and synthesizes a urease that degrades urea-generating ammonia that may protect against acidity. Many strains produce a cytotoxin that promotes infection. The organism can be eliminated from infected patients by combinations of tetracycline, metronidazole, and bismuth.

Food Poisoning

Staphylococcus aureus *S. aureus* is the agent most frequently encountered in cases of **food poisoning**. It is present in the nasopharynx of 10% to 50% of adult populations. The incidence is much higher in the nasal passages of children. It is inevitable that through negligence, unsanitary practice, or accident some *S. aureus* cells will get into food during preparation. Two conditions must be met, however, before food poisoning can occur:

- The contaminated food must be suitable for growth and toxin production by the bacterium.
- The food must stand at a temperature that will permit growth of the bacterium.

For *S. aureus*, this would be room temperature or warmer. *S. aureus* effectively produces enterotoxin during growth on such foods as ham or chicken salad, cream-filled pastry, custards, salad dressing, and mayonnaise. To prevent enterotoxin production, food should be refrigerated before preparation, kept cold during preparation, and refrigerated immediately afterward. Any foods that contain creams or mayonnaise and are exposed to room temperature for any length of time should be discarded. It is foolhardy to take foods such as those listed on a picnic and permit them to sit in the sun. Refrigeration is an absolute necessity.

Staphylococcal food poisoning generally occurs within six hours after ingestion of toxin-containing food. The toxin is heat and acid stable. Severe nausea, vomiting, and diarrhea are the symptoms, and they rarely last longer than 24 hours. However, food poisoning can be a serious problem for individuals weakened by other health problems. Treatment of severe cases is by administration of intravenous fluids to prevent dehydration. Antibiotic therapy is of no value, because the causative agent is the toxin.

Bacillus cereus

B. cereus is an endospore-forming, gram-positive bacillary organism that is a significant source of food poisoning. It is commonly present in soil and can be introduced into food on dust particles. Food poisoning by this organism occurs with various foods but particularly with rice. Rice is often cooked gently and left at room temperature. The heating induces endospores to germinate, and while sitting at room temperature the organism grows extensively and produces enterotoxin. Ingestion of large numbers of bacteria may also result in growth of the *B. cereus* in the gut, where it produces enterotoxin. When this occurs, it might also be considered a food infection. Creamy foods and meat have also been implicated in *B. cereus* food poisoning. Diagnosis is by symptoms and examination of suspected food. A count of 10^5 *B. cereus* cells per gram of food is a strong indicator that this organism is involved. *B. cereus* produces two different enterotoxins:

- Vomiting occurs in 1 to 6 hours after eating contaminated food and is caused by a choleralike enterotoxin that stimulates adenyl cyclase.
- Abdominal pain and profuse diarrhea 4 to 16 hours after ingestion of contaminated food is indicative of yet another enterotoxin, but the action of the toxin is not clear.

Clostridium botulinum

C. botulinum is the etiological agent of the dreaded disease **botulism**. It is widely distributed in soil, on lake bottoms, and in decaying vegetation. The endospores can contaminate vegetables, meat, and fish. Historically, botulism occurred following the ingestion of home-canned, low-acid vegetables such as peas, beans, corn, mushrooms, and meats. The endospores of *C. botulinum* are quite heat resistant and survive unless the processing is complete (120°C for 15 minutes). Botulinum produces a series of toxins that are among the most potent toxins known, and some of the genes for the toxin are coded in lysogenic bacteriophages that attack only this organism. Botulinum toxin is a neurotoxin that attacks nerve cells leading to paralysis. The toxin binds to a receptor on nerve synapses and enters the nerve cell. The release of acetylcholine

from the nerve cell is blocked; muscles cannot contract, resulting in a flaccid paralysis as illustrated in **Figure 28.10**. The toxin is quite heat labile and is destroyed by boiling foods for 10 minutes.

Clostridium tetani

C. tetani is present in soil and in the feces of many animals. It has no invasive ability and enters tissue through punctures, gunshots, or other wounds.

The tetanus toxin is composed of two peptide chains linked by a disulfide bond, and the larger of the chains has a specific receptor that binds to neuronal gangliosides. The tetanus toxin blocks the function of a nerve factor that normally allows muscles to relax. In some unknown way, tetanus toxin prevents the release of inhibitory transmitters (glycine and gamma aminobutyric acid) responsible for muscle relaxation. As a result, opposing muscles that permit movement by alternating contraction/relaxation are in a constant state of contraction (Figure 28.10). This results in a painful spastic paralysis. The spasms can involve muscles of the jaws, and consequently, the disease has been termed *lockjaw*. The toxin can also cause spasms in the respiratory muscles that can result in suffocation.

Clostridium perfringens

C. perfringens is a major agent of food poisoning in the United States. The disease is most often caused by ingestion of meat products, particularly rewarmed meat that has been cooked in bulk at fast-food restaurants. Quite high numbers ($>10^8$) are required for infection, as low numbers of *C. perfringens* are a normal inhabitant of the human intestinal tract. When large numbers of the organism are ingested, they form endospores in the intestine, and during sporulation intertoxin is released. Symptoms occur in 8 to 16 hours and include diarrhea, cramps, nausea, and vomiting. The symptoms last for about 24 hours.

Clostridium difficile

An anaerobic microorganism that forms endospores, *C. difficile* is present in the intestinal tracts of some adults, the number of organisms is quite low, and *C. difficile* causes no apparent infection. If antibiotics such as clindamycin or ampicillin are administered to an individual in an amount sufficient to destroy the normal flora, *C. difficile* can proliferate and be a problem. Overgrowth by the organism can cause a severe necrotizing (death of tissue) process. The infection is termed *pseudomembranous colitis*. It is interesting to note that 50% to 75% of neonates are colonized by *C. difficile* but are completely asymptomatic.

Food- and Water-Borne Infections

There are a number of bacterial species causing disease that are transmitted by contaminated food or water. Among these diseases are salmonellosis, cholera, shigel-

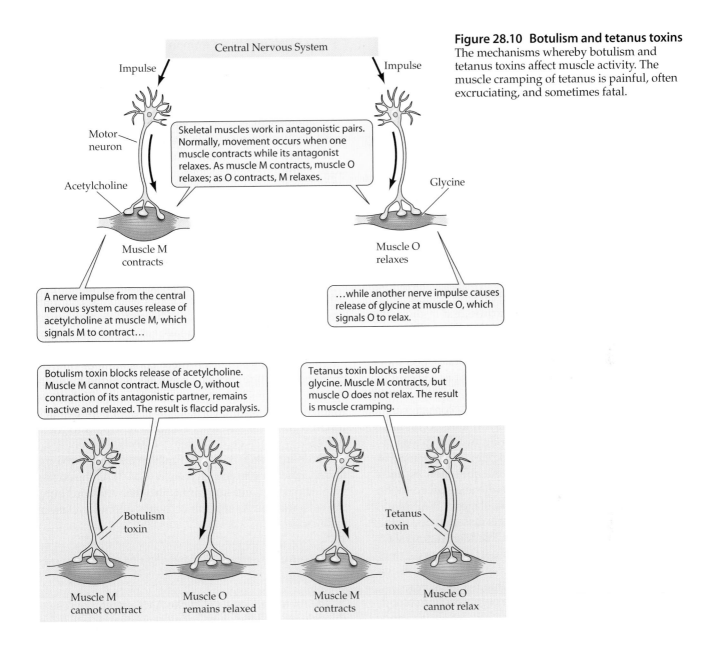

Figure 28.10 Botulism and tetanus toxins
The mechanisms whereby botulism and tetanus toxins affect muscle activity. The muscle cramping of tetanus is painful, often excruciating, and sometimes fatal.

losis, and various forms of enterocolitis. The agents of these diseases, symptoms, diagnosis, and possible treatment will be discussed.

Salmonellosis is an infection fostered by microorganisms in the genus *Salmonella*. There are over 2000 serological types within this genus. These serological types differ by one or more cellular (O) or flagellar (H) antigens. The species most commonly associated with the human infection is *S. typhimurium*.

Various species of salmonellae can be present in the intestinal tracts of animals and birds. As a result, poultry, eggs, and beef can be contaminated with the organism. People who handle these foods, as well as asymptomatic carriers, are a source of food contamination. Given the widespread distribution of salmonellae, all foods, particularly hamburger and poultry, should be treated as contaminated. Hands contaminated by handling an infected animal such as a dog or cat can be a source of disease. Water polluted with human or animal waste is another source of the disease. Some years ago small pet turtles were available in variety stores and pet shops. These turtles were *Salmonella* carriers and were responsible for an estimated 300,000 cases of human salmonellosis per year. Import and sale of these animals is now banned.

There are several million cases of salmonellosis in the United States each year. Most cases of salmonellosis follow ingestion of uncooked egg products such as custards, meringues, or eggnog. Undercooked meat, particularly chicken, is also a source of infection.

Enterocolitis is the most often observed manifestation of salmonellosis. The symptoms occur 10 to 30 hours after infection and include abdominal pain, nausea, vomiting, and diarrhea. The symptoms usually last from 2 to 5 days but may last longer. The organism multiplies in the intestine and produces an enterotoxin with choleralike activity and a cytotoxin that destroys epithelial cells in the intestinal tract. Feces from an enterocolitis patient can contain one billion salmonellae per gram. Diagnosis is by isolation and identification of the infectious agent. Salmonellosis can induce a loss of fluid that may be a serious complication in children, elderly, and debilitated individuals. Treatment is by replacement of lost electrolytes through intravenous fluids.

Salmonella typhi is the etiological agent of **typhoid fever**. There are about 500 cases of typhoid fever in the United States each year and an estimated 2,000 carriers. Typhoid fever epidemics have occurred throughout human history (see Box 30.3). The sole reservoir for *S. typhi* infections is humans, so carriers and improperly treated sewage are the sources of infection. *S. typhi* is passed from feces to hands to food, and public health measures have been established to eliminate food handlers from being carriers. Typhoid infections are caused by ingesting contaminated food, by putting contaminated objects in one's mouth, or by drinking contaminated water. After infection, the onset of symptoms occurs after a 10- to 14-day incubation period, and the disease is marked by headaches, abdominal pain, a high temperature, and rose-colored spots on the abdomen. *S. typhi* cells can penetrate the small intestine and spread to the lymphoid system. During the active stages of the disease, *S. typhi* is present in the blood and is disseminated to the spleen, liver, bone marrow, and gallbladder. Victims shed the bacterium for 3 months or more, whereas carriers often have an infected gallbladder that constantly sheds *S. typhi* into the bile duct and on into the intestines.

Diagnosis is by isolation of the organism from blood, urine, or feces and through serological identification. Patients are treated with chloramphenicol, which must be continued for at least 2 weeks to clear the organism from the gallbladder.

Vibrio cholerae

The etiological agent for **cholera** is *Vibrio cholerae*, a gram-negative, comma-shaped organism with a single polar flagellum. The organism adheres to epithelia in the small and large intestine and secrets **choleragen**, the cholera enterotoxin.

Cholera is a fearful disease spread by fecal–oral transmission through water and food. It is inevitable that cholera will occur in areas that have breakdowns in sewage disposal systems or where human waste pollutes water supplies. Cholera is endemic in India, Pakistan, Bangladesh, and other Asian countries where sewage disposal is inadequate. Serious outbreaks of cholera occurred in Peru and Brazil in recent years, killing over 10,000 people.

The incubation time from ingestion to disease symptoms is only hours to 3 or more days and is dose dependent. One must ingest 10^8 to 10^9 organisms in water to acquire an infection because the stomach acidity can wipe out smaller doses. Ingestion of fewer organisms with contaminated food can result in the disease, because food tends to protect *V. cholerae* from stomach acidity.

Cholera has one major symptom, an acute diarrhea with excretion of 8 to 15 liters of liquid per day. This liquid contains vibrio cells, epithelial cells, and mucus. If untreated, 60% of the patients succumb to dehydration. Treatment by intravenous electrolyte and liquid replacement lowers the death rate to less than 1%.

Cholera is diagnosed symptomatically and by the isolation of the organism from feces. Serology is employed to determine the strain of organism involved. Treatment with streptomycin or other antibiotics may shorten the course of infection. Cholera can be prevented by implementing adequate sewage treatment procedures and by purifying drinking water.

The cholera toxin is a protein consisting of three or more polypeptide chains. Components of the toxin are A_1 and A_2 chains linked covalently by a disulfide bond to form a dimer and a variable number of peptides designated B. The B subunits are loosely linked to the A dimer.

The B subunit(s) have one function and that is to bind to a specific ganglioside glycolipid (GM-1) on the epithelia of the ileum and large intestine. The A chains enter the cell, and the A_1 subunit activates adenyl cyclase as follows:

$$ATP \xrightarrow[\substack{\text{adenyl} \\ \text{cyclase}}]{} \text{cyclic AMP}$$

The increased levels of cyclic AMP effect a release of chloride ion and bicarbonate from mucosal cells that line the intestine. These inorganic ions accumulate in the lumen (inner space) of the intestine, and the normal passage of sodium ion into the mucosal cells is blocked. This creates an osmotic imbalance such that water passes through the intestinal epithelia into the lumen (**Figure 28.11**). This influx of water to the lumen causes the severe diarrhea associated with the disease.

Shigella dysenteriae

Another organism responsible for intestinal disease is *Shigella dysenteriae*. It is the causative agent of **shigellosis,** a form of dysentery. Humans are the only known hosts for *S. dysenteriae,* and the infectious organism is transmitted via the fecal–oral route. Food and water are the major routes of transmission, whereas direct feces-to-mouth transmission is probably responsible for the prevalence of this disease in preschool children. Most cases of the disease are found among children in day-care centers.

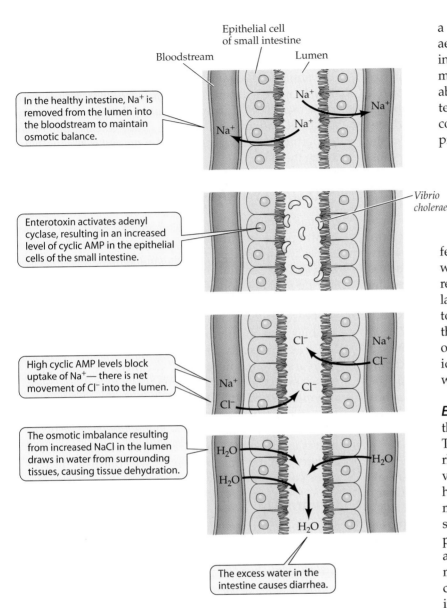

In the healthy intestine, Na$^+$ is removed from the lumen into the bloodstream to maintain osmotic balance.

Enterotoxin activates adenyl cyclase, resulting in an increased level of cyclic AMP in the epithelial cells of the small intestine.

High cyclic AMP levels block uptake of Na$^+$— there is net movement of Cl$^-$ into the lumen.

The osmotic imbalance resulting from increased NaCl in the lumen draws in water from surrounding tissues, causing tissue dehydration.

The excess water in the intestine causes diarrhea.

Figure 28.11 Cholera toxin
The mode of action of cholera toxin.

S. dysenteriae remains localized in intestinal epithelial cells and effects a loss of fluid and ulceration of the colon wall. The shigella toxin kills absorptive epithelial cells, and diarrhea results from this interference with liquid absorption. Isolation and identification of *S. dysenteriae* in feces is one means of diagnosis, but a direct swab of lesions in the colon and subsequent identification of the organism serologically is better. Treatment is by intravenous fluid replacement and with the antibacterials sulfamethoxazole and trimethoprim.

Campylobacter jejuni The etiological agent for a major diarrheal disease termed **campylobacteriosis** is *Campylobacter jejuni*. *C. jejuni* is a gram-negative curved cell with a single polar flagellum. It is microaerophilic. *C. jejuni* is an inhabitant of the intestinal tracts of wild and domestic animals. The organism can cause sterility and abortion in cattle and sheep. It is transmitted to humans via the fecal–oral route in contaminated food and water, and underprocessed chicken and raw milk are common sources of campylobacteriosis. Children can get the infection from domestic dogs. The incubation period after ingestion (fewer than ten organisms can cause infection) is 2 to 4 days. Symptoms include a high fever, nausea, abdominal cramps, and watery, bloody feces. The enterotoxin responsible for campylobacteriosis is similar in activity to the cholera toxin. Another toxin produced by *C. jejuni* is a cytotoxin that causes colitis. Diagnosis is by isolation of the organism from feces and serological identification. The disease can be treated with fluoroquinolone.

Escherichia coli A normal inhabitant of the human digestive tract is *Escherichia coli*. There are strains of *E. coli* that cause diarrheal disease, particularly evident in individuals traveling between countries— hence the name "travelers' diarrhea" (and many other names). These pathogenic strains have virulence factors, carried by plasmids. One of these virulence factors is a surface antigen, the K antigen, that permits attachment and invasion of mucosal cells. Pathogenic *E. coli* strains attach to the intestinal mucosa by pili or fimbriae and produce an enterotoxin. In 1993 an outbreak of a food-borne *E. coli* infection killed four children in the Seattle area and alerted public health officials to the dangers of *E. coli* O157:H7. This pathogenic strain can be transmitted in meat products, particularly ground beef.

Yersinia enterocolitica The etiologic agent for **enterocolitis** in humans is *Yersinia enterocolitica*. The organism is present in the intestinal tracts of cats, dogs, rodents, and domestic farm animals. It can also be isolated from lake and well water. As with many of the previously mentioned enteric diseases, enterocolitis is a fecal–oral transmitted infection. *Y. enterocolitica* adheres to and invades intestinal epithelial cells. The symptoms include fever, diarrhea, and abdominal pain. A form of arthritis can occur several days after onset of acute enteritis. The drug of choice is trimethoprim-sulfa methoxazole.

Listeria monocytogenes An infection called **listeriosis** is caused by *Listeria monocytogenes,* a small gram-positive diphtheroid organism. The organism is unusual in that it has a complex lipopolysaccharide on the cell surface similar to the toxins associated with gram-negative bacteria. *L. monocytogenes* is mainly an animal pathogen and is transmitted to humans through dairy products. Milk that has been improperly pasteurized and the cheese manufactured from such milk are the major causes of infection.

L. monocytogenes may infect pregnant women, and the organism may pass through the placenta to the fetus. In the newborn, the disease is generally pneumonia at birth or shortly thereafter. About 2 to 4 weeks after birth, the listeriosis infection appears as meningitis. This disease also occurs in immunosuppressed persons (such as AIDS patients). Listeriosis can be treated with ampicillin and gentamicin.

Vibrio parahaemolyticus A marine bacterium, *V. para-haemolyticus,* grows only in the presence of moderate salt concentrations. It is the etiologic agent of a food poisoning that follows ingestion of uncooked seafood. The microorganism produces a toxin that causes a diarrhea of rapid onset.

URINARY TRACT INFECTIONS

Infections of the urinary tract are second only to respiratory infections in number of cases annually. Urinary tract infections are by far the prevalent nosocomial infection, causing about 40% of all hospital-acquired (nosocomial) infections. These infections can occur in the kidneys (*pyelonephritis*), in the bladder (*cystitis*), or as an inflammation of the urethra (*urethritis*). Infection of the kidneys generally will lead to infection of the entire urinary tract, but more often infections spread by ascending from the urethra. Proper treatment of urethritis with antibiotics (amoxicillin or trimethoprim) will prevent the spread of the infectious agent to the bladder and kidneys. Because the urethra in females is much shorter than that in males, the incidence of urinary tract infections in females is much higher.

The organism most often associated with urinary tract infections is *Escherichia coli,* the causative agent in 80% of the cases. Other organisms involved are *Proteus mirabilis, Pseudomonas aeruginosa,* and other facultatively aerobic gram-negative rods and cocci. Usually a single strain is involved in an infection, and diagnosis is made by determining the number of organisms present in urine.

The bladder is generally sterile, but bacteria colonize epithelial cells that line the lower end of the urethra. Urine commonly has 1,000 to 10,000 bacteria organisms per ml. Urine is a good growth medium, and accurate counts require prompt processing of the urine specimens. Greater than 10,000 organisms per ml implies an infection. However, in some cases of kidney infection the number of organisms present in urine is relatively low. Diagnosis depends on the presence of irritation and pain especially during urination. Catheterization of hospital patients is a leading cause of infections by opportunistic pathogens generally present in the urethra or on the skin.

SEXUALLY TRANSMITTED DISEASES

Sexually transmitted diseases (STDs) are a growing health problem throughout the world. These diseases affect an alarming portion of the young adult population and may be transmitted to the newborn by transfer during fetal development or during birth. The etiological agents for most STDs are bacteria and viruses. We will discuss the major bacterial diseases: gonorrhea, syphilis, and chlamydial infections. All three bacterial STDs are eminently curable and have been for the last 40 years. Genital herpes and acquired immune deficiency syndrome (AIDS) are the leading viral infections (Chapter 29). STDs have become as much a social problem as a medical concern. The greatest incidence of STDs is in young people ages (15 to 35), the age group that is least likely to seek treatment. Increased sexual activity and multiple partners make control of the diseases particularly difficult.

Gonorrhea

The etiological agent for **gonorrhea** is *Neisseria gonorrhoeae,* a gram-negative generally diplococcal bacterium. Drying, sunlight, and UV light readily kill the organism, but these would have little effect on the organism in the real world. *N. gonorrhoeae* is transmitted in virtually every case by direct contact. Sexual activity between infected and uninfected individuals leads to dissemination of the pathogen. The pili on pathogenic strains of *N. gonorrhoeae* attach to microvilli of mucosal cells of the genitourinary tract. The bacterium is then phagocytosed by cells on the mucosal surface and transported to intercellular spaces and submucosal tissue (**Figure 28.12**). Phagocytosed gonococci within cells are protected from host defenses.

The bacterium can infect the eyes of newborns during birth. Untreated, this infection may lead to serious eye infections and is responsible for blindness in much of the developing world. Tetracycline, erythromycin, or silver nitrate is applied to the eyes of neonates at birth as a preventive measure in the United States. This has essentially eliminated gonorrhea-related eye infections and subsequent blindness.

The incidence of gonorrhea in the United States is alarming, with perhaps as many as one million new

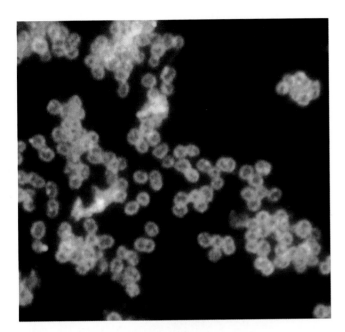

Figure 28.12 Gonorrheal infection
A positive fluorescent antibody test for *Neisseria gonorrhoeae.*
Courtesy of the Centers for Disease Control.

cases occurring annually. This figure would indicate that 1 of every 250 Americans is infected each year, and most are in the 15 to 35 age group. Considering that individuals in this group represent about one-third of the population, the incidence would be about 1 in 80, a public health concern. The probability of infection per contact is about 25% for males and 85% for females. The incidence of gonorrhea increased over fivefold in the period following the widespread use of oral contraceptives. A higher level of sexual activity and decreased use of condoms in the period from 1975 to 1978 is considered responsible for this marked increase. Fear of AIDS has now led to an increased awareness of the dangers of "unsafe" sex and has caused some lowering of the incidence of gonorrhea over the last several years.

Infection with *N. gonorrhoeae* does not confer immunity, and many of the cases observed are reinfections. There is some evidence that chronic gonorrhea in females can induce antibody formation, but with the large number of gonococcus serotypes (at least 16) present in the population, reinfection with another strain is likely. The high incidence of gonorrhea is primarily due to the absence of distinct symptoms in most females. These females are unaware of the infection, and the symptoms of mild vaginitis and minor discharge are attributed to other causes. These chronic carriers serve as a significant reservoir of infection. The gonococcus can infect the urethra, vagina, cervix, and fallopian tubes of females before it is detected and cause pelvic inflammatory disease leading to sterility.

Infected males are rarely asymptomatic. Two to eight days after exposure they experience a painful urethritis and discharge of pus that contains numerous leukocytes. These leukocytes bear intracellular gonococci, and a microscopic examination of stained exudate is an effective diagnostic tool.

A major problem in treating gonorrhea is the increasing incidence of antibiotic resistance. Penicillin has been the treatment of choice, but penicillinase-producing resistant organisms have appeared. Tetracycline and erythromycin have been employed as alternatives. Recently, penicillin-resistant strains have appeared that have a chromosomal-mediated resistance to penicillin that also confer some resistance to tetracycline and erythromycin. Ceftriaxone is the drug of choice. However, unless public health measures or individual responsibility lowers the incidence of this disease, it is likely that antibiotic resistant strains will become more of a threat in the future.

Syphilis

Treponema pallidum is the causative agent of the historically significant disease, **syphilis**, which reached epidemic proportions in Europe in the sixteenth century, as it spread rapidly and caused much suffering. Apparently, *T. pallidum* was considerably more contagious in earlier times than it is today. About one in ten people who have a onetime exposure to syphilis actually will contract the disease. There are 25,000 to 35,000 new cases diagnosed each year in the United States. The lower incidence as compared with gonorrhea is attributed to the absence of asymptomatic female carriers. The highest incidence of syphilis is among homosexual males.

Syphilis is transmitted by sexual contact. The organism enters through mucous membranes or through minor breaks in the skin during sex. About one case in ten occurs in the oral region. Pregnant women can pass the spirochete in utero to the fetus, a terrible disease that affects 300 children per year.

There are three recognizable stages in a syphilitic infection: primary, secondary, and tertiary.

Primary Stage Within 10 days to 3 weeks after exposure, a small, painless reddened ulcer occurs at the site of infection. This ulcer is called a **chancre** and may appear on the penis or in the vaginal or oral regions (**Figure 28.13**). The chancre contains spirochetes, and at this stage the infection is readily transmissible through intimate contact. In about one-third of the cases, the symptoms disappear and the disease progresses no further. In others, the spirochete enters the bloodstream and is distributed throughout the body. The primary stage lasts from 2 to 10 weeks. The disease then enters the secondary stage.

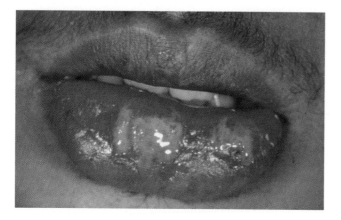

Figure 28.13 Syphilis infection
A primary lesion from a syphilis infection. Photo
©Nussenblatt/Custom Medical Stock Photo.

Secondary Stage After the chancre(s) heal(s), skin
lesions occur about the body including the soles of the
feet and palms of the hand. They also appear on mucus
membranes. The patient is generally feverish. Individuals in this stage of the infection are all serologically
positive and can transmit the disease. After 4 to 8 weeks,
the disease enters a latent period and is not normally
infectious except maternally to the fetus. At this stage,
about one-fourth of the cases are essentially cured, one-
fourth remain latent with relapses to the secondary
stage, and about one-half of these enter the tertiary
stage.

Tertiary Stage Untreated tertiary
syphilis is a degenerative disease with
lesions occurring on skin, in bones,
and in nervous tissue. Lesions frequently occur in the central nervous
system, leading to mental retardation,
blindness, and insanity. Paralysis often
occurs, and those afflicted with tertiary syphilis walk with a stumbling
gait. Lesions in the cardiovascular system can be fatal. There are actually few
spirochetes in the lesions, and most of
the damage is attributed to hypersensitivity to the organism and its products.

Diagnosis can be made when chan-
crous lesions are examined by dark-
field microscopy of the exudate where
the spirochetes can be observed.
Serological tests are widely employed
for diagnosis and depend on the presence of antitreponemal antibodies to
indicate the presence of *T. pallidum*.
Penicillin is an effective treatment and
is particularly effective in primary

stages of the disease. More drugs and prolonged treatment are necessary in the later stages. There is no cure
for symptoms of tertiary syphilis.

Chlamydiae

Chlamydia trachomatis is an obligate intracellular parasite
that causes several different infections in humans. It is
an intracellular parasite and has an exceedingly small
genome, reflecting their limited synthetic ability. The
incidence of these venereal infections greatly outnum-
bers gonorrhea, and there are an estimated 3 to 10 million new cases in the United States each year. *Chlamydia*
reproduce when an **elementary body** attaches to a cell
surface and is phagocytosed into the cell (**Figure 28.14**).
The elementary body then forms an **initial body**, and
these form inclusions containing reticulate bodies. The

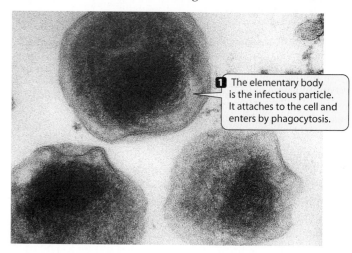

1 The elementary body
is the infectious particle.
It attaches to the cell and
enters by phagocytosis.

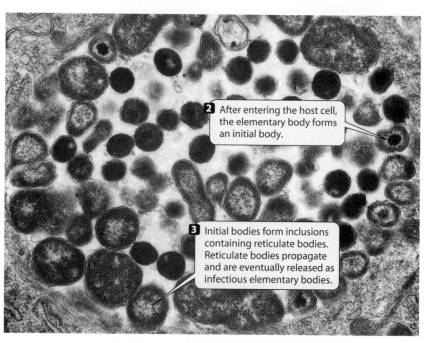

2 After entering the host cell,
the elementary body forms
an initial body.

3 Initial bodies form inclusions
containing reticulate bodies.
Reticulate bodies propagate
and are eventually released as
infectious elementary bodies.

Figure 28.14 Life cycle of *Chlamydia trachomatis*
Micrographs showing the stages in the *Chlamydia* life cycle.
Photos ©D. M. Phillips/Visuals Unlimited.

reticulate bodies reform to elementary bodies that are infectious. These elementary bodies are released on host-cell lysis. The cycle from elementary body to reticulate body and release of infectious elementary bodies requires 48 to 72 hours.

C. trachomatis is also the etiological agent of **trachoma**, an eye infection that occurs worldwide. Over 10 million cases of blindness, principally in developing countries, are attributed to this disease. The organism is passed to the newborn during birth and causes conjunctivitis (eye infection) and respiratory distress. Between 8% and 12% of pregnant women in the United States have chlamydial infections of the cervix, and one-half of babies born to these infected individuals have conjunctivitis. Application of tetracycline to the eyes prevents development of the disease. This organism causes much of the pneumonia in infants.

Chlamydial nongonococcal urethritis (NGU) is probably the most often acquired sexually transmitted disease. In males, *C. trachomatis* causes urethritis, and in females it causes urethritis, cervicitis, and pelvic inflammatory disease. An estimated 50,000 women become sterile each year from NGU. Inapparent infections are common in sexually active adults. Diagnosis is difficult, and until recently it was diagnosed by excluding other diseases. Now it is identified by staining material, taken by swab, with fluorescent antibody. Treatment with tetracycline and erythromycin is effective. Penicillin is ineffective, as *C. trachomatis* lacks a peptidoglycan cell wall. As a significant number of gonorrhea patients are also infected with chlamydiae, it is essential that both diseases be treated.

Another strain of *C. trachomatis* is responsible for a disease in males called *Lymphogranuloma venereum*. This is a sexually transmitted infection that causes swelling of lymph nodes about the groin. It is diagnosed by the presence of swollen lymph nodes and treated with tetracycline or sulfonamides.

VECTOR-BORNE DISEASES

Vectorborne diseases are transmitted from one infected animal to another by means of an intermediary host (Table 28.6). This intermediary host is termed a **vector** and is most often a biting insect (arthropod or louse) that transmits the pathogen while feeding on an uninfected host. The animal that is a constant source of the infectious agent is called a **reservoir**. Most reservoirs are mammals that may or may not be adversely affected by the infectious organism. Some tick and mite vectors also serve as reservoirs. This occurs in cases where the infectious bacterium is transovarially (by infected eggs) transferred from parent to insect progeny.

The bacteria involved in vector-borne diseases are **obligate parasites** that cannot reproduce or survive in nature outside a living host. Those that cause a fatal infection in the reservoir animal must be carried by the vector to another living host. The etiological agents for many of these diseases are in the order *Rickettsiales*, and in the genera *Rickettsia*, *Rochalimaea*, and *Coxiella*. Fleas,

Table 28.6	**Vector-borne bacterial infections of humans**			
Disease	**Etiological Agent**	**Vector**	**Reservoir**	**Transmission**
Rocky Mountain spotted fever	*Rickettsia rickettsii*	Wood tick (western U.S.) Dog tick (eastern U.S.)	Mammals	Tick to human
Q fever	*Coxiella burnetti*	Animal tick	Ticks, cattle	Tick to human, carcass to human
Tularemia	*Francisella tularensis*	Wood tick	Ticks, rabbits	Tick to human, animal other mammals to human
Relapsing fever	*Borellia recurrentsis*	Body louse	Humans	Human to louse to human
Lyme disease	*Borrelia burgdorferi*	Deer tick	Deer, mice, voles	Flea to human
Typhus (epidemic)	*Rickettsia prowazekii*	Human louse	Humans, flying squirrels	Louse to human
Typhus (endemic)	*Rickettsia typhi*	Rat flea	Rats, ground squirrels	Flea to human
Plague	*Yersinia pestis*	Rat flea	Rats	Flea to human, human to human
Scrub typhus	*Rickettsia tsutsugamushi*	Mite	Mites	Mite to human
Trench fever	*Rochalimaea quintana*	Body louse	Lice, humans	Louse to human

lice, or ticks transmit them. The reservoir is generally a mammal.

Tick-borne diseases can be prevented by proper sanitation and by ridding the human environment of rats and other vermin that carry fleas and ticks. For diseases such as Lyme disease and Rocky Mountain spotted fever, one should dress properly when entering outdoor areas potentially infected with ticks. Wear long-sleeved shirts and long pants. Pants should be tucked into socks, and shoes should be worn. Use of insect repellants is of value, and one should check for ticks after leaving an area that is potentially infected. These are summertime diseases of high incidence when ticks are feeding.

Plague

The etiological agent for **plague** is *Yersinia pestis,* a gram-negative nonmotile coccobacillus. Historically, the plague has killed more humans than any other bacterial disease. In the sixth century millions died from the disease, and in the fourteenth century a plague epidemic killed more than one-fourth of the people in Europe (**Box 28.1**).

Y. pestis normally afflicts rodents and is endemic in the southeastern United States in prairie dogs, ground squirrels, wood rats, chipmunks, and mice. These animals are relatively resistant to plague, and the disease resides in them during interepidemic periods. If domestic rats interact with these reservoir animals, the domestic rat becomes infected (**Figure 28.15**). Fleas infesting these infected animals can cause an epidemic of plague in the domestic rat populations. Plague is a fatal disease in domestic rats. When the rats die, the fleas travel to other rats. If rats are not available, the fleas move to and feed on humans, thereby transmitting the disease. Killing rats after the local population of fleas is infected can actually drive the fleas to humans and increase the incidence of plague.

The blood of a domestic rat infected with plague contains a large number of the *Y. pestis* organism. The flea ingests this blood, and the bacterial coagulase induces a clot in the proventriculus of the tick. This prevents food from passing on to the stomach of the flea. The flea becomes hungry and feeds ravenously but regurgitates due to the gut blockage. The regurgitated blood contains

---

| **Milestones Box 28.1** | **A Sad Game** |

There is a children's game where the youngsters stand in a circle holding hands and sing:

> *Ring-a-ring o'rosie*
> *A pocketful of posie*
> *A-tishoo! A-tishoo!*
> *We all fall down.*

When the last line is recited the children drop to the ground. There are a number of versions of this poem which may have originated in the days of the Great Plague of London (1665). Thus, this pleasant little diversion may well have had a macabre origin, as evidenced by the following:

"Ring-o-rosie"— referred to the rosy rash associated with the symptoms of plague.

"Posies of herbs"— were carried during epidemics to ward off the plague.

"A-tishoo"— referred to sneezing, which was considered a final fatal symptom.

"All fall down"— meant that everyone dropped over dead.

Illustration from *Mother Goose,* by Kate Greenaway (1881). The Granger Collection, New York.

> *Ring-a-ring-a-roses,*
> *A pocket full of posies;*
> *Hush! hush! hush! hush!*
> *We're all tumbled down.*

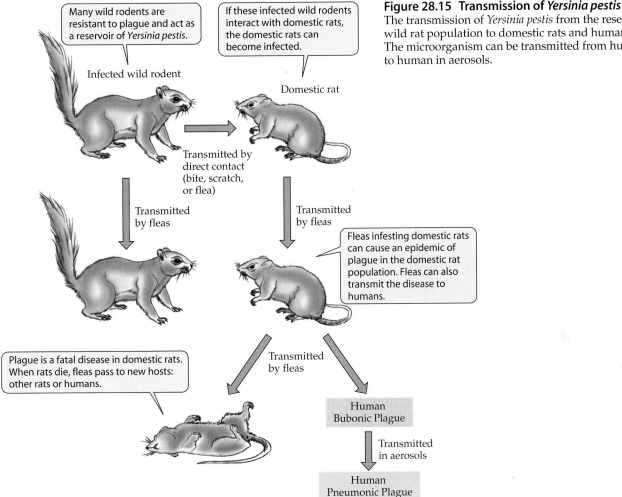

Figure 28.15 Transmission of *Yersinia pestis*
The transmission of *Yersinia pestis* from the reservoir wild rat population to domestic rats and humans. The microorganism can be transmitted from human to human in aerosols.

large numbers of infectious bacteria that enter the bite site, thus infecting the host.

The plague bacillus moves from the entrance site to regional lymph nodes. These lymph nodes become enlarged and tender due to growth of the bacterium. The enlarged node is called a **bubo** and is the origin of the name bubonic plague (**Figure 28.16**). The organism can be engulfed by polymorphonuclear leukocytes and destroyed, whereas those phagocytosed by macrophages are not. Within the macrophage, *Y. pestis* is maintained at a favorable temperature (37°C) and in a low Ca^{++} environment. Low Ca^{++} promotes the expression of virulence factors. The bacilli thrive in macrophages and travel to regional lymph nodes. From lymph nodes they are carried via the bloodstream to the spleen, liver, and other organs. They cause subcutaneous (below the skin) hemorrhages that appear as dark areas on the body surface. These darkened areas and the high level of mortality are the reason for calling plague "Black Death." Movement of the bacilli to the lungs results in pneumonic plague, a disease that is transmitted via the

respiratory route. Untreated, pneumonic plague has 100% fatality rate, and bubonic plague has a fatality rate of 75%.

The plague bacillus produces a number of toxins, and among these is a protein/lipoprotein complex that prevents phagocytosis. There is an exotoxin termed **murine toxin** produced by virulent strains of *Y. pestis* that blocks cellular respiration in laboratory mice and probably has a similar activity in human plague.

Plague is diagnosed by symptoms and by isolation of the organism from aspirates of the buboes. Sputum can be the source of isolates in pneumonic plague, or the organism from either source can be identified serologically. Treatment with streptomycin, chloramphenicol, or tetracycline is effective if given early in the infectious process.

Lyme Disease

Lyme disease results from infection by the spirochete *Borrelia burgdorferi*. The tick-borne disease was first recognized in individuals in Lyme, Connecticut, hence its

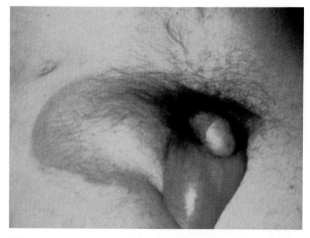

(A)

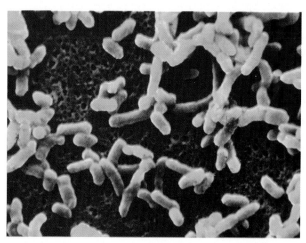

(B)

(C)

Figure 28.16 Plague
Manifestations of plague infection in humans. (A) A bubo (swelling) in the groin area. (B) Gangrene in hand tissue. (C) A scanning electron micrograph of *Yersinia pestis*. A, Photo © C. Stratton/Visuals Unlimited; B, photo ©Biophoto Associates/Photo Researchers, Inc.; C, photo ©CNRI/SPL/Photo Researchers, Inc.

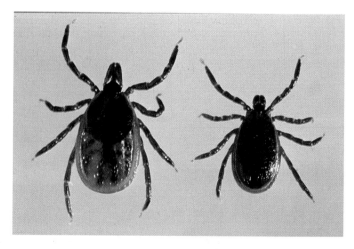

Figure 28.17 Vector of Lyme disease
The deer tick (*Ixodes dammini*) is the major vector of Lyme disease in the United States. Shown here are a male and the much larger female. The female is about 3 mm long. Photo ©Phr/Rocky Mt. Lab/Photo Researchers, Inc.

name. The disease first appeared in 1975 as cases of arthritis, and in 1982 the etiologic agent was identified. Lyme disease is the most rapidly spreading of the tick-borne diseases, with nearly 15,000 cases reported each year. Although Lyme disease is primarily diagnosed in the Northeastern and Middle Atlantic states, it has now spread to virtually every state in the United States.

The major reservoirs for *B. burgdorferi* are the deer and white-footed mouse. The tick involved is mainly the deer tick, *Ixodes dammini* (**Figure 28.17**), although other tick species may also be infected with the spirochete. The deer tick feeds on birds and other animals, rendering them potential sources of Lyme disease. A higher percentage of deer ticks are infested with *B. burgdorferi* than is the case for other ticks carrying tick-borne diseases. The spirochete is transferred transovarially, so the deer tick can be a reservoir as well as the vector.

The deer tick is quite small and not readily seen on human skin. The first manifestation of infection is a skin lesion at the site of the insect bite, which spreads to form a large rash area (**Figure 28.18**). At this stage of the disease it is readily treatable with tetracycline. Disease symptoms include headaches, chills, backache, and general malaise. Untreated Lyme disease progresses to a chronic stage, and the symptoms at this stage are arthritis and numbness in the limbs. The organism may attack the central nervous system, affecting vision and causing paralysis. The best preventive measure is to avoid tick habitats. But should you get bitten and develop a rash, headache, or chills, see a physician immediately.

Rocky Mountain Spotted Fever

The etiological agent for **Rocky Mountain spotted fever** (**RMSF**) is *Rickettsia rickettsii*, a disease first recognized

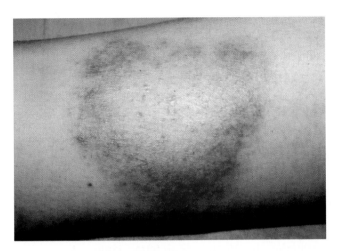

Figure 28.18 Lyme disease
The characteristic rash of Lyme disease at the site of the tick bite. The rash area increases over several days, and effective treatment should start at this stage. Photo ©K. Greer / Visuals Unlimited

in Bitterroot Valley, Montana. The organism is named for the microbiologist H. T. Ricketts, who died of typhus while investigating that disease, also caused by a rickettsiae. RMSF is caused by the bite of an infected tick and is acquired in humans from feces deposited by the tick in and about the bite site.

In the western United States the rickettsiae is carried by the wood tick (*Dermacentor andersoni*) and in the east by the dog tick (*D. variabilis*). *R. Rickettsii* is passed from parent to tick progeny by transovarial passage. *Dermacentor andersoni* is a slow-feeding, hard-shelled tick quite distinct from the fast-feeding ticks that are the vectors of Lyme disease. Consequently, removal of the RMSF tick up to three hours after it initiates feeding can prevent infection. There are about 1,000 cases of RMSF in the United States each year, with more cases in North Carolina and adjacent states than in the Rockies. Dogs are readily infected with *R. rickettsii,* and organisms are present in the blood for two weeks after infection. A dog may serve as a reservoir.

The symptoms of RMSF include a fever, severe headache, and a rash that appears first on the palms and soles of the feet. The organism proliferates in the endothelial linings of small vessels and capillaries. The resultant hemorrhage and necrosis are responsible for the observed rash. Gastrointestinal problems can also occur. RMSF is generally diagnosed via symptoms and by establishing a history of tick bite(s). Tetracycline is an effective treatment, provided it is administered early in the infection.

Typhus

There are two types of the disease typhus, and they are caused by different rickettsial species that share com-

mon antigens. The two types are epidemic and endemic typhus. The etiological agent for epidemic typhus is *Rickettsia prowazekii,* and that for endemic typhus is *Rickettsia typhi.*

Epidemic typhus occurs in areas of substandard living conditions and poor sanitation. The major reservoir for *R. prowazekii* is the human, and the vector is the louse. Flying squirrels can be infested with lice carrying the rickettsia. Passage to humans from the flying squirrel has occurred when these animals move into the attic of a home. There is no transovarial transfer of *R. prowazekii* in the louse, and the infected insect will die within several weeks.

Infection of humans occurs while the louse is feeding, as the infectious organisms are present in feces or regurgitated blood of the insect. The disease is marked by a general bacteremia, with the rickettsia growing in the endothelial cells of blood vessels. Symptoms include a high fever (104°F) with headaches, chills, and delirium. During the fifth or sixth day of illness, a rash develops. Recovery or death occurs 2 to 3 weeks after the onset of symptoms.

Typhus is diagnosed during epidemics by symptoms and, in sporadic cases, by the presence of a circulating antibody. The Weil-Felix test can be employed in diagnosis. This test is based on the fact that strains of the genus *Proteus* share common antigens with *R. prowazekii.* If antibody against *R. prowazekii* is present in serum, it will agglutinate *Proteus* cells. Tetracycline and chloramphenicol are effective in treating typhus. Vaccines are also available for use in areas where typhus is of widespread occurrence.

Endemic typhus, also called **murine typhus**, occurs sporadically in areas throughout the world, including the Gulf Coast states in the United States, but, rarely in epidemics. Immunity to epidemic typhus (*R. prowazekii*) will confer immunity to endemic typhus (*R. typhi*). The primary reservoir for *R. typhi* is the rat or ground squirrel. Neither the rat nor the rat louse is adversely affected by the infection. The disease symptoms, diagnosis, and treatment are the same as for epidemic typhus.

Scrub Typhus

Scrub typhus is also a rickettsial disease, and the etiological agent is *Rickettsia tsutsugamushi.* This disease was of some consequence during World War II and the Vietnam War. The reservoir is the rodent and the vector is a mite. The mite transmits the disease from rodent to rodent. It is transferred transovarially, and the mites lay eggs in soil that hatch into larvae (chiggers) that feed on animals, including humans. The disease symptoms are similar to those encountered with typhus. A primary lesion, called an **eschar,** occurs at the location of the bite. Diagnosis is by the Weil-Felix reaction or by means of serological tests. Tetracycline and chloramphenicol are effective, but treatment should be prolonged because relapses do occur. A successful vaccine has not been developed for scrub typhus.

Q Fever

Q fever received the name Q (Query) because the etiological agent for the disease was long unknown. We now attribute the disease to *Coxiella burnetii*, a rickettsial organism remarkable for its ability to survive outside a host. Arthropods are both a reservoir and vector of Q fever and transmit it from animal to animal. The organism forms an endospore-like body that permits survival on animal carcasses and hides and in feces. Outbreaks in humans occur among animal handlers and through consumption of unpasteurized milk. Q fever causes an influenzalike illness with prolonged fever, headaches, and chills. Pneumonia and chest pains are also symptoms. The disease is insidious in that an endocarditis (inflammation of the heart) can occur months or years after infection. Immunological tests are employed in diagnosing Q fever, and treatment is with tetracycline.

Relapsing Fever

Humans are the reservoir for *Borrelia recurrentis*, the causative agent of **relapsing fever**. *B. recurrentis* is a spirochete that forms coarse irregular coils that are 0.5 μm wide and over 20 μm in length. The disease is transferred from human to human by the body louse. A bite will transfer the spirochete into the bloodstream, which carries it to the spleen, liver, kidneys, and gastrointestinal tract. Multiple lesions occur in these organs. After a fever and 4 to 5 days of illness, the disease symptoms disappear. In most cases, a relapse (hence, relapsing fever) occurs after about 10 days, with a return of the manifestations of disease. These bouts of wellness and relapse occur three to ten times before full recovery occurs. The relapse is attributed to the development of strains with different surface proteins on the outer membrane that are unaffected by the original immune response. A period of time is necessary before the host can generate antibodies to combat each new strain.

The disease can be diagnosed by clinical symptoms and by the presence of *B. recurrentis* spirochetes in the blood. Dark-field microscopy is effective in visualizing spirochetes. The treatment of choice is penicillin.

Tularemia

The etiological agent for **tularemia** is *Francisella tularensis*, an extremely pleomorphic, small gram-negative bacterium. The wood tick is both a reservoir and vector of *F. tularensis* in the Northwest and Midwest and is transmitted in the Southwest by a deer fly. It occurs worldwide mostly from handling of infected animals, particularly rabbits.

F. tularensis is an exceedingly virulent organism, but the reason for this is unclear. The organism grows intracellularly in PMNs. There are three major manifestations of tularemia:

- **Ulceroglandular.** Occurs from the bite of an infected tick or from contact with an infected animal through a scratch, and a lesion occurs at the point of entry.
- **Pneumonic.** Generally spread by the respiratory system during skinning or handling of infected animals.
- **Typhoidal.** Occurs after ingestion of contaminated food or water.

Tularemia can be diagnosed by microscopy of smears obtained from lesions and visualized with fluorescent antibodies; treatment is with tetracycline.

Trench Fever

Trench fever is a disease that apparently occurs mostly during the unsanitary conditions associated with war. Trench fever first occurred in 1915 (World War I), and over 1 million military personnel contracted the disease. The etiological agent is *Rochalimaea quintana*. The primary reservoir is the human, and body lice are the vectors. Disease symptoms include chills, fever, headache, and muscle pain. These symptoms last for 5 days; hence, the name *quintana*. Patients are incapacitated for up to 2 months. Relapses often occur, and antibiotics such as chloramphenicol must be given for an extended time to eliminate the disease.

Cat Scratch Disease

There are over 24,000 cases of cat scratch disease in the United States each year. The bite or a scratch generally by a kitten results in cat scratch fever and occurs most often in young children. Papules (red swellings) are evident 7 to 12 days after injury, and lymph nodes in the vicinity become tender. As the disease progresses, the lymph nodes swell and become pus filled. Enlarged nodes may require surgical drainage, but the disease symptoms are generally mild and localized. Recovery requires a few weeks. The etiological agent is *Rochalimaea henselae*, a small, curved gram-negative rod that can be isolated on blood agar under a CO_2 atmosphere. There may be other causes of this human disease, but no other agent has been isolated.

PROTOZOA AS AGENTS OF DISEASE

Protozoa are the causative agents of many diseases worldwide. Tropical areas near the equator are particularly subject to protozoan diseases, and malaria is considered to be the world's leading health problem. Malaria infects 150 million humans each year and is responsible for over 2 million deaths. Although malaria has been essentially eradicated in the United States,

Research Highlights 28.2   The Water We Drink—Safe or Not

A severe outbreak of a diarrheal disease occurred in Milwaukee, Wisconsin, in April of 1993. This outbreak affected one-fourth of the population of Milwaukee and served as a warning to all Americans that:

1. Our water supplies are under great stress.
2. Disease is lurking and can create havoc with the least breakdown in our defenses.
3. We must not take our water supplies for granted.

The pathogen involved in Milwaukee was a protozoan of the genus Cryptosporidium. Cryptosporidium is a common protozoan in mammals and birds that should be filtered from drinking water during routine water treatment. Ingestion of Cryptosporidium or cysts (Cryptosporidia) in drinking water gave 370,000 citizens of Milwaukee acute diarrhea, nausea and stomach cramps. Undoubtedly there were fatalities from dehydration and other complications, but the exact number of these will never be known.

Why did this happen? The epidemic occurred in early spring, and runoff from the spring thaw may have overstressed the water treatment plant. Dairy farms abound around Milwaukee, and cattle harbor the protozoan. Water from feedlots may have carried the parasites into catchments that supply the city with potable water.

In routine treatment of water, coagulants are added, prior to sand filtration, that remove particles from the water. Alum (aluminum potassium sulfate dodecahydrate) was the coagulant employed in Milwaukee until the fall of 1992, when it was replaced with polyaluminum hydroxychloride. This replacement was due to an EPA concern that alum can leach lead from water pipes, and Milwaukee water supplies contained unacceptable levels of lead.

During March of 1993, employees at one of the water treatment facilities noted that the treated water coming through the plant was unusually turbid. The spring thaw runoff water was evidently not clarified by the replacement coagulant employed to settle particulates. Chlorination does not kill the Cryptosporidia cysts of those of other pathogenic protozoa. The water treatment plant in question was closed down, and the city returned to alum as coagulant.

A report on drinking water quality was released in April of 1993 by the General Accounting Office. The report concluded that water quality assurance programs across the United States are "flawed and underfunded." The conclusion was that 198,000 water supplies are inadequately monitored. This is a problem we cannot ignore.

conditions exist for the reintroduction of the disease. Two other protozoan diseases, amebiasis and giardiasis, are a concern in the United States, and these and malaria will be discussed briefly. Another pathogenic protozoan, *Cryptosporidium*, recently caused a health crisis in Milwaukee, Wisconsin (**Box 28.2**).

Amebiasis

Entamoeba histolytica is the etiologic agent for amebiasis, also known as amebic dysentery. Worldwide there are an estimated 400 million active cases that cause over 100 thousand deaths each year. There are 4,000 to 5,000 new cases in the United States annually. The disease is most prevalent in warmer climates with inadequate water treatment and sanitation.

The infection occurs when the mature cysts of *E. histolytica* are ingested. The cyst moves through the stomach, and the cyst coat is removed in the small intestine yielding the amoeboid form. The amoeboid cell called a **trophozoite** moves to the large intestine and penetrates the intestinal mucosa. In acute amebiasis, lesions are formed in the mucosa, and these lead to severe ulceration. The trophozoites can multiply and invade other intestinal mucosa while feeding on erythrocytes and bacteria. The lesions become infected with bacteria present in the intestine. *E. histolytica* may penetrate the large intestine and move to and produce lesions in the liver, lungs, or brain.

The trophozoites may establish a mild or asymptomatic form of the disease called **chronic amebiasis**. Individuals with this form of the disease become carriers and constantly shed cysts into the environment. They are a concern because they serve as a constant reservoir for infective cysts. Amebiasis is a health problem among homosexual males.

The symptoms of amebiasis are severe dysentery or colitis. The diarrhea is accompanied by blood and mucus shed from the extensive intestinal lesions. Amebiasis can be diagnosed by the presence of trophozoites in the feces filled with ingested red blood cells. Treatment is different for carriers than for those with active infections. Iodoquinol is the drug of choice for asymptomatic carriers, and Aralen phosphate is given in acute amebiasis. The best preventive measure is quality water, although chlorination of water does not destroy the cysts of *E. histolytica*.

Research Highlights Box 28.3

Pathogenic Microorganisms as Weapons of Biological Warfare and Bioterrorism

Infections are most often the result of an accidental encounter of a susceptible individual with a pathogen. Epidemics occur when the infection source is not controlled and the reservoir (or the infected patients) contributes to the spread of the pathogens within the community. The ultimate aim of research in medical microbiology is to understand the basis of disease and the host response, with the explicit aim of eradicating the pathogen from the infected individual. Moreover, one of the missions of public-health services is to determine the sources of infections and develop strategies for limiting the spread of microorganisms. Means of controlling infections, or the vectors that promote dissemination of the pathogen, have historically proven to be extremely effective in controlling disease and potentially saving thousands of lives.

A more sinister side of research in pathogenic microbiology is the

attempt to exploit microorganisms as weapons of mass destruction. Biological weapons are natural or genetically engineered organisms (or their toxic products) in formulations that are intended to cause maximum harm to an infected individual or a targeted population. Biological weapons can be also used to destroy the livestock or agricultural products of the attacked population, leading to secondary effects such as famine. Even before the many concepts of transmissibility of disease were known, biological weapons, in their crudest forms, were part of medieval warfare, wherein the practice of catapulting the cadavers of infected individuals into besieged cities was not uncommon. Similarly, water supplies of the enemy country were routinely poisoned using animals or humans that had succumbed to infections. Evidence for a more sophisticated form of biological warfare can be found in the records of French and Indian War (1754–1767), where British forces used smallpox to initiate epidemics among Native American tribes. Blankets and a handkerchief from a smallpox hospital were distributed to Native American tribes in the Ohio River

valley, with subsequent fatal epidemics, as would be expected in an immunologically naïve population.

Although there are a large number of potential biological weapons of mass destruction that could be propagated and distributed in a deliverable form ("weaponized"), the most effective countermeasures against biological weapons are the results of advances during the last century in understanding the origins and spread of infection. These advances, coupled with the availability of a large repertoire of antibiotics and vaccines against many pathogens, have made much disease, whether naturally occurring or deliberately introduced, preventable or treatable. Therefore, only a handful of pathogens are suitable for use as biological weapons. The leading candidates that may be exploited by countries for the purpose of biological weapons include now only a handful of pathogens. These include the bacterial pathogens Bacillus anthracis, Yersinia pestis, Franceilla tularensis, Clostridium botulinum (or its potent toxin), and Coxiella burnetii. Some viruses have also been tested as biological weapons, and some still can be considered to have a weapon potential today. These

(A)

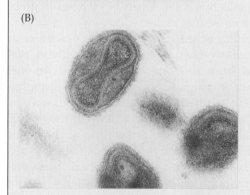

(B)

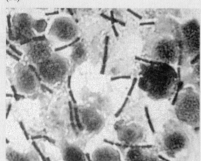

(C)

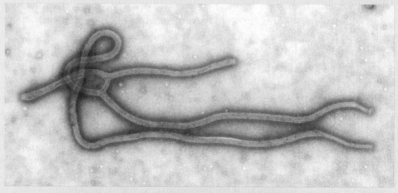

The rogues' gallery. Some potential agents of biological warfare or bioterrorism. (A) *Bacillus anthracis*. (B) Smallpox virus. (C) Ebola virus. Courtesy of F. Murphy, S. Whitfield, and C. Goldsmith/Centers for Disease Control.

include agents of smallpox (or the related monkeypox), Venezuelan equine encephalitis virus, and Ebola virus.

In order to function as an effective biological weapon, the pathogen has to have certain properties, which include ease of preparation and dispersion, stability, and absence of immunological protection in the targeted population. Microorganisms responsible for anthrax and smallpox meet these criteria, and therefore they have to be considered the most likely weapons for use in biological warfare attacks or in acts of bioterrorism. Virulent strains of B. anthracis are easy to isolate from sporadic infections of sheep or cattle. The bacteria grow to large numbers in most routine laboratory media in simple fermenters and form spores, which are stable during storage or

subsequent dissemination. Spores can be readily weaponized by drying them in the presence of organic compounds that prevent clumping and yield fine, highly dispersible particles. These can be then delivered in aerosol form using modified aircraft, such as crop dusters, or in projectiles (missiles and bombshells). The disease has a rapid onset, and when delivered by inhalation route it has a 90% fatality. Large-scale vaccination against anthrax is not routinely carried out in any country at this time.

Although propagation of the smallpox virus requires more sophisticated laboratory facilities and its dissemination is more difficult, it remains on top of the list of potential viral bioweapons. Smallpox virus is highly infectious, with 30% fatality, and currently no treatment is available. One possible scenario for dissemination of smallpox would be a

suicide attack, where an infected person or persons produce an aerosol by excessive sneezing and coughing while attending events involving crowds, such as sporting events, concerts, conventions, or classroom lectures. Virtually everyone is susceptible to smallpox, because vaccination stopped worldwide in the late 1970s.

The recent occurrence of acts of bioterrorism in United States, where anthrax spores were disseminated by mail and resulted in infection of targeted populations as well as mail-facility workers, showed that the threat of deliberate dissemination of pathogens is very real. Public health measures and means for preventing and controlling what may be rare diseases have to be implemented even in communities where natural forms of the disease have been successfully eradicated.

Giardiasis

Giardia lamblia is the agent of an intestinal disease called **giardiasis**. *G. lamblia* is an ancient eukaryote that lacks a mitochondrion, and its rRNA has many bacterialike features. Many domains in *G. lamblia* rRNA resemble that in the *Archaea*. Giardiasis is the most prevalent waterborne diarrheal disease in the United States. It occurs most often in young children, and there are about 30,000 cases per year. Giardiasis can be endemic in child day-care centers.

About 7% of the population in the United States are healthy carriers of giardiasis and shed the cysts in their feces. Transmission of the cysts to humans generally occurs through ingestion of contaminated food or water. Outbreaks often occur in campers or backpackers in wilderness areas when they drink seemingly clear, fresh water that has been contaminated by domestic or wild animals, including beavers, rodents, or deer.

Acute cases of giardiasis are marked by severe diarrhea, cramps, flatulence, and dehydration. The asymptomatic carriers may have intermittent mild cases of diarrhea. Ingestion of the cysts is followed by uncoating and release of trophozoites. These trophozoites inhabit

the lower small intestine where they attach to the mucosal wall through small disks and apparently feed on mucosal material. Diagnosis of giardiasis is by examining for cysts in mucosal material shed in feces. The mucus sheath in the intestinal tract is shed normally, and the trophozoites or cysts would be present in this material. A commercial ELISA test (see Chapter 30) is available for detection of the *G. lamblia* antigens in fecal material. Treatment is with Atabrine or metronidazole. As with *E. histolytica*, the cysts of *G. lamblia* are resistant to chlorination. Campers should boil water to prevent infection. The disease can be prevented in populations at large by proper water treatment.

Malaria

The etiologic agents of malaria are several species of protozoa in the genus *Plasmodium*. Nearly all adults in the middle regions of Africa and in India have been infected. Although the disease has essentially been eradicated in the United States since 1945, it has made a comeback through infected immigrants. There are about 1,000 cases per year recorded in the United States.

Table 28.7	Zoonoses that are transmitted to humans			
Disease	**Organism**	**Transmitted by**	**Host**	**Symptoms**
Anthrax	*Bacillus anthracis*	Endospore inhalation or animal contact	Cattle, goat, sheep	Cutaneous lesions
Bovine tuberculosis	*Mycobacterium bovis*	Milk	Cattle	Lesions in bone marrow of hip, knee
Brucellosis	*Brucella abortus*	Milk, skin/eye contact, or bite	Cattle	Fever, headache
	B. melitensis		Goat	
Leptospirosis	*Leptospira interrogans*	Urine, contaminated water, skin contact	Rodent, dogs, cattle	Meningitis, headaches, fever
Listeriosis	*Listeria monocytogenes*	Raw milk	Cattle	Meningitis
Yaws	*Treponema pallidum* subsp. *pertenue*	Bite	Fly	Skin ulcers, lesions

Malaria results from the bite of a female mosquito of the genus *Anopheles*. The female injects anticoagulants while sucking blood, and the saliva that carries the anticoagulants transmits the malarial parasite into the human bloodstream. Male mosquitoes do not suck blood and do not carry the protozoan.

There are two phases in the life cycle of the protozoan *Plasmodium vivax*, asexual phase in human red blood cells (RBCs) and a sexual stage in the mosquito. When *P. vivax* enters the bloodstream, it passes to the liver, where the sporozoites undergo an asexual fission, and in about 7 days there is the release of **merozoites.** These merozoites invade red blood cells where they reproduce. The red blood cells rupture and release more merozoites that infect other red blood cells. *P. vivax* forms microgametocytes and macrogametocytes in the infected human red blood cells, and these can be taken into the stomach of a female mosquito while feeding on an infected human.

Within the mosquito, the microgametocytes form macrogametocytes and microgametocytes that fuse to form a diploid zygote. Sexual reproduction in the midgut of the mosquito leads to the production of sporozoites that travel to the salivary glands of the mosquito. The sporozoites and anticoagulants are then injected into another victim during feeding.

The symptoms of malaria—chills followed by fever and sweating—are caused by the presence of merozoites and debris from the synchronized rupture of blood cells. A victim may suffer several attacks, and these are followed by remission that can last for days to months. Relapses occur as dormant parasites become activated and emerge from the liver. Malaria cannot be spread by human-to-human contact, as it requires the mosquito vector for transfer, a condition somewhat unique in infectious diseases.

Malaria is diagnosed by finding the parasites in red cells. Serological tests are also available. Treatment is with chloroquine or other quinine derivatives. The only preventive measure is to destroy the breeding grounds for *Anopheles* mosquitoes. DDT was employed extensively for this purpose during the period from 1946 to the 1960s, but the use of this pesticide has been curtailed in the United States.

ZOONOSES

Zoonoses are infections that normally occur in animals but can infect humans through contact with infected animals, animal products, or bites. A number of vector-borne diseases have already been discussed. There are a number of others, and some of these are presented in **Table 28.7.** Many of these diseases occur rarely in the United States, but outbreaks of diseases such as listeriosis do occur when milk is improperly handled. Brucellosis has been a concern in the past, but widespread testing of cattle has eliminated this disease from dairy herds in the United States. Anthrax is mainly a concern at the present time as a germ warfare agent (**Box 28.3**).

SUMMARY

▶ Humans are subject to an array of **skin infections,** of both bacterial and fungal origin. Common cutaneous infections caused by bacteria are staphylococcal infections and by fungi, athlete's foot or ringworm.

▶ **Respiratory infections** are difficult to control because they are generally transmitted in minute air droplets.

▶ **Diphtheria** was once a leading cause of death in children but has been virtually eliminated in the United States by a massive immunization program. It continues to be a concern in economically depressed areas of the world. Whooping cough is another disease that has been controlled in the United States by vaccination. Worldwide it is responsible for 500,000 deaths annually in young children.

▶ **Pneumonia** is an inflammation of the lungs that results from infection by viruses, mycoplasma, bacteria, or fungi. The leading etiologic agent *Streptococcus pneumoniae* is often a secondary infection that follows a viral infection. *Streptococcus pyogenes* is a potent pathogen and can infect the meninges or cause scarlet or rheumatic fever

▶ **Meningitis** is an inflammation of the meninges that can be caused by *S. pyogenes, Haemophilus influenzae,* or *Neisseria meningitidis.* Meningitis is a severe, often fatal, disease.

▶ **Tuberculosis** is an infectious disease, generally occurring in the lungs. It can be cured with antibacterial agents, but new antibiotic resistant strains are developing at an alarming rate, especially among AIDS patients. An estimated 10 million new cases of tuberculosis occur worldwide each year.

▶ **Leprosy** is an ancient disease that has not yet been eliminated, but treatment is possible.

▶ **Fungal diseases** are a problem in part because there are few compounds that will cure infections, and those available are quite toxic. Histoplasmosis, coccidiomycosis, and aspergillosis are lung diseases caused by fungi.

▶ **Gastrointestinal tract diseases** are of two major types—food poisoning from ingestion of food on which toxin producing microorganisms have grown and food infections from drinking liquids or ingesting food contaminated with disease-causing bacteria.

▶ Recent evidence indicates that **peptic ulcers** in humans are caused by growth of a bacterium, *Helicobacter pylori,* in the stomach.

▶ **Botulism** is caused by ingesting food on which *Clostridium botulinum* **has grown.** This microorganism produces one of the most potent toxins known.

▶ A common food infection is **salmonellosis,** caused by various strains of the genus *Salmonella. S. typhi* is the etiological agent of typhoid, a serious human infection. Cholera is a disease caused by drinking water or eating food that has been contaminated with fecal waste.

▶ The incidence of **urinary tract infections** is second only to respiratory infections in the United States in total cases. These are of frequent occurrence in hospital patients.

▶ **Sexually transmitted diseases** (STDs) are a major health concern. These diseases are preventable/curable and are a societal problem.

▶ **Gonorrhea** is an STD that occurs most often in the 15 to 35 age group and is of considerable concern because the organism responsible for the disease is gaining resistance to the antibiotics that are employed in treatment.

▶ **Syphilis** is an STD that occurs less frequently than gonorrhea but is more of a menace because of the severity of the disease.

▶ **Vector-borne diseases** are transmitted from one infected animal to another via an intermediate host. Most of these diseases are caused by obligate parasites such as the rickettsia. Lyme disease and Rocky Mountain spotted fever are vector-borne diseases that are rapidly increasing in incidence. Among the vector-borne diseases are typhus, Q fever, relapsing fever, tularemia, and trench fever.

▶ **Plague** has historically been a major killer of humans. It is carried to humans by rat fleas.

▶ The protozoan disease **malaria** is considered to be the leading health problem in the world. The disease infects 150 million humans each year and is responsible for 2 million deaths. Malaria is passed to humans by the bite of the Anopheles mosquito.

▶ **Amebiasis** and **giardiasis** are protozoan diseases of the intestinal tract. Inadequate treatment of water is responsible for the transmission of these diseases.

▶ **Zoonoses** are infections that are passed to humans through contact with infected animals.

REVIEW QUESTIONS

1. Staphylococcus is a versatile pathogen. What are some of the infections caused by this organism? How can they be controlled?

2. Why are respiratory diseases so difficult to control? What influence did this have on the development of vaccines to prevent these diseases? Consider some of the organisms that attack the different areas of the respiratory system.

3. What is pneumonia? Which organisms cause this disease?

4. Streptococcus pyogenes is the etiological agent for a number of infections. Cite some of these and the symptoms. What factors give this organism invasive ability?

5. What is meningitis? It is caused by several organisms, which is the most dangerous?

6. How does tuberculosis affect an individual? How is it diagnosed? Why is it making a comeback?

7. What are the symptoms of leprosy? Why has it been difficult to study and treat?

8. Fungi are the causative agents of respiratory diseases in humans. Cite examples.

9. What factors are important in lowering the incidence of gastrointestinal tract infections?

10. How does "food poisoning" differ from "food infection"? Give examples. How can either of these be prevented?

11. What are the major sexually transmitted diseases (STDs) that are caused by bacteria? Are they curable? Are they more a societal problem or a medical problem—give reasons for your answer. What is a major health concern with gonorrhea?

12. Do STDs affect the fetus or neonate? If so, how?

13. Define vector and reservoir. Cite examples. Can an organism be both?

14. What is the disease cycle in plague? How can it be prevented? Is plague a concern in the United States?

15. Where did Lyme disease get its name? Why is this disease a danger?

16. What are the symptoms of Rocky Mountain spotted fever? If one is bitten by a dog tick and later has such symptoms; what course should be followed?

17. The remark has been made that General Typhus defeated General Napoleon. Why?

18. How does one get tularemia? How might it be prevented?

19. Protozoan pathogens are a problem in many areas of the world. Where is amebiasis of greatest concern? Giardiasis?

20. Malaria is one of the world's greatest health problem. Why? How can it be prevented?

SUGGESTED READING

Brogden, K. A., J. A. Roth, T. B. Stanton, C. A. Bolin, F. C. Minion and M. J. Wannemuehler. 2000. *Virulence Mechanisms of Bacterial Pathogens*. Washington, DC: American Society for Microbiology.

Cossart, P., P. Boquet, S. Normark and R. Rappuoli. 2000. *Cellular Microbiology*. Washington, DC: American Society for Microbiology.

Fischetti, V. A., R. P. Novik, J. J. Feretti, D. A. Portnoy and J. I. Rood. 1999. *Gram-positive Pathogens*. Washington, DC: American Society for Microbiology.

Miller, V. L. (ed.). 1994. *Molecular Genetics of Bacterial Pathogenesis*. Washington, DC: American Society for Microbiology.

Roth, J. A., C. A. Bolin, K. A. Brogden, F. C. Minion and M. J. Wannemuehler. 1995. *Virulence Mechanisms of Bacterial Pathogens*. Washington, DC: American Society for Microbiology.

Salyers, A. A. and D. D. Whitt. 2002. *Bacterial Pathogenesis: A Molecular Approach*. 2nd ed. Washington, DC: American Society for Microbiology.

Sheld, W. M., W. C. Craig and J. M. Hughes. 2001. *Emerging Infections 5*. Washington, DC: American Society for Microbiology.

Chapter 29

Viral Diseases
of Humans

. . . and hell itself breathes out contagions to this world.
— SHAKESPEARE, HAMLET

A virus (or virion) is a bit of DNA or RNA that can enter and take over the normal functions of a specific host cell. The virus can selfishly supersede the host-cell genomic DNA, forcing the synthetic machinery of the host to function solely in the synthesis of viral particles. The viral genetic material is considered to have originated from ancestors of the organism that they attack (see Chapter 14) and therefore is attuned to the synthetic/genetic machinery of that host. Viral assault on a specific host cell generally leads to its disintegration. However, there are temperate viruses that do establish a latent or dormant infection in a eukaryotic cell (Chapter 14). Active proliferation of viruses, the infectious cycle, may result in cell destruction. This is the mechanism that has evolved to disseminate viruses, and even latent viruses must periodically have an infectious cycle for survival.

Chemotherapeutic agents that can cure most of the viral infections that afflict humans do not now exist. The nature of viral replication is such that administering an agent to rid the body of a viral infection, as we do routinely for bacterial infections, is not feasible. The physiological explanation for this limitation rests on the marked disparity between the components of a bacterium and a eukaryotic cell in contrast to the similarities between a virus and its host. Bacteria generally have peptidoglycan in their cell wall, and an antibiotic (such as penicillin or vancomycin) can interfere with cell wall synthesis without harm to the eukaryotic host that lacks peptidoglycan. In addition, differences between the protein-synthesizing machinery in prokaryotes and eukaryotes can be exploited, and inhibition of protein synthesis in a bacterial invader with antibiotics (such as tetracycline or streptomycin) will not so adversely affect the host. Thus, antibiotics are effective in destroying selected bacterial invaders and "curing" disease.

Viruses, on the other hand, present a different scenario—their reproductive machinery is quite similar to that of the cells they infect. A virion can attach to, invade, and commandeer the synthetic functions of the host, and exploitable differences between virus and host are not available. A chemotherapeutic agent that would interfere with virus replication would likely have a comparable adverse effect on equivalent reactions in a healthy cell. Experience has affirmed that this is essentially the case. There are a limited number of compounds that are administered for specific viral infections: adamantine for influenza A, acyclovir for herpesvirus infections, combined therapy for AIDS virus; but only adamantine has a curative effect.

This chapter is devoted to a discussion of major viral infections in humans. The transmission, entry, and symptoms of viral infections will be discussed along with mechanisms for prevention and/or control of the viral pathogen. For convenience, these infections will be discussed by portal of entry.

VIRAL INFECTIONS TRANSMITTED VIA THE RESPIRATORY SYSTEM

Respiratory diseases are readily transmitted from person to person via airborne droplets and are therefore difficult to control. Normal, everyday human contacts can lead to a broad distribution of infectious particles. Highly contagious diseases, such as influenza, that are spread via the respiratory route can be transmitted rapidly and can reach localized epidemic proportions among human populations that have not previously been exposed.

Because many of the infections transmitted via the respiratory system are so contagious, much emphasis has been placed on developing immunization programs for them. Immunization is favored because quarantine or isolation of the ill has not proven effective. In some cases (smallpox), immunization has been totally effective, and in other cases (HIV), it has so far been a failure. In the case of influenza, vaccination has been somewhat successful by using a combined vaccine made up of antigens from previous flu outbreaks. Consequently, the development of immunity to several strains that passed through the population previously confers limited immunity to a strain that might be introduced at a later date.

Respiratory viral infections affect 85% to 90% of the American people each year. The two major causes of these infections are the flu (influenza virus) and the common cold (Rhinovirus family). All other infectious diseases *combined* affect fewer than 20% of the population. However, there was a marked surge in herpes simplex (HSV) in the 1970s and in human immunodeficiency virus (HIV) in the early 1980s due to increased unprotected sexual activity that altered the picture somewhat. This section presents some of the major viral infections that are spread via the respiratory route: chicken pox, rubella, measles, smallpox, mumps, influenza, and common colds (**Table 29.1**).

Chicken Pox

Chicken pox is caused by the varicella-zoster virus (VZV), a member of the **herpesvirus** group, and is a highly contagious infection that occurs mainly in children. Prior to the development of an effective vaccine, there were over 4 million cases of chicken pox in the United States annually. The eruptions on the skin occur after an incubation period of about 10 to 20 days in the mucosa of the upper respiratory tract. During an infec-

tion the virus is carried to the bloodstream and lymphatics, where replication continues. Chicken pox is characterized by a fever and the small reddish vesicles that erupt on the skin (**Figure 29.1**,A). The rash may occur over much of the body but is more severe on the trunk and scalp than on the extremities. The contagious period begins a few days before the appearance of lesions and subsides a few days after fever ends. The vesicles are painful and itchy and occur over 2 to 4 days as the virus replicates. Scratching by young victims of the disease can cause infection and minor scarring. The virus can move to the lower respiratory system and gastrointestinal tract with more severe consequences. Damage to lungs and to the blood vessels in essential organs can be fatal.

A primary infection confers lifetime immunity to chicken pox but can lead to the presence of a dormant virus in the nuclei of sensory nerve roots. These dormant viruses can be activated by physiological or mental stress or by immunosuppressants, and this activation leads to an active infection commonly called **shingles** (Figure 29.1B). When activated, the virus moves to sensory nerve axons, where it replicates and damages sensory nerves. Skin eruptions occur in the area of nerve damage and are quite painful and itchy. Generally, these lesions appear about the trunk, hence the name *zoster* (Greek for *girdle*), and occur more frequently in adults over 60 years of age.

A vaccine has been employed in Japan for many years, and in 1995 a vaccine was approved for distribu-

(A)

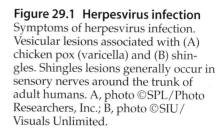

(B)

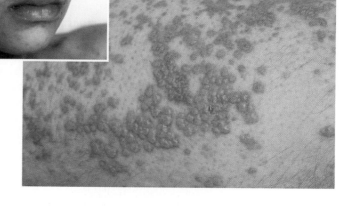

Figure 29.1 Herpesvirus infection Symptoms of herpesvirus infection. Vesicular lesions associated with (A) chicken pox (varicella) and (B) shingles. Shingles lesions generally occur in sensory nerves around the trunk of adult humans. A, photo ©SPL/Photo Researchers, Inc.; B, photo ©SIU/ Visuals Unlimited.

Table 29.1	Droplet or airborne viral diseases of humans		
Infectious Disease	**Viral Agent**	**Control by Vaccination**	**Type**[a]
Chicken pox	Herpesvirus	+	E dsDNA
Rubella	Togavirus	+	E ssRNA
Measles	Paramyxovirus	+	E ssRNA
Smallpox	Vacciniavirus	Eradicated	E ssRNA
Mumps	Paramyxovirus	+	E ssRNA
Influenza	Orthomyxovirus	±	E ssRNA
Common colds[b]	Rhinovirus	–	N ssRNA
	Coronavirus	–	E ssRNA
	Adenovirus	–	N dsDNA

[a]Type E = enveloped, N = naked capsid.
[b]Also spread by contact and fomites.

tion in the United States. The chicken pox vaccine is a live attenuated varicella virus and is administered to children on the same schedule as the MMR (measles, mumps, and rubella) series. Adults who have not been exposed to chicken pox (or have not been vaccinated) should be vaccinated, as the disease can be severe in adults and lead to complications.

Rubella

The infectious agent for **Rubella (German measles)** is a **togavirus** disseminated by respiratory secretions from infected individuals. It is moderately contagious, and infection results in a rash of red spots (**Figure 29.2**) and

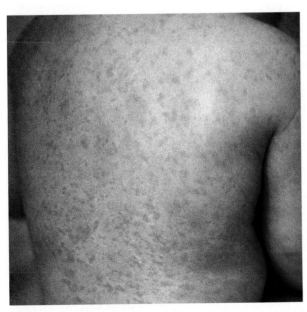

Figure 29.2 Symptoms of rubella
Rash of red spots caused by rubella (German measles). The spots are not raised. Courtesy of the Centers for Disease Control.

a mild fever. The illness is of short duration, and complete recovery occurs within three to four days.

The rubella virus infects the upper respiratory tract and spreads to local lymph nodes. Replicated viruses move throughout the body via the bloodstream. The incubation period varies from 2 to 3 weeks, but the virus is present in the upper respiratory tract for more than a week before the onset of the rash. Transmittable virus is present in the respiratory tract for a period of at least 7 days after the disappearance of the rash. This is another case where an infected individual can transmit a disease prior to the onset of symptoms and after symptoms disappear.

Rubella can be tragic for pregnant females who become infected with the virus. The virus can cross the placenta and infect the fetus during the first few months of pregnancy. This fetal infection may result in congenital defects in the newborn. These defects include hearing loss, microcephaly, cerebral palsy, heart defects, and other abnormalities. Fetal death or premature birth may also occur. A neonate that has been exposed to rubella in utero can shed (transmit) the virus for 1 to 2 years after birth. An estimated 20% of infected babies do not survive the first year.

A live attenuated virus vaccine is administered in the MMR series to preschool children in the United States. It is about 95% effective, and widespread use may develop a herd immunity that will ultimately eliminate the disease. This would substantially reduce the possibility of rubella infections in at-risk pregnant women. There were 152 cases of rubella in the United States in the year 2000 and 7 cases of congenital rubella syndrome. Health officials are concerned that outbreaks will occur among immigrants that were not vaccinated as infants.

Measles

Measles, also known as "common measles" or **rubeola,** is a highly contagious viral disease caused by a **paramyxovirus.** It is an acute systemic infection marked by nasal discharge, cough, delirium, eye pain, and high fever. A rash (**Figure 29.3**) can appear over the entire body, and lesions in the oral cavity are rather common. The measles virus enters through the respiratory tract or conjunctiva (eye). The incubation period is about 11 days after exposure, and the rash appears 3 days after the onset of fever, cough, and other symptoms. These symptoms persist for 3 to 5 days and are followed by the rash, which spreads from the head downward over a subsequent 3- to 4-day period.

Measles is a severe disease and is frequently followed by secondary infections. Pneumonia, inner ear infec-

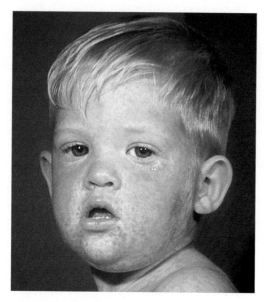

Figure 29.3 Symptoms of rubeola
Characteristic rash associated with rubeola (measles) in a young child. The rash begins on the face and moves downward to the trunk. Courtesy of the Centers for Disease Control.

tions, and infrequently, encephalitis may occur. This encephalitis has nearly a 25% mortality rate and can cause neurological damage in survivors.

Today there are few cases of common measles in the United States due to a comprehensive vaccination program (MMR) that was initiated in the mid 1960s. The vaccine employed is an attenuated live virus and induces immunity in over 95% of recipients. Infants from 12 to 18 months old are vaccinated, a mandatory practice in most states. There have been some outbreaks in recent years among poor inner-city children, immigrants, and illegal aliens who have not been immunized. Each year there are more than 50 million cases of measles worldwide, causing the death of about 1 million people. Occasional outbreaks occur on college campuses and are a concern because adults generally have a more severe infection than that observed in children.

Smallpox

Smallpox, a deadly disease, has caused untold misery and death throughout recorded human history (see **Box 29.1**). The smallpox virus (variola) entered a potential victim through the respiratory tract with an incubation period of 12 to 16 days. Regional lymph nodes became infected. The virus then entered the bloodstream (viremia) and was disseminated throughout the body. The

first symptoms were fever, chills, headache, and exhaustion. The virus continued multiplying in mucous membranes, in internal organs, and on the skin. Lesions occurred over the entire body and were filled with fluid. The lesions often erupted and became infected. These lesions were a considerable concern under the unsanitary conditions that existed in much of the world.

There were two strains of the smallpox virus: (a) variola major, which produced a severe infection, and (b) variola minor, which caused a mild infection. The death rate from variola major infections approached 50%, whereas, variola minor had less than a 1% mortality rate. Infection with variola minor resulted in a lifelong immunity to both strains. In fact, during the years when smallpox was rampant, individuals exposed themselves to patients infected with the mild form in an effort to gain immunity to the more deadly strain.

The World Health Organization initiated a worldwide program for smallpox immunization in 1966. The effort was funded by the United States, Russia, and other industrialized nations. All available humans were inoculated. Because humans are the sole host for the virus, the virus was eradicated, and the program was a great success. The last known case of smallpox occurred in 1977 in Somalia (**Figure 29.4**A and the virus is depicted in 29.4B). The individual received treatment and survived. The disease was declared eradicated in 1979. There are concerns that terrorists might obtain access to stocks of smallpox remaining and spread the disease to susceptible populations (see **Box 29.2**).

(A)

(B)

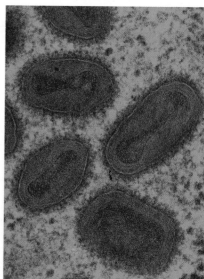

Figure 29.4 Human smallpox
(A) The last known victim of naturally acquired smallpox was Ali Maow Maolin, in Somalia in October 1977. The pustules are characteristic of smallpox. Maolin received treatment and survived. (B) Electron micrograph of the smallpox virus. A, photo ©WHO/J. Wickett; B, photo ©H. Gelderblom/Visuals Unlimited.

Milestones Box 29.1   Smallpox Epidemics in the New World

Infectious diseases have played a major role in shaping the course of human evolution. Viral and bacterial infections have destroyed armies and devastated entire civilian populations. In reality, prior to large-scale vaccination and the use of antibiotics, infectious disease played a major role in the control of human populations, and many of these deadly diseases were due to viruses. One such virus was **smallpox,** a constant threat to Old World inhabitants. Smallpox became a threat to New World inhabitants when explorers, fortune seekers, and immigrants brought the disease to the Americas. Because native populations had no previous exposure to smallpox, they had no resistance to the infection. As a consequence, it was exceedingly deadly to the indigenous peoples of the Americas.

The effect of smallpox epidemics on the Native American populations is well documented by historical records. Two examples are presented to illustrate this point. One was the purposeful employment of the smallpox virus against Native Americans and is a cruel story of "germ warfare" to destroy a perceived enemy. Another is the effect of an accidental exposure to the Aztecs and their subsequent inability to defend themselves against the conquistadors.

During pre-Revolutionary days, Sir Jeffery Amherst was the commander-in-chief of the British forces in North America. At the time, General Amherst was preoccupied with containing a coalition of Native American tribes that had been harassing the western frontiers of Pennsylvania, Maryland, and Virginia. This coalition was under the leadership of Chief Pontiac of the Ottawa tribe. The natives, under Chief Pontiac, had captured several forts along the western frontier in defiance of the colonists who wanted to move westward. To counter this feat, Sir Jeffery Amherst proposed the use of "germ warfare." In a 1763 letter to Colonel Henry Bouquet, commander of the western forces, he offered the following advice:

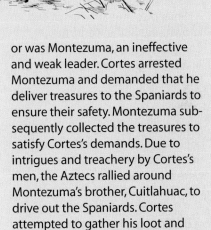

could it not be contrived to send the smallpox among those disaffected tribes of Indians? We must on this occasion use every stratagem in our power to reduce them.

During July of that year, Colonel Bouquet replied to Amherst,

I will try to inoculate the _____ with some blankets that may fall in their hands, and take care not to get the disease myself.

The results of their diabolical plan are unknown, but it is a disgraceful episode in our historical dealings with Native Americans.

On another occasion, a Spanish conquistador inadvertently brought smallpox to Mexico. In June of 1520, Hernando Cortes (1485–1547) and his army marched into Tenochtitlan, which later became Mexico City. The Aztec emperor was Montezuma, an ineffective and weak leader. Cortes arrested Montezuma and demanded that he deliver treasures to the Spaniards to ensure their safety. Montezuma subsequently collected the treasures to satisfy Cortes's demands. Due to intrigues and treachery by Cortes's men, the Aztecs rallied around Montezuma's brother, Cuitlahuac, to drive out the Spaniards. Cortes attempted to gather his loot and escape the island with his men.

As they were attempting their escape, the Aztecs attacked and blocked the Spaniards' passage to the mainland. Many Aztecs and Spaniards were killed. Unbeknownst to the Aztecs, one Spaniard killed in battle was infected by the smallpox virus. The Aztecs looted the dead, including the victim with smallpox. Within two weeks, smallpox infected the Aztecs and killed over one-fourth of the native population. Because Aztecs had never been exposed to smallpox prior to arrival of the Spaniards, they had no resistance to it. Many in the Aztec army succumbed, including the leader, Cuitlahuac.

A few months later, Cortes returned and easily defeated the weak and demoralized Aztec army, looting the fortress city of Tenochtitlan.

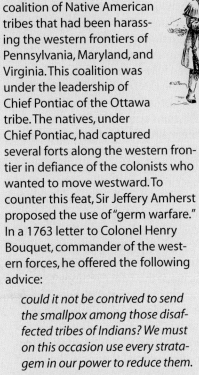

Drawings by Elizabeth Perry

Milestones Box 29.2   **History to Preserve or to Destroy?**

In 1798 Edward Jenner, an English physician, developed an effective vaccination procedure to protect against smallpox (see Chapter 2). Despite this discovery, smallpox continued to kill humans throughout the 19th and well into the 20th century. A worldwide immunization program initiated in 1966 led to the eradication of the disease by 1977. There has not been a documented case of smallpox in the last 24 years.

Smallpox virus survives at two known sites in the world: the Communicable Disease Center in Atlanta, Georgia, and the Institute for Viral Preparations in Moscow, Russia. There are three major strains of the smallpox virus, and the Russian scientists were to establish a genetic map of two strains and the United States the other. All strains were to be destroyed when this task was accomplished.

Scientists are divided on the issue of destroying all of the viral strains now maintained. Some favor destruction, others preservation. Those for preservation point out that genetic maps have not revealed some of the information that only the intact virus can reveal, and so questions remain. Among these questions—How did smallpox kill people? Can the virus yield clues about other diseases? Are there similar diseases that could potentially emerge? Has the virus been fully exploited?

The CDC maintains about 400 vials of smallpox virus and the Russians about 200 vials. These vials are maintained in liquid nitrogen and guarded closely. The fear of biological warfare has led to smallpox vaccination of the Russian, Canadian, Israeli, and U.S. armies. The rest of the world is essentially unprotected.

Should all remaining cultures of the virus be destroyed? This would achieve the goal of complete eradication of the dread disease. Should it be maintained for scientific reasons? Once destroyed, it would not be available to the scientific world or to potential terrorists. The World Health Organization (WHO) in Geneva has recommended that the smallpox stocks should be eradicated, but this has not yet occurred.

During the Cold War period, from the early 1950s to the late 1970s, there was an active development of biological warfare weapons in Russia. One suspected germ warfare agent was anthrax, and there is speculation that smallpox might have also been developed as a military weapon. Since September 11, 2001, and the anthrax scare in the United States, there is understandable fear that the smallpox virus may have fallen into the hands of rogue nations. Release of this virus would be catastrophic, as vaccination against this dread disease ended about 30 years ago, and the level of protection available even to those vaccinated prior to the cutoff date is an unknown. Certainly there is an array of viral and bacterial agents that an amoral despot might loose onto the world. This is a reality in the 21st century.

Mumps

Mumps was a common disease in children prior to the development of an effective vaccine. Mumps is caused by a **paramyxovirus** and is spread via the respiratory route. Disease symptoms occur 18 to 20 days after exposure. The mumps virus multiplies in the upper respiratory tract and progresses to local lymph nodes. The virus moves from the local lymph nodes to the bloodstream, and the resultant viremia disseminates the infection throughout the body. The obvious symptom of the disease is inflammation and swelling of the parotid glands on one or both sides of the neck (**Figure 29.5**). The parotids are salivary glands located below and in front of the ear. Enlargement of these glands can lead to difficulty in swallowing. Infection can develop in the meninges, pancreas, ovaries, testes, or heart. Infection of the pancreas may lead to juvenile diabetes. In males that have reached puberty the viral infection can become localized in the testes, a painful condition that in some cases results in sterility. Encephalitis can also occur as an aftermath of mumps infections, but is quite rare.

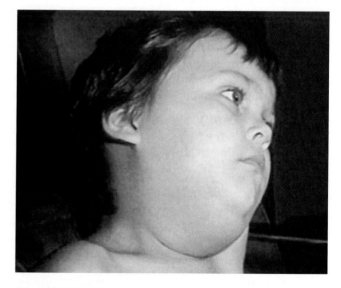

Figure 29.5 Mumps
Enlargement of the parotid gland, typically present in an individual with mumps. Courtesy of Barbara Rice/NIP/Centers for Disease Control.

There is only one antigenic type of the mumps virus. Once the virus is contracted, immunity is apparently life long. A live, attenuated virus vaccine was developed in 1967 that has proven to be at least 95% effective. This vaccine is part of the MMR series and has reduced the number of cases in the United States to fewer than 500 per year. A mumps vaccination prior to entering school is mandatory in most states. There have been mumps outbreaks in states that do not have a mandatory vaccination program.

Influenza

Viral influenza is caused by an **orthomyxovirus.** It is an enveloped virus with an RNA genome surrounded by a matrix protein, a lipid bilayer, and glycoprotein (**Figure 29.6**). Influenza epidemics have occurred throughout recorded history, and pandemics (worldwide epidemics) have occurred with some frequency. The 1918–1919 pandemic, which was caused by a particularly severe and deadly strain of the influenza virus, killed at least 20 million people. In the 1918 pandemic, the United States lost over a half million military personnel; mostly these were otherwise healthy young men. The 1957 pandemic, which was the most recent, originated in central China and is referred to as the Asian flu. The virus involved with this pandemic was a recombinant virus of considerable virulence. The viral strain originated in central China in February 1957 and moved to Hong Kong in April. Air and naval traffic carried the infectious agent worldwide as depicted in **Figure 29.7**. During 2 weeks in October of that year, a peak of 22 million cases was reported. There was also a Hong Kong flu scare in

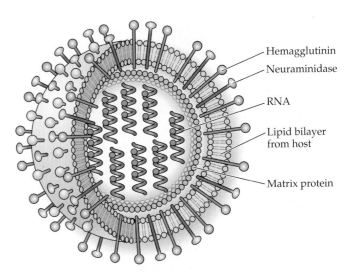

Figure 29.6 Influenza virus
Diagram of the influenza virus, depicting the location of the hemagglutinin and neuraminidase involved in invasion.

the late 1990s, and this led to the slaughter of thousands of chickens to prevent spread of the disease. Domestic chickens were a potential reservoir for the virus.

Once the influenza virus enters human populations, it does not affect any other host. It is mainly transmitted through the air on droplets expelled by sneezing or coughing (**Figure 29.8**). The flu symptoms (chills, fever, headache, malaise, and general muscular aches) appear after a 1- to 3-day incubation period and last for 3 to 7 days.

Once inhaled, the influenza virus settles on the mucous membranes of the upper respiratory tract. The

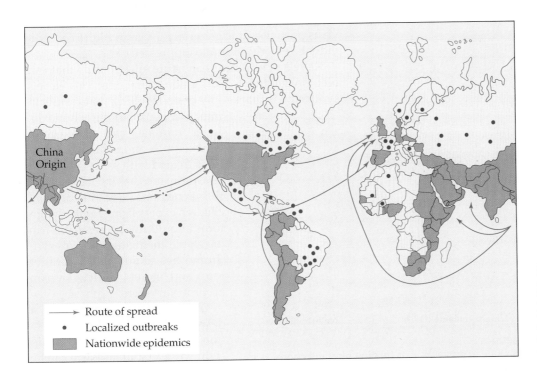

Figure 29.7 Influenza pandemic
Route followed by the influenza pandemic that originated in the interior of China in 1957 and spread throughout the world.

Figure 29.8 A sneeze
High-speed photograph of a sneeze. Photo ©M. R.
Grapes/Michaud/Photo Researchers, Inc.

neuraminidase structure on the viral envelope
hydrolyzes the mucus that lines the epithelial cells in the
respiratory system. This allows the virus to reach the
epithelial cell surface, where the hemagglutinin spikes
attach to the cell. Penetration occurs by endocytosis. The

virus moves from the upper to the lower respiratory
tract, destroying epithelial cells.

The destruction of epithelial cells in the respiratory
system can lead to secondary infections such as bacterial pneumonia caused by *Staphylococcus aureus*, *Streptococcus pneumoniae*, or *Haemophilus influenzae*. These infections are a significant problem in infants, elderly, and
otherwise debilitated individuals. Secondary infections
are the leading cause of death during influenza epidemics, and prior to availability of antibiotics these
deaths generally occurred from pneumonia.

The genome of the influenza virus is segmented into
eight distinct fragments. When more than one virus
infects a cell, there can be a reassortment of genes, which
yields progeny with recombinant genomes. These progeny may then synthesize an altered surface glycoprotein
that is unaffected by any antibodies that are present
from a previous bout with the influenza virus. This
change in the antigenic character of the viral particle is
called **antigenic shift.** It is the major reason why
influenza virus epidemics do occur periodically.

Vaccination against influenza is difficult because the
antigenic shift yields novel strains. Polyvalent vaccines
are generally administered that are composed of a number of viral strains obtained from previous epidemics.
Presently worldwide influenza outbreaks are monitored
so that the virus involved can be isolated and vaccines
to new strains prepared and made available. It is the
goal of these efforts to forestall a pandemic. The drug
amantadine (**Figure 29.9**) is effective in treating influenza A if administered after exposure or early in the infection. Amantadine, given before symptoms appear, can
reduce the incidence of infections by 50% to 75%. The
drug blocks penetration by an uncoating of influenza
virus particles. It is generally given to high-risk individuals (elderly, immunosuppressed).

Reye's syndrome is an acute encephalitis that can
occur in children following some viral infections. Reye's
syndrome is marked by brain swelling that results in
injury to neuronal mitochondria.
Liver damage can also occur.
Chicken pox and influenza infections treated with salicylates, such
as aspirin, increase the incidence of
this syndrome. For this reason,
children experiencing flulike symptoms should not be given aspirin
or related medicinals. Reye's syndrome has about a 40% fatality
rate, and mental deficiency can
occur in survivors.

Guillain-Barré syndrome is a
delayed reaction to influenza infection that generally occurs within
eight weeks. Guillain-Barré can be

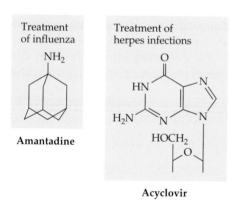

Figure 29.9 Viral chemotherapeutics
Some chemotherapeutic agents of value in preventing or treating viral infections.

a reaction to the virus or to a vaccine. This disease damages the cells that myelinate peripheral nerves. This demyelination results in limb weakness and sensory loss. Recovery is complete in virtually every case.

Cold Viruses

The common cold affects 40% to 45% of Americans each year and with varying severity. It is estimated that Americans miss more than 200 million work or school days annually due to symptoms of the common cold. Colds are caused by several different viruses. **Rhinoviruses** are involved in about 75% of all cases and are the most common etiological agent (**Figure 29.10**). Colds caused by the rhinoviruses generally occur in spring and fall.

Another group of viruses associated with the common cold are the **coronaviruses** (**Figure 29.11**), so named because the envelope projections resemble a solar corona. These viruses generally are involved with colds that occur in midwinter and constitute about 15% of all cases. About 10% of all colds are caused by adenoviruses and various other viruses.

The cold viruses may be transmitted in airborne droplets or by direct contact with infected individuals. Airborne transmission in volunteers has been difficult to demonstrate. The cold viruses can survive on inanimate objects for hours, and fomites (objects that harbor pathogens) are considered to be involved in transmission. Touching inanimate objects such as doorknobs that are contaminated with viral particles through handling by an infected person is considered a major route of transmission.

Cold symptoms include inflammation of the mucous membranes, nasal stuffiness, "runny" nose, sneezing, and a scratchy throat. The symptoms last for about 1 week. Clinical symptoms such as these develop in only about one-half of infected individuals, and those without apparent symptoms are a constant source of infection to the general population. Treatment includes aspirin, antihistamines, nasal sprays, and plenty of fluids.

There are over 110 known serotypes of the rhinovirus, effective antibodies are not generally produced, and any immunity developed is short lived. There is little hope for a successful immunization program for the common cold since there are too many distinctly different viruses (rhinovirus, coronavirus, adenovirus, and others) that cause colds and too many serotypes.

VIRAL PATHOGENS WITH HUMAN RESERVOIRS

There are a number of highly contagious difficult-to-control viral diseases that are transmitted directly from human to human (**Table 29.2**). The human serves as **reservoir,** and the viruses are transmitted in various ways,

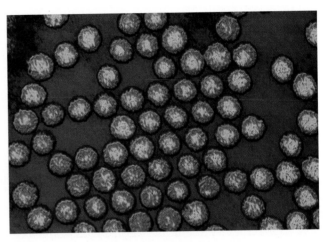

Figure 29.10 Rhinovirus, a cold virus
Electron micrograph of a rhinovirus, an etiologic agent of the common cold. Photo ©O. Meckes/c.o.s./ Gelderblom/Photo Researchers, Inc.

including through contaminated blood, by sexual contact, or intrauterine to a fetus. Some of these viruses are also transmitted through kissing, touching, or via the fecal-oral route through contaminated water. Ingestion of raw or poorly cooked shellfish that has been exposed to raw sewage is a danger. Swimming in septic water can also promote the transmission of selected viruses. Common warts are caused by a virus that is passed from human to human by direct contact, not by touching frogs.

This section presents information on major viral diseases that are transmitted via human contact. These diseases are common in the United States, and these and others are prevalent in other areas of the world as well.

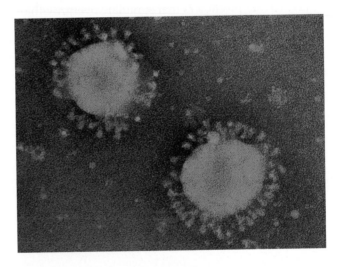

Figure 29.11 Coronavirus, a cold virus
Electron micrograph of a coronavirus, another causative agent of human colds. Note the projections from the viral envelope that give the virus its name. Photo ©R. Dourmashkin/SPL/Photo Researchers, Inc.

Table 29.2	Viral diseases with humans as reservoir and spread directly or indirectly through human contact
Disease	**Mechanism of Transmission**
Human T cell leukemia, serum hepatitis, AIDS	Contaminated blood, exchange of needles between IV drug abusers, sexual contact
Genital herpes, venereal warts	Sexual contact
Infectious mononucleosis, Herpes simplex	Mouth-to-mouth contact
Infectious hepatitis A, gastroenteritis	Fecal-to-oral route, raw shellfish
Polio	Water, food, swimming pools
Warts	Direct contact

Acute T Cell Lymphocytic Leukemia

There is a growing awareness that viruses can transform mammalian cells, and this can lead to sarcoma, leukemia, lymphoma, or other cancers (**Table 29.3**). One virus that can transform cells is the **human T-lymphotropic virus (HTLV),** a retrovirus (see Chapter 14). There are two types of HTLV: HTLV-1 and HTLV-2.

These HTL viruses are transmitted through transfusion with contaminated blood, between intravenous (IV) drug users who exchange needles, or via sexual contact. Breast-feeding may also transmit the oncogene-bearing viruses. The HTLV-1 infects the CD4$^+$ lymphocytes (T cells) and is not cytolytic.

The leukemia virus (HTLV-1) causes cancer after long latent periods by promoting uncontrolled growth of infected cells in two ways. The virus may (1) integrate into the host genome near gene sequences that are involved in growth control and activate their expression or (2) activate promoters that stimulate overgrowth of cells. Whichever the cause, uncontrolled cell growth can transform the lymphocytes, inducing the proliferation of a clone of aberrant T cells. This clone of HTLV growth stimulated lymphocytes reproduces with the accumulation of the aberrant (leukemia) cells. HTLV-1 infection is generally asymptomatic but can progress to acute T cell leukemia in about 5% of those infected. The disease

Table 29.3	Some viruses known to cause tumors in humans
Virus	**Cancer**
Hepatitis B virus	Liver cancer
Human T cell Lymphotrophic virus	T cell leukemia
Epstein-Barr virus	Burkitt's lymphoma, Nasopharyngeal carcinoma
Papilloma virus	Cervical cancer

is generally fatal within a year after diagnosis, and there is no treatment.

The second of these, HTLV-2 is responsible for hairy-cell leukemia. The name is derived from the membrane-derived projections that give leukocytes a "hairy" appearance. The development of this malignancy follows a similar pathway as occurs with the HTLV-1 disease.

The viruses known to cause cancer in humans may have exceedingly long latency between infection and development of the disease. This latency can be 30 or more years. They will not transform cells in vitro, which makes experimental investigation difficult. The presence of the virus can be detected by the identification of virus-specific antibodies in the patient's serum. There is no cure, and prevention is by preventing transmission of the viral particle. At the present time, human blood supplies are being examined for both HTLV and HIV.

Acquired Immune Deficiency Syndrome (AIDS)

Acquired immune deficiency syndrome is caused by an RNA retrovirus, the **human immunodeficiency virus (HIV).** A brief discussion of HIV and the clinical effects of the AIDS infection follows. For a discussion of the effect of HIV on the immune system, see Chapter 27; for a discussion on transmission and development of the virus, see Chapter 14.

The HIV virus is transmitted when the body fluids (blood, saliva, semen, vaginal secretions) of an infected individual come into contact with an open blood vessel (e.g., an open cut or a needle puncture) in an uninfected person. This passage of contaminated body fluids to the bloodstream of an uninfected individual inevitably leads to infection.

In the United States, the population most at risk of acquiring HIV are homosexual and bisexual males, IV drug users, and heterosexuals who have sexual intercourse with IV drug users, prostitutes, or bisexual males. Transmission between heterosexuals who have multiple partners is a growing concern. HIV can also be transmitted by blood transfusions or from infected mothers to the newborn, either directly from the mother's blood to that of the fetus or via mother's milk to the infant.

More than one million individuals in the United States are HIV positive. In other countries, AIDS is not limited to discrete groups but affects men, women, and children throughout the population. Worldwide, over 22 million people have died of AIDS since 1981, and 40 million people now alive are infected. It is estimated that each day the virus spreads to another 16,000 individuals in the world.

There is a stage in the infection, often spanning many years, during which antibodies to HIV are present in blood, but symptoms of AIDS are minimal and generally go unnoticed. The antibodies do not eliminate the virus because the viral genetic information integrates into the genome of CD-4$^+$ T lymphocytes in an infected individual. During this period, diagnosis is accomplished by screening a blood sample for the presence of antibodies against HIV, and the viral infection at this point is known as ARC (AIDS-related complex). Symptoms of ARC include fever, malaise, headache, and weight loss. The symptoms generally occur within weeks of infection and last 1 to 3 weeks. Untreated, the

ARC form of infection develops into full-blown AIDS within a 10-year period. It is rare for an HIV-positive individual to survive for more than 10 years without chemotherapy.

Symptoms of AIDS occur when the bulk (80%–90%) of one's CD-4$^+$ T lymphocytes lose function, due to HIV. Because these lymphocytes are a major part of the normal immune response, individuals with AIDS do not generate an effective immune response and thus are vulnerable to attack by opportunistic infections not generally observed in individuals with a normal immune response (**Figure 29.12**; **Table 29.4**).

(A)

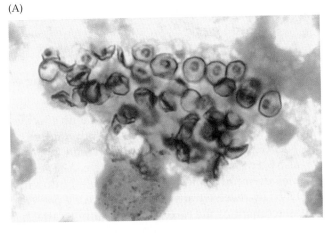

(B)

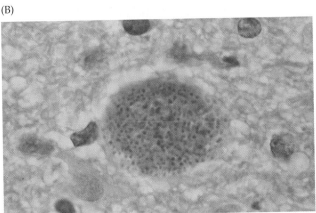

(C)

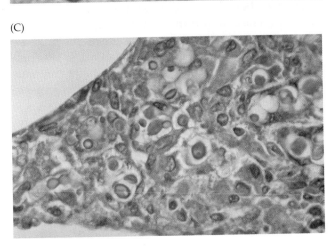

Figure 29.12 Causative agents of diseases in people with AIDS

Microorganisms causing the diseases associated with a fully developed case of AIDS. (A) *Pneumocystis carinii*, from a fungal pulmonary infection. (B) *Toxoplasma gondii*, affecting the brain and central nervous system. (C) *Cryptococcus neoformans*, the cause of liver damage and meningitis. (D) *Candida albicans*, cause of mouth and systemic infections. (E) *Mycobacterium*, cause of intestinal and lung infections. A, courtesy of R. K. Brynes/Centers for Disease Control; B, photo ©G. W. Willis/Visuals Unlimited; C, courtesy of E. P. Ewing, Jr./Centers for Disease Control; D, photo ©Dennis Kunkel Microscopy, Inc.; E, courtesy of E. P. Ewing, Jr./Centers for Disease Control.

(D)

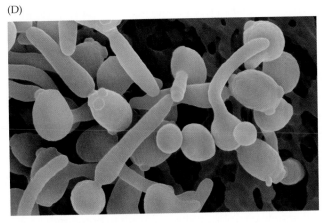

(E)

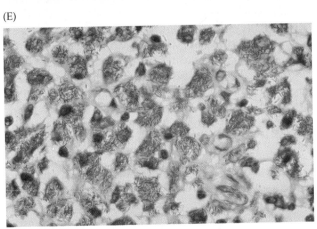

Table 29.4 — Major opportunistic pathogens associated with active cases of AIDS

Organism	Symptoms
Eucaryotic	
Protozoan	
Cryptosporidium	Intestinal infections
Toxoplasma gondii	Brain, central nervous system
Fungi	
Pneumocystis carinii	Pulmonary pneumocystosis
Cryptococcus neoformans	Meningitis, liver damage
Candida albicans	Oral thrush, systemic infections
Histoplasma capsulatum	Systemic infections
Coccidiodes immitis	Pneumonia
Helminths	
Strongyloides stercoralis	Enteric infections
Viral	
Herpes	Ulcerative lesions
Cytomegalovirus	Pneumonia, blurred vision
Procaryotic	
Mycobacterium tuberculosis	Tuberculosis
Mycobacterium sp.	Lymph node infections
Salmonella sp.	Septicemia
Streptococcus pyogenes	Pneumonia

An active case of AIDS can be diagnosed by identification of an array of opportunistic secondary infections including *Pneumocystis carinii* pneumonia. *Candida albicans* is present in the mouth and alimentary tract of healthy individuals. In the immune-compromised AIDS patient it can cause severe mouth infections. Another yeast infection, cryptococcosis (caused by *Cryptococcus neoformans*) can cause a damaging infection in meninges and brain. Toxoplasmosis (*Toxoplasma gondii*) and cryptosporidiosis (*Cryptosporidium*) are protozoan diseases that afflict AIDS patients. The former causes brain damage and the latter a severe form of long-lasting diarrhea. Most of these secondary infections are caused by eukaryotic microbes and are difficult to bring under control. *Toxoplasma* is a protozoan; *Cryptococcus*, *Candida*, and *Pneumocystis* are fungi.

Viral infections are also a concern, including the cytomegalovirus that can cause eye retinal damage and blindness. The herpes simplex virus can cause chronic ulcers or bronchitis. Other viral infections can be carried to the brain tissue by macrophages, causing overproduction of nerve cells that surround the neurons and resulting in a loss of mobility and brain function.

Another symptom of AIDS is Kaposi's sarcoma, a form of cancer that causes purplish blotches on the skin of the legs and feet. In immunocompetent individuals, this cancer is not aggressive and is rarely fatal. In AIDS patients the cancer can spread to the lungs, lymph nodes, and brain. Other malignancies common in AIDS are Burkitt's lymphoma, primary brain lymphoma, and immunoblastic lymphoma. Non-Hodgkins lymphoma and other malignant lesions that are normally removed by immune host defenses can occur in the HIV-infected.

At the present time, treatment of AIDS is limited to efforts to delay symptoms; there is no cure for the disease once it is contracted. The drugs now available are of three types:

- Nucleoside analogs that interfere with the transcription of the single-strand RNA genetic information that otherwise leads to viral double-stranded DNA
- Reverse transcriptase inhibitors that are not nucleoside analogs
- Protease inhibitors that prevent processing of viral polypeptides

The first **nucleoside analog** that was an effective inhibitor of transcription was azidothymidine or AZT (Figure 29.9). Dideoxyinosine (Figure 29.9) has also been employed but is less effective. AZT resembles thymidine and once incorporated into viral DNA by reverse transcriptase the analog prevents attachment of the next base. Inability to attach a base to the nucleotide chain results in **chain termination.** Unfortunately, AZT becomes ineffective quite rapidly in a patient because the high mutation rate results in resistance to the drug. A limited number of mutations are sufficient to render HIV resistant to the nucleotide analogs.

Reverse transcriptase inhibitors are drugs designed to alter the conformation of the catalytic site on reverse transcriptase. Nevirapine (Figure 29.9) is a widely administered reverse transcriptase inhibitor. Even fewer mutations in the viral genome are necessary to overcome the nonnucleoside reverse transcriptase inhibitors. Nevirapine is given to pregnant HIV-positive women during labor since it reduces the transfer of HIV to the neonate in about 50% percent of the cases.

The **protease inhibitors** are computer-designed peptide analogs that bind to the active site on the protease in HIV that processes the viral polypeptides. This prevents virus maturation. HIV is able to generate mutants that retain virulence and are unaffected by the protease inhibitors.

The standard treatment for AIDS is the administration of at least three drugs at the same time. This is termed HAART (highly active antiretroviral therapy). Under such a multidrug protocol, the virus would need to develop several types of resistance simultaneously. This is unlikely even in a virus as mutable as HIV. The HAART treatment usually includes a protease inhibitor combined with nucleoside analogs and/or a reverse transcriptase inhibitor. The patient is monitored periodically for viral load by the sensitive reverse transcriptase-polymerase chain reaction (RT-PCR) assay. Changes in the viral load indicate that the combination of drugs used in treating the patient should be adjusted.

There is a concerted effort underway to prevent HIV infections by immunizing healthy individuals. In addition, chemotherapeutics are being developed that would combat HIV and prevent it from progressing to full-blown AIDS.

Indeed, parts of the HIV retrovirus appear to be vulnerable, including surface glycoproteins (gp120) that could be incorporated into a vaccine. Antibody formed against these envelope proteins could prevent the CD-4$^+$ glycoprotein interaction and prevent infection. Unfortunately the gene encoding gp120 is very mutable, so antibody against one form of the glycoprotein may not recognize another.

Vaccines that induce antibody against other envelope proteins have been more promising. The genes for the envelope proteins are being incorporated into harmless viruses where they are expressed and serve as a vehicle for the HIV envelope protein antigen. Several of these antigen-producing systems have been developed and are in clinical trial.

The detection of anti-HIV antibody blood is an effective diagnostic procedure. It can be applied to blood from individuals or for blood donated by volunteers. The immunological procedure that has proven effective for large-scale screening of blood or in diagnosis of potential victims is the ELISA test (see Chapter 30). However, a positive test in any individual would be confirmed by a highly specific immunoblot procedure (see Chapter 16).

Development of rapid diagnostic tests to screen individuals for HIV infection is underway. Generally these tests are rapid and simple to apply but are not as sensitive nor are they as accurate as tests that are more time-consuming.

Cold Sores (Fever Blisters)

Herpes simplex virus type 1 (HSV-1) is the causative agent of **cold sores** or **fever blisters.** This virus is a member of the **herpesvirus,** a group that includes common animal pathogens. The herpes viruses are noted for their ability to cause latent infections that may be activated days or years after the initial infection. Zoster, the etiological agent of shingles, is one such virus and was discussed previously (see Chicken Pox). Herpes simplex is a double-stranded DNA virus with a genome that is surrounded by an icosahedral nucleocapsid. The capsid is enclosed in a membrane envelope with glycoproteins projecting from this envelope.

Active HSV-1 infection results in blisters around the mouth, lips, and face (**Figure 29.13**). The blister results

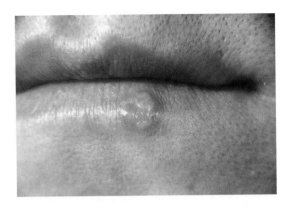

Figure 29.13 Cold sores
Cold sores, also known as fever blisters, are caused by herpes simplex type 1. Courtesy of Dr. Herrmann/Centers for Disease Control.

from host- and virus-mediated tissue destruction. The lesions heal in 1 to 3 weeks without treatment. It is estimated that up to 90% of adults in the United States possess antibodies against HSV-1.

The incubation period is 3 to 5 days. After a primary infection the virus moves to trigeminal nerve ganglia and remains in the dormant state. An active viral infection can occur sporadically throughout the life of an infected individual. Clinical symptoms are triggered by stresses such as excessive sunlight, emotional upset, fever, trauma, immune suppression, or hormonal changes. When the latent virus is activated, it travels down peripheral nerves to the lips, mouth, or facial epidermis, resulting in a reoccurrence of the fever blister(s).

Acyclovir (see Figure 29.9), a guanine analogue, is effective in treating herpes infections. When acyclovir is phosphorylated in cells, it resembles deoxyguanine triphosphate and inhibits the viral DNA polymerase. Acyclovir would be selectively activated by the thymidine kinase in an infected cell, as the virus codes this enzyme. This would not occur in a normal uninfected cell. Dormant viruses would not be affected by acyclovir, and consequently it is not a cure.

Genital Herpes

Genital herpes is generally caused by **herpes simplex virus type 2 (HSV-2),** with about 10% of clinical cases the result of HSV-1 infection. Genital herpes is associated with the anogenital region and is transmitted by direct sexual contact. A primary infection occurs after an incubation period of 1 week, but many primary infections are asymptomatic. The virus causes painful blisters on the penis of a male. In females, the lesions appear on the cervix, vulva, and vagina. The blisters are a result of cell lysis and a local inflammatory response. They contain fluid and infectious viral particles. During this stage, the virus is readily transmitted to unprotected sex partners. The blisters resulting from a primary infection heal spontaneously in 2 to 4 weeks. Recurrent genital herpes infections are generally shorter and less severe than the initial episode. These lesions that occur after primary infection heal in 7 to 8 days. As with HSV-1, the HSV-2 virus travels to nerve cells and remains dormant. Genital herpes can reoccur every few weeks, a few times per year, or not at all. An active case can be activated by a fever. An estimated 50% of those infected with HSV-2 never have symptoms and are unaware of a primary infection. These individuals may shed the virus and can be significant transmitters of the disease.

The virus can be transmitted to an infant during birth, and over 2,000 babies are born in the United States annually with HSV-2. The disease in the neonate ranges from a latent inapparent infection to brain damage and death. To avoid infecting the newborn, cesarean birth is advised if there are virus-infected lesions in the genital area. Unfortunately, the virus can be present in genital secretions without overt symptoms, thereby exposing a vaginally delivered infant.

There is no known cure for herpes virus infections. Oral and topical administration of acyclovir can lessen the symptoms of the disease and can prevent recurrences of HSV if taken prophylactically. Genital herpes has become a significant sexually transmitted disease, with one in five teens infected. According to American Social Health statistics, 50 million Americans have genital herpes, and there are about 1 million newly diagnosed cases each year.

Infectious Mononucleosis

Infectious mononucleosis is caused by a human herpes virus called **Epstein-Barr virus (EBV).** The virus is named for its codiscoverers and was originally isolated when T. Epstein and Y. Barr were seeking the etiologic agent for a malignancy of lymph nodes called Burkitt's lymphoma.

The EBV can be present in the mouth and throat area and is commonly called "kissing disease" because it occurs most frequently among those of college age. The symptoms include enlarged lymph nodes and spleen, sore throat, nausea, and general malaise. These symptoms may be accompanied by a mild fever. Fatigue is a common complaint of mononucleosis patients. The symptoms last for 1 to 6 weeks and rest is the only treatment.

Once contracted, EBV enters lymphatics where the virus multiplies and infects **B lymphocytes (B cells).** The B cells proliferate rapidly and are altered in appearance. These atypical lymphocytes are the earliest indication of a mononucleosis infection. The presence of antibody against EBV is another, more reliable indicator of the disease.

Burkitt's lymphoma is a malignancy that affects children in Uganda and other central African countries. The EBV virus has also been implicated in nasopharyngeal carcinoma that occurs more frequently among the Chinese than people in the Western Hemisphere. EBV is associated with Burkitt's lymphoma but is not the causative agent. Geographic areas where the malignancy occurs have a high incidence of malaria, a disease known to suppress the immune system. Suppression of the immune system might permit EBV to proliferate in B cells. Individuals that have AIDS, and consequently are immunosuppressed, are often victims of an EBV infection.

Poliomyelitis

Poliomyelitis has virtually been eliminated in the United States by an intensive vaccination program. Efforts are being made by the World Health Organization to eradicate the disease worldwide, but it continues to be a concern in developing countries.

Poliomyelitis, also known as **polio** or **infantile paralysis,** is caused by a picornavirus that may survive in food or water for lengthy periods of time. The virus is transmitted by ingestion of contaminated food or water. Public swimming pool water was considered a source of polio infection when the disease was of widespread occurrence in the United States (prior to vaccine development, polio affected 15 of every 1,000 in the United States annually) (**Box 29.3**).

The incubation period after exposure is generally 7 to 14 days. The poliovirus can multiply in the oropharynx and may be transmitted through respiratory droplets during the early stages of infection. Multiplication occurs in tonsils and intestinal mucosa, and the virus moves from the intestinal tract to lymph nodes. It then invades through the bloodstream and is disseminated throughout the body.

There are three forms of clinical illness that can result from invasion by the poliovirus: an abortive infection, aseptic infection, or paralytic polio. The most common form is an **abortive infection.** This infection is asymptomatic or causes some fever, sore throat, nausea, and vomiting. Recovery is rapid, and the symptoms are rarely attributed to the poliovirus. **Aseptic** infection results in similar symptoms except that some of the virus enters the central nervous system, resulting in a stiff neck and back. In these infections, full recovery occurs in about one week. **Paralytic polio** is marked by destruction of motor neurons in the anterior horn of the spinal cord. Because these cells transmit impulses to the motor fibers of the peripheral nerves, their destruction results in paralysis. If the infection involves neurons in the medulla, bulbar poliomyelitis or loss of respiratory function results. Destruction of other neurons can result in loss of motor function and muscle paralysis.

Milestones Box 29.3   Conquering Polio

During the first half of the twentieth century, paralytic polio was among the most dreaded diseases in the industrialized world. It struck randomly in human populations, and the epidemiology was poorly understood. We now know that improved sanitation conditions in the more advanced countries were actually a significant factor in the spread of paralytic polio. The virus normally survived in the intestinal tracts of infants and conferred immunity. Exposure at a very early age was the rule in populations where sanitation was less favorable. If a child was not exposed at an early age but randomly contracted the disease at an older age, paralysis was more likely. Consequently, the chance of contracting paralytic polio increased in populations with an improved standard of living. During the period when polio was rampant, the March of Dimes was organized. This was an organization that collected dimes to support research on this feared disease.

The development of viral vaccines was stymied by the inability to obtain quantities of viral particles. Until the middle of the twentieth century, there was no available technique for propagating virus outside a living host. It was then (1946) that John Enders, Thomas Weller, and Frederich Robbins began a collaboration at the Infectious Disease Laboratory at Boston Children's Hospital. Together they developed techniques for propagating poliomyelitis virus in tissue culture. This was the essential step in the development of the polio vaccine and a major breakthrough in the study of viruses in the laboratory. These three scientists worked together from 1946 to 1952, and the tissue culture methods they developed could be used for selection of strains with altered pathogenicity—a prerequisite for developing effective vaccines for any viral disease. Polio, measles, and mumps vaccines resulted from the techniques developed in John Enders's laboratory.

Enders, Weller, and Robbins received the 1954 Nobel Prize for Medicine for their contribution. Although the polio vaccine has been named for the developers, Salk and Sabin, the Nobel went to the Enders group because they made the conceptual contribution that led to a successful vaccine.

The Salk vaccine for polio was introduced in 1954 and was an inactivated virus administered by injection. Albert Sabin developed an attenuated viral vaccine in 1962 that is given orally. The oral vaccine is more effective than the inactivated virus because it induces an infectious cycle that establishes antibody protection in the intestinal mucosa. As a consequence, vaccinated individuals apparently do not serve as a reservoir of infectious particles.

It is now feared that the poliovirus might become established among immigrants and poor children in the inner city because many of these youngsters are not properly immunized. As we depend on herd immunity in most immunization programs, it is important that all individuals be immunized. No vaccine is effective in 100% of the individuals inoculated, and the goal is to have 90% to 95% of the potential victims immune and thus remove reservoirs. Without sufficient reservoirs, the disease will disappear. In the absence of this herd immunity, a new outbreak of polio could occur and have decidedly disastrous effects on those that did not gain immunity by vaccination.

Hepatitis

The word *hepatitis* means inflammation of the liver, and five distinct viruses have been implicated in hepatitis infections. It is probable that other viral agents of hepatitis are yet to be discovered (**Table 29.5**). These diseases are generally transmitted by exchange of blood/body fluids or via the oral–fecal route.

Hepatitis A (infectious hepatitis) infections are acquired by ingesting raw oysters, clams, or mussels harvested from fecally contaminated water or handled by an infected carrier. Person-to-person transmission can occur in day-care centers and other institutions where sanitary conditions are difficult to maintain. The infectious hepatitis virus is exceedingly stable and can tolerate heating at $56°C$ for 30 minutes. It is also unaffected by many disinfectants. An estimated 40% of all acute cases of hepatitis are caused by hepatitis A.

Hepatitis A has an incubation period of 15 to 50 days. The virus multiplies in the intestinal epithelia and moves via the bloodstream to the liver. It generally causes a mild, self-limiting infection. In severe infections, nausea, vomiting, and fever are the symptoms. Hepatitis A causes a degenerative inflammation of the liver that can lead to liver enlargement and possible blockage of biliary excretions. This blockage leads to jaundice (release of bile pigments into the bloodstream giving a yellowish color to the skin). Damage to the liver is immune-mediated in that antibodies to the virus attack certain hepatic cells.

Recovery takes about 12 weeks, and permanent liver damage is possible. Infections in children are often asymptomatic. Hepatitis A virus appears in the feces of

Table 29.5	Characteristics of hepatitis viruses				
Hepatitis Virus	Common Name	Virus Type	Envelope	Transmission	Incubation Period (Days)
Hepatitis A	Infectious	Picornavirus (RNA)	No	Fecal-oral	15-50
Hepatitis B	Serum	Hepadnavirus (DNA)	Yes	Blood, sexually	45-160
Hepatitis C	Post-transfusion Non-A, Non-B	Flavivirus (RNA)	Yes	Blood, sexually	14-180
Hepatitis D	Delta	Unknown (RNA)	Yes	Parenteral, sexually	15-64
Hepatitis E	Enteric Non-A, Non-B	Calicivirus (RNA)	No	Fecal-oral	15-50

infected individuals 10 days before onset of symptoms and for several weeks after recovery. An active case results in immunity, and about 40% of the people in the United States have antibodies to infectious hepatitis. As symptoms can be more severe in adults than in children, a prophylactic treatment (immunoglobulin) is available to individuals exposed to the virus. A vaccine is now available for at risk individuals.

Hepatitis B virus (HBV) (serum hepatitis) is a more serious threat to human populations than is hepatitis A. An estimated 300 million individuals worldwide are carriers of the virus. Of these, 40% will probably succumb to liver disease. There are about 300,000 humans in the United States infected each year with HBV, with about 5,000 deaths annually.

Individuals infected with hepatitis B have three different antigenic particles present in their serum: Dane particles, spherical, and tubular particles. The Dane particle is the infective form of the virus. The two incomplete particles (spherical and tubular) have the hepatitis B surface antigen as part of the particle and are present in greater quantity than is the infective Dane particle. The presence of these incomplete antigenic particles in serum is an effective indicator of an HBV infection. Their presence is also important in screening for hepatitis B in blood that might be used in transfusions.

The hepatitis B virus is transmitted by needle sharing, acupuncture, tattooing, and potentially by body or ear piercing. The virus can also be transferred during blood transfusions and in semen and is present in saliva and sweat. The virus traverses the placenta, and infants born to infected mothers generally become carriers.

The incubation period for HBV virus varies from 45 to 160 days. The first symptoms include fever, anorexia, and general malaise. This is followed by nausea, vomiting, abdominal pain, and chills. Hepatitis B viral infections can cause extensive liver degeneration resulting in jaundice and release of liver enzymes into the bloodstream. HBV is also associated with primary hepatocellular carcinoma, and this disease accounts for many of the deaths that occur worldwide.

There is no cure for HBV infections, but vaccines for prevention of the disease have been developed. These vaccines are recommended for administration to infants at birth with follow-up doses at 2, 4, and 6 months of age. Hospital workers and others who work in high-risk situations might also benefit from vaccination. Passive immunity with HBV immunoglobulin can be of value in individuals exposed to HBV, but it must be administered within a week of exposure.

The **hepatitis C** virus was formerly called Non-A, Non-B because all known cases not caused by A or B were believed to result from infection by this virus. Hepatitis C virus is a concern in blood transfusion because 5% to 10% of transfusion recipients are victims of this form of hepatitis. Intravenous drug users that share needles also have a high incidence of hepatitis C infections. Generally the epidemiology of hepatitis C is similar to that of hepatitis B virus, and the symptoms are equivalent to those of A or B but generally milder. Prevention is by careful screening of blood used in transfusions.

The **hepatitis D** virus is a defective satellite virus that was formerly known as the **delta agent**. Hepatitis D virus infections occur only in individuals infected with the hepatitis B virus. It can be cotransmitted with HBV or acquired after HBV infection. HBV must be actively replicating for the D virus to generate complete virions. A coinfection with hepatitis D and HBV leads to more severe liver damage and a higher mortality rate than occurs with an HBV infection alone. The coinfection occurs mostly in high-risk individuals such as intravenous drug abusers. The vaccine for HBV can protect against this disease.

The **hepatitis E** virus is a problem in developing countries, and the symptoms and course of infection are similar to that for hepatitis A virus. However, the mortality rate for hepatitis E is about ten times higher than that for hepatitis A viral infections. The mortality rate in pregnant women infected with hepatitis E virus is about 20%, and the reason for this high rate of mortality is unclear.

Warts

Papilloma viruses cause a variety of skin warts and epithelial lesions in humans. Most common warts are benign and disappear or are readily removed by minor surgery or a chemical agent. There are 100 different strains of human papilloma virus, and 15 of these cause genital warts. Genital warts can cause cervical carcinoma, and two strains (strains 16 and 18) account for 95% of these cancers. Transmission of genital warts among human populations is a growing concern. It is estimated that between 10 and 30 million individuals in the United States may be infected with genital papilloma virus or carry the virus in a latent state. The number of new cases each year is estimated to be in excess of 1.5 million with a high prevalence among teenagers.

The human papilloma viruses are nonenveloped, double-stranded DNA viruses. The nucleocapsid is composed of protein. Warts are transmitted by direct contact and occur mainly on skin and mucous membranes. They are not highly contagious. Virtually everyone has had or seen a wart. They are spread by scratching and most often occur in children or young adults. The projections from the skin are a result of skin cell proliferation and/or a longer life span for infected skin cells that accumulate. Common warts can be treated by applying dry ice or liquid nitrogen to the wart or destroying it with a laser beam. Application of podophyllin over an extended time is effective in removing most warts. Podophyllin is a resin extracted from the dried roots of the mayapple.

Viral Gastroenteritis

Several virus types are the causative agents of **viral enteritis,** an inflammation of the stomach or intestines. **Rotaviruses, caliciviruses,** and **astroviruses** are among the viruses involved and are transmitted via the fecal–oral route. These diarrheal diseases are the leading cause of childhood deaths in developing countries. Rotaviruses are the cause of over 4 million cases of infectious diarrhea in the United States annually, with several hundred deaths.

VIRAL DISEASES WITH NONHUMAN RESERVOIRS

There are a considerable number of viruses that infect humans that reside primarily in nonhuman reservoirs. Most are vector-borne and transmitted by the bite of an arthropod, such as a tick or mosquito. The arthropod carrier that bites an infected primary reservoir animal and then bites a human host transfers the infectious virus in the process. Some viruses that reside in nonhuman reservoirs are contracted directly through animal bites, dust, or skin abrasions. A list of natural hosts and vectors is presented in **Table 29.6.** It is evident from the number and diversity of primary hosts involved that prevention of these diseases through the control of the reservoir would be virtually impossible. Reservoirs such as squirrels, rodents, and birds cannot and should not be exterminated. Controlling the infectious agents in a wild animal population is highly unlikely. Ridding the environment of mosquitoes is a worthy goal but not attainable. The practical solution to this problem would be vaccination of domesticated animals and/or human hosts. Unfortunately, effective vaccines are not available for many of these infectious diseases.

This section discusses some of the many viral infections that are transmitted to humans from nonhuman reservoirs.

Colorado Tick Fever

Colorado tick fever is one of the more common tick-borne diseases in the United States. The disease is caused by a reovirus (orbivirus), and the vector is *Dermacentor andersoni.* Colorado tick fever affects

---

Table 29.6	Viral Infections with nonhuman primary hosts and transmitted to humans by vectors		
Disease	**Viral Agent**	**Natural Host(s)**	**Vector**
Colorado tick fever	Arbovirus	Squirrels, chipmunks	Tick (*Dermacentor andersoni*)
Encephalitis			
California	Arbovirus	Rats, squirrels, horses, deer	Mosquitoes (*Aedes* sp.)
St. Louis	Arbovirus	Birds, cattle, horses	Mosquitoes (*Culex* sp.)
Eastern equine	Arbovirus	Birds, fowl	Mosquitoes (*Aedes*)
Venezuelan	Arbovirus	Rodents	Mosquitoes (*Aedes* sp., *Culex* sp.)
Western equine	Arbovirus	Birds, snakes	Mosquitoes (*Culex* sp.)
Lymphocytic choriomeningitis	Arbovirus	Mice, rats, dogs	Dust, food
Rabies	Rhabdovirus	Skunks, raccoons, bats	Dometic dogs and cats
Yellow fever	Togavirus	Monkeys, lemurs	Mosquitoes (*Aedes* sp.)

several hundred people each year, mostly in the western and northwestern United States and in western Canada. The number of cases is considered much higher, but most go unreported.

Symptoms include chills, headache, muscle aches, and fever. The symptoms appear after a 3- to 6-day incubation period and last for 7 to 10 days. The virus infects red blood cells and may persist for 20 weeks after symptoms subside. Fatalities are rare but can occur in young children. The virus apparently traverses the placenta in pregnant females and can cause abnormalities or death of a fetus. A viral vaccine has been developed but is impractical for general use.

Small animals, including rodents, are the primary reservoir for Colorado tick fever (**Figure 29.14**). The virus is passed transovarially, and the tick can be both reservoir and vector. It is a reservoir because it passes the virus to other ticks, and it is a vector when it passes the virus to humans. The disease occurs most often among campers in tick-infested areas. Prevention is by avoiding areas where ticks might be present and by using tick repellants.

Figure 29.14 Encephalitis viruses
Reservoirs and vectors of viruses that are the etiologic agents of human encephalitis infections. The vectors are *Aedes* or *Culex* mosquitoes and the tick *Dermacentor andersoni*.

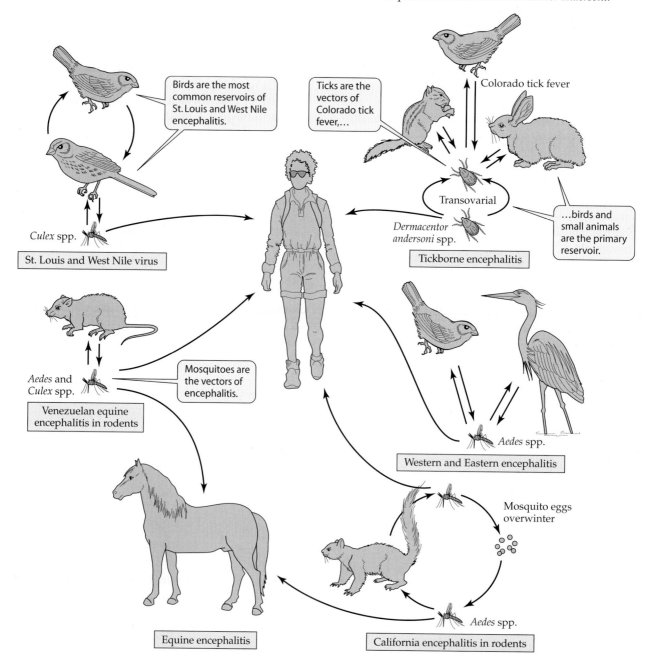

Encephalitis

The term **encephalitis** describes several diseases that are characterized by inflammation of the brain and can lead to neurological damage or fatalities. The major etiological agents of encephalitis in the United States are **arboviruses** that are transmitted by arthropod vectors. Transmission of the virus to humans is illustrated in Figure 29.14.

St. Louis and West Nile encephalitis are caused by flaviviruses, and epidemics of these diseases occur sporadically in the South and Southeast. Birds are the most common reservoirs. Western, Eastern, and Venezuelan equine encephalitis are caused by an **alphavirus** and are transmitted to humans and horses by mosquitoes, although the horse is not an important reservoir.

Symptoms of infection occur 3 to 7 days after exposure. The first bout of infection results in a systemic infection with fever, chills, headache, and flulike symptoms. This is followed by a second phase where the virus moves to the central nervous system and multiplies in the brain. Eastern equine encephalitis can be a severe infection in humans and has a high mortality rate (over 50%). Neurological disorders are a potential consequence for survivors of the disease.

The elimination of breeding sites for vectors might be an effective control measure for these diseases. Vaccines are available for individuals working with the Eastern and Western equine encephalitis virus. A live vaccine against the Venezuelan equine virus is available for use in domestic animals.

Rabies

The causative agent of **rabies** is a bullet-shaped virus of the **rhabdovirus** group (see Table 14.3). The virus is maintained in nature in carnivorous animals. Skunks and raccoons are the principal reservoir, with significant occurrences in bat and fox populations (**Table 29.7**).

Table 29.7	Major sources of the rabies virus	
Animal Group	**Percent**	
Wild Animals		
Skunk	43.0%	
Raccoon	27.7%	
Bat	13.3%	
Fox	2.5%	
Others	1.7%	
Domestic Animals		
Dog	3.6%	
Cat	3.5%	
Cattle	3.7%	
Others	1.0%	

Feral dogs and cats or wild animals carry the disease to domestic animals. Humans are generally exposed through bites or via contact with diseased vectors. The rabies virus is present in saliva of rabid animals and consequently aerosols in bat caves, where rabid bats may nest, are a potential source of human infection.

When transmitted in aerosols, the virus infects the epithelial tissue that lines the upper respiratory system. Generally, rabies is passed among animal reservoirs through a bite or exposure to virus-laden saliva. Although a young skunk infected with rabies will eventually succumb to the disease, it may survive to reproductive age. The progeny then, become infected and pass the disease along to their offspring. Foxes do not survive long after a rabies infection, and though not an important reservoir, they tend to be aggressive and will attack humans.

Rabies is transmitted to humans through aerosols or entry of virus-laden saliva into an open wound, and this can occur by a bite or during handling of an infected animal. The virus multiplies in tissue at the site of inoculation and can remain localized for days or months. The length of time required for the virus to move from the site of infection to the brain depends on several factors: (a) the size of the inoculum, (b) the proximity of the wound to the brain, and (c) the host's age and immune status. The rabies virus travels from the wound site to the peripheral nerve system, farther to dorsal ganglia, and on through the spinal cord to the brain. Infection occurs in the spinal cord, brain stem, and cerebellum. The virus can move from the brain to the eye, salivary glands, and other organs. Viral replication in the brain results in the presence of cytoplasmic inclusions in the affected neurons. These inclusions are called **Negri bodies** and have been a principal diagnostic test for the disease in animals. However, diagnosis is now based on a fluorescent antibody test on nerve tissue.

Symptoms of rabies do not appear until the virus reaches the brain. The initial symptoms include fever, malaise, headache, gastrointestinal upset, and anorexia. After 2 to 10 days, neurological symptoms associated with rabies appear. **Hydrophobia** occurs in 25% to 50% of patients and is characterized by jerky contractions of the diaphragm when the victim attempts to swallow water. Generalized seizures and hallucinations follow, ending in coma and death.

Humans bitten or otherwise exposed to **potential rabies** carriers are given postexposure prophylactic treatment. These treatments are initiated unless it can be confirmed that the suspected animal involved **does not** have a rabies infection. The victim is immunized with a vaccine plus a dose of equine antirabies serum or human rabies immunoglobulin. This passive immunization provides protective antibodies until antibodies are produced in response to the vaccine.

The control of human rabies depends on the effective elimination of the disease in the wild animal populations that serve as reservoirs. Domestic dogs and cats are routinely vaccinated against the disease. Vaccination of wild animals by capture/release is out of the question. Administration of an oral vaccine by placing it on bait is one possible mechanism for widespread inoculation of wild animal populations.

Yellow Fever

Yellow fever, caused by an **arbovirus,** was the first human infection shown to be caused by a virus (1901). It is also the first viral infection demonstrated to be vectorborne. Yellow fever is transmitted by a mosquito, *Aedes aegypti.* The last epidemic in the United States occurred in 1878 and killed over 13,000 people. There has not been a reported case in the United States in the past 70 years, but the potential for reintroduction is real because the vector is present in the southern United States. The reservoir for the disease in Central and South America is the monkey. There are 30,000 to 40,000 deaths worldwide each year from this disease.

The yellow fever virus enters the human body through a mosquito bite and is carried by lymph to local lymph nodes. The virus multiplies in lymph nodes and the spleen and results in a systemic infection. Damage to liver, kidneys, and heart and hemorrhaging of bloody vessels follows the systemic infection. The name *yellow fever* originated because the liver damage results in jaundice. Fever, chills, headache, and nausea are also symptoms of the disease.

Where yellow fever is endemic, it is spread by distinct epidemiological patterns that are either urban or sylvatic. The urban cycle involves human-to-human transmission, with the mosquito serving as vector. In the sylvatic cycle, mosquitoes transmit the virus between monkeys and occasionally to humans.

Elimination of breeding grounds for the vector is an important and effective control measure. A vaccine is available that, when administered intradermally, leads to lifetime immunity.

Ebola and Marburg

Outbreaks of **hemorrhagic fevers** have occurred in recent years in Central and West Africa. Two viruses that are known to be causative agents of these diseases are termed **Ebola** and **Marburg**. Both are ss RNA-enveloped viruses (*Filoviridae*). The viruses are passed among wild vertebrates, including monkeys, and these are the apparent reservoir. Arthropods are believed to be the vector that carries the viruses animal to animal. Transmission mechanisms for these viruses are uncertain at the present time.

There was an outbreak of hemorrhagic fever among scientists in Germany in the late 1960s, with several deaths. The scientists were experimenting on monkeys imported from Uganda that were infected with the Marburg virus. The Marburg virus was transmitted to the victims via saliva or mucus.

In 1976, there was an outbreak of Ebola in Zaire and Sudan that killed over 500, including Belgian physicians and nurses that were treating patients. Another Ebola outbreak occurred in Uganda in the year 2000 and caused the death of 224, including health workers, and another outbreak of Ebola in Gabon is a concern at present with 11 deaths.

The incubation period for Ebola is 4 to 10 days before flulike symptoms occur, followed by vomiting, diarrhea, and severe internal bleeding. Ebola is a highly virulent disease causing death in 50% to 90% of all clinical cases. The disease kills victims faster than it can spread and tends to "burn out" before it reaches epidemic proportions.

Because Ebola is so readily transmitted by bodily fluids, such as mucus, saliva, and blood, it is a deadly threat. For this reason, the World Health Organization monitors it and makes every effort to isolate and treat victims.

Emerging Viruses

The human immunodeficiency virus (HIV), the etiological agent of AIDS, was first described in 1981. Over the last 20 years, this newly emerged virus has had a devastating effect on humans worldwide. There are other viral infections that have emerged in the last 20 years, and these may be a serious threat to human populations. A hantavirus outbreak in the Southwest in the 1900s caused over 30 fatalities and has a 70% to 75% death rate among victims. The virus is apparently disseminated in dust containing rodent urine. Whether this virus may gain other mechanisms of transmission is unknown. Other viral diseases that have emerged include mosquitoborne infections such as **dengue fever** and **chikungunya.** How much of a threat these viruses will be to humans is difficult to assess (see Chapter 14).

SUMMARY

▸ **Viruses** can infect bacteria, plants, and animals. Infections by viruses are more **difficult** to **control** than infections caused by bacteria because the metabolic machinery of the virus is quite equivalent to that of healthy cells.

▸ Infections spread in **airborne droplets** such as colds and influenza are difficult to control because the **daily** life of humans involves person-to-person contact.

▸ **Chicken pox** is a highly contagious disease that generally affects children. It can be a serious infection, and a vaccine is available to prevent the disease. **Shingles** is the result of a dormant viral infection and can occur in later life.

- **Rubella (German measles)** is a viral disease that was particularly dangerous if infections occurred in **pregnant** women, as the virus causes considerable harm to the fetus. Rubella can be controlled by **vaccination** of children. Another measles, **rubeola,** was a concern because it caused permanent neurological damage to some children. Vaccinating infants also can prevent it.

- A deadly disease that has been **eliminated** throughout the world is **smallpox.**

- **Mumps** was a common childhood disease that caused **inflammation** of the **parotid glands** and spread to other areas of the body with unfortunate consequences. Vaccinating young children before they enter school can control this disease.

- **Influenza** can occur in **worldwide epidemics** (pandemics). It is a respiratory disease that can lead to pneumonia. Vaccines lower the risk and/or severity of the disease.

- **Common colds** affect 40% to 45% of Americans each year and result in an estimated **200 million lost school/workdays annually.** They can be neither prevented nor cured.

- Some viruses can **transform** normal human cells to cancerous cells. **Human T cell leukemia** virus is one such and is transmitted human to human in contaminated blood, exchange of IV needles by drug users, or sexual contact.

- **Acquired immune deficiency syndrome (AIDS)** has become a leading killer of young adults. Transmission can occur through transfusions with contaminated blood but most often is transmitted by **sexual contact.** There is **no cure** or prevention except **"safe sex."**

- **Herpes simplex virus type 1** is the etiological agent of **cold sores** that appear on lips and mouth. Another type of herpes (type 2) causes **genital herpes,** a growing problem in the United States. **Acyclovir** is a chemotherapeutic agent that can relieve symptoms of herpes infections but is not a cure.

- **Infectious mononucleosis** is a disease caused by herpesvirus that occurs often in college age individuals. It causes fatigue, and rest is the only cure.

- Another virus disease that can be prevented by vaccination is **poliomyelitis.** It has not been eliminated because there has not been worldwide use of the vaccine. **Polio** can cause **paralysis** by destruction of motor neurons.

- There are a number of **encephalitis** diseases that are characterized by **inflammation** of the **brain.** They are transmitted to humans by vectors that include **arthropods** and **mosquitoes.** Elimination of vector breeding sites is the only preventive measure now available.

- **Rabies** is generally spread by the bite of a domestic animal and wild animals such as **skunks** and **raccoons** that serve as **reservoir.** Feral dogs and cats can carry the virus from the reservoir to domestic animals. Treatment is with antiserum and vaccination of exposed individuals.

- **Yellow fever** is caused by an arbovirus carried by a **mosquito** (*Aedes aegypti*). The disease causes severe liver damage but has been eliminated in the United States at the present time.

REVIEW QUESTIONS

1. Why are viral infections so difficult to control? How might this relate to the origin of virus? In considering this, would one assume that fungal or protozoan infections might also be a problem?

2. Why were the viral diseases transmitted via the respiratory route of such interest to the vaccine producers?

3. What are the symptoms of "shingles," and how does one get this disease?

4. Cite examples of viral diseases that can be transmitted by individuals that show no obvious symptoms.

5. Where have measles outbreaks occurred in recent years?

6. How was smallpox eradicated? Why was this possible from an economic standpoint?

7. What is "herd" immunity? How does this relate to mandatory vaccination? What are some dangers when preschool children do not receive adequate immunization programs?

8. Why is the influenza virus such a problem? Periodically there are epidemics and pandemics of this disease. Why?

9. One might say that the common cold will always be with us. Why is this true?

10. Donated blood is not examined for human T cell leukemia virus at the present time. Should it be?

11. Why would an individual, once infected with HIV, have it for life despite chemotherapy? Elimination of a retrovirus from an infected individual is not feasible. Why?

12. AIDS patients have a major problem with infections by viruses and eucaryotic opportunists. What are some of these, and why is an AIDS victim so vulnerable to them?

13. Herpesviruses cause latent infections. Discuss this process.

14. Genital herpes is a disease that is increasing in frequency. What are the dangers?

15. The Epstein-Barr virus is of some concern in the United States. Why? Why is it more of a problem in other countries?

16. Polio may be eliminated from the world, as was smallpox. What are some problems in achieving this goal?

17. What are some of the diseases caused by hepatitis viruses? How can they be controlled?

18. Can warts cause cancer?

19. What are some of the major viral diseases in humans that have nonhuman reservoirs? How can these diseases be controlled?

20. What animals serve as major reservoirs for rabies?

SUGGESTED READING

Cann, A. J. 2001. *Principles of Molecular Virology*. 3rd ed. San Diego: Academic Press.

Dalgleish, A. G. and R. A. Weiss. 1999. *HIV and the New Viruses*. San Diego: Academic Press.

Granoff, A. and R. G. Webster. 1999. *Encyclopedia of Virology*. 2nd ed. San Diego: Academic Press.

Specter, S., R. L. Hodinka and S. A. Young. 2000. *Clinical Virology Manual*. 3rd ed. Washington, DC: American Society for Microbiology.

Strauss, J. and E. G. Strauss. 2001. *Viruses and Human Disease*. San Diego: Academic Press.

White, D. and F. Fenner. 1994. *Medical Virology*, 45th ed. San Diego: Academic Press.

Chapter 30

Epidemiology and Clinical Microbiology

Alas! regardless of their doom
The little victims play;
No sense have they of ills to come
Nor care beyond today.
– THOMAS GRAY, 1716–1771

The health and well-being of human populations relies, in large measure, on the control of communicable infectious diseases. Human history is a record of devastation by infectious diseases, and up to the early years of the twentieth century, such diseases were the major cause of death worldwide (**Box 30.1**). Infectious diseases continue to take a toll. In the developing nations of Asia, Africa, and South America, they account for nearly 50% of all deaths. In the developed countries, including the United States, Canada, and Western Europe, fewer than 8% of all deaths are from infectious disease. The decreased death rate in the developed nations was brought about by controlling both individual infections and communitywide factors, such as clean water, that would otherwise contribute to the spread of disease.

Worldwide, very few diseases have been truly defeated, but one major affliction that caused much human misery has been eradicated, and that is smallpox (see Box 30.6). The periodic spread of infectious diseases such as influenza to virtually every nation is a reminder that communicable diseases must be a concern. The current catastrophic spread of the human immunodeficiency virus (HIV) confirms that uncontrolled highly communicable diseases can still leave their mark on society.

Two major aspects of infectious disease control are: (1) having an accurate determination of the etiological agent involved in infections and (2) the incidence of this infectious disease in the population. This is where epidemiology and the clinical laboratory come together. Epidemiology is the science concerned with the distribution and prevalence of communicable diseases in populations. The modern clinical laboratory has diagnostic capabilities that can determine precisely the strain of an agent involved in a disease outbreak and accomplish this in a short period of time. This aids the epidemiologist in pinpointing the origin and course of an epidemic and to take measures that might curtail the spread of the infection.

EPIDEMIOLOGY

Epidemiologists work worldwide to monitor infectious diseases in order to institute measures for their control. The role of epidemiologists in pinpointing the outbreak of infectious disease is illustrated by the discovery of AIDS. In June of 1981, the Centers for Disease Control and Prevention in Atlanta compiled data indicating that there were an unusually large number of deaths among immunodeficient men in the Los Angeles area. These deaths were caused by opportunistic pathogens that usually infected people who were immunodeficient, and the deaths involved homosexual males. This discovery was announced a full 2 years before the retrovirus responsible for the

---

Milestones Box 30.1   **The Major Causes of Death in the United States in the Twentieth Century**

In 1900 there were 318 deaths per 100,000 population caused by various infectious diseases and 215 due to heart disease and cancer. Public health measures instituted in the

first half of the twentieth century lowered the death rate from these infectious diseases dramatically, and the death rate is now even lower. Public health measures, antibiotics,

and immunization have combined to protect people in the United States from the scourge of infectious disease. Heart disease and cancer continue to be leading causes of death.

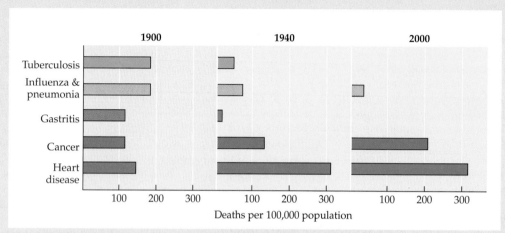

Leading causes of death in the United States in 1900; in 1940, before antibiotics were widely used; and in 2000, by which time antibiotics were widely available.

---

disease was isolated and identified. More important, the relation of the disease to the exchange of bodily fluids was almost fully defined. This is a prime example of the effectiveness of epidemiologists in determining the frequency of a disease even before its etiological agent is identified. Since the identification of HIV, the virus has spread rapidly and is now a concern in all of the continents except Antarctica.

Following is a discussion of some of the terminology that is commonly used in epidemiology, including types of epidemics, incidences, prevalence, mortality, and morbidity.

Terminology

The word *epidemic* appears ominous, as it suggests that a disease may be spreading that will affect a massive number of people. In reality, it means that a disease is occurring in the population at a higher than normal frequency (see **Box 30.2**). Most infectious diseases are endemic; they occur at a low but constant frequency in the population. A significant increase in the frequency of a disease over continents or throughout the world is termed a *pandemic.* Major pandemics caused by the influenza virus occurred in 1918 and 1957, and HIV has clearly reached pandemic proportions.

Epidemics fall into three categories: sporadic, seasonal, or contact. A **sporadic** disease affects a certain percent

of the population throughout the year. Endemic diseases occur sporadically. The incidence may increase slightly during the year, but overall the number of cases is quite constant. A **seasonal** disease is one, such as Lyme disease or Rocky Mountain spotted fever, that is transmitted only in those months when the arthropod vectors are active. A **contact** disease is one, such as colds, that is spread via contact with an infected individual.

The **incidence** of an infectious disease is calculated by the following formula:

$$\text{Incidence} = \frac{\text{number of newly reported cases per a given time}}{100,000 \text{ people at risk}}$$

The **prevalence** of an infectious disease is the percent of the total population that has the disease at a given time. For example, a seasonal disease is more prevalent during a limited number of months of the year. The term *outbreak* is also used and indicates that there are a relatively large number of cases in a limited area at a certain time.

Mortality and Morbidity

There are infectious diseases that cause a limited number of fatalities and others that are fatal to a considerable percentage of those infected. Botulism is generally fatal,

Milestones Box 30.2   The Great Plague in London

The Great Plague in London (1665) was a reasonably well-documented epidemic. The population of London at the time of the Great Plague was estimated to be about 400,000 inhabitants. The number that succumbed in the plague epidemic was somewhere between 25% and 40% of the population.

However, the number of deaths reported is a conservative figure, since many went unrecorded. The numbers for this drawing are mostly from Burial Registers of Churches. Many, including Quakers, were buried in their gardens, and massive numbers of others were buried without record.

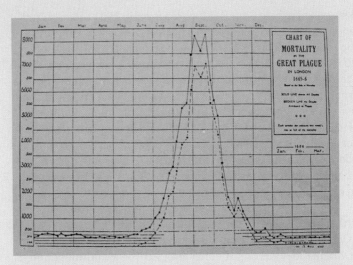

The dramatic increase in death rate in London during the plague epidemic of 1665–66. From Bell, *The Great Plague in London* (1665), p. 238.

but few succumb to staphylococcal food poisoning. The coming of the antibiotic age resulted in a marked change in the number of deaths from diseases such as pneumonia and tuberculosis. The **mortality rate** is the number of deaths from an infectious disease relative to the total number that contract it:

$$\text{Mortality rate} = \frac{\text{deaths due to a disease}}{\text{total infected}}$$

The mortality rate generally varies with the availability of health care. Cholera has a low mortality rate when adequate medical care is available but increases significantly if intravenous fluids and other health care measures are unavailable.

Morbidity is the incidence of infectious disease, both fatal and nonfatal, in a population:

$$\text{Morbidity} = \frac{\text{new cases in a selected period}}{\text{total population}}$$

The morbidity rate is a definitive figure and serves as an accurate measure of the general health of a population. In the developed nations, morbidity is considerably higher than the fatality rate, although both are relatively low for most diseases. The low fatality rate reflects the availability of better health care, antibiotics, and other chemotherapeutics in developed versus developing countries. Fatality figures, in general, are not an accurate measure of health.

CARRIERS AND RESERVOIRS

A significant problem in controlling communicable infectious diseases is the presence of **carriers** in the population. Carriers are individuals with asymptomatic or subclinical infections that are not serious enough to require treatment or curtail a person's activities. Often carriers continue normal duties while "toughing it out" with their disease symptoms and tend to deny to themselves and others that they are ill with a transmissible disease. Unfortunately, these diseased people can expose everyone they encounter to the infectious microorganism they bear. Carriers can also be those convalescing from infectious diseases that return to work or school while harboring and shedding large numbers of infectious organisms. This is particularly a concern in enteric-type diseases, where convalescent patients continue to shed infectious organisms for a considerable length of time after symptoms disappear. In the case of typhoid, some patients (2% to 5%) become **chronic carriers.** A classical case of a chronic typhoid carrier was Mary Mallon, known as Typhoid Mary (see **Box 30.3**), who was well enough to continue working after bouts with typhoid fever.

A **reservoir** is the animate or inanimate site where an infectious disease is maintained in nature between outbreaks. If a disease agent survives only in living hosts, it must have a means of escaping from one host and of traveling and gaining entry into another host. Humans would be the reservoir for diseases transmitted via fecal contamination with food or drinking water as the intermediate (see **Box 30.4**).

Milestones Box 30.3   Typhoid Mary

Mary Mallon, known as Typhoid Mary, has attained the dubious honor of having much of her life story discussed in virtually every microbiology text. Hers is a classical case of the chronic carrier of infectious disease wreaking havoc wherever she traveled. Tracking her down as the source of a typhoid outbreak tested the epidemiological investigatory abilities of the Health Department in New York City in the early 1900s. Yet, the Health Department's handling of the case parallels issues we face today in the AIDS dilemma, where the rights of individuals are pitted against the perceived welfare of the populace.

Mary Mallon was a Swiss immigrant who had a serious case of typhoid fever in 1901 that resulted in a gallbladder permanently infected with *Salmonella typhi*. The infection generated large numbers of the typhoid bacillus that entered her gastrointestinal tract and were shed in her feces. Unfortunately, Ms. Mallon was a cook and housekeeper in New York City and worked in several homes. As she moved from position to position, she left behind 28 cases of typhoid fever.

The New York Health Department, headed by Dr. George Soper, tracked down Mary Mallon and had her arrested. The authorities offered to remove her gallbladder to effect a cure. She refused and was released after 3 years of imprisonment on a pledge to never cook or handle food and to report periodically to the Health Department. Mary Mallon immediately disappeared, changed her name, and became a cook in hotels, hospitals, and sanitaria. She apparently recognized that the disease was caused by her presence, as she quit her job when an outbreak occurred and moved on in order to evade authorities.

After 5 years, Mary Mallon was intercepted during a typhoid outbreak at a hospital. She spent her remaining 23 years on North Brother Island in New York City's East River, where she died in 1938. Typhoid Mary was the source of an estimated 200 cases of typhoid fever, much suffering, and several deaths.

There are a number of diseases termed *zoonoses* that cause either human or animal infections. The animal is the reservoir; a number of these diseases are discussed in Chapter 28.

MODES OF TRANSMISSION OF INFECTIOUS DISEASES

Survival of infectious diseases depends on their transmission from host to host. In most cases, disease-causing organisms do not propagate outside a host or a reservoir. As noted earlier, many infectious diseases are endemic in the general population. There are two general types of localized epidemics (**Figure 30.1**). A **common source epidemic** (disease transmitted from a single source) occurs if a considerable number of humans eat from a large batch of contaminated food. A **host-to-host** epidemic (one human to another) occurs, for example, when susceptible children are brought together during the first days of school. Historically, childhood diseases, such as measles, were epidemic in primary schools during September of each year. Major bacterial diseases that occur in humans are presented in **Table 30.1**, along with potential measures for their control. Many of these diseases were discussed in Chapter 28.

Human Contact and Respiratorily Contracted Diseases

Human contact and respiratory diseases are passed directly by human interaction. Transfer may occur by touching, kissing, coughing, sneezing, or in some cases, on dust particles. Streptococcal and staphylococcal skin infections, such as impetigo, are often transferred among children by scratching. An unimpeded sneeze expels infectious droplets that are about 10 μm in diameter and contain one to several bacteria (see Figure 29.8). The number of droplets per sneeze can be in the thousands. A hearty sneeze can travel at 100 meters/second and

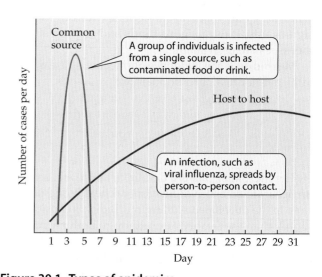

Figure 30.1 Types of epidemics
The different time courses of common-source and host-to-host epidemics.

John Snow and Cholera

John Snow (1813–1858), a British physician, was the first to recognize that humans were a reservoir for cholera. He realized that the feces of cholera patients were highly infectious and surmised correctly that human waste in drinking water could transmit the disease. Snow followed the incidence of cholera from 1853 to 1855 in a wide area of London. At that time, there were two major water systems supplying water to homes in that area: the Southwark & Vauxhall Company and the Lambeth Company. John Snow followed the cholera epidemic in homes of equivalent living standards but supplied by one or the other of these companies. It was obvious that inhabitants of houses supplied by Southwark & Vauxhall had a markedly higher incidence of cholera than those supplied by the Lambeth Company.

In the first 7 weeks of the epidemic, 315 people per 10,000 had died in houses whose water was

supplied by the Southwark & Vauxhall Company compared to 37 per 10,000 people of those whose water was supplied by the Lambeth Company. Snow looked at the water supply for the two companies and found that the Lambeth Company obtained its water from the Thames

above the city, whereas the Southwark & Vauxhall Company drew water from the Thames in an area where untreated sewage from London entered the river. He concluded that the source of the cholera epidemic was the sewage-contaminated water.

The incidence of cholera in the vicinity of the Broad Street pump among a population that drew water from the Thames in an area contaminated by human waste. From Cosgrove *History of Sanitation*, p. 2.

---

contain up to 100,000 bacteria or virus particles. Tuberculosis is a disease that is transmitted in respiratory droplets and is a constant threat.

There are 10,000 to 15,000 new cases of tuberculosis (TB) each year in the United States, and the mortality rate is 10% to 15%. Clinical tuberculosis is generally a slowly progressing, chronic disease, and infected individuals can be asymptomatic for months or even years. These individuals are carriers and may transmit the disease to those they encounter. At the present time, TB infects one-third of the world's population, and 8 million people develop disease symptoms each year. In contrast to TB, leprosy is a poorly transmitted contact disease, and individuals that are infected are predisposed to contract it. There are 50 to 100 new cases annually. Vaccination for leprosy is not available, and treat-

ment of infected individuals is the only known method for control. Vaccination is an effective control for some of the respiratorily transmitted bacterial diseases, such as whooping cough, diphtheria, and such viral diseases such as mumps, chicken pox, rubeola, and measles (see Chapter 29).

Water-, Food-, and Soil-Borne Infections

Many water- and food-borne diseases are caused by pathogens that survive solely in the gastrointestinal tract. Humans are a natural host for many of these infectious microbes, including those that cause cholera, typhoid fever, and shigellosis. Because these microorganisms can survive for a limited time in raw sewage, they can be transmitted through contaminated water supplies. Proper sewage disposal and effective water

Table 30.1	Major bacterial diseases of humans, sources of infection, and potential control	
Disease	**Primary Reservoir**	**Potential Means for Control**
Human Contact and Respiratorily Contracted		
Streptococcal infections	Humans	Antibiotics; vaccine for pneumonia
Staphylococcal infections	Humans	Antibiotics; antiseptics
Meningitis	Humans	Specific antibiotics
Tuberculosis	Humans	Test and treat infected persons
Whooping cough	Humans	Vaccinate infants
Diphtheria	Humans	Vaccinate infants
Leprosy	Humans	Obtain proper treatment; vaccinate in endemic areas
Pneumonic plague	Humans	Eliminate rats and fleas
Water-, Food-, and Soil-borne		
Cholera	Humans	Treat sewage and water; observe proper sanitation
Typhoid fever	Humans	Pasteurize milk; proper treatment of sewage; inspect food handlers
Shigellosis (dysentery)	Humans	Observe proper sanitation
Salmonellosis	Beef, poultry	Cook meat and eggs properly
Campylobacter	Animals, poultry,	Pasteurize milk; thorough cooking of food and water
Tetanus	Soil	Vaccinate
Brucellosis	Cattle	Immunize cattle and pasteurize milk
Botulism	Soil	Properly can and cook food
Staph food poisoning	Humans	Refrigerate food
Legionnaire's disease	Aquatic environments	Clean misting equipment or do not use
Pseudomonas infections	Dust	Clean air in burn wards
Sexually Transmitted		
Gonorrhea	Humans	Eliminate carriers; practice safe sex
Syphilis	Humans	Eliminate carriers; practice safe sex
Chlamydia	Humans	Eliminate carriers; practice safe sex
Louse-borne, Human to Human		
Trench fever	Humans	Proper sanitation; control lice
Relapsing fever	Humans	Control ticks and lice
Typhus (epidemic)	Humans	Proper sanitation; vaccinate
Vector-borne		
Rocky Mountain spotted fever	Mammals, birds	Wear protective clothing and examine body for ticks
Tularemia	Rodents, rabbits	Observe proper care when cleaning wild rabbits
Lyme disease	Deer	Wear protective clothing
Bubonic plague	Rats	Control rats, proper sanitation
Typhus (endemic)	Rodents	Control rats, vaccinate
Scrub typhus	Mites	Control mites
Animal Contact		
Leptospira	Vertebrates	Control rodents, vaccinate domestic animals
Anthrax	Soil	Sterilize wool, hair, other animal products
Psittacosis	Birds	Control bird imports
Q fever	Cattle	Vaccinate animal handlers

treatment are important in control of gastrointestinal diseases. Cases of cholera and typhoid fever are rare in the United States. Shigellosis, a form of dysentery, is quite common and can be transmitted from infant to infant in child-care centers. Outbreaks also occur in elementary schools. Proper sanitation is the best mechanism for control of shigellosis. Pasteurization of milk, proper cooking practices, and immunization of cattle are important in preventing the transmission of food-borne diseases. As a result of such measures, brucellosis in cattle and bovine tuberculosis are no longer a concern in developed nations.

Legionnaire's disease is associated with mists created by cooling towers for air-conditioning systems and vegetable misters in grocery stores and in various aquatic environments. The control of *Legionella pneumophila* is difficult, and the disease occurs sporadically across the United States.

Anthrax has been an important soil-borne disease because the causative organism (*Bacillus anthracis*) can form endospores that survive in soil. Cattle killed by anthrax have been a source of the endospores. Better health practices in the animal industry have curtailed the spread of this disease. Probably the greatest threat from anthrax is its use by a mad despot as a germ warfare agent.

Sexually Transmitted Diseases

The sexually transmitted bacterial diseases are a significant societal problem that can be addressed only when people assume responsibility for their behavior (Table 30.1). All can be treated and cured, yet millions of new cases of gonorrhea and chlamydial infection occur each year. An increasing incidence of antibiotic resistant strains, particularly with the gonococcus, poses a substantial threat to public health. These diseases can be controlled only by public education, public responsibility, and a greater awareness of the problem. The use of condoms and other safe sex practices are the only control now available. Immunization to prevent the sexually transmitted diseases including HIV is not on the horizon.

Three sexually transmitted viral infections are epidemic in the United States. These are genital herpes (HSV-2), genital warts, and the human immunodeficiency virus (HIV). There are an estimated 30 million individuals in the United States infected with HSV-2, and about 1,000,000 new cases are diagnosed each year. Herpes in the neonate is one of the most common life-threatening infections, as there are about 2,000 babies born each year afflicted with this disease. The incidence of genital warts is increasing at an alarming rate and is now one of the more prevalent sexually transmitted diseases, particularly in promiscuous young adults.

HIV has reached pandemic proportions with 920,000 HIV positive individuals in North America and an estimated 40 million infected worldwide (**Figure 30.2**). Since the epidemic began about 20 years ago, 57 million people have been infected with HIV, and 21 million have died because of these infections. HIV infection in sub-Saharan Africa is of great concern, and the United Nations has instituted programs for controlling the disease and treatment of victims. South and Southeast Asia

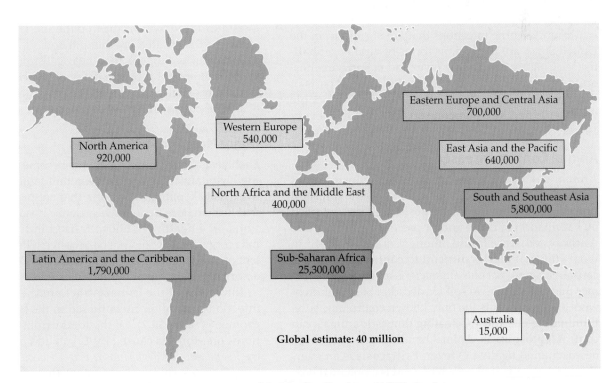

Figure 30.2 Worldwide distribution of HIV infection
The estimated distribution of adult HIV infection through late 2000.

are also experiencing a troubling increase in the incidence of HIV.

Louse-Borne, Human-to-Human Diseases

Louse-borne diseases are those that are passed among human populations by lice bites and are best controlled by proper sanitation. **Epidemic typhus** and **trench fever** are louse-borne diseases and are generally associated with substandard living conditions. Sanitation and elimination of lice can effectively control the spread of these diseases. **Relapsing fever** is a louse-borne disease that occurs mainly among campers who spend time in areas infested with rodents. Campsites in locales where this disease occurs should be selected with care.

Vector-Borne Infections

Two vector-borne infections of concern in the United States are Lyme disease and Rocky Mountain spotted fever. The number of cases of both are increasing in some areas of the United States, and the sole control measure is avoidance of the ticks that serve as vectors. Protective clothing and insect repellent are essential when entering areas potentially occupied by ticks. One should check for ticks on one's body after leaving such areas. Rat control has been effective in the control of vector-borne diseases such as **plague** and **endemic typhus**. **Tularemia** cannot be controlled in wild animal populations. The only way humans can avoid infection by this organism is to not dress freshly killed wild rabbits or, if doing so, to wear rubber gloves and proceed with care.

Mosquito-borne infections that are of concern in the United States are the various forms of **viral encephalitis** (see Chapter 29). **Yellow fever** and **malaria** are mosquito-borne diseases and a concern in various parts of the world. Malaria affects about 500 million people worldwide and kills someone on the average of every 10 to 15 seconds.

Animal Contact Diseases

Animal contact diseases, such as brucellosis, are of decreasing incidence in the United States. The spread of **brucellosis** and **bovine tuberculosis** has been controlled by vaccination of cattle and/or pasteurization of milk. **Anthrax,** a disease found among workers exposed to wool and goat hair, is difficult to control because the causative agent, *Bacillus anthracis*, is an endospore former that can survive in soil for decades. Sterilization of wool and goat hair is the only known control measure. **Eliminating rats and vaccinating domestic animals can control leptospirosis.** Animal handlers can and should be vaccinated against Q fever. **Psittacosis** is an ever-present danger in domestic birds and is a potential hazard for workers in poultry slaughterhouses. Pigeons are also a reservoir for the disease.

NOSOCOMIAL (HOSPITAL-ACQUIRED) INFECTIONS

To effectively invade a host, the pathogen must counteract the host's defenses. These normal defenses (Chapter 26) are generally quite effective in combatting invading organisms. Unfortunately, hospitals bring together patients whose normal defenses are compromised by one circumstance or another, and infections by opportunistic or accidental invaders can occur. Infections acquired in a hospital are called **nosocomial infections.**

Nosocomial infections occur for several reasons:

1. Illness or chemotherapeutics may compromise the immune system.
2. Abrasions or openings in epithelial barriers caused by surgery, catheters, syringes, respirators, or instruments employed by the physician to examine the inner body offer the opportunistic pathogen a site to establish infection.
3. Exposed tissues of burn or wound patients offer access to airborne microorganisms.
4. Cross infection from patient to patient in crowded wards as well as from hospital workers or physicians is possible.
5. The overuse of antibiotics in hospitals has resulted in a hospital environment where drug-resistant microorganisms are ubiquitous.
6. The hospital environment selects for pathogens, as few hospitals have isolation wards; a virulent organism can find a reservoir in patients. This pathogen can be transmitted to other patients.
7. Hospital pathogens may bear plasmids that carry information for multiple drug resistance. As a result, drug resistance can spread rapidly among a population of pathogenic microorganisms.

Major Nosocomial Infections

Each year about 2 million patients become infected during hospitalization, and the resultant fatalities number over 100,000 patients. Bacteria that are part of the normal human flora frequently cause the nosocomial infections. In a healthy nonhospitalized individual these microorganisms would not be invasive. The most prevalent organism in nosocomial infections is *Escherichia coli*, a normal inhabitant of human intestines.

Infections of the urinary tract are the nosocomial infection most often encountered in the hospital environment (**Table 30.2**). Of the total urinary tract infections, one-third is caused by *E. coli* and another third by *Pseudomonas aeruginosa, Enterococcus faecalis,* or *Streptococcus epidermis.* Most of the remainder are caused by other gram-negative bacteria. Infection is generally a consequence of urinary catheterization of

Table 30.2	The relative frequency of nosocomial infections by body site	
Site of Infection	**Percent of Total Infections**	
Urinary tract	40%–42%	
Respiratory	16%–18%	
Surgical wound	17%–20%	
Bacteremias	6%–7%	
Skin infections	6%–7%	
Other	12%	

immobilized patients. Respiratory infections that appear as a form of pneumonia are often encountered and may be caused by *Pseudomonas aeruginosa, Staphylococcus aureus,* or *Klebsiella* sp. These nosocomial infections, which can cause death, result from respiratory devices and the inability of the patient to clear the lungs.

Staphylococcus aureus and pyogenic streptococci are a major concern in surgical patients. An estimated 7% to 12% of all surgical patients have postoperative infection problems. When the gastrointestinal or genitourinary tracts are involved in the surgery, the number of postoperative infections is two to three times higher.

PUBLIC HEALTH MEASURES

The general health of people in the United States has improved dramatically since the time of our Founding Fathers. Life expectancy for a baby born today is about double that in the late eighteenth century. Nutrition has improved markedly, with fresh fruits and vegetables available year around. Housing has also improved, and working conditions are generally less stressful. Potable water is available to virtually all people in the United States, and we have effective treatment systems for human waste. We have eliminated the threat of many deadly diseases such as yellow fever, and antibiotics have alleviated dreaded diseases such as tuberculosis. A combination of these factors has tended to make our lives longer and healthier. How has this happened? By diversified efforts termed **public health,** which refers to the overall health of populations and the efforts of local, state, and the federal public health officials to maintain reasonable health standards (**Box 30.5**).

Each state in the United States has a publicly funded organization generally known as the state health department. The name may vary, but the responsibilities are much the same, and a major function of this organization is to monitor the incidence of infectious diseases in the population. In many states, the health department provides a diagnostic laboratory for infectious organisms

---

Milestones Box 30.5   Public Health Measures and Human Well-Being

The effectiveness of public health measures on the curtailment of infectious diseases in New York City is apparent from this chart. Cholera, a disease transmitted by fecal contamination, was a considerable cause of death prior to the establishment of the Board of Health and Health Department in 1866. Smallpox, yellow fever, and typhus were also significant health problems. Chlorination of water was initiated in 1910 and pasteurization of milk in 1912, and these measures resulted in better control of infectious disease. Note the number of deaths from the influenza pandemic in 1918.

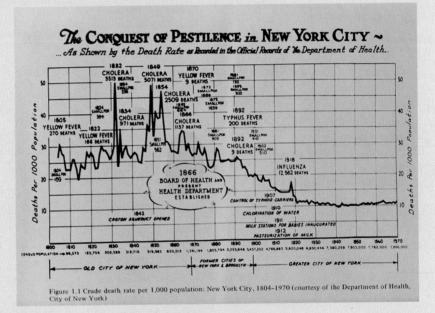

Figure 1.1 Crude death rate per 1,000 population: New York City, 1804–1970 (courtesy of the Department of Health, City of New York)

The effect of public health measures on the incidence of disease. Courtesy of the Department of Health, New York City.

unidentifiable by local clinical laboratories. State agencies follow the incidence of disease in the state and communicate this information to the national agency, the Centers for Disease Control and Prevention (CDCP) in Atlanta, Georgia. The CDCP is a subunit of the United States Public Health Service (USPHS) within the Department of Health and Human Services.

The epidemiology unit of the state health department requires that licensed physicians report cases of selected communicable diseases that occur in patients under their care. Listed in **Table 30.3** are the communicable diseases reported to the North Carolina Division of Epidemiology. Other states have similar requirements. Many of these diseases are then reported to CDCP. It is also required that restaurants and other food or drink establishments report outbreaks of food-borne disease in employees and customers to the local Health Department. Many diseases, such as cholera, plague, and hepatitis, must be reported within 24 hours so that immediate control measures can be implemented.

The effectiveness of our public health institutions is very evident. Potential epidemics are tracked, and in many cases, spread of disease is curtailed. An example of this was the potential for measles (rubeola) outbreaks on college campuses during recent years. The vaccine employed during the infancy of most college students

at that time did not confer a long-lasting immunity. There was the risk of an epidemic in this population, and after a few cases were documented, it was made mandatory that all students lacking proof of the proper immunization report for vaccination. This prevented a potential harmful outbreak, as rubeola can be a serious infection in young adults.

CONTROL MEASURES FOR COMMUNICABLE DISEASES

As we examine diseases and their reservoirs (see Table 30.1), it is evident that some diseases are more readily controlled than others. Diseases that can be prevented by immunization during infancy are controllable but only when adequate vaccination programs are followed. The same applies to most water-borne infectious diseases where adequate water treatment can virtually eliminate the disease. Diseases transmitted in milk can be controlled by maintaining healthy dairy herds and the pasteurization of milk before consumption or prior to conversion to cheese or other milk products.

Many of the respiratory infections, particularly those passed orally, are quite difficult to control. A viral influenza outbreak may spread unchecked around the world. Cold viruses pass rapidly through human pop-

---

Table 30.3	Communicable diseases reported to the epidemiology division of a state health department (North Carolina) and to the Centers for Disease Control and Prevention

AIDS	Granuloma inguinale	Q fever
Anthrax	*Haemophilus influenza* (invasive)	Rabies, human
Botulism	Hemolytic-uremic syndrome	Rocky Mountain spotted fever
Brucellosis	Hepatitis A	Rubella
Campylobacter infection	Hepatitis B	Rubella congenital syndrome
Chancroid	Hepatitis B carriage	Salmonellosis
Chlamydial infection	Hepatitis C	Shigellosis
Cholera	HIV infection	Streptococcal infection, group A
Cryptosporidiosis	Legionellosis	Syphilis
Cyclosporiasis	Leptospirosis	Tetanus
Diphtheria	Lyme disease	Toxic shock syndrome
Dengue	Lymphogranuloma venereum	Toxoplasmosis
E. coli O157:H7 infection	Malaria	Trichinosis
Ehrlichiosis	Measles (rubeola)	Tuberculosis
Encephalitis	Meningococcal disease	Tularemia
Enterococci (vancomycin resistant)	Meningitis, pneumococcal	Typhoid
Food-borne diseases	Mumps	Typhoid, carrier
Clostridium perfringens	Nongonococcal urethritis	Typhus, epidemic
Staphylococcal	Plague	Vibrio infection
Bacillus cereus	Polio, paralytic	Whooping cough
Gonorrhea	Psittacosis	Yellow fever

Source: Information provided by the North Carolina Health Services.

ulations. Humans are gregarious, and infectious droplets resulting from talking, coughing, or a partly stifled sneeze are hazardous to everyone coming in contact with the infected individual. The influenza virus itself can be exceedingly virulent and constantly changes such that immunity to past epidemics does not ensure protection against the next. There is no immunization or treatment for the common cold.

In the following section we will discuss some potential procedures for the control of infectious disease. Difficulties one might encounter are also considered.

Reservoir Control

Control of the reservoirs that sustain infectious microorganisms is effective in some cases but not in others. Domestic cattle have been a reservoir for human infections, but this source has been substantially eliminated by rigorous control measures. *Brucella abortus* causes spontaneous abortion in cattle, and vaccination prevents the spread of the disease among animals and to humans. Cattle are also immunized for bovine tuberculosis to prevent transmission of this infectious microorganism to humans through milk. There are strict rules on the extermination of all cattle that may become infected with lumpy jaw, a disease caused by *Actinomyces bovis*. The elimination of diseased sources is a practical mechanism for preventing the spread of an infectious disease. It is, however, essential that **all infected sources** in the potential reservoir be eliminated, otherwise the disease can erupt and travel rapidly through a nonimmune susceptible population.

The reservoir for a number of significant diseases is rodents, so domestic rats must be destroyed or driven from populated areas to control diseases spread by them. Rats are one wild animal that have evolved with a unique ability to live among humans, and the rats' destructive habits, appetite for grain, and disease-causing potential are all a threat to human welfare. During 1993, a viral infection caused a number of deaths in the southwestern United States that were traced to a hantavirus that was spread by inhalation of aerosols of rat urine and feces. The living conditions of the victims were a primary contributor to the infection.

Quarantine

Animals are frequently quarantined before they are transported from region to region or country to country. Great Britain has eliminated rabies from the isles and has imposed strict regulations on the import of animals to ensure that the disease will not return. The United States enforces a quarantine on cattle and other animals before import to ensure that they are free from infectious diseases. Quarantine was once commonly practiced in the United States at the local level. If a family had a member with scarlet fever or certain other transmissible diseases, a notice would be posted outside the door of their home. The notice warned visitors that an infected individual resided there, and they should not visit lest they carry the disease elsewhere. In cases where the patient was a youngster, all siblings were barred from attending school.

Prior to the availability of antibiotics, tuberculosis patients were confined to a state-controlled sanitarium. Hawaii and other tropical areas had leprosy colonies where individuals infected with the disease were confined. Quarantine of humans in the United States, except in isolation wards of hospitals, has been discontinued. However, by international agreement, some diseases are quarantinable. These diseases include cholera, plague, yellow fever, typhoid, and relapsing fever. Those infected with one of these diseases may be barred from moving from country to country.

Food and Water Measures

Food inspection is a common practice, and among the foodstuffs subject to government inspection are abattoirs and meat processing plants. Milk pasteurization follows mandated procedures and has curtailed the spread of brucellosis, bovine tuberculosis, typhoid, listeriosis, and other diseases. Breakdowns in the pasteurization process have led to common source epidemics. In recent years, there were two listeriosis (*Listeria monocytogenes*) epidemics: one in Massachusetts (1983) resulting from ingestion of insufficiently pasteurized milk and the other in California (1985) from ingestion of Mexican-style cheese, which was apparently made from contaminated milk. There were 104 deaths in these two episodes. Cases such as this are testimony to the need for strict quality control in the food we consume. Foods sold to the general public rightfully must meet reasonable public health standards for quality and sanitation. Despite these precautions, we continue to have periodic incidents that cause considerable human misery.

Water treatment and purification (in the developed nations) have virtually eliminated the threat of cholera, typhoid, and other waterborne diseases. Sewage treatment plants in our municipalities have been effective in processing domestic sewage and in preventing the spread of diseases from this source. A major improvement has been in separating storm runoff that enters street sewers from the domestic sewage systems. In years past, raw domestic sewage and storm water bypassed the sewage disposal plant during heavy storms. This resulted in significant local pollution from raw sewage.

Human and Animal Vaccination

Vaccination has been successfully employed in controlling a number of fearful diseases. There are mandatory laws among the various states that require proof of vaccination before a child can enter school. The basic immunization program is initiated in an infant at about 2 months of age, and children in the United States are immunized against 11 different diseases (**Table 30.4**). These immunizations continue for the first 6 years of life.

There is growing concern that immunization programs are not begun at an early-enough age, particularly among the disadvantaged. The concern is that this delay might establish a significant pool of children up to 5 years of age who are at risk of contracting and disseminating communicable diseases. Extensive immunization of children has been successful to date in virtually eliminating many childhood diseases, such as smallpox, diphtheria, pertussis, and poliomyelitis, and this may be giving the adult population a false sense of security. Some adults are inadequately immunized and potentially can contract childhood diseases. Outbreaks in uninoculated preschoolers can potentially cause limited outbreaks among adults.

In recent years, there has been an increased incidence of autism among young children. Autism is a mental disorder that originates in infancy and is characterized by an inability to develop socially, language dysfunction, and other problems. The basis for the malady is not known. An English physician has made the claim that autism is due to the immunization programs established for infants. He maintained that the MMR series overburdens the immune system and results in autism. Examination of data collected worldwide affirms that the disorder is not related to immunization programs.

Herd immunity is the immunity engendered to a communicable infectious disease when a major segment of the total population is immune to that disease. The greater the percent of a total population immune to an infectious agent, the greater the chance that the entire population will be protected. It is not necessary that 100% of a population be immune for a disease to be eliminated. In practice, few vaccines actually confer immunity to all the individuals vaccinated. Generally, herd immunity is conferred if 70% or more of the population is immune. For highly contagious diseases, such as viral influenza, the percentage of the total susceptible population that must be immune to confer herd immunity is close to 95%.

Vaccination of pets for rabies has been effective in controlling this disease in domestic animals, and generally they are not a threat to humans. Unfortunately, there is a large reservoir of the disease in skunks, raccoons, and other carnivores. Rabid animals among these species continue to be a threat to humans.

Antibiotic Resistance

The development of antibiotic resistance in pathogenic organisms is a major public health concern. One somewhat controversial area in this regard is feeding antibiotics to cattle to increase their growth rate. About one-half of the annual production of antibiotics in the United

Table 30.4	The recommended immunization schedule for infants and young children in the United States
Age	**Vaccine Employed**
Birth	Hepatitis B
2 months	Diphtheria; pertussis; tetanus (DPT)
	Hemophilus B (Hib)
	Poliomyelitis (OPV)
4 months	DPT; OPV; Hib
	Hepatitis B
6 months	Hepatitis B
	DPT; OPV; Hib
12–15 months	DPT; Hib; chicken pox, measles, mumps, rubella (MMR)
4–6 years	OPV; DTP; MMR

States is incorporated into feed for poultry, swine, and cattle. Tetracycline and neomycin are two antibiotics that have been added to animal feed. As the animals consuming these antibiotics generally harbor potential pathogens, such as *Salmonella* sp., the indiscriminate use of antibiotics in feed could select for antibiotic-resistant strains. The animal and pharmaceutical industries do not consider this a danger, and it is not restricted. One antibiotic, penicillin, that was added to animal feed in the past has been restricted and is no longer used for this purpose.

The Centers for Disease Control and Prevention has estimated that 70% of the salmonellosis outbreaks that occurred in the years 1971 and 1983 involved resistant strains that came from food animals. In a documented case, a unique strain of *Salmonella newport*, identified by plasmid profile, was involved. Eighteen cases of salmonellosis were traced to one farm in South Dakota where the antibiotic-resistant strain apparently originated. Studies in Denmark clearly showed that resistance to four antimicrobials declined following bans on their incorporation into animal feed. For example, resistance to the antibiotic vancomycin in *Enterococcus faecium* strains isolated from chickens declined from 72.7% in 1995 to 5.8% in 2000. Incorporation was banned in 1995.

The antibacterial triclosan is a common ingredient of soap and lotions and has also been impregnated into cutting boards, toys, and clothing. The goal is to create bacteria-free materials. However, the result of such widespread nonmedical use is to make us more vulnerable, not less, to infectious agents. It has now been established that bacteria have developed resistance to triclosan, and this in turn engendered resistance to antibiotics such as tetracycline and erythromycin.

WORLD HEALTH

The World Health Organization (WHO) was established in 1948 through the United Nations and is now head-quartered in Geneva, Switzerland. The WHO is committed to the control of diseases and promoting health for the people in over 100 member countries and was instrumental in the eradication of smallpox (**Box 30.6**). The organization is involved with population control, availability of food, and efforts to curtail disease through education. The WHO provides information on developing safe drinking water sources. Worldwide immunization programs for combatting such diseases as diphtheria, poliomyelitis, and tuberculosis are among the goals of the WHO. It is the intention of the WHO that means be found to bring diseases such as malaria and leprosy under control.

Problems in Developing Nations

Diseases that were once prevalent in the Western World but are now controlled by drugs or immunization remain a concern in developing countries. As previously mentioned, infectious diseases cause nearly 50% of the deaths in developing nations, but fewer than 8% of deaths in developed countries. This higher death rate in the developing countries is due to inadequate sanitation, nutritional deficiencies, and substandard housing. The lack of medical care and immunization programs also contribute to a high mortality rate.

Many of the diseases in developing nations result from a lack of adequate supplies of clean water (**Table 30.5**). Cholera, a serious problem in Southeast Asia and a threat in some South American countries, could be eliminated by providing populations with safe drinking water. Better sanitation practices and safe food handling would also decrease the incidence of salmonellosis and

Table 30.5	A number of the diseases that are a major concern in developing nations
Diseases	**Region Where Disease Occurs**
Cholera	Southeast Asia, Central Africa
Salmonellosis	Most
Shigellosis	Most
Yellow fever	Central & South America
Encephalitis	Most
Typhus	Middle East
Typhoid fever	Middle East, Central & South America
Tuberculosis	Most

shigellosis. Typhoid fever is another infectious disease that would be virtually eradicated by the availability of safe water supplies. It is estimated that up to 10% of the population in Latin America are typhoid carriers. Inadequate treatment of sewage is a potential source of typhoid infection from the ever-present carrier reservoir.

Problems in Developed Nations

The developed countries are not without health problems caused by infectious agents. The problems with foot-and-mouth disease in England and the other European countries may well come to the United States. Mad cow disease is also a growing concern (see Chapter 14). In both developed and developing countries, HIV is increasing at an alarming rate and is now considered a pandemic (Figure 30.2). Gonorrhea persists despite the availability of an effective cure for the disease; an estimated 1 of 35 young adults will contract gonorrhea each year. Legionnaire's disease and shigellosis are endemic, and shigellosis is particularly contagious among young

---

Milestones Box 30.6	A Success Story: The Eradication of Smallpox

Smallpox is a dreaded disease that was eradicated by a successful worldwide immunization program. Smallpox is an ancient disease, and the millions of deaths caused by this disease are engraved in the 3,000-year record of human history. Through the first quarter of the twentieth century, thousands of cases occurred annually in the United States. The disease was eliminated in the United States by about 1960 through a long-term extensive vaccination program in preschool children. The World Health Organization (WHO) initiated a worldwide eradication program in 1966. As the disease was eliminated in the developed nations, the campaign's thrust turned to India, Africa, South America, and other developing countries. By 1977, the world appeared free of smallpox. In 1980 the WHO declared the disease eradicated. As a result, children are no longer vaccinated for this disease.

There are several reasons why the eradication of smallpox was so successful and that equivalent programs for other diseases might be less so. Among the advantages of the smallpox vaccine were that the vaccine was low cost, the vaccine raised a high level of immunity, the vaccine did not require refrigeration, and inoculation was by pinpricks with no need for syringes. Thus, individuals in remote villages could be vaccinated. Not all immunization programs are so amenable.

Table 30.6	Immunization recommended or required for travel to developing countries	
Disease	**Traveling to**	
Vaccination Required		
Cholera	Southeast Asia, Albania, Malta, Central Africa, South Korea	
Yellow fever	Central and South America, African countries	
Vaccination Recommended		
Plague	Rural areas of Africa, Asia, and South America	
Serum hepatitis	Africa, Indochina, Russia, Central and South America	
Typhoid fever	Africa, Asia, Central and South America (Specific areas)	
Diphtheria; polio; tetanus; measles; mumps; rubella	Most U.S. citizens already immunized	

children in child-care centers. Lyme disease and Rocky Mountain spotted fever continue to be a concern with no adequate control measures in sight. Vigilance is essential, and the epidemiology programs in the various states and CDCP are doing a commendable job in monitoring communicable diseases in the United States.

Problems for Travelers

When people travel from developed to developing countries, they can encounter potential health problems. Immunizations beyond those routinely given are required or recommended for travel by United States citizens to many areas of the world (**Table 30.6**). Diseases for which there are no immunizations, such as dengue fever, malaria, and typhus are also prevalent in some developing countries. Travelers should check with health authorities before journeying to countries where communicable diseases not encountered in the United States are a threat.

CLINICAL AND DIAGNOSTIC METHODS

Early identification of the causative agent of an infectious disease in individuals and the population at large is of utmost importance. Epidemics can be tracked effectively only where the etiological agent is correctly identified. Therapeutic agents, such as antibiotics, can be chosen more efficiently when the infectious agent involved is known and antibiotic resistance/susceptibility have been determined. Identifying etiological agents can minimize the severity of a disease and shorten the recovery time if proper treatment is immediate and appropriate. In many epidemiological studies, the characterization of the microbe must be carried beyond species to the strain involved. Recent technological advances have provided the clinical laboratory with instruments and techniques that improved the accuracy and shortened the time required for the identification of pathogens.

Searching for Pathogens

There are a number of factors that predispose an individual to the diseased state. Significant among these are the general health of the host, previous contact with the microbe, past medical history, exposure to toxic agents or chemicals, and traumatic or other insults not of microbial origin. These and other elements have considerable bearing on whether an individual will contract a particular disease.

The term *pathogen* can be applied to few organisms if one considers a pathogen to be an organism that **always** causes clinical symptoms. The indigenous microbiota of a human is a varied population, and many of these organisms may cause symptoms of disease under appropriate conditions. Lowered resistance can result in a clinical infection, and it is often difficult to determine which of many organisms present in the diseased state is the one responsible. Clinical microbiologists are trained to sort through the organisms present in a specimen and make decisions on the most likely source of clinical symptoms.

Obtaining Clinical Specimens

The accuracy of a diagnosis is quite dependent on the quality of the specimen delivered to the clinical laboratory. The specimens generally obtained from a patient would be one or more of the following: blood, urine, feces, sputum, biopsy tissue, cerebrospinal fluid, or pus/exudate from a wound. Throat or nasal swabs and fluid aspirated from an abscess could also be submitted to the clinic for diagnosis. Care must be taken to prevent contamination of the specimen by extraneous microorganisms after it is taken. In addition, the specimen must be sufficiently large that all desired tests can be accomplished on that single specimen.

Blood Presence of bacteria in blood (bacteremia) is generally an indication that there is a significant focus of infection somewhere in the patient's body. If a few bacteria enter the bloodstream, these organisms will probably be cleared quickly by natural host defenses. However, when the number of microorganisms present in blood is significant (several per ml) an acute infection is probable somewhere in the body and is shedding microorganisms into the bloodstream. This would lead to a general septicemia, which is rapid propagation of pathogens in the bloodstream. Septicemia results in a fever, chills, and shock. Blood samples for analysis are always taken aseptically with a sterile syringe and the blood should be delivered into a bottle containing a suitable medium and anticoagulant to prevent clots. Clots may entrap bacteria and make isolation of the microbe difficult.

One part of the blood sample is placed in the culture medium to be incubated aerobically, another anaerobically. These inoculated cultures would be placed in an incubator that automatically monitors CO_2 production. Should growth occur, as indicated by generation of CO_2, either aerobically or anaerobically, the microorganism is isolated in pure culture, it is identified, and antibiotic sensitivity is determined. The organisms most commonly associated with blood infections are *Staphylococcus aureus*, *Streptococcus pyogenes*, *Pseudomonas aeruginosa*, and enterics.

Urinary Tract Urinary tract infections are frequently caused by gram-negative bacteria similar to those that are part of the natural human microbiota. Because urine itself will support the growth of bacteria, care must be taken to ensure that once a specimen is taken it must not stand for any length of time at room temperature before analysis. If not analyzed immediately, the sample should be refrigerated. Bacteria that are generally involved in urinary tract infections are gram-negatives such as *Escherichia coli* and species of *Klebsiella*, *Proteus*, or *Enterobacter*.

Urine samples can be analyzed by a direct count of organisms present or by spreading an aliquot over the surface of a MacConkey agar plate and a blood agar plate employing a calibrated loop. The number of colonies that develop on the blood agar plate is a measure of the number of organisms in the urine specimen. If there are 10^5 or more organisms per milliliter of urine,

a urinary tract infection is indicated. MacConkey agar is a selective medium for gram-negative bacteria. It contains bile salts and crystal violet that inhibit growth of gram-positives, the sugar lactose, and neutral red as a dye indicator. *Enterobacter* and *Escherichia* ferment lactose, and the colonies take up neutral red, imparting a reddish color. *Proteus*, *Salmonella*, and *Shigella* do not ferment lactose, and they form white or clear colonies.

Several different media can be employed in characterizing gram-negative bacteria encountered in the clinical laboratory (**Table 30.7**). These differential media would effectively identify the microorganism involved in the infection. The presence of gram-positive staphylococci or streptococci in the original sample would be determined by examining the blood agar plate. These organisms are less frequently present than are the gram-negatives in urinary tract infections.

Other procedures are followed when the infection might be caused by a sexually transmitted infectious microorganism. Gonorrhea is one of the most common sexually transmitted diseases (STDs) among young adults, and the etiologic agent is *Neisseria gonorrheae*. A Gram stain of the purulent urethral discharge in males can be a rapid and reasonably accurate diagnostic procedure. If the symptoms are less obvious, the clinical specimen from males and females must be cultured on selective and nonselective media. A primary medium is the Thayer-Martin medium, which contains the antibiotics vancomycin, nystatin, and colistin. These antibiotics inhibit the growth of many microorganisms commonly present in such specimens but not the growth of *N. gonorrheae*.

Some pathogenic strains of *N. gonorrheae* may be inhibited by vancomycin, so nonselective chocolate agar should also be inoculated. Chocolate agar is prepared from heated blood and is a source of growth factors for *N. gonorrheae*. It also absorbs toxic material that may be present in a rich medium that would otherwise inhibit the growth of the gonococcus. The inoculated plates

Table 30.7	Some agar base media that can be employed to differentiate clinically significant gram-negative bacteria by colonial appearance					
Medium	*Enterobacter aerogenes*	*Escherichia coli*	*Salmonella* sp.	*Shigella* sp.	*Proteus* sp.	Gram+
Bismuth sulfate	Mucoid silver sheen	Little growth	Black with metallic sheen	Inhibited to brown	Green	No growth
Eosin methylene blue	Pink	Purple with black centers	Colorless	Colorless	Colorless	No growth
Salmonella - shigella	Cream to pink	No growth	Colorless	Colorless	Colorless	No growth
Sodium azide agar	No growth	No growth	No growth	No growth	No growth	Growth
MacConkey agar	Pink to red	Pink to red	Colorless	Colorless	Colorless	No growth

should be incubated in an atmosphere of 5% CO_2. Commercial DNA probes have been developed for a rapid diagnosis of gonorrhea.

Specimens obtained in suspected cases of syphilis would be the exudates from open lesions or material from lymph nodes in the affected regions. The presence of spirochetes as determined by dark-field microscopy is a rapid and generally effective diagnostic tool. Staining is not generally effective, as *Treponema pallidum* stains poorly and is only about 0.2 μm in diameter—near the limit of light resolution. The organism cannot be cultured on laboratory media. Identification can be by fluorescein-labeled antitreponemal antibodies or a slide flocculation test (see later discussion). The flocculation tests are based on the presence of antibody to a specific cardiolipin antigen in the serum of syphilis patients.

The common sexually transmitted pathogen *Chlamydia trachomatis* can be identified by inoculating exudate into cell culture and monitoring for growth of the intracellular pathogens. There are several rapid inexpensive tests for detecting *Chlamydia* in urine samples.

Tissue/Abscess   Tissues from biopsies and material from skin lesions or wounds are streaked on blood agar and other rich media. Duplicates should be prepared for incubation aerobically and anaerobically. Care must be taken in obtaining and handling such specimens, as infections of this sort are frequently caused by strict anaerobes that may be adversely affected by contact with atmospheric oxygen. Exudate from infected areas can be collected in a syringe by aspiration and taken directly to the clinic for diagnosis. Such samples may be examined by microscopy after staining with fluorescent antibody. The type and the appearance of lesions and the history of the patient may suggest possible etiological agents and indicate which fluorescent antibody should be applied. Anthrax, plague, buboes, and tularemia lesions have distinct characteristics, and confirmation can be made quickly by serological techniques.

Fecal   The pathogens most often associated with fecal specimens are food- or water-borne microorganisms, and among these are *Vibrio cholerae*, *Campylobacter jejuni*, and species of *Salmonella* and *Shigella*. Careful handling of fecal specimens is essential, for when left at room temperature for a brief period of time, the specimens can quickly become acidic. This acidity will kill pathogens such as *Salmonella* or *Shigella*. To overcome this, fecal samples are generally collected in a buffered medium and delivered quickly to the clinic for diagnosis. The selective medium of choice for fecal specimens is MacConkey or eosin-methylene-blue agar (Table 30.7). Samples should also be streaked on blood agar to deter-

mine whether gram-positive pathogens, such as staphylococcus or streptococcus, might also be present.

Sputum   Sputum and material obtained from the upper respiratory tract should be streaked on blood or chocolate agar. Blood agar is useful for culturing *Streptococcus pyogenes*, *Streptococcus pneumoniae*, and *Staphylococcus aureus*. *Neisseria meningitidis* and *Haemophilus influenza* can be detected on chocolate agar. Acid-fast staining of smears and serological tests can be applied where *Mycobacterium tuberculosis* might be involved in the infection.

Cerebrospinal   Cerebrospinal fluid specimens are cultured on rich media such as blood and chocolate agar. Organisms involved in meningitis (*Neisseria meningitidis*, *Streptococcus pneumoniae*, and *Haemophilus influenzae*) grow on these rich media. Because meningitis is a potentially fatal infection, an immediate diagnosis is mandatory. ELISA tests (discussed on p. 784) are a major tool in determining the nature of the pathogen that might be involved in infections of this type because this permits an immediate diagnosis and start of treatment.

IDENTIFYING PATHOGENS

The major goal of a clinical microbiology laboratory is a prompt, precise identification and characterization of the pathogen involved in an infection. In addition, antibiotic resistance/sensitivity of the causative agent is a major element in selecting a proper treatment. If the infecting strain is resistant to the antibiotic generally employed in such cases, successful treatment depends on administering an alternative antibiotic.

Specimens delivered to the clinical laboratory for microbiological analysis may follow one or more routes:

- The specimen may be examined microscopically following staining or examined directly with a dark-field or phase-contrast microscope.
- A specimen can be streaked or cultured on an enrichment, selective, or differential medium (**Table 30.8**).
- The specimen may be subjected directly to serological, immunofluorescence, ELISA, or other direct diagnostic procedures.

We will discuss each of these in turn. Direct microscopic examination of stained clinical material can be of value with selected specimens. This will give a preliminary diagnosis that can then be confirmed by isolation and identification of the pathogen. In suspected cases of tuberculosis, sputum is subjected to the acid-fast stain, a diagnostic tool for bacteria that have a waxy coat. For suspected gonorrhea, cervical scrapings from females and urethral discharge from males may be Gram stained to visualize the typical gonococcus. There may also be

Table 30.8	Basic types of media that would be employed in the clinical microbiology for isolation of bacterial cultures from specimens

Characteristics media—These media test bacteria for specific metabolic activities, enzymes, or growth characteristics.

Citrate agar—Tests ability to utilize sodium citrate as sole carbon source.

TSI (Triple-sugar-iron) agar—Three sugars are lactose, glucose, and sucrose together with sulfates and a pH indicator. Used in slants to determine relative use of each sugar and generation of sulfide.

SIM (Sulfide, indole, motility) agar—Tests for production of sulfide from sulfate, indole from tryptophan and motility.

Differential and **Selective media**—Contain selected chemicals.

Brilliant green agar—A dye that inhibits gram-positive bacteria, thus selecting for Gram negatives.

Sodium tetrathionate broth—An inhibitor for normal inhabitants of intestinal tract, favors *Salmonella* and *Shigella* species.

MacConkey agar—Contains bile salts and crystal violet that inhibit gram-positives and many fastidious gram-negative organisms, favoring Enterobacteria.

Mannitol salt agar—Contains 7.5% NaCl that is inhibitory to most organisms and favors growth of staphylococci.

Eosin methylene blue—Partly inhibits gram-positives. Eosin gives *Escherichia coli* a metallic greenish sheen. *Enterobacter aerogenes* colonies are pink. Other species are less pigmented.

Enrichment media—Contains blood, serum, meat extract or other nutrients that favor the growth of fastidious bacteria, particularly when present in low numbers. Often employed for clinical specimens such as cerebrospinal fluid.

observable polymorphonuclear leukocytes that contain the characteristic diplococcal *Neisseria gonorrheae* cells. If leprosy is suspected, *Mycobacterium leprae* is identified by direct observation of acid-fast stained specimens from leprous lesions because the organism cannot be grown outside the host. A preliminary diagnosis of syphilis is feasible by dark-field microscopy. Examinations for animal parasites in blood, feces, and so on are generally done by direct microscopic examination of properly collected specimens.

Staining and viewing of some specimens microscopically as they are received in the clinical laboratory is not necessarily productive. The specimen from a given source will generally contain an array of microorganisms, and microscopic examination will not indicate which is the agent of infection. In such cases, samples are inoculated directly into or onto an appropriate growth media. The medium employed will depend on the source of the specimen.

After examining growth on selective or differential media (Table 30.8), an experienced clinical microbiologist can select the best pathway to follow in identification. This would be influenced by microscopic examination, colonial morphology, and specimen source. Microorganisms present in colonies from primary enrichment, selective, or differential media can be subjected to growth-dependent tests for identification (**Tables 30.9** and **30.10**). The media listed have proven to be very accurate in identifying various potential pathogenic microorganisms. These organism would also be tested for antibiotic sensitivity/resistance.

Growth-Dependent Identification

Identification of various pathogens can be achieved by growth-dependent methods, and the procedure

Table 30.9	Growth and colonial characteristics of some commonly isolated organisms on differential agar media

Organisms	Eosin-Methylene-Blue		Mannitol Salt	
	Growth[a]	Color of Colony	Growth of Colony	Color
Enterobacter aerogenes	++	Pink, no sheen	Inhibited	
Escherichia coli	+++	Purple-green metallic sheen	Inhibited	
Klebsiella pneumoniae	++	Green metallic sheen, mucoid colony	Inhibited	
Salmonella typhimurium	++	Colorless	Inhibited	
Staphylococcus aureus	Inhibited		+++	Yellow
Staphylococcus epidermidis	Inhibited		++	Red

[a] ++ good growth; +++ excellent growth

Table 30.10	Major diagnostic tests employed to differentiate pathogenic bacteria in a clinical laboratory	

Test	Purpose	Potential Application
Acid-fast stain	Tests for organisms with high wax (mycolic acid) content in cell surface that stain with hot carbolfuchsin and cannot be decolorized with acid alcohol.	*Mycobacterium tuberculosis* is acid-fast.
Catalase	Enzyme decomposes hydrogen peroxide $H_2O_2 \rightarrow H_2O + O_2$.	*Staphylococcus* from *Streptococcus*.
Citrate	Transports and utilization as carbon source.	Classification of enteric bacteria.
Coagulase	Causes clotting of plasma.	*Staphylococcus aureus* from saprophytic staphs.
Decarboxylases	Tests for ability to decarboxylate amino acids such as lysine, ornithine, or arginine.	Classification of enteric bacteria.
Esculin	Tests for cleavage of a glycoside.	Separate streptococci.
β-galactosidase	Demonstrates an enzyme that cleaves lactose → glucose + galactose.	Separates enterics and identifies pseudomonads.
Gelatin liquefaction	Enzymatic hydrolysis of gelatin.	Identify clostridia and others.
Gram stain	Used as a first and primary differential test.	
Hemolysis	Hemolysis of red blood cells; α-hemolysis, indistinct zone of hemolysis, some greenish to brownish discoloration of medium; β-hemolysis, clear, colorless zone around colonies.	Pathogenic streptococci.
Hydrogen sulfide	Demonstrates H_2S formation by sulfate reduction or from sulfur-containing amino acids.	Identify enterics.
Indole	Determines hydrolysis tryptophan to indole.	Separate enterics.
KCN growth	Tests for ability of microorganisms to grow in the presence of cyanide (inhibits electron transport).	Aerobes/anaerobes.
Lipase	Presence of the enzyme that cleaves ester bonds of fats yielding fatty acids + glycerol.	Separate clostridia.
Methyl red	Checks for acid production from glucose via mixed-acid fermentation.	Separate enterics.
Motility	By microscopic examination or diffusion through soft agar, shows the ability to move.	Motile strains.
Nitrate reduction	Reduction of $NO_3 \rightarrow NO_2$ (nitrate employed as terminal electron acceptor).	Enterics and others.
Oxidase test	Demonstrates presence of cytochrome *c*, which oxidizes an artificial electron acceptor.	Enterics from pseudomonads.
Phenylalanine deamination	Deaminates the amino acid to phenylpyruvic acid.	Proteus group.
Protease	Tests for ability to digest casein.	Separate *Bacillus* species.
Sugar utilization	Growth on pentoses, hexoses, or disaccharides, producing acid and gas.	Many.
Urease	Detects an enzyme that splits urea to $NH_3 + CO_2$.	Separate enterics.
Voges-Proskauer	Detects acetoin as product of glucose fermentation and indicates neutral fermentation.	Differentiate bacillus species and enterics.

followed is generally based on preliminary identification of the microorganism involved.

Commercial diagnostic packets containing various media and selected differential tests have been developed. Among the manual rapid test systems is the API 20E system for the identification of enteric bacteria. These systems (**Figure 30.3**) are compact, they are simple to

inoculate, and many clinical isolates can be analyzed in a short period of time. The sugars and diagnostic tests in these manual systems can be varied and used to confirm the identity of a microorganism based on its presumptive identification. Compact test kits have replaced many of the individual tube tests (**Table 30.11**) that were routine in clinical laboratories in years past and are still a part of

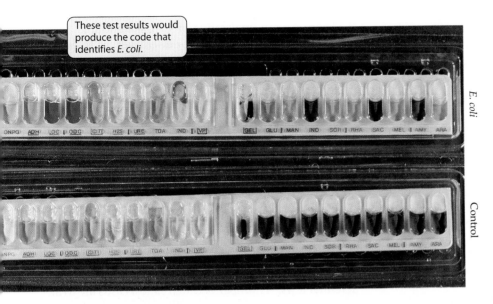

These test results would produce the code that identifies *E. coli*.

E. coli

Control

Figure 30.3 Identification of pathogens

The API 20E system for rapid identification of enteric organisms. There are 21 tests and each would yield a positive or negative result. A number is calculated for each group of three tests, and the resulting seven-digit code will identify most enteric bacteria. Photo ©E. Chan/Visuals Unlimited.

the laboratory section of most general microbiology courses.

Serologic Identification

Culturing of microorganisms including viruses and selected bacteria is not always possible or practical in the clinic laboratory. For rapid diagnosis of many infectious agents, there are commercial test kits that can detect presence of an antigen or antibody. These **serologic** or **immunologic systems** can identify pathogens in clinical specimens rapidly and accurately without culturing of the infectious agent. Rapid immunologic test kits are available to detect: *Haemophilus influenzae* type b, *Neisseria meningitidis* from cerebrospinal fluid, *Streptococcus pneumoniae*, *Helicobacter pylori*, *Bacteroides fragilis*, respiratory syncytial virus, herpes type 1 and 2, HIV, and other suspected pathogens.

| **Table 30.11** | **Growth-dependent tests that differentiate members of the enterobacteria** |

Tests	*Citrobacter freundii*	*Edwardsiella tarda*	*Enterobacter aerogenes*	*Escherichia coli*	*Klebsiella pneumoniae*	*Proteus vulgaris*	*Providencia alcalifaciens*	*Salmonella paratyphi*	*Salmonella typhi*	*Serratia marcescens*	*Shigella dysenteriae*	*Yersinia pestis*
Indole	−	+	−	+	−	+	+	−	−	−	±	−
Methyl red	+	+	−	+	−	+	+	+	+	−	+	+
Voges-Proskauer	−	−	+	−	+	−	−	−	−	+	−	−
Ornithine decarboxylase	−	+	+	±[a]	−	−	−	+	−	+	−	−
Motility	+	+	+	+	−	+	+	+	+	+	−	−
Gelatin liquefaction	−	−	−	−	−	+	−	−	−	+	−	−
KCN (growth in)	+	−	+	−	+	+	+	−	−	+	−	−
Glucose acid	+	+	+	+	+	+	+	+	+	+	+	+
Glucose gas	+	+	+	+	+	±	±	+	−	±	−	−
Lipase	−	−	−	−	−	+	−	−	−	+	−	−
NO$_3^-$→NO$_2^-$	+	+	+	+	+	+	+	+	+	+	+	+
Lactose utilization	±	+	±	+	+	−	−	−	−	−	−	−
H$_2$S on TSI	±	−	−	−	−	+	−	−	+	−	−	−
Citrate	+	−	+	−	+	−	+	−	−	+	−	−

[a]± most strains positive.

In some cases, diagnosis may rely on the presence of antibody in the serum of a patient, but the detection of antibody does not necessarily distinguish between an active and a previous infection. There must also be sufficient time between the onset of the disease and application of serological tests to permit the formation of circulating antibody, a period of 3 to 7 days.

Automated Identification Systems

Fully automated systems have been developed for the identification of many infectious microorganisms. These systems provide a constant monitoring of growth dependent reactions and the susceptibility of the microorganism involved to antimicrobial agents. Such identification systems have many advantages over the manual manipulations that have been employed for many decades. Among these are:

1. There is a minimalization of sample handling.
2. The chance of human error decreases, as there are fewer manipulations.
3. The system delivers results continuously and each test specimen is read and recorded separately. There are no batch readings, so results of individual tests are available at the earliest time possible.
4. These systems have broad applicability and can accommodate updating as new infectious agents, chemotherapeutics, and technical capabilities emerge.
5. They require very little space.

There are a number of systems available, and the one depicted in **Figure 30.4A** has been selected as an example. The basic identification module for the system is a clear plastic card (Figure 30.4B) that has 30 wells containing dried medium and/or reagent chemicals for differential tests. A number of different test cards are avail-

(A)

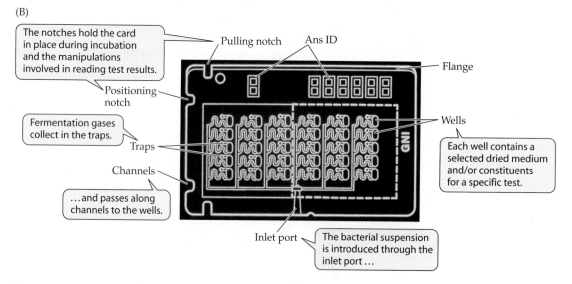

(B)

The notches hold the card in place during incubation and the manipulations involved in reading test results.

Pulling notch   Ans ID

Flange

Positioning notch

Fermentation gases collect in the traps.

Traps

Wells

Each well contains a selected dried medium and/or constituents for a specific test.

Channels

...and passes along channels to the wells.

Inlet port

The bacterial suspension is introduced through the inlet port ...

GNI

Figure 30.4 Automated identification system for pathogens
(A) The automated Vitek system for identification of clinical isolates. The monitor records visual results, and the printer records chartable patient reports. (B) Test card that is placed in the automated Vitek system. The card is clear plastic, measuring 2.25 × 3.5 inches and with 30 individual wells. Courtesy of bioMerieux VITEK, Inc.

Table 30.12	Identification cards available for automated identification of commonly encountered organisms	
Test Card	**Range**	**Genera Detected (speciation within the genus is generally determined)**
GNI	Enterobacteriaceae, Vibrionaceae, glucose nonfermenters	*Achromobacter, Acinetobacter, Citrobacter, Eikenella, Enterobacter, Escherichia, Flavobacterium, Hafnia, Klebsiella, Pasteurella, Proteus, Providencia, Pseudomonas, Salmonella, Serratia, Shigella, Vibrio, Yersinia*
GPI	Coagulase positive and negative, staphylococci, enterococcus, β-hemolytic streptococci, *Corynebacterium, Listeria, Erysipelothrix*	*Corynebacterium, Enterococcus, Erysipelothrix, Listeria, Staphylococcus, Streptococcus* (viridans group)
YBC	Clinically significant yeasts	*Candida, Cryptococcus, Geotrichum, Hansenula, Pichia, Prototheca, Rhodotorula, Saccharomyces, Sporobolomyces, Trichosporon*
ANI	Anaerobic bacteria	*Actinomyces, Bacteroides, Bifidobacterium, Capnocytophaga, Clostridium, Eubacterium, Fusobacterium*
NHI	Fastidious bacteria	*Actinobacillus, Branhamella, Cardiobacterium, Eikenella, Gardnerella, Haemophilus, Kingella, Moraxella, Neisseria*
UID	Detect, identify, and enumerate most common urinary tract pathogens	*Citrobacter, Escherichia, Enterococcus, Klebsiella/Enterobacter, Proteus, Serratia, Staphylococcus, Pseudomonas, yeast*
EPS	Enterics	*Salmonella, Shigella, Yersinia*

able that differ in the substrates that are present in the wells. The selection of a test card is based on the presumptive identification of the primary isolate (**Table 30.12**). Ten test cards can be inoculated in about 3 minutes by drawing in a suspension of an organism that has been **isolated** in **pure culture** by primary enrichment or on a selective/differential medium (**Figure 30.5A**). Following inoculation, the test card fits into a carousel that can hold up to 30 test cards (Figure 30.5B). The carousel containing the inoculated test cards is then placed in the reader incubator module at a selected incubation temperature.

Each carousel is automatically read photometrically once every hour. Turbidity or color changes are recorded and transmitted to a computer system that analyzes the data. The data points that are collected each hour are

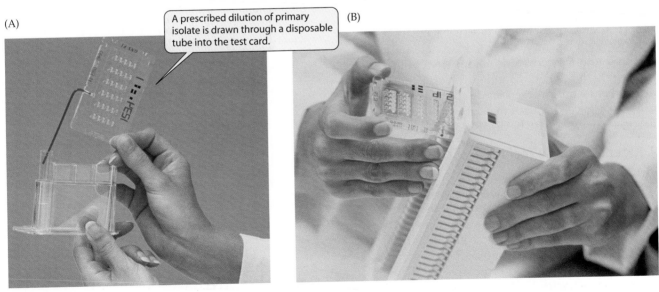

(A)

A prescribed dilution of primary isolate is drawn through a disposable tube into the test card.

(B)

Figure 30.5 Test card filler module
Test card filler module for the Vitek system. (A) Introducing the bacterial suspension under aseptic conditions. (B) Placing the inoculated test card in the incubation chamber. Courtesy of bioMerieux VITEK, Inc.

electronically compared with a known database. When sufficient data is available, the system will print out a most likely and next most likely identification of the microorganism present in the original inoculum. The probability for each is given. The data may also be displayed on a computer screen for reference.

The time required to complete the identification of a clinical isolate varies. Analyzing for the enterobacteria takes from 4 to 6 hours, depending on species. The glucose nonfermenting gram-negative bacteria require 4 to 18 hours. The gram-positives such as staphylococci require 4 to 15 hours. Identification of yeasts is accomplished by incubating the test card for 24 hours off-line before placing it in the reader/incubator. A blood culture organism can be identified in 4 to 6 hours, and susceptibility to antimicrobials is run simultaneously with the other identification tests. By use of automated systems of identification and susceptibility, an appropriate treatment can generally be initiated in less than 8 hours.

Susceptibility cards are available to determine the antimicrobial sensitivity of a microorganism isolated in primary culture. Test cards contain a standard growth medium with graded levels of different antimicrobials. The instrument is programmed to provide information on the antimicrobial that would be most effective. The effectiveness of the antimicrobial would be based on the identification of the organism, origin of the specimen, and other considerations. The effective minimal inhibitory concentration (MIC) of an appropriate antimicrobial is printed out in 6 hours or less. There are at present nine different standard test cards for gram-negative bacteria, two for gram-positive, and four for general antibiotic susceptibility. Each test card has 11 to 12 antimicrobials. A nonautomated or manual method for determining the sensitivity of clinical isolates is the Kirby-Bauer disk diffusion assay, which will be discussed later.

Antigen/Antibody Techniques

Virtually all microbial species and, in many cases, a strain of a pathogenic microorganism will have at least one antigenic component that is unique. Analytical techniques are now available for the recovery and purification of these unique antigens. A purified antigen can be employed to generate a *monoclonal antibody* that will react specifically with that antigen (Chapter 27). Monoclonal antibodies to specific antigens are an effective, highly specific, and rapid diagnostic tool.

One of the diagnostic tests that employs monoclonal antibodies is the **direct ELISA** test (**Figure 30.6**). ELISA is an acronym for Enzyme-Linked Immuno Sorbent

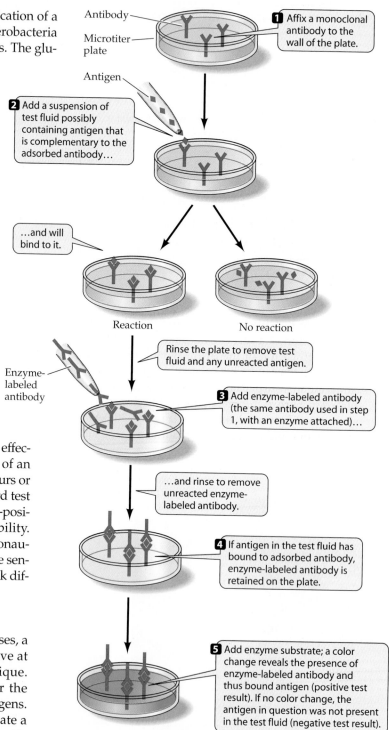

Figure 30.6 Direct ELISA
Direct enzyme-linked immunosorbent assay (ELISA) employing monoclonally produced specific antibodies for detection of an antigen.

Assay, and there are both direct and indirect ELISA tests. The **direct ELISA** test requires a monoclonal antibody that can be chemically adsorbed onto the walls of a well in a microtiter plate (Step 1). A suspension of serum or fluid that one suspects may contain a **specific antigen** is added. If present, the antigen will react with the antibody (Step 2) adsorbed to the wall of the well. Binding of the antigen is both specific and strong enough to withstand rinsing that removes unreacted antigen.

A suspension of an enzyme labeled monoclonal antibody would then be added (Step 3). This added antibody is equivalent to the antibody adsorbed to the well in Step 1 and will react with the same antigen. This antibody has one modification in that a reporter enzyme is conjugated to it without affecting the capacity of the antibody to react with the specific antigen. Enzymes that are conjugated to antibodies are generally those that will give a discernible visual color when the enzyme reacts with its substrate (Step 4).

If the antigen is retained by the antibody at Step 2, the antibody-enzyme conjugate (Step 3) will also be retained specifically and not removed by rinsing. Addition of a substrate (Step 4) for the conjugated enzyme will then result in a readily observed color change—a positive test. No color change indicates that no antigen was absorbed at Step 2 and that the specimen did not have an antigen present that could react with the antibody— a negative test.

The **indirect ELISA** test is one that determines whether a **specific antibody** is present in a specimen such as serum (**Figure 30.7**). HIV is one infection that can be diagnosed by the indirect ELISA assay. Because blood transfusion has a potential for spreading HIV, it is essential that the presence of the virus be determined in donated blood. HIV infection induces the host to generate circulating antibody to specific antigens on the HIV particle. These antibodies are formed during the early stages of HIV infection before the immune system is compromised. Detection of these antibodies in the serum of a patient confirms exposure to HIV.

The procedure for an indirect ELISA test is outlined in Figure 30.7. In this case, the HIV antigen itself is adsorbed to the wall of a microtiter plate (Step 1). Serum suspected of containing antibody against this antigen is added to the microtiter well (Step 2). Rinsing removes any antibodies not specifically attached to the antigen adsorbed to the well. However, if an antibody attaches to the antigen, it can be detected at Step 3 by addition of an antibody-enzyme conjugate suspension. The antibody in this case is one that specifically reacts with human immunoglobulins (generally anti IgG). Presence of the antibody (Step 2) to the antigen results in the formation of a complex (Step 4). Addition of a substrate for the enzyme gives a visual indication that the antibody against the original antigen is present. The indirect

ELISA can also be employed in diagnosing infections such as salmonellosis, plague, brucellosis, syphilis, tuberculosis, leprosy, and Rocky Mountain spotted fever.

Detection by Immunofluorescence

Antibodies tagged by chemically bonding them to fluorescent dyes are commonly employed in the clinical

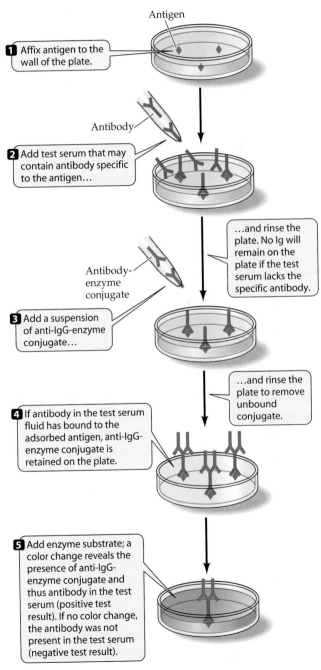

Figure 30.7 Indirect ELISA
Indirect ELISA test to determine the presence of a specific antibody in human serum.

diagnostic laboratory. A fluorescent dye absorbs light of one wavelength, becomes excited, and emits light at another, generally longer, wavelength. This technique can be employed, for example, in diagnosing the genital tract infection caused by *Chlamydia trachomatis*. To accomplish this, a smear of cervix scrapings would be placed on a slide, then a specific antichlamydial antibody-dye conjugate is applied. The antibody employed in detection can be polyclonal or monoclonal and is usually obtained from a selected animal species by injecting it with elementary bodies of *Chlamydia trachomatis* as antigen. Unreacted antibody-dye conjugate is then removed by rinsing. The slide is viewed with a fluorescence microscope, and presence of the pathogen is indicated by a green fluorescence against the dark background of counterstained cells. In place of microscopy, a fluorometer can be employed that electronically detects the total amount of light emitted at the appropriate wavelength.

A problem in identification of pathogens by immunofluorescence is that some species may share antigens. This is particularly the case with enteric bacteria. Employing monoclonal antibodies that interact solely with an antigen unique to a pathogen lessens the problem with cross-reacting antibodies. Immunofluorescence can be a useful technique in identifying pathogens that grow slowly or are difficult to isolate. For example, mycobacterial, chlamydial, and brucellosis infections are diseases where isolation of the organism is time consuming, but early treatment is essential. Other specimens that might be examined by immunofluorescence procedures are genital exudates for gonorrhea, feces for cholera, smears from bubonic plague buboes, tularemia lesions, and legionellosis aspirate. The immunofluorescence technique is applicable where clinical symptoms indicate that one of these pathogens may be responsible for the observed symptoms.

Agglutination

When antigens and antibodies interact, a complex develops that is, quite often, large enough to be visible with the naked eye. This clumping of antigen/antibody is called **agglutination.** Agglutination occurs because antibodies and antigens have two or more binding sites and can form a network of linked antigen and antibody molecules (Chapter 27). Agglutination has been adapted to the clinical laboratory for the rapid detection of either antigens or antibodies and is a relatively inexpensive and specific test. The agglutination test can be augmented by employing antigen- or antibody-coated latex beads. The beads employed are spherical and about 0.8 μ meters in diameter (**Figure 30.8**). Proteins (antigen or antibody) will adhere tightly to the polystyrene surface of the latex beads. When an antigen or antibody is fixed to the bead and a complementary antibody or antigen is added, the agglutination reaction can occur in 30 seconds or less (if positive). A drop of urine, serum, spinal fluid, or other specimen in suspension can be added to the bead-protein complex to determine the presence therein of a specific complementary antigen/antibody. The agglutination test has been employed to identify *Staphylococcus aureus*, *Neisseria gonorrheae*, and *Haemophilus influenzae*. Yeast infections caused by organisms such as *Cryptococcus neoformans* can be diagnosed through agglutination tests.

Plasmid Fingerprinting A plasmid is extra chromosomal DNA that replicates autonomously in a bacterial cell. Plasmids are present in many bacterial genera, including pathogenic microorganisms. Bacterial **strains** that are related generally contain the same number of

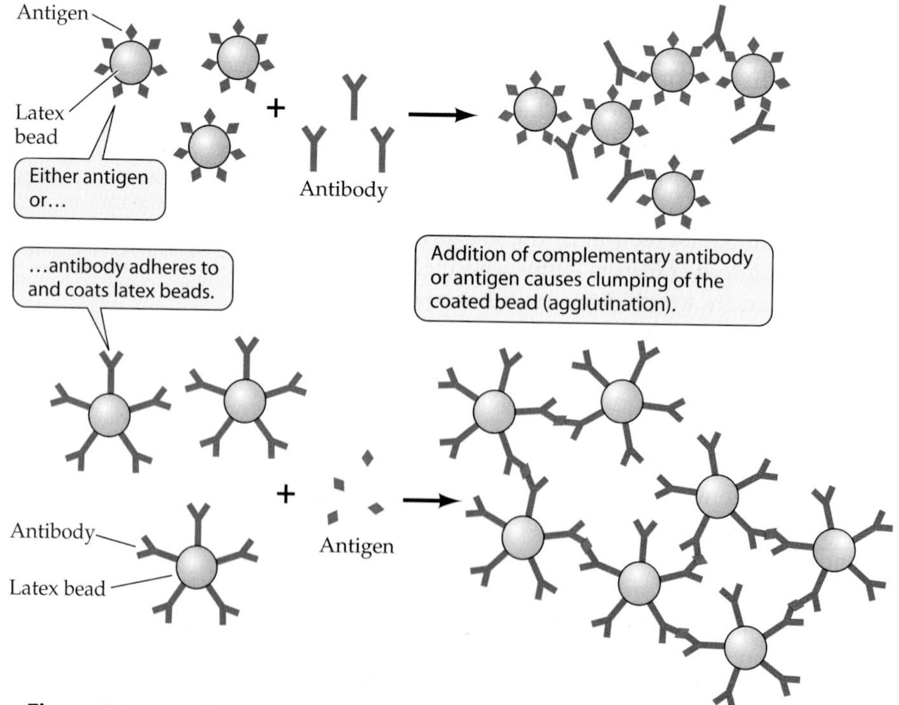

Figure 30.8 Agglutination tests
Latex bead agglutination procedure for the rapid detection of antigens or antibodies.

plasmids of equivalent molecular weight. Determining the number and size of plasmids in strains of a species can be a measure of the relatedness. To analyze for plasmids, a cell mass must be lysed to free intracellular components. The plasmid DNA is then separated from chromosomal DNA by density gradient centrifugation, and the plasmid DNA is applied to an agarose gel and subjected to electrophoresis. The migration rate of each plasmid in the gel is inversely proportional to its molecular weight. The DNA can be visualized in the agarose gel by staining with ethidium bromide. The greater the number of bands with equivalent migration rates in the gel, the more accurately the strain can be identified. The number and migration of bands forms what is called, a **plasmid profile** or **plasmid fingerprint.** Plasmid fingerprinting is not a definitive test because one strain of a species may have several plasmids and another strain may have few or none. However, if strains have several plasmids in common, it is strong evidence that they are related.

Plasmid profiles are of value in tracing the origin of specific strains that may be involved in a local epidemic such as a nosocomial infection. If a hospital has a series of *Staphylococcus* infections and the plasmid profile of the isolated organisms is similar, this can be indicative of a potential common origin. If the organisms involved have significantly different profiles, it would suggest that the infections do not have a common origin. Food poisoning cases caused by enterotoxin from coagulase-positive staphylococci may be traced by this technique. The *Staphylococcus aureus* strain in the nasal passage of a food handler can be subjected to the profile test. If a strain of *S. aureus* from a victim of food poisoning has an identical plasmid profile to that from the food handler, this is evidence that the food handler was a source of the staphylococcal infection.

A DNA analytical procedure, based on digestion of the entire chromosome by restriction enzymes, has been employed successfully in tracing the origin of common-source nosocomial outbreaks. The restriction enzymes yield characteristic DNA fragments that are separated by gel electrophoresis. Patterns of identity based on migration of the fragments indicate relatedness (or unrelatedness).

Phage Typing

Bacterial viruses (bacteriophages) are quite specific to the species or, in many cases, the bacterial strain they can infect. This specificity is based on the presence of viral receptors on the surface of the bacterium where the virus attaches to initiate the infectious process. Susceptibility of bacterial strains to viruses can be an indicator of genus, species, and possibly a strain.

Virus susceptibility can be determined by spreading a selected bacterium over an agar surface. The plate is squared off by marking with a wax pencil on the bottom of the plate. Then a drop of different viral suspensions is applied to each marked square. After suitable incubation for bacterial growth, the plate is examined for plaques, indicating which of the viruses was capable of attacking that bacterium. Phage typing is useful in tracing localized epidemics such as a case of food poisoning.

Nucleic Acid–Based Diagnostics

Advances in molecular biology have provided techniques for a rapid identification of infectious agents. These techniques take advantage of the fact that the nucleotide sequences in one genus or species differ in a measurable way from those in another. A pathogenic microorganism that generates toxins, synthesizes enzymes involved in invasion, or produces virulence factors has unique genetic information. The information for these specific virulence factors resides in the organism's DNA. DNA diagnostics depend on the identification of unique nucleotide sequences that are present in one species or strain but not present in others.

Nucleic acid probes are better than immunological procedures in many ways. Antibodies are sensitive to temperature or other physical conditions that might be necessary to remove interfering material. These harsh conditions that would inactivate an antibody do not harm nucleic acids from pathogens or those in probes.

Diagnostics based on hybridizing or amplifying nucleic acids are gradually being implemented in clinical laboratories. Practical application of these molecular methods has, to date, had some limitations, as they can be time consuming, be expensive, and require the services of experienced technologists. However, these techniques can detect pathogens that available tests cannot and are gaining acceptance, particularly in larger clinical laboratories.

Amplification of target nucleic acids by PCR is necessary in using probe assays for the identification of many pathogens. If the disease agent is present in relatively low numbers or isolation of the pathogen is impractical, such as with a viral or rickettsial infection, amplification is a must.

Amplification assays have been developed for many infectious agents including Epstein Barr virus, enteroviruses, *Ehrlichia, Rickettsia,* herpes simplex, and hepatitis B and C. An amplification assay is also employed to determine the HIV levels in AIDS patients, particularly for those undergoing chemotherapy.

Pneumonia is an example of a disease where probe assays are impractical. Many microorganisms that are agents of pneumonia are carried in the respiratory tract of asymptomatic individuals. A sensitive amplification assay would be overloaded with organisms that may not be involved in the disease process.

Clinical application of DNA technology relies on the tight binding (hybridization) that occurs between complementary single strands of DNA. A single strand of DNA of a known sequence can be employed as a **probe.** If DNA from a clinical isolate has sequences that are complementary to those of a probe, they can hybridize, and these hybrids are detectable.

The DNA from a microorganism obtained from a specimen can be released by treatment with strong alkali and the DNA rendered single stranded (ss DNA). The source of ss DNA may also be a bacterial colony treated with detergent to release the cellular DNA (ds DNA), and this can be rendered single stranded by heating. The ss DNA are affixed to a filter (**Figure 30.9**), and the probe with a reporter molecule attached is added. The temperature and time allowed is adjusted to one at which stable duplexes are formed. This is dependent on the length of the nucleic acid probe and its base composition. After an appropriate time, the unhybridized probe DNA is rinsed through the filter, and the amount of reporter remaining is determined. The reporter molecule can be a fluorescent dye, an enzyme, or radiolabel. A positive test can be determined by fluorescent microscopy, an enzyme dye color change, or radioactivity. There are available DNA probes that will bind to complementary strands of ribosomal RNA (r RNA). A hybrid of rRNA:DNA is more sensitive than the DNA:DNA hybrids, and fewer microorganisms are required for an accurate determination. DNA:DNA probes have been effective in diagnosis of rickettsial and cytomegalovirus infections among others.

A dipstick probe has been developed for the identification of infectious agents in food and crude clinical specimens (**Figure 30.10**). In this method, a clinical specimen is treated with alkali to release DNA and generate single strands from any microorganism present. The specific single-stranded nucleic acid probe is added. This specific ss DNA probe has incorporated in it a sequence of bases (poly T or poly A) that serve as a capture probe. A low molecular weight molecule such as dioxygenin would also be chemically linked to the nucleic acid probe (D in Figure 31.10). Following incubation of the alkali-treated clinical specimen with the nucleic acid probe, one would add nucleases to destroy any single-stranded probe that is unhybridized. A dipstick that bears the capture probe base sequence would be inserted into the hybridization solution. The capture probe with any hybridized DNA would be exposed to antibodies to the dioxygenin reporter, and these antibodies would be coupled to a fluorescent dye or a detectable enzyme. Whether hybridization occurred would be determined by observing an enzyme reaction or fluorescence by microscopy.

Probes are of particular value in diagnosing *Chlamydia trachomatis* and organisms that are involved in respiratory infections. Where crude specimens are to be analyzed (sputum, feces, or discharges), a probe is an effective technique because isolation of the organism is not essential. They are also a method for rapid diagnosis for viruses and other organisms that may not be grown in culture.

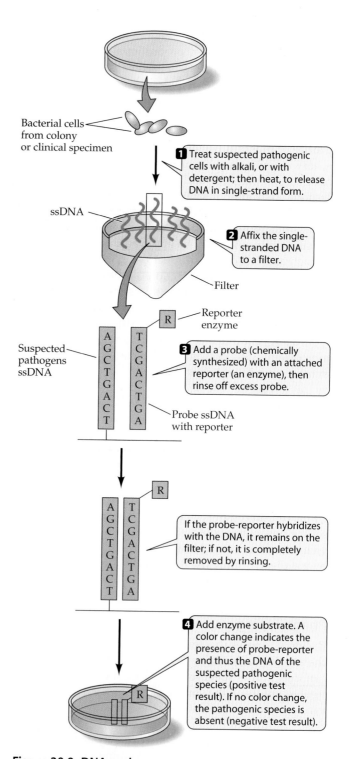

Figure 30.9 DNA probe
DNA probe for detection of a specific microorganism in a clinical specimen.

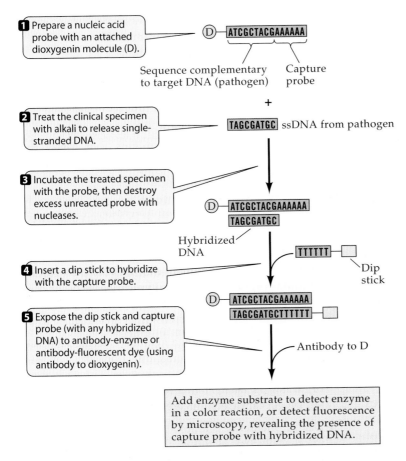

1 Prepare a nucleic acid probe with an attached dioxygenin molecule (D).

Ⓓ—⟨ATCGCTACGAAAAAA⟩

Sequence complementary to target DNA (pathogen)   Capture probe

+

2 Treat the clinical specimen with alkali to release single-stranded DNA.

TAGCGATGC ssDNA from pathogen

3 Incubate the treated specimen with the probe, then destroy excess unreacted probe with nucleases.

Ⓓ—ATCGCTACGAAAAAA
   TAGCGATGC

Hybridized DNA

TTTTTT ☐ Dip stick

4 Insert a dip stick to hybridize with the capture probe.

Ⓓ—ATCGCTACGAAAAAA
   TAGCGATGCTTTTTT ☐

5 Expose the dip stick and capture probe (with any hybridized DNA) to antibody-enzyme or antibody-fluorescent dye (using antibody to dioxygenin).

Antibody to D

Add enzyme substrate to detect enzyme in a color reaction, or detect fluorescence by microscopy, revealing the presence of capture probe with hybridized DNA.

Figure 30.10 Dip stick probe
Dip stick assay procedure for recovery of single-stranded DNA.

ANTIBIOTIC SENSITIVITY

Antibiotics or chemotherapeutics are the treatments available for many bacterial and mycotic infections. As previously mentioned, resistance to these inhibitors is a major concern in the treatment of many infectious diseases. Resistance to one or more antibiotics is now encountered with many infectious microorganisms and is a considerable concern in nosocomial infections. The earlier a patient is treated with an appropriate antibiotic the better.

The susceptibility of a clinical isolate to an antibacterial agent can be determined by the automated system described earlier. However, where this technology is not available, a disk diffusion assay can be employed. The disk assay widely used was developed by W. Kirby and A. W. Bauer and is termed the Kirby-Bauer method. In this procedure, a filter disk is impregnated with a measured amount of antibiotic, and the disk is placed on an agar surface. The disk will absorb moisture from the agar, effecting the

diffusion of antibiotic into the agar medium. The distance that the antibiotic moves outward from the disk depends on the solubility, concentration, and other characteristics of the antibiotic. As the antibiotic diffuses outward, a concentration gradient is established with the highest level in the immediate area of the disk.

In practice, one would inoculate a liquid growth medium by touching an inoculating loop to a colony of bacteria growing on a primary isolation plate and transferring the inoculum to a liquid medium. The inoculated liquid medium would be incubated, and when turbidity is evident, an aliquot would be added to melted agar and poured into a Petri plate. After solidification, antibiotic-containing disks are then placed on the surface of the agar and the plate incubated. After a 12- to 18-hour incubation period, **the zones of inhibition** can be measured (**Figure 30.11**).

The susceptibility of an organism to an antibiotic is directly related to the diameter of the zone of inhibition. The zone diameters can be interpreted by employing a table prepared by Kirby-Bauer (**Table 30.13**). This table was developed by determining minimum inhibitory concentrations (MIC) of antibiotics

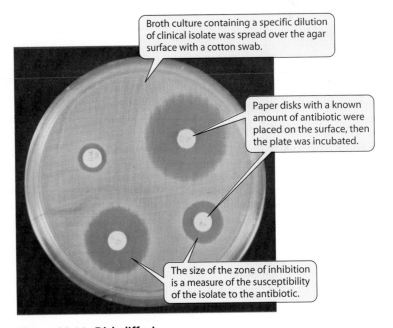

Broth culture containing a specific dilution of clinical isolate was spread over the agar surface with a cotton swab.

Paper disks with a known amount of antibiotic were placed on the surface, then the plate was incubated.

The size of the zone of inhibition is a measure of the susceptibility of the isolate to the antibiotic.

Figure 30.11 Disk diffusion assay
Kirby-Bauer procedure for determining the sensitivity of a clinical isolate to antibiotics. Charts are available for estimating the susceptibility of the clinical isolate from the size of the zone of inhibition. Courtesy of J. J. Perry.

| Table 30.13 | Diameter of the zone of inhibition as a measure of susceptibility to selected antibacterial agents |||

| Antibacterial | Amount on Disk (μg) | Zone Diameter (mm) ||
		Resistant	Sensitive
Ampicillin	10	28 or less	29 or more
Erythromycin	15	13 or less	18 or more
Gentamycin	10	12 or less	15 or more
Tetracycline	30	14 or less	19 or more

effective against many pathogenic bacteria and relating the MIC to the diameter of an inhibition zone.

Minimum inhibitory concentration (MIC) values can also be determined as follows: A series of tubes containing an appropriate medium and graded concentrations of antibiotic are inoculated with equivalent amounts of a suspension of the clinical isolate. The tubes are examined for growth after a suitable incubation period. The lowest concentration of antibiotic that inhibits growth completely is considered the MIC.

Clinical diagnostics has made remarkable advances in the past few years. Automation and rapid specific tests permit accurate identification of an infectious agent in a short period of time. Early treatment saves lives, shortens the illness and convalescence time, and saves money. The epidemiologist and clinical laboratory in concert contribute much to our well-being.

SUMMARY

▶ **Infectious diseases** were the major cause of death worldwide before antibiotics were available. They now are responsible for about 8% of the deaths in the developed countries, but 50% of the deaths in developing countries are due to infectious diseases.

▶ The science concerned with the prevalence and distribution of diseases in the population is **epidemiology.**

▶ An **epidemic** occurs when a disease occurs in the population at a higher than normal frequency. A constant low frequency of disease is called the **endemic** level. A worldwide epidemic is a **pandemic.**

▶ The **incidence** of a **disease** is the number of cases in a susceptible population of 100,000. **Mortality** rate is deaths per total infected. **Morbidity** is the number of new cases during a period divided by the total population.

▶ A **carrier of infection** is an infected individual that transmits a disease to others. A **chronic carrier** is an individual permanently infected and shedding an infectious disease. A **reservoir** is the site where an infectious disease-causing microorganism is maintained between outbreaks.

▶ **Human contact** and **respiratory** diseases are passed by human interaction such as touching, coughing, or sneezing. Among these are **colds, influenza,** and **tuberculosis.**

▶ **Water-** and **food-borne** diseases are generally caused by organisms that infect the human gastrointestinal tract. They are transmitted through fecal contamination.

▶ A major public health concern is **sexually transmitted diseases. HIV, chlamydial infections,** and **herpes** are all on the increase, as are **genital warts.**

▶ Epidemics of **vectorborne** diseases such as **Rocky Mountain spotted fever** and **Lyme disease** occur during summer months, and incidence of Lyme disease in the United States is increasing at an alarming rate.

▶ An **infection acquired** in a hospital is termed a **nosocomial infection.** Nosocomial infections affect an estimated 2 million patients each year. The prevalent organism is *Escherichia coli.* **Staphylococcal** and **streptococcal infections** are also a problem.

▶ Disease **incidence** is followed in most states by requiring that physicians report about 60 different diseases to the state division of epidemiology. They in turn are reported to the Centers for Disease Control and Prevention in Atlanta, Georgia.

▶ Many diseases are controlled in the United States by **vaccination** of babies and young children. Among the vaccinations are those for **mumps, measles, rubeola, diphtheria, pertussis (whooping cough), tetanus, polio, Haemophilus influenza b, chicken pox,** and **hepatitis B. Smallpox** was **eradicated** worldwide by vaccination.

▶ **Reservoirs** that sustain infectious agents and transmit them to humans are a problem in **epidemiology,** and their elimination is important in controlling disease. This is not possible in all cases.

▶ **Food inspection** and **water sanitation** are important public health measures.

▶ A real problem in controlling infectious diseases is development of **drug resistance** among the microbes that cause diseases.

▶ Rapid identification of pathogens involved in an infection leads to better treatment. A clinical laboratory is charged with this responsibility. Much of the identification has been automated.

▶ A specimen is the material delivered to the clinic. It must be handled carefully and may consist of blood, urine, sputum, and so forth.

▸ Automated procedures are now available that can identify clinical isolates in less than 20 hours and also determine resistance/sensitivity to antibiotics.

▸ An ELISA test (Enzyme-Linked-Immuno Sorbent Assay) is a diagnostic procedure that determines whether a specific antigen or antibody is present in a suspension.

▸ Antibodies labeled with fluorescent dyes are an important diagnostic tool to identify pathogens. A specimen is placed on a slide and dye-labeled antibody is applied. Unreacted antibody is rinsed away and the slide is viewed with a fluorescence microscope. Agglutination tests with antibody- or antigen-coated beads, plasmid profiles, and phage typing are other available diagnostic tools.

▸ **Nucleic acid probes** are a technique for detecting pathogens. The methods take advantage of the complementarity of the nucleotides in a probe and equivalent unique base sequences in the genome of a pathogen. Probes are 60 to 80 nucleotides long and generally are a product of chemical synthesis.

▸ When small numbers of a pathogen might be present in a clinical specimen the **polymerase chain reaction (PCR)** can be applied to increase the level of DNA present to detectable levels.

▸ A determination of **antibiotic sensitivity/resistance** in an infectious microorganism is necessary in designing a proper treatment.

REVIEW QUESTIONS

1. Define epidemic, endemic, and pandemic.

2. How does a sporadic disease differ from one that is seasonal? Give examples of some seasonal diseases.

3. Two terms employed in disease reporting are *mortality* and *morbidity*. In fact, the Centers for Disease Control publishes the *Morbidity and Mortality Weekly Report*. What is the significance of these two terms?

4. What is a disease reservoir? Why are diseases that spread from animal reservoirs such a problem?

5. How are human contact diseases spread? Give examples of such diseases. How can they be controlled?

6. Describe how sexually transmitted diseases can be controlled. Which are treatable?

7. Why are nosocomial diseases such a concern? Can they be readily controlled?

8. What are some of the reasons for the doubling of life expectancy in the United States over the last two centuries?

9. We have lessened the incidence of the childhood diseases by vaccination. What are some of these? Why are they still a concern?

10. Name the agency that is responsible for control of disease worldwide. What major success has it had?

11. What are some major disease problems in developed countries? Developing countries?

12. Why is it important that a disease-causing microbe be isolated and identified?

13. What types of specimens are sent to the clinical laboratory? How are some of these treated once there? What identification procedures are generally employed?

14. What is an ELISA test? How is it performed?

15. How is immunofluorescence employed? Agglutination?

16. How are plasmid fingerprinting and phage typing used in the clinic?

17. Nucleic acid probes are of growing importance in the clinical lab. How are they used?

18. What is PCR? How is it employed in virus identification, and why is it necessary?

SUGGESTED READING

Centers for Disease Control and Prevention. *Morbidity and Mortality Weekly Reports*. Atlanta: Centers for Disease Control.

Centers for Disease Control and Prevention. (Annual) *Surveillance*. Atlanta: Centers for Disease Control.

Isenberg, H. D. 1998. *Essential Procedures for Clinical Microbiology*. Washington, DC: American Society for Microbiology.

Murray, P. R., editor-in-chief. 1999. *Manual of Clinical Microbiology*. 7th ed. Washington, DC: American Society for Microbiology.

Rose, N. R., R. G. Hamilton and B. Detrick. 2002. *Manual of Clinical Laboratory Immunology*. 6th ed. Washington, DC: American Society for Microbiology.

Salyers, A. A., and D. D. Whitt. 2001. *Bacterial Pathogenesis: A Molecular Approach*. 2nd ed. Washington, DC: American Society for Microbiology.

Sussman, M., ed. 2001. *Molecular Medical Microbiology*. San Diego: Academic Press.

Truant, A. L. 2001. *Manual of Commercial Methods in Clinical Microbiology*. Washington, DC: American Society for Microbiology.

Tulchinsky, T. H., and E. A. Varavikova. 2000. *The New Public Health*. San Diego: Academic Press.

Applied Microbiology

PART VIII *Applied Microbiology*

Previous page
Bread, cheese, and wine are all products of applied microbiology.
©Guido A. Rossi/Photo Researchers, Inc.

Chapter 31

Industrial Microbiology

There is no field of human endeavor, whether it be in industry or in agriculture, whether it be in the preparation of foodstuffs or in connection with problems of shelter and clothing, whether it be in the conservation of human and animal health and the combating of disease, where the microbe does not play an important and often a dominant part.

— SELMAN A. WAKSMAN, 1943

*I*ndustrial microbiology traces its origins to prehistoric times. Fermentation technology was born when early civilizations unwittingly took advantage of the capacity of microorganisms to produce alcoholic beverages as well as leavened bread and cheese. Early Sumerian city–states (fourth millennium B.C.) produced beer and wine from fermented barley and grapes, respectively. Ancient civilizations also made cheese and salted and fermented meat, thereby preserving more perishable foodstuffs. Modern innovations in food preservation began with the Frenchman François Appert, who developed canning methods in 1809.

During the first half of the twentieth century, wine and beer manufacture, vinegar, and bread yeast production moved from the realm of an ancient art to an established science. Large-scale microbial processes for the manufacture of citric and lactic acid were developed. Acetone, butanol, and ethanol were produced in commercial quantities by fermentation, but are now mostly products of chemical synthesis.

The second half of the twentieth century was an era of dramatic change in the fermentation industry. Much of this change originated with the discovery that antibiotics can be an effective weapon against disease. The search for and production of novel antibacterial, antifungal, antitumor, and antiviral agents became major industries in Japan, the United States, and Western Europe. Other products generated by microbes, including vitamins, sterols, organic acids, amino acids, flavoring agents, enzymes, and fermented foods/beverages, gained commercial importance.

Knowledge gained over the past 20 years has markedly broadened the potential for the production of useful compounds by microorganisms. This new field of endeavor is broadly known as **biotechnology.** Biotechnology is mainly based on the use of living systems (usually microbial) to solve technical problems or achieve technological goals. For example, by genetic engineering, human genes for insulin biosynthesis have been introduced into bacteria, thereby enabling bacteria to synthesize this hormone. Thus, diabetics can receive human insulin for their treatment.

Microorganisms can have an impact on the economic strength of a country. An estimated 5% of Japan's gross national product stems from the broad capabilities of the microbe.

THE MICROBE

The **fermentation industry** is a collective term applied to commercial processes that exploit the capacity of microbes to make useful products. Many of the organisms used commercially, such as the actinomycetes, do not grow anaerobically and carry out fermentations. In industry, fermentation refers to the large-scale production of biological materials in tanks or containers. Every microorganism employed in the fermentation industry is a unique one and is carefully selected for its ability to perform a desired task. Whether a strain is newly isolated from the environment or selected from a stock culture collection, the potential producer of useful products is not significantly different from related species. However, careful development of strains from the original culture is needed before a profitable commercial product is made available.

Metabolites such as amino acids, B vitamins, and nucleic acid bases are overproduced by some microorganisms and excreted into the proximate area of their microenvironment. Other microorganisms in the environment can assimilate these compounds. One strain may overproduce one metabolite but require for its growth a metabolite released by another strain. Such harmonious associations occur widely in nature. Microorganisms that overproduce in nature may, when isolated in pure culture and subjected to laboratory growth, produce even greater quantities of a given metabolite. Unbalanced growth in the laboratory may also lead to the expression of genetic information and the enhanced production of metabolites such as antibiotics.

Microbial strains that are valued for the synthesis of commercial products are highly selected specialists and probably would not survive outside the laboratory setting. Chosen for their genetic stability during the fermentation process and consistency in production, they are fast-growing organisms that generate useful products in a relatively short period of time. Such strains generally must be harmless to humans and to the environment because containment of industrial microbes involved in large-scale fermentations would be difficult and expensive.

Two standard procedures are employed to improve microbial strains that can generate useful metabolites: mutation and selection. Scientists subject a producer strain to mutagenic agents and select the most prolific producer among the progeny. Through repeated and rigorous application of such regimens, phenomenal increases in yield can be realized. For example, the original strains of *Penicillium chrysogenum* that yielded 1.2 milligrams of penicillin per liter have been genetically manipulated to produce over 50,000 milligrams per liter.

Primary and Secondary Metabolites

Some metabolites are synthesized and appear in the medium during active growth, and others appear as the organism completes the growth cycle and enters the stationary phase. **Primary metabolites** are those produced during active growth, and those synthesized after the growth phase nears completion are designated **secondary metabolites (Figure 31.1)**.

A primary metabolite is generally a consequence of energy metabolism and necessary for the continued growth of the microorganism. Recall the discussion of anaerobic metabolism (Chapter 8) where a product of glycolysis serves as electron acceptor for the reduced pyridine nucleotide generated by glyceraldehyde-3-phosphate dehydrogenase (Figure 8.4). In the absence of oxygen, the terminal electron acceptor was pyruvate and the product of anaerobic energy generation was lactic acid. The microorganism involved could not generate ATP (or grow) without the concomitant production of the primary metabolite, lactic acid.

The production of secondary metabolites is a result of complex reactions that occur during the latter stages of primary growth. The hallmark of secondary metabolites is that they are produced by a single or relatively

(A)

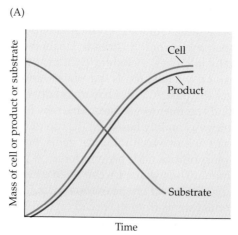

(B)

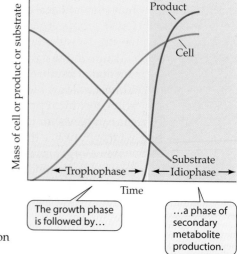

Figure 31.1 Cellular production of primary and secondary metabolites
Time course for substrate utilization, increase in cell mass (growth), and production of (A) primary and (B) secondary metabolites.

few species. In microorganisms that produce secondary metabolites, the growth stage is called the **trophophase.** The phase involved in production of secondary metabolites is the **idiophase.** In some cases, enzymes involved in trophophase growth shift to the production of metabolite during the idiophase. Citric acid fermentation by *Aspergillus niger* is an example of a product resulting from a metabolic shift. During the trophophase, a substrate such as glucose is mostly used for synthesis and as a source of energy. In generating energy, much of the substrate is oxidized to CO_2. During idiophase, glucose is mostly converted to the secondary metabolite, citric acid, with little respiration (CO_2 production). In the production of antibiotics, which are also secondary metabolites, a different scenario prevails. An actinomycete that produces an antibiotic synthesizes a battery of novel enzymes during late exponential and early stationary phase growth. These are responsible for the synthesis of the antibiotic. The synthesis of these enzymes can be a major undertaking. For example, more than 300 genes are involved in the synthesis of enzymes that generate chlortetracycline by *Streptomyces aureofaciens*. At least 72 intermediates are formed, and only 27 have been isolated and characterized. Only a few enzymes involved in chlortetracycline synthesis are required for cell growth.

MAJOR INDUSTRIAL PRODUCTS

The diverse industrial products generated by microorganisms are listed in **Table 31.1**. Bread, cheese, pickles, olives, and mushrooms are common foodstuffs in our diet. Each of these is a direct consequence of microbial activity. Although the role of microbes in the production of cheese and beer is widely known, few are aware that virtually all soft drinks contain citric acid, a product of microbial synthesis.

Flavor enhancers, such as monosodium glutamate (MSG), inosinic acid, aspartate, alanine, and glycine are products of microbial metabolism. Vinegar is a commonly used flavoring agent and preservative. Many of the B vitamins can be produced in significant quantity by selected microbial cultures, but only riboflavin and B_{12} biosynthesis compete economically with chemical synthesis that is the source of the others (biotin, thiamin).

The conversion of low-cost biologically inactive sterols to an activated form is readily accomplished by microorganisms. These are important in maintaining human health. Antibiotics synthesized by microbes have had a profound effect on human health and longevity. In the expanding field of biotechnology, microbes are genetically engineered to produce a variety of proteins and peptides in addition to insulin, such as human growth factor, interferon, blood clotting factors, and vaccines. Production of rhizobia to use as inocula to increase the

Table 31.1	**Major commercial products obtained from microbes**

I. Foods, flavoring agents and food supplements, and beverages
  Foods
    Fermented meat
    Cheeses and milk products
    Edible mushrooms
    Leavened bread-baker's yeast
    Coffee
    Pickles, olives, sauerkraut
    Single-cell protein
  Flavoring agents and food supplements
    Vinegar
    Nucleosides
    Amino acids
    Vitamins
  Beverages
    Wines
    Beer, ale
    Whiskey
  Vitamins
    B_{12}
    Riboflavin
II. Organic acids
  Citric acid
  Itaconic acid
III. Enzymes and microbial transformations
  Commercial enzymes
  Sterol conversions
IV. Inhibitors
  Biocides
  Antibiotics
V. Products of genetically engineered microbes
  Insulin
  Human growth factor

symbiotic fixation of N_2 in legumes (peanuts, peas) is important in food production.

Some of the various major commercial products generated by microorganisms are discussed in the following sections.

FOODS, FLAVORING AGENTS AND FOOD SUPPLEMENTS, VITAMINS, AND BEVERAGES

The microorganism itself may serve as food, but most often they are employed as agents to convert basic foods such as milk or meat to a more desirable product. Microorganisms can also synthesize flavoring agents or flavor enhancers, and these can contribute to the quality and acceptance of foods. Food supplements obtained from fermentations can serve as preservative agents or as supplements to improve the nutritional value of a food.

Foods

Single-*Cell Protein* Bulk microbial biomass, termed *single*-cell protein, or SCP, can be used as a source of human or animal food. SCP could be valuable where climatic conditions cause agricultural crops to fail, where arable land is scarce, or when populations are under threat of starvation. A potential source of food could be the bulk harvesting of bacterial or yeast cells grown on substrates that in themselves would not support human sustenance (**Box 31.1**).

Microorganisms are attractive food sources for a number of reasons:

- Rapid growth
- Growth largely unaffected by climate
- Low requirement for land
- Inexpensive substrates for growth

The potential sources of SCP are yeast or bacterial cell mass. Yeasts and filamentous fungi show greater promise as potential food sources for human consumption than bacteria.

Yeasts grown on a low-nitrogen high-carbon substrate will be rich in lipid, but conversely when grown on high nitrogen will have a higher protein content. Yeast cells are rich in essential B vitamins. However, yeast protein is low in sulfur-containing amino acids and is not sufficiently balanced to provide for human or animal needs. Yeast cells also have a high nucleic acid content, and when consumed in quantity, may cause gout or other ailments. However, bulk yeast cells can be added to wheat or corn flour to provide B vitamins and selected amino acids and thus yield a product that is more nutritionally balanced.

Mushrooms are filamentous fungi and several species are already important as human food. The annual consumption of mushrooms (the fruiting bodies of mycelial fungi) in the United States exceeds 300,000 tons. They are typically grown on decaying organic matter in the soil or on rotting wood. The mycelium grows beneath the surface and under favorable conditions will develop the familiar stalked structure. The stalk and cap are actually compact fungal hyphae, and the gills are the site of reproductive spore formation.

Agaricus bisporus and *Lentinus edulus* are two of several mushrooms grown commercially in the United States and commonly sold in grocery stores (**Figure 31.2**). A generalized flow sheet for growth of *A. bisporus* is presented in **Figure 31.3**. None of the materials used in preparing the compost on which the mushrooms grow is expensive or has many other uses. The mushrooms are grown at a controlled temperature and humidity and in the dark. The mushrooms appear about 3 weeks after the spawn is inoculated into the trays, and production continues for about 6 weeks. The soil remaining in the trays after growth can be used as topsoil or sterilized, aged, and some added back to compost. The shiitake fungus, *L. edulus*, is a more flavorful mushroom that originated commercially in Japan and is now gaining popularity elsewhere. *L. edulus* is a cellu-

(A) *Agaricus bisporus*

(B) *Lentinus edodes*

Figure 31.2 Food mushrooms
Two mushroom species that are produced commercially: *Agaricus bisporus* and *Lentinus edodes*. A, photo ©D. Richter/Visuals Unlimited; B, photo ©Science VU/Visuals Unlimited.

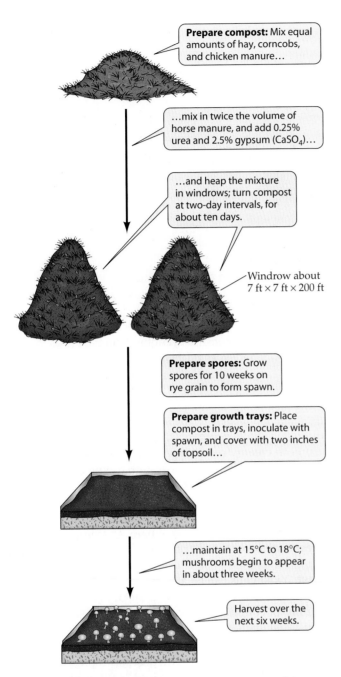

Prepare compost: Mix equal amounts of hay, corncobs, and chicken manure…

…mix in twice the volume of horse manure, and add 0.25% urea and 2.5% gypsum (CaSO$_4$)…

…and heap the mixture in windrows; turn compost at two-day intervals, for about ten days.

Windrow about 7 ft × 7 ft × 200 ft

Prepare spores: Grow spores for 10 weeks on rye grain to form spawn.

Prepare growth trays: Place compost in trays, inoculate with spawn, and cover with two inches of topsoil…

…maintain at 15°C to 18°C; mushrooms begin to appear in about three weeks.

Harvest over the next six weeks.

Figure 31.3 Commercial production of *Agaricus bisporus* Flow chart for commercial production of *Agaricus bisporus*. After the mushrooms are harvested, the compost is pasteurized and some of it recycled.

lose decomposer that grows best on waste lumber from hardwood trees. Logs or waste lumber are soaked in water and inoculated by placing spawn in holes drilled in the wood. The mushrooms fruit from these holes.

Mushrooms have little nutrient value and are eaten mostly for their distinct flavor and texture. They are low in protein, are essentially fat free, and provide only low levels of B vitamins.

Some bacteria have been grown primarily for use in animal feed. For example, using methanol as an inexpensive nutrient, methanol-oxidizing bacteria have been produced commercially in large fermentors in Britain for use as SCP in chicken feed.

Cheese and Yogurt

Yogurt and other fermented milk products are common foodstuffs throughout the world. Commercial yogurt is made from pasteurized low-fat milk thickened by the removal of water or addition of powdered milk. The thickened milk is inoculated with a mixture of *Streptococcus thermophilus* and *Lactobacillus bulgaricus* and incubated at 45°C for about 12 hours. Growth of *S. thermophilus* produces lactic acid, and *L. bulgaricus* contributes lactic acid and aromatics that give the product a distinct taste.

Cheese is an ancient food whose use, in all likelihood, precedes recorded history. Cheese making originated in the Middle East. From there it traveled to Greece and Rome, on to Northern Europe and England, and then to the rest of the world. Over 400 varieties of cheese are made, although many are similar and differ essentially in name only. Cheeses are generally grouped according to texture, as outlined in **Table 31.2**.

The manufacture of cheese can be divided into four distinct phases:

- Coagulation of milk to form a curd
- Separation of the curd from the whey
- Shaping of the curd
- Ripening to achieve flavor and texture

The processing of the material at each of these steps and the nature of the milk source have a marked influence on the finished product. Most cheese is made from cow's milk, but many of the varieties produced worldwide are made from the milk of goats, sheep, mares, camels, and water buffalo.

The first step in cheese production is the coagulation of the milk proteins to form a semisolid curd. This curd entraps fat globules and some water-soluble materials.

Table 31.2	The major types of cheese		
Soft	**Semisoft**	**Hard**	**Very Hard**
Unripened	Blue	Cheddar	Parmesan
Cottage	Roquefort	Gouda	
Mozzarella	Limburger	Swiss	
Cream	Muenster		
Ripened			
Brie			
Coulommiers			
Camembert			

Milestones Box 31.1   Bacteria as Food

The consumption of microbes as food has generally been limited to mushrooms (filamentous fungi) and yeasts. Yeasts are eaten as single-cell protein or as a food supplement (see text). Bacteria—as food—have been collected and eaten since ancient times by tribes in the Central African republic of Chad. The organism is a spiral-shaped cyanobacterium of the genus *Spirulina* (Figure 21.21D), *Spirulina platensis*, and is collected from the bottom of seasonally dried-up ponds and shallow waters around Lake Chad. The cyanobacterial mats are dried in the sun and cut into small cakes. The cakes are called **Dihe.** Dihe is rich in protein (62% to 68% of dry matter) and an important supplement to diets of the desert nomads. *Spirulina* is grown in Mexico and France. The potential yield of 10 tons of protein per acre dwarfs the yield obtained from wheat (0.16 tons/acre) and beef (0.016 tons/acre). *Spirulina* is now sold as a food supplement in health food stores in the United States. It is marketed as dried cakes or powdered product.

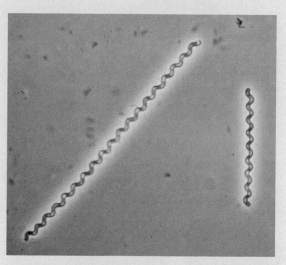

***Spirulina platensis,* a cyanobacterium used as food.** Courtesy of J. T. Staley.

Coagulation is effected by adjusting the pH of the milk to the isoelectric point of casein or adding milk-clotting enzymes. The clotting enzymes (curd-producing) are acid proteases such as rennin, an enzyme originally obtained from calves' stomachs, but now produced by bioengineered microorganisms. Acidification of milk can be accomplished by adding selected cultures of bacteria. Fresh unpasteurized milk has a population of bacteria that will acidify the milk if it is allowed to stand at room temperature. Bacteria that ferment lactose to lactic acid are added to pasteurized milk. These acidifying microorganisms are strains of *Streptococcus lactis, Lactobacillus cremoris, Lactobacillus bulgaricus,* or *Streptococcus bulgaricus.* Some of these bacteria are trapped and remain inside the cheese and become involved in the aging/ripening process.

After the curd is separated from the liquid whey, it is shaped. This process involves adjusting the pH, salting, and physically shaping the curd to ensure proper moisture content and acid production, all of which contribute to the characteristics of the finished cheese.

The ripening process, with or without added microbes, also affects the properties of the cheese product. Fungi or bacteria may be added to the ripening curd, and these add flavors or consistency not contributed by the initial microorganisms. For example, the cheese commonly termed *blue cheese* is inoculated with spores of *Penicillium roqueforti.* If one examines Roquefort cheese in cross section, the "blue veining" where the fungal spores were inoculated are quite evident. For ripening Camembert cheese, *Penicillium camemberti* is added to the surface of the cheese. This fungus produces proteases that transform the cheese to a smooth creamy consistency. Soft cheeses are ripened for a few weeks, semisoft for several months, and a hard cheese like Parmesan for a year or more. Some of the unique features of cheese varieties are the result of bacterial growth. The propionic acid bacteria produce the characteristic holes in Swiss cheese, and propionic acid is responsible for the sharp taste of this cheese.

Yeast Yeast cells can be a source of single-cell protein and may also be employed as a dietary supplement. A number of commercial biochemicals are also obtained from yeast, including enzymes, nucleosides, and natural vitamins. A common component of media for microbial growth is yeast extract, the water-soluble fraction of autolyzed yeast. It is a source of B vitamins, amino acids, and other growth factors.

The production of bakers' yeast annually in the United States exceeds 100,000 tons. The most common use of yeast cells is in the baking industry. Most leavened breads that are marketed throughout the world are products of normal metabolic reactions in yeasts (**Box 31.2**). Yeast is added to bread dough to cause "rising" or leavening prior to baking. During kneading, the dough is manipulated to aid in the yeast fermentation. Kneading also ensures that the CO_2 bubbles in the bread will be small. As yeast ferments, it converts sugar to ethanol and CO_2. The escaping gas (CO_2) that is trapped in wheat gluten results in an expansion of dough generally referred to as the "dough rises." During baking, the CO_2 and ethanol are driven off, leaving the odd-shaped holes in the baked product.

Bulk yeast used in baking or as a food supplement is grown in large fermentors. A flowchart for yeast growth is presented in **Figure 31.4**. A small inoculum is transferred batchwise to larger and larger tanks as shown. Before inoculation into the largest fermentor (S-3), the yeast cells are concentrated and added as a thick slurry. Beet or cane sucrose is the source of carbon and energy. The fermentor is vigorously aerated for rapid growth. The addition of the sugar is carefully monitored to ensure that the yeast uses it quickly and to ensure that ethanol is not produced at the expense of more yeast. Yeast is sold for baking as a compressed block, as moist yeast, or as activated dry yeast. Yeast cake is about 70% moisture and must be refrigerated to maintain activity. Dry yeast is produced by lyophilization to ensure a long shelf life.

Fermented Meat Specialty meat products are manufactured by bacterial fermentation. Among these are summer sausage, Lebanon bologna, salami, and cervelat. The organisms generally involved in meat fermentations are strains of *Pediococcus cerevisiae* and *Lactobacillus plantarum*. Commercial inocula of these microorganisms are available to add to the ground meat. Country cured hams are also a food that gains flavor and texture from fungi that are natural surface contaminants. These fungi are members of the genus *Penicillium* and *Aspergillus*.

Other Food Products Other foods that are products of microbial fermentation are listed in **Table 31.3**. Acidification of cabbage by a lactic acid fermentation to produce **sauerkraut** is an age-old method for long-term storage. Traditionally it is prepared by placing shredded cabbage in a crock or other deep container. Salt (NaCl) is added at 2.2% to 2.8% to restrict the growth of most gram-negative bacteria and draw out water for the fermentation. The crock quickly becomes anoxic, and the natural lactic acid microbiota of the cabbage predominate under these conditions and carry out the fermentation. In today's food industry, *Lactobacillus mesenteroides/Lactobacillus plantarum* are used as inocula in making sauerkraut. The lactic acid produced lowers the pH to a level where little decomposition occurs and the finished product is stable for extended periods of time. A typical fermentation requires about 30 days.

Pickles are produced by placing small cucumbers in a salt brine along with dill or other flavoring agents. The salt (NaCl) concentration, initially about 5%, is increased to 15.9% over a 6- to 9-week period. The high salt concentration extracts sugar from the cucumbers, and these sugars are fermented to lactic acid. Home production of pickles is a result of fermentation by the indigenous lactic acid biota of cucumbers. Commercial production is accomplished by sterilizing the pickles and adding cultures of *Pediococcus cerevisiae* and *L. plantarum* to ensure a uniform product.

Table 31.3	Foods produced by microbial fermentation
Product	**Fermenting Microorganism**
Cocoa beans	*Candida* spp., *Geotrichum* spp., *Leuconostoc mesenteroides*, others
Coffee beans	*Erwinia* spp., *Saccharomyces*
Sauerkraut	*Lactobacillus mesenteroides*, *Lactobacillus plantarum*
Soy sauce	*Rhizopus oligosporus*, *Rhizopus oryzae*
Olives	*Leuconostoc mesenteroides*, *Lactobacillus plantarum*
Pickles	*Pediococcus cerevisiae*, *Lactobacillus plantarum*

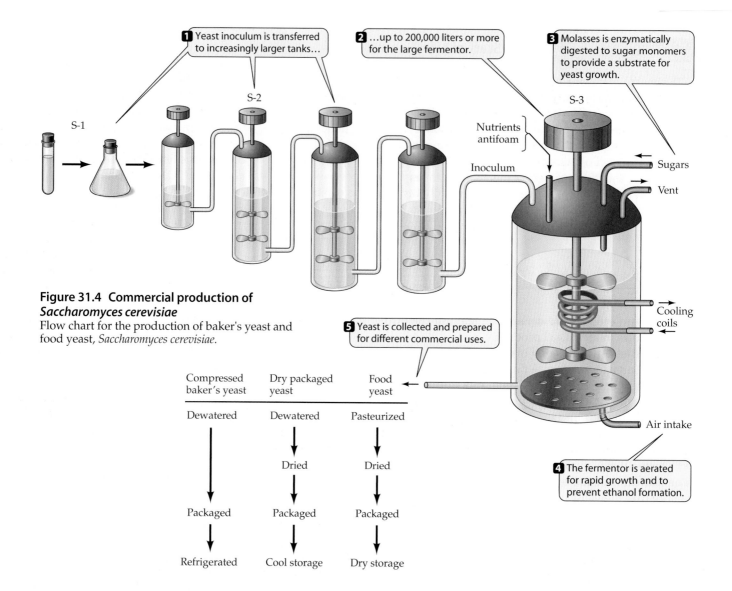

Figure 31.4 Commercial production of Saccharomyces cerevisiae
Flow chart for the production of baker's yeast and food yeast, *Saccharomyces cerevisiae*.

1 Yeast inoculum is transferred to increasingly larger tanks…

2 …up to 200,000 liters or more for the large fermentor.

3 Molasses is enzymatically digested to sugar monomers to provide a substrate for yeast growth.

S-1 S-2 S-3

Nutrients antifoam
Inoculum
Sugars
Vent
Cooling coils
Air intake

5 Yeast is collected and prepared for different commercial uses.

4 The fermentor is aerated for rapid growth and to prevent ethanol formation.

Compressed baker's yeast	Dry packaged yeast	Food yeast
Dewatered	Dewatered	Pasteurized
	Dried	Dried
Packaged	Packaged	Packaged
Refrigerated	Cool storage	Dry storage

Coffee beans develop as berries with an outer pulpy and mucilaginous envelope that must be removed before roasting. The pulp can be removed mechanically, but the pectin that contains the mucilage is removed best by the pectinolytic bacterium *Erwinia dissolvens*. Fungi are also involved in this process. Apparently microbes contribute little to the flavor or aroma of the finished coffee bean.

Chocolate is derived from cacao, a bean that is enclosed in the fruit of a plant that grows in parts of Asia, Africa, and South America. The beans are removed from the fruit and fermented in piles or boxes for several days. The fermentation occurs in two phases: (a) the sugars from the pulp are converted to alcohol and (b) the alcohol is oxidized to acetic acid. The yeasts involved in producing alcohol during the first phase are essential for the development of the chocolate flavor and aroma.

Fermentation of **olives** occurs after the olives have been treated with 1.6% to 2.0% lye at 24°C for 4 to 7 hours. This treatment removes the bitter principal (oleu-ropein) from the green olive. The lye is removed by soaking, and the olives are placed in oak barrels and 7.0% salt brine is added. The brined olives are inoculated with *Lactobacillus plantarum* and fermented for 8 to 10 months.

Soy sauce is prepared by mixing soaked and steamed soybeans with crushed wheat and inoculating with *Aspergillus oryzae* or *A. soyae*. This combination is incubated for several days, and during this period the beans release fermentable sugars, peptides, and amino acids. The fungal-coated beans are then placed in liquid, and salt (sodium chloride) is gradually added to a final concentration of 18%. The total incubation period is 1 year at room temperature. The addition of salt leads to a replacement of the filamentous fungi by lactic acid bacteria and yeast. The lactic acid bacteria are predominantly strains of *Lactobacillus delbrueckii*, and the yeast is *Saccharomyces rouxii*. The liquid resulting from the long-term fermentation is soy sauce. The solids are pressed and used as food for humans and animals.

Flavoring Agents and Food Supplements

Under appropriate conditions, selected bacteria and fungi produce organic compounds that give a desirable flavor or aroma to food. Many of these organics originally served as preservatives. However, with the advent of refrigeration and other means of preservation these flavorful compounds are no longer necessary as preservatives but are valued as flavoring agents. Sauerkraut, mentioned previously, was originally prepared to preserve cabbage for consumption over the long winter. It is now sold commercially because buyers like the flavor imparted by the bacterial fermentation. The following discussion concerns flavoring agents, flavor enhancers, and supplements such as selected amino acids that are added to food to enhance their nutritional value.

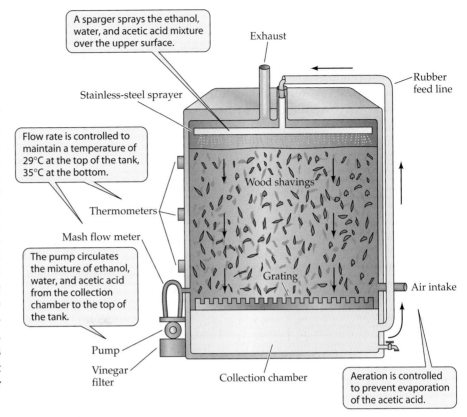

A sparger sprays the ethanol, water, and acetic acid mixture over the upper surface.

Exhaust

Rubber feed line

Stainless-steel sprayer

Flow rate is controlled to maintain a temperature of 29°C at the top of the tank, 35°C at the bottom.

Wood shavings

Thermometers

Mash flow meter

The pump circulates the mixture of ethanol, water, and acetic acid from the collection chamber to the top of the tank.

Grating

Air intake

Pump

Vinegar filter

Collection chamber

Aeration is controlled to prevent evaporation of the acetic acid.

Figure 31.5 Manufacture of vinegar by continuous fermentation
Frings-type trickling generator for the commercial production of vinegar.

Vinegar Vinegar making is at least 10,000 years old and originated with the production of wine. The name *vinegar* comes from the French words *vin* (wine) and *aigre* (sour), as it is essentially sour (acetic acid) or spoiled wine. Vinegar has long been valued as a flavoring agent, preservative, medicine, and consumable beverage. The Romans and Greeks drank dilute vinegar as a beverage and produced it by introducing air into wine casks. An early method of vinegar production was to place an air inlet at the top of a wine cask and draw out vinegar as needed. Fresh wine was then added to replace the vinegar removed.

The production of vinegar in the United States amounts to about 160,000,000 gallons per year. Two-thirds of the vinegar produced is used in commercial products such as in sauces, in dressings, and in the manufacture of pickles and tomato products. The remainder is used unaltered for domestic purposes. The basic reaction in vinegar formation is as follows:

$$CH_3\text{—}CH_2OH \xrightarrow[\text{NADP}^+ \quad \text{NADPH}]{} CH_3CHO \xrightarrow[H_2O]{} CH_3\text{—}\overset{\displaystyle OH}{\underset{\displaystyle H}{C}}\text{—}OH \xrightarrow[\text{NADP}^+ \quad \text{NADPH}]{} CH_3COOH$$

Ethanol   Acetaldehyde   Acetaldehyde hydrate   Acetic acid

The organisms employed in commercial vinegar production are *Acetobacter aceti* and *Gluconobacter oxydans*. Hundreds of bacterial subspecies or strains grow in a commercial fermentor, and mixed cultures of these along with those shown are involved in vinegar production. Two major systems are used for the commercial production of vinegar, the trickling generator and submerged fermentation.

Trickling generator One of the commercial systems for producing vinegar is the Frings-type trickling generator (**Figure 31.5**). The generator is constructed of cypress or redwood and packed with curled beechwood shavings. The collection chamber at the bottom holds about 3,500 gallons. A pump circulates the ethanol-water-acetic acid from the collection chamber and sparges it evenly over the upper surface. The temperature is controlled by flow rate at 29°C at the top and 35°C at the bottom. The system is aerated by a blower at the bottom, and the aeration rate is carefully controlled to prevent evaporation of the acetic acid. Ethanol must be fed constantly to sustain the bacterial population. The maximum acetic acid in solution is 13% to 14%. The life of a well-packed and maintained generator is about 20 years. This is an example of a nonstop **continuous fermentation**.

Batchwise fermentation The other major system for vinegar production is the Frings acetator, which is a submerged, **batchwise fermentation** process (**Figure 31.6**). The Frings acetator is composed of a stainless-steel tank with a high-speed mixer that constantly stirs the microbes, air, ethanol, and nutrients to provide a favorable environment for bacterial growth. Small air bubbles are introduced at the bottom to provide aeration. Because aeration can cause foaming, foam produced is removed by a foam destroyer at the top of the tank. The tank is operated at 30°C, and this temperature is maintained by circulation of cooling water. Ethanol, bacterial inoculum, and selected nutrients required for bacterial growth are added to the acetator in a batch. An electronic instrument monitors the ethanol content. When the alcohol level decreases to 0.2% by volume, about one-third of the finished product is removed, and fresh nutrients and ethanol are added. It takes about 35 hours to make 12% vinegar. The acetator is a more efficient system than the trickling generator.

Several types of vinegar are manufactured and are classified by the origin of the ethanol that the bacterium oxidizes to acetic acid. The basic flavor component in vinegar is acetic acid, but distinct flavors are contributed by the original source of the ethanol. The most widely used types of vinegar include the following:

- *Distilled white vinegar*—made from distilled ethanol. The origin of the ethanol may be fermentation ethanol or obtained by chemical synthesis
- *Cider vinegar*—made from the fermented juice of apples
- *Wine vinegar*—made from a low-quality wine that has been subjected to aerobic oxidation. Either white or red wine can be converted to vinegar
- *Malt vinegar*—produced from the alcohol obtained by fermenting corn or barley starches that have been treated with enzymes to release sugars for fermentation

Nucleotides For centuries the Japanese have used dried seaweed, dried bonito, or other dried fish as flavoring agents. Two of the major ingredients in these dried products that enhance the flavor of food are monosodium glutamate and the histidine salt of inosinic acid. Inosinic acid is a nucleotide (Figure 3.18), and the production of this and other nucleotides will be discussed here. Monosodium glutamate fermentations will be discussed in the section along with other amino acids. As flavor **enhancers,** the descending order of effectiveness among the nucleotides is guanylic acid, inosinic acid, and xanthylic acid. These nucleotides are added to enhance the flavor of soups and sauces and are effective at very low concentrations (0.005% to 0.01%). Production of 5' IMP and 5' GMP is about 4,000 tons/year.

Originally nucleotides were produced by enzymatic hydrolysis of yeast RNA. *Candida utilis*, a yeast, is rich in RNA and was one of the sources. Now, over half the commercial production of nucleotides is by fermentation. The production of 5' IMP is by direct fermentation or through the microbial production of inosine that can be chemically phosphorylated. Strains of *Brevibacterium ammoniagenes* that are auxotrophs for adenine and guanine will produce about 30 grams/liter inosine. Some microorganisms can produce 5' IMP by direct fermentation but must have an altered membrane permeability. Nucleosides (no phosphoryl group) can traverse a normal cytoplasmic membrane but nucleotides (phosphorylated) cannot. Selected permeability mutants of *B. ammoniagenes* can produce about 27 grams/liter 5' IMP.

The production of 5' GMP is generally by a combination of fermentation and chemical synthesis. A purine auxotroph of *Bacillus megaterium* is employed that accumulates an

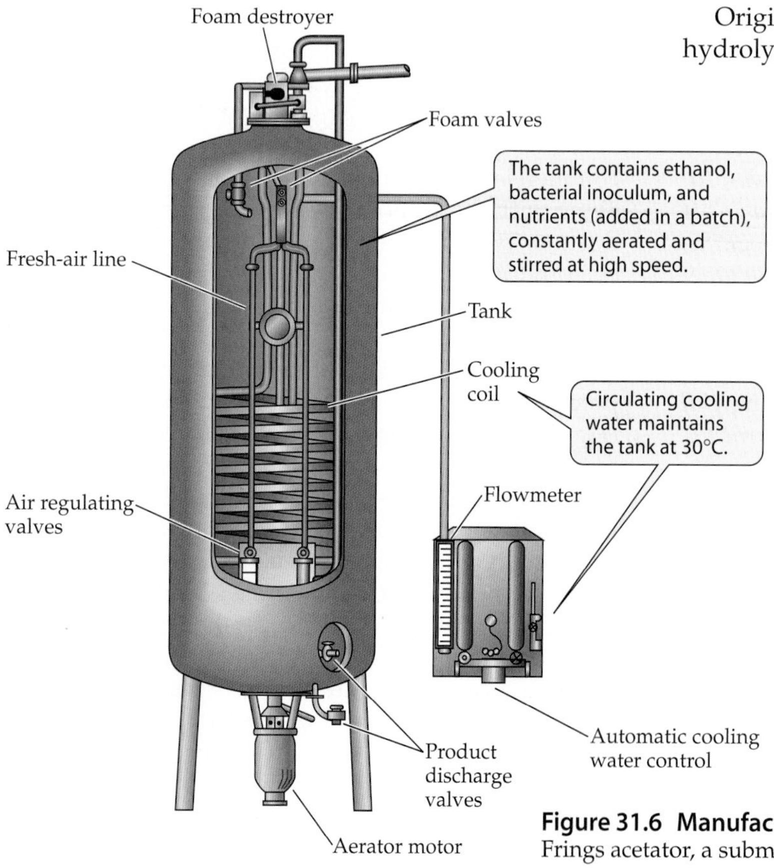

Foam destroyer

Foam valves

The tank contains ethanol, bacterial inoculum, and nutrients (added in a batch), constantly aerated and stirred at high speed.

Fresh-air line

Tank

Cooling coil

Circulating cooling water maintains the tank at 30°C.

Air regulating valves

Flowmeter

Automatic cooling water control

Product discharge valves

Aerator motor

Figure 31.6 Manufacture of vinegar by batchwise fermentation
Frings acetator, a submerged system for vinegar production.

intermediate of purine biosynthesis, aminoimidazole carboxamide ribose (AICAR). The *B. megaterium* strain will produce 16 grams/liter AICAR (**Figure 31.7**), and this is chemically transformed to 5' GMP. Guanosine-excreting mutants are also employed in commercial production of 5' GMP.

Amino Acids Amino acids are a major industrial product with an annual worldwide production of over 400,000 tons. Some of the major uses of amino acids are listed in **Table 31.4**. Amino acids are used to enhance the quality of food, in medicine, or as a starting material in chemical synthesis. The major amino acid produced is glutamic acid at 80% of the total and lysine is second in volume at 10% of the total. The biologically active form of amino acids is the L-form and is the isomer produced by fermentation. Chemical synthesis yields a mixture of the D and L isomers.

Amino acids are synthesized in families by microorganisms based on the precursor metabolite that serves as the starting material. The synthesis of individual amino acids is tightly regulated, and generally the amount synthesized corresponds to the amount essential for the synthesis of cell protein. Thus, the initial reactions in the biosynthesis of a particular amino acid may be inhibited by feedback inhibition (Chapter 13). Repression is also used that curtails the synthesis of an enzyme by an external compound, the repressor. If one wishes to obtain a culture that will produce excessive amounts of an amino acid, then both feedback inhibition and repression must be overcome. Through mutation and selection, microorganisms have been developed that are efficient producers of most of the common amino acids.

Most amino acid manufacturing processes originated in Japan. Some strains that are used commercially are shown in **Table 31.5**. The major glutamic acid-producing organisms have a requirement for biotin and lack significant levels of α-ketoglutarate dehydrogenase. When provided with low levels of biotin, the organism synthesizes leaky and inadequate cytoplasmic membranes (biotin is involved in synthesis of fatty acid). These leaky membranes permit the overproduced glutamic acid to be excreted into the medium. L-lysine can also be produced in high yield by direct fermentation using a mutant of *Brevibacterium flavum* that requires homoserine and leucine. Lysine is excreted into the medium by active transport.

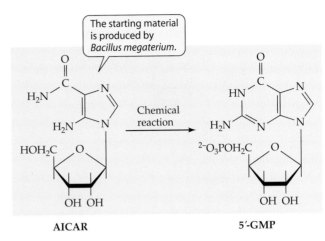

Figure 31.7 Microbiological/chemical production process Example of combined microbiological and chemical processes in generating a commercial product. Aminoimidazole carboxyamide ribose (AICAR)—an intermediate in purine biosynthesis—is produced microbiologically and transformed chemically to 5'-guanosine monophosphate (5'-GMP).

Table 31.4	Uses of amino acids
Amino Acid	**Use**
Sodium Glutamate	Flavor enhancer
L-lysine	Added to plant-derived foods to improve nutritional quality
L-methionine	
L-threonine	
L-tryptophan	
L-tryptophan	Antioxidant in powdered milk
L-histidine	
Aspartate	Improve flavor of fruit juices
Alanine	
L-aspartyl-l-phenylalanine methyl ester (aspartame)	Low-caloric sweetener
Mixed amino acids	Infusion solutions for surgery patients

Table 31.5	Commercially produced amino acids and the microbes involved in their synthesis		
Amino Acid	**Organism**	**Approximate Yield (Grams/Liter)**	**Carbon Source**
L-glutame	*Corynebacterium glutamicum*	>100	Glucose
L-lysine	*Corynebacterium*	39	Glucose
L-lysine	*Brevibacterium flavum*	75	Acetate
L-threonine	*Escherichia coli* K$_{12}$	55	Sucrose

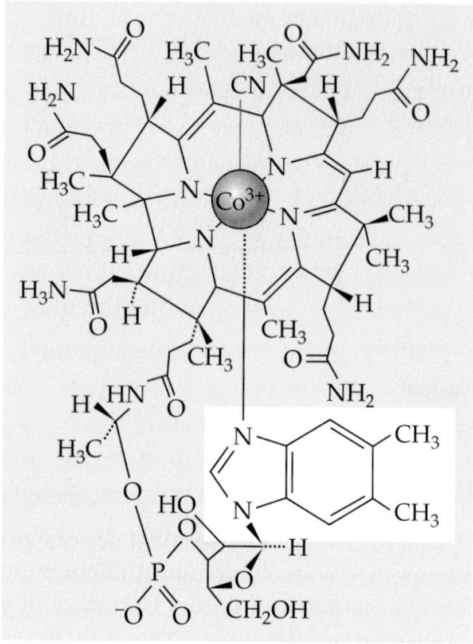

Vitamin B$_{12}$ (cyanocobalamin)

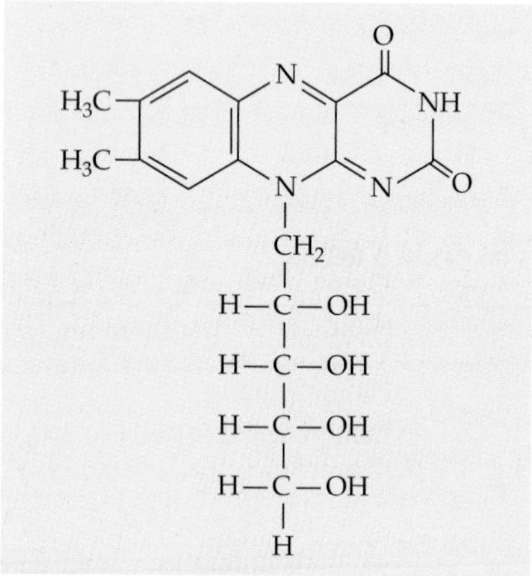

Riboflavin

Figure 31.8 B complex vitamins
Structures of vitamin B$_{12}$ and riboflavin, two members of the vitamin B complex produced commercially by fermentation.

Vitamins Mutation and selection can lead to microbial strains that overproduce almost all of the B vitamins. However, only two of the B vitamins, riboflavin and B$_{12}$, are produced by fermentation, because chemical synthesis of the others is less costly. The structures of vitamin B$_{12}$ and riboflavin are presented in **Figure 31.8**. vitamin B$_{12}$ is a complex molecule that, interestingly, is only synthesized by prokaryotic microorganisms. It is required by humans and supplied in food or is produced by the normal intestinal microbiota. Pernicious anemia can occur in humans that obtain insufficient levels of dietary B$_{12}$.

The annual current production of vitamin B$_{12}$ is about 10,000 kg. Unpurified B$_{12}$ is added to swine and poultry feed at about 12 mg/ton, and this figure affirms that B$_{12}$ is active at exceedingly low supplement levels. Vitamin B$_{12}$ is synthesized in bacteria from intermediates common to heme or chlorophyll synthesis. The organisms involved in vitamin B$_{12}$ production are *Propionibacterium shermanii* and *Pseudomonas denitrificans*. The latter produces about 60 mg/liter under appropriate growth conditions.

Riboflavin occurs naturally in milk, eggs, meat, and other food. It is added as a supplement to bread, milk, and other food to improve their nutritional quality. Yeasts such as *Ashbya gossypii* are employed in the commercial production of riboflavin, and the yield can be in excess of 7 grams/liter.

Vitamin C (ascorbic acid) is produced by a combination of chemical and microbiological synthesis. It is also used as an antioxidant. World production exceeds 35,000 tons per year. The starting substrate, D-glucose, is reduced chemically to D-sorbitol, and a microbial oxidation step leads from D-sorbitol to L-sorbose, which is chemically converted to ascorbic acid (**Figure 31.9**). The microbe involved in the dehydrogenation is *Acetobacter suboxydans*.

Figure 31.9 Manufacture of vitamin C
Production of vitamin C (ascorbic acid) by a combination of fermentation and chemical reactions.

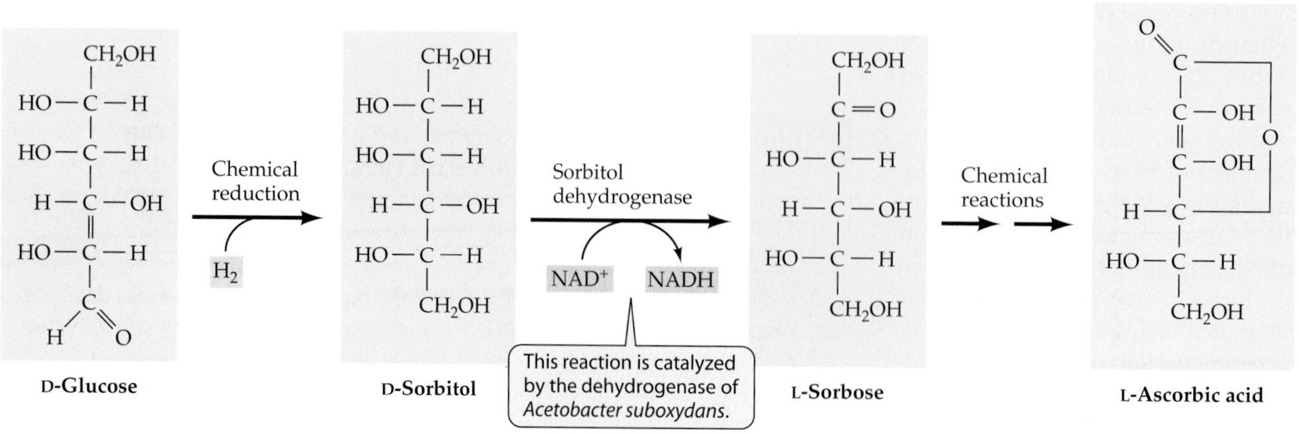

Beverages Containing Alcohol

The alcoholic beverage industry is economically the most significant of all commercial processes that involve microorganisms. Alcohol-containing beverages such as beer and wine are as old as civilization itself. Mead is considered to have been the first consumable alcoholic beverage. Mead is the name given to the product of honey fermentation and is the oldest word associated with drinking; it has been traced to the early Sanskrit language. Quite likely, it originated when honey, which was gathered by the earliest recorded civilizations, became diluted with water, allowing a natural fermentation to take place.

Alcoholic beverages are now produced worldwide from a variety of plant materials. Fruit sugars or grain polysaccharides are the major sources of fermentable substrates (**Table 31.6**). Wine, beer, and distilled beverages are the major alcoholic beverages produced commercially. Yeasts of the genus *Saccharomyces* are the organisms most involved in alcohol production. A discussion of the processes involved in production of major alcoholic beverages follows.

Wine

There are many types of wine (such as grape, peach, pear, dandelion), but by far the most favored is that made from grapes. The leading countries in wine production are Italy, Spain, France, Portugal, United States, Australia, South Africa, and Argentina. Italy and Spain combined produce about 60% of the world total. The best area for wine-producing vineyards is between 30° and 50° North or 30° and 40° South of the equator. In these areas of warm summer and mild winter climate, the grapes ripen slowly and yield superior wines. In the United States, California, New York, Washington, and Oregon have a favorable climate and soil for growth of wine-producing grapes.

Wine occurs in two distinct types—white and red. White wine is made from white grapes or can be made from red grapes provided the skins are removed from the **must** (must is crushed grapes ready for fermentation) prior to fermentation. Red wines gain the red pigmentation (and flavor), as they are partly fermented (3 to 10 days) in the presence of the red skins. During this period, the alcohol formed extracts and solubilizes anthocyanin pigments present in the grape skin. Rosé (pink) wines are made from pink grapes or more often from red grapes but the time of exposure to the skins is for 12 to 36 hours, and much less of the anthocyanin pigments are solubilized. A dry wine is one in which most of the sugar has been fermented to ethanol. A sweet wine has more residual sugar. A fortified wine such as sherry or port contains higher levels of ethanol, and this is attained by supplementing them with ethanol that has been distilled from other wine or from fermented grain.

Yeasts are part of the natural microbiota of grapes. Crushed grapes will undergo a natural fermentation process that eventually produces wine. Natural fermentations are not favored commercially because the natural microbiota can be inconsistent and will not produce sufficient amounts of ethanol. Much of the wine produced by a natural fermentation is not potable. In commercial processes, the grapes are crushed and the must is treated with sulfur dioxide (SO_2) or sometimes pasteurized to destroy the natural microbiota. The must is then inoculated with a proprietary strain of *Saccharomyces ellipsoideus* to bring about the desired fermentation. Laboratory-bred yeasts are favored because they have been selected to tolerate 12% to 15% ethanol, whereas the natural microbiota will cease fermenting when the ethanol content reaches 3.5% to 4%. Quality wine contains 8% to 12% percent ethanol.

A flow chart for red and white wine fermentations is presented in **Figure 31.10**. White wine fermentations are quite direct. The grapes are destemmed, crushed, and skins and solids (pomace) removed. The must is treated with SO_2 and yeast is added. After fermentation, the wine is aged in casks, tanks, or bottles. During a process called racking, red wine is drawn from the primary fermentation vat, placed in casks, and stored at a low temperature.

The aging process is typically much longer for red than white wines. Whereas white wines are ready for consumption within a year of production, most red wines require further aging. During the lengthy aging process used for some of the quality red wines, such as cabernet sauvignon, a second fermentation occurs. This is termed the malo-lactic fermentation. Residual malic acid in the wine is fermented by lactic acid bacteria and converted to lactic acid and CO_2. This lowers the pH of the wine and improves the quality. After aging wines are clarified in a process called fining. Fining can be accomplished by filtering through casein, diatomaceous earth, or bentonite. A few wines are clarified by centrifugation.

Wine fermentations may be accomplished in tanks from 50 to 50,000 gallons, but it is essential that the temperature of the fermentation is maintained at 20°C to

Table 31.6	Major alcoholic beverages and sources of sugars for fermentation
Wine	Fruit sugars
Beer	Barley (sometimes rice or corn)
Whiskey	Rye, corn, barley
Tequila	Agave cactus
Rum	Sugar cane molasses
Mead	Honey
Sake	Rice

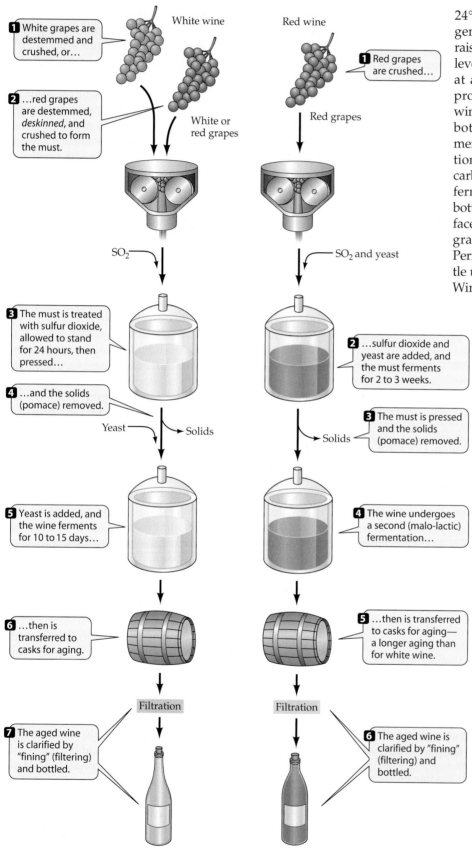

Figure 31.10 Manufacture of wine
Flow charts for the production of red and white wines.

24°C to produce quality wine. Heat generated during metabolism can raise temperatures above the tolerance level of the yeast. Aging is carried out at a lower temperature, as this improves flavor and aroma. Sparkling wines such as champagne are aged in bottles with a secondary yeast fermentation that generates the carbonation (bubbles). Some champagne is carbonated by CO_2 injection after the fermentation. Champagne is aged in bottles that are slanted so that the cork faces down. This allows sediments to gravitate to the neck of the bottle. Periodically the neck is frozen, the bottle uncorked, and sediment removed. Wine lost during this process is replaced before recorking. All wines, including champagne, that are stoppered with a cork should be stored with the bottles lying on their side. This retains moisture in the cork and prevents air from entering. Presence of air allows growth of vinegar bacteria that convert ethanol to acetic acid, thus "spoiling" the product.

Beer The basic raw material for brewing of beer worldwide is barley. The barley must be malted, a process whereby the barley starch is converted to a substrate (sugar) that is amenable to fermentation. Malt beverages are made by fermenting the malted barley, and these include beer, stout, porter, and malt liquor. Sake, an alcoholic beverage produced in Japan, is a product of malted rice. The starch present in barley, rice, and other grains is not available for fermentation unless the polymer is depolymerized to low molecular weight sugars.

The malting process is a series of manipulations that depolymerize starch to free sugars. In the malting process, barley is soaked (steeped) in cold water so that the grains absorb sufficient water to germinate

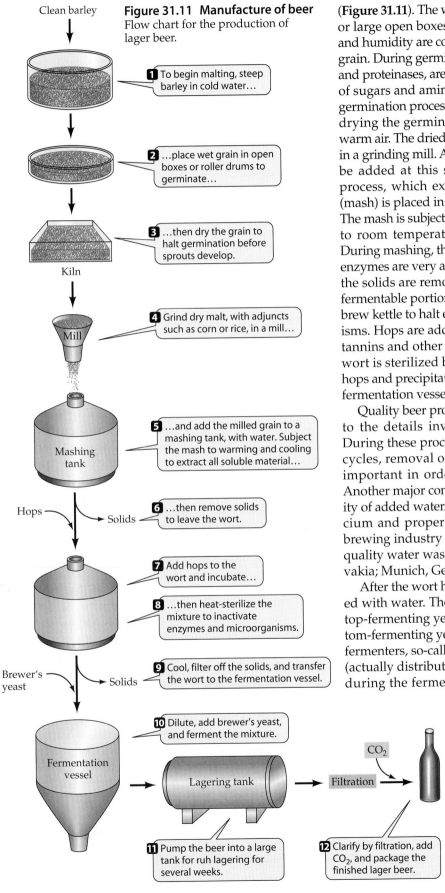

Clean barley

Figure 31.11 Manufacture of beer
Flow chart for the production of lager beer.

1 To begin malting, steep barley in cold water…

2 …place wet grain in open boxes or roller drums to germinate…

3 …then dry the grain to halt germination before sprouts develop.

Kiln

4 Grind dry malt, with adjuncts such as corn or rice, in a mill…

Mill

5 …and add the milled grain to a mashing tank, with water. Subject the mash to warming and cooling to extract all soluble material…

Mashing tank

Hops → Solids

6 …then remove solids to leave the wort.

7 Add hops to the wort and incubate…

8 …then heat-sterilize the mixture to inactivate enzymes and microorganisms.

Brewer's yeast → Solids

9 Cool, filter off the solids, and transfer the wort to the fermentation vessel.

10 Dilute, add brewer's yeast, and ferment the mixture.

Fermentation vessel

Lagering tank

CO_2

Filtration

11 Pump the beer into a large tank for ruh lagering for several weeks.

12 Clarify by filtration, add CO_2, and package the finished lager beer.

(**Figure 31.11**). The wet grain is then placed in rolling tubs or large open boxes in the presence of air. Temperature and humidity are controlled to initiate germination of the grain. During germination, enzymes, including amylases and proteinases, are activated, and these cause the release of sugars and amino acids from the malting grain. The germination process is halted before sprouts develop, by drying the germinated grain in a kiln under a flow of warm air. The dried grain is then broken into small pieces in a grinding mill. Adjuncts such as corn grits or rice may be added at this stage and subjected to the milling process, which exposes the starch. The milled grain (mash) is placed in a mashing tank, and water is added. The mash is subjected to periods of warming and cooling to room temperature to extract all soluble material. During mashing, the proteinases and starch-hydrolyzing enzymes are very active. Following the mashing process, the solids are removed and the liquid remaining is the fermentable portion called wort. The wort is cooked in a brew kettle to halt enzymatic action and kill microorganisms. Hops are added, and the wort is heated to extract tannins and other flavoring agents from the hops. The wort is sterilized by live steam and filtered to remove hops and precipitates. The wort is then transferred to the fermentation vessel.

Quality beer production depends on careful attention to the details involved in the preparation of wort. During these processes, pH, length of heating/cooling cycles, removal of precipitates, and other factors are important in order to produce a desirable product. Another major concern in beer manufacture is the quality of added water. It must be low in carbonates and calcium and properly balanced in other minerals. The brewing industry developed historically at sites where quality water was available, such as Pilsen, Czechoslovakia; Munich, Germany; and Dublin, Ireland.

After the wort has been cooled and filtered, it is diluted with water. The fermentation is initiated either by a top-fermenting yeast (*Saccharomyces cerevisiae*) or a bottom-fermenting yeast (*Saccharomyces carlsbergensis*). Top fermenters, so-called because they are carried to the top (actually distributed throughout) by the CO_2 produced during the fermentation, are used to produce **ale.** In contrast, the bottom fermenters, which tend to settle to the bottom, are used to produce **beer.** The top fermentation is carried out at 14°C to 23°C for 5 to 8 days. The ale is aged at 4° to 8° C.

The bottom fermentation for beer is carried out at 6°C to 12°C for about the same length of time as the ale. Following fermentation, the beer is pumped into a large tank for maturation. Yeast is added at about 1 million cells per ml, and the beer is allowed to "rest" for several weeks at 0.1°C. This process is called the *ruh* (German, meaning "rest") or lagering.

The finished beer is clarified by centrifugation or filtration, CO_2 is added, and the beer is packaged. Draft beer (kegs, bottles, or cans) is generally passed through membrane filters to remove all microorganisms. Prior to the availability of filter systems, bottled beer was pasteurized to prevent spoilage by acetic acid bacteria, and this pasteurization tended to produce off-flavors. Draft beer was considered superior because it was unpasteurized but was kept cold to prevent spoilage. Most European and American beer is lager beer, although ales are increasingly being produced locally by microbreweries in the United States

Distilled Beverages Distilled spirits are extensions of the yeast brewing process just discussed. The fermented liquid is placed in a closed vessel with attached cooling coils. It is then heated, and the volatiles are condensed and collected. Ethanol has a boiling point of 78.5°C, so it can be readily separated from the fermentation medium by distillation. The distillate obtained has a very high ethanol content and must be diluted before aging and bottling.

Gin, vodka, whiskey, rum, and brandy are examples of distilled alcoholic beverages. Gin and vodka are essentially the distillate from an alcoholic fermentation of grain. Gin is produced when the alcohol is refluxed over juniper berries to extract the distinct aroma and flavor of the berries. Rum is manufactured by distilling fermented cane sugar molasses, and brandy is made from distilled wine. Distilled alcoholic beverages are 40% to 43% ethanol by volume (80 to 86 proof). Gin may be somewhat higher.

Whiskey (spelled whisky in Scotland) is defined by the grain employed in the fermentation. Rye whiskey must have 51% or more rye grain in the malting process, and bourbon must have at least 51% corn. There are three types of Scotch whisky: **malt** whisky, **grain** whisky, and **blended** whisky. Malt whisky (single malt) is made from 100% malted barley. Grain whisky is produced from a variety of malted cereals (wheat, maize) that may or may not include malted barley. Blended whisky is a mixture of malt and grain whisky. Most Scotch whisky sold in the United States is blended, but single malt Scotch is growing in popularity. If a Scotch contains any whisky not made from malted barley, it cannot be labeled or sold as **malt** whisky. It is interesting to note that there are over 300 different single-malt Scotches available in Scotland.

The manufacture of whiskey apparently originated in Ireland or Scotland and moved to the United States (bourbon), Canada, and Japan. Scotch and some bourbons are made by a **sour mash** fermentation. A sour mash is produced by introducing a homolactic bacterium into the mash, and the growth of this bacterium lowers the pH of the mash to about 4.0. The distillate from a malted grain fermentation contains volatile products other than ethanol that are generated during the mashing process or fermentation. These volatile products are responsible for the distinct flavors of the various whiskeys. The distillate from grain fermentation is colorless, and the brownish color of the final product results from the aging process. Whiskey is aged in wooden barrels that have been charred on the inside. Scotch is now aged in barrels previously used for aging bourbon or sherry. During aging, desirable flavors are enhanced and the level of some less desirable substances such as fusel oils may be decreased. Fusel oils are mixtures of propanol, 2-butanediol, amyl alcohol, and other alcohols. Whiskey manufacture is an ancient art that involves little science.

ORGANIC ACIDS

Citric acid is an important **organic acid** that is synthesized by a microbial fermentation. Over 130,000 tons are produced worldwide annually. Originally, citric acid was extracted from citrus fruit (a lemon is 7% to 9% citric acid). The market controlled by cartels established in citrus-growing areas making the price prohibitively high. In 1923 a microbial fermentation was developed with high yields of citric acid, and the price plummeted as availability increased. About two-thirds of the citric acid produced commercially is used in foods and beverages including soft drinks, desserts, candies, wines, frozen fruit, jams/jellies, and ice cream. In the pharmaceutical industry, iron citrate is a source of dietary iron. Citric acid is also used as a preservative for stored blood, tablets, and ointments. Citrate has also been replacing polyphosphates in detergents due to the banning of phosphates, as they are an environmental hazard.

Most citric acid comes from a submerged fermentation in large, stainless-steel fermenters using strains of the fungus *Aspergillus niger.* Sucrose is generally the substrate and the citric acid production occurs during the idiophase, as it is a secondary metabolite, and the fermentation follows the pattern presented in Figure 31.1. During the trophophase, mycelium is produced and CO_2 is released. During the idiophase, glucose and fructose are metabolized directly to citric acid. Little CO_2 is produced. Under optimal conditions, about 70% of the sugar is converted to citric acid. A critical factor in the fermentation is the level of iron and other minerals that are inhibitory. The fermentor must be made of stainless steel or be glass lined, and copper fittings must be avoided.

A variety of other organic acids can be synthesized by microbes. One example is **itaconic acid,** which is used in methacrylate resins as a copolymer. Adding 5% of this product to printing ink increases the ink's ability to adhere to paper. It also has potential as a detergent. The organism involved in production of itaconic acid is a strain of *Aspergillus terreus*. In submerged fermentation, the yield from beet molasses is 85% of the theoretical (**Figure 31.12**).

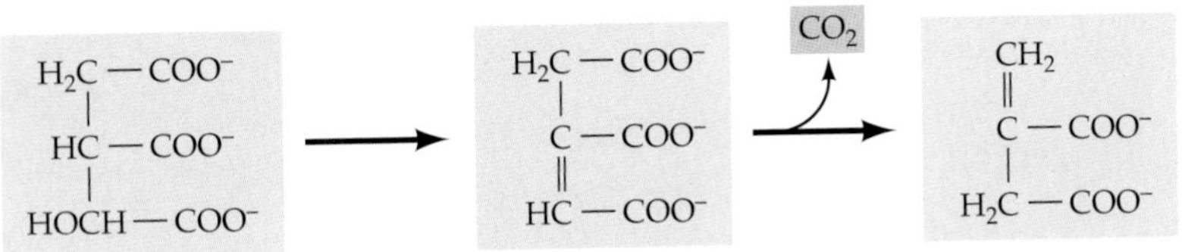

Figure 31.12 Synthesis of itaconic acid
Biochemical reactions in the synthesis of itaconate.

ENZYMES, MICROBIAL TRANSFORMATIONS, AND INOCULA

In a previous section we discussed the enzyme rennin, a coagulant employed in cheese manufacture. This is but one of the many applications of commercially produced enzymes. Selective enzymes from various microorganisms can hydrolyze macromolecules that are of little value, to monomers of greater value, at low cost, and under favorable conditions of temperature and pH.

Enzymes

In nature, microorganisms utilize enzymes to disassemble proteins, starch, pectin, lipids, and other insoluble large molecules, reducing them to monomers that can serve as a carbon or energy source for themselves or others in the environment (see Chapter 12). These hydrolytic enzymes are generally extracellular or on the cell surface and are produced by both bacteria and fungi. As extracellular enzymes, they can be recovered quite readily from the growth medium. Several commercial applications of microbe-generated enzymes are presented in **Table 31.7**. The production of bacterial proteases is a leading industrial process both in bulk and value. Over 500 tons of these enzymes are produced each year and are mostly utilized in the manufacture of detergents and cheese (rennin). The enzymes used in detergents are mostly proteases from selected strains of *Bacillus amyloliquefaciens*. They are added to detergents to promote the solubilization of stains present on the material being washed. A significant problem with the addition of proteases to detergents has been the allergic response to the bacterial protein. This has been overcome by microencapsulation and other techniques that produce dustless protease preparations.

Amylases are enzymes that hydrolyze starch. Various microbial amylases hydrolyze starch in different ways to generate short-chain polymers (dextrins) and maltose. Other enzymes hydrolyze the dextrins and maltose to glucose. The α-amylases cleave internal α-1,4 glycosidic bonds and are produced commercially by members of the thermophilic, endospore-forming genus *Bacillus*. Fungi such as *Aspergillus* species also produce these enzymes. Glucoamylases split glucose from the nonreducing end of starch. These enzymes are in commercial demand for the production of fructose syrups. Fructose is sweeter than glucose and is the favored sugar in soft drinks and in syrups. *Aspergillus niger* is a producer of glucoamylase. Conversion of glucose to fructose can be accomplished with a glucose **isomerase** from bacteria. The most important of the glucose isomerase producers are *Streptomyces* species and *Bacillus coagulans*.

The enzyme Taq polymerase is employed in the **polymerase chain reaction** (see Chapter 16). This enzyme is derived from *Thermus aquaticus*, a thermophilic bacterium that has an optimal growth temperature of 70°C. The polymerase from this organism is heat stable. The enzyme is widely employed in research, diagnostics, and forensic medicine.

Table 31.7	Microbial enzymes and their commercial application	
Enzyme	**Genus of Producer**	**Use**
Bacterial proteases	*Bacillus, Streptomyces*	Detergents
Asparaginase	*Escherichia, Serratia*	Antitumor agent
Glucoamylase	*Aspergillus*	Fructose syrup production
Bacterial amylases	*Bacillus*	Starch liquefaction, brewing, baking, feed, detergents
Glucose isomerase	*Bacillus, Streptomyces*	Sweeteners
Rennin	*Alcaligenes, Aspergillus, Candida*	Cheese manufacture
Pectinase	*Aspergillus*	Clarify fruit juice
Lipases	*Micrococcus*	Cheese production
Penicillin acylase	*Escherichia*	Semisynthetic penicillins
Taq polymerase	*Thermus aquaticus*	Polymerase chain reaction

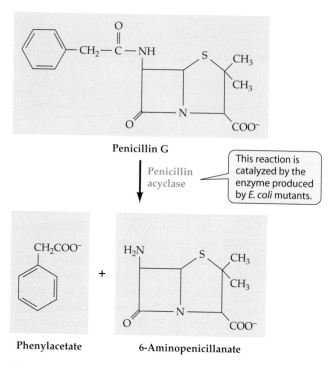

Figure 31.13 Synthesis of precursor of synthetic penicillins
Cleavage of penicillin G (generated by fermentation) to phenylacetate and 6-aminopenicillanate. The latter can be chemically modified to produce effective antibacterials.

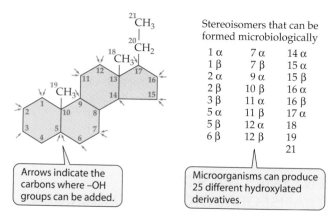

Figure 31.14 Hydroxylation of steroids
Positions at which selected microorganisms can hydroxylate the sterol nucleus. Two stereoisomers are possible for hydroxylation at many of the positions, designated α and β.

The enzyme *asparaginase* is employed medically in the treatment of some leukemias and lymphomas. The metabolism of these tumor cells requires the amino acid L-asparagine. Asparaginase cleaves asparagine to aspartic acid and ammonia, thereby starving the tumor cell. Commercial production of the enzyme is by selected mutants of *Escherichia coli, Erwinea, Carotovora,* and other species.

Penicillin acylases are produced by various bacteria and fungi, but commercial production uses *E. coli* mutants. These enzymes cleave penicillin to 6-amino penicillanic acid and phenylacetic acid (**Figure 31.13**). Then, the free amine on 6-aminopenicillanic acid can be chemically modified to produce various semisynthetic penicillins (see later discussion of Antibiotics).

Microbial Transformations

A dramatic event in the course of medicinal chemistry was the discovery that microorganisms can effectively hydroxylate the steroid nucleus without adverse changes. The anti-inflammatory activity of a steroid compound depends on the presence of an oxygen or hydroxyl in the 11-position (**Figure 31.14**). Naturally occurring steroids produced in significant quantities, such as those from yeasts or plants, lack the 11-hydroxylation. Prior to the discovery that microbes could

hydroxylate the steroid nucleus and with a virtual stoichiometric yield, the hydroxylation was accomplished chemically and the yield was very low. The simple, low-cost microbial hydroxylation therefore had a marked effect on the availability of biologically active sterols.

An outline of the steps involved in the synthesis of biologically active sterols is presented in **Figure 31.15**. The starting material, diosgenin, is from a yam grown in Mexico.

Microbial species are available that can hydroxylate the sterol nucleus at any position (Figure 31.14). Naturally occurring hormones such as progesterone, testosterone, estrone, and those from the adrenal cortex differ in number, type, and location of substituent groups or double bonds on the steroid nucleus. These can now be synthesized inexpensively for medical uses. Other physically active sterols can also be made (**Table 31.8**). Among the useful products are those that are exceedingly effective in preventing the inflammatory response in asthmatics and other allergic individuals. Some are effective as antitumor agents and for ocular diseases. Oral contraceptives would not have been available at low cost without the microbe-based synthetic process for sterols.

Table 31.8	Uses of synthetic sterols
Anti-inflammatory agents	
Sedatives	
Antitumor agents	
Dermatology	
Ocular disease	
Cardiovascular therapy	
Oral contraceptives	

Inocula

Bulk microbial biomass is commercially available as inocula for a number of processes. Lactic acid bacteria are among the microorganisms grown in bulk and sold to fermentation companies involved in the production of cheese and fermented meats. Use of pasteurized milk as a starting material for cheese production is necessary to prevent the spread of milk-borne diseases. However, pasteurization lowers the natural populations of lactic acid bacteria and inoculation with desirable strains to promote cheese production is necessary. Manufacture of summer sausage and other fermented meats is best achieved by inoculating with the appropriate bacterium.

Soil in which leguminous crops are planted often have a low population of symbiotic nitrogen-fixing bacteria. Therefore, planting leguminous crops in these soils may not lead to significant symbiotic nitrogen fixation. Inoculation of seeds with a compatible strain of *Rhizobium* or *Bradyrhizobium* prior to planting can significantly enhance root nodulation and nitrogen fixation. The symbiotic nitrogen fixers are grown commercially and sold as inocula. Field evidence has proven that this is a worthwhile practice.

INHIBITORS

Microbes can produce catabolites that are antagonistic to other biota. As discussed following, these may be a **biocide** (toxin) that kills selected insects or other eukaryotes or an **antibiotic** that inhibits other prokaryotic species.

Biocides

The most successful commercial product generated by a microorganism for insect control is a protein synthesized by *Bacillus thuringiensis*. This protein appears as a parasporal crystal during sporulation and is released along with the spore on disintegration of the sporulating bacterium (**Figure 31.16**). The vegetative cell of *B. thuringiensis* lacks toxicity, but the protein crystal is effective in killing lepidopteran insects. It is most effective in insects that have an alkaline pH in their midgut, as the toxic protein is solubilized at the higher pH. The proteins from ingested spores are cleaved by gut proteases, and the polypeptide toxins destroy gut epithelial cells.

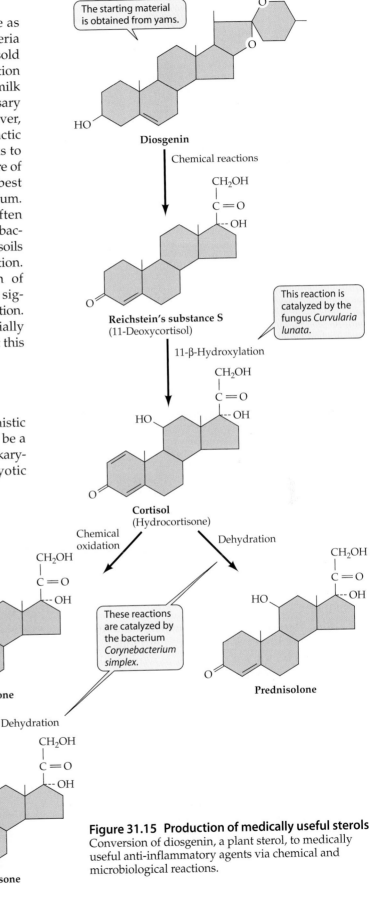

Figure 31.15 Production of medically useful sterols Conversion of diosgenin, a plant sterol, to medically useful anti-inflammatory agents via chemical and microbiological reactions.

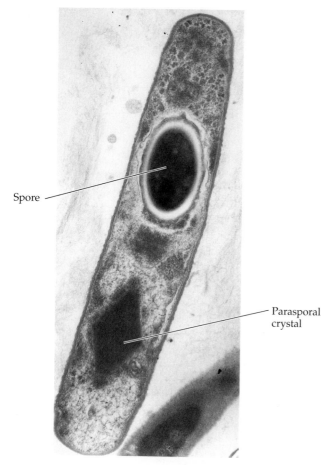

Spore

Parasporal crystal

Figure 31.16 Bacterial insecticide
Spore and diamond-shaped parasporal crystal produced by
Bacillus thuringiensis. The crystal is an effective insecticide.
Photo ©G. B. Chapman/Visuals Unlimited.

The gut contents are released into the blood system of
the insect, resulting in paralysis and death. The toxin is
harmless to other animals because of the neutral or acid
pH of their intestinal tract.

Bacillus thuringiensis grows well in submerged fermentation and under appropriate conditions will produce spores and the parasporal body within 30 hours.
After inoculation of the fermenter, growth, and sporulation, the spore/crystals are recovered and incorporated into an inert carrier and marketed as an insecticide
to be dusted on plants. The active compound is naturally occurring and therefore degraded by soil microbes.

Antibiotics

The use of antibiotics in controlling microbial growth has
been discussed (Chapter 7). Approximately 10,000 different **antibiotics** have been characterized, and about 160
of these are of value and in commercial production.
Antibiotics are produced mainly by fungi and *Bacteria*.
The useful antibiotics are mostly produced by a large

group of filamentous soil bacteria known collectively as
the actinomycetes. **Streptomyces** is the one genus that
has proven to be a most prolific producer, and the search
for commercially valuable antibiotics has centered on it
and related genera. It is a possibility that other microbial
species, not now under investigation, might produce
useful antibiotics. Experience, however, has shown that
microorganisms that have a life cycle, ergo spore formers (both endospores and other spores), are the most
effective in producing useful antibiotic substances.
Antibacterial agents were discussed in Chapter 7, but it
should be emphasized that agents have also been discovered that are active against some fungi, tumors, and
parasites (**Table 31.9**). Many of the antitumor agents are
produced by *Streptomyces* sp., as are agents effective as
antifungals. *Avermectin* is used in veterinary medicine for
treatment of intestinal worms.

Many antibiotics are now available for the medical
profession, but the search for antibiotics continues.
Antibiotics against bacteria, fungi, viruses, and tumors
are actively sought. This search includes antimicrobials
that are effective against disease-causing microbes now
seemingly under control. One of the key problems in the
use of antibiotics and chemotherapeutics is the
pathogen's development of resistance to the agent.
Antibiotic-producing organisms generally bear genetic
information that renders them resistant to the antibiotic they produce. This genetic information (an antibiotic
resistance gene) is thus available in nature and can pass
by horizontal gene transfer through various microbes
and, ultimately and inevitably, to disease-causing organisms. Resistance to antimicrobial agents also occurs
through mutation and selection (see Chapter 13). Hence,
antibiotics are constantly being sought that are not cross-
resistant to those in current use. Cross-resistance is the
phenomenon that occurs when a bacterium becomes

Table 31.9	Antibiotics that are effective against eucaryotic cells	
Producing Organism	**Substance Produced**	
Antifungals		
Streptomyces griseus	Cycloheximide	
Streptomyces noursei	Nystatin	
Penicillium griseofulvum	Griseofulvin	
Streptomyces nodosus	Amphotericin B	
Antitumor		
Streptomyces peucetius	Daunorubicin	
Streptomyces antibioticus	Actinomycin C	
Streptomyces caespitosus	Mitomycin C	
Streptomyces verticillus	Bleomycin	
Antihelminth		
Streptomyces avermitoles	Ivermectin	

resistant to one antibiotic that spontaneously gives resistance to another. It is clear that new agents effective against bacteria, fungi, viruses, or tumors would be of great value.

Several new approaches have been developed for screening for new antibiotics based on an increased understanding of the unique characteristics of pathogenic microbes. Antibacterials targeted against specific virulence factors would minimize the emergence of resistant mutants. This would be appropriate for pathogens that develop virulence factors through quorum sensing.

Molecular models of the proteins involved in adherence would permit the design of specific agents that would prevent this important step in establishing infections (Chapter 26). Sequencing the genomes of pathogens will ultimately identify the gene sequences that code for virulence factors, adherence factors, and toxin production. This can lead to the design of specific chemical agents to counter their activity. Designer drugs such as antisense molecules that react with m RNAs that direct synthesis of proteins involved in pathogenesis are one possibility.

The following discussion presents the industrial approach to the discovery of new and useful antibiotics. Much of the methodology can be applied with some adaptation to the search for biocides, growth factors, or other useful compounds.

Search for New Antibiotics

The indispensable factor in any search for novel compounds is the development of an **assay** system. The search is termed a screen, and the detection system for desired activity is the assay. The assay must give accurate, rapid results at a low cost. Whether the screen is for inhibitors, novel enzymes, or organic or amino acids, an efficient assay is indispensable. A potential procedure that could be employed in the search for antibiotics is as follows:

- **Identify a source of potential producers.** In searching for antibiotics, all past experience affirms that actinomycetes are among the best microbes to screen. The best sources for them are neutral to slightly alkaline soils, composts, and peat.
- **Isolation of organisms.** Isolation is usually accomplished by spreading suitable dilutions of the soil on the surface of an agar medium. Petri plates with a medium made up of hydrolyzed casein (animal), soytone (plant), and yeast extract (microbe) are suitable. Sugar content is best kept low to limit the overgrowth of slower-growing actinomycetes by rapidly growing pseudomonads or bacilli. The leathery, compact dry colonies of the actinomycetes are readily apparent to an experienced technician.
- **Preliminary plate test.** The capacity of the isolated

microorganism to produce an antibiotic substance can be ascertained by using the plate test method. The actinomycete is inoculated across an agar plate as depicted in **Figure 31.17** *Streptomyces* and other actinomycetes form a compact continuous colony when streaked across a plate. After a few days, the test organisms are streaked perpendicular to the actinomycete. Usually a gram-positive and gram-negative bacterium is employed to select for potential agents. A mycobacterium would be used to select for antituberculosis antibiotics and a yeast for antifungals. Rarely is an antibiotic effective against all four of these test organisms unless it is cytotoxic. In the case illustrated, the antimicrobial is effective against *Staphylococcus epidermidis*. If a significant zone of inhibition against any of the test organisms is apparent, the actinomycete passes to the next phase.

- **Assay for novel activity.** With the vast number of known antibiotics, most activity discovered will likely be due to an antibiotic already described. It is of utmost importance that producers of known antibiotics be eliminated from the screen quickly. This is accomplished by following several possible protocols. First, one grows the selected actinomycete in liquid culture and subjects it to a paper disk plate assay to ensure that antimicrobial activity is repeatable (**Figure 31.18**). Various tentative identification procedures then might be applied to the growth medium from the initial fermentation including paper chromatography, electrophoresis, pigment production,

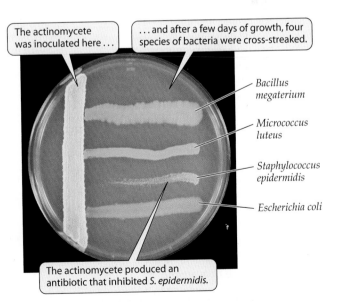

The actinomycete was inoculated here . . .

. . . and after a few days of growth, four species of bacteria were cross-streaked.

Bacillus megaterium

Micrococcus luteus

Staphylococcus epidermidis

Escherichia coli

The actinomycete produced an antibiotic that inhibited *S. epidermidis*.

Figure 31.17 Plate test for antibiotic
Production of an antibiotic substance by a *Streptomyces* sp. isolated from soil. The agar plate contains a medium that supports growth of the actinomycete and the four test bacterial species. Courtesy of J. J. Perry.

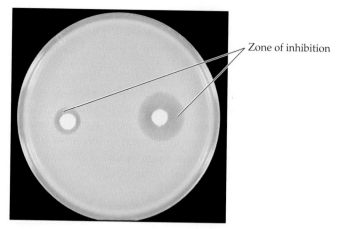

Figure 31.18 Paper disk plate assay for antibiotic
Detection of antibiotic activity by an agar plate assay.
Graded amounts of fermentation broth are added to filter-
paper disks and placed on the surface of an inoculated agar
plate. The size of the zone of growth inhibition is proportional to the amount of antibiotic present. Courtesy of J. J. Perry.

the antimicrobial is the same as that from the original
isolate. Favorable results will lead to production in
laboratory-scale fermentors. Chemists are then
recruited in efforts to chemically identify the antimicrobial substance. If the compound continues to
appear novel, animal testing would be warranted.

- **Animal testing.** Mouse tests tell whether the compound is toxic or destroyed by mammalian enzymes.
Such tests determine whether the antimicrobial is
effective in vivo. If all goes well, the antibiotic proceeds to higher animal tests, human tests, and in the
rare case, ultimately to the pharmacy.

Commercial Production The commercial production of
antibiotics is accomplished in huge fermentors (**Figure
31.19**) that vary in size from less than 40,000 to more
than 200,000 liters. Size is limited by the ability to provide sufficient aeration for maximum rates of growth.
The heat generated during growth is removed by cool-

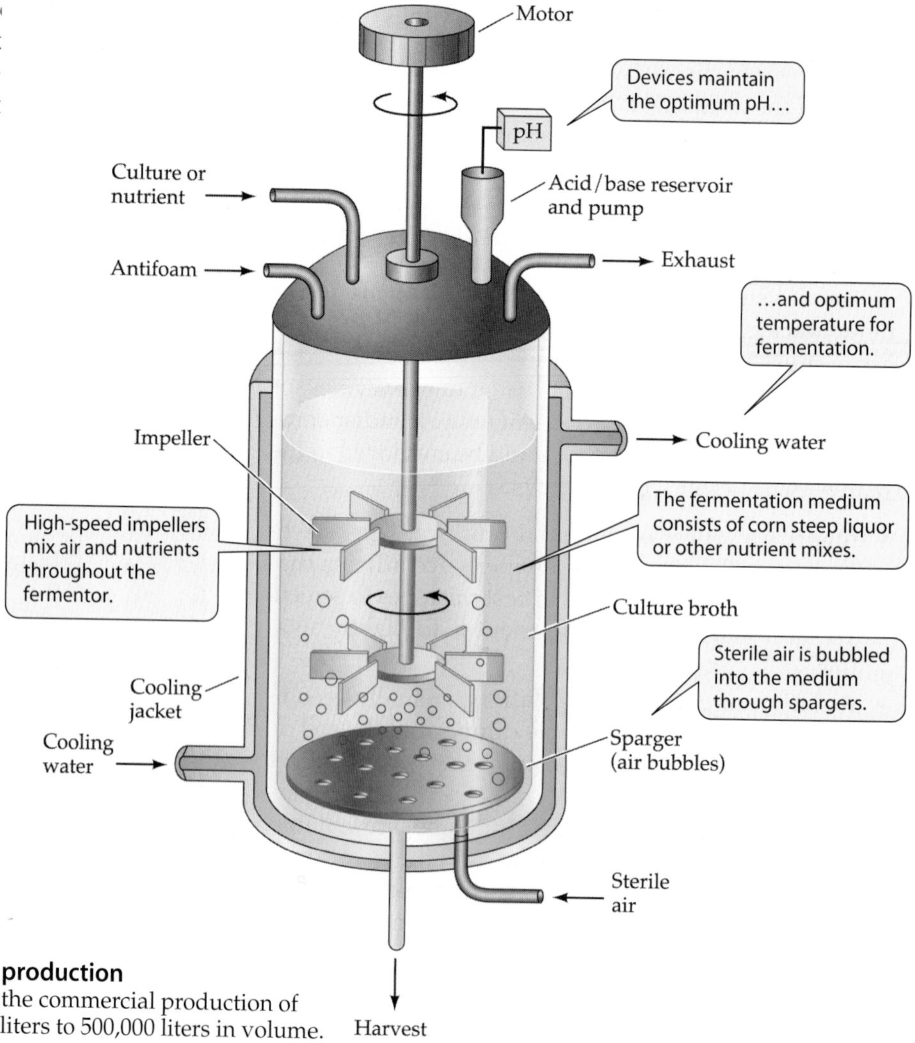

absorption spectra, antimicrobial spectrum, and others. Activity against microbes kn[ ]
to antibiotics now in use can als[ ]
the antimicrobial substance appe[ ]
be more extensively characterized

- **Advanced culture screening.**
At this stage, if the activity
appears novel, the isolated
organism is subjected to a series
of procedures to increase the
amount produced to levels that
permit isolation and characterization. Experience tells us that
newly isolated organisms produce only a few micrograms per
milliliter of antibiotic and generally synthesize a series of chemically related antimicrobials so
that no single one of these
would be present in significant
amounts. Various parameters
are tested for their positive
effects on antibiotic production:
pH, aeration, growth substrate,
length of fermentation, and
other conditions. These efforts to
increase the amount of antibiotic produced are monitored by
assay techniques to ensure that

Motor

Devices maintain
the optimum pH…

pH

Culture or
nutrient

Acid/base reservoir
and pump

Antifoam

Exhaust

…and optimum
temperature for
fermentation.

Cooling water

Impeller

The fermentation medium
consists of corn steep liquor
or other nutrient mixes.

High-speed impellers
mix air and nutrients
throughout the
fermentor.

Culture broth

Sterile air is bubbled
into the medium
through spargers.

Cooling
jacket

Cooling
water

Sparger
(air bubbles)

Sterile
air

Figure 31.19 Fermentor for antibiotic production
Basic components of a fermentor used in the commercial production of
antibiotics. Fermentors range from a few liters to 500,000 liters in volume.

Harvest

ing coils, as the optimal temperature for a typical *Streptomyces* fermentation is generally 26°C to 28°C. Aeration is a significant factor and varies somewhat, but generally requirements are for 0.5 to 1.5 volumes of air/volume of the fermentor/minute. This requires that huge amounts of sterile air be introduced into the fermentor through spargers. Oxygen is poorly soluble in water, and the impellers must turn at over 100 rpm to ensure adequate air saturation of the growth medium. The pH is monitored to maintain the optimal pH for antibiotic production. Antifoam agents are added to control foaming under these high aeration/mixing conditions. Powerful motors drive the impellers, and the shaft of the impeller passes from the outside to the inside of the fermentor. Contamination of the growth medium is a considerable concern and can cause serious economic loss. The fermentor must be constructed so that it can be sterilized completely, usually with live steam passed through the cooling coils, and every outlet must be constructed to preclude contamination.

The medium in a fermentor varies but generally consists of corn steep liquor, a molasseslike substrate, or various combinations of sugar/yeast extract/soytone. The addition of substrates as the fermentation progresses is sometimes desirable. Precursors of antibiotics obtained by chemical synthesis may also be added as the fermentation progresses. The elapsed time for a fermentation run is about 4 days.

The fermentation process is outlined in **Figure 31.20**. The antibiotic-producing stock cultures are carefully maintained either by lyophilization, under liquid nitrogen, or another suitable storage condition. To initiate the fermentation process, the organism is cultivated in a series of fermentation steps, or scale-ups, and at each step the culture is monitored for purity. The final scale-up inoculum is sizable to ensure that the final fermentation occurs quickly and without a lag phase. After growth (trophophase) and antibiotic formation (idiophase), the fermentation medium is filtered to remove mycelium, and the antibiotic is then recovered.

Antibiotics can be classified according to their antimicrobial spectrum, mechanism of action, producer strain, manner of biosynthesis, or chemical structure. A discussion follows on some major antibiotics now employed in the clinic.

Penicillin   Penicillin is produced by *Penicillium chrysogenum* and contains the biologically unusual β-lactam ring. This antibiotic has been in use for over 50 years, and microbes resistant to it have become a serious concern. Semisynthetic penicillins are more effective against microbes resistant to the natural product and are antibiotics of choice clinically. The role of microbial acylase in generating 6-aminopenicillanic acid was shown in Figure 31.13. The structures that can be added to this moiety to form effective antibacterials are illustrated in **Figure 31.21**.

Cephalosporin   This is also a β-lactam antibiotic that is produced by the fungus *Cephalosporium acremonium*. The cephalosporins have lower toxicity than penicillin and a somewhat broader antimicrobial spectrum. They are also resistant to the penicillinase produced by penicillin-resistant microbes. The cephalosporins are produced by fermentation, and some semisynthetic derivatives are generated by chemical modification. The basic structure

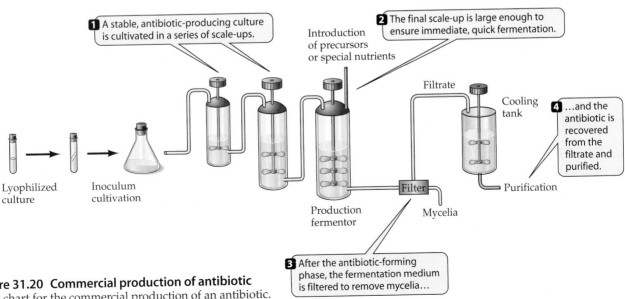

Figure 31.20 Commercial production of antibiotic
Flow chart for the commercial production of an antibiotic.

GENERAL STRUCTURE OF A PENICILLIN

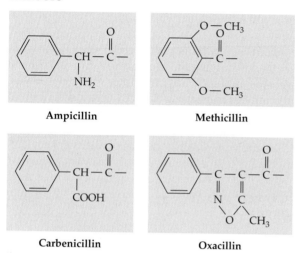

β-Lactam thiazole

R GROUPS

Ampicillin

Methicillin

Carbenicillin

Oxacillin

Figure 31.21 Synthetic penicillins
Chemical derivatives of 6-aminopenicillanate, shown as the R groups substituted in the basic penicillin structure.

GENERAL STRUCTURE OF CEPHALOSPORIN

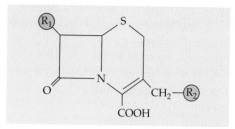

β-Lactam

R GROUPS

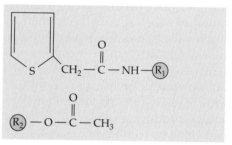

Cephalosporin C

Cephalotin

Cephalexin

Figure 31.22 Cephalosporin and derivatives
Cephalosporin C and some semisynthetic derivatives employed in treatment of infections, shown as the R groups substituted in the basic cephalosporin structure.

of cephalosporin and some semisynthetic derivatives are illustrated in **Figure 31.22**.

Streptomycin Many antibiotics are derivatives of sugars, and a major antibiotic in this group is streptomycin (**Figure 31.23**), which is produced by *Streptomyces griseus*. Streptomycin is composed of amino sugars linked through glycosidic bonds to other sugars. Its discovery was of great medical importance because it was the first effective drug against the scourge of tuberculosis. Resistance to streptomycin has been a serious problem in tuberculosis therapy, and combinations of drugs are now administered over a period of time.

Macrocyclic Lactones Macrolides are antibiotics such as erythromycin that have large lactone rings bonded to sugars. Rifamycin is a macrocyclic lactone antibiotic produced by *Nocardia mediterranei*. Rifampin is a semisynthetic macrocyclic lactone derivative synthesized from rifamycin (**Figure 31.24**). Rifampin is a specific inhibitor of the bacterial DNA-dependent RNA polymerase and is employed in tuberculosis treatment along with isoniazid and pyrazinamide.

Tetracyclines The tetracylines are a major group of antibiotics effective against both gram-positive and gram-negative bacteria. They are also effective against *Rickettsia, Mycoplasma, Leptospira, Spirochetes,* and *Chlamydia*. Some semisynthetic derivations have also been developed to counteract the bacterial drug resistance problem. The tetracylines have a naphthacene core with a number of added R-groups (**Figure 31.25**).

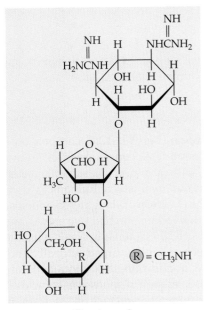

Streptomycin

Figure 31.23 Streptomycin
Structure of streptomycin, an aminoglycoside antibiotic.

GENERAL STRUCTURE OF TETRACYCLINE

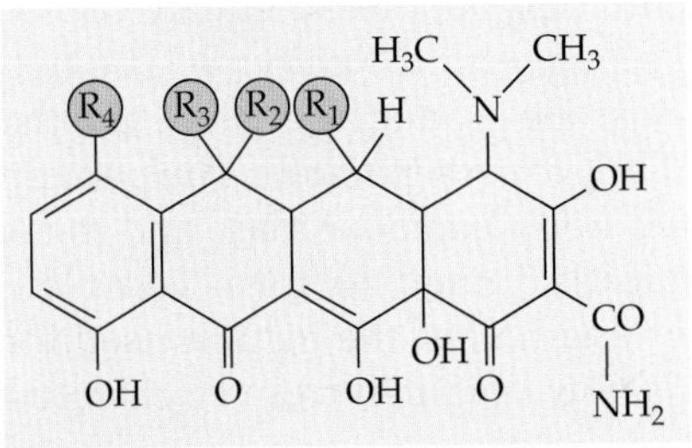

Rifampin

1-Amino-4-methyl piperazine

Isoniazid

Pyrazinamide

$$CH_3-CH_2-\underset{\underset{CH_2OH}{|}}{CH}-NHCH_2CH_2-NH-\underset{\underset{CH_2OH}{|}}{CH}-CH_2-CH_3$$

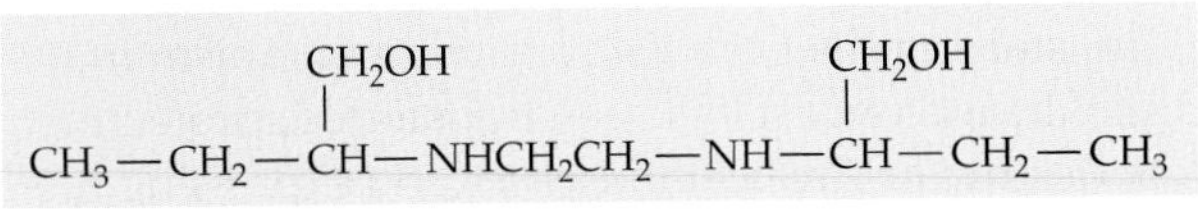

Ethambutol

Figure 31.24 Rifampin and other antituberculosis drugs
Structures of some drugs used in treating tuberculosis. Rifampin is synthesized from the antibiotic rifamycin by addition of 1-amino-4-methyl piperazine. Isoniazid, pyrazinamide, and ethambutol are administered in combination with rifampin or streptomycin in tuberculosis treatment.

Tetracycline	R_1	R_2	R_3	R_4	Production strain or method
Tetracycline	H	OH	CH₃	H	*Streptomyces aureofaciens* (in chloride-free medium); or chemical modification of chlortetracycline
7-Chlortetracycline (Aureomycin)	H	OH	CH₃	Cl	*S. aureofaciens*
5-Oxytetracycline (Terramycin)	OH	OH	CH₃	H	*Streptomyces rimosus*
6-Demethyl-7-chlortetracycline (Declomycin)	H	OH	H	Cl	*S. aureofaciens* (+ inhibitor)
6-Deoxy-5-hydroxy-tetracycline (Doxycycline)	OH	H	CH₃	H	Semisynthetic
7-Dimethylamino-6-demethyl-6-deoxytetracycline (Minocycline)	H	H	H	N(CH₃)₂	Semisynthetic
6-Methylene-5-hydroxytetracycline (Methacycline)	OH	=CH₂		H	Semisynthetic

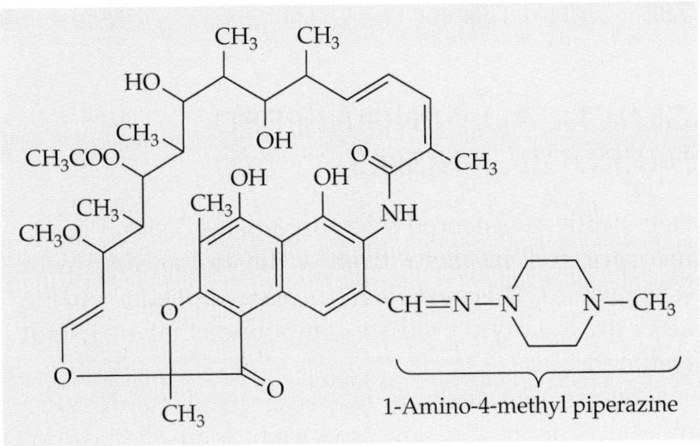

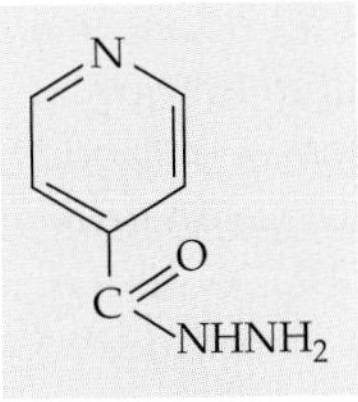

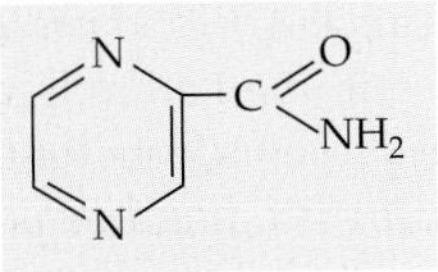

Figure 31.25 Tetracyclines
Structures of the tetracyclines. Some are natural products of *Streptomyces* strains and others are semisynthetic chemical derivatives.

GENETICALLY ENGINEERED MICROORGANISMS

Genetically engineered microorganisms contain genes incorporated in them from a dissimilar organism (Chapter 15). This approach has been applied to enable other organisms to produce compounds that are useful commercially.

Genetic engineering is being used to develop microorganisms that can produce human-related proteins. Proteins are difficult to synthesize, and extracting clinically useful proteins from natural material is a costly and inefficient process. Blood proteins and hormones are generally present in an animal in limited amounts, and purification of proteins from tissue, glands, or blood is expensive and difficult. Engineered bacteria can grow rapidly on simple nutrients and are potential sources of blood proteins, hormones, and so forth in unlimited amounts.

Several challenges are encountered in the course of producing pharmaceuticals through genetic engineering. Transferring the gene into a foreign organism may be difficult, and once it has been transferred, protecting it from destruction by host nucleases and expressing the gene is not always a certainty. One problem in expressing eukaryotic proteins in bacteria occurs when specific posttranslational modification is required, such as glycosylation. Retrieving the product from the organism and purifying it can also be problematic. The Food and Drug Administration must approve drugs before they can be released for clinical use. Their regulations must be met to gain approval for human use. Proving efficacy through animal and human trials and safety assurance can take years.

Human Insulin

Insulin is a small protein (**Figure 31.26**) composed of two peptide chains: peptide A, consisting of 21 amino acids, and peptide B, with 30 amino acids. Until recently, the insulin used in the treatment of human diabetes was extracted from the pancreas of slaughtered animals. Porcine- and bovine-derived insulin had two significant shortcomings. The insulin from these animals was not precisely the same as human insulin structurally and hence not as effective. In addition, they contained animal proteins that caused allergic responses in some diabetics. By 1984, commercial production of human insulin using genetically engineered *Escherichia coli* was realized.

To engineer a bacterium to produce insulin, a synthetic gene was made for each of these polypeptide chains. The gene for each was incorporated separately into two strains via a plasmid vector. The insulin gene was linked to the end of the gene encoding β-galactosidase so that the insulin polypeptide would be coproduced and secreted along with the enzyme. The two gene-engineered strains, one producing chain A and the other chain B, would be grown separately. The peptide produced is separated from β-galactosidase and the A and B chains chemically joined to form complete human insulin. Over one-half of the insulin used for diabetes treatment is now obtained from engineered

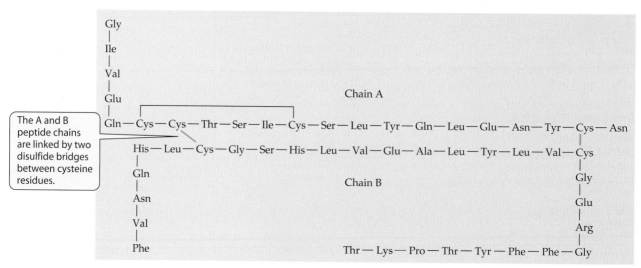

Human insulin

Figure 31.26 Insulin
Amino acid sequence of the hormone insulin. In commercial production of insulin, the two chains are synthesized separately by genetic information cloned into microorganisms and joined chemically to form the active hormone.

bacteria, and the remainder is recovered from slaughtered animals.

Human Growth Hormone

Human growth hormone, HGH, used to treat pituitary deficiencies that result in dwarfism, was previously obtained from the pituitary glands of cadavers. Obtaining enough hormone was both difficult and costly, which severely limited its use. Growth hormone has been recovered from animals but is not effective in humans.

The gene for HGH synthesis has also been engineered into *E. coli*, and this is now the commercial source of the hormone. There were 30,000 children taking the hormone in the United States in 1998.

Vaccines

Genetically engineered antigens are now a reality due to advances in biotechnology. These antigens have many advantages over antigens extracted from pathogenic bacteria or viruses and are generally superior to attenuated bacterial or viral preparations. Antigens generated by bacteria are less costly, more readily purified, and free from contamination by other proteins. One vaccine now obtained from a genetically engineered yeast is that for hepatitis B. Hepatitis B is often transferred person to person in contaminated blood and is a problem for IV drug users, dialysis patients, and individuals that require multiple transfusions. The hepatitis B virus cannot be grown except in the blood of intentionally infected primates. The blood of individuals chronically ill with hepatitis B virus contains a protein particle called HBs Ag. The particle itself is not harmful but can be employed as an effective inducer of anti-hepatitis B antibody. The protein has been cloned into *Saccharomyces cerevisiae* and is now the approved source of the antigen for human immunization.

This is a promising technique for obtaining antigens related to other infectious agents, particularly when propagation of the causative agent in vitro is difficult.

Other Cloned Genetic Systems The potential for developing cloned genetic systems that will serve as a ready source of medically important compounds is endless. As discussed previously, several of these engineered systems are now available or are in development.

Interferons Interferons are proteins normally synthesized by cells that interfere with viral propagation. These may also be effective as anticancer agents.

Blood Proteins Gene engineering for the production of important blood proteins is under development. **Tissue plasminogen** activator is involved in dissolving blood clots, especially during wound healing processes. This protein is useful in dissolving blood clots in the heart or embolisms in other areas of the body. Blood-clotting factors are also essential for hemophiliac patients. These factors are now obtained from pooled blood. With the constant worry about AIDS and other viruses in pooled blood, a gene-engineered source would be most welcome.

Bone Growth Factor Bone growth protein, found in the intracellular bone matrix, is a hormone used for the treatment of patients with osteoporosis or fractured bones. It induces undifferentiated cells to differentiate into bone-forming cells. As the level of this protein declines with age, it is necessary to provide this hormone to elderly individuals with the factor synthesized by gene-engineered microorganisms.

Epidermal Growth Factor This protein stimulates wound healing. A ready source of this factor from genetically engineered bacteria would be a boon to burn and surgery patients.

Potential for Future Developments The future of genetically engineered microorganisms capable of generating a wide array of useful substances seems limitless. The purity of proteins, essential in human well-being, is considerably higher than in those extracted from animal tissues. Fear of contamination by known or unknown human viruses is also eliminated. In addition, the cost of production is minimized, as microbes can be grown inexpensively and in unlimited amounts. The future for this area of industrial microbiology is indeed very bright.

SUMMARY

▶ **Industrial fermentations** are used for the commercial production of antibiotics that cure disease, yeast employed in bread making, vitamins, flavoring agents, organic acids, enzymes, and fermented foods/beverages.

▶ Microorganisms that generate useful industrial products are derived from naturally occurring populations by **mutation** and **selection.**

▶ **Biotechnology** is a field of research and development that is based on studies with microorganisms.

▶ Metabolites generated during active growth are considered **primary metabolites.** Those produced after exponential growth are **secondary metabolites,** each of which is produced by a select group of organisms. The **trophophase** is active growth, and the **idiophase** is involved in stationary growth and results in the production of secondary metabolites.

▶ Microorganisms may be consumed as food or, more important, may **convert foods** such as meat, milk, cabbage, cucumbers, or olives to a desirable product.

Mushrooms are actually microbes (fungi) that are consumed directly.

▶ **Cheese** is an ancient food that has played a role in the movement of human populations to new territories.

▶ Bulk **yeast** is used in the baking industry and as a food supplement.

▶ A lactic acid fermentation of meat produces **Lebanon bologna** and of cabbage yields **sauerkraut. Chocolate** and **soy sauce** are products of fermentation, as are olives and pickles.

▶ **Vinegar** production in the United States amounts to about 160,000,000 gallons per year. Much of this is used in commercial products (sauces, dressings, etc.). Vinegar is a product of ethanol oxidation to acetic acid.

▶ **Flavor enhancers** are compounds that add little flavor by themselves but enhance the flavor of the foodstuff to which they are added. Monosodium glutamate and nucleosides are flavor enhancers that are products of industrial fermentation.

▶ Riboflavin and B_{12} are vitamins produced commercially by microorganisms. Other B vitamins are obtained from chemical synthesis. Some steps in the commercial conversion of glucose to vitamin C are carried out by microorganisms.

▶ **Wine, beer,** and other **alcoholic beverages** are products of fermentation. The starting material in wine fermentation is grapes and for beer and other alcoholic beverages it is grain. The starch in grain must be depolymerized to yield sugars that are then fermented by yeasts such as *Saccharomyces cerevisiae*.

▶ **Distilled beverages** (gin, vodka, whiskey) are essentially distillate from grain alcohol. The flavor is imparted by the volatiles that distill over with the alcohol.

▶ **Citric acid,** a secondary metabolite produced by *Aspergillus* species, is added to soft drinks, candy, and other foods and is also used in detergents.

▶ **Proteases** produced by bacteria such as *Bacillus amyloliquefaciens* are an additive in detergents as stain removers. **Amylases** are employed commercially to convert starch to free sugars.

▶ **Enzymes** have a short half-life and are stabilized by **immobilization.** Cross-linking functional groups to reagents, bonding to inert material, or encapsulation are effective in prolonging the shelf life of enzymes.

▶ **Biotransformation** of low-cost inactive plant sterols to active forms can be accomplished with microorganisms. Sterols are important medicinally as anti-inflammatory agents.

▶ A **natural biocide** from a bacterium is the protein crystal formed during sporulation of *Bacillus thuringiensuis*. The product is sold as an **insecticide.**

▶ About 10,000 **antibiotics** have been characterized with fewer than 100 in commercial production. Most useful antibiotics are produced by actinomycetes. Among the many antibiotics are the **penicillins, cephalosporins, streptomycin,** and **tetracyclines.**

▶ **Microorganisms** have been genetically engineered to produce **insulin, human growth hormone, vaccines,** and other clinically useful compounds.

REVIEW QUESTIONS

1. What events occurred in the second half of the twentieth century that had a marked effect on the fermentation industry?

2. How does an industrial microbe differ from a relative that grows in the environment? What are some characteristics of microorganisms that render them potential producers of commercial commodities?

3. Define primary and secondary metabolite and cite examples of each. What do trophophase and idiophase mean in terms of industrial products?

4. What are some foods that are products of microbial fermentations? The nature of the microbe involved?

5. There are four major steps involved in cheese manufacture. What is the influence at each of these phases on the final product? How does one cheese differ from another?

6. What is the major use of bulk yeast in the United States? Why is yeast not a quality food for humans?

7. How are microbes utilized in the production of: (a) fermented meat, (b) sauerkraut, (c) pickles, (d) coffee beans, (e) chocolate, (f) olives, and (g) soy sauce?

8. Outline the major processes in vinegar manufacture. How do the major types of vinegar differ?

9. How does a flavoring agent differ from a flavor enhancer?

10. What are some of the major uses of amino acids?

11. How does red wine differ from white wine? Outline the steps in the production of these types of wine.

12. In beer manufacture what are: (a) organisms involved, (b) malting, (c) wort, (d) top fermentation, (e) bottom fermentation, and (f) lagering?

13. What are distilled spirits? How does bourbon differ from Scotch and rye? How does whiskey differ from whisky?

14. List some uses of citric acid. What organisms are involved in the citric acid fermentation?

15. Name three industrial-type enzymes. What are their uses? Why do microorganisms produce these enzymes?

16. What is a biotransformation? How does it apply to steroid manufacture?

17. Outline the steps in the search for a new antibiotic.

18. Genetically engineered organisms are involved in production of several medical products. What are some of these, and what is their use?

19. Why is microbial production superior to the isolation of products such as insulin or vaccines from natural sources?

SUGGESTED READING

Alcamo, E. I. 2001. *DNA Technology: The Awesome Skill.* 2nd ed. San Diego: Academic Press.

Crueger, W. and A. Crueger. 1990. *Biotechnology: A Textbook of Industrial Microbiology.* 2nd ed. Sunderland, MA: Sinauer Associates.

Demain, A. L. and J. E. Davies. 1999. *Manual of Industrial Microbiology and Biotechnology.* 2nd ed. Washington, DC: American Society for Microbiology.

Doyle, M. P., L. R. Beuchat and T. J. Montville. 2001. *Food Microbiology: Fundamentals and Frontiers.* 2nd ed. Washington, DC: American Society for Microbiology.

Glick, B. R. and J. J. Pasternak. 1998. *Molecular Biotechnology: Principles and Applications of Recombinant DNA.* 2nd ed. Washington, DC: American Society for Microbiology.

Jay, J. M. 1991. *Modern Food Microbiology.* 4th ed. New York: Van Nostrand Reinhold Co.

Table 32.1	Partial list of water-borne pathogenic microorganisms
Microorganism	**Disease**
Bacteria	
Salmonella typhosa	Typhoid fever
Vibrio cholera	Cholera
Shigella dysenteriae	Bacterial dysentery
Protozoa	
Entamoeba histolytica	Amoebic dysentery
Giardia lamblia	Giardiasis
Naegleria	Meningoencephalitis
Viruses	
Infectious hepatitis	Hepatitis

cannot be used for contact sports because of the possibility of disease. Commercial shellfish fisheries may be adversely affected as well because clams and oysters, which are filter feeders, ingest particulate materials, including bacteria and viruses, resulting in high concentrations of this matter in their digestive tracts (**Figure 32.2**). Regulations restrict the numbers of bacteria in shellfish tissues as well as in the waters where they are raised. Although the shellfish themselves may not be affected by ingested pathogenic microorganisms, the contaminated shellfish cannot be marketed for human consumption.

Figure 32.2 Warning
Sign at a Seattle, Washington, city beach near the discharge point from a sewage treatment plant. The sign indicates that fish and shellfish might be contaminated with microorganisms. Courtesy of J. T. Staley.

Because of these problems with raw sewage disposal, municipalities in the United States now are required to treat their wastewaters *before* they are discharged into receiving waters. The sewage is collected using a system of sewer pipes, usually by gravity flow, and is sent to the wastewater treatment facility that is located near a river or the marine environment where the treated waste can be discharged. Although wastewater treatment does not completely restore the water to its natural state, it reduces the levels of organic matter and the concentrations of bacteria. There are three degrees of wastewater treatment: primary, secondary, and tertiary; they are discussed next.

Primary Wastewater Treatment

The initial step in sewage treatment is termed **primary wastewater treatment** because it involves the first step, namely the settling of the particulate materials and their separation from the dissolved material.

In primary treatment, the sewage coming into the plant is first passed through a screen to remove sticks, plastic bags, and other large pieces of material that might interfere with treatment. The wastewater then flows into the primary treatment tank that acts as a settling basin to allow the heavier particulate material (the **primary sludge**) to settle to the bottom. In addition, a skimming bar removes oils that float on the surface of this tank. Thus, primary treatment is simply a physical process whereby the liquid portion of the wastewater (which comprises over 99% of the waste) is separated from the solids or sludge as well as the floatables.

Secondary Treatment

The typical treatment systems used in the United States are **secondary wastewater treatment** systems. They are called secondary treatment systems because two sequential processes are used in the treatment, a primary treatment stage followed by a secondary treatment stage (**Figure 32.3**).

The **secondary treatment** process is a microbiological process. The liquid portion from the primary tank is passed into another tank (the secondary wastewater tank) and is then aerated. Two different types of aerators have been used. An older type is referred to as a **trickling filter.** In this system, the fluid is sprinkled over a bed of rocks that are about the size of a fist. The wastewater is aerated while it passes from the sprinkler and percolates through the bed of rocks. Microorganisms grow as a biofilm on the surface of the rocks and degrade the organic materials that are in the wastewater. Microorganisms, which include protozoa, algae, bacteria, and viruses, that grow on the surface of the rocks are aerobic, but those underneath are anaerobic fermentative bacteria.

Another other type of secondary treatment process is referred to as the **activated sludge** process (Figure 32.3).

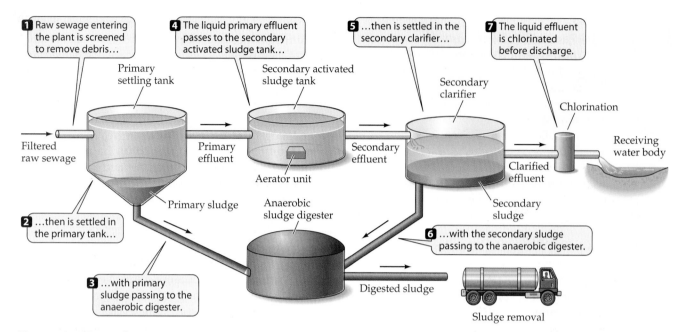

1 Raw sewage entering the plant is screened to remove debris...

4 The liquid primary effluent passes to the secondary activated sludge tank...

5 ...then is settled in the secondary clarifier...

7 The liquid effluent is chlorinated before discharge.

Primary settling tank

Secondary activated sludge tank

Secondary clarifier

Chlorination

Filtered raw sewage

Primary effluent

Secondary effluent

Clarified effluent

Receiving water body

Aerator unit

Primary sludge

Anaerobic sludge digester

Secondary sludge

2 ...then is settled in the primary tank...

3 ...with primary sludge passing to the anaerobic digester.

6 ...with the secondary sludge passing to the anaerobic digester.

Digested sludge

Sludge removal

Figure 32.3 Secondary wastewater treatment
Diagram of a typical secondary activated sludge wastewater treatment plant. The end products are clarified, disinfected water and digested sludge.

In this type of treatment, the effluent from the primary treatment tank is aerated by bubbling air through it in the secondary treatment tank. Aerobic microorganisms, particularly bacteria, grow in colonial aggregates or consortia called **flocs** and degrade the organic material in the waste. Among the bacteria that occur here is the genus *Zoogloea*, a bacterial species that produces gel-like, flocullant microcolonies due to its capsule.

The next step in the treatment consists of settling out the particulate materials from the secondary treatment tank in an additional tank, called the **secondary clarifier.** The liquid portion of this material has much lower levels of organic material in it and can be satisfactorily discharged into a receiving stream or other body of water. However, before final discharge, the fluid is chlorinated to kill many of the potentially pathogenic microorganisms that might be present.

There is also sludge from this secondary clarifier. As in the primary sedimentation tank, this settles out and must also be treated. In typical treatment plants, the primary and secondary sludges are combined and treated as a whole in the **anaerobic sludge digestor.** This is a large-capacity tank in which anaerobic bacteria carry out the final steps of anaerobic degradation of organic materials. These tanks contain fermentative bacteria that produce organic acids, alcohols, carbon dioxide, and hydrogen gas. Methanogenic bacteria grow on the acetic acid and on the hydrogen and carbon dioxide in these reactors and produce methane gas. The methane produced can be reclaimed and used as an energy source

for heating in the treatment plant or elsewhere.

Anaerobic sludge digestion is a slow process and does not result in a complete conversion of sludge to gases. The undigested sludge that remains after treatment is rich in nutrients and must be disposed of elsewhere. It has been used as a fertilizer for agricultural plants that are not used for human consumption because there is concern about viruses and other microbes that might have survived the treatment process.

The organic material in wastewater imparts what sanitary engineers refer to as a **biochemical oxygen demand,** or **BOD,** to the waste. The BOD test is used to measure the amount of oxygen demand, or the amount of oxygen required to decompose the organic material in 5 days at 20°C in a wastewater sample. Initially, raw sewage has a very high BOD, but if treatment has been successful, the BOD level is reduced significantly. For example, trickling filters working satisfactorily can reduce the BOD content of raw sewage by about 75%. Activated sludge treatment units are even more effective when working properly and can reduce the BOD content by about 85%.

Thus, secondary treatment effectively removes much of the organic material from wastewater. In fact, the actual BOD reduction is effected by the bacteria and other microorganisms that grow in the secondary treatment units. These microorganisms are doing what they would normally do in nature, but by doing it in a wastewater treatment plant, the process can be monitored and controlled. The microbes carry out these processes effectively and without any cost to society.

Not all microorganisms are removed by settling in the clarifier unit. Thus, some will leave in the secondary effluent. Most of these microorganisms have grown on the organic materials of the wastewater. However, some of them are coliform bacteria, such as *Escherichia coli,* which come from human feces being treated in the sewage. Because some of the coliform bacteria are water-borne pathogens, it is desirable to kill them. Thus, the effluent from the secondary clarifier is disinfected, usually with chlorine, before it is finally discharged into the receiving water. Although this process is not 100% effective in disinfection of the effluent, it does dramatically reduce the numbers of coliform bacteria and other pathogenic microorganisms that leave the wastewater treatment plant.

Microbial Treatment Problems

The normal degradative activities occurring in the secondary treatment system depend on the types and status of the microorganisms growing in the reactors. Although it is true that the bacteria carry out their activities "for free," it is essential to ensure that conditions in the bioreactors are satisfactory for the growth of the specific organisms that are best suited for the biodegradative processes needed. Thus, in addition to providing the necessary bioreactor tanks, it is essential that the secondary activated sludge unit has sufficient aeration to ensure that the organic materials can be degraded as completely as possible.

One microbiological problem that may occur in the secondary clarifier units is **bulking.** Bulking refers to the poor settleability of sludge. Ideally, the particulate organic material should be completely removed prior to discharge into a receiving stream or other outlet. Thus, it is important that microorganisms and other debris settle to the bottom of the tank where they can be removed and transferred to the sludge digestor. Some species of bacteria are undesirable in secondary clarifiers because they interfere with settling. Filamentous organisms such as *Sphaerotilus natans* and *Thiothrix nivea* are examples of bacteria that interfere with normal settling. Treatment conditions need to be modified to remove them, and this is sometimes difficult to accomplish.

Anaerobic digestors may also encounter problems. Under some circumstances, they "go sour"—they become acidic and cannot support the normal populations of fermenters and methanogens. When this occurs they need to be reseeded with inocula from other digestors.

Tertiary Wastewater Treatment

Although secondary treatment of wastewaters is effective in removing organic matter from the wastes, it does not remove the inorganic by-products of the microbial activity. As can be seen from the following overall treatment formula, inorganic substances such as ammonia and phosphate are produced.

$$\text{Organic material} \rightarrow XO_2 + NH_3 + SO_4^{-2} + PO_4^{-3}$$
$$+ \text{ trace elements}$$

The inorganic products of organic degradation are excellent nutrients for the growth of algae. Discharging these nutrients into a receiving body of water can cause algal blooms. Such excessive algal growth can lead to eutrophication (see **Box 32.1**). Thus, it is possible to have successfully treated a wastewater by secondary treatment, only to have it cause enrichment in the receiving water.

The major nutrients of concern for enrichment of receiving waters are phosphate and ammonia. These nutrients can be removed from secondary effluent by additional treatment referred to as **tertiary wastewater treatment.** Both chemical and microbiological tertiary treatment processes have been developed. Chemical processes are very expensive and will not be discussed further here.

Ammonia removal is effected biologically through a two-step process. The first step is nitrification, the aerobic oxidation of ammonia to nitrite and nitrate, by nitrifying bacteria (see Chapter 19). Although these are chemolithotrophic bacteria, they grow well in properly treated secondary effluent. During this first step, the

---

Research Highlights Box 32.1 **Eutrophication of Lake Washington**

A classical case of lake eutrophication occurred in Lake Washington in the late 1960s. Due to the inflow of secondary effluent from sewage treatment plants around the lake, the lake becomes increasingly enriched. Dr. W. T. Edmondson of the University of Washington noted that phosphate was limiting to algal growth in the lake. Thus, even though the organic material was being satisfactorily removed, the increased growth of algae, due to influx of phospate, resulted in increased primary production and lake eutrophication. Based on his recommendation, the metropolitan area diverted all of the secondary effluent away from the lake. It is now discharged into a much larger body of water, Puget Sound. Since the late 1960s, Lake Washington has returned to its normal mesotrophic status.

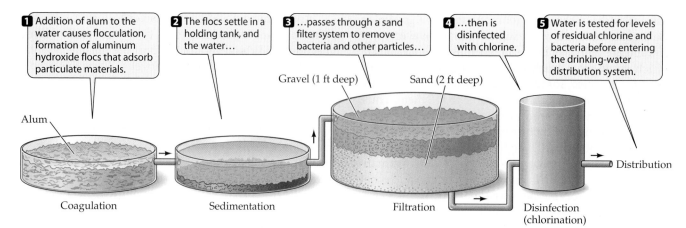

1 Addition of alum to the water causes flocculation, formation of aluminum hydroxide flocs that adsorb particulate materials.

2 The flocs settle in a holding tank, and the water...

3 ...passes through a sand filter system to remove bacteria and other particles...

4 ...then is disinfected with chlorine.

5 Water is tested for levels of residual chlorine and bacteria before entering the drinking-water distribution system.

Gravel (1 ft deep)   Sand (2 ft deep)

Alum

Distribution

Coagulation   Sedimentation   Filtration   Disinfection (chlorination)

Figure 32.4 Drinking-water treatment system
Diagram of a drinking-water treatment system. Microbiological tests are routinely carried out on the product to ensure potability.

ammonia is converted to nitrate. In order to remove the nitrate, it is necessary to carry out denitrification. Denitrification results in the anaerobic conversion of nitrate to form N_2O and N_2 gases, which will dissipate into the air. Therefore, nitrogen is removed from the water as a gas. It should be noted that denitrification by *Pseudomonas* species requires organic carbon. An effective and inexpensive carbon source, methanol, can be added to serve this purpose.

This two-step process for nitrogen removal can be carried out in normal secondary wastewater treatment systems if they are closely monitored. A period of aeration in the secondary clarifier to enhance nitrification is followed by a period of anaerobic incubation that favors denitrification.

Phosphate removal occurs by using bacteria that carry out the uptake of phosphate into their cells. These unidentified bacteria accumulate polyphosphate granules in excess of normal metabolic needs during periods of active growth. They can then be settled out along with the phosphate in a settling tank.

In practice, very few plants in the United States use tertiary treatment. However, this may change as the need for protecting our receiving waters increases.

DRINKING-WATER TREATMENT

As discussed earlier, if a city receives its drinking water from a river, the water may be contaminated with effluent from wastewater treatment plants from cities upstream. In fact, even if there is no city upstream, it is possible that the water can be contaminated from wild or domestic animals that may harbor infectious waterborne microbial agents such as *Giardia lamblia* or enteric viruses or bacteria. In addition, septic tanks from rural areas may also discharge wastewater into the river. Therefore, it is essential for the city to treat its drinking water before it is distributed to its citizens.

Depending on the source of water, the treatment may consist only of disinfection by chlorination or the treatment may be much more extensive. Consider the case

of a large city located on a major river such as the Mississippi. Furthermore, consider that the city is downstream of many other cities. In this example, it is important that the city use an extensive treatment system involving four steps:

- Coagulation
- Sedimentation
- Filtration
- Disinfection

These steps are accomplished in a drinking water treatment plant or facility (**Figure 32.4**).

Coagulation is performed by adding alum (a salt of aluminum sulfate) to the water. When the alum is added to the water, it forms aluminum hydroxide flocs that adsorb particulate materials. The flocs are then settled out by sedimentation in a holding tank. The next step is filtration. The water is passed through a sand filter system to remove the bacteria. Either rapid or slow sand filters are used, and both are effective at removing particulate material, including bacterial cells. If the treatment plant is operating effectively, about 99% of the bacteria present in the raw, untreated water are removed after sand filtration. The final step is **disinfection** of the water. This is usually accomplished by chlorine. The dissolved gas forms a hypochlorite solution (as in household bleach), which effectively disrupts the cell membranes of bacteria, thereby killing them. Disinfection does not kill all of the bacteria, however, so small numbers of bacteria remain in the water.

The drinking water must then be passed through a distribution system before it is consumed. In order to be assured that the water is safe for drinking (**potable**), it needs to be tested. Two approaches have been taken to ensure that drinking water is safe or potable. The first approach is to analyze for the **residual chlorine** level. If chlorine is still present in the water at the tap, at a level of 1 ppm, then it may be considered potable. However, as we have stated, even if there is residual chlorine, bacteria may still be present. Therefore, microbiological tests are routinely performed to determine water potability.

Current microbiological testing uses an **indicator bacterium.** The standard indicator bacterium used in the United States is *Escherichia coli.* Because *E. coli* is found in the intestinal tracts of all humans and some other warm-blooded animals, its presence in a drinking water is *indicative* of fecal contamination. Why use an indicator bacterium? Why not just look for a particular pathogen? The reason for this is that if a pathogen were present, it would be expected to occur in very low concentrations relative to *E. coli.* This is because the pathogen may be from only one infected individual or carrier out of hundreds or thousands of people who live in the city upstream. Therefore, it is extremely difficult to detect the pathogenic bacterium because so few of them would be released into the sewer system in relation to the numbers of *E. coli,* because *all* individuals carry *E. coli.* Furthermore, even if a test is devised for one pathogenic bacterium, what about all the other pathogens? Because there are many different pathogens, separate tests would have to be performed for each one. In contrast, everyone harbors *E. coli* in his or her intestinal tract. If *E. coli* is found in a drinking water sample, it indicates that the water has been contaminated with fecal material, and this finding alone indicates that it is unsafe for drinking.

Although *E. coli* is the indicator bacterium of choice, it is not simple to identify *E. coli* in natural samples because many other bacteria closely resemble it. It belongs to a group of enteric bacteria called the **coliform bacteria.** *Coliform bacteria are defined as non-spore-forming gram-negative rods that ferment lactose to form acid and gas in 24 to 48 hours at 35°C.* Thus, coliform bacteria are defined by experimental conditions. Indeed, several different types of tests are used to identify coliform bacteria. A brief description follows of some of the more common tests.

Total Coliform Bacteria Analyses

One group of tests, the **total coliform bacteria** tests, analyze for the concentration of coliform bacteria in a drinking water sample. There are several tests for total coliform bacteria including the most probable number test and membrane filter tests.

Most Probable Number (MPN) Test This test is the oldest test for total coliform bacteria and is still used as a standard test. It is performed in three different stages:
- Presumptive
- Confirmed
- Completed

The first stage, called the **presumptive test,** is designed to determine whether or not coliform organisms might be present in a sample. In this test, a series of lactose broth tubes (lauryl tryptose medium) are inoculated with the drinking water to be tested (**Figure 32.5**). Typically, fivefold replicates at each of several dilutions are prepared. The first set of five tubes, which are the lowest dilution, are inoculated with 10 ml portions of the water sample to be tested. Because 10 ml is such a large volume of inoculum, 10 ml of the medium is made up double strength, so that it ends up as single strength after addition of the 10 ml inoculum. In the next dilution, the five replicate single-strength tubes are inoculated with 1 ml inocula. Higher dilutions are attained using dilution blanks. The tubes contain lactose, a pH indicator, bromothymol blue, and an inverted vial to detect gas production. They are incubated at 35°C for 24 hours and then read for acid and gas production. If they are positive for gas, they are regarded as presumptive positive. The remaining tubes are incubated an additional 24 hours, and any additional positive tubes are also considered positive presumptive.

The MPN test is quantifiable. Each positive tube is scored at each dilution. Then, the number of positive tubes at each dilution can be used to determine the quantities of presumptive total coliform bacteria. For example, assume that all five tubes from the first two dilutions were positive, three from the third were positive, and only one from the fourth was positive. The results are recorded as 5, 3, and 1 for the final three dilutions. These results can be converted into a quantifiable number taken from a statistical most probable number table (see the American Public Health Association's Standard Methods for the Examination of Water and Wastewater). In this particular instance, the number of total coliform bacteria is 1,100 per 100 ml of the original sample. Note that this concentration is given as *per 100 ml* and is called the **coliform index.**

The presumptive test is followed by the **confirmed test,** which provides additional supportive information about the presence of coliform bacteria. At 24 and 48 hours, each positive presumptive tube is used to inoculate a tube of another broth medium, brilliant green lactose bile broth (BGLB). These BGLB tubes are incubated at 35°C for another 24 to 48 hours to *confirm* acid and gas production. If these are positive, they are considered to be confirmed. At this point, it is still not possible to say that coliform bacteria or *E. coli* are present in the sample, only that the presence of coliform bacteria has been confirmed.

However, a confirmed positive is considered adequate for concern about the potability of the water. For drinking water analysis, even a single positive confirmed test is considered serious. Additional testing is required to verify what the bacterium is. This is

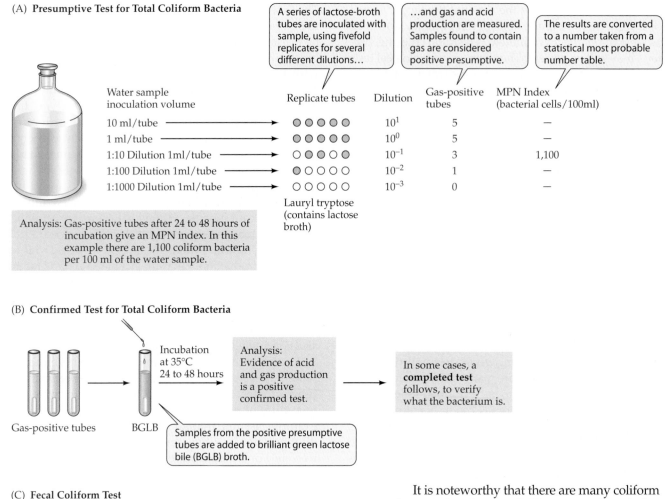

Figure 32.5 Total and fecal coliform analysis
Diagram showing the steps used for total and fecal coliform analysis by the most probable number (MPN) test. (A) Presumptive test, (B) confirmed test, and (C) fecal coliform test. See the text for further details.

accomplished by the completed test. In the **completed test,** a loopful of culture from a positive confirmed BGLB tube is streaked on eosin methylene blue (EMB) plates. If coliform bacteria are present, they will produce colonies with a typical metallic green sheen following incubation at 35°C for 24 hours. Cells from these colonies must be Gram stained to determine if they are gram-negative, non-spore-forming bacteria. If gram-negative, non-spore-forming bacteria are found, then all the criteria have been met for fulfilling the definition of a coliform bacterium (Figure 32.5).

It is noteworthy that there are many coliform bacteria besides *E. coli*. Thus, even though this lengthy procedure has been undertaken, it is still not possible to state that *E. coli* is present in the sample. Further tests are needed to verify this. Moreover, some of these coliform bacteria, such as *Enterobacter aerogenes*, are common soil bacteria and do not reside in the intestinal tracts of warm-blooded animals. Therefore, even a positive completed coliform test would not necessarily mean that fecal coliform bacteria are present in a water sample. Thus, the drinking water tests err on the side of safety. Recurring positive results would require that additional tests be performed to identify the identity and source of the coliform organisms that are detected by the completed tests.

Membrane Filter Analyses Membrane filter tests have also been developed for the identification of total coliform bacteria. These tests entail utilizing a sterile 0.45 μm pore size membrane filter to collect cells from the water sample to be tested. These filters are then placed on an absorbant pad that contains nutrients for the growth of bacteria. They are incubated for 24 hours at 35°C, and the colonies of total coliform bacteria are identified by their characteristic metallic sheen.

Fecal Coliform Bacteria Analyses

One disadvantage of the total coliform bacteria tests is that they select for a variety of bacteria, such as *Enterobacter aerogenes*, that are not indigenous to the intestinal tracts of animals and are therefore not indicative of fecal contamination. Therefore, more specific tests have been designed for fecal coliform bacteria that are more likely to be *E. coli*.

Most Probable Number Analysis Research has shown that *E. coli* strains can grow at elevated temperatures when compared to soil species such as *E. aerogenes*. Therefore, elevated temperature tests have been developed for identifying fecal coliform bacteria.

For the MPN test for fecal coliform bacteria, positive tubes from the MPN total coliform bacteria test described previously are used to inoculate a selective broth medium, which is then incubated at 44°C for 24 hours (Figure 32.5C). Cultures that grow on this medium and produce gas are regarded as fecal coliform bacteria, and most of them, if identified, turn out to be *E. coli*. As with MPN total coliform bacteria, the actual numbers of fecal coliform bacteria in a sample can be quantified by this technique.

Membrane Filter Analysis A membrane filter test has also been developed for fecal coliform bacteria. This test is usually performed directly on water samples. Therefore, the cells are exposed to 44°C on a selective medium immediately after filtration when the membrane filter is placed on a selective *E. coli* (EC) medium and incubated at 44°C. Colonies of fecal coliform bacteria produce a characteristic blue-colored colony on this medium.

All viable counting procedures for coliform testing have limitations. Among the most serious of these is the fact that the tests take so long to perform. If a city were really concerned about an outbreak of some intestinal disease, it would be at least 2 to 4 days before it would be confirmed that coliform bacteria were present. And, as we have already mentioned, even though coliform bacteria may be found, because they cannot be identified as *E. coli* using any of the tests described earlier, it is really not known whether the water sample has fecal contamination. Thus, new research is aimed at developing more rapid and specific tests for *Escherichia coli* and other enteric bacteria.

Alternative Microbiological Techniques for Drinking-Water Analysis Several different approaches are being taken to develop better procedures for testing the potability of drinking waters. One new approach is the presence/absence test.

Presence/Absence Analysis The presence/absence test, or P/A test, looks for total and fecal coliform bacteria using a single large volume of the drinking water sample (100 ml) as an inoculum rather than using a series of tubes or membrane filters. In this test, only a positive or negative is recorded for a given sample. Thus, it is not possible to quantify the numbers of total or fecal coliform bacteria, but only to say they are present or absent in a particular 100 ml sample. This type of test is very simple to perform and thereby enables small municipalities serving from 100 to 10,000 citizens to analyze their own water.

Colorimetric and Fluorogenic Analyses Another approach taken to simplify the analysis of total and fecal coliform bacteria in drinking waters, uses **colorimetric** or **fluorogenic compounds.** These compounds, when cleaved enzymatically, release colored (colorimetric) end products that can be seen visually or fluorescent end products that can be detected when illuminated with short wavelength radiation (ultraviolet radiation), respectively. Rapid tests have been developed for total and fecal coliform bacteria using these compounds. For example, one colorimetric test for total coliform bacteria uses orthonitrophenol galactoside (ONPG) as a substrate for the lactose-splitting enzyme, β-galactosidase, found in *E. coli* and other enteric bacteria that ferment lactose. When this enzyme is present, it cleaves the galactoside, freeing the orthonitrophenol, which has a characteristic yellow color. As a result, a yellow color is indicative of the presence of coliform bacteria.

An example of the use of a fluorogenic substrate is 4-methylumbelliferone-β-D-glucuronide, or MUG (**Figure 32.6**). Almost all strains of *E. coli*, and few other enteric bacteria, produce glucuronidase, an enzyme that cleaves this substrate to release the 4-methylumbelliferone-β-D-glucuronide compound that fluoresces when illuminated with ultraviolet radiation. As in the other tests, these tests require that the bacteria grow in the vials and that they be metabolically active. However, these latter tests do not require such lengthy incubations and follow-up inoculations.

Perhaps the future for detection of *E. coli* and waterborne pathogens lies in the development of probe pro-

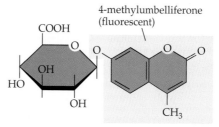

MUG (4-methylumbelliferyl-β-D-glucuronide)

Figure 32.6 Fluorogenic substrate
Chemical structure of 4-methylumbelliferyl-β-D-glucuronide (MUG). This fluorogenic compound is not itself fluorescent, but when cleaved by glucuronidase it produces the fluorescent product 4-methylumbelliferone.

cedures to look for specific organisms directly in natural samples (see Chapter 17). In addition, procedures using the polymerase chain reaction (PCR) could greatly accelerate the detection of fecal contamination.

LANDFILLS AND COMPOSTING

Microbes are also involved in the degradation of garbage and in the process of composting. This is not surprising because these contain large amounts of organic material, and heterotrophic microorganisms require organic materials for growth.

Landfills

All cities collect garbage, called **solid waste,** from private homes and take this to a location away from the city where it is placed in the ground or **landfill.** This is then covered with earth and allowed to decay naturally (**Figure 32.7**).

Landfills contain not only potato peels, tin cans, and discarded cardboard boxes, but also plastic bags, Styro-

foam packing materials, broken appliances, and hazardous chemicals. As a result, during the 1960s and 1970s, most old-fashioned garbage dumps became contaminated sites.

More recently, new regulations have been imposed to improve the treatment of solid waste materials. Recycling is practiced widely in the United States. Thus, cardboard, aluminum, and glass are usually treated separately from garbage. Likewise, hazardous materials are also handled in a separate way. So landfills are beginning to look more like old-fashioned garbage dumps again.

Microbial degradation is the major way in which the organic materials in landfills are degraded. Because the landfilled material is covered with soil, conditions become anoxic. Therefore, anaerobic microorganisms are very important. Methanogens and fermenters, such as *Clostridium* spp. involved in cellulose decomposition, grow in such systems. As a result of the activity of the

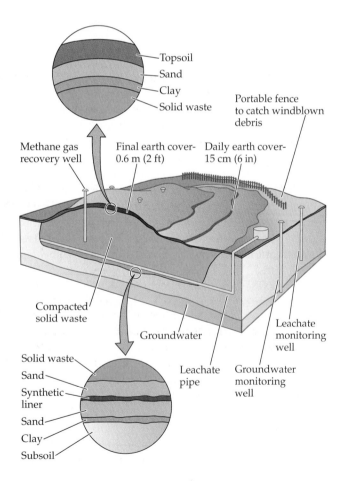

Figure 32.7 Landfill
Diagram of a typical landfill. Solid waste (garbage) is unloaded into the open earth pit. When the pit is full, it is covered with soil.

(A)

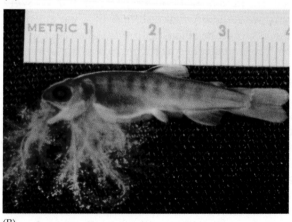

(B)

Figure 32.8 Leachate effects from a landfill
(A) The small stream (entering from the right) contains organic leachate from a landfill. (B) The gills of this salmon fingerling from a hatchery downstream of the leachate entry show heavy bacterial and fungal growth due to the enrichment effects of the effluent. Courtesy of J. T. Staley and James Huff.

Figure 32.9 Composting
A composting container with plant materials from a garden.
Photo ©G. Büttner/Okapia/Photo Researchers, Inc.

anaerobic microorganisms, methane and carbon dioxide are major gases released. In addition, organic acids and alcohols are produced. These and heavy metals derived from corroding materials, may be leached from the landfill and cause deleterious effects on receiving streams (**Figure 32.8**). Heavy metals tend to accumulate in the environment and have adverse effects on plant and animal life. Thus, it is important to monitor effluents from landfills to see that they do not adversely impact the watershed.

Composting

Composting is the process whereby plant materials are decomposed. Because plant materials (leaves and stems) are very high in organic content but low in nitrogen, they are not as readily degraded as most garbage. In composting, the material is placed in a pile or windrow in which the composting process is allowed to occur (**Figure 32.9**). This may be in a container or may simply be in a compost heap on the ground.

Composting typically occurs over a period of weeks. The most important microbial groups involved are the bacteria, in particular the actinomycetes, as well as fungi. The process undergoes several stages including a period of heat generation in which mildly thermophilic microorganisms are selected. Eventually, much of the material being composted is degraded so that the volume is depleted. This material can be used as "mulch," an organic amendment to soils used for gardening. However, it is important to recognize that heavy metals and pesticides or other hazardous materials used in gar-

dening may be concentrated by composting.

The practice of commercial composting is becoming more popular. In this practice, grass clippings, tree trimmings, and other yard waste are collected and composted in large windrows under controlled conditions. This practice has certain advantages over landfills, in that the material can be sold back to the public as mulch.

PESTICIDES

Pesticides are chemical substances manufactured by the chemical industry for control of insects (insecticides) or weeds (herbicides). Many of these are manufactured organic compounds, not previously present on Earth, termed **xenobiotic compounds.** The term **xenobiotic** is derived from Greek and means literally "stranger (xeno) to the biotic environment."

DDT is perhaps the best-known pesticide (**Figure 32.10**). It interferes with the synthesis of chitin, the tough polymeric material composed of β 1-4 glucosamine found in the exoskeletons of insects and crustaceans. Because it is such an effective insecticide, DDT has been used widely all over the world. DDT is an example of a persistent pesticide. Unlike natural organic materials, such as chitin, that are readily degraded by

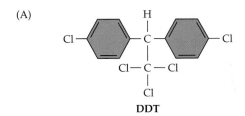

(A)

DDT

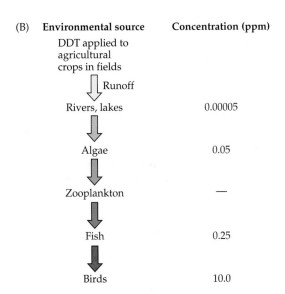

Figure 32.10 DDT and its concentration in a food chain
(A) Chemical structure of the pesticide DDT. (B) An example of biomagnification of DDT in an aquatic ecosystem.

microorganisms in the carbon cycle, some xenobiotic compounds, such as DDT, are not. DDT is recalcitrant to degradation and persists in the environment for long periods of time (see Chapter 12).

Furthermore, DDT is concentrated in the food chain in a process called **biomagnification** (Figure 32.10B). Although it is usually found in relatively low concentrations in soils and aquatic habitats, it is concentrated in fatty tissues. Thus, as it passes through the food chain, it accumulates in the tissues of animals with longer life spans and that occupy the higher trophic levels. The highest consumers in the food chain — fish, birds, humans, and other mammals — can accumulate high concentrations in their tissues. Indeed, some humans have been found with levels greater than 50 ppb (parts per billion) in their tissues, much higher than the allowed concentration for animals, such as beef, used for human consumption.

DDT has proven to be especially harmful to birds that eat insects or fish that have high concentrations of the pesticide. High levels of DDT in birds interfere with normal eggshell formation. The weak eggs are frequently broken in the nest prior to hatching, affecting bird reproduction rates. In the United States, eagles, ospreys, and other birds of prey have been particularly adversely impacted. For this reason, DDT has been banned for use as a pesticide in this country. Banning DDT has resulted in increased eagle populations in North America.

Another example of an herbicide that is quite persistent in the environment and has also been banned is 2,4,5-T (**Figure 33.11**). Although it contains only one additional chlorine atom, it is much more resistant (2 to 3 years in soil to degrade) compared with 2,4-D (3 months to degrade). 2,4-D was also referred to as "agent orange" when it was used on a large scale as a defoliant in the Vietnam War.

Because of the problems associated with some persistent xenobiotic compounds, only readily degraded pesticides are now permitted for use in the United States. Most of these break down through microbial activity in a period of days or weeks following application.

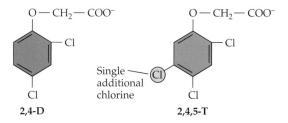

Figure 32.11 Herbicides
Chemical structures of the herbicides 2,4-D and 2,4,5-T, identical except for an additional chlorine in 2,4,5-T.

Alternatives to Use of Persistent Pesticides

Another completely different approach has been taken to control pests: the use of **biological insecticides.** The best example of this is the use of the **Bt protein** produced by *Bacillus thuringiensis* (see Chapters 20 and 31). These compounds are produced normally by microorganisms in the environment and therefore can be readily degraded. Furthermore, they are highly specific in their activity, compared with DDT. Bt specifically affects the development of larval stages of insects without having any effect on birds and other animals that might eat the insects. Of course, Bt can have adverse effects due to killing other insects that are not specific pest targets. More recently Bt genes have been genetically engineered into some agricultural crops so that undesirable insects that begin to eat plants are killed.

Bt is not only used for the control of plant-eating insects, but for other insects as well. For example, some species of *Bacillus* produce effective proteins for the *Anopheles* mosquitos that carry malaria.

BIOREMEDIATION

The rapid growth of the chemical industry during the last 100 years has significantly altered our lifestyle and improved our standard of living. The "chemical age" has also created problems, including monumental environmental pollution. Today, in the United States alone, over 50,000 hazardous waste sites have been identified. Of these, the Environmental Protection Agency (EPA) has designated 1,200 as Superfund sites. These particular sites are so hazardous that the federal government has set aside billions of dollars for their cleanup.

Furthermore, it has been estimated that 15% of the 5 to 7 million underground storage tanks containing toxic chemicals in the United States are leaking. Most of these tanks, such as those at gasoline stations or for home heating oil, contain petroleum products, but some contain considerably more hazardous chemicals.

Moreover, groundwaters in the United States are often dangerously polluted with an array of toxic chemicals, chiefly commercial solvents, such as trichloroethylene (TCE) used for dry cleaning or cleaning machinery and high-technology electronic parts. Groundwater is the source of drinking water for about half of the people in the United States, and it serves as the drinking water supply for over 95% of rural populations. Among the compounds present in groundwater are those listed in **Table 32.2.**

In addition, the mismanagement of pesticides and fertilizers, largely in agricultural areas, has created environmental conditions even in the most remote rural areas. Not only are our fish and wildlife at risk, but hazards exist for humans as well.

Table 32.2	Organic compounds that have been detected in drinking-water wells	
Compound	**Highest Level Reported Mg/Liter**	
Trichloroethylene	27.30	
Toluene	6.40	
1,1,1-Trichloroethane	5.44	
Acetone	3.00	
Methylene chloride	3.00	
Dioxane	2.10	
Ethyl benzene	2.00	
Tetrachloroethylene	1.50	
Cyclohexane	0.54	
Chloroform	0.49	
Di-*n*-butyl-phthalate	0.47	
Carbon tetrachloride	0.40	
Benzene	0.33	
1,2-dichloroethylene	0.32	

What can be done? A number of chemical and or physical procedures can be used for the removal of hazardous chemicals from environments. These chemicals can then be concentrated and stored safely. They could be chemically oxidized and thereby detoxified by incineration, but this causes problems in air pollution. Potential pollutants from waste streams can be concentrated by absorption on a solid phase, such as activated charcoal. Certain other combinations of absorption and extraction methodologies are also available. However, these processes are usually very expensive and not always applicable, particularly for contamination problems in soils and other natural environments. A more practical method for remediating environments contaminated by hazardous waste is by **bioremediation.**

Bioremediation is defined as the use of living organisms to promote the destruction of environmental pollutants. The applications of bioremediation methodology are, in many ways, extensions of the technology that has been so effective in the treatment of urban sewage and industrial wastewater. Bioremediation has been applied to the treatment of unusual contaminations including munitions such as TNT (**Box 32.2**) and even the treatment of radioactive wastes (**Box 32.3**).

Biodegradative Organisms

Evolution has resulted in a vast array of microbes that have broad and flexible biodegradative capacities. They can survive and destroy or detoxify chemicals in a variety of environmental niches (hot, cold, low pH, with or without oxygen, and so on). Often the organisms best suited for bioremediation are the species that are indigenous to a particular polluted habitat or one that is similar to it. The indigenous microbes have shown that they can survive and grow with the toxic substances. Also, mixed populations are superior, in many cases, to axenic cultures in the biodegradation of chlorinated aromatic hydrocarbons.

Although some microbial strains have been selected for bioremediation capabilities, they often tend to be less stable and effective than native populations. Furthermore, it is actually *illegal* to introduce genetically *engineered* bacteria into natural environments because of concern about the unknown effects on natural populations and the difficulty of controlling their dispersal to other sites. Therefore, although genetically engineered bacteria may be used for treatment in bioremediation processes in containers, they cannot be inoculated into the polluted environment to degrade the toxicant compounds.

Hazardous environments are often formidable in that they usually contain more than one type of toxic substance. For example, it is not uncommon that a site will contain heavy metals such as cadmium, lead, or mercury as well as chlorinated organic compounds. It is therefore essential that the microbes selected for such sites be able not only to remediate one group of compounds, but also to survive and grow in the presence of other toxic substances. Increased tolerance to acid, base, temperature, salinity, or heavy metals may be essential.

Advantages of Bioremediation

Bioremediation has a number of advantages, such as broad applicability and low cost, over other processes for remediation. Incineration requires that the toxicant be in a burnable form, and combustion itself may produce toxic smoke. This hazardous smoke problem can be eliminated by applying rigorous conditions during incineration, which is often expensive. Another advantage to bioremediation is the low risk of exposure to hazardous chemicals during clean up. Chemical/physical excavation methods of remediation often require the handling of the hazardous material. Ideally, sediments in riverbeds, harbors, and lakes should be remediated without disturbing the sediments. When sediments are disturbed, this increases the chance of enlarging the area polluted or of transporting them downstream or to adjacent environments. When bioremediation is successful, there is minimal danger to the environment, and the products produced, such as CO_2, H_2O, and fatty acids, are innocuous.

Problems Associated with Bioremediation

Most chemically synthesized compounds are biodegradable, although many are biodegraded at an unacceptably slow rate. Hence, they remain in the environment

Research Highlights Box 32.2   Some Bacteria Get a Charge from TNT

Military sites are often contaminated by a variety of toxic compounds. Among the most problematic are munitions such as TNT (trinitrotoluene). This well-known explosive is a common contaminant of soils near munitions storage areas. In some instances, contaminated soils may contain more than 5% TNT!

A group of microbiologists at the University of Idaho led by Ron Crawford has recently discovered that TNT is readily biodegradable by

Clostridium bifermentans (Figure A). All that is needed is to add starch to the contaminated soils, and the TNT is degraded anaerobically to harmless compounds. This is an example of cometabolism—the bacteria cannot use TNT as an energy source, but they can degrade it if starch is available as a carbon source for their growth (Figure B). Researchers at the University of Idaho have developed a patented treatment process in which contaminated soils are mixed in large reac-

tor containers with starch and an inoculum of the bacterium. This anaerobic treatment process is now commercially available and is used successfully for the removal of TNT from contaminated sites (Figure C).

(C)

(A)

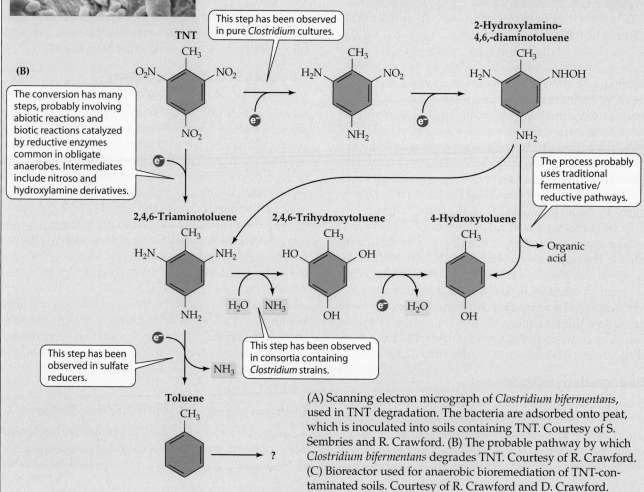

(A) Scanning electron micrograph of *Clostridium bifermentans*, used in TNT degradation. The bacteria are adsorbed onto peat, which is inoculated into soils containing TNT. Courtesy of S. Sembries and R. Crawford. (B) The probable pathway by which *Clostridium bifermentans* degrades TNT. Courtesy of R. Crawford. (C) Bioreactor used for anaerobic bioremediation of TNT-contaminated soils. Courtesy of R. Crawford and D. Crawford.

Research Highlights Box 32.3   Role of Bacteria in Concentration of Radioactive Waste

One of the surprising things about microorganisms is that some can grow in the presence of moderately high concentrations of radioactivity. Of course they are mutated by radiation, but their high population densities, rapid growth rates, and radiation repair mechanisms permit many species to survive and grow. Therefore, they may in the future play important roles in the management of radioactive wastes. One interesting example concerns the reduction of uranium compounds.

Often, radioactive chemicals occur in very low concentrations and are mixed with other chemicals. Microbiologist Derek Lovley and his colleagues developed and patented a process for concentrating radioactive uranium using bacteria. They use metal-reducing species that oxidize organic compounds while reducing uranium from the U(VI) state to the U(IV) state. Whereas

U(VI) is soluble, the reduced form, U(IV), is insoluble and therefore precipitates out from the other wastes. This process thereby concentrates and removes the radioactive uranium from the other waste material (see figure). One of the bacteria

capable of carrying out uranium reduction is the common sulfate-reducing bacterium, *Desulfovibrio desulfuricans*, which uses lactate as a carbon source.

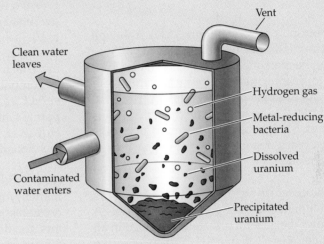

Diagram showing a reactor vessel in which radioactive wastes are concentrated by bacterial reduction and precipitation. Courtesy of Derek Lovley.

---

for long periods of time. To enhance bioremediation rates, it is essential that the proper microbe is available along with appropriate nutrients and other conditions, such as pH and O_2, that favor its growth and activity. Contaminated soil and water sites may have to be amended with fixed nitrogen, phosphorous, and other nutrients that may be limiting microbial growth. In many cases, an alternate energy source must be added in order for rapid biodegradation to occur.

Mineralization to CO_2 is the goal in bioremediation. This does not always occur, and often the pollutant is transformed to a secondary product. This is acceptable when the secondary product is less toxic than the parent compound and utilized by other organisms present in the environment. It is unacceptable when the product is more toxic than the parent pollutant and not readily biodegraded.

Methodology of Bioremediation

Bioremediation can be accomplished in situ (without excavation and recovery) or by methods that require recovery and aboveground treatment. In situ treatment of contaminated soil involves adding nutrients or microbes to the otherwise undisturbed soil. Nutrients include sources of nitrogen, phosphorous, or an alternate energy source. This augmentation encourages the

growth of the indigenous microbes that can catabolize the target pollutant(s). It may also be necessary to sparge air into the polluted area to promote growth of aerobic microorganisms. Microbes that can catabolize selected contaminants may also be inoculated into the environmental niche for in situ bioremediation processes.

There are a number of bioremediation strategies that involve removing or recovering the polluted environment above ground, followed by remediation and replacement after restoration. Some of these methodologies of environmental restoration, which are described next, are:

- Pump and treat
- Bioreactors
- Mound or heap
- Land farming

Pump and Treat   These procedures can be employed effectively in bioremediation of polluted water such as groundwater. The groundwater is pumped to the surface, nutrients are added, and the water is reinjected into the contaminated zone. For a groundwater pollutant such as TCE, the nutrients added would be methane and O_2, inasmuch as the principal organisms responsible for degradation of this compound are the methanotrophic bacteria. The biodegradation of TCE is carried

Figure 32.12 Action of methanotrophic bacteria
Overall action of methane-utilizing bacteria in trichloroethylene (TCE) degradation. Intermediates are not shown. In
the mixed populations of the groundwater environment, a
variety of other heterotrophic bacteria are also involved.

out by methanotrophic bacteria (**Figure 32.12**). Although
the methanotrophs do not use TCE as a carbon and
energy source, the methane monooxygenase they produce is able to dechlorinate and degrade this compound. This is an example of **cometabolism.** Thus, the
methanotrophs need to be supplied with methane during this process. The process of treatment may take
months or even years, depending on the degree of contamination of the site.

Bioreactors Bioreactors are often employed to bring
together the pollutant and the biodegrading microbe. A
bioreactor follows the general principles involved in an
industrial fermentor (see Chapter 31). A slurry of soil or
groundwater is placed in the fermentor, and an inoculum is added. This inoculum may be activated sludge
from a sewage treatment plant, an appropriate pure cul-

ture or a mixed microbial culture, or an inoculum from
a contaminated site. The pure or mixed cultures may be
added as a suspension or on a solid support. Supports
employed are activated carbon, plastic spheres, glass
beads, or diatomaceous earth. The attachment of bacteria to solid support systems was discussed in Chapter
31. Bioreactors can be operated as a continuous culture
system where about 80% of the material is removed
periodically. The remaining 20% serves as inoculum for
the added contaminated material.

Batch reactors may be placed in sequence so that the
microbial population in each has the capacity to biodegrade selected chemicals in the slurry or water. For
example, consider the situation in which the toxicant
material to be treated contains both chlorinated phenolic compounds and TCE. Bacteria such as *Pseudomonas
putida* are able to degrade chlorinated phenolic compounds and use them as a carbon source for growth.
However, this species is not able to degrade TCE. As
mentioned previously, the methanotrophic bacteria are
the preferred organisms for TCE degradation. Furthermore, the conditions necessary for the growth of
methanotrophs (methane gas as energy substrate) are
not necessary for *P. putida*. Thus, a sequence of two reactors, the first containing *P. putida* with conditions ideal
for its growth, followed by a separate reactor for
methanotrophs would result in the complete remediation of both chlorinated phenols and TCE.

The sequential reactor arrangement is particularly

Research Highlights Box 32.4   Degradation of polychlorinated biphenyl compounds (PCBs)

PCBs are double-ringed compounds
that are chlorinated (Figure A).
Depending on the position and
number of chlorine atoms, over 200
different varieties of PCB, called congeners, exist (Figure B). Like DDT,
which is similar chemically, PCBs
accumulate in the fatty tissues of
animals that consume them. Until
recently, the most highly chlorinated
congeners of the PCBs were not
known to be degraded by microorganisms.

Some of the chemical manufacturers that produced PCBs are located on the Hudson River, which
became contaminated with PCBs. In
the 1980s, it was discovered that
contaminated sediments had lower

concentrations of the higher chlorinated PCBs than expected, suggesting they were being degraded in the
environment. Subsequent laboratory studies at Jim Tiedje's laboratory
at Michigan State University confirmed that microorganisms were
degrading PCBs. However, this
breakdown occurred only under
anoxic conditions. This was a truly
exciting discovery because it proved
that these compounds could be
degraded by bacteria and therefore
could be removed from the environment. Furthermore, it indicated that
anaerobic conditions were necessary for degradation of these and
possibly other highly chlorinated
compounds.

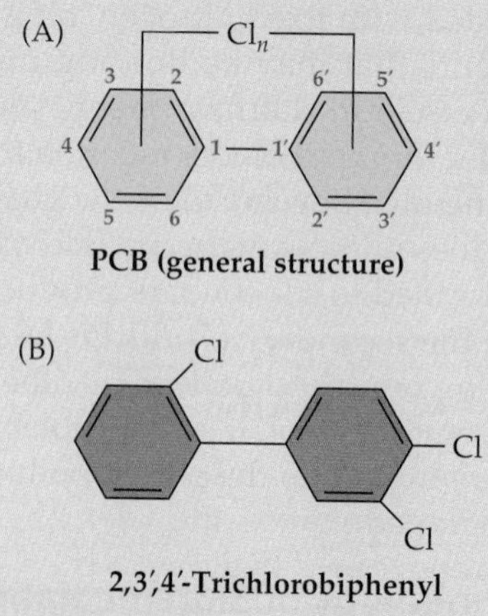

PCB (general structure)

2,3',4'-Trichlorobiphenyl

(A) General structure of a PCB. Some
200 different congeners are possible,
with various locations of the chlorine
atoms. (B) One possible PCB con-
gener, 2,3',4'- trichlorobiphenyl.

important when the contaminated material contains a chemical that is generally toxic to microbes. Removal of this chemical by a selected microbe renders the rest more amenable to bioremediation. Sequential bioreactors also permit the operation of one anaerobically and the following one aerobically. This latter arrangement is particularly useful when highly chlorinated compounds are present, such as PCBs. It is known that the most highly chlorinated forms of the PCBs are degraded only anaerobically. Thus, if the first-stage reactor is anaerobic, the more highly chlorinated compounds will be degraded and the less-highly-chlorinated products from that reactor can be transferred to an aerobic reactor where they can be degraded (**Figure 32.4**).

Mound or Heap Method In this procedure, soil is placed in mounds on a plastic liner. Water, microbes, and nutrient are trickled over the soil. The biodegradation of many pollutants including chlorinated aromatics is augmented by also adding an energy source that is readily utilizable. The effluent from the process can be collected and recycled and the process operated until the pollutants are degraded.

Land Farming

Selected pollutants can be removed from waste streams by running the waste into a confined soil basin. Nutrients may be added to the soil, and the soil can be tilled to increase mixing and aeration. This method is effective where fertile soil lies over a firm, relatively impermeable clay base. The clay base impedes penetration of the hazardous waste into the groundwater.

The Future of Bioremediation

Bioremediation has a promising future, although it is not the sole answer to our huge problem of hazardous waste in the United States. The low cost and broad applicability to many hazardous waste problems make bioremediation a method of choice. However, more research will be required to develop bioremediation processes so that they can be more widely applied in the field.

ACID MINE DRAINAGE AND ACID RAIN

Strip mining practices, in which the surface soil is removed to expose the underlying minerals to be mined, may result in the problem of acid mine drainage. This problem occurs when the minerals that are mined contain significant amounts of sulfide minerals such as pyrite, FeS_2. Strip mining exposes these reduced sulfides to oxygen and rainwater. As a result, sulfur-oxidizing members of the *Bacteria* and *Archaea* flourish and oxidize the pyrite to sulfuric acid. The iron can also be oxidized to form iron oxides.

In areas in Appalachia and in certain midwestern states such as Indiana, coal deposits contain large quantities of pyrite. Strip mining has exposed the sulfides and iron to oxidative activity by thiobacilli. Runoff waters leached from these sites can have pH values as low as 2.5 to 4.5—so low that fish are killed, as are the aquatic plants that live in the stream or receiving waters (**Figure 32.13**).

Recently a novel acidophilic archaeon has been isolated from mine drainage in California. This filamentous organism, which grows as streamers, is named *Ferroplasma acidarmarnus*. It obtains energy by the oxidation of pyrite found in the mine sediments. *Ferroplasma acidarmarnus* is capable of growth at pH 0 (**Figure 32.14**).

There is no simple remedy for acid mine drainage. Therefore, EPA now requires that strip mining areas be reburied after the mining operation is completed. In this

(A)

(B)

Figure 32.13 Acid mine runoff
Acid mine drainage area, showing (A) exposed sulfur called "yellow boy" and (B) dead plants adjacent to a lake of pH 2.5. Courtesy of A. E. Konopka.

Figure 32.14 Ferroplasma acidarmanus
Flowing streamers of *Ferroplasma acidarmanus* growing in acid runoff from a mine containing pyrite. Photo ©K. J. Edwards, Woods Hole Oceanographic Institution.

manner, the sulfides are again removed from exposure and oxidation by the bacteria.

Acid rain is also an environmental problem. This happens when coal and oil that contain large amounts of sulfur are burned as fuels. When these energy sources are burned, the gas sulfur dioxide is released into the atmosphere. When this reacts with light and water, it leads to the production of sulfuric acid and forms acid rain (see sulfur cycle in Chapter 24). Thus, it is desirable to remove the sulfur from coal and oil before burning. One way in which this can be accomplished is by use of thiobacilli to oxidize the sulfur and sulfides to sulfate, which can be removed in solution before the fuel is burned.

SUMMARY

▶ Because of their diverse capabilities of degrading organic materials, microorganisms, particularly bacteria, are used in the treatment of **wastewater** (sewage). Typical sewage treatment plants are called **secondary wastewater treatment** facilities because they entail two successive treatment systems, **primary treatment**, involving the physical separation of particulate material (**sludge**) from the fluid portion of sewage, and **secondary treatment**, in which organic material is degraded by microbial communities. Most secondary wastewater treatment facilities use the **activated sludge process**, an aerobic process, for reduction in organic materials, a major component of the biochemical oxygen demand (BOD) in wastewaters. **Anaerobic sludge digestors**, in which methanogenesis occurs, are used to reduce the organic content of sludge. **Biological tertiary treatment** is used to further treat secondary effluent in some facilities to remove nitrogen (by nitrification followed by denitrification) and phosphorus by Acinetobacter spp.

▶ **Drinking water treatment** is necessary to ensure that the public's health is protected from pathogenic bacteria that may contaminate drinking water supplies. Drinking water is treated at filtration plants which use several process steps including **coagulation** with alum addition, **sedimentation** to remove particulate materials, including bacteria, adsorbed to the alum, and **filtration** through sand filters. The final treatment is **chlorination**, the addition of chlorine gas, which forms hypochlorite solution, to disinfect the water before it is distributed to households and other consumers in the drinking water supply system.

▶ The safety of drinking water is tested by assaying for **total** and **fecal coliform** bacteria. *Escherichia coli*, which is both a total and fecal coliform bacterium, is used as the **indicator** organism, whose presence indicates the contamination of a water supply by mammalian feces. Unfortunately, positive total and fecal coliform tests can be caused by other bacteria, such as Enterobacter aerogenes, which is a soil organism; so additional tests are needed to validate the presence of fecal contamination by E. coli.

▶ **Landfills** (garbage dumps) are used to treat **solid waste** materials. Anaerobic bacteria degrade the organic materials placed in landfills. **Composting** involves the degradation of organic materials such as leaves and grass using microorganisms.

▶ **Bioremediation** is the process whereby toxic materials are degraded by microorganisms which are capable of degrading almost all organic substances including halogenated compounds such as PCBs, almost all pesticides, aromatic and aliphatic hydrocarbons, and even munitions. Bioremediation processes are being used to clean up **EPA superfund sites** scattered about the U. S.

▶ **Acid mine drainage** occurs when mining operations expose reduced sulfur compounds such as pyrite to air and water. *Thiobacilli* oxidize the sulfides to produce sulfuric acid and may lower the pH to 2 to 3. This acid mine drainage kills fish and plants in the vicinity. The only solution to this problem is to cover up the mine after the minerals have been removed so that air and water do not reach the sulfides.

REVIEW QUESTIONS

1. Differentiate between primary and secondary wastewater treatment.

2. Compare the trickling filter with activated sludge treatment.

3. What is the microbiology involved in tertiary treatment for nitrogen removal? For phosphate removal?

4. What do you believe should be done to with the undigested sludge from anaerobic digestors?

5. Why is it so difficult to identify *Escherichia coli* in water samples?

6. Some strains of *Escherichia coli* are pathogenic. Is it still possible to justify the use of this species as an indicator organism?

7. What would be the ideal microbiological test for water potability?

8. Explain the phenomenon of biomagnification and discuss the risk factors for humans living in ecosystems contaminated with DDT.

9. Define cometabolism and provide an example of it.

10. Describe how prokaryotic organisms cause acid mine drainage.

SUGGESTED READING

Bitton, G. and C. P. Gerba, ed. 1984. *Groundwater Pollution Microbiology.* New York: John Wiley & Sons.

Jacobson, M. C., R. J. Charlson, H. Rodhe and G. H. Orians, eds. 2000. *Earth System Science.* New York: Academic Press.

Guthrie, F. E. and J. J. Perry, eds. 1980. *Environmental Toxicology.* New York: Elsevier North Holland.

Mitchell, R. 1992. *Environmental Microbiology.* New York: John Wiley & Sons.

Hurst, C. J., G. R. Knudsen, M. J. McInerney, L. Stetzenbach and R. L. Crawford, eds. 2001. *Manual of Environmental Microbiology.* Washington, DC: American Society for Microbiology.

Clesceri, L. S. and A. D. Eaton, eds. 1999. *Standard Methods for the Examination of Water and Wastewater,* 20th ed. Alexandria, VA: Water Environment Federation.

Omenn, G. S. ed. 1988. *Environmental Biotechnology: Reducing Risks from Environmental Chemicals through Biotechnology.* New York: Plenum.

Glossary

Accessory pigment Pigments that trap light energy and transfer it to chlorophyll molecules in reaction centers. Carotenoids and phycobiliproteins are examples.

Accidental pathogen A microorganism that does not generally cause disease in its normal life cycle. Can cause disease if introduced in an unusual way.

Acetyl-CoA pathway An autotrophic CO_2 fixation pathway utilized by selected anaerobic bacteria. The assimilation of two molecules of CO_2 results in formation of acetyl-CoA.

Acid-fastness A property of some organisms, such as mycobacteria, that have a high lipid content in their outer wall. These organisms retain hot carbolfuchsin stain when rinsed with acid alcohol.

Acidophile An organism that preferentially grows at a pH below 5.4. *Sulfolobus* is one example.

Acquired immunity A specific immunity acquired by exposure to an antigen or by transfer of immune antibodies to a nonimmune individual.

Activation energy Energy introduced that renders molecules capable of chemical interactions.

Active immunity An immune state generated by challenge with a specific antigen.

Active site The specific area of an enzyme where substrate is bound, forming an enzyme-substrate complex.

Active transport Movement of ions or molecules across the cell membrane at an expenditure of energy.

Acute stage The stage in the infection cycle where symptoms are most pronounced.

Adenosine-5′-triphosphate (ATP) The energy carried in all living cells.

Aerobe An organism that utilizes molecular oxygen as terminal electron acceptor in aerobic respiration.

Aerosol Liquid droplets suspended in air.

Aerotolerant An organism that does not utilize molecular oxygen as terminal electron acceptor but is not harmed by O_2.

Agar A sulfur-containing polysaccharide of marine algal origin that is used as a solidifying agent in culture media.

Agglutination Sticking together of microbes, blood cells, or antigen-antibody causing an observable clump.

AIDS Acquired immune deficiency syndrome. An infection caused by HIV, the human immunodeficiency virus.

Akinetes Resting bodies formed by some cyanobacteria.

Alcoholic fermentation Anaerobic metabolism where alcohol (ethanol, butanol, etc.) and CO_2 are major products.

Algae Photosynthetic eukaryotic aquatic organisms.

Alkaliphiles Organisms that inhabit alkaline lakes and grow at a pH up to 11.5.

Allergy Inappropriate immune response to an antigen (allergen).

Allostery Change in the conformation of a protein resulting from the attachment of a compound at a site other than the reactive site. Generally a reversible inactivation of a protein, such as an enzyme.

Ames test A procedure for determining the mutagenic potential of a chemical.

Anabolism Series of reactions involved in the synthesis of macromolecules from monomers. Energy is generally required.

Anaerobe An organism that does not employ oxygen as terminal electron acceptor.

Anaerobic respiration Electron transport oxidation where sulfate, sulfur, nitrate, CO_2, or other oxidized compounds are utilized as terminal electron acceptor.

Anamnestic response Rapid immune response to antigens to which the host has previously been exposed.

Anaphylactic shock A destructive reaction between an antigen and antibody. Results in smooth muscle contraction.

Anaplerotic reactions Replacement reactions that permit the tricarboxylic acid cycle to continue. The replenishing of oxaloacetic acid by carboxylation of pyruvate is a key anaplerotic reaction.

Anion A negatively charged atom.

Anoxygenic photosynthesis Type where O_2 is not produced. Employs electron donors other than H_2O. H_2S, H_2, and reduced organic compounds would donate electrons in anoxygenic photosynthesis.

Antibiotic A metabolite produced by a microorganism that inhibits or destroys other microorganisms.

Antibody A glycoprotein present in body fluids that can bind specifically to an antigen. An immunoglobulin.

Anticodon The three-base sequence on tRNA that is complementary to the three-base codon of mRNA.

Antigen A substance recognized by the body as nonself and can elicit the immune response. An immunogen.

Antigen-presenting cell (APC) Phagocytic cells that process antigens and present them to T cells in the immune response.

Antigenic determinants Areas on an antigen that elicit an immune response. Also termed *epitopes*.

Antigenic drift Antigenic variation resulting from alterations in genes producing an altered antigen that can evade a specific antibody.

Antigenic shift A change in antigenic character resulting from movement of segments of DNA in the genome to other sites.

Antimetabolite   Compound that inhibits metabolism in microorganisms. General structure resembles a natural metabolite and competes for active sites on enzymes.

Antimicrobial agent   Anything that inhibits or kills microbial cells.

Antiparallel   The strands in double stranded DNA are oriented in opposing directions—one $3' \rightarrow 5'$ the other $5' \rightarrow 3'$.

Antiseptic   A chemical generally applied externally that destroys microbes without serious harm to the tissue.

Antiserum   A serum that contains specific antibodies.

Antitoxin   An antibody specific for a toxic substance.

Apoptosis   Elimination of aberrant cells by programmed cell death.

ARC   *Aids Related Complex* is a series of symptoms associated with an HIV infection and may lead to an active case of AIDS.

Archaea   One of the three domains of living organisms; formerly termed *Archaebacteria*.

Aseptic technique   The manipulation of microorganisms such that contamination by undesirable organisms is prevented.

Assimilation   Uptake of nutrient and conversion to cellular components.

Attenuation   Weakening of a pathogen to be employed in immunization or a genetic regulation mechanism that terminates transcription.

Autecology   Relationship of individual organisms or species and their environment.

Autochthonous   Indigenous microbial populations.

Autoimmune   Disorder in which individual becomes immune to self.

Autoimmune disease   State in which antibodies are formed against self-antigens.

Autolysin   Enzymes in bacteria that form breaks in peptidoglycan, permitting incorporation of newly formed wall units during growth.

Autolysis   Process in which a peptidoglycan is disrupted, causing cell lysis caused by autolysins.

Autotroph   An organism that can utilize CO_2 as sole source of carbon.

Auxotroph   A mutant that has lost the ability to synthesize a metabolite normally synthesized by the microbial strain.

Axenic culture   A culture free of contaminating organisms. A pure culture.

B cell (lymphocyte)   An antibody-bearing cell that can differentiate during the immune response to secrete antibody (plasma cell) or to form a memory cell.

Bacillus (bacillary)   An oblong bacterium; also called a rod.

Bacteremia   Presence of bacteria in the bloodstream; generally transient.

Bacteria   One of the three domains, along with *Archaea* and *Eukarya*.

Bactericide   Agent that kills bacteria. Mercuric chloride is a bactericide.

Bacteriocin   Protein produced by a bacterial strain that kills related strains.

Bacteriophage   Virus that infects *Eubacteria* or *Archaea*.

Bacteriorhodopsin   A carotenoid in the purple membranes of halophiles that resembles rhodopsin in the visual cells of animals. It is involved in light-driven ATP synthesis.

Bacteriostatic   Inhibitor of bacterial reproduction or growth but does not kill.

Bacteroid   An osmotically fragile bacterium that lives within protected areas such as nodules or the intestinal tract.

Balanced growth condition   When all metabolic intermediates required for growth are available at the appropriate level.

Banded-iron formations (BIFs)   Bands of ferric iron deposited in geologic strata deposited during the era when oxygenic photosynthesis evolved.

Barophiles   Organisms that thrive at high hydrostatic pressure.

Barotolerant   Can grow under high pressure but generally grows better at normal pressure.

Base composition   Relative percent of the total cellular DNA that is guanine-cytosine (G + C). Remainder is adenine-thymine (A + T).

Batch culture   Growth of a microorganism in a closed vessel, on a suitable medium, at an appropriate temperature, and for a selected time.

Binary fission   Simple division of a bacterium to yield two cells.

Bioconversion (biotransformation)   Biologically induced modification of a substrate to generate a more useful product. The substrate converted is not utilized for growth.

Biochemical oxygen demand (BOD)   Oxygen required by microorganisms to catabolize organic matter present in water.

Biodegradation   Destruction of organic compounds by microorganisms.

Bioluminescence   Production of light by living organisms.

Biomagnification   Accumulation of substances, generally deleterious, by consumer organisms.

Biomarkers   A chemical substance produced by an organism that serves as a signature for the presence or activity of that organism or group of organisms.

Biomass   Total living cellular material in an environment.

Bioremediation   Removal of toxic or undesirable environmental contaminants by microorganisms.

Biosynthesis (anabolism)   Synthesis of macromolecules from monomers during cell growth.

Biotechnology   Use of living organisms to generate useful industrial products. Usually involves genetic manipulation.

Broad-spectrum antibiotic   Antimicrobial that is effective against both gram-positive and gram-negative bacteria.

Budding   Asexual reproduction in which progeny arise from protuberances on the surface of the parent cell.

Calorie   Amount of heat energy required to raise a gram of water from 14.5°C to 15.5°C.

Calvin cycle   The major pathway for CO_2 fixation during photosynthesis in plants, algae, cyanobacteria, and many photosynthetic bacteria. Some chemolithotrophs also utilize this pathway.

Capsid   The protein coat that surrounds viral nucleic acid.

Capsomers   Protein subunits that together form an icosahedral capsid.

Capsule   An organized layer of biosynthetic origin that surrounds the outer wall or envelope of the bacterium.

Carboxysomes   Inclusion bodies present in autotrophs comprised of ribulose-1,5-bisphosphate carboxylase. This enzyme is involved in CO_2 fixation via the Calvin cycle.

Carcinogen   An agent, generally a mutagen, that causes cancer.

Carotenoid   Red to yellow pigments found in many bacteria. Can serve as accessory pigments in photosynthetic organisms. Also a pigment in many non-photosynthetic microorganisms.

Carrier   An infected individual that transmits disease to others. Usually a carrier has a subclinical infection.

Catabolism Reactions involved in reduction of larger molecules to smaller molecules, generally with the release of energy.

Catabolite repression Curtailment of the synthesis of selected enzymes by the availability of glucose or other metabolites as substrate.

Catalyst A compound that increases the reaction rate without itself being altered.

Cationic Having a positive charge.

CD4 Specific antigen on selected cells such as T cells. $CD4^+$ cells are host for HIV.

Cell Basic unit of living matter.

Cell cycle Sequence involved in growth/division of a cell.

Cell-mediated immunity process Whereby T cells destroy foreign or infected cells.

Cell wall A tough structure surrounding the bacterial cytoplasmic membrane that gives the cell shape and protection.

Chaperones Specific proteins that are involved in the folding and assembly of other proteins.

Chemical oxygen demand (COD) Amount of chemical oxidant necessary to oxidize organics in water to CO_2.

Chemiosmosis Development of a chemical potential across a cytoplasmic membrane that can drive ATP synthesis. Protons are driven outward by electron carriers or light.

Chemolithotroph (chemoautotroph) A microorganism that obtains its energy by oxidation of inorganics and utilizes CO_2 as carbon source.

Chemoorganotroph (heterotroph) A microorganism that obtains its energy by oxidation of organic compounds and utilizes carbonaceous substrates for growth other than CO_2.

Chemoreceptors Proteins in the cytoplasmic membrane or periplasm that bind chemicals involved in chemotaxis.

Chemostat A continuous culture apparatus that feeds medium, with one substrate at limiting concentration, into a culture vessel and permits removal of cells at a rate that maintains steady state growth.

Chemotaxis Movement of a microorganism toward a chemical attractant or away from a repellent chemical.

Chemotherapy Treatment of disease with chemicals that destroy the agent without serious harm to the host.

Chlorophyll A tetrapyrrole with a molecule of magnesium in the center that captures light energy during photosynthesis.

Chloroplast The chlorophyll containing organelle in photosynthetic eukaryotes.

Chlorosome An oblong protein-bound structure in green sulfur bacteria that contains the light-gathering pigments. It is located on the inner surface of the cytoplasmic membrane.

Chromosome The site of cellular DNA. The *Archaea* and *Bacteria* generally have one circular chromosome. One microorganism (*Borrelia burgdorferi*) is known that has a linear chromosome. Eukarya are multichromosomal.

Chronic infection Long-term infection.

Classification Placing organisms in groups based on phylogenetic relatedness.

Clonal selection Theory that clones of specific B or T cells are stimulated to reproduce when an appropriate antigen binds to their surface.

Clone Genetically identical progeny of a single parent.

Cloning vector A bit of replicative DNA that can transport inserted foreign DNA into a recipient cell.

Coccus (coccoid) A spherical bacterium.

Codon A sequence of three bases (purines or pyrimidines) in messenger RNA that codes for a specific amino acid.

Colicin Plasmid encoded proteins produced by enterics that harm other enteric bacteria.

Coliform Gram-negative facultative aerobic bacteria that ferment lactose with formation of gas within 48 hours at 35°C .

Colony A visible assemblage of microorganisms growing on a solid surface. Generally from reproduction of a single cell.

Colony forming unit (CFU) Single cell that can form a colony on a solid medium.

Cometabolism Transformation of a nongrowth substrate by microorganisms grown or growing on a utilizable substrate.

Commensalism A symbiotic association where one organism benefits and the other(s) is unaffected.

Common-source epidemic One in which all victims are infected from a single source. Food poisoning is an example.

Community A mixed population of microorganisms in a natural habitat or microcosm.

Compatible solute Organic or inorganic source that increases the internal osmotic pressure inside a bacterium to the level outside.

Competent A bacterial cell that can take up DNA fragments and be transformed.

Competitive inhibitor A chemical analogue that replaces a natural substrate and competes for active sites.

Complement A series of proteins in the blood that act sequentially (cascade) and are involved in removal of antigen/antibody complexes.

Complex medium (undefined) A culture medium in which the chemical composition of the ingredients is not precisely determined.

Concatemer Viral genomic material in which individual genomes are linked end-to-end. The linear structure is split to individual genomes on viral assembly.

Congenital Disease contracted maternally during gestation or birth.

Conjugation Transfer of genetic information by cell-to-cell contact.

Consortium An assembly of several bacteria in which all benefit to some degree from the association.

Constitutive enzyme An enzyme not subject to regulation; always expressed during growth.

Convergent evolution Development of similar characteristics by unrelated organisms because they adapt similarly to an environmental challenge.

Corepressor A low molecular weight compound that functions in repression of an enzyme.

Cortex Dense area between the endospore coat and core in endospore forming bacteria.

Covalent bond A chemical bond resulting from sharing of electrons.

Cross-resistance The development of resistance to one antimicrobial agent effects resistance to another.

Crossing-over In meiosis, where adjacent DNA strands are exchanged.

Crustose Form of a lichen that is compact and grows close to a substrate.

Culture A strain growing on a laboratory medium.

Culture medium A liquid or solid nutrient on which microorganisms can be grown.

Cyanobacteria The *Bacteria* that can grow photosynthetically via oxygenic photophosphorylation. Morphologically diverse.

Cyclic photophosphorylation Cyclical movement of electrons in photosystem I.

Cytochrome Heme proteins involved in electron transport.

Cytokinesis Division of the cytoplasm of a dividing cell.

Cytoplasm The liquid contents of a cell. Surrounded by a cytoplasmic membrane.

Cytoplasmic inclusion An intracellular storage granule in a bacterium.

Cytoskeleton Network of microtubules and microfilaments that gives a eucaryotic cell its shape and the ability to arrange organelles and move.

Cytotoxin A toxic material that destroys cells. Some pathogens produce these compounds.

D value Decimal reduction time. Time required to reduce a population to one-tenth the original.

Dark-field microscopy A process for indirect illumination such that the specimen is illuminated against a dark background.

Death phase The phase in batch culture in which the viable population is in decline.

Decomposition The breakup of complex material to constituent parts.

Defined medium One in which the composition and quantity of all components are known.

Deletion Excision of a part of a gene causing mutation.

Denaturation Change in a protein or other macromolecule that destroys activity.

Dendritic cell An antigen-presenting cell; present in lymph nodes and spleen.

Denitrification Reduction of oxidized forms of nitrogen to nitrogen gas via anaerobic respiration.

Deoxyribonucleic acid (DNA) A nucleotide polymer composed of deoxyribonucleotides joined by phosphodiester bonds. The genome of an organism is composed of DNA.

Diauxie Biphasic growth that can occur when two substrates are available in a culture medium.

Differential media Culture media that permit growth of one bacterial type and inhibit others or permit a microorganism to demonstrate specific biological properties.

Differentiation Changes in the structure of a microorganism during the growth cycle.

Dimer Compound formed by joining of two monomers.

Disease Disturbance of normal structural or functional capacity of an organism.

Disinfect A way to destroy all microorganisms.

DNA library (gene library) Cloned DNA fragments that represent the genes in the entire genome of an organism.

Domain The highest level of classification of all life based on rRNA analysis. The three domains are *Bacteria*, *Archaea*, and *Eucaryota*. A domain is also used to describe an area of a macromolecule.

Doubling time Time required for a population to double in number.

Early message Messenger RNA formed immediately after infection, which is translated to catalytic proteins that disrupt host cell function.

Ecosystem Total organismic community and associated abiotic components of a selected environment.

Ectomycorrhizae Association between fungi and root tip, in which the fungi form a sheath about the root.

Ectosymbiont Lives outside but in close proximity to the host.

Electron acceptor The component that accepts electrons during an oxidation-reduction reaction.

Electron donor The component that donates electrons during an oxidation-reduction reaction.

Electron transport chain Series of electron acceptors and donors that transfer electrons from an electron carrier, such as NADH, to a terminal acceptor, such as O_2. Also functions during photophosphorylation.

Electrophoresis Separation of charged molecules, such as protein or DNA, in an electrical field.

Electroporation Use of an electrical pulse to alter the cytoplasmic membrane promoting uptake of DNA fragments.

Embden-Meyerhof (glycolysis) An anaerobic pathway that catabolizes glucose to two pyruvic acid molecules and generates two molecules of ATP.

Endemic disease One that is present at a constant low level in a population.

Endergonic reaction One that requires energy input to proceed.

Endocytosis Uptake of a particle, such as a virus, by enclosing via membrane extension and pinching off the vesicle formed.

Endogenous pyrogen A protein that induces fever in a host.

Endomycorrhizae A fungus–plant root association in which the fungus penetrates into the root cells.

Endophyte A microorganism, often a cyanobacterium, that lives within a plant.

Endospore A heat-resistant dormant cell that forms within the cell of selected bacteria.

Endosymbiosis Growth of a microorganism within another organism that is generally beneficial to both.

Endosymbiotic theory The theory that the mitochondrion, chloroplast, and other organelles in eukaryotes arose through an endosymbiotic association of bacteria with eukaryotic ancestors.

Endotoxin The lipopolysaccharide portion of the outer envelope of gram-negative bacteria released by cell lysis or during growth, and is toxic to animal hosts.

Energy Capacity to do work.

Enrichment culture Technique which involves addition of a selected substrate to a culture medium to isolate an organism that can grow on that substrate. Physical conditions may also be adjusted to obtain a specific type of organism.

Enteric Intestinal; often used to describe microorganisms associated with the intestinal tract.

Enterotoxin A toxin that adversely affects the intestinal tract.

Entner-Doudoroff A nonglycolytic pathway of glucose catabolism that produces pyruvate and glyceraldehyde-3-PO_4.

Envelope The outer structure in gram-negative bacteria of the membranous layer that surrounds the capsid of some viruses.

Enzyme A highly specific protein catalyst.

Enzyme-linked immunosorbent assay (ELISA) A technique to detect specific antibodies, using enzymes as the detector system.

Epidemic A marked increase in the number of individuals affected by an infectious disease during a given period of time.

Epidemiology The study of prevalence, incidence, and transmission of infectious diseases.

Epilimnion Aerobic warmer layer of water in a stratified lake; lies above the thermocline.

Episome A plasmid that can integrate into the host genome or function separately.

Epitope A specific area on an antigen that elicits an antibody response.

Etiologic agent The specific cause of a disease.

Eukarya One of the three domains. Composed of organisms that have a membrane bound nucleus and division of DNA by mitosis.

Eutrophic   An environment, usually aquatic, overly rich in nutrient.

Exergonic   A reaction that liberates energy.

Exon   Sequences in a split gene that code for messenger RNA.

Exotoxin   A toxin released by cells, generally during growth.

Exponential growth   Balanced growth in which each microbial cell is dividing during a fixed period of time.

Extracellular enzyme   An enzyme excreted by a microorganism to cleave large molecules to a size transportable into the cell.

Extreme   Employed with thermophiles or halophiles to indicate that they grow at the highest temperature or salt concentration of microorganisms presently described.

Facilitated diffusion   Carrier mediated transport across a cytoplasmic membrane.

Facultative   Term applied to aerobes and autotrophs to indicate that the organism can grow either aerobically/anaerobically or autotrophically/heterotrophically under appropriate conditions.

Fastidious   Complex growth requirements generally due to the inability of a microorganism to synthesize requisite monomers.

Feedback inhibition   Condition in which the end product of a biosynthetic pathway can curtail activity of enzymes involved earlier in the sequence of synthetic reactions leading to that end product.

Fermentation   Production of ATP via catabolic sequences in which organic compounds serve as electron donor and organic intermediates as electron acceptor.

Filamentous   A microorganism that, in normal growth, forms long strands.

Fimbriae   Filamentous structures on bacteria that play a role in adherence or in formation of pellicles or masses of cells.

Flagella   Structures involved in motility of bacteria. They are mainly composed of a protein called flagellin.

Flavoproteins   Riboflavin derivatives that are electron carriers.

Fluid mosaic model   The accepted structure of cytoplasmic membranes with the phospholipid bilayers forming a double track. Proteins are embedded through the membrane or peripherally at either surface.

Fluorescence   Emission of light of distinct wavelength after activation with light of a different wavelength.

Fluorogenic compound   A chemical substance that becomes fluorescent when cleaved enzymatically.

Foliose   Lichens that have a leaflike appearance.

Fomites   Inanimate objects that can harbor pathogens and transmit them to hosts.

Food infection   An infection caused by ingestion of food contaminated with disease-causing pathogens.

Food poisoning   Illness caused by toxins present in food due to growth of microorganisms on the foodstuff prior to ingestion.

Food web   Interlinked food chains.

Fragmentation   Reproduction in actinomycetes in which filaments break into individual cells.

Frameshift mutation   Mutation that arises when the RNA polymerase begins producing mRNA at a base other than the proper codon. One base off results in a protein of different amino acid sequence.

Free energy   Total energy available to do useful work.

Fruiting body   A reproductive structure in some bacteria, such as myxobacteria or actinomycetes. Fungi commonly produce fruiting structures.

Fruticose   A lichen that is bush- or treelike.

Fungi   Heterotrophic eukaryotes that have rigid walls.

Fungicide   An agent that kills fungi.

G + C   The DNA of an organism has guanine paired with cytosine and adenine with thymine. G + C refers to the percent of the total DNA that is guanine-cytosine pairs.

Gas vesicles   Organelles with protein membranes that fill with gas in aquatic bacteria; serve as flotation devices.

Gastroenteritis   Inflammation of the stomach lining or intestines; often caused by food poisoning or infections.

Gene   A segment of the genome that codes for a specific polypeptide, protein, tRNA, or rRNA.

Generalized transduction   Transfer of any part of a bacterial genome when packaged randomly in a phage capsid.

Generation time   Time required for a population to double in number.

Genetic code   Triplet nucleotide sequences that specify a specific amino acid in a protein chain.

Genetic engineering   Modification of the genome of an organism.

Genetic map   The precise sequence of genes in the genome.

Genome   The complete genetic information in a cell or virus.

Genotype   Heritable genetic information in a cell.

Genus   A grouping of organisms that are closely related phylogenetically.

Geosmin   Organic molecules produced by actinomycetes and some cyanobacteria that give soil or water a distinct aroma.

Germicide   An agent that kills bacteria.

Germination   Loss of dormancy in an endospore.

Glider   An organism that is motile without aid of flagella. Gliders move only when on a semi-solid surface.

Glycocalyx   Equivalent to capsule, the layer outside the cell envelope or wall.

Glycogen   A branched polysaccharide composed of glucose; used as a storage granule.

Glycolysis   Anaerobic conversion of glucose to pyruvate via the Embden-Meyerhof pathway, generating ATP.

Glycosidic bonds   Covalent bonds between sugars in polysaccharides.

Glyoxylate cycle   A modification of the tricarboxylic acid cycle in which isocitrate is cleaved to form succinate and glyoxylic acid. The latter condenses with acetate to form malate. Functional in organisms growing on two-carbon compounds, such as acetate.

Gram stain   A differential stain that separates bacteria on retention of the dye crystal violet. Those with an outer envelope are decolorized by an ethanol wash (gram-negative), whereas those without an outer envelope retain the dye (gram-positive).

Growth factor   Low molecular weight compounds that must be added to growth media for selected organisms because they cannot synthesize them.

Growth rate   The rate at which bacteria reproduce.

Growth yield   A measure of cellular mass generated per ATP produced from a substrate.

Halophile   An organism that requires very high salt (sodium chloride) for growth.

Hapten   A molecule (generally low molecular weight) that cannot elicit an antibody response against self unless coupled with a larger molecule.

Heat shock proteins Proteins produced under stress, particularly heat that protects the cell.

Helix A helical structure present in DNA and some proteins formed by hydrogen bonding.

Helper T cell A T lymphocyte that cooperates with a B cell in initiating the antibody response.

Hemagglutination Coagulation of red blood cells.

Hemolysins Bacterial toxins that can disrupt cytoplasmic membranes of cells. Generally hemolysins are assayed by use of red blood cells. Important in tissue invasion to release substrates for growth of pathogens.

Herd immunity Immunity to a pathogen in the majority of potential hosts; results in the inability of the pathogen to cause disease.

Heterocyst Specialized cells in cyanobacteria that are sites of nitrogen fixation.

Heterofermentation Fermentation of sugars to a mixture of products.

Heterotroph An organism that utilizes preformed organic substrates as major source of carbon.

Hexose monophosphate shunt (HMS) Metabolism of glucose through 6-phosphogluconate leading to 5-carbon sugars for synthesis of nucleotides and other cellular components.

High-energy compound One that yields free energy on hydrolysis.

Holdfast Material produced by some sessile microorganisms that aids in attachment to solid surfaces.

Homofermentation (homolactic) Fermentation of sugars, mostly to lactic acid.

Horizontal evolution Gene transfer between distantly related organisms.

Host An organism on which a parasite can grow.

Human immunodeficiency virus (HIV) The retrovirus that is the etiological agent of AIDS.

Humoral immunity An immune response involving antibodies.

Hybridoma Fusion of a cancer cell to a specific B cell to generate monoclonal antibodies.

Hydrogen bond A weak, but important, bond between a hydrogen atom and an electronegative atom, such as oxygen or nitrogen. Important in helix formation and other macromolecular interactions.

Hydrolysis Cleavage of a structure by addition of water.

Hydrophilic Affinity for, and solubility in, water.

Hydrophobic Lacking affinity for water; mostly insoluble in water.

Hypersensitivity Harmful, exaggerated immune response.

Hyperthermophile An *Archaeum* or *Bacterium* that has an optimal growth temperature above 80°C.

Hypolimnion The layer in a stratified lake that has a uniform cold temperature and a low level of oxygen.

Icosahedral A virus capsid having 20 equilateral triangular faces and 12 corners.

Idiophase End of the exponential growth phase, when secondary metabolites are synthesized in selected organisms, such as antibiotic producing actinomycetes.

Immobilized One attached to a solid support. May effectively convert substrate to product.

Immune Resistance to infectious agents generally due to specific antibodies.

Immune response Bodily response to the presence of an antigen.

Immunization Eliciting of an immune response by introduction of specific antigen.

Immunodeficiency Inability to generate normal antibody responses to specific antigens.

Immunogen (antigen) Substance that will elicit an immune response.

Immunoglobulin (Ig) Blood protein fraction that is composed of antibodies.

Immunosuppressant An agent that decreases the immune response.

Incidence Number of cases of a disease in a subset of the general population.

Indicator organism A bacterial species that is employed to determine whether an environment is contaminated by human waste. An organism that does not survive for extended periods, but generally outlives enteric pathogens, would be an effective indicator.

Inducible enzyme Enzyme that is synthesized in the presence of a specific substrate (inducer).

Infection Presence and growth of an organism within a host.

Infection thread A minute tunnel through the roots of a legume, in which

Rhizobium cells migrate to reach a tetraploid cell, in which they establish a nodule.

Inflammation A localized response to tissue injury. Characterized by pain, swelling, and redness. The host's mechanism for getting leukocytes and other protective cells to the site.

Initiation site Area of the genome where replication originates.

Insertion Placing of a piece of DNA into the interior of a gene.

Integration Incorporation of a DNA sequence into the genome.

Interference Resistance in a lysogenicized bacterial cell to invasion by another virus.

Interferons Glycoproteins that simulate virus-infected cells to produce antiviral proteins inhibit viral nucleic acid synthesis.

Intron Noncoding intervening sequences in a split gene. Does not appear in the ultimate RNA product.

Invasiveness Innate ability of a pathogen to produce substances that aid in dissemination of the organism within the host.

In vitro Outside a living organism, such as a test-tube experiment.

In vivo In a living organism.

Ionizing radiation High-energy radiation that causes loss of electrons from atoms.

Ionophore A compound that disrupts cytoplasmic membranes, resulting in leakage of cytoplasm.

Isotopes Elements with an increased number of neutrons, but normal electron and proton complement.

Joule A unit of energy.

Kilobase pairs 1,000 base pairs in a fragment of DNA.

Kirby-Bauer An antibiotic disk diffusion assay for susceptibility of a clinical isolate to antibiotics or chemotherapeutics.

Koch's postulates Expression of rules for proving the relationship between a specific microorganism and disease.

Lactic acid fermentation Generally glycolytic with lactic acid as major product.

Lag phase Period after inoculation in which there is no increase in population.

Late message Messenger RNA produced sometime after infection coding for proteins that are involved in virion synthesis.

Latent virus One present in a host without causing detectable symptoms.

Leaching Release of minerals from ore by microbial action.

Lectin Surface protein of a plant cell where microorganisms, such as nitrogen fixation bacteria, can attach.

Legume Plant that can develop nodules for nitrogen fixation.

Lethal dose 50 (LD$_{50}$) Number of microorganisms or level of toxin that will kill 50% of a test population within a fixed time.

Leukocidin A microbial toxin that will destroy phagocytes.

Leukocyte A white blood cell.

Lichen A beneficial symbiotic association between a fungus and a cyanobacterium or alga.

Lipopolysaccharide (LPS) Complex structure containing fatty acids and sugars present on the outer envelope of gram-negative bacteria.

Log phase Exponential phase in growth curve.

Lophotrichous Tuft of flagella at one or both poles of a rod-shaped bacterium.

Lymph Clear, yellowish fluid that bathes tissue and flows through the lymphatic system.

Lymphocyte A leukocyte involved in antibody formation or in cellular immune responses.

Lyophilization Rapid dehydration of frozen material in a vacuum.

Lyphokine Protein secreted by lymphocytes such as activated T-cells. Mediators in the immune response that transmit signals from cell to cell.

Lysis Physical disintegration of a cell.

Lysogenicized cell An archaeal or eubacterial cell bearing a prophage.

Lysosome An organelle in eukaryotic cells that contains digestive enzymes.

Lysozyme Enzyme that disrupts the b(l-4) bond in peptidoglycan causing cell disruption.

Lytic cycle Life cycle of a virus that effects lysis of the host cell.

Macromolecule A large molecule formed by polymerization of small molecules.

Macrophage Phagocytic cell present in blood, lymph, and tissue. These cells destroy pathogens, and some are involved in the immune response.

Magnetosomes Magnetite (Fe$_3$O$_4$) particles that are present in selected bacteria. They are tiny magnets that align the bacterium in a magnetic field.

Magnetotactic Bacteria that orient themselves in a magnetic field.

Maintenance energy Energy required by a cell for remaining viable.

Major histocompatability complex (MHC) Cell surface antigens present in all individuals that are the unique marker of self. They are encoded by a family of genes.

Mast cell Cells that line blood vessels and secrete histamine and other active products in the inflammatory response and hypersensitivity.

Medium The nutrient employed in the growth of microorganisms.

Meiosis Division of a diploid cell to form two haploid daughter cells.

Memory cell Differentiated specific B cells formed during the immune response that can convert to plasma cells when activated by the presence of a specific antigen.

Mesophile Microorganisms that grow optimally at temperatures between 18°C and 40°C.

Mesosome Cytoplasmic membrane invagination in bacteria.

Messenger RNA (mRNA) Single-stranded RNA complementary to template DNA formed via transcription. mRNA generally carries information for one polypeptide and is translated by ribosomes.

Metabolism The sum of all biochemical events in a cell.

Metachromatic granules Polyphosphate polymers stored by some bacteria.

Methanogen Archaea that generate methane in anaerobic environments.

Methylotroph Aerobic *Eubacteria* that utilize methane, methanol, and other one carbon compounds as carbon and energy source.

Microaerophile Microorganisms that require O$_2$ but at lower levels than atmospheric pressure.

Microbiota Microspic organisms collectively present in an environmental niche.

Microcyst Resting spherical structures formed by some *Cytophaga* and *Azotobacteria*.

Microtubules Fibrous protein elements that give structure to eukaryotes.

Mineralization Complete biodegradation of organic mater to CO$_2$ and inorganics.

Minimum inhibitory concentration (MIC) Smallest amount of an antimicrobial agent that inhibits growth of microorganisms.

Mitosis The division of nuclear material in a eukaryote yielding two nuclei of identical chromosomal composition.

Mixed acid fermentation One that generates mixed organic acids (lactic, acetic, etc.) as products.

Mixotroph A microorganism that assimilates organic carbon sources while using inorganic energy sources.

Monoclonal antibody An antibody of a specific type produced by a clone of identical plasma cells.

Monomer A low molecular weight intermediate utilized in the synthesis of cellular macromolecules.

Monotrichous A bacterium with a single flagellum.

Morbidity rate Incidence of a particular disease in a population during a specified time period.

Mordant A substance that increases the affinity of a cell for a dye.

Mortality rate Ratio of deaths from a particular disease relative to the total number infected.

Most probable number (MPN) Measure of the number of microorganisms by dilution. End point is highest dilution yielding growth.

Motility Purposeful movement of a microorganism.

Mutagen A physical agent (radiation) or chemical that induces mutation.

Mutation Heritable change in the genetic make up of a species.

Mutualism A symbiotic association where the partners will gain.

Mycorrhiza Fungal-plant root associations.

Natural killer cells (NK) Lymphocytes present normally (nonimmune) that nonspecifically destroy aberrant host cells or invaders.

Neurotoxin One that harms nerve tissue.

Niche A habitat with all factors necessary for growth of a species.

Nitrification Oxidation of ammonia to nitrate in the environment.

Nitrogen fixation Reduction of atmospheric nitrogen to ammonia. A property present in some *Bacteria* and *Archaea*. The enzyme involved is nitrogenase.

Nonsense codon One that does not code for an amino acid but has a signal to terminate protein synthesis.

Northern blot A technique of hybridization of single-stranded RNA or DNA to nucleic acid fragments attached to a matrix.

Nosocomial infection An infection acquired in a health-care facility or hospital.

Nucleic acid probe A labeled single strand of nucleic acid that can hybridize with a complementary strand in a crude mixture. Employed in pathogen identification.

Nucleocapsid The basic unit of a virion; nucleic acid surrounded by protein capsid.

Nucleotide A monomeric unit that can polymerize to form nucleic acid. Composed of sugar (ribose or deoxyribose), phosphate, and a purine or pyrimidine base.

Nutrient Any substance that is assimilated by a microorganism during growth.

Obligate A term used by microbiologists to indicate an absolute requirement. Obligate aerobe, obligate autotroph, etc.

Okazaki fragments Short fragments of DNA involved in discontinuous replication of DNA on the lagging strand.

Oligotroph A microorganism that can live under low nutrient conditions.

Oncogene A gene that when expressed can convert a normal cell to a tumor cell.

Operator A specific segment of DNA at the start of a gene where a protein (repressor) can bind to control mRNA synthesis.

Operon A sequence in DNA that is controlled by an operator and contains one or more structural genes.

Opines Derivatives of the amino acid arginine or sugars that are synthesized in plants transformed by *Agrobacterium tumefaciens*. Utilization of these compounds is restricted to the crown gall inducing pathogenic agrobacteria.

Opportunist A microorganism that is generally harmless but can cause disease under certain conditions or in an immunocompromised host.

Opsonins Proteins that promote phagocytosis.

Organelle A membrane-bound functional structure in a cell.

Osmophiles Microorganisms that grow best in media of high solute concentration.

Outbreak Sudden high incidence of disease in a given population.

Oxidation Donation of electrons.

Oxidation-reduction (redox) reaction Coupled reaction where one partner donates electrons and the other accepts them. One is oxidized, the other reduced.

Oxidative phosphorylation Employment of the electron transport system to generate a proton motive force that provides energy for ATP synthesis.

Oxygenic photosynthesis Cyanobacterial or plant-type photosynthesis where water serves as electron donor resulting in oxygen production.

Pandemic An epidemic worldwide. HIV is now a pandemic.

Parasite A symbiotic association where one organism benefits and the other is harmed.

Passive diffusion Movement of a molecule from an area of high concentration to one of lower concentration.

Passive immunity A short-lived immunity gained by transferring specific immune antibodies to a nonimmune individual.

Pasteurization Heating a fluid to temperatures that destroy spoilage or disease causing organisms.

Pathogen A disease-causing organism.

Pathogenicity A relative term indicating the disease-causing potential of a microorganism.

Pentose phosphate pathway A pathway that oxidizes glucose-6-phosphate to ribulose-5-phosphate. A source of 5-carbon sugars for DNA and RNA synthesis.

Peptide bond A covalent bond between the carboxyl of one amino acid and the amine of another formed by dehydration.

Peptidoglycan (murein) The polymeric cell wall structure present in most *Bacteria*. It is formed by alternating units of *N*-acetylglucosamine and *N*-acetylmuramic acid. The *N*-acetylmuramic acids are cross-linked by amino acids.

Peptones Enzymatically digested proteins that are used in preparation of culture media.

Periplasm (periplasmic space) A space between the cytoplasmic membrane and peptidoglycan layer (gram-positives) and the outer envelope (gram-negative).

Peritrichous Having flagella distributed over the surface of a bacterium.

Permease A protein in the cytoplasmic membrane that transports material inward.

Phagocyte A blood cell that can capture and digest foreign material.

Phagocytosis Engulfment of a foreign particle or bacterial cell.

Phagosome A membrane-bound vacuole formed in phagocytes by invagination of the cell membrane about a foreign particle.

Phenol coefficient Relative strength of a disinfectant compared with phenol.

Phenotype Expressed properties of a microorganism.

Phospholipid The component part of a cytoplasmic membrane composed of fatty acids, glycerol phosphate, and generally with a polar molecule linked to the phosphate.

Photoautotroph An organism that can utilize light as an energy source and CO_2 as a carbon source.

Photoheterotroph A microorganism that utilizes light energy while assimilating organic compounds as a carbon source.

Photophosphorylation Generation of a proton motive force by use of light energy. This energy drives ATP synthesis from ADP and inorganic phosphate in ATP synthase.

Photosynthesis Use of light energy to generate chemical energy for cell maintenance and CO_2 assimilation.

Phototaxis Purposeful movement of an organism toward favorable wavelengths of light.

Phototroph Organism that utilizes light as a source of energy.

Phycobiliproteins Phycoerythrin (red) and phycocyanin (blue) pigments that trap light in phycobilisomes.

Phycobilisomes Specialized structures on cyanobacterial membranes involved in photophosphorylation.

Phylogeny The evolutionary history and genetic relationships between organisms.

Pili Protein filaments present on bacterial cells that are involved in conjugation.

Pinocytosis A process in eukaryotes for uptake in which the cytoplasmic membrane encloses material and transports it inward.

Plaque A clear area in a lawn of cells on an agar surface resulting from lysis of cells by a virus; or a microbial colony attached to and growing on a tooth.

Plasma cell A short-lived differentiated B cell that synthesizes and secretes quantities of specific antibody.

Plasmid A double strand of circularized DNA that exists and replicates independently in a cell; may integrate into the chromosome. Can carry information for specialized catabolic enzymes or drug resistance.

Plasmolysis A bacterial cell placed in high solute may lose water, causing shrinkage of the cytoplasmic membrane and potential death.

Pleomorphic Bacteria that are variable in shape.

Plus strand Viral nucleic acid that is of a base sequence that can serve as mRNA.

Point mutation Affecting a base pair in a specific location of a genome.

Polar flagellum The end or both ends of a bacillary bacterium.

Poly-β hydroxybutyrate (PHB) A bacterial storage product consisting of a linear polymer of β-hydroxybutyric acid.

Polymerase chain reaction (PCR) Amplification of DNA in vitro by synthesis of specific nucleotide sequences from a small amount of DNA. This technique employs oligonucleotide primers complementary to sequences in the DNA and heat-stable DNA polymerases.

Polymorphonuclear leukocyte (PMN) Motile white blood cells that specialize in phagocytosis.

Porin Proteins that form channels in the outer envelope of gram-negative bacteria.

Porters Membrane proteins involved in transport, both inward and outward.

Posttranslational modification Processing of the RNA transcript after formation to generate mRNA.

Precursor metabolites Twelve intermediates that originate in glycolysis, pentose phosphate pathway, or tricarboxylic acid cycle from which all cellular constituents are synthesized.

Prevalence Total percent of the population infected with a disease at a given time.

Pribnow box A base sequence located about 10 base pairs upstream from the transcription start site. It is the binding site for RNA polymerase.

Primary metabolites Products secreted during the growth phase. Lactic acid and ethanol are examples.

Primary producer Autotrophic organisms that fix atmospheric CO_2 thus providing sustenance for the ecosystem.

Primer A polynucleotide to which the DNA polymerase attaches during DNA replication.

Prion An infectious particle in which no nucleic acid has been detected.

Prokaryote A name previously applied to all microorganisms that lacked a nuclear membrane. Now replaced by *Archaea* and *Bacteria*.

Promoter The region on DNA at the start of a gene where RNA polymerase binds to initiate transcription.

Prophage State in which a temperate viral genome is integrated into the host and replicates in concert with the host genome.

Prostheca Extension of the wall and cytoplasmic membrane to form hyphae, stalks, or unusual shaped bacteria.

Protease An enzyme that cleaves amino acids from a protein.

Protomer Subunit of a viral capsid.

Proton motive force (PMF) An energized state of a cytoplasmic membrane created when the electron transport system extrudes protons. The positivity on the exterior and negativity on the interior creates a charge separation that can drive ATP synthesis.

Protoplast An osmotically sensitive bacterial or fungal cell resulting from removal of the cell wall. The cytoplasmic membrane remains intact.

Prototroph (wild type) Parent organism from which an auxotrophic mutant has been derived.

Pseudomurein Modified peptidoglycan that is present in the cell walls of some *Archaea*.

Psychrophile A cold-loving organism that can grow optimally at 12°C to 15°C and at temperatures below 0°C.

Psychrotroph An organism that grows slowly at 0°C but optimally at around 20°C.

Pure culture (axenic) A culture containing a single strain.

Purple membrane Areas in cytoplasmic membranes of halophilic bacteria containing bacteriorhodopsin.

Pyogenic Pus-forming infection.

Pyrogen A fever-inducing molecule.

Quarantine Restriction of movement of individuals to prevent spread of a contagious disease; to isolate animals to ensure that they do not have a disease.

Quellung reaction Enlargement of a capsulated microorganism in the presence of antibodies to capsular antigen.

Racking Removal of sediments formed in wine bottles during fermentation.

Radioimmunoassay A technique that employs radioisotope-labeled antibody or antigen to detect presence of specific material in body fluids.

Radioisotope An element with a surplus of neutrons that spontaneously decay with emission of detectable radioactive particles.

Reaction center Complex containing multiple bacteriochlorophyll or chlorophyll molecules where photophosphorylation occurs.

Recalcitrance A relative term indicating the resistance of a molecule to microbial attack. Generally measurable by the half-life of the compound in the environment.

Recombinant DNA technology Genetic engineering involving the introduction of a gene into a vector such as a plasmid and introducing this into another species to yield a recombinant molecule.

Recombination Combination of genetic material from two separate genomes.

Reduction potential Tendency of a reduced molecule to donate electrons or an oxidized molecule to accept electrons.

Refraction Bending of light as it passes from one medium to another.

Regulation Sum of the cellular processes that control enzyme function to ensure balanced growth.

Replication Copying of genomic DNA to generate a duplicate copy.

Replication fork A Y-shaped structure in which double-stranded DNA separates, permitting each of the single strands to replicate.

Repressible enzymes The synthesis of these enzymes can be curtailed by the presence of an external intermediate termed the repressor, a specific protein.

Repressor protein A protein that can bind to an operator to prevent transcription.

Reservoir The site or host in nature where pathogenic microorganisms reside and act as a source of infection for humans or other species.

Respiration Energy-yielding catabolic reactions that utilize organic or inorganic compounds as electron donors and acceptors.

Restriction endonucleases   Enzymes that cleave DNA sequences at specific sites. Probably originated to protect cells against viruses and is now employed in genetic engineering.

Retrovirus   Viruses that have a single-stranded RNA genome that is replicated by reverse transcription to form a single-stranded complementary DNA copy that is duplicated to form double-stranded DNA. This double-stranded DNA can be integrated into the genome of the host.

Reverse transcriptase   An RNA-dependent DNA polymerase that generates DNA from RNA.

Rhizosphere   The area surrounding a plant root.

Rho (ρ) protein   A protein that functions to dissociate RNA polymerase after transcription is complete.

Ribonucleic acid (RNA)   A polymer composed of ribonucleotides joined by phosphodiester bonds. The bases in RNA are uracil, adenine, guanine, and cytosine.

Ribosomal RNA (rRNA)   The RNA present in a ribosome and involved in protein synthesis.

Ribosome   The cellular organelle in which protein synthesis occurs.

Ribozyme   RNA that can function as an enzyme.

RNA polymerase   The enzyme that synthesizes mRNA from the DNA template.

Root nodule   Enlarged structure on a leguminous plant root where nitrogen-fixing endosymbionts live.

R plasmids   Plasmids that carry antibiotic resistance genes.

Rumen   The large vat in a ruminant (herbivore) where cellulosic material is digested anaerobically, generating organic acids that sustain the animal.

Saprophyte   An organism that generally grows on decaying organic matter.

Scale-up   Gradual increase in size of industrial fermentation to large fermentors for production of useful chemicals.

Secondary metabolite   Products that are synthesized near the end of the exponential growth phase and during early stationary phase.

Selection   Establishing conditions where organisms with a predetermined desired trait are favored.

Selective medium   One that favors growth of selected microorganism(s). The constituents may inhibit others.

Sense strand   The strand of DNA that RNA polymerase copies to produce mRNA, rRNA, or tRNA.

Septicemia   Infection of the bloodstream with microorganisms or bacterial toxic products.

Serum   Fluid portion of blood resulting from removal of blood cells and fibrinogen.

Sheath   A tubelike structure that encloses chains of bacterial cells.

Shine-Dalgarno sequences   A series of nucleotides on bacterial mRNA that binds to a specific sequence on 16S rRNA in order to properly orient the mRNA on the ribosome.

Siderophore   A low molecular weight compound that complexes with ferric iron and makes it available for transport into a bacterium.

Sigma   A protein that aids RNA polymerase in recognizing the promoter at the start site of a gene.

Signal sequence   Hydrophobic amino acids on the lead end of proteins that move through cytoplasmic membranes carrying proteins outward that function in the periplasmic space.

Single-cell protein   Microbial cell biomass that is primarily a source of protein in animal or human nutrition.

S-layer   A structured layer composed of protein or glycoprotein that covers the surface of selected bacterial species.

Slime layer   Diffuse polymeric material exterior to the cell wall of microorganisms.

Solid waste   The portion of sewage or industrial waste that settles out from the liquid phase.

SOS response   A repair system that is induced in microorganisms that have been damaged.

Southern blot   Hybridization of single strands of DNA or RNA to single-stranded DNA immobilized on a matrix.

Species   Closely related strains that differ from all other strains.

Spheroplast   Removal of the rigid cell wall peptidoglycan from a gram-negative bacterium yields an osmotically fragile cytoplasmic enclosed body. A spheroplast may contain the outer membrane.

Spike   Projection from the outer envelope of a virus.

Spontaneous generation   The long-disproved hypothesis that living organisms can arise from inanimate matter.

Stalk   An elongated structure that emanates from a bacterial cell often used in attaching the cell to a solid surface.

Starter culture   An inoculum consisting of favored microorganisms employed in initiating industrial fermentations.

Stationary phase   Phase in batch culture when growth ceases.

Stem cell   One capable of extensive proliferation, generating more stem cells, and a large clone of differentiated progeny cells.

Sterile   Free of all living things.

Sterilization   Destruction of all living things.

Stock culture   A carefully maintained, stored culture from which working cultures are obtained.

Strain   Descendants of a single organism.

Streak plate   Isolation of colonies by spreading a culture over an agar surface with an inoculating loop.

Strickland reaction   An ATP-generating reaction employed by some clostridial species where one amino acid is oxidized and another serves as electron acceptor. The products are fatty acids and ammonia.

Stromatolite   A fossilized microbial mat often of cyanobacterial origin. Can be a rounded microbial mat of photosynthetic microorganisms.

Substrate   The substance on which an enzyme acts. Also the nutrient(s) on which a microorganism grows.

Substrate level phosphorylation   Generation of a high-energy phosphate bond by reacting an inorganic phosphate with an activated organic compound.

Supercoiled DNA   Twisted double strand of circular DNA.

Suppressor mutation   A mutation that overcomes the effects of another mutation.

Svedberg unit   A measure of the sedimentation rate of a macromolecule on centrifugation; dependent on mass and density.

Symbiosis   Living together or interaction of dissimilar organisms.

Synecology   A branch of ecology that is involved with the development, structure, and distribution of ecological communities.

Synthetic medium   A defined medium, one of known composition, qualitatively and quantitatively.

Synthrophy A species interaction where metabolic capabilities are shared to permit both to thrive.

T lymphocyte (T cell) A lymphocyte that matures in the thymus. T cells are involved in many of the cell-mediated immune responses and in activating specific B cells in the immune response.

T suppressor (TS cell) A cell involved in turning off the antibody production.

Taxonomy Science of identification, classification, and nomenclature.

Teichoic acids Glycerol or ribitol polymers joined by phosphates present in cell walls of gram-positive bacteria.

Temperate virus One that can infect a host without effecting a lytic cycle. The viral genome may be integrated into the host and be replicated along with the host genome.

Template A strand of nucleic acid (DNA or RNA) that can specify the base sequence in a complementary strand.

Thermocline The layer of water in a thermally stratified lake where the temperature changes with depth. The thermocline is located between the warmer epilimnion and the colder hypolimnion.

Thermophile A microorganism that grows optimally at temperatures above 45°C with an upper limit of 75°C to 80°C.

Thylakoid A series of flattened photosynthetic membranes that are impermeable to ions and whose function is to pump out protons to establish a proton gradient essential for ATP synthesis. These membranes contain chlorophyll, electron transport chains, and specific proteins required for photosynthesis.

Ti plasmid A plasmid present in the gall forming *Agrobacterium tumefaciens*. The plasmid has been used as a vector in transferring genes to plants.

Titer The highest dilution of antiserum that reacts visibly with an antigen.

Tolerance Generally refers to the nonresponse of antibodies to self.

Toxemia Effects of toxins present in the bloodstream.

Toxic shock syndrome A potentially fatal massive inflammatory response to the presence of the exotoxin of *Staphylococcus aureus* in blood.

Toxigenicity The relative ability of a microorganism to produce toxins.

Toxin A potentially injurious substance, generally protein or lipopolysaccharide, produced by microorganisms.

Toxoid An exotoxin, such as tetanus toxin, that has been modified so that it does not cause damage but will elicit an antibody response.

Tracer A substrate that is radiolabeled in order to follow its incorporation into a living organism.

Transamination Transfer of an amine from one intermediate to a keto group of another.

Transcription Synthesis of a complementary strand of RNA from a single strand of double-stranded DNA; the enzyme involved is a DNA-dependent RNA polymerase.

Transduction Transfer of genetic information from one bacterium to another by a bacterial virus.

Transfer RNA (tRNA) A small RNA molecule that binds to a specific amino acid and delivers it to a ribosome during translation.

Transformation Gene transfer in microorganisms in which free DNA is taken up by a cell and selected portions are integrated into the genome.

Transgenic Transfer of foreign genetic information to plants or animals by recombinant DNA.

Translation Synthesis of a protein with the aid of a ribosome using the information delivered by mRNA.

Transpeptidation Formation of the peptide bridge between the peptides that extend from *N*-acetyl muramic acid of peptidoglycan.

Transposable element Genetic material that can move from one genomic site to another.

Tricarboxylic acid cycle (Krebs cycle, citric acid cycle) A series of reactions in which a molecule of acetate is completely oxidized to CO_2 generating ATP. Intermediates in the cycle are some of the precursor metabolites involved in biosynthesis.

Trichome A series of bacterial cells in a filament that have close contact with one another. The cells lie parallel lengthwise.

Ultraviolet radiation Wavelengths of light (170–397 nm) that disrupt DNA nucleic acid sequences at the 5′ side of a DNA or RNA molecule.

Vaccine A material that can elicit beneficial antibodies to immunize against disease. Employed in vaccination.

Vector Any of various arthropods, animals that carry pathogenic organisms from one host to another, often from reservoir to humans.

Vehicle (fomite) An inanimate object involved in transmission of infectious organisms.

Vesicle A membrane-bound body. The membrane may be protein or lipid bilayer.

Viable Alive, but in microbiological terms, able to reproduce.

Viable count Determination of total viable cells in a sample.

Virion A mature virus particle that is the extracellular form of the virus.

Viroid An infectious agent of plants that is a very small single-stranded RNA. Not associated with protein nor is it transcribed to generate protein.

Virulence A relative measure of the invasiveness and pathogenicity of a microorganism. Often measured by LD_{50}.

Virus An acellular infectious agent composed of DNA or RNA and able to commandeer the synthetic machinery of a host to generate virus.

Wastewater The liquid portion of sewage or output from a manufacturing process.

Water activity (a_w) A quantitative measure of water availability in an environment.

Wild type Parental type as isolated from nature.

Winogradsky column A glass cylinder packed with mud bearing photosynthetic bacteria. The lower area is anaerobic and upper relatively aerobic. Placed in light, all types of photosynthetic *Bacteria* can thrive.

Xenobiotic A synthetic chemical.

Xerophile A microorganism that grows optimally under low a_w conditions.

Yeast A unicellular fungus.

Zoonoses A disease generally associated with animals other than humans but can be transmitted to humans.

Index